Methods in Enzymology

Volume XXVII
ENZYME STRUCTURE
Part D

METHODS IN ENZYMOLOGY

EDITORS-IN-CHIEF

Sidney P. Colowick   Nathan O. Kaplan

Methods in Enzymology

Volume XXVII

Enzyme Structure

Part D

EDITED BY

C. H. W. Hirs

DIVISION OF BIOLOGICAL SCIENCES
INDIANA UNIVERSITY
BLOOMINGTON, INDIANA

Serge N. Timasheff

GRADUATE DEPARTMENT OF BIOCHEMISTRY
BRANDEIS UNIVERSITY
WALTHAM, MASSACHUSETTS

1973

ACADEMIC PRESS   New York and London

A Subsidiary of Harcourt Brace Jovanovich, Publishers

ACADEMIC PRESS, INC.
111 Fifth Avenue, New York, New York 10003

United Kingdom Edition published by
ACADEMIC PRESS, INC. (LONDON) LTD.
24/28 Oval Road, London NW1

LIBRARY OF CONGRESS CATALOG CARD NUMBER: 54-9110

PRINTED IN THE UNITED STATES OF AMERICA

Table of Contents

Section I. Molecular Weight Determinations and Related Procedures

Section II. Interactions

Section III. Conformation and Transitions

Section IV. Conformation: Optical Spectroscopy

Section V. Resonance Techniques: Conformation and Interactions

Addendum. *Enzyme Structure, Part B—Sequence Determination*

Contributors to Volume XXVII, Part D

Article numbers are in parentheses following the names of contributors.
Affiliations listed are current.

GARY K. ACKERS (15), *Department of Biochemistry, University of Virginia, Charlottesville, Virginia*

ALICE J. ADLER (27), *Graduate Department of Biochemistry, Brandeis University, Waltham, Massachusetts*

BARKEV BABLOUZIAN (32), *Graduate Department of Biochemistry, Brandeis University, Waltham, Massachusetts*

GIORGIO BERNARDI (18), *Laboratoire de Genetique Moleculaire, Institut de Biologie Moleculaire, Paris, France*

J. L. BETHUNE (2), *Department of Biological Chemistry, Harvard Medical School, Boston, Massachusetts*

JOHN R. CANN (12), *Department of Biophysics, University of Colorado Medical Center, Denver, Colorado*

S. P. COLOWICK (17), *Department of Microbiology, Vanderbilt University School of Medicine, Nashville, Tennessee*

JOHN W. DONOVAN (21, 22), *Western Regional Research Laboratory, United States Department of Agriculture, Albany, California*

BURTON P. DORMAN (30), *Department of Chemistry, University of California, Berkeley, California*

STUART J. EDELSTEIN (1, 4), *Division of Biological Sciences, Cornell University, Ithaca, New York*

JOHANNES EVERSE (3), *Department of Chemistry, University of California (San Diego), La Jolla, California*

GERALD D. FASMAN (27, 32), *Graduate Department of Biochemistry, Brandeis University, Waltham, Massachusetts*

DAVID FREIFELDER (8), *Graduate Department of Biochemistry, Brandeis University, Waltham, Massachusetts*

G. A. GILBERT (11), *Department of Biochemistry, University of Birmingham, Birmingham, England*

LILO M. GILBERT (11), *Department of Biochemistry, University of Birmingham, Birmingham, England*

WALTER B. GOAD (12), *Los Alamos Scientific Laboratories, University of California, Los Alamos, New Mexico*

NORMA J. GREENFIELD (27), *Merck, Sharp and Dohme Research Laboratories, Rahway, New Jersey*

GUIDO GUIDOTTI (10a), *The Biological Laboratories, Harvard University, Cambridge, Massachusetts*

FRANK R. N. GURD (34), *Department of Chemistry, Indiana University, Bloomington, Indiana*

WILLIAM F. HARRINGTON (13), *McCollum-Pratt Institute, and Department of Biology, The Johns Hopkins University, Baltimore, Maryland*

JOHN E. HEARST (6, 30), *Department of Chemistry, University of California, Berkeley, California*

KUE HUNG CHAU (28), *Department of Biochemistry and Biophysics, and Cardiovascular Research Institute, University of California, San Francisco, California*

JAMES B. IFFT (7), *Department of Chemistry, University of Redlands, Redlands, California*

PHILIP KEIM (34), *Department of Chemistry, Indiana University, Bloomington, Indiana*

GERSON KEGELES (13, 16), *Section of Biochemistry and Biophysics, University of Connecticut, Storrs, Connecticut*

DAVID L. KEMPER (3), *Department of Chemistry, University of California (San Diego), La Jolla, California*

O. KRATKY (5), *Institut für Physikalische Chemie der Universität, Graz, Austria*

THOMAS F. KUMOSINSKI (9), *Eastern Marketing and Nutrition Research*

Division, Agricultural Research Service, United States Department of Agriculture, Philadelphia, Pennsylvania

JAMES C. LEE (10), Graduate Department of Biochemistry, Brandeis University, Waltham, Massachusetts

H. LEOPOLD (5), Institut für Physikalische Chemie der Universität, Graz, Austria

MARCOS F. MAESTRE (30), Space Sciences Laboratory, University of California, Berkeley, California

MARIO A. MARINI (25), Department of Biochemistry, Northwestern University Medical School, Chicago, Illinois

CHARLES J. MARTIN (25), Department of Biochemistry, The Chicago Medical School/University of Health Sciences, Chicago, Illinois

THOMAS H. MOSS (35), IBM Thomas J. Watson Research Center, Yorktown Heights, New York

HUGH D. NIALL (36), Endocrine Unit, Massachusetts General Hospital, Boston, Massachusetts

YASUHIKO NOZAKI (20), Department of Biochemistry, Duke University Medical Center, Durham, North Carolina

ROBERT M. OLIVER (26), Clayton Foundation Biochemical Institute, and Department of Chemistry, University of Texas, Austin, Texas

HELMUT PESSEN (9), Eastern Marketing and Nutrition Research Division, Agricultural Research Service, United States Department of Agriculture, Philadelphia, Pennsylvania

W. D. PHILLIPS (33), Central Research Department, E. I. du Pont de Nemours and Co., Inc., Wilmington, Delaware

EUGENE P. PITTZ (10), Graduate Department of Biochemistry, Brandeis University, Waltham, Massachusetts

JOHN A. RUPLEY (23), Department of Chemistry, University of Arizona, Tucson, Arizona

H. K. SCHACHMAN (1, 4), Department of Molecular Biology, University of California, Berkeley, California

CARL W. SCHMID (6), Department of Chemistry, University of California, Berkeley, California

ALLAN S. SCHNEIDER (29), Laboratory of Neurobiology, National Institute of Mental Health, Bethesda, Maryland

VICTOR E. SHASHOUA (31), McLean Hospital, Belmont, Massachusetts

ALKIS J. SOPHIANOPOULOS (24), Department of Biochemistry, Emory University, Atlanta, Georgia

H. STABINGER (5), Institut für Physikalische Chemie der Universität, Graz, Austria

H. SUSI (23), Eastern Marketing and Nutrition Research Division, Agricultural Research Service, United States Department of Agriculture, Philadelphia, Pennsylvania

DAVID C. TELLER (14), Department of Biochemistry, University of Washington, Seattle, Washington

SERGE N. TIMASHEFF (9, 10, 23), Graduate Department of Biochemistry, Brandeis University, Waltham, Massachusetts

ROBERT TOWNEND (10), Eastern Marketing and Nutrition Research Division, Agricultural Research Service, United States Department of Agriculture, Philadelphia, Pennsylvania

ELLIOTT L. UHLENHOPP (19), Chemistry Department, University of California (San Diego), La Jolla, California

F. C. WOMACK (17), Department of Microbiology, Vanderbilt University School of Medicine, Nashville, Tennessee

JEN TSI YANG (28), Department of Biochemistry and Biophysics, and Cardiovascular Research Institute, University of California, San Francisco, California

BRUNO H. ZIMM (19), Chemistry Department, University of California (San Diego), La Jolla, California

Preface

This is the second of two volumes of "Enzyme Structure" devoted to physical methods. (Part C, Volume 26 of "Methods in Enzymology," appeared recently.) Although coverage of the various techniques is not exhaustive, it is hoped that the intent of presenting a broad coverage of currently available methods has been reasonably fulfilled.

These volumes present not only techniques that are currently widely available but some which are only beginning to make an impact and some for which no commercial standard equipment is as yet available. In the latter cases, an attempt has been made to guide the reader in assembling his own equipment from individual components and to help him find the necessary information in the research literature.

In the coverage of physical techniques, we have departed somewhat in scope from the traditional format of the series. Since, at the termination of an experiment, physical techniques frequently require much more interpretation than do organic ones, we consider that brief sections on the theoretical principles involved are highly desirable as are sections on theoretical and mathematical approaches to data evaluation and on assumptions and, consequently, limitations involved in the applications of the various methods.

The division of the material between the two parts is arbitrary. Thus, there is a considerable amount of overlap between general categories, and, at times, the descriptions of closely related techniques are found divided between Parts C and D. We do not believe, however, that this should hinder the reader in his use of these volumes for, in every case, each chapter is a completely self-contained unit.

We wish to acknowledge with pleasure and gratitude the generous cooperation of the contributors to this volume. Their suggestions during its planning and preparation have been particularly valuable. We also wish to thank the staff of Academic Press for their many courtesies.

C. H. W. HIRS
SERGE N. TIMASHEFF

METHODS IN ENZYMOLOGY

EDITED BY

Sidney P. Colowick and Nathan O. Kaplan

VANDERBILT UNIVERSITY
SCHOOL OF MEDICINE
NASHVILLE, TENNESSEE

DEPARTMENT OF CHEMISTRY
UNIVERSITY OF CALIFORNIA
AT SAN DIEGO
LA JOLLA, CALIFORNIA

METHODS IN ENZYMOLOGY

EDITORS-IN-CHIEF

Sidney P. Colowick   Nathan O. Kaplan

Section I

Molecular Weight Determinations and Related Procedures

[1] Ultracentrifugal Studies with Absorption Optics and a Split-Beam Photoelectric Scanner[1]

By H. K. SCHACHMAN and STUART J. EDELSTEIN

I. Introduction

As in other areas of research on the ultracentrifuge during the past 50 years, we have witnessed remarkable progress in the development, adaptation, and application of a variety of optical methods for viewing sedimentation processes. Emphases and goals have changed markedly as new problems in biology were recognized, the demands of research workers became more exacting, and the developments in technology opened new avenues for further explorations. In describing the present use and application of the photoelectric scanning absorption optical system, it behooves us to note that the first optical system employed by Svedberg and his colleagues 50 years ago was based on the absorption

[1] This research was supported in part by U.S. Public Health Service Research Grants GM 12159 to H.K.S. from the National Institute of General Medical Sciences and HL 13591 to S.J.E. from the National Heart and Lung Institute, and by National Science Foundation Research Grants GB 4810X to H.K.S. and GB 8773 to S.J.E.

of light by the sedimenting macromolecules.[1a,2] Their absorption optical system, which seems inconvenient, inaccurate, and unwieldy by today's standards, was replaced within 15 years by the schlieren optical system.[3-5] This latter system provided direct viewing of the movement and distribution of molecules in a centrifugal field. But this extraordinarily convenient schlieren optical system gave way in part about 15 years ago because of the pressing demands for enhanced accuracy. Hence, many sedimentation experiments, and particularly sedimentation equilibrium studies, are analyzed today by means of interference optics.[6] Meanwhile, the requirements for greater sensitivity and the need of biochemists to distinguish among the various chemical species present in solutions led to the rebirth of the light absorption optical system which had been discarded prematurely and ignored too long.[7,8]

Accompanying the renewed and widespread use of absorption optics for the study of nucleic acids was a growing frustration with a system which had been denounced variously as "inflexible," "inaccurate," "inconvenient," "laborious," "time-consuming," and even "impossible." Hence efforts were initiated in the late 1950's to incorporate into the optical system some of the products of the technological revolution which had occurred since Svedberg and his co-workers developed and employed absorption optics. The resulting photoelectric scanner has been used widely during the past decade for many types of sedimentation studies. Meanwhile the requirements of the workers have increased again, and the scanner in the form used in most laboratories is no longer considered satisfactory. Thus major changes in it are occurring. In this article we first review the advantages of the absorption optical system in relation to the schlieren and interference systems. This comparison in the next section highlights the principal defect, insufficient accuracy, of absorption optics. Following that, we consider the basic principles of split-beam scanners and the virtues and deficiencies of instruments based on the use of double-sector ultracentrifuge cells. Later sections deal with the applications of existing techniques for a host of sedimentation velocity and equilibrium studies. Both interacting and noninteracting systems are

[1a] T. Svedberg and J. B. Nichols, *J. Amer. Chem. Soc.* **45**, 2910 (1923).

[2] T. Svedberg and K. O. Pedersen, "The Ultracentrifuge." Oxford Univ. Press, London and New York, 1940.

[3] J. St. L. Philpot, *Nature (London)* **141**, 283 (1938).

[4] H. Svensson, *Kolloid-Z.* **87**, 181 (1939).

[5] H. Svensson, *Kolloid-Z.* **90**, 141 (1940).

[6] E. G. Richards and H. K. Schachman, *J. Phys. Chem.* **63**, 1578 (1959).

[7] K. V. Shooter and J. A. V. Butler, *Trans. Faraday Soc.* **52**, 734 (1956).

[8] V. N. Schumaker and H. K. Schachman, *Biochim. Biophys. Acta* **23**, 628 (1957).

illustrated. Experimental aspects, and particularly pitfalls and remedies, are treated in the following section. Finally we discuss the recent development of scanners connected to on-line computers which, though not widely tested as yet, show considerable promise in yielding greatly enhanced accuracy.

II. Comparison of Absorption Optics with Other Optical Systems

The ideal optical system for the ultracentrifuge should be sensitive, convenient, discriminating, versatile, and accurate. All these demands cannot as yet be met by any single system, but the absorption optical system shows considerable promise in fulfilling satisfactorily most of the criteria which research workers would agree upon.

A. Sensitivity

Sensitivity was apparent even in the original optical system devised by Svedberg and his co-workers.[2] Since many biological macromolecules absorb appreciable amounts of light in the near or far ultraviolet region of the spectrum, their migration or redistribution in a centrifugal field can be measured readily by an absorption optical system equipped with a monochromator.[9] For nucleic acids the absorbance at 260 nm is so great that solutions containing only a few micrograms per milliliter can be analyzed readily.[7,8] Comparable absorbances with protein solutions can be achieved with light of wavelength about 220 nm, with the result that proteins can be studied now at these same great dilutions.[10,11] These same macromolecules when added to dilute aqueous solutions cause such small increments in refractive index that neither schlieren optics nor interference optics can rival the absorption method in terms of sensitivity. For some biopolymers, such as polysaccharides, this sensitivity does not prevail since there is little absorption of light by the polymer in a wavelength range which is readily accessible for experimentation. Thus sensitivity must be gauged in terms of the spectral properties of the macromolecules and the solvent. Although some substances could be detected and analyzed readily with infrared light, the experimentation may not be feasible because the solvent itself may absorb most of the light.

[9] H. K. Schachman, L. Gropper, S. Hanlon, and F. Putney, *Arch. Biochem. Biophys.* 99, 175 (1962).
[10] H. K. Schachman, *in* "Ultracentrifugal Analysis in Theory and Experiment" (J. W. Williams, ed.), p. 171. Academic Press, New York, 1963.
[11] H. K. Schachman and S. J. Edelstein, *Biochemistry* 5, 2681 (1966).

B. Convenience

Convenience has been achieved only recently with the development of the photoelectric scanning system.[12-18] Prior to the construction of the scanner, the absorption system was woefully inadequate. Not only were the operations time-consuming and laborious but there was, in addition, the overwhelming psychological drawback that the research worker was unable to observe the sedimentation process during the experiment. The tedium and the delay in analyzing experiments were eliminated when the photoelectric scanner replaced the photography and the required densitometry.[2] Even in its earliest, primitive form the scanner produced rapidly and directly plots of concentration (really absorbance) and concentration gradient versus position in the cell. Subsequent developments which permit multiplexed operations have yielded even greater convenience since many different samples can be studied in a single ultracentrifuge experiment. Since the electrical pulses from the photomultiplier are digitized and interfaced conveniently to dedicated computers[19-21] the scanner has the added convenience of automation. Developments in this area are just beginning, but already the results with on-line computer operations are so promising that the absorption optical system compares favorably with the schlieren and interference optical systems.

C. Discrimination

The absorption optical system has the great advantage of discrimination since different components can be distinguished one from another by way of variations in their absorption properties. In contrast, the schlieren and interference optical systems are inadequate since these methods are responsive to changes in refractive index only and since most solutes cause approximately equal increments in refractive index. Hence schlieren and interference optics afford no possibility for distinguishing or identifying different chemical species in solution. By judicious choice of the wavelength of light with the absorption system the research worker

[12] H. K. Schachman, *Brookhaven Symp. Biol.* **13**, 49 (1960).
[13] J. G. T. Aten and A. Schouten, *J. Sci. Instr.* **38**, 325 (1961).
[14] S. Hanlon, K. Lamers, G. Lauterbach, R. Johnson, and H. K. Schachman, *Arch. Biochem. Biophys.* **99**, 157 (1962).
[15] K. Lamers, F. Putney, I. Z. Steinberg, and H. K. Schachman, *Arch. Biochem. Biophys.* **103**, 379 (1963).
[16] J. C. Deschepper and R. Van Rapenbush, *C. R. Acad. Sci.* **258**, 5999 (1964).
[17] S. P. Spragg, S. Travers, and T. Saxton, *Anal. Biochem.* **12**, 259 (1965).
[18] W. L. Van Es and W. S. Bont, *Anal. Biochem.* **17**, 327 (1966).
[19] S. P. Spragg and R. F. Goodman, *Ann. N.Y. Acad. Sci.* **164**, Art. 1, 294 (1969).
[20] R. Cohen, private communication, 1971.
[21] R. H. Crepeau, S. J. Edelstein, and M. J. Rehmar, *Anal. Biochem.* **50**, 213 (1972).

can "look" selectively at a specific component. In multicomponent systems containing large amounts of a solute such as urea, the net migration or redistribution of a protein can be analyzed unambiguously without complications since the third component can be rendered "invisible" by the appropriate selection of the wavelength of incident light. The discrimination of the absorption system equipped with a monochromator[9] affords unusual advantages for the analysis of interacting systems since "constituent" sedimentation coefficients[22-24] and molecular weights are readily measured by suitable variation of the wavelength of light.[25]

D. Versatility

Absorption optics with a photoelectric scanner afford versatility in both sedimentation velocity and sedimentation equilibrium experiments. A broad concentration range is accessible for direct experimentation since the sensitivity of the recording system can be varied readily by suitable variation of the wavelength of the incident light. The data are in a convenient form for use of transport equations.[25] Although the patterns produced by the recording system are plots of concentration versus distance in the cell, differentiating circuits are available which yield satisfactory plots of concentration gradient versus distance.[11] These derivative patterns are particularly useful for cursory examination of the sedimentation velocity pattern of a given macromolecule since they show directly the number, shapes, and positions of boundaries. However, they are less useful than the integral curves for measurement of concentrations which are obtained rapidly, simply, and accurately. Together the integral and derivative patterns are particularly revealing in detecting the presence of slowly and rapidly sedimenting species in samples. With absorption optics, concentrations are measured directly in absolute terms without the need for the integration which is required with schlieren optics or for the labeling of fringes produced by interference optics.

E. Accuracy

It is in the area of accuracy that absorption optical systems are still inadequate. Although the conversion of the photographic method to the split-beam photoelectric scanner was accompanied by substantial gains

[22] J. W. Williams, K. E. van Holde, R. L. Baldwin, and H. Fujita, *Chem. Rev.* **58**, 715 (1958).
[23] H. K. Schachman, "Ultracentrifugation in Biochemistry." Academic Press, New York, 1959.
[24] H. Fujita, "Mathematical Theory of Sedimentation Analysis." Academic Press, New York, 1962.
[25] I. Z. Steinberg and H. K. Schachman, *Biochemistry* **5**, 3728 (1966).

in precision and accuracy, there are still significant defects for which remedies are needed. An evaluation of the limitations of the accuracy of the scanner must be based, of necessity, on the purpose of the sedimentation experiment. For sedimentation equilibrium studies the accurate determination of concentrations throughout the cell is required; here the photoelectric scanner is clearly inferior to the interference optical system and comparable to schlieren optics. On other occasions the research is aimed at determining the relative amounts of different migrating species and the positions and shapes of the corresponding boundaries. For such studies the use of the photoelectric scanner is warranted since the accuracy compares favorably with that achieved with the schlieren optical system, and the potentially greater accuracy of the interference system can seldom be exploited. In the analysis of the shape of a single sedimenting boundary in terms of artificial sharpening, diffusion, and heterogeneity the interference system is the optical method of choice. Frequently, however, the concentrations required for optimal use of the interference system (or schlieren optics) are so high that appreciable thermodynamic or hydrodynamic nonideality results. Under these circumstances, with nucleic acids for example, the results are hardly interpretable. Therefore, a sacrifice in accuracy is mandatory and recourse to absorption optics with its great sensitivity is necessary.

In balance the absorption optical system has many advantages. Its drawbacks are readily recognized, and it seems likely that suitable remedies can be devised.

III. Principles of Split-Beam Scanner

A. Requirements

In the design of the early versions of the photoelectric scanner several requirements had to be satisfied. First, the photomultiplier used as a sensing element had to be so sensitive to light over a broad range of wavelengths that even with an extremely narrow slit in front of the photocathode there would be sufficient light to produce significant electrical signals.

Second, the photomultiplier had to be mounted in an appropriate housing containing the narrow, adjustable slit and this unit had to be part of a drive assembly so that the movement of the photomultiplier-slit combination at the plane where the cell was imaged provided a measure of the light intensity as a function of radial position in the cell. Since the column heights (in a radial direction) varied from distances of 12 mm in sedimentation velocity experiments to only 3 mm in sedimentation equilibrium studies, the drive assembly had to be linear and reproducible so that distances on the recorder traces were related in an accurately known way to distances in the cell. In addition, it was desirable to be able to vary the "magnification" in the radial direction when steep concentration gradients were present in the ultracentrifuge cell. Appropriate safety switches had to be incorporated in order to prevent damage to mechanical parts in case of malfunctioning of the drive unit.

Third, the photomultiplier receiving a burst of light each time the cell traversed the light beam had to be supplemented with appropriate circuitry so that the amplified electric signal could be converted into an analog trace representing optical density versus distance. The electronics had to be accurate and versatile with appropriate circuits for the measurement of the intensity of transmitted light, the logarithm of intensity (optical density) and their derivatives with respect to distance. Suitable calibrating circuits had to be incorporated in order to provide rapid and simple tests for the reliability of the recording system.

Fourth, the rate of scanning, whether by moving the photomultiplier housing[14] or an oscillating mirror which reflected successive regions of the image onto a stationary photomultiplier,[17] had to be adjustable so that the instrument could be used for both sedimentation velocity and sedimentation equilibrium experiments. For the former, scan times of about 30 seconds (or less) were appropriate, since movement and spreading of most boundaries during these intervals did not interfere seriously with the goal of high resolving power. When a steady-state distribution was obtained in a sedimentation equilibrium experiment, longer scan times of about 6 minutes were employed[11] so that electronic filtering could be used to minimize "noise" without a detrimental effect on resolving power.

Fifth, the inflexibility resulting from the use of a series of light filters to isolate the desired wavelength region had to be eliminated by the incorporation of a monochromator[9] into the optical system. Moreover, procedures had to be developed for the alignment of the system and the focusing of lenses for different wavelengths.[9]

Sixth, the optical system itself had to be improved in order to realize the full potentiality of the photoelectric recording system. This modification required eliminating the limitations stemming from fluctuations in light intensity, nonuniform illumination in a radial direction, reflections from optical surfaces, and the variable scattering of light from oil and dust depositing on the optical components during prolonged and repeated operation of the ultracentrifuge. Attempts to minimize these imperfections led to the development of the split-beam scanner in which the solution (at all radial levels) was continuously compared to a reference liquid (solvent) at conjugate radial distances. By this means most of the optical defects were compensated for automatically.[10,11,15]

Virtually all these requirements were satisfied by the split-beam photoelectric scanner which, though using only a single light beam, simulated double-beam operation by employing the rotor as a light chopper causing first one compartment of the cell containing the solvent to appear in the light path and then the other, which was filled with solution. By storing the voltage generated by the first light burst and then comparing it with that produced by the light transmitted through the second compartment, the scanner produced directly the optical density of the solution relative to the solvent. This was done continuously while the photomultiplier-slit assembly was moved progressively across the image of the ultracentrifuge cell. All signals were converted by the appropriate circuitry to their logarithmic values so that optical densities were produced directly by the recording system.

B. Use of Double-Sector Cells

A reliable method is required for detecting which cell (or compartment) was responsible for each burst of light as the cells move across the optical path. Most of the earlier scanners, though differing slightly in details, used double-sector cells which permitted the scanning photomultiplier to be used directly for the required switching signals. Identification of the reference and sample cells is deduced unambiguously from the time dependence of the light bursts. As the rotor turns, the photomultiplier receives no light until the double-sector cell crosses the light beam. Then two short bursts of light strike the photomultiplier in quick succession. Following that is a long dark period as the rotor makes a complete revolution. By appropriate filling of the compartments of the double-sector cell in terms of the direction of movement of the rotor the first light burst after the long dark period can be made to represent the transmittance of the solvent (or the solution) and the second, after the short dark period, to correspond to the solution (or the solvent). Thus the two light bursts were not only distinguished, but the signals from them were routed unambiguously to the proper holding circuits as the logarithmic values of the respective signals.

Double-sector cells not only permit the use of a simple switching system which is reliable but also they produce flat base lines when solvent is placed in each compartment since the different parts of the windows are very similar in terms of their transmittance of incident light. Comparing different parts of the same windows is preferable to comparing entirely different windows, as would be required if two separate, single-sector cells were employed. However, a split-beam scanner which is based on the use of double-sector cells exclusively possesses several serious disadvantages. First, the two compartments are very close and some light traversing the solvent appears at the photomultiplier even when the solution compartment is directly in the optical path and is being imaged by the camera lens on the slit in the photomultiplier housing. This stray light appearing in the image of the solution compartment is particularly serious in experiments on solutions of high optical density (about 1.5) and constitutes the principal limitation in the linearity of the recording system. Adjustable light-limiting apertures on the collimating and condensing lenses reduce somewhat this "contaminating" light but do not eliminate it completely.[11] Second, the potential signal at the photomultiplier is reduced significantly when double-sector cells are employed. Ideally a long slit at the photomultiplier should be used since the signal generated by the photomultiplier is proportional to the total amount of light incident on the photocathode. However, the slit length at the photomultiplier must be sufficiently short that the photomultiplier cannot "see" both cells at the same time. With double-sector cells this is a serious limitation, for the slit length must be no greater than the image of the central rib which separates the two compartments. As a consequence the length of the slit at the photomultiplier must be about 2 mm. This is a costly restriction since the photocathodes have a diameter of 10 mm. If a full-length slit could be employed in the photomultiplier assembly, the resulting signal would be increased almost 5-fold for the same prevailing light level in the image. With such a slit arrangement the scanner could be operated with much smaller photomultiplier voltages, thereby leading to traces more nearly noise-free and a consequent gain in precision. Alternatively, accurate measurements could be made at lower light levels in the far-ultraviolet region of the spectrum, thereby leading to a substantial enhancement in the sensitivity of the recording system for proteins.

C. Separate Reference and Sample Cells

Both of the drawbacks described above can be circumvented by the use of single-sector cells, with solvent in one and solution in the other, placed in different holes in the rotor. If, for example, the cells are separated by 180°, the dark period between solvent and solution pulses would be of long duration (500 μsec for a rotor turning at 60,000 rpm contrasted to 5 μsec for a double-sector cell). Hence separation of the light pulses by appropriate switching would be simplified. Moreover, by the time the solution cell moved into the optical path, the solvent cell would be completely out of the light beam and there would be no risk that the photomultiplier could "see" both cells simultaneously. Hence stray light would be reduced markedly. Two sets of cell windows would be required for this arrangement (or alternative versions with the cells placed in the rotor at 60° or 90° relative to one another). This should not be a drawback since the transmittance of various windows has been found to be uniformly high and appropriate "base lines" can be established. With separate cells some external signals are required to identify the different cells and to activate the required switches to direct the pulses to appropriate holding circuits.[11] Various methods are available for this purpose, and multiplexing devices are now commonplace.

D. Digitalized Scanner

Until recently all photoelectric scanners were analog instruments which produced recorder traces of optical density as a function of radial level in the cell. The optical densities were obtained through the use of holding circuits which stored the respective signals (after conversion to their logarithmic values). These systems involved electronic filters to reduce noise and they sacrificed potential accuracy by not employing the individual light pulses (in pairs) directly for measurements. In effect, each optical density value at a given radial distance represented many turns of the rotor and the electronic circuitry with its inherent limitations performed the "averaging." Hence efforts have been expended in several laboratories[19-21] to redesign the scanner so that measurements are made directly on the individual light pulses. This modification requires an on-line computer and has the significant advantage of producing data in a digital form. Thus far all workers have used analog-to-digital converters. Experience with these modified instruments has been limited to only a few laboratories, and each instrument has features that differ from the others. Nonetheless it is clear that the use of pairs of individual light pulses in concert with a dedicated computer constitutes a considerable advance over the older split-beam photoelectric scanner, which electronically averages the signals from many pulses by means of holding circuits. In addition, the inadequate electronics used for converting voltages into their logarithmic equivalents are replaced by much more reliable computer calculations.

IV. Applications

A. Sedimentation Velocity Studies

The photoelectric scanner is particularly useful in sedimentation velocity studies aimed at assessing the purity of samples, determining the relative amounts of different components, analyzing interacting systems of large and small molecules, and measuring sedimentation coefficients over a broad range of concentrations. For systems requiring measurements at great dilution it is unrivaled.

1. Measurement of Sedimentation Coefficients. Most determinations of sedimentation coefficients involve measurement of the rate of movement of boundaries, such as those illustrated in Fig. 1. These patterns, obtained with aspartate transcarbamylase at a concentration of 1.5 mg/

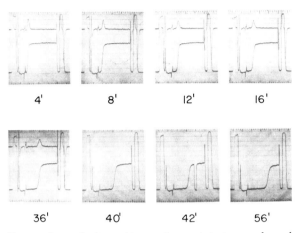

Fig. 1. Sedimentation velocity patterns of aspartate transcarbamylase. The times (in minutes) after attaining a speed of 50,740 rpm are shown below each pattern. Sedimentation is to the right, and increasing concentration (optical density) is in the upward direction. The optical density at 280 nm (in a 1-cm cell) was 0.9 and the slit width on the monochromator of the ultracentrifuge was 2 mm. The two small vertical spikes at the base-line correspond to the menisci of the solvent (pointing in a negative direction) and the solution (in the positive direction). The traces at the left and right of each pattern represent light passing through the inner and outer reference holes of the counterbalance cell. The horizontal grid lines with each fifth line (representing spacings of 10 mm) being slightly heavier than the others are ruled automatically by the recorder as the trace is produced. The vertical lines at the top and bottom of each pattern are drawn by two of the recording galvanometers which receive impulses from the timing generator. The timing marks provide an unambiguous measure of the movement of the photomultiplier in terms of the fractional rotation of the lead screw. In five of the patterns, both the integral and derivative traces are shown, being produced by the recording system simultaneously. After the 36-minute picture was obtained, the derivative pattern was turned off-scale so that the traces corresponding to the reference holes could be seen more clearly. In the 42-minute trace after the photomultiplier had moved into the image of the plateau region of the cell, the null switch on the console was depressed to show the actual base line corresponding to zero optical density. It should be noted that this null position actually is at a level slightly below that corresponding to the supernatant region. This shows that the supernatant liquid in the solution compartment absorbs a small amount of light presumably due to some contaminant that was not present in the reference solvent. For all of these traces, the scanning time was 30 seconds and the trace amplitude was set at 400. The buffer was 0.1 M Tris·HCl at pH 8 containing 10^{-3} M mercaptoethanol and 10^{-4} M EDTA.

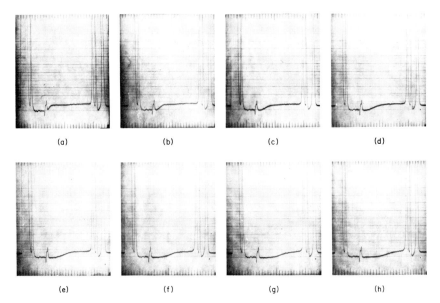

FIG. 2. Sedimentation velocity patterns of oxyhemoglobin. The times after attaining a speed of 60,000 rpm are shown below each pattern. Only integral traces could be obtained for this extremely dilute solution, 5 μg/ml. Extreme sensitivity was achieved by using light with a wavelength of 405 nm. In this buffer, 0.1 M phosphate at pH 7, and protein concentration there was substantial dissociation of the tetramers into smaller molecular weight species and the measured sedimentation coefficient was only 2.4 S.

ml, show both the integral traces representing concentration (actually optical density) as a function of distance and the derivative traces corresponding to the concentration gradient as a function of position in the cell. For this experiment[11] the scanner was operated at relatively low amplification and both the integral and derivative traces were relatively free of "noise" without any sacrifice in resolution in the radial direction. Accurate measurements of boundary positions can be obtained even with much more dilute solutions, as shown in Fig. 2. For this experiment on hemoglobin at a concentration of 5 μg/ml, the incident light had a wavelength of 405 nm in order to permit exploitation of the high extinction coefficient of the protein in that region of the spectrum. When proteins do not contain chromophores, such as heme, extreme sensitivity can be achieved by using light in the far ultraviolet, since peptide bond absorption is so great. Patterns for an extremely dilute solution, 3 μg/ml, of the catalytic subunit of aspartate transcarbamylase are shown in Fig. 3. These traces on an XY-recorder attached to the scanner were obtained with light having a wavelength of 218 nm. For this experiment, additional

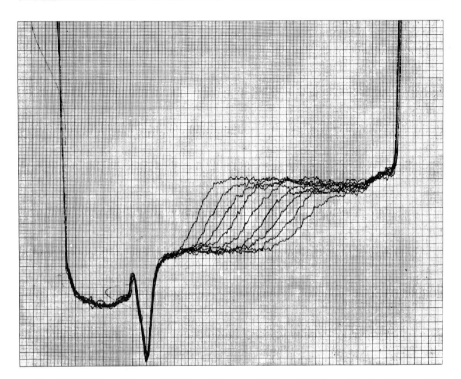

Fig. 3. Sedimentation velocity patterns of the catalytic subunit of aspartate transcarbamylase. These traces were obtained on a Honeywell 520 XY recorder. The scan time was 30 seconds and the individual patterns were obtained at 8-minute intervals. The rotor speed was 60,000 rpm, and the protein concentration was 3 $\mu g/ml$. Extreme sensitivity was achieved by using light of wavelength 218 nm. No dissociation of the oligomeric protein was detected as revealed by the sedimentation coefficient of 5.9 S. (Springer, Yang, and Schachman, unpublished observations.)

sensitivity was achieved by using two separate 7° cells (one for solvent and the other for solution) and a longer slit (10 mm) in the photomultiplier housing along with a multiplexer to identify the two pulses.[11] Experiments at such low concentrations are particularly useful in studies of interacting systems involving association–dissociation equilibrium.

Sedimentation coefficients are readily calculated from both integral and derivative traces. In principle, the sedimentation coefficient is determined from the rate of movement of the square root of the second moment of the gradient curve,[26] but for most proteins with molecular weights greater than 10^4 the position of the maximum ordinate of the gradient curve is a sufficiently accurate measure of the true boundary

[26] R. J. Goldberg, J. Phys. Chem. 57, 194 (1953).

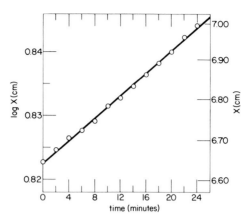

FIG. 4. Determination of the sedimentation coefficient of aspartate transcarbamylase at a concentration of 3 μg/ml. The ordinate represents the logarithm of the distance in centimeters from the axis of rotation (corresponding to the half-height of the integral curve) and the abscissa gives the time in minutes. On the right are shown the actual distances of the boundary from the axis of rotation. For this experiment light of 230 nm was used, the solvent and solution were placed in separate 7° cells, and the slit used in front of the photomultiplier was 10 mm in length. The operating speed was 59,780 rpm, and traces were recorded at 2-minute intervals. The buffer was 0.1 M phosphate at pH 7 containing $10^{-4}\,M$ EDTA.

position. Hence most measurements involve the peaks of the derivative curve or the 50% level in the integral traces. A typical plot of the logarithm of the boundary position as a function of time is shown in Fig. 4. The data for this plot[11] were obtained from the integral patterns by measuring the positions corresponding to the 50% level of the concentration in the plateau region. Such measurements are accurate and can be made rapidly. If greater precision is desired, all the data in the traces can be used in conjunction with the transport equation for the measurement of weight average sedimentation coefficients.[25] Such calculations are required for small molecules (where a complete boundary is not formed) or for systems exhibiting unsymmetrical boundaries. The transport equation can be written

$$s = -\frac{1}{2\omega^2 t}\ln\left[\frac{2\int_{r_m}^{r_p} crdr}{r_p^2 c_0} + \frac{r_m^2}{r_p^2}\right] \tag{1}$$

where r_m and r_p are distances from the axis of rotation to the meniscus, r_m, and to a level, r_p, in the plateau, ω is the angular velocity in radians/second, t is the time of centrifugation, c_0 is the original concentration, c is the concentration at the level, r, and s is the sedimentation coefficient.

Equation (1) can be used with a single trace in which the time, t, must include allowance for the equivalent time of acceleration of the rotor as well as the elapsed time at speed; alternatively Eq. (1) can be used with a series of traces from a plot of the logarithm of the term in brackets in Eq. (1) versus time. For the latter calculation the absolute time is not required. Although calculations with the transport equation are tedious and time-consuming, they are readily performed through the use of desk-type computers and can be made routine. Scanner traces are particularly suited for such measurements since c_0 is available from the initial pattern and the various levels for r_p in the different traces are easily selected. It should be noted that the use of Eq. (1) is directly equivalent to the calculation of sedimentation coefficients from second moments of gradient curves.

2. *Advantage of Combining Integral and Derivative Patterns.* Most of the sedimentation velocity patterns presented above included only integral traces. It should be recognized that the derivative traces are obtained by electronic differentiation of the integral traces and, therefore, are less precise. When the scanning rate is high (as in a 6-second trace) good derivative patterns are produced, but, in contrast, the integral patterns tend to be "noisy," since electronic filtering cannot be performed without a concomitant loss in resolution. Hence some compromise is frequently required. Since the primary data yield the integral curve it is preferable to adjust the scanning rate, slit width at the photomultiplier, and electronic controls to give good integral patterns. For the patterns represented by Figs. 1 and 2 the scanning time was 30 seconds, whereas that for Fig. 3 was 60 seconds. Having both integral and derivative traces in the same experiment is frequently advantageous.[27] Such a combination is shown in Fig. 5. The derivative trace for a ribonucleic acid preparation shows very clearly the presence of two sedimenting components; in contrast, these two principal components are not so readily discernible in the integral trace. The principal advantage of the latter is seen by the slope of the curve in both the supernatant and the plateau regions. The trace shows clearly that a large fraction of the sample is composed of polydisperse, slowly sedimenting material; in addition, there is considerable aggregated material. Neither of these "components" is detectable in the derivative patterns. Figure 5 also shows the scanner trace and a schlieren pattern obtained on a preparation of bushy stunt virus. Whereas the schlieren pattern shows a single, sharp, symmetrical boundary, the integral trace from the scanner shows, in addition, a large amount of aggregated material. This slope in the so-called plateau

[27] H. K. Schachman, *Biochemistry* 2, 887 (1963).

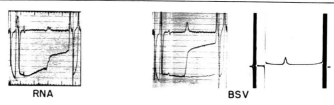

RNA   BSV

FIG. 5. Patterns illustrating the advantages of both the integral and the derivative curves. The pattern on the left is from a sedimentation velocity experiment on a preparation of partially degraded ribonucleic acid (RNA) isolated from tobacco mosaic virus. The derivative pattern shows clearly the presence of two components while the integral curve shows that there is also polydisperse slowly sedimenting and rapidly migrating material. On the right are two patterns from a sedimentation velocity experiment with bushy stunt virus (BSV). The absorption pattern obtained with the single-beam scanning system shows clearly (in the integral curve) the presence of a considerable amount of aggregated material in the preparation. Neither the schlieren pattern on the right nor the derivative pattern from the scanning system shows the presence of this material so vividly.

region in Fig. 5 should be contrasted with the real plateau in the patterns in Fig. 1. Obviously the availability of both the integral and derivative traces simultaneously is a special virtue of the photoelectric scanner.

3. Concentration Determinations at Different Wavelengths. The scanner has special advantages for the analysis of solutions in terms of the concentrations of the various sedimenting components. Amounts of different species, both on a relative and absolute scale, are determined directly and unambiguously in terms of optical density merely by measuring the pen deflection on the recorder from one plateau to another (or from supernatant region to plateau). If the various components have identical extinction coefficients the pen deflections give directly the concentrations. When the spectral properties of the components differ, the scanner can provide additional information about the composition of the solution if the pattern is recorded with light at two or more wavelengths. This virtue of the scanner is illustrated by the traces for aspartate transcarbamylase which was partially dissociated into subunits as a result of reaction with p-hydroxymercuribenzoate.[28] On the left of Fig. 6 is the trace obtained with light of wavelength 280 nm. The absorbance at this wavelength is due predominantly to the protein with very little contribution from the bound mercurial. When the pattern was obtained with light having a wavelength of 248 nm, however, strikingly different results were obtained. As seen in the right-hand pattern in Fig. 6, the "principal" component was that with a sedimentation coefficient of 2.8 S. From measurements of the relative pen deflections at the two wavelengths it

[28] J. C. Gerhart and H. K. Schachman, *Biochemistry* 7, 538 (1968).

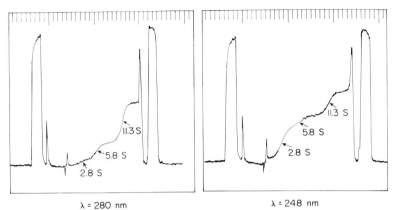

λ = 280 nm   λ = 248 nm

Fig. 6. Sedimentation velocity patterns for aspartate transcarbamylase (ATCase) treated with *p*-hydroxymercuribenzoate (PMB). At a molar ratio of PMB/ATCase equal to 6, some of the enzyme (11.3 S) was dissociated into catalytic subunits (5.8 S) and regulatory subunits (2.8 S). The patterns at 280 and 248 nm were obtained in quick succession after about 50 minutes of sedimentation at 60,000 rpm. Light of wavelength 280 nm measures predominantly protein with very little contribution from the mercurial. In contrast, the light at 248 nm is particularly sensitive to the mercurial bound as a mercaptide complex. Spectral ratios, $A_{280}:A_{248}$, for the three species showed essentially pure protein (about 1.8) for the catalytic subunit and intact enzyme. The value for the regulatory subunit was only 0.18, indicating that the mercurial was bound exclusively to that subunit. This ratio is consistent with a value of 4 mercaptide-bound mercurials per subunit.

was possible to show that the mercurial was bound virtually exclusively to the sulfhydryl groups of the smaller subunit.[28] The spectral ratios, $A_{280}:A_{248}$, for the 5.8 S subunit and the undissociated enzyme (11.3 S) corresponded to protein devoid of mercurial bound to cysteinyl residues. By judicious use of the wavelength of the incident light the nature of the reaction of the enzyme with mercurial was clarified in a way which is not possible with either schlieren or interference optics.

4. *Analysis of Interacting Systems.* In view of the direct relationship between sedimentation coefficients and molecular weights the velocity method is uniquely suited for the detection and quantitative measurement of the binding of small molecules to macromolecules. Such an application of the scanner is illustrated by the patterns shown in Fig. 7 for various mixtures of DPNH and chicken heart lactic dehydrogenase.[27] The pattern for DPNH shows that all the light-absorbing material (at 340 nm) migrated slowly with a sedimentation coefficient about 0.2 S. Upon the addition of increasing amounts of enzyme, progressively more of the light-absorbing material (DPNH) sedimented with a coefficient

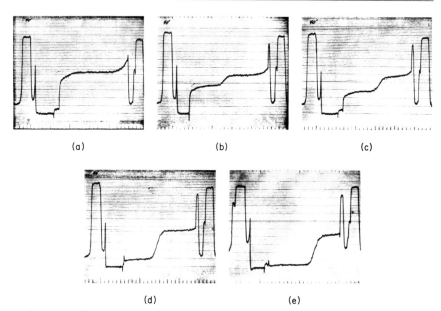

(a) (b) (c)

(d) (e)

FIG. 7. Sedimentation velocity patterns from an interacting system involving an enzyme [lactic dehydrogenase (LDH)] and a coenzyme [reduced diphospho-pyridine nucleotide (DPNH)]. The wavelength of the light was 334 nm, and the slit width on the monochromator was 1.0 mm. The concentration of DPNH in all experiments was $1.3 \times 10^{-4} M$ in $0.1 M$ phosphate buffer at pH 7. The concentration of LDH was varied in the different experiments to give the following molar ratios in the various experiments: (a) DPNH with no LDH; (b) DPNH/LDH = 12; (c) DPNH/LDH = 8; (d) DPNH/LDLH = 4; (e) DPNH/LDH = 2. Sedimentation is from left to right. All experiments were at 59,780 rpm and the patterns were taken 40 minutes after reaching speed.

of 7 S, the value for the pure enzyme. By appropriate treatment of the data in terms of the constituent sedimentation coefficient of the small molecule,[22-25,29,30] it is possible to analyze such patterns in terms of the chemical equilibria describing the interaction between the enzyme and the cofactor. This treatment yields the number of binding sites and the dissociation constants characterizing the various equilibria.

5. *Band Centrifugation.* In the past 10 years the technique of zone centrifugation[31,32] used as a preparative method for the separation and purification of viruses has been adapted to the ultracentrifuge so that

[29] J. R. Cann, "Interacting Macromolecules." Academic Press, New York, 1970.
[30] L. W. Nichol and D. J. Winzor, "Migration of Interacting Systems." Oxford Univ. Press (Clarendon), London and New York, 1972.
[31] E. G. Pickels, *J. Gen. Physiol.* **26**, 341 (1943).
[32] V. N. Schumaker, *Advan. Biol. Med. Phys.* **11**, 245 (1967).

each of the macromolecular species can be examined during its migration through the cell.[33,34] Small lamellae or bands containing the macromolecules are layered onto more dense solvents in specially constructed ultracentrifuge cells.[35,36] This technique, now known as band centrifugation, has proved especially useful for the study of nucleic acids since less material is required, the resolved components are physically separated, and the slow components remain behind the moving bands. As seen in Fig. 8,[37] the photoelectric scanner has been very useful in following the migration of different types of polyoma DNA molecules. The density

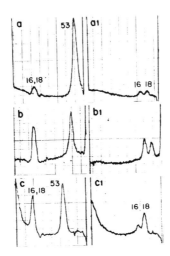

Fig. 8. Band sedimentation of polyoma DNA in alkaline CsCl. The left and right patterns were obtained about 30 minutes and 90 minutes after sedimentation began through an alkaline CsCl solution (pH 12.5) of density 1.35 g/ml. Resolution of the slower components, 16 S and 18 S, is observed in the right-hand patterns. By that time the faster species, 53 S, had already migrated to the bottom of the cell. Component I isolated in a sucrose gradient experiment is shown as a control in (a, a1). The same material treated with pancreatic deoxyribonuclease is shown in (b, b1). The patterns for the DNA which had been subjected to heat denaturation followed by treatment with *Escherichia coli* phosphodiesterase are given in (c, c1). From J. Vinograd, J. Lebowitz, R. Radloff, R. Watson, and P. Laipis, *Proc. Nat. Acad. Sci. U.S.* **53**, 1104 (1965).

[33] J. Rosenbloom and V. N. Schumaker, *Biochemistry* **2**, 1206 (1963).
[34] J. Vinograd, R. Brunner, R. Kent, and J. Weigle, *Proc. Nat. Acad. Sci. U.S.* **49**, 902 (1963).
[35] J. Vinograd, R. Radloff, and R. Bruner, *Biopolymers* **3**, 481 (1965).
[36] J. Vinograd and R. Bruner, Fractions, Beckman Instr. Inc. No. 1, 2 (1966).
[37] J. Vinograd, J. Lebowitz, R. Radloff, R. Watson, and P. Laipis, *Proc. Nat. Acad. Sci. U.S.* **53**, 1104 (1965).

gradient caused by the salt (or D_2O) which is used to provide a positive stabilizing density gradient causes no complications with absorption optics; with schlieren or interference optics the desired optical registration of the moving band is not so direct since all components contribute to the optical patterns in direct proportion to their weight concentration. Since the solute used to provide a stabilizing density gradient is generally at a much higher concentration than the macromolecules, the former dominate in the observed schlieren or interference patterns. In contrast, the photoelectric scanner records the distribution of the light-absorbing macromolecules without "seeing" the stabilizing solute molecules.

6. *Study of Enzymatically Active Species.* Recently the scanner has found application in the identification of active enzyme molecules by the optical registration of the formation of the product of a catalyzed reaction or the disappearance of the substrate.[38-42] This sedimentation velocity method permits the identification of the enzymatically active species in an association–dissociation system and even in impure solutions by examining the formation (or disappearance) of a light-absorbing material (such as DPNH) as the very dilute enzyme molecules migrate through a solution containing substrate. As yet the scanner has been used only rarely for such studies, but it seems likely that this type of application will become increasingly popular.

7. *Difference Sedimentation Velocity.* Although the scanner has considerable potential for difference sedimentation velocity experiments,[43,44] it has been used only rarely for this purpose.[27] This technique, which involves the direct measurement of the small difference in sedimentation coefficients between two similar solutions, is especially useful in detecting conformational changes in proteins which result from the binding of stereospecific ligands. The method is based on the direct subtraction of the concentration-distance curves produced by two samples contained in separate compartments of a double-sector ultracentrifuge cell. Thus far this technique has been used principally with the interference optical system, but it is likely that improvements in the scanner resulting from the incorporation of an on-line computer will lead to increased use of the scanner for this purpose. At present these experiments with the scan-

[38] R. Cohen, *C. R. Acad. Sci.* **256**, 3513 (1963).
[39] J. Rosenbloom, Ph.D. Dissertation, University of Pennsylvania (1965).
[40] R. Cohen, B. Giraud, and A. Messiah, *Biopolymers* **5**, 203 (1967).
[41] R. Cohen and M. Mire, *Eur. J. Biochem.* **23**, 267 (1971).
[42] R. Cohen and M. Mire, *Eur. J. Biochem.* **23**, 276 (1971).
[43] E. G. Richards and H. K. Schachman, *J. Amer. Chem. Soc.* **79**, 5324 (1957).
[44] M. W. Kirschner and H. K. Schachman, *Biochemistry* **10**, 1900 (1971).

ner are performed with some difficulty since the switching circuit does not function satisfactorily if both solutions contain large amounts of light-absorbing macromolecules.

B. Sedimentation Equilibrium Studies

One of the principal attributes of the photoelectric scanner is its versatility for nearly all types of sedimentation equilibrium experiments. Compared to the tedious labor required for plate reading in analyzing interference patterns, the effort required in measuring scanner traces is minimal. Whereas subsidiary information is required for identifying fringes in low-speed sedimentation equilibrium experiments,[45-47] no additional data are needed for evaluating the patterns produced by the scanner. The pen deflections of the scanner recorder can be used directly as an absolute measure of concentration. Since the wavelength of the incident light can be varied at will, both dilute and concentrated solutions can be studied readily. This feature has been exploited in investigations of hemoglobin where the very high absorbance in the Soret region of the spectrum (about 410 nm) permits studies at concentrations of only a few μg/ml.[11,48-50] Alternatively a wavelength can be selected for which the absorbance is low; under these conditions solutions of higher concentrations are readily studied. Accessibility to a broad concentration range is especially useful in studies of association–dissociation equilibria for proteins, such as hemoglobin.[48-50] For most proteins and nucleic acids, wavelengths of light can be found which permit experiments at such low concentrations that corrections for nonideality can be safely ignored. Only for the very large DNA molecules in density gradient sedimentation equilibrium experiments is the nonideality so great that even the scanner is not sufficiently sensitive to permit measurements corresponding to infinitely dilute solutions.[51] As in sedimentation velocity experiments, judicious selection of the wavelength of the incident light permits measurements of the distribution of macromolecules without complications stemming from the presence and redistribution of large amounts of a third component, such as urea or cesium chloride. While the advantages of the scanner for measurements on very dilute solutions are generally

[45] E. G. Richards and H. K. Schachman, *J. Phys. Chem.* **63**, 1578 (1959).
[46] E. G. Richards, D. C. Teller, and H. K. Schachman, *Biochemistry* **7**, 1054 (1968).
[47] D. C. Teller, this volume [14].
[48] S. J. Edelstein, M. J. Rehmar, J. S. Olson, and Q. H. Gibson, *J. Biol. Chem.* **245**, 4372 (1970).
[49] G. L. Kellett and H. K. Schachman, *J. Mol. Biol.* **59**, 387 (1971).
[50] G. L. Kellett, *J. Mol. Biol.* **59**, 401 (1971).
[51] C. W. Schmid and J. E. Hearst, *J. Mol. Biol.* **44**, 143 (1969).

emphasized, it should be noted also that the scanner is preferable to schlieren or interference optics for studies of materials which absorb appreciable amounts of visible light. Thus, for example, studies of sickle cell hemoglobin at concentrations about 150 mg/ml were conducted readily with scanner optics at 750 nm (and with cells having 2-mm optical paths), whereas the other optical systems did not produce useful patterns.[52] Since the scanner permits accurate concentration measurements for many types of macromolecules (not polysaccharides, for example), it is useful for both low-speed and high-speed sedimentation equilibrium experiments.[46,47,53–55] For density gradient sedimentation equilibrium experiments[56,57] the photoelectric scanner is the optical system of choice. Similarly for interacting systems involving dissimilar components, $(A + B \rightleftharpoons C)$, the scanner has unusual potential since it permits measurements of effective constituent molecular weights.[25]

1. Low-Speed Experiments. Figure 9 illustrates various scanner patterns obtained in a sedimentation equilibrium experiment on myoglobin at a concentration of 25 μg/ml. On the left is the pattern observed shortly after the rotor attained the desired speed. Such a pattern is useful for determining the initial concentration, c_0, in centimeters of recorder deflection. Appropriate calibration of the recording system with solutions of known absorbance permits a direct evaluation of the protein concentration; but it should be noted that this conversion of the pen deflection to protein concentration is not required for calculating molecular weights. In general, patterns such as the initial one should be obtained at several different amplifications of the recording system (trace amplitude) for possible use with the equilibrium patterns corresponding to different amplifications. The initial pattern shown in Fig. 9 was obtained at relatively low gain so that the deflections corresponding to the reference holes in the counterbalance (infinite optical density) could be measured if desired. Since the cell was only partially filled (about a 3-mm liquid column), most of the trace shows zero optical density corresponding to the air spaces above the two liquid columns in the double-sector cell. The second pattern recorded after 18 hours of centrifugation shows the redistribution of the protein at sedimentation equilibrium. Distances and

[52] S. J. Edelstein and R. H. Crepeau, unpublished.

[53] D. A. Yphantis, *Biochemistry* **3**, 297 (1964).

[54] D. E. Roark and D. A. Yphantis, *Ann. N.Y. Acad. Sci.* **164**, Art. 1, 245 (1969).

[55] D. C. Teller, T. A. Horbett, E. G. Richards, and H. K. Schachman, *Ann. N.Y. Acad. Sci.* **164**, Art. 1, 66 (1969).

[56] M. Meselson, F. W. Stahl, and J. Vinograd, *Proc. Nat. Acad. Sci. U.S.* **43**, 581 (1957).

[57] J. Vinograd and J. E. Hearst, *in* "Progress in the Chemistry of Organic Natural Products" (L. Zechmeister, ed.), Vol. 20, p. 372. Springer-Verlag, Vienna, 1962.

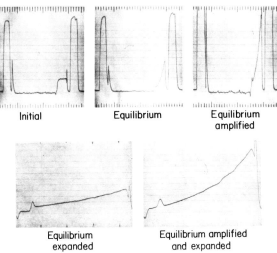

Fig. 9. Sedimentation equilibrium patterns of sperm whale myoglobin. The upper left pattern was recorded at zero time, while the remaining equilibrium patterns were obtained after 18 hours of centrifugation at a speed of 31,410 rpm. The upper center pattern was recorded at a low trace amplitude (400), selected to contain the deflections of the reference holes within the trace. The upper right pattern was recorded at the maximum trace amplitude (1000). The three upper patterns were recorded with a scan time of 30 seconds with the recorder paper moving at 5 mm/sec. The lower patterns were produced with a scan time of 6 minutes with the chart speed the same as above, although only the solution column was recorded. The left and right lower patterns were obtained at trace amplitudes of 400 and 1000, respectively. The traces were recorded with light of 405 nm. The solvent was 0.1 M phosphate buffer at pH 7 and the initial protein concentration was 0.025 mg/ml.

the magnification factor are readily evaluated from the images corresponding to the edges of the reference holes in the counterbalance. Increasing the gain in the amplifier gave the third pattern (labeled equilibrium amplified). Measurements of the recorder deflection as a function of distance were facilitated by obtaining an expanded trace (labeled equilibrium expanded). This expansion of the trace corresponding to the solution column was achieved by decreasing the scanning rate of the photomultiplier-slit assembly without altering the chart drive of the recorder. Only the relevant part of the trace is shown. A complete pattern of the entire cell would entail a 6-minute scan which permits the elimination of much high-frequency photomultiplier noise by suitable adjustment of the electronic filter. This smoothing of the trace can be achieved without sacrificing resolving power if the scanning rate is low. The actual traces of the 3-mm liquid column are about 20 cm in length, thereby permitting accurate measurements of recorder deflection as a function of position in the cell. The last trace shown in Fig. 9 represents the amplified, expanded pattern from which the actual measurements were made.

Patterns such as those in Fig. 9 from a low-speed sedimentation equilibrium experiment can be used directly for the calculation of the molecular weight.[2] Since the concentration is so low, corrections for nonideality are unnecessary and the molecular weight, M, can be calculated from

$$M = \frac{2RT}{(1 - \bar{v}\rho)\omega^2} \cdot \frac{d \ln c}{dr^2} \tag{2}$$

or

$$M = \frac{2RT}{(1 - \bar{v}\rho)\omega^2(r_b{}^2 - r_m{}^2)} \cdot \frac{c_b - c_m}{c_0} \tag{3}$$

In Eqs. (2) and (3), R is the gas constant, T is the absolute temperature, $\bar{v}$ is the partial specific volume, ρ is the density of the solution, ω is the angular velocity in radians/second, c is the concentration with c_b and c_m representing the concentrations at the cell bottoms and meniscus, respectively, and r is the radial distance from the axis of rotation with r_b and r_m corresponding to the position of the bottom of the liquid column and the meniscus, respectively.

2. High-Speed Experiments. In studies of heterogeneity of macromolecules high-speed sedimentation equilibrium experiments have proved to be very powerful since they provide data for the calculation of different molecular weights, such as M_n, M_w, and M_z, corresponding, respectively, to the number-average, weight-average and z-average molecular weights.[47] For these high-speed experiments the scanner is also useful, as shown in Fig. 10, which presents data from an experiment on bovine plasma albumin. The agreement between the measurements with the scanner and the interference patterns is excellent; it should be noted, however, that the accuracy in measuring the concentration distribution from the interference fringes is substantially greater than that obtained with the scanner. The scanner has the advantage, as mentioned earlier, of being more sensitive. Thus at concentrations which are very close to zero in terms of fringe displacements, meaningful data are obtained readily with the scanner. This feature is illustrated in Fig. 11, which shows how the accessible concentration range can be expanded merely by changing the wavelength during the experiment. If light of 280 nm only were available, the measurements of recorder deflection would have indicated that the concentration was virtually zero at values of r^2 less

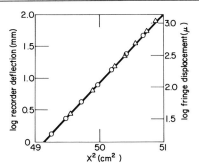

Fig. 10. Molecular weight determination of bovine plasma albumin from sedimentation equilibrium experiments. Both Rayleigh interference (△) patterns and traces from the photoelectric scanner (○) were used for the measurements illustrated in the figure. The ordinate gives the logarithm of the recorder deflection (mm) on the left and the logarithm of the fringe displacement (μ) on the right. The abscissa shows the square of the distance (cm²) from the axis of rotation. Sedimentation was for 22 hours at 24,630 rpm. The solvent was 0.1 M NaCl and 0.01 M acetate at pH 5.4. The initial protein concentration was 0.6 mg/ml.

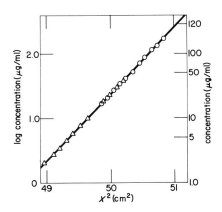

Fig. 11. Molecular weight determination of extremely dilute solutions by sedimentation equilibrium experiments. On the left ordinate is the logarithm of the concentration in micrograms per milliliter, and on the right are the actual concentrations in micrograms per milliliter. These are plotted against the square of the distance (square centimeters) from the axis of rotation. The data were obtained from traces recorded 18 hours after reaching the speed of 24,630 rpm. Two 7° cells were used in a single rotor with one cell containing the solvent and the other the solution. With this combination a long slit (10 mm) was used in the photomultiplier slit assembly. The switching circuit was activated by light pulses passing through the reference hole in the side of the rotor. Traces were produced with light of 280 nm (○) and also with light of 230 nm (△). In both cases the slit width on the monochromator was 2 mm. Recorder deflections were converted to absolute concentrations with data obtained from independent calibration measurements.

than 50 cm². However, a change in wavelength to 230 nm increased the sensitivity so much that appreciable recorder deflections were observed. Hence accurate measurements could be made in the upper region of the liquid column corresponding to concentrations about 0.1 that are measurable with light of longer wavelength.

3. *Interacting Systems.* a. SELF ASSOCIATIONS. The scanner has been used extensively for sedimentation equilibrium studies of self-associating systems such as hemoglobin.[11,48-50] For most of this work, the solutions were so dilute that complications stemming from nonideality were negligible. Various treatments are available for the evaluation of the relevant equilibrium and a detailed comparison of various approaches in terms of their advantages and disadvantages is given elsewhere.[47] Some of these treatments require very accurate experimental data such as those obtained with interference optics; hence their potential cannot as yet be fully exploited for use with data produced by the scanner. The latter has been particularly advantageous in a study of the self-association of the B protein of *Escherichia coli* tryptophan synthetase.[58] Most of the data for the protein could be fit by a monomer–dimer equilibrium along with the presence of higher oligomers in the solution. When the wavelength of the incident light was adjusted to 415 nm, corresponding to the absorption maximum of the cofactor, pyridoxal phosphate, the solution appeared monodisperse with a molecular weight corresponding to the dimer.[58] This result, indicating that the monomers do not bind the cofactor, could not have been deduced from the interference patterns and the application of the scanner and monochromator for this type of system illustrates the power of the absorption optical system.

b. MIXED ASSOCIATIONS. Mixed associations involving a protein and small ions have also been studied by sedimentation equilibrium.[25] For these experiments it has been useful to consider the constituent effective molecular weight, $\bar{M}_{e,A}$, of the small molecule, A, which can be written as

$$\bar{M}_{e,A} = \frac{1}{[A_0]} \left\{ [A]M_{e,A} + \sum_{i=0}^{n} i[PA_i]M_{e,PA_i} \right\} \tag{4}$$

In Eq. 4, $[A_0]$ is the total molar concentration of component, A, which interacts with the protein, P, to form a series of complexes, PA, PA_2, . . . , PA_i, and PA_n. The effective molecular weights of the various species, A and PA_i, are written as $M(1 - \bar{v}\rho)$ with the appropriate subscripts. When the interaction involves a protein and a *small* molecule (for which $M_{e,A}$ is negligibly small compared to $M_{e,P}$), the treatment[25] of multiple

[58] G. M. Hathaway and I. P. Crawford, *Biochemistry* 9, 1801 (1970).

equilibrium[59] coupled with the equations describing sedimentation equilibria for all species[60-63] gives

$$\bar{M}_{e,\mathrm{A}} = M_{e,\mathrm{P}} \frac{r[\mathrm{P}_0]}{[A_0]} \tag{5}$$

In Eq. (5), r is the number of moles of bound A per mole of total protein whose molar concentration is $[\mathrm{P}_0]$.

The measurement of $\bar{M}_{e,\mathrm{A}}$ requires an optical system which provides data for the concentration distribution of the A constituent regardless of the form in which it exists. Some of the A constituent would exist as free A, some would be in complexes with protein of the type PA, some would redistribute as PA_2, some as PA_i, and some as PA_n. With the scanner and the system bovine plasma albumin-methyl orange, it is possible to find an isosbestic point which permits direct measurement of the concentration of the dye constituent throughout the cell. This distribution which, in effect, provides a measurement of $\bar{M}_{e,\mathrm{A}}$, reveals directly the extent of binding of the small dye molecules to the much larger protein molecules. If the dye is largely unbound, the concentration of the A constituent would hardly vary throughout the cell as long as the speed is appropriate for measuring the redistribution of the protein. Alternatively, the A constituent would have the same concentration distribution as the protein if the binding was very large. Binding to an extent between these two extremes would, of course, produce an intermediate concentration distribution, as shown in Fig. 12. Since the concentration of methyl orange on a weight–volume basis is very low, it contributes only negligibly to the refractive index of the solution; hence the interference pattern provides a measure of the distribution of protein which in turn is used for calculating the molecular weight of the protein, $M_{e,\mathrm{P}}$, needed in Eq. 5. The scanner trace gives the data required for evaluating $\bar{M}_{e,\mathrm{A}}$ which is then used in Eq. 5 to evaluate the extent of binding in terms of the parameter r. This type of experiment, involving sedimentation equilibrium of a variety of species as well as chemical equilibrium at all levels in the cell, was shown to be fully the equivalent of a multicompartment equilibrium dialysis experiment[25] in which a series of dialysis bags were suspended in the same vessel containing the dialyzable

[59] I. M. Klotz, in "The Proteins" (H. Neurath and K. Bailey, eds.), Vol. I, Part B, p. 727. Academic Press, New York, 1953.

[60] E. T. Adams, Jr. and H. Fujita, in "Ultracentrifugal Analysis in Theory and Experiment" (J. W. Williams, ed.), p. 119. Academic Press, New York, 1963.

[61] E. T. Adams, Jr., Biochemistry 4, 1646 (1965).

[62] E. T. Adams, Jr. and J. W. Williams, J. Amer. Chem. Soc. 86, 3454 (1964).

[63] L. W. Nichol and A. G. Ogston, J. Phys. Chem. 69, 4365 (1965).

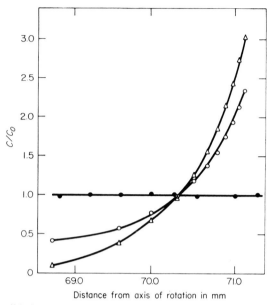

FIG. 12. Equilibrium distribution of the different constituents in a sedimentation equilibrium experiment on a mixture of bovine plasma albumin and methyl orange. The solution contained initially bovine plasma albumin at 0.4 g/100 ml and methyl orange at $7.8 \times 10^{-5} M$ in $0.1 M$ phosphate at pH 5.66. The rotor speed was 12,590 rpm and the distributions shown were obtained after 18 hr of sedimentation at 4°. The wavelength used with a monochromator and the photoelectric scanner was 440 nm. A 3-mm double-sector cell was employed and 0.025 ml of solution was placed in one compartment with solvent in the other. The ordinate represents the concentration, c, relative to the initial concentration, c_0, plotted as a function of distance from the axis of rotation in millimeters. The redistribution of the protein as measured with the Rayleigh interferometer is represented by $\triangle$, and the redistribution of the methyl orange constituent as measured with the photoelectric scanner is signified by $\bigcirc$. Both of these were obtained in the same experiment. The control, indicated by $\bullet$, was obtained in an analogous experiment under identical conditions with a solution containing methyl orange devoid of protein.

component. Each bag contained a solution of different protein concentration, and the concentration of unbound dye was the same in each bag. Similarly at each radial level in the ultracentrifugal cell there was a different protein concentration and a constant concentration of unbound dye. Because of the interaction of the dye with the protein the total dye concentration increased with distance from the axis of rotation since a large amount of dye is bound at the higher protein concentration. Cross-plotting the data of Fig. 12 to give $[A_0]$ versus $[P_0]$ at each radial level yielded a straight line of slope, r, and an intercept (at zero protein concentration) representing the concentration, [A], of the free A component.[25]

This study using the photoelectric scanner for a sedimentation equilibrium experiment on a model system illustrated the potential of the method. As long as wavelengths can be found which permit the measurement of constituent quantities (such as sedimentation coefficient or molecular weight) the amount of information deduced from an experiment on an interacting system will far surpass that gleaned from the use of a nondiscriminating optical system. Extension of this method[25,64] to the study of interactions between dissimilar macromolecules should increase its utility.

4. *Density Gradient Sedimentation Equilibrium.* One of the widest applications of the scanner is in the analysis of the position and shape of the concentration distributions produced in density gradient sedimentation equilibrium experiments involving macromolecules in a multicomponent solvent of density about that of the macromolecules.[56,57] Such studies, especially on nucleic acids, viruses, and ribosomes, are extremely valuable in providing quantitative information as to the buoyant density, homogeneity, molecular weight, and preferential interactions of the macromolecules.[57] Due to the redistribution of the low molecular weight solute a density gradient is produced with the result that (assuming the correct range of densities is achieved) macromolecules at the top of the cell will migrate in a centrifugal direction and those at the bottom will migrate in a centripetal direction. This process leads to a band of macromolecules located at a position related to their density and preferential interaction with solvent. The band has a width which is related inversely to the molecular weight of the macromolecules[57,65-71] and directly to the heterogeneity in their densities.[72,73] Figure 13 shows scanner traces of density gradient sedimentation equilibrium patterns obtained in a study of mitochondrial DNA from human leukemic leukocytes.[74] Each of the solutions contained a reference macromolecule, crab d(A, T), which served as a known density marker for the determination of the density of the

[64] E. T. Adams, Jr., *Ann. N.Y. Acad. Sci.* **164**, Art. 1, 226 (1969).

[65] J. E. Hearst and J. Vinograd, *Proc. Nat. Acad. Sci. U.S.* **47**, 999 (1961).

[66] J. E. Hearst and J. Vinograd, *Proc. Nat. Acad. Sci. U.S.* **47**, 825 (1961).

[67] J. E. Hearst and J. Vinograd, *Proc. Nat. Acad. Sci. U.S.* **47**, 1005 (1961).

[68] J. Vinograd, Vol. 6, p. 854 (1963).

[69] G. Cohen and H. Eisenberg, *Biopolymers* **6**, 1077 (1968).

[70] E. F. Casassa and H. Eisenberg, *Advan. Protein Chem.* **19**, 287 (1964).

[71] H. Eisenberg, *Ann. N.Y. Acad. Sci.* **164**, Art. 1, 25 (1969).

[72] R. L. Baldwin, *Proc. Nat. Acad. Sci. U.S.* **45**, 939 (1959).

[73] N. Sueoka, *Proc. Nat. Acad. Sci. U.S.* **45**, 1480 (1959).

[74] D. A. Clayton, R. W. Davis, and J. Vinograd, *J. Mol. Biol.* **47**, 137 (1970).

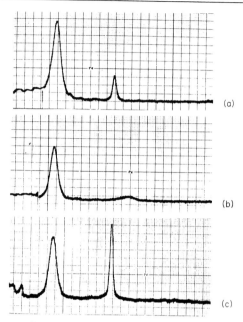

FIG. 13. Density gradient sedimentation equilibrium patterns of mitochondrial DNA from human leukemic leukocytes. The patterns were obtained after 24 hours of sedimentation at 44,770 rpm. In each experiment the band corresponding to the low-density material corresponded to a reference material, crab d(A, T), which had a density of 1.669₆ g/ml. (a) Annealed mixture of approximately equal amounts of heavy dimer strand and light monomer strand plus crab d(A, T). (b) Self-annealed light dimer strand plus crab d(A, T). (c) Annealed mixture of denatured monomer and dimer DNA plus crab d(A, T). The mixture contained 85% dimer and 15% monomer. From Clayton, Davis, and Vinograd, *J. Mol. Biol.* **47**, 137 (1970).

other species. From knowledge of the density gradient in the CsCl solution and the positions of the bands the buoyant densities of the various species produced in the annealing species are readily determined.[74] Moreover, the widths of the bands provide a direct indication of the molecular weights of the different species. An evaluation of the anhydrous molecular weight of the DNA and its partial specific volume requires corrections for preferential interactions[57,65,71] and for nonideality due to the finite concentration of the macromolecules.[51] This technique, which has been of great value in the study of high molecular weight DNA, has been used also for proteins[75,76] even though the bands are much wider because of the lower molecular weights.

[75] J. B. Ifft and J. Vinograd, *J. Phys. Chem.* **70**, 2814 (1966).
[76] J. B. Ifft, "A Laboratory Manual of Analytical Methods of Protein Chemistry" (P. Alexander and H. F. Lundgren, eds.). Pergamon Press, London, 1969.

V. Experimental

The scanner provides a wider range of experimental options than any of the other optical systems. Not only can various wavelengths be selected, but numerous other important adjustments, such as monochromator slit width, photomultiplier slit width, electronic settings and scanning rate, must be made. Most decisions in the choice of operating parameters are based on the purpose of the experiment and the nature of the material. Thus, for example, slow scanning rates, although appropriate for sedimentation equilibrium experiments, would seriously jeopardize the potential resolution of moving boundaries and bands in sedimentation velocity studies. Selection of the scanning rate limits, to some extent, the choice of settings for the electronic filter since excess electric filtering aimed at decreasing "noise" causes a serious deterioration in resolution if the scanning rate is high (as required for a sedimentation velocity experiment). Electric filtering is dependent also upon the nature of the material and the wavelength of the incident light used for the study. If the macromolecules contain chromophores which absorb light in the visible region of the spectrum, such as hemoglobin or proteins which bind pyridoxal phosphate,[58] the light levels employed will be very high and the monochromator slit width should be very small; in addition with visible light the signal-to-noise ratio is sufficiently high that the photomultiplier voltage and electrical filtering can be minimal. For studies in the far ultraviolet region of the spectrum where light levels are small (below 235 nm) the monochromator slit must be wide open and gain on the photomultiplier must be very high with attendant "noise"; under these circumstances increased electrical filtering is necessary. Frequently, compromises are required in order to obtain optimum conditions. In this section the individual components of the photoelectric scanner are considered in terms of various experimental demands. Some discussion of these components and their alignment is available in the manuals supplied by the manufacturers and hence our consideration will deal principally with features which in our experience have proved of special significance.

A. Monochromator

1. Alignment of Light Source. For most experimental work the amount of available light is crucial for attaining optimal accuracy. Hence maximizing light levels is essential. The adjustment of the high intensity light source and the mirrors in the light source housing of the monochromator is performed both visually and with the use of the photomultiplier in conjunction with an oscilloscope to monitor the light pulses.

Not only must the greatest effective intensity of light be obtained, but the illumination across the entire ultracentrifuge must be nearly uniform. Preliminary adjustment of the light source and the mirrors which direct the light into the monochromator can be achieved readily by using visible light and a very narrow slit width (0.05 mm) on the monochromator which is moved forward from under the ultracentrifuge. For ultraviolet light this adjustment with a 2-mm slit opening can be performed by using canary glass over the exit mirror which directs the light vertically. Slight modifications of the knob settings (at the front of the light source housing) for the vertical and left-to-right positions of the light source should be made until the rectangular image on the canary glass is uniformly bright.

The exit mirror of the monochromator housing should be adjusted so as to direct the light beam vertically. This setting, both front-to-back and left-to-right, is performed by examining the rectangular image (with visible light) on the ceiling or on a large board cantilevered from the front of the ultracentrifuge. A plumb line is indispensable for this adjustment. After this is performed, the monochromator is returned to its normal operating procedure and the light level, both in terms of intensity and uniformity of illumination, is checked by examining the pulses produced on the oscilloscope with an empty cell in the spinning rotor. Usually light levels can be enhanced through "fine-tuning" of the various adjustments. The light levels should be checked throughout the length of the cell and the inner and outer reference holes in the counterbalance. Recorder traces for various settings are generally useful in this final positioning. It is important at this stage to bypass the logarithmic circuits so that intensity of transmitted light is recorded. In this way enhanced sensitivity in the positioning of the light source and mirrors is achieved.[9,15] Also it is important to disconnect the feedback circuit which regulates the photomultiplier voltage; otherwise this operation cannot be performed in this fashion.[11] An indication of the light level is provided even with the feedback circuit operating if one monitors the meter indicating the volts per stage; when that reading rises appreciably, the light source should be adjusted.

Even though this feedback regulation of the photomultiplier voltage in response to its output (which in turn is proportional to the light level) corrects for the effect of nonuniform illumination, it is unwise to use it to compensate for a misaligned light source. This servo mechanism is especially useful for high-intensity light sources after several hundred hours of operation since they exhibit wide fluctuations in light intensity. On some occasions the intensity during a scan falls so low that the switching signal derived from the first light pulse is inadequate. For such circumstances, as well as to correct for slight nonuniformity of light across the cell, this feedback circuit is indispensable. It is valuable also for experiments with solvents that absorb some of the incident light. The operation of the feedback circuit ensures that the output from the photomultiplier is constant even though the transmittance of the solvent is significantly less than the air space above the liquid column. The feedback regulation also corrects for nonuniform illumination due to scattering from oil and dust deposits on the collimating and condensing lenses.

2. *Selection of Slit Width.* Although the intensity of light is high over a broad range of wavelengths, it should be recognized that there are peaks of intensity at a variety of wavelengths including 248 nm, 265 nm, 280 nm, 292 nm, 330 nm, 365 nm, 405 nm, 435 nm, and 540 nm. The light source operates at relatively high pressure and thus there is broadening of the emission spectra. Since the intensity of the different bands varies greatly and the dispersion of the monochromator is a function of wavelength, the selection of slit width should be made judiciously. In the ultraviolet region of the spectrum where there is relatively less light, a slit width of 2 mm is necessary; otherwise the light level is so low that the photomultiplier must be operated at high gain with attendant "noise." Studies with visible light, in contrast, should be performed with much narrower slit widths on the monochromator. For light of about 405 nm, a slit width of only 0.3 mm is sufficient to give good light levels; moreover, at such slit widths for this region of the spectrum the dispersion of the instrument is such as to produce reasonably monochromatic light. If much larger slit widths are used in experiments on hemoglobin, nonlinear responses to concentration are observed because of deviations from Beer's law due to the nonmonochromatic light. Hence it is advisable to examine the intensity of light in the region of the desired wavelength so as to find a peak in the intensity and then adjust the slit width to narrow the wavelength band of the incident light. In practice the slit width should be reduced while the ultracentrifuge is operating. In this way the pulses and the recorder deflection can be measured until the maximum optical density is indicated. If the resulting light level is so low as to require an abnormally large photomultiplier voltage a compromise must be made. Experience has shown that a carefully calibrated scanner gives optical densities in good agreement with those measured with a bench-top spectrophotometer.

Some light sources have been found to produce greater light intensities in the ultraviolet (below 265 nm) when they are cooled by the fan incorporated in the light source housing. The adjustments of the fan and vent for controlling the air flow are readily performed while the ultracentrifuge is in operation by examining the change in the light pulses on the oscilloscope. Extra effort in maximizing the light intensity is warranted since the photomultiplier voltage can be adjusted accordingly to give an optimum signal-to-noise ratio.

B. Alignment of Optical System

Compared to the schlieren and interference optical systems the absorption system with a scanner is much easier to align. The monochromator has to be positioned both front-to-back and left-to-right. Apertures must be located properly on both the collimating and condensing lenses. The 45° mirror at the top of the optical track must be oriented correctly. The camera lens must be positioned vertically and horizontally on the optical axis and various positions along the track must be determined for proper focusing at different wavelengths. Finally the photomultiplier-slit assembly must be positioned vertically and the travel of the photomultiplier must be in a horizontal plane. These different adjustments are to some extent interdependent; some can be made visually and others are performed best with the ultracentrifuge and scanner in operation so that the effect of changes in the orientation of the components can be monitored.

1. Aperture Masks for Collimating and Condensing Lenses. Since the photomultiplier scanning assembly is mounted to provide movement in a horizontal plane the basic reference for the alignment procedure is dictated by that orientation. Accordingly, the light-limiting apertures in the vacuum chamber must be aligned relative to that reference. It has proved to be convenient to mount a viewing screen in the plane of the photomultiplier slit. For the instrument developed in Berkeley,[9,11] a ground glass with two perpendicular ruled lines crossing at its center was mounted in a metal tube having the same dimensions as the photomultiplier. This viewing screen was indispensable for adjustment of the 45° mirror and for the placement of the light-limiting apertures on the collimating and condensing lenses. Adjustment of the 45° mirror was performed through the use of a special radial tool that was coupled to the drive shaft of the ultracentrifuge.[57] This tool had a radial arm with a perpendicular cross-bar fastened at a distance of 6.50 cm from the center of the coupling (the screw coupling was taken from the aperture mask aligning tool supplied by the manufacturer for the alignment of the schlieren and interference optical systems). Stability and centering of this radial arm were achieved by fastening a symmetrical brass weight (about 3.5 kg) to the coupling screw. A hole was drilled in the center

---

[57] E. G. Richards, D. C. Teller, V. D. Hoagland, Jr., R. H. Haschemeyer, and H. K. Schachman, *Anal. Biochem.* **41,** 215 (1971).

of the intersection of the cross-bar and the radial arm in order to provide an image at a distance 6.50 cm from the axis of rotation.

The radial arm, coupled to the drive shaft and with the vacuum chamber partially opened, was rotated until the images of the arm and cross-bar were parallel, respectively, to the horizontal and vertical lines on the viewing screen. With the camera lens removed from the optical track the 45° mirror was adjusted horizontally and vertically so that the hole in the radial tool was imaged in the center of the viewing screen. The camera lens was then returned to the track and appropriate adjustment of the mount (vertically and horizontally) was made until the image was again centered on the ground glass screen. It is advisable to slide the camera lens along the optical track as a test for the proper orientation of the track. If the image moves appreciably the track must be adjusted accordingly. If different camera lenses are to be used for different regions of the spectrum, their mounts should be adjusted at this time.

In order to reduce problems arising from stray light, we have incorporated adjustable, rectangular aperture masks which are mounted on the collimating and condensing lenses. These masks could be oriented rotationally so that their edges were parallel to the radial arm. In addition each knife edge was adjustable laterally. The alignment of the lower mask was performed by examining the light on either side of the image of the radial arm. Appropriate scales were mounted on the masks so that settings could be recorded corresponding to the proper positions. The upper mask was oriented relative to the lower mask by examining the light pattern impinging on the upper mask after the lower one had been positioned correctly (for this adjustment the radial tool was removed). Experience has indicated that adjustable masks are preferable to fixed masks which are centered about the lens mounts since the proper reference dictated by the plane of movement of the photomultiplier may preclude the use of the center of the lenses (a detailed discussion of this problem in terms of the schlieren and interference optical systems is given elsewhere[77]). These masks play a crucial role in minimizing stray light by reducing reflections and scattering from the lenses. Without these limiting apertures in the vacuum chamber the image of the first compartment of the double-sector cell was "seen" by the photomultiplier even after it had passed the optical axis of the system and the second compartment was centered in the optical path. At the end of the optical alignment the rectangular openings in the masks were narrowed progressively until the pulses observed on the oscilloscope were affected noticeably. Then each knife edge was moved slightly to increase the opening. This procedure has been found to increase the linearity of the entire optical system, presumably by reducing the stray light that affects adversely studies on solutions of high optical density.

2. *Position of Monochromator.* The mount for the monochromator has slides and calibrated knobs which permit ready and reproducible adjustment both front-to-back and left-to-right.[9] However, there is no simple adjustment available for varying the distance from the exit slit of the monochromator to the collimating lens. That distance is about 81 cm, corresponding to the focal length of the collimating lens for only one wavelength, 265 nm. Hence departures from that wavelength cause the light through the vacuum chamber to be nonparallel. This nonparallelism is slight, however, and not much greater than that which results from the off-axis rays coming through the relatively wide (2 mm) exit slit of the monochromator when ultraviolet light is used. If solutions of low concentration are being studied and there are only small refractive index gradients in the cell, the deviation of this nonparallel light probably causes little difficulty in terms of the present resolution of the absorption optical system. When more intense light sources become available so that very narrow slits can be used, it would be advisable to separate the source-slit combination from the mirror which directs the light vertically through the chamber. In this way the source could be moved to the correct focal position for each wavelength. The left-to-right adjustment of the monochromator should be adjusted so that the light beam is centered about the collimating lens (viewed from beneath). A similar procedure is followed for preliminary adjustment of the monochromator in a front-to-back direction. Final positioning of the monochromator is achieved by examining the distribution of light around the meniscus while the ultracentrifuge is in operation with H_2O or a nonabsorbing buffer in the cell.[9,78] If the camera lens is not focused on the mid-plane of the cell, uniform illumination on either side of the air–liquid meniscus will be observed only if the light source is on the optical axis. With the monochromator incorrectly positioned, there is a bright band of light (due to reflections from the air–liquid interface) on one side or the other. The adjustment involves moving the camera lens several centimeters from its correct focal position and then scanning the region of the cell near the meniscus for a series of positions of the monochromator. This test is sensitive, and the proper front-to-back location can be determined rapidly.[9] Typical patterns from the corollary experiment for determining the correct focal position for the camera lens are illustrated in the next section. When the correct position is obtained, there should be a symmetrical distribution of light around the meniscus;

[78] R. Trautman, *Biochim. Biophys. Acta* **28,** 417 (1958).

with slow scanning rates the fine structure of the meniscus is observed as a pair of vertical pen deflections representing abrupt changes in optical density. Increased sensitivity is achieved if the scanner records intensity of light rather than its logarithm.

3. *Focal Position of Camera Lens.* Focusing of the lenses for different wavelengths is performed by the same method as that used for positioning the monochromator. After the correct position of the latter is determined, the monochromator is moved toward the front of the ultracentrifuge about 2 cm and scans are obtained for different lens positions. That location corresponding to uniform illumination on both sides of the meniscus indicates the correct position of the camera lens in imaging a plane halfway through the cell onto the plane of the photomultiplier slit. For this test to work satisfactorily it is necessary that the solution in the cell be transparent to the incident light and that refractive index gradients be small so as to reduce the deviation of light. Accordingly, the focusing procedure should be performed with a transparent solvent (or H_2O) and at relatively low rotor speeds (about 20,000 rpm). As the camera lens is moved along the track, the band of reflected light will be seen to move from one side of the meniscus to the other. Representative patterns are presented in Fig. 14. At the correct focal position there will be no reflection and the meniscus (when scanned at a slow rate) will appear as two abrupt, closely separated deflections of approximately equal magnitude representing decreased light intensity. A lens can be focused to about ± 0.5 mm in a few minutes with the use of the scanner. It is advisable, therefore, to focus a lens for a variety of wavelengths so that the lens can be positioned properly whenever the occasion arises for studies at a particular wavelength. Moving the camera lens to different positions affects markedly the size of the image at the photomultiplier; hence for each focal position a magnification factor must be determined in order to relate distances on the recorder traces to those in the ultracentrifuge cell. This factor is evaluated from the traces by measuring the distance between the reference holes in the counterbalance cell and comparing it to the actual distance in the cell measured on a microcomparator.

Since ultracentrifuge studies frequently require light varying from 215 nm to 580 nm, it is necessary to have lenses capable of encompassing this broad range of the spectrum. This is achieved through the use of three different lenses. One is used for the far ultraviolet, one for the near ultraviolet, and one for the visible. For convenience each should have its own lens mount. In this way a change can be made from one to another in less than a minute. Recently mirror optics (Beckman Instruments, Inc.) were developed which function for all wavelengths.

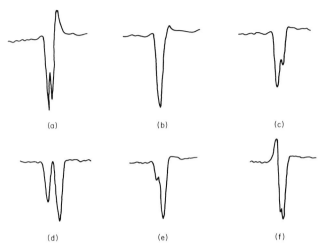

(a) (b) (c)

(d) (e) (f)

FIG. 14. Focusing of the camera lens. This set of experiments was performed with a water-filled cell spinning at 20,000 rpm and with light of wavelength 280 nm. The monochromator was moved off the optical axis about 2 cm toward the front of the ultracentrifuge. The traces represent intensity of light (increasing upward) versus distance from the axis of rotation. Only a small part of the image in the vicinity of the meniscus is shown for each camera lens position. The air space above the liquid column is on the left. These patterns were obtained at a very low scanning rate, corresponding to a 6-minute scan for a trace of the whole image. The photomultiplier slit width was 0.13 mm. Camera lens positions were: (a) 4″; (b) 4¼″; (c) 4½″; (d) 4⁹⁄₁₆″; (e) 4¾″; (f) 5″. The correct focal position is between 4½″ and 4⁹⁄₁₆″.

This unit, consisting of a special holder for a front-surfaced spherical mirror and a similarly coated flat mirror, replaces the three camera lenses ordinarily needed to encompass the entire accessible spectrum. Since the light is reflected from the front surface of the mirrors the system, after being focused for one wavelength, works effectively for all wavelengths. The entire optical system is not completely independent of wavelength, however, since the condensing lens is part of the cell-focusing camera. The wavelength dependence introduced by the condensing lens is very small and has been neglected, especially since the alignment of the mirrors is not, as yet, sufficiently precise to permit a detailed evaluation of the effect of changes in wavelength. In one important respect this combination of mirrors is far superior to the camera lens. Reflecting optics with mirrors produce images of constant size regardless of the wavelength. Thus the magnification needs to be determined for only one wavelength whereas the use of transmittance optics with lenses requires different magnification factors for each wavelength. Also, the focal length of the spherical mirror has been selected to give as large an image as is consistent with the available movement of the photomultiplier. In contrast, with lenses the correct focal position for a specific wavelength may be located far along the optical track toward the photomultiplier; under such circumstances the image at the photomultiplier is small and accuracy is thereby reduced.

4. *Photomultiplier Slit.* Some enhancement of the signal-to-noise ratio can be achieved by increasing the width of the slit in front of the photomultiplier. A continuous range of values varying from zero to slightly over 0.2 mm is available in the commercial instrument. Adjustment is controlled by turning a knob on the face of the ultracentrifuge. The instrument developed in Berkeley[11] has a rotary turret containing fixed slits varying from 0.037 mm to 0.15 mm. Increasing the slit width, though producing a larger signal for a fixed light intensity, leads to a decrease in resolving power. The effect of a change in the width of the photomultiplier slit is seen readily in the region of the trace representing the meniscus. Representative patterns are shown in Fig. 15. With a slit width of 0.037 mm, the recorder trace shows a large, sharp deflection corresponding to the abrupt change in transmitted light at that radial level. In contrast, an analogous trace with a slit about 0.15 mm wide will hardly show a deflection as the photomultiplier moves across the region of the image containing the meniscus. For sedimentation equilibrium experiments in which the concentration gradients are not too large, this loss of resolving power is not serious. However, increasing the width of the photomultiplier slit can lead to serious distortions both in sedimentation velocity experiments with sharp boundaries and in density gradient sedimentation equilibrium experiments with narrow bands. In general the slit width should be as small as is possible for the prevailing light level. Special precautions should be taken to check the length of the photomultiplier slit. If it is longer than the image of the width of the center rib of the double-sector cell, the photomultiplier will "see" parts of both sectors simultaneously and there will be no dark period between the sectors as the cell traverses the optical path. Careful monitoring of the pulses on the oscilloscope reveals this defect, which is readily remedied by reducing the slit length. Having too short a slit is wasteful since the resulting output from the photomultiplier will be below the optimal value.

For studies of extremely dilute protein solutions with ultraviolet light of wavelength about 218 nm (Fig. 3) it has proved to be necessary to increase the sensitivity of the recording system by using longer slits at the photomultiplier.[11] This was accomplished readily by having several 10-mm long slits of different widths in the rotatory turret. These lengths

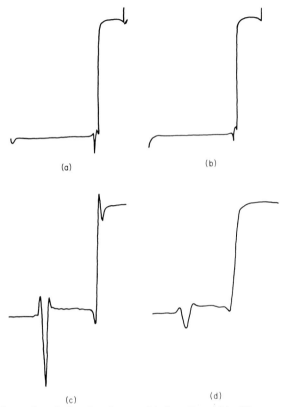

(a)   (b)

(c)   (d)

Fig. 15. Effect of variation in photomultiplier slit width. The patterns of optical density versus distance were obtained on a solution of pyridoxal phosphate in one compartment of the cell and buffer in the other. The speed was 60,000 rpm and the incident light had a wavelength of 410 nm. The patterns at the top were obtained in a 30-second scan so that the whole cell is shown on the trace, whereas those at the bottom were highly expanded by changing the scanning rate to a 6-minute scan while maintaining the same chart rate. Only the regions corresponding to the menisci are shown in the bottom traces. For the patterns on the left the slit width was 0.035 mm while those on the right were obtained with a photomultiplier slit width of 0.13 mm. The increase in resolution with the small slit width is evident from examination of the traces for the solvent and solution menisci.

correspond to the diameter of the photocathode. With such slits (assuming they were fully illuminated) and the same prevailing light the sensitivity is enhanced 5-fold. In order to exploit this potential increase in sensitivity we used special single-sector cells having a large (7°) sector angle. They were made from standard double-sector cells by removing the dividing rib (on a milling machine) of the centerpieces and the window-holders. In addition the apertures in the vacuum chamber were

opened considerably to permit light to pass through the entire cell opening when it was centered on the optical axis. With this arrangement the
solvent and the solution were placed in separate cells located 180° from
each other. Identification of the cells in this symmetrical arrangement
was achieved with a multiplexer triggered by an auxiliary light pulse.[11]
Even though this arrangement involved two different sets of cell windows
(rather than different regions of the same pair of windows) the base
lines with solvent in each cell were satisfactory. Apparently the quartz
windows do not vary appreciably in their transmittance of light. When
a long slit at the photomultiplier is used along with a cell having a large
sector angle the resolution is less than that obtained with the shorter
slit and a double-sector cell. This loss of resolution is attributable to
the increased "effective" width of a long slit (as compared to a short
slit of the same width) in scanning an image having cylindrical (and
not planar) surfaces of equal optical density. Replacement of the
straight, long slit by a curved slit (with a radius corresponding to the
distance from the axis of rotation to the center of the cell) leads to an
increase in resolution. This experiment illustrates one of the potential
pitfalls of the scanning system. Both the width and the length of the
photomultiplier slit must be considered carefully in terms of the experimental arrangement and the magnitude of the concentration gradients
within the ultracentrifuge cell.

C. Electronics

Assessing the performance of the recording system in response to
variations in optical density requires tests of many individual components. The switching circuit must work flawlessly without introducing
spurious voltages into either of the holding circuits (for solution or solvent pulses). Each of the amplifiers must be linear. The logarithmic
amplifier which converts the output from the photomultiplier into its
logarithmic value must operate with fidelity over a broad range. Nulling
the electronics to indicate zero optical density must be performed while
the pulses are observed on an oscilloscope so that the electrical zero
corresponds faithfully to zero optical density. Electrical filtering to
eliminate high frequency noise must not introduce serious distortions in
the curves of optical density versus distance. These individual parts of
the complete system should be checked periodically both separately and
as an integrated unit. In addition the entire scanner including optics and
electronics should be tested frequently with known samples.

1. Oscilloscope. In our experience an oscilloscope is indispensable. Individual pulses representing light intensity can be observed directly, as
shown for two double-sector cells in Fig. 16A, and the signals correspond-

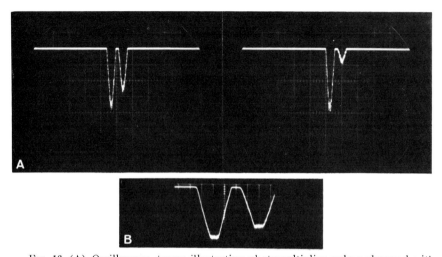

FIG. 16. (A) Oscilloscope traces illustrating photomultiplier pulses observed with two double-sector cells in a single experiment. For each cell the reference compartment contained distilled water and the sample compartments were filled with tryptophan solutions having optical densities of 0.15 and 0.62, respectively. Intensity of light increases in a downward direction. The two double-sector cells were placed at 180° from each other and a multiplexer based on light striking a stationary photomultiplier mounted on the schlieren optical system was used to identify the cells for the electronic circuitry. The logarithm of the ratio of the heights of the pulses (in each pair) yields values for the optical densities which are in good agreement with those measured independently with a spectrophotometer and with the calibrated scanner. (B) Oscilloscope traces showing the trapezoidal shape of the oscilloscope traces. The first pulse shown on the left corresponds to the intensity of light transmitted by the solvent and the second, on the right, represents the transmittance of the solution. Light intensity increases in a downward direction, with each centimeter representing 0.5 V on the photomultiplier output. Each centimeter in a horizontal direction represents 10 μsec.

ing to each sector can be used for calculations of approximate optical densities. Hence the performance of the logarithmic amplifier can be tested directly by comparing the recorder deflection of the scanner with the optical density calculated from the measured ratio of the pulses for the solvent and solution. As shown in Fig. 16B the pulses should be trapezoidal, separated by a short dark period. If the slit at the photomultiplier is too long, the pulses tend to merge with a loss of the dark period. Pulses which do not have a clear trapezoidal shape frequently indicate that light is being scattered by oil or dust deposits on the condensing lens. For ultraviolet light, in particular, oil films on the lenses cause difficulty since the transmittance of the optical system is reduced markedly. Many of the electronic components, such as the circuit for converting signals into logarithmic values, operate optimally when the primary signal is within a certain range. This range is achieved readily if the size of the initial signal at the photomultiplier is optimal. The oscilloscope is the primary tool for securing the proper conditions.

2. *Linearity of Amplifier.* In many sedimentation studies the gain on the amplifier is altered during an experiment in order to provide accurately measurable recorder deflections. As a consequence it is necessary to test the linearity of the recording system for different settings of the amplifier. Such a test is shown in Fig. 17 for three solutions of different optical density. The results show that the recorder deflection is directly proportional to the amplitude setting (this experiment was performed on the instrument built in Berkeley,[15] but an analogous set of data should be obtained on other instruments as well). As seen in Fig. 17, a solution having an optical density of 0.125 produced a deflection of 25 mm with full gain on the amplifier. Since deflections of 2–4 mm are ample for certain types of sedimentation experiments, successful studies can be conducted with solutions having optical densities of only about 0.02. This sensitivity is affected markedly by the type of recorder. High-gain *XY* recorders produce a much larger pen deflection per unit optical density (see Fig. 3).

3. *Proportionality between Recorder Deflection and Optical Density.* Although most scanners contain internal calibrating circuits which are

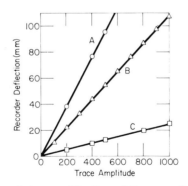

Fig. 17. Performance of the amplification of the recording system. The recorder deflection in millimeters is plotted on the ordinate as a function of the trace amplitude on the abscissa. The units for the trace amplitude represent the divisions of the helipot, 1000 corresponding to maximum gain. Three different experiments were performed in each of which the reference compartment was filled with distilled water and the solution compartments with tryptophan solutions of different optical densities. The rotor speed was about 24,000 rpm, and the wavelength on the monochromator was 280 nm with a slit opening in the monochromator of 2 mm. Curve A represents a solution having an optical density of 0.95; B corresponds to 0.54; and C to 0.125.

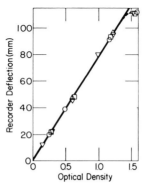

Fig. 18. Linearity of the recording system as a function of optical density. Solutions of known optical density were examined at various wavelengths in a double-sector cell with distilled water as the reference liquid. The recorder deflection in millimeters is plotted on the ordinate as a function of optical density on the abscissa. All optical densities for the solutions were measured in a Zeiss spectrophotometer and converted to a thickness of 12 mm to correspond to the optical path of the ultracentrifuge cells. All measurements were made at a trace amplitude of 500 and the rotor speed was 59,780 rpm. Measurements were made with solutions of tryptophan at 280 nm ($\triangledown$); BrCTP at 298 nm ($\triangle$); and maleate at 235 nm ($\square$) and 240 nm ($\bigcirc$).

very useful, it should be recognized that these units provide tests of the electronics only. In effect they introduce attenuators which reduce the electrical signal in a known way and thereby provide recorder deflections for different electrical signals which simulate known optical densities. The calibration steps of 0.2 optical density unit produced by the Spinco scanner (Spinco Division, Beckman Instruments, Inc.) are extremely convenient and furnish a reference for each trace. With this unit the recorder deflections for the solutions are not directly proportional to optical density; rather the recorder deflections are referred to the calibration stair-steps which in turn yield values of optical density. Other instruments[11,15] do not provide such calibration data on each trace, but instead are designed so that the measured recorder deflection is proportional to the optical density. That such linearity can be achieved between recorder deflection and optical density is shown in Fig. 18. These results show that the scanner responds linearly with optical density up to values about 1.4. Similar behavior is observed with the Spinco scanner if the recorder deflections are first converted into optical densities by means of the calibration stair-steps. These values of the optical density are in good agreement with those measured with a spectrophotometer.[79]

[79] W. Bauer and J. Vinograd, *J. Mol. Biol.* 33, 141 (1968).

As seen in Fig. 18, the response of the recording system was no longer proportional to the concentration of the solute when the solutions had optical densities about 1.5. Since independent tests of the various electrical components showed that this deviation from linear behavior could not be attributed to their malfunctioning, the limitations must have arisen from optical problems such as stray light. Removal of the light-limiting apertures mounted on the collimating and condensing lenses leads to a decrease in the range of linearity of the recorder response. A similar decrease in the linear range occurs when the lenses are heavily coated with oil droplets and dust which cause much of the light to be scattered. These observations indicate that the departure from linearity for highly absorbing solutions results from stray light appearing in the image of the solution compartment as it passes by the photomultiplier-slit assembly. This pitfall represents one of the principal difficulties with the split-beam scanning system. As a consequence it is essential that the linearity of the scanner be checked frequently.

The linearity of the recording system can be checked in a single experiment by the combined use of the scanner and interference optics. By judicious selection of the rotor speed in terms of the molecular weight of the solute an appreciable concentration gradient is produced throughout the liquid column. This concentration distribution as a function of position in the cell is measured both with the interference optical system and the photoelectric scanner. Cross-plotting of the data at corresponding positions gives directly the relationship between recorder deflection (or optical density) and fringe displacement (or solute concentration) as shown in Fig. 19. Bovine plasma albumin, because of its relatively low extinction coefficient and large molecular weight, is an ideal substance for simultaneous measurement of its concentration by interference optics and by the scanner. The centrifugal field should be sufficiently high that the concentration at the meniscus is virtually zero. In the experiment illustrated by Fig. 19 sedimentation equilibrium was attained, but it should be noted that the attainment of equilibrium is not required. Figure 19 shows clearly that the recorder deflection is proportional to protein concentration. This test demonstrates that stray light directed in a radial direction (from a region of low optical density to one of high absorbance) must be negligibly small in magnitude. Otherwise the measured optical density at the bottom of the cell would be less than the true value.

All experiments with the scanner require balancing the electronics so that the recorder deflection is zero in a region of the image corresponding to zero optical density. The oscilloscope is indispensable for this procedure because there is always a risk that the light pulses might not be

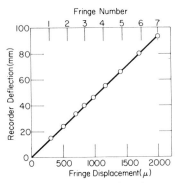

Fig. 19. Linearity test by combined use of interference optics and photoelectric scanner. The ordinate represents the recorder deflection in millimeters and the abscissa the fringe displacement in microns on the lower scale and as fringe number on the upper scale. The data were obtained from a sedimentation equilibrium experiment on bovine plasma albumin, at an initial concentration of 0.6 mg/ml in 0.1 M NaCl–0.01 M acetate buffer, pH 5.4. The rotor speed was 24,630 rpm. To obtain the data for this graph the recorder deflections and the fringe displacements were measured at corresponding values of the distance from the axis of rotation and then cross-plotted to produce the results shown above.

equal in magnitude even though the transmittances should be the same. Frequently the nulling can be performed with the photomultiplier receiving light passing through the air space above the liquid columns. However, it is observed sometimes that the light pulses there are not of equal magnitude, presumably because the windows in the solution compartment transmit less light after having been wet with a solution of high optical density. Hence it is important to use a specially designed counterbalance cell which can be used for this nulling procedure. This counterbalance cell serves not only as a reference for zero optical density, but also as a reference for known radial distances (where the recorder deflection changes abruptly) and for determination of the magnification factor relating distances in the cell to those on the trace.

4. Recorder. Various recorders have been used satisfactorily for registration of the data produced by the scanner. It is essential to have rectilinearity, fast response time, different chart speeds and variable sensitivities to permit expansion of the scale for optical densities. In addition, several pens should be available so that both integral and derivative patterns can be recorded simultaneously. Since the lead screw which drives the photomultiplier across the image and the recorder have independent drives it is necessary to incorporate a timing generator which rules the recorder paper with transverse reference lines which indicate the exact movement of the photomultiplier.[14] This has been achieved by mounting on the end of the lead screw a slotted disk which in conjunction with a small light and a photosensitive element generates marker pulses during the revolution of the lead screw. Through this means an automatic check is provided for possible slippage of the two independent drives for the photomultiplier and the recorder. The instrument built in this laboratory[11,14,15] uses a Honeywell 906B Visicorder oscillograph as a recorder. This unit contains 14 plug-in mirror galvanometers which direct ultraviolet light onto photosensitive paper and can be used to record data for several ultracentrifuges operating at the same time. The writing speed is extremely high, and the traces of optical density can be made to encompass the full width, 15 cm, of the paper. In contrast, the maximum pen excursion of the Offner recorder used in the commercial instrument is only about 7 cm. As a consequence of this limited range, various workers[47,79] have employed XY recorders. Typical traces from such a unit are shown in Fig. 3. With this Honeywell 520 XY recorder, slightly modified in this laboratory (M. Springer and H. K. Schachman, unpublished), the optical density range covers 25 cm and the distance in the cell can be expanded to 38 cm. For sedimentation velocity experiments the patterns can be superimposed onto one sheet of paper; this practice not only provides convenience, but also it reveals trends in the experiment very vividly. The signal for the X readings (corresponding to radial levels in the cell) were obtained from a calibrated, linear potentiometer mounted on the housing for the photomultiplier. With such a device a nonlinear movement from the rotating lead screw or intermittent pauses because of a faulty drive introduce no difficulty. In addition, the uncertainties stemming from backlash in the screw and coupling are eliminated.

If an XY recorder is incorporated into the scanner, it is important that it have a time for the X scale. It is useful to be able to register optical densities at some fixed position as a function of time. These plots can be extremely useful in detecting vibration or precession of the rotor due to malfunctioning ultracentrifuge drives. For this test the photomultiplier should be positioned in the image at a place where small movements in a radial direction will lead to large changes in optical density. The position corresponding to the meniscus serves well for this purpose. Alternatively one can use a position near the bottom of the cell corresponding to a large gradient of optical density with distance (as in a sedimentation equilibrium experiment). Quantitative assessments of the movement of the axis of rotation can be made in this way (a detailed discussion of this problem is presented elsewhere[47]).

VI. Analysis of Data

The transformation of the scanner patterns into meaningful data for the analysis of sedimentation velocity and equilibrium experiments is straightforward. Concentrations are measured in absolute terms as recorder deflections and distances on the traces are converted into actual radial levels (in the cell) by using the magnification factor and the position corresponding to the images of the reference holes in the counterbalance cell.

For sedimentation velocity experiments, boundary positions frequently can be taken as the radial level at which the concentration is one-half that in the plateau region. Alternatively (and more accurately) the equivalent boundary position can be determined in a rigorous manner. This procedure, which is equivalent to using the transport equation [see Eq. (1)], is more laborious, but it can be programmed readily for desk-type computers.[80] If the system is polydisperse, the resulting value is a weight average sedimentation coefficient (assuming all species have equal extinction coefficients). The use of Eq. (1) is mandatory if the sedimenting species are so small that the concentration at the meniscus is greater than zero. Equation (1) is required also for the calculation of constituent sedimentation coefficients[25]; in this case the integration is performed across all boundaries to the plateau containing all species.

The data from the scanner are particularly appropriate for sedimentation equilibrium calculations since no auxiliary information other than the equilibrium trace is required for most computations. For low-speed experiments (where the concentration ratio from the bottom to the top of the solution column is less than 10) the concentrations are evaluated directly from the recorder deflections. This feature constitutes one of the principal advantages of the scanner in comparison to interference or schlieren optics. Similarly the concentrations are evaluated directly in high-speed experiments (where the concentration ratio is larger than 100). Some precautions need be observed to correct for the possible presence in the solution of low molecular weight light-absorbing species. Therefore, after the equilibrium pattern is obtained, the speed of the rotor should be increased about 2-fold to check for such impurities. If a nonzero optical density trace is obtained after overspeeding, that trace should be used as a baseline for the equilibrium pattern. Correction for such contaminants is performed by direct subtraction of the corresponding base-line readings from the measured recorder deflections in the equilibrium pattern. These corrections are especially important in high-speed sedimentation equilibrium experiments where the concentration is virtually zero in the upper region of the solution column. Because of the marked sensitivity of the scanner we have observed frequently that dialysis of protein solutions, even in boiled and scrupulously rinsed

[80] R. Trautman, *Ann. N.Y. Acad. Sci.* **164**, Art. 1, 52 (1969).

dialysis tubing, leads to the presence of nonsedimenting, ultraviolet light-absorbing material. Such material contributes only negligibly to the refractive index of the solution and, as a consequence, is not detected by the use of interference optics. In a way, then, the major advantage of the scanner, its sensitivity, is not without its drawbacks. Small amounts of light-absorbing contaminants can easily complicate determinations of molecular weight by the high-speed sedimentation equilibrium method. Hence extreme care should be exercised in nulling the instrument properly and in determining the correct base line if such contaminants are detected.

VII. Future Developments

Although the photoelectric scanner has been very useful for a wide variety of ultracentrifugal studies, it is clear that major improvements in it can be anticipated and that these in turn will lead to new applications. Fruitful developments are already emerging from the use of the on-line computer for automatic scanner operation. With computer systems utilizing signals from individual light pulses and the conversion of voltages to digital values a number of problems associated with the electronics of the scanner, such as drift, limited range, and nonlinearity, can be eliminated. Moreover, analyses of large numbers of pulses provide precise statistical indices regarding the reliability of the data. Improvements in data collection permit the extension of the range of optical densities for which accuracy can be attained. Of course one of the principal advantages of computer systems is the availability of data in a digital form for immediate processing in programs suitable for analyzing the sedimentation experiment.

A number of approaches to automatic scanner operation have been reported.[19-21,81] In the simplest system the electronic circuits of the existing scanner are still employed for conversion of intensities of transmitted light to optical densities, and these in turn are digitized.[81] Although this system reduces the tedium in collecting data, the weaknesses inherent in the electronic processing of pulses are still retained and the potential of the computer is not fully exploited. The other systems,[19-21] although differing substantially in design and extent of development, analyze individual light pulses which are converted to digital form, transfer the values to an on-line computer, and calculate optical densities numerically. Such systems afford unusual opportunities for in-

[81] A. H. Pekar, R. E. Weller, R. A. Byers, and B. H. Frank, *Anal. Biochem.* **42**, 516 (1971).

creasing accuracy by altering data acquisition at the command of the computer. Thus, for example, where light levels are low, as in the bottom of the solution column in a sedimentation equilibrium experiment, the scanning rate could be reduced to ensure the collection of signals over a longer time interval. Such interactions between the computer and the scanner have not yet been achieved, but the possibilities for attaining these features are already available.

One instrument, which has been described fully only recently,[21] is now in routine use. Since it shows much potential already and can be improved readily by further modification, it is outlined here to indicate future directions. Its basic design is shown in Fig. 20. The system centers around a triggering circuit which responds to the rise in voltage for each photomultiplier pulse and generates a convert command to the analog-to-digital converter. The convert command is delayed by a variable setting to coincide with the plateau level of the light pulse. In addition to the

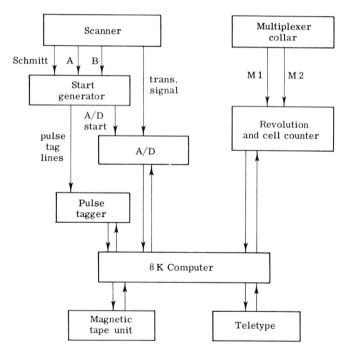

Fig. 20. Block diagram of basic hardware. Information from pulse tagger (which indicates sample, reference, or base line) and bits from the A/D converter are read simultaneously into computer and correlated with information from the revolution and cell counter derived from the rotor collar of the Beckman multiplex unit. Programs and data are stored on a cassette-type magnetic tape unit. Interaction with the system is through the teletype.

convert signal for each pulse in the pair for solvent and solution, a third convert signal is used to sample the base-line voltage in the dark space. The digital value (12 bits) for each conversion is then read into the on-line computer along with information on the origin of each pulse in terms of cell and sector. Since the computer word length is 16 bits and only 12 bits are used for the digital voltage, 4 bits remain for identification purposes. In this way all the incoming data are tagged with identifying information to guard against errors in assigning the pulses within the computer. A Nova computer (Data General Corp.) with 8 K of memory was used for minimal operation; upon the inclusion of 4 K additional memory, operations could be performed with a FORTRAN compiler requiring 12 K. As data enter the computer they are stored for 100 revolutions of the rotor, then sorted, averaged, and analyzed to give one value of the optical density (and its standard deviation) for each radial level in each cell. The values of optical density are stored along with an index given by the median rotor revolution number. When the scan is completed the computer sorts the optical density values and calculates the radial position for each point by interpolating on the basis of the revolution number corresponding to the disappearance and appearance of the outer and inner reference holes in the counterbalance cell. The absolute radial distances for these references are stored in the computer and thus conversion of revolution numbers to real distances is direct. Since the computer operation is rapid, the data can be collected from each pair of compartments in multicell rotors during each revolution of the rotor. In this way data from different cells enter the computer in a string and the identifying bits are used to sort the data into appropriate storage buffers.

When the collection of data is completed, the values of optical density and radius are stored on magnetic tape, displayed in a variety of ways, or analyzed directly in a fitting program. Such programs are available for sedimentation equilibrium experiments on both homogeneous and interacting (self-associating) systems. The data can be examined prior to fitting through the use of a high-speed line printer, an XY plotter, or a cathode ray tube (CRT) display.

If tabular display of the data is desired, it is essential that a high-speed line printer (135 characters per second) be incorporated. With multicell rotor a single scan will produce hundreds of points which are indexed, recorded in terms of radial position and the square of the radial distance from the axis of rotation. For each point there is also tabulated the optical density and the standard deviation in the optical density. The squares of the radial positions are useful for plotting of the results by hand. Suitable programming of the computer directs the printing of only

POINT	R(CM)	Y(OD)	SIGMA Y(OD)	R^2
+00001	+6.13774	+2.17581	+.00338	+37.67197
+00002	+6.13362	+1.86527	+.00488	+37.62134
+00003	+6.12943	+1.56517	+.03044	+37.56999
+00004	+6.12531	+1.34290	+.01852	+37.51944
+00005	+6.12124	+1.14688	+.02085	+37.46969
+00006	+6.11718	+1.00179	+.01516	+37.41998
+00007	+6.11306	+.88580	+.01377	+37.36952
+00008	+6.10899	+.77473	+.01134	+37.31987
+00009	+6.10493	+.69426	+.01033	+37.27025
+00010	+6.10087	+.60651	+.01074	+37.22066
+00011	+6.09681	+.53923	+.01109	+37.17111
+00012	+6.09281	+.47244	+.00805	+37.12235
+00013	+6.08868	+.41892	+.00840	+37.07211
+00014	+6.08456	+.37056	+.00840	+37.02189
+00015	+6.08037	+.32354	+.00750	+36.97095
+00016	+6.07631	+.28460	+.00811	+36.92156
+00017	+6.07212	+.25081	+.00756	+36.87070
+00018	+6.06793	+.21905	+.00737	+36.81986
+00019	+6.06381	+.20238	+.00737	+36.76981
+00020	+6.05974	+.17855	+.00676	+36.72056
+00021	+6.05556	+.15705	+.00655	+36.66983
+00022	+6.05149	+.14014	+.00775	+36.62064
+00023	+6.04743	+.12586	+.00640	+36.57149
+00024	+6.04331	+.11199	+.00640	+36.52161
+00025	+6.03912	+.09948	+.00662	+36.47102
+00026	+6.03512	+.09341	+.00676	+36.42272
+00027	+6.03118	+.08747	+.00609	+36.37521
+00028	+6.02699	+.07360	+.00676	+36.32472
+00029	+6.02293	+.06185	+.00704	+36.27577
+00030	+6.01893	+.06171	+.00655	+36.22760
+00031	+6.01487	+.05844	+.00625	+36.17871
+00032	+6.01081	+.04779	+.00569	+36.12986
+00033	+6.00662	+.04750	+.00676	+36.07953
+00034	+6.00249	+.04507	+.00724	+36.03000
+00035	+5.99837	+.03702	+.00640	+35.98049
+00036	+5.99431	+.03212	+.00585	+35.93177
+00037	+5.99024	+.03765	+.00647	+35.88309
+00038	+5.98606	+.02817	+.00724	+35.83294
+00039	+5.98199	+.02587	+.00662	+35.78432
+00040	+5.97793	+.02472	+.00676	+35.73573
+00041	+5.97381	+.02437	+.00617	+35.68643
+00042	+5.96968	+.02322	+.00560	+35.63716
+00043	+5.96562	+.02787	+.00655	+35.58867
+00044	+5.96143	+.02342	+.00625	+35.53873
+00045	+5.95737	+.01521	+.00640	+35.49031
+00046	+5.95324	+.01903	+.00632	+35.44117
+00047	+5.94924	+.01994	+.00617	+35.39357
+00048	+5.94512	+.01882	+.00750	+35.34450
+00049	+5.94099	+.01837	+.00594	+35.29547
+00050	+5.93681	+.01913	+.00683	+35.24574
+00051	+5.93268	+.00876	+.00647	+35.19677
+00052	+5.92862	+.01362	+.00617	+35.14859
+00053	+5.92456	+.01073	+.00625	+35.10043

FIG. 21. Output of data collection program on high-speed line printer. Data taken from an experiment on human carboxyhemoglobin S in 0.1 M phosphate, pH 7. The light had a wavelength of 405 nm, and the data were recorded after 18 hours of centrifugation at 30,000 rpm.

a certain fraction of the data or that falling between predetermined limits. This feature is especially useful in the subsequent utilization of fitting programs for analyzing concentration distributions in sedimentation equilibrium experiments. A typical page of output, printed in about 1 second, is shown in Fig. 21.

For graphical presentation of the data both the XY plotter and the CRT have proved to be valuable. The former is illustrated in Fig. 22A as a plot of optical density versus radial position. Although not shown in the diagram, the values along both the X and Y axes are printed automatically (in tabular form) by the high-speed printer. This plot is controlled through a graphical output monitor program which selects axes for optical densities and radial levels so as to give maximum spacing of the data. Since many of the data are analyzed in terms of the logarithm of concentration, the values of optical density in Fig. 22A are converted to logarithms for plotting, as in Fig. 22B, as a function of the square of distance from the axis of rotation. The equivalent CRT displays (Figs. 23A and 23B) have the advantage that changes in the plots can be made readily to permit enlargement of those regions that require closer examination. No additional instructions need be given to the computer.

In the displays of the logarithm of optical density versus the square of radial distance there is distinct curvature, as expected for a heterogeneous or self-associating system. Near the top of the solution column the concentration is very low, and there is considerable scatter in the plot representing the logarithm of concentration versus the square of distance. For systems exhibiting curvature of the type shown in Figs. 22B and 23B, computer programs are employed which attempt to fit the concentration data (Figs. 22A and 23A) as a series of exponentials. Each term represents a different molecular species, and the equilibria among them are expressed in terms of appropriate dissociation constants.

It appears likely that the on-line computer system will be very valuable for the analysis of interacting systems, where the savings in time and labor provided by the computer will permit calculations that otherwise might not have been undertaken. Heretofore the precision of the data produced by the scanner has not been sufficient to warrant such extensive calculations. Clearly additional accuracy is sorely needed. In particular, "noise" levels have to be reduced. This goal can be accomplished in part by the collection of more pulses of light for each average value of the optical density. At high optical densities where there are only small numbers of photons reaching the photomultiplier per revolution of the rotor, longer scan periods are required. It seems desirable to have a scanning device (perhaps with a step motor instead of a con-

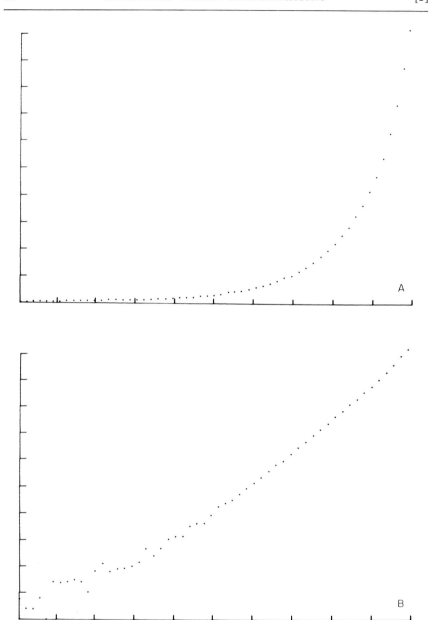

FIG. 22. Data display on XY plotter. The data of Fig. 21 are plotted on an XY recorder in two forms: (A) optical density versus distance from the axis of rotation; (B) logarithm of optical density versus the square of the radial distance from the axis of rotation. The actual values of the scales, both ordinate and abscissa, are tabulated independently on the high-speed printer.

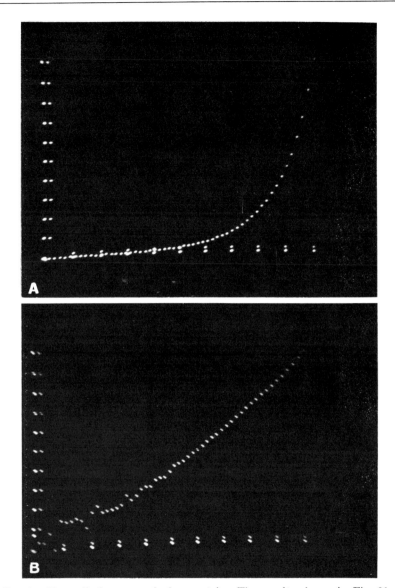

Fig. 23. Data display on cathode ray tube. The results shown in Fig. 21 are displayed as: (A) optical density versus distance from the axis of rotation; (B) logarithm of optical density versus square of the radial distance.

tinuous drive) which will provide automatically for adjustments in rate according to the prevailing light intensity at the photomultiplier.

Most scanners use only a fraction of the light traversing the cell at any given radial level as the cell sweeps across the optical axis. This occurs because these systems are based on measuring the peak height of the light pulses. Such scanners are simple in design and construction; moreover, their performance is independent of cell geometry. However, a substantial fraction of the potential signal at a given radial level is wasted since *all* of the light reaching the photomultiplier is not used. Hence further investigations are needed to assess techniques which involve integration of the signal across the entire pulse.[20] This arrangement requires two identical openings in the cell assembly and high-precision switching to permit gathering all the light in the appropriate channels of the recording system. These requirements can be met with existing techniques. Doubtless the integration of the respective signals for the solvent and solution compartments will provide greater signal-to-noise ratios than mere peak detection and, as a result, enhanced accuracy can be expected.

Clearly the use of the individual light pulses in conjunction with an on-line dedicated computer has provided a considerable gain in accuracy already. This approach requires the incorporation of an analog-to-digital converter so that the data can be obtained in a tractable form for subsequent manipulation by the computer. Simplicity could be achieved if the light pulses could be utilized directly by modern photon counting techniques. Such photon counting techniques have many advantages over the dc analog method, especially at very low light levels. However, the existing light levels, although not as great as we would like, are already too great for the present capacity of photon counters. Hence, if this approach is to prove viable there must be a substantial gain in the counting rate over present-day instrumentation. Where light levels are extremely low, as in the far ultraviolet region of the spectrum, the photon counting technique has much to commend it. Attractive as this approach appears, it seems, nonetheless, that gains in light intensity are likely to be obtained sooner than counting rates can be increased. If this proves to be the case, photon counting may not prove to be as accurate as the use of photomultipliers with analog-to-digital conversion. Despite the present limitations in photon counting, it is so amenable for direct split-beam operation that the technique should be explored further.

Considerations of the photon counting approach have been very fruitful in showing that the accuracy of any scanner is limited statistically by the number of events that can be measured. Light levels for solutions of zero optical density are too high for present counters; but for solutions of optical density above 2, the number of photons is not very great.

Hence, for such solutions, statistically reliable signals could be obtained only if each optical density measurement was based on many individual light pulses. In principle this could be achieved by very slow scanning rates in the region of the image where the optical density is high. Unfortunately, however, the scanning times may prove prohibitively large for certain types of sedimentation experiments. Thus it is useful to consider electronic scanning with a television camera as an alternative to mechanical scanning with a moving photomultiplier tube. Television cameras with a vidicon tube have great potential since the image of the entire cell is "seen" as it sweeps by the light beam (different radial levels are not visualized sequentially as with a moving photomultiplier). Moreover, this image on the vidicon tube is scanned electronically in about 30 msec. The data are obtained directly in digital form through the use of an optical multichannel analyzer. Vidicon tubes sensitive in the ultraviolet region of the spectrum are now available and fairly good signal-to-noise ratios can be obtained at the present state of the art. This approach has one major drawback in that it cannot be used directly as a split-beam system in which the solvent and solution cells follow one another rapidly in crossing the light beam. Hence, the two light beams must be separated physically by appropriate optical techniques, or, alternatively a gating system must be devised. Although the former route can be made to work satisfactorily, the introduction of additional optical components is likely to bring with it new problems. However, developments can be expected so that image intensifiers can be turned off and on in a fraction of a microsecond. Such devices should permit the use of the vidicon tube with double-sector cells. Obviously developments of this type are highly desirable. When they occur the merits of this approach can be assessed in relation to the existing (and improved) mechanical photoelectric scanners.

[2] Determination of Molecular Weights in the Ultracentrifuge Using Time-Lapse Photography

By J. L. BETHUNE

The analytical ultracentrifuge is probably the most widely used instrument for accurate determination of molecular weights of biological macromolecules. Highest accuracy is obtained by employment of the technique of low speed sedimentation equilibrium,[1] in which a dissolved

[1] H. K. Schachman, in "Ultracentrifugation in Biochemistry." Academic Press, New York, 1959.

macromolecular substance is sedimented in an appropriate field until an equilibrium concentration gradient is established, the field being chosen to give, at equilibrium, a finite concentration of the substance at the air–liquid meniscus.

The fundamental equation for sedimentation equilibrium of a homogeneous ideal solute is[2]

$$C(r) = C(r_m) \exp[M(1 - \bar{v}\rho)\omega^2(r^2 - r_m^2)/2RT] \tag{1}$$

where $C(r)$ is the concentration at radical position r; $C(r_m)$ is the concentration at the air–liquid meniscus, r_m; M is the molecular weight; $\bar{v}$ is the partial specific volume; ρ is the density; ω is the rotor velocity, rpm $\times 2\pi/60$; R is the gas constant; T is the absolute temperature. This may be rewritten as

$$\ln \frac{C(r)}{C(r_m)} = \frac{M(1 - \bar{v}\rho)}{RT} \omega^2 \frac{r^2 - r_m^2}{2} \tag{2}$$

Thus, M can be obtained by plotting $\ln(C)$ against r^2. It is necessary, however, that either $C(r)$ or some quantity proportional to it be known throughout the liquid column. While the absorption optical system will yield this directly,[3] the more accurate Rayleigh interferometric system gives directly only the difference in concentration between any point r and the meniscus, r_m. Although a number of methods of determining meniscus concentration are employed (for a general review, see Creeth and Pain[4]), these generally are most readily applicable to a single solvent or dilute buffer, and serious experimental and/or theoretical problems may be encountered if they are employed for solvents containing, e.g., high concentrations of guanidine hydrochloride, salt, or urea.

We have recently employed the technique of time-lapse photography to determine meniscus concentrations in an approach which is not subject to these limitations.[5,6]

Principle of the Method

Either of two approaches may be employed. (1) After sedimentation equilibrium is attained at low speed, the rotor is slowly speeded up until the meniscus is depleted of solute. The process is monitored continuously by time lapse photography, and the fringe shift at the meniscus (displayed by the Rayleigh interferometer) from that obtaining at equilib-

[2] W. J. Archibald, *J. Phys. Colloid Chem.* **51**, 1204 (1947).

[3] H. K. Schachman and S. J. Edelstein, *Biochemistry* **5**, 2681 (1966).

[4] J. M. Creeth and R. H. Pain, *Progr. Biophys. Mol. Biol.* **17**, 217 (1967).

[5] J. L. Bethune, *Biochemistry* **9**, 2737 (1970).

[6] R. T. Simpson and J. L. Bethune, *Biochemistry* **9**, 2745 (1970).

rium to that after depletion is complete, represents the meniscus concentration. This procedure is analogous to that of LaBar.[7]

(2) The reverse process is employed. The meniscus is first depleted at high speed, then the speed is dropped to that required for attainment of equilibrium, and the fringe rise at the meniscus is monitored. The fringe rise is the result of diffusion of the solute to the meniscus in the lower field. After equilibrium is attained, the meniscus concentration is directly equal to this fringe rise.

Instrumentation

This is described for the Spinco Model E analytical ultracentrifuge.

The optical system is aligned,[8] by means of a Kodak 77A Walten filter and a parallel slit mask. Certain physical modifications must be made in the ultracentrifuge to permit the installation of a time-lapse camera unit. The metal covering at the camera end of the centrifuge is cut out to allow mounting of the components described. A square hole, 5.0 cm on a side and centered on the optic axis is cut in the cassette holder to allow direct observation of the fringe pattern. An aluminum plate with a similar opening is made to fit over the cassette holder opening, and a covering plate of opaque plastic, drilled to accept the nosepiece extension of the camera, is fitted to this. Slotted screw holes allow horizontal movement of the aluminum plate and vertical movement of the plastic plate, to properly position the camera at any desired position in the fringe pattern.

An L-shaped table is fabricated from $3/8$-inch aluminum stock to support the camera behind the focal plane of the instrument. This support is fixed to the top of the cassette holder, so that the entire assembly is fixed with reference to the optical axis. No significant deflection of the optical tube should occur on addition of the whole mass of the assembly to the end of the tube.

The camera employed is a 16 mm Bolex Paillard modified by Zeiler Instruments, Inc. (Boston, Massachusetts), for time-lapse photography, at intervals from 0 to 5 minutes, with exposure times up to 150 seconds. No ancillary camera lens was employed; rather, a nosepiece fabricated from 2.9-cm diameter aluminum tubing, 1.7 cm long, was positioned to fix the film at the focal plane of the optical system.

Alignment of the camera is accomplished by placing the holes in the movable plates to center the fringe pattern both horizontally and vertically in the aperture. The camera assembly is then positioned with-

[7] F. E. LaBar, *Proc. Nat. Acad. Sci. U.S.* **54**, 31 (1965).

[8] K. E. van Holde and R. L. Baldwin, *J. Phys. Chem.* **62**, 734 (1958).

out disturbing the movable plates and is clamped to the support table. The camera is not moved during the course of a centrifuge run.

Photography and Evaluation

Motion pictures are taken on 16 mm Plus −X negative motion picture film, Type 7231, generally with exposure times of 100–150 seconds. The film is developed in a manual tank-type developer with continuous agitation for 10 minutes at ∼20° using either a 1:3 dilution of Kodak HC-110 developer or Kodak D-11 developer. After one wash with water, the films are fixed with a 1:3 dilution of Kodafix acid fixer for twice the time required for clearing, and then washed with tap water and air dried. (No distortion of the emulsion, compared with that obtained on Kodak IIG spectroscopic plates, has been noted.)

The data are evaluated using a modified Kodak Analyst 16 mm projector. This projector permits automatic, flicker-free projection at speeds down to 1 frame per second, and is equipped with blower cooling so as to allow full light intensity even when stopped for single-frame projection. The pictures are projected onto a ground-glass screen with a raster to allow quantification of movement of the interference fringes.

Additional photographs on Kodak Spectroscopic IIG plates are utilized for evaluation of equilibrium patterns. The plates are read on a Mann two-dimensional comparator, utilizing previously noted techniques.[9,10]

Cell and Window Selection

We have used single cells, with 12 mm filled Epon center pieces and sapphire windows. The cells are checked for leakage and the windows for irreversible distortion by first loading them with water (see below for loading technique) and then spinning them at 50,000 rpm for 8 hours. The speed is then dropped to 5000 rpm and maintained at this level for another 16 hours. The whole process is monitored by the time-lapse camera. Leakage is immediately apparent from changes in position of the menisci, while for satisfactory results in application of the techniques all window distortion should be relaxed in 2000 seconds.[5] The criterion for such relaxation is that the fringes should reach a constant position in this time interval.

Loading Technique

Since we have used a 16 mm camera, the maximum column height that can be completely observed is of the order of 7 mm (the magnifica-

[9] J. L. Bethune, *Biochemistry* 4, 2691 (1965).
[10] E. G. Richards, D. C. Teller, and H. K. Schachman, *Biochemistry* 7, 1054 (1968).

tion factor of the camera lens is 2.19). While it is not necessary to observe the whole column to determine the fringe change at the meniscus, such observation allows facile detection of any problems arising in use of the ultracentrifuge.

In general, the column heights used have been of the order of 2–3 mm. For example, when Hamilton microsyringes are used, 0.01 ml of hydrocarbon FC-43 (Spinco Co.) is placed in the reference channel of the double sector interference cell, and 0.02 ml in the solution side. Then 0.11 mm of solvent is placed in the reference channel and 0.096 ml of solution in the solution side. This assures that the solvent column is slightly longer than the solution column and overlaps it at both ends. Thus the entire solution column is visualized and any mismatch due to redistribution of solvent components is minimized.

Selection of Rotor Speeds

In general, some estimate of the molecular weight, however rough, is available. This will dictate the rotor velocity for the low speed equilibrium portion of the run, where a concentration ratio of three or four to one between the ends of the solution column is preferable.[11] From the fundamental Eq. (2), then

$$\ln 4 = \frac{M(1 - \bar{v}\rho)\omega^2}{RT} \frac{r_b{}^2 - r_m{}^2}{2}$$

For $\bar{v} = 0.75$ ml/g; $\rho = 1$ g/ml; $T = 293°$K; $r_b = 7$ cm; $r_m = 6$ cm; speed (rpm) $= (2.446 \times 10^6)/\sqrt{M}$. The concentration distribution should be examined visually after a few hours of sedimentation. An incorrect choice of the rotor speed will generally be obvious by that time, and adjustment can be made.

For the meniscus depletion portion of the run, we have routinely used a speed of 42,000 rpm, except in cases of molecules such as ribonuclease in guanidine hydrochloride, where the speed chosen was 52,000 rpm. In terms of the above criterion, this corresponds roughly to a concentration ratio of fifty.

Experimental

In the first approach indicated, after the attainment of low-speed sedimentation equilibrium, the rotor is accelerated slowly (at a driving current of about 3.9 ± 0.1 Å) to a speed which ensures depletion of the kinetic unit at the meniscus. Usually this rate of acceleration requires

---

[11] T. Svedberg and K. O. Pedersen, "The Ultracentrifuge." Oxford Univ. Press, London and New York, 1940.

about 8–10 hours to accelerate the rotor from the low speed condition (10,000–20,000 rpm) to the high speed condition (45,000–60,000 rpm). As the rotor accelerates, the meniscus is gradually depleted of protein, resulting in a fringe drop at that position. The total fringe drop then represents J_m for the low-speed run. Alternately, if the record is continuous, the first fringe drop, i.e., that occurring from the time of initial imposition of the field to equilibrium, when added to J_m, gives J_0, the initial concentration.

Although in most cases buffer base lines show no deviation of the fringe across the column greater than 0.1 fringe, the *entire* fringe pattern as a unit may shift up or down by as many as six fringes. The extent of this fringe shift is a function of the centerpiece and windows employed, the refractive index of the solvent, and the technique of assembly of the cell. The shifting is relatively minor until speeds greater than 30,000 rpm are attained but becomes increasingly important above this speed. This artifact may be due to wedging of the cell centerpiece. It is fully reversible on deceleration in an essentially instantaneous fashion. A blank is run every time the cell is assembled, and appropriate corrections are made to the fringe shift observed in the experimental run.

In the second approach, the solution of protein is centrifuged at a high speed until the meniscus is fully depleted. Then, the rotor is decelerated to an appropriate speed for the equilibrium run, and allowed to equilibrate by diffusion. The appearance of protein at the meniscus, reflected as a fringe rise, rather than its disappearance, is followed in the film record. Since the artifactual shifting of fringes under the centrifugal field is reversed rapidly upon deceleration whereas diffusion of protein to the meniscus is a much slower process,[5] no correction is necessary for wedging in this type of experiment. Moreover, the field-imposed distribution of the solvent components at the high speed relaxes to that characteristic of the lower speed much more rapidly than does the protein. Thus the protein is diffusing through a much more homogeneous solvent than is at first apparent, allowing full definition of the kinetic unit.

To cover a range of protein concentrations across the liquid column, the rotor may be run at several speeds, allowed to attain equilibrium at each, and thereby provide a significantly wider range of protein concentration within a single experiment.

In both approaches, attainment of equilibrium is checked by evaluation of photographs taken on spectroscopic IIG plates at intervals of a few hours. The positions of a few fringes or the total fringe count along the liquid column can be checked quickly; constancy within experimental error can then afford an operational criterion for attainment of equilibrium.

Evaluation of the Records

The motion picture is projected at high speed, and the position of the menisci and constancy of the final equilibrium pattern are rapidly checked. Then, for the first approach, one fringe at the meniscus in the initial equilibrium pattern is fixed on a raster line, using the highest magnification compatible with film quality, the projector is run at a slow speed and the number of fringes passing the reference point in the raster constitutes the fringe drop. The fractional fringe can be estimated using a ruler. The blank record is treated in the same way, the total deviation being added algebraically to the experimental record.

In the second approach, one flat fringe at the meniscus is selected and fixed on a raster line, and the fringe rise is counted. A blank is similarly evaluated.

The meniscus is defined operationally as that position most centripetal in the liquid column at which a fringe can be distinguished. Since the absolute fringe number at this position is known from the film record, the information is immediately transferable to the patterns recorded on plates. For evaluation of the plates, they are first aligned in a two-dimensional comparator, using the center reference fringe (since there must be an odd number of fringes, the center fringe in both reference patterns is immediately identified). Using the wire edge as a known reference position, the radial positions of successive fringes or fractions of fringes in the liquid column are identified.

Calculations

The absolute concentrations in terms of fringes are obtained by identifying the number of fringes determined from the film record, J_m, with the first fringe recorded from the plate. Successive fringe positions then represent concentrations of $J_m + 1$, $J_m + 2$, etc. The molecular weight can then be determined using Eq. (2) or, in its differential form,

$$M_{app} = \frac{2RT}{(1 - \bar{v}\rho)\omega^2} \frac{d \ln J}{dr^2}$$

where it is assumed that

$$C \alpha J$$

Data analysis is facilitated through utilization of two computer programs written for the Scientific Data Systems 940 time-sharing system. The first of these programs calculates the parameters of the least square equation for the plot of $\ln J$ vs. r^2 across the cell, yielding molecular weights averaged over the entire cell volume. In addition, the program includes as output molecular weights calculated between each pair of comparator readings, to detect both nonlinearity in the plot and erroneous comparator readings.

The second program is particularly applicable for determinations in solvents containing high concentrations of denaturing agents, where the marked concentration dependence of M_{app}, if there is a significant concentration change across the cell, may give a high degree of error when the molecular weight is evaluated over the column height to yield a M_{app} at one concentration, i.e., C_0, or $(C_m + C_b)/2$.[12] This program calculates a least-square fit of nine readings centered around every fourth reading through the cell. Thus, if the displacement across the cell is 15 fringes and readings are made at half-fringe intervals, the output would correspond to M_{app} calculated at the 2nd, 4th, 6th, 8th, 10th, and 12th fringes, i.e., at $J_m + 2$, $J_m + 4$, etc., where J_m is the fringe number at the meniscus. In addition, molecular weights calculated between each pair of readings are included in the output to detect erroneous comparator readings. In systems where a high dependence of M_{app} on concentration is observed, this approach provides fuller utilization of the information inherently present in the data, resulting in smaller errors than does evaluation of the distribution across the column as a whole.

Applications

These approaches were first applied to ribonuclease and α-chymotrypsinogen in guanidine hydrochloride,[6] where the molecular weights determined were, for ribonuclease, 13,250 ± 500 (amino acid analysis 13,683) and for α-chymotrypsinogen 25,500 ± 400 (amino acid analysis 26,000). Further applications include alkaline phosphatase from *Escherichia coli* and alcohol dehydrogenase from horse liver in a variety of denaturing solvents,[13,14] and human placental 17β-estradiol dehydrogenase in 40% glycerol.[15]

Several comments concerning the relative advantages and disadvantages of the two methods presented are appropriate. The second method, that of initial meniscus depletion, followed by attainment of equilibrium through diffusion in a low centrifugal field, would appear to lead to significantly higher precision, in large measure due to elimination of corrections for cell wedging and solvent redistribution. However,

[12] J. W. Williams, K. E. van Holde, R. L. Baldwin, and H. Fujita, *Chem. Rev.* **58**, 715 (1958).

[13] R. T. Simpson, Doctoral Dissertation, Harvard University, Cambridge, Massachusetts, 1969.

[14] D. B. Pho and J. L. Bethune, *Biochem. Biophys. Res. Commun.* **47**, 419 (1972).

[15] D. J. W. Burns, L. L. Engel, and J. L. Bethune, *Biochem. Biophys. Res. Commun.* **44**, 786 (1971).

certain requirements in the protein-solvent system under study must be met for this method to be applicable. Thus, the occurrence of a very slowly reversible precipitation of protein at the base of the cell at high speeds would lead to impractically long times for the attainment of equilibrium at the lower speed. Second, in the study of large molecules, characterized by low values of the diffusion coefficient, inconveniently long times again would be necessary for attainment of equilibrium.

The first method, i.e., sedimentation to equilibrium, followed by meniscus depletion on imposition of a higher field also suffers from operational disadvantages. Thus, when high speeds are necessary for depletion to occur fully, problems of cell wedging and solvent redistribution necessitate significant corrections in the determination of the concentration of sedimenting material present at the meniscus at equilibrium, or redefinition of the kinetic unit, the concentration distribution of which is determined.

For systems where it is applicable, the second method would seem to offer more advantages. Fortunately, as noted above, the use of time-lapse photography as a monitor during the ultracentrifuge experiment allows the facile detection of various problems in the use of the instrument. Thus, the problems possibly encountered in these methods of molecular weight determination, i.e., baseline fringe shifts, nonattainment of equilibrium, and loss of solute due to irreversible aggregation, are all readily detectable on simple viewing of the motion pictures taken during the run.

[3] Active Enzyme Centrifugation

By David L. Kemper and Johannes Everse

René Cohen[1] first demonstrated the feasibility of determining sedimentation and diffusion constants of an enzyme in its catalytically active form. The technique involves the layering of a small amount of enzyme on an appropriately buffered solution that contains all substrates normally used to observe the enzyme-catalyzed reaction. This may be done by the use of a synthetic boundary cell. The movements of the enzyme in the cell under the influence of the centrifugal force may thus be followed spectrophotometrically or by observing the change in refraction of the solution due to the disappearance of a substrate or the formation of a product. Cohen termed this method *"Active Enzyme*

[1] R. Cohen, *C. R. Acad. Sci.* **256**, 3513 (1963).

Centrifugation," since only the polymeric form of the enzyme that possesses catalytic activity is observed with this technique. Although the method as described by Cohen yields satisfactory results, the manipulations are laborious and time consuming, and they lack a certain degree of quantitation. The method described here has been developed in our laboratory during the course of our investigations on the action of enzymes as related to structure; it consists mainly of a number of modifications of the Cohen method in order to achieve an increased simplicity in the execution of the centrifugation and improvements in its accuracy and quantitation.

"Active Enzyme Centrifugation," as described in this section, requires an analytical ultracentrifuge, equipped with a monochromator and a photoelectric scanning system. A 12 mm double sector cell is used, equipped with sapphire windows and a Vinograd-type double sector centerpiece.

Active enzyme centrifugation is based on the principle that an enzymatic reaction occurring in the cell can be followed by some physical method during centrifugation. In many cases enzymatic reactions may be monitored spectrophotometrically, which is convenient as well as highly accurate. This technique can be used in the centrifuge by utilizing a photoelectric scanning system. The method is not necessarily limited to reactions in which optical density changes occur in either substrates or products during catalysis. Enzymatic reactions may also be monitored, if the enzymatic reaction can be coupled to a secondary reaction which involves a change in optical density. For example, an enzymatic reaction that releases or takes up hydrogen ions may be measured if an appropriate indicator has been added to the weakly buffered mixture. Another example would be a reaction in which ketone formation occurs; addition of phenylhydrazine to the reaction mixture will result in the formation of a hydrazone wherever the ketone has been formed owing to the presence of the enzyme. A method using coupled systems that use more than one enzyme has not been fully worked out as yet.

Since only the displacement of the active form of the enzyme is observed with this method, any inactive forms of the enzyme which may be present in the preparation are not observed. For this reason it is not always necessary to use highly purified enzymes in order to obtain meaningful results; in fact, successful experiments may be done with relatively crude dialyzed tissue extracts (Fig. 1).

Active enzyme centrifugation may thus be applied to many enzyme systems. This paper describes the theoretical as well as the practical aspects of the procedure in some detail; however, the discussion of the manipulations involved in the technique has been limited to a thorough

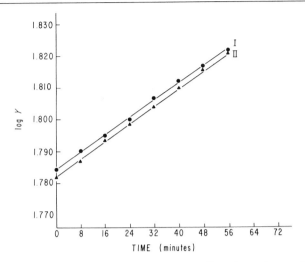

Fig. 1. Comparison of the active enzyme centrifugation of a purified enzyme (●——●) and a crude extract (▲——▲). Enzyme: chicken H_4 LDH (I), homogenized chicken heart supernatant (II). Reaction mixture: 0.1 mM Tris·HCl, pH 6.5, 0.35 mM DPNH; 1.3 mM pyruvate; 0.04 mM phenol red, 10% glycerol. Speed: 59,780 rpm.

description of some of the procedures that have been used in our laboratory. It must be left to the individual to adapt this method to his particular enzyme system.

Theoretical Considerations

The method of active enzyme centrifugation, as described here, combines the techniques of ultracentrifugation and spectrophotometry. Thus the theoretical aspects involved in these two techniques apply directly to active enzyme centrifugation; they are covered in a variety of textbooks on the subjects and will not be repeated here. This section will be restricted to a theoretical consideration of the problems resulting from the combination of ultracentrifugation and spectrophotometry.

Boundary Formation

The amount of enzyme needed for active enzyme centrifugation is proportional to its turnover number, and is usually a few micrograms, dissolved in a small amount of an appropriate buffer. During acceleration of the centrifuge, this amount of enzyme will be layered as a thin film on the contents of the centrifuge cell. It is imperative that the formed boundary is sharp, and that no enzymatically active material is present at any point in the cell other than at the meniscus before the desired ultracentrifugal speed has been reached. This implies that the density of the substrate solution in the cell has to be somewhat higher than that of the enzyme solution. In general, this may be achieved by dissolving the enzyme in $0.01\,M$ buffer, whereas the substrates are contained in $0.1\,M$ buffer. In some cases, however, it may be necessary to increase the density of the substrate solution by the addition of NaCl, glycerol, or any other suitable compound. The suitability of such an additive is dependent mainly on three criteria: (1) it should in no way affect the enzymatic system to be tested; (2) it should not react with substrates or any other products that might shift the equilibria from normal assay conditions; and (3) the molecular weight of the additive should be small enough to prevent sedimentation or gradient formation in the centrifugal field where it will be used. For the latter reason, sucrose is not suitable, if high ultracentrifuge speeds are to be employed.

For relevant results to be obtainable, the density and the viscosity of the substrate solution should remain constant throughout the cell during the test and should be accurately known (see Data Analysis). Too high a density and/or vicosity may adversely affect the accuracy of the observed rate of sedimentation because of the application of large correction factors, and should be avoided.

Substrate and Cofactor Limitations

Similar considerations as outlined above are applicable with respect to the substrates and cofactors present in the cell. Cohen[2] has published a detailed analysis of the conditions that apply to the various constituents of the cell and of how these conditions affect the final results. He concluded that the contribution of the cofactors and substrates to the density and viscosity of the reaction medium is negligible at the concentrations generally used for enzymatic assays. However, in certain cases when high concentrations of substrate are required, such contributions may become significantly large, and the observed data may need to be corrected for these effects. Furthermore, it is inherent to the successful use of active enzyme centrifugation, that the reactants (cofactors, substrates, products) do not sediment to any appreciable extent during the experiment. This consideration is necessary since a change in the distribution of reactant concentration will affect the equilibrium conditions of the enzyme reaction. We have found that such a change in distribution will affect the rate of sedimentation. Cohen found that DPNH has a sedimentation coefficient (s) of 0.22 ± 0.03 S; such a small value will not affect the results obtained with most pyridine nucleotide-linked enzymes. Prob-

[2] R. Cohen, B. Giraud, and A. Messiah, *Biopolymers* 5, 203–255 (1967).

lems may arise, for example, when lipids are used as a substrate for certain lipases. Such problems must be dealt with individually for each enzyme system and are beyond the scope of this article.

Enzyme Concentration

The amount of enzyme required for active enzyme centrifugal analysis is dependent on its activity and its molecular weight. During the centrifugation the enzyme is located in the cell in a relatively narrow band, which displaces itself slowly. The turnover of the substrate, at the point where the enzyme is located, is thus many times higher than when the same amount of enzyme is evenly distributed over the entire solution, as is the case in a spectrophotometric assay. Nevertheless, the relationship between these two conditions may be experimentally evaluated. It has been our experience that an amount of enzyme that promotes a change in optical density of between 0.001 and 0.005 per minute in a 3-ml cuvette is an appropriate amount to be used in active enzyme centrifugation. This amount of enzyme, dissolved in 10 μl of buffer, will yield satisfactory results, provided that sufficient amounts of the substrates are present in the reaction medium. The overall change in optical density during centrifugation is then sufficiently large to produce a reasonable deflection on the scanner tracing at a low noise level. The amount of enzyme needed for a reasonable deflection may fall outside of the given ranges if the protein is very large or small. The empirical ranges are based upon the relative mobilities of medium-sized molecules, since a small protein will remain in a given region longer than an enzyme of larger size under comparative conditions. Our experience has shown that the indicated amounts of enzyme are very satisfactory for proteins with molecular weights from 30,000 to 150,000.

Enzyme Distribution

Ideal conditions for active enzyme centrifugation with respect to the amount of enzyme to be employed are illustrated in Fig. 2A. The upper part of Fig. 2A represents the change in optical density promoted by the enzyme, observed on the recorder tracings of the scanner at some time during centrifugation. From this tracing one calculates the position of the enzyme at that particular time, and from a series of such tracings one calculates the rate of displacement of the enzyme. The lower part of Fig. 2A indicates the distribution of the active enzyme in the cell. The concentration distribution represents a bell-shaped curve as a result of diffusion. Under ideal conditions the curve of the scanner tracing is related to the distribution of the enzyme as indicated by the lines A, B,

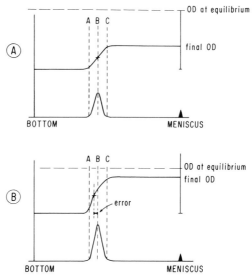

Fig. 2. (A) Relation between recorder trace and enzyme distribution under ideal conditions. Top: recorder trace; bottom: enzyme concentration distribution. The maximum in enzyme concentration coincides with the midpoint of the recorder deflection. (B) Relation between recorder trace and enzyme distribution, using too high enzyme concentration. Most of the reaction has been catalyzed by the enzyme, represented by the left half of the enzyme distribution curve; as a result, the maximum in enzyme concentration no longer coincides with the midpoint of the recorder graph.

and C. The midpoint of the change in optical density coincides with the highest concentration of enzyme, that is, with the top of the bell-shaped curve (line B). Furthermore, the optical density curve is symmetrical in shape, reflecting the distribution of the enzyme at any one point between A and C. These conditions can only be met when at any one point between A and C the amount of enzyme is rate-limiting. If a substrate concentration becomes rate-limiting somewhere between the points B and C, the observed curve no longer reflects the true distribution of the enzyme, as shown in Fig. 2B. This figure illustrates a recorder tracing that is obtained when the enzyme concentration is too high and the equilibrium of the reaction is being approached at point C. The figure clearly indicates that serious errors may be encountered under these conditions, resulting in values for s that are too high. Meaningful results may generally be obtained if care is taken that no more than 30–50% of the available substrates have been converted to products after the enzyme band has passed; this estimate is obviously based upon the assumption that the available amount of substrate is initially much higher than its K_m.

Similar pitfalls may be encountered if the enzyme under investigation is inhibited by one of the products, or activated by one of the substrates, since there is a gradual change in the concentrations of these entities between the points A and C. In order to obtain meaningful information from active enzyme centrifugation, it is necessary for the investigator to be thoroughly familiar with the kinetic properties of the enzyme and to be able to adjust the conditions during centrifugation so that no changes in the kinetic properties of the enzyme occur during centrifugation.

Technical Considerations

Type of Centerpiece

The technique described here uses a Vinograd type double sector centerpiece (Beckman part No. 331359), shown in Fig. 3. The sector at the right, which has a small hole adjacent to it, contains the substrate solutions, and the left sector contains the reference solution. The enzyme is contained in the small hole at the right, and is quantitatively transferred through small grooves to the meniscus of the substrate under influence of the centrifugal force. Changes in optical density occurring in the sample sector are registered by the scanner recorder, using the reference compartment as a blank. The scanner thus acts as a dual-beam spectrophotometer, automatically compensating for any variations in the cell that are not related to the action of the enzyme. The results obtained with this technique are reproducible with an accuracy of better than 95%.

Fig. 3. Vinograd-type double sector centerpiece (Beckman-Spinco, part No. 331359).

Cell Assembly

A properly assembled cell will prevent the enzyme from coming into contact with the reaction mixture prior to the time that the centrifugal speed reaches approximately 10,000 rpm. At this time the enzyme solution is forced through the capillary channels into the cell cavity and will form a sharp, unperturbed boundary with the reaction mixture.

The amount of enzyme to be used is contained in a volume of 10 μl, which is placed in the small hole to the upper left of the centerpiece cutouts. The enzyme is added with a fine-tipped micropipette. The tip is held at the bottom of the hole to avoid capturing an air bubble under the enzyme solution. If an air bubble is trapped under the enzyme solution while the hole is being filled, the enzyme will be forced out when the cell is torqued down, owing to the pressure differential created in the enzyme hole with respect to the substrate compartment. The premature displacement of the enzyme can be determined by inspecting the centerpiece through the window before filling the sectors themselves. If small beads of solution are seen along the sector wall after cell assembly, one can be assured that some of the enzyme has been forced out prematurely. The cell must then be disassembled and cleaned.

Another problem that frequently occurs is a movement of the enzyme solution along the wall of the small hole, due to adhesion, which may also result in a precentrifugation displacement of part of the enzyme solution into the cell cavity. Since the densities of the two solutions are different the problem is recognized by the appearance of density gradients near the meniscus of the sector containing the reaction mixture. This problem may be successfully prevented by applying a light coating of silicone to the upper part of the enzyme hole wall. This prevents the enzyme solution from coming in contact with the window until the cell has been completely assembled and placed under a centrifugal force. Care needs to be exercised in applying the silicone coating to avoid a clogging of the capillary channels from the enzyme hole to the cell cavity with excess silicone. The coating is best applied by a sparingly coated Q-Tip.

Composition of Solutions

The total amount of enzyme to be used has to be measured accurately, for reasons discussed in the theoretical considerations. We have established experimentally that the amount of enzyme to be used may be determined from the results of a regular assay under the conditions to be used during the centrifugation.

The composition of the reaction mixture is subject to individual needs and may be varied over a wide range, provided that its density exceeds the density of the enzyme solution in order to assure proper layering. Any compounds used in the enzyme solution for the purpose of stabilizing the enzyme, e.g., mercaptoethanol, glycerol, EDTA, should also be present in the reaction mixture. Losses in enzymatic activity during centrifugation may be indicated by a slanted line between the meniscus and the enzyme boundary. The results of such experiments are more complicated to calculate, and are apt to contain significant errors. It is advisable to repeat such experiments, using more favorable conditions.

Reference Solution

The preparation of the assay mixture and the reference solution, which will be used to fill the cell, are also subject to some consideration. Two principally different situations may be encountered during active enzyme centrifugation; in one case the optical density in the sample cell *increases* as a result of the enzyme action, whereas in the other case the optical density *decreases*. In the systems in which an increase in optical density is expected, both sections of the cell are filled with the same solution, i.e., the initial assay mixture is used as the reference solution. The wavelength is usually set at the absorption maximum of the formed product. In the systems in which the optical density decreases as a result of the enzyme action, the reference solution contains the same ingredients as the assay solution, but without the absorbing substrate.

Scanning Interval

The initial adjustments of the scanning system are performed during acceleration of the centrifuge. This task is greatly simplified if an oscilloscope equipped with an external trigger circuit is attached to the scanner. A final adjustment of the slit over the image of the reference holes in the counterbalance is done just before the desirable speed is reached, in order to be able to start taking recordings at constant time intervals as soon as possible after the centrifuge has attained the desired speed. This procedure allows for a maximum number of recordings during the early part of the centrifugation, when the enzyme band has not yet been substantially broadened by diffusion.

Recordings may be made at 4-minute intervals, using medium scanning speed and 5 mm/sec recorder speed. If a multiplex is available, and more than one sample is centrifuged simultaneously, recordings may be made using the fast scanning speed and 25 mm/sec recorder speed.

Specific Methods

A. *A Dehydrogenase Reaction Involving Oxidation of a Reduced Pyridine Coenzyme*

Enzyme: H_4 Lactate dehydrogenase from chicken heart, in $0.01\,M$ phosphate buffer, pH 7.5
Reaction Mixture
  Phosphate buffer, $0.1\,M$, pH 7.5
  Pyruvate, 1 mM
  DPNH, 0.15 mM
Reference Solution: Same as reaction mixture; however, DPNH is omitted
Speed: 59,780 rpm
Temperature: regulated at 20°
Wavelength: 340 nm

The amount of enzyme needed is contained in 10 μl, and promotes a change in optical density of 0.005 per minute at 340 nm in a 3-ml cuvette. Recordings are taken at intervals of 8 minutes.

B. *A Dehydrogenase Reaction Involving Reduction of an Oxidized Pyridine Coenzyme*

Enzyme: H_4 lactate dehydrogenase from chicken heart, in $0.01\,M$ Tris·HCl buffer, pH 8.9
Reaction Mixture
  Tris·HCl buffer, $0.1\,M$, pH 8.9
  L-Lactate, 20 mM
  DPN$^+$, 10 mM
Reference Solution: Same as reaction mixture
Speed: 59,780 rpm
Temperature: Regulated at 20°
Wavelength: 340 nm

The amount of enzyme required is contained in 10 μl and promotes a change in optical density of 0.001 per minute at 340 nm in a 3-ml cuvette. The recordings are taken at intervals of 4 minutes.

C. *Enzyme Reactions That Promote an Increase in pH*

$$\text{Pyruvate} + \text{DPNH} + \text{H}^+ \rightarrow \text{L-lactate} + \text{DPN}^+$$

Enzyme: M_4 Lactate dehydrogenase from chicken muscle
Reaction Mixture
  Tris·HCl buffer, $1 \times 10^{-4}\,M$, pH 6.5

Phenol red, $4 \times 10^{-5} M$
Pyruvate, $1 \times 10^{-3} M$
DPNH, $1.4 \times 10^{-4} M$
Reference Solution: Same as reaction mixture
Speed: 59,780 rpm
Temperature: Regulated at 20°
Wavelength: 560 nm

The amount of enzyme needed is the same as that required for B. Recordings are taken at intervals of 4 minutes. The reaction mixture is yellow at pH 6.5, and becomes red at more basic pH. The pK of phenol red is 7.3. The change in optical density with pH is reasonably linear between pH 6.5 and 8.0, and the total change in pH occurring in the cell should be kept within these limits. This can be accomplished by adjusting either the enzyme concentration or the concentration of the buffer. In certain enzyme systems the buffer may be omitted if one of the substrates has a pK around neutrality, thus allowing for a reasonably accurate adjustment of the pH. It should be realized, however, that in such cases complications may arise if the buffering substrate is present in relatively high concentrations and its pK is near that of phenol red.

In this type of experiment, the course of the enzymatic reaction is coupled with an indicator system which absorbs in the visible region of the spectrum. This variation is useful if no change in absorption occurs during the reaction at any accessible wavelength, or when the changes in absorption are obscured by the presence of other absorbing substances. Any suitable indicator system may be employed, provided the conditions as stated in the Theoretical section are met. These are: (1) the stoichiometry of the indicator reaction is known; (2) no appreciable sedimentation of the indicator occurs at the centrifugal force used; and (3) the indicator system does not interfere with the enzymatic activity.

D. Enzyme Reactions That Promote a Decrease in pH

$$\text{DPN}^+ + \text{H}_2\text{O} \rightarrow \text{adenine diphosphoribose} + \text{nicotinamide} + \text{H}^+$$

Enzyme: DPNase from *Neurospora* in 0.1 mM Tris·HCl buffer, pH 7.5
Reaction Mixture
   Tris·HCl buffer, $1 \times 10^{-4} M$, pH 7.5, in 0.1 M NaCl
   Phenol red, $4 \times 10^{-5} M$
   DPN$^+$, $1 \times 10^{-3} M$
Reference Solution: Reaction mixture, adjusted to pH 6.0

Speed: 59,780 rpm
Temperature: Regulated at 20°
Wavelength: 560 nm

The amount of enzyme needed to catalyze the hydrolysis of DPN⁺ is 0.5 unit (Volume II [114]), contained in 10 μl of buffer. Recordings are taken every 8 minutes. See also the remarks under C.

Other Spectrophotometric Assays

Any spectrophotometric assay can basically be applied to studies of active enzyme centrifugation, provided several points are given due consideration. In the first place, the centrifugal mobility of the substance that is being monitored spectrophotometrically is of importance. If the sedimentation rate of this substance is such that it may form a concentration gradient or display a movement either upward or downward under the influence of the centrifugal force, such that it does not maintain a homogeneous distribution during the experiment, serious errors may be introduced into the calculations. Second, one needs to assure oneself that none of the other compounds involved in the reaction are displaced owing to the centrifugal force, with the exception of the enzyme. In the third place, it needs to be ascertained that no interaction, other than those inherent to the catalytic activity, occurs between the protein and the material that is to be monitored.

Data Analysis and Interpretation

The most critical point in the analysis of active enzyme centrifugation is the determination of the enzyme distribution within the sedimenting zone from the obtained scanner tracings. To obtain this enzyme distribution it is necessary to apply the mathematical equations defining zone centrifugation in a sectored cell. Active enzyme centrifugation is governed by the following equation:

$$\frac{d[\text{substrate}]}{dt} = k[\text{product}] \, E + D \, \frac{d^2[\text{substrate}]}{dx^2}$$

If the obtained scanner trace curve is symmetrical as evidenced by equal areas at A and C of Fig. 4, and the line AC is straight, then $d[\text{substrate}]/dt$ is constant and the second derivative is equal to zero. Therefore, at any point where $d[\text{substrate}]/dt$ has a positive value, active enzyme must be present.

To determine the sedimentation behavior of the active enzyme the mean distribution of the protein must be determined. This distribution

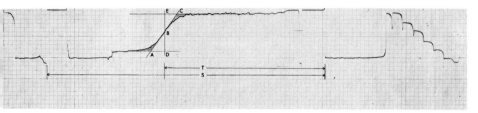

FIG. 4. Calculation of the midpoint B, and its distance to the outer reference line. The distance between the two reference lines is indicated by the distance S, and is 16.1 mm.

in a sectored cell is defined by the second moment which is calculated from the following equation:

$$\log x = \frac{\int \log xd \ (\text{mass})}{\int d(\text{mass})}$$

However, in order to calculate the second moment, an accurate baseline must be determined. As with other centrifugation analyses, the desired accuracy in the baseline determination is very difficult to obtain. When the peak is symmetrical, however, the second moment is closely approximated by the first moment, which is the half height of the scanner trace. For this reason the first moment is more easily defined than the second moment and can be utilized for calculating the sedimentation coefficient of the active enzyme. The error introduced by approximating the second moment under these conditions is not greater than 2%. When the peak becomes distorted or asymmetrical the error introduced by this approximation usually does not exceed 15%. It is therefore reasonable to calculate the sedimentation coefficient of an active enzyme by measuring the rate of migration utilizing the scanner trace half height.

The data reduction utilizing the half height is identical to the extraction of the data from scanner tracings of a conventional sedimentation experiment, and is outlined in Fig. 4. Parallel lines are drawn indicating the absorption before and after the enzymatic reaction. The distance between the two parallel lines presents the overall change in absorption that has occurred as a result of the enzymatic action. Point B indicates the point on the curve that is equidistant from the parallel lines. This point thus coincides with the point of maximal enzyme concentration (see Fig. 2A). A line is then drawn tangent to the curve and passing through point B. This line intersects the two parallel lines at point A and C. Furthermore, a vertical line is drawn through point B, intersecting the parallel lines at points D and E. The shaded areas in Fig. 4 should be of approximately equal areas. If this is not the case, further calculations are useless, and the experiment should be repeated, using less enzyme. If the abnormality is persistent, with changes in enzyme concentration, then the possibility exists that more than a single active species is present.

In order to determine the distance from the center of rotation to the enzyme band one measures the distance from point B to the inside edge of the right reference line (T), and then the distance between the

DATA WORKSHEET

Enzyme: Mannitol-1-phosphate dehydrogenase Enzyme conc.: $\Delta OD_{340} = 0.001/min/10 \ \mu l$
Reaction concentrations: 0.1 M Tris buffer, pH 7.5; 1.7 mM fructose-6-PO$_4$; 0.13 mM DPNH;

Speed: 59,780	Temp: 5570 (22°)	Sample Interval: 8 min
Magnification	27.7 /16.1 mm	

Time (min)	T (mm)	T/mag (mm)	73-T/mag(= Y) (mm)	log(73 − T/mag) (= log Y)
0	17.2	10.0	63.0	1.800
8	16.6	9.65	63.35	1.802
16	16.0	9.30	63.70	1.805
24	15.2	8.85	64.15	1.807
32	14.6	8.50	64.50	1.810
40	13.8	8.05	64.95	1.812
48	13.2	7.70	65.30	1.815
56	12.6	7.35	65.65	1.818
64	12.0	7.00	66.00	1.820
72	11.3	6.60	66.40	1.822

$$s_{obs} = \frac{3.50}{(rpm)^2} \times slope$$

$$s_{20,w} = s_{obs} \left(\frac{\eta_t}{\eta_{20}}\right) \left(\frac{\eta}{\eta_0}\right) \left(\frac{1 - \bar{V}\rho_{20,w}}{1 - \bar{V}\rho_t}\right)$$

inside edges of the two reference lines (S). The distance between the inside edges of the reference holes in the counterbalance is 16.1 mm. From these data the magnification factor (mag) is calculated:

$$\text{mag} = \frac{S}{16.1}$$

The actual distance from the enzyme to the outer reference hole is thus T/mag. Since the outer reference hole is 73.00 mm from the center of rotation, the actual distance from the center of rotation to the center of the enzyme band is:

$$Y = 73.00 - T/\text{mag}$$

This value is determined for every scanning trace obtained in the experiments, as outlined in the table.

The subsequent treatment of the data in order to obtain the sedimentation coefficient is analogous to that of a centrifugation experiment using schlieren or interference optics. A graph is made, plotting values of log Y vs. time; this plot should result in a straight line. Nonlinear or broken lines are indicative of the fact that complications arose during the centrifugation. Such complications may result from leakage of the cell as well as from denaturation of the protein during the experiment or from secondary kinetic phenomena such as product inhibition. A plot of the log Y values, obtained in the table, with time is presented in Fig. 5. The slope of this line (the time is expressed in minutes) is

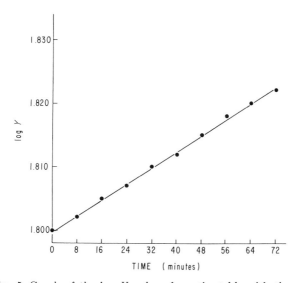

Fig. 5. Graph of the log Y values from the table with time.

related to the observed sedimentation coefficient s_{obs} by the following equation:

$$s_{obs} = \frac{3.50}{(rpm)^2} \times slope$$

In order to determine $s_{20,w}$ corrections for the viscosity and the density of the substrate solution must be made. The following equation may be used for this purpose:

$$s_{20,w} = s_{obs} \left(\frac{\eta_t}{\eta_{20}}\right) \left(\frac{\eta}{\eta_0}\right) \left(\frac{1 - \bar{v}\rho_{20,w}}{1 - \bar{v}\rho_t}\right)$$

in which η/η_0 is the principal correction factor corresponding to the relative viscosity of the substrate solution to that of water, η_t/η_{20} is the viscosity of water at the operating temperature relative to that at 20°, and $\rho_{20,w}$ and ρ_t are the densities of water at 20° and of the substrate solution at $t°$, respectively.

We have found that a worksheet similar to that of the table is very helpful in performing the calculations as well as to maintain a permanent record of the evaluation of the experiment. The data presented in the table and in Fig. 5 were taken from an actual centrifugation experiment performed in our laboratory. The table and Fig. 5 are given to illustrate the total calculations needed for the interpretation of an active enzyme centrifugation experiment. Since the amount of enzyme in the experiment is very small, no significant change in $s_{20,w}$ is to be expected when the enzyme concentration is extrapolated to zero.

Acknowledgment

This work was supported in part by grants from the National Cancer Institute (CA-11683) and the American Cancer Society (P-77M) to Dr. N. O. Kaplan.

[4] Measurement of Partial Specific Volume by Sedimentation Equilibrium in H₂O–D₂O Solutions[1]

By STUART J. EDELSTEIN and H. K. SCHACHMAN

Advances in ultracentrifugal techniques, particularly the development of the photoelectric scanning absorption optical system[1a] (see

[1] This research was supported in part by U.S. Public Health Service Research Grants HL 13591 to S.J.E. from the National Heart and Lung Institute and GM 12159 to H.K.S. from the National Institute of General Medical Sciences, and by National Science Foundation Research Grants GB 8773 to S.J.E. and GB 4810X to H.K.S.
[1a] H. K. Schachman and S. J. Edelstein, *Biochemistry* 5, 2681 (1966).

also article [1] in this volume), have permitted molecular weight measurements to be conducted with very small amounts of material. However, in all cases, determinations of molecular weights by sedimentation methods require values for the partial specific volume, $\bar{v}$, of the protein or other material examined. For proteins in particular, accurate values of $\bar{v}$ are necessary since errors in $\bar{v}$ are magnified about 3-fold in the subsequent calculations of molecular weights. The magnification of errors occurs because $\bar{v}$ for proteins is about 0.75 ml/g and the buoyancy term, $(1 - \bar{v}\rho)$, appearing in all basic equations[2-4] has a value of 0.25 for dilute buffer solutions of density, ρ, about 1.0 g/ml. The classical methods for determining $\bar{v}$ require large amounts of material, both for dry weight measurements of concentrations and for accurate density measurements by pycnometry.[5] Where density gradient columns[6] or floats[7-9] are used, smaller amounts of material can be employed for the density measurements. However, relatively large amounts of purified protein are required for determining dry weight concentrations, and these amounts frequently exceed the quantities needed for the sedimentation experiments themselves. In order to eliminate the requirements and limitations of these techniques, a method was developed that permits measurement of $\bar{v}$ along with the molecular weight by performing parallel sedimentation equilibrium experiments in solutions of H_2O and D_2O.[10] The method has the advantage of requiring only small amounts of material, and it involves operations that are only a minor extension of the sedimentation equilibrium experiments themselves. In addition, the method can be applied to multicomponent systems, such as protein solutions in guanidine hydrochloride.[11]

It should be emphasized that, while the D_2O method is convenient and is readily performed with small amounts of material, it does require

[2] T. Svedberg and K. O. Pedersen, "The Ultracentrifuge." Oxford Univ. Press, London and New York, 1940.

[3] H. K. Schachman, "Ultracentrifugation in Biochemistry." Academic Press, New York, 1959.

[4] H. Fujita, "Mathematical Theory of Sedimentation Analysis." Academic Press, New York, 1962.

[5] N. Bauer, in "Physical Methods of Organic Chemistry" (A. Weissberger, ed.), p. 253, Vol. 1, Part 1. Wiley, New York, 1949.

[6] K. O. Linderstrøm-Lang and H. Lanz, C. R. Trav. Lab. Carlsberg Ser. Chim. **21**, 315 (1938).

[7] M. O. Dayhoff, G. E. Perlmann, and D. A. MacInnes, J. Amer. Chem. Soc. **74**, 2515 (1952).

[8] D. V. Ulrich, D. W. Kupke, and J. W. Beams, Proc. Nat. Acad. Sci. U.S. **52**, 349 (1964).

[9] M. J. Hunter, J. Phys. Chem. **70**, 3285 (1966).

[10] S. J. Edelstein and H. K. Schachman, J. Biol. Chem. **242**, 306 (1967).

[11] J. O. Thomas and S. J. Edelstein, Biochemistry **10**, 477 (1971).

very accurate sedimentation equilibrium data; hence multiple determinations are required in order to obtain enhanced precision. Although the D_2O method may be essential where material is limited, it is not likely to supplant the classical methods of densitometry, where extreme accuracy is required in the determination of $\bar{v}$. A considerable gain in the precision of the sedimentation equilibrium method can be achieved[10] through the use of $D_2{}^{18}O$ in place of D_2O, but the expense of the former may be restrictive. In this article the basic principles of the sedimentation equilibrium method are presented along with illustrations of its application. The analyses of both two-component and three-component systems are described along with the use of both the photoelectric scanner and interference optics.

Theory

Two-Component Systems. The determination of $\bar{v}$ involves two sedimentation equilibrium experiments in H_2O and D_2O and the solution of two simultaneous equations for the unknown quantities, $\bar{v}$ and M, where M is the molecular weight of the macromolecules.

The behavior of macromolecules in an ideal solution at sedimentation equilibrium in an aqueous solvent can be described by Eq. (1).

$$M(1 - \bar{v}\rho_{H_2O}) = \frac{2RT}{\omega^2}\left(\frac{d \ln c}{dr^2}\right)_{H_2O} \tag{1}$$

where the subscript, H_2O, refers to normal water, R is the gas constant, T is the absolute temperature, ω is the angular velocity of the rotor in radians per second, c is the concentration, and r is the distance from the axis of rotation in centimeters. When the macromolecules are dissolved in D_2O, their molecular weight is increased as a result of deuterium exchange and their partial specific volume is decreased by the same relative amount. Accordingly Eq. (1) takes the form

$$kM\left(1 - \frac{\bar{v}}{k}\rho_{D_2O}\right) = \frac{2RT}{\omega^2}\left(\frac{d \ln c}{dr^2}\right)_{D_2O} \tag{2}$$

where k is the ratio of the molecular weight of the macromolecules in the deuterated solvent to that in the nondeuterated solvent. The subscript D_2O refers to the deuterated solvent. These two equations can be solved simultaneously for $\bar{v}$ to give

$$\bar{v} = \frac{k - [(d \ln c/dr^2)_{D_2O}/(d \ln c/dr^2)_{H_2O}]}{\rho_{D_2O} - \rho_{H_2O}[(d \ln c/dr^2)_{D_2O}/(d \ln c/dr^2)_{H_2O}]} \tag{3}$$

The value of k can be estimated reliably from knowledge of the composition and structure of the macromolecules by computing the number of exchangeable hydrogens. For all proteins k is relatively constant since the exchangeable hydrogens reside principally in the one amide hydrogen per amino acid residue in the polypeptide backbone and to a relatively minor extent in the amino acid side chains. The value of $k = 1.0155$ has been determined for a number of proteins from studies of deuterium exchange in pure D_2O (see references in 10). Although less extensive data are available for the deuterium exchange into other macromolecules, the appropriate values of k can be calculated through the use of the related model compounds. When the solvent contains D_2O at concentrations significantly below 100%, the value of k is reduced proportionately. Thus for 1:1 mixtures of H_2O and D_2O, k would be only about 1.00775.

Three-Component Systems. For three-component systems it is convenient to adopt the nomenclature of Casassa and Eisenberg,[12] where the macromolecules are designated as component 2, and components 1 and 3 refer to water and the low molecular weight solute, respectively. The buoyancy term, $(1 - \bar{v}\rho)$, for such systems must be corrected for possible preferential interactions between the macromolecules and either of the other components. As shown by Casassa and Eisenberg,[12] this correction can be achieved by defining an apparent specific volume, ϕ'_2, in terms of the density increment, $(\partial\rho/\partial c_2)_{\mu^0}$, and the parameter, ξ_1, which describes the preferential interaction. Their expression can be written as

$$(1 - \phi'_2\rho) = (\partial\rho/\partial c_2)_{\mu^0} = (1 - \bar{v}_2\rho) + \xi_1(1 - \bar{v}_1\rho) \qquad (4)$$

where the subscript μ indicates constancy of chemical potentials of components diffusible through a semipermeable membrane and the superscript zero refers to vanishing concentration of the macromolecules. In Eq. (4) the preferential interaction of the macromolecules is written as the binding of component 1 (hydration) where ξ_1 is $(\partial w_1/\partial w_2)_{\mu^0}$; this term expresses the weight of component 1 in grams which must be added to the solution, per gram of component 2, in order to maintain constancy of μ_1 and μ_3.

The concentration gradient of macromolecules in such three-component systems is described by the equation

$$\frac{d \ln c_2}{dr^2} = \frac{M_2\omega^2(1 - \phi'_2\rho)}{2RT} \qquad (5)$$

The interaction term in Eq. (4) can also be measured using the H_2O–D_2O technique[11] by defining a volumetric interaction parameter q_1, where

[12] E. F. Casassa and H. Eisenberg, *Advan. Protein Chem.* **19**, 287 (1964).

$$q_1 = (\partial V_1/\partial w_2)_{\mu^0} = \xi_1/\rho_1^0 \tag{6}$$

Here V_1 is the volume and ρ_1^0 the density of component 1 which is bound to or excluded from the macromolecules. The apparent specific volume, ϕ'_2, can be expressed in terms of q_1 by

$$\phi'_2 = \bar{v}_2 + q_1(1 - \rho_1^0/\rho) \tag{7}$$

Two simultaneous equations can now be formulated, analogous to the derivation for a two-component system in H_2O and D_2O. The concentration gradient at sedimentation equilibrium is given by

$$\left(\frac{d \ln c_2}{dr^2}\right)_{H_2O} = \frac{M_2\omega^2[(1 - \bar{v}_2\rho_{H_2O}) + q_1(\rho_1^0 - \rho_{H_2O})]}{2RT} \tag{8}$$

For D_2O solutions the corresponding equation becomes

$$\left(\frac{d \ln c_2}{dr^2}\right)_{D_2O} = \frac{kM_2\omega^2[(1 - \bar{v}_2\rho_{D_2O}/k) + q_1(\rho_1^0 - \rho_{D_2O})]}{2RT} \tag{9}$$

It should be noted that ρ_1^0 in Eqs. (8) and (9) refer to pure H_2O and

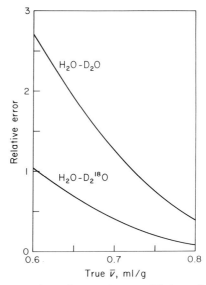

FIG. 1. Effect of errors in sedimentation equilibrium data on the calculated value of $\bar{v}$. The relative error, i.e., the error in $\bar{v}$ per unit error in the ratio $(d \ln c/dr^2)_{D_2O}/(d \ln c/dr^2)_{H_2O}$ is presented on the ordinate as a function of the true $\bar{v}$ on the abscissa for hypothetical experiments in H_2O and D_2O as well as H_2O and $D_2^{18}O$. From S. J. Edelstein, Ph.D. Thesis, University of California, Berkeley, 1967.

D_2O, respectively. These two equations can then be solved simultaneously to yield a value of q_1.

$$q_1 = \frac{\bar{v}_2(\rho_{D_2O} - F\rho_{H_2O}) + F - k}{k(\rho^0_{D_2O} - \rho_{D_2O}) - F(\rho^0_{H_2O} - \rho_{H_2O})} \tag{10}$$

In Eq. (10) F is the ratio of the concentration gradients $(d \ln c_2/dr^2)$ in D_2O and H_2O, $\rho^0_{D_2O}$ and $\rho^0_{H_2O}$ refer to the density of pure D_2O and

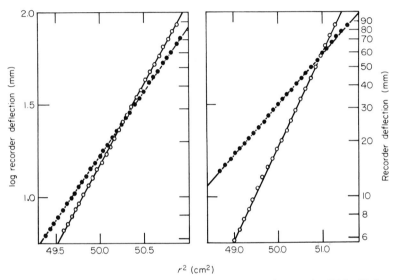

FIG. 2. Sedimentation equilibrium of α-chymotrypsinogen in H_2O, D_2O, and $D_2^{18}O$ solutions. In the experiment represented by the plots at the left, the solution (containing 20 μg of α-chymotrypsinogen in 0.1 ml of 0.1 M phosphate at pH 7 in H_2O) was placed in one compartment of a double-sector cell with solvent in the other compartment. A second cell contained the analogous solution of protein in D_2O in one compartment and solvent in H_2O in the other. The speed was 36,000 rpm. In the second experiment (speed 28,000 rpm) (plots at the right), the solution in cell 2 contained 90% $D_2^{18}O$, and that in cell 1 contained the analogous solution and solvent in H_2O. Measurements were made from expanded traces of the type illustrated by the bottom patterns in Fig. 3, and the data were plotted as the logarithm of the recorder deflection against the square of the distance, r, from the axis of rotation. $\bigcirc$, results in H_2O; $\bullet$, those in D_2O (or $D_2^{18}O$). The light used with the photoelectric scanner had a wavelength of 280 nm. Since the $D_2^{18}O$ itself had a slight absorbance at this wavelength due to an unidentified impurity, a corrected baseline was established after the equilibrium patterns were recorded. This was accomplished by acceleration of the rotor to 60,000 rpm and measurement of the recorder deflection in the supernatant liquid [H. K. Schachman and S. J. Edelstein, *Biochemistry* **5**, 2681 (1966)]. Data from S. J. Edelstein and H. K. Schachman, *J. Biol. Chem.* **242**, 306 (1967).

H_2O D_2O^{18}

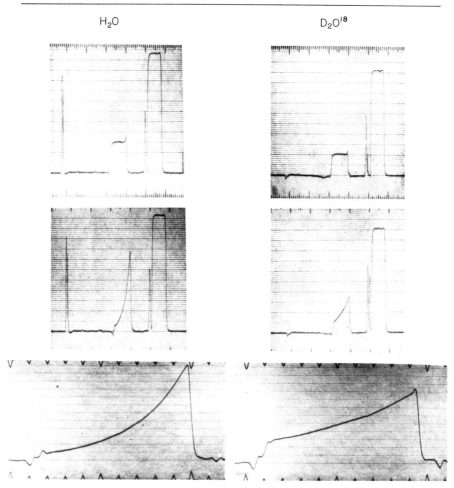

FIG. 3. Sedimentation equilibrium patterns for myoglobin in H_2O and $D_2^{18}O$ solutions. In the experiment, two double-sector cells were used with multiplex operation of the photoelectric scanner [H. K. Schachman and S. J. Edelstein, *Biochemistry* **5**, 2681 (1966)]. Cell 1 contained 4 μg of myoglobin in 0.1 ml of 0.1 M phosphate, pH 7, in H_2O. The traces representing this solution are at the left. Cell 2 contained 2.7 μg of myoglobin in the same buffer in 90% $D_2^{18}O$, and the corresponding traces are at the *right*. The $D_2^{18}O$ solution was prepared by gravimetric dilution of 0.1 ml of a myoglobin solution (containing 27 μg of myoglobin in 1.0 M phosphate at pH 6.7) with 0.9 ml of $D_2^{18}O$. About 0.1 ml of fluorocarbon FC 43 was added to each compartment of the double-sector cells in order to produce a transparent region at the cell bottom and thereby permit accurate measurements of the absorbance throughout the liquid columns. The traces were recorded, with light of wavelength 405 nm; the traces at the top were obtained immediately after the rotor attained the equilibrium speed of 28,000 rpm. The time required for a scan of the image of the entire cell was 30 seconds, and the trace amplitude was 400 (1000 is

pure H_2O, respectively, and ρ_{D_2O} and ρ_{H_2O} correspond to the densities of the solutions of the macromolecules in D_2O and H_2O, respectively. When the value of q_1 is substituted into Eq. (6), we obtain

$$\phi'_2 = \frac{\bar{v}_2(kJ - \rho_{D_2O}) + k - F}{kJ + F_{\rho_{H_2O}}} \tag{11}$$

where

$$J = \frac{\rho_{H_2O}(\rho^0_{D_2O} - \rho_{D_2O})}{\rho_{H_2O} - \rho^0_{H_2O}} \tag{12}$$

It should be noted that the notation used here differs somewhat from that presented earlier.

Experimental Determinations

Two-Component Systems. Although the equations presented above show that it is possible to obtain the partial specific volume for two-component systems, it is important to recognize the requirement for precision in the experimental data. An experimental error of 1% in the determination of the slopes of plots of $\ln c$ vs r^2 may lead to errors of several percent in the estimation of $\bar{v}$ by the H_2O–D_2O method. The exact dependence of errors in $\bar{v}$ on errors in primary data will depend on the value of $\bar{v}$ itself—the higher the $\bar{v}$, the greater the perturbation in the equilibrium concentration distribution caused by D_2O and the smaller the multiplication of errors. A summary of the dependence of the errors in the determination of $\bar{v}$ is presented in Fig. 1. As indicated by the lower line in the figure, errors are substantially reduced if $D_2^{18}O$ is employed instead of D_2O. With $D_2^{18}O$, perturbations of the concentration distribution are greater and the ratio of slopes $(d \ln c/dr^2)_{D_2O}/(d \ln c/dr^2)_{H_2O}$ can be obtained more accurately. This point is illustrated by a comparison of H_2O–D_2O and H_2O–$D_2^{18}O$ determinations of $\bar{v}$ for α-chymotrypsinogen. As shown in Fig. 2, the slopes representing the concentration distributions differ much more widely for the H_2O–$D_2^{18}O$ system than

---

the setting for maximum amplification). In the center are the traces after 16 hours of centrifugation. At the bottom are expanded and amplified traces showing only the regions of the cells corresponding to the myoglobin solutions. For these traces the scanning period was increased to 6 minutes so as to give an improved signal-to-noise ratio through the use of an electronic filter at the slower scanning rate. In these expanded traces the 3-mm column of the solution appeared as 20 cm on the final traces. In order to facilitate measurements of the concentration distribution the recorder deflection was amplified by adjustment of the helipot control to 600 for cell 1 and to 1000 for cell 2. From S. J. Edelstein and H. K. Schachman, *J. Biol. Chem.* **242**, 306 (1967).

for the H_2O–D_2O experiment. The dramatic perturbation introduced by $D_2{}^{18}O$ can also be seen in Fig. 3 which presents the actual scanner traces obtained for myoglobin in H_2O and $D_2{}^{18}O$ buffers.

Three-Component Systems. For measurements of interactions in multicomponent systems, the multiplications of errors due to inaccuracies in the primary data are even larger. For example, in the analysis of molecular weights of aldolase in H_2O and D_2O an experimental error of 3% in a determination of the quantity $d\ln c_2/dr^2$ is magnified to an error of about 100% in the estimation of q_1.[11] However, because of the nature of the equations involved, the errors are reduced when the actual molecular weight is estimated. For three-component systems the multiplication of errors is complex, depending on both the value of q_1 and the densities of the solvents, which will vary markedly, depending on the amounts of the third component. Therefore, some estimation of the errors involved at all stages of the calculation should be performed, particularly for multicomponent systems to ensure that the accuracy obtained is sufficient to permit satisfactory interpretation of the data. When such precautions are taken, adequate results can be obtained, as in the case of aldolase in guanidine-HCl (Gu-HCl). Figure 4 summarizes

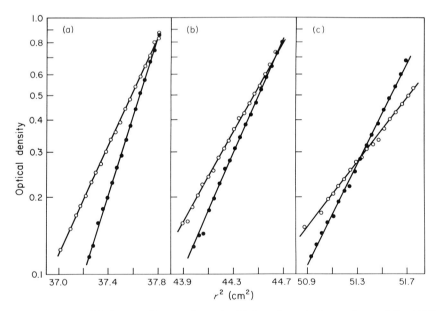

Fig. 4. Equilibrium sedimentation of aldolase in solutions of guanidine-HCl (Gu-HCl) and guanidine-DCl. Solutions contained 0.2 mg/ml of aldolase in protonated (●) or deuterated (○) solvents. (a) 3 M Gu-HCl, (b) 5 M Gu-HCl, and (c) 7 M Gu-HCl. The rotor was maintained at 20°; the speed was 40,000 rpm. From J. O. Thomas and S. J. Edelstein, *Biochemistry* **10**, 477 (1971).

TABLE I

QUANTITIES DERIVED FROM DENSITY PERTURBATION ULTRACENTRIFUGATION
OF ALDOLASE IN GUANIDINE (Gu)-HCl-0.01 M MERCAPTOETHANOL AT 20°[a]

Concentration of Gu-HCl or GU-DCl	3 M	5 M	7 M
ρ_{H_2O}	1.075	1.120	1.164
ρ_{D_2O}	1.173	1.213	1.246
$M_2(1 - \phi'_2\rho) \times 10^{-3}$			
$\quad$ H$_2$O	9.6 ± 0.1	7.4 ± 0.2	5.8 ± 0.2
$\quad$ D$_2$O	7.3 ± 0.3	5.4 ± 0.2	4.0 ± 0.1
$M_{2,app} \times 10^{-3}$			
$\quad$ H$_2$O	47 ± 1	44 ± 1	43 ± 1
$\quad$ D$_2$O	51 ± 2	48 ± 2	45 ± 1
q_1	−0.6 ± 0.6	−0.4 ± 0.4	−0.1 ± 0.1
ϕ'_2	0.70 ± 0.04	0.70 ± 0.04	0.73 ± 0.01
$M_{2,true} \times 10^{-3}$	39 ± 6	34 ± 6	39 ± 4

[a] Each set of data is the average of four experiments; the errors represent average deviations. For details of calculations, see J. O. Thomas and S. J. Edelstein, *Biochemistry* **10**, 477 (1971).

three sets of measurements, comparing the sedimentation equilibrium distributions of aldolase in Gu-HCl–H$_2$O solutions and deutero-guanidine-DCl–D$_2$O solutions at 3 M, 5 M, and 7 M Gu-HCl (or deutero-guanidine-DCl). As seen in Table I, apparent molecular weights about 45×10^3 were obtained, but corrections for guanidine binding, as revealed by values of $q_1 < 0$ and $\phi'_2 < \bar{v}_2$, yield values close to 39×10^3. The results indicate that analysis of the data without correction for preferential interactions would lead to an erroneous molecular weight for the polypeptide chains and a misleading conclusion about the number of subunits in the native protein. However, the data corrected for preferential interactions show that aldolase contains four subunits, a conclusion consistent with other information.[13] This same approach with Gu-HCl and deutero-Gu-DCl has also been applied in studies of DNA polymerase[11] and the M-line protein of skeletal muscle.[14]

Comparison of Scanner and Interference Methods. Measurements of $\bar{v}$ for two-component systems can be performed with either the photoelectric scanner or with interference methods. The results with the scanner in its commercial form (Beckman Instruments, Inc.) or in the model developed at Berkeley[1a] are generally less accurate than corresponding experiments with the Rayleigh interference optical systems. A

[13] E. Penhoet, M. Kochman, R. C. Valentine, and W. J. Rutter, *Biochemistry* **6**, 2940 (1967).
[14] K. Morimoto and W. F. Harrington, *J. Biol. Chem.* **247**, 3052 (1972).

TABLE II

MOLECULAR WEIGHT AND PARTIAL SPECIFIC VOLUME OF BOVINE PLASMA ALBUMIN[a]

Speed (rpm)	Optics	Solution	$d \ln c/dr^2$	$M_w \times 10^{-4}$	$\bar{v}$ (ml/g)
10,000	Interference	H_2O	0.2025 ± 0.00086	7.98	0.743 ± 0.0040
		D_2O	0.1558 ± 0.00060	7.84	
	Absorption	H_2O	0.2099 ± 0.0016	8.27	0.740 ± 0.0098
		D_2O	0.1624 ± 0.0019	8.17	
26,000	Interference	H_2O	1.152 ± 0.0044	6.72	0.726 ± 0.0033
		D_2O	0.9112 ± 0.0035	6.78	
	Absorption	H_2O	1.163 ± 0.016	6.78	0.727 ± 0.0110
		D_2O	0.9205 ± 0.0058	6.85	

[a] Bovine plasma albumin (0.88 mg/ml) was examined at sedimentation equilibrium in 0.2 M NaCl, 0.01 M acetate, pH 5.4, in H_2O and D_2O. Each analysis was performed with 0.1 ml of solution. The experiments were conducted with a 12-mm multichannel cell [D. Yphantis, *Biochemistry* **3**, 297 (1964)] with sapphire windows and open window holders to achieve compatibility with interference and absorption optical systems. A Rayleigh aperture mask was included in the centrifuge chamber [E. G. Richards, D. C. Teller, and H. K. Schachman, *Biochemistry* **7**, 1054 (1968)]. Data from the absorption optical system were obtained with light of 280 nm. Final solutions were prepared by gravimetric dilution of stock solutions of protein and buffer, D_2O and H_2O of known density. After dilution the concentration of D_2O was 91.0%. The data were obtained after 20 hours of centrifugation at 10,000 rpm at 20° and after an additional 20 hours at 26,000 rpm. Interference plates were read with a Nikon microcomparator. Labeling of fringes at the lower speed was achieved by the method of F. E. LaBar (*Proc. Nat. Acad. Sci. U.S.* **54**, 31 (1965)] since concentrations were below the level required for synthetic boundary cell determinations [E. G. Richards, D. C. Teller, and H. K. Schachman, *Biochemistry* **7**, 1054 (1968)]. Labeling of fringes at the higher speed was achieved by the method of D. Yphantis [*Biochemistry* **3**, 297 (1964)]. Values of $d \ln c/dr^2$ were obtained by linear least squares analysis. Numbers following ($\pm$) represent standard deviations. Values of M_w were calculated on the basis of $\bar{v} = 0.734$ ml/g and are corrected for deuteration where appropriate. Calculations of $\bar{v}$ were performed with a value of $k = 1.0140$, the measured value, $k = 1.0155$, corresponding to pure D_2O, corrected to correspond to 91.0% D_2O. Errors in $\bar{v}$ ($\pm$) are the product of the root-sum-square of the individual fractional errors in $d \ln c/dr^2$ and the ratio $(d \ln c/dr^2)_{D_2O}/(d \ln c/dr^2)_{H_2O}$ corrected by 0.9, the reduction factor obtained from Fig. 1.

direct comparison of the two methods has been conducted,[15] and the results for experiments with bovine serum albumin are summarized in Table II. As can be seen from these results, the standard deviations for the slopes of ln c vs. r^2 are consistently smaller with data from interference measurements, and these lower error levels are reflected in the

[15] S. J. Edelstein, Ph.D. Thesis, University of California, Berkeley, 1967.

more accurate determinations of $\bar{v}$. It is possible that some of the recent developments with the scanner system involving on-line computer operation[16] would enhance the levels of precision to the range found with the interference optical system (see also article in this volume on photoelectric scanner). The schlieren optical system has been employed for determination of the $\bar{v}$ of ferredoxin,[17] but the accuracy with this optical system is likely to be below the level available with the scanner or interference techniques.

In studies where interaction parameters are measured, as in sedimentation equilibrium experiments with concentrated solutions of guanidine HCl, the scanner has particular advantages. Since wavelengths can usually be found where the third component does not absorb light, the operations with multicomponent systems are only slightly more difficult than with two-component systems. However, with interference optical system measurements, the third component creates several problems. If the solutions in the reference and sample sectors are not perfectly matched in terms of guanidine-HCl concentration or solution volume, the entire fringe pattern may be displaced, giving inaccuracies in the estimation of concentration or possibly blurring of the fringes if the refractive indices of solution and the reference liquid (solvent) differ appreciably. Therefore, measurements of this type should be performed with the scanner if at all possible and efforts should be made to reduce the level of error by repetition of experiments. In the studies with aldolase (Table I) and DNA polymerase[11] in guanidine hydrochloride a satisfactory level of errors was achieved by averaging four determinations for each individual point. The multiplexing capability of the scanner[1a] permits multiple determinations to be performed simultaneously very conveniently.

Density Determination. The values of solution densities needed for calculations can be substantially less precise than would be needed for $\bar{v}$ measurements by pycnometry. Therefore smaller volumes can be used in density determinations, consistent with the requirements for the sedimentation equilibrium measurements (about 0.1 ml). Adequate density measurements can be performed with a calibrated micropipette possessing a constriction which is used as a reference for reproducible filling. In this way densities can be measured with a few tenths of a milliliter in conjunction with a precision balance such as the Mettler microanalytical balance. Moreover, at the concentrations of protein usually employed (<1 mg/ml), the contribution of the protein to the density is negligible

[16] R. H. Crepeau, S. J. Edelstein, and M. J. Rehmar, *Anal. Biochem.* **50**, 213 (1972).
[17] T. Devanathan, J. M. Akagi, R. T. Hersh, and R. H. Himes, *J. Biol. Chem.* **244**, 2846 (1969).

(<0.003 g/ml) and the density of the solutions needed for the calculations can be measured with the solvent, which is readily available in larger amounts.

Discussion

Since its introduction in 1967, the H_2O–D_2O sedimentation equilibrium method for determining $\bar{v}$ in the ultracentrifuge has been applied to a wide range of materials. The scope of applications is illustrated by Table III which summarizes the results obtained with the method for small molecules (MW $< 10^3$), such as adenosine and enterobactin, and a wide range of macromolecules up to large aggregates, such as 80 S ribosomes (MW about 3.5×10^6). Values of $\bar{v}$ range from near 0.6 ml/g for ribosomes, nucleosides, and metalloproteins, such as ferredoxin, to 0.88 ml/g for the photosynthetic reaction center, which is rich in lipoprotein. A few proteins have values of $\bar{v}$ below 0.7 ml/g, but the majority fall in the range 0.72–0.74 ml/g. In a small number of cases $D_2{}^{18}O$ has been employed as the heavy water.

Prior to the development of the H_2O–D_2O sedimentation equilibrium method for measurement of $\bar{v}$, a similar approach had been employed based on sedimentation velocity experiments.[18-23] Although the velocity method has been used in a small number of recent studies, the equilibrium technique appears likely to replace it. The preference for the equilibrium method stems from several advantages: (1) smaller quantities of material are required; (2) viscosity data on the solutions are not needed;. (3) temperature control can be less stringent; (4) no ambiguities arise due to possible changes in shape (frictional factor) of the macromolecules in D_2O; and (5) analysis can be performed by equilibrium measurements with materials which are too small for accurate determination of sedimentation coefficients.

While few serious pitfalls exist with the H_2O–D_2O method for measurement of $\bar{v}$ at sedimentation equilibrium, special caution must be exercised in studies on systems involving association–dissociation equilibria. D_2O has been found to affect subunit interactions in a number of proteins, and in the case of hemoglobin attempts have been made to correct $\bar{v}$ measurements for D_2O effects on the absolute molecular weight dependence on concentration (S. J. Edelstein, unpublished re-

[18] T. Svedberg and I. B. Eriksson-Quensel, *Nature* (*London*) **137**, 400 (1936).
[19] D. G. Sharp and J. W. Beard, *J. Biol. Chem.* **185**, 247 (1950).
[20] P. Y. Cheng and H. K. Schachman, *J. Polym. Sci.* **16**, 19 (1955).
[21] S. Katz and H. K. Schachman, *Biochim. Biophys. Acta* **18**, 28 (1955).
[22] W. G. Martin, W. H. Cook, and C. A. Winkler, *Can. J. Chem.* **34**, 809 (1956).
[23] W. G. Martin, C. A. Winkler, and W. H. Cook, *Can. J. Chem.* **37**, 1662 (1959).

TABLE III

SUMMARY OF $\bar{v}$ MEASUREMENTS BY SEDIMENTATION EQUILIBRIUM
STUDIES IN H_2O–D_2O SOLUTIONS[a]

Material	$\bar{v}$ (ml/g)	M	Method	Reference[b]
80 S Ribosome (*Paramecium*)	0.601	3.47×10^6	I	(1)
Ferredoxin	0.61	5,800	P	(2)
Ferredoxin	0.62	6,000	S	(3)
Adenosine	0.62	254	P	(2)
Deoxynucleotidyl transferase	0.65	32,360	I*	(4)
Vitamin B_{12} binding protein	0.679	66,000	P*	(5)
Nonheme iron protein	0.684	13,000	P	(6)
Fructose diphosphatase (spinach chloroplast)	0.697	73,000	I	(7)
Surface protein (*Paramecium*)	0.702	301,000	I	(8)
Ovoinhibitor C	0.707	48,700	P	(9)
Human plasminogen	0.714	81,000	P	(10)
Chymotrypsinogen A	0.715	25,700	I	(11)
Glucose-6-phosphate dehydrogenase	0.717	103,000	I	(12)
α-Chymotrypsinogen	0.72	25,000	P*	(2)
Molybdoferredoxin	0.72	140,000	P	(13)
D-Serine dehydrogenase	0.727	45,800	I	(14)
Human serum high density lipoprotein	0.728	27,500	I	(15)
Bovine serum albumin	0.73	68,000	I + P	(16)
Galactosyl transferase	0.73	23,000	P	(17)
Citrate lyase	0.73	550,000	I	(18)
Phosphomannose isomerase	0.73	45,000	I	(19)
M-line protein	0.73	88,000	I	(20)
Vitamin D binding protein	0.73	52,800	I	(21)
Canine immunoglobulin A	0.731	277,000	P	(22)
Citrate synthetase	0.733	100,000	I	(23)
Fructose diphosphatase	0.733	127,000	I	(24)
Aspartate amino transferase	0.734	79,000	P	(25)
Insect flight muscle phosphorylase	0.735	97,000	P	(26)
Adenovirus type 2 hexon	0.736	333,000	P	(27)
Urocanase	0.737	110,000	P	(28)
Spermidine dehydrogenase	0.738	76,000	I*	(29)
DNA polymerase	0.74	115,000	P	(30)
Myoglobin	0.74	17,100	P*	(2)
Type B toxin of *Clostridium botulinum*	0.74	167,000	I	(31)
Liver aldolase	0.741	158,000	I	(32)
Human thyroxine-binding prealbumin	0.742	59,700	I	(33)
Chymotrypsin inhibitor I	0.743	39,900	I	(34)
Glutathione reductase	0.744	124,000	I	(35)
Ovomacroglobulin	0.744	637,000	P	(36)
Phosphoglucose isomerase	0.745	125,000	P	(37)

(*Continued*)

TABLE III (*Continued*)

Material	$\bar{v}$ (ml/g)	M	Method	Reference[b]
Aspartokinase	0.75	116,000	I	(38)
Rat liver arginase	0.75	118,500	I	(39)
Iron-sulfur protein I	0.75	21,000	I	(40)
Kidney bean erythroagglutin	0.75	140,000	P*	(41)
Human serum high density lipoprotein	0.776	30,500	P*	(42)
Protein of human erythrocyte membrane	0.78	49,900	I	(43)
Photosynthetic reaction center	0.88	153,000	C	(44)
Enterobactin	—	670	P	(45)

[a] The name of the material along with $\bar{v}$ and molecular weight is presented, as well as the method employed. I indicates interference, S represents schlieren, P is an abbreviation for photoelectric scanner, and C refers to camera optics in conjunction with a densitometer; an asterisk after the symbol for the method indicates that ^{18}O-enriched heavy water was used.

[b] Key to references:

1. A. H. Reisner, J. Rowe, and H. Macindoe, *J. Mol. Biol.* **32**, 587 (1968).
2. S. J. Edelstein and H. K. Schachman, *J. Biol. Chem.* **242**, 306 (1967).
3. T. Devanathan, J. M. Akagi, R. T. Hersh, and R. H. Himes, *J. Biol. Chem.* **244**, 2846 (1969).
4. L. M. S. Chang and F. J. Bollum, *J. Biol. Chem.* **246**, 909 (1971).
5. R. Grasbeck, K. Visuri, and V. H. Stenman, *Biochim. Biophys. Acta* **263**, 721 (1972).
6. Y. I. Shethna, *Biochim. Biophys. Acta* **205**, 58 (1970).
7. B. B. Buchanan, P. Schurman, and P. P. Kalberer, *J. Biol. Chem.* **246**, 5952 (1971).
8. A. H. Reisner, J. Rowe, and R. W. Sleigh, *Biochemistry* **8**, 4636 (1969).
9. J. G. Davis, J. C. Zahnley, and J. W. Donovan, *Biochemistry* **8**, 2044 (1969).
10. G. H. Barlow, L. Summaria, and K. C. Robbins, *J. Biol. Chem.* **244**, 1138 (1971).
11. R. F. DiCamelli, P. D. Holohon, S. F. Basinger, and J. Lebowitz, *Anal. Biochem.* **36**, 470 (1970).
12. C. Olive and H. R. Levy, *J. Biol. Chem.* **246**, 2043 (1971).
13. H. Dalton, J. A. Morris, M. A. Ward, and L. E. Mortenson, *Biochemistry* **10**, 2066 (1971).
14. W. Dowhan, Jr. and E. E. Snell, *J. Biol. Chem.* **245**, 4618 (1970).
15. A. Scanu, W. Reader, and C. Edelstein, *Biochim. Biophys. Acta* **160**, 32 (1968).
16. S. J. Edelstein, Ph.D. Thesis, University of California, Berkeley, 1967.
17. D. Romeo, A. Hinckley, and L. Rothfield, *J. Mol. Biol.* **53**, 491 (1970).
18. T. J. Bowen and M. G. Mortimer, *Eur. J. Biochem.* **23**, 262 (1971).
19. R. W. Gracy and E. A. Noltmann, *J. Biol. Chem.* **243**, 3161 (1968).
20. K. Morimoto and W. F. Harrington, *J. Biol. Chem.* **247**, 3052 (1972).
21. P. A. Peterson, *J. Biol. Chem.* **246**, 7748 (1971).
22. H. Y. Reynolds and J. S. Johnson, *Biochemistry* **10**, 2821 (1971).

TABLE III (Continued)

23. J.-Y. Wu and J. T. Yans, J. Biol. Chem. **245**, 212 (1970).
24. C. L. Sia, S. Traniello, S. Pontremoli, and B. L. Horecker, Arch. Biochem. Biophys. **132**, 325 (1969).
25. S. C. Magu and A. T. Phillips, Biochemistry **10**, 3397 (1971).
26. C. C. Childress and B. Sacktor, J. Biol. Chem. **245**, 2927 (1970).
27. R. M. Franklin, U. Petterson, K. Akervall, B. E. Strandberg, and L. Philipson, J. Mol. Biol. **57**, 383 (1971).
28. D. J. George and A. T. Phillips, J. Biol. Chem. **245**, 528 (1970).
29. C. W. Tabor and P. D. Kellogg, J. Biol. Chem. **245**, 5424 (1970).
30. J. O. Thomas and S. J. Edelstein, Biochemistry **10**, 477 (1971).
31. W. H. Beers and E. Reich, J. Biol. Chem. **244**, 4473 (1969).
32. R. W. Gracy, A. G. Lacko, and B. L. Horecker, J. Biol. Chem. **244**, 3913 (1969).
33. L. Rask, P. A. Peterson, and S. F. Nilsson, J. Biol. Chem. **246**, 6087 (1971).
34. J. C. Melville and C. A. Ryan, J. Biol. Chem. **247**, 3445 (1972).
35. R. D. Mavis and E. Stellwagen, J. Biol. Chem. **243**, 809 (1968).
36. J. W. Donovan, C. J. Mapes, J. G. Davis, and R. D. Hamburg, Biochemistry **8**, 4190 (1969).
37. K. T. Tsuboi, K. Fukunaga, and C. H. Chervenka, J. Biol. Chem. **246**, 7586 (1971).
38. C. Biswas, E. Gray, and H. Paulus, J. Biol. Chem. **245**, 4900 (1970).
39. H. Hirsch-Kolb and D. M. Greenberg, J. Biol. Chem. **243**, 6123 (1968).
40. D. V. Dervartanian, Y. I. Shethna, and H. Beinert, Biochim. Biophys. Acta **194**, 548 (1969).
41. T. H. Weber, H. Aro, and C. T. Nordman, Biochim. Biophys. Acta **263**, 94 (1972).
42. V. Shore and B. Shore, Biochemistry **7**, 3396 (1968).
43. S. A. Rosenberg and G. Guidotti, J. Biol. Chem. **243**, 1985 (1968).
44. F. Reiss-Husson and G. Jolchine, Biochim. Biophys. Acta **256**, 440 (1972).
45. J. R. Pollack and J. B. Neilands, Biochem. Biophys. Res. Commun. **38**, 989 (1970).

sults). However, the corrections are quite complex, and a much safer and direct remedy is to find conditions where association–dissociation equilibria are not present.

Summary

Measurements of $\bar{v}$ can be performed conveniently in the ultracentrifuge with the H_2O–D_2O technique if some sacrifice in precision compared to classical methods can be tolerated. The methods are applicable to both two-component and three-component systems, and in all cases relatively small amounts of material are required. Experiments can be performed with either the photoelectric scanner or the interference optical system. The interference optical system is to be preferred in terms of the higher levels of precision available, although automated methods with the scanner are likely to reduce errors with that technique. In studies of three-component systems containing large amounts of guanidine-HCl or other low molecular weight solutes, the photoelectric scanner is to be preferred since the third component is not likely to introduce complications with absorption optics; in contrast, such components affect markedly the interference patterns. For two-component systems, $\bar{v}$ can also be estimated from the amino acid composition if available,[24] although complications for metalloproteins and enzymes with prosthetic groups may limit applications. An enhancement in the precision of the measurements can be achieved by use of $D_2^{18}O$ in place of D_2O.[10] While the cost of the $D_2^{18}O$ is likely to restrict usage, it is possible to recover the solvent by appropriate techniques and thereby reduce the expense.

[24] T. L. McMeekin and K. Marshall, *Science* **116**, 142 (1952).

[5] The Determination of the Partial Specific Volume of Proteins by the Mechanical Oscillator Technique

By O. KRATKY, H. LEOPOLD, and H. STABINGER

The partial specific volume, $\bar{v_1}$, of a solute is a characteristic parameter that can be used in investigations of protein associations and changes in conformation, as well as in studies on protein solvent interactions and various other intermolecular interactions. It provides also information needed for the determination of particle mass (molecular weight) by means of ultracentrifugation and small-angle X-ray scattering.

The value of the apparent partial specific volume $\bar{v_1}(c)$, measured at a single solute concentration, c, is related to three characteristic parameters of the system by Eq. (1):

$$v_1(c) = \frac{1}{d_2}\left(1 - \frac{d - d_2}{c}\right) \tag{1}$$

where d is the density of the solution in g/cm^3, d_2 is the density of the solvent in g/cm^3, and c is the concentration in g/cm^3. This relation suggests that partial specific volume may be determined by means of a precision differential density measurement $(d - d_2)$. If a certain error in the partial specific volume, $\Delta\bar{v_1}/\bar{v_1}$, is allowed, the allowable tolerance in the determination of the density difference and the concentration is given by the following relations:

$$\Delta(d - d_2) = -d_2 c \bar{v_1} \frac{\Delta\bar{v_1}}{\bar{v_1}} \tag{2}$$

for the absolute error in density difference, and

$$\frac{\Delta c}{c} = \frac{\overline{v_1}d_2}{1 - \overline{v_1}d_2}\frac{\Delta\overline{v_1}}{\overline{v_1}} \tag{3}$$

for the relative error in concentration.

The precision requirements can best be illustrated by a typical example. Assuming reasonable values for a protein solution: $\overline{v_1} = 0.75$ cm^3/g, $d_2 = 1.0$ g/cm^3, $c = 0.005$ g/cm^3, if the concentration measurement has an error of 1.0%, and we wish to measure $\overline{v_1}$ within 0.15%, application of Eqs. 2 and 3 results in a maximal permissible error in the density difference of

$$\Delta(d - d_2) = 4.10^{-6} \text{ g/cm}^3$$

The measurements of d and d_2 then must be carried out with maximal absolute errors of $\pm 2 \times 10^{-6}$ g/cm^3. The tolerable temperature error, ΔT, can be calculated from: $\Delta T\gamma = \Delta d$, where γ is the cubic coefficient of expansion.

For the example used above of an aqueous protein solution, γ has a value of ca. 2×10^{-4} cm^3 deg^{-1}. Therefore, the temperature must be kept constant to within $\pm 1 \times 10^{-2}$°C. The requirement for the precision in the density measurements can be reduced by increasing the protein concentration, as can be seen from Eq. 2. An increase in concentration is, however, limited by the following factors: (1) in many cases, it is not possible to make highly concentrated solutions; (2) often the amount of sample available is insufficient; and (3) $\overline{v_1}$ may be concentration dependent.

Specifications of a Suitable Method of Density Determination

The critical quantity used in measurements of molecular weights by ultracentrifugation and small-angle X-ray scattering (see this volume [1], [9] and [14]) is $(1 - \overline{v_1}d_2)$ resp. $(1 - \overline{v_1}\rho_2)$. Then for a typical situation in which $\overline{v_1}\rho_2 = 0.70$–$0.85$, a 1% error in M requires that $\overline{v_1}$ be known to ± 0.001. For a protein concentration of 10 mg/ml, this requires that the quantity $d - d_2$ be measured with an accuracy of approximately 7×10^{-6} g/cm^3.

From the above we may conclude that a density measuring device meeting the requirements of the determination of partial specific volumes of proteins should meet the following specifications: (1) The differential density $(d - d_2)$ should be determinable with an uncertainty less than 10^{-5} g/cm^3. This requires thermostating of the sample to ± 0.01°C. (2) The measuring time should be short (minutes), since it is difficult to obtain long-term temperature stability of this order with commercial thermostats. (3) The results of the measurement should not be affected by the viscosity and the surface tension of the sample. (4) Volatility of the solvent should not cause any concentration deviation during the measuring process. (5) The density determinations should be possible with a small amount of solution, since, in case of substances of biological interest, often only very small amounts are available. We believe that 1 cm³ should be reasonable.

During the study of this problem, it became evident that there was no need to develop an absolute method, since air, water, and mercury are available as internationally approved standards for calibration. It should be pointed out, furthermore, that the numerical value of the density of water is fixed by an international convention and, for this reason, any densimeter will deliver only relative results. Low-accuracy densimeters can be calibrated during manufacture to read in absolute values. In such instruments, it is impossible to readjust calibration constants, which may change because of aging, etc. In the required range of accuracy, a permanent calibration in absolute values is completely impossible. Therefore, in developing our instrument, we chose differential density determination, using water as a reference, since it is easily available everywhere. As the densities of most samples of biological interest are similar to that of water, this method also should lead to the most accurate results.

In examining various approaches, we found that the natural frequency of a U-shaped glass tube filled with sample would be an easily measurable quantity.[1] The natural frequency is a function not only of the mechanical properties of the vibrating tube, but also of the density of the filling substance. Although this principle was not new, a device based on this principle with an accuracy higher than that obtained with an ordinary weighing pipette had not been described anywhere. The resulting device met all the requirements presented above. It had the additional advantage that the measuring time was of the order of 1 minute or better, permitting constant temperature during the measurement, and afforded a variety of additional applications.

Principle of the Technique

The precision density measurement in the apparatus described herein is based on the determination of the natural frequency of an electronically excited, mechanical oscillator, its effective mass being composed of its own unknown mass and the well-defined, but also unknown, volume of the sample under investigation. In order to assure that this volume be well defined, the oscillator is made of a hollow, U-shaped glass tube which can be filled with the liquid sample. A schematic drawing is shown in Fig. 1.

[1] H. Stabinger, O. Kratky, and H. Leopold, *Monatsh. Chem.* **98**, 436 (1967); O. Kratky, H. Leopold, and H. Stabinger, *Z. Angew. Phys.* **27**, 273 (1969); H. Leopold, *Elektronik* **19**, 297, 411 (1970).

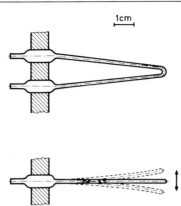

Fig. 1. Schematic of the vibrator.

The mode of vibration is that of a bending-type oscillator. The positions of its vibrating nodes, which in fact determine the limits of the volume of sample taking part in the motion, is kept stable by the abrupt change in the cross section of the glass tubes. In addition, the elemental volumes of the sample in the vicinity of the nodes will make a decreasing contribution to the effective total mass of the sample, since, close to the nodes, the amplitudes decrease. Therefore, small changes in the positions of the nodes do not affect the effective volume of the sample. The U-shape was chosen to define a plane of vibration, thereby eliminating the risk of an elliptic vibration with an ambiguous resonant frequency. Furthermore, the U-shape facilitates the filling and rinsing procedures and permits density determinations on flowing samples.

Over a small frequency range (sufficiently large for the determination of the partial specific volume), the motion of a bending-type oscillator may be described by the simple mass spring model of Fig. 2, where c

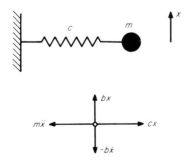

Fig. 2. Analytic model of the vibrator.

is the constant of elasticity of the spring, x is the elongation, and m is the total mass, which in our case is

$$m = M_0 + dV$$

M_0 is the effective mass of the empty vibrator, d is density of the sample, and V is the volume of the sample taking part in the motion. Assuming that the mass, m, performs an undamped oscillation (the elastic force, cx, and the dynamic force, $m\ddot{x}$, are in equilibrium), its resonant frequency, f, is given by

$$2\pi f = \sqrt{\frac{c}{m}} = \sqrt{\frac{c}{M_0 + dV}} \tag{4}$$

From this equation one can derive the density of the sample, d, which is

$$d = A(T^2 - B)$$

where T is the period $(1/f)$ of oscillation and A and B are constants which contain d, M_0, and V. For a density difference between two samples $(d_1 - d_2)$, this equation may be written as

$$d_1 - d_2 = A(T_1^2 - T_2^2), \text{ or } \Delta d = A\Delta T^2 \tag{5}$$

T_1 and T_2 are the respective period readings of the two samples. The two constants, A and B, can be determined by two calibration measurements using air and very clean water as standards of known density. Accurate data on the densities of these two substances are available over a wide range of temperatures.

The excitation system is designed in such a way as to excite and maintain undamped oscillation of the filled vibrator. In order to accomplish this, it must impose a force, $-b\dot{x}$, to the vibrating mass which, in the stationary case, should compensate completely for the damping force, $b\dot{x}$. Figure 3 shows a schematic diagram of the excitation system. A

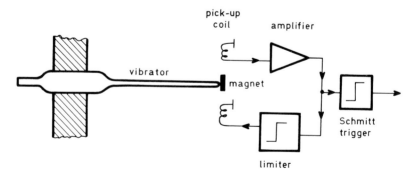

FIG. 3. Excitation system.

magnetic rod induces in the pick-up coil a voltage proportional to the velocity of the vibrator. This voltage is fed back to the excitation coil after being amplified and limited. The gain of the loop is adjusted in such a way that the amplitude of the oscillator filled with various samples reaches quickly the level given by the limiter. As is evident from the vectorial diagram of Fig. 2, any phase shift in the excitation system will cause the force, $-b\dot{x}$, to have a component in the direction of the dynamic force, thereby altering the effective mass of the vibrator. A constant contribution of this induced mass could, of course, be included in the calibration constants via M_0 but in order to keep the effect of a deviation of the phase shift negligible, the absolute phase shift is made very small by setting a large overall band width (3–100,000 Hz) for the excitation system relative to the natural frequency of the oscillator (approximately 500 Hz). The limiter keeps the amplitude to a level of about 0.01 mm in order to allow motion to be performed in the elastic region of deformation. The Schmitt-trigger circuit squares the excitation waveform, so that it can be handled in a digital period-measuring circuit.

To determine the density of the sample with an accuracy in the sixth decimal place, it is necessary to measure the period of the oscillator with approximately 10-fold accuracy. This factor, which contains a margin of safety, is based on the ratio between the mass of the sample and that of the empty vibrator. On the other hand, it is desirable to keep the measuring time as short as possible, in order to avoid effects of long-term temperature fluctuations.

The shortest measuring time is determined by the uncertainty in detection of the zero crossing of the velocity amplitude by the excitation system and the Schmitt trigger. According to the band width of the system, this uncertainty amounts to 10 μsec.

Therefore, the total measuring time must be at least 10^7 times 10 μsec; this results in 100 seconds, which compares favorably to the 6 hours necessary to count 10^7 periods of the oscillator. Figure 4 shows the electronic implementation of a period-measuring circuit which makes use of the above considerations.

The start command sets flip-flop 1, allowing the output of the Schmitt trigger to pass gate 1. The first positive zero crossing of the velocity amplitude of the vibrator after the start command will, therefore, set flip-flop 2 and initiate the counting of the clock pulses of the 100 kHz crystal oscillator. After $n + 1$ pulses of the Schmitt-trigger output, the divider will reset the flip-flop, thus blocking the 7-digit counter. The time-lapse of n full periods of the vibrator will show up on its display (in units of 10 μsec). In practice, the quantity n can be varied from 100 to 10,000, depending on the desired accuracy.

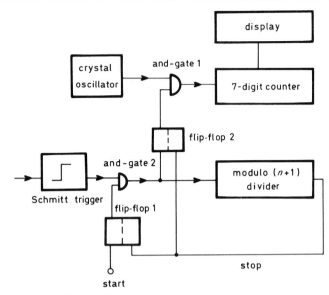

FIG. 4. Block diagram of period-meter.

In addition to this basic circuit, many devices can be added such as a digital printer or a small computer for direct read out of the density. For measurements of the density of streaming liquids, the circuit of Fig. 4 was modified in such a way that a continuous series of measurements could be carried out, the result of each being stored in an intermediate memory until completion of the next measurement. Fig. 5 shows the photograph of a digital densimeter for laboratory use which incorporates the features of Fig. 4. Figure 6 depicts the vibrator of this instrument. This instrument is manufactured by A. Paar, Graz, Austria.

FIG. 5. Digital densimeter for laboratory use.

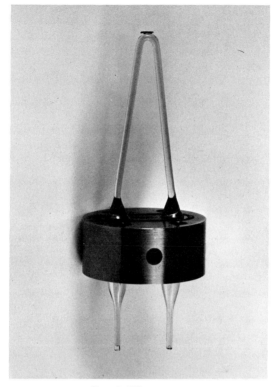

FIG. 6. Vibrator.

Procedure for a Differential Density Measurement

General

The determination of a differential density is based on the simplified equation

$$\Delta d = A \Delta T^2$$

As a first step, the constant, A, is determined by measuring the differential density of two known samples. The best choice for these samples should be air and water, on account of the significant and well reproducible difference in their densities. If the constant, A, is determined once by means of this procedure, any differential density can be evaluated from the corresponding period readings according to Eq. (5). It should be pointed out that any relative error in A affects Δd to the same extent. As a consequence, the constant, A, is not a critical figure in the determination of a partial specific volume (where Δd is usually very small).

Before a measurement is made, the vibrator must be rinsed with a suitable solvent and dried by streaming air. The pump built into the digital oscillator densimeter requires about 5 minutes for the drying procedure. After an additional interval of 3 minutes, needed to establish thermal equilibrium, a period measurement is carried out. This gives T for air. Then, the vibrator is filled with distilled water. Since the heat capacity of the liquid is higher, about 15 minutes is required to attain thermal equilibrium in this case. This is usually checked by taking several succeeding period measurements. At thermal equilibrium, the values remain constant. The period of the water-filled oscillator is determined, and from the two measurements the constant, A, is calculated. The differential density between any two substances (e.g., solvent and solution) may be evaluated now using this constant A.

Possible Errors

Temperature. In the determination of the differential density between two liquids with approximately equal coefficients of thermal expansion (e.g., water and an aqueous solution), it is necessary only to keep the temperature constant during the measuring process, the absolute value of the temperature being of secondary interest. This is not true if the two substances have essentially different coefficients of expansion, and especially if it is desired to measure the absolute density of an unknown liquid from the density difference between it and water. It is very unlikely that such a case will be encountered in protein work.

Residues within the Vibrator. The constant A is determined by the temperature-dependent constant of elasticity of the measuring oscillator, by its volume, and by its mass. As a consequence, residues within the vibrator, as well as a deviation of the measuring temperature, will alter the constant A and through it the results of a differential density measurement. Therefore, when measurements are carried out at a new temperature, or if insoluble residues remain within the vibrator, it is necessary to redetermine the constant A by redoing measurements on air and water. In the case of soluble residues within the vibrator, only careful and patient rinsing will eliminate these sources of error. The increasing degree of cleanness can be checked easily by taking frequent period measurements on the empty vibrator. The readings should tend asymptotically to a constant value which is identical with the original reading for an air-filled vibrator if all the residues were soluble, or to a new but constant value if only the insoluble fraction of the residues is left. When the measuring temperature is below the dew point, atmospheric air should be replaced by a dried gas as calibrating substance, since, otherwise, condensing vapor will cause unstable readings.

Special Comments on the Determination of the Partial Specific Volume. A determination of $\bar{v}_1$ from a differential density measurement requires that very small differences of absolute densities be measured. Therefore, any differences in density between solution and solvent which are not due to the dissolved substance will lead to a large error in $\bar{v}_1$. In addition to the requirement of constant temperature, it is absolutely essential that the solvent used in making up the solution have exactly the same density as the reference solvent in the differential density measurement. For example, extreme care must be exercised that no additional salt be introduced with the protein.

Gases dissolved in the sample have only a negligible effect on the density. However, the measurement becomes completely worthless if gas bubbles appear in the measuring cell due to an increase in temperature. There is, however, no danger that the formation of bubbles will be overlooked, since these render stable period readings impossible. The simplest way to avoid the problem of gas bubbles is to introduce the sample at a temperature a few degrees higher than that of the experiment.

Some Applications

Test Measurements

The accuracy of the instrument was tested by measurements on sodium chloride solutions; their density values are available in the Landolt-Börnstein tables.[2] This system was chosen because it is possible to determine densities up to six decimals on these solutions, which can be prepared at exactly known concentrations and are available in unlimited amounts.

The density difference was calculated from the equation

$$T_1^2 - T_2^2 = A(d_1 - d_2)$$

The constant A was determined from measurements on air and water. The time lapse for 2×10^4 periods was measured by a crystal-oscillator; this quantity, T, was given in units of 10^{-5} sec. The measuring temperature was 19.00°C. The results were

$$T_{air} = 4,378,260$$
$$T_{H_2O} = 5,110,602$$

The density of the air was $d_{air} = 0.001173$ g/cm^3, that of water, $d_{H_2O} = 0.998597$ g/cm^3, giving $A = 6,970,039 \times 10^6$. Measurements of a 5.674% NaCl solution gave $T_{sample} = 5,138,397$. The density difference between

[2] Landolt-Börnstein: "Physikalisch-Chemische Tabellen," 1, Ergänzungs-Band, p. 202. Springer, Berlin, 1927.

TABLE I

DENSITIES d AND PARTIAL SPECIFIC VOLUMES OF BOVINE SERUM ALBUMIN IN 0.2 M NaCl AND IN 6 M GuHCl-0.1 M β-MERCAPTOETHANOL AT 25° [a]

Solvent (M)	d	$\bar{v}_1$
NaCl (0.2)	1.0053	0.734
GuHCl (6)	1.1415	0.729
GuHCl (6)	1.1407	0.727

[a] According to E. Reisler and H. Eisenberg, *Biochemistry* 8, 4572 (1969). Each set of data consists of four experimental points in the concentration range 4.6–11.2 mg/ml. No trend with concentration was observed.

the NaCl solution and water, thus, amounts to $\Delta d = 0.040888$ g/cm³. This value agrees to the last decimal with the values given in Landolt-Börnstein.[2]

This example shows that density differences can be determined exactly up to six decimals using only 1 cm³ of solution.

Examples of Applications

The digital oscillator densimeter based on the mechanical oscillator technique was developed in our group only a few years ago and has not yet been very widely applied. It seems desirable, therefore, to point out as examples some typical measurements from several laboratories.

Reisler and Eisenberg[3] tested the instrument itself in the course of their studies on proteins in guanidine hydrochloride solutions and stated: "Experimetal precision, checked with specially prepared salt samples, was limited to 2×10^{-6} g/ml in density due to the thermostat performance (temperature constancy around 20 and 25° was ± 1–2×10^{-2}) and small fluctuations in room temperature. Every run of experiments (four protein samples at different concentrations) was preceded and followed by a check of the calibration constant of the instrument. In the absence of protein, the density difference between GuHCl solution inside and outside the dialysis bags was found to be negligible and no correction for membrane assymmetry was required."

Table I gives the results of their measurements on bovine serum albumin (using our nomenclature). The partial specific volume was found to decrease from 0.734 in 0.2 M NaCl to 0.728 in 6 M GuHCl in the presence of 0.1 M β-mercaptoethanol.

For rabbit muscle aldolase in water, the same authors found $\bar{v}_1$ to be 0.737 and 0.739 at 20° and 25°, respectively. In 3, 4, 5, and 6 M

[3] E. Reisler and H. Eisenberg, *Biochemistry* 8, 4572 (1969).

TABLE II

PHYSICAL PROPERTIES OF INDIVIDUAL IMMUNOGLOBULINS FROM
BOVINE SERUM AND COLOSTRUM[a]

Type	Source	$\bar{v}_1$ ml/g, 25°C
Gs	Colostrum	0.735 ± 0.003
Gs	Bovine serum	0.734 ± 0.001
1	Bovine serum	0.735 ± 0.001
2	Bovine serum	0.736 ± 0.002

[a] According to B. Kickhofen, D. K. Hammer, and D. Scheel, *Hoppe-Seyler's Z. Physiol. Chem.* **349**, 1755 (1968).

GuHCl at 25°, $\bar{v}_1$ was found to be practically unchanged, with values of 0.733, 0.732, and 0.733, respectively.

In the course of a very detailed study on the isolation and characterization of γG immunoglobulin from bovine serum and colostrum, Kickhofen et al.[4] reported on the measurement of the partial specific volume of four bovine γ-globulins (γGs from serum and colostrum, γ_1 and γ_2): the samples were exhaustively dialyzed against phosphate-buffered saline. The values of $\bar{v}_1$ measured by these authors are presented in Table II. Their data on the limits of error are noteworthy. Furthermore, there is full agreement with our measurements, in which we found for human γGl myeloma protein (Eu) a value of $\bar{v}_1$ of 0.733.[5]

Skerjanc et al.[6] have reported an interesting study on the influence of urea on the partial specific volume of chymotrypsinogen A. The partial specific volumes have been found to be independent of protein concentration in the concentration range studied, i.e., up to 4%. The same solutions were used for dilatometric measurements as enough sample was available. A comparison of the results of the two methods can be seen in Fig. 7. The partial specific volume of the protein at first increases with urea concentration, reaches a maximum, decreases, reaches a minimum and then increases again. The authors estimate the error in $\bar{v}_1$ to be ± 0.001 ml/g.

We have measured $\bar{v}_1$ of *Helix pomatia* hemocyanin.[7] Using the measured value, $\bar{v}_1 = 0.732$, we obtained a molecular weight of 8.95×10^6 by means of small-angle X-ray scattering, in good agreement with ultracentrifugal values.

[4] B. Kickhofen, D. K. Hammer, and D. Scheel, *Hoppe-Seyler's Z. Physiol. Chem.* **349**, 1755 (1968).
[5] I. Pilz, G. Puchwein, O. Kratky, M. Herbst, O. Haager, W. E. Gall, and G. M. Edelman, *Biochemistry* **9**, 211 (1970).
[6] J. Skerjanc, V. Dolocek, and L. Lapanje, *Eur. J. Biochem.* **17**, 160 (1970).
[7] I. Pilz, O. Kratky, and I. Moring-Claesson, *Z. Naturforsch.* **25b**, 600 (1970).

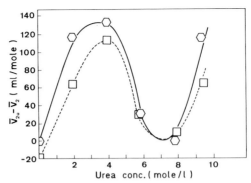

FIG. 7. Plots of the differences of the partial molar volumes of chymotrypsinogen A in urea solutions and water, respectively, as a function of urea concentration. Solid line ($\bigcirc$): values obtained from density measurements; dashed line ($\square$): values obtained from dilatometric measurements (according to J. Skerjanc, V. Dolecek, and L. Lapanje, *Eur. J. Biochem.* **17**, 160 (1970).

We have also found the partial specific volume of the apoenzyme of yeast glyceraldehyde-3-phosphate dehydrogenase to be: $\bar{v}_1 = 0.745 \pm 0.005$ cm³/g.[8] The uncertainty of this result is mainly due to a 2% error in concentration. This value for $\bar{v}_1$ is in good agreement with the data of Jaenicke *et al.*[9]: $\bar{v}_1 = 0.747$ cm³/g at 40°. With our experimental value for $\bar{v}_1$, the molecular weight from small-angle X-ray scattering is: $M = 141,000$. This agrees sufficiently well with the value of 144,700, given by Jaenicke *et al.*[9] It also agrees with the value of approximately 142,000 calculated from the known amino acid composition.[10]

Finally, mention should be made of the investigation of the clinical use of the apparatus,[11] of measurements on fatty acid synthetase,[12] on the rabbit muscle glycogen phosphorylase,[13] and on bacteriophages fr and R17.[14]

[8] H. Durchschlag, G. Puchwein, O. Kratky, I. Schuster, and K. Kirschner, *Eur. J. Biochem.* **19**, 9 (1971).

[9] R. Jaenicke, D. Schmid, and S. Knof, *Biochemistry* **7**, 919 (1968).

[10] G. M. T. Jones and J. I. Harris, *Abstr. Commun., 5th Meeting Fed. Eur. Biochem. Soc.*, No. 740, 185 (1968).

[11] G. P. Tilz, H. Leopold, R. Heschl, and S. Sailer, *Klin. Wochenschr.* in press.

[12] I. Pilz, M. Herbst, O. Kratky, D. Oesterhelt, and F. Lynen, *Eur. J. Biochem.* **13**, 55 (1970).

[13] G. Puchwein, O. Kratky, C. F. Golker, and E. Helmreich, *Biochemistry* **9**, 461 (1970).

[14] P. Zipper, O. Kratky, R. Herrmann, and T. Hohn, *Eur. J. Biochem.* **18**, 1 (1971).

[6] Density Gradient Sedimentation Equilibrium

By JOHN E. HEARST and CARL W. SCHMID

In 1957, Meselson, Stahl, and Vinograd[1] introduced the remarkable technique of sedimentation equilibrium in a density gradient for the study of the properties of high molecular weight DNA in solution. Because of the very high molecular weights of DNA, conventional sedimentation equilibrium cannot be used; for even at the lowest centrifugal fields which are practical, the force on these molecules is too large and they become distributed in extremely narrow distributions. Sedimentation equilibrium in a density gradient avoids this problem because the experiment is performed in a solution which is almost precisely buoyant for the DNA, reducing the centrifugal force on the molecules by several orders of magnitude. The band or distribution of DNA in the centrifugal cell is wide enough to be accurately measurable because the force on the DNA molecule is so small. All procedures described in this chapter are equally applicable to proteins, although generally resolution is poorer because of the lower molecular weights.[1a]

The centrifugal field also generates an equilibrium density gradient which is caused by the redistribution of the salt and the compression of the solution. Fortuitously, the buoyant density of DNA is a function of base composition, so the technique also provides a method for separating and identifying different varieties of DNA in a mixture.[2] Meselson and Stahl[3] proved in 1957 that the replication of DNA in *Escherichia coli* is semiconservative by following the history of density labeled (^{15}N) DNA which had been transferred to light medium. The density gradient method provided something unique in labeling techniques, for it made possible the physical separation of isotopically labeled material from unlabeled material.

We present herein the methods for standardizing buoyant density data for DNA. The source of the different density scales is discussed and recommendations are made as to the choice of a standard scale. Factors

[1] M. Meselson, F. W. Stahl, and J. Vinograd, *Proc. Nat. Acad. Sci. U.S.* **43**, 581 (1957).
[1a] J. B. Ifft, "A Laboratory Manual of Analytical Methods of Protein Chemistry" (P. Alexander and H. P. Lundgren, eds.), Vol. 5, p. 151. Pergamon, New York, 1969.
[2] C. L. Schildkraut, J. Marmur, and P. Doty, *J. Mol. Biol.* **4**, 430 (1962).
[3] M. Meselson and F. W. Stahl, *Proc. Nat. Acad. Sci. U.S.* **44**, 671 (1958).

TABLE I

$$G \frac{\text{g-sec}^2}{\text{cm}^5} \text{ AT } 20°\text{C}$$

Salt	Cs formate	CsCl	CsTFA	CsBr	Cs_2SO_4	$CsCl-OH^a$
$G \times 10^{+10}$	4.02 ± 0.08	8.21 ± 0.17	11.46 ± 0.02	16.4 ± 0.1	19.0 ± 0.36	8.08 ± 0.08

[a] The value of G for alkaline CsCl defines neutral DNA as $Cs3n/2DNA$, where n is the number of nucleotides [C. W. Schmid and J. E. Hearst, *Biopolymers* **10**, 1901 (1971)]. All other values refer to Cs_nDNA.

TABLE II

G VERSUS TEMPERATURE[a,b]

Salt	Empirical relation
Cs formate	$G = 4.308 \, (1-0.00334 \, t°C) \times 10^{-10}$
CsCl	$G = 9.849 \, (1-0.00802 \, t°C) \times 10^{-10}$
CsTFA	$G = 13.83 \, (1-0.00855 \, t°C) \times 10^{-10}$
Cs_2SO_4	$G = 23.01 \, (1-0.00824 \, t°C) \times 10^{-10}$

[a] C. W. Schmid and J. E. Hearst, *Biopolymers* **10**, 1901 (1971).
[b] G for CsCl and Cs_2SO_4 was extensively calibrated in the range 8° to 33°. For CsTFA and Cs formate the above relations are based on two temperatures 8° and 20°.

influencing resolution are discussed. Finally, most of the data required for the use of this elegant tool are compiled.

The review starts with a discussion of the factors governing band shape or the distribution with the intent that this section will provide the intuitive knowledge required for an understanding of the complications.

Molecular Weight and the Equilibrium Concentration Distribution

The equilibrium distribution of DNA in the density gradient is readily calculated from thermodynamics.[4-8] At infinite dilution the distribution of a homogeneous DNA will be Gaussian [Eq. (1)].

$$C = C_0 \exp\left(-\frac{\delta^2}{2\sigma^2}\right) \tag{1}$$

where C_0 is the concentration at band center, C is the concentration at a distance δ from band center, and σ is the standard deviation of the Gaussian band.

In calculating a molecular weight from the distribution in the band an effective density gradient must be used.[4] This gradient can be measured experimentally by observing the spacing between [^{14}N]DNA and [^{15}N]DNA.[9-11] The parameter G or $(1 + \Gamma')/\beta_{eff}$ is calculated from Eq. (2) and has been tabulated in Table I for several salts at 20° and in Table II as a function of temperature.

[4] J. E. Hearst and J. Vinograd, *Proc. Nat. Acad. Sci. U.S.* **47**, 999 (1961).
[5] J. E. Hearst and J. Vinograd, *Proc. Nat. Acad. Sci. U.S.* **47**, 1005 (1961).
[6] J. E. Hearst, J. B. Ifft, and J. Vinograd, *Proc. Nat. Acad. Sci. U.S.* **47**, 1015 (1961).
[7] E. F. Casassa and H. Eisenberg, *J. Phys. Chem.* **65**, 427 (1961).
[8] G. Cohen and H. Eisenberg, *Biopolymers* **6**, 1077 (1968).
[9] C. W. Schmid and J. E. Hearst, *Biopolymers* **10**, 1901 (1971).
[10] J. Vinograd and J. E. Hearst, *Fortschr. Chem. Org. Naturst.* **20**, 372 (1962).
[11] H. Eisenberg, *Biopolymers* **5**, 681 (1967).

$$G \equiv \frac{1 + \Gamma'}{\beta_{\text{eff}}} = \frac{\Delta m \rho_{s,0}}{m \Delta r \omega^2 r_0} \tag{2}$$

Δm is the change in mass per nucleotide upon isotopic substitution of ^{15}N for ^{14}N, $\rho_{s,0}$ is the buoyant density, m is the mass DNA per Cs nucleotide, Δr is the distance between the two peaks. ω is the angular velocity, and r_0 is the average position of the bands relative to the center of the rotor. The quantity, Γ', is the thermodynamic net hydration and the effective gradient is related to β_{eff} by Eq. (3).

$$\left(\frac{\partial \rho}{\partial r} \right)_{\text{eff}} = \frac{\omega^2 r_0}{\beta_{\text{eff}}} \tag{3}$$

An apparent molecular weight of dry Cs DNA, $M_{3,\text{App}}$, is calculated with Eq. (4)

$$\frac{1}{M_{3,\text{App}}} = G \left(\frac{\omega^4 r_0{}^2}{RT\rho_{s,0}} \right) \sigma^2{}_{\text{App}} \tag{4}$$

where R is the gas constant and T is the absolute temperature. The true molecular weight, M_3, must be determined experimentally by extrapolation of the apparent molecular weight to zero concentration.[12] There are two methods to analyze the bands. The concentration distribution may be numerically integrated to find the apparent molecular weight which is then extrapolated to zero concentration by plotting log M_{App} against the average concentration, $\langle C \rangle$.[12] Equations (5)–(7) define the numerical integrations and operations required for this procedure. Figure 1 presents an example of such an extrapolation.

The variance:

$$\langle \delta^2 \rangle = \frac{\int_{-\infty}^{\infty} C \delta^2 d\delta}{\int_{-\infty}^{\infty} C d\delta} \tag{5}$$

The average concentration

$$\langle C \rangle = \frac{\int_{-\infty}^{\infty} C^2 d\delta}{\int_{-\infty}^{\infty} C d\delta} \tag{6}$$

The extrapolation

$$\ln M_{\text{App}} \equiv \ln \frac{\rho_{s,0}RT}{\langle \delta^2 \rangle G \omega^4 r_0{}^2} = \ln M_3 - B'\langle C \rangle \tag{7}$$

An alternate method is to measure the band half-width, σ_{App}, directly

[12] C. W. Schmid and J. E. Hearst, *J. Mol. Biol.* **44**, 143 (1969).

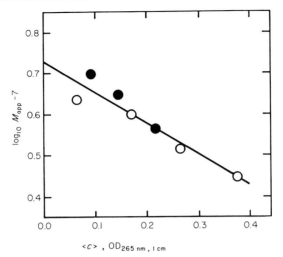

Fig. 1. Molecular weight of D3 DNA. ● 25,000 rpm, ○ 35,000 rpm. Extrapolation to infinite dilution using the moments analysis (Eq. 7).

at the point where the concentration equals 0.606 of that at the maximum concentration.[12] Equation (8) describes the form of this extrapolation, and Fig. 2 presents a representative plot.

$$\ln M_{\text{App}} = \ln \frac{\rho_{s,0} RT}{(\sigma_{\text{App}})^2 G \omega^4 r_0^2} = \ln M_3 - B C_0 \tag{8}$$

In the absence of density heterogeneity, the first of these methods provides a number average molecular weight[2] and the second provides a

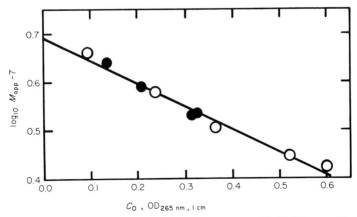

Fig. 2. Molecular weight of D3 DNA. ● 25,000 rpm, ○ 35,000 rpm. Extrapolation to infinite dilution using a direct measure of the band width (Eq. 8).

TABLE III
MOLECULAR WEIGHTS OF HOMOGENEOUS PHAGE DNA's[a]

	Phage			
Method of analysis	T7	T5	T4	D3
Moments, $M \times 10^{-6}$ dalton, Na_nDNA	24.8	68.7	113	40.6
Direct measure of σ, $M \times 10^{-6}$ dalton, Na_nDNA	24.1	80.1	106	36.9

[a] Data from C. W. Schmid and J. E. Hearst, *J. Mol. Biol.* **44**, 143 (1969); *Biopolymers* **10**, 1901 (1971); and unpublished results.

weight average molecular weight.[12] The numerical analysis has the advantage of greater statistical significance and applicability to non-Gaussian bands. It has the disadvantages of requiring more labor and being more sensitive to small baseline errors. This is the cause for the larger scatter in Fig. 1 than in Fig. 2.

Molecular weights for four different phage DNA's have been found by these methods and are presented in Table III.

Density Heterogeneity

The considerations of the last section apply only to a sample that is homogeneous in buoyant density. Fragmented DNA's, in particular, may be heterogeneous in base composition and therefore in buoyant density as well. This density heterogeneity increases the band width and decreases the apparent molecular weight.[13] The most general class of heterogeneity possible is one in which the density and the molecular weight of the DNA may in some way be correlated. Such a system is difficult to deal with experimentally. Most of the density heterogeneity one must deal with is fortunately the result of breakage of very high molecular weight DNA from higher organisms. For such a case there should not be any correlation between molecular weight and density in the sample, and a simplifying assumption is possible.

In the absence of correlation between density heterogeneity and molecular weight heterogeneity, the variance of the distribution may be assumed to be the sum of a contribution from diffusion and a contribution from density heterogeneity. We may write[14,15]

[13] N. Sueoka, *Proc. Nat. Acad. Sci. U.S.* **45**, 1480 (1959).
[14] J. E. Hearst, Ph.D. Thesis, California Institute of Technology, 1961.
[15] C. W. Schmid and J. E. Hearst, *Biopolymers*, in press (1973).

$$\frac{1}{M_{App}} = \frac{G\omega^4 r_0^2 \langle \delta^2 \rangle}{RT\rho_{s,0}} = \frac{1}{M_3} + \frac{G\beta_B^2 a^2}{RT\rho_{s,0}} \langle (GC)^2 \rangle \tag{9}$$

where $\langle (GC)^2 \rangle$ is the variance of the GC composition, β_B is related to the buoyancy gradient by an equation analogous to Eq. (3) and is defined in the following section, and the parameter, a, is the slope of the plot of buoyant density against GC content for a given salt solution. The derivation assumes a linear dependence of buoyant density versus GC and also assumes that all heterogeneity in density arises from GC heterogeneity and therefore there can be no odd bases or glucosylated bases[16] if the analysis is to succeed. From Eq. (9) we can conclude that changing the speed of the centrifuge does not aid in separating the effects of diffusion and density heterogeneity,[17] but changing the salt solution in which the DNA is studied enables one to make this separation. The parameters, G, β_B, and a, are all functions of the salt used. Generally, salts which result in weak gradients such as Cs formate at any arbitrary speed have a high resolving power with respect to density heterogeneity, and salts with large gradients such as Cs_2SO_4 have poor resolving power. Figure 3 shows the distribution of λ half-molecules in three different salts. We see that the bands are best resolved in Cs formate, next best in CsCl, and worst in Cs_2SO_4.

Table IV summarizes the data obtained in these three salt solutions on λ half molecules. The method should prove to be of great value, in future years, in obtaining molecular weights and evaluating the amount of density heterogeneity in the DNA sample.[18,19] The Cs_2SO_4 estimate of $\langle (GC)^2 \rangle$ is different from that for the other two salt solutions.

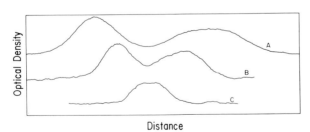

Distance

Fig. 3. Equilibrium concentration distribution of mechanically sheared halves of λcI857 DNA in various solvents. All traces are at the same magnification. Trace A is cesium formate at 35,000 rpm; trace B is cesium chloride at 35,000 rpm; trace C is cesium sulfate at 30,000 rpm.

[16] W. Szybalski, Vol. 12B, p. 330.
[17] J. B. Ifft, D. E. Voet, and J. Vinograd, *J. Phys. Chem.* **65**, 1138 (1961).
[18] C. L. Schildkraut and J. J. Maio, *J. Mol. Biol.* **46**, 305 (1969).
[19] H. Yamagishi, *J. Mol. Biol.* **49**, 603 (1970).

TABLE IV

MOLECULAR WEIGHT AND COMPOSITION HETEROGENEITY OF
SHEARED LAMBDA DNA[a]

Salt	$M_{App} \times 10^{-6}$ dalton[b]	$\beta_B a \times 10^{-8\,c}$	$\dfrac{\beta_B{}^2 a^2 G}{RT\rho_{s,0}} \times 10^{+4}$	$\langle GC^2 \rangle^{1/2\,d}$
Cs formate	1.17	2.166	4.254	0.0435
CsCl	1.88	1.167	2.544	0.0435
Cs$_2$SO$_4$	7.08	0.218	0.246	0.0616

[a] C. W. Schmid and J. E. Hearst, *Biopolymers*, in press (1973).
[b] The apparent molecular weight of Cs$_n$DNA is defined by Eq. (3). For Cs$_2$SO$_4$ a small virial correction was performed by Eq. IIa; for CsCl and Cs formate no correction was used as the M_{app} is essentially a measure of density heterogeneity (Fig. 2).
[c] CsCl is taken as the standard [C. L. Schildkraut, J. Marmur, and P. Doty, *J. Mol. Biol.* **4**, 430 (1962)], and Cs formate and Cs$_2$SO$_4$ are calibrated to it as explained in the text. For Cs$_2$SO$_4$ $\beta_B a$ varies by a factor of approximately 2 over the range 25% to 70% GC (C. W. Schmid and J. E. Hearst, *Biopolymers*, in press (1973); W. Szybalski, Vol. 12B, p. 330).
[d] Calculated with Eq. (3) assuming for λ halves $M_3 = 21 \times 10^{+6}$ daltons. For λ NaDNA MW $= 32 \times 10^6$, and for CsDNA MW $= 42 \times 10^6$.

We believe this disagreement comes from the strong nonlinear dependence of buoyant density on base composition in Cs$_2$SO$_4$ solutions.[16] This test system also suffers from a certain correlation between density and molecular weight in λ halves,[20] but the results are encouraging.

The Buoyant Density

The buoyant density of a DNA sample may be measured with precision. Under favorable circumstances buoyant density differences of 0.0005 g/ml are significant. Since the buoyant density depends on such things as the base composition,[2,16] the extent of density labeling,[3,21] and the presence of unusual bases,[16] it is possible to measure these quantities from a buoyant density determination.

By far the most satisfactory way of determining buoyant density is to determine band position relative to a standard band.[2] The standard most frequently used is *E. coli* DNA, and if it has a density too near that of the DNA of unknown density, a secondary standard which has been standardized to *E. coli* should be used. The density difference between the standard DNA and the unknown DNA is then calculated by Eq. (10) and the calibration data in Table V.

[20] W. Doerfler and D. S. Hogness, *J. Mol. Biol.* **33**, 635 (1968).
[21] R. L. Baldwin, P. Barrand, A. Fritsch, D. A. Goldthwait, and F. Jacob, *J. Mol. Biol.* **17**, 343 (1966).

$$\rho_1 - \rho_0 = \frac{\omega^2}{2\beta_B} (r_1^2 - r_0^2) \qquad (10)$$

where r_1 and r_0 are the positions of the two bands relative to the center of rotation, ω is the angular velocity in radian/second, $\rho_1 - \rho_0$ are the two buoyant densities which are banded in the same solution, and β_B is related to the buoyancy gradient by an equation similar to Eq. (3).

The values of β_B in Table V are among the more commonly used values in the literature. The values for CsCl and Cs formate were measured in this laboratory,[9] and we believe them to be the most precise available. The value for Cs_2SO_4 is 13% lower than the value determined in this laboratory, but since almost all buoyant densities for DNA in Cs_2SO_4 in the literature are based on this number,[16] it seems foolish at this point to recommend a change, although such a change may become desirable in the future.

While density differences can be determined with great accuracy, absolute buoyant densities are far less accurate. For example, both 1.7100 g/ml[2] and 1.7035 g/ml[16] have been used as the standard density of *E. coli* DNA. By accident, the first of these numbers is probably close to the 1 atm pressure buoyant density of *E. coli* DNA[22] and therefore serves as a good standard for buoyant densities at 1 atm. The second number is the buoyant density of *E. coli* DNA which is banded in the center of a cell at an angular velocity of 45,000 rpm where the pressure is about 160 atm.[16] We favor the use of the first number, that is 1.7100, as the standard density of *E. coli* since 1 atm numbers are independent of liquid column height, angular velocity, and position of the band. The

TABLE V
BUOYANCY GRADIENT CONSTANTS

Salt	Cs_2SO_4[a]	CsCl[b]	Cs formate[c]
$1/\beta_B \, \dfrac{\text{g-sec}^2}{\text{cm}^5} \times 10^{+10}$	14.6	9.35	4.21

[a] This value is calculated from a buoyancy gradient (W. Szybalski, Vol. 12B, p. 330), assuming that value refers to $r_0 = 6.60$ cm.

[b] Several calculations and measurements of this quantity are in good agreement: J. E. Hearst, J. B. Ifft, and J. Vinograd, *Proc. Nat. Acad. Sci. U.S.* **47**, 1015 (1961); J. Vinograd and J. E. Hearst, *Fortschr. Chem. Org. Naturst.* **20**, 372 (1962); W. Bauer, F. Prindaville, and J. Vinograd, *Biopolymers* **10**, 2615 (1971); C. W. Schmid and J. E. Hearst, *Biopolymers* **10**, 1901 (1971).

[c] C. W. Schmid and J. E. Hearst, *Biopolymers* **10**, 1901 (1971).

[22] W. Bauer, F. Prindaville, and J. Vinograd, *Biopolymers* **10**, 2615 (1971).

disadvantage of this density scale is that the CsCl solution which one makes up to band the DNA in the center of the cell must be 0.003–0.006 g/ml lower in density than the stated buoyant density in order to correct for the fact that the band is at high pressure.[22] It would clearly be counterproductive to consider still other standard densities at this point.

The other method for determining buoyant density is to band the DNA in two solutions of different initial density and interpolate the band to the root mean square position of the cell, assuming this is the position where the density of the solution is that of the starting solution.[10] This procedure is far less accurate than using a marker, and it is not recommended unless one is using a new salt solution or a substance other than native DNA where standards are not available.

The GC differences in two DNA's can be calculated directly from their spacing in a gradient by Eq. (11).

$$\Delta GC = \frac{\omega^2(r_1{}^2 - r_0{}^2)}{2a\beta_B} \tag{11}$$

The important parameter, $a\beta_B$, is tabulated in Table IV. For CsCl the GC data of Schildkraut, Marmur, and Doty were employed.[2] These data were selected because they are clearly presented and done with care. The ratio of the quantity $a\beta_B$ to the same quantity in another salt can be obtained from the relative spacing of two DNA bands in the two solutions. This method was used to obtain the values of $a\beta_B$ for Cs formate and Cs_2SO_4[15] in Table IV. This method assumes a linear dependence of buoyant density on GC, and, while we trust the results for Cs formate, we believe that the absence of linearity in Cs_2SO_4[16] makes the figure for that salt very unreliable.

The buoyant properties of single-stranded DNA have been studied under alkaline and neutral conditions. In neutral CsCl the buoyant density of single-strand DNA is about 0.02 g/ml denser than native DNA. A larger difference is observed in Cs_2SO_4.[16] The buoyant density of single stranded DNA is dependent on its base composition.[23] Under alkaline denaturing conditions the buoyant density is dependent on the GT composition.[24]

For preparative and analytical studies, it is often desirable to enhance density differences. Synthetic polynucleotides have been bound to single-strand DNA's to increase the density differences between com-

---

[23] S. Riva, I. Barrai, L. Cavalli-Sforza, and A. Falaschi, *J. Mol. Biol.* **45**, 367 (1969).
[24] J. Vinograd, J. Morris, N. Davidson, and W. F. Dove, Jr., *Proc. Nat. Acad. Sci. U.S.* **49**, 12 (1963).

plementary strands.[25,26] Intercalating dyes produce a density difference between DNA's of different superhelical density.[27] Heavy metal (Hg^{2+} and Ag^+) ions enhance density differences arising from differences in base composition.[28-30] By using different salts, base compositional differences may be enhanced (Cs formate) or diminished (Cs_2SO_4) relative to the differences observed in CsCl (Fig. 3). Little use has been made of the greater resolution of such salts as Cs formate. The resolution of λ halves in Cs formate is comparable to the separation in $Hg^{2+} - Cs_2SO_4$.[31]

A nonequilibrium method for separating components that would ordinarily be difficult to resolve has been considered. This technique employs high speed centrifugation which results in the formation of very sharp bands, followed by slow speed centrifugation so that the components sediment away from each other.[32]

Finally, since there still is some experimental error in the β_B's for many salt systems and since they are unknown for other salts and macromolecules (denatured DNA, viruses, etc.) it would be very beneficial if the primary data were presented in the literature as well as the buoyant density differences. Such data should be presented as the distances between bands multiplied by $\omega^2\bar{r}$ to eliminate angular velocity and the position of the bands in the gradient as parameters. We suggest the symbol Δ to represent this quantity; its definition is contained in Eq. (12), where $\bar{r}$ is the arithmetic mean of the position of the two bands.

$$\Delta \equiv \omega^2\bar{r}(\Delta r) \qquad (12)$$

The value of Δ between calf thymus DNA and $E.$ $coli$ DNA calculated from the data of Schildkraut, Marmur, and Doty[2] is -1.31×10^7 $cm^2/$ sec^2, where the negative indicates that calf thymus DNA is less dense then the reference $E.$ $coli$ DNA.

The Approach to Equilibrium

The approach to equilibrium is a transport process in a centrifugal field and may be used to measure the sedimentation coefficient. The ap-

[25] W. C. Summers, *Biochim. Biophys. Acta* **182**, 269 (1969).
[26] H. Kubinski, *Anal. Biochem.* **35**, 298 (1970).
[27] R. Radloff, W. Bauer, and J. Vinograd, *Proc. Nat. Acad. Sci. U.S.* **57**, 1514 (1967).
[28] G. Corneo, E. Ginelli, and E. Polli, *J. Mol. Biol.* **48**, 319 (1970).
[29] R. H. Jensen and N. Davidson, *Biopolymers* **4**, 17 (1966).
[30] U. S. Nandi, J. C. Wang, and N. Davidson, *Biochemistry* **4**, 1687 (1965).
[31] J. C. Wang, U. S. Nandi, D. S. Hogness, and N. Davidson, *Biochemistry* **4**, 1697 (1965).
[32] R. Anet and D. R. Strayer, *Biochem. Biophys. Res. Commun.* **34**, 328 (1969).

proach to equilibrium has all the thermodynamic complications[33] mentioned for sedimentation equilibrium in addition to the problem of an accurate description of the salt distribution. To simplify the description of the salt distribution, sedimentation may be studied in either a preformed gradient[34] or a centrifugally generated equilibrium gradient.[35] It is most convenient in analytical ultracentrifugation to employ an internally generated equilibrium density gradient. This may be accomplished by low speed centrifugation and by waiting about 24 hours before taking data.[34,35]

A complete description of the nonequilibrium concentration distribution is very difficult. To avoid this problem the sedimentation coefficient, DM_3, can be related to moments of the concentration distribution as shown in Eqs. (13) and (14).[33,35]

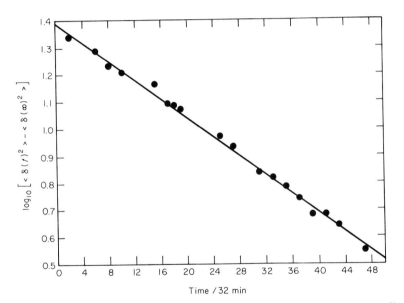

FIG. 4. Sedimentation velocity of phage α DNA in a buoyant density gradient. The moments are conveniently expressed in arbitrary chart dimensions. The initial point was taken 40 hours after the run was started. The average concentration increased through the range 0.11–0.21 $OD_{265\ nm,\ 1\ cm}$ during the time sequence followed.

[33] J. E. Hearst, *Biopolymers* 3, 1 (1965).
[34] R. L. Baldwin and E. M. Shooter, "Ultracentrifugal Analysis in Theory and Experiment" (J. W. Williams, ed.), p. 143. Academic Press, New York, 1963.
[35] M. Meselson and G. M. Nazarian, "Ultracentrifugal Analysis in Theory and Experiment" (J. W. Williams, ed.), p. 131. Academic Press, New York, 1963.

$$\langle \delta(t)^2 \rangle - \langle \delta(\infty)^2 \rangle = [\langle \delta(0)^2 \rangle - \langle \delta(\infty)^2 \rangle] \exp - \frac{2Dt}{\sigma^2} \qquad (13)$$

$$DM_3 = \left[\frac{RT\rho_{s,0}}{G\omega^4 r_0^2} \right] \frac{D}{\sigma^2} \qquad (14)$$

The proportionality constant between M_3 and σ^2 is identical to the quantity appearing at equilibrium [Eq. (4)]. Use of G from Tables I or II defines M_3 for dry Cs DNA. D is the diffusion constant for the temperature and solvent employed and $\langle \delta(t)^2 \rangle$ is the time-dependent variance of the band.

Transport data demonstrating the use of Eq. (2) are presented in Fig. 4. The slope of this plot, $2D/\sigma^2$ is listed in Table VI together with the necessary corrections to a solvent of the viscosity and density of water at 20°.[36,37] The value is also corrected to the molecular weight and partial specific volume of Na DNA. DM_3/RT which equals M_3/f is multiplied by 0.75[9] to correct to Na DNA and by $(1 - \bar{v}\rho)$, 0.444,[8,38] to form the sedimentation coefficient.

This method of measuring the sedimentation coefficient eliminates the problems of speed effects on high molecular weight DNA's and avoids problems of convection that are experienced in the absence of density stabilization. The sedimentation coefficients found by this procedure are in reasonable agreement with $s_{20,w}$'s found by conventional boundary sedimentation (Table VI). The moment analysis, however, is lengthy, and it is not likely that this method will receive much use except in special circumstances.

The Buoyant Medium

When banding DNA it is prudent to check the solvent density before starting the run. This is conveniently done by measuring the refractive index of the solution. Table VII contains the parameters which relate the index of refraction to the solution density by the equation $\rho^{25°C} = a\,n_D^{25°C} - b$. The a's, b's, and approximate buoyant densities of DNA are tabulated for numerous salt solutions. The relations summarized in Table VII and those below refer to a salt dissolved in water.

For CsCl a more exact relation between refractive index and density is $\rho^{25°C} = 1.1584 - 10.2219\,n_D^{25°C} + 7.5806\,(n_D^{25°C})^2$ which covers the density range 1.1 g/ml $< \rho <$ 1.9 g/ml.[39] For cesium sulfate a more exact relation is[40]

[36] R. Brunner and J. Vinograd, *Biochim. Biophys. Acta* **108**, 18 (1965).
[37] P. A. Lyons and J. F. Riley, *J. Amer. Chem. Soc.* **76**, 5216 (1954).
[38] J. E. Hearst, *J. Mol. Biol.* **4**, 415 (1962).
[39] J. B. Ifft, W. R. Martin, III, and K. Kinzie, *Biopolymers* **9**, 597 (1970).
[40] D. B. Ludlum and R. C. Warner, *J. Biol. Chem.* **240**, 2961 (1965).

TABLE VI

SEDIMENTATION VELOCITY IN A BUOYANT DENSITY GRADIENT[a]

DNA[b]	r_0	$\langle c \rangle$ optical density 265 nm, 1 cm	$d \ln[\langle \delta^2 \rangle - \langle \delta(\infty)^2 \rangle]/dt \times 10^{+5}$ sec^{-1}	$\dfrac{M_{\text{CsDNA}}{}^c}{f_{20,\text{w}}} \times 10^{+13}$ sec^{-1}	$\dfrac{M_{\text{NaDNA}}(1 - \bar{v}\rho)^c}{f_{20,\text{w}}} \times 10^{+13}$ sec^{-1}	$s_{20,\text{w}}$ from boundary sedimentation (S)
Phage α	6.647	0.16	2.092	107	35.6	34.5 ± 0.7
Phage α	6.670	0.055	1.95	99	33.0	34.5 ± 0.7
Phage D3	6.503	0.165	2.059	114	38.0	40.3 ± 0.5
Phage D3	6.602	0.100	2.066	111	37.0	40.3 ± 0.5

[a] C. W. Schmid and J. E. Hearst, *Biopolymers*, in press (1973).

[b] The speed was 24,960 rpm, and buoyant densities of 1.697 and 1.722 were used for α and D3 DNA's, respectively.

[c] $f_{20,\text{w}}$ is a molar friction factor corrected to standard solvent conditions.

$$\rho = 0.9954 + 11.1066(n - n_0) - 26.4460(n - n_0)^2$$

where n_0 is the refractive index of water.

Several observations on the applicability of different salts are summarized below. Cs acetate, Cs formate, and mixtures of these two salts provide shallow density gradients with very good resolving power. Equilibration in these salt solutions is slow and the initial density must be adjusted very close to the buoyant density of the DNA. These salts are ideally suited for very high molecular weight DNA. In Cs acetate solutions it is difficult to obtain discernible bands with anything other than very high molecular weight DNA. In Cs_2SO_4 and Cs_2SeO_4 the steep density gradient produces poor resolution, rapid equilibration, and a medium for banding low molecular weight samples. CsBr and CsI are nearly saturated at the densities required for banding DNA and are of little use.

The trifluoroacetate ion is a denaturant for DNA and either potassium or cesium trifluoroacetate serves as a buoyant solvent for DNA.[41] So-

TABLE VII

Approximate DNA Buoyant Densities and Refractive Index
Equations for Several Salts

Salt	$\rho_{s,0}$	a	$-b$
Cs acetate	1.959^b	10.1428^b	12.5548^b
	1.946^a	10.7527^a	13.4247^a
50–50 acetate–formate	1.864^d	—	—
Cs formate	1.767^a	13.7363^a	17.4286^a
CsCl	1.7100^c	$—^f$	$—^f$
Cs trifluoroacetate	1.576^d	—	—
CsBr	1.63^a	9.9667^a	12.2876^a
CsI	1.55^a	8.8757^a	10.8381^a
Cs_2 oxalate	1.508^b	10.5904^b	13.2769^b
Cs_2SeO_4	1.479^a	12.0919^a	15.1717^a
Cs_2SO_4	1.4260^c	$—^f$	$—^f$
K trifluoroacetate	1.52^e	—	—
Li silicotungstanate	1.138^a	—	—

[a] T4 DNA [J. E. Hearst and J. Vinograd, *Proc. Nat. Acad. Sci. U.S.* **47**, 1005 (1961); J. Vinograd and J. E. Hearst, *Fortschr. Chem. Org. Naturst.* **20**, 372 (1962); J. E. Hearst, Ph.D. Thesis, California Institute of Technology, 1961].

[b] Calf thymus DNA [L. Zolotor and R. Engler, *Biochim. Biophys. Acta* **52**, 145 (1967)]; coefficients a and b refer to $\rho^{23°}$ and $n^{23°}$.

[c] 50% GC DNA (W. Szybalski, Vol. 12B, p. 330).

[d] 50% GC DNA [M. J. Tunis and J. E. Hearst, *Biopolymers* **6**, 1345 (1968)].

[e] T7 DNA at 25°C [M. J. Tunis and J. E. Hearst, *Biopolymers* **6**, 1325 (1968)].

[f] See text.

[41] M. J. Tunis and J. E. Hearst, *Biopolymers* **6**, 1325 (1968).

dium iodide serves as a inexpensive salt with better resolution than cesium chloride.[42]

Finally, several attempts have been made to calculate and measure the composition density gradient of many salts.[13,39,40] While this is useful for the study of new materials in density gradients, analytically this quantity should not be used for the determination of buoyant density differences between two bands or for the calculation of molecular weights using band widths unless corrections are made.

Experimental Procedures

All DNA studies utilize the ultraviolet optical system of the analytical ultracentrifuge. In recent years this means that the experimenter has flexibility with respect to the wavelength of light being utilized because most instruments are now equipped with a monochromator. The most common wavelength used for DNA work is 265 nm because of the absorption maximum of nucleic acids near this wavelength and because the high pressure mercury light source has a strong emission band at that wavelength. Most instruments are also equipped with a photoelectric scanner which eliminates the need for photography, and with double-sector centerpieces, makes double-beam optics possible so that baselines are indeed flat even in very strong salt gradients. A major time-saving device is the multiplexer which makes it possible to run as many as six cells at a time and record the cells independently. It is sometimes necessary to use a $-1°$ wedge as the top window of the cell to compensate for the refractive index gradient generated by the salt solution. The gradient, if sufficiently high, bends the light beam so far that partial obstruction of the light beam occurs.

It is advisable to avoid the use of aluminum or aluminum-filled Epon centerpieces because of the corrosive effects of salt solutions. The most satisfactory centerpieces are fabricated from titanium,[43] charcoal-filled Epon, or Kel-F.

To avoid leaks, it is advisable to clean all cell parts before assembly and to lubricate the screw ring and upper Bakelite gasket. The cell should be tightened with a torque of 120–125 inch-pounds. About 30 seconds should be allowed for the flow of the centerpieces and gaskets during tightening. Scratched windows should be avoided. For DNA work, syringe needles must not be used to fill the cells because the DNA is broken by the shear in passing through the needles. This laboratory fills the partially assembled cells from the top using glass pipettes.

[42] R. Anet and D. R. Strayer, *Biochem. Biophys. Res. Commun.* **37**, 52 (1969).
[43] J. E. Hearst and H. B. Gray, Jr., *Anal. Biochem.* **24**, 70 (1968).

Knowledge of the density of the homogeneous solution is necessary. Densities are most conveniently determined by weighing a known volume of solution in a 0.3-ml calibrated micropipette. No attempt is made to control the temperature of the sample during the measurement, which is performed at room temperature. The resulting densities are accurate to ± 0.001 g/ml. Data for the refractive index vs. density of salt solution greatly simplify the setting up of runs. The data for many salts can be obtained from the International Critical Tables,[44] and when not available are readily measured. Certain of these data are given in Table VII.

It is usually desirable to buffer the solution. The addition of buffer is readily accomplished during mixing of the concentrated salt solution with water and macrospecies. Changes in pH up to 0.5 pH unit have been observed for buffers diluted into 7 molal CsCl solutions. Buffers in low concentration do not interfere with refractive index measurements. For Cs salts, additive volumes may be assumed in preparing solutions of desired density from concentrated stock solutions and water or buffer solutions. The additive mixing relation is discussed in more detail by Vinograd and Hearst.[10]

It is necessary to estimate the amount of DNA or protein to add to the cell in order to end the run at equilibrium with a measurable band. If C_0 is the polymer concentration at band center and C_i is the initial concentration uniformly distributed over the cell a helpful relationship is[10]

$$\frac{C_0}{C_i} = 0.40 \frac{L}{\sigma}$$

where L is the length of the liquid column and σ is the standard deviation of the equilibrium band. Generally DNA increases in concentration by a factor of 10–20 in a full centrifuge cell so it is common to start a run with a DNA concentration about $OD_{1cm}^{260} = 0.03$.

Acknowledgment

We wish to acknowledge that this work was supported in part by Grant GM 11180 of the U.S. Public Health Service and that C. W. S. was supported by a predoctoral fellowship No. 1-F01-GM 46,314-01.

[44] "International Critical Tables." McGraw-Hill, New York, 1933.

[7] Proteins in Density Gradients at Sedimentation Equilibrium

By JAMES B. IFFT

The preceding chapter[1] provides an excellent introduction to the theory of density gradient sedimentation equilibrium. The techniques required to obtain accurate molecular weights, including the effects of density heterogeneity, are discussed. The buoyant medium and methods employed in the measurement of the buoyant density are described. A brief description of experimental procedures is given. All this material is directed specifically toward the analysis of nucleic acids in density gradients.

The purpose of the present chapter is to provide additional information on experimental techniques appropriate to the investigation of proteins. There are several reasons for this separate discussion of proteins. Proteins are much smaller than DNA. Molecular weights range between about 20,000 and a few hundred thousand as opposed to several million for nucleic acids. Because the standard deviation of the macromolecular band in the density gradient is inversely related to the square root of the molecular weight of the polymer, protein bands are much broader than nucleic acid bands and in fact occupy most of the length of the centrifuge cell. Figure 1 demonstrates such a band for the average size protein, bovine serum mercaptalbumin (BMA).[2] This behavior places several constraints both on experimental techniques and subsequent analyses. It is impossible to band two proteins in the same analytical cell and obtain the density of an unknown protein from the distance it bands away from a known marker, as described by Hearst and Schmid.[1] Absolute densities must be obtained from a thermodynamic analysis of the salt distribution alone. The wide band also dictates that an experimental baseline is mandatory if the area under the band is to be integrated to obtain a molecular weight.

The second difference between banding proteins and nucleic acids is the monomer composition of the two species. Proteins generally contain about 20 different amino acids in contrast to the four bases of DNA. This means that an analysis of monomer composition of proteins will be considerably more complicated than the straightforward plot of GC

[1] J. E. Hearst and C. W. Schmid, this volume [6].
[2] J. B. Ifft, *in* "A Laboratory Manual of Analytical Methods of Protein Chemistry" (P. Alexander and H. P. Lundgren, eds.), Vol. 5, p. 151. Pergamon, Oxford, 1969.

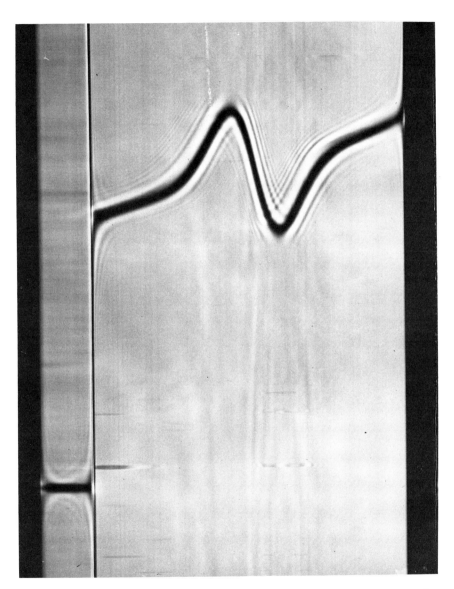

FIG. 1. Bovine serum mercaptalbumin, 0.1%, in CsCl of $\rho_e = 1.279$ g/ml. pH = 5.08, acetate buffer ($\mu = 0.01$) at 56,100 rpm and 25° at sedimentation equilibrium. Single sector, 4°, Kel-F centerpiece. J. B. Ifft, *in* "A Laboratory Manual of Analytical Methods of Protein Chemistry" (P. Alexander and H. P. Lundgren, eds.), Vol. 5, p. 151. Pergamon, Oxford, 1969.

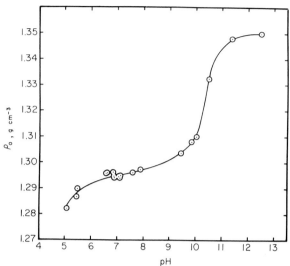

Fig. 2. Buoyant density, ρ_0, of bovine serum mercaptalbumin in CsCl as a function of pH. All runs were performed in 4°, single sector, Kel-F centerpieces at 56,100 rpm at 25°. Diameters of circles represent maximum experimental uncertainties.[3] A. E. Williams and J. B. Ifft, *Biochim. Biophys. Acta* **131**, 311 (1969).

content of a DNA vs. the buoyant density. Additionally, there are seven ionizable amino acid residues which make proteins zwitterions. Thus, the total charge on the protein varies significantly with pH. This charge variation causes appreciable changes in the ion binding and hydration of the protein, resulting in significant changes in buoyant density. Each protein experiment must therefore be conducted at a definite pH, and this pH must be noted along with every buoyant density result. Figure 2 shows the effect of pH variation on the buoyant density of BMA.[3] Another consequence of the differences in monomer composition is that the densities of proteins are much lower than DNA, 1.3 instead of 1.7. This means that a much wider variety of solvents are available to band proteins.

A third distinction between experiments involving these two biopolymers is the photographic system employed. The very large extinction coefficients of nucleic acids at 265 nm indicates the use of the absorption optical system. Proteins have much lower values of ϵ at 280 nm. This results in an increase in sensitivity, of a factor of only 2, of the absorption system over the schlieren system. In general, the exact center of the broad protein band is located more readily with the latter

[3] A. E. Williams and J. B. Ifft, *Biochim. Biophys. Acta* **131**, 311 (1969).

optical system. This requires, however, that manual analyses be used because no equipment similar to the photoelectric scanner has been developed for the schlieren system.

This chapter then will treat only those aspects of experimental procedure and analyses that differ from the discussion presented in the previous chapter as determined by the basic differences between DNA and protein behavior as noted above. After this discussion, a few applications of studies of proteins and polypeptides in density gradients will be discussed briefly.

Experimental Procedures

Selection of Solvent. The buoyant densities of all proteins measured to date fall in the range 1.20–1.35 g/ml. This permits a much wider flexibility in the selection of the two component solvents because the densities of saturated solutions of almost all the halides of the alkali metals range up to at least 1.25 g/ml.[4] This is fortunate for several reasons. Because of the ion-binding properties of proteins, it is often useful to measure the buoyant density,[5] ρ_0, of a given protein in a wide variety of solvents to determine the effects of an anion series (CsCl, CsBr, CsI) or a cation series (KBr, RbBr, CsBr). Also, it has been shown[6,7] that the resolution of two proteins in a density gradient in a preparative centrifuge is directly proportional to the density gradient proportionality constant, β, as defined by the density gradient relation:

$$\frac{d\rho}{dr} = \frac{\omega^2 r}{\beta} \tag{1}$$

The correct procedure in such cases is to select salts having the largest β values[4] which have densities extending beyond that of the polymer under study. Implicit in both of the above criteria is the assumption that the protein is both soluble and stable in the solvent selected.

Highly purified rubidium and cesium salts may be obtained from the following sources: Apache Chemicals, Rockford, Illinois; Alfa Inorganics Inc., Beverly, Massachusetts; Harshaw Chemical Company, Cleveland, Ohio; K&K Laboratories, Inc., Plainview, New York; Pierce Chemical Company, Rockford, Illinois. Other alkali metal and am-

---

[4] J. B. Ifft, W. R. Martin III, and K. Kinzie, *Biopolymers* **9**, 597 (1970).

[5] The subscript zero will be used to denote the properties of that variable at the center of the protein band. The subscript "e" is used to denote the properties of the initial solution. The superscript zero will describe the properties of the variable at atmospheric pressure.

[6] A. S. L. Hu, R. M. Bock, and H. O. Halvorson, *Anal. Biochem.* **4**, 489 (1962).

[7] J. B. Ifft and J. Vinograd, *J. Phys. Chem.* **66**, 1990 (1962).

monium salts can be purchased from regular chemical supply sources at low cost and in satisfactory purity (>99.5%).

Selection of Initial Solution Density. If a density at any pH is available for the protein under investigation, the initial solution density can be adjusted up or down from that depending upon the pH noting that ρ_0 increases with increases in pH for all proteins studied in CsCl gradients to date. If no data are available, a safe assumption to make is that if one is dealing with a pure protein having no prosthetic groups, a recognizable band can be obtained using an initial solution density, ρ_e^0, of 1.27–1.30 g/ml. This is somewhat surprising in view of the wide diversity of densities of the amino acids which range between 1.1 and 1.7 g/ml. It indicates fairly homogeneous amino acid compositions for the proteins measured to date. An examination of densities of model compounds indicates that the densities of lipoproteins will be less, and the densities of gluco- and nucleoproteins will be greater, than that of pure proteins.

Selection of Buffer. Virtually any buffer can be used in density gradient experiments provided of course that the pK_a of the buffer is within ±1 pH unit of the desired pH. One precaution has recently been noted,[8] however, in precise work where the density changes rapidly with pH. This is the effect of pressure on the equilibrium between the ionized and nonionized forms of the buffer.

The thermodynamic dependence of the equilibrium constant on pressure is

$$\frac{d \ln K}{dP} = -\frac{\Delta V}{RT} \tag{2}$$

or if ΔV is independent of pressure:

$$pK_P - pK_1 \doteq \frac{P \Delta V}{2.3RT} \tag{3}$$

The quantity ΔV is the volume change in milliliters per mole which occurs upon ionization. An approximate value for acetic acid–acetate in CsCl of $\rho_e = 1.3$ g/ml is 10.0 ml/mole. Insertion of the appropriate values for this solution at full speed at 25° indicates that the pH will be 4.724 during the centrifuge experiment as opposed to 4.757 when the solution is at rest. This change of 0.033 pH unit may or may not be significant depending upon the type of experiment. However, amines have ΔV values of 30 ml/mole. This leads to a pH shift of 0.1 unit, which is often too large a shift. Some buffer systems which cover almost the entire pH range and their ΔV values are given in Table I. pH values

[8] J. B. Ifft, *C. R. Trav. Lab. Carlsberg* **38**, 315 (1971).

TABLE I
BUFFERS AND THEIR VOLUME CHANGES UPON IONIZATION

Buffer	pK_a	ΔV, ml/mole (0.1 M KCl)[a]
Acetate	4.74	10.9
Phosphate, secondary	7.12	25
Glycine, α-NH$_3^+$	9.6	21.8
Carbonate, secondary	10.33	27.8
Phosphate, tertiary	12.32	—

[a] Data are from H. H. Weber, *Biochem. Z.* **218**, 1 (1930).

above and below these limits can be reached with dilute solutions of HCl and CsOH if the salt in use is CsCl.

If no titration studies are envisioned, the value of ρ_0 at pI, the isoelectric point of the protein in that salt solution, should be used to minimize charge and ion-binding effects. Unfortunately very few pI's are known in solutions of 3–4 M salt. Generally, extrapolations from low salt concentration data will have to be employed until these data become available.

Selection of Centrifuge Parameters. In most cases, the highest angular velocity possible for the rotors available should be employed. This is necessary to contain all the band in the 1.2 cm of the gradient column, especially for smaller proteins, and it shortens the time required to reach equilibrium. For a protein of molecular weight of 10^5, a useful order of magnitude relation is:

$$\omega^2 \simeq 7 \times 10^6 \beta_0^0 \qquad (4)$$

in order to band it with a standard deviation of 1 mm in a solution of $\rho_e = 1.3$ g/ml near the center of the cell at 25°. This requires a velocity of 55,000 in CsCl but 67,000 in RbCl.

One precaution must be observed in using high angular velocities. As we have extended our studies to other salt systems, we occasionally have found a loss of light at the camera, usually at the bottom but sometimes in the middle of the cell as well. This loss has turned out to be due to gradients so large that light was lost at the schlieren phase plate. Table II gives the maximum values of $\omega^2/\beta_e^0 \times 10^2$ which the Spinco optical system can accommodate without loss of a portion of the image. Of course, increasing the amount of negative wedging reverses this effect.

Three rotors are used most of the time in density gradient studies with proteins. Table III provides the necessary data to select a particular

TABLE II

MAXIMUM VALUES OF $\omega^2/\beta_e^0 \times 10^2$ WITHOUT LIGHT LOSS AS A
FUNCTION OF WINDOWS EMPLOYED

Max. value of $\omega^2/\beta_e^0 \times 10^2$	Windows employed
1.85 ± 0.21	$+1°$
2.42 ± 0.08	Flat
3.45 ± 0.35	$-1°$

rotor for a run. It also lists the possible wedge combinations so that two, three or four solutions can be run simultaneously. Because the images are recorded simultaneously on the 2×2 inch plate, the full four-cell capacity of the An-F rotor is seldom realized. A counterbalance fills the fourth hole and is used to obtain real distances in the cells. The reference hole in the other two rotors serves this purpose when two cells are run in them.

The 4°, 12 mm, single-sector Kel-F centerpiece is satisfactory and convenient for most purposes. No gaskets are required, and it distorts only after several hundreds of hours at full speed. Because the schlieren image plunges so steeply at band center (see Fig. 1), buoyant density values accurate to ± 0.001 g/ml can be obtained without the aid of an experimental baseline. If an accurate baseline is required, the double-sector, capillary-type synthetic boundary centerpiece made of filled Epon works nicely.[7]

All values of β and density versus refractive index data have been calculated or measured at 25.0°. Thus, unless data is determined by the investigator for other temperatures, most studies are restricted to this one temperature.

For reasons given above, the schlieren optical system is used in most cases. Metallographic Kodak film has a high sensitivity to the green light which passes through the Kodak Filter No. 77A and provides a permanent record of each run.

TABLE III

SPECIFICATIONS FOR SPINCO ANALYTICAL ROTORS

	An-F	An-D	An-H
Maximum speed	52,640	59,780	67,770
Maximum number of cells	4	2	2
Wedge combinations	$+1°, 0°, -1°, -2°$	$+1°, 0°$	$+1°, 0°$
	$0°, -1°, -2°$	$0°, -1°$	$0°, -1°$

Preparation for the Run. In contrast to the DNA work described earlier,[1] proteins are not sheared on passage through a syringe needle, again a property of their modest size. Proteins are subject to surface denaturation, however, and protein solutions should be passed through these needles at a minimum velocity. Any needle between No. 22 and No. 26 gauge may be employed. The tips should be ground flat to avoid scratching the centerpiece. If a multiple-cell run is made, care should be taken in filling the cells to ensure that three distinct menisci will result.

About 2 ml of solution should be prepared for each run. The final protein concentration in the cell should be about 0.1% (1 mg/ml), and the buffer concentration should be 0.01 M. The density ρ_e should be adjusted to match the expected ρ_0 of the protein. The volumes of concentrated salt solution, V_1 of density ρ_1, and of water, V_2 of density ρ_2, which need to be mixed to give a final density of ρ_e, are given by the additive volume relationship: $V_1/V_2 = -(\rho_e - \rho_2)/(\rho_e - \rho_1)$. A trial value of V_1 is selected to give approximately $V_1 + V_2 = 2.0$ ml.

The concentrated salt solution can be prepared from the data given in the Merck Index or by empirically adding salt to 5 or 10 ml of water until a saturated solution is obtained. The refractive index and pH of this solution should be measured. The protein stock solution should be about 1% and should be obtained directly from an ion exchange column or by exhaustive dialysis versus deionized water. At least one other experimental technique such as sedimentation velocity should be employed prior to the density gradient run to demonstrate qualitatively the purity and homogeneity of the protein. The stock buffer solution of the desired pH should be 0.1 M.

Serological measuring pipettes are used to measure out the following quantities of solutions:

$$V_{\text{solution}} = V_1(\text{conc. salt}) + 0.1V_2(\text{protein}) + 0.1V_2(\text{buffer}) + 0.8V_2(\text{H}_2\text{O}) \quad (5)$$

The solution should be thoroughly mixed and its pH and ρ_e measured. If either quantity differs significantly from the desired values, adjustments can be made by adding small amounts of 6 M HCl or 6 M CsOH and/or concentrated salt solution or water.

The most convenient technique for measurement of densities is refractometry. A number of linear and two quadratic relationships between refractive index, n_D^{25}, and ρ^{25} are given in the preceding chapter. Additional quadratic and linear relationships are given in Tables IV and V, respectively. Care should be taken to occasionally restandardize the refractometer with the test piece and to routinely check the calibra-

TABLE IV

DENSITY-REFRACTIVE INDEX RELATIONS[a] FOR AQUEOUS SALT SOLUTIONS AT 25°

$$\rho^{25} = a + b(n_D^{25}) + c(n_D^{25})^2$$

Salt	Coefficients of equation			Valid density range (g/ml)
	a	b	c	
CsCl	1.1584	−10.2219	7.5806	1.00–1.90
CsBr	2.7798	−12.2102	8.1615	1.05–1.50
CsI	4.5245	−13.5833	8.2067	1.05–1.55
RbCl	21.7661	−39.2834	17.7843	1.05–1.35
RbI	−7.1472	4.8363	0.9561	1.05–1.75

[a] All relations are from J. B. Ifft, W. R. Martin III, and K. Kinzie, *Biopolymers* **9**, 597 (1970).

tion with distilled water. The prisms must be maintained at 25.0 ± 0.1°C and a small correction be applied for the protein concentration.[2]

The pH of the solution must also be measured before and after the run. Because of the small quantities involved, especially at the conclusion of the run, a miniature combined electrode in conjunction with a precision pH meter is required.

Analysis of the Run. A standard procedure for the analysis of plates from density gradient runs is available.[2] Two modifications of it are recommended. One is the use of the newer quadratic density–refractive index relationships given above. The other is in the determination of the isoconcentration position, r_e. The use of the following equation permits much greater freedom in the selection of the salt and angular velocity providing only that values of $\beta^0(\rho)$ are available.

TABLE V

DENSITY-REFRACTIVE INDEX RELATIONS FOR AQUEOUS SALT SOLUTIONS AT 25°

$$\rho^{25} = a(n_D^{25}) - b$$

Salt	Coefficients of equation		Valid density range (g/ml)
	a	$-b$	
NaCl	4.23061	4.64125	1.00–1.19[a]
KBr	6.4786	7.6431	1.10–1.35[b]
RbBr	9.1750	11.2410	1.15–1.65[b]

[a] J. B. Ifft, *in* "A Laboratory Manual of Analytical Methods of Protein Chemistry" (P. Alexander and H. P. Lundgren, eds.), Vol. 5, p. 151. Pergamon, Oxford, 1969.
[b] J. B. Ifft and J. Vinograd, *J. Phys. Chem.* **70**, 2814 (1966).

$$(r'_e)^2 - (r_e)^2 = D(r_b{}^2 - r_a{}^2)^2/48 \qquad (6)$$

The quantity r'_e is the root mean square of the radial position of the meniscus, r_a, and the radial position of the cell bottom, r_b. The value $D = g\omega^2/\beta_e{}^2$ where g equals the slope of the $\beta^0(\rho)$ plot at $\rho_e{}^0$ and β_e is the density gradient proportionality constant at $\rho_e{}^0$.

A computer program utilizing a sliding 5-point fit to the $\beta(\rho)$ curve has been written[9] to determine the slope of the curve at every point. Values of $g(\rho)$ for the important case of CsCl are given in Table VI.

Applications

The use of density gradient sedimentation equilibrium in the study of proteins was reviewed in 1969.[2] Thus, only three recent applications of this technique to proteins will be presented here.

Buoyant Titration of BMA. The term buoyant titration was coined by Professor Vinograd to describe the measurement of the buoyant density of a polymer as a function of pH. If changes in density occur as the state of ionization of the biopolymer changes, an inflection in the buoyant density curve should appear paralleling the usual sigmoidal potentiometric titration curve. Figure 2 presents the results of the first study on a protein, BMA.[3]

These data were analyzed in terms of changes in water, anion and cation binding as the several residues titrate utilizing the amino acid composition of BMA, published values of the pK_a, and the assumption that the molecule must remain electrically neutral. This procedure yielded values of the number of Cs^+ and Cl^- bound as a function of pH. The results indicated that a majority of the histidine and lysine residues

TABLE VI

SLOPE OF $\beta(\rho)$ PLOT, g, AS A FUNCTION OF DENSITY FOR CsCl AT 25°

Density	Slope of $\beta(\rho) \times 10^{-9}$	Density	Slope of $\beta(\rho) \times 10^{-9}$
1.05	-89.718	1.50	-0.752
1.10	-44.987	1.55	-0.540
1.15	-15.475	1.60	-0.414
1.20	-8.045	1.65	-0.311
1.25	-4.929	1.70	-0.233
1.30	-3.225	1.75	-0.172
1.35	-2.207	1.80	-0.109
1.40	-1.392	1.85	-0.040
1.45	-0.996	1.90	-0.041

[9] R. Almassy, Senior Thesis, University of Redlands, 1971.

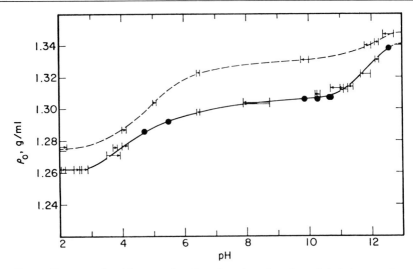

FIG. 3. Buoyant densities, ρ_0, of native (——) and carbamylated (———) ovalbumin in CsCl as a function of pH. All runs were performed in 4°, single-sector, Kel-F centerpieces at 50,740 or 52,640 rpm at 25°. The arrows reflect the pH shifts with time, pH initial → pH final. Runs involving no pH shifts are indicated by a filled circle (●).

release a chloride ion as the pH is increased and additional cesium ions are bound as the pH passes the pK_a of the tyrosines in this protein.

Buoyant Titration of Native and Chemically Modified Ovalbumin. The first complete titration over the entire pH range from pH 2 to 12.5 for a protein has been reported for ovalbumin and its selectively carbamylated derivative.[8] The data in Fig. 3 indicate that the buoyant behavior of BMA and ovalbumin are quite similar in behavior between pH 6 and high pH values. The lower plateau at pH 2 indicates that no further changes in ρ_0 occur after protonation of all the carboxyl groups. The buoyant density of the carbamylated protein in which 18 out of the 20 lysine residues had been converted to neutral homocitrulline residues was 0.025 g/ml higher than that of the native protein.

These data suggest that the low pH inflection is due to the ionization of the carboxyl residues and the high pH inflection is due to the deprotonation of both the lysine and tyrosine residues. The magnitudes of the $\Delta\rho_0$'s were used to show that the density change occurring at the inflection pH of 4.2 was due to a loss of Cl⁻ while the $\Delta\rho$ at pH 11.7 was due to a combination of Cl⁻ loss and Cs⁺ gain.

The effects of the volume change upon ionization on the observed changes in density were computed from the relation:

$$\Delta\rho = \frac{-\rho\Delta V}{V} \tag{7}$$

where ΔV is the volume change upon ionization and V is the volume of the polymer. These changes in density constitute an appreciable fraction of the observed changes.

Buoyant Titrations of the Ionizable Homopolypeptides. In order to more fully understand the above titration behavior of proteins, we have measured the buoyant behavior of seven ionizable homopolypeptides in CsCl as a function of pH.[10] Figure 4 presents these data. All these polypeptides display plateau regions at low and high pH. In every instance, ρ_0 increases with pH. The inflection pH occurs very close to the pK_a, which has been measured for these polymers by potentiometric titrations. The densities range from 1.08 to 1.700. This is somewhat sur-

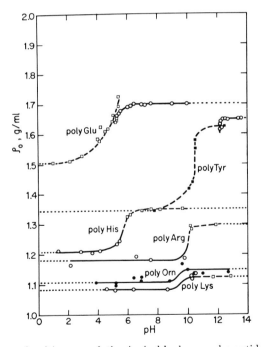

FIG. 4. Buoyant densities, ρ_0, of the ionizable homopolypeptides in CsCl as a function of pH. All runs were performed in 4°, single-sector, Kel-F centerpieces at 25°. Angular velocities varied between 44,770 and 59,780. ○, Soluble; □, precipitate.

[10] R. Almassy, J. S. V. Zil, L. Lum, and J. B. Ifft, unpublished observations.

prising in that proteins display the rather narrow range of densities mentioned earlier.

Acknowledgment

The author wishes to acknowledge that this work was supported in part by Grant GM 18871 from the U.S. Public Health Service.

[8] Zonal Centrifugation

By DAVID FREIFELDER

Zonal centrifugation through preformed concentration gradients is widely used both in the determination of hydrodynamic properties and in the purification of macromolecules. For the most part, the greatest application has been in nucleic acid research, although it has in recent years proved to be a valuable tool in the study of proteins. The method in outline is as follows: A small volume of a solution of a macromolecule is layered on top of a preformed concentration gradient contained in a centrifuge tube. The tube is centrifuged and after some period of time, the tube is removed from the centrifuge and the contents of the tube are fractionated. Then the amount of the particular macromolecule contained in each fraction is assayed in some way. In preparative work the fractions containing the substances(s) of interest are retained and usually dialyzed to remove the gradient material; in this way the centrifugation actually constitutes a stage in purification. In analytical work one obtains from the shape of the sedimentation profile the sedimentation coefficient of the substance(s) and often the degree of heterogeneity.

Types of Concentration Gradients

The purpose of a concentration gradient is to stabilize against convective and mechanical disturbances during layering, centrifugation, and fractionation. In addition, the precise form of the gradient can provide other advantages, as will be seen below.

The material most commonly used to form a concentration gradient is sucrose, since it is inexpensive and relatively easy to handle. However, although this has never been clearly documented, commercial sucrose appears sometimes to be contaminated with various enzymes from the sugar cane. The most notable substances are ribonucleases, deoxyribonucleases, and proteases. For this reason some workers in RNA research substitute glycerol for sucrose. Glycerol is useful in that it can be obtained in highly pure form, it mixes readily with water thus avoiding the long periods of time required to make concentrated sucrose solutions, and, most important, it has a stabilizing effect on many enzymes. On the other hand, a disadvantage of glycerol is that its high viscosity increases the length of time required for sedimentation.

A useful system that has seen few applications is a gradient made of increasing concentrations of D_2O. This has the advantage that no interfering substance is introduced and dialysis of the fractions obtained from the gradient is often unnecessary. In fact, there is one type of material for which an H_2O–D_2O gradient is the method of choice—that is, when the substance of interest is osmotically sensitive. For example, many subcellular particles maintain their integrity only in solutions of high osmotic strength, e.g., in the presence of 20% sucrose. Clearly a sample suspended in 20% sucrose can be layered onto a sucrose gradient only if the top of the gradient has a sucrose concentration significantly greater than 20%. One could use a 25–50% or 25–60% gradient, but this might result in many problems due to the high buoyancy and viscosity and to potential osmotic shock when the very high sucrose concentration is reduced. An H_2O–D_2O gradient containing 20–25% sucrose avoids these problems. This system has been used to characterize hormone-containing particles from rat hypothalamus (Dorothy Freifelder, unpublished work); the particles suspended in 20% sucrose were centrifuged through a gradient varying from 22% sucrose in H_2O to 25% in 90% D_2O.

It is also possible to incorporate into concentration gradients a substance to reduce secondary structure. Some examples are the use of alkali to study single-strand DNA and 70% dimethyl sulfoxide in sucrose to eliminate internal hydrogen bonding of RNA. Presumably other denaturants, such as formamide, dimethyl formamide, or guanidinium chloride, could be used instead.

Steepness of Concentration Gradients

The literature abounds with concentration gradients of various degrees of steepness, although 5–20% sucrose is certainly the most common. The reasons for the various choices are usually not made clear. As mentioned above, the primary purpose of the gradient is to protect against mechanical and convective disturbance so that the steeper the gradient, the more stable it is. However, the steepness of the gradient is determined both by the concentrations of the two solutions used to generate the gradient and by the total length of the liquid column, i.e., the length of the centrifuge tube so that the concentration difference between the two starting solutions must be greater for longer tubes. This fact is not commonly considered and the "standard" 5–20% sucrose gradient is used in all manner of centrifuge tubes; indeed several workers have detected instability when a 5–20% gradient is used in the more recently developed high resolution long centrifuge tubes (e.g., for the Spinco SW40 and SW41 rotors).

The steepness of the gradient is important for a second, usually forgotten, purpose. Since the centrifugal force increases linearly with the distance from the center of rotation, sedimenting particles move increasingly more rapidly with time. In a 5–20% sucrose gradient contained in a Spinco SW39 rotor (equivalent to the SW50 and SW65), the increasing buoyancy and viscosity of the sucrose gradient with distance from the center of rotation compensate for the dependence of centrifugal force on radius so that the sedimenting particles move linearly (instead of logarithmically) with time. This arrangement has the advantage that measurement of sedimentation coefficients is greatly simplified since s is proportional to D, the distance traveled in a particular period of time. It must be emphasized that (to the author's knowledge) for no other rotor has there been a determination of the composition of a gradient that would yield sedimentation linear with time. Hence only with a 5–20% sucrose gradient in a SW39, SW50, or SW65 rotor can s or even relative s be determined with precision. In fact, with long tubes, the error may be quite large.

Preparation of Concentration Gradients

The simplest type of apparatus for the preparation of a linear concentration gradient is shown in Fig. 1. The left container is filled with the lower density liquid, and the right with the higher. A mixer of some kind (either a magnetic stirrer or a standard laboratory mixer) is put in the denser liquid. As liquid drains from the mixing chamber, the liquid of lower density flows in to maintain equal hydrostatic pressure (not equal heights since the densities are different). The outflowing liquid is allowed to run down the side of a slightly inclined centrifuge tube in which the gradient is formed. If the liquid flow is slow enough, a homogeneous concentration gradient will form in the tube; if it is too fast, mixing will occur in the centrifuge tube. A safe flow rate is $\frac{1}{3}$ ml per minute. The channel connecting the two chambers should be filled with the lighter liquid; otherwise the first few tenths of a milliliter will be at constant concentration. The flow rate can be determined either by gravity (i.e., the vertical distance from the exit part of the mixing tube to the end of the tubing) or by a peristaltic pump. If gravity is used, the flow rate will not be constant since the pressure head decreases); however, other than slight inconvenience, this causes no

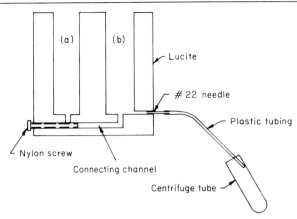

FIG. 1. Apparatus for preparation of linear density gradient. The nylon screw is open so that there is free access from chambers (a) to (b) via the connecting channel. The liquid of lower density is put in (a). The connecting channel is closed by turning the nylon screw, and the lower density liquid which has flowed into (b) is removed and added to (a). Liquid of higher density is added to (b); the amounts of liquid in (a) and (b) are chosen to equalize the hydrostatic pressure head. A mixer is placed in (b), the nylon screw is turned to open the connecting channel, and the liquid is allowed to flow through the needle and tubing into the centrifuge tube.

difficulty. If a peristaltic pump is used, it should have as little pulsation as possible. The author has used both Sigmamotor pumps (Sigmamotor, Middleport, New York) and the Polystaltic pump (Buchler Instruments, Fort Lee, New Jersey) with success.

For a smooth flow of liquid down the wall of the centrifuge tube, the end of the tubing is placed against the wall of the centrifuge tube. It is wise to use some type of holder for the tubing so that it does not flip out. Figure 2 shows a simple holder constructed in the author's laboratory for holding several tubes.

If polyallomer tubes are used, most aqueous solutions form small droplets on the wall of the centrifuge tube rather than a continuous stream of fluid. These droplets move rapidly down the wall of the tube, splash at the bottom and make uneven, unsatisfactory gradients. In general, polyallomer tubes are to be avoided but if, for some reason, they are necessary (e.g., if very high centrifugal speeds are required), concentration gradients can be made by one of two simple variations of the standard procedure. In the first, 0.01% sodium dodecyl sulfate or sodium sarcosinate is incorporated into the gradient solutions. The addition of the detergent enables the polyallomer tube to be wet. If the detergent is undesirable, a second method may be used, i.e., the gradient

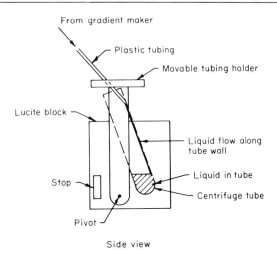

Side view

Fig. 2. Holder for centrifuge tubes while gradient is being formed. Plastic tubing from the gradient marker is pushed through the holes in the movable tubing holder. The holder is swung aside (to the left in the figure) to rest against the stop. The centrifuge tubes are placed in the holes in the Lucite block, the movable tubing holder is swung back, and the tubing is adjusted so that it touches the upper wall of the centrifuge tube. The figure is a side view. The author's holder has positions for six centrifuge tubes.

is formed in reverse. Here the solutions of lower and higher concentration are interchanged in the apparatus of Fig. 1 so that the low density part of the gradient is produced first. The exit part of the tubing is then placed at the very *bottom* of the centrifuge tube. In this way the solution being added, whose density continually increases, at the bottom of the gradient slowly pushes the gradient upward. When the gradient is complete, the exit tubing must be removed slowly and carefully.

Since gradients are usually used in groups of three to six (because of rotor design), it is advantageous to be able to make several at one time. There are numerous commercial instruments that can do this. These are probably all adequate, but they are rather expensive ($500–1200). The least expensive system that is still automatic is that shown in Fig. 3. Whereas fabrication of this also comes to about $500 including labor, the major component, i.e., the peristaltic pump, can also be used for fractionation (see below) and often already exists in many laboratories.

Recipes for Gradient Solutions

Sucrose Gradients. The sucrose solutions are usually prepared in a dilute buffer such as 0.02 M Tris[hydroxymethyl]aminomethane. Salts,

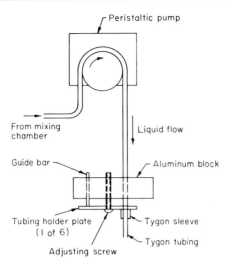

Fig. 3. Apparatus for controlling the flow of liquid to several centrifuge tubes. Liquid is pumped from a part of the mixing chamber to the tube. By adjustment of the tubing holder plate with the adjusting screw, the tubing is stretched. The degree of strength determines the internal diameter of the tubing and thereby the flow rate. The aluminum block is normally mounted on the pump body.

such as $MgCl_2$, or chelating agents, such as ethylenediaminetetraacetate, are added if necessary for stability of the material to be studied. NaCl can be incorporated up to $1\,M$ if high ionic strength is necessary. $X\%$ sucrose means X g per 100 ml of solution. For short centrifuge tubes holding ca. 5 ml, a 5–20% sucrose gradient can be used. For long tubes, such as the Spinco SW41 (13 ml) or SW25.1 (34 ml), a 5–30% gradient is needed to ensure stability, although 5–20% gradients are in common use. In general, when high ionic strength is used, since the gradient is flatter, it may be necessary to increase the steepness by increasing the density of the denser sucrose solution. This can also be accomplished by superimposing on the sucrose gradient a salt gradient—for example, 5% sucrose in $0.5\,M$ NaCl to 20 or 30% sucrose in $1\,M$ NaCl. It is wise but not absolutely necessary to boil or autoclave solutions to inactivate enzymes present in the sucrose and to eliminate growth of molds. *Note:* Sucrose and NaCl cannot be autoclaved together for more than 5 minutes at 15 psi or the sucrose will char; solutions should be autoclaved separately and then mixed. If sucrose solutions are to be kept for long periods of time, they should be stored out of the light to avoid growth of algae.

Glycerol Gradients. The low and high density solutions are usually

10% and 30% glycerol, respectively, in an appropriate buffer. These solutions should be kept sterile to avoid bacterial growth.

H_2O–D_2O Gradients. The low density solution is an aqueous buffer; the high density solution is the same buffer prepared in 95% (or higher purity) D_2O. Normally the buffer in the D_2O is adjusted to be at a slightly higher concentration than in the H_2O to avoid flattening the gradient by inadvertently having the H_2O buffer slightly more concentrated. It is good practice to superimpose on an H_2O–D_2O gradient a small concentration gradient to improve stability. Addition of 0.1 M NaCl to the D_2O is quite adequate.

Centrifugation Procedures. Zonal centrifugation is best carried out using a centrifuge with a rotor stabilizer, i.e., some system which prevents wobbling of the rotor during acceleration and deceleration. An example of such a system is that found in the Spinco Model L2 series. The older type Model L centrifuges lack this and when this machine is used, gradients often show broad peaks indicative of some type of disturbance. It is not meant to imply that zonal centrifugation is not possible in these older instruments, but only that the rate of failure will be higher.

The braking system of the machine must be set so that during deceleration the brake is turned off when the speed is reduced to a few thousand revolutions per minute, since at very low angular velocity the braking system frequently produces rotor wobble and therefore mechanical disturbances in the gradient. Furthermore, if the brake is improperly set, the rotor may come to a full stop and then accelerate slightly in the reverse direction. In most modern instruments the brake is satisfactorily adjusted, but it is wise to consult the service man.

Normally, swinging buckets are used for zonal centrifugation. However, it is possible to use angle rotors if the gradient is steep enough to provide the necessary stability.

Gradients are frequently disturbed while placing the rotor on the drive shaft or while removing the rotor after a run is completed. This is usually a result of sticking. To avoid this the drive spindle or drive shaft and the rotor base should be kept clean and well lubricated with the appropriate material provided by the manufacturer.

A destructive disturbance often results if the centrifuge tube sticks in the bucket of the rotor. This is usually caused by dirt, sucrose, or water in the bucket—which should be kept clean and dry. If a tube cannot be removed from the bucket, the gradient can be saved by fractionation from the bottom through the top (see below).

On occasion, a centrifuge tube collapses during the run. This is usually due to one of three causes: (1) The tube was defective. Tubes should always be carefully examined for cracks, deformation, and dents. (2) The tube was not sufficiently filled; tubes should be filled to within 3 mm of the top. (3) The bucket had water in it. Keep buckets clean and dry.

The temperature used for sedimentation depends primarily on the stability of the material being studied. Centrifugation itself can be carried out over a wide range of temperatures. The one precaution is that a gradient to be centrifuged at a particular temperature should be prepared at that temperature since any temperature change after preparation of the gradient can cause serious convective disturbance.

Layering the Sample on the Gradient

As in the preceding section the sample to be layered should be at roughly the same temperature as the gradient to avoid convection.

Clearly it is also necessary to avoid mechanical disturbance. Layering is normally done by allowing the solution to run slowly down the wall of the centrifuge tube. If it is at a lower density than the top of the gradient, it will form a thin layer. If the flow down the wall of the tube is too rapid, the layering will be poor and there will be mixing with the upper part of the gradient. Therefore, the liquid should be touched to the wall very near the surface of the gradient—preferably not more than 2 mm away. If the solution does not wet the wall, the wall can first be wet with a drop of the lower density solution used to make the gradient. Pasteur pipettes are very useful for layering.

The maximum volume of the sample to be layered depends upon the size of the centrifuge tube. As the volume of the layer increases, the band width after centrifugation will be greater. However, there is not a linear relation between band width and sample volume since the macromolecule sediments more rapidly in the low density layer than anywhere in the gradient, resulting in sharpening prior to entering the gradient in much the same way that sharpening occurs in disc electrophoresis. Nonetheless, it is wise to restrict the sample volume to 0.1 ml for tubes of narrower diameter (e.g., 0.5-inch tubes) and 0.5 ml for the tubes of greatest diameter (1–1.25 inches).

Fractionation of Gradients

The most common method of fractionation is to puncture the bottom of the tube and collect the drops. This has been done in three ways. In the first, the rate of dripping is uncontrolled and is determined only by gravity. In the second, a rubber stopper with a needle through it is placed in the top of the tube and the needle is attached via tubing to a glass syringe. The flow rate is then controlled by pressing the syringe. In the third, tubing is attached to the needle and air is pumped into the sealed tube by a peristaltic pump. In general, fractionation is best when the flow rate is controlled and kept to less than 0.5 ml per minute. At high flow rates the gradient may be disrupted. Two valuable tube holders are shown in Fig. 4.

It is also possible to fractionate the gradients (a) from the bottom but through the top or (b) directly from the top. For (a) a needle (usually No. 22) is slowly lowered to the bottom of the tube. After 1 minute to reestablish equilibrium, the gradient is either pumped out by a peristaltic pump operating on tubing attached to the outer end of the needle or the tube is sealed and a light oil is forced onto the top of the gradient. For (b) a rubber stopper containing two needles is inserted into the tube. One needle is flush with the inner surface

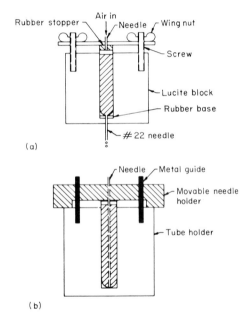

FIG. 4. Two tube holders used in fractionation. (a) The tube is pressed by the metal plate against the rubber base in which the lower needle is mounted. This pressure makes a seal that ensures that all liquid leaving the tube enters the needle. The tube is sealed with a rubber stopper through which a needle passes which is connected to a syringe in order to control flow of the liquid. A No. 27 needle is inserted into the lower No. 22 needle and the tube bottom is punctured. (b) A long needle is inserted through the gradient to the tube bottom. The liquid is then pumped out using a peristaltic pump. The metal guides hold the movable needle holder which is removed to insert the tube into the tube holder. The movable needle holder serves to prevent lateral movement of the needle while it passes through the gradient.

of the stopper and the other needle extends to the bottom of the tube. A heavy oil is pumped to the bottom of the gradient, thus raising the level of the liquid until it passes through the other needle. There are various commercial units available for fractionation of centrifuge tubes.

For most purposes the various methods of fractionation give similar results. Methods (a) and (b) above have an advantage over dropping from a hole in the bottom of the tube only in that the tube is not destroyed. However, there are a few situations in which top and bottom fractionation offer particular advantages. For instance, if the centrifuge tube contains a gelatinous pellet, top fractionation is better since with bottom fractionation one never knows when the pellet will be dislodged. On the other hand, if after centrifugation there is any viscous material on the top of the gradient, bottom fractionation is clearly preferable.

Assaying Fractions

The fractions obtained can be assayed by content of radioactivity, chemical tests, enzymatic activity, fluorescence, etc. Most assays are straightforward and require little or no adaptation to gradient techniques. One need only ascertain that the gradient materials do not have a strong effect on the assay.

Determination of Sedimentation Coefficient (s)

The absolute determination of a sedimentation coefficient from a zonal centrifugation run is a formidable task because of the numerous corrections that must be made. However, with a few precautions *relative s* values are easily obtained. This is normally accomplished by mixing with the substance of unknown s at least *two* other substances of known s. It cannot be overemphasized that the common practice of adding a single marker can lead to significant errors since, as pointed out in a previous section, the relative distances sedimented are time independent only in a 5–20% sucrose gradient run in any SW39 or SW50 rotor. With two or preferably three markers, one can empirically relate all the distances sedimented.

There are two principal limitations in the determination of relative s. (1) Since one measures only the relative distances traveled, one must know where to choose the sedimentation origin. This is not trivial because the original sample layer has finite thickness and because also there is accumulation of material at the sample–gradient interface. Clearly the uncertainty in the origin will depend upon the volume of the sample layered and the relative densities of the sample and the top of the gradient. In practice this probably introduces an error of not more than 2–3% in s-value. (2) A more serious problem arises from the finite size of the fraction taken. This limits the accuracy of identification of the fraction containing a "peak" and probably introduces an error of about ±5%.

The usual reason for measuring s is to determine the molecular weight (M). This is accomplished by use of either theoretical or empirical equations relating s and M. It is essential to realize that (a) the s used in theoretical equations is *not* the s measured by zonal centrifugation and (b) most empirical equations have used the s measured with the analytical ultracentrifuge. The crux of the problem is that the s measured by zonal centrifugation in a preparative centrifuge is always less than the s-value from the analytical centrifuge. The reason is that in the preparative centrifuge the tubes are not sector shaped and a significant fraction of the molecules are sedimented toward the walls. This gives rise to aggregation and localized convection. Furthermore, in the derivation of the theoretical equations the assumption is made that there are no wall effects. To date, there is no rigorous theoretical analysis of wall effects but the observation that the s-values obtained in tubes with parallel walls differs from those from sector-shaped walls is well known.

In order to avoid these problems, it is essential that one use empirical relations that are derived from measurements using zonal centrifugation. If such a relation does not exist for the materials being studied, it is recommended that three or four substances of known M be cosedimented with the substance of unknown M. The theoretical equations should probably not be used until a relation between the two types of s-values is worked out.

Advantages and Disadvantages of Zonal Centrifugation

Advantages are (1) small amount of material needed; (2) little dependence of s on concentration; (3) sample sediments free of interfering substance.

Disadvantages are (1) inability to measure absolute s; (2) poorer resolution than analytical ultracentrifuge.

[9] Small-Angle X-Ray Scattering

By HELMUT PESSEN, THOMAS F. KUMOSINSKI, and SERGE N. TIMASHEFF

I. Introduction

Among the methods available for the characterization of globular proteins, small-angle X-ray scattering (SAXS) is particularly powerful. This method is capable of yielding the radius of gyration and, when used on the absolute-intensity scale, the molecular weight, hydrated volume, surface-to-volume ratio, and degree of hydration of a particle in solution.[1-6] To obtain this information, it is necessary to measure two auxiliary parameters: the concentration and the partial specific volume of the protein.

In addition to the molecular parameters, one may obtain thermodynamic parameters of interacting systems,[4,5] such as association constants of aggregating subunit systems and the degree of preferential interaction of proteins with components of mixed solvent systems. Furthermore, in the case of highly concentrated solutions (>100 mg/ml), in which there are strong long-range intermolecular interactions, SAXS can be used to determine the radial distribution function of the interacting (or even partly immobilized) macromolecules in solution.[1,5,7,8] This, in turn, yields the interaction potentials characteristic of the operative forces.[9] At still higher concentrations, distinct bands may appear, as the system gradually becomes ordered and X-ray scattering passes over into small-angle X-ray diffraction.[10] The recent development of absolute intensity apparatus[6,11,12] (see Section III, A) has rendered

[1] A. Guinier and J. Fournet, "Small-Angle Scattering of X-Rays." Wiley, New York, 1955.

[2] W. W. Beeman, P. Kaesberg, J. W. Anderegg, and M. B. Webb, *in* "Handbuch der Physik" (S. Flügge, ed.), p. 321. Springer-Verlag, Berlin and New York, 1957.

[3] V. Luzzati, *Acta Crystallogr.* **13**, 939 (1960).

[4] S. N. Timasheff, *in* "Electromagnetic Scattering" (M. Kerker, ed.), p. 337. Pergamon, Oxford, 1963.

[5] S. N. Timasheff, *J. Chem. Educ.* **41**, 314 (1964).

[6] H. Pessen, T. F. Kumosinski, and S. N. Timasheff, *J. Agr. Food Chem.* **19**, 698 (1971).

[7] G. Fournet, *Acta Crystallogr.* **4**, 293 (1951).

[8] G. Fournet, *Bull. Soc. Fr. Mineral. Cristallogr.* **74**, 37 (1951).

[9] J. G. Kirkwood and J. Mazur, *J. Polym. Sci.* **9**, 519 (1952).

[10] P. Saludjian and V. Luzzati, *in* "Poly-α-amino Acids" (G. D. Fasman, ed.), p. 157. Dekker, New York, 1967.

[11] V. Luzzati, J. Witz, and R. Baro, *J. Phys. (Paris)*, Suppl. **24**, 141A (1963).

practical such a characterization of macromolecules, and aggregates of molecules, including enzymes, in solution.

Fundamentally, the method of X-ray scattering differs little from that of light scattering, the theoretical principles being essentially identical. The differences that exist arise from differences in the wavelengths of the radiations in the two cases. In light scattering, the wavelength is of the order of 4000 Å; in SAXS it is ~1.5 Å. Both techniques are based on concentration fluctuations of the solution under examination. In light scattering an auxiliary parameter required is the refractive index increment of the macromolecular solute. This can be measured directly. In SAXS such a measurement is not possible, since the refractive index is practically indistinguishable from unity. In order to express the concentration fluctuations, it becomes necessary, therefore, to calculate a corresponding quantity, which for SAXS is the electron density, i.e., the number of electrons per unit volume. This can be done from the chemical compositions of the solution components.

The electromagnetic theory basic to SAXS is also basic to X-ray diffraction, and the two techniques are founded on the same phenomena. They differ, however, in the nature of the observations. X-ray diffraction results from destructive and constructive interferences in scattered radiation, evidenced by discrete spots or bands that correspond to characteristic repeat distances within an ordered structure, such as a crystal. In SAXS the scattered radiation is diffuse and a generally monotone function of angle. X-ray diffraction reflections usually correspond to small interatomic distances and thus are found at higher angles; SAXS corresponds principally to molecular dimensions and is concentrated mostly within a cone a few degrees from the incident beam. There is an intermediate region (2–5°) in which the internal order of macromolecules begins to manifest itself. This leads to the appearance of secondary maxima and minima superimposed on the scattering curve, resulting in a wavelike appearance of the angular dependence of scattering at these higher angles. As will be shown (see Section II, B), the positions of these fluctuations in the scattering curve are very useful in assigning structural models to particular macromolecules.

II. Principles

A. Nature of the Phenomenon

When a beam of electromagnetic radiation strikes an electron, some of the energy is momentarily absorbed and the electron becomes dis-

---

[12] O. Kratky and H. Leopold, *Makromol. Chem.* **75**, 69 (1964).

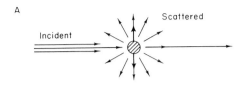

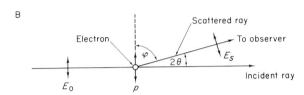

FIG. 1. Fundamentals of small-angle X-ray scattering. (A) Basic "scattering" event. (B) Definition of terms.

placed from its unperturbed position due to the force exerted on it by the electric field. As a result, the electron is set into periodic motion with a frequency equal to that of the exciting radiation. Since, according to the Maxwell equations, any accelerating or decelerating charge must radiate an electromagnetic wave in all directions and since the radiation reemitted by the electron has the same frequency as the exciting radiation, the experimental observation gives the impression that the incident radiation is scattered in all directions by the electron, as depicted schematically in Fig. 1A. This observation is the origin of the terms "X-ray scattering" as well as "light scattering." In the quantum view the incident X-ray protons undergo perfectly elastic collisions with the electron, leaving their energy (i.e., their frequency) unchanged. Hence this type of scattering is called "elastic," "unmodified," "coherent," and, because it obeys Bragg's law, "Bragg scattering."[12a]

While a detailed derivation of the theory of X-ray scattering can be found in various monographs,[1,13] we will outline here briefly the arguments followed. The action of an electric field of strength E on a polarizable particle quite generally induces in it a dipole moment, p, whose magnitude is

$$p = \alpha E \tag{1}$$

[12a] Another distinct phenomenon is "inelastic" or "modified scattering," also termed "incoherent" because of its lack of definite phase relationships, and "Compton scattering." It becomes more pronounced the higher the energy of the radiation (e.g., gamma rays), but is negligible at small angles. SAXS is concerned solely with elastic scattering.

[13] W. H. Zachariasen, "Theory of X-Ray Diffraction in Crystals." Wiley, New York, 1945.

where α is a proportionality constant known as the polarizability and is a measure of the induced distortion of the molecule. For an electromagnetic wave, E, the amplitude of the electric field vector can be expressed by

$$E = E_0 \cos 2\pi(\nu t - x/\lambda) \tag{2}$$

where E_0 is the maximum amplitude, ν is the frequency, λ is the wavelength, t is the time, and x is the location along the line of propagation. Here, the cosine function represents the phase angle, the determination of which is one of the main problems in structure determination by X-ray diffraction. Combining Eqs. (1) and (2), the amplitude, E_s, of the re-emitted electric field, which is proportional to d^2p/dt^2, results in:

$$E_s = -\frac{4\pi^2\nu^2}{c^2 r} p \sin \varphi \tag{3}$$

where c is the velocity of light, r is the distance between the scattering particle and the observer, and φ is the angle between the dipole axis and the line joining the point of observation to the dipole, as shown in Fig. 1B. The intensity, I, of the radiation is equal to the product of the amplitude and its complex conjugate[13] (the conjugate of a complex quantity $a + ib$ being defined as $a - ib$), $I = EE^*$, whose magnitude is[14]

$$I = |E_s|^2 \tag{4}$$

If we take now a source of unpolarized radiation and resolve the scattered radiation into components parallel and perpendicular to the electric vector of the incident radiation, and sum up the total, having first combined Eqs. (1), (2), (3), and (4), and remembering that $c/\nu = \lambda$, we obtain the familiar Rayleigh equation, which is fundamental for light scattering (see this volume [10]).

$$I_{\text{scat}} = |E_s|^2 = \frac{8\pi^4\alpha^2}{r^2\lambda^4} I_0(1 + \cos^2 2\theta) \tag{5}$$

where I_{scat} is the intensity of the scattered radiation, I_0 is that of the incident radiation and 2θ is the angle between the directions of the incident and scattered rays, as defined in Fig. 1B.[15]

B. Scattering from an Electron

For radiation whose frequency is high compared to the natural frequency of the dipole (as is the case for X-rays, though not for visible light), the polarizability, α, can be expressed as

[14] This is the source of the loss of phase angle information in X-ray diffraction.
[15] The angle 2θ is equal to twice the reflection, or Bragg, angle which is familiar with X-ray diffraction.

$$\alpha = \frac{-e^2}{4\pi^2\nu^2 m} \tag{6}$$

where e is the charge and m is the mass of the scattering element, the electronic charge and electronic mass in the case of an electron. Combination of Eqs. (5) and (6) results in the Thomson equation, which is the fundamental equation of X-ray scattering:

$$I_{\text{scat,e}} = \frac{e^4}{m^2 c^4 r^2} I_0 \left(\frac{1 + \cos^2 2\theta}{2} \right) \tag{7}$$

where $I_{\text{scat,e}}$ is the scattered intensity from a single, independent electron. The product $e^4 m^{-2} c^{-4}$ is known as the electron-scattering factor (also termed electron-scattering cross section and Thomson's constant; equal to the square of the so-called electron radius), and the quantity in parentheses is known as the polarization factor.[16] Introducing the numerical values of e, m, and c, into Eq. (7) results in

$$\frac{I_{\text{scat,e}}}{I_0} = 7.90 \times 10^{-26} r^{-2} \left(\frac{1 + \cos^2 2\theta}{2} \right) \tag{8}$$

This indicates that the intensity of radiation scattered by a single electron is more than 25 orders of magnitude smaller than the intensity of the incident beam. Now, the mass of a proton is 1840 times greater than that of an electron. Since mass appears to the second power in the denominator of Eq. 7, scattering by a proton will be ca. 3.4×10^6 times weaker than that by an electron. As a result, in X-ray scattering, as well as in X-ray diffraction, essentially only the electrons in matter are detected. This has given rise to the term "electron density"[17] and to the practice in X-ray scattering of expressing all mass units in numbers of electrons.

While the scattering from a single electron is extremely weak, in real systems, such as a protein in aqueous medium, we measure the total scattering from all the electrons in the irradiated volume (which is of the order of 0.1 ml). This result in a measured scattered intensity only 10^4 to 10^5 times weaker than the incident radiation. Such an in-

[16] Comparison of Eqs. (5) and (7) reveals a notable difference between the wavelength dependences of light scattering and small-angle X-ray scattering. The Rayleigh equation (Eq. 5) contains the well-known inverse fourth power of wavelength dependence of light scattering. In the Thomson equation (Eq. 7) no such relationship appears; i.e., X-ray scattering intensities are not wavelength dependent. This difference derives from the circumstance that in the Rayleigh equation the polarizability, which appears squared in the numerator, is not a simple function of wavelength; for X-rays, however, the polarizability is proportional to the square of the wavelength (see Eq. 6), allowing all the wavelength terms to cancel from the equation.

[17] This is also the origin of the term "electron density map" used in X-ray diffraction.

tensity is measurable, but the four to five orders of magnitude difference in intensities between the incident and scattered radiations lead to the extremely stringent instrument collimation requirements which will be discussed below (see Section III, A).

C. Scattering from Particles

1. Scattering Envelope

In molecules such as those of proteins the electrons are not independent, since the relative positions of the atoms are fixed in space. It is reasonable, therefore, to expect interactions between the scattering of individual electrons. Since the dimensions of macromolecules are always large relative to the wavelength of the incident X-radiation, interference occurs between the radiation scattered from individual scattering elements (in the case of X-ray scattering, these are individual electrons) within the macromolecule, with the result that the intensity of scatter is a strong function of the angle of observation, 2θ. The reason for this is shown in Fig. 2A. Here we have a particle which is large with respect to the wavelength of the radiation. Consider scattering from elements n and m observed at points P and Q. We find that when radiation scattered by elements n and m reaches point P (in the forward direction), if angle 2θ is small, the difference between the pathlengths of the two rays ($nm + mP - nP$) is small, so that they are not greatly out of phase with each other and interference is small. However, when the radiation scattered from n and m reaches point Q (in the backward direction) the total distance traveled by the ray from m is much greater than that from n (greater by $nm + mQ - nQ$), with the result that the two rays can become completely out of phase, leading to destructive interference. In the forward direction, i.e., along the incident beam, scattered radiation from n and m is fully in phase, there is no destructive interference and the total scattering is the sum of the scatterings from all elements within the particle. As a result, the scattering envelope (i.e., the angular dependence of the scattering) has an asymmetric shape such as shown in Fig. 2B. (For clarity, the envelope is shown much less elongated than should actually be the case.) Since scattering in the forward direction falls on top of the much stronger incident beam, it cannot be measured directly. This results in the requirement of extrapolation to zero angle. The shape of a typical recording of the angular dependence of scattering of X-rays obtained from a protein solution is shown in Fig. 2C. At very low angles, the angular dependence of the scattering is essentially gaussian. At increasing values of 2θ (above 1°), the intensity drops to very low values, decreasing asymptotically to a

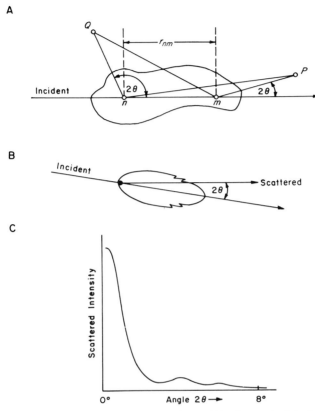

Fig. 2. Angular dependence of scattering. (A) Internal interference. (B) Schematic drawing of scattering envelope. (C) Schematic representation of typical recording of data.

constant background value. In this higher-angle region, secondary maxima and minima become superimposed on the weak and diminishing radiation.

a. *Debye Equation.* In 1915 Debye[18] showed that the angular dependence of the scattering from a particle of any shape, averaged over all orientations, is given by

$$I_{scat}(s) = \sum_{m=1}^{N} f_m \sum_{n=1}^{N} f_n \frac{\sin 2\pi s r_{nm}}{2\pi s r_{nm}} \tag{9}$$

$$s = (2/\lambda) \sin \theta$$

where N is the total number of scattering elements in the particle, f_m

[18] P. Debye, *Ann. Phys.* (*Leipzig*) **46**, 809 (1915).

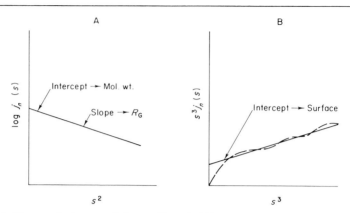

FIG. 3. Types of plots used in small-angle X-ray scattering. (A) Guinier plot. (B) Soulé-Porod plot.

and f_n are the scattering factors of any pair of scattering element, r_{nm} is the distance between elements n and m, λ is the wavelength of the radiation, and 2θ is the angle between the incident and scattered beams. From this equation, the angular dependence of the scattering of variously shaped bodies can be calculated by introducing specific expressions for r_{nm}, characteristic of the geometry of the particular body. The approximate shape of a scattering particle can be determined by comparing the experimental scattering envelope with envelopes calculated for various geometric models. It is not necessary, however, to know the shape of a particle to obtain certain information about its structure.

b. *Guinier Equation; Radius of Gyration.* In 1939 Guinier[19] showed that scattering yields a characteristic geometric parameter of any particle which is independent of any assumption regarding its shape, namely the radius of gyration, R_G, i.e., the root-mean-square of the distances of all the electrons of the particle from its center of electronic mass. Expanding Eq. (9), Guinier showed that in the case of isotropic particles for a point source radiation, and for small values of the product R_Gs,

$$i_n(s) = i_n(0) \left(1 - \frac{4}{3}\pi^2 R_G^2 s^2 + \cdots\right) \approx i_n(0) \exp\left(-\frac{4}{3}\pi^2(R_G)^2 s^2\right) \quad (10)$$

where $i_n(s)$ is the scattered intensity at angle 2θ corresponding to a given value of s (normalized to the energy of the incident beam, i.e., referred to the scattering produced by a single electron under the identical conditions; see Sections III, A and IV, A); $i_n(0)$ is the normalized intensity extrapolated to zero angle. Thus, at very low angles, a plot of log $i_n(s)$ versus s^2 gives a straight line (Fig. 3A), the slope of which is $(4/3)\pi^2 R_G^2$.

[19] A. Guinier, *Ann. Phys. (Paris)* **12**, 161 (1939).

As will be shown below, the intercept, $i_n(0)$, is proportional to the square of the molecular weight.[20]

As a practical matter, it is rarely possible to utilize a point source, on account of its insufficient intensity. The geometry generally chosen for the source is one defined by a narrow slit (see Section III, A). If the slit is long, so that its height exceeds the angular range, measured at the detector, at which observable scattering occurs, it is said to be an "infinitely high" slit. In analogy to Eq. 10, the normalized scattering intensity $j_n(s)$ from an infinite-slit source is given by[1,3]

$$j_n(s) = j_n(0) \exp\left(-\frac{4}{3}\pi^2 R_a^2 s^2\right) + \phi(s) \tag{11}$$

where $j_n(0)$ is $j_n(s)$ extrapolated to zero angle, R_a is the apparent radius of gyration, i.e., that referring to a finite concentration of solute, and $\phi(s)$ is a residual function expressing the difference between the gaussian portion of Eq. (11) and the scattering actually observed.

c. Deconvolution. The theoretical point-source scattering curve can be constructed from the experimental infinite-slit data by an appropriate mathematical transformation,[1] which is fairly simple in principle but, in practice, is attended with considerable difficulties. It is usually carried out numerically on a digital computer. Using the Luzzati[3] form of this transformation, $i_n(s)$ and $j_n(s)$ are related by

$$i_n(s) = -\frac{1}{\pi}\int_0^\infty \frac{dj_n(s^2+t^2)^{1/2}}{d(s^2+t^2)^{1/2}}\frac{dt}{(s^2+t^2)^{1/2}} \tag{12}$$

2. Molecular Weight

If in Eq. (9) each scattering element is taken as one electron in the particle, and the particle contains m electrons, then it can be shown that, for a single particle in vacuum, $i_n(0) = m^2$. For J noninteracting particles per unit volume, the total scattering at $2\theta = 0°$ is Jm^2. Expressing the concentration in mass per volume units, c, $J = c/m$, in vacuum,

$$\frac{i_n(0)}{c} = m \tag{13}$$

Now, passing to the real case of macromolecules in solution, the fluctuation theory of scattering[21,22] gives for a two-component system,

[20] The notation adopted in this presentation is that of Luzzati.[3] A variety of other notations have been used in the literature. These, in general, refer to intensity normalized in different ways and to different ways of expressing the angular function. The most common alternate notations either use the Bragg angle θ as such, or the symbol h, defined as $h = (4\pi \sin \theta)/\lambda$.

[21] W. Kauzmann, "Quantum Chemistry." Academic Press, New York, 1957.

[22] C. Tanford, "Physical Chemistry of Macromolecules." Wiley, New York, 1961.

e.g., an enzyme dissolved in water, the following relation between the excess scattering of solution over solvent per unit volume $\Delta i_n(0) = i_n(0)_{\text{solution}} - i_n(0)_{\text{solvent}}$, and the fluctuations of the electron density, $\Delta\rho_s^2$, in a volume element δV:

$$\Delta i_n(0) = \delta V \overline{\Delta\rho_s^2} \tag{14}$$

Here, $\overline{\Delta\rho^2} \equiv \overline{\rho^2} - \bar{\rho}^2$ and ρ_s is the electron density of the solution (number of electrons per unit volume). Since

$$\overline{\Delta\rho_s^2} = \left(\frac{\partial\rho_s}{\partial c_e}\right)_{T,p} \overline{\Delta c_e^2} \tag{15}$$

where c_e is the solute concentration expressed as the ratio of the number of electrons of solute to that of solution (essentially a weight fraction), application of the thermodynamic relations for concentration fluctuations[22] results in[22a]

$$\left(\frac{\partial\rho_s}{\partial c_e}\right)_{T,p}^2 \frac{c_e(1-c_e)^2}{\Delta i_n(0)\rho_s^2} = \frac{1}{m_{\text{app}}} = \frac{1}{m}\left[1 + \frac{c_e}{RT}\left(\frac{\partial\mu^e}{\partial c_e}\right)_{T,p}(1-c_e)\right] \tag{16}$$

where R is the gas constant, T is the thermodynamic temperature, μ^e is the excess chemical potential of the solute, and m_{app} is an apparent mass (in electrons) of the particle calculated for each finite concentration of protein at which scattering measurements are made.[23] Since the electron density increment

$$\left(\frac{\partial\rho_s}{\partial c_e}\right)_{T,p}$$

is not a directly measurable quantity, it must be replaced in a working equation by one that is readily measurable. At constant temperature and pressure,

$$\frac{d\rho_s}{dc_e} = \rho_s\frac{(1-\rho_s\psi_2)}{(1-c_e)} \tag{17}$$

where ψ_2 is the electron partial specific volume of the solute. In studies on enzymes, the measurements are performed in solution; thus the intensity value used is the excess scattering of solution over solvent, $\Delta i_n(s)$. For the sake of simplicity, we will drop the symbol Δ, taking note that, in what follows, $i_n(s)$ and $j_n(s)$ refer to the excess scattering. Combining Eqs. (13) through (17), expressing the derivative of the excess chemical potential, μ^e, of solute with respect to concentration as the usual virial expansion, such as used in light scattering and osmometry,[22] and setting

---

[22a] S. N. Timasheff, *Advan. Chem. Ser.* (1973), in press.

[23] This equation is analogous to the light-scattering equation, with electron density increment replacing refractive index increment.

$\rho_s \approx \rho_1$ (since at low solute concentrations, the electron densities of solvent and solution will be the same within experimental error), leads to

$$m_{app} = i_n(0)(1 - \rho_1\psi_2)^{-2}c_e^{-1} \qquad (18a)$$

and

$$m = m_{app} + 2Bm^2c_e \qquad (18b)$$

where B is the second virial coefficient. Extrapolation to zero concentration of a plot of $1/m_{app}$ versus c_e leads, then, to m from the ordinate intercept and, with m known, to B from the slope. The molecular weight, M, is readily obtained from m, since

$$M = mN_A/q \qquad (19)$$

where q is the number of electrons per gram of the particle, calculated from its chemical composition, and N_A is Avogadro's number.

3. Other Parameters

For an isotropic particle of uniform electron density, at large values of s, and using slit optics,[1,24]

$$\lim_{s \to \infty} s^3 j_n(s) = \lim_{s \to \infty} s^3 j^*_n(s) + \delta^* s^3$$
$$j^*_n(s) \equiv j_n(s) - \delta^* \qquad (20)$$

where $\lim_{s \to \infty} s^3 j_n(s) \equiv A$ and δ^* are constants and $j^*_n(s)$ is a corrected normalized scattering intensity defined by the equation. A plot of $s^3 j_n(s)$ versus s^3 follows the form shown in Fig. 3B: as s^3 increases, the product $s^3 j_n(s)$ first increases rapidly in nonlinear fashion; at sufficiently high values of s^3 this function assumes a linear form, with weak fluctuations superimposed on it. The intercept of the straight line portion of this plot is A and its slope is δ^*. As has been shown by Luzzati et al.,[24] δ^* reflects the internal structure of the macromolecule. Knowledge of A and $j^*_n(s)$ permits the calculation of several other molecular parameters.[24]

The external surface area, S, of the particle in solution, is given by[3,25,26]

$$S = 16\pi^2 A(\rho_2 - \rho_1)^{-2} \qquad (21)$$

where ρ_2 is the mean electron density of the hydrated particle. The hydrated volume, V, can be obtained by integration under the scattering curve,

[24] V. Luzzati, J. Witz, and A. Nicolaieff, *J. Mol. Biol.* 3, 367 (1961).

[25] G. Porod, *Kolloid-Z.* 124, 83 (1951).

[26] J. L. Soulé, *J. Phys. Radium Phys. Appl.*, Suppl. 18, 90A (1957).

$$V = \frac{i_{n(0)}}{\int_0^\infty 2\pi s j^*_n(s)ds} = \frac{m(1 - \rho_1\psi_2)^2 c_e}{\int_0^\infty 2\pi s j^*_n(s)ds} \tag{22}$$

It may be shown that the surface-to-volume ratio is

$$\frac{S}{V} = \frac{8\pi A}{\int_0^\infty s j^*_n(s)ds} \tag{23}$$

The excess electron density of the hydrated particle over that of solvent, $\Delta\rho = \rho_2 - \rho_1$, can be calculated from

$$\Delta\rho = \frac{\int_0^\infty 2\pi s j^*_n(s)ds}{c_e(1 - \rho_1\psi_2)} + \rho_2 c_e(1 - \rho_1\psi_2) \tag{24}$$

The degree of hydration, H, expressed as the ratio of the number of electrons of water of hydration to the number of electrons of the dry particle, is

$$H = \frac{\rho_1(1 - \rho_2\psi_2)}{\Delta\rho} \tag{25}$$

The exact values of the parameters of Eqs. 22–25 are obtained by extrapolation to zero protein concentration.

With a knowledge of a number of molecular parameters—namely, M, R_G, V, and S/V—the possible overall geometry of the unknown particle becomes highly restricted. Further information on the particle shape may be obtained from scattering at higher angles. At these angles ($2\theta > 2°$), the X-ray scattering curves develop maxima and minima superimposed on the Guinier relationship. The positions of these are well defined for different geometric models, and scattering curves for various models have been calculated (see Section IV, E). Comparison of the experimental curves with those calculated for various likely models then suggests the choice most compatible with the data.

4. Multicomponent Systems

All the foregoing relations are rigorously valid only for true two-component systems, such as a macromolecular solute immersed in pure solvent, e.g., an enzyme in water. In biological systems we normally deal with thermodynamically more complicated multicomponent systems, the additional components being buffer salts, dispersing agents, or other perturbants. The proper interpretation of data requires, therefore, the application of multicomponent thermodynamic theory. In practice, it is found that in dilute buffers ($\leqslant 0.2\,M$) the multicomponent effects are negligibly small and in most cases the above two-component equations may be used directly. In the case of some enzymes, however, the solution properties of the macromolecules are such that a dispersing agent must be added at high concentration (e.g., concentrated salt, urea, or organic solvents). In such a case, just as in light scattering and equilibrium sedimentation, the parameters measured contain a contribution from the interactions between the macromolecule and solvent components. Multicomponent fluctuation theory yields for small-angle X-ray scattering an equation for the molecular weight similar to that obtained in light scattering (see this volume [10]). Defining[27,28] principal solvent as component 1, enzyme as component 2, and the additive as component 3, we have[5,29]

$$\left(\frac{\partial \rho_s}{\partial c_{e,2}}\right)^2_{T,p,c_{e,3}} \frac{c_{e,2}(1 - c_{e,2})^2}{\Delta i_n(0)\rho_s^2} = \frac{1}{m_2(1+D)^2}\left[1 + \frac{c_{e,2}}{RT}B'\right]$$

$$D = \frac{(\partial \rho / \partial c_{e,3})_{T,p,c_{e,2}}}{(\partial \rho / \partial c_{e,2})_{T,p,c_{e,3}}}\left(\frac{\partial c_{e,3}}{\partial c_{e,2}}\right)_{T,p,\mu_3} = \frac{(1 - c_{e,2})}{(1 - c_{e,3})}\frac{(1 - \rho_s\psi_3)}{(1 - \rho_s\psi_2)}\left(\frac{\partial c_{e,3}}{\partial c_{e,2}}\right)_{T,p,\mu_3} \quad (26)$$

where the symbols have their previous meaning, and B' is a complicated function of interactions between the various components, identical to that found in light scattering (see this volume [10]). In order to obtain the true molecular weight in such a system, it becomes necessary to measure the extent of preferential interaction, i.e., $(\partial c_{e,3}/\partial c_{e,2})_{T,p,\mu_3}$, between the enzyme and solvent components in an auxiliary experiment, for example by equilibrium dialysis. The various geometric parameters also become complex, since solvent interactions result in the fluctuating unit being no longer isotropic; as a result, the values obtained from Eqs. (11), (21), (22), and (23) can serve only as qualitative estimates. Equations taking multicomponent effects into account can be developed, however, for the shape parameters.

5. Polydispersity

If the particles present vary in molecular weight and size, the observed scattering is the sum of contributions from all components. The Guinier equation then becomes

$$\frac{i_{n(s)}}{Kc_e} = \frac{\Sigma i_n(s)_i}{K\Sigma c_{e,i}} = \frac{\Sigma c_{e,i}m_i}{\Sigma c_{e,i}}\left(1 - \frac{4}{3}\pi^2 s^2 \frac{\Sigma c_{e,i}m_i R_{G,i}^2}{\Sigma c_{e,i}m_i}\right) \quad (27)$$

[27] G. Scatchard, J. Amer. Chem. Soc. 68, 2315 (1946).

[28] W. H. Stockmayer, J. Chem. Phys. 18, 58 (1950).

[29] S. N. Timasheff and R. Townend, in "Physical Principles and Techniques of Protein Chemistry" (S. J. Leach, ed.), Part B, p. 147. Academic Press, New York, 1970.

The values of the molecular weight and radius of gyration obtained in such systems are, then, the weight average molecular weight and a radius of gyration of an average which is a function of the particle shape.[30]

D. General Remarks

1. Resolution

Similarly to other methods that measure the size and shape of macromolecules, the resolution of small-angle X-ray scattering can be defined as the limit of molecular dimensions that can readily be measured. In discussing this question, it is expedient to compare small-angle X-ray scattering to the related technique of light scattering. Resolution can be expressed essentially in terms of the highest value of the radius of gyration that can be measured. Since the Guinier equation (Eq. 10) is an approximation based on the assumption that $R_G^2 s^2 \ll 1$, in practical terms this reduces the question to that of the lowest value of the angular parameter, s [see Eq. (9)], attainable. In light scattering, with radiation of ~ 4000 Å wavelength and the usual limits of angular measurements, ca. $2\theta = 20°$, the limit of attainable s^{-1} is $\sim 1.2 \times 10^4$ Å. The Guinier equation then imposes an upper value of ~ 1200 Å for the radius of gyration experimentally accessible. Similarly, in small-angle X-ray scattering, the wavelength is 1.5 Å, while the smallest angle that can be readily reached is ca. 10′; thus, the limit of attainable s^{-1} is ~ 540 Å. This sets 55 Å as an upper limit for the radius of gyration that can be reasonably measured. For a spherical protein, $R_G^2 = \frac{3}{5} r^2$ (where r is the radius of the sphere); this sets 65 Å as the maximal value for the measurable radius, which corresponds to a molecular weight of ca. 300,000. The converse of this relation is the limit of the lowest molecular dimensions that can be detected. In light scattering this is of the order of $\lambda/10$, or ca. 400 Å; in small-angle X-ray scattering, since the wavelength is very small, there is, in principle, no similar limit. In practice, however, the low intensities of scattering from biological macromolecules require a minimal molecular dimension of ~ 10 Å.

When the overall dimensions of a particle are large compared to the maximal resolution attainable, i.e., when $R_G^2 s^2$ is not small, the angular dependence of scattering no longer obeys the Guinier equation. Relations can be developed, however, for the angular dependence of the scattered intensity for particular structural models. For example, in the case of long asymmetric structures, such as those encountered in myosin, col-

[30] E. P. Geiduschek and A. Holtzer, *Advan. Biol. Med. Phys.* **6**, 431 (1958).

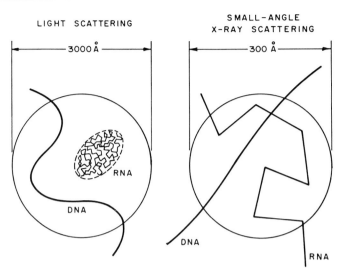

LIGHT SCATTERING

SMALL-ANGLE
X-RAY SCATTERING

Fig. 4. Comparison of resolutions obtained with light scattering and small-angle X-ray scattering.

lagen, or in the associated form of glutamate dehydrogenase,[31] the parameters which are readily obtainable are the cross-section radius of gyration, R_c, i.e., the radius of gyration for rotation about the long axis, and the mass per unit length, m/l, of the equivalent rigid rod. The corresponding scattering equation is[32]

$$j_n(s) = \frac{m}{2l} c_e(1 - \rho_1\psi_2^2) \exp(-\pi^2 R_c^2 s^2) K_0(\pi^2 R_c^2 s^2) \qquad (28)$$

where m is mass in electrons, l is length in Å, and the other symbols have their previous meaning; K_0 indicates a kappa function.

A striking example of the differences in resolution between light scattering and small-angle X-ray scattering is shown in Fig. 4, where a comparison is given of the structural features observed by the two techniques for DNA and ribosomal RNA. In small-angle X-ray scattering, DNA appears as a rigid rod with cross-sectional dimensions characteristic of a Watson–Crick double helix.[32] In light scattering, where ten times greater dimensions are viewed, DNA appears as a stiff wormlike chain.[33,34] The heavy (32 S, 1.9×10^6 molecular weight) component

[31] H. Sund, I. Pilz, and M. Herbst, Eur. J. Biochem. 7, 517 (1969).
[32] V. Luzzati, A. Nicolaieff, and F. Masson, J. Mol. Biol. 3, 185 (1961).
[33] V. Luzzati and H. Benoit, Acta Crystallogr. 14, 297 (1961).
[34] C. Sadron and J. Pouyet, Proc. 4th Int. Congr. Biochem., Vienna, Vol. 9, p. 52. Pergamon, Oxford, 1958.

of ascites tumor cell ribosomal RNA is seen in light scattering as an asymmetric globular structure with a radius of gyration of 355 Å,[4,35] while small-angle X-ray scattering at ten times greater resolution shows that this molecule is actually a zigzag chain composed of double-helical segments ca. 85 Å in length linked by flexible joints.[4,36,37] It should be pointed out, furthermore, that if the small-angle X-ray scattering experiments on the RNA had been limited to angles greater than 1° (i.e., $s^{-1} < 85$ Å), the angular dependence of scattering would have been that characteristic of a rigid double-helical rod. This example shows how two related techniques which view molecular dimensions at different resolutions can be used to great advantage together to obtain rather detailed structural information on macromolecules in solution and to compare the geometric features of two structurally related molecules.

2. Limitations

The method of small-angle X-ray scattering is almost unique in its ability to give simultaneously geometric and thermodynamic parameters of macromolecules in solution; it also has great limitations and is best used late in a study, i.e., on already well defined systems. The principal limitation stems from the low intensity of scattering and the necessity to perform measurements at angles very close to the incident beam. This imposes the instrumental requirement of an extremely fine degree of collimation, which will be discussed below (see Section III, A). The second difficulty, related to the low intensities, is the necessity to use slit sources of radiation. The theoretical equations describing the angular dependence of the scattering from structures of various shapes, however, have been derived for a point source. This requires the use of mathematical transformations such as Eq. (12) to "desmear" the data (see Section IV, B); such operations are not feasible without the use of computers. Related also to the low intensity of scattering are two additional limitations. One is the need to use lengthy scans ($\sim$24 hours per concentration point) in order to record a sufficient number of counts over the angular range normally covered (from $+8°$ to $-8°$, see below, Section III, B). The other is the necessity to use high protein concentrations (5–70 mg/ml). Therefore, in a highly nonideal system, such as an enzyme at a pH far from its isoionic point, the product $2Bmc_e$ of Eq. 18, could become very large, introducing great uncertainty into the measured molecular weight. In an associating system, the data would

[35] M. J. Kronman, S. N. Timasheff, J. S. Colter, and R. A. Brown, *Biochim. Biophys. Acta* **40**, 410 (1960).
[36] S. N. Timasheff, J. Witz, and V. Luzzati, *Biophys. J.* **1**, 526 (1961).
[37] S. N. Timasheff, *Biochim. Biophys. Acta* **88**, 630 (1964).

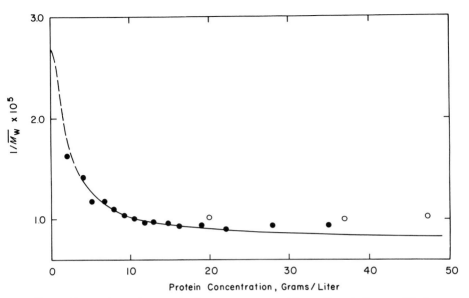

FIG. 5. Comparison of concentration ranges covered by sedimentation equilibrium (– – –), light scattering (●), and small-angle X-ray scattering (○).

fall essentially into the concentration range where the aggregated species predominate. In such systems it is best to combine the application of small-angle X-ray scattering with that of other thermodynamic techniques, such as light scattering and sedimentation equilibrium, in which the measurements are performed over progressively decreasing concentration ranges. Such a comparison is shown in Fig. 5 for the association of β-lactoglobulin A. As can be seen, good agreement can be obtained between sedimentation equilibrium,[38] which gives the details of the reaction in the low concentration range, light scattering, which is used to characterize the middle and high concentration ranges,[39,40] and small-angle X-ray scattering.[41]

The fact that the small-angle X-ray scattering data are available at high concentrations has, however, the advantage that, in an associating system, the radius of gyration measured is essentially that of the enzyme polymer. This follows directly from the types of averages that are measured. Just as in light scattering, small-angle X-ray scattering

[38] D. E. Roark and D. A. Yphantis, *Ann. N.Y. Acad. Sci.* **164**, 245 (1969).
[39] S. N. Timasheff and R. Townend, *J. Amer. Chem. Soc.* **83**, 464 (1961).
[40] T. F. Kumosinski and S. N. Timasheff, *J. Amer. Chem. Soc.* **88**, 5635 (1966).
[41] J. Witz, S. N. Timasheff, and V. Luzzati, *J. Amer. Chem. Soc.* **86**, 168 (1964).

yields the weight-average molecular weight and a higher-order average radius of gyration

$$\overline{R_G^2} = \frac{\Sigma c_i M_i R_{G,i}^2}{\Sigma c_i M_i} \tag{29}$$

Since the relation of the radius of gyration to the molecular weight is a function of the shape of the particle, this average can be complicated. The proper relations for a variety of geometric models have been tabulated elsewhere.[29,30]

3. The Role of Dust

The higher-order average of the radius of gyration measured in small-angle X-ray scattering, together with the form of the Guinier equation, an exponential in $-s^2 R_G^2$, leads to an experimental simplification. Namely, the "dust problem," familiar in light scattering (see this volume [10])[42,43] is not found in small-angle X-ray scattering. "Dust" particles are so large that their scattering is almost fully in the forward direction. Thus, when angles of $>5'$ are reached, i.e., the angular range in which small-angle X-ray scattering measurements usually begin, there is no significant contribution remaining from "dust" scattering. As a result, as a practical matter, small-angle X-ray scattering does not require the elaborate solution clarification techniques normally used in light scattering.

III. Experimental

A. Apparatus

1. Basic Requirements

Despite the similarity, or even basic identity, of small-angle X-ray scattering (SAXS) theory to the theory of light scattering, instrumentation for measuring the corresponding phenomena has taken quite different forms.

In principle, both techniques require the same basic apparatus components: (a) a radiation source, (b) means for selecting desired wavelengths from this source, (c) a collimating system for ensuring a desired geometry for the irradiating beam, (d) a sample holder or, in the case of samples in solution, a sample vessel or cell, (e) some type of goniometer arrangement for allowing observation of the scattered radia-

---

[42] M. Bier, Vol. 4, p. 165.
[43] K. A. Stacey, "Light Scattering in Physical Chemistry," Chapter 3. Academic Press, New York, 1956.

tion over a range of precisely defined angles with respect to the incident radiation, (f) a radiation detector mounted on the movable arm of the goniometer, and (g) a data readout with optional ancillary data-processing equipment. (In an earlier stage of development, and occasionally still in use today for certain purposes, the function of the last three components is performed by appropriately arranged photographic film. These are subsequently evaluated with a densitometric device to relate the intensities of the photographic record to angular positions; hence the apparatus is referred to as a "scattering camera.")

As a practical matter, the special characteristics of X-rays have made it necessary for X-ray scattering apparatus to assume very particular forms. Furthermore, the slow development of the apparatus over the past four decades has reflected the special interests and personal approaches of individual workers in the field, many of whom attempted designs of their own, few of which ever have become commercialized.

In the following, the main features, as well as the most prominent instruments illustrative of a given type will be described briefly. Since scattering is a general phenomenon of which diffraction may be considered to constitute a particular case, it will not be surprising that many of the less specialized apparatus components used for X-ray scattering are the same as those familiarly used in X-ray diffraction equipment. In fact, the term "diffractometer" is frequently used to refer to a SAXS apparatus comprising a goniometer, as contrasted to one comprising a camera. In view of the variety of relevant design features and, at the same time, the scarcity of models in general use, the principal special requirements, many of which are interrelated, will first be considered in some detail.

2. Special Requirements

a. Stability of Source and Apparatus. Scattering is a quantum process, hence a random event, and acceptable counting statistics presuppose a certain minimum number of counts.[44] Since dilute solutions of biological macromolecules are weak scatterers (typical scattered intensities at 1° scattering angle might be of the order of 20–100 counts/second), the accumulation of the requisite number of counts, whether measured continuously or discretely at a number of points over the angular range of interest, involves relatively long total counting times, of the order of 12–24 hours, and sometimes more. During this entire time, the intensity of the primary beam is expected to remain constant,

---

[44] H. P. Klug and L. E. Alexander, "X-Ray Diffraction Procedures," p. 270. Wiley, New York, 1954.

as is every other parameter of the instrument, such as its geometric relationships and the detector gain. These considerations, which are of no concern in the case of a camera, are matters of great consequence with a detector instrument.

Stability of the detector and of the signal-processing electronics may be achieved by the use of stabilized power supplies and by suitable design and selection of electronic components to assure negligible drift. Physical stability to assure maintenance of the relative positions of all the apparatus elements, such as source, slits, sample, and detector, is achieved by proper mechanical design, including choice of dimensions, materials of construction, balance, methods of support, and fastenings of components. Source and goniometer are usually mounted on a rigid steel or marble plate, often the top of a commercial X-ray generator. (In the construction of our instrument we have made use of the rigidity afforded by an 8-inch-thick granite slab.) In addition, dimensional changes due to temperature fluctuations should be controlled by ambient air conditioning to within $\pm 0.5°C$, with special avoidance of exposure to transient temperature extremes; i.e., the apparatus should not be directly exposed to the stream of air issuing from the conditioning system, and it is desirable to shield it by air locks from less well controlled air from adjacent rooms. Variations in barometric pressure and humidity, because of their effects on proportional detectors and on static charges which may affect the measuring system, have also been found to be detrimental.[45,46]

Stability of the X-ray source has presented much more of a problem. It is necessary to start with an X-ray generator incorporating a high degree of voltage regulation and tube-current stabilization. Diffraction tubes generally employ water cooling to protect the life of the targets. Close temperature control (to $\pm 1°C$) of this cooling water, though not usually important in diffraction work, is essential in scattering work to maintain the dimensional stability of the target, as otherwise the focal spot will tend to wander, compromising the delicate alignment of the scattering apparatus. Instruments employing crystal monochromators (to be discussed below), which image the narrow focal spot of a fine-focus tube (typically, 0.4×8.0 mm), are more sensitive to even slight focal-spot wandering (of the order of a fraction of a micrometer) than are pure slit-collimating instruments. In these, the slits eliminate a relatively large portion of the much larger focal spot of a

[45] O. Kratky, in "Small-Angle X-Ray Scattering" (H. Brumberger, ed.), p. 64. Gordon & Breach, New York, 1967.

[46] T. W. Baker, J. D. George, B. A. Bellamy, and R. Causey, Advan. X-Ray Anal. 11, 361 (1968).

regular tube in order to produce the narrow primary beam required, and a slightly shifted focal spot will still completely illuminate the slits. Since higher intensity primary beams will yield higher scattered intensities and thus allow shorter counting times or scans, high-power tubes have been employed using currents in excess of 80 mA, as compared with currents nearer 25 mA for a fine-focus tube. However, to dissipate the heat generated by the electrons bombarding the anode at this rate, rotating anodes have to be used. The rotating motion and the slight vibration involved again aggravate the problem of focal spot wander.[47]

Despite all measures taken to obtain stability of the entire system, the primary beam will still be found to have some tendency to drift over the relatively long experimental times involved. Detection of this drift, and possible correction for it, has been attempted by monitoring the beam intensity. Practical difficulties inherent in this approach have prevented it from being adopted to any great extent in scattering work, although a recent design of a monitor system has been described by Kratky et al.[48] The application of signal averaging, a technique increasingly used to improve signal-to-noise ratio (e.g., in nuclear magnetic resonance work), has been suggested,[49,50] though again not widely adopted.

In the absence of a monitoring device, it is not possible to establish directly whether the primary beam intensity had remained constant throughout an extended experiment. An indirect check on constancy, however, can be obtained by virtue of the near-perfect symmetry of the scattered intensities on the two sides of zero angle which is a characteristic of a well aligned instrument. This is accomplished by folding a chart record of intensity vs. angle at the zero angle position and examining by means of an illuminator whether the two branches of the recorded curve are coincident. If instability due to any cause, in any part of the total system, has occurred during an experiment, it is virtually certain to manifest itself here as a lack of symmetry, since the likelihood of one kind of disturbance being precisely compensated by another is vanishingly small. It goes without saying that the chart paper may not be opaque, and also that this valuable check is not available unless the scattering instrument is so designed as to allow measurements on both sides of the direct beam.

[47] W. W. Beeman, in "Small-Angle X-Ray Scattering" (H. Brumberger, ed.), p. 197. Gordon & Breach, New York, 1967.

[48] O. Kratky, H. Leopold, and H.-P. Seidler, Z. Angew. Phys. 31, 49 (1971).

[49] C. R. Peters and M. E. Milberg, Rev. Sci. Instrum. 37, 1186 (1966).

[50] M. Berman and S. Ergun, Rev. Sci. Instrum. 40, 1144 (1969).

b. Quality of Source. Beyond the aspects of stability and intensity, mentioned in the preceding section, the cross section of the primary beam, and hence the focal spot, is required to be homogeneous, and, for the slit optics discussed below, it must be of rectangular rather than of trapezoidal or other shape. Particularly with crystal monochromatization, the effect of any deviation from these requirements becomes readily noticeable.

The spectral purity of the radiation obtained from the tube is of great importance, especially if the primary beam is not monochromatized. In the absence of other monochromatization, even the use of a pulse-height analyzer does not yield sufficiently fine energy resolution to prevent smearing of the scattering curve due to the energy distribution.[51]

Even an uncontaminated target and tube producing pure copper radiation, the radiation most often chosen for SAXS, does not yield radiation of a single wavelength since, depending on the relation of the exciting potential to the Duane–Hunt short-wavelength limit,[52] varying amounts of the copper continuum and of characteristic radiation other than the desired K_α line are present. A pulse-height analyzer can eliminate the continuum and the unwanted characteristic lines, including nearly all of the K_β, though not the higher harmonics of the K_α line if the potential is high enough to excite them. The use of balanced filters, i.e., a combination of Ni and Co filters of the proper thickness whose absorption edges bracket the Cu K_α line, will accomplish the same result, but in either case the K_α line will still consist of the K_{α_1} – K_{α_2} doublet. A well aligned high quality crystal monochromator, on the other hand, can separate this doublet and yield the K_{α_1}, the larger component, free from all but traces amounting to a few percent of the K_{α_2}.[53,54]

Monochromator crystals may be flat, but they yield higher intensities when bent. In the configuration of Johann,[55] a thin crystal plate, elastically or plastically bent to a radius of curvature R (Fig. 6), focuses rays from the source S, diffracted by an angle θ obeying Bragg's law, approximately to point F, where S, F, and the center of the crystal face lie on a circle (the Rowland circle) having a diameter equal to the radius of curvature of the bent crystal. This crystal is tangent to the circle,

[51] See Guinier and Fournet,[1] p. 85.
[52] See reference 44, p. 81.
[53] D. R. Chipman, *in* "Methods of Obtaining Monochromatic X-Rays and Neutrons" (F. Herbstein, ed.), pp. 55–58. Union Crystallogr., Utrecht, Netherlands, 1967.
[54] J. Witz, *Acta Crystallogr.* **A25**, 30 (1969).
[55] H. H. Johann, *Z. Phys.* **69**, 185 (1931).

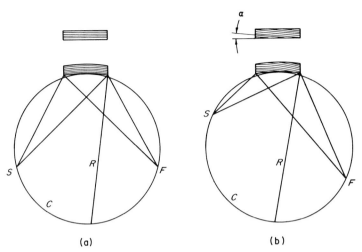

(a)   (b)

Fig. 6. Johann bent-crystal monochromators: (a) symmetrical; (b) asymmetrical.
S, source; *F*, focus; *C*, circumference of Rowland circle; *R*, radius of curvature of crystal. From H. H. Johann, *Z. Phys.* **69**, 185 (1931).

since the radius of the circle is only one-half the radius of curvature of the crystal. For exact focusing, the face of the crystal must be ground to be coincident with, rather than just tangent to, the circle according to Johansson.[56,57] This latter technique is more exacting, the crystals are much more costly, and the anticipated increase in intensity due to more perfect focusing is frequently not realized because of other deviations from ideality, so that the Johann type crystals are the ones more generally used. For Cu K_α radiation, quartz laminae, elastically bent, have been found most suitable. The 1011 lattice planes are the ones utilized; by cutting the crystal faces at an angle α (typically, 8°) to these planes, a desirable asymmetry results which allows a more favorable geometry for the instrument. With this method, the intensity of the focal spot is better utilized, and a longer optical path is made available for the slit system, because the monochromator may be located much closer to the source, than would be possible with a symmetrically cut crystal.[58] Since for a given spacing d and angle θ, Bragg's law, $n\lambda = 2d \sin \theta$, is satisfied by only one value of $n\lambda$, true monochromatization (except for the admission of harmonics) is achieved.

 c. Slit Geometry. Scattering theory has been derived for a point source. Point sources or, more realistically, spot sources are so weak in

[56] T. Johansson, *Naturwissenschaften* **20**, 758 (1932).
[57] T. Johansson, *Z. Phys.* **82**, 507 (1933).
[58] A. Guinier, *C. R. Acad. Sci.* **223**, 31 (1946).

intensity that they have been employed in few instruments. Customarily, one resorts to a source having a negligible extension in only one dimension, i.e., an illuminated slit, which produces a beam of narrow rectangular cross section. Such a slit or line source can be conceived of as consisting of a large number of spot or point sources along a straight line, and the scattering curves produced by such a slit are derived from those produced by a point source by the "smearing," or superposition, of a large number of the latter, in a manner corresponding to the mathematical operation of convolution. In order to be evaluated, such so-called "slit-smeared" experimental curves generally must be "desmeared" or deconvoluted. In practice, this is not always a simple matter. Numerous methods, graphical, analytical, and numerical, have been described.[59-66]

For this deconvolution, the actual intensity distribution of the longitudinal beam profile must be known. The mathematics become somewhat simpler if the assumption of an infinite slit height may be made, i.e., if the height of the slit is greater than the arc length over which the detector can see any appreciable scattering.

The earliest slit geometry still in common use employs four slits (see 3, a, below), the first close to the source, the second close to, and just ahead of the sample, and two more slits between sample and detector. The first two are collimating slits whose function is to remove all but a thin, nearly parallel bundle of rays from the primary beam coming from the source. The third is the receiving slit; it defines the angular position at which the detector reads the intensity. The fourth is an antiscatter slit which prevents the detector from seeing any scattered or parasitic radiation from directions other than the irradiated volume of the sample.

Use of a curved-crystal monochromator leads to a second slit geometry (see 3, b, below). The monochromator described in the preceding section, besides selecting the desired wavelength, also performs a focusing function. Whereas the four-slit geometry selects from the widely diverging beam issuing from the X-ray tube window a very thin and only very slightly diverging beam, the monochromator focuses a widely

[59] V. Gerold, Acta Crystallogr. 10, 287 (1957).
[60] W. Ruland, Acta Crystallogr. 17, 138 (1964).
[61] J. Mazur and A. M. Wims, J. Res. Nat. Bur. Stand. Sect. A 70, 467 (1966).
[62] P. W. Schmidt, Acta Crystallogr. 8, 772 (1955).
[63] S. Heine, Acta Phys. Austr. 16, 144 (1963).
[64] B. Chu and D. M. Tan Creti, Acta Crystallogr. 18, 1083 (1965).
[65] J. A. Lake, Acta Crystallogr. 23, 191 (1967).
[66] O. Kratky, G. Porod, and Z. Skala, Acta Phys. Austr. 13, 76 (1960).

diverging beam, and thus concentrates a converging beam on the sample and the detector. There is little, if any, net gain in intensity, since various losses due to the monochromator are large enough to outweigh the collimation losses due to the first two slits of the pure four-slit system. The monochromator still requires the use of four slits; however, the first two slits in this system no longer have a collimating function. The purpose of the first slit is solely to eliminate parasitic radiation produced by scattering from the monochromator holder and elsewhere, while the second slit removes radiation scattered from the edges of the first. The quality of these two slits and their adjustment (and particularly that of the second slit) are exceedingly critical in this system, as is the fine adjustment of the monochromator itself to the exact focusing position. The number of elements which must be in precise alignment makes this system very dependent on effective provisions to facilitate these potentially very time-consuming adjustments.

A third geometry, that of Kratky (see 3, c, below), is particularly successful in eliminating parasitic radiation originating from slit edges and, thus, is capable of very high resolution. In place of the initial two slits it uses a precisely machined asymmetric system of steel blocks to produce a region virtually free of parasitic scattering very close to one side of the direct beam, albeit at the price of making inaccessible to observation one half of the scattering region, namely, that on the other side of the direct beam.

Still other, less generally used, collimating systems will be described together with the specific instruments that use them (see 3, d, below).

d. Resolution and Angular Range. The problems of the geometric definition of the beam and of the elimination of parasitic scattering are central to the design of any SAXS instrument. Aside from the geometry of the goniometer, the precision of its position readout and any associated gearing, and the width of the receiving slit, the resolution of the instrument depends on the width of the primary beam at the receiving slit. In all the systems mentioned, narrowness of the primary beam is achieved to a great extent at the expense of its intensity, so that a compromise has to be reached between the precision with which data may be obtained and the concentration at which the sample may be studied, or its inverse, the extended time required for the study, with its attendant drawbacks. Nevertheless, the monochromator system, because of its converging beam, and the Kratky system, because of its freedom from parasitic scattering, have inherent advantages over the four-slit system with its diverging beam and substantial parasitic scattering from its collimating slits.

Some of the information of interest in SAXS is contained in the data at the smallest angles and, because of the inverse relationship between d and θ indicated by the Bragg law, the larger the molecules (or particles) studied, the smaller these angles. Yet, the smallest angles (forward scattering, close to zero angle) are experimentally inaccessible because they are occupied by the primary beam which, owing to its finite width, must extend some distance on either side of the zero angle. Being of the order of 10^5–10^6 times more intense than the adjacent scattered intensities produced by all but the strongest scatterers, it will mask these as soon as they overlap. It is true that, to protect the detector, the primary beam is invariably blocked in X-ray work by a heavy-metal beam stop or some equivalent device, but this provision itself leads to some degradation of data. If the beam stop, or its equivalent, is slightly too wide, it will block also the measurement of some otherwise detectable scattered radiation; if it is slightly too narrow, it will allow some of the direct beam to spill over and swamp scattered radiation. Furthermore, the primary beam is not truly a sharp beam entirely confined to a limited angular region. The so-called rocking curve, the curve of beam intensity vs. angle, has tails extending out to fairly large angles (except in the case of some double-crystal instruments, where these tails may be very sharply limited). The tails are exceedingly faint compared to the peak intensity, but they are by no means negligible on the millionfold smaller scale of the scattered radiation.

It is clear then that the collimation or narrowness of the beam, besides influencing the general precision of the data, determines the small-angle limit. It can be readily appreciated that the aforementioned requirements are particularly sensitive to the problems of stability, discussed above (see Section 2, a). In fact, it may be said that the greatest single difficulty in designing a SAXS instrument resides in this complex of requirements.

Other information of value is obtained from scattering at relatively high angles, up to about 8°. Here the scattered intensities are very low. To obtain measurements within reasonable counting times it is desirable to work with as much scattering material as possible, i.e., with relatively high concentrations. There are, however, practical upper limits to the usable concentrations of protein solutions, and thus the counting times are inevitably lengthened. Again, the requirement for long-term stability makes itself felt here.

e. Absolute-Scale Intensity Measurements. Absolute-scale measurements allow the calculation of particle parameters not readily obtained otherwise, as discussed in the section on principles (see Section II, C). Absolute-scale intensity or, for short, absolute intensity, or absolute-unit measurements, may be defined fundamentally as the intensity scattered by the sample at any angle in terms of the intensity scattered by a single classical electron under the same conditions, i.e., expressed in electron units.[67] For working purposes it has been defined in terms of the ratio of the scattered intensity to that of the incident beam.[68,69] These two definitions, and yet others, may not be entirely equivalent, the second one being not necessarily independent of the collimation system; a more detailed discussion may be found elsewhere.[70]

The chief experimental difficulty consists in measuring the intensity of the primary beam. The intensity of the unattenuated primary beam, as mentioned above, is so much greater than the scattered intensity that the two cannot be directly measured on the same scale by any practical apparatus. In fact, the unattenuated direct beam is much too intense for the counting speed of even modern X-ray detecting devices with few exceptions.[71] Hence, for any direct comparison with the scattered radiation, it must be attenuated or sampled in some precisely defined manner.

Attenuation by calibrated filter foils,[11,68,72–74] by utilization of the Bragg reflection from a perfect Si crystal,[75] or by fractional-time sampling of the beam by means of a rotating disk with a calibrated hole[69,76] have all been employed with some success, although each method has its own difficulties. The filter method requires flaw-free foils of very uniform thickness, and the calibration of a set of filters to the required accuracy is a very time-consuming procedure, again dependent on stable radiation. The anomalous transmission method and the rotating-disk method are, unlike the foil method, experimentally awkward and not well adapted to routine use. Indeed, the rotating-disk method has been used essentially in only one laboratory, where it has served in the calibration of a large number of secondary standards, which in turn are employed, there and elsewhere, in measurements where scattered intensities are in this way indirectly referenced to their respective primary beams.[45]

[67] D. P. Riley, in "X-Ray Diffraction by Polycrystalline Materials" (H. S. Peiser, H. P. Rooksby, and A. J. C. Wilson, eds.), p. 439. Inst. Phys., London, 1955.
[68] V. Luzzati, Acta Crystallogr. 13, 939 (1960).
[69] O. Kratky and H. Wawra, Monatsh. Chem. 94, 981 (1963).
[70] R. W. Hendricks, J. Appl. Crystallogr. 5, 77 (1972).
[71] H. Witte and E. Wölfel, Rev. Mod. Phys. 30, 51 (1958).
[72] D. L. Weinberg, Rev. Sci. Instrum. 34, 691 (1963).
[73] G. Damaschun and J. J. Müller, Z. Naturforsch. A 20, 1274 (1965).
[74] H. Pessen, T. F. Kumosinski, S. N. Timasheff, R. R. Calhoun, Jr., and J. A. Connelly, Advan. X-Ray Anal. 13, 622, 626 (1970).
[75] B. W. Batterman, D. R. Chipman, and J. J. DeMarco, Phys. Rev. 122, 68 (1961).
[76] O. Kratky, Z. Anal. Chem. 201, 161 (1964).

Indirect comparison with the primary beam can be accomplished by reference to a standard scatterer which may be a primary or a secondary standard. Use of a secondary standard, a sample calibrated by means of one of the direct methods just mentioned, has been the method chosen by Kratky and collaborators, employing samples of polyethylene.[77-79] It is an absolute requirement that the standard possess long-term physical, chemical, and radiation stability; the last problem, in particular, appears not to have been completely solved.

The second indirect method of comparison is by establishing a primary standard by calculation of its scattering properties from basic data.[72] This approach is feasible for gases,[80-82] gold sols,[83,84] and silica gels.[85] The number of materials suitable for such standards is limited, and the measurement of their scattering is apt to take very long. Thus, this method also is not suitable for routine use. In addition, the further the physical state of the standard differs from that of the samples to be studied (protein solutions, in our case), the more will the geometry of the setup be inevitably different, and comparability of the two types of measurements becomes questionable.

f. Alignment Provisions and Other Apparatus Features. In designing SAXS apparatus, overriding consideration must be given to adequate provisions for adjusting the many elements (monochromator, slits, sample holder, scanning arm pivot, and detector) through which the incident and the scattered beams are required to pass. Each has a number of degrees of freedom and the proper alignment of several of these is quite critical. Moreover, these alignment provisions must also assure stability.

Regarding precision and repeatability, these provisions may range in various instruments from simple manual shifting of entire apparatus subassemblies on their supports, and their maintenance in position simply by gravity, to motions in dovetailed or ball-bearing tracks, controlled by precision or differential screws. Position indications, if provided, may be by simple scales, with or without vernier, by dial indicator or, most often, by micrometer heads. Part of the reason for the popularity of

[77] O. Kratky, I. Pilz, and P. J. Schmitz, *J. Colloid Interface Sci.* **21**, 24 (1966).
[78] I. Pilz and O. Kratky, *J. Colloid Interface Sci.* **24**, 211 (1967).
[79] I. Pilz, *J. Colloid Interface Sci.* **30**, 140 (1969).
[80] L. Katz, "Absolute Intensity Measurements of Small-Angle X-Ray Scattering." Ph.D. Thesis, University of Wisconsin, 1959.
[81] L. B. Shaffer, "Absolute X-Ray Scattering Cross Sections of Liquids and Solutions." Ph.D. Thesis, University of Wisconsin, 1964.
[82] L. B. Shaffer and W. W. Beeman, *J. Appl. Crystallogr.* **3**, 379 (1970).
[83] O. Kratky, G. Porod, and L. Kahovec, *Z. Elektrochem.* **55**, 53 (1951).
[84] P. H. Hermans, D. Heikens, and A. Weidinger, *J. Polym. Sci.* **35**, 145 (1959).
[85] I. S. Patel and P. W. Schmidt, *J. Appl. Crystallogr.* **4**, 50 (1971).

micrometer heads is the fact that they combine the positioning and indicating functions in one relatively compact device. Where fairly large adjustment forces as well as great positional accuracy is required, a separate drive in conjunction with a dial indicator may be more suitable.

If the scattering apparatus is intended for one type of application only, flexibility and general accessibility of the apparatus are not essential factors. If studies are to extend to a variety of materials, temperatures, concentrations, states of aggregation, or particle sizes, an apparatus permitting various adjustments and the use of possible accessories becomes desirable. Almost inevitably, there must be some sacrifice of stability, because an adjustable element, unless especially well constructed, is never quite as rigid as a similar fixed one, and an accessible apparatus layout will tend to be more spacious and not as rigid as a more compact one. Good mechanical design, however, can minimize any detrimental consequences of such a compromise.

One factor in this respect is the spatial orientation of the apparatus. The great majority of designs have utilized a horizontal layout, i.e., one in which the scanning motion is in a horizontal plane (confusingly, this is sometimes referred to as "vertical," because the motion is about a vertical axis). Some major designs (those of Kratky and Skala[86] and of Luzzati et al.[11]) use a vertical motion (and, consequently, a horizontal axis), as do some successful standard X-ray diffractometers (e.g., that manufactured by Phillips[87]).[87a] The advantages of the vertical design follow from its inherent compactness: savings in floor space, and possibly (but not necessarily) increased rigidity. In diffraction work it is customary to use an X-ray generator whose top surface fulfills the function of a work table, with the X-ray tube mounted, tower fashion, on its center. If the tube is thus mounted vertically, a four-window tube housing allows the use of as many as four instruments (diffractometers, cameras, or other accessories) at the same time. This results in a highly efficient utilization not only of the available space, but also of the operating capacity and of the limited life of the tube.[88]

Because of the generally increased compactness, with the apparatus,

[86] O. Kratky and Z. Skala, Z. Electrochem. Ber. Bunsenges. Phys. Chem. **62**, 73 (1958).

[87] W. Parrish, E. A. Hamacher, and K. Lowitzsch, Philips Tech. Rev. **16**, 123 (1954).

[87a] From this configuration, the occasional use of the alternative designation of "low-angle" scattering for SAXS is readily understood.

[88] If the tube focus is rectangular, as is usually the case, one pair of opposing windows will present horizontal line sources; the other pair, because of foreshortening, will present square or spot sources. While only the former are usable for scattering instruments requiring slit sources, the other two ports remain available for other apparatus.

so to speak, extending into the air, flexibility and general accessibility are somewhat diminished compared to a horizontal layout. If, in order to permit an extended angular range of observation, the goniometer is mounted on a vertical base plate, the various subassemblies must be mounted and various adjustments must be performed in this generally cramped vertical plane, in which even temporarily placed components need to be clamped to remain in place. Similar steps are simpler and more convenient on a horizontal table. A horizontal instrument can be modified much more readily, with fewer problems of maintaining mechanical balance and original rigidity. It must be understood that a horizontal instrument, having vertical slit requirements, requires a horizontally mounted tube, either or both of whose vertical line-source windows will be thus available. One of the spot-source windows will be pointing downward and will be unusable, and the other will be pointing upward and will not be conveniently usable. This is hardly a sacrifice, since the spot sources are rarely utilized in an apparatus primarily devoted to scattering work.

3. Survey of Existing Instruments

a. Slit-Collimation Instruments. Early designs of slit instruments incorporated three slits, two collimating slits and one guard slit ahead of the sample. The choice of the various dimensions for optimum results has been discussed extensively,[89-91] and a number of cameras have been built on this basis (e.g., Hosemann).[92]

By way of digression, it may be noted here that camera-type instruments almost invariably have provisions for a vacuum path, for two reasons. The first, which applies to goniometer instruments as well, is the reduction of absorption losses by air, which increase the intensity problem. The second reason lies in the parasitic scattering produced in the air volume between sample and detector. As pointed out by Luzzati,[11] this scattering is serious in the case of a camera because the entire emulsion is subject to it during the whole period of exposure, but is less serious in the case of a goniometer. Here the presence of a third and a fourth slit between sample and detector limits the direction from which parasitic scattering can impinge on the detector at any one time to the very small amount arising along the line of the scattering angle at that time, since a goniometer involves a sequential method of detection. As a

[89] O. E. A. Bolduan and R. S. Bear, J. Appl. Phys. 20, 983 (1949).
[90] K. L. Yudowitch, J. Appl. Phys. 20, 1232 (1949).
[91] See Guinier and Fournet,[1] pp. 86 ff.
[92] R. Hosemann, Ergeb. Exakt. Naturwissenschaften 24, 142 (1951).

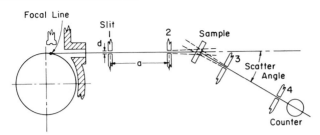

FIG. 7. Schematic top view of X-ray tube and four-slit scattering geometry. From W. W. Beeman, in "Small-Angle X-Ray Scattering" (H. Brumberger, ed.), p. 197. Gordon & Breach, New York, 1967.

result, vacuum chambers are frequently not used with goniometers, and if used, they are usually found only between the sample and the detector. The addition of another slit results in the four-slit geometry (Fig. 7) favored by Beeman[93] and his collaborators[81,94] and still currently used. In order to obtain sufficient intensity, rotating X-ray anodes are frequently used with this type of instrument. Absolute intensity work, when undertaken, has utilized comparison with scattering from a standard gas (see 2, c, above). A commercial form of this type of apparatus (but without special provisions for absolute measurements, and without symmetrically adjustable slits) is the instrument manufactured by Rigaku Denki, Ltd., Tokyo, distributed in the U.S. by Engis, Inc. Instruments of this kind have been used in the study of biological solutions by Anderegg et al.,[94] Ritland et al.,[95] and Brierre.[96]

    b. *Crystal Monochromator Instruments.* A goniometer instrument with a monochromator consisting of a flat crystal has been described by Kahovec and Ruck.[97] Most designs, however, have taken advantage of the observations of Guinier[58] that (a) curved crystals yield both greater intensity and the advantages of focusing, and (b) asymmetrically cut curved crystals, following a suggestion by Fankuchen,[98] have considerable practical advantage (see 2, b, above). The resulting geometry (Fig. 8) shows the angle of convergence of the primary beam, ω, and the region b, subject to parasitic scattering, which limits the definition of the primary beam. The relative advantages of this system have been

[93] See Beeman,[47] p. 198.
[94] J. W. Anderegg, W. W. Beeman, S. Shulman, and P. Kaesberg, *J. Amer. Chem. Soc.* **77**, 2927 (1955).
[95] H. N. Ritland, P. Kaesberg, and W. W. Beeman, *J. Chem. Phys.* **18**, 1237 (1950).
[96] R. T. Brierre, "Small-Angle X-Ray Scattering Investigation of Proteins in Solution." Ph.D. Thesis, Duke University, 1965.
[97] L. Kahovec and H. F. Ruck, *Z. Elektrochem.* **57**, 859 (1953).
[98] I. Fankuchen, *Nature (London)* **139**, 193 (1937).

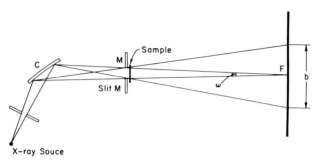

FIG. 8. Schematic diagram of system employing bent-crystal monochromator. C, crystal; F, film plane. From A. Guinier and J. Fournet, "Small-Angle Scattering of X-Rays." Wiley, New York, 1955, p. 102.

discussed by Guinier and Fournet,[99] as have those of various more complicated systems using double crystal monochromatization. Because of increased complexity, the latter, while having certain uses for X-ray diffraction, are not practical for routine SAXS work.

A method of using a Guinier-type instrument, without vacuum chamber, in such as manner as to cancel various constants peculiar to the instrument and its geometry has been devised by Luzzati and co-workers.[11,68] Whereas all instruments mentioned so far have been horizontal, that of Luzzati, which makes use of the commercial Philips goniometer, is vertical. It uses a fine-focus tube in conjunction with a curved-quartz monochromator, calibrated nickel-foil filters for absolute measurements, and a Geiger–Müller tube or proportional counter for detection. A further development of this type of instrument has been described by Pessen et al.[6,100] (Fig. 9). It uses the same method for absolute measurements as that of Luzzati, but differs from it by the use of a horizontal goniometer, a sealed-window proportional detector in conjunction with a pulse-height analyzer, and various instrument refinements. These two instruments have found extensive use in studies of biological materials in solution. An attempt along similar lines has been made by Renouprez et al.,[101] whose instrument has been used in the study of solid catalysts. Another diffractometer utilizing crystal monochromatization and aiming at high resolution, ease in alignment and rigidity of construction is that of Kavesh and Schultz,[102] who used it in

[99] See Guinier and Fournet,[1] pp. 100 ff.
[100] See Pessen et al.,[74] pp. 618–631.
[101] A. Renouprez, H. Bottazzi, D. Weigel, and B. Imelik, J. Chim. Phys. 62, 131 (1965).
[102] S. Kavesh and J. M. Schultz, Rev. Sci. Instrum. 40, 98 (1969).

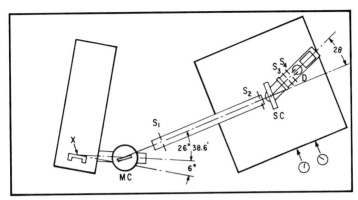

FIG. 9. Schematic top view of scattering apparatus with monochromator and four slits. X, X-ray source; MC, monochromator; S_1, S_2, beam-defining slits; SC, sample cell; S_3, receiving slit; S_4, antiscatter slit; D, detector. From H. Pessen, T. F. Kumosinski, and S. N. Timasheff, *J. Agr. Food Chem.* **19**, 698 (1971).

studies of crystalline polymers. No SAXS diffractometers of these various types are commercially available, although the components for the Luzzati instrument are available from commercial sources.

c. Block-Collimation Instrument. A highly original solution to the parasitic scattering problem which limits ultimate resolution is embodied in the Kratky instrument.[86] As the diagrams (Fig. 10) of the collimating system show, parasitic scattering from slit edges is suppressed to an extraordinary degree by means of a special arrangement of blocks with highly finished surfaces, which replace a more conventional slit system. As mentioned above (see Section 2, c), it is an unavoidable consequence of this system that one-half of the primary beam is blanked out before it reaches the sample, and only one side of a scattering curve is observable. For absolute measurements, secondary standards (polyethylene samples standardized against an intensity-attenuating rotating disk) are routinely employed (see Section III, A, 2, e, above). Like the Luzzati instrument, this one has a vertical goniometer. It should be noted that, unlike the Luzzati-type instruments described in the preceding section which are constructed to satisfy the assumptions of the "infinite slit," the Kratky instrument, depending on the construction of the X-ray tube used with it, may require that a weighting function descriptive of the primary-beam longitudinal profile be used in the deconvolution calculations. This instrument is particularly well constructed with respect to compactness and rigidity. Manufactured by Anton Paar, KG (Graz, Austria), and distributed by the Siemens Group and by Seifert and Co. (Ahrensburg, West Germany) around the world,

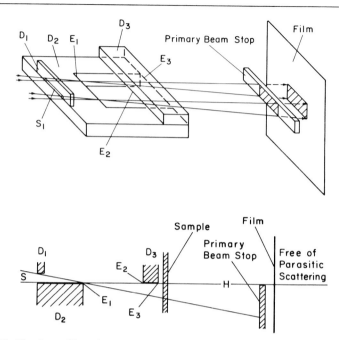

FIG. 10. Kratky collimation system. D_1, entrance block; D_2, U-shaped block; D_3, bridge block; S, entrance opening slit; E_1, E_2, E_3, edges; H, principal section. From Anton Paar, KG, promotional literature.

it is at present the SAXS instrument in widest use. It has been employed extensively in studies of biological solutions.[31,103–105]

d. *Other and Special Instruments.* The problem of slit-smearing can be avoided by use of a point focus. Simple pinhole collimation to obtain a point focus, however, results in unacceptably severe intensity losses. These losses can be somewhat alleviated by using focusing crystals. A single, spherically bent quartz crystal has been used in a scattering instrument constructed by Hagström and Siegbahn.[106] Henke and Du-Mond[107] have constructed an instrument in which a monochromatic point-focus beam is produced by total reflection from an ellipsoidal mirror. Combinations of two crossed cylindrically bent crystals have

[103] O. Kratky and W. Kreutz, *Z. Electrochem. Ber. Bunsenges. Phys. Chem.* **64**, 880 (1960).

[104] I. Pilz, O. Kratky, F. von der Haar, and F. Cramer, *Eur. J. Biochem.* **18**, 436 (1971).

[105] H. Durchschlag, G. Puchwein, O. Kratky, I. Schuster, and K. Kirschner, *Eur. J. Biochem.* **19**, 9 (1971).

[106] S. Hagström and K. Siegbahn, *J. Ultrastruct. Res.* **3**, 401 (1960).

[107] B. L. Henke and J. W. M. DuMond, *J. Appl. Phys.* **26**, 903 (1955).

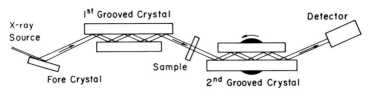

Fig. 11. Multiple reflection diffractometer according to Bonse-Hart. Each grooved crystal contains five Bragg reflections. From U. Bonse and M. Hart, in "Small-Angle X-Ray Scattering" (H. Brumberger, ed.), pp. 121–130. Gordon & Breach, New York, 1967.

also been used. Shenfil et al.[108] used two quartz crystals in reflection; Furnas[109] used a mica crystal in transmission, crossed with a quartz crystal in reflection. Franks[110] used total reflection from two crossed bent glass plates. The Franks camera has been built commercially and distributed in the United States by the Jarrell-Ash Co.

Such instruments may be adapted to meet the requirements of high resolution, as well as irradiation of a very limited area of the specimen of interest, for instance, where a fine-grained polycrystalline material is to be examined. Still, their intensity is generally too low to allow the study of weakly scattering solutions of biological materials.

A variation on the ellipsoidal mirror consists in the more efficient toroidal mirror of Elliot.[111] Cameras allowing the optional use of double Franks mirror optics or Elliot toroid optics, developed by G. D. Searle, Ltd., are available from Elliot Automation Radar Systems, Ltd. in England, represented by Picker Corporation in the United States. They have been applied to studies of polymers, tissues, and cell constituents and to biological macromolecules in the solid form, such as nucleic acids, conjugated proteins, and polypeptides.

Among other special designs is a high-resolution camera built by Brumberger and Deslattes,[112] which utilizes the Borrmann effect, in which a germanium crystal of fairly high perfection is used in anomalous transmission to give an astigmatic image of the source. Another effect utilized for monochromatization is that of multiple total reflections from opposite sides of a groove in a single silicon crystal, described by Bonse and Hart[113] (Fig. 11). As a consequence of the repeated reflec-

[108] L. Shenfil, W. E. Danielson, and J. W. M. DuMond, J. Appl. Phys. 23, 854 (1952).
[109] T. C. Furnas, Jr., Rev. Sci. Instrum. 28, 1042 (1957).
[110] A. Franks, Brit. J. Appl. Phys. 9, 349 (1958).
[111] A. Elliot, J. Sci. Instrum. 42, 312 (1965).
[112] H. Brumberger and R. Deslattes, J. Res. Nat. Bur. Stand. 68C, 173 (1964).
[113] U. Bonse and M. Hart, in "Small-Angle X-Ray Scattering" (H. Brumberger, ed.), pp. 121–130. Gordon & Breach, New York, 1967.

tions, there occurs a progressive enhancement of the peak-to-background ratio of the reflection curve. A second grooved crystal, located between sample and detector, is rotated to perform the scanning. The result is probably the sharpest peak available from any instrument, having negligible tails, and suitable for extremely high-resolution work. An instrument described by Koffman[114] which uses this principle is manufactured by AMR and distributed in the U.S. by Philips Electronic Instruments. Its performance compared to that of the Kratky instrument has been evaluated by Kratky and Leopold.[115] It appears that, although producing more intensity than point-focus instruments, the repeated reflections result in such losses that the intensity left is insufficient for the study of weak scatterers, such as proteins in solution.

B. Procedure

1. Introduction

SAXS is a versatile method that has been applied to the study of a wide variety of systems in different fields, among which are particle and pore sizes in catalysts, grain sizes and clustering in alloys, ceramics and glasses, critical phenomena, colloidal micelles, crystallinity in polymers, order in tissue constituents, and biopolymers in solution. It is to be expected that each application dictates practical aspects peculiar to its requirements. Here we are concerned only with the study of macromolecules (predominantly globular proteins) in solution. In what follows, we will confine ourselves mainly to the procedures which are currently being used in our laboratory and which are further developments of the methods of Luzzati. While some of the methods may not be strictly applicable to a different apparatus setup, this material should be illustrative of the general approach taken in this type of work.

The experimental procedure may be divided into preliminary steps, namely those performed only once (such as filter calibration), or only once for a series of runs (such as apparatus alignment, if called for by a test for alignment), and operational steps, which have to be performed individually for each sample run. These may be either preparatory (preparation of sample, determination of protein concentration, determination of partial specific volume, and measurement of cell thickness), or they may be the actual data gathering, i.e., measurements of

---

[114] D. M. Koffman, Advan. X-Ray Anal. 11, 332–338 (1968).
[115] O. Kratky and H. Leopold, Makromol. Chem. 133, 181 (1970).

primary beam intensities and of scattered intensities as a function of scattering angle. They will be discussed in this order.

2. Preliminary Steps

a. *Filter Calibration.* As indicated above, measurements of incident-beam intensity require that a set of calibrated attenuating filters be at hand. The basic construction requirement is freedom for pinholes. In addition to careful selection of the nickel foils used, the chances of inhomogeneities being present are further reduced by building up each filter to the requisite thickness from layers of thinner foils, so that very slight imperfections will tend to average out. The most efficient design of a set of filters is one in which the filter factors are so related that each of the denser filters is approximately equivalent to exactly one combination of the less dense ones. (One or two of the lower values may be constructed in duplicate, to facilitate arranging various combinations.) Thus, a series of factors such as 2^1, 2^2, 2^4, 2^8, . . . is suitable. A further requirement is that the densest filter be adequate to get the count rate at the peak of the primary beam down to a value which is not only measurable with the equipment available, but which is also below the range where the counting-system dead-time correction and the peak shift of the pulse-height distribution become of importance. Practical count rates of less than 20,000 counts per second at a peak rate of perhaps 10 million photons per second (without filter) require a filter factor of over 500.

With a filter of that approximate value (although initially the value is not known precisely) in place, the attenuation factors of each of the less dense filters in the series may be determined by measuring in turn the intensity of a stable beam without, and with, the unknown less dense filter in place, and taking the ratio of the respective intensities. When a sufficient number of the less dense filters have been calibrated in this manner to approximately equal, in combination, the attenuation factor of the unknown densest filter, the combination of the less dense filters is placed in the filter holder and the beam intensitites without, and with, the densest filter added are determined in turn. Their ratio gives the filter factor of the latter. Because of the propagation of errors, it is evident that the attainment of a desired precision in the attenuation value of the densest filter requires a very much higher precision of each of the less dense filters. Fortunately, this tends to be the case anyway, since during equal counting intervals a less dense filter will accumulate larger counts, resulting in lower relative counting errors. In each determination, the counting interval must be long enough to accumulate a total count sufficiently large to give a low relative error, as determined by counting statistics.[44]

During the time of filter calibration, constancy of the beam intensity must be ascertained by periodic checks. If a change has been detected by monitoring and can be well defined, it may be possible to correct for it. Corrections for counting-system dead time, although much smaller for proportional counters than for Geiger–Müller counters, also must be applied separately for each filter. The subject is discussed in some texts on X-ray diffraction and elsewhere.[11,75,116,117]

b. Apparatus Alignment. Alignment is a particularly exacting process. It starts with the monochromator. The bending press, which clamps the crystal lamina to give the required curvature, is mounted on a horizontally rotatable platform in such a way that the center of rotation coincides with the center of the concave front face of the elastically deformed crystal. With the X-rays on and the shutter open (suitable shielding precautions having been taken) manipulation of, first, the coarse, and finally, the fine rotational adjustment (the latter by means of a sensitive differential tangent screw) will bring the monochromatized beam corresponding to the K_α line into view on a fluorescent screen in the dark. The monochromator platform is supported so as to permit a variety of adjustments, axial, transverse vertical and horizontal, and rotary in a vertical plane. All these adjustments may have to be applied in an iterative fashion until near-perfect alignment is accomplished. The criterion for perfect alignment of a good crystal is the presence of a rectangular and homogeneous beam cross section and its sudden appearance and disappearance, without shifts in position, upon a slight change of the horizontal rotatory fine adjustment in either direction.[118]

Next, the goniometer table as a whole is adjusted to align the first two slits with the monochromatized beam. These slits (which are continuously adjustable and symmetrically opening and closing) must be optically aligned beforehand, so that their median lines and the goniometer axis of rotation lie in the same vertical plane. The goniometer table is mounted on two superimposed platforms, each of which allows a mode of adjustment. The lower platform allows translation transverse to the optical axis of the monochromatized primary beam; this motion is actuated by a differential screw and may be read on a dial in-

[116] See Klug and Alexander,[44] p. 281.
[117] D. R. Chipman, *Acta Crystallogr.* **A25**, 209 (1969).
[118] A. Guinier, "Théorie et Technique de la Radiocristallographie," p. 192. Dunod, Paris, 1964.

dicator. The upper platform is designed to be rotatable about a pivot which can be adjusted to coincide with the median line of the first slit; this rotation is actuated by a two-speed screw, allowing both coarse and fine adjustment, and is similarly dial indicated. The primary beam, visualized by a fluorescent screen, is first threaded through the first slit by translation of the lower platform. Next, the second slit is opened wide, and the sample holder is replaced with an auxiliary slit mounted and aligned in such a way that its median line coincides with the goniometer axis. The primary beam is then threaded through the auxiliary slit by rotation of the upper platform, the first slit remaining essentially in place.

The detector is now aligned approximately. The first and the auxiliary slits define the zero-angle position and, with the third and fourth slits removed, the goniometer readout is adjusted to zero, using the detector in conjunction with a strip-chart record. The third and fourth slits are replaced and adjusted so that they just admit the primary beam to the detector. The second slit is replaced last and is adjusted until it just fails to graze the primary beam. Its purpose is to eliminate edge-scattering produced by the first slit; its edges must not be irradiated by the primary beam, leading to further parasitic scattering. The quality of this slit and its alignment are exceedingly critical. The final test is a constant-speed scan from about 0.5° on one side of zero to 0.5° on the other, with the intensity versus time recorded on a strip chart. Until a symmetrical scan is obtained, some, or all, of the preceding adjustments may have to be repeated one or more times. The first two slits are then opened to the desired width, which is defined by the smallest angles to be measured in a given experiment. Since frequently it is necessary to carry out in sequence runs with different slit openings, it is advantageous to have the slit openings controlled by micrometer heads; it should become, then, a matter of routine to open and close these slits as required.

3. Preparatory Steps

a. *Sample Preparation.* Since all molecular parameters must be evaluated by extrapolation to infinite dilution (see Section II), it is necessary in every case to study a concentration series. Hence a series of protein concentrations, obtained by dilution from a stock solution, must be so chosen that the points are as far separated as possible in order to yield well defined concentration plots for the various parameters. The upper limit to the concentrations may be set by the availability of a scarce material, by its limited solubility, or by excessive viscosity of more concentrated solutions; it rarely exceeds 100 g/liter. The lower limit is imposed by the technique itself. In the case of biological macromolecules, even moderately dilute solutions (much under 10 g/liter) produce so little excess scattering that, given the random nature of the radiation process, the difference between solution and solvent scattering is not sufficient to yield statistically meaningful results.

The scattering we are concerned with is the excess scattering, i.e., the scattering due to the macromolecular solute of interest alone. As shown in Section II (see also this volume [10]), this means that the solvent used for the blank measurements must have a chemical potential identical to that of the solvent as it exists in the solution. Hence, ideally the solvent used as blank for each protein dilution must be the dialyzate of that particular dilution. In practice, when the concentrations of the nonaqueous components of the solvent are low, e.g., 0.1 M salt, this is sufficiently approximated by using the dialyzate of the stock solution as reference solvent and as diluent.

These considerations differ little from those applicable to any other thermodynamic technique, for example, light scattering. Fortunately, as pointed out above (see Section II, D) the very troublesome problem encountered in light scattering, namely the removal of every trace of dust particles, need not concern us here.

It may be added that in the choice of solvents (usually buffers or other dilute salt solutions), in SAXS one is limited to fairly low salt concentrations. Since the scattering intensity is a function of electron density, the solvent scattering at high salt concentrations could easily mask the scattering of the sample. Furthermore, since X-ray scattering is a function of electron concentration, light ions are preferable to heavy ones; for example, fluoride should be used preferably to chloride.

b. Protein Concentration and Partial Specific Volume Determinations. Protein concentrations must be known with high precision, since accurate extrapolation to zero concentration is required, and concentration enters into the expression for the measurement of the molecular weight (see Eq. 16), even at extrapolation to zero concentration. As in other protein work, careful ultraviolet absorbance measurements (see this volume [21]) are the method of choice, provided the absorptivity at some given wavelength is known. Otherwise, dry weight measurements might be necessary.

The partial specific volume must also be known precisely, since it appears as the square in the molecular weight equation (Eq. 18). At present, instrumentation is available to carry out such measurements to a precision of better than ±0.2% (see this volume [5] and Kupke and Beams[119]).

c. *Cell Thickness Measurement.* Since the cell thickness determines the number of scatterers within the irradiated volume seen by the detector, it needs to be accurately known to permit expressing scattered intensities in terms of the scattering of a single electron; and, indeed, the cell thickness appears in the expression for the normalized intensities (see Section IV, A). Inasmuch as this thickness measurement is related to the construction of the cell, we shall digress briefly to describe the kinds of cell or sample container used in work with dilute solutions.

A cell must satisfy certain requirements with respect to optical path length, volume, and windows. For an optimal signal-to-noise ratio, the path length should be so chosen as to result in a maximum ratio of scattering relative to absorption. This is a criterion generally taken into consideration in diffraction work.[119a] For dilute protein solutions, the optimum path length works out to about 1 mm. The sample volume must be balanced between the geometric requirements of the X-ray beam (e.g., the infinite slit-height assumption) and the need to use the minimum amount of sample, biological material frequently being in scarce supply. The window material must be radiation stable and transparent, should not contribute disturbing scattering of its own, and should have sufficient rigidity to maintain a given geometric shape.

These requirements can be met by quartz capillaries, as used in diffraction work, and by assembled cells with flat windows, which may be either demountable or cemented to an appropriate spacer and frame. The preferred window material is mica; mylar film has also been used but lacks the rigidity necessary to maintain a nearly flat configuration. Because round capillaries present problems in establishing the precise path length and irradiated volume, we have adopted the use of flat windows in a demountable cell of about 0.35 ml volume. The frame elements, windows and 1-mm spacer (preferably of Teflon) may be clamped together by machine screws. For uniformity in tightening, to assure leak tightness without causing undue distortion of the windows, we have adopted a screw-ring assembly similar to that of standard infrared absorption cells. Suitable changes in dimensions take into account the optical requirements of the thickness-measuring device to be described below, and the fact that the thick, rigid windows used in infrared work are here replaced by exceedingly fragile sheets of mica (ca. 0.0175 mm thick).

[119] D. W. Kupke and J. W. Beams, see Vol. 26 [5].
[119a] See Klug and Alexander,[44] p. 205.

If, in addition to the concentration, the X-ray absorption coefficient of the solution is known from previous experiments, the sample thickness may be calculated by the use of Beer's law from intensity measurements taken on the same sample container filled with sample and solvent, in turn. However, an X-ray system cannot be relied upon to retain constant intensity between these two measurements without special precautions, nor is it always easy to obtain a reliable value for the absorption coefficient. A further complication stems from the fact that the cell windows are somewhat elastic. This precludes an exactly reproducible path length for consecutive fillings with solutions of slightly different properties, such as density, viscosity and surface tension. For these reasons the cell thickness is preferably determined in an independent auxiliary masurement.

This may be done with an instrument of a type first used in Luzzati's laboratory which comprises two opposed microscopes with fine-focusing adjustments between which the filled sample cell is placed. The microscope optics are chosen to give a very shallow depth of field (below 10 μm), so that the position of the microscope tube, as measured by a suitable indicator, is a precise indication of its focal plane at any particular setting. One microscope remains fixed and serves to define a reference plane in space. One face of one window of the sample cell is brought into coincidence with this plane by means of an adjustable sample cell holder. The other microscope which had been previously zeroed by focusing on this same plane is then adjusted to focus on the appropriate face of the second cell window. The difference between the two positions of this microscope gives the sample thickness, after refractive index corrections for solution and window material.

4. Data Gathering

a. Primary Beam Intensities. If the stability of the X-ray source as well as that of the slit and detection systems could be absolutely relied on, there would be a need to take only a single measurement of the primary beam intensity; this single value could then be applied to the difference between the observed values of solution and solvent scattering. As a matter of fact, however, even the best of systems must be expected to undergo some fluctuations (see Section III, A). That is why the defining equation for excess normalized scattering (Eq. 32, below) shows the normalization involving the direct beam intensities performed separately for solution and for solvent, before one is subtracted from the other. This procedure, although it is not universally followed, is proper inasmuch as the scattered intensities for solution and solvent may be determined during widely separated time periods, when the beam intensity cannot be assumed to have been identical.[120]

Ideally, the beam intensity should be recorded concurrently with that of the scattered radiation. In the absence of a monitor (see Section III, A), this is not possible. All that one can do is to sample the beam at times when it is practical. If step scanning is chosen, it is possible to intercalate beam intensity measurements between steps of the scattered intensity measurement. As it is not practical to do this too often, the adopted practice, is to do it either at a few predetermined positions during the step scan, or else routinely at certain times of the day (e.g., at the beginning and end of the work day, and at noon). If continuous scans are chosen, it is not practical to interrupt a scan in progress, and one is limited to making beam measurements before and after each scan. In our practice, which uses primarily continuous scanning, the integrated intensity of the direct beam is measured between $\pm 0.3°$, at a scanning speed of $\frac{1}{8}°$ per minute, using an appropriate filter to limit the count rate, as discussed in Section III, B.

b. *Scattered Intensities.* Disregarding possible hybrid systems, there are essentially two ways of scanning, referred to above: by steps, and continuously. Step scanning has certain advantages that have made it the method chosen in the majority of laboratories. With the prevalence of modern digital data processing equipment, it appears logical to acquire data in digital form, as is done in step scanning. Since this is a discrete sampling procedure, in which certain angular positions (usually, but not necessarily, equidistant) are preselected for counting for a fixed time or a fixed count, it would also seem to offer some saving in time.

However, some finite time during which no counting can take place is expended while the scanning arm slews from one position to the next. Furthermore, the number of points required for good definition of a curve is quite large (usually at least 100), and the counting time at each point needs to be quite long. For a relative error of 1%, a total count of 10,000 is required, giving a standard deviation of $\sqrt{10,000} = 100$, or 1%.[44] Along the tail of a scattering curve, which accounts for perhaps 75% of the entire scan, a count of, typically, 25 counts per second may be expected, implying a counting time of $10,000/25 = 400$ seconds,

---

[120] This is the basis for the difference between our statement (see Section I) that a number of parameters may be obtained only from absolute measurements and occasional statements in the literature which imply that, except for molecular weight, other molecular parameters beyond radius of gyration are obtainable from relative measurements alone. Correct as this may be in principle, it is not realistic in view of what has been said above regarding the operational Eq. (32) and the possibility of long-term shifts in the intensity of the primary beam.

or nearly 7 minutes, per point, exclusive of the time required for slewing. Along the steeper portions of the curve, counting times may be much shorter, but rapid changes in curvature would benefit from more closely spaced points. Flexible arrangements, which can take into account the requirements suggested by the shape of the curve and thus make more efficient use of time, have been used by Kratky and Kratky[121] and others.[122]

Regarding the digital character of the data, it should be borne in mind that, because of the random errors characteristic of scattering data, it is not possible to subject them to the required slit-smearing correction (Section IV) without prior smoothing; without smoothing, computational artifacts are prone to arise and the data become severely degraded. Despite continual attempts at developing computer-adaptable smoothing routines (e.g., Oelschlaeger,[123] Damaschun et al.[124]), these efforts have not been successful enough to induce those in this field to abandon manual smoothing. This, of course, largely negates the potential advantages of automatic data processing.

Pending the development of more reliable smoothing methods, we have found it practical, for the most part, to retain the older practice of continuous scanning with strip chart recording, followed by manual smoothing of the graphic record, and finally to digitization. The strip chart is required in any case, to check on the symmetry of the scan.

The readability of the record depends to a great extent on the choice of the time constant of the ratemeter whose output is recorded.[125] Too short a time constant causes a jittery and cluttered trace, obscuring the trends to be looked for; too a long a time constant will distort the record. Choice of time constant is often regarded more as an art than a science, and certain rules of thumb are appealed to. It should be apparent, however, that scanning speed and receiving slit opening are the determinant variables, as discussed by Klug and Alexander.[126] In consequence of the very weak observed intensities, scanning speeds in SAXS need to be much slower than is customary in diffraction work, speeds of 0.5 degree per hour being typical. At speeds as low as this (approximately 60-fold slower than would be typical in diffraction), much higher time

[121] C. Kratky and O. Kratky, Z. Instrumentenkunde 72, 302 (1964).
[122] H. Leopold, Z. Angew. Phys. 25, 81 (1968).
[123] H. Oelschlaeger, Acta Phys. Austr. 30, 323 (1969).
[124] G. Damaschun, J. J. Müller, and H.-V. Pürschel, Acta Crystallogr. A27, 11 (1971).
[125] Similar remarks would apply to the gating times of an electronic counter used in conjunction with a digital-to-analog converter, optionally used in place of the more conventional ratemeter.
[126] See Klug and Alexander,[44] p. 310.

constants are permissible. Whereas time constants between 0.5 and 16 seconds are customary in diffraction, and commercial diffraction apparatus rarely provides a time constant as high as 40, we have concluded that 300 seconds would not be excessive for SAXS. With appropriate instrument modification, such values can be realized. We have found routinely that a time constant of 200 seconds is very satisfactory, since it produces a quiet, undistorted, and interpretable record.

IV. Data Evaluation

It is evident from the discussion of the previous sections that the problem of data evaluation can be quite tedious. Due to the low count rates, it is necessary to use as many data points as possible. We have found it advantageous to utilize the continuous method of scanning rather than the discrete point method, which also allows a more efficient smoothing of the data. The only requirement for the continuous scanning method is the use of a sufficiently long time constant, i.e., of the order of 200 seconds.

A. Ancillary Calculations

Since SAXS measures the scattering from the electrons in a particle (see Section II, A), it follows that all parameters must be expressed in the corresponding electron-$Å^3$, and not gram-milliliter units. This unit transformation can be easily accomplished with the use of the following expressions.

The partial specific volume, $\bar{V}$, in units of milliliters per gram, can be transformed to ψ, expressed in $Å^3/el$, by

$$\psi = \bar{V}10^{24}/q_p \tag{30}$$

Here, q_p is the number of electrons per gram of the protein. This can be readily calculated from the amino acid composition or an elemental analysis.

The protein concentration, c_e, in units of electrons protein per electron solution, is given by

$$c_e = \frac{gq_p}{q_s - (q_s - q_p)g} \tag{31}$$

where q_s is the number of electrons per gram of solvent and g is the gram fraction of protein in units of gram protein per gram solution.

Finally, the excess normalized scattering function, $j_n(s)$ is obtained from[3]

$$j_n(s) = \left(\frac{I(s)}{\lambda^2 t \rho_s (7.9 \times 10^{-26}) E_0 (ds/dt) f}\right)_{\text{solution}}$$

$$-\left(\frac{I(s)}{\lambda^2 t \rho_1 (7.9 \times 10^{-26}) E_0 (ds/dt) f}\right)_{\text{solvent}} \quad (32)$$

where $I(s)$ is the scattered count rate at each value of s in counts per second, which is averaged on both sides of the primary beam for each s, λ is the wavelength of the Cu K_{α_1} peak in Å,[126a] t is the thickness of the filled cell in cm; 7.9×10^{-26} is the value of the scattering of a single electron derived from Thomson's equation (see Eq. 8); E_0 is the total number of counts under the primary beam while scanning at a set rate of ds/dt; f is the product of the filter factors used for attenuation purposes when measuring the energy of the primary beam, and ρ_s and ρ_1 are the densities of the solution and solvent, respectively, in $el/\text{Å}^3$. These can be calculated from

$$\rho_i = \frac{d_i q_i}{10^{24}} \quad (33)$$

where q_i is the number of el/g and d_i is the density in the usual units of grams per milliliter of material i, here solution and solvent.

When some seventy or more $j_n(s)$ values, ranging from approximately 0.25 to 5°, 2θ, are calculated, the data evaluation can begin. First a plot of $s^3 j_n(s)$ versus s^3 is generated. A typical plot is shown in Fig. 12 for the scattering of lysozyme at various concentrations. The limiting slope of the curve at high s^3 gives δ^* while the intercept gives the A value (see Eq. 20). The concentration dependences of these parameters are reflected in Fig. 12 by the change in the slope as well as the intercepts when going from 36 g/liter to 22 g/liter. The A parameter is used in the calculation of the surface, S, and the surface-to-volume ratio, S/V, of the scattering particle, while δ^*, a parameter which measures the contribution due to internal atom diffraction, must be subtracted from each $j_n(s)$ value (see Eq. 9) to give the quantity $j^*_n(s)$ which is used for the rest of the calculations (Eqs. 20, 21, 23).

B. Molecular Parameters

A Guinier plot is then constructed by plotting $\log_e j^*_n(s)$ versus s^2 (Fig. 13). Here, it is important to recall that, by definition, this plot is linear at small values of s for a homogeneous substance. The appearance of nonlinearity or multiple-linearity in this region implies that the

---

[126a] The presence of λ^2 in the denominator of Eq. (32) does not mean that the scattering intensity is a function of wavelength, seemingly contradicting Eq. (7). In Eq. (32), λ^2 appears trivially as a conversion factor between $(2\theta)^2$ and s^2.[126b]

[126b] See Guinier and Fournet,[1] p. 18.

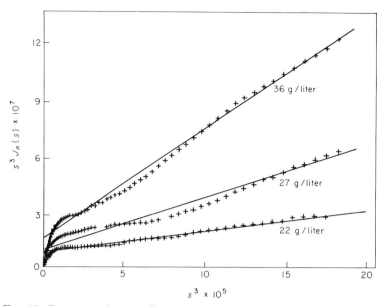

FIG. 12. Computer plots of $s^3 j_n(s)$ vs. s^3 for lysozyme at various protein concentrations. From H. Pessen, T. F. Kumosinski, and S. N. Timasheff, *J. Agr. Food Chem.* **19**, 698 (1971).

system being measured is nonhomogeneous. With a homogeneous system, the slope of the linear region generates the apparent radius of gyration, R_a (Eq. 11), while the intercept yields the $j_n(0)$ value. Figure 13 shows the large concentration dependence of $j_n(0)$, which is reflected

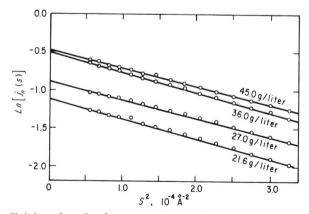

FIG. 13. Guinier plots for lysozyme at several protein concentrations. From H. Pessen, T. F. Kumosinski, and S. N. Timasheff, *J. Agr. Food Chem.* **19**, 698 (1971).

in the intercepts of the Guinier plots at concentrations from 45 to 21.6 g/liter, while the R_G values, proportional to the slope of a Guinier plot, are only mildly concentration dependent.

The nonlinear portion of the curve gives rise to the residual function $\phi(s)$ (Eq. 11) by subtracting the calculated Guinier function from each $j^*_n(s)$ value and fitting these results to a polynominal in s^2, such that $\phi(s)$ is now defined as

$$\phi(s) = \sum_{i=0}^{N} a_i s^{2i} \tag{34}$$

The integral

$$Q = \int_0^\infty s j^*_n(s) ds \tag{35}$$

which has been termed the invariant, Q, by Porod[127] must now be evaluated before proceeding to the final calculations. This is accomplished by substituting Eq. (34) into Eq. (11), which yields

$$j^*_n(s) = j^*_n(0) \exp(-4/3\pi^2 R_a^2 s^2) + \sum_{i=0}^{N} a_i s^{2i} \tag{36}$$

Then, integrating according to Eq. (35), the first term analytically and the second term from zero to a predetermined high angle, s_L, we find that

$$Q = \sqrt{3/\pi}\, j^*_n(0)/4R_a + \sum_{i=0}^{N} \frac{a_i}{2} s_L^{2(i+1)} \tag{37}$$

Here it is important to discuss the problems associated with the numerical analysis of the residual function and the invariant. In the first place, care must be taken when fitting the residual function to a polynominal. Only the best polynominal least-squares routine should be employed, since the data in this region are very imprecise. Second, the roots of the polynominal must be calculated in order to obtain a good value for s_L. It is possible to pick an erroneous value for which the residual function would already be negative. This would result in a value for the invariant that is much too small and would lead to too large values of the hydrated volume, the surface-to-volume ratio, and the degree of internal hydration and to a too small value of the electron density difference.

Using the same type of argument as was used in the calculation of

[127] G. Porod, *Kolloid-Z.* **124**, 83 (1951); **125**, 51 (1952).

the invariant, it is possible to derive from Eqs. (12) and (36) the following expression for the deconvoluted scattering function,[3]

$$i_n(s) = 2\sqrt{\pi/3}\,j^*{}_n(0)R_a \exp\left[-\frac{4}{3}\pi^2 R_a^2 s^2\right]$$
$$-\frac{1}{\pi}\sum_{i=0}^{N} 2ia_i \int_0^\infty (s^2 + l^2)^{(i-1)}dl \quad (38)$$

The integral in the second term can be evaluated analytically by using a simple recursion formula (derived from integration-by-parts) which can be found in any mathematical table of integrals. However, since the numerical calculation of this term is rather lengthy, it would necessitate the use of a computer program. These desmeared scattering values are then used in a second Guinier plot for calculating the true radius of gyration, R_G, from the slope and $i_n(0)$ from the intercept.

In his original work, because of the lack of computer facilities, Luzzati[128] derived the expressions for $i_n(0)$ and R_G using a Maclaurin expansion of Eq. (11). They are

$$i_n(0) = 2\sqrt{\pi/3}\,j^*{}_n(0)R_a - \frac{1}{\pi}\int_0^\infty s^{-2}\phi(s)ds \quad (39)$$

and

$$R_G^2 = \frac{R_a^2 + \dfrac{9\sqrt{3\pi}}{16\cdot4}\dfrac{1}{j^*{}_n(0)R_a}\displaystyle\int_0^\infty s^{-4}\phi(s)ds}{1 - \dfrac{\sqrt{3\pi}}{2\pi^2}\dfrac{1}{j^*{}_n(0)R_a}\displaystyle\int_0^\infty s^{-2}\phi(s)ds} \quad (40)$$

By making use of the fact that the lim $s^3 j^*{}_n(s) = A$ (a constant), the integrals in Eqs. (39) and (40) were approximated[24,41] by

$$\int_0^\infty s^{-2}\phi(s)ds \approx \sum_0^a s^{-2}\phi(s)\Delta s + \lim_{s\to\infty} s^3 j^*{}_n(s)\int_a^\infty \frac{ds}{s^5} \quad (41)$$

and

$$\int_0^\infty s^{-4}\phi(s)ds = \sum_0^a \frac{\phi(s)}{s^4}\Delta s + \lim_{s\to\infty} s^3 j^*{}_n(s)\int_a^\infty \frac{ds}{s^7} \quad (42)$$

where a is the value of s at which the scattering function $j^*{}_n(s)$ reaches for all practical purpose, a constant limit. At very low values of s,

[128] V. Luzzati, *Acta Crystallogr.* **11**, 843 (1958).

$\phi(s)/s^2$ is obtained by interpolation between the measurable range of s and $s = 0$, since $\lim \phi(s)/s^2 = 0$. The function $\phi(s)/s^4$ attains a constant value at low s, and thus can be calculated between $s = 0$ and $s = a$.

It is interesting to note that the contribution of the integrals in Eqs. (39) and (40) is usually of the order of 1–3% for normal globular protein. This contribution increases, however, as the concentration of the protein solution increases, or as the molecules become larger and more asymmetric. Although the above expressions can be calculated on a desk calculator, it is preferable to use Eq. (38) when a computer is available, since Eqs. (39) and (40) were derived from an expansion in which all but the first two terms were dropped. It is quite possible that with larger, less globular proteins large errors will develop in Eqs. (39)–(42); these might easily escape the cognizance of the investigator.

From the calculated values of $i_n(0)$, Q, and A, the rest of the structural parameters, namely, the molecular weight, M, the hydrated volume, V, the surface-to-volume ratio, S/V, the electron density difference, and the degree of internal hydration, H, can be calculated by direct use of Eqs. (18)–(25) at each protein concentration. Each parameter, in turn, is extrapolated to zero protein concentration in order to cancel virial effects [see Eqs. (16)–(19)].

C. Typical Examples

Typical examples of results obtained using this type of analysis are given in Table I for ribonuclease, lysozyme, and α-lactalbumin. The values of the various molecular parameters extrapolated to zero protein concentration show the close overall structural similarity between lyso-

TABLE I

STRUCTURAL PARAMETERS

Parameter	Ribonuclease	Lysozyme	α-Lactalbumin
R_G, Å	14.8	14.3	14.5
M	12,700	13,600	13,500
V, Å^3	22,000	24,200	25,100
S/V, Å^{-1}	0.29	0.25	0.24
H, g_{H_2O}/g_{Prot}	0.46	0.33	0.37
a/b from $\left(\dfrac{3V}{4\pi R_G^3}\right)^a$	1.87	1.42	1.43
a/b from $\left(R_G\dfrac{S}{V}\right)^a$	3.70	2.92	2.82

[a] Assuming a prolate ellipsoid of revolution.

zyme and α-lactalbumin,[6,129] which is in essential agreement with the expectations raised by Browne et al.,[130] who have postulated that the secondary and tertiary structures of these two proteins should be similar on the basis of their homologous amino acid sequences.[131] The ribonuclease data for R_G, M, V, S/V, and H, on the other hand, show the sensitivity of the small-angle X-ray scattering technique for distinguishing between globular proteins of the same general size but different conformations. The R_G values for the three proteins are almost the same, but the S/V and V parameters are significantly different. Thus, it can be concluded that lysozyme and α-lactalbumin are very similar proteins in their overall structures, while ribonuclease is different from both of them. Furthermore, the crystallographic radii of gyration have been calculated for ribonuclease by Kartha as 13.5 Å[132] and for lysozyme by Blake et al. as 13.8 Å.[133] The solution values measured by SAXS are slightly higher. This could well be the result of the fact that SAXS gives the geometric parameters of the hydrated protein in solution, in which the surface side chains have much more freedom of motion[134] than in the crystalline state.

Another parameter which is useful in correlating SAXS data with other solution or crystallographic data is the axial ratio of an equivalent ellipsoid of revolution, a/b, where a is the major axis and b is the minor axis. This parameter can be calculated[24] from the products $(3V/4\pi R_G{}^3)$ and $R_G(S/V)$, respectively. A working graph, derived by Witz et al.[24] relating these quantities to the axial ratio is shown in Fig. 14. One merely calculates the values for $(3V/4\pi R_G{}^3)$ and $(R_G(S/V))$ from small-angle X-ray scattering data extrapolated to zero protein concentration, and reads from the curves the corresponding values of the axial ratio, a/b, for a prolate and an oblate ellipsoid of revolution. Such a calculation was performed for ribonuclease, lysozyme, and α-lactalbumin and the results, assuming prolate ellipsoids, are shown in Table I.

It is interesting to note that the value for lysozyme calculated from $(3V/4\pi R_G{}^3)$ was found to be 1.42, whereas the crystallographic value reported by Blake et al. was 1.5.[133] This agreement is quite good. The

[129] H. Pessen, T. F. Kumosinski, and S. N. Timasheff, Fed. Proc., Fed. Amer. Soc. Exp. Biol. 30, 1180 (1971). (Abstract.)

[130] W. J. Browne, A. C. T. North, D. C. Phillips, K. Brew, T. C. Vanaman, and R. L. Hill, J. Mol. Biol. 42, 65 (1969).

[131] K. Brew, T. C. Vanaman, and R. L. Hill, J. Biol. Chem. 242, 3747 (1967).

[132] G. Kartha, J. Appl. Crystallogr. 4, 417 (1971).

[133] C. C. F. Blake, D. F. Koenig, G. A. Mair, A. C. T. North, D. C. Phillips, and V. R. Sarma, Nature (London) 206, 757 (1965).

[134] K. O. Linderstrøm-Lang and J. A. Schellman, in "The Enzymes" (P. D. Boyer, H. Lardy, and K. Myrbäck, eds.), Chapter 10. Academic Press, New York, 1959.

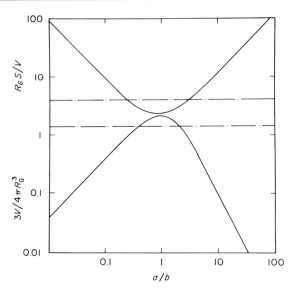

Fig. 14. Nomogram relating the geometric parameters of ellipsoids of revolution to the axial ratio, a/b. From V. Luzzati, J. Witz, and A. Nicolaieff, *J. Mol. Biol.* **3**, 367 (1961).

value for α-lactalbumin is 1.43, which again shows the close structural similarities between the two proteins. Ribonuclease, on the other hand, has an axial ratio of 1.8 which is much higher than the other two proteins. This, however, is consistent with the crystallographic results, which show a greater degree of asymmetry and consequently a larger axial ratio.[135] It is striking to note that the axial ratios calculated from $(R_G(S/V))$ are all considerably larger for all three enzymes. This apparent anomaly is due to the fact that proteins are not solid geometric entities but do indeed consist of surfaces with many holes and clefts. Thus, for a given volume, the surface is considerably larger than would be predicted for a corresponding ellipsoid of revolution. In this way it can be seen that SAXS is very sensitive to the surface topology of the molecules. This is brought out even more strikingly by the fact that the difference between the axial ratios calculated from the volume and the surface is larger for ribonuclease than for lysozyme and α-lactalbumin. This is a direct result of the fact that the ellipsoidal model is even more incorrect for ribonuclease than for the other two proteins, since in ribonuclease a single polypeptide chain does indeed protrude somewhat from the main body of the molecule.

[135] G. Kartha, J. Bello, and D. Harker, *Nature (London)* **213**, 862 (1967).

D. Precautions

When the calculations which lead to the molecular parameters are carried out with a computer, great care must be taken in programming. Since the residual function is not theoretically defined in an analytic fashion and since the Guinier region varies in s position and magnitude with the type and concentration of material to be investigated, a program must be written with a large amount of operator decision-making. A self-contained program could lead to errors in the values of various s limits arising from the small scattering values at high angles and at low concentrations of protein.

It is also necessary to consider in detail the problem of polydispersity which up to now has been only mentioned. Since in a nonhomogeneous system the Guinier plot shows a double linear or nonlinear character, one attempts to fit the data to a double gaussian function,[24] assuming a two-component system,

$$j^*_n(s) = A \exp\left[-\frac{4}{3}\pi^2\alpha^2s^2\right] + B \exp\left[-\frac{4}{3}\pi^2\beta^2s^2\right] + \phi(s) \qquad (43)$$

where A, B, α, and β are adjustable parameters that are related to $j_n(0)$ and the R_G values of the two components and $\phi(s)$ is the normal residual function. It has been shown[24] that $i_n(0)$ and R_G values can then be obtained from

$$i_n(0) = 2\sqrt{\pi/3}\,(A\alpha + B\beta) - \frac{1}{\pi}\int_0^\infty s^{-2}\phi(s)ds \qquad (44)$$

and

$$R_G = \left(\frac{3}{4\pi^2}\left[\frac{8\pi^2}{3}\sqrt{\pi/3}\,(A\alpha^3 + B\beta^3) + \frac{3}{2\pi}\int_0^\infty S^{-4}\phi(s)ds\right][i_n(0)]^{-1}\right)^{1/2} \qquad (45)$$

These parameters are weight–average values and must be used in conjunction with the protein concentration in order to find the intrinsic values of the two species. These are obtained from

$$X^2 = \frac{C_1X_1^2 + C_2X_2^2}{C_t} \qquad (46)$$

$$C_t = C_1 + C_2$$

where X is the structural parameter (G_G, $i_n(0)$, etc.) at C_t, the total concentration, X_1 and X_2 are the structural parameters of the individual species, and C_1 and C_2 are their mass concentrations. Since these equations are not analytic in their solutions for X_1 and X_2, it becomes necessary to use a curve–fitting routine or a series of tables in X_1 and X_2, C_1 and C_2 which are calculated at each X and C_t. This, however, is extremely cumbersome and not very precise. Therefore, a large number of experiments at various concentrations must be performed when dealing with a polydisperse system. In such systems, it is very advantageous to employ other methods, such as sedimentation velocity or sedimentation equilibrium, to aid in finding the concentration distribution, i.e., C_1 and C_2 values at every C_t.

E. High-Angle Region

The final discussion deals with the calculations at high protein concentrations (>100 g/liter) for the particle shape, which is reflected in the positions and magnitudes of maxima and minima that appear at high angles ($2\theta > 4°$).[1] Two alternative methods for handling these calculations are in use currently. These will be described in turn, allowing the reader to choose his preference.

The first method has been used primarily by the Kratky school.[105] In this method the experimental $j_n(s)$ values are deconvoluted, according to Eq. (12), to give $i_n(s)$ (which is normalized to an intercept of unity) at each angle and compared with theoretical curves calculated for various models.[136–139] Such a comparison is shown in Fig. 15, in which the normalized $i_n(s)$ values (expressed by the symbol Φ) are plotted as a double logarithmic plot as a function of (sR_G) for experimental data obtained on yeast glyceraldehyde-3-phosphate dehydrogenase and compared with theoretical curves for various models. It is interesting to note that only a small maximum and minimum appear on the experimental curve. This is due to the low scattering intensities at high angles. The deconvolution integral [see Eq. (12)], first derivatizes then integrates a function, tending in the process to smear details of a curve, especially when the precision of the data is not maximal.

In the second method, proposed by Luzzati, the theoretical curves calculated for point source optics are transformed to slit optics.[41] This approach makes it possible to compare the experimental scattering points, $j^*_n(s)$, directly with the theoretical curves for various models. The only

---

[136] O. Kratky and G. Porod, *Acta Phys. Austr.* **2**, 133 (1948).
[137] G. Porod, *Acta Phys. Austr.* **2**, 255 (1948).
[138] P. Mittelbach and G. Porod, *Acta Phys. Austr.* **14**, 185, 405 (1961); **15**, 122 (1962).
[139] P. Mittelbach, *Acta Phys. Austr.* **19**, 53 (1964).

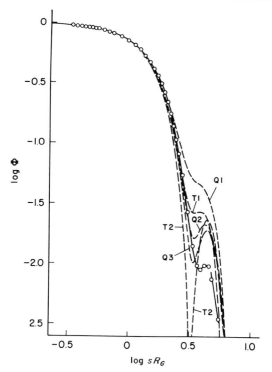

FIG. 15. Log Φ vs. log (sR) plot for comparison of the experimental scattering curve for the apoenzyme of yeast glyceraldehyde-3-phosphate dehydrogenase in 50 mM sodium pyrophosphate, 5 mM Na EDTA, and 0.2 mM dithiothreitol at pH 8.5 with theoretical scattering curves of model bodies built up from four rotation ellipsoids. T1, T2 tetrahedral configuration of the subunits; Q1, Q2, Q3, quadratic configuration of the subunits. From H. Durchschlag, G. Puchwein, O. Kratky, I. Schuster, and K. Kirschner, *Eur. J. Biochem.* **19**, 9 (1971).

transformation necessary in $j^{*}{}_{n}(s)$ is to normalize the function so that $i_{n}(0) = 1$. This is accomplished by

$$j(s) = \frac{j^{*}{}_{n}(s)}{c_{e}m(1 - \rho_{1}\psi)^{2}} \tag{47}$$

The quantity $(j(s))$ is then plotted as function of (sR_{G}) on a double logarithmic plot and compared with the convoluted, or smeared, theoretical curves for various geometric models. Some of the convoluted scattering curves may be constructed from available tables[140]; others may be calculated by convoluting the $i_{n}(s)$ geometric model func-

---

[140] P. W. Schmidt, *Acta Crystallogr.* **8**, 772 (1955).

TABLE II
COMPARISON OF SAXS RESULTS WITH THOSE OF OTHER TECHNIQUES

Protein	Molecular weight			Stokes radius (r)		
	Small-angle X-ray scattering	Amino acid composition	Small-angle X-ray scattering[a]	Sedimentation velocity[b]	Titration[c]
Ribonuclease	12,700[d]	13,800	19.1[f]	22.4	20.7
Lysozyme	13,600[f]	14,300	18.5[f]	18.8	17.9
α-Lactalbumin	13,500[f]	14,500	18.5[f]	18.8	—
Bovine serum albumin	81,200[g]	77,000[e]	39.5[g]	38.3	38.2
β-Lactoglobulin dimer	36,600[h]	36,300	27.7[h]	27.0	27.0
β-Lactoglobulin A octamer	(144,000)[h]	(145,200)	44.4[h]	43.3	43.8
α-Chymotrypsin	22,000[i]	25,200	23.3[i]	23.5	—

[a] Calculated from: $r = \left(\dfrac{5}{3}\right)^{1/2} R_G$

[b] Calculated from: $r^{1/3} = \dfrac{M(1 - \bar{v}_2\rho)}{6\pi\eta N_A s_{20,w}^\circ}$

(η: solution viscosity, $s_{20,w}^\circ$ = sedimentation coefficient); see C. Tanford, "Physical Chemistry of Macromolecules." Wiley, New York, 1961.

[c] Calculated from: $\dfrac{1}{r} = \dfrac{2DkTw}{e^2} + \dfrac{\kappa}{1 + \kappa a}$

(D, dielectric constant; k, Boltzmann constant; T, absolute temperature; e, electronic charge = 4.8×10^{-10} esu, w = electrostatic work function derived from the titration curve; X, Debye-Hückel screening parameter; a = center-to-center distance of closest approach between protein and small buffer ion); see C. Tanford, "Physical Chemistry of Macromolecules." Wiley, New York, 1961.

[d] Recalculated from H. Pessen, T. F. Kumosinski, and S. N. Timasheff, J. Agr. Food Chem. **19**, 698 (1971).

[e] The molecular weight of bovine serum albumin is 69, 000; however, for the sake of comparison with the $\bar{M}_w$ value obtained in small-angle X-ray scattering, the contribution from 5% dimer, normally present, has been taken into account.

[f] H. Pessen, T. F. Kumosinski, and S. N. Timasheff, J. Agr. Food Chem. **19**, 698 (1971).

[g] V. Luzzati, J. Witz, and A. Nicolaieff, J. Mol. Biol. **3**, 379 (1961).

[h] J. Witz, S. N. Timasheff, and V. Luzzati, J. Amer. Chem. Soc. **86**, 168 (1964).

[i] W. R. Krigbaum and R. W. Godwin, Biochemistry **7**, 3126 (1968).

tions[136-139] with the use of Eq. (12). An example is shown in Fig. 16 for β-lactoglobulin at pH = 5.7 in 0.1 M acetate buffer.[41] Here, the experimental $j(s)$ data are compared with various convoluted curves, namely those for the sphere, the two-sphere and the parallelepiped models. The experimental curve shows large amplitudes of the maximum and the minimum, which have not been diminished by convolution of the data. It is obvious, however, that none of the geometric models fit the experimental scattering curve. This is in great part due to the nature of protein structure. Proteins, in general, are not smooth geometric bodies with uniform internal structures. Therefore, an exact fit of the experimental points to a geometric model should not be expected. Differences should indeed occur between these normalized theoretical and experimental curves and information only on the overall gross structure,

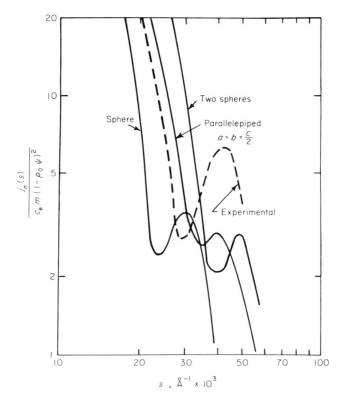

Fig. 16. Normalized scattering of β-lactoglobulin A and B in the higher angle range in 0.1 M sodium acetate at pH 5.7. The dashed line represents the experimental curve, and the solid lines represent the convoluted theoretical curves calculated for various models. From J. Witz, S. N. Timasheff, and V. Luzzati, *J. Amer. Chem. Soc.* **86**, 168 (1964).

e.g., whether the protein structure lies between a single sphere and a two sphere model, should be expected.

Using the shape factor parameters and high-angle data, together with calculated axial ratios at various conditions, such as pH, temperature, solvent composition, an extremely large amount of information may be obtained on protein tertiary and quaternary structural changes and related to biological function.

V. Conclusion

In conclusion, it seems desirable to give some comparisons of the values of parameters measured with those obtained by other techniques. This can be readily done for the molecular weight and the radius of gyration. In Table II such a comparison is provided of the molecular weights and the Stokes radii calculated for equivalent spheres for several radius of gyration can, in general, be obtained with other physical and proteins. It is evident that good agreement of the molecular weight and chemical methods. This result gives confidence as well in the validity of the other molecular parameters which are measured, namely the hydrated volume, the surface to volume ratio and the degree of hydration, since these are derived from the same raw data.

Acknowledgment

This work was supported in part by NIH Grant GM-14603 (to S.N.T.) and NSF Grant GB-12619 (to S.N.T.).

[10] Light Scattering and Differential Refractometry

By EUGENE P. PITTZ, JAMES C. LEE, BARKEV BABLOUZIAN,
ROBERT TOWNEND, and SERGE N. TIMASHEFF

I. Introduction

Among the techniques available for the characterization of macromolecules in solution, a particularly useful one is light scattering. With this method, it is possible to determine at relatively low protein concentration the molecular weight, degree of association, interactions with solvent components, and, if the macromolecule is large enough, its size and general shape.

There are a number of excellent reviews of this technique in the literature.[1-7] These reviews cover in detail the theory and methods in-

[1] P. Doty and J. T. Edsall, *Advan. Protein Chem.* **6**, 35 (1951).
[2] E. P. Geiduschek and A. Holtzer, *Advan. Biol. Med. Phys.* **6**, 431 (1958).
[3] M. Bier, see Vol. 4, p. 165.

volved in the application of light scattering to two component systems, with little[7] or no discussion of multicomponent systems, and the reader is referred to them for a detailed treatment of two-component experiments. The purpose of this chapter is to present the basic principles, to discuss the problems and to describe the practical techniques involved in performing light scattering and differential refractometry measurements on multicomponent systems, as well as to describe some of the instrumentation available for angular measurements. Frequently, the techniques described will be those adopted in this laboratory and with which the authors are most familiar.

II. Theory

A. Thermodynamics

1. Multicomponent Systems

Multicomponent theory was first treated in 1915 by Zernicke[8,9] and later developed in detail by Ewart et al.,[10] Brinkman and Hermans,[11] Kirkwood and Goldberg,[12] and Stockmayer.[13] More recently, it has received an extensive treatment by a number of investigators.[13-27] In

[4] G. Oster, "Physical Methods of Organic Chemistry" (A. Weissberger, ed.), 3rd ed., Part III, 2107. Wiley (Interscience), New York, 1960.

[5] K. A. Stacey, Light scattering in physical chemistry. In "A Laboratory Manual of Analytical Methods of Protein Chemistry" (P. Alexander and R. J. Block, eds.), Vol. 3, p. 245. Butterworth, London, 1961.

[6] S. N. Timasheff, in "Electromagnetic Scattering" (M. Kerker, ed.), p. 337. Pergamon, Oxford, 1963.

[7] S. N. Timasheff and R. Townend, in "The Physical Principles and Techniques of Protein Chemistry" (S. Leach, ed.), Part B, p. 147. Academic Press, New York, 1970.

[8] F. Zernicke, Dissertation, University of Amsterdam, 1915.

[9] F. Zernicke, Arch. Neerl. Sci. III A 4, 74 (1918).

[10] R. H. Ewart, C. P. Roe, P. Debye, and J. R. McCartney, J. Chem. Phys. 13, 159 (1946).

[11] H. C. Brinkman and J. J. Hermans, J. Chem. Phys. 17, 574 (1949).

[12] J. G. Kirkwood and R. J. Goldberg, J. Chem. Phys. 18, 54 (1950).

[13] W. H. Stockmayer, J. Chem. Phys. 18, 58 (1950).

[14] H. Shogenji, Busseiron Kenkyu 62, 1 (1953).

[15] C. M. Kay and J. T. Edsall, Arch. Biochem. Biophys. 65, 354 (1956).

[16] T. Ooi, J. Polym. Sci. 28, 459 (1958).

[17] A. Vrij, Doctoral Dissertation, University of Utrecht, 1959.

[18] S. N. Timasheff and M. J. Kronman, Arch. Biochem. Biophys. 83, 60 (1959).

[19] D. Stigter, J. Phys. Chem. 64, 842 (1960).

[20] E. F. Casassa and H. Eisenberg, J. Phys. Chem. 64, 753 (1960).

the present chapter, we shall briefly present the theory as it is applied in differential refractometry and light scattering measurements.

a. Differential Refractometry. In treating three-component systems, the generally accepted notation is that of Scatchard[28] and Stockmayer[13] in which the protein is designated as component 2, water as component 1, and the nonaqueous solvent as component 3. Expressing the concentration in the proper thermodynamic units, i.e., on the molal scale, the preferential interaction of component 3 with the protein is measured directly by the difference between the refractive index increments measured at conditions at which the chemical potential of the third component and its molality are, in turn, kept identical in the solution and in the reference solvent.

At constant temperature and pressure, the total change in refractive index, dn, with a change in solution composition, is given by Eq. (1):

$$dn = \left(\frac{\partial n}{\partial m_2}\right)_{T,p,m_3} dm_2 + \left(\frac{\partial n}{\partial m_3}\right)_{T,p,m_2} dm_3 \tag{1}$$

where n is the refractive index, m_i is the molal concentration of component i (moles per 1000 g of water), T is the thermodynamic temperature, and p is the pressure. Taking the derivative of Eq. (1) with respect to m_2 at constant chemical potential, μ, of component 3, we get

$$\left(\frac{\partial n}{\partial m_2}\right)_{T,p,\mu_3} = \left(\frac{\partial n}{\partial m_2}\right)_{T,p,m_3} + \left(\frac{\partial n}{\partial m_3}\right)_{T,p,m_2}\left(\frac{\partial m_3}{\partial m_2}\right)_{T,p,\mu_3} \tag{2}$$

Rearranging Eq. (2) gives the preferential interaction,

$$\left(\frac{\partial m_3}{\partial m_2}\right)_{T,p,\mu_3}$$

of component 3 with component 2:

$$\left(\frac{\partial m_3}{\partial m_2}\right)_{T,p,\mu_3} = \left\{\left(\frac{\partial n}{\partial m_2}\right)_{T,p,\mu_3} - \left(\frac{\partial n}{\partial m_2}\right)_{T,p,m_3}\right\} \bigg/ \left(\frac{\partial n}{\partial m_3}\right)_{T,p,m_2} \tag{3}$$

The preferential interaction of component 1 with component 2 is given by

[21] E. F. Casassa and H. Eisenberg, *Advan. Protein Chem.* **19**, 287 (1964).
[22] C. Strazielle and H. Benoit, *J. Chem. Phys.* **58**, 675 (1961).
[23] C. Strazielle and H. Benoit, *J. Chem. Phys.* **58**, 678 (1961).
[24] A. Vrij and J. Th. G. Overbeek, *J. Colloid Sci.* **17**, 570 (1962).
[25] M. E. Noelken and S. N. Timasheff, *J. Biol. Chem.* **242**, 5080 (1967).
[26] H. Inoue and S. N. Timasheff, *J. Amer. Chem. Soc.* **90**, 1890 (1968).
[27] S. N. Timasheff and H. Inoue, *Biochemistry* **7**, 2501 (1968).
[28] G. Scatchard, *J. Amer. Chem. Soc.* **68**, 2315 (1946).

$$\left(\frac{\partial m_1}{\partial m_2}\right)_{T,p,\mu_3} = -\frac{m_1}{m_3}\left(\frac{\partial m_3}{\partial m_2}\right)_{T,p,\mu_3} \tag{4}$$

b. Light Scattering. A rigorous development of the molecular theory of light scattering will not be presented here, but the reader is referred to the presentation of Fixman[29] and the later analysis of Kerker[30] (see also Timasheff and Townend[7] and this volume [9]). Here, we shall limit ourselves to a brief description of the basic phenomena and then proceed to the presentation of the pertinent equations of light scattering from macromolecules in multicomponent systems.

Light will pass through any medium undeflected as long as its polarizability and density are uniform. Whenever variations in these parameters take place, light is scattered in all directions, the shape of the scattering envelope depending on the size and shape of the particles and on the wavelength of the electromagnetic radiation used. If we assume that particles are immersed in a medium of polarizability α_0, the observed increase in scattering when the particles are introduced into the medium is the result of the excess polarizability of the particles over that of the medium. Since all the molecules in the solution are in constant thermal motion, there will be constant concentration and density fluctuations in a volume element, δV, if it is small enough. If this volume element is examined over a period of time, its total polarizability will be found to fluctuate. The scattering in solution is proportional to the time average of the fluctuations in the polarizability within a volume element. The total scattering from all volume elements due to concentration fluctuations of the macromolecules is

$$R_\theta = \frac{8\pi^4 V \delta V \overline{\Delta C^2}}{\lambda_0}\left(\frac{\partial \alpha}{\partial C}\right)^2_{T,p} \tag{5}$$

where R_θ is the Rayleigh ratio which is related to the turbidity by $\tau = 16\pi R_\theta/3$, C is the concentration of the macromolecule in grams per milliliter, α is the polarizability, and λ_0 is the wavelength of the light *in vacuo.*

The polarizability may be related to the refractive index, n, by the Maxwell and Lorenz equations

$$\alpha = (n^2 - 1)/4\pi \tag{6a}$$

and Eqs. (5) and (6a) combined with the thermodynamic relationship

$$\delta V \overline{\Delta C^2} = \frac{-kTC_2\bar{v}_1}{\left(\dfrac{\partial \mu_1}{\partial C_2}\right)_{T,p}} \tag{6b}$$

[29] M. Fixman, *J. Chem. Phys.* **23**, 2074 (1955).
[30] M. Kerker, "The Scattering of Light and Other Electromagnetic Radiation." Academic Press, New York, 1969.

where k is Boltzmann's constant, $\bar{v}_1$ is the partial molal volume of the solvent, and μ_1 is the chemical potential. In this way, in dilute solution, we obtain the light scattering equation of Debye,[31] shown here in three commonly used forms

$$\frac{KC_2}{R_\theta} = \frac{1}{M_2}\left[1 + \left(\frac{\partial \ln \gamma_2}{\partial C_2}\right)C_2\right] \tag{7a}$$

$$\frac{KC_2}{R_\theta} = \frac{1}{M_2}\left[1 + \frac{C_2}{RT}\frac{\partial \mu_2^{(e)}}{\partial C_2}\right] \tag{7b}$$

$$\frac{HC_2}{\Delta\tau} = \frac{KC_2}{R_\theta} = \frac{1}{M_2}[1 + 2BC_2 + 3CC_2^2 + \cdots] \tag{7c}$$

where $K = 2\pi^2 n^2 (\partial n/\partial C_2)^2_{T,p}/N_A\lambda^4$, $H = 16\pi K/3$, γ_2 is the activity coefficient of the macromolecules, $\mu_2^{(e)} = RT \ln \gamma_2$ is its excess chemical potential, B, C, etc. are the second, third, and higher virial coefficients, $\Delta\tau$ is the excess turbidity of solution over pure solvent, and N_A is Avogadro's number.

When KC_2/R_θ or $HC_2/\Delta\tau$ is plotted as a function of C_2, a curve is obtained, the intercept of which is the reciprocal of the weight average molecular weight and whose slope is $2B$, the second virial coefficient. The experimental quantities which are required are $(\partial n/\partial C_2)$, the refractive index increment of the macromolecule, and the turbidities of the solvent and of solutions of different concentrations of the macromolecule.

Equations (7a) through (7c) deal with two-component systems. It is possible to write out a generalized equation for multicomponent systems.[7] In such systems, preferential interactions between the macromolecule and solvent components necessarily take place; the magnitude of their effect on light scattering depends on the particular system. We shall now extend Eq. 7 to three-component systems. Explicitly, when the scattering intensities of protein solutions are measured in a water–nonaqueous solvent mixture, keeping the molality of the nonaqueous solvent identical in the solvent and in the solution, multicomponent theory results in the equation

$$H'\left(\frac{\partial n}{\partial C_2}\right)^2_{T,p,m_3}\frac{C_2}{\Delta\tau} = \frac{1}{(1+D)^2}\left(\frac{1}{M_2} + 2B^0C_2\right) + O(C_2)^2 \tag{8a}$$

where

$$H' = \frac{32\pi^3 n^2}{3N_A\lambda^4}$$

$$D = \frac{(\partial n/\partial m_3)_{T,p,m_2}}{(\partial n/\partial m_2)_{T,p,m_3}}\left(\frac{\partial m_3}{\partial m_2}\right)_{T,p,\mu_3} \tag{8b}$$

[31] P. Debye, J. Phys. Colloid Chem. **51**, 18 (1947).

B^0 is an apparent second virial coefficient involving interaction constants between solute and solvent and other thermodynamic parameters.[7]

In such systems, a plot of $H'(C_2/\Delta\tau)$ as a function of concentration extrapolates not to the true molecular weight of the macromolecule, but to the product of the molecular weight and a function of preferential interaction with solvent components. The deviation of this extrapolation, $[(1 + D)^2 M_2]^{-1}$ from the reciprocal of the true molecular weight, M_2^{-1}, is a measure of the extent of this interaction. Using Eq. (8b), the amount of preferential interaction, at times also referred to loosely as "preferential binding," of component 3 with component 2 can be calculated.

Protein associations frequently occur in multicomponent systems, altering the weight-average molecular weight from the true molecular weight. A measurement of the true weight-average molecular weight in the given solvent system is, therefore, required. This can be accomplished by performing the light scattering and differential refractometry experiments under conditions at which the chemical potentials of the solvent components are identical in the solution and the reference solvent. This can be accomplished by establishing dialysis equilibrium first, and then by using the dialyzed solution and the dialyzate in the experiments. Under such conditions the system reduces to a pseudo-two component one and the intercept of the light scattering plot gives the true weight-average molecular weight.[16,17,19-21,26] Thus, introducing Eq. (2) into Eqs. (8) results in[7]

$$\frac{H'C_2}{\Delta\tau}\left(\frac{\partial n}{\partial C_2}\right)^2_{T,\mu_1,\mu_3} = \frac{1}{M_2}[1 + 2B'C_2] \qquad (9)$$

where B' is the correspondingly complex second virial coefficient.[7]

From the above presentation, it follows, therefore, that in multicomponent systems, the light scattering and differential refractometry experiments should be carried out in two ways, namely with and without prior dialysis against the reference solvent. It should be emphasized that each protein dilution must be dialyzed individually for these experiments.

Preferential interaction can be measured by differential refractometry alone; it is reasonable to ask, then, why should these experiments be done by light scattering? The advantages of light scattering are two: first, the measurements after dialysis afford a control on changes in molecular weight; second, in light scattering, the preferential interaction is obtained from the experiments without dialysis, i.e., from the deviation of the apparent molecular weight from M_2 in a "standard" light-scattering experiment; thus, no procedures involving membranes are used, with possible adsorption effects. The dialysis step in light scattering is needed only to control the molecular weight.

2. Protein–Protein Interactions

While interactions between protein molecules may be of a great variety, involving various attractive and repulsive forces, we shall limit our discussion to the actual formation of complexes. Referring to Eq. 7, the general rule is that attractive forces make negative contributions to

$$\left(\frac{\partial \mu_2^{(e)}}{\partial C_2}\right)_{T,p}$$

while repulsive forces make positive contributions to this term. Thus, in the case of attractions, apparent molecular weights at finite concentrations are high, while in the case of repulsions, they are low. Since in the case of charged macromolecules, such as enzymes, electrostatic repulsion may result in strong deviations from the true molecular weight at finite concentration, it is preferable to extend the measurements to high dilution and to work in the presence of a reasonable amount of supporting electrolyte (at an ionic strength of 0.1–0.2).[32]

If multicomponent effects are neglected, or the solutions are first dialyzed against the reference solvent, we may examine these systems in terms of Eqs. (7). Three cases will be examined: a limited association, progressive polymerization and a heterologous association.

   a. *Limited Association.* The stoichiometry is

$$nP \overset{k}{\rightleftharpoons} P_n$$

In this case, the contribution to the second term of Eq. (7) is[32a]

$$C_2\left(\frac{\partial \mu_2^{(e)}}{\partial C_2}\right)_{T,p} = -\frac{Kn(n-1)f_2C_2^{n-1}}{M_m^{n-1}f_2 + Kn^2f_2^nC_2^{n-1}} = \left(\frac{M_m - \bar{M}_w}{\bar{M}_w}\right) \quad (10)$$

where n is the degree of association, M_m is the monomer molecular weight, f_2 is the fraction of protein not aggregated at concentration C_2, and K is the equilibrium constant given by

$$K = \frac{(1 - f_2)M_m^{n-1}}{nf_2^nC_2^{n-1}} \quad (11)$$

The light-scattering equation for this case reduces to

$$\frac{HC_2}{\Delta \tau} = \frac{1}{\bar{M}_w} + \frac{2B_0C_2}{M_m} \quad (12)$$

[32] S. N. Timasheff, Polyelectrolyte properties of globular proteins. *In* "Biological Macromolecules" (A. Veis, ed.), Vol. III. Dekker, New York, 1967.
[32a] R. Townend and S. N. Timasheff, *J. Amer. Chem. Soc.* **82**, 3168 (1960).

where $2B_0$ is the second virial coefficient of the monomer. Thus, if we set $2B_0 = 0$ or account for it in a specific way, the light scattering at any concentration gives the weight-average molecular weight of the equilibrium solution. Using this value of $\bar{M}_w$ in Eqs. (10) and (11), it becomes possible to calculate the equilibrium constant, if the degree of association n is known. Frequently n can be obtained from the asymptotic value of $\bar{M}_w$ at high protein concentration, when the aggregated species is predominant.

b. *Progressive Association.* The stoichiometry is

$$2A \underset{}{\overset{k_2}{\rightleftharpoons}} A_2$$

$$A_2 + A \overset{k_3}{\rightleftharpoons} A_3$$

$$A_3 + A \overset{k_4}{\rightleftharpoons} A_4$$

$$\vdots \qquad \vdots$$

$$A_{n-1} + A_n \overset{k_n}{\rightleftharpoons} A_n$$

This case has been treated by Steiner[33] and by Doty and Myers.[34] Using the presentation of Steiner, at any concentration of macromolecule, C,

$$\bar{M}_w = M_m \left(1 + \frac{d \ln X}{d \ln C}\right)^{-1} M_m X[1 + 4k_2(XC/M_m) + 9k_2k_3(XC/M_m)^2 + 16k_2k_3k_4(XC/M_m)^3 + \cdots] \quad (13)$$

X is defined by the integral

$$\ln X = \int_0^C \left(\frac{M_m}{\bar{M}_w} - 1\right) d \ln C \quad (14)$$

In practice, integral (14) is first evaluated by integration of the light-scattering data with respect to concentration. Then, the limiting slope of a plot of $\bar{M}_w/XM_m$ as a function of XC/M_m gives k_2; the slope of the plot of

$$\left(\frac{\bar{M}_w}{XM_m} - 4k_2XC/M_m\right)$$

as a function of $(XC/M_m)^2$ gives k_3; the values of all the higher equilibrium constants are obtained from such a sequence of plots.

c. *Heterologous Association.* This is the case found in frequently encountered interactions, such as enzyme–inhibitor complex formation, antigen-antibody reactions and catalytic subunit-regulatory subunit interactions of allosteric enzymes. The stoichiometry is

[33] R. F. Steiner, *Arch. Biochem. Biophys.* **39**, 333 (1952).
[34] P. Doty and G. E. Myers, *Discuss. Faraday Soc.* **13**, 51 (1953).

$$nA + mB \underset{}{\overset{k}{\rightleftharpoons}} A_nB_m$$

The problem may be reduced to the determination of the interaction constant $(\partial m_4/\partial m_2)_{T,p,\mu_4}$ of Eq. 8, where components 4 and 2 are proteins [Eq. (8) may be used directly, replacing subscript 3 by 4 where it appears, since in the Scatchard notation all macromolecules are numbered even; odd numbers are reserved for dialyzable components]. In order to eliminate the effect of interactions with small molecules or ions present in the system, all the protein solutions should be first dialyzed against the solvent and the dialyzate used as reference solvent. A series of light-scattering experiments must be carried out in order to obtain M_2, M_4, $(\partial\mu_2^{(e)}/\partial C_2)$ and $(\partial\mu_4^{(e)}/\partial C_4)$. The required experiments are: light scattering of a mixture of the two proteins under interacting conditions; light scattering of each protein individually in the same solvent. The latter experiments give M_2, M_4, $(\partial\mu_2^{(e)}/\partial C_2)$ and $(\partial\mu_4^{(e)}/\partial C_4)$. Combination of these values with the experimental data on the interacting mixture gives $(\partial m_4/\partial m_2)_{T,p,\mu_4}$ from which n/m and k may be calculated. The details of these calculations are available in the literature.[6,35]

B. Size and Shape of Macromolecules

1. Radius of Gyration

As pointed out in this volume [9], when the dimensions of a particle became comparable to the wavelength of the incident radiation, interference occurs between the radiation scattered from individual elements within the particle. In the case of light scattering, this effect becomes significant when the maximal dimension of a particle becomes of the order of $\lambda/10$. Since the wavelength of the incident radiation is of the order of 4000 Å, this means that particle dimensions must attain 400 Å in order to be resolved by light scattering. While most enzyme molecules are smaller than that and are not amenable to such studies of their dimensions, the geometric aspects of light scattering become quite useful in the case of large extended proteins, such as myosin[36] or large molecular aggregates.[37]

The general case of scattering from large particles has been treated by Debye[31] and extended by Guinier[38] to the measurement of radii of gyration. The principles involved are discussed in this volume [9]. The

[35] G. A. Pepe and S. J. Singer, J. Amer. Chem. Soc. 81, 3878 (1959).
[36] A. Holtzer and S. Lowey, J. Amer. Chem. Soc. 81, 1370 (1959).
[37] H. Eisenberg and E. Reisler, Biopolymers 10, 2363 (1971).
[38] A. Guinier, Ann. Phys. (Paris) [11] 12, 161 (1939).

results of these calculations give the following relation between the angular dependence of the scattering intensity, $I(h)$, and the radius of gyration, R_G

$$I(h) = K'M_2^2\left(1 - \frac{h^2}{3}R_G^2 + \cdots\right) \tag{15}$$

$$h = \frac{4\pi \sin(\theta/2)}{\lambda}$$

where M_2 is the molecular weight of the macromolecule, R_G is its radius of gyration, and θ is the angle formed between the directions of the incident and scattered rays and $\lambda = \lambda_0/n$, the wavelength measured in a medium of refractive index n. This equation, which is expressed here in the accepted notation of light scattering, is identical with Eq. (10) of this volume [9], in which the angular function is expressed by the symbol s, defined in the customary notation of small angle X-ray scattering.

If in a volume V there are J non-interacting particles, the total scattering is then the scattering from a single particle multiplied by $J = C_2N_A/M_2$, where C_2 is the concentration of the macromolecule in grams per milliliter, and N_A is Avogadro's number. Combining this with Eq. (15), remembering that, for small x, $(1 - x)^{-1} \cong 1 + x$, and introducing the proper optical and instrumental constants, gives

$$\frac{KC_2}{R_\theta} = \frac{1}{M_2}\left(1 + \frac{h^2R_G^2}{3} - \cdots\right) \equiv \frac{1}{M_2P(\theta)} \tag{16}$$

Thus, if the particle is sufficiently large, its scattering intensity will vary with the angle of observation. In Eq. (7), it has been shown that the scattering intensity is also a function of concentration, expressed through the second virial coefficient $2B$.

a. Zimm Plot. It is clear, therefore, that the intensity of the light scattered from solutions of large molecules depends both on the solute concentration and on the scattering angle. In order to determine the second virial coefficient B, the scattering at each concentration must be extrapolated to zero angle, while the determination of R_G demands the extrapolation of the angular variation of scattering to zero concentration at each angle. Calculation of the molecular weight requires both extrapolations. These two requirements may be expressed in a single equation, which combines Eqs. (7) and (16)

$$\frac{KC_2}{R_\theta} = \frac{1}{M_2}\left(1 + 2BC_2 + \frac{16\pi^2 \sin^2(\theta/2)}{3}\frac{R_G^2}{\lambda^2}\right) \tag{17}$$

This is the basis of the well known Zimm plot.[39] In practice, KC_2/R_θ

[39] B. H. Zimm, *J. Chem. Phys.* **16**, 1099 (1948).

is plotted against $\sin^2(\theta/2) + kC_2$, where k is an arbitrary constant chosen to space the data conveniently and to facilitate the extrapolations to zero angle and zero concentration. The weight-average molecular weight, reciprocal particle-scattering factor $[P^{-1}(\theta)]$, and second virial coefficient, $2B$, are then determined by the following extrapolations:

$$\frac{1}{M_2} = \left(\frac{KC_2}{R_\theta}\right)_{\substack{\theta=0 \\ C_2=0}} \tag{18}$$

$$P^{-1}(\theta) = \left(\frac{KC_2}{R_\theta}\right)_{C=0} M_2 \tag{19}$$

$$2B = \left[\frac{d\left(\frac{KC_2}{R_\theta}\right)}{dC_2}\right]_{\theta=0} \tag{20}$$

The radius of gyration, R_G, may be obtained from $P^{-1}(\theta)$ by

$$R_G = \frac{\lambda}{4\pi}\left(3M_2 \left(\frac{d\left(\frac{KC_2}{R_\theta}\right)}{d\sin^2\theta/2}\right)_{C_2=0}\right)^{1/2} \tag{21}$$

b. Yang Plot. An alternate way of plotting Eq. (17) has been described by Yang,[40] who plots $KC_2/[R_\theta \sin^2(\theta/2)]$ as a function of $1/\sin^2(\theta/2)$, with concentration, C_2, as a parameter. Such a plot gives a series of straight lines with a common intercept. Extrapolation of the data at each angle to zero concentration gives the weight-average molecular weight from the reciprocal of the slope. The radius of gyration is obtained from the common intercept on the ordinate, i.e., by extrapolation to $(\sin^2(\theta/2))^{-1} = 0$. The Yang plot affords a very good check of the Zimm plot, when the precision of the extrapolated values is in doubt. In particular it helps in detecting overlooked experimental errors at low angles, since in the Yang plot these points are well separated and far away from the origin. The Yang method requires a second plot for the determination of the second virial coefficient, $2B$. In this, KC_2/R_θ is plotted as a function of $1/C_2$, and the data at each concentration are extrapolated to zero angle. The common intercept gives the value of $2B$; and the slope at zero angle, the reciprocal of the molecular weight. Typical Zimm and Yang plots are shown on Fig. 1.

2. Molecular Shape

a. Direct Calculation. Using the general Debye relation[31] (see Eq. (9) of this volume [9]), it is possible to calculate the function $P^{-1}(\theta)$

[40] J. T. Yang, *J. Polym. Sci.* **26**, 305 (1957).

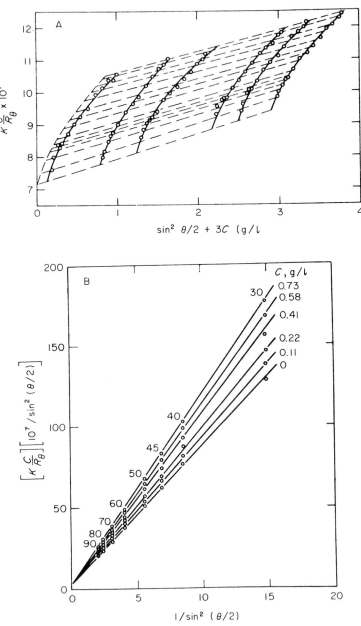

FIG. 1. Light scattering of ascites tumor cell H (ribosomal) RNA. (A) Zimm plot. (B) Yang plot. From M. J. Kronman, S. N. Timasheff, J. S. Colter, and R. A. Brown, *Biochim. Biophys. Acta* **40**, 410 (1960).

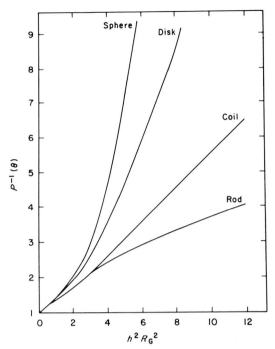

Fig. 2. The reciprocal particle scattering factor, $P^{-1}(\theta)$, of a sphere, disk, random coil, and rod. From E. P. Geiduschek and A. Holtzer, *Advan. Biol. Med. Phys.* **6**, 431 (1958).

for molecules of any given shapes, using characteristic geometric parameters. Geiduschek and Holtzer[2] have compiled the function $P^{-1}(\theta)$ for molecules of various shapes. The dependences of $P^{-1}(\theta)$ on the product $h^2 R_G^2$ for spheres, disks, random coils, and rods are shown in Fig. 2, and the expressions for the scattering functions of various molecular shapes are summarized in Table I. In general, the $P^{-1}(\theta)$ versus $h^2 R_G^2$ curve of an asymmetric particle will lie below that of a more compact (spherical) particle. In conjunction with this analysis, Geiduschek and Holtzer[2] have also presented the variation of $P^{-1}(\theta)$ with changes in the stiffness of coiled molecules. This is shown in Fig. 3. For very large particles, the number-average molecular weight, $\bar{M}_n$, can be obtained from light scattering. Benoit[41] and Yang[40] have shown that extrapolation from the asymptotic region of the high-angle range gives for random coils

[41] H. Benoit, *J. Polym. Sci.* **11**, 507 (1953).

TABLE I
SCATTERING FUNCTIONS FOR VARIOUS STRUCTURES

Structure	$P(\theta)$	Radius of gyration	$\langle R_G^2\rangle_{av}$
Sphere	$\left[\dfrac{3}{x^3}(\sin x - x\cos x)\right]^2$ $\left\{x = \dfrac{hD}{2}; D = \text{diameter of sphere}\right.$	$\left(\dfrac{3}{5}\right)^{1/2}\dfrac{D}{2}$	$\left(\dfrac{1}{\bar{M}_w}\Sigma C_i M_i^{5/3}\right)^{3/2}$
Thin rod	$\dfrac{1}{x}\displaystyle\int_0^{2x}\dfrac{\sin w}{w}\,dw - \left(\dfrac{\sin x}{x}\right)^2$ $\left\{x = \dfrac{hL}{2}; L = \text{length of rod}\right.$	$L/\sqrt{6}$	$(z+1)$-average
Random coil	$\dfrac{2}{x^2}[e^{-x}+x-1];\ x = \dfrac{h^2\langle r^2\rangle_{av}}{6}$ $\{\langle r^2\rangle_{av}^{1/2} = \text{root-mean-square end-to-end distance}\}$	$(\langle r^2\rangle_{av}/6)^{1/2}$	z-average
Wormlike chain	$e^{v/3A}\left\{F\left[\dfrac{vx}{3A}\right] - \dfrac{v}{3A}F\left[\left(\dfrac{v}{3A}+1\right)x\right]\right.$ $\left.+ \dfrac{v}{29A}F\left[\left(\dfrac{v}{3A}+2\right)x\right] - \cdots\right\}$ $v = \dfrac{h^2AL^2}{x^2};\ A = \dfrac{x}{3} - 1 + F(x)$ $= \left[\dfrac{x}{3} - 1 + \dfrac{\langle r^2\rangle_{av}}{L^2}\right]$ $F(y) = \dfrac{2}{y^2}(y - 1 + e^{-y})$ $\left\{\begin{array}{l}L = \text{stretched out chain length}\\ \langle r^2\rangle_{av}^{1/2} = \text{root-mean-square end-}\\ \text{to-end distance}\\ x = \text{stiffness parameter}\\ x = \dfrac{L}{a}\\ a = \text{persistence length}\end{array}\right.$	$\dfrac{L}{x}\sqrt{A}$	z-average
Thin disk	$\dfrac{2}{x^2}\left[1 - \dfrac{1}{x}J_1(2x)\right]$ $\left\{x = hR; R = \text{radius of disk}; J_1 = \text{Bessel function of order 1}\right.$	$R/\sqrt{2}$	z-average

$$\left(\frac{KC_2}{R_\theta}\right)_{\substack{C_2=0 \\ hR_G \to \infty}} = \frac{1}{2}\left[\frac{1}{\bar{M}_n} + h^2\frac{b^2}{6}\right] \tag{22}$$

where b is the length of a statistical chain segment. For rods of uniform cross section[42]

$$\left(\frac{KC_2}{R_\theta}\right)_{\substack{C_2=0 \\ hR_G \to \infty}} = \frac{1}{\bar{M}_n}\left[\frac{2}{\pi^2} + h\bar{L}_n\right] \tag{23}$$

where $\bar{L}_n$ is the number-average length of the rods. For rods of nonuniform cross section,[43] this is

$$\left(\frac{KC_2}{R_\theta}\right)_{\substack{C_2=0 \\ hR_G \to \infty}} = \frac{2}{\pi^2}\frac{\Sigma M_i w_i L_i^2}{(\Sigma M_i w_i/L_i)^2} + \frac{h}{\Sigma M_i w_i/L_i} \tag{24}$$

where $w_i = N_i M_i/N$, and L_i is the length of the rod.

b. *Dissymmetry Method.* The construction of a Zimm or a Yang plot requires that measurements be carried out at ten to fifteen angles for at least four concentrations. The labor involved is considerable. In

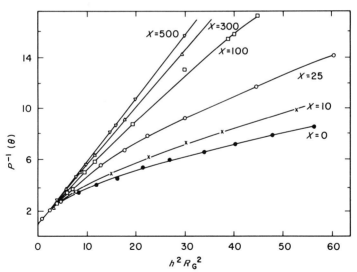

Fig. 3. The reciprocal particle scattering factors of coils with persistence, for various values of the stiffness parameter, x. (x is the ratio of the stretched-out length, L, of the molecule, to the persistence length a, or the number of persistence units in the chain.) From E. P. Geiduschek and A. Holtzer, *Advan. Biol. Med. Phys.* **6**, 431 (1958).

[42] A. Holtzer, *J. Polym. Sci.* **17**, 432 (1955).
[43] E. F. Casassa, *J. Chem. Phys.* **23**, 596 (1955).

cases where the functional form of $P(\theta)$ for the particles in solution is known and is singled-valued, or where the molecular size is sufficiently small so that particles of various shapes have essentially identical dispersions, the amount of labor can be reduced by using an alternative method which requires the measurement of KC_2/R_θ at only three angles as a function of concentration. This is known as the dissymmetry method.[2] The scattering intensity is measured at angles of 45°, 90°, and 135° at a series of concentrations, and values of KC_2/R_{90} and the dissymmetry, Z, are calculated, where $Z = R_{45}/R_{135}$. A plot of $1/(Z - 1)$ versus C_2 is extrapolated linearly to $C_2 = 0$, and $[Z]$ is obtained from the intercept, $1/([Z] - 1)$, where $[Z]$ is readily related to Z by

$$\frac{1}{Z - 1} = \frac{1}{[Z] - 1} + \frac{2BC_2MP(45)}{[Z] - 1}$$

From the value of $[Z]$ and the standard graph of R_G/λ versus $[Z]$ for the appropriate model, R_G/λ is obtained and hence R_G. To determine molecular weight, KC_2/R_{90} is plotted against C_2 and linearly extrapolated to $C_2 = 0$. The intercept is $1/M_2P(90)$ and can be converted to molecular weight by using another standard graph of $1/P(90)$ versus $[Z]$ for the same model.

The disadvantage in using this method is its requirement that a model of known $P(\theta)$ be assumed. In the usual experimental situation, the particle shape is not known, nor is it known to what extent the system is polydisperse. This renders uncertain the exact functional form of $P(\theta)$ which must be used and will frequently result in incorrect molecular weight and size. However, in cases where the molecular size is known to be small enough so that $P(\theta)$ is essentially the same for all shapes in the measurable angular range, the dissymmetry method may be employed with confidence. In the usual range of angles and wavelengths used, this corresponds to dissymmetries less than about 1.4.

c. Conclusion. In conclusion, angular measurements in light scattering usually result only in a knowledge of the radius of gyration. However, in those cases in which the molecular shape is already known or can be deduced from comparison of $P^{-1}(\theta)$ with theoretical curves for various structures, R_G can be transformed into a more meaningful shape parameter, such as the length of a rod, the radius of a sphere, the root-mean-square end-to-end distance of a coil, etc.

III. Experimental

A. Instrumentation

The methods for preparing solutions for light scattering and differential refractometry are essentially the same. For light scattering it is necessary to subject the carefully prepared solutions to special filtration or ultracentrifugation treatments in order to remove glass dust and other sources of extraneous turbidity from the system. In this section, we will describe in detail the methods of preparation of solutions, in particular for multicomponent systems and the methods, instrumentation, etc., involved in general light scattering measurements.

1. Light Scattering

Compared to other instruments used to measure molecular weights and solute–solvent interactions, light scattering instruments and differential refractometers are rather simple.

A typical light scattering photometer consists of a light source, collimation system to obtain a parallel beam of light, a transparent cell for the material under study, a collimation system for receiving the scattered light, and a means of comparing the intensities of the incident and scattered beams. The measurement of Rayleigh's ratio involves the comparison of the intensities of the incident and the scattered beams, the latter in protein systems being many times weaker than the incident beam. Stacey[5,44] and Kratohvil[45,46] have prepared excellent reviews of a variety of light scattering instrumentation, and the reader should refer to these authors for a comprehensive discussion of the subject. In this manuscript we will discuss several typical light scattering and differential refractometry instruments, in particular those used in our laboratory.

a. The Brice Photometer. Figure 4 represents the essential features of the Brice-Phoenix Universal Light Scattering Photometer, 2000 series.[47,47a] This instrument has the flexibility necessary to measure fluorescence, reflectance, and depolarization, as well as light scattering. These various techniques can be used with a minimum amount of rearrangement of the components of the instrument. Here we shall discuss the instrument only with respect to turbidimetric measurements.

The Brice photometer[47b] contains as a light source a high-pressure mercury vapor lamp which is stabilized by a special compensating ballast transformer and constant-current lamp regulating unit. Standardized color filters are used to isolate the mercury lines at 546 and 436

[44] K. A. Stacey, "Light Scattering in Physical Chemistry." Butterworth, London, 1956.

[45] J. P. Kratohvil, *Anal. Chem.* **36**, 458R (1964).

[46] J. P. Kratohvil, *Anal. Chem.* **38**, 517R (1966).

[47] B. A. Brice, M. Halwer, and R. Speiser, *J. Opt. Soc. Amer.* **40**, 768 (1950).

[47a] Mention of companies or products is for the convenience of the reader and does not constitute an endorsement by the U.S. Department of Agriculture.

[47b] This instrument is manufactured by the Phoenix Precision Instruments Co., a division of VirTis Co. Inc. Gardner, New York.

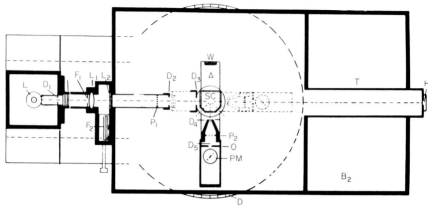

Fɪɢ. 4. Diagrammatic sketch of Brice photometer. L, mercury lamp; F_1, monochromatic filters; S, camera shutter; F_2, four neutral filters mounted on a sliding carriage; L_1, achromatic lens ($f = 122$ mm); L_2 cylindrical lens ($f = 200$ mm) with axis horizontal. SC, scattering cell (40×40 mm) on fixed table; D, graduated disk attached to rotatable arm A, carrying removable working standard, W, and the receiving system (shown in $90°$ position, $0°$ position dotted); limiting rectangular diaphragms in primary beam, D_1 (13 mm wide $\times$ 11 mm high), D_2 (12×15 mm); limiting rectangular diaphragms in receiving system; D_4 (3.1×6.4 mm) and D_5 (7.4×22 mm); O, opal glass depolarizing diffusor close to the multiplier phototube PM; P_1 and P_2 positions of demountable polarizer and analyzer; T, blackened removable tube serving as a light trap and as a means of aligning parts; H, covered peephole; B_1, scattering compartment; B_2, power supply compartment. From B. A. Brice, M. Halwer, and R. Speiser, *J. Opt. Soc. Amer.* **40**, 768 (1950).

nm in sufficient purity and intensity. As in most light scattering equipment, the scattering sample is fixed and a photomultiplier moves around it on a turntable. A variety of side-window photomultiplier tubes can be used in this instrument, such as selected 931-A, 1P21, 1P22, and 1P28. Since the measurements require the instrument to read the transmitted and scattered lights in sequence with the same photomultiplier tube, a high level of stability is required both of the light output and of the high voltage for the photomultiplier tube, as significant fluctuations cannot be permitted between the two readings.

In most laboratories the usual method of readout from the photomultiplier is to record the fluctuations in the current of the photomultiplier tube on a sensitive galvanometer.[5,44,45,46] The limiting factor in such measurements is the precision with which the data can be read. These limitations are due mainly to noise created by thermal fluctuations in electronic components, high voltages on the photomultiplier tube and momentary fluctuations in light output. This noise creates a wander of one galvanometer division, which can introduce an error of 5% when the reading is only 20 as can be the case in dilute aqueous solutions of proteins. This error plus the compounded errors of taking readings of 0° and 90° for both sample and solvent may result in even greater uncertainty. In our system, a significant change in the readout device has been made.[48] The current from the PM tube is passed through a 14 kΩ resistor. The voltage appearing across the output terminals is fed into a Hewlett-Packard Model 412A vacuum tube voltmeter (VTVM) which contains a highly stable dc amplifier, used to drive the meter movement. This meter serves the same purpose as the galvanometer. The output of the VTVM is applied to the input of a voltage to frequency converter (Dymec Model 2210, manufactured by Hewlett-Packard Corp.), and the ac output from this is counted for a 1.000- or 10.00-second period by an electronic counter with a crystal-controlled time base (Hewlett-Packard Model 5512A). The counts are displayed on "Nixie" tubes. Counting for a fixed period of time averages out most random noise. The photomultiplier voltage is kept between 1200 and 1300 V (using an external transistorized power supply), and the output reading is adjusted with the neutral filters to give a $\theta = 0°$ meter reading of between 0.06 and 0.08 V. Under these conditions, the anode current from the 1P21 is of the order of 0.4–0.6 μA and four significant figures, with variation in the last one only, can be obtained on successive 1-second counts.

b. Wippler-Scheibling Instrument. An instrument which is widely used and is available commercially is that of Wippler and Scheibling.[49,50] This instrument, shown in Fig. 5, has a distinct advantage in angular measurements. The cells are fabricated of thin glass tubing with a hemispherical bottom, and when in use they are immersed in a thermostated bath of benzene. Furthermore, the wall of the benzene bath behind the back (the surface of the cell opposite the light source) of the cell has the shape of a horn. With a black background, the horn-shaped structure absorbs the scattered light transmitted through the back surface of the cell. The combination of the submersion of the cell in a benzene bath and the special designs minimize strong reflections and refractive index corrections (see below, "Reflection correction"). The photomultiplier, analyzer, and prism move as a unit around the cell for angular measurements.

c. Low-Angle Instrument. Equations (16) and (17) are valid for

[48] J. A. Connelly, unpublished experiments, 1964.
[49] C. Wippler and G. Scheibling, *J. Chim. Phys.* **51**, 201 (1954).
[50] This instrument is manufactured by the Société Française d'Instruments de Contrôle et d'Analyses (SOFICA), France.

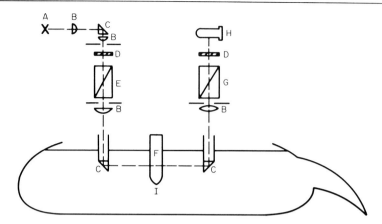

Fig. 5. Schematic of a SOFICA light scattering photometer. A, light source; B, lens; C, prism; D, filter; E, polarizer; F, light scattering cell; G, analyzer; H, photomultiplier; I, benzene bath; J, horn-shaped part of benzene bath for light trap.

only small values of $h^2R_G^2$. Therefore, when the ratio R_G^2/λ^2 is large, higher terms in the expansion become significant, rendering it necessary to extend the experiments to progressively lower angles in order to minimize the product $h^2R_G^2$ and reach the angular region in which the plot is linear in $\sin^2(\theta/2)$. In most commercially available photometers the smallest angle of observation is about 25°–30°. In the case of molecules of very high axial ratio, such as DNA, serious errors are introduced by the linear extrapolation of light-scattering data from 30°.[4,51] For this reason, special attempts have been made to extend light scattering measurements to lower angles.[52]

Harpst, Krasna, and Zimm[53] have constructed a low-angle light scattering photometer capable of measurements at angles as low as 10°. Figure 6 shows schematically the optical system employed. The major features of the instrument, which render it capable of low-angle measurements, are the use of an intense light source (Osram HBO-200 or high-pressure mercury arc lamp) and of a long, rectangular sample cell in which the scattered light can be measured at low angles with no interference from the incident beam. The rapid changes in the arc position of the mercury lamp result in large intensity changes in the main beam. In order to compensate for such fluctuations, it is essential to monitor all parts of the beam by placing a beam splitter into the con-

[51] C. Sadron and J. Pouyet, *Proc. Int. Congr. Biochem. 4th 1958*, Vol. 9, p. 52. Pergamon, Oxford, 1959.

[52] S. Katz, *Nature (London)* **191**, 280 (1961).

[53] J. A. Harpst, A. I. Krasna, and B. H. Zimm, *Biopolymers* **6**, 585 (1968).

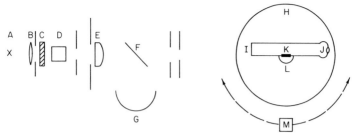

FIG. 6. Schematic of the low-angle light scattering photometer and cell: (A) Osram lamp; (B) condensing lens; (C) interference filter; (D) vertical polarizer; (E) collimating lens; (F) beam splitter of glass cover slips; (G) monitor photo-tube; (H) sample table; (I) incident light beam; (J) horn-shaped light trap of rectangular glass cell; (K) rectangular viewing slit located at center of sample table and at center of circular window; (L) circular viewing window; (M) photo-multiplier-containing detector. From J. A. Harpst, A. I. Krasna, and B. H. Zimm, *Biopolymers* 6, 585 (1968).

verging beam; this reflects a constant fraction of all parts of the beam into the monitoring phototube. The scattered light is monitored by the photomultiplier-containing detector, which is mounted on a rod revolving around the central support of the cell assembly. The outer cell walls are painted with an opaque solvent-resistant black paint, except for a viewing slit for observing the scattered light. The cell is cemented to a brass plate in such a way that the slit is at the center of rotation of the photometer. A semicircular piece of Pyrex glass tubing is cemented to the outside of the cell, rendering the center of the viewing slit equidistant from all portions of the circular window. A liquid of the same refractive index as the solution within the cell is pipetted into the circular viewing chamber to eliminate multiple reflections and the need for refractive index corrections.

Using this photometer, Harpst, Krasna, and Zimm[53] have measured the molecular weights of calf-thymus and T7 DNA at both high (above 30°) and low (below 30°) angles. The low-angle light scattering molecular weights were found to be about twice as large as those obtained from the extrapolation of data taken above 30°. These molecular weights are now in agreement with estimates obtained by other techniques.[54,55]

Aughey and Baum[56] have also constructed a light scattering photometer which is capable to measure the light scattered by large particles, e.g., lycopodium powder at angles between 0.05° and 140° from the

[54] J. A. Harpst, A. I. Krasna, and B. H. Zimm, *Biopolymers* 6, 595 (1968).
[55] D. M. Crothers and B. H. Zimm, *J. Mol. Biol.* 12, 525 (1965).
[56] W. H. Aughey and F. J. Baum, *J. Opt. Soc. Amer.* 44, 833 (1954).

incident beam. This specialized instrument, however, seems to lack general availability and applicability to the study of biomacromolecules.

2. Light Scattering Cells

Cells can be essentially of two types: those used in light scattering measurements at 0° and 90°, and those for angular measurements. Figure 7 is a photograph of the cells used in our laboratory for light scattering measurements.

a. 90° Cells. Light scattering measurements, using the Brice photometer, result in absolute turbidities if optical cells 30 or 40 mm² are used.[57] The light beam impinging on the cell is 1.2 cm high and 1.2 cm wide. The capacity of these cells is at least 25 ml, requiring several hundred milligrams of low molecular weight protein. As a result, Dintzis[58] developed the small cell technique, which requires less than 2.2 ml of solution to fill above the light beam. This cell which is essentially a regular 1-cm cuvette is shown on the left of Fig. 7. The closely fitting Teflon caps, shown in the figure, are constructed in such manner that the cell contents can be mixed by inversion. The cell is placed in a special holder (Fig. 7) which centers it properly on the sample turret. A special set of narrower slits accompanies these cells. In order to measure absolute turbidity, it is necessary to calibrate each 1-cm cell against the standard 3-cm cell. Such a calibration is accomplished by measuring in succession in the two cells the light intensity due to fluo-

Fig. 7. Cells and holders for the Brice photometer.

[57] B. A. Brice, M. Halwer, and R. Speiser, *J. Opt. Soc. Amer.* **40**, 768 (1950).
[58] S. N. Timasheff, H. M. Dintzis, J. G. Kirkwood, and B. D. Coleman, *J. Amer. Chem. Soc.* **79**, 782 (1957).

rescence and scattering at 90°, I, from fluorescein solutions of several concentrations (5×10^{-7} to 5×10^{-5} g/liter). This results in a proportionality constant, k, for each small cell:

$$k = \frac{I_{(3\text{-cm cell})}}{I_{(1\text{-cm cell})}} \tag{25}$$

During the calibration, the large cells are used with 1.2-cm slits, while the narrow slits are employed with the 1-cm cells. The use of the 1-cm cells requires that these be placed into the photometer in a reproducible manner relative to the light beam. This is accomplished by slight adjustments in the cell position until its reflection falls symmetrically on slit D_3 of the photometer (Fig. 4).

b. *Angular Cells.* The most common and commercially available angular cells are designed with a cylindrical viewing region and flat faces for the entrance and exit windows of the incident beam.

i. WITNAUER-SHERR CELL.[59] The cell is designed specifically for the Brice photometer. The body of the cell consists of standard-wall Pyrex tubing with a pair of lengthwise parallel windows, with flat inner and outer faces. One half of the tube is frosted over its full inside length to reduce multiple reflections of stray light. Some investigators prefer to have the rear of the cell coated with a black paint.[60] Painting either the outside or inside surface is not satisfactory due to poor reproducibility, incomplete absorption of light and reflection at the glass–paint interface. The frosted side, furthermore, seems to introduce glass particles into the system rendering difficult accurate measurements at low angles. This cell requires at least 15 ml of solution, setting as a requirement the availability of large quantities of samples. A nonfrosted cylindrical cell is also available.

ii. HELLER-WITECZEK CELL.[61] The special feature of this cell is the presence of two Rayleigh horns which trap all stray light completely. One Rayleigh horn is situated at the exit end of the scattering cell. The backward reflection is eliminated by another horn with a very wide and curving aperture covering the angles $\theta = 30°$–$150°$. The wide-angle horn is made from dull-black epoxy resin reinforced with fiberglass.

With these two types of cells, samples of the macromolecule solution have to be clarified and then transferred into the cells. It is very easy to introduce dust particles during the transfer process, even when sample solutions are filtered directly into these cells.

[59] L. P. Witnauer and H. J. Sherr, *Rev. Sci. Instrum.* **23**, 99 (1952).
[60] Y. Tomimatsu and K. J. Palmer, *J. Phys. Chem.* **67**, 1720 (1963).
[61] W. Heller and J. Witeczek, *J. Phys. Chem.* **74**, 4241 (1970).

iii. KRONMAN-TIMASHEFF CELL.[62] In an effort to minimize the volume of solution required and to eliminate sample transfer, Kronman and Timasheff[62] designed a filter-cell combination which enables the solution under study to be cleaned and measured in the same vessel. This cell and the cell mounted in a special bracket designed to hold it in the Brice photometer are shown on the right side of Fig. 7. These cells are generally usable down to 20° and require 8–10 ml of solution before filtration. The great advantages of these cells are the speed and ease with which solutions can be clarified. The cell consists of two cylindrical compartments joined by a connecting tube at the bottom. The optical side is essentially a Witnauer-Sherr cell, the bottom of which is a sintered-glass filter pad. The solutions are introduced into the non-optical side and forced to the "clean" optical side up through the filter disk by applying a small pressure of nitrogen gas. In this way, all transfers of the clean solution are eliminated and no new air solution interfaces are formed during filtration, minimizing the chance of introducing dust particles. One filtration is generally sufficient to provide a dust-free solution for angular measurements, as no sparkles are detected at low angles with a high intensity light source. In addition, there is no loss of sample during clarification, as is found sometimes with other methods of solution clarification, e.g., Millipore filtration and centrifugation. It is known that Millipore filters (type GS and HA) retain biological macromolecules.[54,63] Therefore at low concentrations there may be significant losses of sample along with the removal of dust particles. Despite these advantages, the Kronman-Timasheff cell is not practical for measurements on large macromolecules, since force-filtration through an ultrafine sintered-glass disk will most likely shear rigid DNA molecules and the comparatively high limiting angle of 20° would prevent these cells from being used to measure the molecular weight of very large particles.[54] These cells are most useful, therefore, in the middle range of molecular dimensions, such as found in ribosomal RNA.[64]

3. Reflection Correction

The absolute calibration of a light scattering photometer involves considerations of suitable corrections for reflection effects.[65-67] The geom-

[62] M. J. Kronman and S. N. Timasheff, *J. Polym. Sci.* 40, 573 (1959).
[63] J. C. Lee, Doctoral thesis, Case Western Reserve University, 1971.
[64] M. J. Kronman, S. N. Timasheff, J. S. Colter, and R. A. Brown, *Biochim. Biophys. Acta* 40, 410 (1960).
[65] B. A. Brice, M. Halwer, and R. Speiser, *J. Opt. Soc. Amer.* 40, 768 (1950).
[66] C. I. Carr, Jr. and B. H. Zimm, *J. Chem. Phys.* 18, 1616 (1950).
[67] Y. Tomimatsu and K. J. Palmer, *J. Polym. Sci.* 54, 527 (1961).

etry of both the incident and scattered beams must be corrected for the fact that light is passing through interfaces at which the refractive index changes. In the case of the incident beam, the order is air-to-glass-to-cell liquid; it is reversed for the scattered beam. These changes in refractive index cause refraction of the rays and reflections from surfaces. As a result, different volumes of liquid are illuminated and viewed by the detector than would be the case if these interfaces did not exist or were "eliminated" by immersing the cell and detector in a liquid with a refractive index approaching that of the cell-wall glass. It is for this reason that the instruments of Wippler and Scheibling[49] and of Harpst, Krasna, and Zimm[53] are designed with the cell immersed in a bath of benzene or a liquid with the same refractive index as the solution within the cell. However, for a Brice-type of photometer in which the cell is sitting in the air and the detector rotates in the air around it, it is essential to correct for these refraction and reflection effects. Tomimatsu and Palmer[67,68] and Kratohvil[69] have derived expressions for reflection corrections, and Tomimatsu and Palmer[67,68] have reviewed the literature on the problem of reflection effects in light scattering measurements. For a detailed discussion and rigorous derivation of the equations, the reader is referred to the above references. The equations for corrected intensities developed by various investigators will be considered now.

a. *Tomimatsu-Palmer Method.* According to these investigators the Rayleigh ratio, R_θ, can be calculated from the observed scattering ratios, G_θ/G_0, by means of the following equation

$$R_\theta = \frac{TDan^2(R_w/R_c)}{1.04\pi h} \left(\frac{\gamma}{\gamma'}\right)\left(\frac{\sin\theta}{1 + \cos^2\theta}\right)\left(\frac{1}{(1 - R)^2(1 - 4R^2)}\right) \quad (26)$$

$$\left\{\left[\left(\frac{G_\theta}{G_0} F\right)_{\text{solution}} - \left(\frac{G_\theta}{G_0}\right)_{\text{solvent}}\right] - 2R\left[\left(\frac{G_{180-\theta}}{G_0} F\right)_{\text{solution}} - \left(\frac{G_{180-\theta}}{G_0} F\right)_{\text{solvent}}\right]\right\}$$

where γ/γ' is the calibration factor relating the geometry of the narrow beam and cylindrical cell to that of the standard beam and standard cell, h is the height of the beam in the cell, G_θ is the intensity of light observed at an angle θ, F is the attenuation factor of neutral filters, R is the fraction of the primary beam reflected at the exit window and is defined by the refractive index of the glass and TD is an optical factor correcting for the diffuse transmittance and imperfections of the working standard diffusor. These terms are discussed in detail below [see

[68] Y. Tomimatsu and K. J. Palmer, *J. Phys. Chem.* **67**, 1720 (1963).
[69] J. P. Kratohvil, *J. Colloid Interface Sci.* **21**, 498 (1966).

Eq. (31)]. Therefore, in order to obtain a Rayleigh ratio at an angle θ by this method, one must also measure the intensity at the supplementary angle. The recent work of Lee[63] has raised the question whether this method accounts for all the necessary corrections for a clear cylindrical cell.

b. *Kratohvil Method*. The equations developed for corrected intensities can be put in the general form:

$$G_\theta = X(G'_\theta - Y G'_{180-\theta}) \tag{27}$$

where X and Y are constants depending only on f_a and f_l for a particular geometrical-optical situation. f_a and f_l are the fractions of light reflected at perpendicular incidence at the glass–air and glass–liquid interfaces, respectively. Table II summarizes the equations for various cases and gives numerical values of X and Y for $f_a = 0.0370$ and $f_l = 0.00255$ (Pyrex and water at $\lambda_0 = 546$ nm), and the values of the ratio $G_\theta/G'_{180-\theta} = X(1 - Y)$. A similar table can be constructed for X and Y at $\lambda_0 = 436$ nm. However, the recent report by Heller and Witeczek[61] has indicated that the data corrected by this method still deviate from theoretical values.

c. *Harpst, Krasna, and Zimm Method*. Other workers have used empirical methods for the same corrections. The procedure of Harpst, Krasna, and Zimm[53] utilizes low molecular weight, random–coil polymers to obtain angular correction factors for the effects of the variations in refraction and reflection at the light scattering cell surfaces and the change in the volume of solution viewed at different angles. The ratio of the excess scattering of the polymers at 90° to their excess scattering at any other angle, θ, is defined as the angular correction (S_{90}/S_θ), where

$$S_\theta = \left(\frac{G_\theta}{G_0}\right)_{\text{solution}} - \left(\frac{G_\theta}{G_0}\right)_{\text{solvent}} \tag{28}$$

The polymers used include various concentrations of polymethacrylic acid and polyacrylamide. Various investigators have shown that polymethacrylic acid in dilute HCl shows no dissymmetry.[70,71] For measurements of the Rayleigh ratio, the instrument is calibrated with samples of known Rayleigh ratio or known turbidity. The calibration constant C of a cell for unpolarized incident light is given by the equation

$$CS_{90} = R_{90} = (3/16\pi)(2.303A/l) \tag{29}$$

[70] A. K. Katchalski and H. Eisenberg, *J. Polym. Sci.* **6**, 145 (1951).
[71] A. Oth and P. Doty, *J. Phys. Chem.* **56**, 43 (1952).

TABLE II

A SUMMARY OF EQUATIONS DEVELOPED FOR VARIOUS GEOMETRICAL–OPTICAL ARRANGEMENTS[a]

Arrangement	X	Y	$G_\theta/G'_\theta = X(1-Y)$ when, $G'_\theta = G'_{180-\theta}$
Cylindrical cell, clear faces	$1/t_a^2 t_{l_1}^2(1-4A^2) = 1.090$	$2A = 0.0789$	1.004
Cylindrical cell, painted outside	$1/t_a^2 t_{l_1}^2[1-(f_l+A)^2] = 1.085$	$f_l + A = 0.0420$	1.039
Cylindrical cell, painted inside; or semioct. cell (45°, 135°)	$1/t_a^2 t_{l_1}^2(1-A^2) = 1.085$	$A = 0.0395$	1.042
Cylindrical cell, clear faces or frosted inside; $f_l = 0$	$1/(1-2f_a) = 1.080$	$2f_a = 0.0740$	1.000
Cylindrical cell, painted outside, or inside, or semioct. cell; $f_l = 0$	$1/(1-2f_a) = 1.080$	$f_a = 0.0370$	1.040

[a] The numerical values are those for $f_a = 0.0370$ and $f_l = 0.00255$. $t_a = 1 - f_a$; $t_l = 1 - f_l$; $A = f_l + t_l f_a$.

where S_{90} is the observed excess instrument reading corresponding to the scattering from a calibrating solution at 90°, R_{90} is the Rayleigh ratio, and A is the optical absorbance measured in a cell of length l at the same wavelength as used in the scattering measurements. The constant C is determined only at 90° and is applied to measurements at other angles by correcting the scattering at these angles with the correction factors determined with the various polymers. The standards with known Rayleigh ratios that can be used are benzene, toluene, and a standard sample of Dow Styron polystyrene.[66,72]

4. Differential Refractometers

Light scattering experiments and measurements of preferential interactions between solvent components and macromolecules require that the refractive index increment of the solution be known with both high precision and high accuracy. For example, the preferential interaction between macromolecules and solvent components is measured directly by the difference between two refractive index increments, as shown in Eq. 3. This difference is frequently small (of the order of 5–10% of the total value). In light scattering experiments [Eqs. (7) and (9)], the refractive index increment is squared, so that any error in this quantity is also squared.

It is not practical to employ an absolute refractometer for such measurements primarily because of the effect of temperature on the absolute refractive index n. Since the differences in n are small, this would require temperature control to better than ±0.001° over considerable periods of time. These complications may be circumvented by employing an instrument which measures directly the difference in refractive index between protein solution and solvent, Δn. By having the solution and solvent in thermal contact in a specially designed cell, differential refractometers currently in use can measure Δn with a precision of 4 to 6 × 10⁻⁶.

a. Brice-Halwer Instrument. The most widely used and most readily commercially available differential refractometer is a visual one, designed by Brice and Halwer[73] and produced by the Phoenix Precision Instrument Company.[74] Similarly to the Brice light scattering photometer, this instrument has a light source, a mercury lamp, with a filter that transmits light at 436 or 546 nm. The light passes through a vertical slit and a square glass cell which is partitioned diagonally. The cell is

[72] J. P. Kratohvil, G. Dezelic, M. Kerker, and E. Matijevic, J. Polym. Sci. 57, 59 (1962).

[73] B. A. Brice and M. Halwer, J. Opt. Soc. Amer. 41, 1033 (1951).

[74] Model BP-2000 V.

placed in a thermostated compartment. The slit image is focused by a double convex lens and its position is determined by measuring the deviation of the slit image with a microscope fitted with a micrometer eyepiece. The two compartments of the partitioned cell are filled with liquid; the cell is rotated by 180° on a turntable, reversing the direction of the deviation. This results in establishing the zero Δn instrument baseline and in doubling the deviation when the two compartments are filled with liquids of different refractive indices.

Several other types of differential refractometers have been constructed; these are described by Timasheff and Townend.[7]

b. Photoelectric Differential Refractometer. In 1957, the late Dr. B. A. Brice of the Eastern Regional Research Laboratory, U.S. Department of Agriculture, designed and constructed a photoelectric differential refractometer[75] which is based essentially on the same principles as those of his visual instrument.[73] We have constructed and utilized successfully an instrument similar to that designed by Brice.[75a] Its principal advantages are relief from eye strain and related speed of operation, since it has been found in practice that owing to visual fatigue an experimenter is not capable of taking more than a few measurements in one day with the visual instrument. The components of this instrument are available commercially, and it can be constructed by a competent laboratory instrument shop.

Just as in the visual differential refractometer, this instrument takes advantage of the linear displacement of a slit image as a function of refractive index. The major difference between the instruments is that commercially available photocells are used to detect the slit image and a null detector microvoltmeter (Keithley Instruments, Model 155) is used to determine the exact position of the slit image relative to the oppositely polarized photocells. Schematics of this instrument are shown in Figs. 8A and B, and Fig. 9 is a photograph of it.

Light from the light source (A) is passed through the filter (blue or green) contained in the filter compartment (B). The light next impinges upon the slit (C), passing through and forming a slit image that is transmitted down the blackened light tube (D). The solvent is introduced first into the front compartment of the divided cell. With the cell positioner (E) pushed "in," the slit image passes through the solvent compartment only of the divided cell (F), and is reflected back through the same compartment by a mirror (G). The image then goes back down the tube and strikes a mirror (H) which reflects it downward by 90° where it impinges on a trolley (I), on which two oppositely

[75] B. A. Brice, unpublished results.
[75a] E. P. Pittz and B. Bablouzian, *Anal. Biochem.*, 1973, in press.

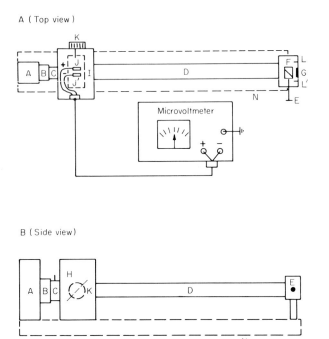

FIG. 8. Schematic of Brice photoelectric differential refractometer (see text).

FIG. 9. Photograph of the Brice photoelectric differential refractometer.

polarized photoelectric cells (J, J') are mounted with a gap between them. The micrometer dial (K) controlling the position of the trolley (I) is then rotated until the slit image falls exactly between the two photoelectric cells, illuminating them equally. This is indicated by nulling the reading on the polarized microvoltmeter scale. The micrometer dial reading is then recorded. The cell positioner (E) is pulled to the "out" position, where the light passes through the solution compartment, now wedge shaped, through the diagonal glass partition and the solvent compartment, now also wedge shaped. In this position, the beam is deflected an amount proportional to Δn. Back reflection from mirror (G) doubles the deviation. The micrometer screw is again rotated until the slit image falls between the photoelectric cells, and the position is again recorded. This procedure is carried out again with solvent in both compartments of the cell, as well as with sample and solvent positions reversed, i.e., with sample in the rear compartment and solvent in the front compartment. This procedure is carried out to cancel cell imperfections. If we allow X' to be the difference between micrometer readings with E in the "in" and "out" positions with solvent in both compartments, and let X be the difference between micrometer readings in the "in" and "out" positions with sample in the rear and solvent in the front, then $\Delta X = X - X'$ is the displacement of the slit image due to sample. The instrument is calibrated by plotting ΔX versus Δn for

TABLE III

REFRACTIVE INDEX DIFFERENCES Δn BETWEEN SODIUM CHLORIDE
SOLUTIONS AND DISTILLED WATER[a]

Concentration		$\Delta n \times 10^6$ (25°C)		
g/100 ml	g/100 g H_2O	589 nm	546 nm	436 nm
0.0938	0.0941	165	166	173
0.1034	0.1037	182	184	190
0.3362	0.3375	587	592	614
0.5602	0.5627	974	982	1019
0.6866	0.6900	1191	1202	1246
0.9090	0.9142	1572	1586	1645
1.1240	1.1311	1939	1956	2028
1.6465	1.6595	2824	2848	2954
2.0327	2.0513	3474	3504	3634
3.7307	3.7854	6293	6347	6583
6.7405	6.9092	11151	11247	11670
10.488	10.895	16996	17144	17800
16.011	16.988	25260	25480	26430

[a] A. Kruis, Z. Phys. Chem. **34B**, 13 (1936).

samples of known refractive index, such as KCl or NaCl of known concentration. The values of Δn for standard solutions are listed in Table III.[76] Experience has shown that Δn is a linear function of ΔX for the instrument in use in our laboratory.

B. Procedures

1. Removal of "Dust"

a. Clarification of Solvents. Since the scattering intensity at $\theta = 0°$ is proportional to the square of the molecular weight and to the product of this with the square of the radius of gyration at other angles [Eq. (15)], trace amounts of large particle contaminants give serious errors in molecular weights and radii of gyration. This is known as the "dust problem."

Light scattering requires a large quantity of particle-free solvent, usually water, for making solutions, rinsing dust and contamination away, etc., and clarification of water is particularly difficult. In our practice, we have been able to obtain "clean" water by the following process. Laboratory distilled water is passed through an ion exchange resin, filtered through a very fine Millipore filter (Type VM, pore size 50 nm)[77] and redistilled in an all-glass still of special design, which is capable of producing 600 ml of particle-free water per hour.

A photograph of our still is shown in Fig. 10. The still pot (A) is a 3-liter Pyrex flask that has been modified to accommodate a filling neck (B) and cap (C) made from a 34/45 taper joint. These help to eliminate glass dust from the still. The springs on the cap (C) eliminate breakage due to occasional bumping.

The introduction of organic contaminants (a source of oily substance) is eliminated by prefiltering the still water through the membrane filter, which removes most debris. Since these membrane filters may contain detergent,[78] each membrane must be washed with boiling water to extract the contaminant before use. In order to prevent bumping while boiling, since this creates glass dust, platinum tetrahedra[79] are introduced into the distilling section as boiling chips, using water free of organic contaminants or oxidants. These are the only boilizers found that do not introduce dust. These tetrahedra remain effective for long periods of time, and can be regenerated by hot HNO_3–H_2SO_4.

During distillation, the steam passes through a spray trap, T, and

[76] A. Kruis, *Z. Phys. Chem.* **34B**, 13 (1936).
[77] Filters and accessories are obtained from Millipore Corp., Bedford, Massachusetts.
[78] R. D. Cohn, *Science* **155**, 195 (1967).
[79] Beckman platinum tetrahedra. Available from A. H. Thomas Co., Philadelphia, Pennsylvania, Catalog No. 8313.

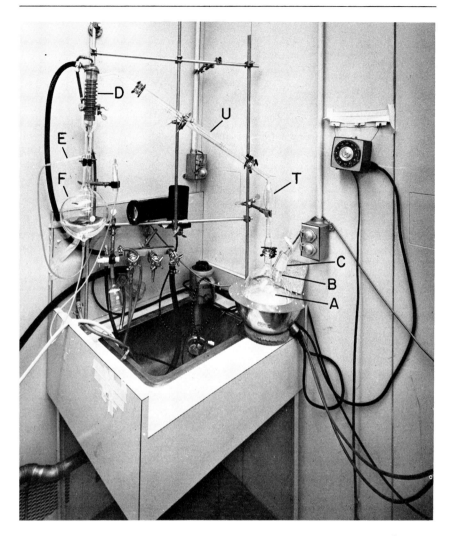

FIG. 10. Photograph of the still used to clarify water for light scattering.

then progresses up along a wide uninsulated tube, U, (27 mm i.d., 50 cm long). This has the effect of preventing dust from migrating to the receiver side of the still by the high reflux rate relative to the upward velocity of the steam. The most important problem with such a still is the generation of glass dust. For example, vibrating joints are a main source of such dust. By using Pyrex O-ring joints[80] with Buna N rings, enough flexibility exists to prevent both breakage and generation of glass

[80] Size 25, Corning Glass Catalog No. 6780.

dust in the system. As can be seen from Fig. 10, the receiver side-con-
denser (D), vent (E), and receiver (F) are in one piece, helping to
prevent the production of glass dust. The Friedrich's condenser (D)
used in our system can be fused to the receiver (F) and the vent (E)
by a competent glassblower. Within the vent section, a drop tip is fused
to the side, letting the distillate flow down the side of the receiver and
preventing bubble formation due to splashing. The vent (E) is constructed for a wide-bore (9 mm) tubing to prevent droplets from being
washed backward. The vent is connected to a ⅛ inch bore gum rubber
tubing, more than 1 foot in length, which leads to a three-way stopcock.
The function of this stopcock is to act as a vent during distillation, or
as a drain of condensate when the still is steam-cleaned. Clean water is
drawn out through the bottom of the receiver. Gum rubber tubing can
be used as an extension from the receiver; other polymers should be
used with great caution, since many (e.g., polyvinyl chloride) contain
plasticizers.

When the still is first set up, or on the rare occasions when it must
be disassembled, a week or more of steaming is required to wash out
the particulate contaminants. If oily substances are accidentally introduced into the still, a water-alcohol solvent is frequently sufficient
to wash these out, without disassembling the still. As a general procedure, the still should be steamed for 10 hours or more as soon as oily
or particulate contaminants are observed in the distillate.

Organic solvents of low viscosity such as 2-chloroethanol, benzene,
toluene, ethanol, etc., are easily cleaned by a single distillation in a
usual laboratory still. Alternative methods that have been used are
centrifugation and filtration.[81-84] Organic solvents of high viscosity, such
as ethylene glycol, glycerol, 2-methyl-2,4-pentanediol, are not easy to
clarify. Filtration through sintered-glass filters, however, will leave these
solvents mote free. Unfortunately, the viscous solvents stabilize microbubbles, and these can often be mistaken for dust particles.

b. *Clarification of Protein Solutions.* In our operations, all protein
stock solutions are prepared from solvents that have been clarified as
described in the preceding section. These stock protein solutions are
then centrifuged for 30–40 minutes in a high speed centrifuge. As a
final clarification step, the protein stock solutions are filtered through
ultrafine Pyrex sintered-glass filters (see Fig. 11).[3,85] These filters have

[81] G. Bernardi, *Makromol. Chem.* **72**, 205 (1964).
[82] G. V. Schulz, H. J. Cantow, and G. Meyerhoff, *J. Polym. Sci.* **10**, 79 (1953).
[83] G. V. Schulz, O. Bodmann, and H. J. Cantow, *J. Polym. Sci.* **10**, 73 (1953).
[84] W. B. Dandliker and J. Kraut, *J. Amer. Chem. Soc.* **78**, 2380 (1956).
[85] J. P. Kratohvil, G. Dezelic, M. Kerker, and E. Matijevic, *J. Polym. Sci.* **57**, 59
(1962).

Fig. 11. System for delivering clarified solutions to light-scattering cells.

the advantage that the solution is pushed upward, thus avoiding splashing, foaming, or bubble-formation that can result in aggregation, denaturation, formation of persistent microbubbles or picking up of dust from the air. Two Bier[3,S5] filters of different size are shown in Fig. 11. The small filter (made from a Pyrex No. 36060 funnel 10-mm disk) on the jack, is used for protein stock solutions while the larger type (20-mm disk funnel) is used for clarification of solvents. The filters are constructed from ultrafine grade sintered Pyrex glass Büchner funnels. These filters should be constructed so that a pipette can be inserted into the "clean" side to remove stock solution. Also a minimum of space is allowed below the sintered disk, since this represents essentially dead volume, i.e., solution that is unavailable for light scattering. The filling tube must be tubulated in order to connect it to a gum rubber pressure hose. These filters are cleaned by boiling in 50% HNO_3–H_2SO_4, followed by exhaustive flushing of the disk with clean water and finally solvent. The protein is then force-filtered up into the clean side and no new protein–air interface is formed. Solvent is cleaned by similar filtration in the larger size filter.

2. Light Scattering Measurements

The solvent, after force filtration through the large Bier filter, can be withdrawn from it by the device shown in Fig. 11. The Bier filter containing clean solvent is placed on a standard laboratory jack (B) and raised carefully to a microburette (C) which is gradually introduced into it, great care being taken to avoid glass-to-glass contact between the filter and the clean microburette which can create glass dust. The microburette is mounted solidly on a ring stand (D). The tip of the microburette which had been siliconed previously is placed just below the surface of the solvent, and solvent is gradually withdrawn, the laboratory jack being adjusted accordingly. The laboratory jack is lowered, and the solvent in the microburette is transferred to the cell and discarded. This process, which effectively rinses the cell with ultra-clean solvent, is repeated several times. Finally, ~2.2 ml of solvent is delivered to the tared Dintzis cell; it is capped, weighed, mounted in the light scattering photometer, oriented, and checked for particles. Particles appear as bright stars in the light beam, and the appearance at angles close to $\theta = 0°$ of any sparkling or otherwise visible particles is an indication that the solvent is not suitably clean for experimentation. The cell can be easily examined at low angles by using a small mirror mounted on a rod.

If no particles appear, the measurement is carried out as follows in the fully calibrated[55,56] Brice instrument. Using the 436 nm (blue) filter, the first three light-attenuating neutral filters (F_2 of Fig. 4) are placed in the "in" position and the photomultiplier detector is set at 0°; the reading, G_0 (F1,2,3), is recorded. The detector is next set at 90°, all neutral filters are pulled out and a reading, G_{90}, is again taken. This process is repeated several times, taking G_0 and G_{90} readings in succession. The ratio of these two readings, multiplied by the proper attenuating filter factor, is the raw data scattering intensity of the solvent. As an example, let the reading G_0 (F1,2,3) be 4655 and that of G_{90} (no filters) be 193. The intensity of scattering relative to that of the incident beam, which is further attenuated by the working standard, is

[86] Y. Tomimatsu and K. J. Palmer, J. Polym. Sci. 35, 549 (1959).

$$I_{\text{solvent}} = \left(\frac{G_{90}}{G_0}\right)_{\text{solvent}} \frac{1}{F4} = \frac{193}{4655 \times 0.0436} = 0.9509$$

Since in the working equation [see below, Eq. (31)], the product of the attenuation factors of all four filters is included, at this point in the calculations, it is necessary to eliminate the attenuation of the filter(s) which had not been used. Since $F1$, $F2$, and $F3$ were used in the present measurement, the G_{90} to G_0 ratio must be divided by the transmittance of neutral filter $F4$. $F4 = 0.0436$ is a typical value for this attenuation.

Next, using the smaller Bier filter and the same procedure for filtering and filling the microburettes, clean protein stock (usually 20–80 mg/ml concentration), is added to the cell contents, making sure to keep the tip of the microburette below the solution surface in the cell. Increments of 0.100 ml are convenient. After the addition, the cell is weighed and the protein concentration in the cell is calculated either by volumetric or gravimetric dilution. The protein concentration, C_{add}, can be calculated by using the equation:

$$C_{\text{add}} = \left(\frac{W_{\text{add}}}{W_{\text{add}} + W_{\text{solv}}}\right) C_T \tag{30}$$

where C_T is protein concentration in the stock solution, W_{add} is the weight of the protein stock solution added, and W_{solv} is the weight of the solvent. While this expression is not exact, since the protein solution and solvent have different densities, their difference is small and the error introduced is negligible. After addition of the protein stock solution, the cell contents are thoroughly mixed by gentle inversion, the sides of the cell, which had been presiliconed, are washed down with clean water, the cell is remounted in the cell holder and is again oriented in the photometer. The solution is again inspected visually for dust. If no dust particles or bubbles are observable after a 10-minute period, G_0 ($F1,2,3$) and G_{90} are again recorded. As protein concentrations increase, G_{90} increases in value and neutral filters are progressively introduced. If now, the values are $G_0 = 4657$ and $G_{90} = 1043$,

$$I_{\text{solution}} = \left(\frac{G_{90}}{G_0}\right)_{\text{solution}} \frac{1}{F4} = \frac{1043}{4657 \times 0.0436} = 5.1368$$

The excess scattering intensity, I, is then

$$I = I_{\text{solution}} - I_{\text{solvent}} = 5.1368 - 0.9509 = 4.1859$$

Knowing the protein concentration, e.g., $C_{\text{add}} = 1.6120 \times 10^{-3}$ g/ml, we obtain the ratio C/I, which in this example is 0.387×10^{-3} g/ml. Multiplication of C/I by a product of the constant H or K (Eq. 7) with various instrument factors gives $H(C/\Delta\tau)$ or $(KC)/R_\theta$. In the Brice method, these constants are combined into a working constant, H'':

$$H'' = \frac{2\pi^3(dn/dc)^2 1.045h}{\lambda^4 N_A TDFaR_W/R_c} \tag{31}$$

where

dn/dc = refractive index increment of the macromolecule in ml/g.

1.045 = Fresnel correction for backward reflection of exit window–air interface. The latter number need not be used when small Pyrex cells are standardized vs. the absolute cell; the difference becomes part of the standardization constant, k.

h = height of beam in cell, 1.20 cm.

λ = wavelength *in vacuo* of light in centimeters.

N_A = Avogadro's number.

TD = An optical factor, correcting for the diffuse transmittance and imperfection of the working standard diffusor, which intercepts the incident beam when the detector is set to $0°$.

F = The product of the transmittances of all four neutral filters. This factor is applied to the denominator of Eq. (31) for convenience, since $0°$ readings are generally taken with most or all of the filters in the beam.

a = The "a constant." This constant compares the working standard (this diffuses the incident beam when the detector is at $0°$) to the standard opal glass diffusor.[57] It is a function of the geometry of the light beam, and changes slightly as the lamp ages. It should be redetermined regularly using the "absolute" 12×12 mm slits; it must be redetermined when a lamp is replaced. Substitution of narrow slits when using 1 cm Dintzis-type cells or Kronman-Timasheff filter cells[7] does not change "a"; the change becomes part of the "standardization constant" k.

R_W/R_c = residual refractive index correction. This factor arises because of the foreshortening of the field of view of the detector "refractive index effect."[66]

If the Dintzis cell is used, H'' is further divided by the cell calibration constant, k, defined by Eq. (25).

A sample calculation is given in Table IV, where the constants used are: $\lambda = 436$ nm, dn/dc is 0.196 ml/g; $R_W/R_c = 1.001$; $TD = 0.260$, $a = 0.0435$; $h = 1.2$, $N_A = 6.02 \times 10^{23}$; F, the product of the neutral filters is equal to 5.176×10^4.

TABLE IV

SAMPLE CALCULATION OF LIGHT-SCATTERING MEASUREMENTS

Cell No. R   Cell k 1.638   Cell wt. 23.76722   Solvent 0.01 M sodium acetate, 0.02 M NaCl

Cell + solvent wt. 26.07934   Solvent wt. 2.31212   Solvent volume 2.31908

Stock concentration (g/ml) 13.16×10^{-3} $H/k = 0.1452$

Protein stock (g)	Total weight (g)	C (g/ml) $\times 10^3$	G_0 (F1,2,3)	G_{90} (no filters)	Ratio	Filter at 90°	$1/F$	G_{90}/G_0 Corr.	I	C/I $\times 10^3$	$HC/\Delta\tau \times 10^5$
Solvent			4666, 4646, etc. Av. 4655	191, 189, etc. Av. 193	0.0415	—	22.936	0.951	—	—	—
0.32452	2.63644	1.6120	4656, 4659, etc.	1042, 1034, etc.	0.2240	—	—	—	—	—	—
			4657 av.	1043 av.	0.2240	—	22.936	5.1368	4.1859	0.3870	5.619
—	—	—	—	—	—	—	—	—	—	—	—
—	—	—	—	—	—	—	—	—	—	—	—
—	—	—	—	—	—	—	—	—	—	—	—
—	—	—	—	—	—	—	—	—	—	—	—
—	—	—	—	—	—	—	—	—	—	—	—
1.637	3.94927	5.4562	4527, 4528, etc.	2819, 2816, etc.		—					
			4519 av.	2817 av.	0.6234	—	22.936	14.2974	13.3465	0.4088	5.936

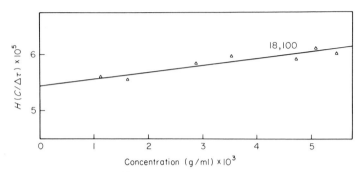

FIG. 12. Standard plot of $H(C/\Delta\tau)$ vs. protein concentration. From E. P. Pittz and S. N. Timasheff, unpublished results, 1972.

Figure 12 shows a plot of the results calculated in Table IV.[87] The reciprocal of the $HC/\Delta\tau$ intercept gives M_w, the weight average molecular weight, which in this case is 18,100. For large macromolecules which require the measurement of an angular envelope, the same procedure is carried out using the appropriate cell and performing the measurements at a number of angles.

3. Differential Refractometry

In describing a differential refractometry experiment, we will refer to the photoelectric differential refractometer (Figs. 5 and 6) since experiments carried out on the Brice-Hawler[73] instrument are well documented.[7,73]

In most respects, the requirements for obtaining good results are identical for the Brice-Halwer[73] instrument and the photoelectric instrument. These requirements are: (a) prevention of liquid from crossing between cell compartments, accomplished by siliconing the divided cell; (b) temperature equilibrium between the cell compartments; 20 minutes is the time usually allowed for equilibrium to be attained, although 10 minutes has been found to be sufficient; (c) the instrument must be calibrated against solutions of known standard refractive index difference,[73] giving a constant $k = \Delta n/\Delta X$ where ΔX is the linear displacement of the drive necessary to place the slit image exactly between the two parts of the split photocell.

An experiment is carried out as follows, with typical instrumental readings given at each step:

1. With the cell holder in the IN position, solvent is placed in both

---

[87] E. P. Pittz and S. N. Timasheff, unpublished results, 1972.

the left-hand (L) and right-hand (R) cell compartments, and the system is allowed to come to thermal equilibrium.

2. The out-of-balance current is zeroed and the reading is taken on the micrometer dial attached to the drive, e.g., 2491. The balance is upset and restored several times to obtain an average. The cell holder is then pulled to the "OUT" position, and the balance is restored by turning the micrometer screw, giving a second reading, e.g., 2686.

3. The difference between the $(OUT)_0$ and $(In)_0$ positions is the instrument blank for this solvent. The "0" subscript here indicates that solvent is in both compartments. In this example this value is $+195$.

4. The above procedure is repeated with sample in the left-hand compartment and solvent in the right. This time, let

$$(OUT)_s - (IN)_s = 2418 - 2492 = -74$$

Here subscript s indicates that sample is present in the left compartment.

5. Then, $\Delta X = [(OUT)_s - (IN)_s] - [(OUT)_0 - (IN)_0]$, in the present examples: $-74 - 195 = -269$.

6. Δn is then calculated from $\Delta n = -k \, \Delta X$. With a typical value of $k = 1.0111 \times 10^{-6}$, in the present example $\Delta n = 272 \times 10^{-6}$.

7. With knowledge of the concentration, C, $\Delta n/C$ is calculated; in the present example, $C = 1.391 \times 10^{-3}$ g/ml and $\Delta n/C$ is 0.196 ml/g.

8. Procedures 4 through 7 are repeated for several samples (usually 5) at different protein concentrations and, if no concentration dependence is noted, an average is taken. Typical results are:

$$0.196 \pm 0.000$$
$$0.196 \pm 0.000$$
$$0.194 \pm 0.002$$
$$0.193 \pm 0.003$$
$$0.195 \pm 0.001$$
$$0.199 \pm 0.003$$

Average: $0.196 \pm 0.001_5$

4. Techniques Involved in Multicomponent Experiments

a. Preparation of Solutions. The methods of preparing solutions containing more than two components for measurement of preferential interactions of solvent components with macromolecules either by differential refractometry or by light scattering are identical. Greater care must be taken, however, to keep the solution mote-free when doing light scattering.

i. MEASUREMENTS AT CONSTANT MOLALITY OF COMPONENT 3, m_3. Three methods are used in our laboratory for preparing solutions for multicomponent measurements at constant m_3.

(a) A water–nonaqueous solvent mixture of known molality is prepared by mixing known weights of the two components. This is used as reference and as dilution solvent. A concentrated stock solution of the protein $(20 - 80 \times 10^{-3}$ g/ml) is prepared in an aqueous medium. Increments of this stock solution are added volumetrically to the solvent, each time adding a calculated amount of component 3 necessary to keep the molality of component 3 in the protein solution equal to that of the solvent. This method requires knowledge of the amount of water bound to the protein when preparing the aqueous solution, since this water must be taken into consideration in determining the molality, m_3, and in keeping it constant.

(b) The protein is dried under vacuum at low temperature $(25–50°)$ in the presence of phosphorus pentoxide. A known weight of water (or buffer) is added to the dried protein, and the protein is completely dissolved in the water. A known weight of nonaqueous solvent is added to the aqueous protein solution to give the desired m_3. Then the solvent is made up accurately by weight to the nearest milligram to the same m_3 as that of the protein solution. The volume of solvent prepared is usually a volume considerably larger than that of the stock solution, so that this solvent can be used for dilution.

(c) If the protein is very soluble in the water–nonaqueous solvent combination, then solvent can be made up to the proper m_3, and then added to protein that had been dried as described in (b) above.

ii. MEASUREMENTS AT CONSTANT MOLARITY OF COMPONENT 3, C_3. One milliliter of an aqueous stock solution of protein is accurately delivered into a 5-ml volumetric flask by a microburette. A predetermined volume of the organic solvent is then added to adjust this solution to the given molarity, C_3, and the flask is filled to the line by adding mixed solvent of the given molarity, C_3. This is then used as the stock for the serial light-scattering or differential refractometry measurements. It should be pointed out, however, that measurements made at contant C_3 rather than m_3, have several disadvantages. For example, it is extremely difficult to deliver accurate volumes of the two solvents. Also, it is difficult to fill the 5-ml flask to the line reproducibly enough to obtain precise results. Second, measurements at constant molarity, C_3, give preferential interaction results that can be misleading. When the preferential interactions are done on the C_3 or gram per milliliter scale, the volume that the protein occupies is not properly considered. It is possible, therefore, to obtain preferential interaction with water on the C_3 scale when the correct thermodynamic interaction is preferentially with component 3, as would be found in experiments on the m_3 scale. Such a measurement of preferential hydration is spurious, as it only represents volumetric dilution of component 3 by addition of protein. It is, therefore, incorrect to say that the concentration units employed are a matter of personal preference.[88,89] The difference between the concentration scales is illustrated in Fig. 13. Comparison of the various solution-reference solvent pairs shows that proper substraction of solvent contribution is attained only on the m scale, while the C scale results in overcorrection.

　　b. *Differential Refractometry of Multicomponent Systems.* Equation (3) can be transformed into

$$\frac{M_3}{M_2}\left(\frac{\partial m_3}{\partial m_2}\right)^0_{T,p,\mu_3} = \left(\frac{\partial g_3}{\partial g_2}\right)^0_{T,p,\mu_3}$$

$$= \frac{1}{(1-\bar{V}_3 C_3)}\left\{\left(\frac{\partial n}{\partial C_2}\right)_{T,p,\mu_3} - \left(\frac{\partial n}{\partial C_2}\right)_{T,p,m_3}\right\} \Big/ \left(\frac{\partial n}{\partial C_3}\right)_{T,p,m_2} \quad (32)$$

where $(\partial g_3/\partial g_2)_{T,p,\mu_3}$ is the preferential interaction of component 3 with component 2 on a gram per gram basis; $\bar{V}_3$ is the partial specific volume of the third component; C_3 is the concentration of the third component in grams per milliliter; $(\partial n/\partial C_3)_{T,p,m_2}$ is the refractive index increment of the solvent, measured by plotting n versus C_3; $(\partial n/\partial C_2)_{T,p,m_3}$ is the refractive index increment of the protein at constant m_3, and is measured as described in Section III, B, 3. The term $(\partial n/\partial C_2)_{T,p,\mu_3}$ is measured by taking the difference in refractive index between a protein solution that has been equilibrated with an "infinite" excess of solvent and that of the solvent. In practice, $(\partial n/\partial C_2)_{T,p,\mu_3}$ is set equal to $(\partial n/\partial C_2)_{T,\mu_1,\mu_3}$, since it has been shown that the difference between these two terms is very small.[90] The dialysis experiment is carried out as follows:

　　1. Prepare a stock solution of protein dissolved in water (or buffer).
　　2. Add component 3 to the right m_3 (if the protein is soluble at the high concentration).
　　3. Make a large excess of solvent of the same m_3.
　　4. Dilute stock with solvent to give 10 solutions with a concentration range of 0.5–5 × 10^{-3} g/ml.
　　5. Put each solution (approximately 2.5 ml) into thoroughly cleaned "No Jax" dialysis tubing (Union Carbide Corp., New York, N.Y.). This tubing is precleaned by boiling several times in sodium bicarbonate and then in distilled water.
　　6. Be sure that a small bubble exists in each tube.
　　7. Add these bags to the solvent of m_3 and let them dialyze for ap-

[88] M. E. Noelken, *Biochemistry* **9**, 4122 (1970).
[89] J. A. Gordon and J. R. Warren, *J. Biol. Chem.* **243**, 5663 (1968).
[90] D. Stigter, *J. Phys. Chem.* **64**, 842 (1960).

proximately 1 week (a kinetic study can be carried out to determine the time necessary to reach equilibrium).

8. Carry out the measurements as described in Section III, B, 3 for differential refractometry on two-component systems.

The preferential interaction results are then calculated using equations, such as Eq. (3) or Eq. (32). The preferential interaction of component 3 with component 2 is related to the preferential interaction of component 1 with 2 on a weight per weight basis by

$$\left(\frac{\partial g_1}{\partial g_2}\right)_{T,\mu_1,\mu_3} = -\frac{g_1}{g_3}\left(\frac{\partial g_3}{\partial g_2}\right)_{T,\mu_1,\mu_3} \tag{33}$$

Measurements made at constant molarity $(\partial C_3/\partial C_2)_{T,\mu_1,\mu_3}$ are related to those made at constant molality (m_3) by

$$\left(\frac{\partial g_3}{\partial g_2}\right)^0_{T,\mu_1,\mu_3} = \frac{g_3}{(\bar{V}_1)_{T,p,m_2}C_3}\left(\frac{\partial C_3}{\partial C_2}\right)^0_{T,\mu_1,\mu_3} + C_3(\bar{V}_2)_{T,p,m_3} \tag{34}$$

where g_3 is the weight of component 3 per gram of component 1, $(\bar{V}_1)_{T,p,m_2}$ is the partial specific volume of component 1 at constant molality of component 2, $(\bar{V}_2)_{T,p,\mu_3}$ is the partial specific volume of the protein, and $(\partial C_3/\partial C_2)_{T,\mu_1,\mu_3}$ is the preferential interaction of component 3 with the macromolecule on the molar scale, and is measured by:

$$\left(\frac{\partial C_3}{\partial C_2}\right)_{T,\mu_1,\mu_3} = \left\{\left(\frac{\partial n}{\partial C_2}\right)_{T,\mu_1,\mu_3} - \left(\frac{\partial n}{\partial C_2}\right)_{T,p,C_3}\right\} \Big/ \left(\frac{\partial n}{\partial C_3}\right)_{T,p,C_2} \tag{35}$$

The essential difference between Eq. (35) and Eq. (32) is the $(\partial n/\partial C_2)_{T,p,C_3}$ term. The measurement of this term is described in the previous section.

Table V[27] shows data for the preferential interaction of β-lactoglobulin with solvent components in water-2-chloroethanol. As can be seen by comparing columns 3 and 4, $(\partial n/\partial C_2)_{T,p,m_3}$ and $(\partial n/\partial C_2)_{T,p,c_3}$ are different, this difference becoming greater at the higher concentrations of chloroethanol. Columns 6 and 7 show how these differences in refractive index increments at constant molality and molarity are reflected in the magnitude and sign of the reported preferential interaction of solvent components with protein. It must be pointed out here that $(\partial g_3/\partial g_2)_{T,\mu_1,\mu_3}$ reflect the thermodynamically correct preferential interactions. This becomes clearer if we consider that, by working at constant molality of the third component, we are judging preferential interactions with respect to a reference point such that the preferential binding $(\partial m_3/\partial m_2)_{T,\mu_1,\mu_3}$ or $(\partial g_3/\partial g_2)_{T,\mu_1,\mu_3}$ is zero when the amount of third component per 1000 g, or per gram, of water is identical on both sides of the membrane at osmotic equilibrium.

TABLE V

PREFERENTIAL INTERACTION OF SOLVENT COMPONENTS WITH β-LACTOGLOBULIN A IN THE WATER-2-CHLOROETHANOL SYSTEM ON THE MOLAL AND MOLAR SCALES, DETERMINED BY DIFFERENTIAL REFRACTOMETRY

Vol % of 2-chloroethanol	$\left[\dfrac{\partial n}{\partial C_2}\right]_{T,\mu_1,\mu_3}$	$\left[\dfrac{\partial n}{\partial C_2}\right]_{T,p,m_3}$	$\left[\dfrac{\partial n}{\partial C_2}\right]_{T,p,c_3}$	$\left[\dfrac{\partial n}{\partial C_3}\right]_{T,p,m_2}$	$\left[\dfrac{\partial g_3}{\partial g_2}\right]_{T,\mu_1,\mu_3}$	$\left[\dfrac{\partial C_3}{\partial C_2}\right]_{T,\mu_1,\mu_3}$	$\left[\dfrac{\partial C_1}{\partial C_2}\right]_{T,\mu_2,\mu_3}$
5	0.192[a]	0.184[a]	0.188	0.103[a]	0.081[a]	0.039	
10	0.195	0.179	0.186	0.103	0.171	0.087	
20	0.198	0.169	0.184	0.100	0.359	0.140	
30	0.202	0.159	0.182	0.094	0.643	0.212	
40	0.192	0.152	0.180	0.092	0.718	0.131	
50	0.175	0.143	0.178	0.089	0.711	−0.034	0.029
60	0.150	0.137	0.177	0.088	0.366	−0.307	0.175
80	0.117	0.131	0.175	0.082	−0.848	−0.717	0.155
					$(0.183)^b$		

[a] The values in these columns are taken from S. N. Timasheff and H. Inoue, *Biochemistry* **7**, 2501 (1968).

[b] The value in parentheses is for $(\partial g_1/\partial g_2)_{T,\mu_1,\mu_3}$.

c. Light Scattering of Multicomponent Systems. It is important to point out that turbidimetric measurements on proteins in multicomponent systems are relatively insensitive to small changes in solvent composition as compared to differential refractometry, where a difference in solvent composition between the dialyzed protein solution inside the dialysis bag and the bulk solvent outside the bag is a direct indicator of preferential solvent interactions. Turbidimetric measurements in multicomponent solvents are sensitive to changes in polarizability and density fluctuations in the local environment of the protein, relative to the bulk solvent. Any one of the methods of preparing protein solutions in multicomponent solvents [Section III, B, 4, a, i (a)–(c)] can be used when turbidity measurements are made.

The experimental procedure for carrying out the light scattering measurements is the same as in Section III, B, 2, if methods (b) or (c) are used for preparing solutions. If method (a) is used, then an additional filtering step and microburette are needed to deliver the pure nonaqueous solvent to the light-scattering cell.

In a three-component system where no protein association takes place, the turbidity measurements at constant chemical potential (μ_3) should differ little from those at constant molality (m_3). It is the refractive index increments at constant m_3 that changes significantly (Eq. 8) when strong preferential interactions of solvent components with proteins take place. Hence, it is important to carry out the equilibrium dialysis experiment indicated by Eq. (11) to obtain the true molecular weight. Again, a large change will take place in $(\partial n/\partial C_2)_{T,\mu_1,\mu_3}$, rather than in $\Delta\tau$, if no association takes place. If there is association taking place, $\Delta\tau$ values, given the same protein concentrations, will increase to a significant extent. Thus, it is important to carry out light scattering, as well as differential refractometry experiments to establish whether the third component induces protein aggregation.

Figure 14 shows $H(C_2/\Delta\tau)$ vs. concentration results for ribonuclease A in 30% (v/v) aqueous 2-methyl-2,4-pentanediol (MPD) under conditions of constant μ_3 and m_3.[87] It is found by the dialysis measurement that the weight average molecular weight is 23,320, indicating that association is occurring. Using Eq. (8a), together with values of M_2 measured with and without dialysis, we obtain D, which is also defined as:

$$D = \frac{(1 - C_3\bar{V}_3)_{m_2}}{(1 - C_2\bar{V}_2)_{m_3}} \frac{(\partial n/\partial C_3)_{T,p,m_2}}{(\partial n/\partial C_2)_{T,p,m_3}} \left(\frac{\partial g_3}{\partial g_2}\right)_{T,p,\mu_3} \tag{36}$$

and

$$\left(\frac{\partial g_1}{\partial g_2}\right)_{T,\mu_1,\mu_3} = -\frac{g_1}{g_3}\left(\frac{\partial g_3}{\partial g_2}\right)_{T,\mu_1,\mu_3} \tag{37}$$

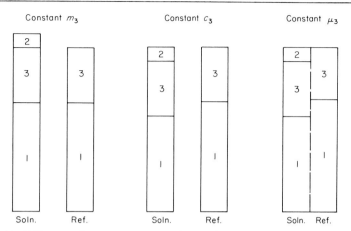

FIG. 13. Difference between concentration scales: 1, H_2O; 2, protein; 3, third component (cosolvent or salt).

In this example, $HC_2/\Delta\tau$, measured at constant m_3, when extrapolated to zero concentration, gives an apparent molecular weight of 19,950. Then, $(1 + D)^2$ (23,320) = 19,950, giving $D = -0.0751$. Since, $\bar{V}_3 = 1.023$ ml/g, $C_3 = 0.281$ g/ml, $C_2 \to 0$, $(\partial n/\partial C_3)_{T,p,m_2} = 0.122$ ml/g and $(\partial n/\partial C_2)_{T,p,m_3} = 0.171$ ml/g, $(\partial g_3/\partial g_2)_{T,p,\mu_3} = -0.148$ g MPD per gram of ribonuclease, and $(\partial g_1/\partial g_2)_{T,p,\mu_3} = 0.377$ g of H_2O per gram of ribonuclease. Hence, we see an example of a system in which a combination

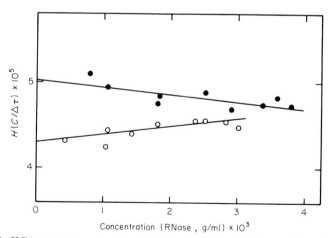

FIG. 14. $HC/\Delta\tau$ vs. concentration plots for ribonuclease A in 30% (v/v) 2-methyl-2,4-pentanediol at constant molality of component 3 (m_3) (●——●, 19,950) and at constant chemical potential (μ_3) (○——○, 23,320). From E. P. Pittz and S. N. Timasheff, unpublished results, 1972.

of light-scattering and differential refractometry, with and without dialysis, result in a determination both of the degree of association of a protein and of the extent of its preferential interaction with solvent components, in this case with water.

Acknowledgment

This work was supported in part by NIH Grant GM-14603 and NSF Grant GB-12619.

[10a] Osmotic Pressure

By GUIDO GUIDOTTI

The first studies on the osmotic pressure of protein solutions began at the turn of the century, and until the advent of the ultracentrifuge these measurements provided the only way of determining the molecular weight of proteins. The true molecular weight of hemoglobin was in fact established first by Adair in 1925 by measurements of osmotic pressure.[1] The method has not been extensively used during the past fifty years for two reasons. The precision necessary in the measurement of a pressure difference between solution and solvent requires the use of large amounts of protein, and in only a few cases have these quantities been available. In the second place, it is of great importance that the time necessary to make the measurements be short, especially when one is dealing with native proteins. Recently, instruments have become available which require small volumes of solution and in which the approach to equilibrium is fast. It is for these reasons that renewed interest in the method has arisen. The method is attractive because it is relatively uncomplicated, and it allows one to obtain the molecular weight of a protein rapidly. Two reviews provide a comprehensive view of the theoretical[2] and of the practical[3] considerations on osmotic pressure and should be consulted for detailed information.

Theory

Consider two solutions separated by a semipermeable membrane, at constant temperature and pressure. At equilibrium the chemical potential of all *permeable* components must be the same on both sides of the

[1] G. S. Adair, *Proc. Roy. Soc. Ser. A.* **108**, 627 (1925).
[2] J. T. Edsall, *in* "The Proteins" (H. Neurath and K. Bailey, eds.), 1st ed., Vol. 1B, p. 549. Academic Press, New York, 1953.
[3] D. W. Kupke, *Advan. Protein Chem.* **15**, 57 (1960).

membrane. If one solution contains a nondiffusible component, like a protein molecule, then the diffusible components will flow across the membrane into the protein solution, where their chemical potentials are lower than on the side containing only diffusible components. This tendency can be counteracted by applying a hydrostatic pressure to the solution into which solvent is flowing in order to counteract the flow. When the excess pressure is just sufficient to counterbalance the flow of diffusible components, equilibrium is attained. The pressure difference is the osmotic pressure. Equilibrium is reached because the chemical potential of the diffusible components becomes equal on both sides of the membrane, and this is so because the chemical potential of any component increases with increasing pressure in proportion to its partial molar volume. The fundamental equation of osmotic pressure is given by van't Hoff's law:

$$\lim_{m_2 \to 0} \frac{\pi}{m_2} = RT$$

where π is the osmotic pressure, m_2 is the number of moles of solute per liter of solvent, R is the gas constant, and T is the absolute temperature. At low concentrations of solute, $m_2 = c_2/M_2$ where M_2 is the molecular weight of the solute and c_2 is the concentration in milligrams per milliliter of solution. Thus if c_2 is known, measurements of π at a given T give the molecular weight of component 2, since

$$\lim_{c_2 \to 0} \frac{\pi}{RTc_2} = \frac{1}{M_2}$$

If the measurements are made at 20° and the osmotic pressure is in centimeters of H_2O, the appropriate value of RT is 2.481×10^4 (cm $H_2O \cdot$ liters)/mole. It is the value of π/c_2 at zero solute concentration that is related to the molecular weight, and this value has to be obtained by extrapolation. A convenient way to do this is to use the relationship:

$$\frac{\pi}{RTc_2} = \frac{1}{M_2} + Bc_2 + Cc_2^2 + \cdots$$

where B, C, ..., etc. are the second, third, ..., etc. virial coefficients and are obtained by experiment. In practice, only the second virial coefficient is important and need be considered in dilute solutions.

Scatchard[4] had derived rigorous and complete equations for the osmotic pressure of dilute protein solutions containing salt (three-component system). The expression for B is:

[4] G. Scatchard, *J. Amer. Chem. Soc.* **68**, 2315 (1946).

$$B = \frac{1000}{2M_2{}^2} \left(\frac{Z_2{}^2}{2m_3} + \beta_{22} - \frac{\beta_{23}{}^2 m_3}{2 + \beta_{33} m_3} \right)$$

where Z_2 is the charge on the protein, m_3 is the concentration of salt in moles per liter, and the β's are derivatives of activity coefficients:

$$\beta_{22} = \frac{\partial \ln \gamma_2}{\partial m_2}, \qquad \beta_{23} = \frac{2\partial \ln \gamma_3}{\partial m_2}, \qquad \text{and} \qquad \beta_{33} = \frac{2\partial \ln \gamma_3}{\partial m_3}$$

This expression indicates that the value of B depends on three terms. The first term in parentheses, $Z_2{}^2/2m_3$, is the Donnan term (for a uni-univalent salt). It is always positive, but it approaches zero as the charge on the protein approaches zero, and it becomes small when the concentration of salt increases. Thus this term will be important when the solution contains low salt concentrations and has a pH different from that of the isoionic point of the protein.

The second term is the only one that would not disappear for a protein at the isoionic point in salt-free water. This term includes the contribution for the excluded volume of the molecule ($B = 4v_2/M_2$), and effects due to protein–protein interactions (both due to charge fluctuation and to specific association). In the latter case, the term will be negative. Finally, the third term depends upon the protein–salt and salt–salt interactions and is almost always negative. It is important when the protein–salt interactions are strong.

It is clear from this very brief analysis of the osmotic pressure equations that a large amount of information can be obtained from the osmotic pressure of macromolecules: not only can the molecular weight be determined, but the shape, charge, and interactions of the molecules can be studied.

Experimental Methods

Instruments[5]

There are two requirements for a good osmometer. One is rapid attainment of equilibrium, which implies that the surface of the membrane be

---

[5] While it may be appropriate in some cases to build an instrument with special features, it should be noted that there are at least two commercial instruments now available which embody the features described above. They are the Melabs CSM-2 membrane osmometer (Wescan Instruments, Inc., 2968 Scott Blvd., Santa Clara, California 95051) and the Hewlett-Packard membrane osmometer (Hewlett-Packard Co., Avondale Division, Route 41, Avondale, Pennsylvania 19311), which was formerly known as the Mechrolab Model 503 membrane osmometer. The Melabs instrument utilizes a pressure transducer on the solvent side of the membrane, and the difference in pressure between the two solutions is measured

large relative to the volume of the solution in contact on both sides of the membrane. The other is the size of the sample chamber, which should be small. An example of such an osmometer is shown in Fig. 1, which depicts the Scatchard modification of a Hepp osmometer.[7]

Preparation of Sample

The sample of protein is dissolved in the appropriate buffer at a high protein concentration, dialyzed for 24–48 hours against the solvent until equilibrium is reached, and then diluted to the required concentration with the buffer.

In most cases this procedure is sufficient to ensure that the chemical composition of the solution on both sides of the membrane remains fixed except for the concentration of protein. In practice it is even feasible to start with a concentrated solution of protein in water or in very dilute buffer, and to add a small amount of this solution to a large volume of solvent (for example: if one wishes to measure the osmotic pressure of a solution containing 10 mg of protein per milliliter one can dilute a solution containing 200 mg/ml to the required concentration with solvent). The small imbalances in ion concentrations that exist will equilibrate rapidly across the membrane.[8]

Dilutions of the protein solution are made so that one has 5–10 solutions which span the concentration range between 1 mg/ml and 20 mg/ml.

---

by the transducer during small changes in volume. The Hewlett-Packard instrument operates to null changes in volume on the solvent side of the membrane by changing the hydrostatic pressure of the solvent on that side of the membrane. Both these instruments require small volumes of solutions to fill the sample chamber (0.3–0.5 ml), and they reach equilibrium in approximately 10–15 minutes. The surface area of the working part of the membrane is approximately 7 cm² and the volume of solution in contact across the membrane is 0.15 ml. The approach to equilibrium can be monitored by displaying the change in hydrostatic pressure (Hewlett-Packard instrument) or the output of the pressure transducer on a strip-chart recorder. In this way a record is obtained of the approach to equilibrium and of the final equilibrium osmotic pressure. A good description of the procedures useful for operation of the Hewlett-Packard instrument has been presented by Paglini.[6]

[6] S. Paglini, *Anal. Biochem.* **23**, 247 (1968).
[7] G. Scatchard, *Amer. Sci.* **40**, 61 (1952).
[8] If the ionic strength of the solution is very low (<0.05) and the charge on the protein is substantial, the simple procedures described above are not adequate to define the protein component. In the latter cases, one should define the protein component in more elaborate ways,[2,4] or one will have to be satisfied with the estimation of a molecular weight that might include contributions from the protein and from "bound" diffusible components.[9]
[9] E. F. Casassa and H. Eisenberg, *Advan. Protein Chem.* **19**, 287 (1964).

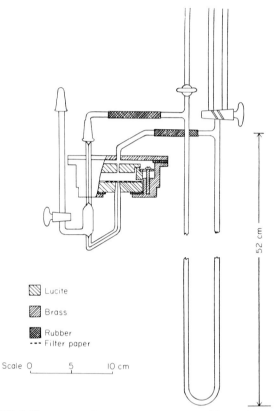

FIG. 1. Modified Hepp osmotic pressure apparatus. From G. Guidotti, *J. Biol. Chem.* **242**, 3685 (1967).

The samples are passed through a millipore filter (0.45 μ pore size) which has been washed with buffer to elute the bound detergent, and then they are kept in a water bath set at the same temperature as that at which the osmotic pressure will be measured. The pH and the protein concentration of each dilution of the protein are measured before it enters the sample chamber and after it is removed from the sample chamber. In this way, one ensures that the concentrations are exact (that is, that dilution has not taken place in the chamber) and that the pH is constant.

Membranes

A good membrane should allow rapid equilibration of water and ions, but it should be impermeable to the protein. It should also be sufficiently rigid to withstand ballooning under pressure. The main types are cellophane[10] membranes and collodion membranes, both of which are available commercially. Collodion membranes can also be prepared by the experimenter himself, who can select the degree of permeability according to the conditions of preparation.[11]

In all cases the membranes should be equilibrated with the solvent for at least 24 hours prior to use and then for an additional 24 hours after being placed in the instrument. In most cases badly fitting membranes and leaky membranes show up during the 24-hour period of equilibration in the instrument, and they must be replaced.

Method of Operation

The osmotic pressure is determined on solutions of protein that vary in concentration from 1 mg/ml to 10–20 mg/ml. In general, the pressures obtained with solutions that contain less than 1 mg/ml are small and unreliable; and the concentration range up to 10–20 mg/ml is sufficient to obtain a reliable molecular weight. These estimates are appropriate for proteins in the molecular weight range from 30,000 to 100,000. With small proteins, concentrations below 1 mg/ml will generate adequate pressures, whereas with large proteins (>100,000) the range of concentrations should be extended above 20 mg/ml. If the amount of material is not limited, it is advisable to use the following routine to obtain repeated measurements of the osmotic pressure of the various solutions. Measurements are made first on the most concentrated protein solution and then in serial fashion on the more dilute solutions. The chamber is not flushed with buffer between measurements. After measuring the pressure of the most dilute protein solutions, one reverses the order and measures the pressures of the successively more concentrated protein solutions. Thus in the first instance one proceeds from high to low con-

---

[10] "Bac-T-Flex" B-19 and B-20 membranes can be obtained from Schleicher and Schuell, Inc., 543 Washington Street, Keene, New Hampshire 03431. The B-19 membranes are more porous than the B-20 membranes. The equilibration of water and of ions across the B-19 membranes is rapid and takes place in less than 2 minutes (i.e., if one places 0.1 M NaCl on one side of the membrane and 0.15 M NaCl on the other side, equilibrium is reached in less than 2 minutes). When the same experiment is carried out with B-20 membranes, the time of equilibration is greater than 20 minutes. Accordingly, there is little leakage of proteins down to a molecular weight of less than 10,000 with the B-20 membranes, but there is measurable leakage of myoglobin (MW 17,800) through the B-19 membranes. Visking Cellophane casing (Visking Co., 6733 West 65th Street, Chicago, Illinois 60638) can also be used; the size 27/32 can be cut into an appropriate shape to fit the apparatus. The permeability of these membranes corresponds roughly to that of the B-20 membranes.

[11] C. W. Carr, D. Anderson, and I. Miller, *Science* **125**, 1245 (1957).

centrations and in the second, from low to high concentrations. This routine has the effect of damping out small systematic errors that arise from incomplete exchange of one protein solution by another in the sample chamber. Alternatively, at each protein concentration measurements can be made on three to four different samples of the solution.

Treatment of the Data

Values of the osmotic pressure are converted to pressure in centimeters of water (with knowledge of the density of the solvent) and the function π/c or π/RTc is plotted against c. In the simplest case, that is, when there are no protein–protein interactions and when the ionic strength of the solvent is sufficiently high, one obtains data that extrapolate in a linear fashion and with a positive slope to give a value of π/c at $c = 0$. This value yields the molecular weight of the protein under

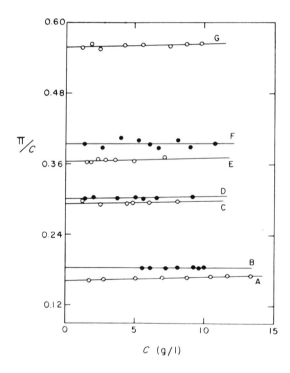

Fig. 2. Plots of π/c vs c for native proteins. (A) Aldolase, (B) lactate dehydrogenase, (C) enolase, (D) alcohol dehydrogenase, (E) serum albumin, (F) methemoglobin, and (G) ovalbumin. From F. J. Castellino and R. Barker, *Biochemistry* **7**, 2207 (1968).

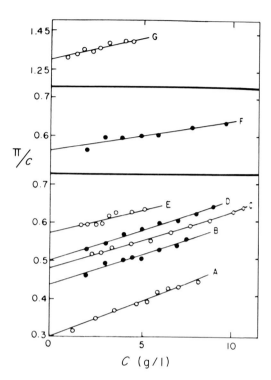

$$C \ (g/l)$$

Fig. 3. Plots of π/c vs c for proteins in GuHCl-MSH (0.5). (A) Serum albumin, (B) ovalbumin, (C) aldolase, (D) alcohol dehydrogenase, (E) lactate dehydrogenase, (F) enolase, and (G) methemoglobin. From F. J. Castellino and R. Barker, *Biochemistry* **7**, 2207 (1968).

consideration.[12] Examples of this type of plot are shown in Figs. 2–4 and in Tables I and II.

The slope of the line gives the value of the second virial coefficient. In principle, one can extract from this value information concerning the

[12] If the system is polydisperse, the value obtained corresponds to that of the number-average molecular weight, defined by

$$\tilde{M}_n = \frac{\Sigma c_i}{\Sigma (c_i/M_i)} = \frac{\Sigma m_i M_i}{\Sigma m_i}$$

where the c_i are the weight concentrations of the species i, the m_i are the molar concentrations of the species i, and the M_i are the molecular weights of each species i. Clearly, the chief contribution of the number-average molecular weight is made by the smallest molecules. Thus osmotic pressure measurements are relatively insensitive to a small fraction of large molecules in the system, in contrast to procedures (e.g., light-scattering measurements) which determine weight-average molecular weights.

TABLE I

The Molecular Weights of Native and Dissociated Proteins and the Number of Subunits as Determined by Osmometry[a]

Protein	Solvent density (g/cm³)	$RT \times 10^{-4}$ (cm 1 mole⁻¹)	π/c (cm 1 g⁻¹)	M_n	Subunits ±3%
Serum albumin	1.012	2.4941	0.365 ± 0.003	68,320 ± 600	1.0
Serum albumin + G[b]	1.150	2.0475	0.302 ± 0.005	67,790 ± 1,000	
Ovalbumin	1.012	2.4941	0.559 ± 0.003	44,620 ± 300	1.0
Ovalbumin + G	1.150	2.0475	0.440 ± 0.006	46,530 ± 600	
Alcohol dehydrogenase	1.012	2.4941	0.290 ± 0.006	86,000 ± 1,750	2.1
Alcohol dehydrogenase + G	1.150	2.0475	0.502 ± 0.003	40,790 ± 300	
Enolase	1.009	2.5015	0.303 ± 0.003	82,550 ± 800	2.3
Enolase + G	1.150	2.0475	0.561 ± 0.003	36,500 ± 200	
Methemoglobin	1.008	2.5040	0.393 ± 0.007	63,720 ± 1,100	4.0
Methemoglobin + G	1.150	2.0475	1.293 ± 0.006	15,840 ± 800	
Lactate dehydrogenase	1.012	2.4941	0.183 ± 0.002	136,290 ± 1,400	3.8
Lactate dehydrogenase + G	1.150	2.0475	0.566 ± 0.013	36,180 ± 800	
Aldolase	1.008	2.5040	0.160 ± 0.001	156,500 ± 1,000	3.7
Aldolase + G	1.150	2.0475	0.483 ± 0.003	42,400 ± 300	

[a] From F. J. Castellino and R. Barker [*Biochemistry* 7, 2207 (1968)].
[b] G = proteins in GuHCl–mercaptoethanol (0.5).

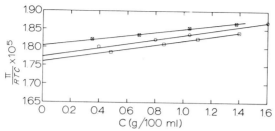

Fig. 4. Osmotic pressure data for hemoglobins in $0.2\,M$ NaCl, pH 7. ⊠, oxy-hemoglobin; ⊙, CO hemoglobin; □, deoxyhemoglobin. From G. Guidotti, *J. Biol. Chem.* **242**, 3685 (1967).

charge on the protein, the shape of the protein, and the protein–salt interaction. In practice, such deductions are very painstaking, and they have been done only rarely.[13,14]

In many cases protein–protein interactions are sufficient to cause the slope of the plot of π/c against c to have a negative value. In such a case one may be interested only in the molecular weight of the smallest molecule. In this case one attempts to obtain measurements of the osmotic pressure at the lowest possible concentration of protein and then extrapolates these values to zero concentration. Since, as is shown in Fig. 5, the curves have a negative slope, one will obtain a virtual intercept on the π/c axis which can correspond to the molecular weight of the smallest species of interest.

Finally, the data may be of the type depicted in Figs. 6 and 7, which show the presence of curvature in the plot of π/c vs. c. These data can yield under appropriate circumstances not only values of the molecular

TABLE II

MOLECULAR WEIGHTS OF HEMOGLOBIN IN $0.2\,M$ NaCl, AT pH 7 AND $20°$[a]

Type of hemoglobin	Molecular weight (g/mole)	B' (mole-ml g^{-2})
Oxyhemoglobin	55,400	4.7×10^{-5}
CO hemoglobin	56,400	6.1×10^{-5}
Deoxyhemoglobin	57,000	5.9×10^{-5}

[a] From G. Guidotti [*J. Biol. Chem.* **242**, 3685 (1967)].

[13] G. Scatchard, A. C. Batchelder, and A. Brown, *J. Amer. Chem. Soc.* **68**, 2320 (1946).

[14] J. T. Edsall, H. Edelhoch, R. Lontie, and P. R. Morrison, *J. Amer. Chem. Soc.* **72**, 4641 (1950).

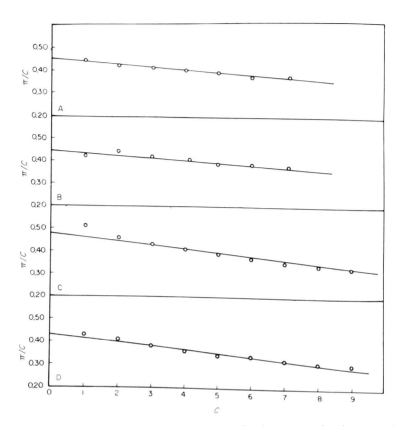

FIG. 5. Effect of pH on osmotic pressure of tobacco mosaic virus protein in phosphate buffer, ionic strength 0.1; π is the osmotic pressure in centimeters of water and c, the protein concentration in milligrams per milliliter. (A) pH 8.0, $\bar{M}_n = 50,500$; (B) pH 7.5, $\bar{M} = 53,400$; (C) pH 7.0, $\bar{M}_n = 49,500$; (D) pH 6.5, $\bar{M}_n = 54,650$. From K. Banerjee and M. A. Lauffer, *Biochemistry* **5**, 1957 (1966).

weights of the aggregated and disaggregated species, but also of the dissociation constant for the system (Tables III and IV). Treatment of these data, however, is somewhat complicated because curvature is the result of two factors. The negative slope is due almost entirely to protein–protein interaction which can be described by an appropriate dissociation constant, K. The positive slope derives from the usual interaction (second virial) coefficient, B. Thus one has to sort out for a given set of data two numbers: one for the value of K and the other for the value of B. There are several ways in which this can be done, and which include simple curve-fitting approaches[15] and more sophisticated computer search

[15] G. Guidotti, *J. Biol. Chem.* **242**, 3685 (1967).

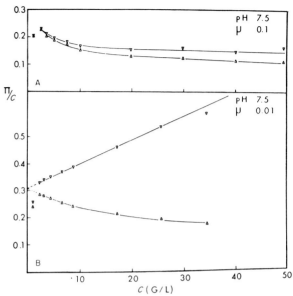

Fig. 6. Plot of π/c vs c for tobacco mosaic virus protein, in phosphate buffer at room temperature (▼). Corrected curve (▲) is the result obtained after allowing for Donnan effect and excluded volume effect. (A) pH 7.5; ionic strength 0.1; (B) pH 7.5, ionic strength 0.01. From H. Stauffer, S. Srinivasan, and M. A. Lauffer, *Biochemistry* 9, 193 (1970).

analyses.[16,17] In any event, it is important to remember that the interaction coefficient can have a dominating role in such cases. The lower panel of Fig. 6 presents a vivid demonstration of the contrasting effects which protein–protein association, and the excluded volume and the Donnan effect have on the osmotic pressures measured with solutions of tobacco mosaic virus protein. The upper curve shows the osmotic pressure data actually obtained: there is no evidence of protein–protein interaction. However, when these data are corrected for the Donnan effect and the excluded volume term (lower curve), then the aggregation of the protein becomes evident.

An interesting example of the use of osmometry in the analysis of protein–protein interactions is shown in Fig. 8. The mixture of horse and human hemoglobins acts under these solution conditions as a single species of hemoglobin. This means that hybrid molecules of the type

$$(\alpha_1\beta_1)^{\text{human}}$$

$$(\beta_2\alpha_2)^{\text{horse}}$$

[16] E. T. Adams, Jr., *Biochemistry* 4, 1655 (1965).
[17] R. F. Steiner, *Biochemistry* 7, 2201 (1968).

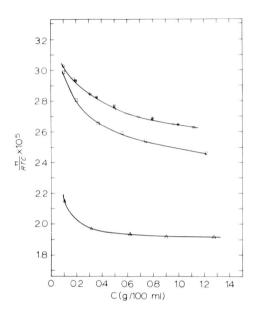

Fig. 7. Osmotic pressure data for hemoglobins in 0.4 M MgCl$_2$, pH 7. $\otimes$, cyanmethemoglobin; $\bigcirc$, oxyhemoglobin; $\square$, CO hemoglobin; $\triangle$, deoxyhemoglobin. From G. Guidotti, *J. Biol. Chem.* **242**, 3685 (1967).

are present in solution or that the contact surfaces that are engaged in holding dimers together in the tetrameric molecules are the same in horse and human hemoglobins.

In conclusion, two main types of information can be obtained by

TABLE III

TETRAMER TO DIMER DISSOCIATION CONSTANTS FOR OXYHEMOGLOBIN IN SOLVENTS AT pH 7 AND 20°[a]

Solvent	K ($M \times 10^5$)	B' (mole-ml g^{-2})
0.4 M MgCl$_2$	60	12.5×10^{-5}
1 M MgCl$_2$	$\sim$1000	
1 M NaCl	2	16.8×10^{-5}
2 M NaCl	12	17.4×10^{-5}
1 M CH$_3$COONa	2	16.8×10^{-5}
1 M NaClO$_4$	$\sim$1000	
1 M NaI	$>$1000	

[a] From G. Guidotti [*J. Biol. Chem.* **242**, 3685 (1967)].

TABLE IV
DISSOCIATION CONSTANTS FOR TREATED HEMOGLOBINS IN 0.4 M MgCl$_2$
AND 2 M NaCl, pH 7 AND 20°[a]

| | K | |
| | ($M \times 10^5$) | |
Type of hemoglobin	0.4 M MgCl$_2$	2 M NaCl
Oxyhemoglobin	60	12
EM-treated oxyhemoglobin		5
CO hemoglobin	30	7.5
Iodoacetamide-treated CO hemoglobin	30	7.5
EM-treated CO hemoglobin	25	5
Cyanmethemoglobin	60	
Iodoacetamide-treated cyanmethemoglobin	60	
EM-treated cyanmethemoglobin	35	

[a] From G. Guidotti [*J. Biol. Chem.* **242**, 3685 (1967)].

measurement of the osmotic pressure of protein solutions: the molecular
weight of the protein, and parameters related to the protein–protein
interactions in the system.

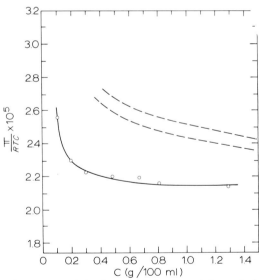

FIG. 8. Osmotic pressure data for an equimolar mixture of six CO hemoglobins
(normal, iodoacetamide-treated, and N-ethylmaleimide-treated human and horse
hemoglobins) in 2 M NaCl, pH 7. The lower dashed line is the theoretical curve for
a mixture of six solutes; the upper dashed line is the theoretical curve for a mixture of nine solutes (shown because horse red blood cells contain two major types of
hemoglobin). From G. Guidotti, *J. Biol. Chem.* **242**, 3694 (1967).

Section II
Interactions

[11] Sedimentation Velocity Measurement of Protein Association

By LILO M. GILBERT and G. A. GILBERT

There are many general accounts[1-8] available covering the subject of this article (see this volume [12]) and numerous original papers devoted to the development of the subject. Those who want general information, or who are specialists in the field, are in fact well catered for, but the enzymologist or the protein chemist who comes across a case of protein interaction and would like to pursue it further needs more specific guidance. This article is therefore addressed to readers with average resources at their disposal who want to get as much information as possible from their sedimentation velocity measurements on proteins. The problem of interpreting such physical measurements is in a sense always an open ended one. A model can be postulated for a system, and a reasonable fit obtained to experimental data, but one can never be sure that another model would not fit the data at least as well. A certain degree of scepticism is helpful, and attention to the evidence to be obtained by the other approaches to be found in this volume. These have their own advantages, just as velocity sedimentation has the particular merit of molecular resolution in the centrifuge cell, making it less sensitive to the presence of impurities or decomposition products.

There are two significant trends in this field. One is the development of computer programs[7,9-20] with which realistic sedimentation velocity

[1] T. Svedberg and K. O. Pedersen, "The Ultracentrifuge," p. 28. Oxford Univ. Press, London and New York, 1940.

[2] H. K. Schachman, "Ultracentrifugation in Biochemistry." Academic Press, New York, 1963.

[3] H. Fujita, "Mathematical Theory of Sedimentation Analysis." Academic Press, New York, 1962.

[4] "Ultracentrifugal Analysis" (J. W. Williams, ed.). Academic Press, New York, 1963.

[5] L. W. Nichol, J. L. Bethune, G. Kegeles, and E. L. Hess, *in* "The Proteins" (H. Neurath, ed.), 2nd ed., Vol. II, Chapter 9. Academic Press, New York, 1964.

[6] J. M. Creeth and R. H. Pain, *Progr. Biophys. Mol. Biol.* **17**, 217 (1967).

[7] J. R. Cann, "Interacting Macromolecules: The Theory and Practice of their Electrophoresis, Ultracentrifugation, and Chromatography." Academic Press, New York, 1970.

[8] T. J. Bowen, "An Introduction to Ultracentrifugation." Wiley (Interscience), New York, 1970.

[9] H. Vink, *Acta Chem. Scand.* **18**, 409 (1964).

[10] D. J. Cox, *Arch. Biochem. Biophys.* **112**, 259 (1965).

schlieren patterns can be simulated for reacting systems on the basis of interaction constants and other physical parameters that are fed into the programs. Notably these parameters include diffusion terms which, with few exceptions,[21] were omitted for simplicity in earlier treatments of the sedimentation of interacting proteins. The other development is the computer-controlled optical scanning of ultracentrifuge cells during the course of a run[22] to obtain a digital record of optical density at closely spaced intervals throughout each cell.

Obviously, a great step forward will have been achieved when these two approaches have been brought together and the theoretical calculation of protein distribution can be matched by statistical methods to the digital record,[23] to discover the stoichiometry and interaction constants of a protein system. Until such time, much can be done with what is already available in most laboratories.

We will begin by showing how one can analyze a set of data obtained in the conventional way for the dependence of the sedimentation coefficient of a substance on its concentration. We will assume a basic knowledge of ultracentrifuge theory and practice, and we will also assume that it will not be necessary to reiterate warnings concerning time-dependent effects where reactions are slow, or other considerations amply dealt with in an earlier review,[5] or in other chapters of this volume.

Simple Association

We do not have to distinguish between the apparently opposite processes of the dissociation of a native protein into subunits or its association into oligomers. Fortunately both processes are encompassed by a single formal description, even though the same protein may be dissociated under some conditions and associated under others. For instance, bovine β-lactoglobulin A dissociates into two equal subunits at

---

[11] D. J. Cox, *Science* **152**, 359 (1966).

[12] D. J. Cox, *Arch. Biochem. Biophys.* **129**, 106 (1969).

[13] D. J. Cox, *Arch. Biochem. Biophys.* **146**, 181 (1971).

[14] M. Dishon, G. H. Weiss, and D. A. Yphantis, *Biopolymers* **4**, 449 (1966).

[15] M. Dishon, G. H. Weiss, and D. A. Yphantis, *Biopolymers* **5**, 697 (1967).

[16] J. R. Cann and W. B. Goad, *J. Biol. Chem.* **240**, 148 (1965).

[17] J. R. Cann and W. B. Goad, *Advan. Enzymol.* **30**, 139 (1968).

[18] W. B. Goad and J. R. Cann, *Ann. N.Y. Acad. Sci.* **164**, 172 (1969).

[19] J. L. Bethune and G. Kegeles, *J. Phys. Chem.* **65**, 1761 (1961).

[20] D. F. Oberhauser, J. L. Bethune, and G. Kegeles, *Biochemistry* **4**, 1878 (1965).

[21] E. O. Field and A. G. Ogston, *Biochem. J.* **60**, 661 (1955).

[22] S. P. Spragg, *Anal. Chim. Acta* **38**, 137 (1967).

[23] R. Trautman, S. P. Spragg, and H. B. Halsall, *Anal. Biochem.* **28**, 396 (1969).

low pH,[24,25] whereas it associates[26-31] into an oligomer containing eight of these elemental subunits under other conditions. To illustrate the effect of dissociation on sedimentation velocity with an important historical example, we give in Table I the numerical data of Townend et al.[25] obtained for the sedimentation of β-lactoglobulin A,B (a mixture of the two genetic forms, A and B) at pH 1.6. It can be seen from Table I that the value of the sedimentation coefficient $\bar{s}$ at first rises with concentration of protein and then falls after passing through a maximum at a constituent concentration $\bar{w}$ of protein of about 0.8 g/dl. This contrasts with the linear negative concentration dependence accepted as the norm for simple noninteracting proteins, expressed by the equation

$$\bar{s} = (s)_0(1 - \mathbf{g}\bar{w}) \tag{1}$$

where $\mathbf{g}$ is a constant. The quantity $(s)_0$ is the sedimentation coefficient of the protein at the limit of zero concentration where molecules sediment uninfluenced by their neighbors. The principal influence of these neighbors is to increase the density of the solution, the viscosity of the solution, and the backflow of solvent. As a consequence, the value of $\mathbf{g}$ depends upon such characteristics as the effective hydrodynamic volume of the protein molecule. Reversible association between the molecules upsets the linear relationship of Eq. (1).[32,33] It would be invaluable to know more about the "normal" value of $\mathbf{g}$ for a molecule of given shape and size because departures from it could then be used as an indication of protein association. We believe a good approximation to this value for uncharged spherical globular proteins to be 0.07 dl/g, based on the value for hemoglobin in neutral solution[34] at concentrations at which hemoglobin is largely in its tetrameric form. The ground for this belief is that normal hemoglobin shows no sign of further association, as evidenced

[24] R. Townend and S. N. Timasheff, J. Amer. Chem. Soc. 79, 3613 (1957).
[25] R. Townend, L. Weinberger, and S. N. Timasheff, J. Amer. Chem. Soc. 82, 3175 (1960).
[26] A. G. Ogston and J. M. A. Tilley, Biochem. J. 59, 644 (1955).
[27] R. Townend, R. J. Winterbottom and S. N. Timasheff, J. Amer. Chem. Soc. 82, 3161 (1960.
[28] R. Townend and S. N. Timasheff, J. Amer. Chem. Soc. 82, 3168 (1960).
[29] H. A. McKenzie, Advan. Protein Chem. 22, 56 (1967).
[30] H. A. McKenzie, W. H. Sawyer, and M. B. Smith, Biochim. Biophys. Acta 147, 73 (1967).
[31] J. M. Armstrong and H. A. McKenzie, Biochim. Biophys. Acta 147, 93 (1967).
[32] K. O. Pedersen, Cold Spring Harbor Symp. Quant. Biol. 14, 140 (1950).
[33] G. W. Schwert, J. Biol. Chem. 179, 655 (1949).
[34] E. Chiancone, L. M. Gilbert, G. A. Gilbert, and G. L. Kellett, J. Biol. Chem. 243, 1212 (1968).

TABLE I

SEDIMENTATION OF β-LACTOGLOBULIN A, B, pH 1.6[a]

Concentration $\bar{w}$ (g/dl)	0.098	0.13	0.21	0.24	0.25	0.56	0.83	1.00	1.07	1.54	2.10
Sedimentation coefficient, s	2.06	2.22	2.16	2.26	2.16	2.27	2.32	2.28	2.30	2.23	2.05

[a] Experimental data from R. Townend, L. Weinberger, and S. N. Timasheff, *J. Amer. Chem. Soc.* **82**, 3175 (1960).

by the absence of crystallization or precipitation, even at the very high concentration at which it is present in the red cell. Until further empirical evidence is brought to bear on the question, this value can be used only as a rough guide. It should be noted that highly asymmetric or random coil proteins are better described by the expression, familiar from polymer solution theory,

$$\bar{s} = (s)_0/(1 + \mathbf{g}\bar{w}) \tag{2}$$

where $\mathbf{g}$ has usually a much higher value than for globular proteins, due to the much larger effective hydrodynamic volume of such molecules.

If the protein molecules are in a state of reversible association, (whether rapidly reversible or not), more than one type of species must be sedimenting in the plateau region of the ultracentrifuge cell. The total flux, $\bar{s}\bar{w}$, of material is the sum of the separate fluxes,[1] $s_i w_i$, of each species i.

$$\bar{s}\bar{w} = \sum_i s_i w_i \tag{3}$$

As Goldberg[35] has shown, $\bar{s}$ is found experimentally by determining the rate of movement of the second moment of the concentration gradient of the boundary, as portrayed by the schlieren pattern, or of the first moment of the concentration profile if the boundary is detected directly by optical density measurements. If the boundary is symmetrical, the rate of movement of the mode of the schlieren pattern is a satisfactory measure of $\bar{s}$.

The concentrations of the various products of association are related to each other by the law of mass action,

$$\begin{aligned}
w_1 &= L_{1,1}w_1 \; ; L_{1,1} = 1 \\
w_2 &= L_{1,2}w_1{}^2 \\
&\;\vdots \\
w_i &= L_{1,i}w_1{}^i \\
&\;\vdots
\end{aligned} \tag{4}$$

(If each w_i is expressed in grams per deciliter and each association constant $L_{1,i}$ in deciliters per gram rather than in liters per mole, inaccuracies in molecular weight assumptions do not affect the values of the association constants.) The constituent concentration $\bar{w}$ is given by

$$\bar{w} = \sum_i w_i \tag{5}$$

Substitution of Eq. (4) in Eq. (3) leads to the equation

[35] R. J. Goldberg, *J. Phys. Chem.* **57**, 194 (1953).

$$\bar{s}\bar{w} = \sum_i s_i L_{1,i} w_1{}^i \qquad (6)$$

The left-hand side of Eq. (6) is composed solely of quantities that can be obtained experimentally with an ascertainable degree of accuracy, while the right-hand side contains the parameters that have to be evaluated by appropriate methods of data analysis. It is a matter of general experience that extremely high precision and reproducibility of measurement is essential if several parameters in an equation of type (6) are to be determined to a reasonable level of accuracy. Here there is the additional complication that the s_i and $L_{1,i}$ are certainly functions of the concentration $\bar{w}$, and allowance must be made for this. The simplest assumption to make, and the most the data will bear, is that each flux $w_i s_i$ is modified by the term $(1 - g\bar{w})$. This is in effect to make the assumption that to a first approximation each s_i and $L_{1,i}$ is modified by a correction factor increasing in proportion to the first power of $\bar{w}$, without apportioning the correction between s_i and $L_{1,i}$. Adopting a single value for g for all species is an undesirable assumption, but also unfortunately an unavoidable contemporary expedient.

In its final form Eq. (6) becomes

$$\bar{s}\bar{w} = (1 - g\bar{w}) \sum_i (s_i)_0 L_{1,i} w_1{}^i \qquad (7)$$

Boundary Shape

Nothing has yet been said about the shape of the boundary, but since this is often informative, and has great importance for future developments, it will now be treated in some detail. The shape of all boundaries is dominated by diffusion which leads to the familiar Gaussian form of schlieren patterns if no interaction is taking place. If, however, there is interaction, some striking effects can be superimposed upon the normal pattern. Chymotrypsin,[36] mercuripapain,[37] and β-lactoglobulin A provide good examples of this. The process which disturbs the Gaussian shape is the lagging behind at the boundary of the smaller molecules in the reaction mixture, thereby setting up a new equilibrium corresponding to a concentration lower than that within the plateau region. Because of the lower concentration, the average size of oligomer is less, the average sedimentation velocity is less, and the lost molecules can never catch up again. Thus the boundary broadens more than it would by diffusion alone. A useful estimate of the magnitude of this effect can be

[36] V. Massey, W. F. Harrington, and B. S. Hartley, *Discuss. Faraday Soc.* **20**, 24 (1955).
[37] E. L. Smith, J. R. Kimmel, and D. M. Brown, *J. Biol. Chem.* **207**, 533 (1954).

obtained by considering what would happen in the absence of diffusion.[38-41] This is not entirely a theoretical exercise as it is easy to show that such a "diffusion-free" boundary, or "asymptotic" boundary as it is often called, can be constructed *experimentally* from measurements on finite difference boundaries. This finite difference approach leads rather directly to the formulas required for describing the asymptotic boundary, and therefore we will deal with it next.

Finite Difference and Differential Boundaries

An initially sharp boundary between two solutions which differ only slightly in concentration spreads by diffusion[3,42] at a rate determined by the mean value of the diffusion coefficient of the solute. In the limit of a vanishingly small difference in concentration, the shape of the boundary becomes perfectly Gaussian and the velocity of the "equivalent sharp boundary"[35,43] becomes the same as that of the mode of the boundary. If one makes velocity measurements on finite difference boundaries of decreasing difference and extrapolates to zero concentration, the velocity of the true differential boundary can be determined experimentally[44,45] (though few such measurements have been done). Familiarity with the idea of such differential boundaries is crucial to an understanding of the effect of interaction on the shape of boundaries. Consider a succession of such differential boundaries set up between solutions of gradually increasing concentration, each solution being separated from the next by a plateau region. Imagine the equivalent sharp differential boundary to be constructed between each solution, and then, since the length of each plateau region is arbitrary, let each such length be reduced to zero. The result is a continuous "integral" boundary made up of infinitesimal steps, the velocity of any point in the boundary being given by the velocity of the corresponding step (i.e., the experimentally determinable differential boundary) at that concentration. This continuous boundary is the boundary that would be seen if diffusion were absent, with one proviso. If any region of the boundary would apparently move at a slower velocity than another region at lower concentration, a physically impossible involuted boundary with two values

[38] G. A. Gilbert, *Discuss. Faraday Soc.* **20**, 68 (1955).
[39] G. A. Gilbert, *Proc. Roy. Soc. Ser. A* **250**, 377 (1959).
[40] G. A. Gilbert, *Proc. Roy. Soc. Ser. A* **276**, 354 (1963).
[41] L. W. Nichol and A. G. Ogston, *Proc. Roy. Soc. Ser B* **163**, 343 (1965).
[42] H. Fujita, *J. Chem. Phys.* **31**, 5 (1959).
[43] L. G. Longsworth, *J. Amer. Chem. Soc.* **65**, 1755 (1943).
[44] R. F. Steiner, *Arch. Biochem. Biophys.* **49**, 400 (1954).
[45] R. Hersh and H. K. Schachman, *J. Amer. Chem. Soc.* **77**, 5228 (1955).

to represent the concentration at a single point would tend to form. Instead, we would then have a self-sharpening or "hypersharp" boundary with the faster region of lower concentration overridden and suppressed by the slower region of higher concentration.[39,46]

To calculate the shape of a diffusion-free interaction boundary, it is therefore sufficient to find the corresponding differential boundary velocity at each concentration, taking care finally to replace any involuted region by an equivalent hypersharp region. As we have said, this is realizable experimentally, but it turns out that the experiments need not be conducted on finite difference boundaries. To show this, we next calculate the velocity of a differential boundary, as Hersh and Schachman[45] have done (see also Miller[47] and Fujita[42]).

Let two parallel planes α and β delimit a volume of solution within which there is a finite difference boundary parallel to α and β between plateau concentration levels $\bar{w}$ and $(\bar{w} + \Delta\bar{w})$. Let the fluxes of the constituents in these plateau regions be $\bar{w}\bar{s}$ and $[\bar{w}\bar{s} + \Delta(\bar{w}\bar{s})]$, respectively. After unit time the boundary will have moved a distance s, if we use s as the symbol for the velocity of a finite difference boundary. We assume that the planes α and β have been so positioned that meanwhile the boundary remains between them. Then for conservation of mass, the decrease $s\Delta\bar{w}$ in the amount of substance between the planes must equal the difference between the fluxes into and out of the planes, i.e.

$$s\Delta\bar{w} = \Delta(\bar{w}\bar{s}) \tag{8}$$

Rearranging and proceeding to the limit gives for the velocity s of the differential boundary, and therefore of the diffusion-free boundary at the level of concentration $\bar{w}$, the equation

$$s = \frac{d(\bar{w}\bar{s})}{d\bar{w}} = \bar{s} + \bar{w}\frac{d\bar{s}}{d\bar{w}} \tag{9}$$

We will call s the profile velocity. According to Eq. (9) measurement of $\bar{s}$, the weight-average velocity of the integral boundary, as a function of $\bar{w}$ to give $d\bar{s}/d\bar{w}$ is an alternative and equally rigorous way of determining *experimentally* the shape of the diffusion-free boundary, s versus $\bar{w}$.

To calculate the profile velocity[40] s as a function of $\bar{w}$ for a reaction boundary for any given model system, one substitutes in Eq. (9) the expression for $\bar{s}$ given in Eq. (7) to obtain

[46] D. DeVault, *J. Amer. Chem. Soc.* **65**, 532 (1943).
[47] L. W. Miller, *Z. Phys. Chem. (Leipzig)* **69**, 436 (1909).

$$s = \frac{d}{d\bar{w}} \left\{ (1 - \mathfrak{g}\bar{w}) \sum_i (s_i)_0 L_{1,i} w_1{}^i \right\} \tag{10}$$

By recalling the definition of the constituent[48] concentration $\bar{w}$ [Eq. (5)] and making use of the relationship

$$d/d\bar{w} = (1/(d\bar{w}/dw_1))(d/dw_1) \tag{11}$$

the differentiation on the right-hand side of Eq. (10) can be carried out to give for s

$$s = (1 - \mathfrak{g}\bar{w}) \frac{\sum\limits_i iw_i(s_i)_0}{\sum\limits_i i(w_i)} - \mathfrak{g} \sum_i w_i(s_i)_0 \tag{12}$$

Incidentally, this equation is identical with the equation for s obtained [Eq. (16) of Gilbert[40]] by considering the effect of reversible reaction on the flux across planes in an "integral" boundary.

Differentiation of Eq. (12) with respect to $\bar{w}$ gives

$$\frac{ds}{d\bar{w}}$$

$$= (1 - \mathfrak{g}\bar{w}) \frac{\left(\sum\limits_i iw_i\right)\left(\sum\limits_i i(i-1)w_i(s_i)_0\right) - \left(\sum\limits_i i(i-1)w_i\right)\left(\sum\limits_i iw_i(s_i)_0\right)}{\left(\sum\limits_i iw_i\right)^3}$$

$$- 2\mathfrak{g} \frac{\sum\limits_i iw_i(s_i)_0}{\sum\limits_i iw_i} \tag{13}$$

The asymptotic schlieren pattern can be constructed by plotting the reciprocal of $(ds)/(d\bar{w})$ as ordinate calculated by Eq. (13) against s as abscissa calculated by Eq. (12).

Practical Considerations

Dimerization

If association is suspected, it is advisable to test whether the data can be adequately represented by a monomer–dimer system. If so, there is not much point in postulating more elaborate systems with further adjustable parameters unless there is independent evidence which re-

---

[48] A. Tiselius, *Nova Acta Regiae Soc. Sci. Upsal.* **7**, No. 4, 1 (1930).

quires this. As an aid to detecting dimerization,[49] we have calculated $\bar{s}$ as a function of $\bar{w}$ for a monomer–dimer system for a range of values of $L_{1,2}$.

The necessary equations are

$$w_1 = \frac{-1 + \sqrt{1 + 4L_{1,2}\bar{w}}}{2L_{1,2}} \tag{14}$$

$$w_2 = L_{1,2}(w_1)^2 \tag{15}$$

$$\bar{s} = \frac{1 - \mathbf{g}\bar{w}}{\bar{w}} \left(w_1(s_1)_0 + w_2(s_2)_0 \right) \tag{16}$$

$$s = (1 - \mathbf{g}\bar{w}) \left(\frac{w_1(s_1)_0 + 2w_2(s_2)_0}{w_1 + 2w_2} \right) - \mathbf{g}(w_1(s_1)_0 + w_2(s_2)_0) \tag{17}$$

$$\frac{d\bar{w}}{ds} = \frac{1}{ds/d\bar{w}}$$

$$= 1 \bigg/ \left\{ (1 - \mathbf{g}\bar{w}) \frac{2w_2(s_2)_0(w_1 + 2w_2) - 2w_2(w_1(s_1)_0 + 2w_2(s_2)_0)}{(w_1 + 2w_2)^3} \right.$$
$$\left. - 2\mathbf{g} \frac{w_1(s_1)_0 + 2w_2(s_2)_0}{w_1 + 2w_2} \right\} \tag{18}$$

In the absence of any change in frictional coefficient, an increase of n times in the size of a particle causes its velocity of sedimentation to increase by the factor $(n)^{2/3}$. For illustration we have therefore taken the relative velocities of monomer and dimer as 1 and $2^{2/3}$, respectively. In special cases there may be evidence, for example, from X-ray studies, which allows an estimate to be made of the approximate value of the ratio of the frictional coefficients of the monomer and dimer, which can be used to modify the factor $2^{2/3}$. We have assumed $\mathbf{g}$ to have the value mentioned above of 0.07 dl/g. The result of the calculations is shown in Fig. 1.

The most significant feature of Fig. 1 is the maximum in each curve and its progressive displacement toward zero concentration at high and low values of $L_{1,2}$. There is no value of $L_{1,2}$ for which the maximum occurs at a concentration above about 1 g/dl (for $\mathbf{g} = 0.07$ dl/g). It can be shown by differentiating $\bar{s}$ in Eq. (6) with respect to $\bar{w}$ and equating to zero that the value of w_1 at the maximum of a curve [and hence the value there of $\bar{w}$ by Eq. (5)] is given for the monomer–dimer system by the cubic equation

$$2rL_{1,2}{}^3w_1{}^3 + (4r + 1)L_{1,2}{}^2w_1{}^2 + 2(r + 1)L_{1,2}w_1 - (r - 1)L_{1,2}/\mathbf{g} + 1 = 0 \tag{19}$$

where r is the ratio $(s_2)_0/(s_1)_0$, set in this instance as $2^{2/3}$, i.e., 1.587.

[49] G. A. Gilbert, *Nature (London)* **186**, 882 (1960).

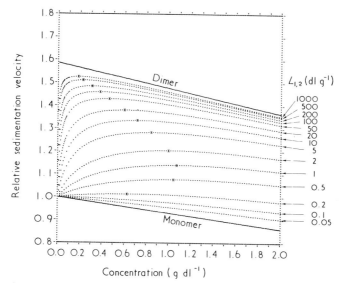

FIG. 1. Chart to demonstrate for a range of values of the association constant $L_{1,2}$ the dependence on constituent concentration of the sedimentation velocity of a monomer–dimer system in reversible equilibrium. Velocities are shown relative to the velocity of monomer at the limit of zero concentration. Maxima in the curves are indicated by crosses.

If the data of Table I are plotted[25] (Fig. 2), it becomes clear that they mark out a curve rather similar in shape to one in the middle region of Fig. 1. What is missing is some way of correlating the relative velocities making up the ordinates of Fig. 1 with the absolute sedimentation coefficients of Table I. For this correlation, a value for either $(s_1)_0$ or $(s_2)_0$ is needed. Figure 2 shows how difficult it is to obtain an estimate of $(s_2)_0$ by extrapolation, and indicates that effort should be put into measurements at very low concentrations to find $(s_1)_0$. Technically it is rather difficult to do this owing to the risk of convection, and controls should be done in parallel with nonassociating proteins to ensure that spuriously low sedimentation coefficients are not being found (a very real danger of frequent occurrence). Alternatively, it may be possible to estimate $(s_2)_0$ or $(s_1)_0$ by using data obtained under other conditions where association is either almost complete or almost absent. Thus in an earlier analysis[50] of this β-lactoglobulin system, we used a value of 2.87 S for $(s_2)_0$ after considering published work on the sedimentation of β-lactoglobulin under less acid conditions.[25] There is X-ray evidence

[50] L. M. Gilbert and G. A. Gilbert, *Nature* (*London*) **192**, 1181 (1961).

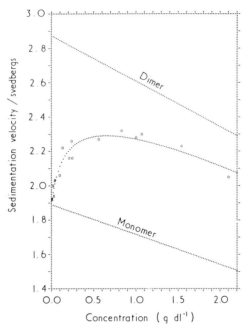

Fig. 2. Sedimentation of β-lactoglobulin A,B, at pH 1.6. $\bigcirc$, Experimental data of R. Townend, L. Weinberger, and S. N. Timasheff, *J. Amer. Chem. Soc.* **82**, 3175 (1960). X, Simulated data with normally distributed random errors. .... Theoretical curve fitted to experimental data by least-squares program. Limiting values for the velocities of monomer and dimer are also shown. Values of the parameters used are those listed in Table II.

from the work of Green and Aschaffenburg[51] that to a first approximation the monomer should be treated as a sphere and the dimer as a prolate ellipsoid of axial ratio 2 and relative frictional coefficient 1.044. The ratio $(s_2)_0/(s_1)_0$ becomes $1.587/1.044 = 1.520$, and $(s_1)_0$ is then $2.87/1.520 = 1.89$ S. By dividing the $\bar{s}$ values in Table I by 1.89 and comparing with Fig. 1, it is found that an interaction constant of about 5 dl/g is roughly consistent with the data.

Least-Squares Analysis

We could pursue this qualitative approach further by constructing a finer net of theoretical lines based on the ratio 1.520 for $(s_2)_0/(s_1)_0$, and then looking at the effect of altering the value of **g** from its initial arbitrary value of 0.07 dl/g, but this would leave us with no measure

[51] D. W. Green and R. Aschaffenburg, *J. Mol. Biol.* **1**, 54 (1954).

of the goodness of fit of the curves to the data, nor with any estimate of the standard errors of the parameters that we would find. We therefore turn to conventional nonlinear least squares fitting procedures by computer program to minimize the sum of squares of the residuals (S.S.R.), using the qualitative fit to form the initial S.S.R. where

$$S.S.R. = (\bar{s}_{exp} - \bar{s}_{calc})^2 \tag{20}$$

and then modifying the parameters to minimize the S.S.R. Computer centers can usually give advice on least squares procedures, and many will carry programs that can be adapted for use with this system.

These programs depend upon iteration and work best when the number of parameters to be found is small and the initial guessed values are not too far from the final optimum values. In the present instance, we first limit the number of parameters to two, namely, $L_{1,2}$ and $\mathbf{g}$, by assuming that $(s_1)_0$ and $(s_2)_0$ have the values 1.89 and 2.87 S, respectively. Table II gives the output of the computer program when it operates on the data of Table I.

It will be noticed that there is a slight bias of the residuals, but that the standard error (0.05 S) for an individual measurement is reasonable. The value of 4.34 dl/g found for the association constant $L_{1,2}$ agrees remarkably well with the value Townend et al.[25] obtained by the entirely independent method of light-scattering.

TABLE II

LEAST SQUARES ANALYSIS OF DATA OF TABLE I[a]

$\bar{w}$	s_{exp} (S)	w_1	w_2	s_{calc} (S)	Residual
0.098	2.06	0.074	0.024	2.110	−0.046
0.13	2.22	0.092	0.037	2.146	0.076
0.21	2.16	0.133	0.077	2.206	−0.042
0.24	2.26	0.147	0.093	2.221	0.040
0.25	2.16	0.151	0.099	2.226	−0.065
0.56	2.27	0.262	0.298	2.287	−0.019
0.83	2.32	0.337	0.493	2.282	0.031
1.00	2.28	0.379	0.621	2.267	0.007
1.07	2.30	0.395	0.675	2.259	0.036
1.54	2.23	0.492	1.05	2.190	0.033
2.10	2.05	0.590	1.51	2.086	−0.047

[a] Minimum S.S.R. = 0.0213; standard error of a single measurement = 0.049; association constant $L_{1,2}$ = 4.34 ± 0.58 dl/g; hydrodynamic factor $\mathbf{g}$ = 0.092 ± 0.008 dl/g; correlation coefficient of $L_{1,2}$ with $\mathbf{g}$ = 0.64. This output was obtained with the limiting sedimentation coefficients of the monomer and the dimer set at $(s_1)_0$ = 1.89 S and $(s_2)_0$ = 2.87 S respectively.

Townend *et al.* found 4.44 dl/g, expressed in the form $K_d = 2.5 \times 10^{-4}$ mole/liter which we have converted to dl/g using the relation

$$L_{1,2} = \frac{20}{18,000} \frac{1}{K_d} \, dl/g \tag{21}$$

in which the number 18,000 is the value they took for the molecular weight of the monomer.

By using the computer program of Cox,[10-13] and by assuming diffusion coefficients for monomer and dimer of 10.2×10^{-7} and 7.77×10^{-7} cm²/second respectively (calculated from the Svedberg equation

$$(D_i)_0 = RT(s_i)_0/(1 - \bar{v}\rho)M_i$$

with $\bar{v} = 0.75$ ml/g), it is possible to generate a theoretical schlieren pattern for any given concentration of lactoglobulin and time of centrifugation to compare with experiment. A generated schlieren pattern is shown in Fig. 3A. We also show the asymptotic schlieren pattern, both patterns being calculated for a plateau concentration of 0.75 g/dl. At this concentration the asymptotic boundary is already hypersharp at its leading edge and we therefore show next how to calculate such composite hypersharp boundaries.

Hypersharp Diffusion-Free Boundaries

An involuted boundary profile obtained from Eq. (10) is shown in Fig. 3B to demonstrate the calculation of the position of the hypersharp

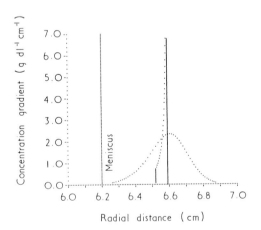

Fig. 3A. Comparison of an "asymptotic" and computer-simulated schlieren pattern for the sedimentation of a monomer–dimer system in reversible equilibrium. Simulation by computer program of D. J. Cox [*Arch. Biochem. Biophys.* **146,** 181 (1971)]. Values of the parameters used are those listed in Table II.

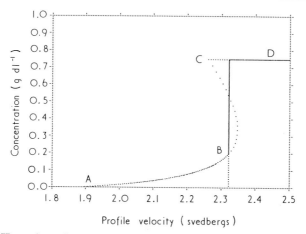

Fig. 3B. Hypersharp "asymptotic" concentration profile for the sedimentation of a monomer–dimer system in reversible equilibrium. Values of the parameters used are those listed in Table II.

region. The lower continuous region AB joins the hypersharp region BC at the coordinate position $(s', \bar{w}')$. This point has to be found by ensuring that the total flux $(\bar{s}°\bar{w}°)$ in the plateau region is equal to the contribution $(\bar{w}° - \bar{w}')s'$ of the hypersharp region added to that of the continuous region $(\bar{w}'\bar{s}')$, where $\bar{s}'$ is the velocity of the equivalent sharp boundary of the continuous region. We have

$$\bar{w}°\bar{s}° = \bar{w}'\bar{s}' + (\bar{w}° - \bar{w}')s' \tag{22}$$

and so

$$s' = \frac{\bar{w}°\bar{s}° - \bar{w}'\bar{s}'}{(\bar{w}° - \bar{w}')} \tag{23}$$

One now proceeds by iteration. For a guessed value of w', one calculates s' for the continuous region by Eq. (12) and then for the hypersharp region using Eq. (23). The two values will not agree at first and therefore the two regions will not join. w' is altered progressively until the two calculated values of s' do agree. With a programmable desk calculator the adjustment of w' to a satisfactory value can be carried out automatically by the machine.

Simulation with Pseudo Random Errors

The reader may have wondered how objective the analysis above was of the ultracentrifuge data for β-lactoglobulin. How valid, in fact, was the choice of values for the sedimentation coefficients of the monomer and dimer, and what would be the effect of a different choice? Could the values of these coefficients have been deduced from the internal evidence provided by the experiments instead of being assumed?

When we attempt to do this, we find in fact that we are not able to estimate limiting sedimentation coefficients from the data of Table I by least-squares analysis even when the sedimentation coefficients are tied by setting $r = (s_2)_0/(s_1)_0 = 1.520$, nor do we believe it to be possible without more extensive data as we show below. In order to find out how the value of $L_{1,2}$ depends upon the choice of the sedimentation coefficients, we have altered the value of $(s_1)_0$ progressively from 1.815 to 2.050 S, keeping r constant at 1.520 so that $(s_2)_0$ varies from 2.760 to 3.117 S. The results for $L_{1,2}$ are plotted in Fig. 4, along with the least squares values for **g** and the corresponding minimum of the sum of the squares of the residuals (S.S.R.). It is clear that the least-squares value found for $L_{1,2}$ is *very* sensitive to the value chosen for the sedimentation coefficients, and that not much weight after all can be put on the astonishingly close agreement between the values of $L_{1,2}$ deduced from the light-scattering and the sedimentation measurements.

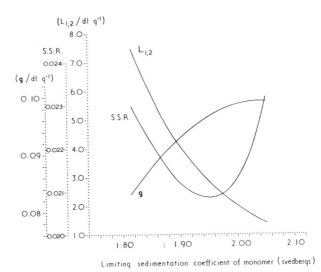

Fɪɢ. 4. Sedimentation of monomer–dimer systems in reversible equilibrium. Dependence of the minimum value of the sum of squares of the residuals (S.S.R.) and of the values found for the association constant $L_{1,2}$ and the hydrodynamic factor **g** on the value assumed for the sedimentation coefficient $(s_1)_0$ of monomer of β-lactoglobulin (see text).

Any similar investigation of an associating protein will give rise to questions of this kind, and underlying them the essential question: What further data or what improvements in precision are necessary to narrow the uncertainties in the derived parameters to prescribed limits? The most effective way of answering such questions lies in the setting up of simulated systems which mirror the actual system in as much detail as possible. These can then be analyzed by the same least squares procedures with the advantage that one knows what the values of the parameters should be because they have been put into the system in the first place. If one fails to extract sensible values of the parameters from the simulated system, one can test different methods of data analysis, and if these fail, one can seek to find how the simulated system needs to be modified, for instance by extending the range of concentration over which measurements are postulated to be made, or by increasing the precision of the data. On the basis of these findings one can get guidance on how to modify the actual experiment, or perhaps be driven to realize that the experiment is not feasible at all. Simulation has the additional advantage of revealing whether two parameters are so highly correlated that a given experimental approach would fail to give their separate values.

To illustrate the use of simulation[52-54] by a specific example, let us suppose that we wish to find out how to modify the experiment we have just analyzed so that we can improve the accuracy of our estimate of $L_{1,2}$. One obvious way is to reduce the uncertainty in the estimated value of $(s_1)_0$. Let us therefore simulate pseudoexperimental points in the region below 0.1 g/dl, and include them in the data for least squares analysis. The parameters on which to base the simulation are those listed in the computer output in Table II. Exact values of $\bar{s}$ are calculated for let us say $\bar{w}$ equal to 0.01, 0.02, 0.03, and 0.04 g/dl. Next we add appropriate random errors which we can calculate by making use of a table of normally distributed random numbers with a mean of zero and a standard deviation of one. To each calculated value of $\bar{s}$ is added one of the random numbers after the number has been multiplied by the "standard error of a single measurement," in this case 0.05 S.

It is advisable to produce several sets of data based on several choices of random numbers. One set is plotted in Fig. 2 with crosses to indicate the new pseudoexperimental points. Least squares analysis for the *three* parameters, $L_{1,2}$ g, and $(s_1)_0$, is now found to be feasible. The

[52] G. A. Gilbert, *Anal. Chim. Acta* **38**, 275 (1967).
[53] J. Myhill, *Biophys. J.* **7**, 903 (1967).
[54] J. Myhill, *J. Theoret. Biol.* **23**, 218 (1969).

set of data illustrated in Fig. 2, for example, leads to estimated values of $L_{1,2} = 4.03 \pm 1.42$ dl/g, $g = 0.093 \pm 0.008$ dl/g, and $(s_1)_0 = 1.90 \pm 0.04$ S. That the values of the estimated parameters are very sensitive to the points in the low concentration range can be demonstrated by supposing the true value of $L_{1,2}$ to be 2.50 dl/g, g to be 0.097 dl/g and $(s_1)_0$ to be 1.97 S instead of 4.34, 0.092, and 1.89, respectively for the simulation of the four new points (see Fig. 4). Least squares analysis with this set of pseudoexperimental points added to the original data leads to $L_{1,2} = 2.43 \pm 0.73$ dl/g, $g = 0.098 \pm 0.008$ dl/g, and $(s_1)_0 = 1.97 \pm 0.03$ S, which shows that the system is dominated by the points at low concentration, and that this is the region in which it is essential to obtain more actual data.

Perhaps enough has been said to show that an appreciation of the numerical significance of experimentally measured interaction constants can hardly be attained without the parallel analysis of carefully varied simulated systems. In general, it is best to begin simulation studies as soon as possible after the first rough values for the parameters have been obtained. An "error-free" curve for $\bar{s}$ versus $\bar{w}$ is then calculated with these parameters, suitable intervals for observations are chosen, and points with normally distributed errors constructed. One begins with minimal errors, say 10^{-6} S, to get the program working. Then the errors are raised to realistic values, gradually if necessary, and a selection of points sought that results in a reasonably small error for the estimated parameters. Unless this can be achieved with simulated data, one cannot expect it to be possible using data from an actual experiment, since the same program has to be used to evaluate the experimental data as is used for the simulated data.

Any identified assumption can be tested. For instance in the present case for lactoglobulin, $(s_2)_0/(s_1)_0$ has been set at $2^{2/3}/1.044$ whereas in other cases, as for hemoglobin, $2^{2/3} \times 1.044$ might be more appropriate. Only by simulation can one determine whether such an assumption is at all critical in the given circumstances.

Oligomeric Systems

Elaboration of a system to include species higher than dimer increases the number of parameters and, much more than proportionately, the difficulty of finding their values accurately. However, encouraging results have been obtained in a qualitative sense from many sedimentation studies of oligomeric substances, and in the classical studies of the reversible association of β-lactoglobulin A even good quantitative agreement was obtained with the results of other methods such as light scattering and ORD.[27-30,55,56] We will in fact take advantage of the great deal that is known about the association of bovine β-lactoglobulin A to illustrate the characteristic effects of reversible aggregation beyond dimer on the sedimentation behavior of a substance. At pH 4.65 in sodium acetate-acetic acid buffer of ionic strength 0.1, and at low temperature, β-lactoglobulin A is predominantly in the form of dimers, tetramers, hexamers, and octamers, so much so that except at very low concentrations of protein, the contribution of monomers can be neglected, as will be seen. We will take the dimer therefore as the associating species, representing it by A_2 in the following set of equilibria:

$$A_2 + A_2 = A_4 \qquad L_{2,4} = w_4/w_2^2$$
$$A_4 + A_2 = A_6 \qquad L_{2,6} = w_6/w_2^3 \qquad (24)$$
$$A_6 + A_2 = A_8 \qquad L_{2,8} = w_8/w_2^4$$

The constitutent concentration $\bar{w}$ is given by

$$\bar{w} = w_2 + L_{2,4}w_2^2 + L_{2,6}w_2^3 + L_{2,8}w_2^4 \qquad (25)$$

In order to carry out the simulation we could have used values for the various parameters which are available from previous studies,[27,28,40,57] but we take this opportunity to indicate briefly the path of present developments by refining these values in the following way. (We must emphasize, however, that a complete analysis would be much more detailed and extensive.)

The principal parameters required are the association constants $L_{2,i}$, the sedimentation coefficients $(s_i)_0$ and the concentration dependence g. (At present we do not see how activity coefficients could be separated in sedimentation velocity experiments from the combined effects of the L's and g, and therefore do not include them explicitly.) We keep to the principle that the model adopted initially must be the simplest, consistent with known biochemical facts, that fits the data. We therefore initially set $L_{2,6} = \frac{3}{4}(L_{2,4})^2$ as if the bonds between dimer and dimer, and dimer and tetramer had the same molar free energy of formation. We also assume a pseudocubical structure for the octamer, and that completion of the octamer by addition of dimer to hexamer proceeds with a free energy of formation greater by the factor $-RT \ln \gamma$ than it would be if the bonds had continued to be similar in energy to the dimer–dimer bond. The appropriate value of $L_{2,8}$ is then given by $L_{2,8} = \frac{1}{2}\gamma(L_{2,4})^3$. It will be assumed, as previously,[46] that the value of the

[55] S. N. Timasheff and R. Townend, *Protides Biol. Fluids Proc. Colloq.* **16**, 33 (1969).

[56] T. T. Herskovits, R. Townend, and S. N. Timasheff, *J. Amer. Chem. Soc.* **86**, 4445 (1964).

[57] L. M. Gilbert and G. A. Gilbert, *Nature* (*London*) **194**, 4834 (1962).

sedimentation coefficient $(s_2)_{0(20,w)}$ can be taken as 2.87, $(s_4)_{0(20,w)}$ as 4.57, $(s_6)_{0(20,w)}$ as 6.09, and $(s_8)_{0(20,w)}$ as 7.37 S. The values of $L_{2,4}$, $L_{2,6}$, $L_{2,8}$, and g remain to be found by the process of simulation, comparison with experiment, and iteration.

We were fortunate in obtaining from Dr. H. McKenzie full details of the sedimentation velocity experiments of Armstrong and McKenzie,[31] together with accurate photographs of the original plates (see, for examples, Fig. 2, a and b, of their paper). Pure bovine β-lactoglobulin A had been run at 1.5 g/dl at 2.5° in 0.1 I sodium acetate–acetic acid buffer, pH 4.65. The correction factor used for calculating $(s)_{20,w}$ from $(s)_{2.5\text{ solution}}$ was 1.711. The successive exposures on the photographs were measured by a microcomparator accurate to ± 10 μm to obtain a digital record for each schlieren pattern containing about 200 co-ordinates. These were punched onto tape. Although no solvent base line was available, it was possible to estimate a very minor correction for a linear sloping base line from the extremes of the schlieren pattern. Using a programmable Wang 700 desk calculator with tape reader, we fitted each pattern over short sections by least squares to quartic polynomials to enable peak height to be related to radial distance by analytical expressions. These equations were then used to obtain the second moment of the radial position of the schlieren pattern.[58] From the second moment the dilution due to sector shape and centrifugal field could be calculated and hence the factor relating the area of the pattern to concentration found, after which integration of the quartics with respect to radial distance gave $\bar{w}$ as a function of radial distance r. This was needed for the following reason.

A serious difficulty in simulating ultracentrifuge schlieren patterns arises from the fact that a boundary leaving a meniscus cannot be formed instantly at full speed, and that its shape is modified by the effects of acceleration and the presence of the meniscus, effects that are very difficult to allow for in a simulation program. Boundaries formed in synthetic boundary cells likewise have initial imperfections and do not escape the effect of acceleration. These difficulties can be avoided entirely, as Cox[10,11] has pointed out in his series of very perceptive papers, by basing simulation on an experimental peak measured on an exposure taken only after full speed has been reached, and after the peak has completely left the meniscus. The values of $\bar{w}$ and r derived from such a peak are entered into the simulation program,[11,13] and the shape of peaks at further times are computed to compare with further exposures in the experiment. We chose the peak from an exposure cor-

---

[58] R. Trautman, *Ann. N.Y. Acad. Sci.* **164**, 52 (1969).

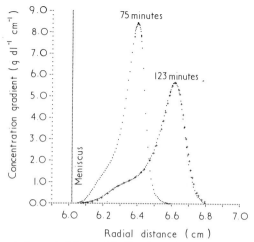

Fig. 5. Sedimentation of β-lactoglobulin A under associating conditions (1.5 g dl^{-1}, 2.5°, sodium acetate–acetic acid buffer ionic strength 0.1, pH 4.65). . . . ., Measurements on the experimental schlieren patterns for exposures at effective running times of 75 and 123 minutes obtained by J. M. Armstrong and H. A. McKenzie [*Biochim. Biophys. Acta* **147**, 93 (1967)]. + + + +, Computed schlieren pattern for 123 minutes based on measurements on the pattern at 75 minutes. Data from the 75-minute pattern were processed by a WANG 700 desk calculator and used as input data in the simulation program of D. J. Cox [*Arch. Biochem. Biophys.* **146**, 181 (1971)]. Parameters used for the simulation were the following. Running time: (123 − 75) = 48 minutes. Speed: 59 780 rpm. Molecular weight of dimer: 36,800. Association contants: $L_{2,4} = 1.20$ dl g^{-1}, $L_{2,6} = 0.837$ dl^2 g^{-2}, $L_{2,8} = 20.7$ dl^3 g^{-3}. Hydrodynamic coefficients: 0.070 dl g^{-1}. Sedimentation coefficients at 2.5°C: $(s_2)_0 = 1.68$ S, $(s_4)_0 = 2.67$ S, $(s_6)_0 = 3.56$ S, $(s_8)_0 = 4.32$ S. Diffusion coefficients at 2.5°C: $(D_2)_0 = 4.05 \times 10^{-7}$ cm^2 sec^{-1}, $(D_4)_0 = 3.22 \times 10^{-7}$ cm^2 sec^{-1}, $(D_6)_0 \times 2.86 \times 10^{-7}$ cm^2 sec^{-1}, $(D_8)_0 = 2.60 \times 10^{-7}$ cm^2 sec^{-1}.

responding to an effective running time at full speed of 75 minutes as the initial peak from which to start simulation, and compared the result with the peak from an exposure made 48 minutes later. After a few attempts, values of the L's and g had been successfully selected which gave the correspondence between experiment and simulation shown in Fig. 5. As a further check, simulation[13] was then carried out as if from an initially sharp boundary for a total of 75 + 48 minutes. The correspondence between experiment and simulation (Fig. 6) is again rather precise.

The values of the parameters arrived at by this iteration procedure were $L_{2,4} = 1.201$ dl g^{-1}, $L_{2,6} = 0.837$ dl^2 g^{-2}, $L_{2,8} = 20.746$ dl^3 g^{-3}, $\gamma = 24$, and g = 0.07 dl g^{-1}. We now use these parameters to calculate s as a function of $\bar{w}$ for this system with the result seen in Fig. 7. Next we

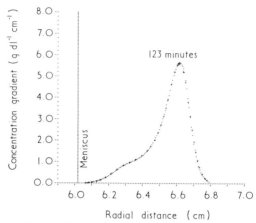

Fig. 6. Sedimentation of β-lactoglobulin A under associating conditions (1.5 g dl^{-1}, 2.5°C, sodium acetate–acetic acid buffer ionic strength 0.1, pH 4.65). . . . ., Measurements on experimental schlieren pattern for an effective running time of 123 minutes obtained by J. M. Armstrong and H. A. McKenzie [*Biochim. Biophys. Acta* 147, 93 (1967)]. + + + +, Computed schlieren pattern assuming a running time of 123 minutes from an initially sharp boundary at the meniscus position. Input data otherwise as for Fig. 5.

calculate the asymptotic schlieren pattern for $\bar{v}^0 = 1.268$ g dl^{-1} and compare it with the corresponding computer-simulated pattern by superimposing the patterns (Fig. 8) in such a way that their second moments correspond in position relative to a common meniscus. Considering that

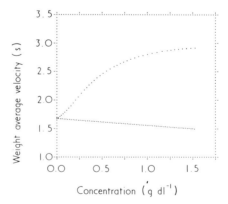

Fig. 7. Characteristic "cooperative" appearance of the curve for sedimentation velocity versus concentration for a reversibly polymerizing substance. The model is based on β-lactoglobulin A and assumes dimer, tetramer, hexamer, and octamer in reversible equilibrium.

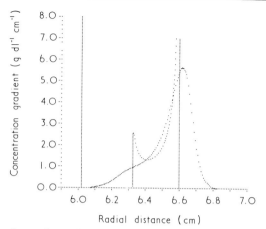

FIG. 8. Comparison of an "asymptotic" hypersharp schlieren pattern with a computer simulated schlieren pattern [D. J. Cox, *Arch. Biochem. Biophys.* **146,** 181 (1971)] for a reversibly polymerizing substance based on the self-association of β-lactoglobulin A, assuming dimer, tetramer, hexamer, and octamer in reversible equilibrium (see text).

the asymptotic pattern ignores diffusion completely, a surprisingly useful degree of correspondence between the two patterns exists, enabling the easily calculated asymptotic pattern to be used as a preliminary to full computer simulation.

Conclusion

In the past, velocity sedimentation has been very much the poor relation of equilibrium sedimentation in respect to its acceptability for determining molecular weight averages and degrees of self-association. We hope that the demonstrations above are convincing that in future the velocity method will be truly complementary to the equilibrium method used in conjunction with simulation by computer. Often this will mean no increase in laboratory work, since velocity sedimentation is usually carried out before there is any thought of proceeding to equilibrium measurements. Similarly, the dependence of sedimentation velocity upon concentration is usually a matter of routine determination. Hitherto much of the information gathered in this way, and as a rule preserved in permanent form, has not been put to use. If one draws a parallel with astronomy and high energy physics, it is not fanciful to forsee a similar development of automated methods of plate measurement and computerized interpretation[58] perhaps in a suitably equipped center. If one believes this, then it seems sensible to take even greater care over the recording and filing of experimental details, such as rotor temperature, magnification factors (or reference cell dimensions), exposure times and durations, correction factors (or actual densities and viscosities) and solute concentrations, to justify the labor of a full analysis by computer simulation. It should be remembered that a special advantage of the velocity method is its power to resolve, and its ability to discriminate between impurity and principal component, thus enabling conclusions to be reached even before the final stages of purification.

Self-association is only one small facet of the general problem of macromolecular interaction. Indeed it will not have escaped the reader's notice that even the elementary example with which we began this article involved a hybrid mixture of β-lactoglobulin A and B, which it was wrong to treat as a single entity. Initially we did in fact intend to pursue the general case in as much detail as the special case of self-association, but it is indicative of the pace of development of the subject that during our preparation of this article it became possible for the first time to deal realistically with experimental data for self-associating systems. It cannot be long before equally detailed treatment is possible for the general case.

Acknowledgments

We are very grateful to Dr. D. J. Cox for sending us full details of his computer programs, and to Dr. H. A. McKenzie for original experimental data concerning β-lactoglobulin A. We have received support from the Science Research Council of Great Britain during the period of preparation of this article.

[12] Measurements of Protein Interactions Mediated by Small Molecules Using Sedimentation Velocity[1]

By JOHN R. CANN and WALTER B. GOAD[2]

The significance of protein–small molecule interactions of the type $P + nHA \rightleftharpoons P(HA)n$ for electrophoresis and chromatography is discussed in Volume XXV [11]. There we were primarily concerned with situations in which P and P(HA)n differ in electrophoretic mobility but not significantly in frictional coefficient; but many of the same considerations can be applied to sedimentation in the event that cooperative bind-

[1] Supported in part by Research Grant 5R01 AI01482 from the National Institutes of Health, U.S. Public Health Service. Contribution No. 411 from the Department of Biophysics and Genetics, University of Colorado Medical Center, Denver, Colorado 80220.

[2] Work done under the auspices of the United States Atomic Energy Commission.

ing of small molecules (or ions) causes sufficient alteration in macromolecular conformation so as to change the sedimentation coefficient measureably. Given the appropriate conditions, the sedimentation pattern of such an interacting system can show a bimodal reaction boundary despite instantaneous establishment of equilibrium. This will be so if the ratio of protein to small molecule concentration is such that concentration gradients of unbound small molecule are produced along the ultracentrifuge cell due to reequilibration during differential transport of P and $P(HA)n$. Suppose, however, that the ratio is sufficiently small that the small molecule concentration cannot be perturbed significantly during transport. In that event, the system effectively collapses to a simple isomerization reaction, $A \rightleftharpoons B$. Three situations are recognized here: (a) If reequilibration is instantaneous, the initial equilibrium composition will be maintained during transport; and the sedimentation pattern will show a single peak with weight average sedimentation coefficient. (b) If the rates of interconversion are so slow that significant reequilibration does not occur during sedimentation, the pattern will show two boundaries corresponding to the isomers. (c) If the half-times of interconversion are of the order of the time of sedimentation, the pattern will show three peaks. These several predictions apply to both moving-boundary and band sedimentation in the analytical ultracentrifuge and to zone sedimentation through a preformed density gradient in the preparative instrument.

While the aforementioned types of protein–small molecule interaction are of considerable interest particularly as they apply to biological control mechanisms, ultracentrifugation is peculiarly sensitive to changes in the state of association of a marcromolecule. Accordingly, our primary concern here is with interactions in which binding of ligand favors macromolecular association or dissociation. The consequences of this class of interaction have recently been explored theoretically.[3-6] Computations have been made for reaction of the type, $mM + nX \rightleftharpoons M_mX_n$, in which a macromolecule, M, associates into an m-mer with the mediation of a small ligand molecule or ion, X, of which a fixed number, n, are bound into the complex, M_mX_n. The conservation equations with cyclindrical divergence have been solved numerically on a high speed computer assuming that local equilibrium attains at every instant. In certain respects the results for ligand-mediated dimerization, $2M +$

[3] J. R. Cann and W. B. Goad, *Advan. Enzymol.* **30**, 139 (1968).

[4] W. B. Goad and J. R. Cann, *Ann. N.Y. Acad. Sci.* **164**, 172 (1969).

[5] J. R. Cann and W. B. Goad, *in* J. R. Cann, "Interacting Macromolecules," Chapters IV and V. Academic Press, New York, 1970.

[6] J. R. Cann and W. B. Goad, *Science* **170**, 441 (1970).

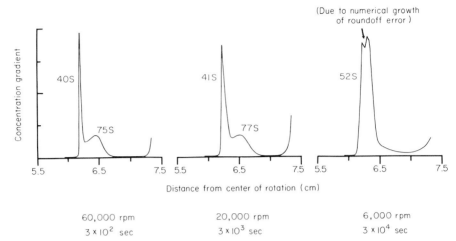

Fig. 1. Theoretical moving-boundary sedimentation patterns for the ligand-mediated dimerization reaction, $2M + 30X \rightleftharpoons M_2X_{30}$, at three different values of the centrifugal field. Rotor speed and time of sedimentation under each pattern. The following values of the macromolecular concentrations and parameters were chosen to approximate the sedimentation of high molecular weight DNA: $C_{10} = 10^{-10} M$, $C_{20} = 5 \times 10^{-11} M$, $s_1 = 40 S$, $s_2 = 80 S$, $D_1 = D_2 = 10^{-9}$ cm^2 sec^{-1}. The small molecule concentration and other parameters were assigned the values: $C_{30} = 4.5 \times 10^{-9} M$, $s_3 = 0.1 S$, $D_3 = 10^{-6}$ cm^2 sec^{-1}. In this and the following figures the subscripts 1, 2, and 3 designate monomer, polymer, and unbound ligand, respectively; C_{10}, C_{20}, and C_{30} initial equilibrium concentration of monomer, polymer, and unbound ligand; and other symbols, their usual meaning. From J. R. Cann and W. B. Goad, *Advan. Enzymol.* **30**, 139 (1968).

$nX \rightleftharpoons M_2X_n$, are the more revealing with regard to fundamental principles and can be summarized as follows:

For appropriate choice of ligand concentration, the computed moving-boundary sedimentation pattern show two well resolved peaks (Figs. 1–4) even when dimerization is mediated by the binding of only a single ligand molecule into the complex. This result is in contrast to the Gilbert theory of sedimentation[7,8] (see also Volume XXVII [11]), which correctly predicts that the sedimentation pattern of the system, $2M \rightleftharpoons M_2$, will show only a single peak for instantaneous establishment of equilibrium. In the case of ligand-mediated dimerization, resolution is dependent upon the production of concentration gradients of unbound ligand by re-equilibration during differential transport of macromonomer and complex. Increasing the initial ligand concentration, holding the percent dimeriza-

[7] G. A. Gilbert, *Discuss. Faraday Soc.* **20**, 68 (1955).

[8] G. A. Gilbert, *Proc. Roy. Soc. Ser. A* **250**, 377 (1959).

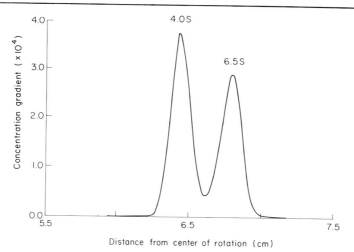

FIG. 2. Theoretical moving-boundary sedimentation pattern computed for the ligand-mediated dimerization, $2M + 30X \rightleftharpoons M_2X_{30}$. The following values of the macromolecular concentrations and parameters were chosen to approximate the sedimentation of a protein of molecular weight 60,000: $C_{10} = 7 \times 10^{-5} M$, $C_{20} = 3.5 \times 10^{-5} M$, $s_1 = 4 S$, $s_2 = 6.35 S$, $D_1 = 6 \times 10^{-7}$ cm^2 sec^{-1}, $D_2 = 4.76 \times 10^{-7}$ cm^2 sec^{-1}. For the small molecule: $C_{30} = 10^{-5} M$, $s_3 = 0.1 S$, $D_3 = 10^{-5}$ cm^2 sec^{-1}. Rotor speed, 60,000 rpm; time of sedimenation, 5540 seconds. From W. B. Goad and J. R. Cann, *Ann. N.Y. Acad. Sci.* **164**, 172 (1969).

tion constant, results in progressive coalescence of the two peaks with concomitant drift in their sedimentation velocities toward the weight average value until resolution disappears entirely at the highest concentration. In the limit of sufficiently high ligand concentration (or low

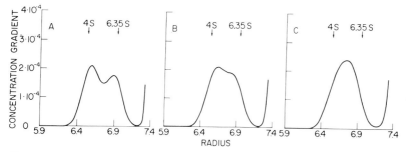

FIG. 3. Theoretical moving-boundary sedimentation patterns for the ligand-mediated dimerization reaction, $2M + 6X \rightleftharpoons M_2X_6$. Dependence of boundary shape at 50% dimerization upon the initial concentration of unbound small ligand molecule: (A) $C_{30} = 5 \times 10^{-5} M$; (B) $10^{-4} M$; (C) $2 \times 10^{-4} M$. Other parameters as in Fig. 2 except for time of sedimentation, 6431 sec. From J. R. Cann and W. B. Goad, *in* J. R. Cann, "Interacting Macromolecules," p. 201. Academic Press, New York, 1970.

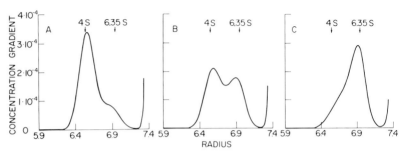

FIG. 4. Theoretical moving-boundary sedimentation patterns for the ligand-mediated dimerization reaction, $2M + 6X \rightleftharpoons M_2X_6$. Dependence of boundary shape upon percent dimerization, $K = 4.57 \times 10^{29}$ M^{-7}: (A) 25% dimerization, $C_{30} = 3.89 \times 10^{-5} M$; (B) 50% $5 \times 10^{-5} M$; (C) 75%, $6.74 \times 10^{-5} M$. $C_{10} + 2C_{20} = 14 \times 10^{-5} M$; time of sedimentation, 6431 seconds; other parameters as in Fig. 2. From J. R. Cann and W. B. Goad, in J. R. Cann, "Interacting Macromolecular," p. 201. Academic Press, New York, 1970.

centrifugal field, Fig. 1), the Gilbert theory is valid because the concentration of ligand cannot be perturbed significantly under these conditions. The patterns for 50% dimerization ($n = 6$) displayed in Fig. 3 illustrate the way in which resolution decreases with increasing ligand concentration as the Gilbert limit is approached. Comparison of patterns for $n = 1$, 6, and 30 shows that resolution can occur at higher ligand concentrations when the interaction is cooperative. In general, the higher the cooperativeness, the wider the range of ligand concentrations over which resolution can occur. Finally, except for very high cooperativeness, the areas under the two peaks do not faithfully reflect the initial equilibrium composition. Nor do the peaks sediment with the sedimentation coefficients of macromonomer and dimer. Under some conditions the slow peak sediments more rapidly than the monomer since it contains dimer, while the fast peak always sediments slower than the dimer since it contains monomer. These are properties which we have come to associate with bimodal reaction boundaries.[8a] The set of

[8a] $Footnote$ $added$ in $proof.$ The theory has recently been extended to take cognizance of the hydrodynamic dependence of the sedimentation coefficients of the reacting species upon concentration (J. R. Cann and W. B. Goad, $Arch.$ $Biochem.$ $Biophys.$ **153**, 603 (1972)). For appropriate ratios of protein to ligand concentration the theoretical sedimentation patterns show bimodal reaction boundaries even when the concentration dependence of the sedimentation coefficients is severe. The boundaries are subject, however, to a Johnston-Ogston type effect which accentuates the discrepancy between the relative areas of the two peaks and the relative, equilibrium concentrations of monomer and dimer. As in the case of constant sedimentation coefficients, the velocities of the peaks per unit centrifugal field are not equal to the sedimentation coefficients of the monomer and dimer. Moreover, these

patterns presented in Fig. 4 for $n = 6$ illustrates how the shape of the reaction boundary depends upon the initial percent dimerization as determined by total ligand concentration at constant macromolecule concentration and fixed equilibrium constant.

Ligand-mediated tetramerization $(4M + 4X \rightleftharpoons M_4X_4)$ and ligand-mediated dissociation $(M_4 + 4X \rightleftharpoons 4MX)$ reactions have likewise been considered. For conditions such that the free ligand concentration is significantly perturbed by reequilibration during differential transport of macromonomer and tetramer, the patterns deviate significantly from Gilbert patterns. In particular, the area of the slow peak is 20 to 50% smaller than predicted by the Gilbert theory as elaborated by Fujita[9] for a sector-shaped cell. This precludes calculation of an apparent association constant from the macromolecular concentration at the position of the minimum in the bimodal reaction boundary and the degree (m) of association or dissociation.

The foregoing discussion underscores the fact that unambiguous interpretation of the sedimentation behavior of associating-dissociating systems in terms of stoichiometry and energetics ultimately depends upon the combined use of at least one other physical method. Thus, classical equilibrium sedimentation or light scattering can be used to distinguish ligand-mediated association-dissociation from other interactions mediated by small molecules and to determine the size of the monomer and polymer and the apparent association constant as a function of ligand concentration. Light-scattering experiments can also provide information on the rates of reaction. Equilibrium dialysis measurements permit direct characterization of the protein-small molecule interaction in terms of the number of binding sites, intrinsic affinity, and possible cooperative binding. Kegeles and his co-workers[10,10a] have applied Archibald molecular weight determinations and temperature jump kinetic analysis using light scattering as an indicator to elucidate the velocity sedimentation behavior of lobster hemocyanin. Dimerization of hemocyanin upon lowering the pH from 9.6 to 9.2 is mediated by the binding of 4–6 Ca^{2+} and 2–4 H^+; the reaction is reversible; both dimerization and dissociation of the dimer are very rapid processes; and the sedimentation pattern shows two peaks as predicted above for such an interaction.

Nor are zonal methods immune to the complications attending ligand-

---

quantities cannot be determined by extrapolation of the velocities of the peaks per unit field to infinite dilution of macromolecule at constant total ligand concentration.

[9] H. Fujita, "The Mathematical Theory of Sedimentation Analysis," Chapter IV. Academic Press, New York, 1962.
[10] K. Morimoto and G. Kegeles, *Arch. Biochem. Biophys.* **142**, 247 (1971).
[10a] M. Tai and G. Kegeles, *Arch. Biochem. Biophys.* **142**, 258 (1971).

mediated interactions.[6] In the case of dimerization mediated by the binding of 6 ligand molecules into the complex, the results for band sedimentation (Fig. 5) are quite analogous to those for moving-boundary sedimentation. Thus, for appropriate choice of ligand concentration, the computed band pattern shows a well resolved, bimodal reaction zone. Increasing the initial ligand concentration, holding the percent dimerization constant, results in progressive coalescence of the two peaks with

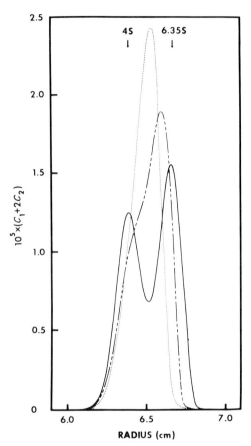

Fig. 5. Theoretical band sedimentation patterns for the ligand-mediated dimerization reaction, $2M + 6X \rightleftharpoons M_2X_6$. ———, $C_{30} = 10^{-6} M$; ———— —— ———, $10^{-5} M$; . . . . . . , $5 \times 10^{-5} M$. Time of sedimentation, 4740 seconds at 60,000 rpm; $s_3 =$ 0.15 S; other parameters as in Fig. 2. Unbound ligand initially distributed uniformly along the centrifuge cell; a bimodal zone of virtually the same shape as shown here for $C_{30} = 10^{-6} M$ was also obtained when the computation was for unbound ligand initially present only in the starting zone. From J. R. Cann and W. B. Goad, *Science* **170**, 441 (1970).

concomitant drift in the sedimentation velocities toward a value close to the weight average until resolution disappears entirely at the highest concentration. The mean sedimentation coefficient of the unimodal zone shown at the highest ligand concentration decreases slightly during the course of sedimentation due to some dissociation of the dimer within the spreading zone.

The latter behavior assumes major proportions in the case of ligand-mediated tetramerization, $4M + 4X \rightleftharpoons M_4X_4$. Consider, for example, 50% tetramerization at equal constituent concentrations of ligand and macromolecule (Fig. 6). The mean sedimentation coefficient of the predicted unimodal zone decreases rapidly and continuously from the weight average value to a value approaching that of the monomer as the zone migrates down the centrifuge cell, i.e., as the zone spreads due to differential transport of M and M_4X_4 and to diffusion. In fact, for all practical purposes it can be said that the system sediments essentially as the monomer. This is certainly so when the ligand is initially present only in the starting zone and approximately so when it is

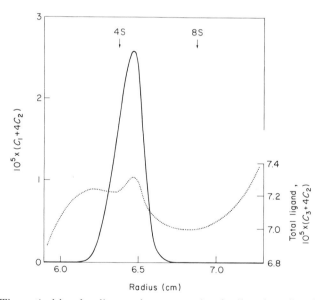

Fig. 6. Theoretical band sedimentation pattern for the ligand mediated tetramerization reaction, $4M + 4X \rightleftharpoons M_4X_4$. ――――, macromolecule concentration; ――――, total ligand concentration. $C_{10} = 7 \times 10^{-5}\,M$, $C_{20} = 1.75 \times 10^{-5}\,M$, and $C_{30} = 7 \times 10^{-5}\,M$; unbound ligand initially distributed uniformly along the centrifuge cell; $s_1 = 4\,S$, $s_2 = 8\,S$, $s_3 = 0.15\,S$; $D_1 = 6 \times 10^{-7}\,cm^2\,sec^{-1}$, $D_2 = 3.8 \times 10^{-7}\,cm^2\,sec^{-1}$, and $D_3 = 10^{-5}\,cm^2\,sec^{-1}$; 4250 seconds at 60,000 rpm. From J. R. Cann and W. B. Goad, *Science* **170**, 41 (1970).

initially distributed throughout the cell. Such behavior reflects the strong concentration dependence of tetramerization; as the macromolecule concentration within the spreading zone decreases, the tetramer dissociates by mass action into slower sedimenting monomer. Even for much stronger interaction (e.g., 90% tetramerization at the same constituent concentrations as in Fig. 6), the theory predicts a centripetally skewed zone whose mean sedimentation coefficient decreases with time. Such an interaction would be detected in practice by the nonlinear dependence of the logarithm of the mean position of the zone upon time and in some instance by comparison of the macromolecule and total ligand patterns (Fig. 6). Whereas the zone of macromolecule is unimodal, the distribution of total ligand is bimodal provided that initially the free ligand is distributed uniformly throughout the centrifuge cell. The bimodal distribution of total ligand can be understood as follows: As the tetramer dissociates it releases ligand which remains behind the advancing zone. The sum of the broad peak of released ligand, the peak of ligand bound to the remaining tetramer and the distorted background of unbound ligand is bimodal. Careful note should be taken of the difference between the distribution of macromolecule and total ligand, since it is not uncommon in practice to follow zone sedimentation by analysis of fractions for specific ligand. This is often the only method available for detecting a specific protein in a partially purified cellular extract; but a bimodal distribution of ligand need not necessarily indicate inherent heterogeneity. Nor would it seem to be unique for ligand-mediated association. Conceivably pressure-sensitive ligand binding without a change in state of aggregation or frictional coefficient would give a similar result if pressure favored dissociation of the complex(es). In that event, however, the sedimentation coefficient of the protein zone would not decrease with time of sedimentation.

In addition to ligand-mediated association reactions, we have examined the ligand-mediated dissociation reaction, $M_4 + 4X \rightleftharpoons 4MX$, for which the theory predicts a bimodal reaction zone at appropriate concentration of unbound ligand initially distributed throughout the cell. At higher ligand concentration the zone is unimodal and sediments like the monomer as described above for ligand-mediated tetramerization; but in contrast to ligand-mediated tetramerization, the total ligand pattern shows a single peak which matches the zone of macromolecule both in shape and sedimentation velocity. When the ligand is initially present only in the starting zone, the theoretical patterns exhibit bimodal zones or zones showing a major fast peak and a broad, intense centripetal shoulder over a wide range of parameters. Generally speaking, the fast peak migrates with a sedimentation coefficient close to that of the tetramer and the slow peak or shoulder, the monomer, but the distribution of material between the two peaks does not correspond to the initial equilibrium composition. The amount in the slow peak, while increasing with increasing degree of dissociation, is in all cases considerably less than reckoned from the initial equilibrium concentration of monomer.

These various results can be applied with only quantitative reservations to zone sedimentation through a density gradient in the preparative ultracentrifuge and to molecular sieve chromatography on Sephadex or other gel-permeation supports. Thus, the several considerations presented above have important implications for the many conventional applications of these methods to the analysis and characterization of biological materials. In particular, bimodal zones need not necessarily indicate inherent heterogeneity. It cannot be overemphasized that unequivocal proof of heterogeneity is afforded only by isolation of the various components. The recommended procedures for distinguishing between ligand-mediated interactions and inherent heterogeneity are much the same as those described, for electrophoresis and chromatography.[11] They include: (a) Analysis of fractions under the same conditions as used in the original separation. (The partition cell can be used for fractionation in the analytical ultracentrifuge.) For interactions, the fraction will behave like the unfractionated material and show multiple peaks, while for heterogeneity a single peak will be obtained. The fractions must be reconstituted to the concentration used in the original separation, and the ligand concentration restored prior to analysis. It is conceivable that an interaction mediated by a small molecule may not even be suspected, but this will not invalidate the fractionation test if the fractions are reequilibrated against the buffer or other supporting electrolyte by dialysis. If these precautions are not observed, fractions from an interacting system might show a single peak thereby leading to an incorrect conclusion of inherent heterogeneity. (b) Sedimentation analyses at progressively lower protein concentration. In certain cases as delineated above, resolution of the reaction boundary into two peaks will decrease with decreasing protein concentration. At sufficiently low protein concentration the small molecule concentration cannot be perturbed significantly during transport, and the pattern will show a single peak. (c) Sedimentation analyses at various rotor speeds. At sufficiently low rotor speed the gradients of small molecule upon which resolution depends in a rapidly equilibrating system cannot be maintained against diffusion, and the protein will sediment as a single peak. It is self evident that the time of sedimentation as governed by rotor speed will effect

[11] J. R. Cann, see Vol. XXV [11].

the sedimentation pattern of a kinetically controlled interaction. (d) Systematic variation of the concentration of the interacting small molecule once it has been identified.

Finally, if circumstances attending an unusual sedimentation behavior of a unimodal zone (e.g., increasing sedimentation coefficient with increasing protein concentration in the known presence of a ligand) indicate ligand-mediated association-dissociation, the system should be examined by band sedimentation; and the instantaneous sedimentation coefficient extrapolated to zero time in order to characterize the interaction quantitatively.

[13] Pressure Effects in Ultracentrifugation of Interacting Systems

By WILLIAM F. HARRINGTON and GERSON KEGELES

In this chapter we will consider the effect of the pressure gradient developed in high and low speed centrifugation on the distribution of species in chemically reacting systems of macromolecules. Current theories of sedimentation equilibrium of interacting systems do not in general take into account the possibility of changes in the partial specific volume on association, and no theory exists for the effect of these changes on transport processes in velocity sedimentation experiments. Since pressures of the order of 100–500 atm are generated at the base of a centrifuge cell at high rotor speeds, even extremely small changes in specific volume can lead to marked effects in sedimentation behavior.

Over the past few years a number of well documented studies demonstrating volume changes on polymerization of proteins have been reported. In general, these aggregation reactions show positive volume increments, in the absence of denaturing conditions (see Table II), suggesting that ionic or hydrophobic bonding plays a dominant role in the polymerization process. The evidence is strong that the formation of such bonds in aqueous systems requires a decrease in ordered water structure[1-4] about the groups involved in bonding leading to a positive volume change of about 10–20 cc/mole bond. Since a large number of ionic or hydrophobic interactions may be involved in the association of

[1] W. Kauzmann, Advan. Protein Chem. 14, 1 (1959).
[2] H. S. Frank and M. W. Evans, J. Chem. Phys. 13, 507 (1945).
[3] G. Nemethy and H. A. Scheraga, J. Chem. Phys. 36, 3401 (1962).
[4] G. Nemethy and H. A. Scheraga, J. Phys. Chem. 66, 1773 (1962).

protein molecules to form high molecular weight aggregates, the result-ing molar volume change can exceed this value by one to three orders of magnitude, and the apparent association constant for polymerization can be dramatically shifted by the pressure gradient developed in an ultracentrifuge cell.

The fact that pressures in the range of 70–400 atm can produce striking transformations in the aggregation state of various polymeric systems of biological interest has been known for many years (for a recent review, see Zimmerman[5]). Brown[6] was the first to observe that the protoplasmic gel of the *Arbacia* egg was liquefied under moderate pressure, and Marsland and Brown[7] later showed that a similar trans-formation occurs in gels prepared from rabbit muscle actomyosin. The early experiments of Brown and Marsland[8] and the later studies of Landau *et al.*[9] relating the concomitant change in structural stability of the protoplasmic gel and the form and activity of *Amoeba proteus* under increasing pressure are most provocative. With increasing pressure in the range of the gel–sol transition of the plasma gel, the *Amoeba* is gradually transformed into a large inert sphere. On releasing the pres-sure (400 atm, maintained 5 minutes) the plasma gel contracts suddenly, leaving a broad, clear zone between the spherical cell membrane and the contracted gel, then spreads peripherally, bulging the membrane in an irregular fashion. Within minutes after decompression, the *Amoeba* com-mence renewed activity. Similar reversible changes in cell morphology and locomotor activity are seen in other protozoa as well. *Blepharisma* and *Paramecium*[10–12] and *Tetrahymena pyriformis*[13–15] undergo changes in structure and activity under pressure which can be directly related to dissociation of the protoplasmic gel. It is of considerable interest, in terms of the type of bonding involved in the associated structures, that the critical pressure required for dissociation is directly linked to the temperature. For example, the critical pressure required to bring about

[5] A. M. Zimmerman (ed.), "High Pressure Effects on Cellular Processes." Academic Press, New York, 1970.

[6] D. E. S. Brown, *J. Cell. Comp. Physiol.* **5**, 335 (1934).

[7] D. A. Marsland and D. E. S. Brown, *J. Cell. Comp. Physiol.* **20**, 295 (1942).

[8] D. E. S. Brown and D. A. Marsland, *J. Cell. Comp. Physiol.* **8**, 159 (1936).

[9] J. V. Landau, A. M. Zimmerman, and D. A. Marsland, *J. Cell. Comp. Physiol.* **44**, 211 (1954).

[10] W. Auclair and D. A. Marsland, *Biol. Bull.* **115**, 384 (1958).

[11] H. Asterita and D. A. Marsland, *J. Cell. Comp. Physiol.* **58**, 49 (1961).

[12] A. C. Giese, *Exp. Cell Res.* **52**, 370 (1968).

[13] R. E. Simpson, Thesis, University of Iowa, Iowa City, Iowa.

[14] L. Lowe-Jinde and A. M. Zimmerman, *J. Protozool.* **16**, 226 (1969).

[15] S. Zimmerman and A. M. Zimmerman, *in* "High Pressure Effects in Cellular Processes" (A. Zimmerman, ed.), p. 179. Academic Press, New York, 1970.

distinct morphological changes as well as solution of the gel structure decreases by 70 atm for each 5° drop in temperature in the case of *Amoeba proteus.*[9]

Another pressure-dependent structural transition thought to be closely linked to gel–sol transformations is the furrowing process by means of which a cell divides itself into two daughter cells. If pressure (400–500 atm) is applied in telophase, just as the cell begins to pinch inward prior to division, the cleavage furrow is reversed and the cell is restored to a single spherical or ovoid structure.[16] Cleavage can proceed only when the temperature–pressure conditions are such that the gel strength does not fall below a certain well-defined minimum. The linked pressure-temperature relationships required for blocking cell division in *Arbacia* eggs are shown in Fig. 1, which is taken from the work of Marsland.[16]

Recently, the important role played by microtubules in these processes has been recognized. Microtubular elements constitute an intrinsic part of a variety of protoplasmic structures (e.g., mitotic spindles, suctorian tentacles, cilia, flagella, many protozoan structures, and the cortical cytoplasm of various animal and plant cells), and the pressure-dependent gel–sol transformations appear to be closely related to depolymerization of the microtubular system of these structures.[15,17,18] At moderate pressures (275–400 atm) the needlelike extensions or axopodia of *Actinosphaerum* become unstable and retract as a result of dissociation of the centrally located microtubular system, which runs from the base to the tip of these structures. On releasing the pressure, repolymerization of the microtubules occurs and the organism resumes its normal form and activity. Microtubules of the mitotic apparatus also appear to be susceptible to high pressure dissociation,[16] and Zimmerman and Philpott[19] have demonstrated that high pressures disorganize the cytoplasmic microtubules of *Arbacia* eggs. The possibility that reversible polymerization–depolymerization reactions of the protoplasmic gel structure are intimately associated with the development of mechanical energy in cytokinesis and cell movement has been emphasized in recent years, particularly by Marsland and Landau.[16,20,21]

[16] D. A. Marsland, *in* "High Pressure Effects in Cellular Processes" (A. Zimmerman, ed.), p. 259. Academic Press, New York, 1970.

[17] L. G. Tilney, Y. Hiramoto, and D. A. Marsland, *J. Cell Biol.* **29**, 77 (1966).

[18] J. A. Kitching, *in* "High Pressure Effects in Cellular Processes" (A. Zimmerman, ed.), p. 155. Academic Press, New York, 1970.

[19] A. M. Zimmerman, *in* "High Pressure Effects in Cellular Processes" (A. Zimmerman, ed.), p. 253. Academic Press, New York, 1970.

[20] J. V. Landau, *Ann. N.Y. Acad. Sci.* **78**, 487 (1959).

[21] D. A. Marsland, *Int. Rev. Cytol.* **5**, 199 (1956).

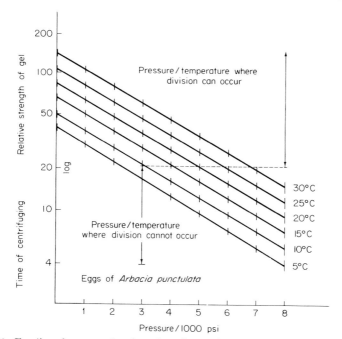

Fig. 1. Family of curves showing the effects of pressure, at different temperatures, on the structural strength of the gelated cortical cytoplasm of the *Arbacia* egg and the relation of these data to the capacity of the egg to perform the work of cleavage. Any combination of temperature and pressure which reduces the cortical gel strength beyond a critical value (about 20% relative to the atmospheric value at 23°) is just adequate to block cell division. Data obtained with a "pressure-centrifuge" designed by D. E. S. Brown [*J. Cell. Comp. Physiol.* 5, 335 (1934)]. Centrifuge times are expressed directly in seconds, and the degree of variation is indicated by the length of the markers. Reproduced from D. A. Marsland [*in* "High Pressure Effects in Cellular Processes" (A. Zimmerman, ed.), p. 259. Academic Press, New York, 1970].

In addition to the transformations observed in the three-dimensional fibrous networks of cells, a number of well defined macromolecular systems have been found to depolymerize under moderate pressure. These systems include the synthetic filaments of myosin,[22] crystals of sickle cell hemoglobin,[23] sea urchin ribosomes,[24] the polymeric form of β-casein,[25] and aggregates of poly-L-valyl-ribonuclease.[26] Moreover, the

[22] R. Josephs and W. F. Harrington, *Proc. Nat. Acad. Sci. U.S.* 58, 1587 (1967).
[23] M. Murayama, *Clin. Chem.* 13, 578 (1967).
[24] A. A. Infante and R. Baierlein, *Proc. Nat. Acad. Sci. U.S.* 68, 1780 (1971).
[25] T. A. J. Payens and K. Heremans, *Biopolymers* 8, 335 (1969).
[26] M. S. Kettman, A. H. Nishikawa, R. Y. Morita, and R. R. Becker, *Biochem. Biophys. Res. Commun.* 22, 262 (1965).

positive molar volume increments reported for a number of other macro-molecular systems suggest that these, too, would be partially or completely depolymerized by moderate pressures (see Table II).

To make progress in understanding these phenomena at the molecular level, it seems essential to investigate the pressure dependence of the isolated, purified macromolecular systems, using techniques that will provide information on the distribution of species at chemical equilibrium. In the following discussion we consider the effect of the pressure gradient developed in velocity and equilibrium centrifugation and show how these techniques can be used to detect pressure dependence and provide data relevant to the fundamental mechanisms underlying association processes.

Detection of Volume Changes in Interacting Systems

Rapidly interacting systems of the nM $\rightleftarrows$ P type, in which a volume change occurs on polymerization, will be affected by hydrostatic pressure to an extent depending on the magnitude of the *molar* volume change of the reaction[27] and the pressure gradient developed in the ultra-centrifuge. We assume that at each level of the cell the species concentrations satisfy the local value of the equilibrium constant of the chemical reaction. Then the pressure dependence of the equilibrium constant of polymerization is

$$\ln K(x) = \ln K_0 - \frac{1}{RT} \int_{x=x_0}^{x=x} \Delta V \left(\frac{\partial P}{\partial x} \right) dx \tag{1}$$

where $K(x)$ is the equilibrium constant at any point x in the liquid column and K_0 is the equilibrium constant at the meniscus position, x_0. The change in molar volume upon formation of 1 mole of polymer is ΔV, and P, R, and T are the pressure, gas constant, and temperature, respectively. In the present discussion we are interested in the effect of moderate pressures on aqueous systems. Assuming that the density of the solution and the density increment for each solute species are not functions of pressure,[28] Eq. (1) can be integrated to give:

$$\ln K(x) = \ln K_0 - \frac{\Delta V \rho \omega^2}{2RT} (x^2 - x_0^2) \tag{2}$$

where ρ is the solution density and ω the rotor velocity. Thus the volume change is readily obtained from a plot of log K against pressure, where the pressure at any depth x in the cell is given by

[27] L. F. Ten Eyck and W. Kauzmann, *Proc. Nat. Acad. Sci. U.S.* **58**, 888 (1967).
[28] P. F. Fahey, D. W. Kupke, and J. W. Beams, *Proc. Nat. Acad. Sci. U.S.* **63**, 548 (1969).

$$P_x = \frac{\rho \omega^2}{2} (x^2 - x_0^2) + P_0 \qquad (3)$$

P_0 is the pressure at the liquid–air meniscus (except in special experiments, $P_0 = 1$ atm).

Because of the large molecular weight of macromolecules such as proteins, a small change in partial specific volume on self-association will result in a large molar volume change, ΔV. Since it is the molar volume change rather than the fractional or percent change which governs the magnitude of the pressure dependence of K, clearly, the larger the molecular weight of the interacting species, the more readily pressure dependence will be detected. Consider three hypothetical polymerization reactions for which the molar volume change on association is $+0.5\%$ and the molecular weights of the polymer species are 10^5, 10^6, and 10^7, respectively.

Table I[29] shows the striking changes in equilibrium constant expected for the three self-association reactions over the range of hydrostatic pressures commonly generated at the base of a rotating ultracentrifuge cell. It should be emphasized that the number of monomeric units associating to form the polymeric species is irrelevant; only the change

TABLE I

EFFECT OF PRESSURE ON INTERACTING SYSTEMS

Pressure (atm)	MOLECULAR WEIGHT OF PRODUCTS[a]		
	MW = 10^5 (K_0/K_x)	MW = 10^6 (K_0/K_x)	MW = 10^7 (K_0/K_x)
1	1	1	1
40	1.9	6.3×10^2	10^{28}
80	3.6	4.0×10^5	10^{56}
120	6.9	2.5×10^8	10^{84}
160	13.2	1.6×10^{11}	10^{112}
200	25.1	9.1×10^{13}	10^{140}
300	126	9.8×10^{20}	10^{210}

[a] K_0 is the equilibrium constant at atmospheric pressure, and K_x is the equilibrium constant at the pressure indicated in the left-hand column. Results give the ratio of K_0 to K_x (i.e., the perturbation of the equilibrium due to pressure) for three reactions in which $\bar{V}$ increases from 0.73 ml/g to 0.73365 ml/g, an increase in molar volume of 0.5% ($\Delta \bar{V} = 3.65 \times 10^{-3}$ ml/g). The molecular weight of the products are 10^5, 10^6, and 10^7. At the maximum speed attainable in the ultracentrifuge (72,000 rpm), the pressure across the cell is 480 atm; the pressure across the cell at 59,780 rpm is 330 atm. Reproduced from W. F. Harrington and R. Josephs [Develop. Biol., Suppl. 2, 21 (1968)].

[29] W. F. Harrington and R. Josephs, Develop. Biol., Suppl. 2, 21 (1968).

in specific volume and the size of the aggregate are pertinent to the calculation. Since small changes in the partial specific volume are common in interacting systems, the effects seen in Table I are likely to be the rule rather than the exception for molecular aggregates.

Dependence of Sedimentation Patterns on Rotor Speed

The radial variation of the equilibrium constant along the centrifuge cell resulting from the pressure gradient can have a profound effect on the sedimentation patterns observed in an interacting system. Thus velocity sedimentation studies at various rotor speeds provide an important and sensitive technique for the detection of pressure effects. The reversible association of myosin in the pH range 8–8.5 at ionic strength 0.1–0.2 M KCl to form a sharply defined polymeric species of molecular weight 50 to 60 $\times$ 10^6 is a good example.[30]

For purposes of the present discussion we assume the simplest model for the interacting myosin system, that of a rapidly reversible n monomer $\rightleftharpoons$ polymer equilibrium. In a later section of this chapter the more realistic model of monomer $\rightleftharpoons$ dimer $\rightleftharpoons$ polymer will be discussed, based on recent sedimentation equilibrium evidence demonstrating the presence of a monomer $\rightleftharpoons$ dimer equilibrium in high and low salt media. The qualitative features of the argument are not expected to be altered by the model chosen when a large number of monomer (dimer) units associates to form polymer.

At low rotor speed (9000 rpm), two well resolved peaks are observed in the sedimentation pattern (Fig. 2)—a broad, slowly sedimenting peak ($s_{20,w}^\circ = 6.5$ S), which we provisionally identify as myosin monomer, and a hypersharp rapidly sedimenting fast peak ($s_{20,w}^\circ = 150$ S). At this low speed, the experimentally observed distribution of mass in the ultracentrifuge cell is consistent with the predictions of the Gilbert theory[31,32] for a system in which the single type of polymer molecule is composed of a large number of monomeric units and the equilibrium constant strongly favors association. Analysis of the concentration profile according to the Gilbert theory reveals[33] that the concentration change across the slow boundary very closely approximates the monomer concentration, whereas the concentration change across the fast boundary closely approximates the polymer concentration. Thus the slow boundary may be identified as monomer and the fast boundary as polymer. The two well resolved peaks constitute

[30] R. Josephs and W. F. Harrington, *Biochemistry* 5, 347 (1966).
[31] G. A. Gilbert, *Discuss. Faraday Soc.* 20, 68 (1955).
[32] G. A. Gilbert, *Proc. Roy. Soc. Ser. A* 250, 377 (1959).
[33] R. Josephs and W. F. Harrington, *Biochemistry* 7, 2834 (1968).

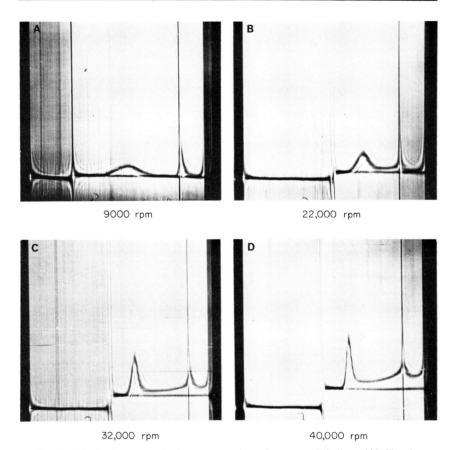

9000 rpm 22,000 rpm

32,000 rpm 40,000 rpm

FIG. 2. Effect of rotor velocity on myosin–polymer equilibrium. (A) Total protein concentration 0.4%; time of centrifugation, 18 hours. The breadth of the slow peak is due to diffusion. (B) Total protein concentration 0.6%; time of centrifugation, 5 hours. (C) Total protein concentration 0.6%; time of centrifugation, 1.5 hours. (D) Total protein concentration 0.6%; time of centrifugation, 1 hour. Each polymer solution was exhaustively dialyzed against $0.18 M$ KCl, $2 \times 10^{-3} M$ Veronal, pH 8.3. Rotor velocity is indicated in the figure. Temperature was 5°. Double sector cell. Reproduced from R. Josephs and W. F. Harrington [*Proc. Nat. Acad. Sci. U.S.* **58**, 1587 (1967)].

a reaction boundary even though the concentration gradient in the region between them is virtually undetectable. These unusual features stem from the large value of n, the number of monomeric units associating to form polymer. Now, when identical myosin solutions are examined at progressively higher rotor velocities the concentration profiles are significantly and continuously altered. Increasing speed results in a

marked elevation of the concentration gradient, both between the monomer and polymer peaks, and in the region centrifugal to the polymer
peaks. The elevation of the base line is seen to increase with rotor
speed, suggesting that the monomer–polymer equilibrium is altered as
a result of the increasing hydrostatic pressure gradient established
throughout the liquid column.

The origin of the most salient features in these sedimentation patterns may be qualitatively understood in terms of the gradients generated by the "individual" sedimentating species. Assume that increasing hydrostatic pressure produces a shift in the chemical equilibrium
myosin $\rightleftarrows$ polymer toward increasing monomer concentration. At constant rotor velocity the pressure gradient $[(\partial P/\partial x) = \rho\omega^2 x]$ increases
with radial distance; hence, as soon as the rotor is brought to speed, and
before any mass transport occurs, the concentration of monomer and
polymer at each level of the liquid column will adjust to satisfy the
value of the equilibrium constant at that level. Thus, the monomer
concentration should increase with increasing depth of the liquid column,
and the polymer concentration should decrease. Since the pressure
gradient increases with the square of the rotor velocity we would anticipate a progressive elevation in the concentration gradient of monomer
and a concomitant and corresponding shift in the (negative) gradient
of polymer with increasing rotor velocity. Since the total concentration
is invariant with respect to radial distance before sedimentation begins,
this phenomenon cannot be detected by the schlieren or interference optical systems of the ultracentrifuge. Nevertheless, in accordance with
Rayleigh's scattering law, the high molecular weight myosin polymer
(MW $\cong 50 \times 10^6$) will scatter light far more strongly than monomer,
and the concentration gradient of the polymer species can be easily monitored by measurement of the optical density at low wavelength. Figure 3
shows the optical density at 365 nm as a function of radial distance at
four different rotor speeds obtained from schlieren photographs immediately after rotor speed was reached. It is clear from these results that,
at sufficiently high rotor speed, the concentration of polymer does in fact
decrease continuously from the meniscus to the base of the cell. Moreover, we would expect the concentration gradients of each species,
established as a result of the pressure gradient, to be completely reversible in the absence of mass transport. Consistent with this thesis,
the results of Fig. 3 are found to be the same irrespective of whether
the rotor speed is raised from 20,000 to 52,000 rpm or first raised to 52,000
rpm and then immediately decreased to 20,000 rpm. This technique—
the recording of radial change in optical density at low wavelength over
a wide span of rotor speeds—offers a sensitive diagnostic test for the

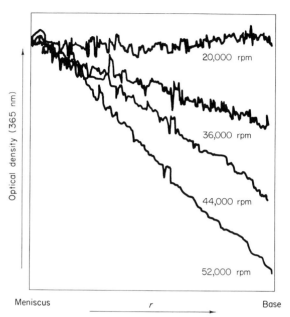

FIG. 3. The effect of rotor velocity on the radial optical density profile of a myosin–polymer equilibrium system. Protein concentration 1% in 0.178 M KCl, Veronal $2 \times 10^{-3} M$, pH 8.3, 30 mm Kel-F single-sector cell. Rotor velocity is indicated in the figure. Temperature, 5°. Optical density is in arbitrary units and was obtained from a microdensitometer tracing of schlieren photographs taken with ultraviolet ($\lambda = 365$ nm) light. At 52,000 rpm the extinction change across the photographic plate ($x_0 \to x_b$) is about 0.9 OD units. Reproduced from R. Josephs and W. F. Harrington [*Proc. Nat. Acad. Sci. U.S.* **58**, 1587 (1967)].

presence of a pressure-dependent associating system. The test presupposes that the aggregate species is of much higher molecular weight than the monomeric unit.

In the case of the myosin–polymer system the large value of n renders c_m, the concentration of "monomer," practically independent of the concentration of polymer. Consequently, the sedimentation of polymer will have a vanishingly small effect on the concentration profile of monomer, and as this species sediments it will uncover the concentration gradient of monomer established when the rotor has reached operating speed. The experimentally observed concentration gradients centripetal to the polymer boundary of Fig. 2 thus clearly reflect the effect of the increase in the pressure gradient, with both radial distance and rotor velocity. At levels centrifugal to the polymer boundary, the factors influencing the gradient profile are more complex. Calculations employing the countercurrent distribution analog of Bethune and Kegeles[34] reveal a very sharp negative concentration gradient between the leading edge of the fast peak and the cell bottom; this gradient acts to provide convective mixing in this region.[35] A detailed consideration of pressure-induced convection and its effect on the gradient profile will be given below.

The technique of varying rotor speed can also be employed to demonstrate pressure-dependent dissociation of macrostructures during zonal centrifugation. Anomalies in the sedimentation patterns of both eukaryotic and prokaryotic ribosomes under a variety of ionic conditions have often been interpreted in terms of conformational changes in the ribosome and ribosomal subunits. The magnitude of the effect that pressure can have on the sedimentation behavior has been appreciated only recently.[24] In these studies, see urchin ribosomes (75 S) were layered onto 15–30%, sucrose gradients in a buffer (0.005 M MgCl$_2$, 0.05 M triethanolamine, pH 7.8) containing 0.24 M KCl and sedimented at various rotor speeds. At 24,000 rpm the sedimentation pattern shows only a single peak with sharp trailing edge; at 30,000 rpm, a single slower-sedimenting major peak with a trailing shoulder; at 41,000 rpm, two distinct peaks. Isolation of the material under each peak followed by a determination of the RNA size, reassociation, and *in vitro* activity in the presence of poly(U) confirmed that the bimodal profile at high speed represented a pressure-induced dissociation of the 75 S free ribosomes into 35 S and 56 S subunits.

The time dependence of the sedimentation pattern at high rotor speed (41,000 rpm) is instructive. In the early stages of centrifugation (2.5 hours) the band traverses the sucrose gradient as though it were composed entirely of 75 S particles. As the band moves deeper into the liquid column (3.7 hours), it broadens and a reduction in sedimentation coefficient occurs (to about 67 S). In the late stages of centrifugation (5 hours), the single broad peak is transformed into two well resolved peaks with apparent sedimentation coefficients of 55 S and 65 S.

The effect of the pressure gradient on the sedimentation pattern in zonal centrifugation is somewhat easier to comprehend than that in velocity sedimentation for two reasons: (1) In zonal centrifugation, the presence of a strong preestablished density gradient of solvent acts to suppress convective disturbances. (2) The shift in equilibrium constant resulting from the pressure gradient occurs progressively as the sedimenting zone of macromolecules moves deeper into the liquid column.

[34] J. L. Bethune and G. Kegeles, *J. Phys. Chem.* **65**, 1761 (1961).
[35] G. Kegeles and M. Johnson, *Arch. Biochem. Biophys.* **141**, 63 (1970).

In the case of velocity sedimentation the pressure-dependent distribution of sedimenting species is established immediately on reaching speed throughout the entire liquid column. Convective disturbances resulting from mass transport may act from the very beginning of sedimentation to minimize detection of pressure dependence or to nullify correct interpretation (see below). Once bands of pure components have been resolved in zonal centrifugation, each substance sediments independently,[36] although it is not always feasible to produce such complete resolution in reacting systems.

The sedimentation behavior of sea urchin ribosomes under zonal centrifugation suggests the existence of a rapidly reversible equilibrium between the parent molecule and its constituent subunit species, in which association of the subunits results in a positive molar volume change. That is, for the reaction

small subunit (35 S) + large subunit (56 S) $\rightleftharpoons$ ribosome (75 S), $\Delta V > 0$

As the ribosome band sediments through increasing levels of hydrostatic pressure, the equilibrium (ribosome $\rightleftharpoons$ subunits) is continuously shifted toward the slower-moving subunits, and the apparent sedimentation coefficient of the broadening but unresolved major peak decreases. When the diffuse band has reached a sufficient depth it resolves into two separate subunit peaks. Infante and Baierlein[24] have shown that the time-dependent changes in sedimentation patterns can be closely simulated from the sedimentation coefficients of the three species assuming a volume change of 500 cc per mole of ribosomes on association of the subunits.

The striking changes in sedimentation patterns observed in this system on increasing the rotor speed emphasize again the importance of the molar volume, which is related to the high molecular weight of the polymeric species (ribosomes), not to the number of monomeric units involved in the association process.

To demonstrate pressure dependence unequivocally, it may sometimes be necessary to compare elution profiles of an interacting system obtained from a separation method, such as gel filtration chromatography (at atmospheric pressure), with those derived from gradient or velocity sedimentation. In studies on the interaction of seryl and leucyl transfer RNA synthetases with their cognate transfer RNA's, Knowles et al.[37] observed significant differences in complex formation when mixtures of

[36] J. Vinograd, R. Bruner, R. Kent, and J. Weigle, *Proc. Nat. Acad. Sci. U.S.* **49**, 902 (1963).
[37] J. R. Knowles, J. R. Katze, W. Konigsberg, and D. Söll, *J. Biol. Chem.* **245**, 1407 (1970).

the enzymes and crude tRNA were examined by gradient sedimentation and by gel filtration. These were ascribed to the effect of the pressure gradient on the enzyme–tRNA complex association constant. In this instance, the pressure gradient favors formation of the complex, indicating a negative molar volume of formation. The findings of Knowles *et al.* may account for the results of Lagerquist and Rymo,[38] who have shown that only one of the two yeast tRNA[val] species forms a stable complex with its cognate tRNA synthetase in Sephadex G-100 chromatography, but in sucrose gradient sedimentation both forms are found to associate with the enzyme.

Detection of pressure dependence can be extremely difficult in interacting systems where a small number of low molecular weight species are involved in chemical equilibrium, and detailed analysis of the velocity sedimentation profiles may be required to demonstrate the pressure effect. From Archibald molecular weight studies on α-chymotrypsin at several different protein concentrations in 0.2 ionic strength phosphate buffer, pH 6.2, Rao and Kegeles[39] concluded that the enzyme exists as a mixture of monomer, dimer, and trimer under these ionic conditions. Calculations based on the countercurrent distribution analog[34] and the Gilbert theory[31,32] show that the velocity sedimentation pattern of this chemically reacting system should exhibit a sizable slow shoulder, yet the observed sedimentation velocity profile clearly shows a single symmetrical peak. It seems likely that this anomaly results from the presence of a pressure effect favoring the monomeric species. When a positive molar volume change for trimerization is included in the countercurrent analog computation, the predicted sedimentation velocity profile corresponds closely to that observed experimentally.[40]

Layering Techniques

Proof of the presence of pressure dependence in an interacting system may often be obtained by overlayering the solution column in an ultracentrifuge cell with an inert medium such as mineral oil. In this way any effect of pressure on the chemical reaction can be established before significant mass transport occurs. This technique was employed to demonstrate pressure dependence in the myosin polymer system.[22] Varying thicknesses of mineral oil, previously equilibrated with solvent, were layered over identical myosin solutions of the same column height, and each of these preparations was centrifuged at 40,000 rpm for 75 minutes to resolve the monomer and polymer boundaries (Fig. 4). The

[38] U. Lagerquist and L. Rymo, *J. Biol. Chem.* **244**, 2476 (1969).
[39] M. S. N. Rao and G. Kegeles, *J. Amer. Chem. Soc.* **80**, 5724 (1958).
[40] G. Kegeles and M. Johnson, *Arch. Biochem. Biophys.* **141**, 59 (1970).

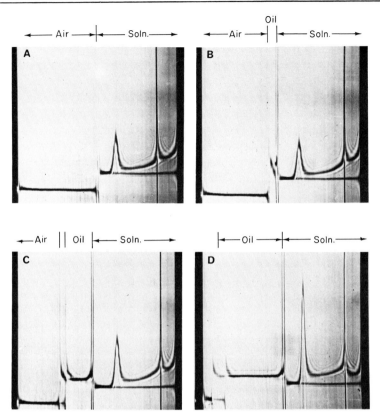

Fig. 4. The effect of hydrostatic pressure on the myosin monomer–polymer equilibrium at constant rotor velocity of 40,000 rpm; temperature, 5°. Varying amounts of mineral oil (density = 0.85 g/ml), previously equilibrated with dialyzate were added to aliquots of 0.66% myosin solution which had been dialyzed against 0.185 M KCl, 2×10^{-3} M Veronal, pH 8.3. The lower (centrifugal) meniscus at the oil–solution interface is that of the protein solution, and the upper (centripetal) oil–air meniscus corresponds to the protein sector. Time of centrifugation for each frame was 75 minutes. Reproduced from R. Josephs and W. F. Harrington [Proc. Nat. Acad. Sci. U.S. 58, 1587 (1967)].

resulting sedimentation patterns show a progressive elevation of the gradient curve between the two peaks with increasing hydrostatic pressure (thickness of mineral oil) analogous to that observed in the experiment depicted in Fig. 2. Moreover, the area of the gradient profile centripetal to the polymer boundary was observed to increase markedly ($\sim$240%) with increasing pressure. Since in the limit of large n and equilibrium constant, K, this area closely approximates the monomer concentration in the beginning of each experiment, it can be concluded that the monomer–polymer equilibrium is a function of hydrostatic pressure.

Similar results would be expected if varying amounts of solvent were layered over a suspected pressure-dependent system in a synthetic boundary cell at low speed and the sedimentation patterns were compared after centrifuging for an equivalent time at constant high rotor velocity. A modification of this technique was employed to establish pressure dissociation of ribosomes.[24] Free ribosomes were layered over an abbreviated sucrose gradient and then the tube was filled to the top with a sucrose solution of lower density to generate a strong hydrostatic pressure head above the sedimenting zone immediately after reaching operating speed. After 2 hours of centrifugation at 41,000 rpm, the single ribosomal band was transformed into two peaks with sedimentation coefficients characteristic of the 35 S and 56 S subunits (Fig. 5). However,

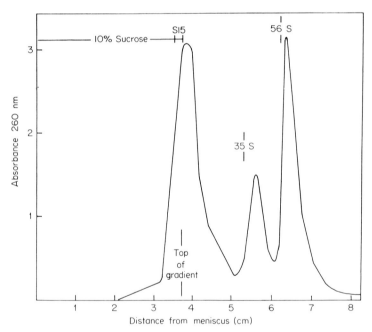

Fig. 5. Dissociation of ribosomes is related to position in centrifuge tube; 0.4 ml of ribosomal extract was placed on a 6.7-ml, linear 15–30% sucrose gradient (0.20 M KCl), and then 5 ml of 10% sucrose was layered above the sample. The system was centrifuged for 2 hours at 41,000 rpm. The reference points (35 S and 56 S) are the positions reached by separated subunits run in parallel in similar abbreviated gradients. Absorbance at the top of the gradient is nonribosomal material in the extract. Reproduced from A. A. Infante and R. Baierlein [*Proc. Nat. Acad. Sci. U.S.* **68**, 1780 (1971)].

when ribosomes were layered onto the top of a conventional 15–30% sucrose gradient and centrifuged 2.5 hours at the same rotor velocity, no significant dissociation was observed.

The layering technique may be essential to distinguish between pressure-dependent reactions and other types of interaction which may show rotor-speed dependence, e.g., kinetically controlled processes and macromolecule–small molecule interactions.[41] As we have noted earlier, changes in the sedimentation pattern during mass transport of an interacting system are often difficult to interpret, even in the absence of pressure effects, and the possible existence of convective disturbance resulting from negative concentration gradients during transport of pressure dependent reactions adds to the complexity. Simple comparative studies which involve only variations in the hydrostatic pressure head above the rotating liquid column may permit detection of changes in the sedimenting patterns which can be directly attributed to pressure-dependent chemical reactions.

Use of the Pressure Chamber

Studies of pressure-dependent macromolecular systems in the ultracentrifuge may be simplified and their scope broadened appreciably through the use of a recently designed device of Schumaker *et al.*[42] which permits loading and sealing of the analytical cell under high pressure. An ultracentrifuge cell containing the solution of interest is placed in the well of a pressure chamber (Fig. 6), and the top section of the chamber, which is fitted with a floating, spring-loaded screwdriver, is bolted firmly in place. The contents of the centrifuge cell are free to equilibrate with their surrounding environment since the brass filling-hole screw is temporarily fastened to the tip of the screwdriver prior to assembly of the pressure chamber. After the pressure chamber is filled with N_2 gas at pressures up to 135 atm, the cell is sealed by rotation of the filling-hole screw, and it can then be employed for investigation of macromolecular systems at low rotor velocity. The elevation in the schlieren base line above the liquid–gas interface permits direct determination of pressure within the cell at the conclusion of the centrifuge experiment. This technique would appear to be a powerful new approach to the detection and analysis of pressure-dependent phenomena.

The pressure chamber technique has been employed by Schumaker *et al.*[42] to investigate the pressure dependence of the myosin–polymer associating system. Solutions of the protein in a solvent comparable

[41] J. R. Cann, "Interacting Molecules," pp. 171 ff. Academic Press, New York, 1970.
[42] V. N. Schumaker, A. Wlodawer, J. T. Courtney, and K. M. Decker, *Anal. Biochem.* **34**, 359 (1970).

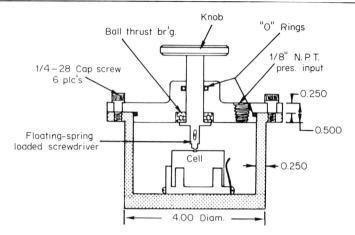

Fig. 6. Pressure chamber allowing solutions in analytical ultracentrifuge cell to be equilibrated with nitrogen under moderate pressures. Reproduced from V. N. Schumaker, A. Wlodawer, J. T. Courtney, and K. M. Decker [*Anal. Biochem.* **34**, 359 (1970)].

to that used in the studies of Fig. 2 were adjusted to 100 atm in the pressure chamber and centrifuged at 9000 rpm. The resulting sedimentation patterns revealed a marked depletion in the size of the polymer peak and enhancement of the slower sedimenting "monomer" peak compared to a conventional centrifuge run at the same speed, but at 1 atm pressure. This same technique has also been used to demonstrate that the well-known rotor speed-dependent aggregation phenomenon of high molecular weight DNA is not a pressure-dependent process.[43] The degree of aggregation at various rotor velocities, as judged from the monomer concentration and the sedimentation coefficient of the monomer species, is virtually identical in the presence and in the absence of high pressure ($\sim$112 atm).

When low-speed Archibald molecular weight measurements at the gas–liquid meniscus were performed at various imposed hydrostatic pressures by use of a modified Schumaker-type pressure loading chamber, it was found possible[44] to evaluate quantitatively the molar volume change for the rapid, reversible hexamer–dodecamer reaction of lobster hemocyanin. While this method is thermodynamically exact for the determination of volume changes, the interpretation of the data depends on knowledge of the process taking place, and is therefore still somewhat subjective.

[43] V. N. Schumaker and P. Poon, *Biopolymers* **10**, 2071 (1971).
[44] V. P. Saxena and G. Kegeles (personal communication, 1972).

Estimation of Molar Volume Changes

The change in volume that can give rise to significant pressure dependence in the ultracentrifuge lies well within the accuracy of routine partial specific volume measurements. For example, the absolute difference between the partial specific volumes of monomeric myosin and its polymer is only $6 \pm 1.2 \times 10^{-4}$ ml/g; that between the constituent 35 S and 56 S subunits and free sea urchin ribosomes, only 2×10^{-4} ml/g. (Specific volume changes, which have been reported for a number of protein association reactions, are given in Table II.) Methods generally utilized for measurement of volume changes in aqueous solutions consist either of determining the partial specific volume, $\bar{V}$, of the species before and after the transition of interest (the difference, $\Delta \bar{V}$, being a measure of the volume change) or more directly by dilatometry. The former method is limited by the inherent accuracy of $\bar{V}$ measurements, which routinely lie in the range ± 0.01 ml/g, whereas employment of the latter is technically extremely tedious and requires large solution volumes and a high degree of temperature control. The problem can be appreciated from the very precise measurements of apparent specific volumes as a function of pressure by Fahey et al.[28] on tobacco mosaic virus, in both the polymerized and unpolymerized state, employing the magnetic densitometer.

Stevens and Lauffer[45] found an increase in volume of 0.00741 ± 0.00003 ml/g at $4°$ when TMV protein was polymerized in a dilatometer by adjusting the pH from 7.5 to 5.5. Since polymerized TMV protein shows an increase in volume of 0.00247 ml/g on titration from pH 6.8 to pH 5.5,[46] the difference between these two values, 0.0049 ± 0.00003 ml/g, represents the increase in partial specific volume when TMV protein undergoes polymerization at the low pH (see Lauffer[47]). In the magnetic densitometer studies of Fahey et al.[28] no net change in apparent specific volume was observed with increasing pressure in the solution of TMV protein at pH 7.5 (340 atm) or polymerized protein at pH 5.5 (400 atm). The extreme deviations about the mean measured apparent specific volume in these experiments were ± 0.002 ml/g and are consequently within the range of the volume change observed by Stevens and Lauffer.

Magnetic densitometer measurements on ribonuclease and turnip yellow virus reveal that the apparent specific volumes of these nonassociating proteins are unchanged within the precision of the measure-

[45] C. L. Stevens and M. A. Lauffer, *Biochemistry* **4**, 31 (1965).
[46] R. Jaenicke and M. A. Lauffer, *Biochemistry* **8**, 3083 (1969).
[47] M. A. Lauffer, in "Subunits in Biological Systems" (S. N. Timasheff and G. D. Fasman, eds.), p. 194. Dekker, New York, 1971.

TABLE II
VOLUME CHANGES FOR SOME MACROMOLECULE ASSOCIATION REACTIONS

System	Molecular weight polymer	n, degree of polymerization	ΔV^a (ml/g)	Reference
Poly-L-Valyl ribonuclease	Indefinite	Indefinite	$1.5 \times 10^{-2\,b}$	d
S-Peptide + S-protein (ribonuclease)	13,700	$A + B \rightleftharpoons C$	2.3×10^{-3}	e
35 S + 56 S subunits (ribosomes)	2.8×10^6	$A + B \rightleftharpoons C$	1.8×10^{-4}	f
Collagen	$>10^8$	$>3 \times 10^2$	8×10^{-4}	g
Myosin	50×10^6	~ 100	6×10^{-4}	h
Lobster hemocyanin	9.4×10^5	$6 - 12$	6×10^{-5}	i
Tobacco mosaic virus	$>50 \times 10^6$	>500	5×10^{-3}	j
Actin	Indefinite	Indefinite	1.5×10^{-3}	k
Flagellin	Indefinite	Indefinite	3.8×10^{-3}	l
Sickle cell hemoglobin	Indefinite	Indefinite	$5.9 \times 10^{-3\,b}$ (<50 atm pressure)	m
β-Casein	1.25×10^6	52	>0	n
Serum albumin	Indefinite	Indefinite	1.2×10^{-3}	o
Lysozyme	28,800	2	-3×10^{-2}	p
tRNA synthetase and tRNA	125,000	$A + B \rightleftharpoons C$	<0	q
α-Chymotrypsin	72,000	3	>0	r
Antibody–Antigen				
DNP + DNP-lysine	155,000	$A + nB \rightleftharpoons C$	2.3–$5.2 \times 10^{-3\,c}$	s
DNP + DNP-BγG	155,000	$A + nB \rightleftharpoons C$	2.6–$5.5 \times 10^{-3\,c}$	s
FAB (DNP) + DNP-lysine	52,000	$A + nB \rightleftharpoons C$	3.1–$3.8 \times 10^{-3\,c}$	s
Poly Glu^{56}Lys^{38}Tyr^6 + poly Glu^{56}Lys^{38}Tyr^6	155,000	$A + B \rightleftharpoons C$	$4.2 \times 10^{-3\,c}$	s
Poly Glu^{66}Lys^{40} and poly Glu^{66}Lys^{40}	155,000	$A + B \rightleftharpoons C$	1.5×10^{-3}	s
Enzyme–Inhibitor				
Lysozyme + N-acetyl-D-glucosamine	14,600	$A + B \rightleftharpoons C$	3.3×10^{-3}	t
Ribonuclease + cytidine 2′(3′)-monophosphate	13,700	$A + B \rightleftharpoons C$	1.7×10^{-3}	u,v

^{a} For the reaction n monomer $\rightleftharpoons$ polymer.

^{b} Activation volume (ΔV^*) obtained from rate measurements.

^{c} ΔV based on molecular weight of antibody. Most, if not all, of molar volume change is associated with this particle (Ohta *et al.*[5]).

^{d} M. S. Kettman, A. H. Nishikawa, R. Y. Morita, and R. R. Becker, *Biochem. Biophys. Res. Commun.* **22**, 262 (1965).

TABLE II (*Continued*)

[e] R. Y. Morita and R. R. Becker, *in* "High Pressure Effects on Cellular Processes" (A. M. Zimmerman, ed.), p. 71. Academic Press, New York, 1970.

[f] A. A. Infante and R. Baierlein, *Proc. Nat. Acad. Sci. U.S.* **68**, 1780 (1971).

[g] J. M. Cassel and R. G. Christensen, *Biopolymers* **5**, 431 (1967).

[h] R. Josephs and W. F. Harrington, *Biochemistry* **7**, 2834 (1968).

[i] V. P. Saxena and G. Kegeles (personal communication, 1972).

[j] C. L. Stevens and M. A. Lauffer, *Biochemistry* **4**, 31 (1965).

[k] T. Ikkai and T. Ooi, *Biochemistry* **5**, 1551 (1966).

[l] B. R. Gerber and H. Noguchi, *J. Mol. Biol.* **26**, 197 (1967).

[m] M. Murayama and F. Hasegawa, *Fed. Proc.* **28**, 536 (1969).

[n] T. A. J. Payens and K. Heremans, *Biopolymers* **8**, 335 (1969).

[o] R. Jaenicke, *Eur. J. Biochem.* **21**, 110 (1971).

[p] G. J. Howlett, P. D. Jeffrey, and L. W. Nichol, *J. Phys. Chem.* **76**, 777 (1972).

[q] J. R. Knowles, J. R. Katze, W. Konigsberg, and D. Söll, *J. Biol. Chem.* **245**, 1407 (1970).

[r] G. Kegeles and M. Johnson, *Arch. Biochem. Biophys.* **141**, 59 (1970).

[s] Y. Ohta, T. J. Gill, III, and C. S. Leung, *Biochemistry* **9**, 2708 (1970).

[t] D. M. Chipman and N. Sharon, *Science* **165**, 454 (1969).

[u] J. P. Hummel, D. A. Ver Ploeg, and C. A. Nelson, *J. Biol. Chem.* **236**, 3168 (1961.)

[v] G. G. Hammes and P. R. Schimmel, *J. Amer. Chem. Soc.* **87**, 4665 (1965).

ments ($\pm 5 \times 10^{-4}$ ml/g) up to pressures of 410 atm. Thus the assumption of $\partial \bar{V}/\partial P = 0$ for each species made in integrating Eq. (1) seems reasonable.

Gerber and Noguchi[48] utilized a Carlsberg-type dilatometer to investigate the volume change associated with the polymerization of flagellin at neutral pH. One arm of the dilatometer was filled with 5 ml of monomer solution (6.9 mg/ml, in $0.2\,M$ KCl-phosphate buffer, pH 7.1) and the other with 5 ml of "seed" solution of the same protein concentration which had been subjected to sonic degradation. The dilatometer was immersed in an accurately controlled bath which was maintained at constant temperature (reported to be $\pm 0.0002°$) by holding the temperature of the room 3° below the desired bath temperature. Replicate dilatometric experiments at 22–28° gave volume changes in the range 150 ml/mole of monomer ($\Delta \bar{V} = 3.8 \times 10^{-3}$ ml/g) within 5%.

Cassel and Christensen[49] found that a thermostated bath controlled to $\pm 0.001°$ was not sufficient to measure the volume change which occurs on aggregation of tropocollagen ($\Delta \bar{V} = 8 \times 10^{-4}$ ml/g) by a conventional dilatometric technique. Two dilatometers were employed, one containing 17 ml of the protein solution, the other filled with the same volume of phosphate buffer. The volume change was estimated by observing the difference in meniscus heights between the sample and dummy dila-

[48] B. R. Gerber and H. Noguchi, *J. Mol. Biol.* **26**, 197 (1967).

[49] J. M. Cassel and R. G. Christensen, *Biopolymers* **5**, 431 (1967).

tometers accompanying the time-dependent polymerization (temperature range 14–25°).

Evaluation of the molar volume change from sedimentation velocity experiments is possible if the equilibrium constant of the polymerization reaction can be estimated at various levels in the cell. In the special case of a rapidly reversible $n\mathrm{M} \rightleftharpoons \mathrm{P}$ reaction in which the number of monomeric units is large and the polymer of a specific size, the concentration of the two species at various levels of the cell can be estimated from a plot of the turbidity as a function of radius determined from the scanning optical system of the ultracentrifuge (see Fig. 3). The method presupposes that the scattering from the monomeric unit will be negligible compared to that from the high molecular weight polymer. Thus, in principle, the concentration distribution of both species can be estimated at various rotor speeds as a function of pressure before mass transport is initiated.

An estimation of the molar volume change can also be obtained from an analysis of the sedimentation patterns during mass transport. Consider the set of sedimentation patterns of the myosin–polymer system shown in Fig. 2. Since we assume chemical equilibrium at all levels of the cell, the equilibrium constant at each level may be readily calculated from the simple mass action expression $\log K = \log c_p - n \log c_m$. The polymer boundary exhibits a very high degree of self-sharpening, and interference fringes recorded by the Rayleigh optical system of the ultracentrifuge stop abruptly at a point immediately centripetal to the hypersharp polymer peak. Thus the radial position of the polymer boundary may be precisely determined. According to arguments presented earlier,[29,33] the large value of n in this system renders c_m practically independent of c_p. Consequently, sedimentation of polymer will have a vanishingly small effect on the concentration profile of the monomer. However, sedimentation of monomer will result in a reduction in monomer concentration at each radial position in the cell, and consequently some polymer will dissociate at each level to maintain the monomer at equilibrium concentration. The monomer concentration at the position of the polymer peak may thus be equated to the total protein concentration corresponding to the area of the gradient profile centripetal to the polymer boundary. The polymer concentration, c_p, at the position of the polymer boundary can be estimated either from the area under the fast peak (at low speed) or from the difference between the original total concentration, corrected for radial dilution, and the monomer concentration. Both methods are subject to considerable error, the first method because of hypersharpening effects, and the second because of convective disturbances in front of the fast boundary in systems where dissociation is strongly favored by pressure. Nevertheless, for a polymerizing system with large n, it is not necessary to know the polymer concentration at all accurately since the monomer concentration is raised to the nth power in the calculation of the equilibrium constant. Calculations of the gradient profiles of monomer and polymer[35] in the myosin–polymer system based on the countercurrent distribution analog of Bethune and Kegeles[34] show, in agreement with these predictions, that the slow boundary is virtually entirely monomer. Moreover, the estimation of K_p according to the method outlined above corresponds closely to that derived from the countercurrent distribution analysis.

Typical plots of the logarithms of the equilibrium constant versus pressure for the myosin–polymer system are shown in Fig. 7 for two

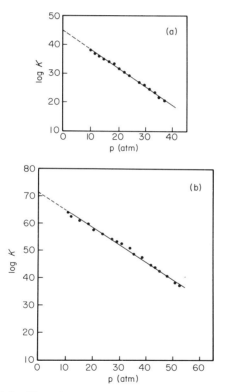

FIG. 7. Plots of $\log K$ against pressure for the myosin–polymer equilibrium. *Top*: Rotor velocity 22,000 rpm, KCl concentration $0.194\,M$; $2 \times 10^{-3}\,M$ Veronal, pH 8.3; protein concentration is 1.0 g/100 ml. *Bottom*: Rotor velocity 32,000 rpm, KCl concentration $0.18\,M$; $2 \times 10^{-3}\,M$ Veronal, pH 8.3; protein concentration 0.6 g/100 ml. Reproduced from R. Josephs and W. F. Harrington [*Biochemistry* **7**, 2834 (1968)].

different salt concentrations and rotor velocities. From the slopes of these plots the molar volume change, ΔV, estimated from Eq. (2), for reversible polymerization of myosin is $+32$ liters per mole of polymer.

Analysis of the pressure dependence of self-association of myosin is based on the assumption that only two species, monomer and polymer, are present in detectable concentration at chemical equilibrium. Recently, high speed sedimentation equilibrium and laser light-scattering studies have demonstrated that the myosin monomer is in rapid reversible equilibrium with a dimer species in high salt solvent systems.[50,51] As the ionic strength is lowered to a level just above the threshold concentration for polymerization, the equilibrium constant for dimerization is elevated appreciably ($K_2 = 20$ dl/g at 0.2 M KCl, pH 8.3).[52] Thus it seems possible that the slowly sedimenting peak in Fig. 1 represents a monomer–dimer reaction boundary and that three species, monomer, dimer, and polymer, coexist in equilibrium at each level of the centrifuge cell. The distribution of species will be governed by the two equations

$$2\mathrm{M_1} \rightleftharpoons \mathrm{M_2} \qquad K_2 = \frac{[\mathrm{M_2}]}{[\mathrm{M_1}]^2} \tag{4}$$

and

$$m\mathrm{M_2} \rightleftharpoons \mathrm{P} \qquad K_p = \frac{[\mathrm{P}]}{[\mathrm{M_2}]^m} \tag{5}$$

where the apparent equilibrium constant is

$$K_{\mathrm{app}} = K_p[K_2]^m = \frac{[\mathrm{P}]}{[\mathrm{M_1}]^{2m}} \tag{6}$$

and $m (\sim 50)$, the number of dimer units associating to form polymer. Evaluation of the equilibrium constant, K_{app}, as a function of pressure requires a knowledge of the pressure dependence of the monomer–dimer equilibrium.

For this system we may evaluate the pressure dependence by observing that the derivative of the logarithm of K_{app} with respect to pressure will be given by

$$\frac{\partial \ln K_{\mathrm{app}}}{\partial \mathrm{P}} = \frac{\partial \ln K_p}{\partial \mathrm{P}} + m \frac{\partial \ln K_2}{\partial \mathrm{P}} \tag{7}$$

From Eq. (2) it will be seen that

[50] J. Godfrey and W. F. Harrington, *Biochemistry* 9, 894 (1970).
[51] T. Herbert and F. Carlson, *Biopolymers* 10, 2231 (1971).
[52] W. F. Harrington, M. Burke, and J. S. Barton, *Cold Spring Harbor Symp. Quant. Biol.* in press (1973).

$$\frac{\partial \ln K_2}{\partial P} = -\frac{2M_1\Delta\bar{V}_2}{RT} \tag{8}$$

and

$$\frac{\partial \ln K_p}{\partial P} = -\frac{2mM_1\Delta\bar{V}_p}{RT} \tag{9}$$

Thus

$$\frac{\partial \ln K_{app}}{\partial P} = -\frac{2mM_1}{RT}[\Delta\bar{V}_p + \Delta\bar{V}_2] \tag{10}$$

where $\Delta\bar{V}_2$ is the partial specific volume change accompanying dimerization and $\Delta\bar{V}_p$, the partial specific volume change which occurs when the dimer is polymerized. If the specific volume change on dimerization is zero, the concentration of monomer can be estimated, as before, by assuming that the gradient profile centripetal to the polymer boundary is composed solely of monomer and dimer species. Thus from a single velocity sedimentation run the concentration of monomer at each position of the polymer boundary is calculated from K_2 (under these ionic conditions) and K_{app} is estimated from Eq. (6). We expect $\Delta\bar{V}_2$ to be greater than zero, but significantly lower than $\Delta\bar{V}_p$ since insertion of dimer into the interior protein environment of polymer will likely release a larger amount of structured water than that desorbed on dimerization. Nevertheless, an accurate estimation of the molar volume change from sedimentation experiments will require direct measurement of $\Delta\bar{V}_2$. With this information, the gradient of the dimerization equilibrium constant through the liquid column can be calculated and the concentration of monomer estimated at each level of the cell. It should be emphasized that the analysis employed here is based on a large value of n, the number of monomeric units in the polymer, and cannot be expected to hold for low degrees of polymerization.

In associating systems where the equilibrium constant of polymerization is affected by salt gradients generated by the high centrifugal field, pressure dependence can also be estimated by layering varying amounts of mineral oil over the protein solution column. Since $\log K$, estimated at the oil–solution meniscus, is not affected by the salt gradient and is unlikely to be influenced by affects arising from transport during sedimentation, its measurement provides a valuable cross check on the procedures employed to evaluate $\Delta\bar{V}$.

In general the incremental volume change associated with polymerization is expected to be relatively insensitive to small changes in the ionic environment. Cassel and Christensen[49] reported $\Delta\bar{V}$ of aggregation of tropocollagen rods to be independent of ionic strength over the range

0.20–0.50 M. The molar volume change estimated for the reversible association of myosin is invariant with KCl concentration between 0.145 and 0.18 M.[33] Infante and Baierlein[24] found a single value of ΔV (500 ± 100 ml) to give a good fit between the simulated and experimental sedimentation patterns of ribosomes over the salt concentration range 0.1–0.5 M KCl. On the other hand, the equilibrium constant of polymerization is often markedly dependent on the ionic strength and pH of the surrounding medium. These features—a relatively constant molar volume term, and a rapidly varying equilibrium constant of polymerization with ionic conditions—lead to an important conclusion relevant to the effect of pressure on interacting systems. It will be seen from inspection of Eq. (2) that the detection of pressure dependence in any system will depend on the relative magnitude of the two terms on the right. Associating systems that show virtually no pressure dependence under one set of ionic conditions may exhibit dramatic effects when the ionic strength, pH, or temperature is altered to bring log $K_{(0)}$ into the range of the molar volume term.

Consider the association of 35 S and 56 S subunits to form ribosomes. Infante and Baierlein observed the equilibrium constant to vary from 10^{+17} at 0.1 M KCl to 10^{+7} at 0.5 M KCl. Since ΔV is virtually constant at 500 ml/mole, the second term of Eq. (2) will vary from 0 to ~10 at the bottom of the liquid column at 30,000 rpm. At low ionic strength (~0.1 M KCl) the pressure gradient will have essentially no detectable effect in dissociating the ribosomes since association of subunits is so strongly favored (log K_0 = 17 in concentration units of moles per liter). At higher salt concentrations, log K_0 decreases into the range of the molar volume term and large pressure effects are expected and observed. Similar results are predicted in the myosin–polymer equilibrium system. The value of log K_{app} increases from 25 (in concentration units of grams per deciliter) at 0.20 M KCl to ~500 at 0.11 M KCl in a pH 8.3 buffer system while the molar volume term varies from 0 to ~56 near the bottom of the ultracentrifuge cell at 40,000 rpm. We expect pressure dependence to be difficult if not impossible to detect in the ultracentrifuge in the low salt environment, whereas at higher ionic strengths the two terms of Eq. (2) are comparable in magnitude, leading to striking pressure-dependent sedimentation behavior. In view of these considerations it seems likely that other large macrostructures, e.g., animal and plant viruses and phage, which are composed of a large number of subunits, although appearing to be perfectly stable in ordinary velocity sedimentation studies, may well undergo dissociation under the pressure gradient developed in ultracentrifugation through appropriate alterations in ionic environment.

The Interpretation of Ultracentrifuge Experiments

Ideally, the ultracentrifuge should now take its place as an exquisitely sensitive tool, not only for estimating the effect of pressure on chemical equilibria between sedimenting species, but in addition for simultaneously defining the species which are involved. As already indicated in the discussion above, certain unexpected peculiarities can arise in sedimentation velocity experiments, as well as in zonal sedimentation velocity and equilibrium experiments and ordinary sedimentation equilibrium experiments. Since it has been shown above that any chemically reequilibrating system can, under the proper circumstances, be expected to show marked sensitivity to the hydrostatic pressures existing in many types of ultracentrifuge experiments, such sensitivity may, in fact, also provide pitfalls for the unsuspecting investigator. This section is intended to point to some pitfalls which have been recognized, and to suggest how they may be at least detected, in real experiments.

Band Sedimentation

It has been demonstrated, for example, that in zonal (band) sedimentation[36] each species travels at its own characteristic rate, once the species have been completely resolved. On the basis of model calculations for the countercurrent distribution of systems undergoing bimolecular complex formation,[53] however, it is also clear that in the case of complex moving faster than reactants, unless the formation constant of a bimolecular complex is very small, such a reacting system cannot be completely separated into the reacting species in a zonal experiment. When this system is separated into two, or possibly three, zones, one major zone must contain complex in equilibrium with its reaction partners and the other major zone must be a zone of that reactant which is present in molar excess in the mixture being resolved, provided that equilibrium is reached at all stages of the separation. A minor zone of the second reactant may possibly also occur. In the case of a polymerization reaction, separation into zones of the oligomers is not to be expected,[54] although the zones may become peculiarly skewed. Such are also the expectations for zonal (band) sedimentation experiments on rapidly reequilibrating reacting systems of these types, *in the absence of hydrostatic pressure effects*. The sensitivity of such reacting systems to hydrostatic pressure in the ultracentrifuge acts to complicate the process of zonal (band) sedimentation, in sometimes startlingly unexpected ways. Thus, it has

[53] J. L. Bethune and G. Kegeles, *J. Phys. Chem.* **65**, 1755 (1961).
[54] J. L. Bethune and G. Kegeles, *J. Phys. Chem.* **65**, 433 (1961).

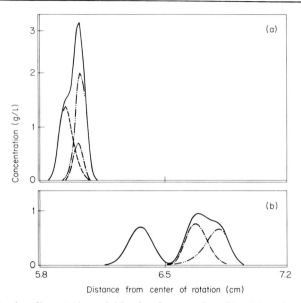

Fig. 8. Band sedimentation of bimolecular complex. Densities of reactants and complex are 1.31 (A), 1.31 (B), and 1.32 (C); molecular weights are 70,000, 140,000, and 210,000; ultracentrifuge frequency is 1000 rps; sedimentation coefficients are 4 S, 6 S, and 8 S; dissociation constant is 10; solution density is 1.0. Key to concentration profiles: total; - - -, (A); — · —, (B); — · · —, (C). Reproduced from G. Kegeles, L. Rhodes, and J. L. Bethune [*Proc. Nat. Acad. Sci. U.S.* **58**, 45 (1967)].

been predicted[55] that when a bimolecular complex is in equilibrium with its reaction partners in a band sedimentation experiment, and a large volume of reaction stabilizes the complex at higher pressures, only one zone moves at the expected (assigned) rate—that of the completely resolved reactant in molar excess. As shown[55] in Fig. 8, after appreciable sedimentation a partially resolved zone of complex now contains the bimolecular complex and *only one of the two reactants* in any appreciable amount. The formation constant of complex, which was only 10 at 1 atm, has become so huge at high pressure that virtually all of the other reactant has either been forced into complex, or squeezed out of the zone, forming a zone moving at its own characteristic sedimentation rate. Moreover, even if one had a specific analytical method to follow the complex and each reactant, one could not obtain the correct sedimentation coefficient of either the complex or the reactant moving with the complex, because the species in this zone *do not travel at their assigned*

[55] G. Kegeles, L. Rhodes, and J. L. Bethune, *Proc. Nat. Acad. Sci. U.S.* **58**, 45 (1967).

rates. Such a prediction may seem to stretch the imagination. However, an example of this general type, already referred to in the section on detection of volume changes, is the study of leucyl- and seryl-tRNA synthetase-cognate tRNA complexes.[37] Under conditions of gel permeation chromatography where resolution between enzyme and cognate tRNA is effected without detection of any complex (because of an extremely small formation constant)[37] zone sedimentation in a deuterium oxide density gradient gives rise to patterns containing two or three zones. When enzyme is present in excess, there is a large zone of free enzyme, an adjacent smaller zone of tRNA, and a large zone of complex containing both species. When tRNA is present in excess, there is a zone of free tRNA containing virtually no enzyme, and a zone of complex containing both species. In both cases, hydrostatic pressure in the ultracentrifuge has stabilized the complex and caused enzyme and tRNA to travel together under conditions where they do not bind together to any appreciable extent at 1 atm pressure. In particular, the enzyme in the presence of cognate tRNA travels much faster than does free enzyme alone in similar density gradient sedimentation experiments, demonstrating unequivocally the formation of complex.

A prediction[55] was made in the case of pressure-dependent dimerization that such a system should show both a time-dependent and a rotor speed-dependent sedimentation coefficient in band sedimentation experiments. This is intuitively acceptable, since the macromolecular system in the band, on entering a region of higher pressure, reacts chemically in the direction of decreasing volume, thereby altering its sedimentation rate to correspond more closely to that of the favored species. In ordinary sedimentation velocity experiments, such a system might possibly exhibit a rotor speed-dependent sedimentation coefficient, but without showing time dependence within an individual experiment, provided that there is a nearly time-invariant region (plateau region) below the moving boundaries.

Moving Boundary Sedimentation (Sedimentation Velocity)

In this standard procedure, one might expect fewer complications due to hydrostatic pressure effects, inasmuch as it could be hoped that a region of constant composition would exist below all the moving boundaries. That this is not necessarily so is clearly demonstrated[22] in Fig. 3, where the nonzero slope of the scattering response curve for myosin at higher rotor velocities, even before any appreciable transport has taken place, makes it clear that one cannot expect any plateau region mixture of constant composition at any later time. At best, the region at the bottom of the cell in high speed experiments might be a plateau region containing only monomer. Since the sedimentation process must transfer mass in such experiments at a rate characteristic of the composition of the plateau region (i.e., monomer), one is immediately confronted with a quandary: How is this possible, when there is a clearly visible boundary moving with a rate characteristic of polymer, and another boundary moving with a rate characteristic of monomer? The answer is that there must be regions of the cell in which negative concentration gradients are present,[56] in order to compensate for an otherwise-too-large rate of transfer of mass to represent the monomer present at the bottom of the cell. This qualitative argument found support in earlier more quantitative predictions of negative concentration gradients[55] for various hypothetical examples.

More recently, it has been possible to simulate fairly precisely[35] the sedimentation behavior of myosin itself under the original assumptions[22] of a monomer–single polymer equilibrium. These calculations are based on the countercurrent distribution analog[34] of the moving boundary sedimentation experiment, but include in addition estimates[22] of the equilibrium constant for polymerization and the effect of pressure on the equilibrium constant. Whereas the original Gilbert theory[31,32] assumes a plateau region below all moving boundaries and is, as a result, fundamentally not equipped to investigate such alterations in composition below the moving boundaries as may be generated by a pressure gradient, the countercurrent analog calculations have no such restriction. The first such set of calculations[35] for myosin sedimentation velocity experiments did indeed reveal a hypersharp inverted concentration gradient generated below the polymer peak. These calculations suffered from the omission of the effect of concentration on the sedimentation of myosin, a very pronounced effect causing hypersharpening of the polymer boundary. When this effect was included in a recalculation,[57] the predicted monomer and polymer concentration gradient peaks appeared very nearly like those of the experiments of Josephs and Harrington,[22,23] except for the prediction of a now grossly emphasized inverted concentration gradient below the polymer peak. These predictions are shown in Fig. 9, the upper pattern indicating total concentration gradients, and the lower pattern the concentration profiles of monomer and polymer, as well as total mass. It is noted that the incidence of the large negative gradient of total concentration also marks the position below which no polymer exists. This is then, in effect, a marker for the critical micelle pressure of the polymer at a specified myosin concentration: when the pressure is raised

[56] G. Kegeles, *Biopolymers* **7**, 83 (1969).
[57] G. Kegeles, *Arch. Biochem. Biophys.* **141**, 72 (1970).

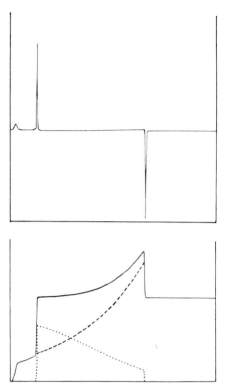

Fig. 9. Computed total concentration gradient pattern (above) and concentration profiles (below) for 0.6% myosin in 0.18 M KCl, 0.002 M Veronal buffer at pH 8.3. Hydrodynamic concentration dependence of monomer and polymer are included. Patterns correspond to 857 seconds of sedimentation at 40,000 rpm. Key to concentration profiles: ——, total; - - -, monomer; · · ·, polymer. Reproduced from G. Kegeles [*Arch. Biochem. Biophys.* **141**, 72 (1970)].

infinitesimally above this value, the remaining polymer melts completely into monomer.

It is obvious that such large negative concentration gradients, and correspondingly large negative density gradients, cannot and do not persist in an ultracentrifuge, unless compensated by superposed extremely steep artificial density gradients. In the present case, there would be no practical way to develop or maintain such steep positive density gradients. Since the effect is constantly self-generating from the instant the sample reaches quite appreciable rotor speeds, the only possibility is that there must be continual convective mixing in the region below and up to the hypersharp polymer boundary. Fortunately, this boundary (Fig. 9) contains a sufficient positive gradient to stabilize the centripetal region of the cell contents, and the only disturbing effect is to mix out the leading edge of the polymer boundary, and thereby somewhat decrease its area, in the experimental schlieren pattern.

Convective mixing effects may be expected in reversibly self-aggregating systems when pressure favors the dissociation process, on the basis of the arguments above.[56] Yphantis[58] has further investigated the conditions for potential convection prior to any transfer of mass, corresponding to the experiments shown in Fig. 3. He has found that, in fact, *initial* instability is independent of the sign of ΔV (whether association or dissociation is favored by increasing pressure). He has also developed a criterion for *initial* instability which indicates, for self-aggregating systems, that the degree of polymerization is a critical factor.

It should be emphasized again that, when convective instability is present, mixing must take place continually to produce a stable column in which the density constantly increases with radius of rotation. This can happen only if some of the schlieren pattern is mixed out. The final result, however, except for small characteristic spikes near the base line, may look like a perfectly acceptable pattern, even though much of the potential resolution and area have been mixed out. One very effective test is to see whether the area under the peaks of the pattern, corrected as need be for radial dilution, is approximately equal to the area under a synthetic boundary peak formed by layering solvent over solution at low speed.[58]

Density Gradient Sedimentation Equilibrium

Equilibrium banding of macromolecules in a self-generating gradient of low molecular weight material has been developed into a very powerful tool,[59] with especially wide application in the field of nucleic acid research.[60] When a chemically reacting system is potentially resolvable in such a density gradient, because of differences in buoyant density between reactants and their products, there is generally an appreciable volume change on reaction. Since, again, it is the *molar* volume change that is effective in changing the equilibrium constant with pressure, any density gradient sedimentation equilibrium experiments which do, in fact, produce resolution between reacting species within the density range

[58] D. A. Yphantis, M. Dishon, G. H. Weiss, and M. Johnson, *Biophys. Soc. Abstr.* **12**, 97a (1972).

[59] M. Meselson, F. W. Stahl, and J. Vinograd, *Proc. Nat. Acad. Sci. U.S.* **43**, 581 (1957).

[60] J. Vinograd and J. E. Hearst, "Equilibrium Sedimentation of Macromolecules and Viruses in a Density Gradient," *Fortschr. Chem. Org. Naturst.* **25**, 372 (1962).

of available gradients[59,60] are likely to lead to extreme sensitivity to hydrostatic pressure.[61]

A previously only very sketchily described modification was made[55] in the countercurrent distribution analogue calculation,[34] in order to permit the prediction of transient states and final states of equilibrium banding of hypothetical chemically reacting macromolecular systems in imposed stabilizing density gradients. The computation consisted basically of equilibrating a train of tubes in which one or more adjacent *pairs* were originally filled with an equilibrium mixture, and then shifting upper phases one tube to the right, and lower phases one tube to the left, at each transfer. To eliminate the "hole-in-the-middle" effect, a pair of tubes was assigned to each "station," and the contents of a whole "station" were equilibrated before transfer. Use was made of the mass-conservation equations for countercurrent distribution[34] to require that at equilibrium, a single pure substance would band as a Gaussian error curve.[59] This stipulation automatically served to define the required shape of the partition coefficient gradient. Then, arbitrarily, for a mixture which was potentially resolvable, the centers of the bands for different species were assigned different locations in the density gradient column by simple shearing of the controlling gradient of partition coefficient. In effect, different buoyant densities were thereby assigned to the species, although the gradient of partition coefficient was for simplicity kept identical for all species. In the simplified calculations, the width of the zone is not correlated with molecular weight. For reacting species, it was then to be required that the concentrations at each stage of equilibration satisfy the specified statement of the law of mass action for the chemical reaction. Finally, the volume of reaction was computed from the assigned densities and molecular weights, and the *local* value of the equilibrium constant was computed (Eq. 2) at each position in the cell [at each value of hydrostatic pressure (Eq. 3)], and this local value was satisfied at each equilibration. The computation was iterated, from several cases of different initial positions and distributions of macromolecular material, until no further shifts in equilibrium distribution could be detected (requiring as many as 20,000 iterations in some cases). As a check, it was required that total mass of macromolecular material be precisely conserved throughout the computation.

Of some practical interest might be the predictions[55] for reactions forming a bimolecular complex. Figure 10 shows an early stage (a) and the final stage (b) in the density gradient banding of a system in which

---

[61] The importance of including the effects of hydrostatic pressure in these calculations was kindly pointed out to one of us (G. K.) by Dr. Lynn Ten Eyck.

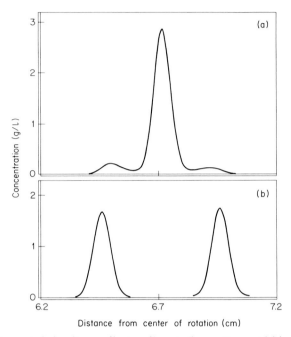

FIG. 10. Computed density gradient sedimentation patterns of bimolecular complex of intermediate density. Volume of reaction is zero. Dissociation constant of complex is 10^{-4}. Upper pattern (a): transient state during approach to equilibrium. Lower pattern (b): final equilibrium state. For details of computation see text. Reproduced from G. Kegeles, L. Rhodes, and J. L. Bethune [*Proc. Nat. Acad. Sci. U.S.* **58**, 45 (1967)].

the complex has a density intermediate between those of the reactants, having a value such as to lead to precisely zero volume of reaction. This would be an unusual condition, but it serves as a reference with which to compare the more likely cases of appreciable volume of reaction. As seen in Fig. 10, the large amount of intermediate density complex present in early stages (a) eventually dissociates completely into its reaction partners (b) at equilibrium. Contrasted with this is the unsymmetrical case where a complex, of buoyant density intermediate between those of its reaction partners, is nevertheless favored strongly by increasing pressure (Fig. 11). In Fig. 11a, an excess of the less dense reactant is present in the initial mixture, and in Fig. 11b, an excess of the denser reactant is present. Both patterns represent final equilibrium conditions, and in both patterns the large zone is for all practical purposes pure complex, now stabilized by a very large pressure effect, while the small zone in either case represents the reactant present in excess. It is noted that

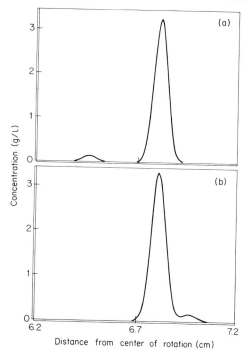

FIG. 11. Computed density gradient sedimentation equilibrium patterns of bimolecular complex of intermediate density, with pressure favoring complex formation. (a) Excess of less dense reactant, (b) Excess of denser reactant. Dissociation constant of complex is 10^{-4} at 1 atm. Volume of formation is -2446 cm^3 per mole of complex. Reproduced from G. Kegeles, L. Rhodes, and J. L. Bethune [*Proc. Nat. Acad. Sci. U.S.* **58**, 45 (1967)].

the dissociation constant for complex is 10^{-4} at 1 atm in both Figs. 10 and 11. Whenever the complex is denser than both reactants, a stabilized zone of complex and a zone of reactant are to be expected at equilibrium.[53] While density gradients representing very high concentrations of strong electrolyte[59,60] would probably strongly dissociate many bimolecular complexes, and perhaps preclude the experimental detection of interactions involving nucleic acids, the use of sucrose gradients for the banding of proteins[62] might not have such drastic effects, and, conceivably, protein–protein interactions might be subject to study in such experiments. Predictions were also made[55] for the equilibrium banding of dimerizing systems, indicating the possibility of an equilibrium distribution containing only a monomer band, only a dimer band, or two bands—one mostly monomer and one mostly dimer, depending largely on

[62] J. B. Ifft and J. Vinograd, *J. Phys. Chem.* **70**, 2814 (1966).

the buoyant densities and the dimerization constant. These calculations assumed very large molar volume changes on dimerization, -3628 cm^3 per mole of dimer.

Probably much more realistic calculations for the distribution *in an analytical cell* of proteins of low molecular weight were made by Ten Eyck and Kauzmann,[27] assuming a molecular weight of 30,000 for monomer, and $\Delta V = -250$ cm^3/mole or $\Delta V = -750$ cm^3/mole for the formation of 1 mole of dimer. Their calculations assumed a linear increase of density with radius of rotation from 1.237 g/cm^3 at the top of the column, 6.3 cm from the axis of rotation, to 1.437 g/cm^3 at the bottom of the cell, 7.2 cm from the axis of rotation. Their final equilibrium calculations were then made directly from the thermodynamic equations for sedimentation equilibrium. Their conclusions differed from those described above for the monomer–dimer system in that the bands filled the major portion of the analytical cell, as would be expected for macromolecules of this molecular weight range,[62] and the bands of monomer and dimer were separated so little, in consequence of the much smaller ΔV values assumed, that schlieren optical patterns for total protein would be expected to reveal only small pressure effects, even on comparison of such experiments with similar ones having an overlay of 3 mm of immiscible liquid of density 1.2. Nevertheless, they concluded that there would indeed be enormous shifts in the dimerization constant with pressure, even in these experiments.

Conventional Sedimentation Equilibrium

This procedure for the direct measurement of molecular weights was one of the first devised by Svedberg and his colleagues.[63] It has the attractive advantage of being based on solid thermodynamic principles.[63,64] For a considerable period, the method was avoided, however, because of the long time (days to weeks) required to establish macromolecular equilibrium in a column of the length of an ordinary cell column (some 10–15 mm). When the advantages were developed of performing such experiments in short columns,[65,66] this again became a method of choice among many investigators of protein systems. The speed of attainment of equilibrium and the possibility of increased precision in short columns has been enhanced by the "meniscus-depletion" procedure of Yphantis,[67] in which samples are centrifuged at much higher speeds than those con-

[63] T. Svedberg and K. O. Pedersen, "The Ultracentrifuge." Oxford Univ. Press, London and New York, 1940.

[64] R. J. Goldberg, *J. Phys. Chem.* **57**, 194 (1953).

[65] K. E. van Holde and R. L. Baldwin, *J. Phys. Chem.* **62**, 734 (1958).

[66] D. A. Yphantis, *Ann. N.Y. Acad. Sci.* **88**, 586 (1960).

[67] D. A. Yphantis, *Biochemistry* **3**, 297 (1964).

ventionally used.[63] Since this high speed technique is often preferred, it seemed important to investigate the possibility of the occurrence of appreciable pressure effects when chemically reacting systems are studied in this way. It has been repeatedly emphasized that the thermodynamic equations for sedimentation equilibrium in differential form must hold for the various species present, under all conditions, including the case of chemical reactions between species,[68,69] and even including the case of a dependence of the equilibrium constant on hydrostatic pressure.[27,70] This is not to say that the actual observable distribution of concentration or concentration gradient of any given species in a reacting system having an appreciable volume of reaction is independent of pressure, however, since the constants of integration which determine the levels of concentration throughout the cell are directly affected by the pressure gradient. The consideration of an extreme case in which a complex AB is formed from reactants A and B with a very large decrease in volume suffices to illustrate this point. Because of the large hydrostatic pressure at the bottom of the cell, A and B are postulated to react there completely to form AB, which cannot redissociate. Suppose that in the original mixture, there was, in all forms, a 2-fold molar ratio of A to B. Then at the cell bottom there will coexist A and AB. Since there is no B present, because the formation constant of AB is effectively infinite, at the cell bottom, this region of the cell acts as a sink for B, continually removing it from the rest of the cell. At equilibrium, there is no B at the bottom of the cell; moreover, there is also no B present anywhere else in the cell. This is a very different situation from that produced if B did not react with A. While it is true that the same differential equation for the distribution of the logarithm of concentration of B is still satisfied, this is now a trivial point in practice; numerically, the equation for B in our extreme case simply states that a concentration gradient of zero *equals* zero. Thus, in practice, one needs to work out the computation of all integration constants for any practical set of parameters (solution densities, centrifuge angular velocity, molecular weights, partial specific volumes, temperature, distances from center of rotation of meniscus and cell bottom, etc.). This was completely developed for the case of a bimolecular reaction[71] and for the case of self-aggregation.[72,73] Although

[68] E. T. Adams, Jr. and H. Fujita, *in* "Ultracentrifugal Analysis in Theory and Experiment" (J. W. Williams, ed.), p. 119. Academic Press, New York, 1963.

[69] L. W. Nichol and A. G. Ogston, *J. Phys. Chem.* **69**, 4365 (1965).

[70] J. R. Cann, "Interacting Macromolecules," pp. 34-40. Academic Press, New York, 1970.

[71] G. Kegeles, S. Kaplan, and L. Rhodes, *Ann. N.Y. Acad. Sci.* **164**, 183 (1969).

[72] G. J. Howlett, P. D. Jeffrey, and L. W. Nichol, *J. Phys. Chem.* **74**, 3607 (1970).

[73] G. J. Howlett, P. D. Jeffrey, and L. W. Nichol, *J. Phys. Chem.* **76**, 777 (1972).

the complete algebraic equations are available, they are, in the case of the bimolecular reaction,[71] a quadratic equation for the concentration of one species at the meniscus, at equilibrium, containing a large number of transcendental factors. In practice, the solution is adequately visualized only by inserting numerical values into this expression and solving for all unknowns with a digital computer. When this was done, the complete solution for the experimentally observable total concentration, or its gradient, could be compared with that expected in a hypothetical similar reacting system having zero volume of reaction.[71] The results are shown for one such calculation, in Fig. 12, for concentration; and Fig. 13, for concentration gradient, the solid curves representing the case of $\Delta V = 0$, and the circles representing the corresponding computations for $\Delta V = -1000$ cm^3 per mole of complex. Table III shows the numerical values computed for the local value of the weight-average molecular weight for these two cases, as well as the ratio of the equilibrium constant at various levels to that at 1 atm. From Figs. 12 and 13 it is also clear, however, that no data are really attainable experimentally in this case, except in the small region between 7.08 cm and somewhat less than 7.20

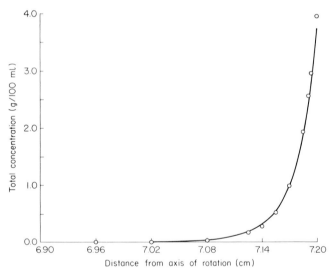

Fig. 12. Computed values of concentration versus position in cell for two reacting systems (A and B) forming a bimolecular complex, at sedimentation equilibrium. Solid curve, $\Delta V = 0$. Circles, $\Delta V = -1000$ cm^3 per mole of complex. Ultracentrifuge frequency, 30,000 rpm. For solid curve, $V_A = V_B = V_{AB} = 0.76$ cm^3/g. For circles, $V_A = 0.75$ cm^3/g, $V_B = 0.786$ cm^3/g, and $V_{AB} = 0.76$ cm^3/g. The initial concentration of each species was taken as 0.1 g/100 ml. The molecular weight of A was 30,000, that of B was 50,000. Reproduced from G. Kegeles, S. Kaplan, and L. Rhodes [*Ann. N.Y. Acad. Sci.* **164**, 183 (1969)].

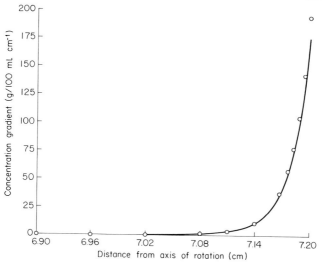

FIG. 13. Computed values of concentration gradient versus position in cell at sedimentation equilibrium for the same two reacting systems as in Fig. 12. Reproduced from G. Kegeles, S. Kaplan, and L. Rhodes [*Ann. N.Y. Acad. Sci.* **164**, 183 (1969)].

cm from the center of rotation. Figure 14 shows the difference, in fringes, which would be expected between Rayleigh interference patterns for the two systems. Where this difference becomes appreciable, above one fringe, the separation on the photograph between fringes in the direction of the

TABLE III

COMPUTED VALUES OF $M_{w,x}$ AND $K(x)/K(x_0)$ AT A ROTOR SPEED OF 30,000 RPM[a]

Distance from axis of rotation (cm)	Weight-average molecular weight $\Delta V = 0$	Weight-average molecular weight $\Delta V = -1000$	$K(x)/K(x_0)$ $\Delta V = -1000$
6.90	30,469	31,151	1.00
6.93	30,702	31,498	1.09
6.96	31,064	31,975	1.18
6.99	31,645	32,673	1.28
7.02	32,622	33,785	1.39
7.05	34,345	35,731	1.51
7.08	37,472	39,316	1.65
7.11	42,942	45,638	1.79
7.14	51,239	54,916	1.95
7.17	60,926	64,851	2.12
7.20	69,135	72,289	2.31

[a] Reproduced from G. Kegeles, S. Kaplan, and L. Rhodes, *Ann. N.Y. Acad. Sci.* **164**, 183 (1969).

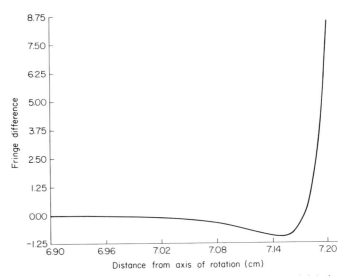

FIG. 14. Number of fringes that would be detected, in a Rayleigh interference optical system, with $\Delta V = -1000$ cm³/mole minus the number of fringes that would be detected with $\Delta V = 0$, for the reacting systems in Fig. 12. Fringe difference = $(1.2)(0.0018)$ $(\Delta C)/5.46(10)^{-5}$, where ΔC = concentration difference in grams per 100 ml. Reproduced from G. Kegeles, S. Kaplan, and L. Rhodes [*Ann. N.Y. Acad. Sci.* **164**, 183 (1969)].

centrifugal field would be only about 5 μm in the usual ultracentrifuge optical system. Thus, although Table III shows a systematic drift between the weight average molecular weights in these two cases, as the bottom of the column is approached, this is reflected in only very small differences in primary experimental data, in the bottom 1 mm of the 3-mm column. Moreover, the experimentalist would actually have access only to one of the two systems; only the one having $\Delta V = -1000$ cm³/ mole would be real. The other system—comparison or reference reacting system having $\Delta V = 0$, but all other parameters identical—is purely hypothetical. Before the experimentalist could construct the reference sedimentation equilibrium curves, he would have to have precise data for the molecular weights of the species, nonideality coefficients, and the equilibrium constant for the reaction at 1 atm pressure, a very considerable demand. The ratio of equilibrium constants does, nevertheless, drift away from unity very markedly at higher pressures in this case, as is shown in Table III. This indicates that an extremely small error in primary data for the molecular weight measurement can produce a very significant effect on the computed equilibrium constant. These calculations have led to several additional conclusions. Since the value of

−1000 cm³/mole of complex is rather extreme for molecular weights in the range of 30,000 to 80,000 assumed here, it is not likely that, *for such a system,* very appreciable errors in molecular weight would derive from pressure effects at rotor speeds up to 30,000 rpm in the "meniscus-depletion" experiments. However, if one wished to estimate equilibrium constants, or nonideality coefficients (assumed to be unity in the calculations which have been discussed), careful tests would have to be made to assure independence of hydrostatic pressure. The calculations discussed[71] lend very little confidence to the hope of achieving accurate measurements (or *any* measurements) of volume of reaction from such experiments on a system of the type postulated. Howlett *et al.*[72] have been more optimistic, and have outlined a scheme for obtaining both the equilibrium constant at one atmosphere and the volume of reaction from such data, for a dimerizing system. They have applied this to the study of the dimerization of lysozyme.[73]

One additional point may be added. A possible method of testing for sensitivity to pressure is to examine the same solution at a series of different rotor speeds. If the equilibrium constant is independent of pressure, and no other disturbing problems exist, a plot of the weight-average molecular weight against total concentration should be a single curve, containing points from all the different experiments.[72,74] If this is not the case, heterogeneity and experimental error may also contribute to the failure to obtain a complete overlap,[75-78] but the problem may be at least partially due to a pressure effect.[72]

[74] D. A. Yphantis, private communication, 1968.
[75] P. G. Squire and C. H. Li, *J. Amer. Chem. Soc.* **83**, 3521 (1961).
[76] P. D. Jeffrey and J. H. Coates, *Biochemistry* **5**, 489 (1966).
[77] J. C. Nichol, *J. Biol. Chem.* **243**, 4065 (1968).
[78] D. A. Albright and J. W. Williams, *Biochemistry* **7**, 67 (1968).

[14] Characterization of Proteins by Sedimentation Equilibrium in the Analytical Ultracentrifuge[1]

By DAVID C. TELLER

I. Introduction

The technique of sedimentation equilibrium is perhaps at present the most common method of characterization of proteins and enzymes. In this article I present some useful methods for the characterization of proteins by this technique. Most of the enzymes that we have studied in this laboratory have been quite pure from a chemical standpoint, but physically paucidisperse owing either to reversible chemical association (or dissociation) or to being irreversibly dissociated. Included are the methods we use in the determination of molecular weights and equilibrium constants for chemically reacting proteins as well as for nonreacting enzymes. A few methods used in other laboratories are also presented.

The ability to characterize a particular system depends to a large degree on the quality of the primary data. Consequently, considerable discussion will be related to such methods as experimental design, optical systems, methodology for baseline determinations, data acquisition.

For the study of chemically interacting systems or otherwise heterogeneous preparations, there are two approaches that can be taken to analyze experimental data. One approach is to take all the observations of a concentration distribution and express these as a sum of exponentials in the constituents. All the data are treated at one time to determine equilibrium constants and stoichiometries. The second approach is to use the molecular weight averages calculated at single points in the concentration distribution. These averages are then used to determine the parameters of the system under investigation at each point. There are advantages to each approach that are not provided by the other. For the whole system approach, it is generally required that the molecular weights of the species be accurately known, presumably from other experiments. Stoichiometry and equilibrium constants are the unknowns. In this approach, the presence of competing equilibria or species not participating in the equilibria are difficult to detect but can lead to significant errors. However, quite complex reaction stoichiometries can

[1] This work was supported by a grant from the National Institute of General Medical Sciences (GM 13401), National Institutes of Health, U.S. Public Health Service.

be solved by this method. In the other method, using molecular weight averages at each point, it is possible to determine both molecular weight and stoichiometry of aggregation, but only for simple systems such as monomer–dimer, monomer–dimer–trimer, and indefinite association. Significant nonideality seriously impairs the accuracy of the calculations.

In this article I shall first approach the problem as one of characterizing an unknown protein in terms of molecular weight and other hydrodynamic properties. In the final sections, associating proteins of known molecular weight are considered in order to determine thermodynamic parameters of protein structure and function.

II. Preliminary Experiments

A. Sedimentation Velocity

In the characterization of a protein we usually begin by performing sedimentation velocity experiments using the schlieren and absorption optical systems in order to construct a graph of the dependence of the sedimentation coefficient as a function of concentration. The buffer system which we normally choose is one in which the enzyme is fully active. Of course, it is necessary to have sufficient electrolyte present to suppress the primary charge effect.[2] For most enzymes this s vs. C graph will be linear and can be characterized by the equation,

$$s = s°(1 - \kappa C) \tag{1}$$

For nonreacting, globular protein systems,[3] κ will be about 9×10^{-3} l/g. An upward trend of sedimentation coefficients with increasing concentration indicate aggregation of some type. Also, values of κ in the range of 0 to 5×10^{-3} l/g may indicate weak aggregation or slight dissociation of the enzyme. In the construction of this graph, it is proper to choose the concentration as that in the plateau region,[4] but the initial concentration loaded into the cell is sufficiently close to the plateau concentration so as to not introduce serious error.

For the measurement of sedimentation coefficients from absorption optics (scanner), we have found it useful to perform two calculations.

[2] The primary charge effect is the suppression of sedimentation coefficients due to the repulsion of similarly charged particles. The charge effect can be decreased by utilizing ionic strengths of 0.1 to 0.5 M electrolyte. It can be eliminated by experiments at the isoelectric point of the protein, but this is not always possible to attain for enzymes due to loss of enzymatic activity, aggregation, etc.

[3] J. M. Creeth and C. G. Knight, *Biochim. Biophys. Acta* **102**, 549 (1965).

[4] R. J. Goldberg, *J. Phys. Chem.* **57**, 194 (1953).

First, we calculate the sedimentation coefficient from the inflection point of the boundary, plotting log r vs. time in the usual way. In this method, the experimenter psychologically pays attention to the sharpest boundary present, essentially ignoring slower or faster moving material.

The second useful calculation which we make on the scanner data is that of the equivalent boundary method. This method[4-6] takes all the material present between the meniscus and an arbitrary position in the plateau and redistributes the material as an infinitely sharp but equivalent boundary. For this method it is necessary to have a baseline from which concentrations can be determined. The best baseline technique which we have found is to continue the sedimentation velocity run until all the material has sedimented (generally, this takes 5 or 6 hours). Equation (2) may be used to calculate the equivalent boundary positions.[5]

$$\bar{r}^2 = r_p{}^2 - \frac{2}{C_p} \int_{r_m}^{r_p} C r \, dr \tag{2}$$

Where $\bar{r}$ is the equivalent boundary position, r_p is the plateau position with concentration C_p. (Since the units of C cancel in this equation, graph inches or centimeters suffice for C.) To obtain an accurate value for the integral, surprisingly few points (approximately 15) need be taken and a trapezoidal integration suffices for the calculation. When the scanner is connected to an x,y plotter as described below, it is quite simple to use a pair of dividers to determine the values of C. By choosing equal increments in r on the scan, the equation can be programmed for a desktop computer quite readily. The advantage of this calculation is that all the sedimenting material is "seen" in the measurement of the sedimentation coefficient. If the sedimentation coefficient calculated by plotting $\ln \bar{r}$ vs. t is the same as that using the half-height method, the protein will probably be homogeneous in sedimentation equilibrium experiments. A difference between the two numbers is indicative of dispersity of some type. In order to determine the error associated with the calculation of the sedimentation coefficients by this method, we back-calculate C_0 from the radial dilution Eq. (3)

$$C_0 = C_p \frac{\bar{r}^2}{r_m{}^2} \tag{3}$$

and it can be shown that

$$(\tilde{s} - s)/s \approx -\epsilon/C_0 \tag{4}$$

[5] H. K. Schachman, "Ultracentrifugation in Biochemistry." Academic Press, New York, 1959.
[6] M. M. Rubin, Ph.D. Dissertation, University of California, Berkeley, 1966.

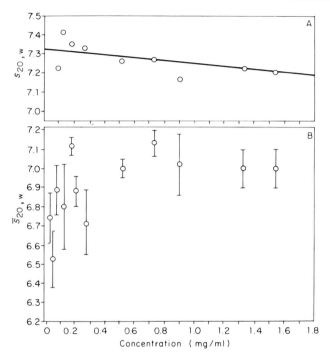

FIG. 1. Determination of the s vs. C dependence of *Acinetobacter* glutaminase by absorption optics. (A) Sedimentation coefficients calculated from the inflection point of the boundary. (B) Equivalent boundary calculations on the same data as (A).

Where $\tilde{s}$ is the true value of the sedimentation coefficient, s is the observed value, ϵ is the error is concentration, and C_0 is the observed initial concentration calculated from the radial dilution. For ϵ we use the standard deviation of the C_0's calculated at each time. If there is a systematic trend in C_0 (as there sometimes is), it means that the physical laws of sedimentation are not being followed: owing to convection, chemical equilibrium, denaturation of the protein, pressure effect, or other causes. In any case, these calculations give an indication of the reliability of the experiment.

To illustrate the utility of these two procedures for the determination of s vs. C, Fig. 1 shows some recent results from *Acinetobacter* glutaminase.[7] In Fig. 1A are shown the results from the calculation by inflection point of the boundary of the scan. A linear extrapolation of

[7] J. S. Holcenberg, D. C. Teller, J. Roberts, and W. C. Dolowy, *Fed. Abstr.*, 1972; *J. Biol. Chem.* **247**, 7750 (1972).

the points given by the straight line in Fig. 1A gives $s = 7.32 \pm 0.04$ S $(1 - 0.013 \pm 0.005$ $C)$ for values of Eq. (1).[8] Figure 1B shows the equivalent boundary calculation of same data together with three more determinations at low concentration. The error bars are calculated from Eq. (4). Note that the sedimentation coefficients are uniformly less those of Fig. 1A, and the shape of a curve drawn through the points would indicate dissociation of the protein.

B. Diffusion Experiments

Free diffusion experiments in a Tiselius cell as described by Schachman[9] are rarely performed at the present time. The method for measurement of the diffusion coefficient which we describe here is theoretically sloppy but suffices as a first approximation to the diffusion coefficient to allow a preliminary determination of molecular weight. If insufficient material is available, a preliminary value of the molecular weight can be calculated from the sedimentation coefficient and the assumption of a spherical molecule as described by Schachman.[9]

For this experiment we chose a relatively high protein concentration (2–10 mg/ml), twice the concentration of a reliable sedimentation velocity experiment. A double-sector synthetic boundary cell of the capillary type is used together with the schlieren optical system. Solution (0.15 ml) is placed in the right sector of the cell (facing B ring, filling holes up), and 0.40 ml of solvent is placed in the left sector. The centrifuge is accelerated to approximately 6000 rpm, and then the voltage is turned back to a small value so that the rotor decelerates slowly but without braking or speed control. The temperature is controlled at the same value as the sedimentation velocity experiment. Photographs are taken at an appropriate schlieren bar angle as a function of time. Initially, photographs are taken at 2-minute intervals, then 4 minutes, 8 minutes, 16 minutes, etc. for 1–3 hours. The photographs are placed on a microcomparator with 10- to 20-fold magnification and the boundary traced on graph paper. The magnification of the microcomparator must be known. The inflection points on the curve are assumed to occur at

$$y_i = \frac{1}{\sqrt{e}}\, y_{\max} \tag{5}$$

where y_i is the height of the inflection points and $y_{\max}$ is the maximum of the schlieren peak (Table I, column 3[10]). Asymmetry is ignored. The

---

[8] Numerous other determinations by this method and at higher protein concentrations gave $s^0_{2,w} = 7.46 \pm 0.08$ S.[7]

[9] H. K. Schachman, see Vol. 4, p. 32 (1957).

[10] To keep the table short, only every second measurement is recorded.

TABLE I
DIFFUSION COEFFICIENT OF YEAST ALDOLASE[a]

Picture No.	Time (min)	y_{max}	$y_{max}/\sqrt{e}$	x_i measured	Correction factor[b]	x_i corrected ($\times 10^2$)	$x_i^2 (\times 10^3)$ (cm²)
1	4	16.95	10.28	1.60	21.733	3.68	1.35
3	8	17.25	10.46	1.70	21.733	3.91	1.53
5	12	16.35	9.92	1.80	21.733	4.14	1.62
7	16	15.10	9.16	2.03	21.733	4.67	2.18
9	20	13.85	8.40	2.05	21.733	4.72	2.23
11	24	13.00	7.89	2.13	21.733	4.90	2.40
13	28	12.35	7.49	2.40	21.733	5.52	3.05
15	32	12.10	7.34	2.45	21.733	5.64	3.18
17	36	11.50	6.98	2.55	21.733	5.87	3.45
19	40	22.25	13.50	5.36	43.466	6.18	3.82
21	44	21.35	12.95	5.50	43.466	6.33	4.00
23	48	20.70	12.56	5.73	43.466	6.59	4.34

[a] Least squares line: $x_i^2 = 11.60 \pm 0.27 \times 10^{-7}t + 0.94 \pm 0.05 \times 10^{-3}$; $D_{obs} = 5.80 \pm 0.13 \times 10^{-7}$ cm²/sec at $C_0 = 12.84$ mg/ml. At 6.42 mg/ml, $s_{obs} = 4.72$ S; $(1 - \bar{v}\rho)_{obs} = 0.2616$; since $T = 293°$, $M_{s/D} = 76 \times 10^3$ g/mole from Eq. (6).
[b] Magnification of centrifuge is 2.1733.

distance, $2x_i$, is measured across the peak at the height y_i (Table I, column 5). The values of x_i are then corrected to real distances in the cell by division by the magnification of the ultracentrifuge optics and the microcomparator optics. The values of x_i^2 are then plotted on graph paper vs. time. The slope of the straight line is approximately $2D$ where D is the apparent diffusion coefficient in square centimeters per time unit. Correction for the time unit to seconds yields an approximate diffusion coefficient.

Thus a preliminary determination of the molecular weight can be calculated from the Svedberg equation,

$$M = \frac{RTs_{obs}}{D_{obs}(1 - \bar{v}\rho)_{obs}} \tag{6}$$

In footnote a of Table I are shown the results of an experiment of this type on yeast aldolase. The height-area method of measurement of diffusion coefficient (see Schachman[9]) may also be used for these calculations, but we have found that this inflection-point method gives more precise results. If the magnification of the microcomparator is unknown, it is necessary to use the height-area method.

The experiments in this section have been designed with two purposes. First, a careful determination of the dependence of the sedimentation coefficient upon concentration can reveal a great deal of information concerning the protein under study. Strong, rapid interactions can be diagnosed using appropriate theories described in this volume (see [11]–[13]). Weak interactions can be detected by anomalies in the dependence of s on concentration and the weight average sedimentation coefficients from the equivalent boundary method may be used for calculation of equilibrium constants, if desired. The diffusion experiment described is "quick and dirty" but is sufficiently accurate (5–15%) to determine experimental parameters for future investigations such as the optimum rotor speed, initial concentrations, and time to equilibrium for the sedimentation equilibrium experiments.

III. Modifications of the Ultracentrifuge Which Facilitate Sedimentation Equilibrium Experiments

Prior to discussion of experimental design and more sophisticated topics in the area of sedimentation equilibrium experiments for the characterization of proteins, I would like to discuss some modifications that can be made to an ultracentrifuge in order to facilitate the determination of molecular parameters using this tool. Also, in this section, some problems with ultracentrifuge drive units and other pragmatic topics will be discussed. The rationale for this discussion is that the experimenter is limited in accuracy of molecular weight determinations by the precision of the tools he uses to make the determinations. Throughout this article it is assumed that the ultracentrifuge facility available is equipped with a Beckman Model E analytical ultracentrifuge (with scanner), a microcomparator (Nikon 6C or equivalent), and programmable desk-top computers.

A. Scanner

Of the three optical methods presently available in the analytical ultracentrifuge, the scanner is the most recently developed.[11-13] The commercially available unit from Beckman can be modified quite simply in a number of ways which are useful to the experimenter. Here we will discuss some modifications performed in this laboratory which have proved to be convenient.

The R-S recorder of the scanner is a y,t recorder, and the ±15 V

[11] S. Hanlon, K. Lamers, G. Lauterbach, R. Johnson, and H. K. Schachman, *Arch. Biochem. Biophys.* **99**, 157 (1962).
[12] H. K. Schachman, L. Gropper, S. Hanlon, and F. Putney, *Arch. Biochem. Biophys.* **99**, 175 (1962).
[13] K. Lamers, F. Putney, I. Z. Steinberg, and H. K. Schachman, *Arch. Biochem. Biophys.* **103**, 379 (1963).

power supply is close to the tolerance level. We have replaced the ±15 V supply with a Hewlett-Packard ±15 V, 1.25 A power supply. Also, we have replaced the unregulated −24 V supply in the scanner with a Hewlett-Packard regulated power supply. In the ±15 V supply we now have sufficient current to drive the x coordinate of an x,y plotter (EAI 1110 variplotter) via a voltage divider. The voltage divider consists of a 10-turn 10 $\kappa\Omega$ helipot (preceded and followed by 10 $\kappa\Omega$ resistors) geared to the transmission screw of the scanner via a 2:1 gear reduction ratio. The output from the terminal board pin 20 of the Beckman system to the x,y plotter allows both the R-S recorder and the external recorder to run in parallel. This device is extremely convenient for sedimentation velocity experiments since all graphs of concentration vs. distance are superimposed (see Fig. 2). For equilibrium work, scans are also superimposed, which is helpful in the determination of error limits.

A second device which is useful for equilibrium work is the use of a log amplifier on the y axis of the recorder. The zero optical density

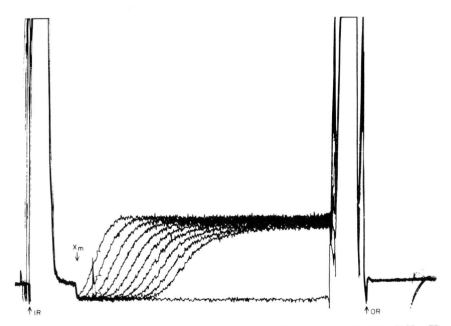

Fig. 2. Sedimentation velocity of bovine serum albumin in 0.2 M Tris·HCl, pH 7.8, at 20° and 60,000 rpm. Scanning intervals are at 8 minutes. The baseline is a 208-minute scan. The ragged inner reference edge is due to plotter difficulties on several of the scans. $s_{20,w}$ = 4.42 S by inflection points 4.77 ± 0.07 S by equivalent boundary. The disparity of these numbers indicates that heavy material is present.

voltage of the scanner is ca. -0.2 V. By means of a Heathkit standard voltage reference source (No. EUW-16A) this -0.2 V can be canceled to ± 2 mV. (This can also be done by an operational amplifier circuit.) A log amplifier module for the plotter then converts optical density to log optical density. Thus the output of the scanner is proportional to log C vs. r. For homogeneous materials such as bovine chymotrypsinogen A, molecular weights can be determined within 2.5% in extremely simple calculations. We have not been very successful in obtaining point-by-point molecular weight averages with this system up to the present time, however.

For specialized systems it is sometimes desirable to modify the gain

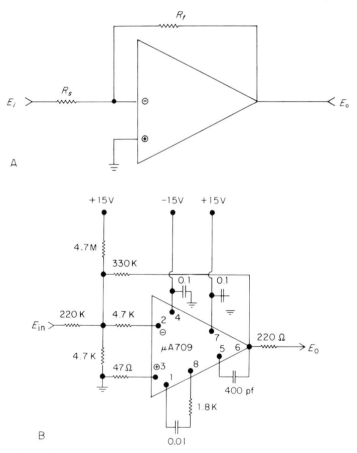

Fig. 3. Operational amplifiers for scaling scanner output. (A) Scheme showing an inverting operational amplifier. The output voltage E_o is given by $-(R_f/R_s)E_i$. (B) The circuit used in this laboratory to scale the scanner output for input to a PDP-12 computer. The circuits between pins 1–8 and 5–6 are for input and output frequency compensation and can be omitted if the Fairchild μA741 amplifier is used.

and/or offset of the scanner output signal. Examples of this are for A/D converters of on-line computers, or special x,y plotters. In this department we wished to scale the output of the scanner so that 0 to -2.5 V corresponded to -1 V to $+1$ V for input to a PDP-12 computer. The best and cheapest way to modify such signals is through operational amplifiers. While circuits for operational amplifiers are common, they may be difficult for biochemists to understand. Figure 3B presents the circuit diagram for an operational amplifier designed to use the ±15 V power supply of the scanner and scale the output voltages to ±1 volt. This circuit is designed about a Fairchild μA709 amplifier and cost about $5 to build. Logarithmic amplifiers can also be built from such components.[14]

B. Rayleigh Optics

Several tools may be constructed which aid the alignment of the Rayleigh optical system. These aids have recently been published in detail and will not be repeated in this article.[15,16]

A device which aids greatly in the plate reading procedure for sedimentation equilibrium experiments is a mask which fits on the swinging gate assembly.[17] This mask (Fig. 4) consists of metal strips which block the light from the cell in one position and from the reference holes in the other position. Monochromatic pictures are taken of the cell with the mask in the up position, and, without moving the plate, the mask is pushed down, the filter is removed, and a white light picture is taken of the reference holes. This device makes it quite simple to align the plate on the microcomparator stage.

D. A. Yphantis (personal communication) has constructed an interference mask with three slots in the region of the reference holes, but the usual two in the region of the cell. With this mask and a single sector counterbalance in the rotor, the fringes are spread about twice as far apart in the reference holes, again allowing easier alignment of the plate on the microcomparator stage.

Modification of the light source slit to produce Rayleigh and schlieren photographs in the same exposure have been described.[18,19] The device

[14] J. N. Giles, "Fairchild Semiconductor Linear Integrated Circuits Applications Handbook." Fairchild Semiconductor, Mountain View, California, 1967.
[15] E. G. Richards, D. C. Teller, and H. K. Schachman, *Anal. Biochem.* **41**, 189 (1971).
[16] E. G. Richards, D. C. Teller, V. D. Hoagland, Jr., R. H. Haschemeyer, and H. K. Schachman, *Anal. Biochem.* **41**, 215 (1971).
[17] T. A. Horbett and D. C. Teller, *Anal. Biochem.* **45**, 86 (1972).
[18] B. B. Massie, E. B. Titchener, and S. Hanlon, *Arch. Biochem. Biophys.* **128**, 753 (1968).
[19] W. D. Behnke, D. C. Teller, R. D. Wade, and H. Neurath, *Biochemistry* **9**, 4189 (1970).

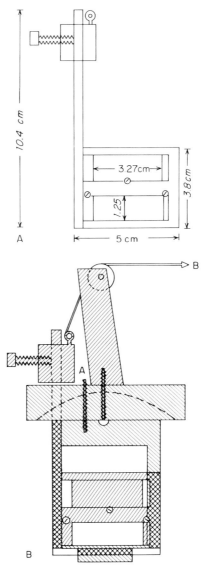

Fig. 4. Modified swinging gate assembly for superposition of white light and monochromatic Rayleigh photographs. (A) Mask with dimensions. The masks are set on a milled edge and the positions adjusted by means of the screws. The brass cylinder with the loop is used to adjust the down (white light of reference holes) position of the mask. A rubber band is stretched from the bolt to swing the gate out of the path for schlieren photographs. (B) Complete assembly with mask in the down position. Right slanted shading shows newly built parts. Left slanted lines or crosshatched lines indicate original pieces of the assembly. The bolt marked A is used to adjust the height of the mask in the up position (monochromatic). The pulley and string are attached to a spring which is tightened for the up position of the mask.

described by Massie *et al.*[18] appears superior to the T-shaped slit developed in these laboratories because the Rayleigh pattern is on the bottom of the photographic plate. With the T-shaped slit, the Rayleigh pattern is at the top of the plate and sometimes is obscured by the schlieren pattern in the presence of steep gradients.

At high centrifuge speeds the distortion of the sapphire windows is often quite severe using normal cell assemblies. We have recently found that this window distortion can be eliminated by using Teflon window liners to replace the usual Bakelite ones.[17] The Teflon which we use (Universal Plastics Co., 650 S. Adams, Seattle, Washington) is 0.015-inch thick and cut to the same width and length as the Bakelite liners. We also place a piece of Bakelite on the top of the window to minimize window rotation during cell tightening. We have not observed window distortion with these window liners. It should be cautioned, however, that materials from different manufacturers or perhaps even different lots from the same manufacturer may not have exactly the same properties of hardness, compressibility, etc. Further, at rotor velocities greater than 36,000 rpm it may be desirable to eliminate the Bakelite clip as this would probably compress the Teflon. Elimination of window distortion is presently being investigated in several laboratories.

C. Ultracentrifuge Drives

Ultracentrifuge drive units purchased from Beckman Instruments are quite variable in their behavior. The height of the rotor varies about 5 mm among various drives necessitating the realignment of the Rayleigh optical system after each drive change in order to maintain the proper focus of the optics. However, what is far more disastrous is the high and low frequency precessions of the rotor induced by the drive unit. High frequency precession destroys or modifies the protein gradient. Low frequency precession limits the Rayleigh fringe resolution in sedimentation equilibrium experiments, and completely prevents the use of the scanner for sedimentation equilibrium experiments; thus far, we have not been able to correct it by leveling the drive unit or other means.

The simplest method of checking drive stability is by means of the absorption scanner. One first positions the scanner in the air space of the counterbalance to obtain the minimum noise level (Fig. 5). One then moves the scanner to the edge of the counterbalance or other steep gradient, stops the scanner and observes the behavior on the y,t display of the R-S Recorder, as shown in Fig. 5A. In Fig. 5A is presented an example of both low and high frequency precession. In the air space, the noise level of the scanner is approximately ±0.5 mm maximum deviation. At the edge of the counterbalance, the high frequency

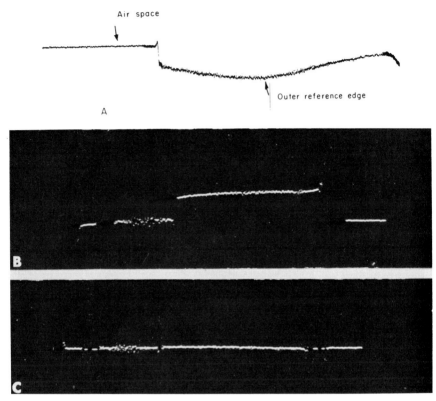

Fig. 5. Rotor movement and its effect on gradients at 52,000 rpm. (A) Scanner recording of the air space and at the edge of the counterbalance. High speed precession is seen by the increased noise at the counterbalance edge. Slow movement of the rotor is seen by the low frequency component (chart speed: 10 mm/sec). (B) The high frequency precession of (A) precluded establishment of an equilibrium gradient of aniline naphthalene sulfonate (ANS) at 1000 minutes. (C) Subtraction of 1257 minute scan of ANS from the scan of (B) demonstrating that there is no change in the pattern with time.

noise is ±1-2 mm. The effect of this high frequency movement on the gradient of aniline naphthalene sulfonate (ANS) is dramatic since virtually no gradient has formed at 1000 minutes (Fig. 5B), while the other two cells in the rotor (at higher initial concentrations) had attained sedimentation equilibrium. Further, there is no change in the concentration distribution with time as shown in Fig. 5C, where we have subtracted a scan at 1257 minutes from that at 1000 minutes. This high frequency precession is present in all drives to some extent and the point at which it becomes harmful is difficult to state. However, it can be observed in almost all published figures of log C vs. r^2 from high speed sedimentation equilibrium experiments where the authors have graphed data below 100 μ fringe displacement. These figures reveal a break in the linearity of the curve at low concentration due to convection. In general, it will be found that the log C value is too small at the point where the curve is discontinuous. At the present time, we feel that this discontinuity may be an indication of convection due to high frequency precession, although this is only a hypothesis.

The low frequency precessions of ultracentrifuge drives do not appear to affect gradients adversely, but strongly limit the ability to calculate precise molecular weights of proteins. Figure 5A shows the slow movement of the rotor as detected by the scanner. Figure 6 demonstrates more dramatically the effect this precession has on the determination of molecular weight. The frequency of the slow precession is speed dependent, having higher frequency at low speeds, thus corresponding to a nutation of the rotor. Clearly, this nutation precludes the determination of point-by-point molecular weight averages from the scanner data. In the Rayleigh photographs, at high speed equilibrium, the fringes are clear at low gradients, blur out at intermediate gradients, and become clear again at steep gradients within the same gradient, although this is not always true since the precession amplitude varies with time over a cycle of 10 minutes to several hours. In this laboratory, we have tried to eliminate the movement by such means as drive leveling, rubber bands attached to various points on the drive, and resupport of the

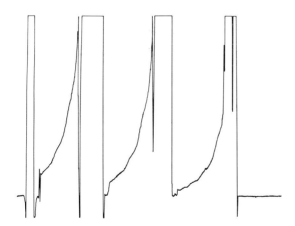

Fɪɢ. 6. Effect of rotor motion on the observed distribution at sedimentation equilibrium. α-Chymotrypsin centrifuged at pH 4.4 with 0.01 M acetate, 0.2 M KCl at 18,000 rpm for 20 hours in a six-channel centerpiece. The rotor movement superimposes a sine wave pattern on the exponential protein distributions.

drives with rubber stoppers. To date, all such measures have been unsuccessful. What is worse is the fact that only about 1 drive in 12 lacks the nutation.

Some workers, including D. A. Yphantis, are attempting to eliminate the effects of the nutation by such means as pulsed lasers to take a photograph of a single rotation of the rotor. Since such apparatus is usually not available in the biochemistry laboratory, we will not consider these methods in detail. However, it is fervently hoped by the author that some means can be found to eliminate these difficulties, since, at the present time, two effects of high frequency precession and low frequency nutation of ultracentrifuge drives are the limiting factors in the determination of molecular weights of proteins in the ultracentrifuge.

IV. High Speed and Low Speed Sedimentation Equilibrium

A. General Considerations

Sedimentation equilibrium experiments at the present time are characterized by the terms "high speed" or "meniscus depletion" and low speed experiments. This subclassification has arisen on the basis of the paper by D. A. Yphantis in 1964.[20] Low speed sedimentation equilibrium experiments[21-23] are performed at low speeds such that C_b/C_m is approximately 3–10 and require measurement of the initial concentration in order to establish the concentration scale when using the Rayleigh optics.[23] High speed sedimentation equilibrium experiments are performed at higher rotor velocities and lower initial concentrations, and C_b/C_m is in the range of 1000–10,000.[20] In the low speed experiments attempts are made to observe all of the material present in the preparation, while usually no such attempt is made in the high speed equilibrium method. In Table II, we have attempted to compare these two methods in some of their properties. Various authors[24,25] have discussed intermediate cases but, for the time being, we will consider only the two extremes presented in the table.

In order to compare the methods of measurement of protein molecular weights, it is necessary to first consider some facts of life concerning the ultracentrifuge. First, the z-average molecular weights do not require measurement of concentration but only of the gradient (as in schlieren

[20] D. A. Yphantis, *Biochemistry* **3**, 297 (1964).
[21] K. E. van Holde and R. L. Baldwin, *J. Phys. Chem.* **62**, 734 (1958).
[22] E. G. Richards and H. K. Schachman, *J. Phys. Chem.* **63**, 1578 (1959).
[23] E. G. Richards, D. C. Teller, and H. K. Schachman, *Biochemistry* **7**, 1054 (1968).
[24] F. E. Labar, *Proc. Nat. Acad. Sci. U.S.* **54**, 31 (1965).
[25] J. E. Godfrey and W. F. Harrington, *Biochemistry* **9**, 886 (1970).

TABLE II
COMPARISON OF HIGH SPEED AND LOW SPEED SEDIMENTATION
EQUILIBRIUM EXPERIMENTS

Property	Low speed equilibrium	High speed equilibrium
C_b/C_m	3–10	1000–10,000
Initial concentration		
Value	2–20 mg/ml	0.1–1.5 mg/ml
Measurement	Necessary[a]	Not necessary
Time to equilibrium	Long	Short
Sectorial cells	Desirable	Unnecessary
Accuracy	High	Lower
Experimental technicalities	Many	Few
Pitfalls	Many	More
Sensitivity		
Small components	Low	High
Large components	High	Low
Molecular weight averages from C vs. r data	M_w, M_z	M_n, M_w, M_z

[a] For a good method of labeling of white light fringes in this method see S. J. Edelstein and G. H. Ellis, *Anal Biochem.* **43**, 89 (1971).

optics). Consequently, most errors in concentration are eliminated. However, the calculation of the z-average requires a second derivative of Rayleigh or absorption optical system data with a consequent increase in the noise to signal ratio. Both low speed and high speed equilibrium experiments suffer equally from this effect. Second, the weight-average molecular weights cannot be calculated without knowledge of the values of the concentration. While the units of concentration are irrelevant, the error in calculation of weight-average molecular weights is[20,26]

$$\frac{M_{w,r}^{\text{true}} - M_{w,r}^{\text{obs}}}{M_{w,r}^{\text{true}}} = \frac{\epsilon}{C_{\text{obs}}} \tag{7}$$

Where $\epsilon = C_{\text{obs}} - \tilde{C}$. C_{obs} is the observed concentration and $\tilde{C}$ is the true concentration. At high concentrations as in low speed equilibrium, $C_{\text{obs}} \gg \epsilon$ so that accuracy is high. In the high speed equilibrium method, ϵ is close to C_{obs} so that accuracy suffers. There are also further consequences of this little formula. One consequence (and the criterion of design of high speed experiments) is that if C_{obs} is less than ϵ, for example at the centripetal meniscus, then an estimate of ϵ can be used

[26] D. C. Teller, T. A. Horbett, E. G. Richards, and H. K. Schachman, *Ann. N.Y. Acad. Sci.* **164**, 66 (1969).

to determine the concentration scale as well as the probable errors in M_w when $C_{obs} > \epsilon$ toward the base of the cell. On the other hand, if instantaneous adsorption of protein on the walls and windows of the cell occurs in the low speed method, it cannot be detected and ϵ may become quite large. Finally, since the fringe count, $J = C - C_a$ where C_a is a point near the meniscus, it is necessary to construct a concentration scale from the fringe data in order to measure weight-average molecular weights in the ultracentrifuge.

Third, the formula for calculation of number average molecular weight is,

$$M_{n,r} = \frac{A^{-1}C}{\int_{r_m}^{r} Cd(r^2) + \dfrac{C_m}{AM_{n,m}}} \tag{8}$$

Where $M_{n,r}$ is the number average molecular weight at the point r, $A = (1 - \bar{v}\rho)\omega^2/2RT$, C_m is the meniscus concentration, and $M_{n,m}$ is the number average molecular weight at the meniscus of the cell. It is not necessary that this constant be determined at the meniscus but, since it is a constant due to an integration, it must be known at some point in the centrifuge cell. It is possible to prove that $C_m/AM_{n,m}$ can be determined for any experiment by using finite-difference calculus,[27] but little practical success has been achieved for high meniscus concentrations. Clearly if $C_m \approx \epsilon$, the measurement error, then this term represents only a small part of the integral and accurate number average molecular weights can be obtained, in principle. This device was first used by Wales et al.[28] to compute number average molecular weights from sedimentation equilibrium experiments. At present, it is not practical to estimate $C_m/AM_{n,m}$ from low speed sedimentation equilibrium experiments.

B. High Speed or Low Speed?

This question is impossible to answer on an a priori basis. The general properties of the enzyme must be assessed from its s vs. concentration dependence and further experiments be planned at that point. Of course, one must take into consideration the other evidence available concerning the enzyme. It should be kept in mind, however, that the low speed method is not well suited for molecular weight determinations of unstable enzymes, for enzymes which tend to aggregate extensively, nor for extremely large ($M > 10^6$ g/mole) enzymes. The high speed method, on

[27] D. C. Teller, Ph.D. Dissertation, University of California, Berkeley, 1965.
[28] M. Wales, F. T. Adler, and K. E. van Holde, J. Phys. Colloid Chem. 55, 145 (1951).

the other hand, is less accurate and extremely sensitive to the presence of low molecular weight material. At some times this is an advantage, but for other systems this sensitivity obscures important data.

V. Methods in Sedimentation Equilibrium Experiments (Rayleigh Optics)

In this section we describe the methods we presently use for high speed sedimentation equilibrium experiments. The methods used for low speed sedimentation equilibrium experiments have been discussed in great detail by Richards et al.[23] The procedure published in that paper is described in sufficient detail that a repetition here would be superfluous. The methods described in this section have been published by Yphantis,[20] Teller et al.,[26] and Horbett and Teller.[17]

A. Preliminary Considerations

For high speed sedimentation equilibrium experiments, it is extremely important that the camera lens be focused at the proper (2/3) plane in the cell.[15,16] In this technique, steep gradients are produced at low values of the total concentration so that errors due to improper camera lens position become very significant.[20] Systems that are heterogeneous can appear homogeneous and conversely, owing to improper lens positions.

A second initial precaution which we have found worthwhile is to flatten the centerpieces used for the experiments. Upon receipt of the centerpieces of the Yphantis[20] design from Beckman Instruments, we flatten them on a lapping plate consisting of a glass telescopic mirror mount. The lapping compound which we use is John Crane Lapmaster No. 1900 (Johns Manville Corporation), which has been fractionated by sedimentation in water. The initial lapping is performed using the fast sedimenting components, and final polishing is accomplished with the very fine fractions. A rubber stopper placed on top of the centerpiece provides the weight for lapping which utilizes a figure-eight motion of the arm holding the rubber stopper lightly. The tolerance we use for all portions of the centerpieces is $<10^{-4}$ inch as measured by a micrometer.

In order to measure leakage by a finished centerpiece, solutions of $1 M$ Na_2CO_3 and phenolphthalein are centrifuged from 1 to 8 hours with cell loading as shown in Fig. 7. A pink color observed in the viewing screen indicates leakage. The reason for this rather elaborate leakage test is that serious errors in molecular weights are observed when leakage occurs during an experiment. Horbett[29] has calculated that it is neces-

[29] T. A. Horbett, Ph.D. Dissertation, University of Washington, Seattle, 1970.

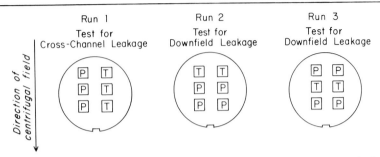

Fig. 7. Cell filling format for leakage test. P represents 0.13 ml of 0.01 g/l pheno-phathalein; T represents 0.13 ml of 1 M Na$_2$CO$_3$, pH 11. Leakage is detected by observation of a pink color in the viewing screen.

sary to detect leakage at a sensitivity of 0.03 μl in order that the effects of such leakage be less than fringe reading errors. The usual test of meniscus movement is not sufficiently sensitive for this measurement.

Finally, we believe that it is worthwhile to keep in the laboratory a single protein sample which has been well characterized and period-ically check its molecular weight distribution. α-Chymotrypsin is very good for this purpose since it reversibly associates at pH 4.4. Since dif-ferent lots of α-chymotrypsin have different association constants, it is desirable to use a single lot for this purpose. In 0.01 M acetate, 0.2 M KCl, pH 4.40, most Worthington Biochemicals α-chymotrypsin prepara-tions have an association constant for dimerization between 26 and 44 $\times$ 10^3 liters/mole at 20°.[30] Occasional experiments with such a protein serve as a control that all aspects of the methodology are correct. It is particularly important to have such a standard available when anoma-lous results are found in experiments on other enzymes.

B. Criteria for Design of High Speed Sedimentation Equilibrium Experiments

In the characterization of a protein system we have an approximate molecular weight based on the sedimentation and diffusion experiments presented in Section II. In this section we will use those data to cal-culate the optimum rotor speed, initial concentration, column height, and time to reach sedimentation equilibrium for high speed experiments.

In order to choose the optimum speed and initial concentration, we may set several criteria: First, the maximum observable concentration gradient at the cell bottom will be limited by the fringe resolution (200–400 fringes/cm in the cell). Second, the meniscus concentration should be sufficiently small that errors in its estimation will not jeopardize the

---

[30] D. D. Miller, T. A. Horbett, and D. C. Teller, *Biochemistry* **10**, 4641 (1971).

validity of the experiment (<0.1 fringe). Third, initial concentrations should be as high as possible so that a stable protein gradient is formed to minimize the danger of convection. Fourth, the time to reach equilibrium should be 24 hours or less, since the enzyme may not be stable for longer times. Fifth, rotor speeds should be kept below 40,000 rpm, if possible, to reduce dangers of cell leakage and window distortion; however, as previously mentioned, window distortion can be eliminated by using Teflon window liners.

The sedimentation of a homogeneous substance is given by the equation

$$C = C_0 H e^{(\Psi+1)} / \sinh H \approx 2 C_0 H e^{\Psi H} \tag{9a}$$

$$H = \frac{(1 - \bar{v}\rho)\omega^2}{4RT} M(r_b^2 - r_m^2) \tag{9b}$$

Where the cell has been assumed sectorial, C_0 is the initial concentration and $\Psi = 2(r^2 - r_b^2)/(r_b^2 - r_m^2)$. r is radial distance in centimeters from the rotation axis, $\bar{v}$ is the partial specific volume, ρ the solution density, ω the angular velocity in radians per second, R is the gas constant = 8.3144×10^7 erg/deg. mole, T is the absolute temperature, and M is the molecular weight in grams per mole. The approximate equality applies to high speed sedimentation equilibrium experiments since $\sinh H \approx e^{H/2}$ when $H > 2$. Further, since column heights are small, the fact that the Yphantis[20] cell is not sectorial can be neglected. To apply the first criterion we differentiate Eq. (9a) and estimate it at the cell base

$$\left(\frac{dC}{dr}\right)_b = \frac{8r_b C_0 H^2}{r_b^2 - r_m^2} \tag{10}$$

To apply the second criterion, we evaluate Eq. (9a) at the meniscus.

$$C_m = 2C_0 H e^{-2H} \tag{11}$$

We combine these two equations with the elimination of C_0 to obtain

$$He^{2H} = \left(\frac{\left(\frac{dC}{dr}\right)_b}{C_m}\right)\left(\frac{r_b^2 - r_m^2}{4r_b}\right) \tag{12}$$

The second term in brackets varies less than 0.1% for a fixed column height with 5.8 cm < r_b < 7.2 cm. Consequently, we can tabulate this term for a variety of column heights as shown in the legend of Fig. 8. Second, a plot of log He^{2H} vs. H is almost linear as shown in Fig. 8. Thus, we can choose $(dC/dr)_b$ at an arbitrary number of 200–400 fringe/cm, depending on the drive, $C_m = 0.04$ fringe and pick the value of H from the graph ($H = 3.2$). From Eq. (9b) we then calculate the rotor speed.

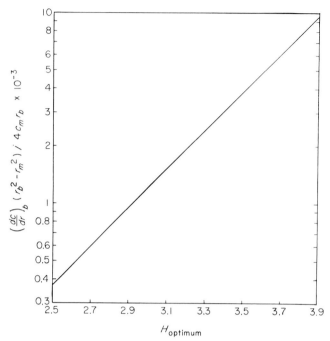

FIG. 8. Determination of the optimum value of H to determine the equilibrium speed. The ordinate is $He^{2''}$ of Eq. (12). The tabulation lists the values of the second bracketed term of Eq. (12).

Column height (mm)	$\dfrac{r_b{}^2 - r_m{}^2}{4r_b}$ (cm)
1	0.0496
2	0.0985
3	0.1466
4	0.1940
5	0.2408

This procedure can be repeated for a variety of column heights, but a simpler method is presented below. We now apply the third criterion. For each column height the optimum value of the initial concentration can be calculated from Eq. (10). At this point, some subjectivity is required, since it is desirable to keep the concentration high yet obtain reliable results. If the meniscus concentration is greater than 0.1 fringe, it cannot be established accurately. If the meniscus concentration is very low, then high concentrations cannot be obtained at the cell base (due to a steep gradient and rotor movement) and so accuracy suffers (recall the error Eq. 7). Because there is uncertainty in the heterogeneity of the preparation as well as the fact that we wish to test for the presence or absence of chemical equilibria, we generally perform experiments on several samples of the enzyme at varying initial concentration. When the Yphantis[20] six-channel centerpiece is employed we generally run a 3-fold range of initial concentrations of enzyme. Yphantis (personal communication) recommends a broader range of concentrations (20-fold) instead of the 3-fold range recommended here for the detection of chemical equilibria; however, I feel that all enzyme preparations contain sufficient impurities to obscure the major interactions if such a broad range of initial loading concentrations is used. In essence, we are closing our eyes to minor contaminants of the preparation.

In order to determine the time required to reach equilibrium, we use the equation of Mason and Weaver [31-33]

$$t_{max}^{(hours)} = \frac{2(r_b - r_m)}{\omega^2 s_{obs}(r_b + r_m) \cdot 1800} \tag{13}$$

The column height dictates (to some extent) the centerpiece to be used for the experiment. It is not practical to use greater than 3 mm column heights in the Yphantis[20] six-channel centerpiece.

A simple way to decide on the column height and equilibrium speed is presented in Fig. 9. Based on the stability of the material, we decide on the length of the experiment (say 24 hours). One then finds the intersection of this time (right ordinate) with the molecular weight of the material (say 15×10^4 g/mole). This point is closest to the 4 mm solution column height on the graph. However, we would like to use a Yphantis[20] centerpiece as well as allow some latitude in time so that we can be certain that equilibrium is reached. Thus, a 3-mm column height with an equilibrium time of about 15 hours is most satisfactory. The right ordinate (solid lines) then dictates that the experiment should be performed at 14,000 rpm. The optimum initial concentration for this experiment is taken from the legend of Fig. 9 as 0.77 mg/ml. We would probably decide to use our three initial concentrations as 1.2, 0.8, 0.4 mg/ml.

These lines were calculated for a protein with $\bar{v} = 0.75$ ml/g and the density of water at 20°. If the density is much greater than that of water

[31] M. Mason and W. Weaver, *Phys. Rev.* **23**, 412 (1924).

[32] W. Weaver, *Phys. Rev.* **27**, 499 (1926).

[33] T. Svedberg and K. O. Pedersen, "The Ultracentrifuge." Johnson Reprint Corp., New York, 1940.

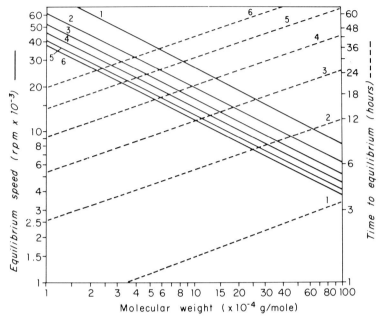

Fig. 9. A simple way to decide column heights and equilibrium speeds for high speed experiments. The first bracketed term of Eq. (12) was taken as 10^4 cm^{-1}. The column height is chosen according to the intersection of the molecular weight (abscissa) and time to equilibrium (right ordinate, dashed lines). The equilibrium speed is determined from the column height and molecular weight on the right ordinate.

Column height (mm)	Approximate C_0 (mg/ml)
1	0.36
2	0.58
3	0.77
4	0.93
5	1.11
6	1.26

(say D_2O or guanidine HCl) then faster speeds must be used. To correct for density, the formula is

$$\text{rpm}_{\text{exptl}} = \text{rpm}_{\text{graph}}/2(1 - \bar{v}\rho)^{1/2}{}_{\text{exptl}}$$

For example, if $\bar{v} = 0.70$ ml/g and $\rho = 1.2$ for our experiment, then $(1 - \bar{v}\rho)^{1/2}{}_{\text{exptl}} = 0.4$ so that the equilibrium speed should be 14 ×

$10^3/0.8 = 17.5 \times 10^3$ rpm. The closest speed on the ultracentrifuge (18,000 rpm) would be chosen. If the temperature was to be controlled at low temperature (4°), then the time to equilibrium is increased by a factor of 1.5 due to the increased viscosity of water. In the example given above, this would increase the time to equilibrium from 15 hours to about 23 hours. To allow a safety factor in the attainment of equilibrium, it might be desired to decrease the column heights from 3 mm to 2.5 mm together with a slight reduction of the C_0 values; alternatively the experiment could be continued for 30 or more hours with a 3-mm solution column.

For the Yphantis[20] six-channel centerpieces, 0.07 ml solution corresponds approximately to 2 mm column height, and 0.13 ml corresponds to 3 mm column height. In the standard aluminum-filled epoxy double-sector centerpieces (part No. 305685, Beckman Instruments) the volume relation[23] is approximately,

$$V = 2.443 - 0.3393r \qquad (14)$$

where V is the volume of solution to be used. The base of the cell occurs at $r_b = 7.2$ cm. Thus from the formula, with $r_m = 6.9$ cm approximately 0.10 ml of solution is required for a 3-mm solution column height, and a 7-mm column requires 0.24 ml of solution. For very short solution columns of 1 mm or less, the Yphantis[34] style eight-channel centerpiece is quite convenient to use. Here, 0.05 ml corresponds to approximately 1 mm column height.

C. At Equilibrium

In this laboratory we usually do not bother to check that the material has reached sedimentation equilibrium at the end of the calculated period of time. While it would be desirable to do this in principle, in practice we are generally very lazy. If the distribution of concentration as seen in the viewer of the ultracentrifuge looks strange we frequently change the speed and continue the experiment for another 24 hours; however, for most proteins, the preliminary experiments are sufficiently accurate so that the preexperimental calculations are correct. We take a photograph of the distribution, develop and fix it prior to changing the speed of the ultracentrifuge. Examination of this photograph with a magnifier is generally adequate to determine that the run is successful.

D. Baseline Patterns

The success of the experiment is totally dependent on the success of this part of the experiment. Yphantis[20] recommended that the base-

[34] D. A. Yphantis, *Ann. N.Y. Acad. Sci.* **88**, 586 (1960).

line used for high speed sedimentation equilibrium experiments be that of a water–water baseline at the end of the protein–solvent run. However, at the present time[17] there are reasons to believe that the proper baseline to use in a photograph of a low speed picture of the resuspended sample taken after the high speed equilibrium experiment. This is true for the Yphantis-style[20] centerpieces. For the externally loaded, six-channel ultracentrifuge cell described recently by Ansevin et al.,[35] this is not true. In brief, the following considerations apply to baseline determinations. Ultracentrifuge cell assemblies suffer a certain amount of hysteresis due to changing speeds. A cell barrel that is somewhat out of round will twist to varying degrees dependent on the rotor velocity. Since the sapphire windows used in these cells are not flat nor symmetrical, baselines will vary as a function of speed due simply to the geometry of the cell housing and other components. These speed dependencies will appear in photographs as rotations of baselines. Also, the sapphire windows, window holders, and centerpiece move from their initial positions to final positions dependent on their mass, density, and frictional resistance to movement. These considerations dictate that baseline experiments should be performed after the equilibrium experiment and without disassembly of the cell since it is virtually impossible to reassemble a cell so that all components resume their former positions.

Thus, for the Yphantis[20] six-channel cell, we recommend that the rotor and cell be examined with a magnifier after the equilibrium experiment to determine its position relative to the scribe lines on the rotor and cell. The cell is then removed from the rotor and vigorously shaken for about 5 minutes to destroy the gradient, replaced in the rotor with the same position as determined previously and taken to low speed (1200–4000 rpm) for the baseline photograph. If the rotor is allowed to spin at low speed for about 0.5 hour, all thermal gradients and concentration gradients should be destroyed. Further, any denatured protein should be sedimented at this time. A picture is then taken for the baseline experiment. For the six-channel centerpiece described by Ansevin et al.,[35] which can be flushed with solvent, it is only necessary to clean the cell without disassembly and return the rotor to the same speed as in the equilibrium experiment.

We occasionally check our cells with respect to baseline reproducibility by performing a water–water baseline experiment at high speed (32,000 rpm) and again at low speed (3200 rpm). This test serves to determine whether the aluminum cell housing has become twisted during use. If these baselines do not reproduce between speeds, we replace the

[35] A. T. Ansevin, D. E. Roark, and D. A. Yphantis, *Anal. Biochem.* **34**, 237 (1970).

cell housing with a new one. For further consideration of baseline reproducibility, see Horbett and Teller.[7]

E. Plate Reading

This is the most tedious part of the equilibrium experiment and causes ultracentrifuge technicians to seek other jobs ("It was a great job except for that plate reading"—D. Trout, a former technician). A vast variety of semiautomatic and automatic plate readers are presently in the process of development which hopefully will alleviate this problem in the future, but at present they are not available in all laboratories. The following description for plate reading of high speed equilibrium experiments is generally satisfactory (modified from Aune and Timasheff[36]).

In this laboratory we position the plate on the microcomparator with the wire in the counterbalance positioned at 5.000 mm on the x-micrometer. In this way it is easy to correct for baselines and compare data between experiments from the micrometer readings alone. The necessary data for calculation include the positions of inner and outer reference edges (outer edge of inner reference and inner edge of outer reference), the meniscus position, the x,y coordinates of a point close to the meniscus and readings from the baseline and equilibrium photographs. In this laboratory, the magnification of the microcomparator is 100×, because we use a modified microcomparator.[37] However, we have recently built the automatic microdensitometer described by D. J. DeRosier et al.[37a] For readings by eye, the best magnification to use is ×50. Below ×50 the individual fringes are too small to read accurately, and above ×50 they are too large.

Three light fringes are read at each x coordinate, and the positions are averaged from the meniscus to the base of the cell at every 100 μ. The three readings are averaged at each point, and if the average differs by more than 5 μ from the value read for the middle fringe, the three readings are repeated. When the fringes curve steeply enough so that they occur at increments of less than 100 μ between light and dark fringes, in this laboratory we keep the y-displacement constant and measure the x-position of the half fringes. When large computers are to be used for calculation of the data, this method is preferable. However, if calculations are to be performed on a small computer, it is best to read equal x increments throughout the cell in order to use orthogonal

[36] K. C. Aune and S. N. Timasheff, *Biochemistry* **10**, 1609 (1971).

[37] D. C. Teller, *Anal. Biochem.* **19**, 256 (1967).

[37a] D. J. DeRosier, P. Munk, and D. J. Cox, *Anal. Biochem.* **50**, 139 (1972).

least squares polynomials in r, as will be discussed in the next section. In this laboratory we read the baseline at 200-μ increments in x. Again, the same number of readings at each x position should be made as were made for the equilibrium experiment. To apply the baseline to the equilibrium data, it is first smoothed. There are several ways of smoothing the data ranging from the very involved smoothing and filtering described by Horbett and Teller[17] to graphing the data and drawing a smooth curve through the points. Intermediate is the use of smoothing formulas of five or seven points at a time.

The smoothing formulas for 5 points with baseline values f_i are[38]

$$y_0 = (f_{-2} + f_{-1} + f_0 + f_1 + f_2)/5$$
$$y_{-1} = (4f_{-2} + 3f_{-1} + 2f_0 + f_1)/10$$
$$y_{-2} = (3f_{-2} + 2f_{-1} + f_0 - f_2)/5 \qquad (15)$$

The formulas for y_{+1} and y_{+2} are obtained by reversing the numbering of the ordinates in the last 2 equations. These smoothing formulas correspond to fitting least squares straight lines through the equally spaced points.

After subtraction of the baseline corrections, one is now prepared to make the calculations to obtain point-by-point molecular weight averages.

F. Calculation of the Data

In this section we shall describe the method of calculation of the data using a desk-top calculator or small programmable computer. Several programs exist for these calculations using large computers. In this laboratory, we strongly recommend that students using sedimentation equilibrium perform these slower calculations on a few experiments in order to learn more about the errors encountered and artifacts that arise in these experiments. Throughout this discussion we will assume that data have been collected at equal increments in x on the microcomparator. Some of the formulas are incorrect if that is not true.

1. Preliminary Estimate of the Meniscus Concentration

In this laboratory, we use units of fringes for concentration rather than microns, as is quite common. To convert the data to fringe units, j, we divide the y-micrometer values less the value of the first point by the fringe separation. To obtain a preliminary estimate of the

[38] F. B. Hildebrand, "Introduction to Numerical Analysis." McGraw-Hill, New York, 1956.

meniscus concentration we plot log j vs. x, where the x values are the microcomparator readings between 0 and about 0.5 fringe ($140\,\mu$).

We extrapolate back from the linear part of the curve to the meniscus position, x_m. The value of the concentration at x_m is then added to all concentrations. This process is repeated a second time. The values of the $\ln j' = \ln(j + C_m)$ are again plotted vs. x and back-extrapolated to x_m. The reason for using x values rather than r^2 values in this procedure is to attempt to minimize the effects of heterogeneity in this region of the cell. This procedure is readily performed on semilog graph paper. At this point after addition of the C_m value to j, we have an estimate of the concentration scale for which we will use the symbol C.

2. Least Squares

In order to further discuss the calculation of the data from sedimentation equilibrium experiments, it is first necessary to discuss general principles of calculation of theoretical functions from experimental sources. In all aspects of the centrifuge experiment, experimental errors are introduced. These errors take such forms as both random errors and systematic errors. The problem is further complicated because we do not know the molecular weight distribution prior to the experiment. We only know that the fringe distribution at equilibrium for an ideal solute will be given by the function

$$C - C_m = \sum_{i=1}^{n} C_{pi} e^{A_i M_i (r^2 - r_p^2)}$$

where C is the concentration in fringes, C_m is the meniscus concentration, the points C_{pi}, r_p represent an integration constant, and the C_{pi} are the fringe concentrations of species i at some arbitrary position r_p. $A_i = (1 - \bar{v}_i \rho)\,\omega^2/2RT$, and M_i is the molecular weight of species i. In addition, the observed concentrations, C, are perturbed by the systematic and random errors. The object of the experiment is to recover the values of M_i and simultaneously determine whether the C_{pi} are dictated by the chemical laws of mass action.

In the attempt to recover this information we will frequently make the assumption that any function, $f(x)$ can be approximated by a Taylor's series over small regions of the independent variable, x; that is

$$f(x) = f(x_0) + \frac{f'(x_0)}{1!}(x - x_0) + \cdots + \frac{1}{n!}f^{(n)}(x_0)(x - x_0)^n + \cdots \quad (16)$$

in which x_0 is a constant. This equation can be reduced to a polynomial of the form,

$$f(x) = a_0 + a_1x + a_2x^2 + \cdots + a_nx^n \tag{17}$$

What we wish to do in curve fitting is minimize the deviations of the observed values $f(x_j)$ from their predicted values y_j determined by the right side of Eq. (17). At least this way we can minimize the random components of the errors in the data. A convenient way to do this mathematically is to minimize the sums of the squares of the deviations of the points (N in number) from their observed values. That is,

$$\sum_{j=1}^{N} [y_j - f(x_j)]^2 = \text{minimum} \tag{18}$$

Minimizing the sums of squares of the values is not necessarily the best way to solve this problem for all data; however, it is the usual method used for two reasons: First, it is convenient to handle mathematically; and second, it allows good error prediction if only random errors are present.

We equate $f(x_j)$ at each point with the right side of Eq. (17) and place this in Eq. (18) to yield,

$$\sum_{j=1}^{N} (y_j - a_0 - a_1x_j - a_2x_j^2 - \cdots - a_nx_j^n)^2 = \text{minimum}$$

To obtain the minimum of this function, we differentiate with respect to the a values since these do not change from point to point; we set the derivatives to zero to find the minimum. This produces the equations

$$\frac{\partial}{\partial a_0} = 0 = \Sigma y_j - Na_0 - a_1\Sigma x_j - a_2\Sigma x_j^2 - \cdots - a_n\Sigma x_j^n$$

$$\frac{\partial}{\partial a_1} = 0 = \Sigma y_jx_j - a_0\Sigma x_j - a_1\Sigma x_j^2 - a_2x_j^3 - \cdots - a_n\Sigma x_j^{n+1}$$

$$\cdots$$

$$\frac{\partial}{\partial a_n} = 0 = \Sigma y_jx_j^n - a_0\Sigma x_j^n - a_1\Sigma x_j^{n+1} - a_2\Sigma x_j^{n+2} - \cdots - a_n\Sigma x_j^{2n}$$

Since x_j and y_j are observed parameters, the only unknowns are the a values. However, we have had to differentiate the summation n times to minimize all n parameters. Thus, we have produced n equations with unknowns $a_0, a_1, a_2, \ldots a_n$. It is a simple matter to solve these equations for the unknowns by computers at the present time. All computer manufacturers give programs capable of solving such linear systems of equations when they sell the computer to a customer. The matrix of these equations is

$$
\begin{bmatrix} a_0 \\ a_1 \\ a_2 \\ \vdots \\ a_n \end{bmatrix} = \begin{bmatrix} N & \Sigma x_j & \Sigma x_j^2 & \cdots & \Sigma x_j^n \\ \Sigma x_j & \Sigma x_j^2 & \Sigma x_j^3 & \cdots & \Sigma x_j^{n+1} \\ \Sigma x_j^2 & \Sigma x_j^3 & \Sigma x_j^4 & \cdots & \Sigma x_j^{n+2} \\ \cdot & \cdot & \cdot & \cdot & \cdot \\ \Sigma y_j x_j^n & \Sigma x_j^{n+1} & \Sigma x_j^{n+2} & \cdots & \Sigma x_j^{2n} \end{bmatrix}^{-1} \begin{bmatrix} \Sigma y_j \\ \Sigma y_j x_j \\ \Sigma y_j x_j^2 \\ \vdots \\ \Sigma y_j x_j^n \end{bmatrix}
\tag{19}
$$

One needs only to use a matrix inversion program on the sums in the right-hand brackets of this equation to solve for the values of a. There are a few technical details which should be observed for such matrix inversions. One is to keep the sums of x_j^k close to unity. This can be achieved by subtraction of or division by a constant in the observed x_j, y_j data. A second pitfall to steer around is that one bad point can strongly influence the predicted curve. A point which is deviant by five or six standard deviations can badly skew the predicted value of the line toward itself. This is particularly common when semilogarithmic functions are used, as in kinetics or high speed sedimentation equilibrium.

A third, more subtle, pitfall is to select the proper amount of data for the curve fitting. As mentioned earlier, this function was selected on the basis of a Taylor's series in which $x - x_0$ must be small for the series to converge. Thus, attempts to pack too much data into the sums will produce systematic deviations of the predicted values of the function; on the other hand, if not enough data are included, the predicted curve may follow the experimental data too closely and lead to nonphysical values for physical parameters. Hildebrand[38] gives an estimate of the efficiency of such curve fitting as

$$
\epsilon = \left(\frac{\Sigma(y_i - \bar{y}_i)^2}{N} \right)^{1/2} \left(\frac{N}{N - (n + 1)} \right)^{1/2}
\tag{20}
$$

Where ϵ is the efficiency, the first term is the root-mean-square deviation of predicted and observed values and $N - (n + 1)$ is the number of degrees of freedom of the polynomial. Choice of the minimum ϵ among a variety of polynomials of varying degrees, n, fitted to the same number of data, N, is one criterion of choice of the best polynomial.

There are other examples of least squares techniques which are worthwhile to consider in this section. The first to be considered is when several variables constitute the x_j values. For example, in the approach to sedimentation equilibrium, the concentration at a point in the centrifuge cell is a function both of distance and time, which are separately measured variables. Another example will appear in the discussion of hybridization reactions of proteins. To perform a least squares curve fitting on such data we can proceed as in the previous development:

$$
f(x,z) = a_0 + a_1 x + a_2 x^2 + \cdots + b_0 + b_1 z + b_2 z^2 + \cdots
$$

By partial differentiation of the observed equation with respect to the parameters a and b, a set of linear equations will again be obtained which may be used to fit the data. Of course, in the example given, a_0 and b_0 will appear as a single term, $a_0 + b_0$.

Returning to the polynomial in x as the single independent variable, a special case occurs when the values of x_j are equally spaced. This condition allows each term of a polynomial series to be calculated independently of the others based solely on point position rather than value. Such a circumstance results in considerable simplification of the formulas and makes calculation of least squares equations very simple and accurate, since matrix inversions are not required. It allows one to write the least squares formulas as single equations for each term. These polynomials are called orthogonal or Gram polynomials and a review of their properties is given in Chapter 7 of Hildebrand.[38]

It is possible and simple to insert weighting functions into least squares analyses. These weighting functions can be inserted in either of two ways depending on whether it is desirable to use strong weighting or weak weighting of the data points. It was mentioned above that a strongly deviant point will affect adversely the least-squares line through all the data. Weighting functions are a simple way to avoid this adverse behavior in curve-fitting calculations. Let us consider that we are smoothing experimental data and that one value of ln C at low C is bad. If we were to apply least-squares fitting of 5 or 7 points over the region, then this bad point would dominate the predicted values of the concentration in this region of the cell. One way to reduce this effect is to weight each point according to how well it fits the predicted curve in the region. If a point fits the predicted curve it should receive a high weight, but a badly fitting point should be discounted so as not to affect the rest of the curve. To achieve this purpose we can write a weighting function for each point,

$$w_j = \frac{1}{|\Delta y_j + \delta|} \tag{21}$$

In this equation, δ is some constant, say the rms fit to the data in the interval considered, Δy_j is the deviation of y_j between observed and predicted values from the least squares fit. If the point y_j fits the curve perfectly, then w_j is $1/\delta$. If it fits badly then $w_j \ll 1/\delta$. Consequently, w_j reflects the goodness of fit of the data to the empirical least-squares formula (18) by requiring that

$$\sum_{j=1}^{n} w_j[y_j - f(x_j)]^2 = \text{minimum} \tag{22}$$

or second, require that

$$\sum_{j=1}^{n} [w_j y_j - w_j f(x_j)]^2 = \text{minimum} \tag{23}$$

These two formulas for weighting functions w_j will predict different curves. In formula (22), the terms w_j will appear in all sums of the matrix as a linear term. In formula (23), it will appear as w_j^2. Clearly, Eq. (23) contains stronger weighting than formula (22). But which is better for obtaining practical results from sedimentation equilibrium data cannot be predicted *a priori*.

3. Data Smoothing

There are many procedures available for smoothing the data, and it is difficult to recommend a single procedure that will be successful in all cases.

The only principles involved in data smoothing are: First, linearize the data as much as possible prior to smoothing; and, second, do not smooth too many times. The simplest smoothing operation on equilibrium data is to use orthogonal polynomials on log C vs. r data (not r^2). If we let f_i represent the values of log C_i then the equations for 5-point linear smoothing are given by Eqs. (15).

Five-point third-degree smoothing formulas are:

$$y_0 = (-3f_{-2} + 12f_{-1} + 17f_0 + 12f_1 - 3f_2)/35$$
$$y_{-1} = (2f_{-2} + 27f_{-1} + 12f_0 - 8f_1 + 2f_2)/35$$
$$y_{-2} = (69f_{-2} + 4f_{-1} - 6f_0 + 4f_1 - f_2)/70$$

Seven-point smoothing formulas are:
Linear,

$$y_0 = (f_{-3} + f_{-2} + f_{-1} + f_0 + f_1 + f_2 + f_3)/7$$
$$y_{-1} = (7f_{-3} + 6f_{-2} + 5f_{-1} + 4f_0 + 3f_1 + 2f_2 + f_3)/28$$
$$y_{-2} = (5f_{-3} + 4f_{-2} + 3f_{-1} + 2f_0 + f_1 - f_3)/14$$
$$y_{-3} = (13f_{-3} + 10f_{-2} + 7f_{-1} + 4f_0 + f_1 - 2f_2 - 5f_3)/28$$

Cubic,

$$y_0 = (-2f_{-3} + 3f_{-2} + 6f_{-1} + 7f_0 + 6f_1 + 3f_2 - 2f_3)/21$$
$$y_{-1} = (-4f_{-3} + 16f_{-2} + 19f_{-1} + 12f_0 + 2f_1 - 4f_2 + f_3)/42$$
$$y_{-2} = (8f_{-3} + 19f_{-2} + 16f_{-1} + 6f_0 - 4f_1 - 7f_2 + 4f_3)/42$$
$$y_{-3} = (39f_{-3} + 8f_{-2} - 4f_{-1} - 4f_0 + f_1 + 4f_2 - 2f_3)/42$$

In all cases, the formulas for y_1, y_2, and y_3 are obtained by changing the sign of the subscripts.

Some general comments on smoothing are in order at this time. The amount of smoothing of the data increases with the number of points used in the smoothing formula and decreases with increasing values of the degree, n, of the smoothing formula. The ideal smoothing operation eliminates the noise from the data, but the essential characteristics of the function are not modified. The reason why the degrees of the polynomials given above are odd is that when n is even, the same centerpoint formulas are calculated as for the odd equation of next higher degree. For example, a seven-point quadratic expression for y_0 is identical to that of the cubic formula given above. The only differences are in the end-point formulas, and it is assumed that enough data have been collected so that these are infrequently applied.

The first time in smoothing the end points should be kept at their original values unless they appear unusual, in which case they should be discarded. The data are smoothed, and the smooth values replace the observed values except at the ends. In the last smoothing operation, the end points generally are discarded entirely and only the zero point smoothing formula used.

Inspection of the graph of ln C vs. r reveals that the points are tightly bunched on the ordinate at high concentrations. Thus, these points are not effectively smoothed in this procedure. In our computer program we circumvent this problem by a second smoothing operation on the raw data. We plot $\int Cd(r^2)$ vs. C and smooth this function 2 times. (The method of computing the integral is given in the next section.) This cannot be done with the above smoothing formulas since the points are not equally spaced in C. However, in this graph, the points at low concentration are bunched and those at high concentration spread out. We then use the ln C vs. r smoothed data from 0 to 2 fringe and the $\int Cd(r^2)$ vs. C data above two fringes.

Yphantis and co-workers (Roark and Yphantis[39] and personal communication) use weighting functions of the type described in the previous section.

Whatever the procedure used for smoothing the observed data, it cannot be used indiscriminately without the introduction of artifacts. On the other hand, to obtain accurate point-by point molecular weight averages, it is absolutely essential at the present time to use smoothing. To test a smoothing procedure, it is recommended that several sets of data be simulated and random numbers of mean zero and standard deviation of 5–10 μ be superimposed on the data to test the procedure. We do not require that the true value of the data be recovered but require that the predicted error bars at each point encompass the true values.

[39] D. E. Roark and D. A. Yphantis, *Ann. N.Y. Acad. Sci.* **164**, 245 (1969).

4. Calculation of Point-by-Point Molecular Weight Averages

Normally, in programs for large computers, one attempts to refine the preliminary estimate of the meniscus concentration at this point. These procedures are probably two tedious to perform on a desk-top computer, however.

The first calculation to be performed for the determination of molecular weights is to obtain values of $\int_{r_m}^{r} C d(r^2)$. In the low concentration region where the gradient is shallow, we use the quadratic formula,

$$\int_{r_i^2}^{r_{i+1}^2} C d(r^2) = \frac{1}{2} (C_{i+1} + C_i)(r_{i+1}^2 - r_i^2)$$
$$- \frac{1}{6} \frac{(r_{i+1}^2 - r_i^2)^3}{(r_{i+2}^2 - r_i^2)} \left(\frac{C_{i+2} - C_{i+1}}{r_{i+2}^2 - r_{i+1}^2} - \frac{C_{i+1} - C_i}{r_{i+1}^2 - r_i^2} \right) \quad (24)$$

We only use this integration procedure for displacements less than 0.75 fringe (200μ). Above this displacement we use the formula

$$\int_{r_i^2}^{r_{i+1}^2} C d(r^2) = \frac{(r_{i+1}^2 - r_i^2)}{\ln C_{i+1}/C_i} (C_{i+1} - C_i) \quad (25)$$

A simple trapezoidal integration (which is the first term of Eq. 24) cannot be used on the data since it introduces a bias. Equation 25 is derived from a trapezoidal integration of $\ln C$ vs. r^2 by assuming that this function is linear between successive points.[26] For the calculation techniques employed by Roark and Yphantis[39] these integrals are not required; however, we feel that there are advantages to be gained from their calculation.

We are now prepared to estimate the number-average molecular weight at the meniscus. Two methods will be discussed for this calculation; the first is theoretically exact, and the second is an approximation which has been used frequently in the literature. The equation for the number-average molecular weight (Eq. 8) can be rearranged to the form

$$\int_{r_m^2}^{r^2} C d(r^2) = - \frac{C_m}{AM_{n,m}} + \frac{C}{AM_{n,r}} \quad (26)$$

Successive differentiation of this formula with respect to C leads to an infinite series[27] which can be shown by finite-difference calculus that a graph of $\int_{r_m^2}^{r^2} C dr^2$ vs. C can be extrapolated to $-C_m/AM_{n,m}$ at $C = 0$. If the material is homogeneous, this is clearly true since the function is a straight line. We plot the values of the $\int C d(r^2)$ vs. C between 0 and 1 fringe, fit this with least-squares polynomials of order 1 through 3 and decide by the efficiency function, Eq. (20), the proper equation to choose. The intercept, a_0, is $-C_m/AM_{n,m}$. This same procedure can be performed graphically by making the extrapolation with a French curve or other drawing tool. The method used by Roark and Yphantis to obtain this constant utilizes weight- and z-average molecular weight estimates; but, remarkably, it works out to be the same procedure.

The second method used to obtain this constant is to assume that the number and weight-average molecular weights are identical at the meniscus, evaluate the weight-average molecular weight and use this number for $M_{n,m}$. The disadvantage of this procedure is that weight-average molecular weights are quite uncertain in the meniscus region so that unless the value of C_m is very small (less than $5\,\mu$), a bias is propagated throughout the M_n data. Nevertheless, some workers have had success using this method.[36]

At this point, all data below 0.35 fringe ($100\,\mu$) are discarded. It is essential to collect these data for the estimates of meniscus concentration and number average molecular weight at the meniscus, but attempts to calculate reliable molecular weight information in this region of the cell have proved futile. Worse, if the data are saved, a bad bias is sometimes propagated into data that would otherwise be reliable.

We now proceed to the estimate of the weight-average molecular weight of each point in the cell. We estimate $M_{w,r}$ by two methods, the slope of $\ln C$ vs. r^2 and by $\int Cd(r^2)$ vs. C. For $\ln C$ vs. r^2 it is simple to use an orthogonal least squares straight line of 5 or 7 points for the estimate of $M_{w,r}$ at the midpoints. For 5 points the equation is

$$2AM_{w,r_0} = \frac{0.1}{r_0\Delta r}\,[2(u_2 - u_{-2}) + u_1 - u_{-1}] \qquad (27)$$

where $u = \ln C$. Aune and Timasheff[36] found that 5 points was sufficient in their calculations. Noisy data require the use of more points in such equations. In our computer programs we use 11-point quadratic equations in this calculation. We also estimate $M_{w,r}$ from the reciprocal of the slope of the plot of $\int Cd(r^2)$ vs. C. As mentioned above, in this graph the data are widely spread at high concentrations so this estimate of $M_{w,r}$ deviates from that calculated from $\ln C$ vs. r^2. An average of the two methods is probably the best estimate of $M_{w,r}$. We then smooth the $M_{w,r}$ data obtained in this average as a function of C.

In order to obtain an estimate of the probable error in $M_{w,r}$, the equation

$$\frac{M_{w,r}^{\text{true}} - M_{w,r}^{\text{obs}}}{M_{w,r}^{\text{true}}} = \frac{\epsilon}{C_{\text{obs}}} \qquad (7)$$

may be used. In our computer programs, we estimate ϵ from the sum of the errors in the various meniscus concentration estimates and the root-mean-square movement of the points in the initial smoothing of C vs. r. For desk-top calculations the value of ϵ could be taken as approximately $2\,C_m$ as a first approximation. It is important that some trial calculations be performed on simulated data containing random errors as well as experiments on well-characterized proteins to establish the reliability of the calculations and estimates of the probable errors in the molecular weights.

We calculate number average molecular weights from the equation

$$M_{n,r} = \frac{A^{-1}C}{\int_{r_m}^{r} Cd(r^2) + C_m/AM_{n,m}} \tag{8}$$

since great care was taken in the calculation of the integrals and the estimate of $C_m/AM_{n,m}$. On the other hand, Roark and Yphantis[39] use the formula

$$M_{n,r} = \frac{A^{-1}C}{\int_{C_m}^{C} \dfrac{dC}{AM_{w,r}} + \dfrac{C_m}{AM_{n,m}}} \tag{28}$$

This method involves the estimation of M_w in the region which we have now discarded, namely between the meniscus and 0.35 fringe. If this method for calculation of $M_{n,r}$ is used, it is probably best to use an integration formula similar to Eq. (24) rather than a trapezoidal integration to calculate the values of the integrals in the denominator.

For the calculation method used in Eq. (8), the error is approximately[26]

$$\frac{M_{n,r}^{\text{true}} - M_{n,r}^{\text{obs}}}{M_{n,r}^{\text{true}}} \approx \frac{\epsilon(r^2 - r_m^2)AM_{n,r}^{\text{obs}}}{C_{\text{obs}}} - \frac{\epsilon}{C_{\text{obs}}} \tag{29}$$

This formula is only approximate since it assumes that all C_m terms are zero. The formula shows, however, that errors in concentration estimates are propagated throughout the solution and do not decrease at high C_{obs} as is true for the errors in $M_{w,r}$. Thus, for example, a bad baseline determination can result in a significant displacement of the $M_{n,r}$ vs. C curve while it has much less effect on the $M_{w,r}$ vs. C curve.[17]

The calculation of the z-average molecular weight is quite sensitive to a variety of errors, although insensitive to errors in concentration labeling. In the large computer, we compute the z-average from the equation of Wales[40]

$$M_{z,r} = M_{w,r}(1 + d\ln M_{w,r}/d\ln C) \tag{30}$$

[40] M. Wales, J. Phys. Chem. **52**, 235 (1948).

and use 21 points in the linear to cubic fit of $\ln M_{w,r}$ vs. $\ln C$. Simulation experiments and experience with real proteins has shown that the reliability of the calculation is only about ± 5–10%. Roark and Yphantis[39] use the equation

$$M_{z,r} = M_{w,r}{}^2 d(1/CM_{w,r})/d(1/C) \qquad (31)$$

in their computer program with $\pm 3\%$ as the reported accuracy.

Since the calculation of the z-average involves taking the second derivative of experimental data, it is not surprising that some difficulty is encountered in its calculation. Aune and Timasheff[36] were unable to reliably compute this molecular weight average by a desk-top computer. It may be possible to calculate this average using the technique of Nazarian[41] for the experimental data.

Some tedium is encountered in these calculations of the experimental C vs. r data. By use of points equally spaced in r, the calculations are considerably simplified. A simplification of the remaining calculations which we have occasionally discussed in this laboratory but never actually tried is to interpolate equal increments in C from the C vs. r data and then treat these as a second set of data for the molecular weight averages which are calculated as functions of C. Clearly, a great deal of investigation into the various methods of calculation of point-by-point molecular weight averages is required at the present time.

VI. Dissociating Proteins—General Considerations

A. Molecular Weights as a Function of Concentration

In a high speed sedimentation equilibrium experiment, the concentration in the meniscus region approaches zero. Since the rotor speed for such an experiment was calculated using the molecular weight of the species in greatest amount based on $M_{s/D}$, any protein in the mixture with molecular weight less than that used to determine the rotor speed will be preferentially observed in the meniscus region. The number-average and weight-average molecular weights reflect this fact by being lower in the meniscus region than throughout the rest of the cell. The degree to which the presence of the lighter components affect the molecular weight averages at each point in the cell is a function of its concentration. Decreased molecular weights near the meniscus will occur for all dissociating systems, whether chemical equilibrium exists or not. A homogeneous, nondissociating protein will not show such molecular weight dependence.

[41] G. M. Nazarian, *Anal. Chem.* **40**, 1766 (1968).

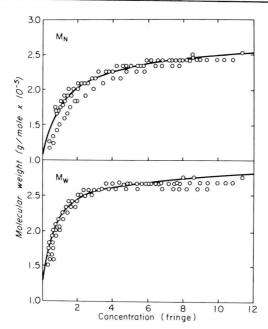

FIG. 10. *Escherichia coli* glutamic decarboxylase in 0.1 M sodium phosphate buffer pH 7.0. The experiment was run for 32 hours at 12,000 rpm. The upper graph is the distribution of M_n as a function of C while the lower graph is for M_w vs. C. The lines are predicted from $M_1 = 104 \times 10^3$ g/mole, $k_2 = 443 \times 10^3$ liters/mole, $k_3 = 1631 \times 14^4$ liters/mole.

For a protein in chemical equilibrium, the ultracentrifuge cell is characterized as a continuous sequence of thermodynamic phases, each in chemical equilibrium.[4] The various molecular weight averages calculated at each point in the cell are functions only of the concentration at which they are calculated as long as pressure effects are negligible.[39,42,43] The values will not vary with the distance from the center of rotation, initial concentration, or loss of material at the cell base due to packing. Therefore, for a system in chemical equilibrium, graphs of the point-by-point molecular weight averages vs. concentration should superimpose. In contrast, a heterogeneous system, not in chemical equilibrium will show a marked dependence of the molecular weight averages with concentration provided different initial loading conditions of the cell are compared.[44]

[42] E. A. Guggenheim, "Thermodynamics." North-Holland Publ., Amsterdam, 1957.

[43] E. T. Adams, Jr., *Proc. Nat. Acad. Sci. U.S.* **51**, 509 (1964).

[44] C. E. Harris, R. D. Kobes, D. C. Teller, and W. J. Rutter, *Biochemistry* **8**, 2442 (1969).

Figures 10 through 14 illustrate varieties of this behavior. Figure 10 depicts the molecular weight distribution for glutamic decarboxylase from *Escherichia coli* at high pH. The molecular weight distributions nicely superimpose in this figure indicating that all species are in chemical equilibrium. Figure 11 is the distribution observed for yeast

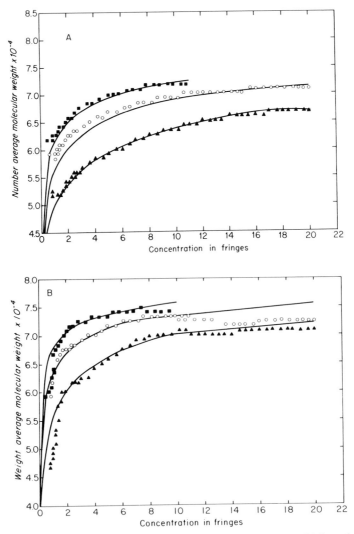

FIG. 11. Molecular weight distributions observed for yeast aldolase in 0.05 M glycylglycine, 0.05 M 2-mercaptoethanol, 0.05 M KCl pH 7.5. (A) Number-average data; (B) Weight-average data. ■ represents an initial loading concentration of 0.42 mg/ml; ○, 0.83 mg/ml; and ▲, 1.24 mg/ml.

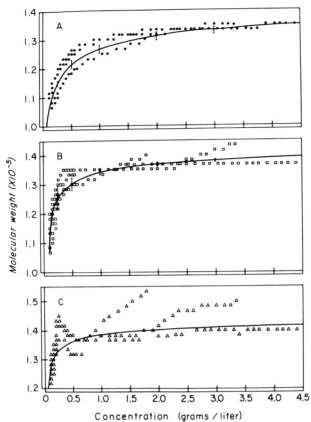

Fig. 12. Molecular weight distributions found for rabbit muscle D-glyceraldehyde-3-phosphate dehydrogenase at three initial loading concentrations in 0.1 M Tris·HCl, 0.001 M EDTA, 0.001 M 2-mercaptoethanol, pH 7.0 and 5°C. (A) Number-average data; (B) weight average data; (C) z-average data. The lines are drawn for $M_1 = 72 \times 10^3$ g/mole, $k_2 = 1.7 \times 10^6$ liters/mole. The error bars represent potential errors in molecular weights due to uncertainty in fringe labeling. The bars are the averages of the three channels and centered on the molecular weights predicted by k_2.

aldolase[44] at three separate loading concentrations. Apparently, chemical equilibrium does not occur in this case. Figures 12 and 13 illustrate other behavior.

In Fig. 12 (rabbit muscle D-glyceraldehyde-3-phosphate dehydrogenase), an anomaly is seen in the weight-average and z-average distributions.[45] The molecular weight averages at high concentrations devi-

[45] V. D. Hoagland, Jr. and D. C. Teller, *Biochemistry* 8, 594 (1969).

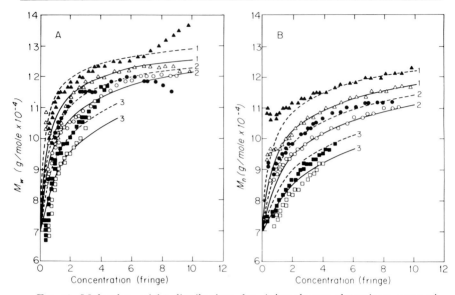

Fig. 13. Molecular weight distributions for *Acinetobacter* glutaminase–asparaginase. (A) M_w data; (B) M_n data. Data represented by open symbols and solid
lines was taken from a photograph at 26 hours at 16,000 rpm and 20°. The filled
symbols and dashed lines were data from a photograph at 43 hours at the same
speed and temperature. The buffer was 0.01 M sodium phosphate, 0.2 M NaCl,
3 mM NaN$_3$, pH 7.0. The numbers on the lines indicate the channel pairs of the
Yphantis [D. A. Yphantis, *Biochemistry* **3**, 297 (1964)] centerpiece; 1 is the centripetal channel pair. The lines were calculated from $M_1 = 69,000$ g/mole with k_2
values and initial concentrations as tabulated below:

Time (hours)	Symbol	Channel number	Initial concentration (mg/ml)	Apparent k_2 ($\times 10^{-3}$) liter/mole
26	- -▲- -	1	0.75	706
	- -●- -	2	0.50	257
	- -■- -	3	0.25	114
43	—△—	1	0.75	369
	—○—	2	0.50	178
	—□—	3	0.25	72

ate from the other data indicating the presence of heavy material which
is not in chemical equilibrium. The reason why such behavior is observed
in the data with dilute initial concentrations can be explained in the
following way. At high initial loading concentrations, material is packed
at the base of the cell and, owing to drive precession, fringes cannot be
resolved in this region. In the most dilute samples, all (or almost all)

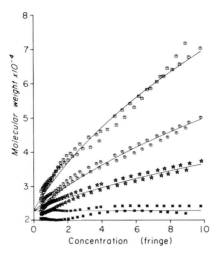

Fig. 14. Point-by-point molecular weight averages for fraction II of bovine procarboxypeptidase A. The solid lines are those predicted for indefinite self-association with $k/M_1 = 0.376$ 1/g and $M_1 = 22.87 \times 10^3$ g/mole. Squares are $M_{z,r}$ data, circles are $M_{w,r}$ data, stars are $M_{n,r}$ data, and crosses are $2M_{n,r} - M_{w,r}$ data. The buffer was $0.038\,M$ KH$_2$PO$_4$, $0.2\,M$ KCl pH 7.5. The temperature was 5°.

of the material can be observed, including the heavy material. The heavy material is obscured at high concentrations since only the lightest materials are seen by the high speed technique. Another anomaly is depicted in Fig. 12C for the z-average molecular weight dependence on concentration. In the region of 0.25–0.50 grams/liter (1–2 fringes), molecular weights are high and decrease at higher concentration. Several factors could cause this behavior: lack of sedimentation equilibrium or convection due to drive precession. At the time of these experiments we were using an overspeeding procedure to attempt to reduce the time to equilibrium, however, this procedure may have actually increased the time to equilibrium,[26] and we have since abandoned it. Convection is unlikely in this example since fringes could be resolved up to 18 fringes (4.5 grams/liter).

Finally, in Fig. 13 a complicated system can be seen. The enzyme is glutaminase–asparaginase from *Acinetobacter*.[7] The distributions of M_n and M_w are close but do not superimpose very well. Further, the highest molecular weights are observed for the highest initial loading concentrations. At the present time, we believe that this is due to a combination of two effects: first, a slight equilibrium dissociation of the enzyme together with a slow time-dependent irreversible dissociation of the tetramer to the dimeric form. The molecular weight distributions represented by the solid points and the dashed lines are from a photograph taken after 26 hours at 16,000 rpm. The distributions given by the open points and solid lines are from data collected at 43 hours at the same speed. At this later time there is clearly more dimeric material present than earlier. This may be seen more readily in the number-average molecular weight data than in the weight-average data.

The four examples above are the type of molecular weight distributions most commonly observed in high speed sedimentation equilibrium experiments. A radically different type of distribution is observed for other protein systems, as shown in Fig. 14 for fraction II of bovine procarboxypeptidase A.[46] In this figure, instead of the hyperbolic distribution of molecular weight averages, M_z is almost linear and so is M_w. We also see that $2M_n - M_w$ (lowest curve) is almost constant. Such molecular weight distributions do not occur as frequently as the hyperbolic ones but are surprisingly common.[47] The continued increase of molecular weight with concentration in Fig. 14 indicates that the interacting molecules are not asymptoting at a single molecular weight value but forming high polymers of the monomeric unit.

The remainder of this article is devoted to the consideration of methods by which molecular weight information such as depicted in Figs. 10–14 can be used in an analysis of (a) constituent molecular weights of proteins, (b) stoichiometry of associations whether reversible or not, and (c) methods of study of hybrid formation of proteins.

Prior to considering methods of calculations on these data it is worthwhile first to consider some factors such as terminology. We shall consider a "monomer" as the smallest species which participates in chemical equilibria or other dissociating reaction. Thus "monomer" corresponds to protomer of the Monod, Wyman, and Changeaux[48] terminology. For illustration consider a protein composed of 4 identical subunits as the active enzymatic species. From studies of the denatured enzyme, a subunit molecular weight is determined. If the protein reversibly dissociates in solution from tetramer to dimer, as far as the equilibrium is concerned, $2E \rightleftharpoons E_2$. We will consider this as a monomer–dimer reaction in spite of the fact that each "monomer" is composed of two subunits. For aggregation of the tetramer to octamer, the tetramer becomes the "monomer" and the octamer is the "dimer." If both reactions can be seen for the same protein, it becomes a "monomer–dimer–tetramer" system. Thus, the terminology is based, not on the physical com-

[46] D. C. Teller, *Biochemistry* **9**, 4201 (1970).

[47] G. Schmer, E. P. Kirby, D. C. Teller, and E. W. Davie, *J. Biol. Chem.* **247**, 2512 (1972).

[48] J. Monod, J. Wyman, and J.-P. Changeaux, *J. Mol. Biol.* **12**, 88 (1965).

position of the protein, but on the stoichiometry of the observed chemical equilibria.

As a final comment in this section, it should be emphasized that we are determining molecular weights of the protein system under investigation rather than studying the chemical equilibria per se. While the two studies are not incompatible, the enzymologist is usually interested in molecular weights rather than free energies of self-association. For this reason, it is worthwhile to consider a principle discussed later in this article: for a given protein we would like to focus our attention on the important information and not be side-tracked by minor contaminants of a protein preparation, either heavy or light. If a protein such as glycogen phosphorylase can exist in two states of known molecular weight ratio, the best way we can determine the molecular weights of the individual species is to mix the two proteins and treat it as a dissociating species of the type $r\mathrm{E} \rightleftharpoons \mathrm{E}_r$ where r is the ratio of molecular weights. In this way, dissociation of E, if slight, will not be observed nor will aggregation of E_r, if slight. Our attention will be focused on the relation between the two (the ratio, r) and hence, the calculated values of molecular parameters will reflect this attention. The minor associations and dissociations can then be detected by separate experiments on E and E_r alone. This approach can be shown to be valid from information theory and is presented in the final section. This realization was made by Seery et $al.$[49] in a qualitative way, but it required further work[44,45] to show its validity.

B. Thermodynamic Nonideality

This section provides a discussion of nonideal behavior of macromolecules which affects their equilibrium distribution in a gravitational field. It includes a discussion of the physical basis of thermodynamic nonideality as well as present methods of treating the problem.

For a system at sedimentation equilibrium, there is no change in any of the potentials.[4] For constant temperature and negligible compressibility of the components, the concentration distribution of the ith component is given by,

$$\frac{C_i M_i (1 - \bar{v}_i \rho_0) \omega^2 r dr}{RT} = dC_i + M_i C_i \sum_{k=1}^{n} B_{ik} dC_k + \sum_{k=1}^{n} \bar{v}_i C_k dC_i \quad (32)$$

This equation results from a thermodynamic derivation of the sedimen-

[49] V. L. Seery, E. H. Fischer, and D. C. Teller, $Biochemistry$ **6**, 3316 (1967).

tation equilibrium distribution.[4,50] For the derivation it is assumed that
the chemical potential is given by $\mu_i = \mu_i^0 + RT \ln \gamma_i C_i$, where γ_i is
the activity coefficient of the ith component, of concentration C_i, and μ_i
is its chemical potential. The units of C_i can be arbitrary, but the
activity coefficient is defined as unity at $C_i = 0$, which corresponds to a
pure solvent standard state for all components. The terms in B_{ik} arise
from a Taylor's series expansion of $\ln \gamma_i$ as a function of total concentration[50]

$$\ln \gamma_i = M_i \sum_k B_{ik} C_k + \mathrm{O}(C_i C_k)$$

The B_{ik} are called virial coefficients. Now, for protein, the $\bar{v}_i$ terms on
the right of Eq. (32) are negligible for protein concentrations less than
10–20 mg/ml. Most globular enzymes also have negligible values of
B_{ik} under conditions of enzymatic activity as long as the ionic strength
is in the range of 0.1–0.5 M.[51,52] The positive term in B_{ik} arises from two
causes: First, all molecules have excluded volume at least four times
their molecular volume.[51] This excluded volume means simply that two
molecules cannot occupy the same space. For neutral, globular proteins,
this space is quite small and hence the B_{ik} are generally small for such
systems. The second cause of nonideality is due to charges on the
proteins. When the proteins are charged, they repel each other and this
function depends on the distance between the particles. If the solution
is concentrated in macromolecules then the repulsion can be quite strong.
The effect of the repulsion is to increase the excluded volume of the
molecule. In general, the charge can be screened by use of high ionic
strength, however.

The term in B_{ik} as I have written it is a "catch-all" term for repulsive
forces. It is always positive since negative values of this term represent
attractive forces which are better described by equilibrium constants.
Thus, we consider that these first virial coefficients are always positive,
never negative.

The presence of the B_{ik} terms in Eq. (32) makes manipulation of
the equation quite complicated since the distributions of the various
species are coupled to one another via the B_{ik} terms. However, Adams
and Fujita[53] pointed out two simple and not unreasonable assumptions

[50] H. Fujita, "Mathematical Theory of Sedimentation Analysis." Academic Press,
New York, 1962.
[51] C. Tanford, "Physical Chemistry of Macromolecules." Wiley, New York, 1961.
[52] D. E. Roark and D. A. Yphantis, *Biochemistry* **10**, 3241 (1971).
[53] E. T. Adams, Jr. and H. Fujita, *in* "Ultracentrifugal Analysis in Theory and Experiment" (J. W. Williams, ed.). Academic Press, New York, 1963.

which allow considerable simplification of the equation, namely, represent the B_{ik} term of this equation as a weight average in the total concentration, C:

$$B_i = \Sigma B_{ik} C_k / \Sigma C_k$$

and further assume that all B_i for a series of oligomers are identical so that a single coefficient, B, suffices for all species. As van Holde $et\ al.$[54] emphasize, this assumption is arbitrary but reasonable. The effect on Eq. (32) is to make the nonideality term depend on the total concentration of like materials and to increase with increasing size of the molecule. With the further assumption of equal partial specific volumes, Eq. (32) becomes

$$C_i M_i A d(r^2) = dC_i + M_i C_i B dC \tag{33}$$

where the substitution $A = (1 - \bar{v}\rho_0)\omega^2/2RT$ has been made and the right terms in $\bar{v}$ of Eq. (32) is included in the B term. All methods of study of associating systems may be derived from this starting equation. It should be kept in mind, however, that some assumptions have been incorporated which may not be universally true.

Goldberg[4] has shown that all possible molecular weight averages of the type

$$M_k = \frac{\Sigma C_i M_i^k}{\Sigma C_i M_i^{k-1}}$$

are available from sedimentation equilibrium experiments, and Teller $et\ al.$,[26] among others, have published a few of them derived from Eq. (32). It is much simpler to use Eq. (33) for these averages, however.

Since the virial coefficients are unknown, the molecular weight averages calculated from the formulas of Section V are not the true values but "apparent" averages. These are related to the true values by the relations

$$M_{na} = M_n/(1 + \tfrac{1}{2}BM_nC) \tag{34a}$$

$$M_{wa} = M_w/(1 + BM_wC) \tag{34b}$$

$$M_{za} = M_z/(1 + BM_wC)^2 \tag{34c}$$

These three equations indicate that molecular weights are always positive. However, both Roark and Yphantis[39] and this laboratory have found negative molecular weight averages under conditions of high charge and low ionic strength ($0.1 M$ acetic acid). This indicates that higher order virial terms are occasionally important, and further, they must be negative terms (see Roark and Yphantis[52]).

---

[54] K. E. van Holde, G. P. Rosetti, and R. D. Dyson, $Ann.\ N.Y.\ Acad.\ Sci.$ **164**, 279 (1969).

For the most part, however, globular proteins can be considered thermodynamically ideal for ionic strengths in the range of 0.1–0.5 M and at concentrations less than 5 mg/ml. The effects of virial coefficients are less than the experimental errors of measurement of point-by-point molecular weight averages.

C. Molecular Weight Space—The "Two Species" Plot

It is possible to think of molecular weight averages (M_n, M_w, M_z, etc.) as existing in a two-dimensional space, only part of which is physically available. Such a concept was first utilized by Sophianopoulos and van Holde[55] in a study of lysozyme and later exploited by several other workers, notably Roark and Yphantis.[39] For this treatment we first consider the nature of the molecular weight averages on a weight per volume scale

$$M_n = \Sigma C_i / \Sigma C_i / M_i \tag{35a}$$

$$M_w = \Sigma C_i M_i / \Sigma C_i \tag{35b}$$

$$M_z = \Sigma C_i M_i^2 / \Sigma C_i M_i \tag{35c}$$

$$M_{z+1} = \Sigma C_i M_i^3 / \Sigma C_i M_i^2 \tag{35d}$$
etc.

If only two species are present with molecular weights M_1 and M_2 and the solution is thermodynamically ideal, then[26,39]

$$M_{k,r} = -M_1 M_2 \left(\frac{1}{M_{k-1,r}} \right) + M_1 + M_2 \tag{36}$$

where $M_{k,r}$ is a weight-average (M_w), z-average (M_z), or other average molecular weight at a point, and M_{k-1} is the next lower average at the same point, M_n for $k = w$, M_w for $k = z$, etc. M_1 and M_2 represent the molecular weights of two species present in the solution. For an ideal system, $M_k \geqslant M_{k-1}$ since molecular weights are never less than unity. If $M_k = M_{k-1}$, then a graph of Eq. (36) is given by the hyperbola shown in Fig. 15. This is the locus of ideal homogeneous molecular weights. As long as the material is ideal and experimental errors of data analysis do not exist, then no molecular weights could be found in the space below and left of the curve. If two species with molecular weight $M_1 = M_1$ and $M_2 = 2M_1$ were present in the solution, then the data points should ideally lie along the line connecting these two points in Fig. 15. On the other hand, if the molecular weights were M_1, $M_2 = 3M_1$, a separate line denoted by $3M_1$, M_1 would be observed. Now consider that $M_1 = 2M_1$ and $M_2 = 3M_1$. The data would lie along this line. For a monomer–

[55] A. J. Sophianopoulos and K. E. van Holde, *J. Biol. Chem.* **239**, 2516 (1964).

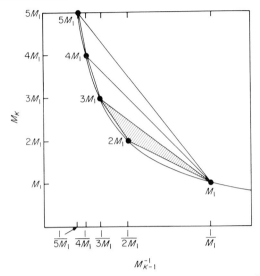

Fig. 15. Allowed space of molecular weight averages. The ordinate, M_k, represents the weight-, z-, or higher average molecular weight. The abscissa is the reciprocal value of the next lower average (number- or weight-average, respectively). The hyperbola is the locus of ideal, homogeneous molecular weight averages. The lines and shaded area are discussed in the text.

dimer–trimer system of M_1, $2M_1$, $3M_1$ the points would all fall in the shaded triangle. Departure from this stoichiometry toward higher polymers moves the points vertically on the graph. On the other hand, thermodynamic nonideality and concentration labeling errors cause points to move downward on the ordinate.

With our unknown protein system, we can draw the limiting hyperbola, but at this point are unable to specify the M_1, $2M_1$, etc., lines on the graph since, in general, these are unknown. We should like now to consider the practical aspects of the distribution of data on this graph.

In this laboratory we plot M_z vs. $1/M_w$, M_w vs. $1/M_n$ and the function, $2M_w - M_z$ vs. $2/M_n - 1/M_w$.[39] Let us first consider the effect of thermodynamic nonideality on these functions and then the errors due to concentration labeling in the original data.

As mentioned in the preceding subsection, thermodynamic nonideality depresses the values of the observed molecular weight averages. Since M_n, M_w, and M_z are affected differently by the B term of Eqs. (34), in various manipulations of these averages these terms in nonideality can either cancel or get larger. In the case of the two-species function, they become larger.

The two-species graphs emphasize the nonideality quite strongly,[39] causing departure from the "allowed space." It is for this reason that the function $2M_w - M_z$ vs. $2/M_n - 1/M_w$ is used. The reader can verify that this function eliminates all C terms, leaving only C^2 terms in B. This is one method of overcoming the nonideality factors.

A second type of problem in such a graph arises from errors in labeling the fringes with true concentrations. Again, these errors add together in the graph of M_k vs. M_{k-1} in a complicated way but, in any case, cause departures of the data below the limiting hyperbola of Fig. 15. These two effects can both be seen in the data of Fig. 16 taken from Fig. 10 for glutamic decarboxylase from $E.$ $coli$. The curvature toward the origin at high M_k is due to thermodynamic nonideality, while similar curvature at low M_k is due to errors in labeling of the meniscus concentration. The lines were drawn at a much later time.

Such a graph is independent of the chemical equilibria which may be occurring in the system, at least as long as only two species are present in the system. For example, yeast aldolase molecular weight data (Fig. 11) graphs nicely along the monomer–dimer line, and all data fol-

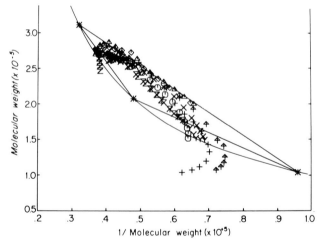

Fig. 16. Allowed space of molecular weight averages for $Escherichia$ $coli$ glutamic decarboxylase. Only each third point of data is graphed. The symbols used are tabulated below.

Channel No.	M_w vs. $1/M_n$	M_z vs. $1/M_w$	$2M_w - M_z$ vs. $2/M_n - 1/M_w$
1	○	△	+
2	×	◇	⋔
3	⊼	Z	Y

low this line in spite of the fact that this is a case of irreversible dissociation.[36]

In spite of the difficulties, such graphs are frequently useful to obtain an estimate of M_1, the molecular weight as well as the stoichiometry of the system. A ruler can be used to draw the limits of M_1 and stoichiometry of the system, using the straightest part of the curves as guidelines for M_1 and M_2.

This graph does not work well with only slightly associated or disassociated systems[26] since it is dominated by experimental error and thermodynamic nonideality in this case. For this reason, we routinely graph the function $2M_{n,r} - M_{w,r}$ as a function of C on the same graph with $M_{n,r}$ and $M_{w,r}$ vs. C. For slightly associated systems,[20] it is simple to extrapolate this function to $C = 0$ in order to obtain M_1. If the system is undergoing indefinite self-association, then $2M_{n,r} - M_{w,r} = M_1$, which makes it quite simple to recognize this mode of association. For monomer–dimer–tetramer and some cases of monomer–dimer–trimer interactions, $2M_{n,r} - M_{w,r}$ will be less than M_1 at high concentrations, thus indicating these types of stoichiometry.

If the monomer molecular weight is known, it is possible to use a more sensitive graphical method to obtain the stoichiometry of interaction[56] (also D. E. Roark and D. A. Yphantis, personal communication; K. E. van Holde, personal communication). Let $M_2 = n\,M_1$ in the two-species Eq. (36), where n is the stoichiometry. The two-species equation is then solved for n, which yields

$$n = \frac{M_{k-1}}{M_1}\frac{(M_k - M_1)}{(M_{k-1} - M_1)} \tag{37}$$

where M_1 is the molecular weight of the monomer. Clearly, for any monomer–n-mer system, the right side of this equation should be constant. For the abscissa, van Holde (personal communication) uses a quantity

$$\frac{M_w}{M_1} - 1 = (n - 1)w_2 = Q$$

where w_2 is the weight fraction of n-mer for a two-species system. For an attempt to eliminate nonideality, M_n, M_w, and M_z can all be used in the function

$$n = \frac{M_{y1}}{M_1}\frac{(2M_w - M_z - M_1)}{(M_{y1} - M_1)} \tag{38}$$

where $M_{y1} = (2/M_n - 1/M_w)^{-1}$.[39]

[56] H. G. Elias and R. Bareiss, *Chimia* **12**, 53 (1967).

In this laboratory we have found it preferable to use the concentration, C, as the abscissa in this graph, since van Holde's Q is affected by nonideality. Figure 17 shows such a graph for $E.$ $coli$ glutamic decarboxylase. In these graphs, $M_1 = 104 \times 10^3$ was assumed. In Fig. 17A, M_k was M_z and M_w. The values of n are seen to be below 3 in the range where the data are reliable (greater than 2 fringes). In Fig. 17B the value of n was calculated from Eq. (38). Here, the errors in molecular weights accumulate so that the data below 4 fringes are unreliable. On the basis of the data in this figure above 4 fringes, a monomer–trimer system is indicated.

The discussion in this section has been devoted to the general determination of preliminary ideas concerning the protein under investigation. In the following sections we consider specific stoichiometries for the simultaneous determination of molecular weight and stoichiometry, since generally M_1 is unknown or uncertain. The principle of the method is to assume a stoichiometry of interaction, find the best fit to that stoichiometry by variation of assumed values for M_1, and then evaluate

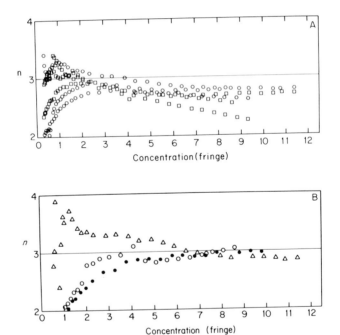

FIG. 17. Graphical attempt to determine the stoichiometry for $Escherichia$ $coli$ glutamic decarboxylase. $M_1 = 104 \times 10^3$ g/mole for the calculation. (A) M_k of Eq. (37) was M_z (circles) and M_w (squares). (B) Determination of the stoichiometry from the ideal Eq. (38).

(from the goodness of fit to the data) whether the correct stoichiometry was assumed.

VII. Determination of M_1 and k_2 for Monomer–Dimer Systems

For this method,[45] we utilize point-by-point equilibrium constants which are calculated from the expressions

$$\frac{2k_{2,n}}{M_1} = \frac{2M_{n,r}(M_{n,r} - M_1)}{C(2M_1 - M_{n,r})^2} \tag{39}$$

$$\frac{2k_{2,w}}{M_1} = \frac{M_1(M_{w,r} - M_1)}{C(2M_1 - M_{w,r})^2} \tag{40}$$

$$\frac{2k_{2,z}}{M_1} = \frac{(M_{z,r} - M_1)(3M_1 - M_{z,r})}{4C(M_1 - M_{z,r})^2} \tag{41}$$

where M_1 is the molecular weight of the smallest species present. C can have any units which are proportional to grams/liter; the k's can later be corrected to molar values from the proportionality factor. The n, w, or z subscript merely indicates that it was calculated from the corresponding molecular weight moment; ideally all $k_{2,i}$ would be identical. Equation (39) was derived by Nichol et al.,[57] and Eq. (40) was given by Rao and Kegeles.[58]

At each point we construct the average of the values of k_2 as,

$$\bar{k}_2 = \Sigma_i w_i k_{2i} / \Sigma_i w_i \tag{42}$$

where the summation in i is over the number-, weight-, and z-determined constants. The weights, w_i which we use are,

$$w_n = |M_{w,r} - \tilde{M}_{w,r}|^{-1} + |M_{z,r} - \tilde{M}_{z,r}|^{-1}$$

$$w_w = |M_{n,r} - \tilde{M}_{n,r}|^{-1} + |M_{z,r} - \tilde{M}_{z,r}|^{-1}$$

$$w_z = |M_{n,r} - \tilde{M}_{n,r}|^{-1} + |M_{w,r} - \tilde{M}_{w,r}|^{-1}$$

The tilde over the symbol indicates that the value of the molecular weight was calculated from $k_{2,i}$, C, and M_1. It does not seem to make any difference if the weights are taken as the reciprocal squares of the deviations; however, if one average is missing (such as z-average molecular weights), it may be desirable to use reciprocal squares as weighting factors. At each point, the reliability of the $\bar{k}_2$ values is estimated from

$$E(\bar{k}_2) = \left[\frac{\Sigma_i w_i k_{2i}^2}{\Sigma_i w_i} - \bar{k}_2^2 \right]^{1/2} \tag{43}$$

This function is the weighted root-mean-square uncertainty of the constant which has been determined at each point, j, in the cell.

[57] L. W. Nichol, J. L. Bethune, G. Kegeles, and E. L. Hess, Proteins 2, 309 (1964).
[58] M. S. N. Rao and G. Kegeles, J. Amer. Chem. Soc. 80, 5724 (1958).

In order to calculate the predicted values of the molecular weight averages at each point, it is necessary to solve a quadratic equation to obtain C_1, the concentration of monomer,

$$C_1 = \frac{-1 \pm (1 + 4K_aC)^{1/2}}{2K_a}$$

where K_a is proportional to $2k_2/M_1$, depending on the units of C chosen. The sign to choose is the positive one. Also, the calculation is performed only for positive values of K_a.

In order to characterize the behavior of all concentrations by a single parameter, another weighted average k_2 is constructed,

$$\hat{k}_2 = \Sigma_j f_j \bar{k}_{2,j} / \Sigma_j f_j \tag{44}$$

where the sum in j is carried over all data points greater than one fringe. The weighting factors, f_j, are the values of $1/E(\bar{k}_{2,j})$ determined by inversion of Eq. (43).

To obtain an estimate of the reliability of $\hat{k}_2$, the root-mean-square uncertainty can be calculated as

$$E(\hat{k}_2) = \left[\frac{\Sigma_j f_j \bar{k}_{2,j}^2}{\Sigma f_j} - \hat{k}_2^2 \right]^{1/2} \tag{45}$$

In practice, this is an overestimate of the uncertainty of $\hat{k}_2$, so we rarely actually use it.

In order to obtain M_1 for an unknown system, we assume a reasonable value for M_1, calculate $\hat{k}_2$ in the procedure described by Eqs. 39–44. From the value of $\hat{k}_2$, M_1, and C, $\tilde{M}_{n,r}$, and $\tilde{M}_{w,r}$ values are predicted both for the data used in the computation (greater than 1 fringe) and for all the data. The average deviations of the molecular weights are then calculated from,

$$E(M_n) = \sum_{j=1}^{N} |M_{n,j} - \tilde{M}_{n,j}|/N \tag{46}$$

where the $\tilde{M}_{n,j}$ are calculated from $\hat{k}_2$ and j is the point number. A similar formula is used for the calculation of $E(M_w)$. The value of M_1 is then incremented and the entire calculation repeated. Finally, a graph of $E(M_n)$ and $E(M_w)$ vs. M_1 (assumed) is constructed. If the system is truly that of monomer–dimer, then a clear, low minimum occurs in this parabolic function as M_1 (assumed) passes through its true value.

This calculation has certain characteristics that should be discussed. First, if M_1 (assumed) is too small then no positive values of $\bar{k}_2$ or $\hat{k}_2$ can be found. As M_1 (assumed) approaches M_1 (true) from the low side, the $E(M_n)$ and $E(M_w)$ functions decrease quite steeply to a minimum.

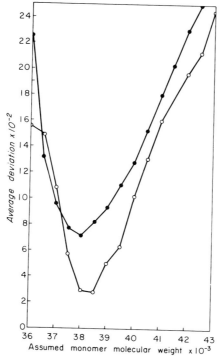

Fig. 18. Determination of the monomer molecular weight of yeast aldolase from one channel of data. M_1 (assumed) was varied from 36×10^3 to 43×10^3 and equilibrium constants, k_2, calculated for each M_1 value. From k_2 and M_1 the values of the point-by-point M_n and M_w were predicted. The figure represents the average deviations of M_n and M_w from the observed data. Open circles are average deviations of M_n data, and filled circles are from M_w data.

The rise from this minimum as M_1 (assumed) increases may not be as steep and may have local maxima, but the general trend is upward. As M_1 (assumed) becomes much larger than its true value, the values of $E(M_n)$ and $E(M_w)$ become quite erratic as individual regions of the M vs. C curves are well fit by the model.

Three examples of the experimental determination of M_1 by this method may be chosen for illustration. Figure 18 was calculated for one channel of data on yeast aldolase.[14] It will be recalled that this protein did not behave as though chemical equilibrium were present (Fig. 11). However, the data from single channels showed distinct minima as shown in this graph. The data depicted in Fig. 19 taken from *E. coli* glutamate decarboxylase yields no minimum in the function over a broad range of M_1 (assumed). In this case, the stoichiometry was assumed as 2

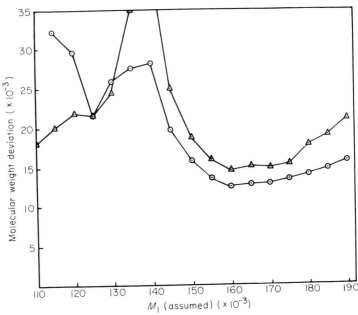

FIG. 19. Search for M_1 for *Escherichia coli* glutamic decarboxylase. M_1 was varied in increments of 5000 g/mole over the range shown and the best monomer–dimer fit to the data calculated for each M_1. No minimum in molecular weight deviations could be found for these data when a monomer–dimer system was assumed. ⊙, M_w; ▲, M_n.

while actually trimers are present in this solution. Because molecular weights were superimposable, all the data from 3 channels was treated at the same time. From this example, we see that when the stoichiometry assumed is incorrect, no minimum will be found. This is the criterion for the correctness of the stoichiometry. The third example is from a calculation on glutaminase from *Acinetobacter*.[7]

The data of Fig. 20 are taken from calculations of the 26-hour data of Fig. 13. This figure shows that occasionally smooth parabolic behavior is found for the molecular weight deviations (top figures, channel 1) but often the curves are ragged and bumpy as in channel 3 data of this figure. The deviations of the number-average molecular weight of channel 2 for all the data show two minima, the one occurring at $M_1 = 81 \times 10^3$ g/mole is lower than the one at $M_1 = 68 \times 10^3$ g/mole. In this case, the latter minimum was taken to obtain the average M_1 of all experiments as discussed in the next paragraph, since most other data indicated the value was near 70×10^3 g/mole. The calculations of Fig. 20 together with data from three other sedimentation equilibrium experiments gave $M_1 =$

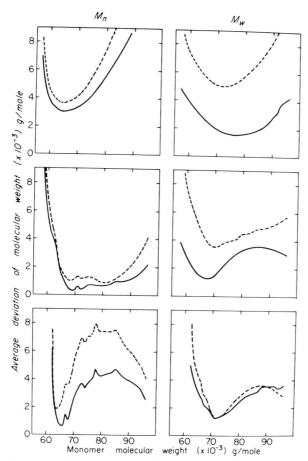

Fig. 20. Search for M_1 from *Acinetobacter* glutaminase–asparaginase data (26-hour data of Fig. 13) at increments of 1000 g/mole. This figure illustrates some of the shapes seen for the deviation function as M_1 is varied. Dashed lines represent deviations based on all the data. Solid lines are the deviations of data from concentrations greater than one fringe. The left figures are number average molecular weight deviations and the right panels are weight average molecular weight deviations.

$69 \pm 3 \times 10^3$ g/mole for this enzyme.[7] The ragged behavior of the deviation function depicted in Fig. 20 is presumably due to errors in the $M_{n,r}$, $M_{w,r}$ and $M_{z,r}$ data.

In order to summarize the values of M_1 as a single number, we use the values of M_1 at the minima from individual experiments as a weighted mean,

$$M_1 = \Sigma M_{1,k}/E(M_x)_k/\Sigma 1/E(M_x)_k$$

where k denotes the kth experiment, $E(M_x)_k$ represents the deviation of M_n or M_w for this experiment (Eq. 46). The root-mean-square uncertainty is calculated from a formula of the type given by Eq. (43).

In general, it will be found that the deviation of M_1 is surprisingly small from one experiment to another. Clearly, a relatively large computer is required for these calculations. If such a computer is not available, it might be possible to choose 10 or 15 $M_{n,r}$, $M_{w,r}$, and $M_{z,r}$ values and perform the calculation on a desk top computer.

A slight amount of thermodynamic nonideality can be tolerated in the calculation. The effect of nonideality is to cause M_1 (minimum) to be less for M_w deviations than for M_n deviations. As a first approximation to correct for such slight nonideality, one might use the function

$$M_1 \approx 2M_1[E(M_n = \text{min})] - M_1[E(M_w = \text{min})]$$

Although generally, the range of values of M_1 are overlapping. Nonideality enters into $k_{2,n}$, $k_{2,w}$, and $k_{2,z}$ in different ways so that the method is probably unreliable under conditions of strong nonideality.

We have attempted to use such a "search" procedure utilizing the ideal moments of Roark and Yphantis[39]; however, these calculations have routinely failed to date. The reason for failure seems to be due to the fact that a large amount of information about dimerization is canceled in formulating the ideal moments.

Clearly, the method can also be used when M_1 is known. Under these conditions, the amount of calculation is considerably reduced but is still larger than for other methods to be described below; however, it does yield constants that fit the data better. Aune and Timasheff[36] used a quite similar method in their study of the dimerization of α-chymotrypsin at low pH.

A simpler, but related, calculation method has recently been designed for use with the PDP-12 computer and may work well with small desk-top computers. In this case we use M_1 as input and guess the value of k_2. The program simply calculates the root-mean-square deviation of the predicted and observed values for all the data from a given molecular weight average (such as $M_{w,r}$ data). At each M_1 (assumed), the value of k_2 is guessed until the minimum deviation is found. A new M_1 is assumed, and a series of k_2 values is guessed for each M_1 (assumed). The M_1 (assumed) and k_2 corresponding to it which shows the minimum deviation is chosen. This method requires a much smaller program and much less molecular weight data than the previous method. The guesswork to find k_2 which has a minimum deviation of the data is quite rapid. There are several disadvantages over the preceding method, however: First, the data are unequally spaced as a function of C so that the data at low concentrations (which are the most unreliable) dominate the *rms* deviations. This disadvantage could be overcome, however, by choosing molecular weight points approximately equally spaced in concentration or by the design of a suitable weighting function. Second, strongly deviant points can dominate the *rms* values leading to "best-fitting" values for parameters which do not describe the data. Again, such points could be discarded prior to the calculation. Third, all the data are not calculated at the same time so that the constraint that $M_{n,r}$, $M_{w,r}$, and $M_{z,r}$ all correspond to the monomer–dimer model is lacking. There is a strong temptation to use the data from only a single set of molecular weight averages that can easily lead to incorrect stoichiometry assumptions.

VIII. Determination of M_1, k_2, and k_3 for Monomer–Dimer–Trimer Systems

As in the case of the monomer–dimer systems, for this calculation we assume that the system is that of monomer–dimer–trimer. Presumably the hypothesis that the system is that of monomer–dimer only has been rejected at this stage due to data such as those of Fig. 19. For this calculation, it is essential that all the averages, $M_{n,r}$, $M_{w,r}$, and $M_{z,r}$ are available since the averages will be used two at a time with the prediction of the remaining one serving as a weighting factor. The equations for calculation of the weight fractions of monomer, dimer, and trimer are given by equations:

$$f_{nw,1} = (M_wM_n + 6M_1^2 - 5M_1M_n)/(2M_1M_n)$$
$$f_{nw,2} = (8M_nM_1 - 2M_wM_n - 6M_1^2)/(M_nM_1)$$
$$f_{nw,3} = (3M_nM_w + 6M_1^2 - 9M_1M_n)/(2M_1M_n) \qquad (47)$$

$$f_{wz,1} = (M_wM_z - 5M_wM_1 + 6M_1^2)/2M_1^2$$
$$f_{wz,2} = (4M_wM_1 - M_wM_z - 3M_1^2)/M_1^2$$
$$f_{wz,3} = (M_wM_z - 3M_wM_1 + 2M_1^2)/2M_1^2 \qquad (48)$$

$$f_{nz,1} = (-19 + 5M_z/M_1 + 30M_1/M_n - 6M_z/M_n)/2(6 - M_z/M_1)$$
$$f_{nz,2} = 2(13 - 4M_z/M_1 - 12M_1/M_n + 3M_z/M_n)/(6 - M_z/M_1)$$
$$f_{nz,3} = 3(-7 + 3M_z/M_1 + 6M_1/M_n - 2M_z/M_n)/2(6 - M_z/M_1) \qquad (49)$$

In these equations, f is the weight fraction of a species and the subscripts nw, wz, and nz refer to the molecular weight averages from which they were calculated. The three weighting factors for the average k_2 and k_3 values are calculated from the reciprocal deviation of the remaining molecular weight. k_2 and k_3 are calculated from the formulas

$$k_{2,x} = M_1 f_{x,2}/2C f_{x,1}^2 \qquad (50)$$

and

$$k_{3,x} = 2M_1 f_{x,3}/3C f_{x,1} f_{x,2} \qquad (51)$$

where the subscript x denotes the molecular weight averages from which these were calculated. If either $k_{2,x}$ or $k_{3,x}$ is less than or equal to zero, the k values are discarded from further calculation. At each point an average k_2 and k_3 are calculated.

In order to compute the weighting factors for the averages, it is necessary to solve the cubic equation,

$$\frac{3k_2 k_3}{M_1^2} C_1^3 + \frac{2k_2}{M_1} C_1^2 + C_1 - C = 0 \qquad (52)$$

for C_1. There is always at least one real root for such an equation. In the computer program we use either of two methods, explicit solution or a Newton–Raphson iteration.[38]

The values of C_2 and C_3 (dimer and trimer) are obtained from

$$C_2 = \frac{2k_2}{M_1} C_1^2 \quad \text{and} \quad C_3 = \frac{3k_2 k_3}{M_1} C_1^3$$

regardless of the method used to calculate C_1. From C_1, C_2, and C_3, all the molecular weight averages can be calculated to use as weighting factors and error estimates of $\bar{k}_2$, $E(\bar{k}_2)$, $\bar{k}_3$, and $E(\bar{k}_3)$ calculated in the manner denoted by Eq. (43). For each M_1 (assumed), the values of $\hat{k}_2$ and $\hat{k}_3$ are estimated by the same method given in Section VII. We then try to predict the entire molecular weight average vs. C curve from M_1 (assumed), $\hat{k}_2$, and k_3. The minimum deviation of the observed values from predicted values is taken as the correct M_1, $\hat{k}_2$, and $\hat{k}_3$, provided that a clear minimum can be found.

The behavior of the deviation graph, unlike the monomer–dimer case, is somewhat strange for perfect (hypothetical) data. Figure 21A shows this strange behavior. Slightly above the true value of M_1, a sharp spike occurs in the magnitude of the deviations of both M_n and M_w data; further, the deviations are smooth for M_1 less than the true value but quite ragged above this value. We would expect to see this behavior reflected to some degree in real data and, indeed, it is as shown in Fig. 21B for glutamic decarboxylase. A rather sharp rise in the deviations of both M_n and M_w occurs between M_1 (assumed) of 104×10^3 and 105×10^3 g/mole; although the minimum deviation of M_w occurs at $M_1 = 102 \times 10^3$ g/mole. It is presumed that the shift in the minimum value is due to thermodynamic nonideality—M_w is more affected by this parameter than is M_n. In any case, it is clear from this graph and the cal-

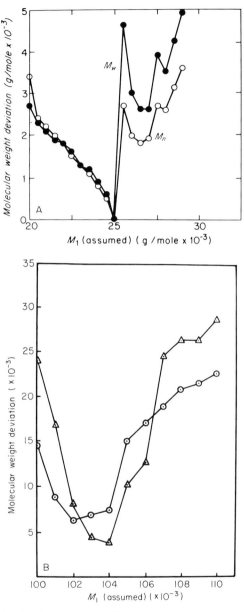

Fig. 21. Search for M_1 in monomer–dimer–trimer system. (A) Test of the procedure with hypothetical data. Filled circles, M_w data; open circles, M_n data. (B) Calculations on the data of *Escherichia coli* glutamic decarboxylase. Circles are M_w data; triangles are M_n data. At $M_1 = 104 \times 10^3$ g/mole, $k_2 = 443 \times 10^3$ liters/ mole and $k_3 = 1631 \times 10^3$ liters/mole.

culations for glutamic decarboxylase that the data are nicely described by a monomer–dimer–trimer system with M_1 approximately 104×10^3 (within 2%) and $k_2 = 44 \times 10^4$ liters/mole and $k_3 = 163 \times 10^4$ liters/ mole. Whether glutamic decarboxylase is a monomer–dimer–trimer system or a monomer–trimer system will be discussed at a later point in this article.

Whether this method of calculation of molecular weights and equilibrium constants will be generally applicable for other proteins is obscure. The molecular weight data from this experiment were exceptionally good, as may be seen in Fig. 10. Further calculations to account for thermodynamic nonideality in this system have been unsuccessful, and it is not clear where the problems lie in these latter calculations. Also, it is not certain that the weighting factors used in the averaging to find the $\hat{k}_i$ are the best ones. The use of the uncertainties in prediction of other averages may not be good weighting functions at all. They work well for the monomer–dimer systems but may not be efficient for multiple parameters. Further work is required on this method.

IX. Indefinite Self-Association

This chemical mechanism corresponds to a polymerization reaction with equal energy for the addition of each monomeric species; that is,

$$2 \, P_1 \rightleftharpoons P_2 \qquad k = (P_2)/(P_1)^2$$
$$P_2 + P_1 \rightleftharpoons P_3 \qquad k = (P_3)/(P_2)(P_1)$$
$$P_3 + P_1 \rightleftharpoons P_4 \qquad k = (P_4)/(P_3)(P_1)$$
$$\cdots \qquad\qquad \cdots$$
$$P_{j-1} + P_1 \rightleftharpoons P_j \qquad k = (P_j)/(P_{j-1})(P_1)$$
$$\cdots$$

This type of chemical equilibrium system is surprisingly common for proteins, and, further, the relations of the molecular weight averages in the chemical equilibrium system are the same if the protein is randomly aggregated or degraded in an irreversible sense. For the chemical equilibrium case the mechanism has been termed "isodesmic" by van Holde and co-workers.[54] Figure 14 is an example of molecular weight distributions corresponding to this mechanism. A great many papers occur in the literature which discuss this type of association.[54,56,59-62] The quality of random addition of monomers makes all the equations de-

[59] K. E. van Holde and G. P. Rosetti, *Biochemistry* **6**, 2189 (1967).
[60] C. E. Smith and M. A. Lauffer, *Biochemistry* **6**, 2457 (1967).
[61] E. T. Adams, Jr. and M. S. Lewis, *Biochemistry* **7**, 1044 (1968).
[62] R. C. Deonier and J. W. Williams, *Biochemistry* **9**, 4260 (1970).

rived by Flory[63] for condensation polymerization or random degradation of a polymer[51] applicable. Rather than a complete discussion of this mechanism, we will confine this discussion to the ideal case, and then consider the effects of nonideality and chain termination.

This mechanism requires two sites on each monomer unit: a donor and an acceptor. Depending on the dihedral angle between these two sites relative to the center of mass of the monomer, a variety of geometries such as linear strings or helices may occur. The outstanding characteristic in the recognition of this mechanism in sedimentation equilibrium is that $2M_n - M_w$ is a constant (M_1). In sedimentation velocity experiments, the leading edge of the schlieren boundary is hypersharp. If we let $K = k/M_1$ then the concentration distribution becomes

$$C = C_1/(1 - KC_1)^2$$

which has been derived by several authors mentioned above. Starting from this equation, many linear relations that involve combinations of M_n, M_w, M_z, and C can be derived.

The relations that we have found useful are

$$M_n = M_1 + KM_1^2C/M_n \tag{53}$$

$$M_w^2 = M_1^2 + 4KM_1^2C \tag{54}$$

$$2M_n - M_w = M_1 \tag{55}$$

$$M_w = M_1 + 2KM_1^2C/M_n \tag{56a}$$

and the rearranged form,

$$\frac{M_wM_n}{C} = 2KM_1^2 + M_1\frac{M_n}{C} \tag{56b}$$

$$\frac{M_zM_w}{C} = 2KM_1^2 + M_w^2/C \tag{57}$$

$$M_zM_w = M_1^2 + 6KM_1^2C \tag{58}$$

$$3M_w - 2M_z = M_1^2/M_w \tag{59}$$

As Howard Cossell might say, such a plethora of relations may lead to a bit of confusion since each of these relations emphasizes different aspects of the data. Owing to the various errors in molecular weight averages, somewhat different values of M_1 and K will be found by each of these equations. In the two-species graph (Fig. 22) the values of M_z vs. M_w and M_w vs. M_n appear to parallel the limiting hyperbola, but $2M_w - M_z$ vs. $2/M_n - 1/M_w$ data cluster about the value of M_1. Usually,

[63] P. J. Flory, "Principles of Polymer Chemistry." Cornell Univ. Press, Ithaca, New York, 1953.

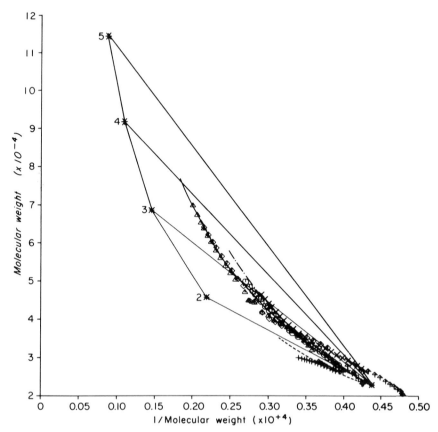

FIG. 22. Two-species graph of the data for fraction II of procarboxypeptidase A. The points labeled by the numbers represent dimer, trimer, etc., of 22.87×10^3 g/mole. The solid curved line is the predicted curve of $M_{z,r}$ vs. $M_{w,r}^{-1}$. The values predicted for $M_{w,r}^{-1}$ vs. $M_{n,r}$ are given by the broken line ($-\cdot-\cdot-$). The equilibrium constant used for the calculation of these curves was 0.376 liter/g. Only data above one fringe displacement are plotted in the figure. The symbols used are tabulated below.

Channel No.	$M_{z,r}$ vs. $1/M_{w,r}$	$M_{w,r}$ vs. $1/M_{n,r}$	$2M_{w,r} - M_{z,r}$ vs. $2/M_{n,r} - 1/M_{w,r}$
1	△	○	+
2	◇	×	⟁

the linear relations will suffice to establish M_1 and K reasonably well provided some assurance of thermodynamic ideality can be established. If the solution is nonideal, several complications arise.

The most serious effect of thermodynamic nonideality is that it becomes difficult (or impossible) to distinguish between nonideality and chain termination.[64] Both effects cause $2M_n - M_w$ to show upward parabolic behavior. Nonideality can be expressed by the equation,

$$2M_{na} - M_{wa} = M_1 + \tfrac{1}{2}BM_1^2(4K - BM_1)C^2$$
$$+ BM_1^2[K^2 - 7BM_1K/2 + 241(BM_1)^2/256]C^3 + \cdots \quad (60)$$

where the subscript a denotes apparent molecular weight averages (that is, those which are observed). The comparable equation for chain termination is rather complicated but produces results in $2M_n - M_w$ quite similar to those seen for nonideality. There are two reasons why chain termination might occur for enzymes: First, either the donor or acceptor sites of a large fraction of monomers may be "crippled"; second, ring formation could occur. The occurrence of rings would probably cause the system to appear as monomer-n-mer since the addition of the last unit to the string of monomers has approximate energy of $-2RT \ln k$ and a large entropy term; however, if the rings are large, this may not be true. Proof of the presence or absence of crippled association sites would be a more difficult experimental problem. For proteins it would appear worthwhile to attempt to block the association by shift of pH or other parameter to show the presence or absence of nonideality.

Van Holde and Rosetti[59] suggested a method for accounting for nonideality in order to determine equilibrium constants for this mechanism of association. The equilibrium constant may be determined by the following procedure: Since $2M_n - M_w = M_1$ is the limiting value (at C=0) of Eq. (60), M_1 can be found. Further, by the assumption that $BM_i = iBM_1$, the equilibrium constant should be a constant. Equation (34b) may be solved for M_w as

$$M_w = M_{wa}/(1 - BM_{wa}C)$$

If this substitution is placed in Eq. (54), the result is

$$\frac{M_{wa}^2}{(1 - BM_{wa}C)^2} = M_1^2 + 4KM_1^2C$$

With the substitution that $R_{wa} = M_{wa}/M_1$, we obtain:

$$K = \left[\frac{R_{wa}^2}{(1 - BM_1R_{wa}C)^2} - 1 \right]/4C \quad (61)$$

In order to calculate K, one merely guesses the value of B until K is a constant when graphed (or computed) as a function of concentration. The value of B which gives rise to zero slope when K is plotted vs. C is taken as the best value.

---

[64] D. C. Teller, *Biochemistry* **9**, 4201 (1970).

As a check on the value of K and B, Eqs. (53) and (34a) can be combined for apparent number average molecular weights

$$K = \frac{R_{na}^2 - R_{na}(1 - \frac{1}{2}BM_1R_{na}C)}{C(1 - \frac{1}{2}BR_{na}M_1C)^2} \qquad (62)$$

where $R_{na} = M_{na}/M_1$. Alternatively, the values of B and K may be iterated from this equation and used to predict M_{wa}.

Of course, it is possible to devise several other methods of calculation of K and B from the linear relations of Eqs. (53)–(57).

X. Analysis of the Concentration Distance Distribution

In this section we depart from the attempt to compute both stoichiometry and molecular weight of the monomeric species and assume that M_1 is known by experiments with blocked self-association. This method is quite powerful for the estimation of equilibrium constants in both self-associating and hybridizing systems.

Equation (33) may be directly integrated from some point C_p, r_p to predict the concentration at any point in the cell,

$$C_i = C_{pi}e^{AM \cdot (r^2 - r_p^2)}e^{-M_iB(C-C_p)} \qquad (63)$$

Where C_i is the concentration of the ith species of molecular weight M_i, $A = (1 - \bar{v}\rho)\omega^2/2RT$, and C is the total concentration. Neglect of the nonideal term in B leads to a much simpler equation describing the concentration distribution of a solute at sedimentation equilibrium which is amenable to simple curve fitting. If we now define the quantity, Γ as,

$$\Gamma = e^{AM_1(r^2 - r_p^2)} \qquad (64)$$

then, provided that M_1 is known, all quantities in Γ are known since r_p is a totally arbitrary position. By choosing r_p^2 as $(r_b^2 + r_m^2)/2$ the values of Γ can be kept small. Thus, we can treat Γ as a set of completely known variables at each point, j, in the cell. The observed concentration is the sum of the concentrations C_i provided that all have the same refractive index increment, so

$$C = \sum_{i=1}^{n} C_i = \sum_{i=1}^{n} \Gamma^i C_{pi} \qquad (65)$$

where we have assumed that $M_i = iM_1$ in this equation; that is, only polymers of the monomer are present. This equation is a simple polynomial in Γ at each point, j, in the cell:

$$C_j = C_{p1}\Gamma_j + C_{p2}\Gamma_j^2 + C_{p3}\Gamma_j^3 + \cdots + C_{pn}\Gamma_j^n \qquad (66)$$

and as such is amenable to least-squares treatment. If we let C_j be the predicted value of the concentration at the jth point from the parameters C_{pi}, the least-squares treatment of Section V gives,

$$\Sigma_j C_j \Gamma_j = C_{p1} \Sigma \Gamma_j^2 + C_{p2} \Sigma \Gamma_j^3 + \cdots + C_{pn} \Sigma_j \Gamma_j^{n+1}$$
$$\Sigma C_j \Gamma_j^2 = C_{p1} \Sigma \Gamma_j^3 + C_{p2} \Sigma \Gamma_j^4 + \cdots + C_{pn} \Sigma \Gamma_j^{n+2}$$
$$\cdots \qquad \cdots \qquad \cdots$$
$$\Sigma C_j \Gamma_j^n = C_{p1} \Sigma \Gamma_j^{n+1} + C_{p2} \Sigma \Gamma_j^{n+2} + \cdots + C_{pn} \Sigma \Gamma_j^{2n} \qquad (67)$$

Due to the large values that $\Sigma \Gamma_j$ can obtain even when $r_p^2 = (r_b^2 + r_m^2)/2$ it is useful to divide C_j by Γ_j producing the equation:

$$C_j/\Gamma_j = C_{p1} + C_{p2} \Gamma_j + C_{p3} \Gamma_j^2 + \cdots + C_{pn} \Gamma_j^{n-1} \qquad (68)$$

in which the value of the deviations C_j/Γ_j are minimized in the least squares procedure. The matrix of this modified equation is exactly that of Eq. (19) with $C_{pi} = a_{i-1}$ and $x_j = \Gamma_j$. In order to determine the stoichiometry of the self-association it is necessary that the calculated values of the C_{pi} be positive since negative values are nonphysical. In order to evaluate the C_{pi}, it is desirable to perform a mass conservation type of average to see whether the values are reasonable.

Mass conservation simply is a type of average over the cell volume; i.e.,

$$\int_{r_m}^{r_b} C dV \Big/ \int_{r_m}^{r_b} dV$$

$dV = \phi a r dr$ for sectorial cells where ϕ is the sector angle, a is the cell thickness. We then may make the statement that initial mass = final mass. If we denote the initial mass of component i by C_{0i}, then

$$C_{0i} \int_{r_m}^{r_b} \phi a r dr = \int_{r_m}^{r_b} C_i \phi a r dr$$

or

$$C_{0i} = \int_{r_m}^{r_b} C_i d(r^2)/(r_1^2 - r_m^2) \qquad (69)$$

since ϕ and a are constants that cancel. Now, if the system is not in chemical equilibrium and r_m and r_b are the true values, then the values of C_{0i} are the same values which actually exist in the remainder of the solution which is sitting in a test tube in the refrigerator. However, suppose that C_b at r_b cannot be seen owing to packing of material at the base of the cell. We now use r_N, the last readable r value, as the value of r_b and perform the integration to determine C_{0i}. In this case, C_{0i} is the average initial concentration of species i that we can see when the system is at sedimentation equilibrium. Similarly, if the system is in chemical equilibrium, such a mass average will not reflect the contents of the test tube but will only be an average concentration of species i at the condition of sedimentation equilibrium and rotor speed chosen for the experiment, but will not reflect the chemical equilibrium.[43] Again, C_{0i} is a hypothetical initial mass, but nonetheless a useful parameter. If we put the value of C_i from Eq. (64) into the numerator of Eq. (69) and perform the integration, we obtain

$$C_{0i} = \frac{C_{pi}[e^{iAM_1(r_b{}^2 - r_p{}^2)} - e^{iAM_1(r_m{}^2 - r_p{}^2)}]}{iAM_1(r_b{}^2 - r_m{}^2)} \tag{70}$$

While I have used the idea of mass conservation to develop this equation, it is unfair to call C_{0i} the initial mass of the species i. It is much better to call it the mass average concentration and always keep in mind that it may be a hypothetical quantity due either to an invalid use of r_b or the existence of chemical equilibrium.

We then take the values of C_{pi} determined in a least squares fashion and calculate their mass average concentrations as in Eq. (70). The criterion for choice of the best stoichiometry is the maximum number of species that have positive values of C_{0i}. However, there are several qualifications that apply to this criterion and are best explained by reference to calculations on hypothetical data.

Consider the model of a monomer–dimer–trimer–hexamer system given in Table III. At values of monomer–dimer–trimer . . . , n-mer below the true value, large negative deviations of the C_{0i} are observed. However, when monomer through hexamers were assumed to be present, C_{04} and C_{05} were only slightly negative while species 1, 2, 3, and 6 were reasonable. By deletion of columns 4 and 5 and row 4 and 5 of the least squares matrix, the data in the last row of Table III were obtained. This procedure gave quite reasonable values for the C_{0i} in Table III

TABLE III
ILLUSTRATIVE CALCULATIONS ON MODEL DATA

Number of components assumed	C_{01}	C_{02}	C_{03}	C_{04}	C_{05}	C_{06}
2	−8.04	23.06				
3	14.07	−22.46	23.40			
4	−3.44	33.70	−37.66	22.40		
5	4.48	−0.72	20.14	−21.64	12.74	
6	3.00	7.37	1.71	−0.06	−0.11	3.09
True values	3.00	7.50	1.50	0.00	0.00	3.00
4*[a]	2.94	7.60	1.43	0.00	0.00	3.03

[a] For the 4* entry in the table, 4-mer and 5-mer were assumed to be absent.

although the exact values were not recovered. Similar behavior is observed for real data.

The values of C_{pi} (but not C_{0i}) can be formulated into equilibrium constants since they represent the concentration of species i at the point r_p,

$$\frac{i}{M_1^{i-1}} \prod_{m=2}^{i} k_m = C_{pi}/C_{p1}^{i} \tag{71}$$

provided the C_p values are converted to grams per liter concentration units, the k values will be in liter per mole units according to this formula. If we now reconsider the ultracentrifuge Eq. (65), it is clear that the values of k merely represent a change in parameters. That is, C_{pi} is given either by the C_{pi} parameter or by the combined parameter of Eq. (71) solved for C_{pi} which depends on C_{p1} and the equilibrium constants. This leads to the important conclusion that a paucidisperse protein at sedimentation equilibrium behaves as if it were in chemical equilibrium as well, as long as the data from a single cell are considered. Only when data from several initial loading concentrations, radial distances, and rotor speeds are combined need one pay attention to chemical equilibria.

Table IV shows data[27] for α-chymotrypsin at pH 6.2 in 0.2 M phosphate buffer. The characteristics of this data are typical for such calculations on real data. When two components are postulated, positive C_{0i} are found, but the data are fit better by three components. When four com-

TABLE IV

ANALYSIS OF MONOMER–POLYMER EQUILIBRIA[a]

Number of components postulated	Concentrations of				$\sum_{r_m}^{r_b} \|\Delta C_r\|$[b]
	Monomer	Dimer	Trimer	Tetramer	
2	16.31	6.22			1.2045
	16.18	6.28			1.1644
3	18.38	2.02	2.13		0.2917
	18.17	2.23	2.05		0.2966
4	16.67	7.34	−3.43	1.95	0.2308
	16.39	7.77	−3.74	2.03	0.2297

[a] α-Chymotrypsin, pH 6.2, $\mu = 0.2$ M phosphate, $C_0 = 22.5$ fringes.
[b] $\Delta C_r \equiv C_r$ (observed) − C_r (predicted). The upper entry of each pair is from mass conservation labeling of fringes, and the lower values are from white light fringe labeling in the low speed sedimentation equilibrium method. E. G. Richards, D. C. Teller, and H. K. Schachman, *Biochemistry* **7**, 1054 (1968).

ponents are postulated, the fit to the data is better, but the C_{0i} are not all real. In a study of bovine serum albumin, Reinhardt and Squire[65] used a computer program which required that all coefficients, C_{pi}, be zero or positive. Such methods require more sophisticated programming than a simple matrix inversion but are probably better for determination of the stoichiometry. This point will be discussed later in this article.

This type of calculation is also well suited for the determination of hybrid formation between two different chemical species. If we let

$$\Gamma_P = \exp[A M_P(r^2 - r_p^2)] \quad \text{and} \quad \Gamma_F = \exp[A M_F(r^2 - r_p^2)]$$

then a paucidisperse mixture of proteins F and P is described by the equation,

$$C = \sum_{k=1}^{n} \Gamma_F{}^k C_{p,F,k} + \sum_{j=1}^{m} \Gamma_{Pj} C_{p,P,j}$$
$$+ C_{p,F_1P_1}\Gamma_F\Gamma_P + C_{p,F_2P_1}\Gamma_F{}^2\Gamma_P + C_{p,F_1P_2}\Gamma_F\Gamma_P{}^2 + \cdots (72)$$

This equation is linear in the C_p parameters so the same least-squares method may be used for its solution. Table V[19] shows an attempt to recover the values of C_p from model data in which $n = m = 2$ and the three hybrids of Eq. 72 were assumed to be present. There are three criteria that must be fulfilled for selection of the model: (1) the calculated C_p values must be positive, (2) the back-calculated mass averages must be consistent with the model, and (3) the back-calculated molecular weight averages must be consistent with the observed data. It may be seen that only 4 of the 14 models tested in this table represent real solutions (Model Nos. 4, 6, 10, and 11 of Table V). Of these, only two (10 and 11) had acceptably low values of molecular weight deviations. As a practical example, Behnke et al.[19] were able to apply such an analysis to data from a mixture of carboxypeptidase A and fraction II of procarboxypeptidase to establish that a 1:1 hybrid was present in the mixture.

Such concentration distance analyses are useful for the determination of stoichiometry and equilibrium constants. Errors in molecular weight of the species involved in the system propagate linearly (to a first approximation) into the values of C_{pi}. Such errors can lead to the conclusion that trimer is present when it actually is absent. Thus, this method of analysis of associating systems has several disadvantages as well as advantages:

First, the advantage is that the primary data of C vs. r are used in

[65] W. P. Reinhardt and P. G. Squire, *Biochim. Biophys. Acta* **94**, 566 (1965).

TABLE V

ILLUSTRATIVE CALCULATIONS ON MODEL HYBRID DATA[a]

Set No.	Composition							Rms fit		
	A(1)B(0)	A(0)B(1)	A(1)B(1)	A(2)B(0)	A(0)B(2)	A(2)B(1)	A(1)B(2)	M_n	M_w	C
1	+	−	−	+	+	−	+	146,223	97,938	142.639
2	+	+	+	−	−	+	0	9,337	156,622	4.637
3	+	+	+	−	−	0	+	3,641	4,282	5.128
4	+	+	+	+	+	0	0	143	136	0.105
5	+	+	+	−	0	−	+	1,000	982	0.701
6	+	+	+	+	0	+	0	122	107	0.082
7	+	+	−	−	+	0	0	3	3	0.004
8	+	+	−	0	−	−	+	7,313	10,627	6.481
9	+	+	+	0	+	0	+	11	22	0.005
10	+	+	+	0	+	0	0	2	2	0.003
*11	+	+	0	0	0	0	0	2	2	0.000
12	−	+	0	0	0	0	+	95,076	502,352	0.097
13	−	+	0	0	0	+	0	19,014	10,492	0.071
14	+	+	0	0	0	+	−	2,028	1,895	0.014

[a] $M_1(A) = 27,000$, $M_1(B) = 40,000$; − = negative coefficients for C_p; + = positive coefficients for C_p; 0 = assumed absent in composition. The asterisk indicates the correct solution.

the analysis; systematic deviations introduced by bad computation procedures are minimal. The second advantage is that quite complex stoichiometries can be established by the method—much more complicated than can be performed by the use of the "search" procedures. Third, the equations are linear so that simple, straightforward linear analyses can be used in the computations. Fourth, it is not necessary to make any extrapolations from the data in the curve-fitting calculations such as are required in the Steiner method given in Section XI.

These advantages are offset by several disadvantages: First, the molecular weight of the monomeric species must be accurately known for the calculation of Γ at each point. Knowledge of M_1 is often difficult to obtain with sufficient accuracy so that stoichiometries can be established or disproved by the method. Second, it is not obvious that weighting functions can be used to advantage in the least-squares equations so that a single bad point or run of bad points can strongly affect the data analysis. Third, data must be processed from a single cell in each calculation; that is, data from several overlapping concentrations cannot be combined unless exactly the same conditions of radial distance and rotor velocity are met. The methods given in the next section are superior in this point.

XI. Analysis of Molecular Weight Distributions When M_1 Is Known

Historically, these methods were the first to be applied practically for associating protein systems.[66-68] As discussed in the introduction, most of these methods use a whole-system approach, fitting the entire molecular weight vs. concentration distribution at one try. At present, the methods are important because they are amenable to analysis of nonideal systems. In the method outlined in the preceding section, only one set of data can be treated at a single time. For the methods in this section, all of the data from several experiments can be combined, provided that chemical equilibrium exists; that is, if molecular weight averages are a single continuous function of concentration (within experimental error). Under these circumstances it is also possible to extract the number average molecular weight from low speed equilibrium experiments[69]

$$\frac{C}{M_n} = \int_0^C \frac{dC}{M_w} \tag{73}$$

[66] R. F. Steiner, *Arch. Biochem. Biophys.* **39**, 333 (1952).
[67] R. F. Steiner, *Arch. Biochem. Biophys.* **49**, 400 (1954).
[68] R. F. Steiner, *Arch. Biochem. Biophys.* **53**, 457 (1954).
[69] E. T. Adams, Jr., *Biochemistry* **4**, 1646 (1965).

so that one may perform the integration denoted by the right of the equation and extract the number-average molecular weight. For non-ideal systems, the apparent number-average molecular weight is obtained.

A. Method of Steiner

In order to evaluate this and other methods, it is desirable to consider some equations that will help clarify later discussion. The total concentration of the solution is given by

$$C = C_1 + \frac{2k_2}{M_1}C_1^2 + \frac{3k_2k_3}{M_1^2}C_1^3 + \cdots + \frac{ik_i}{M_1^{i-1}}C_1^i + \cdots \tag{74}$$

This expression follows directly from Eq. (71) of the preceding section. For simplicity of notation let $K_2 = 2\ k_2/M_1$, $K_3 = 3\ k_2k_3/M_1^2$, etc. This equation now reads,

$$C = C_1 + K_2C_1^2 + K_3C_1^3 + \cdots + K_iC_1^i + \cdots \tag{75}$$

The principle of the methods discussed in this section is to compute C_1 from experimental data at each total concentration, C.

Steiner in 1952[66] showed by differentiation of Eq. (75) together with the definition of the weight-average molecular weight (Eq. 35b) that

$$\frac{M_w}{M_1} = \frac{d \ln C}{d \ln C_1} \tag{76}$$

from which the weight fraction of monomer, x_1, can be extracted since $C_1 = x_1C$:

$$\ln x_1 = \int_{C=0}^{C} \left(\frac{M_1}{M_w} - 1\right) d \ln C \tag{77}$$

In order to obtain the value of x_1 at each point, the values of $[(M_1/M_w) - 1]/C$ are graphed as a function of C and the area shown by the shading in Fig. 23 is measured. The value of $(M_1/M_w) - 1)/C$ at $C = 0$ can be shown to be $-2k_2/M_1$ for ideal systems.

Since k_2 is unknown at this point, it is necessary to have molecular weight averages at low values of C so that the extrapolation to $C = 0$ is not wildly inaccurate. If a big gap exists between $C = 0$ and the first data, the value of the integral between the first points dominates the calculation. This is the biggest weakness of this method since a non-linear extrapolation to $C = 0$ is required.

After the extrapolation is made, it is a simple matter to calculate the integral from a formula such as Eq. (24). From the values of x_1

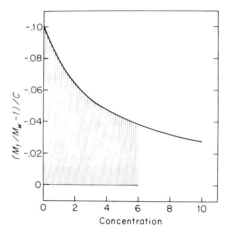

Fig. 23. Illustration of the method of Steiner for a monomer–dimer system. The association in this case is weak since $2k/M_1 = 0.1$ concentration^{-1}. The units of concentration are arbitrary. The shaded area is the value of ln x_1 at $C = 6$.

determined by the integral, there are several ways to proceed: Adams[70] eliminates the equilibrium constants starting with the highest until a single analytical expression exists with only one unknown. We prefer to use a slight modification of the original method of Steiner, which uses least-squares curve fitting. Since the Adams procedure is well documented[70] and since we have not had experience with it, we will only explain the procedure used in this laboratory.

Since $C_1 = x_1C$, and x_1 has been calculated at each experimental point, j, we can calculate C_1 values at each of these points. Equation (75) written for each point becomes

$$C_j = C_{1,j} + K_2 C_{1,j}^2 + K_3 C_{1,j}^3 + \cdots \qquad (78)$$

if this equation is divided by $C_{1,j}$ and 1 is subtracted from it, it becomes

$$\left(\frac{C}{C_1} - 1 \right)_j = K_2 C_{1,j} + K_3 C_{1,j}^2 + \cdots \qquad (79)$$

The least-squares equations for this function are the same as given by Eq. (67). By fitting a polynomial through the data, the equilibrium constants are determined from the coefficients of the polynomial. For example, if the system is that of monomer–dimer, a straight line will occur for ideal systems (Fig. 24). Upward curvature in such a graph

[70] E. T. Adams, Jr., *Fractions*, No. 3, (1967). This article contains references to the work on self-associating systems as well as a clear explanation of the methods used by Adams.

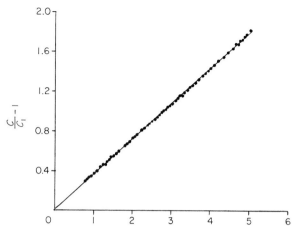

FIG. 24. Graph of $C/C_1 - 1$ vs. C_1 from Steiner analysis of M_w data for α-chymotrypsin (Nutritional Biochemicals Lot 6397) in 0.2 M KCl, 0.01 M potassium acetate, pH 4.40 at 20°. The initial concentration was 0.75 mg/ml. $k_2 = 18.0 \pm 0.1 \times 10^3$ liters/mole; intercept = 0.009 ± 0.010.

indicates the presence of higher polymers. When nonideality is significant, downward curvature results. If a particular coefficient in the polynomial is negative or small, the appropriate row and column of the matrix is deleted and the matrix inversion repeated. In this way the stoichiometry of the interaction can be estimated.

The number average molecular weight data can also be used in such an analysis.[63] The weight fraction of monomer is

$$\ln x_1 = \frac{M_1}{M_n} - 1 + \int_{C=0}^{C} \left(\frac{M_1}{M_n} - 1\right) d \ln C \qquad (80)$$

from these values of x_1 at each point, the same procedure for determination of equilibrium constants can be used.

Equation (79) represents a polynomial with zero intercept. There are three causes of nonzero intercepts in graphs such as Fig. 24: (1) Systematic concentration labeling errors which affect M_n and M_w differently; (2) the use of incorrect monomer molecular weight in the calculations; and (3) systematic errors in the numerical integration procedures, such as a bad extrapolation to $C = 0$.

The effect of nonideality is to decrease the values of the molecular weights calculated from the observed data according to Eqs. (34). When such apparent molecular weight averages are used in this Steiner type of analysis, the concentration of the monomer which is calculated is incorrect.[70] The perturbation due to the viral coefficient is given by the expression,

$$C_{1,\text{app}} = C_1 e^{BM_1 C} \tag{81}$$

Equation (78) then becomes,

$$C = C_{1,\text{app}} e^{-BM_1 C} + K_2 C_{1,\text{app}}^2 e^{-2BM_1 C} + K_3 C_{1,\text{app}}^3 e^{-3BM_1 C} + \cdots \tag{82}$$

To illustrate the complexity of this system, we may expand the exponents by Taylor's series, restrict it to two components (monomer and dimer) to yield the approximate result,

$$\frac{C}{C_{1,\text{app}}} - 1 = (K_2 - BM_1)C_{1,\text{app}} - 3BM_1 K_2 C_{1,\text{app}}^2 - 2BM_1 K_2^2 C_{1,\text{app}}^3 \tag{83}$$

(In the expanded exponentials we have assumed that $C_{1,\text{app}} = C_1$.) This equation demonstrates downward curvature for monomer–dimer systems. If trimer and higher species are present, the equations are very complicated, since positive and negative terms compensate. It is for this reason that Adams[70] uses a term-by-term elimination of unknowns.

While it has not been done in the literature, it is possible to solve Eq. (82) in a least-squares manner for values of both B and K_i. In this equation, the values of C and $C_{1,\text{app}}$ are known at each point, j. If B were known, the equation would take the form,

$$C_j = C_{1,\text{app},j} \delta_j + K_2 C_{1,\text{app},j}^2 \delta_j^2 + K_3 C_{1,\text{app},j}^3 \delta_j^3 + \cdots \tag{84}$$

where $\delta_j = e^{-BM_1 C_j}$. So one way to approach the problem would be to choose a series of reasonable values of B and compute the least-squares equivalent of Eq. (84) for each value of B. The deviations of

$$\sum_{j=1}^{N} (C_j{}^{\text{obsd}} - C_j{}^{\text{calc}})^2$$

could be graphed vs. B (assumed) and the minimum deviation on the graph taken as the best value of B.

A better way to perform such curve fitting of a nonlinear equation is known as the Newton–Raphson method. The detailed procedure is given in the Appendix; but basically it is an automatic way to perform the calculation described in the preceding paragraph. By this technique, values of the K's and B may be found which fit the data best.

One final calculation in the use of this method which is essential is to compute predicted values of M_n and M_w as a function of concentration from the equilibrium constants. For this method it is not sufficient to simply calculate root-mean-square deviations of the predicted and observed data. We have observed that often the predicted curve from this calculation does not describe the observed data due to various errors committed during the calculation. If such predicted curves do not describe the data, the sources of error must be determined.

B. Method of van Holde, Rosetti, and Dyson

This method uses concentration and weight-average molecular weight data in a unique way. The method is amenable to calculation of ideal systems on small computers and perhaps can be utilized for nonideal systems on such computers. A limited number of copies of a FORTRAN program to carry out the calculations are available on request from R. Dyson.[54] As with the Steiner method, we will first consider ideal systems and then extend the analysis to nonideal cases.

The method begins from the equation of Steiner[67] (Eq. **76**),

$$\frac{d \ln C}{d \ln C_1} = \frac{M_w}{M_1} \tag{76}$$

Instead of integration from $C = 0$, the lower limit of integration is an arbitrary point, C_a. C_a may be taken as a middle point in the data or the lowest concentration point. In the latter case, then,

$$C_1 = C_{a1} \exp\left(M_1 \int_{\ln C_a}^{\ln C} \frac{1}{M_w} d \ln C\right) \tag{85}$$

The only unknown value in this equation is C_{a1}, the concentration of monomer at $C = C_a$. Since C_a is greater than zero, the limiting value of the exponent as $C \to 0$ is of no concern in the equation. Note, however, that the value of the exponent can be determined from the known values of M_1 and $M_{w,j}$ and C_j just as in the Steiner method. No extrapolation to $C = 0$ is required, however. This is a distinct advantage over the Steiner method.

We now define

$$I(C) = \int_{\ln C_a}^{\ln C} \frac{1}{M_w} d \ln C$$

for simplicity of notation. Also, since $C_{ai} = K_i C_{a1}{}^i$ we may rewrite Eq. (**75**) as

$$\frac{C_{a1}e^{M_1 I(C)}}{C} + \frac{C_{a2}e^{2M_1 I(C)}}{C} + \frac{C_{a3}e^{3M_1 I(C)}}{C} + \cdots = 1 \tag{86}$$

Also, the equation for C_{ai} may be substituted into the definition of M_w to yield,

$$M_w = M_1 \frac{C_{a1}e^{M_1 I(C)}}{C} + \frac{2M_1 C_{a2}e^{2M_1 I(C)}}{C} + \frac{3M_1 C_{a3}e^{3M_1 I(C)}}{C} + \cdots \tag{87}$$

This equation is divided by M_1, and 1 is subtracted from both sides to produce

$$\frac{M_w}{M_1} - 1 = -1 + \frac{C_{a1}e^{M_1 I(C)}}{C} + \frac{2C_{a2}e^{2M_1 I(C)}}{C} + \frac{3C_{a3}e^{3M_1 I(C)}}{C} + \cdots \quad (88)$$

In this equation, the negative of Eq. (86) may be substituted to eliminate the term in C_{a1},

$$\frac{M_w}{M_1} - 1 = \frac{C_{a2}^2 e^{M_1 I(C)}}{C}$$

$$+ \frac{2C_{a3}e^{3M_1 I(C)}}{C} + \cdots + \frac{(i-1)C_{ai}e^{iM_1 I(C)}}{C} + \cdots \quad (89)$$

This equation may be used for curve fitting of ideal systems.

If we define $X_j = (i-1)e^{M_1 I(C_j)}/C_j$ at each datum point, j, the equation may be seen to be a polynomial in this variable X_j

$$\frac{M_{w,j}}{M_1} - 1 = C_{a2}X_j^2 + C_{a3}X_j^3 + \cdots + C_{ai}X_j^i \quad (90)$$

The polynomial is linear in the unknown parameters, C_{ai}, so that least-squares techniques may be applied to it. In order to determine C_{a1}, it is only necessary to use the constraint that C_a (total) is the sum of C_{ai}. Since data may be taken either from a single cell or from several experiments, this method seems very powerful for the analysis of all monomer–polymer systems, regardless whether chemical equilibrium exists or not.

The problem of thermodynamic nonideality has also been considered by van Holde et al.[54] for this method of analysis. The result is

$$\frac{M_{wa}/M_1}{1 - BM_{wa}C} - 1 = \sum_{i=2}^{n} C_{ai}X_i e^{-iM_1 B\Delta C} \quad (91)$$

where $\Delta C = C - C_a$ and M_{wa} is the apparent molecular weight which is observed. As discussed in the previous subsection, it should be possible to guess values of B and solve the now linear equations to obtain the best value of B. The procedure followed by Dyson is to solve the ideal system by least squares and then iterate the nonlinear Eq. (91) by a Newton–Raphson refinement procedure (Appendix) to obtain the best values of C_{ai}.

Needless to say, a similar relation in Eq. (91) can be derived for the number average molecular weight data (Dyson, personal communication). One begins with the equation

$$d \ln C_1 = \frac{1}{C} d \left(\frac{CM_1}{M_n} \right) \quad (92)$$

If we consider the right side of this equation, it may be integrated by parts,

$$\int \frac{1}{C} d\left(\frac{CM_1}{M_n}\right) = \frac{M_1}{M_n} + \int \frac{M_1}{M_n} d\ln C \tag{93}$$

Let

$$H(C) = \frac{1}{M_n} + \int_{\ln C_a}^{\ln C} \frac{1}{M_n} d\ln C$$

then

$$C_1 = C_{a1}e^{M_1 H(C)} \tag{94}$$

proceeding as with the weight-average molecular weight, the result is

$$1 - \frac{M_1}{M_n} = \frac{1}{2}\frac{C_{a2}e^{2M_1 H(C)}}{C} + \frac{2}{3}\frac{C_{a3}e^{3M_1 H(C)}}{C} + \cdots$$
$$+ \frac{i-1}{i}\frac{C_{ai}e^{iM_1 H(C)}}{C} \tag{95}$$

Which may be treated in exactly the same fashion as discussed for Eq. (90). The nonideal analog of Eq. (91) is

$$1 - \frac{M_1}{M_{na}} - \frac{1}{2}BCM_1 = \sum_{i=2}^{n} \frac{1}{i} C_{ai} Y^i e^{-iM_1 B\Delta C} \tag{96}$$

where $Y = (i-1)e^{M_1 H(C)}/C$. In principle, Y should equal X of Eq. (91); but since $H(C)$ and $I(C)$ are subject to different calculation errors, they will not be identical for real data.

C. Method of Deonier and Williams

This subsection presents two methods which have been used for nonideal monomer–dimer systems with good success.[62] The simplest method for monomer–dimer systems is to solve the nonideal form of the weight-average molecular weight for the equilibrium constant, K_2,

$$K_2 = \frac{1}{4C}\left\{\frac{R_{wa}^2}{[2(1 - BM_1 R_{wa}C) - R_{wa}]^2} - 1\right\} \tag{97}$$

where $R_{wa} = M_{wa}/M_1$ and K_2 has reciprocal C units. By assumption of a variety of values of B the values of $K_2(B)$ are fit to a linear least-squares equation in C and when the slope is zero, the value of B has been found. $K_2(B)$ means the values of K_j at C_j, M_{wj} for an assumed value of B. This is clearly a simple procedure to follow for testing of monomer–n-mer hypotheses. For $n \neq 2$ it is necessary to solve the appropriate equation for K_n.

The second method of Deonier and Williams[62] begins with the explicit formulation of the apparent weight-average molecular weight as a function of concentration[53]

$$\frac{2M_1}{M_{wa}} - 1 = (1 + 4K_2C)^{-1/2} + 2BM_1C \tag{98}$$

Now, the term in $(1 + 4K_2C)^{-1/2}$ can cause the left side of the equation to vary rapidly with concentration as C approaches zero as discussed in Section XI, A. To generate a more slowly varying function, the authors eliminated the radical by means of an integration procedure. Following their procedure, the following definitions are presented: Let,

$$a = 4K_2$$
$$b = BM_1$$
$$F(C) = \frac{2M_1}{M_{wa}} - 1$$

so that Eq. (98) becomes

$$F(C) = (1 - aC)^{-1/2} + 2bC \tag{99}$$

$G(C)$ is then defined as the integral of $F(C)$,

$$G(C) = \int_0^C F(C)dC = \frac{2}{a}[(1 + aC)^{1/2} - 1] + bC^2 \tag{100}$$

$G(C)$ in this equation has physical significance. It is,

$$G(C) = C\left[\frac{2M_1}{M_{na}} - 1\right]$$

This equation is solved for the radical, $(1 + aC)^{1/2}$ to combine it with Eq. (99) with the result

$$G(C) - bC^2 = \frac{2}{a}[F(C) - 2bC]^{-1} - \frac{2}{a} \tag{101}$$

The equation is then rearranged to produce

$$\frac{F(C)G(C)}{1 - F(C)} = \frac{2}{a} + b\frac{C^2F(C) - 2CG(C)}{1 - F(C)} - 2bC\frac{bC^2 - 2/a}{1 - F(C)} \tag{102}$$

which is written as,

$$U = \frac{2}{a} + bV - W \tag{103}$$

The significance of U, V, and W can be obtained by comparison with Eq. (102). While it is not obvious from symbol comparison, U and V are the dominant variables in this equation. W is much smaller than either term. This is important because this term contains unknown values while U and V can be determined from experimental data of M_{wa} and C at each point.

In order to apply Eq. (103) to experimental data, the term in W is assumed zero. A graph of U vs. V is constructed to determine $2/a$ and b by least-squares. From these two parameters, W can be calculated at each point. A new linear least-squares equation is used with the values of U_j, V_j, and W_j at each point; $2/a$ and b are again estimated. Repetition of this calculation with new values of W_j leads to rapid convergence.[62]

This method is much more complicated than the first method. There do not seem to be very many advantages to it, but it is presented because it represents a method that is mathematically stable.

D. Method of Roark and Yphantis

Roark and Yphantis[39] suggested a method that should be quite powerful for highly associated, nonideal systems. The method can be used to determine the stoichiometry of nonideal systems. These authors use thermodynamically ideal moments of molecular weight averages. We will consider only the use of M_{na}, M_{wa}, and M_{za} for the calculation of the ideal moments. Roark and Yphantis[39] have considered several other possibilities that are not included here.

The basis of this approach is to combine point-by-point apparent molecular weight averages in such ways that the linear B terms of Eqs. (34) are eliminated. The first two such moments (and the easiest to calculate) are

$$M_{y1} = (2/M_{na} - 1/M_{wa})^{-1} = (2/M_n - 1/M_w)^{-1} \qquad (104)$$

$$M_{y2} = M_{wa}{}^2/M_{za} = M_w{}^2/M_z \qquad (105)$$

These new averages are not simple summations of concentrations and molecular weights but nonetheless do reflect the association or dissociation of the material, although sometimes in weird ways. If we consider only monomer–n-mer systems for the moment and let α be the weight fraction of n-mer. Then,

$$\left(\frac{M_{y1}}{M_1}\right)^{-1} = \frac{2}{1 - \alpha + \alpha/n} - \frac{1}{1 - \alpha + n\alpha} \qquad (106)$$

and

$$\frac{M_{y2}}{M_1} = \frac{(1 - \alpha + n\alpha)^3}{1 - \alpha + n^2\alpha} \qquad (107)$$

from the definitions of M_n, M_w, and M_z. Graphs of M_{y1}/M_1 and M_{y2}/M_1

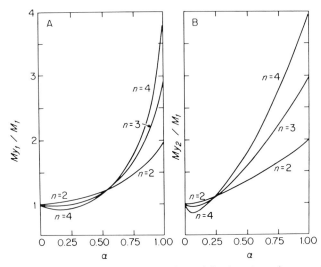

Fig. 25. Ideal moments as a function of the weight fraction of n-mer for various stoichiometries, n. (A) M_{y1}/M_1 vs. α; (B) M_{y2}/M_1 vs. α.

vs. α are presented in Fig. 25A and B for various assumed monomer–n-mer stoichiometries. Both functions are always greater than unity for monomer–dimer systems, but are less than 1 for weak association with monomer–trimer and monomer–tetramer associations. The figures indicate that, if the amount of higher polymer is greater than 25%, M_{y2}/M_1 may be useful if the amount of n-mer is greater than 50%.

For a monomer–n-mer reaction, the equilibrium constant, K_n as defined by Eq. (75) is given as

$$K_n\, C^{n-1} = \alpha/(1 - \alpha)^n \tag{108}$$

A graph of $\alpha/(1 - \alpha)^n$ vs. C^{n-1} should be linear for the correct stoichiometry of a monomer–n-mer reaction. For an assumed stoichiometry, one need only construct the values of M_{y1}/M_1 and M_{y2}/M_1, calculate the values of n, and look for the most linear plot of Eq. (108).

In order to determine whether this procedure was practical, I performed this calculation on the data from glutamic decarboxylase data of Fig. 10, assuming $M_1 = 104 \times 10^3$ g/mole. The results indicate that the method may be quite useful for the determination of the stoichiometry of associating systems and complimentary to the Dyson method for analysis of associating systems.

The results are shown in Figs. 26A, B and 27A, B. I first assumed that $n = 3$ since most of the M_{y2}/M_1 values were greater than 2. From graphs such as those of Fig. 25, values of α were computed for each of

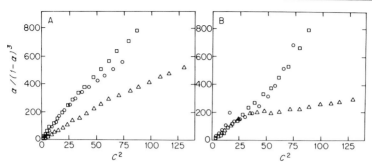

Fig. 26. Test for the stoichiometry of association for *Escherichia coli* glutamic decarboxylase. (A) M_{y1} data with monomer–trimer assumed; (B) M_{y2} data for the same stoichiometry. Symbols: ◯ channel 1, △ channel 2, ☐ channel 3 data. The units of C are fringes. Method of D. E. Roark and D. A. Yphantis, *Ann. N.Y. Acad. Sci.* **164**, 245 (1969).

58 values of M_{y1}/M_1 and M_{y2}/M_1 at approximately ½ fringe increments of C in the data.

The values of $\alpha/(1 - \alpha)^n$ were then graphed vs. C^{n-1} as indicated in Figs. 26 and 27. It is clear from the figures that it is not possible to determine precise equilibrium constants from the slopes of the data. However, it is quite clear that tetramers need not be considered in any

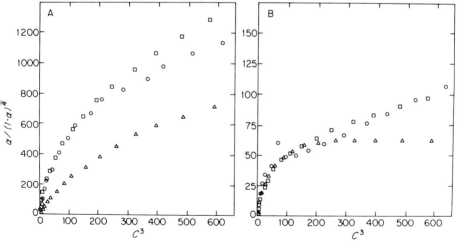

Fig. 27. Test for the stoichiometry of association for *Escherichia coli* glutamic decarboxylase. (A) M_{y1} data for $n = 4$; (B) M_{y2} data for the same stoichiometry. The symbols are the same as in Fig. 26. The units of C are fringes. Note the difference in ordinate values between (A) and (B). Method of D. E. Roark and D. A. Yphantis, *Ann. N.Y. Acad. Sci.* **164**, 245 (1969).

future calculations. An eyeball estimate of K_3 from the initial slope of Fig. 26B gives 6 ± 2 fringe^{-2}, and approximately the same value would satisfy the data of Fig. 26A. A remarkable inconsistency occurs when the data of Fig. 27 are considered. When the stoichiometry was assumed to be $n = 4$, the M_{y1} data indicate K_4 as approximately 6 fringe^{-3}, but the M_{y2} data indicate that 0.5 fringe^{-3} should be the value, taking the slope to be ordinate/abscissa at $C^3 = 100$ fringe^3. The extreme curvature observed in Fig. 27, in spite of the fact that the data do not superimpose well, suggests that this can be used as a general method for indicating the stoichiometry of self-associating systems, provided that α of Fig. 25 is large. This type of technique is important since it indicates the number of components which should be assumed in methods such as that used by van Holde *et al.*[54] or the Steiner type of method. These appear to be the best methods of calculating equilibrium constants for nonideal systems.

This method has some disadvantages that are obvious from Fig. 25. M_{y2} (Fig. 25B) is rather imprecise due to the difficulty of obtaining M_{za}. Thus M_{y2} is useful for determination of stoichiometry only if $\alpha > 0.5$, as was the case for the glutamic decarboxylase data. The situation is even worse for M_{y1} data (Fig. 25A). Only at values of $\alpha > 0.8$ is it possible to uniquely distinguish among the curves for n. In the case used for example (Figs. 26 and 27), α was generally greater than 0.5 and so was well determined. When the association is weak (as in the case of lysozyme[62]), it may be impossible to determine n by this method.

E. Conclusions and Future Directions

In this subsection it would be desirable to prescribe a single method of approach to the study of self-associating systems; however, it is not possible to do this at present. Each of the methods I have presented has been successfully used with data from experiments on real systems of biological molecules. But the available methods are not satisfactory for the biochemist who isolates an enzyme and wishes to study its molecular properties. On the other hand, they are sufficient for the physical chemist who desires to learn the thermodynamics of interaction of well-characterized (M_1 known exactly) proteins.

The immediate problems to solve are quite clear, and the methods of solutions are also known: It is necessary to develop a method for determination of M_1 and all k_i for any associating protein system. The methods of Hoagland and Teller[45] are very primitive in this respect, but they work for simple stoichiometries and ideal systems. Variants of the method of Dyson[54] appear most promising in this respect. Specifically, it is possible to vary M_1 in this technique and search

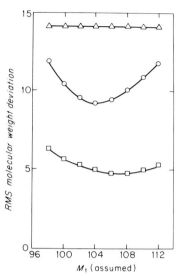

FIG. 28. Search for M_1 for *Escherichia coli* glutamic decarboxylase data. A monomer–trimer stoichiometry was assumed as described in the text. At $M_1 = 104 \times 10^3$ g/mole $k_3 = 102 \times 10^{10}$ liter^2/mole^2 and $B = 1.08 \times 10^{-6}$ ml mole/g^2. Triangles are M_n data, squares are M_w data, and circles are M_z data. Method of K. E. van Holde, G. P. Rosetti, and R. D. Dyson, *Ann. N.Y. Acad. Sci.* **164**, 279 (1969).

for the minimum in the deviation of the expected parameter (C, M_{wa}, or M_{na}).

However, there are difficulties with this approach, and these are illustrated in Fig. 28 for the glutamic decarboxylase data. The M_w data were used as described in Section XI, B. When the system was fit by a monomer–dimer–trimer system, no positive values of C_{a2} could be found for molecular weights less than 108×10^3. A monomer–trimer system was then attempted, and the results are shown in the figure. The minimum in the fit to the M_w data occurred between 106 and 108×10^3. The minimum deviation in M_z occurs at 104×10^3. No minimum deviation could be found in the number average data; predicted values of M_n were uniformly smaller than the observed values. This behavior is predicted if $M_{n,m}$ is estimated too large. This is probably why a monomer–dimer–trimer system was observed in the search for M_1, k_2, and k_3 of Section VIII.

Probably the best way to proceed to determine all molecular parameters of self-associating systems is to use a combination of the methods of Scholte[71] and Dyson.[54] Specifically, the procedure would be to generalize Eq. (86) to the form

[71] T. G. Scholte, *Ann. N.Y. Acad. Sci.* **164**, 156 (1969).

$$C = \sum_{i=1}^{n} C_{ai} e^{iM_1 I(C)} e^{-iM_1 B \Delta C}$$

where $I(C) = \int_{\ln C_a}^{\ln C} 1/M_{wa} \, d \ln C$. The number of components, n, should probably be about six. This equation can then be solved for variations of M_1 with the added condition that the C_{ai} are greater than or equal to zero. Only Reinhardt and Squire[65] and Scholte[71] have incorporated this important restriction in the computer programs. The reason for choosing this function is that data from several cells can be combined if chemical equilibrium exists. By graphing the deviations of $C^{\text{obs}} - C^{\text{pred}}$ as a function of M_1, it should be quite straightforward to obtain all values of C_{ai} and B as well as determine the stoichiometry. For this proposal it would be important to use data points equally spaced in C, in order to obtain a good fit to the whole curve.

XII. Data Simulation

In this brief section, I would like to discuss the method used in this laboratory to compute simulated interacting systems. Such procedures are necessary to test computer programs prior to calculation of real data. By incorporation of systematic and random errors into the simulated data, it is possible to determine whether a proposed calculation method will work with experimental data.

The equation for the distribution of a single, ideal solute at sedimentation equilibrium is given by the equation

$$C_i = C_{0i} H_i e^{z H_i} / \sinh H_i \tag{9a}$$

as discussed in Section V. Here

$$H_i = \frac{(1 - \bar{v}\rho)\omega^2}{4RT} M_i (r_b^2 - r_m^2) \tag{9b}$$

and $z = (2 \ r^2 - r_b^2 - r_m^2)/(r_b^2 - r_m^2)$. The C_{0i} are defined by the integrals

$$C_{0i} = \int_{r_m^2}^{r_b^2} C_i d(r^2)/(r_b^2 - r_m^2) \tag{69}$$

for sectorial cells. To compute the distribution of a paucidisperse mixture, it is only necessary to specify the values of C_{0i}, M_i, r_b, r_m, etc. and sum Eq. (9a) for all components at various positions of r.[27] Frequently, however, it is desired to compute systems at chemical equilibrium. From thermodynamics we know that

$$K_i = C_i / C_1^i$$

where K_i has units of $C^{-(i-1)}$. For self-associating systems, $H_i = iH_1$ so that K_i becomes

$$K_i = \frac{C_{0i}}{C_{01}{}^i} \frac{iH_1}{H_1{}^i} \frac{(\sinh H_1)^i}{\sinh iH_1}$$

from Eq. (9a). From this equation, C_{0i} can be calculated from C_{01} and K_i.

If high speed sedimentation equilibrium experiments are to be simulated, then an important simplification of Eq. (9a) can be made: for $H_i \geqslant 2$, $\sinh H_i \approx \frac{1}{2}e^{H_i}$ and the equation becomes

$$C_i = 2C_{0i}H_i e^{\Psi H_i} \tag{109}$$

where $\Psi = Z - 1 = 2(r^2 - r_b{}^2)/(r_b{}^2 - r_m{}^2)$ as in the right approximation of Eq. (9a) given earlier (p. 365). To show how this equation may be used for simulation, the example of indefinite self-association may be used.[19] Here,

$$C_i = iK^{(i-1)}C_1{}^i \tag{110}$$

where $K = k/M_1$ where k is the molar equilibrium constant, and so k/M_1 has units of liters per gram. Equations (110) and (109) can be combined to produce a relation for C_{0i}

$$C_{0i} = (2KC_{01}H_1)^{i-1}C_{01} \tag{111}$$

Now, C_0 is the sum of the individual C_{0i} so that this equation becomes

$$C_0 = C_{01} + 2KC_{01}H_1C_{01} + (2KC_{01}H_1)^2C_{01} + \cdots$$

which is a series that may be condensed to the form

$$C_0 = C_{01}/(1 - 2KC_{01}H_1)$$

This may be calculated for C_{01} with the result

$$C_{01} = C_0/(1 + 2KC_0H_1)$$

Thus, one may specify H_1 and C_0. C_{01} is obtained from this equation. The other C_{0i} are then obtained from Eq. (111).

The C_{0i} concentrations calculated by these procedures for chemically reacting systems never can actually exist. They are defined as the total mass average of monomers, dimers, etc., at sedimentation equilibrium. They represent the initial concentrations of a hypothetical solution which is not in chemical equilibrium, but which would have been observed if a mixture of proteins had been combined and centrifuged under the same conditions of r_m, r_b, ω, etc.

This same approach may be used for the simulation of nonideal chemically reacting mixtures, but it is easier to use other methods. The nonideal expression for any C_i is

$$C_i = C_{pi}e^{AM_i(r^2-r_p^2)}e^{-M_iB(C-C_p)} \tag{63}$$

So that by choosing a point in the cell, C_p, r_p, and C_{p1} the values of C_i can be generated by a few iterations, since $C_{pi} = C_{p1}{}^iK_i$.

XIII. The Amount of Information Required to Establish a Stoichiometry of Associating Systems

In this section, I would like to consider two problems that have been discussed in the literature: First, how much information is sufficient concerning a particular reacting system to establish its stoichiometry? Second, is it worthwhile to obtain molecular weight averages other than the weight average from the same C vs. r data?

In order to attempt to answer these questions, I would like to approach the problem by the method outlined by Weber[72] and improved by Deranleau.[73] This method utilizes information theory for the solution.[74]

In order to answer these questions, we define a saturation fraction as

$$S = (C_2 + C_3 + C_4 + \cdots + C_n)/(C_1 + C_2 + C_3 + \cdots + C_n) \tag{112}$$

where the C_i represent concentrations of the individual species. Clearly S is C polymer/C total; and $1 - S = x_1$ is the weight fraction of monomers. S represents the saturation of monomeric units by other monomers. It can be solved for measured parameters using the method of Steiner

$$S = 1 - \exp\left[\int_{C=0}^{C}\left(\frac{M_1}{M_w} - 1\right)d\ln C\right]$$
$$= 1 - \exp\left[\frac{M_1}{M_n} - 1\right]\exp\left[\int_{C=0}^{C}\left(\frac{M_1}{M_n} - 1\right)d\ln C\right]$$

Now S is directly proportional to the probability of reaction, and $(1 - S)$ represents the lack of reaction. The information in a system of independent probabilities, p_i, is

$$I(p) = -K\sum_{i=1}^{n} p_i \ln p_i \tag{113}$$

where $I(p)$ is the information that can be extracted from the probabilities p_i. These probabilities must sum to unity. If $K = (\ln 2)^{-1}$ the units of information are "bits." Since S was defined according to the probability of a reaction or no reaction $(1 - S)$, then Eq. (113) can be cast into the form[73]

[72] G. Weber, *in* "Molecular Biophysics" (B. Pullman and M. Weissbluth, ed.). Academic Press, New York, 1965.
[73] D. A. Deranleau, *J. Amer. Chem. Soc.* **91**, 4044 (1969).
[74] L. Brillion, "Science and Information Theory." Academic Press, New York, 1962.

$$I(S) = -K[S \ln S + (1 - S) \ln (1 - S)] \tag{114}$$

Now, the maximum amount of information concerning a system is[73]

$$I_{\max} = \int_{S=0}^{1} I(S)dS \tag{115}$$

and the information at any intermediate value is

$$I = \int_{S=0}^{S} I(S)dS \tag{116}$$

Assuming that data are collected from the point $C=0$. The fractional information from an experiment calculated from $C=0$ becomes

$$I/I_{\max} = S - S^2 \ln S + (1 - S)^2 \ln (1 - S) \tag{117}$$

Now, in any self-associating system it is not possible to obtain $I_{\max}$ owing to mass action. However, it is possible to use the rate of accumulation of information as a criterion of experimental information[73]

$$\frac{I}{SI_{\max}} = 1 - S \ln S + \frac{1}{S} (1 - S)^2 \ln (1 - S) \tag{118}$$

This function has a maximum at 83% of the total possible information corresponding to $S = 0.76$ (Fig. 29). This point represents the optimum in accumulation of information concerning self-associating systems and suggests that the stoichiometry of a self-associating system can be specified when the weight fraction of monomer has been reduced to 0.24.

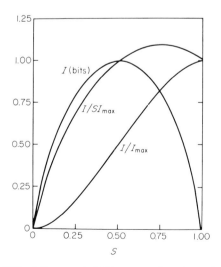

FIG. 29. Values of $I(S)$, $I/I_{\max}$, and $I/SI_{\max}$ as a function of the saturation fraction, S. At the maximum in $I/SI_{\max}$, $I/I_{\max} = 0.83$ and $S = 0.76$.

For complex formation where it is possible to add one component independently of the others, this criterion of $S = 0.76$ is often easily fulfilled.[73] For the self-associating systems, this can be accomplished only by performing experiments at a variety of dilutions.

Having this criterion for discussions concerning stoichiometry is not complete by itself. Many proteins are sufficiently insoluble so that $x_1 = 0.24$ cannot be attained. However, when $x_1 = 0.24$ can be achieved, then this seems sufficient to establish a thermodynamic stoichiometry.

Deonier and Williams[62] have expressed doubt that the calculation of molecular weight averages other than the weight-average contributes additional information for imprecise data. van Holde and Rosetti[59] and van Holde et al.[54] feel strongly that such data can be put to good use. It is clear that Roark and Yphantis[39] and Hoagland and Teller[45] agree with the latter view since their methods of analyses are based on combinations of these averages.

The problem that we will consider is to decide whether a system is a monomer–dimer system or an indefinite self-association for weakly associating system. For a monomer–dimer system, the saturation fraction, S is simply the fraction of dimer. Figure 30A presents the fractional information which can be obtained from each of M_n, M_w, and M_z; 50% of the information corresponds to $M_n/M_1 = 1.33$, $M_w/M_1 = 1.50$, $M_z/M_1 = 1.67$. For an indefinite self-association (Fig. 30B), $S = 1 - (1 - KC_1)^2$. Values of I/I_{max} are graphed only to $S = 0.75$; that is, $I/I_{max} = 0.83$. From this graph, it is clear that a monomer–dimer system would not be considered if 50% of the information was available, since $M_z/M_1 = 2.5$. However, if only M_w had been calculated, such a model might be considered. Aside from the fact that $2M_n - M_w = M_1$ would allow easy recognition of the indefinite association, this graph demonstrates the utility of several averages.

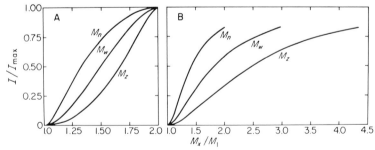

Fig. 30. Fractional information as a function of molecular weight averages, M_x/M_1, for self-association systems. (A) Monomer–dimer system; (B) indefinite self-association.

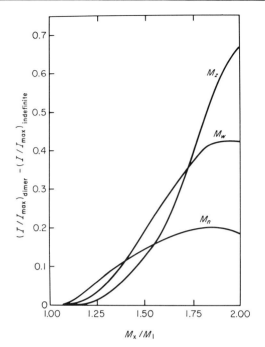

FIG. 31. Fractional information difference between monomer–dimer and indefinite self-association models as a function of molecular weight averages. At values of M_x/M_1 greater than 1.4, it should be possible to distinguish these models.

Figure 31 demonstrates the desirability of obtaining molecular weight averages in the region of high saturation fractions. Here, the ordinate is the difference in fractional information between the dimerizing system and indefinite association. It represents the degree to which one can distinguish between the two models. Let us assume that the system is truly monomer–dimer, and we are attempting to distinguish this model from an indefinite self-association. Table VI demonstrates how great a saturation is required to distinguish these models. Several points are clear from the figure and the table. First, it is worthwhile to calculate the Z-average molecular weights, and higher averages, if possible. Second, from the table it seems probable that a distinction between the two models cannot be made at saturation fractions reaching only 0.35 unless the data for M_z are very precise. Third, as S exceeds 0.45, the models become distinguishable for the usual errors encountered in M_n, M_w, and M_z. Certainly, at $S = 0.55$ one would have little difficulty in selecting the proper model.

Table VII demonstrates that it is possible to distinguish an indefinite

TABLE VI
SATURATION FRACTION REQUIRED TO DISTINGUISH A MONOMER–DIMER
SYSTEM FROM AN INDEFINITE ASSOCIATION MODEL

	$\Delta(I/I_{max})^a$		
S^b	M_n data	M_w data	M_z data
0.25	0.02	0.03	0.07
0.35	0.05	0.08	0.13
0.45	0.08	0.15	0.23
0.55	0.11	0.23	0.36

^{a} $\Delta(I/I_{max})$ is $(I/I_{max})_{dimer} - (I/I_{max})_{indefinite}$.
^{b} In this table, S is the weight fraction of dimer.

self-associating system from a monomer–dimer model on the basis of
M_z data, provided that the saturation fraction exceeds 0.25 and the
z-average data are reliable.

The figures and tables presented in this section have been calculated
for thermodynamically ideal systems. For nonideal proteins, since M_z
is depressed by the nonideality more than M_w which, in turn, is more
affected than M_n, it seems necessary to attain higher saturation fractions
than with ideal systems. For example, Deonier and Williams[62] found
$S \simeq 0.35$ for the monomer–dimer model and approximately 0.40 for
the indefinite association model but were unable to distinguish the mechanisms using M_w data alone. In this case, the z averages would have been
of assistance.

It is probably clear at this point that the distinction between a
monomer–dimer–trimer model and an indefinite self-association for
weakly associating systems is even more difficult to analyze[54] requiring
quite accurate values of molecular weight averages to distinguish among

TABLE VII
SATURATION FRACTION REQUIRED TO DISTINGUISH AN INDEFINITE
ASSOCIATION SYSTEM FROM A MONOMER–DIMER MODEL

	$\Delta(I/I_{max})^a$		
S^b	M_n data	M_w data	M_z data
0.15	0	0.01	0.03
0.25	0.02	0.07	0.26
0.35	0.05	0.18	0.60

^{a} $\Delta(I/I_{max})$ is the same as Table VI.
^{b} In this table, $S = 1 - (1 - KC_1)^2$.

postulated stoichiometries. Nonetheless, if a sufficiently large value of the saturation fraction is obtained for a particular model, it seems reasonable to use that model for further thermodynamic calculations.

Many proteins are highly associated assemblies of a discrete number of identical subunits, but dissociate slightly in solution. As an example, consider triose phosphate dehydrogenase.[45] In order to establish its stoichiometry of dissociation, we could proceed as previously and ask that the saturation fraction exceed 0.76. However, here the problem is to show dissociation rather than association. Consequently it makes more sense to reverse the analysis and demonstrate that S is less than 0.24. However, in this case no dissociation of tetramer to monomeric subunits can be detected unless NAD is removed from the enzyme. In

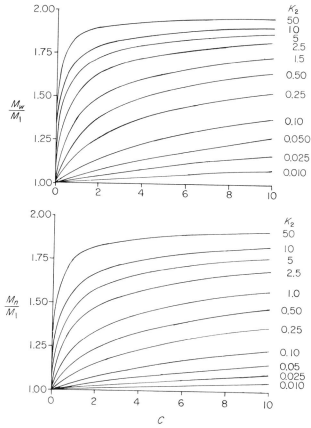

Fig. 32. Molecular weight averages as a function of concentration for various dimerization constants in a monomer–dimer system. The units of K_2 are reciprocal concentration units.

the presence of NAD, the dimer is the smallest species that can be detected by present ultracentrifuge techniques. In such cases of slight dissociation of an oligimer, it makes sense to me to demonstrate a small value of the saturation fraction, S, where S is calculated from the smallest observable oligimer.

That these stoichiometric considerations of weak association and slight dissociation of oligomeric enzymes are important can be illustrated by reference to Fig. 32. This figure shows M_w/M_1 and M_n/M_1 vs. C for a variety of dimerization constants. Now, consider that the molecule is tetrameric. Observation of molecular weight averages vs. concentration alone cannot be used to determine whether the tetramer is weakly associating ($K_2 = 0.1$–0.25) or dissociating ($K_2 \geqslant 50$) since the errors in measurement of the averages are very uncertain at low concentration. For this reason it is desirable to find perturbants of the enzyme (such as dilute urea) which increase the dissociation or association and lead to S values in the region of 0.5.

XIV. Why?

Why should one wish to study protein–protein interactions? The usual answer that one encounters is that it is important to determine these thermodynamic parameters for their own sake. In this section I would like to go one step further and present my own reasons for interest in protein–protein interactions. I shall present these reasons as problems and hypotheses concerning protein structure and function.

Problem 1: Do all self-associating systems involve a shift in pK of ionizable groups? I raise this question because Aune and Timasheff fit the dimerization of α-chymotrysin at low pH by a shift in pK according to the linked functions of Wyman.[75] At the same time, Horbett and I had fit the curve according to classical electrostatic theory, which does not require pK shifts. The difference between these two models is that a pH shift should occur upon dimerization if the linked function theory is correct, whereas no pH shift would be seen in the electrostatic model. Horbett measured the shift in pH upon dilution of very concentrated solutions of α-chymotrypsin. He found a change in hydrogen ion binding very close to that predicted by the shift in pK's given by Aune and Timasheff.[36] Thus, we have concluded that their model of the dimerization of this protein is correct. There are also reasons for believing that this is a general phenomenon. First, most reversible equilibria of interaction between proteins are variable functions of pH, and frequently the change of ΔF with pH is too steep to

[75] J. Wyman, *Advan. Protein Chem.* **19**, 224 (1964).

be attributed to the ionization of a single group; however a shift in pK of a single group is sufficient to fit the data.

Second, if you think about it, if any ionizable groups are involved in self-association or complex formation between proteins, there must be a shift in pK due to the change in the electrostatic environment.

Problem 2: Is the repulsive force due to the surface potential negligible in all cases of protein–protein interaction? There are several reasons for asking this question. First, the pK values given by Aune and Timasheff[36] for the ideal association of α-chymotrypsin fit the observed hydrogen ion binding data better than the pK's given which included long-range electrostatic interactions. Second, when an ionizable dye is attached to a protein,[76] the values do not give flattened titration curves as expected from electrostatic theory,[51] even when sodium dodecyl sulfate is used to increase the protein charge. Third, the Verwey–Overbeek[77] theory of the repulsive surface potential is designed for large particles with small charge. If one includes their attractive potential for proteins together with the repulsive term, one concludes that all proteins should not be soluble. At the distances separating interacting proteins (5–10 Å), this surface potential may be negligible in most cases.

Problem 3: Do all protein–protein interactions involve polypeptide chain termini? This problem is more important biologically than the preceding two and much more difficult to establish. The reason for presenting the question is that there are indications that it may be at least partially true. It has been established for α-chymotrypsin that the C-terminal tyrosine 146 is essential for dimerization at low pH.[78,79] In hemoglobin, the C-terminal residues play an important role in subunit interactions.[80] The N-terminal residues may be involved in subunit binding in lactic dehydrogenase.[81] Since bovine procarboxypeptidase A (a three-subunit structure) dissociates upon activation to carboxypeptidase A, the amino terminal residues are implicated in this case also.[82]

[76] I. M. Klotz and J. Ayers, *J. Amer. Chem. Soc.* **79**, 4078 (1957).

[77] E. J. Verwey and J. Th. G. Overbeek, "Theory of the Stability of Lyophobic Colloids." Elsevier, Amsterdam, 1948.

[78] J. A. Gladner and H. Neurath, *J. Biol. Chem.* **206**, 911 (1954).

[79] T. A. Horbett and D. C. Teller, Joint Conference of the Chemical Institute of Canada and the American Chemical Society, Toronto, Canada, May, Coll. 14, 1970.

[80] M. F. Perutz, *Nature (London)* **228**, 726 (1970).

[81] M. J. Adams, G. C. Ford, R. Koekoek, P. J. Lentz, Jr., A. McPherson, Jr., M. G. Rossman, I. E. Smiley, R. W. Schevitz, and A. J. Wonacott, *Nature (London)* **227**, 1098 (1970).

[82] J. R. Brown, R. N. Greenshield, M. Yamashaki, and H. Neurath, *Biochemistry* **2**, 867 (1963).

The final reason why I think this may be a general phenomenon arises from a consideration of evolution: an easy way to convert an enzyme from a single polypeptide chain to an oligimer would be the addition of a sticky "tail" on one or the other terminus.

This hypothesis suggests that a great deal about the mechanism of protein–protein interactions might be deduced by digestion of the protein with exopeptidases to change the polypeptide termini.

There are numerous questions of a similar type which might be asked about specific ions and solvents such as D_2O. For example, does D_2O always enhance self-association? Does a high concentration of $CaCl_2$ depress such reactions? Hopefully, the answers to these and other questions will begin to appear in the near future.

Appendix: Newton–Raphson Method of Calculating Virial Coefficients

To illustrate the Newton–Raphson method of fitting nonlinear polynomial functions, we will use the nonideal equations for the concentration such as are found in Sections X and XI. These equations may be written in the form

$$C = \sum_{i=1}^{n} a_i x^i e^{iBM_1C}$$

In this equation, the a_i are coefficients and the x^i values vary in significance according to the method being used. The way to proceed in the curve fitting is to calculate the ideal equations with $B = 0$ to determine the first approximation for the values of a_i. To show the form of the future iterations, we make the function

$$D_j = -C_j + \sum_{i=1}^{n} a_i x_j^i e^{iBM_1C_j} \tag{A.1}$$

where the subscript j refers to the point number. The derivatives with respect to all parameters of the D_j are

$$\frac{\partial D_j}{\partial a_i} = x_j^i e^{iBM_1C_j} \qquad i = 1, 2, \cdots n \tag{A.2a}$$

$$\frac{\partial D_j}{\partial B} = \sum_{i=1}^{n} a_i i M_1 C_j x_j^i e^{iBM_1C_j} \tag{A.2b}$$

and we now solve the matrix of equations for the sums over the data points $j = 1, 2, \ldots N$

$$
\begin{bmatrix}
\sum_j \dfrac{\partial D_j}{\partial a_1} D_j \\[4pt]
\sum_j \dfrac{\partial D_j}{\partial a_2} D_j \\[4pt]
\vdots \\[4pt]
\sum_j \dfrac{\partial D_j}{\partial a_n} D_j \\[4pt]
\sum_j \dfrac{\partial D_j}{\partial B} D_j
\end{bmatrix}
=
\begin{bmatrix}
\sum_j \left(\dfrac{\partial D_j}{\partial a_1}\right)^2 & \sum_j \dfrac{\partial D_j}{\partial a_2}\dfrac{\partial D_j}{\partial a_1} & \cdots & \sum \dfrac{\partial D_j}{\partial a_n}\dfrac{\partial D_j}{\partial a_1} & \sum \dfrac{\partial D_j}{\partial B}\dfrac{\partial D_j}{\partial a_1} \\[6pt]
\sum \dfrac{\partial D_j}{\partial a_1}\dfrac{\partial D_j}{\partial a_2} & \sum \left(\dfrac{\partial D_j}{\partial a_2}\right)^2 & \cdots & \sum \dfrac{\partial D_j}{\partial a_n}\dfrac{\partial D_j}{\partial a_2} & \sum \dfrac{\partial D_j}{\partial B}\dfrac{\partial D_j}{\partial a_2} \\[6pt]
\cdots & \cdots & & \cdots & \cdots \\[6pt]
\sum_j \dfrac{\partial D_j}{\partial a_1}\dfrac{\partial D_j}{\partial a_n} & \sum \dfrac{\partial D_j}{\partial a_2}\dfrac{\partial D_j}{\partial a_n} & \cdots & \sum \left(\dfrac{\partial D_j}{\partial a_n}\right)^2 & \sum \dfrac{\partial D_j}{\partial B}\dfrac{\partial D_j}{\partial a_n} \\[6pt]
\sum \dfrac{\partial D_j}{\partial a_1}\dfrac{\partial D_j}{\partial B} & \sum \dfrac{\partial D_j}{\partial a_2}\dfrac{\partial D_j}{\partial B} & \cdots & \sum \dfrac{\partial D_j}{\partial a_n}\dfrac{\partial D_j}{\partial B} & \sum \left(\dfrac{\partial D_j}{\partial B}\right)^2
\end{bmatrix}
\begin{bmatrix}
\Delta a_1 \\[4pt]
\Delta a_2 \\[4pt]
\vdots \\[4pt]
\Delta a_n \\[4pt]
\Delta B
\end{bmatrix}
$$

$$(A.3)$$

The values of the D_j and the partial derivatives are calculated from Eqs. (A.1) and (A.2). The values of Δa_i and ΔB are used to update the estimates from the previous calculations

$$a_i{}^{new} = a_i{}^{old} - \Delta a_i$$

and

$$B^{new} = B^{old} - \Delta B$$

The iteration is repeated until the values of Δa_i and ΔB are sufficiently small.

Further discussion of nonlinear curve fitting is given by Worsley and Lax,[83] Haschemeyer and Bowers,[84] and Dyson and Isenberg.[85]

[83] B. H. Worsley and L. C. Lax, *Biochim. Biophys. Acta* **59**, 1 (1962).
[84] R. H. Haschemeyer and W. F. Bowers, *Biochemistry* **9**, 435 (1970).
[85] R. D. Dyson and I. Isenberg, *Biochemistry* **10**, 3233 (1971).

[15] Studies of Protein Ligand Binding by Gel Permeation Techniques

By Gary K. Ackers

General Considerations

Interactions between proteins and small ligand species can be effectively studied by a group of techniques based on the distribution of solute within porous gel networks. Frequently the distribution process is utilized in a chromatographic experiment and the technique is variously called gel filtration, gel permeation chromatography, molecular sieve chromatography, or simply gel chromatography. In other applications the differences between solute distribution of the various molecular species are utilized in a purely equilibrium situation and measured by direct optical scanning or by batch equilibration procedures. Because of the similarity in basis between these approaches, they will be described collectively in this chapter as *gel permeation techniques*. We will be concerned with experimental methods that utilize the principles of gel permeation for the study of macromolecule–ligand interactions. Attention will be focused on practical applications rather than theoretical analyses or details of instrumentation. For these the reader is referred elsewhere to original sources or to selected review articles.[1-3] Only certain results of the theoretical treatments will be presented as they form the basis of experimental procedure or interpretation of experimental results. The reader is also referred elsewhere[3] for detailed descriptions of molecular sieve media, including their chemical and physical properties, ranges of porosity, etc. It should be noted that the most commonly employed gels of dextran, polyacrylamide, or agarose are frequently sufficiently inert in moderate ionic strength buffers so that adsorptive interactions do not interfere with determinations down into the microgram per milliliter ranges of concentration. In cases where this is not true, appropriate corrections can be made. Using molecular sieve techniques, it is possible for the experimenter to obtain a great deal of information with only very limited apparatus and very small quantities of protein.

Several procedures have been developed in recent years which take advantage of the equilibrium dialysis properties of gel partitioning systems for the study of ligand binding. The two methods to be described here have been chosen because they offer convenience and precision. The simplest method in terms of required equipment is the Hummel–Dreyer technique (see below, under heading "The Hummel–Dreyer Method") carried out by elution chromatography.[4,5] Basically the only equipment required is a chromatographic column, a small amount of gel, and a spectrophotometric monitoring device. For many applications to enzyme systems the last of these is either replaced by or used in a sensitive biological assay procedure. Such activity assays permit study of

[1] G. K. Ackers, *Advan. Protein Chem.* **24**, 343 (1970).
[2] D. J. Winzor, *in* "Physical Principles and Techniques of Protein Chemistry" (S. J. Leach, ed.), Part A, p. 451. Academic Press, New York, 1969.
[3] L. Fischer, "An Introduction to Gel Chromatography." Wiley (Interscience), New York, 1969.
[4] J. P. Hummel and W. J. Dreyer, *Biochim. Biophys. Acta* **63**, 530 (1962).
[5] G. F. Fairclough and J. S. Fruton, *Biochemistry* **5**, 673 (1966).

enzyme systems in impure form and may extend the concentration range down into the nanogram per milliliter region. Because of these advantages and the ready accessibility in any biochemical laboratory of necessary equipment, the Hummel–Dreyer elution method will be described in detail.

A generally more precise technique is the Brumbaugh–Ackers method (see below, under heading "The Brumbaugh–Ackers Method"), which utilizes direct optical column scanning and measures ligand binding at many points within a single column.[6,7] Although this approach requires considerably more sophisticated instrumentation and expense, the advantages of speed and precision certainly warrant the additional investment for any extensive studies of macromolecular interaction. The necessary instrumentation is now commercially available (Schoeffel Inst. Co., Westwood, New Jersey) so that time-consuming efforts of instrument building are no longer required. In addition, the direct optical-scanning approach allows one to carry out experiments of the Hummel–Dreyer type with improved convenience, speed, and precision.[7]

Definitions: Quantities of Interest to Binding Studies

Before describing the techniques, a few comments are necessary regarding the type of molecular system to be studied and the experimental quantities of interest. The type of system to be studied is represented by the general scheme for reaction equilibria involving the successive binding of small ligand species, L, to a macromolecular component, P.

$$P + L \rightleftharpoons PL$$
$$PL + L \rightleftharpoons PL_2$$
$$PL_{i-1} + L \rightleftharpoons PL_i \tag{1}$$
$$PL_{n-1} + L \rightleftharpoons PL_n$$

The successive binding constants for the system are represented by

$$K_i = \frac{[PL_i]}{[PL_{i-1}][L]} \qquad i = 1, \cdots n \tag{2}$$

In a system of this type, the experimental quantity required for determination of successive binding constants is the binding ratio, r, given by

$$r = \frac{[L_0] - [L]}{[P_0]} \tag{3}$$

In Eq. (3) $[P_0]$ is the total (constituent) concentration of macromolecule, defined by the relationship

[6] E. E. Brumbaugh and G. K. Ackers, *J. Biol. Chem.* **243**, 6315 (1968).
[7] E. E. Brumbaugh and G. K. Ackers, *Anal. Biochem.* **41**, 543 (1971).

$$[P_0] = \sum_{i=0}^{n} [PL_i] \qquad (4)$$

The quantity $[L]$ is concentration of free (unbound) ligand and $[L_0]$ is the total (constituent) concentration of ligand,

$$[L_0] = [L] + \sum_{i=1}^{n} i[PL_i] \qquad (5)$$

Although a complete discussion on the interpretation of binding ratios in terms of molecular models is beyond the scope of this chapter, consideration of the simplest case provides an illustration of the types of conditions under which binding ratios should be determined. If all of the n binding sites are identical and have identical intrinsic binding constants K, then $K_i = [(n - i + 1)/i]K$. The Scatchard equation relates the experimental quantities r and $[L]$ to the stoichiometry n, and binding constants, K.

$$\frac{r}{[L]} = K(n - r) \qquad (6)$$

This relationship illustrates the fact that the binding constants can in principle be determined from measurement of the binding ratio r as a function of free ligand concentration $[L]$. Such data are usually analyzed by plotting $r/[L]$ versus r. For cases where Eq. (6) is applicable, a straight line results with slope K and abscissa-intercept of n. If a non-linear curve is found, more complex forms of analysis must be attempted which are beyond the scope of present discussion.[8-11] It is important to note that binding ratios of extremely high precision are required for unequivocal resolution of any but the simplest cases,[8,9] The remainder of this chapter will deal with ways of accurately determining the binding ratio for protein systems to which ligand species are reversibly bound.

The Hummel–Dreyer Method

Principle

The determination of binding ratios can be efficiently carried out using a procedure first developed by Hummel and Dreyer.[4] A gel column

[8] I. Klotz and D. L. Hunston, *Biochemistry* **10**, 3065 (1971).
[9] G. Weber, *in* "Molecular Biophysics" (B. Pullman and M. Weissbluth, eds.). Academic Press, New York, 1965.
[10] D. A. Deranlau, *J. Amer. Chem. Soc.* **91**, 4044 (1969).
[11] D. A. Deranlau, *J. Amer. Chem. Soc.* **91**, 4050 (1969).

is equilibrated with solution containing the ligand at a desired concentration [L]. A small sample of protein solution in which the total ligand concentration equals that of the column-saturating solution is then added to the column. If the protein binds ligand, the solvent of this sample will be depleted with respect to ligand. When the sample is chromatographed on the column, the protein will be separated from the ligand-depleted solvent, moving ahead with its bound ligand. The resulting elution profile will exhibit a peak in ligand concentration above the ligand-saturation baseline. This peak represents the excess ligand bound. A corresponding trough will follow, representing the depletion that resulted from the bound ligand which was carried ahead with the protein. These effects are illustrated in Fig. 1. In principle, the area of the trough and peak are equal. However, measurement of the trough area is usually simpler and more reliable. From the area of the trough and known amount of protein applied, the binding ratio is determined corresponding to the ligand saturation level used.

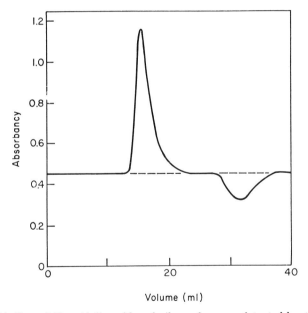

Volume (ml)

FIG. 1. Binding of 2'-cytidylic acid and ribonuclease as detected by the Hummel and Dreyer procedure on a Sephadex G-25 column equilibrated with 0.1 M acetate buffer, pH 5.3. Absorbancy measurements at 285 mm (ordinate) indicate a positive peak (left) and negative trough (right) with respect to the (horizontal) baseline absorbancy. From J. P. Hummel and W. J. Dreyer, *Biochim. Biophys. Acta* **63**, 530 (1962).

Experimental Procedures

Equipment Setup

Since maximum separation between ligand and macromolecule is desired, a gel should be chosen from which the macromolecule species is totally excluded. For most protein systems, Sephadex G-25 or G-50 are appropriate. In cases where ligands bind to Sephadex (e.g., aromatic ligands), an appropriate polyacrylamide gel should be used instead (e.g., Bio-Gel P-50). Usual procedures of soaking and swelling of gels are to be employed as recommended by the manufacturers. Most commercial preparations no longer contain "fines" that must be decanted after swelling.

Small glass columns (10–50 ml bed volume) should be used which have a length:diameter ratio of at least 10—e.g., a 1 cm × 25 cm column. The column is carefully packed by continuous addition of a slurry containing the desired gel bed material. Settling of the beads can be observed by eye. It is important that the column be packed in a continuous bed with no "layering discontinuities" present. The column should be fitted with porous polyethylene or Teflon disks at each end or other fittings which ensure well-defined extremities of the column bed. In some cases glass wool may be used at the bottom. However a clearly defined top of the column bed is of crucial importance for analytical determinations. The column should be jacketed so that the temperature may be regulated by an external thermostated bath. It is desirable that this bath also contain space for thermal equilibration of samples prior to their addition to the column. A reservoir containing buffer or the desired saturating solution of ligand allows flow into the column from the top (upward flow is of no advantage in these determinations and in fact can cause serious problems if the column bed shifts). The reservoir need only be a few centimeters above the top of the column. This height is not critical for tightly cross-linked gels such as G-25 or G-50, which are relatively insensitive to pressure effects. After the column has been packed it should be allowed to flow for several hours at its maximum "natural" flow rate and the porous disk at the top is adjusted if necessary after this period.

The outflow tubing at the bottom of the column is then connected to a precisely regulated pump (such as the LKB Varioperpex) which may be set for any flow rate less than the maximum "natural" rate mentioned above. It is desirable to calibrate the pump while connected to the column by passing the outflow tube into a burette and making successive volume readings corresponding to known time intervals. Calibration of the pump, although not usually critical, is of importance in systems where the establishment of the plateau between peak and trough is flow-rate dependent. In those cases it is highly desirable to be able to make reproducible flow-rate settings from one experiment to another.

Provision should be made for the measurement of effluent concentration profiles either by collecting small increments of effluent in a fraction collector, or by connecting the outflow tubing to a flow cell so that continuous monitoring by a spectrophotometric instrument is effected. The flow cell may be placed either before or after the pump. However, it is desirable to have the monitoring device as close to the column as possible to minimize possible distortion artifact that might result from flow characteristics of long tubing sections.

Procedure for a Binding Experiment

After the column has been equilibrated at the desired temperature, it is saturated by passing 1.5–2 column volumes of solution containing ligand at the desired concentration. This concentration will be the free ligand concentration [L] referred to in the analysis. The rigorous criterion of saturation for the column is that ligand concentration emerging from the bottom of the column must equal that being introduced at the top. If there is any doubt, these concentrations may be checked, e.g., spectrophotometrically, and compared. Usually it is sufficient to record effluent ligand concentration until it has achieved a constant value.

A 1-ml sample containing a mixture of ligand and protein is made up from stock solution (the best procedure is to weigh solutions to be mixed) so that (1) the concentration of protein in the sample is known and (2) the total concentration of ligand in the sample equals the concentration with which the column has been saturated. This sample is equilibrated for 30 minutes in a thermostated bath at the same temperature as the column to which it will be applied.

After the sample has equilibrated, it is applied to the top of the column as cleanly as possible and elution is subsequently carried out with the original saturating ligand solution. The elution profile is determined either by continuous recording of absorbance during elution, or by graphical plotting of the concentrations of collected fractions versus time or, equivalently, volume. It is essential that the elution profile exhibit a distinct peak and trough, and that a *plateau exist between them* of constant ligand concentration equal to the saturating level applied. If this plateau is not present it is likely that the binding equilibria are slow and the experiment should be repeated at a lower flow rate. In cases where no plateau can be established the experiment cannot be used for a valid determination of binding ratio and another method should be used.

Calculations of Amount Bound and Binding Ratio

The immediate object of the experiment described above is to determine the amount of ligand bound by the macromolecule sample applied to the column. Although in principle one can use either the peak or trough area to determine this binding ratio, the trough is usually the most reliable and convenient of the two. This is especially the case if concentrations of ligand are monitored spectrophotometrically, since there is no interference due to absorbance or light scattering of the protein which migrates with the excess ligand peak of the measured elution profile. Determination of the binding ratio is carried out as follows.

For Automatically Recorded Elution Profiles

From the strip chart recording of absorbance versus time the area of the trough is first determined (in square inches) by graphical integration or planimetry. The number of micromoles of ligand bound by the protein sample is then calculated from the formula

$$\mu\text{moles bound} = \frac{\text{area of trough}}{ab(\epsilon_L \times 10^{-3})} \tag{7}$$

In this formula a is the recorder pen displacement corresponding to unit absorbance and b is the chart movement (in inches) corresponding to a 1-ml increment of column effluent.

For Concentrations Measured from Collected Fractions

For data obtained as a series of absorbances corresponding to collected fractions, the amount of ligand is calculated from the formula

$$\mu\text{moles bound} = \frac{\sum_i (\Delta A_i)(\text{ml}_i)}{\epsilon_L \times 10^{-3}} \tag{8}$$

In this formula ΔA_i is the difference between the absorbance of fraction i and the baseline absorbance (i.e., that of saturating ligand solution). The volume of fraction i is ml_i, and ϵ_L is the molar extinction coefficient of ligand. The summation is carried out for all fractions within the trough region of the elution profile. For other types of ligand concentration determinations (e.g., biological activity or radioactive assay), a procedure similar to that of Eq. (8) is used.

Once the number of moles of bound ligand have been determined, the binding ratio r is calculated as the number of micromoles of ligand bound divided by the number of micromoles of protein in the sample applied.

Variation of Conditions

Since the binding ratio is a function *only* of [L], it is not useful only to carry out a series of experiments in which the amount of protein in the sample is varied without changing the saturating level of ligand. This necessitates a somewhat tedious process of completely saturating the column with solution at each desired concentration [L]. However, it should be noted that it is necessary only to use two different saturations in order to determine k and n for a system with n identical binding sites.[12] Substituting the defining relation for r (Eq. 3) into Eq. (6) yields, after rearrangement

$$\frac{1}{[L_0] - [L]} = \frac{1}{[P_0]} \left\{ \frac{1}{n} + \frac{1}{nK([L_0] - [L])} \right\} \tag{9}$$

Therefore a plot of reciprocal of ligand bound versus reciprocal of protein concentration yields a straight line of slope

$$\frac{1}{n} \left\{ 1 + \frac{1}{K([L_0] - [L])} \right\}$$

If two such plots are constructed, from data of different saturating levels [L], the two slopes enable calculation of the two unknowns n and k. In principle then, only one protein concentration [P_0] need be used plus two saturating levels [L]. However, it is desirable to obtain data from more [P_0] values in order to estimate precision.

The Brumbaugh–Ackers Method

Principle

This method is based on the direct optical scanning of a small chromatographic column which has been saturated with solution containing a desired mixture of ligand and macromolecule. In the simplest case, conditions are arranged so that the macromolecule, P, and all complexes, PL_i, are totally excluded from the gel. The system is then similar to an equilibrium dialysis experiment in which the mobile phase (void spaces) of the column correspond to the "inner" compartment, containing macromolecular species, complexes, and unbound ligand. The solvent space of the stationary phase (internal volume) acts as the

[12] I thank Dr. C. Huang for pointing out this method for analysis.

"outer" compartment containing only free ligand. When such an equilibrated column is optically scanned, the measurement of absorbances yield, at each point within the column, a determination of the binding ratio. Since the absorbance measurements are recorded digitally at several hundred points along the column, the experiment is equivalent to a large collection of replicate equilibrium dialysis experiments and appropriate statistical analysis can be used to evaluate the accuracy of results. In addition to the high precision attainable, the method is simple and convenient and requires only small quantities of material.

We will first describe the column scanning experiment and its use in measuring the phenomenological parameters necessary to the characterization of equilibrium partitioning within a gel column. We will then consider how these measurements can be used to determine the free ligand concentration and binding ratio for a system of the type represented by Eq. (1).[13]

The Equilibrium Saturation Technique

The scanning gel chromatograph uses precision-bore quartz columns (1 × 10 cm bed volume) which are packed with gel in the normal way and scanned at any desired wavelength of the visible or ultraviolet regions. Details of instrument design and operation are beyond the scope of the present discussion. It is assumed that a suitable instrument of commercial or home-made origin is available[6,7] so that accurate measurements of absorbance versus distance along the axis of the column can be made. We will discuss the conditions under which these measurements are to be carried out for binding studies and their interpretation in terms of binding parameters.

The equilibrium saturation experiment is shown diagrammatically in Fig. 2. First the column is equilibrated with buffer and optically scanned at the desired wavelength to establish a baseline. Subsequently the column is completely saturated with a solution of desired composition with respect to solute species of interest. After equilibration has taken place throughout all penetrable volume elements of the column, a second scan is performed in order to determine the apparent solute concentration as a function of distance along the column axis. There are seven regions of interest in the scan, shown in Fig. 2: (1) the air space above the column, (2) a spike at the interface between air and the top of the column, (3) a region of air space within the column, (4) a meniscus at the interface of air and solution above the column bed, (5) liquid

---

[13] The method of analysis also applies rigorously to systems of greater complexity in which any interactions between macromolecular species are also present, i.e., it applies to all cases of ligand-mediated protein subunit interactions.

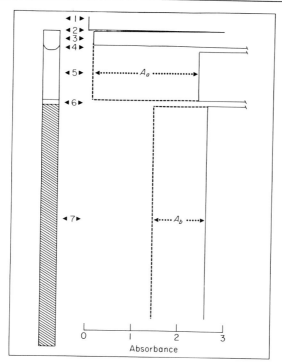

Fig. 2. Schematic diagram of column and idealized scan. Regions of interest are (1) air above the column, (2) spike corresponding the air–column interface, (3) air inside the column, (4) meniscus at the air–liquid interface, (5) liquid above gel bed, (6) porous disk at top of gel bed, (7) gel bed. Dashed lines represent baseline absorbance of buffer-saturated column. Solid trace represents absorbance of column saturated with solution. A_a is absorbance of solution above column bed, and A_b is corresponding absorbance within column bed. From E. E. Brumbaugh and G. K. Ackers, *Anal. Biochem.* **41**, 543 (1971).

above the gel bed, (6) a spike corresponding to the porous disk which separates the column bed from inflowing solution above it, and (7) the column bed itself. A comparison of absorbance values (after baseline subtraction) in region 5 (constant throughout the region) and any point within region 7 yields a measure of the *partition cross section* ξ pertaining to that point in region 7:

$$\frac{A_b}{A_a} = \xi \tag{10}$$

where A_b is absorbance at a point within the gel bed and A_a is absorbance in free solution above the column bed. The partition cross section is the fundamental quantity which characterizes the equilibrium partitioning of solute. This quantity represents the solute distribution volume per unit distance along the column axis, and is directly determined by optical scanning of the column. For a solute species, j, the partition cross section is related to three other experimental quantities

$$\xi_j = \alpha + \beta\sigma_j \tag{11}$$

where α is the void volume cross section of the column, β is the internal volume cross section, and σ_j is the partition coefficient for species j. For most purposes values of α, β, and ξ_j are normalized to unit column cross-sectional area and thus represent fractions of the column cross-section. It is this normalized ξ which is determined by the column-scanning procedure. Because the packing characteristics of gel columns are invariably nonuniform, the quantities α and β (and consequently ξ) will vary from point to point along the axis of the column. The partition coefficient, on the other hand, is constant to within very narrow limits throughout the column.[14]

The scanning chromatograph provides data in the form of several hundred absorbance values corresponding to positions along the column axis. Calculations from a scan are best performed by computer, and the data are provided in digital form (e.g., on punched paper tape) for this purpose. The absorbance measurements corresponding to region 5 are first averaged providing a single value of A_a for comparison with each value of A_b within region 7. The values of A_b are determined by subtraction of corresponding baseline absorbances at each point from the absorbances measured in the saturation scan. Subsequently the ratio $A_b{:}A_a$ is calculated at each point to obtain corresponding values of ξ.

Procedure for Binding Studies

In order to obtain binding ratios from measured values of ξ, it is necessary to know α and β at each point within the column. The saturation experiment described above is first carried out using a totally excluded solute species (e.g., tobacco mosaic virus) for which $\sigma = 0$. The values of ξ obtained from these scans thus represent the void fraction α as a function of distance along the column. Corresponding values of $\alpha + \beta$ are obtained by subsequently carrying out the experiment with a totally nonexcluded solute (e.g., glycylglycine-scanned at 220 nm) for which $\sigma = 1$. The partition cross sections thus obtained are equal to $\alpha + \beta$ at each point within the column, and β is obtained at each point by subtraction. Once α and β values have been determined, correspond-

[14] H. S. Warshaw and G. K. Ackers, *Anal. Biochem.* **42**, 405 (1971).

ing saturation experiments are carried out with solutions comprised of mixtures of ligand and protein. Each solution is made up from stock solutions by weight so that total concentration $[P_0]$ and $[L_0]$ are accurately known. From scans of these saturation experiments the free ligand concentration and binding ratio is calculated at each point within the column bed as described below.

Calculation of Free Ligand Concentration and Binding Ratio

Since the total concentrations of both ligand, $[L_0]$, and macromolecule $[P_0]$ with which the column has been saturated are known, it is evident from Eq. (3) that the binding ratio can be determined if the free ligand concentration $[L]$ can be found. The calculation of $[L]$ follows from the measurement of A_a and A_b at each point within the column, according to the relationships described below. The absorbance above the column bed is written in terms of species concentrations and extinction coefficients

$$A_a = \epsilon_P[P] + \epsilon_{PL}[PL] + \epsilon_{PL_2}[PL_2] \cdots + \epsilon_{PL_n}[PL_n] + \epsilon_L[L] \quad (12)$$

The absorbance at any point within the column bed is

$$A_b = \epsilon_P\xi_P[P] + \epsilon_{PL}\xi_{PL}[PL] + \epsilon_{PL_2}\xi_{PL_2}[PL_2] \cdots + \epsilon_{PL_n}\xi_{PL_n}[PL_n] + \epsilon_L\xi_L[L] \quad (13)$$

Since all macromolecular species are excluded,

$$\xi_P = \xi_{PL} = \xi_{PL_2} \cdots = \xi_{PL_n} = \alpha \quad (14)$$

and with

$$A_b = \alpha A_a - \epsilon_L\beta[L] \quad (15)$$

so that the free ligand concentration is calculated from

$$[L] = \frac{A_b - \alpha A_a}{\epsilon_L\beta} \quad (16)$$

Since the extinction coefficient, ϵ_L, of free ligand can easily be known independently, the determination of α, β, A_a, and A_b in this experiment enables one to calculate free ligand concentration regardless of the complexity of the system. The calculation is done at many points within the column, yielding constant values of $[L]$ and consequently r.

Method of Correction for Binding to the Gel

If ligand also binds to the gel, it is necessary to determine the concentration dependence of this binding in order to be able to make corrections at any desired free ligand concentration. This function, the ligand-gel binding isotherm, is determined by measuring the apparent partition cross section $\xi'_L = A_b/A_a$ for the column saturated with different ligand concentrations in the absence of protein. In these measurements $A_a = \epsilon_L[L]$ and $A_b = \epsilon_L \xi_L[L] + \epsilon_L Q_L$ where Q_L is the apparent concentration of ligand bound to the gel. At each point within the column this quantity can be determined as

$$Q_L = [L](\xi'_L - \xi_L) \tag{17}$$

When ξ'_L is measured at a series of different saturating concentrations $[L]$ of ligand, the binding isotherm $Q_L(L)$ is determined. Note that a comparison of ξ'_L with ξ_L provides a test for binding to the gel. If $\xi'_L = \xi_L$ then $Q_L = 0$ and the absence of binding is established. When binding is present ξ'_L will be greater than ξ_L and will approach ξ_L in the limit of zero ligand concentration. Accuracy of the binding isotherm rests on a proper determination of ξ_L. For Sephadex gels the use of glycylglycine is recommended for this purpose since it has been shown to occupy the same distribution volume within the gel as water (i.e., $\sigma = 1$). If the ligand binds to the gel, Eq. (13) must include an $\epsilon_L Q_L$ term with the result

$$[L] + \frac{Q_L([L])}{\beta} = \frac{A_b - \alpha A_a}{\epsilon_L \beta} \tag{18}$$

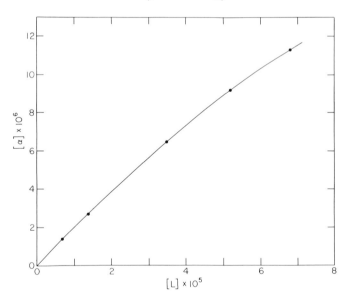

FIG. 3. Methyl orange–Sephadex binding isotherm. Ordinate represents concentration of methyl orange bound to gel. Abscissa is corresponding free ligand concentration. Data are averages over a 30-point segment of gel bed. From E. E. Brumbaugh and G. K. Ackers, *Anal. Biochem.* **41**, 543 (1971).

Once $Q_L([L])$ has been determined, the left-hand side of Eq. (18) becomes a known function from which the free ligand concentration $[L]$ may be calculated for each measured value of A_b/A_a. Equation (18) defines a completely general method for determination of $[L]$. The form of the function Q_L will depend on the particular system studied and must be determined experimentally. An example of such an isotherm is shown in Fig. 3, representing the binding of methyl orange to Sephadex G-75. Data of this kind can usually be fitted to a polynomial function of low degree which provides a convenient way of expressing Q_L for use of Eq. (18). A second-degree polynomial

$$Q_L = a_0 + a_1[L] + a_2[L]^2 \tag{19}$$

is frequently adequate for this purpose. From data of the type shown in Fig. 3, the constants a_0, a_1, and a_2 are determined by polynomial regression.

DETERMINATION OF BSA-MO BINDING BY EQUILIBRIUM SATURATION METHOD[a]

Expt. no.	$[P_0] \times 10^5$	$[L_0] \times 10^5$	$[L] \times 10^5$	$SD \times 10^5$	No. of points	r
1	1.4412	7.6552	4.128	0.0176	138	2.448
2	0.7146	7.6839	5.483	0.0314	245	3.080
3	0.1320	7.7077	7.151	0.0086	156	4.216
4	1.5503	5.1448	2.658	0.0358	249	1.604
5	0.7670	5.0953	3.531	0.0237	240	2.039
6	0.3779	5.1702	4.266	0.0188	246	2.392
7	0.0607	5.1693	4.973	0.0122	237	3.224
8	1.4927	1.0398	0.510	0.0089	241	0.355
9	0.7409	1.0517	0.713	0.0066	228	0.457
10	0.1476	1.0456	0.970	0.0080	204	0.510

[a] E. E. Brumbaugh and G. K. Ackers, *Anal. Biochem.* **41**, 543 (1971).

An example of results obtained using this method is shown in the table for the binding of methyl orange to bovine serum albumin. The standard deviations pertain to final calculated values of $[L]$ after correction for the binding isotherm and provide a measure of the experimental precision obtainable. In cases where such correction is not necessary, the precision can be as much as an order of magnitude higher. A careful selection of gels is therefore warranted. For studies of binding with aromatic ligand species, it is recommended that polyacrylamide gels be used instead of Sephadexes.

[16] Determination of Equilibrium Constants by Countercurrent Distribution

By Gerson Kegeles

Introduction and General Principles

One of the basic ideas behind the use of countercurrent distribution methods to study chemical equilibria is that *if one knows* the stoichiometry of the reactions and the values of the equilibrium constants and of the individual partition coefficients, one can make a *unique* prediction of the detailed shape of the countercurrent distribution pattern.[1-3] One can then perform appropriate experiments on the reacting mixture, and make detailed comparisons between experimental results and predictions.[4,5] Thus in this approach, the experimenter is basically performing a somewhat sophisticated curve-fitting procedure by adjusting in turn one of a series of unknown parameters.[4,5] As in all such procedures, the choice of a model to represent a real system may be highly appropriate but, because of real experimental error, the correctness of the choice of a model is never absolutely certain.[5]

A second basic idea is that, if one performs a countercurrent experiment analogous to sedimentation or descending-boundary free electrophoresis, the rate of transport of mass in the constant-composition (plateau) region is a weight-average of the rate of mass transport for all reacting species at the plateau region composition.[6] By complete analogy to the general techniques of sedimentation and electrophoresis,[7] one can then use such countercurrent distribution data, plus those for the individual species, to determine equilibrium constants. This technique presupposes that it is possible to isolate individual reacting species and measure partition coefficients for each, which may be most nearly true in the case of reactions between dissimilar species. This leads to a procedure whose basic principle is essentially synonymous with that of single extraction, but in which it is possible to attain higher experimental accuracy.

A variation of this idea is to perform countercurrent distribution

[1] J. L. Bethune and G. Kegeles, *J. Phys. Chem.* **65**, 433 (1961).
[2] J. L. Bethune and G. Kegeles, *J. Phys. Chem.* **65**, 1755 (1961).
[3] J. L. Bethune and G. Kegeles, *J. Phys. Chem.* **65**, 1761 (1961).
[4] V. P. Saxena and G. Kegeles, *Arch. Biochem. Biophys.* **139**, 206 (1970).
[5] R. C. Williams, Jr. and L. C. Craig, *Separ. Sci.* **2**, 487 (1967).
[6] G. Kegeles, *Arch. Biochem. Biophys.* **144**, 763 (1971).
[7] L. W. Nichol, J. L. Bethune, G. Kegeles, and E. L. Hess, *in* "The Proteins" (H. Neurath, ed.), 2nd ed., Vol. 2, p. 305. Academic Press, New York, 1964.

experiments in which, somewhere in the middle of the apparatus, there is an abrupt shift in total concentration, providing a boundary between two plateau regions of constant composition. The movement of this "concentration-boundary" in countercurrent distribution is completely analogous to the studies made in electrophoresis,[8] chromatography,[9] and sedimentation.[10-12] As indicated,[12] such so-called "differential" measurements in the case of a self-interacting solute (a single thermodynamic component) will characterize a Z average of the quantity being studied (in the present case this is the rate factor $K/(1 + K)$). When only one reaction step is taking place, the combination of the weight-average and Z-average values leads quite directly to the equilibrium constant.[12] Even if many species are present, however, the combination of Z-average and weight-average data can be used in principle[12] to evaluate several equilibrium constants without the need to measure values for every isolated individual species.

Procedures

Zonal Countercurrent Distribution

The application to reacting systems of the usual, zonal procedure, in which solute is placed in only one, or a very few tubes in the apparatus[13] depends on the possibility of affecting appreciably the shape of the pattern by partial decomposition of complexes, or polymers, during separation.[1,2] An example[4] is the study of the reaction between $HgBr_2$ and $HgCl_2$ to form $HgClBr$. This reaction had been studied by Raman methods[14,15] and reported to have a formation constant in methanol of 2.0 ± 0.2. Subsequent studies by Marcus,[16] however, reported a probable value of 100, using single extraction of radiomercury from water into benzene. Spiro and Hume[17] reported a probable value of 13.8 in water, using differential spectrophometric methods. Subsequently, Marcus and Eliezer[18] revised the interpretation of the original data of

[8] L. G. Longsworth, *J. Amer. Chem. Soc.* **65**, 1755 (1943).
[9] E. Chiancone, L. M. Gilbert, G. A. Gilbert, and G. L. Kellett, *J. Biol. Chem.* **243**, 1212 (1968).
[10] H. K. Schachman and W. F. Harrington, *J. Polym. Sci.* **12**, 379 (1954).
[11] R. T. Hersh and H. K. Schachman, *J. Amer. Chem. Soc.* **77**, 5228 (1955).
[12] G. Kegeles, *Proc. Nat. Acad. Sci. U.S.A.* **69**, 2577 (1972).
[13] L. C. Craig and D. Craig, *in* "Technique of Organic Chemistry" (A. Weissberger, ed.), Vol. III. Wiley (Interscience), New York, 1950.
[14] M. Delwaulle and F. François, *C. R. Acad. Sci.* **208**, 999 (1939).
[15] M. Delwaulle and F. François, *C. R. Acad. Sci.* **208**, 1002 (1939).
[16] Y. Marcus, *Acta Chem. Scand.* **11**, 610 (1957).
[17] T. G. Spiro and D. N. Hume, *J. Amer. Chem. Soc.* **83**, 4305 (1961).
[18] Y. Marcus and I. Eliezer, *J. Phys. Chem.* **66**, 1661 (1962).

Marcus[16] to give a value of 16. Although the difference between 100 and 16 at first sight seems large, it actually represents only 1.59 kcal for the formation of 2 moles of HgClBr. It is apparent from the work of Marcus[16,18] that single extraction does not provide the requisite experimental leverage to measure this small an amount of binding energy with any appreciable precision. The difference between the spectra of the species is also so small[17] as to make very precise measurement extremely difficult. Moreover, the mixed compound HgClBr has not been isolated, and its physical properties can only be inferred.[4,16,17]

In the case of a very weak complex, however, the study of the shape of the countercurrent distribution pattern is particularly useful, especially if, as in the present case, the product formed by reaction, HgClBr, travels between the reactants. Thus in a low-resolution zonal countercurrent experiment, the complex which is initially formed is observed to decompose progressively, and to disappear in about 50 transfers.

In Fig. 1, taken from Saxena and Kegeles,[4] is shown a comparison between predictions and experimental results (open circles) for 40 transfers between water and benzene. The ordinates are total mercury per tube, experimentally estimated spectrophotometrically with dithizone at 623 nm, or predicted, on the basis of formation constants, K_U, of 2, 15, and 100 (in benzene) for 2 moles of HgClBr. The partition coefficients used in the calculation were $K_A = 1.05$ for $HgCl_2$, $K_B = 0.080$ for $HgCl_2$ and $K_C = 0.39$ for HgClBr. Even though this value chosen for K_C was

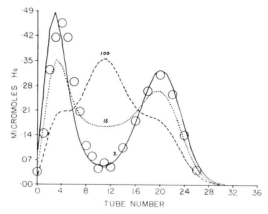

Fig. 1. Countercurrent distribution of 2.49 μmoles of $HgCl_2$ plus 2.90 μmoles of $HgBr_2$ for 40 transfers. Circles are experimental points, with a radius of 0.025 μmole, representing experimental error expected. Curves are computed from an assigned value of $K_C = 0.39$, and $K_U = 2$, 15, and 100 as labeled (see text). From V. P. Saxena and G. Kegeles, *Arch. Biochem. Biophys.* **139**, 206 (1970). Reprinted by permission of Academic Press.

not yet optimized, this figure, which represents essentially a first examination of the problem, shows that an unequivocal selection can be made for the formation constant K_U in the neighborhood of 2, rather than 15 or 100. Further work proved[4] that the results come out to be in agreement with those of Delwaulle and François.[14,15]

In order to make the predictions shown as continuous curves in Fig. 1, it was necessary to compute the contents of each tube in the train after each transfer, which can be done quite effectively with the aid of a digital computer.[1-4] In the present case, moles as well as grams of solute are conserved. When $A°$ moles of $HgCl_2$, $B°$ moles of $HgBr_2$ and $C°$ moles of HgClBr are originally placed in a tube containing equal volumes of water and benzene, the number of moles of $HgCl_2$ or $HgBr_2$ lost through chemical reaction is given for $K_T \neq 4$ by

$$X = \frac{1}{2(K_T - 4)} [K_T(A° + B°) + 4C°] - \frac{1}{2(K_T - 4)} [K_T^2(A° - B°)^2$$
$$+ 16K_TA°B° + 4K_TC°\{C° + 2(A° + B°)\}]^{1/2} \quad (1)$$

and for $K_T = 4$ by

$$X = [4A°B° - (C°)^2]/4(A° + B° + C°) \quad (2)$$

In Eqs. (1) and (2), the quantity K_T is an overall equilibrium constant, for equal volumes of the both phases in the tube, given in terms of the quantities K_U, K_A, K_B, and K_C defined in the text above, by

$$K_T = K_U(1 + K_C)^2/(1 + K_A)(1 + K_B) \quad (3)$$

Assuming that one has placed predetermined quantities $A°$ and $B°$ in one tube, numbered zero, one calculates the quantities, A, B, and C at equilibrium in tube zero by use of Eqs. (1), (2), and (3). The amounts in the upper phase A_U, B_U and C_U are computed by multiplying the total quantities in the tube by the fractions $K_A/(1 + K_A)$, $K_B/(1 + K_B)$ and $K_C/(1 + K_C)$, respectively. At this point these quantities, A_U, B_U, and C_U can be transferred symbolically to the next tube, numbered one, and the entire operation reiterated for tubes zero and one; this process is repeated over the number of transfers performed in the actual experiment.

A description of the general calculation procedures has been given,[1-3] and many complete FORTRAN II programs as used in IBM 709 and 7090 digital computers have been described by Bethune.[19] The programs for an IBM 1620 computer for the present case have been described by Saxena.[20]

When a bimolecular complex travels either faster or slower than

[19] J. L. Bethune, Ph.D. Dissertation, Clark University, 1961.
[20] V. P. Saxena, Ph.D. Dissertation, Clark University, 1965.

both reactants, it does not decompose completely, because it must always move together with both reactants in order to satisfy the law of mass action.[2] It should be perfectly possible to study the formation constant of such a complex by zonal countercurrent distribution, but experience is lacking for practical information as to the accuracy attainable.

Moving-Boundary Countercurrent Distribution

In these experiments the whole train is initially filled with a uniform composition of solution, and in each transfer pure solvent is added to one end of the train (tube zero). A moving-boundary system ensues, below which there remains a region of initial constant composition (the plateau region). In the plateau region, each solute species, whether or not a reaction is taking place, is characterized by its total amount per tube, T_i, and its partition coefficient, K_i. The amount of the ith solute being moved forward per transfer in this plateau region is simply $T_i K_i / (1 + K_i)$. If T_i is given as a mass, then the total mass of solute moved forward per transfer in the plateau region is just

$$Q = \sum_i T_i K_i / (1 + K_i) \tag{4}$$

If this is divided by the total solute mass per tube, T,

$$Q/T = \frac{\sum_i T_i K_i / (1 + K_i)}{\sum_i T_i} = \overline{\left(\frac{K}{1 + K}\right)}_w \tag{5}$$

Thus the quantity Q/T represents the weight-average value of the rate factor $K/(1 + K)$, averaged over all solute species. However Q/T is also simply the fraction of the mass in the tube which is moved forward in one transfer, so that it is identical to the fractional amount of solute extracted in a single extraction experiment. By methods similar to those for electrophoresis[8] and sedimentation,[21,22] it was shown[6] that Q/T can also be measured from the detailed shape of the boundary system. If the total amount of solute in the rth tube after n transfers is denoted, in the boundary region, by $S_{n,r}$, then it has been shown[6] that

$$\sum_{r=1}^{n} (S_{n,r+1} - S_{n,r}) r = n \overline{\left(\frac{K}{1 + K}\right)}_w T \tag{6}$$

[21] R. J. Goldberg, *J. Phys. Chem.* **57**, 194 (1953).
[22] R. Trautman and V. N. Schumaker, *J. Chem. Phys.* **22**, 551 (1954).

Thus, one analyzes all the tubes up to the plateau region, and multiplies the first difference between successive tubes by the tube number, and sums the products to obtain the weight-average rate factor $\overline{\{K/(1 + K)\}}_w$ in the plateau region. This procedure, to be useful for studying equilibria, conveniently assumes that all the nonideality (change of an average K value across the boundary region) can be ascribed to the reaction, i.e., that the K_i value for a given solute species is a constant which is a reasonable expectation, holding exactly for thermodynamic ideality. If this is not so, the dependence of each K_i on such factors as concentration must be separately evaluated—usually a formidable task. However, it is to be noted that such dependence is thermodynamic, rather than hydrodynamic, in origin, for countercurrent distribution. If the individual K values are known or can be estimated, then a series of experiments at different initial loading concentrations will provide different values for the weight-average value of $K/(1 + K)$, from which it it possible to obtain equilibrium constants, just as with the use of molecular weight or sedimentation data.[7]

Differential Moving-Boundary Countercurrent Distribution

If the apparatus is loaded so that part of the train (say approximately half) is uniformly filled with one composition or concentration of solute, and the remainder is filled with a slightly different composition or concentration, then a "concentration-boundary" or "differential boundary" will result. In the case where there is no chemical reaction, and no dependence of the individual partition coefficients on concentration or composition for any other reason, this boundary will, in fact, remain stationary in the apparatus during the countercurrent process. If a difference in the transfer rate of solute mass on the two sides of this boundary exists then the boundary will move. If the net difference in transfer of solute mass on the two sides of the boundary can be ascribed to chemical reactions, the movement of this concentration-boundary will provide additional data concerning the values of the equilibrium constants for the reactions. By analogy with electrophoresis,[8] gel permeation chromatography,[9] and sedimentation,[10–12] it is a straightforward matter to show that the rate of movement, R_D, of this boundary, measured in units of tubes per transfer is given by

$$R_D = \frac{\overline{\left(\dfrac{K}{K+1}\right)}_w' T' - \overline{\left(\dfrac{K}{K+1}\right)}_w T}{T' - T} \tag{7}$$

Here T' and T represent total solute mass per tube in the plateau regions below and above the differential boundary (or resolved boundaries), where the weight-average rate factors are $\{\overline{K/(1+K)}\}'_w$ and $\{\overline{K/(1+K)}\}_w$, respectively. When the difference in concentration $T' - T$ becomes very small, Eq. (7) may be replaced by the differential equation

$$R_D = \frac{d(RT)}{dT} \tag{8}$$

where R now represents the weight-average rate factor $\{\overline{K/(1+K)}\}_w$.

In the case of sedimentation, it has been demonstrated[12] that, for the case of a single thermodynamic component as solute, the corresponding differential sedimentation coefficient (neglecting hydrodynamic effects) is, in fact, the Z-average sedimentation coefficient for the solute. By analogy, we find that the quantity R_D is given by

$$R_D = \bar{R}_Z = \overline{\left(\frac{K}{1+K}\right)_Z} \tag{9}$$

If the symbol $\bar{R}_w$ is chosen to represent the weight-average rate factor $\{\overline{K/(1+K)}\}_w$, then *if the only reaction is the formation of a single polymer* (*m-mer*) *from a monomer, the weight fraction of monomer is given*[12] *by*

$$T_1/T = \frac{m}{m-1} \frac{\bar{R}_Z - \bar{R}_w}{\bar{R}_Z - R_1} \tag{10}$$

Here R_1 is the rate factor $K_1/(1+K_1)$ for monomer. From Eq. (10) the overall equilibrium constant is obtained by a straightforward application of the law of mass action:

$$K_T = \frac{T - T_1}{T_1{}^m} \tag{11}$$

The equilibrium constant K_U in a single phase (the upper phase) is related to K_T by an equation[1] analogous to Eq. (3)

$$K_U = K_T \frac{R_P}{R_1{}^m} \tag{12}$$

Here R_P is the rate factor for polymer, which can be obtained from the relationship[6]

$$\bar{R}_w = T_1 R_1 + (1 - T_1) R_P \tag{13}$$

The equilibrium constant K_L in the lower phase is then[1] given by

$$K_L = K_U K_1{}^m / K_P \tag{14}$$

where K_P is the partition coefficient of polymer (*m*-mer).

In order to obtain the quantity R_D from experimental observations, one should again take the "first-moment" of the concentration-gradient across the differential moving boundary:

$$R_D = \frac{\sum_{r_0}^{n} (S_{n,r+1} - S_{n,r})(r - r_0)}{\Delta T} \qquad (15)$$

Here the position r_0 is the starting tube number for the differential boundary, and ΔT is the (small) difference in total solute mass per tube across the initial boundary.

This procedure makes available an additional average, the Z-average rate factor, and thus potentially helps to fix more closely the value of the equilibrium constant, *in this special case*. Whether this will be useful in practice will depend on the accuracy attainable.

Summary

Three procedures have been outlined which may be used to obtain chemical equilibrium constants from countercurrent distribution experiments. The first, which has been used experimentally and is likely to be the most sensitive, depends on fitting predictions computed from a model to the experimental zonal countercurrent distribution pattern. The fit is optimized by variation in succession of the unknown parameters, such as equilibrium constants. For this procedure, it is necessary to perform iterative computations with the aid of a digital computer, to make the necessary predictions.

The second procedure consists of measuring directly the weight-average rate factor $\{\overline{K/(1 + K)}\}_w$ from the countercurrent distribution pattern in moving boundary experiments. This is done at different loading concentrations or compositions, and the data are combined with individual rate factors for the species present, to obtain equilibrium constants.

The third procedure is to study the movement of differential moving boundaries between two large regions of constant composition. Such data, in the case of only one reacting solute, which is a single thermodynamic component, gives rise to the Z-average rate factor. In the case of a monomer–single polymer equilibrium, the use of this quantity, together with the weight-average rate factor and the rate factor for monomer, leads very directly to the equilibrium constant for the reaction.

There is an important potential advantage in the use of the countercurrent distribution technique rather than continuous transport methods in such determinations. Even if the reactions are quite slow, one can allow the system to come to equilibrium at each stage in countercurrent distribution with no loss in resolution whatsoever, and one can also test quite carefully to assure that equilibrium has been achieved. However, due to the effects of diffusion in continuous transport methods, resolution is impaired by decreasing the field responsible for migration. Moreover, if the speed of reaction is slow enough so that appreciable migration takes place before chemical equilibrium is locally reestablished, a wide variety of artifacts can also occur, which seriously complicate the possibility of achieving any meaningful equilibrium measurements.[23]

[23] J. R. Cann, "Interacting Macromolecules." Academic Press, New York, 1970.

[17] Rapid Measurement of Binding of Ligands by Rate of Dialysis

By F. C. WOMACK and S. P. COLOWICK

This procedure was originally developed for rapid measurement of binding of sugars and nucleotides to yeast hexokinases.[1,2] The method has since been applied successfully in other laboratories, e.g., for binding of nucleotides,[3-5] pyrophosphate,[6] calcium ions,[7] and amino acids.[8] The method is especially useful in cases where the long period required for establishing equilibrium across a membrane, as in classical equilibrium dialysis, would result in destruction of labile enzymes or degradable substrates. It has the additional advantage that a series of measurements can be made on the same sample, to study effects of varying ligand concentration, or of adding other reagents which influence ligand binding. It can be applied whenever the establishment of chemical equilibrium between a ligand and a macromolecule is a reasonably rapid process.

[1] S. P. Colowick and F. C. Womack, *J. Biol. Chem.* **244**, 774 (1969).
[2] S. P. Colowick, F. C. Womack, and J. Nielsen, *in* "The Role of Nucleotides for the Function and Conformation of Enzymes" (Alfred Benzon Symposium 1968), (H. M. Kalckar, ed.), p. 15. Munksgaard, Copenhagen, 1969.
[3] N. C. Brown and P. Reichard, *J. Mol. Biol.* **46**, 39 (1969).
[4] J. G. Nørby and J. Jensen, *Biochim. Biophys. Acta* **233**, 104 (1971).
[5] C. Hegyvary and R. L. Post, *J. Biol. Chem.* **246**, 5234 (1971).
[6] L. Butler, *J. Biol. Chem.* **244**, 777 (1969).
[7] K. G. Walton, "Energized Processes of Mitochondria: Ion-supported Swelling and Calcium Uptake." Dissertation, Vanderbilt University, August, 1970.
[8] R. B. Loftfield and A. Pastuszyn, Wenner-Gren Symposium on Structure and Function of Oxidation Reduction Enzymes, pp. 585–594 (1972).

Principle

When a ligand forms a complex with a macromolecule, the concentration of free ligand in the equilibrium mixture can be monitored by the rate of dialysis across a suitable membrane. By using radioactive ligands of high specific activity, the rate can be monitored quickly, without affecting appreciably the total amount of ligand in the reaction mixture. By making additional rate measurements after successive additions of unlabeled ligand, all the data for a Scatchard-type plot for determination of the number of binding sites and the binding constant(s) can be collected in about 20 minutes with a single enzyme sample.

The rate of dialysis can be monitored conveniently by use of the simple device illustrated in Fig. 1. The apparatus consists of a dialysis cell with an upper chamber, containing the macromolecular component and the labeled ligand, separated by a membrane from the lower chamber, through which buffer is pumped at a constant rate and from which the effluent is sampled for measurement of radioactivity. After a sufficient volume of buffer has passed to permit a steady state to be reached, i.e., when the rates for isotope entering the chamber by diffusion and leaving by flow of buffer are practically equal, the concentration of labeled ligand in the effluent becomes a true measure of the concentration of free labeled ligand in the upper chamber. The volume of flow required to reach greater than 98% of this steady state value is about four times the volume of the lower chamber (for theory, see Colowick and Womack[1]). Thus, a *single measurement* in the steady state provides a measure of the rate of dialysis and hence of the concentration of free labeled ligand in the

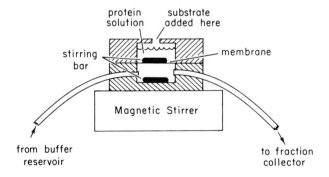

Fig. 1. Diagram of the apparatus for measuring substrate binding by rate of dialysis. Reprinted by permission of the *Journal of Biological Chemistry* [S. P. Colowick and F. C. Womack, *J. Biol. Chem.* **244**, 774 (1969)]; and Munksgaard Publishing Company [S. P. Colowick, F. C. Womack, and J. Nielsen, *in* "The Role of Nucleotides for the Function and Conformation of Enzymes" (H. M. Kalckar, ed.), p. 15. Munksgaard, Copenhagen, 1969].

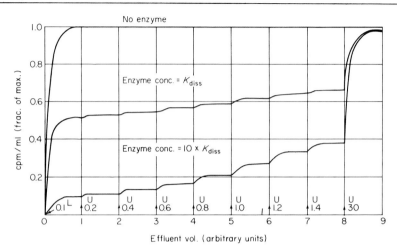

Feffluent vol. (arbitrary units)

Fig. 2. Theoretical curves for binding measurements by rate of dialysis. The ordinate shows calculated steady-state radioactivity of effluent when a labeled compound (L) is added to the upper chamber in the presence of different concentrations of enzyme. The total concentration of ligand present initially and after successive additions of unlabeled compound (U) is shown at the arrows and is expressed as fraction of the total enzyme sites present. The counts per minute in the effluent, when expressed as fraction of maximum, indicate directly the fraction of ligand which is free in the upper chamber.

reaction mixture. If a part of the labeled ligand is bound to enzyme, a lower steady-state value will be found in the effluent, the level depending on enzyme concentration and K_{diss} (see Fig. 2). Subsequent additions of unlabeled ligand result in a progressively larger fraction of the isotope in the free state, and hence in corresponding increases in the steady-state concentration of isotope in the effluent. Finally, excess unlabeled ligand is added, and a maximum value for isotope concentration in the effluent is reached, which corresponds to that expected when no appreciable fraction of the labeled ligand is bound.

From these data, one can calculate the concentration of ligand in the free (F) and bound (B) state and then estimate the total concentration of enzyme sites (B_T) as well as the dissociation constant K_{diss} from intercept and slope of a Scatchard-type plot of B vs. B/F, using the following equation[1]: $B = B_T - K \cdot B/F$.

For examples of results obtained, the original papers cited may be consulted.

Details of Apparatus

The device used in this laboratory was derived from a 1-ml continuous-flow dialysis cell (Bel-Art Products, Pequannock, New Jersey) simply by drilling a 5-mm hole in the enzyme compartment to permit addition of substrate or other agents to the solution being dialyzed, and increasing the depth of the upper compartment from 4 mm to 9 mm in order to accommodate a 1.5-ml volume without overflow during mixing. It should be pointed out that these cells, which have a relatively large membrane surface (inner diameter of cell 19 mm) are not ideally designed for the present purpose. They were designed for rapid removal of small molecules, whereas the purpose here is to bleed off the minimal amount of ligand which will permit accurate monitoring of radioactivity of effluent. In practice, the loss of ligand with this device, as modified, has not exceeded 3% over a 20-minute run in the cold when sugars or nucleotides are used.

The lower chamber (19 mm × 10 mm) has a capacity of 2.8 ml and is completely filled with buffer solution, which is pumped through at a constant rate of 8 ml per minute. With this rate, a steady state is reached in 1.5 minutes. Both chambers are mixed by means of Teflon-covered magnetic stirring bars.

The membrane, a square cut from cellophane dialysis tubing (Visking No. 36, Union Carbide Co.) is clamped between the Lucite blocks, which are held together with stainless steel screws. Membranes are cleaned in boiling 1% acetic acid and rinsed with distilled water before use. For procedures for altering membrane porosity if desired, see Craig.[9]

The inner diameter of the tubing leading from the lower chamber to the fraction collector is 1 mm and the length is 40 cm. The diameter has been increased over that originally recommended[1] in order to avoid excessive pressure in the lower chamber at high flow rates. Excessive pressure may lead to flow of fluid from the lower to the upper chamber. The "dead space" between lower chamber and fraction collector remains negligible (0.3 ml).

Some useful modifications of the apparatus have been made by Brown and Reichard.[3] They have reported the use of cells with a 13-mm inside diameter, and a lower chamber volume of 0.3 ml or 1.5 ml. The flow rates used with these cells were 2.5 ml/minute and 2.5 to 10 ml/minute, respectively. The volume of enzyme sample in the upper chamber was 1 ml in both cases. This design lowers the ratio of membrane surface to upper chamber volume, thereby decreasing the loss of isotope from the upper chamber during an experiment. The small volume in the lower chamber decreases the dilution of isotope in that chamber and also permits a more rapid approach to the steady state. A shop drawing of this type of cell, but with an inner diameter of only 10 mm and a

[9] L. C. Craig, see Vol. XI, p. 870.

lower chamber volume of 0.16 ml, has been kindly supplied by Dr. Reichard (Fig. 3).

Walton[7] has described two ways to minimize the ratio of membrane surface to upper chamber volume. One of these is to use an upper chamber which is tapered to give a larger diameter at the top. The other is to sandwich the membrane between microscope slides with a small hole (3 mm in diameter) drilled in each.

The measurements in this laboratory are usually carried out in the cold room ($T = 5°$) but have also been done at room temperature. We have noted an increase of approximately 3-fold in the rate of dialysis of ADP at 25° vs. 5°. The effect of temperature on the rate of dialysis of various solutes has been discussed in detail by Craig.[9] The lower temperature is, of course, preferable for minimizing action of enzymes on added substrates[2,3] and for minimizing loss of ligand in dialysis. For more adequate temperature control, the buffer reservoir,

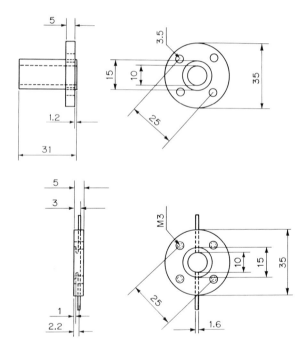

Fig. 3. Modified flow-dialysis apparatus of N. C. Brown and P. Reichard [*J. Mol. Biol.* **46**, 39 (1969)] for substrate-binding measurements. Shop drawing kindly supplied by Dr. Reichard. The views below are for the lower chamber, those above for the upper chamber of the apparatus, which was constructed of tubular Plexiglas. Dimensions shown are in millimeters.

inlet tube, and dialysis may all be thermostated,[3,4] but this is not essential if all parts are equilibrated at the ambient temperature.

The rate of flow of buffer through the lower chamber must be constant in order to reach a steady concentration of isotope in the effluent. Dr. Kenneth Walton (personal communication) found more reproducible results by use of gravity flow from a buffer reservoir than by use of a peristaltic pump.[7] Similarly, Nørby and Jensen[4] have used a Mariotte bottle for maintaining a constant pressure head of the perfusing buffer.

The measurement of radioactivity in the effluent is ordinarily run on aliquots from the collected fractions.[10] In experiments with dialysis of ^{125}I-labeled insulin, Crofford et al.[11] have monitored the count rate continuously by passing the effluent through a coil located in the well of a gamma counter.

Details of Procedure

The first step in carrying out a binding experiment with this method is to study the behavior of the radioactive ligand in the apparatus when no enzyme is present. Ideally, the radioactivity found in the effluent in the steady state should be maximal, i.e., not influenced by subsequent additions of unlabeled ligand (see Fig. 2). In some cases, these ideal results have not been found, especially when dealing with charged ligands in buffers of low ionic strength. Thus, in experiments with $^{45}Ca^{2+}$, in 10 mM Tris buffer, the addition of excess unlabeled Ca^{2+} increased the radioactivity in the effluent. However, when the concentration of the buffer was raised to 40 mM, this anomaly was no longer observed (Dr. Kenneth Walton, personal communication; see also dissertation[7]). In our measurements of sugar or nucleotide binding, we ordinarily use 50 mM phosphate or 50 mM Tris·HCl buffers to avoid such anomalies.

The control experiment without enzyme also serves as a check on the proper functioning of the apparatus. A constant value for radioactivity in the effluent means that the buffer flow rate is constant and that there is no appreciable dilution of the upper chamber by net fluid flow from below.

---

[10] Radioactivity measurements were done on 0.5–1.0-ml aliquots of the effluent, diluted in 10 ml of Bray's solution for liquid scintillation counting. In measurements of ^{3}H-labeled ATP and ADP in the effluent, which contained $10^{-2} M$ $MgCl_2$, variability was encountered in counts per minute measurements of a given sample at different times, presumably owing to precipitation of the Mg salts of the nucleotides in Bray's solution. This can be overcome either by adding Cab-o-sil thixotropic powder to each sample prepared for counting, or by substituting 10 ml of Aquasol (New England Nuclear) for the Bray's solution.

[11] O. B. Crofford, N. L. Rogers, and W. G. Russell, Diabetes 21, Suppl. 2, 403 (1972).

The control experiment with labeled ligand alone also serves to check on the radiopurity of the labeled material. Dr. William Jakoby (personal communication) has informed us of a case in which the tritium in a so-called ^{3}H-labeled nucleotide was largely in a form which was rapidly exchangeable with water. (See also Bradbury and Jakoby.[12]) The nucleotide nevertheless appeared "radiochemically pure" by paper chromatography. However, in the dialysis apparatus the exchangeable tritium appeared rapidly in the effluent as a large peak of radioactivity which declined rapidly because of the extremely fast dialysis of ^{3}H ions out of the upper chamber. Similarly, an appearance of a transient peak of radioactivity in the effluent on addition of ^{125}I-labeled insulin to the upper chamber has been noted and tentatively ascribed to iodide.[11]

Once it is established by control experiments with labeled ligand alone that the system is reasonably free of the above artifacts, the actual binding measurement can be undertaken. For this purpose, enzyme at a molar concentration of binding sites estimated to be equal to or greater than the expected dissociation constant of the enzyme–ligand complex (see Fig. 2), is added to the upper chamber in the selected buffer, with other additions, such as metal ions, metal chelators, sulfhydryl compounds, as required. The buffer flowing through the lower chamber should be supplemented with all dialyzable components other than the ligand to be tested.

The radioactive ligand is then added to the upper chamber at a molar concentration about one-tenth that of the anticipated enzyme sites, and the collection of fractions of effluent is begun. With a flow rate of 8 ml per minute in our apparatus, we collect 2-ml samples every 15 seconds. Since the counts per minute per milliliter in the effluent reaches a constant value after 1.5 minutes, additions of unlabeled ligand can then be made every 2 minutes, covering the range from 0.2 to 1.4 times the molar concentration of enzyme sites expected, as indicated in Fig. 2. These additions are made in small volumes (5–10 μl) in order to avoid excessive dilution. The final addition to give 30 times the enzyme concentration, or more when feasible, should, when corrected for dilution of the enzyme solution and loss of isotope from the upper chamber, yield a value essentially equal to that obtained in a control sample with labeled ligand and no enzyme. This final value, therefore, serves as an internal control, so that ordinarily there is no need to run control samples without enzyme, once the preliminary tests for artifacts have been carried out.

[12] S. L. Bradbury and W. B. Jakoby, J. Biol. Chem. 246, 6929 (1971).

The actual concentrations of enzyme and ligand used in a particular case will depend on the dissociation constant of the system. The radioactivity added to the upper chamber has to be high, of the order of 10^6 to 10^7 cpm, because of the high dilution factor (ca. 10^3 to 10^4) during flow dialysis. Nevertheless, the high specific activities of labeled compounds make it feasible to attain the required number of counts per minute at very low molarities. Thus, with ^{32}P-labeled ATP (4–10 Ci/mmole), dissociation constants as low as $10^{-7} M$ have been readily measured with this method.[4,5] With [^{14}C]glucose (0.15 Ci/mmole), a K_{diss} around $10^{-5} M$ is easily measurable.[1,2] In cases where dissociation constants are too low for measurement by this method, the procedure would nevertheless be useful as a means of detection of ligand-binding proteins with such characteristics.

[18] Chromatography of Proteins on Hydroxyapatite

By GIORGIO BERNARDI

The purpose of the present article is to discuss our present knowledge of the mechanism of adsorption of proteins on hydroxyapatite (HA[1]), particularly as far as the interaction of amino acid side groups with adsorption sites on HA crystals is concerned. The main conclusions arrived at in investigations carried out in the author's laboratory[2–4] can be summarized as follows. Two different types of adsorbing sites exist on the surface of HA crystals: calcium sites and phosphate sites. The former appear to bind acidic groups, carboxyls, and phosphates; the latter bind basic groups. This picture fits with the known amphotheric character of HA crystals.[5] Elution is caused by anions (usually phosphates), which compete with the carboxyl or phosphate groups of proteins for the calcium sites of HA; or by cations (Na⁺, K⁺ or, more effectively, Ca²⁺ or Mg²⁺), which compete with the basic groups of proteins for the phosphate groups of HA.

---

[1] Abbreviations: HA, hydroxyapatite; MW, molecular weight; KP, potassium phosphate buffer; NaP, sodium phosphate buffer; if no pH indication is given for KP or NaP, the pH is 6.8.

[2] G. Bernardi and T. Kawasaki, *Biochim. Biophys. Acta* **160**, 301 (1968).

[3] G. Bernardi, see Vol. 22, p. 325.

[4] G. Bernardi, M. G. Giro, and C. Gaillard, *Biochim. Biophys. Acta* **278**, 409 (1972).

[5] S. Mattson, E. Kontler-Anderson, R. B. Miller, and K. Vantras, *Kgl. Lantbruks-Hoegsk. Ann.* **18**, 493 (1951), quoted by S. Larsen, *Nature (London)* **212**, 212 (1966).

From a practical point of view, our investigations have shown that: (1) elution of proteins from HA columns can be obtained using a number of eluents other than the usual phosphate buffer, pH 6.8; this leads to a remarkable increase in the potentialities of the method; (2) the chromatographic behavior of basic, neutral, and acidic proteins on HA columns operated with different elution systems can be predicted to a considerable extent, thus permitting a less empirical approach to separation problems; in turn, the elution molarities of proteins in different solvent systems can be used to identify the nature of the interacting amino acid side groups.

Other aspects of chromatography of proteins on HA columns have been reviewed in Volume 22 (p. 325).

Chromatography of Proteins

Basic Proteins. The chromatographic behavior of basic proteins is characterized by relatively high elution molarities when the KP elution system is used; as shown in the table, lysozyme, cytochrome c, ribonuclease A, α-chymotrypsin, and spleen acid DNase are all eluted in the 0.12–0.23 M KP range; a lysine-rich histone (30% lysine) was eluted at about 0.55 M KP.

NaCl or KCl molarity gradients can be used to elute basic proteins: the elution molarities, in this case, are equal to about twice the eluting molarities of phosphate buffer, pH 6.8 (see the table). In the case of the lysine-rich histone, gradient elution was not tried, but a 3 M KCl step completely removed the protein from the column.

As indicated in the table, all basic proteins are removed from the columns by very low molarities of $CaCl_2$. The proteins having the lowest KP elution molarities, lysozyme, and ribonuclease A, are eluted by the initial 0.001 M $CaCl_2$ step; those having increasingly higher KP elution molarities are eluted by the $CaCl_2$ molarity gradient below 0.025 M $CaCl_2$. $MgCl_2$ behaved very similarly to $CaCl_2$ as an eluent of basic proteins.

As shown in the table, elution by KP, pH 7.8, takes place at lower molarities than at pH 6.8, the opposite being true for elution at pH 5.8. The ratio of the eluting phosphate molarities at pH 5.8 and pH 6.8 was constant for all basic proteins investigated; the ratio of the eluting molarities at pH 7.8 and pH 6.8 showed some small fluctuations.

Acidic and Neutral Proteins. These proteins show a distinctly different chromatographic behavior on HA columns compared to basic proteins, being eluted at generally lower molarities by the KP system. When using the KCl system, acidic and neutral proteins are either eluted by very high KCl molarities compared to KP or they are not eluted at

ELUTION MOLARITIES OF SOME PROTEINS FROM HA COLUMNS[a]

Protein	Isoelectric point	Eluting solvents				
		KP 6.8	KP 7.8	KP 5.8	KCl	$CaCl_2$
Lysozyme	10.5–11.0[f]	0.12	0.08	0.15	0.25	0.001
Cytochrome c	9.8–10.1[g]	0.23	0.20	0.30	0.48	0.007
RNase A	9.7[h]	0.12	0.09	0.15	0.23	0.001
α-Chymotrypsin[b]	8.1[i]	0.16	0.10	0.20	0.32	0.01
Spleen acid DNase	10.2[j]	0.22	0.115	0.32	0.44	0.02
Spleen acid exo-nuclease[c]	—	0.125	0.065	0.195	0.25	>3.0
Myoglobin	7[g]	0.12	0.08	0.17	0.80	>3.0
Snail acid DNase	5.9[k]	0.11	0.04	0.20	0.54	>3.0
Pancreatic DNase	4.7[l]	0.04	0.01	0.12	0.4	>3.0
Bovine serum albumin[d]	4.7[m]	0.06	0.01	0.17	>3.0	>3.0
Pepsin[e]	1[n]	0.03	0.01	$\begin{cases} 0.08 \\ 0.12 \end{cases}$	>3.00	>3.0

[a] Chromatographic experiments were performed on 1×18–23 cm columns of HA prepared as described in Volume 22, p. 325. KCl gradients were in 0.01 M KP. In the $CaCl_2$ chromatograms, after loading the proteins on HA columns equilibrated with 0.001 M NaP, columns were washed with 0.001 M NaCl; elution was then performed, in succession, with a 0.001 M $CaCl_2$ step, a 0.001 to 0.05 M $CaCl_2$ molarity gradient, and steps of increasing concentrations (up to 3 M) of $CaCl_2$. When even 3 M $CaCl_2$ could not desorb proteins, columns were washed again with 0.001 M NaCl and proteins were eluted with a 1 M NaP step. For other experimental details see G. Bernardi, M. G. Giro, and C. Gaillard, *Biochim. Biophys. Acta*, in press.

[b] 19–23% of the material was not retained by the columns.

[c] Isoelectric point is not known. Chromatographic data suggest that isoelectric point is close to neutrality.

[d] Bovine serum albumin was eluted in one main peak followed by a shoulder, corresponding to the dimer.[o] The elution molarity given is that of the main peak.

[e] 21–25% of the material was not retained by the columns. Two peaks were eluted by the KP molarity gradient at pH 5.8.

[f] P. Jollès, *in* "The Enzymes" (P. D. Boyer, H. Lardy, and K. Myrbäck, eds.), 2nd ed., Vol. 4, p. 431. Academic Press, New York, 1960.

[g] E. G. Young, *in* "Comprehensive Biochemistry" (M. Florkin and E. H. Stotz, eds.), Vol. 7, p. 25. Elsevier, New York, 1963.

[h] L. B. Barnett and H. B. Bull, *Arch. Biochem. Biophys.* **89**, 167 (1960).

[i] E. A. Anderson and R. A. Alberty, *J. Phys. Colloid Chem.* **52**, 1345 (1948).

[j] G. Bernardi, E. Appella, and R. Zito, *Biochemistry* **3**, 1419 (1965).

[k] J. Laval, Thesis, University of Paris, 1970.

[l] U. Lindberg, *Biochemistry* **6**, 355 (1967).

[m] G. I. Loeb and H. A. Scheraga, *J. Phys. Chem.* **60**, 1633 (1956).

[n] G. E. Perlmann, *Advan. Protein Chem.* **10**, 23 (1955).

[o] D. B. Menzel and E. G. Richards, private communication.

all by KCl molarity gradients up to 3 M. None of the neutral and acidic proteins can be removed by the columns by $CaCl_2$ molarities as high as 3 M. After washing the columns with 0.001 M NaCl, all these proteins can, however, be totally eluted by a 1 M NaP step. Spleen exonuclease (a protein whose unknown isoelectric point is probably close to neutrality) resembled neutral and acidic proteins in being not eluted by $CaCl_2$.

For acidic and neutral proteins, elution by KP, pH 7.8, or by KP, pH 5.8, takes place at lower and higher molarities, respectively, compared to KP, pH 6.8. These effects are, therefore, qualitatively the same as for basic proteins; in the case of acidic proteins, they are, however, much stronger than in the case of basic proteins.

Phosphoproteins. Phosphoproteins represent a special case in that they have a very high affinity for HA. It has been shown[6] that the two egg-yolk phosphoproteins, α- and β-lipovitellin, which are identical in amino acid and lipid composition, but different in their protein phosphorus contents can be easily separated on HA columns. When elution was performed with a molarity gradient, instead of the stepwise, technique originally used, it could be shown[2] that β-lipovitellin, the electrophoretically slow component, was eluted by 0.4 M KP, whereas α-lipovitellin, the fast component, was eluted by 0.75 M KP. The third egg-yolk phosphoprotein, phosvitin, a protein in which almost 50% of the amino acid residues are phosphoryl serines, was eluted by an exceptionally high phosphate molarity, 1.2 M KP.

The Chromatographic Behavior of Proteins on HA

This may be summarized and interpreted as follows.

1. The finding that KCl elution molarities of all basic proteins tested are systematically twice as large as the corresponding KP elution molarities may be interpreted as indicating that the elution of basic proteins is caused by the cations of the eluents, since the eluting K$^+$ concentration in the KCl system is roughly the same as that in the KP system.[7]

2. The fact that the eluting molarities of basic proteins by $CaCl_2$ are 20–200 times lower than those of NaCl or KCl is in keeping with the suggestion made above that, in the chromatography of basic pro-

---

[6] G. Bernardi and W. H. Cook, *Biochim. Biophys. Acta* 44, 96 (1960).

[7] The eluting concentration of K$^+$ in the KCl system is, in fact, higher by about 30% than that of the KP, pH 6.8, system. This difference may, however, be due to a difference in the activity coefficients of the two salts and/or to the fact that the pH of the KCl solution, at the eluting molarity, is lower than 6.8 (the eluting K$^+$ concentration in the KCl system is practically the same as that in the KP, pH 5.8, system).

teins, elution is caused by the cations, not by the anions of the eluent. The fact that cations having a very strong affinity for phosphate ions like Ca^{2+} and Mg^{2+} are endowed with a much greater eluting power than cations having a weak affinity for phosphate, like K^+ and Na^+, suggests two important points: that the adsorbing sites for basic proteins are to be identified with phosphate groups at the surface of the crystals and that elution of basic proteins takes place because of a competition between the cations of the eluent and the basic amino acid side groups of proteins for phosphate sites on HA, or, in other words, that elution of basic proteins is not simply a ionic strength effect.

3. Acidic and neutral proteins behave similarly to acidic polypeptides and nucleic acids [2,4,8,9] in that their adsorption on HA is little, or not at all, affected by NaCl, KCl, or $CaCl_2$. An explanation for this behavior is that the adsorption of acidic and neutral proteins, like that of acidic polypeptides, phosphoproteins, and nucleic acids, is due only, or partly, to the interaction of their acidic groups with calcium sites at the surface of HA crystals. Elution is, therefore, expected to be caused by anions able to compete with the macromolecules for the calcium sites on HA. It is not surprising that anions having little affinity for calcium, like chloride, are poor eluents compared to the phosphates normally used as eluents.

4. Basic proteins are eluted from HA columns at relatively high molarities of phosphate, pH 6.8. The five proteins listed in the table are all eluted in the 0.12–0.23 M phosphate range; the lysine-rich histone is eluted at an even higher molarity, 0.55 M. In contrast, neutral and acidic proteins seem to be eluted at low phosphate molarities, in the 0.03–0.12 M range (phosphoproteins are an exception to this rule, see point 7, below). This different behavior may be understood in the following terms: the usual eluents, NaP or KP, pH 6.8, while very effective, because of their phosphate ions, in competing with the carboxyl groups of proteins for the calcium sites on HA, are much less effective in competing with the basic groups of the proteins for the phosphate sites on HA because of the low affinity of Na^+ or K^+ for such groups. Therefore, the usual phosphate buffers are good eluents for acidic proteins, as well as for acidic polypeptides, phosphoproteins, and nucleic acids and poor eluents for basic proteins. Conversely, $CaCl_2$ is an excellent eluent for basic proteins and a very poor one for acidic proteins.

5. An unexpected finding obtained with spleen exonuclease, myoglobin, snail acid DNase, and pancreatic DNase is that these proteins,

[8] G. Bernardi, *Biochim. Biophys. Acta* **174**, 423, 435, 449 (1969).
[9] G. Bernardi, see Vol. XXI, p. 95.

which are eluted by KCl molarities in the 0.2–0.8 M range cannot be eluted by $CaCl_2$ molarities as high as 3 M. It is interesting to ask whether in these cases Ca^{2+} binds to the carboxyl groups of these proteins and strengthens the adsorption by forming bridges to phosphate groups on HA.

6. The data of the table show that the elution molarity of proteins by KP, appears to be increased by a constant factor at pH 5.8 and decreased (also by a constant factor, but with larger fluctuations) at pH 7.8, compared to the usual pH 6.8. Both effects appear to be much greater for acidic than for basic proteins. A satisfactory explanation for this phenomenon is not yet available.

7. The very high eluting molarities required by phosphoproteins in comparison with nucleic acids may be due, in part at least, to the mono-esterified phosphate groups in contrast to the diesterified groups of nucleic acids. Another consideration is that phosphoproteins have runs of phosphorylserines, which form areas of very high density of groups able to interact with HA.

Effect of Secondary and Tertiary Structure of Proteins on Their Chromatographic Behavior on HA Columns.

Investigations by Bernardi and Kawasaki[2] showed that proteins are much less retained or not retained at all by HA columns equilibrated with 0.001 M KP when they are in their denatured state. It appears that the disruption of the secondary and tertiary structures of proteins by 8 M urea or heat causes a strong reduction in their interaction with HA. This phenomenon can be explained by the fact that the random coil configuration of the denatured protein causes a decrease in the number of amino acid side groups able to interact per unit of protein surface in contact with HA, and therefore a decrease in the elution molarity of denatured proteins. This decrease may be due to the following reasons: (1) acidic or basic groups present at the surface of the native protein will, in part, disappear from the "surface" of the denatured protein, which is known to have a random coil configuration; (2) local clusters (due to the existence of secondary and tertiary structures) of acidic and basic groups will disappear in the denatured state in favor of a more random distribution over the entire protein "surface."

Chromatography of Synthetic Polypeptides

Acidic Polypeptides. These polypeptides show a strong affinity for HA; poly-L-glutamate and poly-L-aspartate are eluted at about 0.25 M and 0.35 M KP, respectively. Elution is not due to the ionic strength of the eluting buffer, but to a specific competition by phosphate ions for

HA sites binding carboxyl groups, as shown by the following findings: (1) $3 M$ KCl does not elute poly-L-glutamate; (2) elution with a linear gradient between $0.001 M$ KP–$1 M$ KCl and $0.5 M$ KP (therefore at a practically constant ionic strength, KP having a ionic strength which is close to twice its molarity) does not change the phosphate eluting molarity of poly-L-glutamate; (3) statistical copolymers of poly-L-glutamate with phenylalanine, lysine, and serine, are eluted by KP at a slightly lower molarity than poly-L-glutamate, the phosphate concentration needed for elution decreasing with decreasing glutamate content. (4) A copolymer of DL-histidine and benzyl-L-glutamate (1:1 molar ratio) was not retained by a column equilibrated with 1 mM KP.

Chromatography of poly-L-glutamate and poly-L-aspartate in the presence of $8 M$ urea caused no change in the phosphate eluting molarity. This result, at variance with what is found in the case of proteins endowed with secondary and tertiary structure (in which case denaturation causes a drastic drop in the elution molarity) is not surprising since both carboxylic polymers already are in a random coil configuration at neutral pH.

Basic Polypeptides. The chromatographic behavior of poly-L-lysine, poly-L-arginine, and poly-L-ornithine is characterized by an even stronger affinity for HA than that of acidic polypeptides. In fact, basic polypeptides are so strongly retained by HA columns that they cannot be eluted by KP gradients reaching a molarity of 1. An exception to this general rule was found with the low-molecular-weight poly-L-lysine samples (MW = 7000); in this case a large aliquot of the retained material was eluted between 0.1 and $0.5 M$ KP as a series of peaks (see Fig. 5 of Bernardi[3]).

All basic polypeptides investigated can, however, be completely desorbed by $3 M$ NaCl, or $3 M$ KCl, or by molarity gradients of these salts. Poly-L-ornithine, for instance, is eluted at about $1.75 M$ by KCl or NaCl molarity gradients in $0.01 M$ KP or $0.01 M$ NaP, respectively. Basic polypeptides can also be eluted by rather weak molarities of $CaCl_2$. This solvent can also elute material which is not eluted by KP.

Neutral Polypeptides. Poly-L-histidine, poly-L-serine, poly-L-tyrosine and poly-L-proline are not retained by HA columns equilibrated with 1 mM KP. The behavior of the latter two polypeptides is in agreement with the lack of interaction of nonpolar amino acids with HA (see below).

It can be concluded that the chromatographic behavior of synthetic polypeptides fits very well with the general ideas derived from the study of proteins. A complication existing in the case of synthetic polypeptides is due to the fact that their elution molarities appear to depend upon their molecular weights, at least when this is below a certain level.

Adsorption of Amino Acids

Tiselius *et al.*[10] found that neutral and dicarboxylic amino acids show very weak or no absorption on HA columns. Basic amino acids were found to have slight affinity, arginine and lysine having an R_f of about 0.4 in 0.001 M NAP,[1] but displaying a considerable tailing. In contrast, Hofman[11] reported that aspartic acid has by far the lowest R_f of 20 amino acids chromatographed on thin layers of HA. Elden and Howell[12] found, using light-scattering measurements, that glycine, tyrosine, arginine and histidine fail to interact with HA at pH 7, whereas lysine, aspartic and glutamic acids, serine, threonine and hydroxyproline do interact.

In order to resolve the apparent conflict of these results, Bernardi *et al.*[4] investigated the adsorption isotherms of amino acids on HA equilibrated with 1 mM KP, pH 6.8 or 7.8.

At pH 6.8, all polar amino acids tested (arginine, lysine, histidine, aspartic acid, glutamic acid, serine) were adsorbed to rather similar extent. Among nonpolar amino acids, alanine, valine, and phenylalanine were not adsorbed at all; glycine, but not glycylglycine, was adsorbed. If 0.01 M KCl was present in the equilibration buffer, arginine was not adsorbed anymore, whereas aspartic acid was slightly more adsorbed; histidine and serine were not affected in their adsorption properties.

At pH 7.8, arginine and lysine were more strongly adsorbed than at pH 6.8. In contrast acidic amino acids were less adsorbed, glutamic acid being not adsorbed at all. The adsorption of histidine and serine varied very little at the two different pH values. All other amino acids, including alanine, valine, leucine, isoleucine, tyrosine, phenylalanine, tryptophan, proline, hydroxyproline, glutamine, asparagine, and glycine, were not adsorbed.

In conclusion: (1) adsorption on HA was only found with polar amino acids; (2) the adsorption of basic amino acids is decreased by 0.01 M KCl, in agreement with the similar effect on basic polypeptides and proteins. Other features, like the adsorption of glycine at pH 6.8 and the increased adsorption of aspartic acid in the presence of 0.01 M KCl, are not easily explainable at the present time. Similarly, the effect

[10] A. Tiselius, S. Hjertén, and O. Levin, *Arch. Biochim. Biophys.* **65**, 132 (1956).
[11] A. F. Hofman, *Biochim. Biophys. Acta* **60**, 458 (1962).
[12] H. Elden and D. S. Howell, *Fed. Proc., Fed. Amer. Soc. Exp. Biol.* **19**, 142 (1960).

of pH on the adsorption of amino acids does not show any clear correlation with the results obtained with proteins. The finding that histidine and serine are adsorbed raises the question of the possible intervention of hydroxyamino acids and histidine in the protein–HA interaction. This seems unlikely, however, in view of the fact that poly-L-histidine and poly-L-serine are not retained by HA equilibrated with 0.001 M KP.

Section III

Conformation and Transitions

[19] Rotating Cylinder Viscometers

By Elliott L. Uhlenhopp and Bruno H. Zimm

Most protein molecules are sufficiently small and spherical for their viscosities to be measured very accurately by classical capillary viscometric techniques. A few protein-containing solutions, however, contain very large, asymmetric molecules or molecular aggregates for which accurate viscosity measurements can be made only by resorting to the low-shear conditions which are found in rotating cylinder viscometers. When large molecules are subjected to a shearing force, the molecules tend to align themselves with the direction of fluid flow. If the tendency toward alignment is greater than the tendency toward a random, equilibrium distribution of orientations due to normal thermal motion, then the viscosity observed for the solution will be less than under zero-shear conditions, and the solution can be described as non-Newtonian. (The viscosity of Newtonian solutions is independent of the force used to shear the solution.) Conversely, flow-induced aggregation can lead to an increase of viscosity with shear rate. To obtain viscosity numbers characteristic of the undisturbed molecules, it becomes necessary to extrapolate to conditions of zero shear. Such extrapolations are much easier and more reliable with rotating cylinder viscometers since the shearing force is more easily varied and is much smaller than the shearing forces generated by capillary viscometers.

The floating rotor viscometer introduced by Zimm and Crothers[1] has found widespread use for studying the viscosity of DNA solutions under conditions of low shear,[2] as well as for examining different types of non-Newtonian behavior, such as the increase or decrease in viscosity with change in shearing force or time of shear. In the last few years this viscometer has also been used to study the shear-dependent viscosity of certain protein solutions.

The viscosity of a solution, measured in poises, is simply the ratio of the shear stress exerted on the solution, in dynes/cm^2, divided by the shear rate, in sec^{-1}. Rotating cylinder viscometers can either fix the shear stress and measure the shear rate, or fix the shear rate and measure the shear stress. The latter type includes viscometers where either the inner[3] or the outer[4,5] cylinder is rotated, and the stress on the oppos-

[1] B. H. Zimm and D. M. Crothers, *Proc. Nat. Acad. Sci. U.S.* **48**, 905 (1962).
[2] B. H. Zimm, *in* "Procedures in Nucleic Acid Research" (G. L. Cantoni and D. R. Davies, eds.), Vol. 2, pp. 245–261. Harper, New York, 1971.
[3] P. J. Gilison, C. R. Dauwalter, and E. W. Merrill, *Trans. Soc. Rheol.* **7**, 319 (1963).

ing cylinder is measured directly. These viscometers are usually rather complicated in construction and are designed for specific applications. Some incorporate a guard ring[3,5-7] for circumventing the problem of surface films when working with protein-containing solutions. Others measure the rheological properties of blood[8] by providing special geometries[7,9,10] so that normal stresses and time-dependent phenomena can be investigated.[11-14] The discussion here will be limited to the former type of rotating cylinder viscometer, where the shear stress is fixed by magnetic interaction of external rotating magnets with the inner cylinder, and the shear rate is measured by observing the rate of rotation of the inner cylinder.

The basic floating-rotor viscometer is shown in Fig. 1. The solution whose viscosity is to be measured is placed in the well created by the outer cylinder (called the "stator"), and the inner cylinder ("rotor") is inserted from above, using an instrument which grasps the inside of the rotor (for example, "Triceps" from Universal Technical Products, Inc., Forest Hills, New York 11375). Enough additional solution is added to make the rotor float and to bring the scribed horizontal line on the rotor up to the scribed line on the stator. If the meniscus is not perfectly clean, additional solution can be added and then removed carefully from the meniscus until the two scribed lines again match exactly. The meniscus from which the rotor is suspended must remain free of any floating particles which might interfere with rotor movement. If a protein surface film is suspected, additional solvent can be added and/or removed from the meniscus to see whether interference can be minimized and to observe any time-dependent formation of the film. The addition of a drop of decyl alcohol sometimes liquefies such a film.

The external rotating magnets are turned on, and the rotor's time of revolution is recorded either manually, by observation through a

[4] H. Eisenberg and E. H. Frei, *J. Polym. Sci.* **14**, 417 (1954).
[5] M. Joly, *Biorheology* **1**, 15 (1962).
[6] L. Dintenfass, *Nature (London)* **213**, 179 (1967).
[7] A. Evans, J. P. A. Weaver, and D. N. Walder, *Biorheology* **4**, 169 (1967).
[8] Y. Nubar, *Biophys. J.* **11**, 252 (1971).
[9] L. Dintenfass, *Biorheology* **6**, 33 (1969).
[10] L. Dintenfass, *Biorheology* **2**, 221 (1965).
[11] M. I. Gregersen, S. Usami, B. Peric, C. Chang, D. Sinclair, and S. Chien, *Biorheology* **1**, 247 (1963).
[12] G. R. Cokelet, E. W. Merrill, E. R. Gilliland, H. Shin, A. Britten, and R. E. Wells, Jr., *Trans. Soc. Rheol.* **7**, 303 (1963).
[13] L. Dintenfass, *Biorheology* **1**, 91 (1963).
[14] M. Kaibara and E. Fukada, *Biorheology* **6**, 73 (1969).

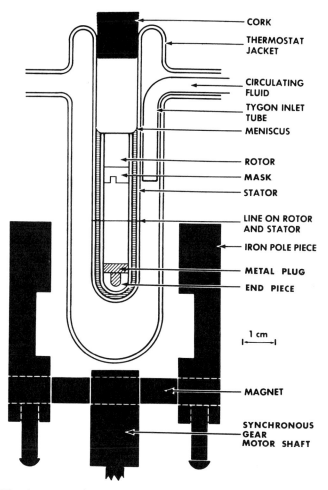

FIG. 1. Floating-rotor viscometer of B. H. Zimm and D. M. Crothers [*Proc. Nat. Acad. Sci. U.S.* **48,** 905 (1962)] with modifications, principally by Hays [J. B. Hays and B. H. Zimm, *J. Mol. Biol.* **48,** 297 (1970)]. Rotor is driven by interaction of the circulating permanent magnets with the aluminum plug. Revolution time is recorded electronically using a light beam focused by lenses on the off-center rotor mask [A. Prunell and J. Heimark, *Anal. Biochem.* **42,** 202 (1971)] and picked up by a photocell on the opposite side of the chamber.

telescope with cross hair, or electronically. Relative viscosity is obtained by comparing revolution times for the solution vs. the solvent alone. Both rotor and stator must be cleaned rigorously with detergent, chromic acid, or alcoholic potassium hydroxide before use. Any condensate or droplets on the inside of the rotor must be removed to prevent imbalance and rotor wobble. Finally, temperature control is critical for accurate viscosity measurements, and a good circulating bath ($\pm 0.05°$) must be employed.

The shear stress for a rotating cylinder viscometer can be found from measurements with a known Newtonian liquid (e.g., water). The average shear stress, $\langle S \rangle$, is given by

$$\langle S \rangle = \pi\Omega\eta \frac{R_1 + R_2}{R_2 - R_1} f(R_1, R_2)$$

where Ω is the rotor speed in revolutions per second, η is the viscosity of the solution in poises, R_1 is the radius of the inner cylinder, R_2 is the radius of the outer cylinder, and $f(R_1, R_2)$ is a correction factor which is near unity and is given by

$$f(R_1, R_2) = \frac{8R_1^2 R_2^2}{(R_1 + R_2)^3(R_2 - R_1)} \ln\left(\frac{R_2}{R_1}\right)$$

If the distance between the inner and outer cylinder is small compared to the radius of the inner cylinder, then the shear stress reduces to the torque times the radius of the inner cylinder divided by the surface area of the inner cylinder, and the shear rate reduces to the tangential velocity divided by the distance between the cylinders.

Since the floating-rotor viscometer was originally described, several refinements have been reported. Two different types of rotor–magnet interaction have been utilized to drive the rotor: hysteresis of induced magnetic moments and eddy-current induction. The former is produced by rotating permanent magnets interacting with the ferromagnetic material in the bottom of a rotor to generate a torque which is almost totally independent of the speed of magnet rotation but which can be varied by changing the field strength.[1,15–19] More recently, the ferromagnetic rotor plug has been replaced by diamagnetic conducting materials such as aluminum,[20] brass, bronze, or gold. The eddy currents induced in these materials by the rotating magnets produce their own magnetic fields, interact with the rotating magnets, and thus drive the rotor around. Such interaction is sensitive to both the strength *and* the speed of rotation of the external field. (The exact shear stress now depends on the rotation rates of both the external field and the rotor, so that to obtain relative

[15] M. Boublík, *Chem. Listy* **59**, 1343 (1965).
[16] H. J. Scherr, H. C. Vantine, and L. P. Witnauer, *J. Phys. E* **3**, 322 (1970).
[17] E. V. Frisman, L. V. Shchagina, and V. I. Vorobev, *Colloid J.* **27**, 102 (1965).
[18] M. Boublík and E. Chloubová, *Chem. Listy* **62**, 1114 (1968).
[19] O. Quadrat and P. Munk, *Coll. Czech. Chem. Commun.* **30**, 3631 (1965).
[20] G. C. Berry, *J. Chem. Phys.* **46**, 1338 (1967).

viscosities the experimental data must be manipulated in a slightly different manner. For hysteresis drive, relative viscosity is obtained simply by dividing the rotation time for the solution by the rotation time for the solvent, but for the eddy-current drive, prior calibration with the solvent alone is required to see exactly how rotor speed depends on magnet speed.[21,22] The shear stress can be altered by selecting several reproducibly different speeds of magnet rotation by use of a synchronous motor with attached variable gearbox. (An Apcor Multiratio Gearmotor, Geartronics Corp., North Billerica, Massachusetts, is presently in use in the authors' laboratory.)

Calculations are simplified if the rotating permanent magnets are replaced by two stationary alternating-current electromagnets surrounding the stator. When supplied with two currents that are mutually out of phase, the magnetic field "rotates" around the stator at a constant speed (60 Hz), and the relative viscosity once again can be calculated as the solution time divided by the solvent time. The current can be precisely varied to change the strength of the rotating magnetic field in order to achieve different shear stresses.[23-26] Eddy-current drive appears to be more stable over a wide range of temperatures.[20] In addition, the solid aluminum plugs can be machined easily to cylindricity in order to obviate rotor wobble. To this end, rotors are now precision machined out of bars of Kel-F rather than being made of glass.[22] As a result, rotors are better balanced and less fragile; furthermore, the aluminum plug can be removed easily and another inserted if necessary (see Fig. 2).

Photoelectric timing has been used to make the recording of revolution times easier and more accurate.[16,19,25,27,28] A light source on one side of the stator shines a beam through the rotor which is interrupted periodically by the rotor revolution and is picked up by a photocell on the opposite side. In one such device described recently the light beam is interrupted by a thin nylon mask in the rotor with two off-center slits so that light passes through the rotor once per revolution.[27] The photosignal is then timed by an automatic timer. Using a similar system in our laboratory, we have utilized a Hewlett Packard 5216A Electronic

[21] S. J. Gill and D. S. Thompson, *Proc. Nat. Acad. Sci. U.S.* **57**, 562 (1967).

[22] J. B. Hays and B. H. Zimm, *J. Mol. Biol.* **48**, 297 (1970).

[23] A. R. Sloniewsky, G. T. Evans, and P. Ander, *J. Polym. Sci. Part A-2* **6**, 1555 (1968).

[24] V. I. Zak, *Polymer Sci. USSR* **10**, 2270 (1968).

[25] O. C. C. Lin, *Macromolecules* **3**, 80 (1970).

[26] R. E. Chapman, Jr., L. C. Klotz, D. S. Thompson, and B. H. Zimm, *Macromolecules* **2**, 637 (1969).

[27] A. Prunell and J. Neimark, *Anal. Biochem.* **42**, 202 (1971).

[28] E. L. Uhlenhopp, Ph.D. Thesis, Columbia University, New York, 1971.

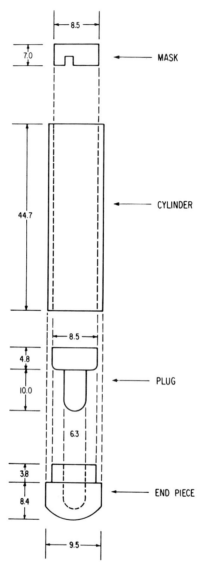

Fig. 2. Exploded view of rotor. The cylinder and end piece are made of precision-machined Kel-F [poly(chlorotrifluoroethylene)], the metal plug of solid aluminum, and the mask of aluminum foil. Specifications are in millimeters.

Counter as a timer to obtain relative viscosities routinely reproducible to better than 0.1%

Other instrumental modifications include a tube for adding or removing solutions from the bottom of the chamber,[20] a tube with a mixing chamber so that reagents can be added while the solution remains in the viscometer,[29] an outer evacuated jacket to decrease heat loss from the water jacket during high temperature runs,[15] and a second viscometer chamber which monitors solvent viscosity from one run to the next to detect any possible fluctuations.[24]

The one modification which cannot be made easily on a floating rotor viscometer involves the problem of solid surface films created by solutions of certain globular proteins. Protective guard rings are not possible for this type of instrument, and, although addition of a solvent or of decyl alcohol to the meniscus can serve as an indication of whether or not a film is interfering with rotor movement, high precision is difficult to obtain under these conditions. To circumvent the problems of solid surface films, protein chemists might well consider using the type of rotating cylinder viscometer described by Gill and Thompson.[21] In this instrument the rotor is inverted; instead of floating from the meniscus, it is submerged into the solution by exerting pressure directly on the solution from above. An air bubble trapped inside the rotor is compressed until the rotor becomes neutrally buoyant and remains suspended in solution. The rotor then acts as a Cartesian diver, and, although centering forces are less powerful than with a floating rotor and somewhat more rotor drift is therefore encountered, film problems are entirely avoided since no surface is under shear. Rotor height can be monitored continually and adjusted by a photoelectric servo system. Further refinements have been incorporated into a "retardimeter,"[26] which is essentially a Cartesian-diver rotating cylinder instrument in which both viscosity and elasticity can be measured. These measurements are useful with DNA solutions. A rotating polarized light beam passes vertically up through the viscometer chamber, and thence through amplification, squaring, and phase-differentiating electronics to a variable speed Esterline Angus recorder which provides a continual trace of rotor movement. Although this instrument is considerably more complex to construct than the original floating rotor viscometer, its capabilities are considerably enhanced also. It is presently being used to study native and denatured DNA in total lysates of bacterial and eukaryotic cells. No film problems have even been encountered, in spite of the complex nature of the solutions involved.

Only within the last few years has the floating rotor viscometer been used to investigate the shear dependence of solutions containing large, asymmetric protein molecules. Typical studies include low-shear viscometric examination of solutions of hyaluronic acid,[30-32] mucoproteins

[29] K. E. Reinert and K. Geller, *Chem. Instrum.* **1**, 391 (1969).
[30] J. H. Fessler and L. I. Fessler, *Proc. Nat. Acad. Sci. U.S.* **56**, 141 (1966).
[31] P. Silpananta, J. R. Dunstone, and A. G. Ogston, *Biochem. J.* **109**, 43 (1968).
[32] F. A. Meyer, B. N. Preston, and D. A. Lowther, *Biochem. J.* **113**, 559 (1969).

of interest in connection with cystic fibrosis,[33,34] and tropocollagen.[35] Another general area where the floating-rotor viscometer has quite logically been employed is the study of DNA-protein interaction systems. For example, the viscosity of DNA–histone complexes has been investigated as a function of temperature,[36-39] irradiation,[40] amount of histone added,[38,41,42] and pH.[43] The instrument has also been used as a sensitive assay for DNase.[44,45] The release of DNA from N4 virus particles has been followed,[46] as well as the labile attachment of fowlpox virus DNA to its protein coat.[47] The interaction of native and denatured DNA with intercalating and ionically binding dyes and antibiotics has also been studied using the floating-rotor viscometer.[48-53] Other systems for which this instrument has proved useful include studies on the acidic protein from bovine adrenal medulla chromaffin,[54] agarose and agaropectin,[55] wheat flour hemicellulose,[56] human serum albumin,[57] an extracellular polysaccharide from *Xanthomonas campes-*

[33] F. K. Stevenson and P. W. Kent, *Biochem. J.* **116**, 791 (1970).

[34] F. K. Stevenson, *Clin. Chim. Acta* **23**, 441 (1969).

[35] P. L. Privalov, I. N. Serdyuk, and E. I. Tiktopulo, *Biopolymers* **10**, 1777 (1971).

[36] S. N. Alam, R. A. Alam, and E. Harbers, *Biochim. Biophys. Acta* **209**, 550 (1970).

[37] P. Henson and I. O. Walker, *Eur. J. Biochem.* **14**, 345 (1970).

[38] M. Boublík, J. Šponar, and Z. Šormová, *Coll. Czech. Chem. Commun.* **32**, 4319 (1967).

[39] H. Bujard and E. Harbers, *Z. Naturforsch.* **20**, 719 (1965).

[40] M. Vogt and E. Harbers, *Strahlentherapie* **133**, 426 (1967).

[41] J. Šponar, M. Boublík, I. Frič, and Z. Šormová, *Biochim. Biophys. Acta* **209**, 532 (1970).

[42] M.-C. Touvet-Poliakow, M. P. Daune, and M. H. Champagne, *Eur. J. Biochem.* **16**, 414 (1970).

[43] I. O. Walker, *J. Mol. Biol.* **14**, 381 (1965).

[44] C. D. Steuart, S. R. Anand, and M. J. Bessman, *J. Biol. Chem.* **243**, 5308 (1968).

[45] H. Kopecká and M. Kohoutová, *Folia Microbiol.* **14**, 54 (1969).

[46] G. Rialdi, P. Profumo, and A. Ciferri, *Biopolymers* **8**, 701 (1969).

[47] L. G. Gafford and C. C. Randall, *Virology* **40**, 298 (1970).

[48] W. Kersten and H. Kersten, *Biochem. Z.* **341**, 174 (1965).

[49] D. S. Drummond, N. J. Pritchard, V. F. W. Simpson-Gildemeister, and A. P. Peacocke, *Biopolymers* **4**, 971 (1966).

[50] W. Kersten, H. Kersten, and W. Szybalski, *Biochemistry* **5**, 236 (1966).

[51] R. L. O'Brien, J. L. Allison, and F. E. Hahn, *Biochim. Biophys. Acta* **129**, 622 (1966).

[52] E. Hirschberg, I. B. Weinstein, N. Gersten, E. Marner, T. Finkelstein, and R. Carchman, *Cancer Res.* **28**, 601 (1968).

[53] H. L. White and J. R. White, *Biochemistry* **8**, 1030 (1969).

[54] A. D. Smith and H. Winkler, *Biochem. J.* **103**, 483 (1967).

[55] T. G. L. Hickson and A. Polson, *Biochim. Biophys. Acta* **165**, 43 (1968).

[56] E. W. Cole, *Cereal Chem.* **46**, 382 (1969).

[57] J. Kirschbaum, *J. Pharm. Sci.* **59**, 854 (1970).

tris,[58] and anionic polysaccharides such as alginate.[59] The list is not meant to be comprehensive, but only to indicate the wide range of problems to which the low-shear floating-rotor viscometer has already been applied.

Acknowledgments

The authors wish to acknowledge support in the form of a grant from the National Institutes of Health (GM-11916) to B. H. Z., and a postdoctoral fellowship from the Damon Runyon Memorial Fund for Cancer Research (DMF-643) to E. L. U.

[58] F. R. Dintzis, G. E. Babcock, and R. Tobin, *Carbo. Res.* **13**, 257 (1970).
[59] O. Smidsrød and A. Haug, *Biopolymers* **10**, 1213 (1971).

[20] Solubility Measurements

By YASUHIKO NOZAKI

Methods of measuring solubility are so numerous and diverse that it is far beyond the scope of this chapter as well as the author's capability to describe even typical ones. We shall, therefore, limit the present discussion to the solubility measurements of amino acids and the related compounds mostly in aqueous solvents, in which the author has some experience.[1]

Solubility is the concentration of the substance in question which is at equilibrium with its solid phase at a certain temperature and pressure. The determination of solubility, therefore, requires the following conditions to be fulfilled.

First, the establishment of equilibrium must be verified. This can be done in several ways. For instance, by assaying the solution after

---

[1] For a more general approach to the subject of solubility measurements, see W. J. Mader, R. D. Vold, and M. T. Vold in "Technique of Organic Chemistry" (A. West, ed.), Vol. I, Part I, p. 655. Wiley (Interscience), New York, 1959.

For a theoretical treatment of solubility in general, see J. H. Hildebrand and R. L. Scott, "The Solubility of Non-electrolytes." Reinhold, New York, 1950.

Numerous solubility data of amino acids and their theoretical interpretation are presented by E. J. Cohn and J. T. Edsall, "Proteins, Amino Acids and Peptides," Chapters 8–11, Reinhold, New York, 1943; and by J. T. Edsall and J. Wyman, "Biophysical Chemistry," Vol. I, Chapter 5, Academic Press, New York, 1958.

An aspect of the application of solubility data to the thermodynamic interpretation of protein denaturation is described by C. Tanford, *Advan. Protein Chem.* **24**, 1 (1970).

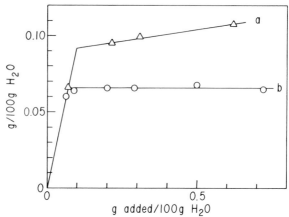

FIG. 1. Solubility of N-carbobenzoxyglycine in water at 25°. Curve a for an impure sample. Curve b after the sample was recrystallized from 30% ethanol.

different periods of equilibration, or reaching the same solubility value both from oversaturation, say, by cooling the solution saturated at higher temperature, and from undersaturation, that is, the normal way of saturation.

Second, the solid sample must be pure. If the sample is not pure, the solubility will normally show a steady increase with increasing amount of solid phase as shown in Fig. 1, curve a, reflecting the decrease in activity coefficient because of the interaction of the impurity with the substance in question. Incidentally, this phenomenon serves as a criterion for the purity of a sample. When a sample is pure, its solubility is constant regardless of the quantity of the saturating body as shown in Fig. 1, curve b.

If these conditions are fulfilled, the solubility must be a uniquely fixed value. However, there have been reported more than one value for the solubility of some amino acids. For instance, the solubility of L-tyrosine at 25° was reported as 0.0479 g in 100 g of water by Dunn, Ross, and Read[2] and 0.0453 g in 100 g of water by Dalton and Schmidt.[3] It is rather likely that many investigators, in the past, have adopted either one of these values and rejected the other. We had done the same until our own two samples of L-tyrosine showed two sets of solubility values, 0.0451 g/100 g of water[4] and 0.0452 g/100 g of water,[5] on the one hand, and 0.0475 g/100 g of water[6] on the other, each being very

[2] M. S. Dunn, F. J. Ross, and L. S. Read, *J. Biol. Chem.* **103**, 579 (1933).
[3] J. B. Dalton and C. L. A. Schmidt, *J. Biol. Chem.* **103**, 549 (1933).
[4] Y. Nozaki and C. Tanford, *J. Biol. Chem.* **238**, 4074 (1963).
[5] Y. Nozaki and C. Tanford, *J. Biol. Chem.* **246**, 2211 (1971).
[6] Y. Nozaki and C. Tanford, *J. Biol. Chem.* **245**, 1648 (1970).

similar to those values given by the two groups of investigators, respectively. Another example is L-tryptophan for which two sets of solubility data, 1.28 g and 1.38 g in 100 g of water, have been reported.[5,7]

Since there is little doubt about the purity of the two solid samples in these cases, the explanation for the two sets of solubility values could be that those amino acids existed in two different forms. The difference could be in crystalline form as in the case of glycine (α, β, and γ forms)[8,9] or due to the formation of a complex. It cannot be predicted that the lower solubility is the correct value, since it corresponds to the stable form. In the case of suspected complex formation between diglycine and guanidine hydrochloride in the solvent, a new solid phase precipitated out to lower the solubility. Therefore, it would be helpful, in order to avoid such ambiguous situations, to analyze the solid phase before and after the solubility measurements, for instance, by X-ray diffraction.[10] Also it is a good practice to keep records of conditions for the preparation of the sample, in particular the solvent used for recrystallization.

In this connection it may be noted that optically active compounds offer a special case, since it is likely that crystal lattice, and hence the lattice energy of an optically active isomer, can be quite different from that of its racemic isomer. Their solubilities may be different, although it is not predictable by how much. For example, the solubility of dl-glutamic acid is 2.643 g in 100 g of water, and that of d-glutamic acid is 0.8878 g in 100 g of water at 25°,[2] while the solubility of dl-alanine is 16.72 g in 100 g of water and that of d-alanine is not very different, being 16.65 g in 100 g of water at 25°.[3]

On the other hand, there is no difference between the solubilities of optically active isomers. For example, an identical solubility was reported for D- and L-tyrosine.[11]

Experimental Procedure

Mixing Chamber and Shaking Apparatus. There are many ways of mixing the solid phase and the solvent. The following is, therefore, only intended to give some idea of how to assemble an apparatus from ordinary laboratory items to begin with.

We have found ground-glass joints convenient to make mixing chambers, as shown in Fig. 2.

[7] J. O. Hutchens and E. P. K. Hade, *in* "Handbook of Biochemistry" (H. A. Sober, ed.), p. B10. Chemical Rubber Company, Cleveland, 1968.

[8] R. Marsh, *Acta Crystallogr.* 11, 654 (1958).

[9] Y. Iitaka, *Acta Crystallogr.* 13, 35 (1960); 14, 1 (1961).

[10] E. P. K. Hade, personal communication, 1972.

P. S. Winnek and C. L. A. Schmidt, *J. Gen. Physiol.* 18, 889 (1934–1935).

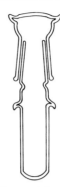

Fig. 2. Glass mixing chamber.

We have also found that a Teflon sleeve inserted between the ground-glass parts was sufficient for sealing, thus eliminating the use of grease. Rubber bands stretched taut are preferred to metal springs to secure tight seal.

The addition of several glass beads helps mixing. ⊥-shaped glass tubes will completely eliminate the possibility of leakage, but we prefer the type described above, which is completely immersible. For larger quantities of mixture, other types of containers, say, Erlenmeyer flasks, will be the choice.

The number of ways to agitate mixtures is also almost limitless. There are commercial rotators, reciprocating as well as rocking shakers, and magnetic stirrers. We made use of an oblong Warburg shaker bath. The manometer was replaced by a sturdy metal stand (Fig. 3a). Two such stands support a Lucite rack (Fig. 3b) which is designed to hold 5 mixing tubes, or a metal cage to hold the Erlenmeyer flasks.

We normally prepare 5 mixtures in such a way that 2 mixtures contain less than sufficient quantities of solid sample and 3 contain more than enough to saturate the solvent. The latter three are to yield the solubility in relation to the quantities of the sample added and the former two serve as a recovery test.

For shaking the Erlenmeyer flasks we use a wire cage attached to the metal supports and a metal plate with 5 holes to keep the flasks in place in the cage when immersed.

Preparation of the Sample and Mixing with Solvent. The amounts of sample should be chosen as described earlier. The free space in the mixing chamber may be flushed with an inert gas, such as nitrogen, if necessary. For samples which are extremely sensitive to oxygen or carbon dioxide, the water to be used can be boiled to expel such gases and then cooled under a slow nitrogen gas flow. One should remember, though,

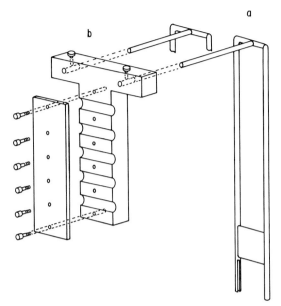

FIG. 3. Shaking rack and metal supports.

that commercial nitrogen is not completely free from carbon dioxide, oxygen, and other gases.

The accuracy of temperature control required depends on the effect of temperature on solubility. Without knowledge of this effect, temperature constant ± 0.05° will be adequate to begin with.

It must be found out by experiment how long the mixing should be carried out. In general, the lower the solubility, the longer the time required to reach equilibrium.[1] The equilibration period is affected by the size of the crystal also. Cohn et al. equilibrated the same sample with several changes of solvent. They judged that the sample was pure and the system had reached equilibrium when all supernatants showed the same concentration (solubility).[12]

If the sample consists of large crystals, their size must be reduced by grinding in a clean glass mortar. But it should be done carefully, because rubbing the crystals with the pestle might cause the distortion of crystal structure by raising the temperature, or otherwise.

Separation of the Solution. A glass filter funnel whose stem has been replaced by a thinner tubing long enough to reach the bottom of the mixing chamber is quite adequate for separating the solutions. When

[12] E. J. Cohn, T. L McMeekin, J. T. Edsall, and J. H. Weare, *J. Amer. Chem. Soc.* **56**, 2270 (1934).

mixing is terminated, the chamber is detached from the rack, its upper portion is held above the bath water surface and wiped with tissue paper. The cap is then removed, the filter stem is inserted in the mixing chamber still kept in the bath, and the supernatant is collected on the fritted glass disc by suction through a glass tube attached to the top of the filter by means of a rubber stopper. An aliquot for assay should be weighed out as soon as possible.

Assay. Again there are many ways of assaying the supernatant, but a few comments may deserve mention.

Probably the simplest and least ambiguous method of assay would be by dry weight. The size of the sample to be assayed depends on the accuracy required and the precision of the balance available, but with a balance with a precision of ± 0.1 mg, an aliquot to give at least 10 mg of residue on drying should be assayed.

Heating in a forced-air oven at 107° has been the routine method for drying protein solutions and the same condition has been adopted successfully to purely aqueous solutions of amino acids with the exception of glutamine which undergoes thermal decomposition. In the case of glutamine, therefore, the weight decrease was recorded with time, and the weight of residue was determined from the plateau region before decomposition set in. When the solvent is volatile, the dry weight method is always a possibility, but the drying conditions have to be chosen so as to suit the system in question. For example, we evaporated most of the solvent at 35° *in vacuo*, when it was a mixture of water and ethanol or ethylene glycol, and then transferred the residue to the oven to complete evaporation.

A micro dry weight method has been developed for determining the concentration of solutions containing about 1 mg of protein dissolved in 1 ml of sample solution.[13] A combination of an electric microbalance and Teflon weighing cups[14] weighing 50–60 mg is used. Silica gel is placed inside the balance chamber to keep it dry, since no cover for the container is available. A polonium[15] cartridge is placed beneath the Teflon container, because static electricity makes weighing impossible otherwise.

The spectrophotometric method is also widely used. Tyrosine presents a typical case because of its large absorptivity and low solubility. It is advantageous to scan the whole spectral range before a particular wavelength is chosen for assay. For instance, the possibility of oxidation could

[13] R. Smith, J. Dawson, and C. Tanford, *J. Biol. Chem.* **247**, 3376 (1972).
[14] Cahn Gram Electrobalance and No. 2034 weighing cups by Ventron Instruments Corporation, Paramount, California.
[15] Staticmaster Model No. 3C500 by Nuclear Products Co., El Monte, California.

be detected in this manner, and a protective measure could be taken, as is the case with tyrosine which undergoes autoxidation.

The use of radioisotope-labeled compounds is another general method.[16] In practice there are no upper or lower limits to the concentrations which may be determined by this method; therefore it may be applied quite generally, particularly when a large body of data is required, provided adequate labeled compounds are available.

[16] K. H. Dooley and F. J. Castellino, *Biochemistry* 11, 1870 (1972).

[21] Ultraviolet Difference Spectroscopy— New Techniques and Applications

By John W. Donovan

Recently, a difference between the conformation of a protein in solution and in a crystal has been demonstrated by use of spectrophotometry. The difference, the position of the side chain of tyrosine residue 248 at the active site of carboxypeptidase A[1] appears to be directly related to zinc binding at the active site of the enzyme, and therefore to its biological activity. It is probable that other conformational differences exist between proteins in solution and in the crystal. Conformational differences should be recognized more easily as techniques for determining local conformation in solution are improved to give higher "resolution"—i.e., significant information about small portions of the surface or interior of a protein. There are now many different ways to use ultraviolet absorption spectroscopy to obtain information about protein conformation in solution at different degrees of "resolution."

Much of the following is applicable also to the visible spectroscopy of proteins. However, because the chromophores of the amino acids do not absorb visible light, studies in the visible region are restricted to conjugated proteins or protein–coenzyme complexes,[2] interaction of proteins with substances (such as transition metal ions) that absorb visible light, and proteins to which "reporter groups" are attached.[3]

No theory, only a very brief background, together with some recent developments and a few results of general interest, will be presented

[1] J. T. Johansen and B. L. Vallee, *Proc. Nat. Acad. Sci. U.S.* 68, 2532 (1971).
[2] R. J. Johnson and D. E. Metzler, see Vol. 18A, p. 433.
[3] M. E. Kirtley and D. E. Koshland, Jr., see Vol. 26 [25].

here. Recent articles which should be consulted are those by: Laskowski,[4,5] Kronman and Robbins,[6] Donovan,[7] Herskovits,[8] and Wetlaufer.[9] Ultraviolet techniques currently in use or being developed can be roughly classified as: (1) those in which the changes in the spectra of chromophores are observed under the dynamic conditions of biological interest (e.g., change of pH, state of aggregation, conformation, interaction with substrate or inhibitor), and (2) those in which a change is introduced by the experimenter under static conditions—those for which the conformation or degree of aggregation of the protein remains constant. These latter techniques can all roughly be classified together as "spectral perturbation" techniques. Perturbations may be produced by nearly any physical or chemical means, e.g., alteration of pH, temperature, pressure, or solvent composition, and are generally useful for the study of the protein in its native conformation. While all these techniques lend themselves to kinetic uses, kinetic measurements, as such, are not considered in this section.

Recent Developments Pertaining to the Interpretation of Difference Spectra

When a chromophore undergoes an alteration in environment which produces a small alteration in its absorption spectrum, ϵ, the spectrum is shifted an amount $\Delta\lambda$ along the wavelength scale and changed in intensity. If the alteration in intensity is ignored, the difference between the altered spectrum, $\epsilon(\lambda - \Delta\lambda)$, and the original one, ϵ, has the appearance of the first derivative of the absorption spectrum,[10] that is, it can be represented adequately by the first term on the right-hand side of Eq. (1).

$$\Delta\epsilon = \epsilon(\lambda - \Delta\lambda) - \epsilon(\lambda) = -\left(\frac{d\epsilon}{d\lambda}\right)\Delta\lambda + \frac{(\Delta\lambda)^2}{2!}\left(\frac{d^2\epsilon}{d\lambda^2}\right) - \cdots \quad (1)$$

When the perturbation is greater, the spectrum is shifted more, and generally, the change in intensity of absorption is sufficiently great that it should not be neglected. Bailey et al.[11] have considered in detail the

---

[4] M. Laskowski, Jr., in "Spectroscopic Approaches to Biomolecular Conformation" (D. W. Urry, ed.), p. 1. American Medical Association, Chicago, Illinois, 1970.

[5] M. Laskowski, Jr., Fed. Proc., Fed. Amer. Soc. Exp. Biol. 25, 20 (1966).

[6] M. J. Kronman and F. M. Robbins, in "Fine Structure of Proteins and Nucleic Acids" (G. D. Fasman and S. N. Timasheff, eds.), p. 271. Dekker, New York, 1970.

[7] J. W. Donovan, in "Physical Principles and Techniques of Protein Chemistry" (S. J. Leach, ed.), Part A, p. 101. Academic Press, New York, 1969.

[8] T. T. Herskovits, see Vol. XI, p. 748.

[9] D. B. Wetlaufer, Advan. Protein Chem. 17, 303 (1962).

[10] C. H. Chervenka, Biochim. Biophys. Acta 31, 85 (1959).

[11] J. E. Bailey, G. H. Beaven, D. A. Chignell, and W. B. Gratzer, Eur. J. Biochem. 7, 5 (1968).

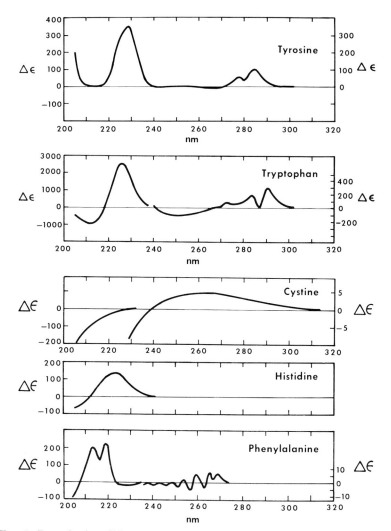

Fig. 1. Perturbation difference spectra of amino acids produced by 20% (v/v) ethylene glycol. From J. W. Donovan, *J. Biol. Chem.* **244**, 1961 (1969), and reproduced by permission of the American Society of Biological Chemists.

intensity changes produced by solvent perturbation, and Fisher *et al.*[12] have shown the necessity of considering the higher terms in the expansion of Eq. (1) when the shift in the absorption spectrum, $\Delta\lambda$, is large.

Difference spectra for perturbation of the chromophores commonly found in proteins are shown in Fig. 1. They resemble the first derivatives

[12] H. F. Fisher, D. L. Adija, and D. G. Cross, *Biochemistry* **8**, 4424 (1969).

TABLE I

ESTIMATED CHANGES IN MOLAR ABSORBANCE PRODUCED BY TRANSFER OF A
CHROMOPHORE FROM THE INTERIOR OF A PROTEIN INTO WATER[a]

Chromophore	λ_{max}[b] (nm)	$-\Delta\epsilon_{max}$	λ_{max}[b] (nm)	$-\Delta\epsilon_{max}$
Phenol	287	700	228	2,000
Indole	292	1600	225	13,800
Benzene	259	36	218	1,260
Disulfide	265	30	—	—
Imidazole	—	—	223	780

[a] Calculated from the observed perturbation with 20% ethylene glycol, assuming that the protein interior is equivalent to 120% ethylene glycol [J. W. Donovan, J. Biol. Chem. 244, 1961 (1969)].

[b] Both λ_{max} and $\Delta\epsilon$ show changes with different proteins and solvents.

of the absorption spectra, $d\epsilon/d\lambda$, and are characteristic of each chromophore—see Herskovits[8] for a more extended discussion. The difference spectra of Fig. 1 have been obtained over a larger wavelength range than is generally used.[13-18] The change in absorption produced by a perturbation is ordinarily proportional to the absorption coefficient of the absorption band.[19] Since the absorption coefficients of these chromophores are greater at shorter wavelength, the changes produced by ionization or solvent perturbation are greater. At shorter wavelength, less protein is needed for equivalent accuracy in determining $\Delta\epsilon$, but corrections for solvent absorption[8] and impurities are more important. Table I gives the changes in molar absorption ($\Delta\epsilon$) at the wavelengths (subject to minor variation for particular proteins) of the maxima in the difference spectra. To a first approximation, the maxima in the difference spectra occur at the wavelengths of greatest steepness ($d\epsilon/d\lambda$) in absorption spectra (Eq. 1).

Interpretation of Difference Spectra. When absorption differences are obtained in the form of difference spectra, the usual procedure is to compute a fit to the observed difference spectrum by summation of perturbation difference spectra of model compounds, such as those in

[13] Spectra and difference spectra for these changes at shorter wavelengths have been presented by Green,[14] McDiarmid,[15] Bailey et al.,[11] Móra and Elödi,[16] Donovan,[7,17] and Dyson and Noltman.[18]

[14] N. M. Green, Biochem. J. 89, 599 (1963).

[15] R. S. McDiarmid, Ph.D. Thesis, Harvard Univ., Cambridge, Massachusetts, 1965.

[16] S. Móra and P. Elödi, Eur. J. Biochem. 5, 574 (1968).

[17] J. W. Donovan, J. Biol. Chem. 244, 1961 (1969).

[18] J. E. D. Dyson and E. A. Noltman, Biochemistry 8, 3544 (1969).

[19] N. S. Bayliss, J. Chem. Phys. 18, 292 (1950).

Fig. 1. When the protein has only one type of chromophore of interest, a comparison can be made with the perturbation spectrum of the completely unfolded protein (usually in a denaturant such as $8\,M$ urea, with protein disulfide bonds reduced). The reader is referred to Herskovits[8] for details of alternate representations of data, corrections for urea, etc. Many recent calculations have been carried out by computer. A spectrophotometer with digital tape or card output (or better yet, direct interface to the computer) reduces the accounting work.

Interpretation and Calculation of Absorption Spectra. Attempts are continually being made to account as completely as possible for all the features of the spectrum of a protein in solution. Preliminary to this effort is the understanding of the spectrum of the chromophores in solution. For calculations of spectra, the wavelength shift, intensity change, and sharpness of vibrational bands are three of the parameters used. The most ambitious attempt to calculate the spectra of model chromophores in different solvents from the vapor spectrum only has been made by Laskowski and co-workers.[4]

The absorption spectrum of ribonuclease in 1:1 glycerol:water at $77°K$ has been analyzed.[20] Model spectra of the phenolic chromophore were determined at $77°K$ in both glycerol–water and organic solvents (methanol–glycerol and ether–pentane–alcohol mixtures), and fitted by sums of vibrational bands of Gaussian shape, separated by the known vibrational frequencies of this chromophore. This fitting procedure gave the intensity ratios of the vibrational bands. The spectrum of ribonuclease could then be fit by sums of vibrational bands with these separations and intensities. The fitting consisted in shifting the vibrational spectra on the wavelength scale so that the absorption spectrum was matched closely. While this procedure does not guarantee a unique fit, the result obtained seems acceptable in terms of the chemical and other spectroscopic experiments previously carried out. The positions of the 0–0 vibrational bands of the six phenolic groups of ribonuclease seem to be in reasonable agreement with the wavelengths observed for the 0–0 vibrational bands of the phenolic chromophores of model compounds dissolved in the aqueous and nonaqueous solvents studied. In ribonuclease, three phenolic groups appear buried—0–0 vibrational bands at 288.5 nm (1 group) and at 286 nm (2 groups)—and three appear exposed to solvent (0–0 vibrational band at 283.5 nm).

However, first derivative spectroscopy[21] indicates that the absorption spectrum of ribonuclease in aqueous solution at room temperature

[20] J. Horwitz, E. H. Strickland, and C. Billups, *J. Amer. Chem. Soc.* **92**, 2119 (1970).
[21] J. F. Brandts and L. J. Kaplan, *Fed. Proc., Fed. Amer. Soc. Exp. Biol.* **30**, 1108Abs (1971).

is not broader than that of N-acetyltyrosine ethyl ester—suggesting that all phenolic groups have equivalent environments. The apparent contradiction between this result and that of Horwitz et al.[20] may be produced by the opposite shifts the absorption spectra of buried and exposed chromophores undergo with changes in temperature (see below).

The circular dichroism spectrum of ribonuclease at 77°K was interpreted[20] in a similar fashion to the absorption spectrum, but since the intensities of the CD bands of a given chromophore cannot be assigned a priori, and the disulfide chromophores make a large contribution to the CD spectrum, reference to the absorption spectrum was necessary for the analysis. Strickland and co-workers have analyzed the vibrational structure of the absorption and CD spectra of tyrosyl, tryptophanyl, and phenylalanyl,[20,22] chromophores at 77°K.

A technique for generating a simulated difference spectrum, termed a "fine structure plot," has been described.[23] A best approximation to the observed spectrum of a protein is computed, with the restriction that only a limited number of absorption bands and adjustable parameters are used in the computation. A smooth fit to the observed spectrum of a protein is obtained from the sum of two absorption bands of lognormal shape (each with four parameters: position, intensity, band width, and skewness).[2] The difference between the observed spectrum and this calculated spectrum contains the "fine structure," or "wiggles" in the observed spectrum, and resembles a difference spectrum obtained experimentally. This procedure has some advantages over the determination of a difference spectrum. A difference spectrum clearly shows changes in absorption, but does not, for example, reveal any abnormalities in absorption of a protein in which all chromophores are buried unless the environment of these chromophores differs in the reference and sample solutions. A "fine structure plot" of such a protein might reveal its abnormalities in the positions of the two computed absorption bands. Although considerably more computation work is required, accurate calculation of the spectra of proteins would, of course, be preferable to both difference spectra and "fine structure plots." At present, even when a good fit to the observed spectrum can be calculated,[20] no indication is obtained that the fit is unique, and the interpretation of the calculated spectrum in terms of the environment of the individual chromophores remains very difficult.[4]

[22] E. H. Strickland, M. Wilchek, J. Horwitz, and C. Billups, J. Biol. Chem. 245, 4168 (1970); E. H. Strickland, J. Horwitz, and C. Billups, Biochemistry 8, 3205 (1969); J. Amer. Chem. Soc. 91, 184 (1969).
[23] D. E. Metzler, C. Harris, I. Yang, D. Siano, and J. A. Thompson, Biochem. Biophys. Res. Commun. 46, 1588 (1972).

Fig. 2. Directions of the transition moments in the plane of the indole chromophore, after Yamamoto and Tanaka.[25]

Abnormal Difference Spectra. Particularly in the 290 nm wavelength region, many examples now exist of difference spectra of proteins that cannot be approximated by combinations of "normal" difference spectra such as those of Fig. 1. A typical abnormality is a peak (often negative) near 300 nm.[24] Also typical is the failure to observe such abnormal spectra for denatured proteins. It appears likely that the greatest proportion of such abnormal spectra are produced by the fixed location of a chromophore with respect to its environment in the native protein. Among the effects that can be expected from the fixed orientation of a chromophore are those listed: (1) The orientation of a perturbant (dipole, charge, or polarizable group in the protein, or in a bound ion or substrate molecule) with respect to the transition moment of the chromophore can determine the magnitude of a perturbation and the energy difference between the excited and the ground state (wavelength of absorption band). (2) When the different electronic transitions of the same chromophore have transition moments oriented in different directions, the fixed position of a perturbant can have unequal effects on the magnitudes and wavelengths of the absorption bands of these transitions. Thus, when any degree of orientation exists, the absorption bands at shorter wavelengths (near 230 nm) may be affected differently from those at longer wavelengths (near 280 nm). (3) For the indole chromophore of tryptophan, the two overlapping absorption bands in the 290 nm wavelength region, ^1L_b and ^1L_a, have transition moments that are at a large angle to one another in the plane of the rings (Fig. 2).[25] Orientation of a perturbant with respect to tryptophan or orientation of one indole chromophore close to another will give "abnormal" spectra and difference spectra in the 290 nm wavelength region.[26] (4) Coupling between adjacent chromophores leads to band splitting, or broadening, depending on the closeness of the coupling.[22]

[24] V. S. Ananthanarayanan and C. C. Bigelow, *Biochemistry* **8**, 3717, 3723 (1969).
[25] Y. Yamamoto and J. Tanaka, *Bull. Soc. Chem. Japan* **45**, 1362 (1972).
[26] L. J. Andrews and L. S. Forster, *Biochemistry* **11**, 1875 (1972); E. H. Strickland, C. Billups, and E. Kay, *Biochemistry* **11**, 3657 (1972).

Solvent Perturbation

This now-classic technique[27] allows the determination, under favorable conditions, of which of a protein's chromophores are exposed to solvent by measurement of the absorption changes produced upon alteration of the composition of the solvent. The rationale of the experiment has been clearly explained by Laskowski[5] and the experimental technique was described in detail by Herskovits.[8] The magnitude of the perturbation is proportional to the product of the polarizability of the solvent and the absorption coefficient of the electronic transition if no specific interaction between solvent and chromophore occurs.[19,28] Tables of the perturbation produced on the phenol and indole chromophores of model compounds by various perturbants are given by Herskovits and Sorensen[29] and by Kronman and Robbins.[6] Difference spectra (220–320 nm) produced by ethanol, dioxane, and urea are given by Bailey et al.[11]

Difference spectra obtained for globular proteins in the 230 nm wavelength region are attributable almost entirely to their aromatic chromophores.[11,16,17] Figure 3 shows experimental and calculated perturbation difference spectra for aldolase in the 230 nm wavelength region. The perturbation difference spectra obtained with ethylene glycol at both longer and shorter wavelengths are fit by the same sums of perturbation difference spectra of model compounds shown in Fig. 1.

Calculated and observed difference spectra, like those of Fig. 3, often differ by a wavelength displacement. This occurs when the chromophores in the model compounds have an environment which is different from that of the same chromophores in the protein. Most of the discrepancy in wavelength of Fig. 3 can be attributed to the absence of the peptide bond and the presence of free carboxyl and amino groups in the model compounds used in the experiments to determine the absorption change produced by ethylene glycol.

Alteration of solvent composition by addition of perturbant must not alter protein conformation, otherwise the other chromophores (those buried in the interior of the protein) would become exposed. Linearity between the magnitude of the perturbation ($\Delta\epsilon$) and concentration of perturbant generally indicates absence of conformational change. However, auxiliary measurements should be carried out to establish constancy of protein conformation; the use of another technique such as

[27] T. T. Herskovits and M. Laskowski, Jr., J. Biol. Chem. 235, PC56 (1960); 237, 2481 (1962); C. C. Bigelow, C. R. Trav. Lab. Carlsberg Sér. Chim. 31, 305 (1960).
[28] N. S. Bayliss and E. G. McRae, J. Phys. Chem. 58, 1002 (1954).
[29] T. T. Herskovits and M. Sorensen, Biochemistry 7, 2523 (1968).

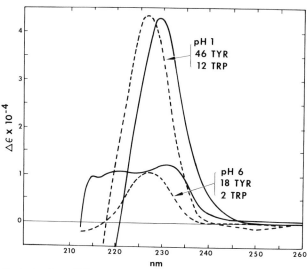

Fig. 3. Perturbation difference spectra of rabbit muscle aldolase (MW 1.58 ×
10⁵) produced by 20% ethylene glycol, in the shorter ultraviolet wavelength region.
The observed difference spectra (————) are compared with calculated perturbation
difference spectra (— — —) for 46 moles of tyrosine plus 12 moles of tryptophan
(at pH 1) and for 18 moles of tyrosine plus 2 moles of tryptophan (at pH 6). From
J. W. Donovan, *J. Biol. Chem.* **244**, 1961 (1969); reproduced by permission of the
American Society of Biological Chemists.

optical rotation allows detection of at least the larger conformational
changes that may accompany introduction of perturbants into the
solvent. The ability of aliphatic alcohols to alter the conformation
of selected globular proteins increases with increasing length of the
carbon chain.[30] Methanol has the least effect on conformation; glycols
have relatively little effect on conformation at the concentrations of 20%
to 40% (v/v) generally used for perturbation experiments.

Although a perturbant may have little or no effect on conformation
at a pH and ionic strength at which a protein has great stability, under
other conditions (particularly extremes of pH) the same perturbant
can stabilize or destabilize the native form of the protein.[31] Accordingly,
if perturbation methods must be used to measure the exposure of chromophores under conditions of marginal stability of protein conformation,
it is essential to demonstrate by other techniques that addition of

[30] E. E. Schrier and H. A. Scheraga, *Biochim. Biophys. Acta* **64**, 406 (1962); E. E.
Schrier, R. T. Ingwall, and H. A. Scheraga, *J. Phys. Chem.* **69**, 298 (1965); T. T.
Herskovits, B. Gadebeku, and H. Jaillet, *J. Biol. Chem.* **245**, 2588 (1970).
[31] T. T. Herskovits and M. Laskowski, Jr., *J. Biol. Chem.* **243**, 2123 (1968).

perturbant does not alter the equilibrium constant for the conformational change.

Allied with conformational changes of the protein are interactions of the perturbant with the protein. Strong interactions make the perturbant a denaturant.[32] Weaker interactions may cause conformational changes. Still weaker interactions are characterized by presence of perturbant adjacent to particular protein side chain groups (chromophoric or nonchromophoric) in greater or lesser amount than that expected from the composition of the solvent mixture. When the local concentration of perturbant is greater than in the bulk of the solvent mixture, "preferential binding of perturbant" occurs. The reverse situation is termed "preferential binding of water" (if that is the solvent present in greatest amount). If preferential binding in the vicinity of a chromophore is produced not by interaction with the chromophore itself, but with the groups in the protein *adjacent* to the chromophore, then an incorrect exposure will be calculated for the chromophore, since the chromophore in the model compound will not be solvated to the same extent. This type of preferential interaction may be detected by the use of a different perturbant which does not bind preferentially (e.g., D_2O or a glycol). In certain instances, the existence and magnitude of preferential interaction may be determined by "double perturbation" or "competitive perturbation." In these experiments, a second perturbant is introduced to determine the degree of interaction of the first perturbant with the exposed chromophores. Interpretation can be difficult.[7] A special case of preferential interaction, the formation of charge transfer complexes, is considered below.

Even when no conformation change takes place and when no preferential binding of perturbant occurs, the apparent degree of exposure of chromophores may still be a function of the perturbant employed. Perturbant molecules of smaller size[33] presumably have greater access to those chromophores that are exposed to solvent, but are close to the irregular surface of the protein. Chromophores that are perturbed by small perturbants but not by larger perturbants may be located in "crevices" in the protein. Experiments must be designed and carried out with care in order to distinguish between "partially exposed" chromophores—those partially embedded in the surface of the protein—and chromophores located in crevices.[5]

Ovomucoid, a glycoprotein of molecular weight 2.7×10^4 which contains six tyrosine residues but no tryptophan residues, provides an

[32] C. C. Bigelow, *J. Mol. Biol.* **8**, 696 (1964).
[33] See Herskovits[8] for a table of sizes of perturbant molecules.

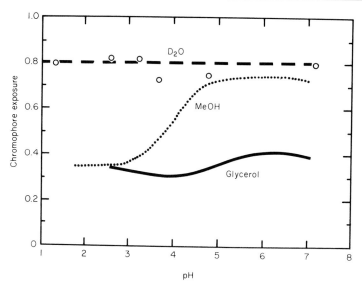

Fig. 4. Fractional exposure of the phenolic chromophores of ovomucoid to three perturbants as a function of pH. The methanol and glycerol data are taken from T. T. Herskovits and M. Laskowski, Jr. [*J. Biol. Chem.* **237**, 3418 (1962)] and the D_2O data from J. W. Donovan [*Biochemistry* **6**, 3918 (1967)]. The latter data have been recalculated in terms of 100% exposure for thioglycolic acid-reduced ovomucoid in 8 M urea, to put all relative exposures in this figure on the same basis. The change in exposure to methanol between pH 3 and pH 5 accompanies a conformation change of the protein.

example of how different degrees of exposure may be obtained by use of different perturbants. As the pH of a solution of ovomucoid is reduced from 5 to 3, the relative exposure, or fractional exposure, of the phenolic chromophores to methanol (Fig. 4) decreases from about 0.7 to 0.3. The fractional exposure (buried = 0, fully exposed = 1.0) to Carbowax (a larger glycol) or glycerol remains constant at about 0.3 as the pH is lowered. The change in exposure of the phenolic groups to methanol is accompanied by a change in optical rotation.[34] Interestingly, relative exposure to D_2O is essentially constant throughout this pH range.[35] The correct interpretation of these data is still uncertain, but it appears likely that a crevice or crevices in which some of the phenolic chromophores are located decreases sufficiently in size during the conformational change to exclude methanol but not D_2O.

When a chromophore is partially embedded at the surface of a protein and partially exposed to solvent, but no crevices are present,

[34] T. T. Herskovits and M. Laskowski, Jr., *J. Biol. Chem.* **237**, 3418 (1962).
[35] J. W. Donovan, *Biochemistry* **6**, 3918 (1967).

interpretation of perturbation experiments is very difficult.[5] X-ray diffraction results suggest that partial exposure of chromophores is common. Williams and Laskowski[36] have presented a method for distinguishing between complete and partial exposure by carrying out solvent perturbation experiments on derivatives of the protein in which differing numbers of chromophores have been chemically modified. The use of different perturbants with these derivatives provide additional information about the solvent-accessibility of the partially exposed chromophores.[36]

Thermal Perturbation

Changes in spectra produced by change in temperature have been used for thermodynamic studies of heat denaturation of proteins and nucleic acids. These absorption changes are produced by an alteration of chromophore environment brought about by the temperature-induced conformational change. When a difference in temperature between sample and reference produces no conformational change, but does produce a change in absorption, then the use of the difference in temperature between sample and reference solutions can be considered to be a perturbation method,[37,38] similar to solvent perturbation. Chromophores in a protein undergo absorption changes because of temperature-induced changes in the medium in which they are located. Unlike the solvent perturbation technique, in which, under ideal conditions, only *selected* chromophores (those in contact with solvent) are perturbed, *all* chromophores are perturbed by change in temperature.

Chromophore Environment. At least in the solvent medium, perturbation can be assumed to be approximately isotropic. The chromophore is surrounded by, on the average, a uniform *liquid* solvent with polarizability a definite (usually linear) function of temperature. However, this is probably a very bad approximation for the "microenvironments" inside the protein. X-ray diffraction measurements show that the interiors of proteins in crystals are characteristically *solid*—all interior side chains are fixed in definite positions with respect to the polypeptide backbone and other side chains. Accordingly, this essential anisotropy means that, for the chromophores within the protein, both the magnitude and spectral character (band shape) of perturbations produced by temperature difference will depend on changes in angle, distance, and relative orientation between the perturbants (the adjoining portions of the protein interior) and the chromophore, rather than on change in

[36] E. J. Williams and M. Laskowski, Jr., *J. Biol. Chem.* **240**, 3580 (1965).

[37] J. Bello, *Biochemistry* **8**, 4542, 4550 (1969).

[38] W. Cane, *Fed. Proc., Fed. Amer. Soc. Exp. Biol.* **28**, 469 (1969).

distance alone. In general, perturbations of chromophores within the protein caused by alteration of temperature should be smaller than those of chromophores exposed to solvent as long as the microenvironments within the protein retain their solid-phase character.

The Effect of Temperature on Molecular Spectra. (1) Lower temperatures sharpen (increase height, decrease width) vibrational bands because the frequency of molecular collisions is decreased, and the population of higher rotational levels decreased. (2) Lower temperatures may cause some vibrational bands of low intensity to disappear (this is particularly true of vapor spectra)—those bands originating from (thermally populated) higher vibrational levels of the ground state.[39] (3) Lower temperature (unlike change of solvent) does not diminish the integrated intensity of any vibrational band (with the exception listed above) because, at least to a good first approximation, the intensity of these bands is determined by the potential function for each electronic level. These potential functions (interatomic energy as a function of interatomic coordinates) are not temperature dependent.[40]

The Role of the Solvent. The polarizability of the solvent produces a "red shift" of the spectrum of a chromophore.[19] Since the polarizability is directly related to solvent density, which normally decreases with temperature, the polarizability and red shift of the spectrum will decrease with increase in temperature, resulting in a shift in the spectrum to shorter wavelengths (blue shift). An exception to this generalization appears with that biologically important solvent: *water.* For water, increase in temperature increases the mean red shift of the spectrum of a chromophore, resulting in a shift in the spectrum to longer wavelengths (red shift). Figure 5 shows the effect of cyclohexane and water on the mean shift of the spectrum of anisole as a function of temperature.[41] Note that (a) both solvents produce red shifts with respect to the vapor spectrum, (b) the red shift for water is smaller, and (c) temperature has opposite effects in the two solvents.

Application to Proteins. The absorption spectra of the chromophores of a protein show temperature perturbations.[37,38,42-44] Those chromophores exposed to solvent water should have their absorption spectra *red-*

---

[39] Thermal energy, kT, at room temperature is equivalent to 200 cm^{-1}. This is the same order of magnitude as the frequencies of some of the vibrational modes of the benzene ring.

[40] Vibrational–electronic interactions which modify the potential function, generally grouped under the heading of the Jahn-Teller effect, are neglected here.

[41] G. L. Tilley, Ph.D. Thesis, Purdue Univ., Lafayette, Indiana, 1967.

[42] J. Bello, *Biochemistry* 9, 3562 (1970).

[43] G. M. Lehrer and R. Barker, *Biochemistry* 10, 1705 (1971).

[44] S. J. Leach and J. A. Smith, *Int. J. Protein Res.* 4, 11 (1972).

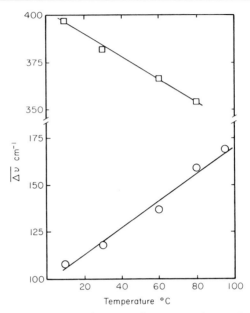

Fig. 5. The temperature dependence of the mean shift, $\Delta\bar{\nu}$ for anisole in cyclohexane (□) and in water (○). The mean shift is one of the kernel parameters of an integral equation devised to measure the overall wavelength shift and band shape changes of an absorption spectrum in solvent with respect to the vapor spectrum. A shift of 400 cm⁻¹ is roughly equal to 3 cm⁻¹ in this wavelength region. Figure by courtesy of G. Tilley.

shifted with increase in temperature. If the environment of the buried chromophores can be assumed to be equivalent to that of a (liquid) organic solvent, their spectrum should undergo a *blue* shift with increase in temperature, similar to that depicted in Fig. 5. According to the usual convention, a red shift produces a positive difference spectrum, and a blue shift, a negative one. Such temperature perturbation spectra for indole and phenol are shown in Fig. 6. If measurements are made at the longest wavelength maximum of the difference spectrum, then (if acetonitrile is a good model for that portion of the interior of a protein in which the buried chromophores are embedded) the exposed chromophores make the greatest contribution to the difference spectrum. Bello[37] has used the magnitude of the longest wavelength maximum in the difference spectrum of ribonuclease to determine the number of exposed phenolic groups from the linear dependence of molar difference absorption coefficient, $\Delta\epsilon$, on temperature. When all the chromophores of a protein are exposed by denaturation of the protein in guanidine

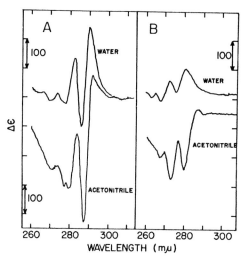

Fig. 6. Thermal perturbation difference spectra of (A) indole, and (B) phenol, in acetonitrile and in water. Sample temperature, 40.4°; reference temperature, 13.4°. From G. M. Lehrer and R. Barker, *Biochemistry* **10**, 1705 (1971), by permission of the American Chemical Society.

hydrochloride, the number of exposed chromophores can be calculated from the thermal perturbation spectrum.[37,42,44] For native proteins containing both tyrosine and tryptophan, the interpretation of the complex difference spectra obtained by temperature perturbation is difficult. The greatest unknown factor is the effect of the protein interior on the buried chromophores.[38] Lehrer and Barker[43] have assumed that acetonitrile is equivalent to the protein interior. Their data accordingly indicate that a tryptophan in the interior is only one-third as affected by temperature as one on the exterior (Table II). Interpretation of phenolic maxima near 280 nm is made more difficult by superimposed tryptophan absorption.

Method. The experimental technique used for thermal perturbation is straightforward. Absorption cells with integral jackets, or standard cells and jacketed cell holders are used. The reference solution is maintained at constant temperature (the temperature range 10–25° is convenient) by circulating water from a constant temperature bath. The sample solution is maintained at a higher or lower temperature. A temperature difference of 10° to 20° between sample and reference produces absorption differences large enough for routine measurements. Thermistors inserted through the cell covers are convenient for determining solution temperature. Since altering the temperature of the sample and

TABLE II

THERMAL PERTURBATION SPECTRA OF PHENOL AND INDOLE—DIFFERENCE
SPECTRA MAXIMA, AND THE CHANGE IN ABSORBANCE PER DEGREE[a]

	$\Delta\epsilon$ deg^{-1} at $(\lambda)^b$		
Phenol		Indole	
Water	Acetonitrile	Water	Acetonitrile
—	+0.9 (285)	+5.9 (290)	+1.8 (292)
+2.4 (281)	−5.1 (280)	+0.9 (282.5)	−5.5 (284)
+0.8 (274)	−5.5 (273)	—	—

[a] Selected data from: G. M. Lehrer and R. Barker, *Biochemistry* **10**, 1705 (1971), reproduced by permission of the American Chemical Society. See Fig. 6 for difference spectra.

[b] Wavelengths in nanometers. A positive sign indicates positive $\Delta\epsilon$ when the temperature of the sample solution is greater than the temperature of the reference solution.

reference compartments may affect the wavelength calibration of the spectrophotometer,[45] it may be necessary to maintain the cell compartment walls at constant temperature for work of highest accuracy. Linearity of perturbation with increase in temperature difference indicates that no conformation change which affects the chromophore has taken place in the temperature range to which the sample has been exposed. If the absorption difference as a function of temperature extrapolates linearly to zero at the temperature of the reference,[44] it may be inferred that the conformation of the protein in the sample solution is the same as that in the reference solution.

Concentration Difference Spectra

Difference spectra obtained between a concentrated solution and a dilute solution of protein have been given the convenient but ungrammatical name, "concentration difference spectra." Equal amounts of absorbing material are placed in the sample and reference light paths by making the path length of the dilute solution equal to the product of the path length of the concentrated solution and the concentration ratio. No absorbance difference is observed unless the nature (e.g., state of ionization) or environment of the chromophores is a function of protein concentration. On association of protein subunits, chromophores may become buried at a newly formed interface, or conformational changes of the subunits may bury previously exposed chromophores. The shape

[45] M. Laskowski, Jr., personal communication.

of the difference spectrum allows identification of the chromophores altered. Solvent perturbation experiments carried out separately on the associated and dissociated systems should show the average number and type of chromophores buried upon association. The same information may also be calculated from the concentration difference spectrum by use of Table I, if the fraction of protein associated can be determined.

Concentration difference spectroscopy has been used to study the association of a number of proteins.[46] The equilibrium constant (or constants, in the case of multiple equilibria) for association can be calculated if the magnitude of the absorption change can be determined over a sufficient concentration range. After the association constant has been determined, the fraction of protein associated at any concentration can be calculated.

Formation of Charge Transfer Complexes

The specific interaction between a chromophore and a "perturbant" in which a charge transfer complex[47] is formed is a very promising technique for probing protein surfaces. Upon excitation, a transfer of charge from one member of the complex (the donor) to the other (the acceptor), which must be in close proximity, produces new absorption bands at longer wavelengths than those at which either the donor or acceptor normally show absorption (Fig. 7). Both donor and acceptor may be covalently attached to the protein, or one may be a distinct solute species. In principle, both may be solutes if at least one is strongly bound to the protein. In most of the charge transfer work reported for proteins, the donor is a side-chain chromophore (usually indole) of the protein.

Charge Transfer Complexes of the Indole Chromophore. "Solvent perturbation" of the indole ring by the *N*-methylnicotinamide cation (MN) shows all the characteristics of the formation of a charge transfer complex specific for the indole chromophore.[47,48] Two broad absorption bands, with maxima at approximately 315 and 360 nm, appear. The former overlaps tryptophan absorption. The complex between MN (acceptor) and indole (donor) is presumed to occur by face-to-face association of the aromatic rings. The magnitude of the charge transfer

[46] These include: glutamic dehydrogenase [D. G. Cross and H. F. Fisher, *Arch. Biochem. Biophys.* **110**, 222 (1965)]; insulin [J. A. Rupley, R. D. Renthal, and M. Praissman, *Biochim. Biophys. Acta* **140**, 185 (1967)]; glucagon [J. C. Swann and G. G. Hammes, *Biochemistry* **8**, 1 (1969)]; methemoglobin [H. Uchida, J. Heystek, and M. H. Klapper, *J. Biol. Chem.* **246**, 6843 (1971)].
[47] S. Shifrin, *Biochim. Biophys. Acta* **81**, 205 (1964); M. Shinitzky, E. Katchalski, V. Grisaro, and N. Sharon, *Arch. Biochem. Biophys.* **116**, 332 (1966).
[48] D. A. Deranleau and R. Schwyzer, *Biochemistry* **9**, 126 (1970).

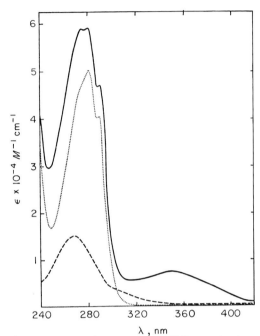

FIG. 7. Absorption spectra of chymotrypsin (ChT, ...), chymotrypsin alkylated with α-bromo-4-nitroacetophenone (ChT-NAP, ——), and α-bromo-4-nitroaceto-phenone (BrNAP, – – –) at 25° in 0.001 M HCl. The spectrum of BrNAP was obtained in a solution containing 5% acetonitrile. The absorption band with maximum at 350 nm, observed with ChT-NAP but not with either ChT or BrNAP, has the characteristics of a charge transfer band. From D. S. Sigman and E. R. Blout, *J. Amer. Chem. Soc.* **89**, 1747 (1967), by permission of the American Chemical Society.

absorption band which appears when MN binds to native lysozyme[49] indicates that only one indole ring in lysozyme is so situated that its face is freely accessible to molecules the size of MN (Table III).

Thermodynamic Parameters for Charge Transfer Formation. Binding of donor to acceptor is weak. Association constants calculated for model systems and for lysozyme[47-49] range from 1 to 5 liters mole⁻¹. For the formation of the 1:1 complex between N-acetyltryptophan and MN, at 25°, $\Delta G°$ is -1 kcal mole⁻¹, $\Delta H°$ is -3.5 kcal mole⁻¹, and $\Delta S°$ is -8.3 e.u.[48] Accordingly, high molar ratios of acceptor to protein (approximately 1000:1) are needed for experiments. In addition, the relatively small molar absorption of the complex requires concentrated solutions. A donor concentration of $7 \times 10^{-4} M$ (10 mg protein/ml)

[49] D. A. Deranleau, R. A. Bradshaw, and R. Schwyzer, *Proc. Nat. Acad. Sci. U.S.* **63**, 885 (1969).

TABLE III

RELATIVE AREAS, BAND MAXIMA, AND ABSORPTION COEFFICIENTS FOR
N-METHYLNICOTINAMIDE CHLORIDE COMPLEXES: GAUSSIAN
APPROXIMATION[a]

Donor	Short-wavelength band		Long-wavelength band		ϵ at 350 nm (liter/ mole cm)
	λ_{max}(nm)	Area (%)	λ_{max}(nm)	Area (%)	
Lysozyme	312	59	357	41	1040
3-Indoleacetic acid	314	65	361	35	950
N-Acetyl-L-tryptophan	313	62	361	38	850
N-Acetyl-L-trypto- phanamide	314	63	361	37	760

[a] D. A. Deranleau, R. A. Bradshaw, and R. Schwyzer, *Proc. Nat. Acad. Sci. U.S.*
63, 885 (1969). Reproduced by permission.

and acceptor concentrations of 0.005 M to 0.9 M were used to determine
the association constant for lysozyme (3.2 liters mole^{-1}).[49]

Experimental Considerations. To determine the association constant for lysozyme,[49] solid MN was weighed directly into the spectrophotometer cell, and absorption measurements were made at a wavelength (350 nm) at which neither donor nor acceptor had significant
absorption. Measurements were made at room temperature. The small
value of the enthalpy of formation of charge transfer complexes[47,48]
indicates that precise control of sample temperature is not necessary.
Use of tandem cells[8] is convenient in some instances.[47,48] Scatchard
plots give values for association constant and molar absorption coefficient.

Since the binding constants observed thus far are so small, saturation of the binding site with acceptor cannot be attained. Determination of the number of donors in a protein thus would appear to require
that both the association constant and the molar absorption coefficient
be obtained. Accordingly, a number of experiments at constant donor
(protein) concentration, and with acceptor concentrations in the range
0.1 M to 1 M are required to obtain data for a Scatchard plot. The molar
absorption coefficient of the charge transfer complex for the protein,
divided by 900 to 1000 (Table III) should give the number of indole
chromophores "exposed" to MN.

*Charge Transfer Complexes of the Phenol, Benzene, and Disulfide
Chromophores.* In addition to the indole chromophore, the other aromatic
chromophores—those of tyrosine and phenylalanine—and the disulfide
bond act as donors in charge transfer complexes with a variety of

acceptors containing a substituted benzene ring.[50,51] In the experiments reported so far with lysozyme and α-lactalbumin,[52] MN has formed charge transfer complexes *only* with the indole group. Each charge transfer complex should be identifiable by its absorption spectrum.[50,51]

Inter- and Intramolecular Charge Transfer Complexes. When a suitable acceptor is covalently attached to a protein, an intramolecular charge transfer complex[47] may form (Fig. 7). The 2,4-dinitro- and 2,4,6-trinitrophenyl haptens appear to form charge transfer complexes when bound by their specific antibodies. The changes in spectrum are characteristic of an interaction with an indole chromophore in the antibody, and fluorescence quenching of indole is observed.[47,48,53] Although the charge transfer complex thus forms part of the site of attachment of hapten to antibody, most of the standard free energy of formation of the antigen–antibody complexes (-6 to -11 kcal mole^{-1})[54,55] must be attributed to other types of interactions.

Instruments, Samples, and Procedures

A double-beam spectrophotometer is most desirable for measuring difference spectra. With this instrument, errors caused by changes in sensitivity of measuring components with time are eliminated by passage of the light beam either simultaneously or alternately through both the sample and the reference solutions. For measuring difference spectra of highly absorbing samples, double-monochromator spectrophotometers, which have low levels of stray light, are preferable. The wavelength scale should be calibrated with a mercury arc of low intensity and/or the hydrogen lines emitted by the hydrogen discharge lamp source. After the wavelength calibration has been carried out, the absorbance scale of the instrument should be checked.[56]

Warm-up and Drift. Particularly when difference spectra of small magnitude are measured, amplifier drift on warm-up can cause significant errors. The instrument should accordingly be allowed time to stabilize. The warm-up time necessary can be judged by allowing the instrument to record the absorption of empty cell compartments (air–air baseline

[50] D. S. Sigman and E. R. Blout, *J. Amer. Chem. Soc.* **89**, 1747 (1967).

[51] J. R. Little and H. N. Eisen, *Biochemistry* **6**, 3119 (1967).

[52] R. A. Bradshaw and D. A. Deranleau, *Biochemistry* **9**, 3310 (1970); F. M. Robbins and L. G. Holmes, *J. Biol. Chem.* **247**, 3062 (1972).

[53] S. F. Velick, C. W. Parker, and H. N. Eisen, *Proc. Nat. Acad. Sci. U.S.* **46**, 1470 (1960).

[54] H. N. Eisen and G. W. Siskind, *Biochemistry* **3**, 996 (1964).

[55] J. R. Little and H. N. Eisen, *Biochemistry* **5**, 3385 (1966).

[56] The standard alkaline chromate solution [G. W. Haupt, *J. Res. Nat. Bur. Stand.* **48**, 414 (1952)] or calibrated filters obtainable from the National Bureau of Standards may be employed for this purpose.

with no cells) until the recorded value is constant. The photomultiplier dynode voltage (or slit program) should be adjusted to the setting desired for subsequent use. Longer warm-up times are required for more exacting work.

Photomultipliers will be noisier when high voltage is first applied to them, and they will exhibit some hysteresis. Long-term hysteresis (longer than sample-reference alternation time as observed by the detector) is not a source of error for instruments which employ only one detector to measure the light transmitted through both sample and reference solutions. However, differences in long-term hysteresis of photomultipliers in instruments which employ separate detectors for sample and reference light paths will produce errors if the instrument has not been allowed time to stabilize.

Signal, Noise, and Bandwidth. With good amplifier design, photomultiplier detectors are "shot-noise" limited. The shot-noise (noise originating from the randomness of emission of electrons from the photocathode) is proportional to the square root of the light intensity received by the photocathode.[57] The photomultiplier signal is proportional to the first power of the light intensity. (Some detectors, e.g., PbS photoconductive cells, are limited by the thermal noise, independent of light intensity, originating in the input resistance of the amplifier.) The optimum absorbance (that absorbance giving the smallest relative error of measurement) with a shot-noise limited system is about 1.0 (Fig. 8). However, samples with absorbances of 2 or 3 are measured nearly as accurately.

The instrumental operating variable which primarily determines whether an absorption spectrum or a difference spectrum has been recorded accurately is not the photomultiplier dynode voltage, so often reported in experimental sections of papers on difference spectra, but, instead, the bandwidth of the spectrum isolated by the monochromator. The bandwidth necessary to obtain an accurate measure of the absorbance is inversely proportional to the curvature, or second derivative of the absorption spectrum, $d^2\epsilon/d\lambda^2$, *not* the slope, or first derivative, $d\epsilon/d\lambda$. It can be shown[58] that when the ratio of the spectral bandwidth[59] (SBW) of the instrument to the natural bandwidth (NBW) of the

[57] V. K. Zworykin and E. G. Ramberg, "Photoelectricity and Its Application," Chapter 13. Wiley, New York, 1949.

[58] See, for example, the Cary Instruments Application Report AR 14-2 (1964), Cary Division of Varian Instruments, Monrovia, California.

[59] Ordinarily, SBW is the width of the band of radiation passed by the monochromator, measured at half its peak intensity. The shape of the spectral band, called the "slit function," is usually triangular. For a fixed slit width, the spectral bandwidth as a function of wavelength is determined by the dispersion curve for the instrument.

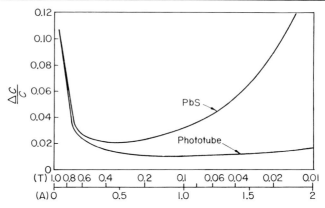

Fig. 8. The relative error in concentration ($\Delta C/C$) as a function of transmittance T and absorbance A, for a thermally noise-limited detector (PbS) and for a shot-noise limited multiplier phototube (Cary Instruments Application Report AR 14-2, 1964). Signal-to-noise (rms) 100:1 at $T = 1.0$. By courtesy of Cary Division of Varian Instruments.

absorption spectrum is 0.5, 0.2, and 0.1, the observed peak height will be 0.9, 0.98, and 0.995, respectively, of the true peak height. Thus, the accurate determination of an absorption spectrum prescribes limits on the slit width employed. For monochromators with entrance and exit slits which are adjusted simultaneously, a decrease in slit width by a factor of two (obtained by an increase in photomultiplier voltage) results in a 4-fold decrease in energy output from the monochromator. This produces a 4-fold decrease in detector output, accompanied by a 2-fold decrease in noise. The overall result is a 2-fold decrease in signal-to-noise ratio. Thus, while it is necessary to narrow the slit so that accurate values of absorbance can be obtained, unnecessarily narrow slits produce an undesirable decrease in signal-to-noise ratio without significantly improving the accuracy of the absorbance measurement.

Recording spectrophotometers now available usually are designed to operate in one of two modes: (a) constant slit width (equivalent to constant bandwidth if the dispersing elements are gratings only), obtained by variation of photomultiplier voltage during scanning; (b) constant photomultiplier voltage. In the latter mode, the slit width is programmed to maintain constant energy at the detector for one beam (usually the reference beam). Although the slit width varies during a wavelength scan, the signal-to-noise ratio remains constant in this mode of operation. A few spectrophotometers offer the choice of either programming mode. Since it seems likely that errors in obtaining difference spectra are more often made through use of excessively large slits than from other causes, the constant slit width mode of operation would appear to be the better choice for those who lack familiarity with sophisticated spectrophotometers. However, the simple test of rescanning a difference spectrum with the slit width of the instrument decreased by a factor of two or more will show up any inaccuracies, regardless of programming mode chosen.

If the difference in absorbance between sample and reference solution is very small, highly absorbing solutions may be used to increase the magnitude of the difference spectrum. The practical maximum absorbance of solutions may be limited by insufficient light output from the monochromator at the maximum slit width allowing good resolution, by stray light, or by fluorescence from the samples. In any case, a very small difference spectrum is best recorded on an expanded scale (e.g., 0–0.1 absorbance).

Some methods for improving the difference spectrum (increasing the signal-to-noise ratio) are as follows: (1) The spectrum may be scanned very slowly, and a long time constant used in the output (recorder) circuit. The improvement in signal-to-noise ratio is proportional to the square root of the reduction in scanning speed. (2) If necessary, resolution may be traded for an increase in signal-to-noise ratio. For difference spectra, this is a particularly good exchange. The resolution required for an accurate determination of a spectrum is determined by the second derivative of the spectrum, not the slope, or first derivative. However, the height of the difference spectrum is proportional to the first derivative of the absorption spectrum, and is maximal at a wavelength at which the second derivative of the spectrum is zero. At a resolution which is less than optimal for the accurate determination of the height of an absorption peak, the absorption on the side of the peak is determined quite accurately. Thus, it should be possible to use a slit width which is three to five times greater than optimal (i.e., SBW/NBW of 0.3 to 0.5, rather than 0.1) without any loss in height of the difference spectrum. The dispersion curve of the instrument should be consulted when selecting the photomultiplier dynode voltage and/or the slit width. (3) Resolution may be compromised even further, by use of still larger slit widths, if necessary. Even if the difference spectrum recorded does not show its maximum height, it will normally still show the proper dependence upon pH, temperature, or other factors. However, it may not obey Beer's law.

Stray Light and Sample Fluorescence. Stray light is light emerging from the monochromator with wavelengths other than those included within the spectral region determined by the nominal wavelength and the slit width. Often not absorbed by the sample, stray light passes on

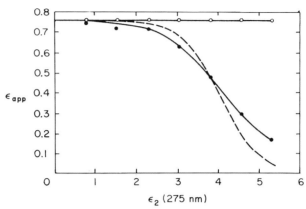

FIG. 9. Apparent differences in absorbance at 275 nm of successive pairs of K₂CrO₄ solutions (ordinate), plotted against the true absorbance of the more concentrated solution (abscissa). ○, Measurements made with a double-monochromator spectrophotometer; ●, measurements made with a single-monochromator spectrophotometer. The dashed line is a theoretical curve calculated using the equation: $\epsilon_{app} = \log [(10^{-\epsilon_1} + A)/(10^{-\epsilon_2} + A)]$, where ϵ_1 is the lower absorbance and ϵ_2 the higher and A (the stray light expressed as a fraction of the total incident light) is taken as 2×10^{-4}. From E. Mihalyi, *Arch. Biochem. Biophys.* **110**, 325 (1965).

to the detector, decreasing the apparent absorption of the sample. It can be measured with special filters, or by using highly absorbing samples. The lower the level of stray light, the higher the absorbance which can be attained before deviations from Beer's law are observed. Fluorescence of the sample has the same effect as stray light. A simple method to determine whether stray light or sample fluorescence is an important factor in measurements of difference spectra has been described by Mihalyi.[60] A series of solutions having equal increments in concentration is used. They are measured one against another, in such a way that the concentration difference between the sample solution and the reference solution is always constant. A plot of the apparent differences in absorbance against the (calculated) true absorbance of the more concentrated solution is prepared. Figure 9 shows such a plot for the same pairs of solutions measured in a single- and in a double-monochromator spectrophotometer. The superiority of a double-monochromator instrument is evident. Similar data obtained with protein solutions and a double-monochromator spectrophotometer give curves similar to the dashed line of Fig. 9. These data indicate that the absorption of a protein solution should not be greater than 3 absorbance units if the

[60] E. Mihalyi, *Arch. Biochem. Biophys.* **110**, 325 (1965).

protein has the usual fluorescence efficiency of about 10%. Thus the detector in double-monochromator spectrophotometers will ordinarily receive more fluorescence from protein solutions than stray light from the monochromator.

Suggested Initial Operating Procedure. Determine a difference spectrum using a convenient protein concentration and, if necessary, an expanded scale (0–0.1 absorbance) for measurement. Decrease the slit width by a factor of two and redetermine the spectrum. If a difference is observed in the spectrum recorded, again decrease the slit width and rescan. Repeat until the changes are not significant. The scanning speed should be slow enough to avoid "override" on the absorption peaks, and the sensitivity of the recorder should be great enough so that the maxima of the absorption peaks are not cut off ("flat-topping"). If no difference is observed when the slit width is reduced, it is possible that the original slit width selected was too narrow. Open the slits in increments until some change in the recorded spectrum is observed, then reduce the slit width by about a factor of two to recover optimum slit width and optimum signal-to-noise ratio. Sometimes interchanging sample and reference cells gives a difference spectrum which is not identical except for sign. This is an indication of photometric error, most likely to be due to fluorescence. However, the absence of a change in the recorded spectrum or difference spectrum when sample and reference are interchanged is not an assurance that such errors are absent.

Prepare a series of sample and reference solutions having different protein concentrations. Measure the difference spectrum for each pair of matched solutions. A plot of the height of the difference spectrum against the protein concentration should be a straight line passing through the origin (Beer's law). If the line curves downward at higher concentrations, choose a protein concentration in the linear portion of the curve for subsequent measurements of the difference spectrum as a function of pH, temperature, or other experimental variables. Criteria for choosing optimal concentrations of solutions for spectrophotometric measurement of binding of absorbing ligands to proteins have been presented by Brill and Sandberg.[61]

Recommended Measurements. Absorption spectra and difference spectra should be determined using a wavelength-scanning spectrophotometer equipped with a chart recorder. Failing this, at least the preliminary experiments and some experiments selected at random should be carried out with this apparatus. Ideally, the ultraviolet absorption should be measured from 350 nm (from longer wavelength if tur-

---

[61] A. S. Brill and H. E. Sandberg, *Biophys. J.* **8**, 669.

bidity is significant, or protein conjugates having absorption in this wavelength region are present) to as short a wavelength as possible, consistent with sample absorption, instrument sensitivity, or absorption by oxygen in the air. There are many reasons for recording the complete spectrum; they include the following: (1) Turbidity of samples, sometimes not easily recognizable by eye, can be detected by the increase in apparent absorption from 350 nm to 310 nm, since in this region nonconjugated proteins are transparent. A flat baseline in this wavelength region, particularly in the case of difference spectra, gives assurance that turbidity is not contributing to the absorption. (2) Any errors in the preparation of solutions are usually immediately apparent from the magnitude of, or distortions in, the spectra or difference spectra obtained. (3) Rescanning a spectrum or difference spectrum after a short period of time shows whether the samples are unstable, precipitation is taking place, chemical reactions involving absorbing materials are occurring—in short, whether any time-dependent process is affecting the measured absorption. (4) When spectrophotometric titrations are being performed by measurement of either direct or difference absorbance, nonideality of absorption produced by other chromophores or by other chemical reactions or ionizations can readily be detected by the failure of the absorption curves to pass through the isosbestic points. For difference spectra, these isosbestic points will normally fall on the baseline. (5) Inspection of the complete spectra and difference spectra often results in the observation of perturbations of other chromophores, and this assists in the interpretation of results. For example, if perturbation of both tyrosine and phenylalanine chromophores occurs simultaneously, the perturbation is more likely to be produced by a conformational change than by changes in hydrogen bonding of the phenolic groups.

An air–air and/or water–water baseline should first be obtained for the instrument and/or instrument plus cells. A single solution of the highest protein concentration to be used should be divided and placed in both the sample and reference cells, and the baseline be redetermined. This last baseline is the reference–reference baseline. A comparison of the water–water baseline with the reference–reference baseline will reveal any deviations in the baseline resulting from changes in the energy received by the phototubes because of slit-width changes or photomultiplier dynode voltage changes produced by the absorption of the sample. If differences are noted, the reference–reference baseline should be chosen for correcting the measured spectra, or the solutions diluted so that the two baselines do not differ significantly.

Sample Preparation and Manipulation. The pH of the sample solution can be adjusted by adding concentrated acid or base with a syringe microburet or a glass rod drawn to a fine tip. Except at extremes of pH or for strongly buffered solutions, the volume change so produced is usually negligible. The solutions can be mixed in 1-cm square absorption cells by using a small magnetic "flea,"[62] or by holding a polyethylene or paraffin film over the top of the cell and inverting it. Microelectrodes which can be placed directly into the cell are convenient for measuring pH. Volumes of reagent added to attain extremes of pH are significant, and the absorption measured must be corrected for dilution. Before the absorption measurement is made, however, the reference solution must be diluted with a volume of water or buffer equal to that of the reagent added to the sample solution. A pH-stat is most convenient for pH adjustment, since a desired pH can be obtained without trial-and-error addition of reagent.

Another procedure for preparing solutions of identical concentration at different pH values is to add an aliquot of protein stock solution (clarified by filtration[63] or centrifugation) to each of a series of tightly stoppered vials. Identical aliquots of a series of buffer solutions are then added, one to each cell. One of the resulting solutions then serves as the reference, the others as samples. Herskovits[8] describes in detail a similar procedure for preparing solutions for solvent perturbation experiments.

Below 240 nm, the anions of some common salts and buffers absorb ultraviolet radiation in the process of charge transfer of electrons to the solvent. The wavelengths (nm) of *maximum absorption* of some common anions are: Cl^-, 181; Br^-, 200; I^-, 226; OH^-, 187; NO_3^-, 194.[64] Perchlorate ion is commonly used as an inert anion at the shortest wavelengths.[65] Absorption spectra of some common anions are given by Gratzer and by Buck *et al.*[66]

Absorption Cells. Cells are available commercially in a large variety of styles and sizes. They can be obtained with integral water jackets for constant temperature work near ambient temperature, with attached

[62] A centrifugal stirrer for rapid mixing in square cuvettes described by R. H. Conrad [*Anal. Chem.* **39**, 1039 (1967)] is available commercially.

[63] Membrane filters usually contain several percent detergent by weight. This detergent is slowly leached out by water solutions. These filters should be water-washed before use if the presence of detergent is objectionable [R. D. Cahn, *Science* **155**, 195 (1967)].

[64] H. L. Friedman, *J. Chem. Phys.* **21**, 319 (1953).

[65] L. J. Saidel, *Arch. Biochem. Biophys.* **54**, 184 (1955).

[66] W. B. Gratzer, *in* "Poly-α-amino Acids" (G. Fasman, ed.), Chapter 5, p. 177. Dekker, New York, 1967; R. P. Buck, S. Singhadeja, and L. B. Rogers, *Anal. Chem.* **26**, 1240 (1954).

Dewar flasks for work at liquid nitrogen temperatures, or with a closed air space so that cells can be evacuated and a known amount of gas admitted, as for oxygen-binding studies of hemoglobin. The most commonly used cells are those of 1-cm path length, both cylindrical and rectangular types. The shorter path length cells of these two styles also find much use, particularly for experiments which require high concentrations of absorbing solutes. An alternative to short path length cells is insertion of a precisely machined silica block into a 1-cm square cell. Blocks are available which will reduce the path length to 0.05 mm. However, insertion of a block produces four wetted surfaces through which the light beam must pass, instead of the two surfaces in an ordinary cell. If adsorption of protein or other light-absorbing material on the cell surfaces occurs, greater errors can be produced by use of the block inserts.

Absorption cells of the ordinary path lengths are usually matched by the manufacturer to ±0.01 mm or less. Matched cells can be tested by measuring a difference spectrum when the same nonfluorescing solution of high absorbance is placed in each cell. The severest demands on quality and material of cell construction and of cleaning methods are presented by absorption measurements in the far ultraviolet wavelength range, and have been discussed by Gratzer.[66] For the wavelength region above 200 nm, the use of the best quality fused silica cells presents no problems with cell matching. Washing cells with a swab of cotton on a wooden splint dipped in a solution of a nonionic detergent, followed by thorough rinsing with distilled water and air-drying, is quite satisfactory. It is a good idea to occasionally measure the absorption of the distilled water against an air path to check for impurities. Similarly, solutions of concentrated acids, bases, or buffers should be measured against distilled water to discover accidental contamination with ultraviolet-absorbing impurities. Many commercial detergents of the alkyl sulfonate type have absorption spectra similar to proteins. If these detergents are used on other laboratory glassware, a little distilled water should occasionally be added to a pipette or beaker, then drained into an absorption cell and the spectra determined in order to check on the thoroughness with which the glassware is rinsed. Remember that cleaning the absorption cells thoroughly is time wasted if ultraviolet-absorbing impurities are introduced from the regular laboratory glassware.

Turbidity. For nonconjugated proteins, turbidity, even if not apparent to the eye, may be present if the absorption increases uniformly with decreasing wavelength from 400 nm to about 310 nm, where the indole chromophore of tryptophan starts to absorb light. If clarification of stock solutions by centrifugation or filtration, alteration of buffer, ionic strength, or pH does not remove the long-wavelength "tail" of the spectrum, the presence of an absorbing impurity should be considered before making corrections for scattering. If scattering is present, a plot of log A vs. log λ will be linear in the 400–310 nm region.[67] The slope of such a plot is rarely 4, as predicted for scattering from particles which are small compared to the wavelength of the light used, but is often 2 to 3. If the "tail" is small and absorbancies near 280 nm only are required, it is usually sufficiently accurate to extrapolate the tail of the spectrum or difference spectrum to the wavelength used for calculations, and subtract the value of absorption so extrapolated. If the scattering is large enough so that the "tail" is steep, the double logarithmic plot is required. An extrapolation of the straight line portion of the double logarithmic plot into the wavelength region of protein absorption allows the determination of log A for scattering at any working wavelength. Log A is then converted to A and subtracted from the observed absorbance at each wavelength. The corrected absorption spectrum may then be plotted.

[67] E. Schauenstein and H. Bayzer, J. Polym. Sci. **16**, 45 (1955); S. J. Leach and H. A. Scheraga, J. Amer. Chem. Soc. **82**, 4790 (1960).

[22] Spectrophotometric Titration of the Functional Groups of Proteins

By JOHN W. DONOVAN

A spectrophotometric titration of a protein is usually understood as the titration of its phenolic groups by measurement of the change in absorption at 295 nm, characteristic of the conversion of the un-ionized phenolic groups to phenolate ions. There are many other uses for the pH dependence of the absorption spectrum. This section reviews less-used spectrophotometric titrations and reconsiders older uses, in particular the interpretation of titration curves obtained at high pH. Fundamental material pertaining to spectrophotometric titrations has been presented by Beaven and Holiday.[1] Descriptions of experimental techniques and interpretation of titration curves of proteins are given by Steinhardt and Reynolds[2] and by Nozaki and Tanford.[3]

Principle. A spectrophotometric titration can be carried out whenever

[1] G. H. Beaven and E. R. Holiday, Advan. Protein Chem. **7**, 319 (1952).
[2] J. Steinhardt and J. A. Reynolds, "Multiple Equilibria in Proteins." Academic Press, New York, 1969.
[3] Y. Nozaki and C. Tanford, see Vol. 11, p. 715.

dissociation of a proton results in a change in the spectrum of one or more chromophores. If the dissociating group is not actually a chromophore, it must interact with some chromophore so that proton dissociation results in a change in the spectrum of the protein. Dissociation constants of metal ions or any ion or molecule bound to a protein which produces a change in spectrum can be obtained in the same manner. Since spectral changes are often small, a reference solution of similar absorption properties is normally used, and measurements are obtained as differences from this reference solution.

Consider the dissociation of a proton from an uncharged compound. At a definite pH, some fraction α of the total number of chromophores is in the ionized form (the chromophores themselves need not be ionized, but must be affected by the ionization), with the spectrum $\epsilon_I(\lambda)$. The remaining fraction, $(1 - \alpha)$, is un-ionized and has the spectrum $\epsilon_U(\lambda)$. The apparent dissociation constant, K', is then

$$K' = a_{H^+}[\alpha/(1 - \alpha)] \tag{1}$$

which can be written in the Henderson–Hasselbalch form

$$pK' = pH - \log[\alpha/(1 - \alpha)] \tag{2}$$

The absorbancy,[4] A, of a sample solution of total chromophore concentration (or total protein concentration) c, measured in a 1-cm cell, is given by the sum of the optical densities of the ionized and un-ionized forms. In addition to being a function of chromophore concentration, it is a function of both pH and wavelength.

$$A_{sample}(c, pH, \lambda) = [\epsilon_I\alpha + \epsilon_U(1 - \alpha)]c \tag{3}$$

When the absorption change is determined by means of the difference method, the solvent is not used as the reference solution in the spectrophotometer. It is most convenient to choose a reference solution of the same compound at the same concentration, in which all the chromophores are either ionized or un-ionized. Choosing as a reference solution the un-ionized compound at the same concentration (pH fixed and $\alpha = 0$)

[4] Absorbancy (absorbance or absorption), A, or optical density, D, is defined as: $A = D = \log I_0/I$, where I_0 is the light entering the sample and I the intensity unabsorbed by the sample. The absorbancy is proportional to concentration, c, of absorber (Beer's law) and path length, l, of absorber (Lambert's law): $A = \epsilon c l$. The proportionality constant ϵ is called the molar absorption coefficient when the absorber concentration is expressed in moles per liter and the absorber path length in centimeters. Units of ϵ are liter mole^{-1} cm^{-1}. When two or more different absorbers (chromophores) are present together, the additional assumption of additivity of absorption is usually made: $A = \Sigma_i \epsilon_i c_i l_i$. Additivity generally holds true when the chromophores do not interact with one another.

$$A_{\text{reference}}(c, \lambda) = \epsilon_U c \qquad (4)$$

If the molar difference absorption coefficient is defined as

$$\Delta\epsilon(\lambda) = \epsilon_I(\lambda) - \epsilon_U(\lambda) \qquad (5)$$

then the absorbancy difference, which is a function of wavelength, pH, and chromophore concentration, can be written in a form resembling Beer's law

$$A_{\text{sample}} - A_{\text{reference}} = \Delta A(\lambda) = \Delta\epsilon(\lambda) \cdot \alpha(\text{pH}) \cdot c \qquad (6)$$

Thus, for a fixed total concentration the following conditions exist.

1. For *any* definite pH of the sample solution, the difference spectrum will have the wavelength dependence of the difference absorption coefficient, $\Delta\epsilon$. This can be seen more clearly by taking logarithms of both sides of Eq. (6).

$$\log \Delta A(\lambda) = \log[\Delta\epsilon(\lambda)] + \log[\alpha(\text{pH})] + \log c \qquad (7)$$

If the logarithm of ΔA is plotted (ordinate) as a function of wavelength (abscissa) for all values of the pH used in the titration, all curves ("log absorption spectra") will have *identical* shape, but will be vertically displaced by the different values of $\log[\alpha(\text{pH})]$. This is an important test which a spectrophotometric titration must pass before any confidence can be placed in the interpretation given it. Fulfillment of this test is assurance that only one type of absorption change is measured throughout the titration—for example, that only one kind of chromophore, or the same proportions of different kinds of chromophores, are being titrated throughout the pH range.

2. At *any* fixed wavelength (excepting isosbestic points, at which the absorption of the chromophore in the ionized form is the same as that in the un-ionized form) and concentration, the absorbancy difference measured as a function of pH will give the titration curve of the chromophore. The absorbancy difference at any wavelength, divided by the molar difference absorption coefficient at that wavelength, gives the *identical* titration curve (Eq. 6).

The molar difference absorption coefficient used may be that of the protein or of the chromophore. These are related by

$$\Delta\epsilon_{\text{protein}} = n\Delta\epsilon_{\text{chromophore}} \qquad (8)$$

where n is the number of chromophores titrated.

Thermodynamic Data. When the titration is reversible, the apparent dissociation constant K' can be obtained from a plot of Eq. (2), if the effect of the charges on the protein can be neglected (see below). Alternatively (and much easier), the pH at which α equals $(1 - \alpha)$, the "half-ionization point," is equal to pK' (Eq. 2). Then

$$\Delta G^0 = -RT \ln K' \qquad (9)$$

Measurement of the temperature dependence of the ionization constant allows the calculation of the enthalpy of ionization[5]

$$-\Delta H^0 = R[d(\ln K)/d(1/T)] \qquad (10)$$

The entropy change is given by

$$\Delta S^0 = (\Delta H^0 - \Delta G^0)/T \qquad (11)$$

The temperature dependence of the absorption difference cannot, in general, be used to determine the enthalpy of ionization, because the molar difference absorption coefficients are a function of temperature.[5,6]

A spectrophotometric titration of a protein with hydrogen ion can be considered: a, the determination of the degree of dissociation of protons from ionizable chromophoric side chain groups; b, the determination of the change in absorption as a function of pH. It is not true that a and b are equivalent statements. They are equivalent only when the proportionality factors[7] relating change in absorption to proton binding remain constant over the pH range investigated. The failure to make this distinction is the essential error committed when spectrophotometric titrations are misinterpreted.

Scope and Limitations. Some chromophores are titrated directly, other chromophores have their spectra altered indirectly, and simultaneously other functional groups which do not absorb are titrated. Accordingly, the consumption of base or acid is not generally measured in a spectrophotometric titration. This is convenient in the alkaline pH range, since special precautions to avoid the absorption of carbon dioxide by the sample are unnecessary. In addition, the spectrophotometric titration can be carried out accurately at extremes of pH, since the large errors which arise in determining the amount of hydrogen ion bound to the protein at extremes at pH^3 do not present themselves in the spectrophotometric titration. Usually measurements are made only of pH and of light absorption. Only those chromophores which are affected by the alteration of pH are observed. Titratable groups which are not in themselves chromophores or do not affect the absorption of chromophores will not be observed. The spectrophotometric titration is selective, and can be used together with a potentiometric titration to

---

[5] J. Hermans, Jr., J. W. Donovan, and H. A. Scheraga, *J. Biol. Chem.* **235**, 91 (1960).

[6] J. W. Donovan, see this volume [21].

[7] The molar difference absorption coefficient, $\Delta\epsilon$, and the effective chromophore concentration, c. If a conformational change of a protein takes place during a titration, for example, *both* these factors may change over the pH range of the titration.

APPROXIMATE CHANGES IN MOLAR ABSORBANCE FOR IONIZATION OF
CHROMOPHORES IN PROTEINS

Group	Short wavelength region		Longer wavelength region	
	λ, nm	$\Delta\epsilon$	λ, nm	$\Delta\epsilon$
Phenolic	243	11,100[a]	295	2330[b]
Sulfhydryl	235	5,000[c]	—	—
Imidazolium	225–240	300[d]	—	—

[a] Average of values reported by J. Hermans, Jr., *Biochemistry* **1**, 193 (1962); A. C. M. Paiva and T. B. Paiva, *Biochim. Biophys. Acta* **56**, 339 (1962); D. S. Eisenberg and J. T. Edsall, *Science* **142**, 50 (1963); J. W. Donovan, *Biochemistry* **3**, 67 (1964).

[b] From G. H. Beaven and E. R. Holiday, *Advan. Protein Chem.* **7**, 319 (1952). In 8 M guanidine hydrochloride, the change in molar absorbancy is 2480 [H. Edelhoch, *Biochemistry* **6**, 1948 (1967)].

[c] A large variation is observed in proteins—see J. W. Donovan, *Biochemistry* **3**, 67 (1964) and J. E. D. Dyson and E. A. Noltman, *Biochemistry* **8**, 3533 (1969).

[d] J. W. Donovan, *Biochemistry* **4**, 823 (1965); J. W. Donovan, M. Laskowski, Jr., and H. A. Scheraga, *J. Amer. Chem. Soc.* **83**, 2686 (1961).

distinguish the titration behavior of groups which ionize in the same pH region, e.g., ϵ-amino groups from phenolic groups.

Since each type of chromophore has its own characteristic difference spectrum, inspection of the difference spectrum of a protein allows the determination of which chromophores are being titrated in a particular pH range. Since the chromophores of nonconjugated proteins absorb in the ultraviolet wavelength region, this wavelength region is employed. All the titratable groups in a protein should be titratable spectrophotometrically, since all absorb somewhere in the ultraviolet region of the spectrum. The carboxyl, amino, and guanidinium groups of proteins, with absorption bands in the region of 200 nm, have never been successfully titrated spectrophotometrically, but the carboxyl groups in a dipeptide have.[8] The sulfhydryl group and the imidazole group, which have absorption maxima near 235 and 220 nm, respectively, can be titrated spectrophotometrically (see below). The table lists the wavelengths of absorption difference maxima and molar difference absorption coefficients for the change in the state of ionization of chromophores of proteins.

It is convenient to make a distinction between a change in the nature of the chromophore when it is titrated (e.g., phenol changes to phenolate ion) and an absorbancy change of a chromophore caused by the ionization of neighboring groups, or by alterations in the conformation of the

[8] L. J. Saidel, *Arch. Biochem. Biophys.* **54**, 184 (1955).

protein, or a change in position, number, or type of solvent molecules in the neighborhood of the chromophore.[9] It is also convenient to divide the pH region in which spectrophotometric titrations can be carried out into three: the acid pH region, below pH 6; the neutral pH region, pH 6–8; and the alkaline pH region, above pH 8.

Acid pH Region

At acid pH, except in proteins containing phosphate or sulfate groups, ordinarily only carboxyl groups are titrated. The titration is usually carried out potentiometrically (electrometrically). The change in ionization state of the carboxyl groups at acid pH may have three major effects on the other chromophores in the protein: (1) the change in charge on one or more carboxyl groups may perturb both ionizable and un-ionizable chromophores; (2) the change in the net charge on the protein produced by titration of the carboxyl groups may induce aggregation, or (3) produce a conformational change of the protein which exposes to solvent chromophores that were previously inaccessible to solvent, or vice versa. The kind of chromophores altered can be determined by the nature of the change in the absorption spectrum, each chromophore having its characteristic perturbation difference spectrum.[6,10]

Interpretation of Acid Difference Spectra. A typical difference spectrum obtained in the acid pH range is that of ovomacroglobulin (MW 6.6×10^5), shown in Fig. 1. A step-by-step interpretation of this difference spectrum illustrates the analytical processes employed.

A. First to be considered is the *sign* of the difference spectrum. A positive difference spectrum usually indicates that a larger number of chromophores are exposed to solvent in the reference solution. Accordingly, Fig. 1 suggests that chromophores become exposed to solvent when the pH of a solution of ovomacroglobulin is reduced from 6.5 to 2.8.

---

[9] These effects on the absorption spectra of chromophores are usually termed "perturbations." Ordinarily, perturbations are much smaller in magnitude than the changes in absorption produced by the ionization of a chromophore [D. B. Wetlaufer, J. T. Edsall, and B. R. Hollingworth, *J. Biol. Chem.* **233**, 1421 (1958)], but exceptions occur [J. W. Donovan, M. Laskowski, Jr., and H. A. Scheraga, *J. Amer. Chem. Soc.* **83**, 2686 (1961)]. The absorption change accompanying the ionization of the phenolic chromophore can also be considered a perturbation by introduction of negative charge on the phenolic oxygen, but it is still useful to make some distinction between "direct" and "indirect" (perturbation) effects.

[10] E. J. Williams and J. F. Foster [*J. Amer. Chem. Soc.* **81**, 865 (1959)] have shown that the pH dependence of absorption changes in acid gives titration curves which, although similar to potentiometric titration curves of the protein, are not identical with them.

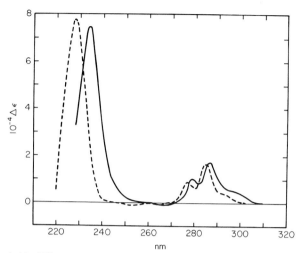

Fig. 1. Acid difference spectrum of ovomacroglobulin, expressed in terms of molar difference absorption coefficient of the subunit (MW 3.3×10^5). Solid line: absorption difference observed when a pH 6.5 solution is measured against a pH 2.8 solution as reference. Dashed line: difference spectrum calculated [J. W. Donovan, *J. Biol. Chem.* **244**, 1961 (1969)] for exposure of 24 phenolic and 2 indole chromophores per subunit. The subunit contains approximately 100 phenolic and 30 indole chromophores, of which 32 and 8, respectively, are exposed to solvent at pH 6.5. From J. W. Donovan, C. J. Mapes, J. G. Davis, and R. D. Hamburg, *Biochemistry* **8**, 4190 (1969). Reproduced by permission of the American Chemical Society.

B. Next, the type of chromophores affected, and their number, must be determined. In the 280–300 nm wavelength region, the peaks in the difference spectrum at 280 and 288 nm, characteristic of perturbation of phenolic chromophores, are much more prominent than the characteristic indole difference peak near 290 nm. Accordingly, the number of phenolic chromophores perturbed is much larger than the number of indole chromophores perturbed. A first approximation to the number of phenolic and indole chromophores producing this difference spectrum was calculated by dividing the height of the 288 nm difference peak by the molar difference absorption coefficient (700) for exposure of the phenolic chromophore (see Table I of this volume [21]), and the height of the 295 nm shoulder by the molar difference absorption coefficient (1600) for exposure of the indole chromophore. This very naive calculation yielded the result: 26 phenolic and 4 indole chromophores. A trial difference spectrum was obtained by summation of 26 times the phenolic and 4 times the indole difference spectra of model compounds. Subsequent trial-and-error adjustments showed that the solid curve of Fig. 1 was fit best over the wavelength range 230–310 nm by a calculated difference spectrum which was a sum of the difference spectra for 24 phenolic and 2 indole chromophores (dashed line in Fig. 1). The calculated difference spectrum is displaced in wavelength from the observed difference spectrum because the model compounds used to obtain the perturbation difference spectra (see Fig. 1 of this volume [21]) were free amino acids, not peptides.

C. An exercise in curve-fitting is not very enlightening unless it produces a result with some correspondence to reality. The important question: Are 24 phenolic and 2 indole chromophores *really* transferred from the interior of the protein to the solvent when the pH is lowered? Solvent perturbation experiments indicate that this is the correct interpretation. Separate ethylene glycol perturbation experiments with ovomacroglobulin indicate an average of 32 phenolic and 8 indole chromophores exposed to solvent (per 3.3×10^5 g) at neutral pH, and an average of 56 phenolic and 10 indole chromophores at acid pH. Accordingly, exposure of chromophores appears to occur when the pH is reduced. Had these additional experiments not been in agreement with the first interpretation of the difference spectrum of Fig. 1, other possibilities, such as charge effects on chromophores, or chromophores exposed to solvent, but not to ethylene glycol,[6] would have to be considered.

The pH dependence of the acid difference spectrum of Fig. 1 is quite different from the typical potentiometric titration curve of a protein in the acid pH range. The difference spectrum, however, closely follows the dissociation of the protein into subunits, as determined by area measurements on schlieren photographs of sedimentation velocity experiments (Fig. 2), and sedimentation equilibrium measurements of molecular weight at neutral and acid pH.

Finally, the interpretation of this change in absorption as resulting from exposure of chromophores is entirely in accord with dissociation of the protein into subunits, since even if no conformational change in the subunits were to take place upon dissociation, those chromophores at the interface between subunits would become exposed to solvent on separation of the subunits.

Intermediate pH Region

Normally only the imidazole chromophore and the α-amino group are titrated between pH 6 and pH 8.[3] Only limited studies of the ionization of the imidazole chromophore in proteins and in model compounds have been made spectrophotometrically. The experimental difficulties are great, since the absorption change of this chromophore is small, and the other chromophores in the protein have large absorption

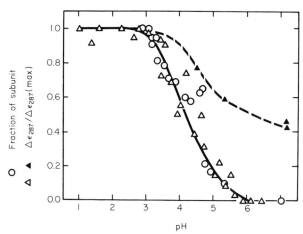

Fig. 2. The pH dependence of the dissociation of ovomacroglobulin: ◯, the fraction of subunit determined from area measurements of schlieren pictures; △, relative change in absorption at 287 nm, vs. reference solution, pH 7; ▲, attempted reversal (from pH 3) of the dissociation, measured by absorption change at 287 nm. From J. W. Donovan, C. J. Mapes, J. G. Davis, and R. D. Hamburg, *Biochemistry* **8**, 4190 (1969). Reproduced by permission of the American Chemical Society.

coefficients in the wavelength region in which the imidazole chromophore has absorption. Nuclear magnetic resonance (NMR) appears to be a more practical technique for following both the ionization[11,12] and the perturbation[13] of the imidazole chromophore in proteins. However, spectrophotometric titration of the imidazole chromophores *can* be carried out. Luckily, since most proteins have greatest conformational stability in this pH range, conformational changes which might give rise to large absorption changes from alteration of the environment of the phenol and indole chromophores are unlikely to occur. However, changes in charge which occur during titration might affect the absorption of other chromophores.

Imidazole Groups in Ribonuclease. When difference spectra of the imidazole groups of ribonuclease were obtained (Fig. 3), an absorption change near 280 nm, characteristic of perturbation of an abnormal (probably buried) phenolic chromophore was observed. The sign and magnitude of the absorption change and the position of the difference peak all indicated that the environment of the phenolic chromophore and

[11] J. H. Bradbury and H. A. Scheraga, *J. Amer. Chem. Soc.* **88**, 4240 (1966).
[12] D. H. Meadows, O. Jardetzky, R. M. Epand, H. H. Ruterjans, and H. A. Scheraga, *Proc. Nat. Acad. Sci. U.S.* **60**, 766 (1968).
[13] D. H. Sachs, A. N. Schechter, and J. S. Cohen, *J. Biol. Chem.* **246**, 6576 (1971).

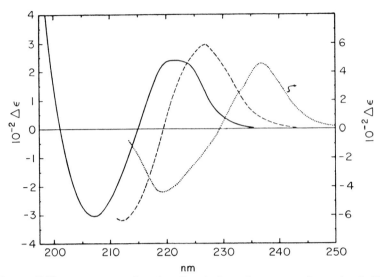

FIG. 3. Difference spectra for the dissociation of a proton from the imidazole chromophore. Solid curve, imidazole, pH 9.5 vs. pH 3.3 (left ordinate); dashed curve, histidine methyl ester, pH 6.6 vs. pH 4.1 (left ordinate); dotted curve, ribonuclease, pH 6.5 vs. pH 5.0 (right ordinate), uncorrected for superimposed phenolic perturbation. From J. W. Donovan, *Biochemistry* **4**, 823 (1965). Reproduced by permission of the American Chemical Society.

its state of ionization were not a function of pH. The pH dependence (the titration curve) of the perturbation indicated that this phenolic chromophore in ribonuclease was perturbed by the change in charge on an imidazole group and on an α-amino group as these groups were titrated.

When an absorption change is observed for the phenolic chromophore in the 280 nm wavelength region, a corresponding, and larger, alteration in its absorption simultaneously occurs at shorter wavelength, near 230 nm. Thus, changes in absorption of ribonuclease, due to perturbation of one of its phenolic groups, take place simultaneously with the ionization of the imidazole groups of ribonuclease, and in the same wavelength region, near 230 nm. Studies of the charge perturbation of the phenolic chromophore showed the relation between the absorption change in the 280 nm region and in the 230 nm region. From the measured perturbation difference spectrum for the phenolic group in ribonuclease at 280 nm, and the difference spectrum for the charge perturbation of the phenolic goup in model compounds,[14] a difference spectrum near 230 nm could be calculated for the phenolic group in ribonuclease.

[14] D. B. Wetlaufer, *Advan. Protein Chem.* **17**, 303 (1962).

Accordingly, the observed difference spectra for the imidazole groups of ribonuclease obtained in the 230 nm wavelength region were corrected for the perturbation of the phenolic group by subtracting the calculated difference spectrum.

In general, absorption changes produced by charge perturbations of chromophores may differ both in magnitude and in sign (direction of wavelength shift) from solvent perturbations. The magnitude of the perturbation depends on the specific nature of the local environment of the perturbed chromophores, and thus cannot be estimated *a priori.*[6] The pH dependence of the difference spectrum of the imidazole groups in ribonuclease proved to be in good agreement with both the potentiometric titration curve[15] and with the titration curves of the individual histidine residues obtained by the NMR technique.[11,12]

High pH Region

In the pH range above 8, the "nonchromophoric" amino and guanidinium groups and the "chromophoric" sulfhydryl and phenolic groups are titrated. The phenolic absorption maximum shifts from approximately 278 nm to approximately 293 nm in the near ultraviolet (difference peak at approximately 295 nm). At shorter wavelength, the difference peak (approximately 242 nm) is of greater magnitude (see the table). The latter difference peak is used in titrating the phenolic groups in proteins containing heme, since absorption of the heme obscures the absorption change of the phenolic groups near 280 nm.[16]

Perturbations. It should be evident that spectrophotometric titrations carried out at alkaline pH are subject to the same perturbations of chromophores as titrations carried out in the other pH ranges. These perturbations may be produced in the absorption spectrum of the chromophore being titrated, either in its ionized or un-ionized form, as well as upon all other chromophores in the protein. For this reason, it is good practice to determine the entire spectrum or difference spectrum at every pH. A plot of the "log absorption difference spectra" can be made,[17] or instead, any one spectrum can be subtracted from any other. If there are no perturbations, then all differences between spectra will have identical shape (but different magnitude), characteristic of the chromophore being titrated.[17] It is strongly suggested that every detail in the recorded spectra be accounted for by actually matching spectra or difference spectra with one another and with model spectra.

An excellent example of a critical study of the spectrophotometric

[15] C. Tanford and J. D. Hauenstein, *J. Amer. Chem. Soc.* **78,** 5287 (1956).
[16] J. Hermans, Jr., *Biochemistry* **1,** 193 (1962).
[17] See the first section: Principle.

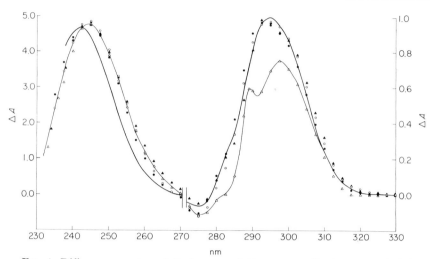

FIG. 4. Difference spectra of fibrinogen solutions, normalized to a peak height of 4.7 at the short-wavelength peak: ○, pH 10.2 native vs. pH 8.9 native; ●, pH 10.9 native vs. pH 10.2 native; ▲, pH 11 denatured vs. pH 11 native; △, pH 13 denatured vs. pH 11 denatured. Heavy line is a calculated difference spectrum for the amino acid tyrosine, pH 13 vs. pH 7, for which the neutral pH spectrum was shifted 3.2 nm to longer wavelength. From E. Mihalyi, *Biochemistry* **7**, 208 (1968). Reproduced by permission of the American Chemical Society.

titration of a protein is afforded by Mihalyi's work on fibrinogen.[18] Figure 4 shows that differences between spectra obtained for fibrinogen from pH 9 to pH 11 are identical and characteristic of tyrosine ionization only. Denaturation at pH 11 also gives a difference spectrum characteristic of tyrosine only. The difference between spectra obtained for protein denatured at pH 13 and protein denatured at pH 11 shows a perturbation (negative peak at 292 nm) which is characteristic of unmasking of indole chromophores. The difference spectrum for the perturbation was obtained by subtracting the normalized pH 13 difference spectrum from the normalized pH 11 difference spectrum. This perturbation difference spectrum was matched by sums of difference spectra obtained by solvent perturbation of tyrosine and tryptophan. The abnormality in the difference spectrum for tyrosine ionization was then attributable to exposure of *both* tyrosine and tryptophan chromophores to solvent on denaturation at pH 13.[18]

A rapid check for abnormalities in the absorption difference spectra for ionization of phenolic groups is the calculation of the ratio of the

[18] E. Mihalyi, *Biochemistry* **7**, 208 (1968). The experimental section of this reference is well worth a careful reading.

height of the difference peak at 242–244 nm to that at 294–296 nm. A ratio of 4.7 corresponds to tyrosine ionization only; a higher ratio is indicative of unmasking of buried chromophores.[18]

Conformational Inferences. Frequently, titrations of phenolic groups are carried out not merely because of intrinsic interest in these groups, but because the investigator wishes to determine from their normal or abnormal titration behavior, by inference, whether the protein has a compact, or folded, structure in its native conformation. Caution must be observed in making such interpretations. The conformation of the protein at alkaline pH may be different from the conformation at neutral pH, so that chromophores which titrate normally may be partly or completely buried at neutral pH.[19] Usually, the conformation of a globular protein will be compact at neutral pH, and less compact, or completely unfolded, at acid or alkaline pH. However, the opposite behavior *may* also occur.[20] Conformational changes of the protein which take place in alkaline solution are often indicated by time-dependent changes in absorption, but sometimes chemical reactions produce these time-dependent changes. It is wise to carry out parallel experiments with other techniques to substantiate conformational changes. Convenient techniques are optical rotation, circular dichroism, light scattering, viscosity, sedimentation velocity, or NMR.

Side Reactions. Because spectrophotometric titrations are usually carried as far as pH 13 to 14 in water solution, special care must be taken to ensure that chemical reactions which affect the absorption are recognized if they occur. In addition to conformational changes, reactions catalyzed by hydroxide ion, or in which hydroxide ion is a reactant, are common.[21,22] Above pH 11, measurements should always be made to determine whether the spectra are time dependent. Such time dependence may be a change in *shape* of the absorption spectrum, or a change in the *magnitude* of the spectrum, for which the shape of the spectrum remains (nearly) constant. The second case would be typical of a time-dependent ionization of phenolic groups, resulting from a conformational change of the protein. A titration curve obtained under these circumstances is not an equilibrium curve, and cannot be used to determine equilibrium constants for ionization. If the *shape* of the absorption spectrum is a function of time, other reactions or processes occur. These may

[19] T. T. Herskovits and M. Laskowski, Jr., *J. Biol. Chem.* **243**, 2123 (1968).
[20] For ovomucoid, exposure of the phenolic chromophores to methanol, dimethyl sulfoxide, and ethylene glycol *decreases* when the pH is changed from pH 7 to pH 2 [T. T. Herskovits and M. Laskowski, Jr., *J. Biol. Chem.* **237**, 3418 (1962)].
[21] C. J. Garratt and P. Walson, *Biochem. J.* **105**, 51c (1967).
[22] J. W. Donovan, *Biochem. Biophys. Res. Commun.* **29**, 734 (1967).

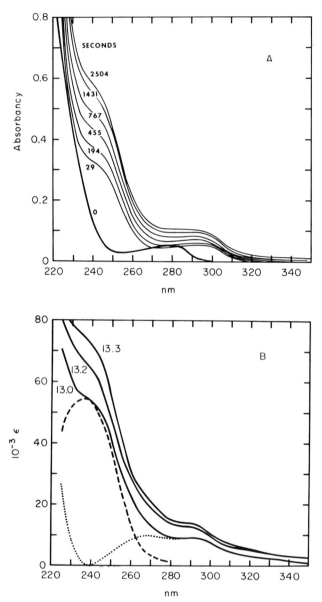

Fig. 5. (A) absorption spectra of a $4.2 \times 10^{-6}\,M$ solution of ovomucoid, before (0) and after addition of KOH to bring the pH to 13.0. The large absorption change which occurs in the first 30 seconds is produced by ionization of phenolic groups. The isosbestic points near 280 nm are not present in spectra obtained after 30 seconds. (B) The total change in absorption per mole of ovomucoid, not including ionization of phenolic groups, observed at three values of pH. These absorption

be separate from, or simultaneous with, the ionization of the phenolic groups. If the time-dependence is due to an unfolding of the protein, perturbations of the spectra of other chromophores may appear at the same rate as the phenolic groups ionize.

Hydrolysis of Disulfide Bonds at High pH. In some instances, entirely new absorption bands are produced as a function of time of holding at alkaline pH. Figure 5 shows new absorption bands produced upon hydrolysis of disulfide bonds.[23,24] Note that successive scans (Fig. 5A) are not superimposable, and that the differences between successive scans (Fig. 5B) are not characteristic of absorption differences of the ionization of phenolic groups (see Fig. 4). The hydrolysis of the disulfide bond liberates sulfhydryl ion in a reaction first-order with respect to time.[22] The reaction is also first order with respect to hydroxide ion activity.[22] The bimolecular rate constants at room temperature are approximately 10^{-2} liter mole^{-1} sec^{-1}. Activation energies for splitting of disulfide bonds have been determined.[25] When reactions such as disulfide bond splitting take place during a spectrophotometric titration, the experiments should be carried out as quickly as possible, and extrapolated to zero time of addition of alkali. Lower temperatures will reduce the rate of disulfide hydrolysis.[25]

Since hydrolysis of disulfide bonds may occur at as low a pH as 11,[24] proteins should not be exposed to alkaline pH for longer than necessary during isolation and purification. After a protein has been exposed

[23] The consumption of hydroxide ion which occurs on hydrolysis of disulfide bonds has been observed in potentiometric titrations of proteins by Y. Nozaki and C. Tanford [*J. Amer. Chem. Soc.* **89**, 736 (1967); see also Vol. 11, p. 715].

[24] Exposure of lysozyme to high pH in guanidine hydrochloride results in change in ultraviolet absorbance. New functional groups were observed on back-titration to neutral pH [R. Roxby and C. Tanford, *Biochemistry* **10**, 3348 (1971)].

[25] J. W. Donovan and T. M. White, *Biochemistry* **10**, 32 (1971).

difference spectra are the differences between the absorption spectrum at infinite time (about 3000–5000 seconds) and the absorption spectrum back-extrapolated to zero time *after* addition of alkali. The pH 13.0 curve is approximately the difference between the two curves labeled 2504 seconds and 29 seconds in Fig. 1A. The dashed curve is an absorption spectrum of 10 moles of sulfhydryl ion (per mole of ovomucoid), calculated with *n*-butylmercaptan in $1 M$ NaOH as model [L. H. Noda, S. A. Kuby, and H. A. Lardy, *J. Amer. Chem. Soc.* **75**, 913 (1953)]. The dotted curve is the residual spectrum obtained when the dashed curve is subtracted from the pH 13.0 difference spectrum. It is presumed to be the absorption spectrum of unknown products formed in side reactions. The small absorption decrease resulting from loss of disulfide bonds is neglected here. From J. W. Donovan, *Biochem. Biophys. Res. Commun.* **29**, 734 (1967). Reproduced by permission of Academic Press.

to high pH, as in the titration of its phenolic groups, its complete ultra-violet absorption spectrum (not just the absorption at 295 nm) should be determined. Careful examination of the spectrum should reveal whether any deleterious changes have taken place.

All spectrophotometric titrations of phenolic groups should include tests for irreversible absorption changes—the simplest one is to determine whether isosbestic points observed on forward titration to higher pH are preserved on the reverse titration. When kinetic measurements of the rate of ionization of phenolic groups are determined, complete spectra, again, must be obtained to ensure that hydrolysis of disulfide bonds is not mistaken for ionization of phenolic groups. If ionization of phenolic groups and hydrolysis of disulfide bonds occur simultaneously, the experimenter will have to determine the relative rates of each process in order to interpret the experimental results, since hydrolysis of disulfide bonds might, in some instances, be required before ionization of phenolic groups can occur.

The rate of hydrolysis of disulfide bonds is proportional to the activity of hydroxide ion in solution.[22,25] Accordingly, although good experimental evidence for hydrolysis of disulfide bonds below pH 11 has not been presented, there is good reason to expect disulfide bond hydrolysis will take place even in weakly alkaline solutions (at pH 9, for example). However, the dependence of the rate of hydrolysis on hydroxide ion activity means that the rate will be 10-fold smaller if the pH of the solution is lowered by one pH unit. In the absence of convincing evidence that a lower limit of pH exists below which hydrolysis of disulfide bonds does not occur, long-term storage of protein in weakly alkaline solution should be avoided.

Titration of Sulfhydryl Groups in Proteins. Although the sulfhydryl groups of many enzymes are essential to their function, generally the availability of these groups is determined by reaction, e.g., with p-mercuribenzoate[26] or with 5,5'-dithiobis-(2-nitrobenzoic acid).[27] However, the ionized sulfhydryl group has an absorption maximum at approximately 235 nm (Fig. 5), while the absorption of the nonionized sulfhydryl group is negligible at this wavelength.[28] Unfortunately, the other aromatic chromophores absorb strongly at this wavelength, the change in absorption accompanying ionization of the phenolic group is large, and changes in perturbation of the aromatic absorption during a conformational change can also be large. If the difference spectrum for the conformational change of the protein can be determined (e.g., by denaturation in

[26] P. D. Boyer, *J. Amer. Chem. Soc.* **76,** 4331 (1954); see Vol. 3, p. 940.
[27] G. L. Ellman, *Arch. Biochem. Biophys.* **82,** 70 (1959).
[28] L. H. Noda, S. A. Kuby, and H. A. Lardy, *J. Amer. Chem. Soc.* **75,** 913 (1953).

acid, or by denaturation in guanidine hydrochloride and correction for the solvent perturbation produced by guanidine hydrochloride[18]), then the number of the sulfhydryl (and phenolic) chromophores ionized at any value of pH can be obtained from the set of equations[29]

$$\Delta\epsilon_{295} = \Delta\epsilon^F_{295} \cdot F + \Delta\epsilon^T_{295} \cdot T$$

$$\Delta\epsilon_{243} = \Delta\epsilon^F_{243} \cdot F + \Delta\epsilon^T_{243} \cdot T + \Delta\epsilon^S_{243} \cdot S \qquad (12)$$

$$\Delta\epsilon_{235} = \Delta\epsilon^F_{235} \cdot F + \Delta\epsilon^T_{235} \cdot T + \Delta\epsilon^S_{235} \cdot S$$

Here, F is the fraction of the conformational change ($0 \leq F \leq 1$), T is the number of phenolic groups ionized, and S, the number of sulfhydryl groups ionized. Three of the eight absorption difference coefficients required are obtained from the difference spectrum for the conformational change, three from the difference spectrum for ionization of the phenolic chromophore,[29] and two from the spectrum of the ionized sulfhydryl group. Solutions of the equations determine the titration curves for the sulfhydryl groups and the phenolic groups, as well as the pH dependence of the conformational change of the protein. If no conformational change takes place during the titration, all terms in F and one of the equations are omitted.

Conformational changes occurred for proteins for which spectrophotometric titrations of sulfhydryl groups have been determined[29,30] so no values of pK' for normal sulfhydryl groups were obtained. In denaturing solvents, pK values obtained spectrophotometrically for the sulfhydryl groups in proteins are: 9.8 (thiolated gelatin[31]), 10.2 (phosphoglucose isomerase in 1% sodium dodecyl sulfate[30]), and 8.7 (aldolase in 4 M urea[29]). The pK of mercaptoethanol is reported to be 9.5[32] and 9.65[33] in water, 9.5 in 4 M urea,[29] and 10.25 in 8 M urea.[33] Accordingly, the pK' obtained for the sulfhydryl groups in denatured aldolase seems abnormally low. Values of pK for the thiol group obtained from potentiometric titration curves are given by Nozaki and Tanford.[3]

Methods

Apparatus. Absorption of carbon dioxide by the sample solution at alkaline pH is not an important factor in spectrophotometric titrations,

[29] J. W. Donovan, *Biochemistry* **3**, 67 (1964).

[30] J. E. D. Dyson and E. A. Noltman, *Biochemistry* **9**, 3533 (1969).

[31] R. Benesch and R. E. Benesch, *in* "Sulfur in Proteins" (R. Benesch, R. E. Benesch, P. D. Boyer, I. M. Klotz, W. R. Middlebrook, A. G. Szent-Györgyi, and D. R. Schwarz, eds.), Chapter I.2, p. 15. Academic Press, New York, 1959.

[32] J. T. Edsall and J. Wyman, "Biophysical Chemistry," Vol. I. New York, Academic Press, 1958.

[33] G. M. Bhatnagar and W. G. Crewther, *Int. J. Protein Res.* **1**, 213 (1969).

since the solution is not exposed to air for very long, the pH change so produced is small, and the actual pH of the solution used for each spectrum is determined separately. Titrations are usually carried out in small open beakers or even in the cuvette in which absorption measurements are made. The titrant should be very concentrated—a $5 M$ to $10 M$ solution is recommended—so that dilution of the sample solution by added titrant will either be negligible or small enough so that corrections for volume changes will be necessary only at higher pH and can be made easily. Most pH meters now commercially available are quite adequate.[34] The meter–electrode combination must always be calibrated against a minimum of two buffer solutions of known pH, preferably buffers in the pH range of the titration. Potassium tetroxalate, $0.05 M$ (pH 1.68 at 25°), $0.1 M$ HCl (pH 1.09, 0–50°), $0.01 M$ borax (pH 9.18 at 25°), and saturated calcium hydroxide (pH 12.42 at 25°) are recommended by Bates.[35] The alkaline pH standards must be prepared with carbonate-free water, and protected from absorbing carbon dioxide from the air.

The reference electrode, normally calomel or silver chloride, usually makes electrical contact with the sample solution through a KCl salt bridge. When titrations at 0° are to be carried out, the concentration of the KCl solution in the reference cell should not be much greater than $4 M$ (saturated with silver chloride or calomel, depending on the internal structure of the electrode) so that crystals of KCl do not precipitate upon cooling. Junctions formed by cracks about wires or metal plugs are likely to become clogged with protein and produce false readings of pH. Junctions formed by asbestos or glass fibers seem more trouble-free as long as the flow rate through the opening is adequate. This is also true of open, narrow capillary junctions.[3] However, if the flow of bridge solution is too great, the ionic strength of the sample solution (as well as its concentration) will be altered. The static (renewable) junction formed in a U-tube of small diameter has also generally been found satisfactory. If temperature changes of the electrodes are frequent, small (so-called "semimicro") electrodes with silver-silver chloride internal elements in *both* the reference and glass electrodes should be used, since these adjust quickly to temperature changes and have, in addition, a small temperature coefficient of EMF because of the identity of the internal elements. An additional bridge of KCl may also be used to keep

[34] When use of a pH meter is undesirable, an indicator added directly to the sample solution allows spectrophotometric determination of the pH from a reading of absorbancy in the visible wavelength region (J. Léonis and C. H. Li. *J. Amer. Chem. Soc.* **81**, 415 (1959)).

[35] R. G. Bates, "Determination of pH." Wiley, New York, 1964.

the reference electrode out of the sample solution, and thus not subject to temperature changes.[3] In special cases, a bridge can be used to prevent undesirable ions from entering the solution. Various types of cells, bridges, and junctions are described by Bates.[35]

The glass electrode should be selected with care.[3] Since a spectrophotometric titration can be carried out accurately to a higher pH than a potentiometric titration, it is important to avoid sodium ion errors of the glass electrode.[35] If the base to be used as titrant has sodium ion as the cation, a special "high pH" glass electrode, insensitive to sodium ion, may be used. It is convenient to avoid the sodium ion error of the glass electrode by using KOH as titrant and KCl as electrolyte, but nomograms for correction of sodium ion error, provided with the glass electrode by the manufacturer, may be used to correct the observed pH. KOH solutions can be prepared by dissolving analytical grade reagent directly in distilled (not deionized)[36] water, unless carbonate-free reagent is necessary.[37] The absorption spectrum of a dilution of the KOH titrant should be determined at shorter wavelengths, to check for ultraviolet-absorbing impurities.

Although a pH meter will almost always give a pH reading when the glass and reference electrodes attached to it are immersed in a solution of protein, the pH indicated by the meter should never be accepted unthinkingly. The experimenter should always remember the many ways (some obvious, some not so obvious) that things can go wrong. Are the electrodes fully immersed in the solution? While one type of glass electrode will give a correct reading when only touching the surface of a solution, another requires immersion to a depth of at least a centimeter. Is there any indication of drift in pH? Drift may merely be caused by poor stirring of the solution, but it may be produced by slow absorption of carbon dioxide by alkaline solutions. Drift is usually the first sign that the reference electrode junction is becoming clogged, or that the glass electrode is coated with a layer of protein gel. After titrating a protein to alkaline pH, is the proper pH obtained when the electrodes are immersed in neutral buffer, e.g., phosphate at pH 6.8? Failure to read the pH of a neutral buffer correctly afterward does not necessarily invalidate a titration of a protein to alkaline pH, since protein adhering to the electrodes may be precipitated by water used to wash the electrodes, or by the neutral pH buffer, clogging electrodes that had been functioning properly when the protein was soluble. But again, failure to check a neutral pH buffer may mean that the electrodes have finally reached the end of their useful life. Washing the electrodes free of protein with alkaline buffer or with 0.01 M KOH before checking with neutral buffer solution can restore confidence in the experiment. Alternatively, the electrodes may be taken out of the alkaline protein solution and immersed directly, without rinsing, in a large quantity of 0.01 M or 0.1 M KOH. The observed pH should be then very close to that determined for the same alkali solution before the titration. Each experimenter will find checks of this sort that will convince him that his electrodes and meter are functioning properly—to some extent, the checks which can be employed depend on the system investigated. Viscous solutions of high molecular weight glycoproteins are particularly nasty because of their tendency to form gel coatings on the electrodes, effectively buffering them against pH changes of the solution.

Procedures. Spectrophotometric titrations may be carried out directly, in which case the spectrum of the sample solution is measured against water or buffer as reference. Complete spectra are measured after each addition of titrant. This procedure is highly recommended for a preliminary investigation. If the titrant absorbs appreciably, or if the solvent absorption is a function of pH, either corrections for the absorption of added titrant or solvent absorption must be made or an equal amount of titrant must always be added to the reference solution before the spectrum is determined.

A spectrophotometric titration (here, as an example, assumed to be the titration of phenolic groups) may be carried out by means of the difference technique. A solution of protein at acid or neutral pH (phenolic groups not ionized) is used as reference solution, and additions of titrant are made to an identical solution placed in the sample cell. The spectra recorded are then difference spectra. Concentration of protein used depends on the number of phenolic groups present in the protein and the wavelength used for measurement. If the volume of titrant added to the sample solution is significant, then an equal amount of water or buffer must be added to the reference solution to keep the concentration of protein in the two cells identical. If only phenolic groups are titrated, and no complications present themselves, then the two isosbestic points near 280 nm, characteristic of phenolic groups, will be present throughout the course of the titration—that is, all difference spectra should pass through two points (of zero absorption) in the neighborhood of 280 nm (Fig. 4). Ordinarily, titrant can be added in successive increments to the same solution, and spectra re-recorded. However, if time-dependent ionization occurs, or if other reactions occur, spectra taken later (at higher pH) will be incorrect. Checks should be made for these errors, and fresh samples should be used at high pH so that accumulated errors from slow reactions or from evaporation of solvent will be absent. The entire spectrum or difference spectrum should be recorded after each addition of titrant. Unusual changes in absorption or shifts in baselines due to turbidity will then easily be recognized.

Determination of Intrinsic pK. Spectrophotometric titration curves are usually presented as the change in molar absorption at a specified wavelength if the molar absorption coefficient of the protein is known, or as the change in absorption for a given concentration of protein (w/v or molarity), as a function of pH (Fig. 6). If the titration curve is reversible, the pH at which the degree of ionization, α, equals one-half is the apparent pK, pK'. When the effect of the net charge of the protein upon ionization behavior is taken into account by assuming simple physical models for proteins, Eq. (2) becomes

$$\text{pH} - \log[\alpha/(1 - \alpha)] = \text{p}K_{\text{int}} - 0.868w\bar{Z} \tag{13}$$

where w can be regarded as an empirical electrostatic factor, and $\text{p}K_{\text{int}}$ is the pK corrected for the effect of the charge on the protein.[3] The net charge on the protein, $\bar{Z}$, can be calculated from the potentiometric titration curve if ion binding is absent. Figure 7 shows examples of plots of Eq. (13) for the spectrophotometric titration of phenolic groups of fibrinogen.[18] The intercept (at $\bar{Z} = 0$) is the intrinsic pK (pK_{int}), and the slope is equal to $-0.868w$.

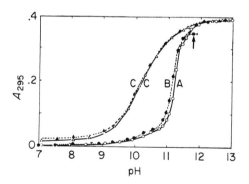

Fig. 6. Spectrophotometric titration of flagellin monomer (MW 37,000), at a concentration of 0.73 mg/ml, in 33 mM phosphate. Titrated from pH 7.0 with NaOH (A, ○); back titration with HCl (C, △); retitration of C with NaOH (C′, ▲). B (●), 0.1 M phosphate buffer was added to flagellin treated with 20 mM NaOH for 1 day (arrow at ■), then the titration was carried out from pH 7 by addition of NaOH. Optical rotation measurements show that a conformational change takes place between pH 10.8 and pH 11.5 on titration with NaOH. From M. Taniguchi, *Biochim. Biophys. Acta* **207**, 240 (1970). Reproduced by permission of Elsevier Publishing Co.

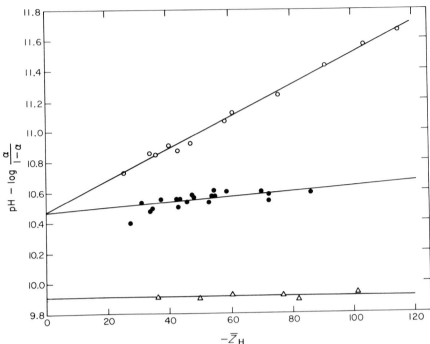

FIG. 7. Spectrophotometric titration curves at 25° of fibrinogen in 0.25 M KCl (native, ○), in 0.25 M KCl plus 5 M urea (●), and in 5 M guanidine hydrochloride (△), plotted according to Eq. (13). $\bar{Z}_H$ was calculated (from the potentiometric titration curve) for a molecular weight of 10^5, approximately the molecular weight of one of the spherical constituents of fibrinogen. From E. Mihalyi, *Biochemistry* **7**, 208 (1968). Reproduced by permission of the American Chemical Society.

Tests for Reversibility. Some tests for reversibility are made by quickly bringing the alkaline solution of protein back to neutrality with acid, allowing the solution to stand for a period of time, and then carrying out the spectrophotometric titration again by adding increments of alkali. This procedure, in general, is not a test for thermodynamic reversibility, since a period of standing at neutral pH may allow a protein which has been unfolded by titration to high pH to refold into the native conformation before the second titration is carried out. Such a reversibility test, however, can be quite useful when carried out in addition to a direct reverse titration, in which increments of acid are successively added to the alkaline pH solution. It is often found that while a direct reverse titration does not superimpose upon the forward titration (i.e., the forward and reverse titration curves enclose a "hysteresis loop"), after a sample has been held again at neutral pH, the original titration curve is reproduced closely (Fig. 6). This behavior, often observed when the titration of phenolic groups occurs at abnormally high pH, is nearly always characteristic of refolding of a protein at neutral pH, as measurements of optical rotation, for example, usually demonstrate. A reasonable inference is that at least part of the abnormality of the phenolic groups is produced by their placement in the native protein. Conformation changes linked to association may have large[38] or small[39] effects on titration behavior.

Abnormal Titration Curves. When the pK' of the ionizable groups, as measured from the half-ionization point of the titration curve, is significantly abnormal,[40] a correction should be carried out for the effect of the net charge of the protein by constructing a plot of Eq. (13). This will require the calculation of the pH dependence of $\bar{Z}$. When calculated from the potentiometric titration curve without corrections for ion binding,[41] $\bar{Z}$ is designated $\bar{Z}_\mathrm{H}$. If the plot of Eq. (13) is linear and pK_int is in the normal range, then the abnormal pK' or apparent steepness of titration curves in the pH ranges 10–11 and 4–3 is due to the charge on the protein. Plotting data in the form of Eq. (13) is one of the best methods for determining abnormalities. Conformational changes generally show up as "breaks," or changes in slope, of these plots, since the electrostatic factor, w, for unfolded proteins is smaller than for globular proteins.[3]

Abnormal groups can be "normalized" by dissolving the protein in a denaturant (e.g., urea or guanidine hydrochloride). The titration curve obtained in the denaturant can serve as a reference curve—groups have normal pK values,[42] electrostatic effects are diminished in urea, essentially absent ($w = 0$) in guanidine hydrochloride (Fig. 7), so that the total number of groups of each type may be calculated more readily.[3] The normal pK_int of a phenolic group in a protein is approximately 9.8.[3] The phenolic groups in native fibrinogen have an abnormal pK_int of 10.47, as shown by the intercept in Fig. 7. In both 5 M urea and in 5 M guanidine hydrochloride, the electrostatic factor is markedly reduced, as is evidenced by the small and zero slope, respectively, of those lines in Fig. 7. At first glance, it would appear that the phenolic groups are still abnormal

[38] M. Taniguchi, *Biochim. Biophys. Acta* **207**, 240 (1970).
[39] W. B. Gratzer and G. H. Beaven, *J. Biol. Chem.* **244**, 6675 (1969).
[40] Normal pK values are given by Nozaki and Tanford.[3]
[41] C. W. Carr, see Vol. 26 [9].
[42] Normal values of pK depend on the concentration of denaturant. Values of pK' for acetic acid, imidazole, *n*-butylamine, and phenol as a function of concentration of KCl, urea, and guanidine hydrochloride are given by J. W. Donovan, M. Laskowski, Jr., and H. A. Scheraga, *J. Mol. Biol.* **1**, 293 (1959).

in 5 M urea, since the intercept is the same as that obtained for the native protein. However, the pK_{int} of the phenolic group of the amino acid tyrosine is 10.38 in 5 M urea and 9.90 in 5 M guanidine hydrochloride.[18] Thus, the intercepts for fibrinogen in 5 M urea and in 5 M guanidine hydrochloride, different as they are, both show normal titration behavior of the phenolic groups in these denaturing solvents.[12]

Temperature Dependence. The temperature dependence of ionizable groups, not often studied, can give information about the nature of the abnormalities that are present. For example, if the enthalpy of ionization is extremely large, a conformational alteration of the protein is likely. Conversely, if the enthalpy of ionization of an "abnormal group" is normal, the group may be subject to perturbation by a charge, the position of which, with respect to the chromophore, is not altered with temperature. In this case, a plot of Eq. (13) might be linear, but w would be considerably larger than expected for the size of the protein.

[23] Difference Infrared Spectrophotometric Titration of Protein Carboxyls

By SERGE N. TIMASHEFF,[1] H. SUSI, and JOHN A. RUPLEY[2]

The pH dependence of the dissociation of ionizable groups in proteins is normally examined by standard acid-base titration methods.[3-7] Potentiometric titrations, however, measure only the total uptake of protons and cannot distinguish between the titration of various types of ionizable groups when there is an overlap in the pH zones of their ionizations.

Although carboxylic acids normally titrate in a pH region outside

[1] This work was supported in part by the National Institutes of Health Grant No. GM 14603 and the National Science Foundation Grant No. GB 12619.
[2] This work was supported in part by the National Institutes of Health Grant No. GM 09410 and the National Science Foundation Grant No. GY 22632.
[3] Y. Nozaki and C. Tanford, see Vol. 11, p. 715.
[4] C. Tanford, *Advan. Protein Chem.* **17**, 70 (1962).
[5] C. Tanford, "Physical Chemistry of Macromolecules," p. 548. Wiley, New York, 1961.
[6] E. J. Cohn and J. T. Edsall, "Proteins, Amino Acids and Peptides," p. 444. Van Nostrand-Reinhold, Princeton, New Jersey, 1943.
[7] S. N. Timasheff, "Biological Polyelectrolytes" (A. Veis, ed.), p. 1. Dekker, New York, 1970.

the zone of ionization of other titratable residues, a serious overlap may occur when the protein contains carboxyls that titrate at abnormally high pH values or other groups with abnormally low values of apparent pK. Typical examples are β-lactoglobulin,[8-10] which contains per monomer subunit one buried carboxyl that becomes ionized only at pH 7.5, i.e., in the region of histidine and α-amino ionization; carbonic anhydrase,[11,12] which contains a number of buried histidines that do not become protonated until pH 4.5, i.e., at the pK of normal carboxyls; and lysozyme,[13] which contains one carboxyl with a pK of 6.5. In such systems it becomes difficult to assign observed pK values to specific groups from potentiometric titration data alone. A way of circumventing such difficulties is to monitor the titration of particular groups by following spectral changes characteristic of these groups. This approach is well known in the ultraviolet spectrophotometric titration of tyrosine hydroxyls.[3,4,14] In the case of carboxylic residues, it is possible to carry out a similar analysis by using infrared absorption to monitor the degree of ionization.[9,13]

Carboxyl groups absorb in the spectral region of 1550–1710 cm^{-1}. The ionized residues have an absorption band around 1570 cm^{-1} while protonated carboxyls are characterized by a band near 1710 cm^{-1}.[15-18] From the intensities of these two infrared bands it is possible to determine directly the fractions of carboxyls which are either ionized on nonionized at any value of pH.

While the use of infrared absorption eliminates the problems inherent in potentiometric titration, it requires the use of special techniques because of two complications: (1) the bands characteristic of carboxyl groups appear as shoulders on the much stronger amide I band of the peptide links; (2) in aqueous solution the frequency region of interest is totally masked by strong H_2O absorption. These two problems are circumvented by using a differential infrared absorption technique

[8] C. Tanford and V. G. Taggart, J. Amer. Chem. Soc. 83, 1634 (1961).
[9] H. Susi, T. Zell, and S. N. Timasheff, Arch. Biochem. Biophys. 85, 437 (1959).
[10] J. J. Basch and S. N. Timasheff, Arch. Biochem. Biophys. 118, 37 (1967).
[11] L. M. Riddiford, J. Biol. Chem. 239, 1079 (1964).
[12] L. M. Riddiford, R. H. Stellwagen, S. Mehta, and J. T. Edsall, J. Biol. Chem. 240, 3305 (1965).
[13] S. N. Timasheff and J. A. Rupley, Arch. Biochem. Biophys. 150, 318 (1972).
[14] J. W. Donovan, this volume [22].
[15] G. Ehrlich and G. B. B. M. Sutherland, J. Amer. Chem. Soc. 76, 5268 (1954).
[16] R. C. Gore, R. B. Barnes, and E. Petersen, Anal. Chem. 21, 382 (1949).
[17] H. Susi, in "Structure and Stability of Biological Macromolecules" (S. N. Timasheff and G. D. Fasman, eds.), p. 575. Dekker, New York, 1969.
[18] H. Lenormant and E. R. Blout, Nature (London) 172, 770 (1953).

and substituting D_2O for H_2O as solvent.[9,13] D_2O can be used as a solvent for proteins without altering their conformation.[19] In the presence of D_2O, however, the pK values of ionizable groups are about 0.5 unit higher on the pD scale than on the pH scale in H_2O.[20,21]

Methods of Procedure

Protein Solutions. In order to carry out the titration in D_2O, care must be taken that the protein amide group hydrogens exchange to the maximal extent with deuterium and that residual water of hydration also be converted from H_2O to D_2O. This is accomplished most easily either by dissolving a batch of protein in D_2O or by suspending a slurry of protein crystals in D_2O; this solution or slurry is then left standing in D_2O for at least 24 hours, after which it is lyophilized. The degree of hydrogen–deuterium exchange can be monitored by the disappearance of the band characteristic of NH bending at ca. 1540 cm^{-1} [17,22] and the appearance of strong ND absorption near 1450 cm^{-1} (see Fig. 4 of Susi[22]).[13] The lyophilized protein is then dissolved in D_2O of the desired pD and salt composition. The protein solution should have a concentration between 30 and 90 g/liter. Aliquots of this solution are then adjusted to the proper pD by addition of concentrated NaOD or DCl and brought to final volume and ionic strength by further addition of D_2O or a solution of the supporting electrolyte in D_2O. In a typical infrared absorption experiment the total volume required is ca. 2 ml; thus, all pD and ionic strength adjustment can be made by a total addition of 0.1 ml of acid, base or salt solution to 2 ml of stock protein solution. These additions are made most easily with micropipettes. Protein concentration may be measured by any standard technique, such as dry weight or absorption in the ultraviolet spectral range.

pD Measurements. The pD measurements may be performed with any standard pH meter equipped with electrodes small enough to do a measurement on 2 ml of solution. It is advantageous to use a combination electrode, since this prevents leakage of H_2O into the solution from the calomel electrode. The electrodes are standardized with any of the usual standard buffers. They are then rinsed with D_2O and immersed into the protein solution. Equilibrium between electrode and D_2O is reached after 15–30 minutes. The value of the pD measured is the reading on the pH meter plus 0.4.[20,21]

[19] H. Susi, L. Stevens, and S. N. Timasheff, *J. Biol. Chem.* **242**, 5460 (1967).
[20] R. P. Bell, "The Proton in Chemistry." Cornell Univ. Press, Ithaca, New York, 1959.
[21] P. K. Glasoe and F. A. Long, *J. Phys. Chem.*, **64**, 189 (1960).
[22] H. Susi, see Vol. 26 [22].

Infrared Measurements. Reference solutions are prepared at selected pD values. A good internal check can be obtained by using reference solutions of various pD values. This provides for overlap of results obtained relative to a number of fixed reference points. The experimental and reference solutions are then introduced into two "matched" sealed cells of conventional design, with a nominal 0.1 mm path length, equipped with CaF$_2$ windows. If variable path length cells are used, it is frequently convenient to adjust the cell thickness of the sample-containing cell until a flat baseline spectrum is given in a region in which the amide groups do not absorb, e.g., at 1800 cm^{-1}. Difference spectra are then recorded between 1400 and 1900 cm^{-1} at instrumental settings proper for the particular spectrophotometer used. Direct spectra are also measured at selected pD values. In all the experiments the spectrophotometer must be purged with dry nitrogen, since atmospheric water vapor introduces insensitivity in the region of 1650 cm^{-1}.

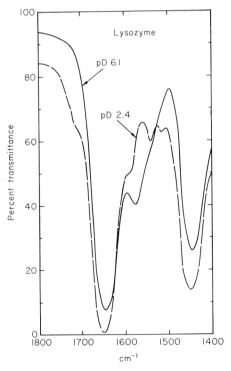

Fig. 1. Infrared spectra of lysozyme in D$_2$O at pD 2.4 and 6.1. Pure D$_2$O in reference beam. The dashed line spectrum has been arbitrarily displaced by 10% transmittance. Reproduced from S. N. Timasheff and J. A. Rupley, *Arch. Biochem. Biophys.* **150**, 318 (1972).

With the number of manipulations involved, it is almost impossible to avoid the introduction into the samples of minute amounts of H_2O, for example from atmospheric water vapor. Such small H_2O impurities in the deuterated solvent could cause complications because of the H—O—H bending frequency which occurs in the 1640 cm⁻¹ region. Fortunately, H_2O exchanges very rapidly with D_2O according to the equation $H_2O + D_2O \rightleftharpoons 2\ HOD$. The HOD bending mode absorbs at considerably lower frequencies, in the vicinity of 1450 cm⁻¹, and, since the equilibrium lies very far on the favorable side, this does not present any serious problems if the amount of initial H_2O is small.

Difference spectra are normally obtained over a broad pD range, e.g., between 1.0 and 10.0, using reference protein solutions at intermediate pD values, such as 7.0 and 4.0. As a spot check, direct spectra are also recorded at selected pD values. Typical direct and difference spectra obtained with lysozyme[13] are shown in Figs. 1 and 2. The direct spectra are characterized by a strong amide I band centered around 1650 cm⁻¹ with shoulders evident at 1560–1575 cm⁻¹ and ca. 1710 cm⁻¹. The essential absence of the amide II band near 1550 cm⁻¹ and the presence of a strong band near 1450 cm⁻¹ testify to the extensive deuteration of the peptide NH groups.[17,22-24] The difference spectra are characterized by pD dependent bands at 1565 cm⁻¹ and 1707 cm⁻¹; these correspond to COO⁻ and COOD absorption, respectively. The detail frequently seen between 1620 and 1670 cm⁻¹ is most probably due to pH perturbation of the backbone amide I band. At the protein concentration used, the amide I band in the direct spectrum is of such intensity that, at its peak, only 1–10% of the total energy is transmitted, i.e., it has an absorbance between 1.0 and 2.0. The apparent intensities of the difference spectral bands, on the other hand, vary with the amount of carboxylic side chains found in the proteins of interest and with the extent of their ionization. Their intensity usually falls in the range between 95 and 60% transmittance, i.e., their absorbance, ΔOD, varies between 0.020 and 0.200.

Baseline controls must be carried out periodically, using protein solutions of identical pD in the two cells, the reference solution having been diluted in a manner identical to the experimental samples. Typical baselines are shown in Fig. 2. As can be seen, with proper matching of pD and protein concentration there is essentially no detectable deviation from a smooth line in the regions of interest, i.e., near 1570 and 1710 cm⁻¹. At 1650 cm⁻¹, i.e., in the region of the amide I band, mismatching by

[23] E. R. Blout, C. de Lozé, and A. Asadourian, *J. Amer. Chem. Soc.* **83**, 1895 (1961).
[24] A. Hvidt and S. O. Nielsen, *Advan. Protein Chem.* **21**, 288 (1966).

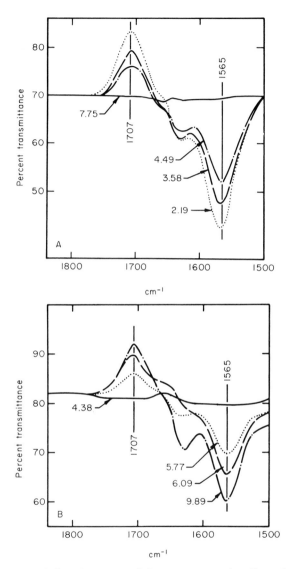

FIG. 2. Difference infrared spectra of lysozyme as a function of pD. (A) pD 7.75 reference; (B) pD 4.38 reference. (The instrument was set so that the matched spectra at 1800 cm⁻¹ were recorded at 65–85% transmittance; this permitted to record both positive and negative differences in a single scan.) The numbers next to the curves indicate the pD at which the spectrum was recorded. Reproduced from S. N. Timasheff and J. A. Rupley, *Arch. Biochem. Biophys.* **150,** 318 (1972).

2% transmittance can occur. Since this corresponds to an absorbance of approximately 0.01–0.02, the mismatching is usually no greater than 1%, while normally much better reproducibility can be obtained.

Interpretation

Data Analysis. The intensities (in units of % transmittance or absorbance) of the difference bands are read relative to 1800 cm^{-1}, at which frequency there should be no detectable difference spectrum, if the cell thickness and protein concentration matchings are done accurately. The values are then normalized to 100% transmittance or zero absorbance for the reference solution, converted to absorbance units if recorded as transmittance and corrected for the deviation of the absorbances in the regions of 1570 and 1710 cm^{-1} from that of 1800 cm^{-1} in the proper baseline. This procedure is necessary, since the baselines, while smooth, usually are not horizontal straight lines, but display a slight frequency dependence, as shown in Fig. 2. The experimental error due to the reading of intensities is of the order of 3–5%.

Calculation of Carboxyl Ionization. The absorbance values at 1707 cm^{-1} relative to the value obtained at total ionization, and at 1565 cm^{-1} relative to the value at zero ionization, are plotted as a function of pD and compared with each other. If A_{1565} and A_{1707} are normalized in such a manner that the maximum values correspond to the total number of ionizable carboxyls, the two curves should be mirror images of each other. Figure 3A and B shows typical data obtained on lysozyme. The curves in this figure are identical because the ordinate (left-hand side of the figure) has been reversed in Fig. 3A. The data are then analyzed in a standard way, as any titration curve, using the Linderstrøm-Lang equation[25] or similar equations.[5,7]

In detecting the presence of abnormally titrating groups and in establishing their pK values, the fraction of ionized carboxyls in the protein, α, is calculated for a distribution of the groups over a number of classes of type i, with characteristic pK$_i$. The overall degree of carboxyl ionization, α, is given by

$$\alpha = \frac{\Sigma n_i \alpha_i}{\Sigma n_i} \tag{1}$$

where n_i is the number of groups of type i. The degree of ionization of each type of group, α_i, is given by, for example, the Linderstrøm-Lang

[25] K. O. Linderstrøm-Lang, *C. R. Trav. Lab. Carlsberg* **15**, No. 7 (1924).

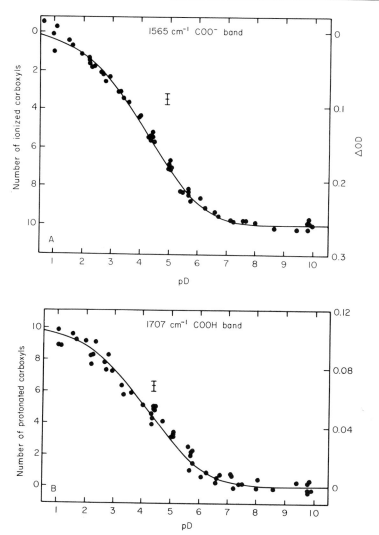

FIG. 3. Typical infrared titration curves as obtained on lysozyme. (A) 1565 cm⁻¹ band; (B) 1707 cm⁻¹ band. The dots are experimental points representing optical density differences from either the fully protonated (A) state or the fully ionized (B) state. The solid lines are calculated curves using best fitting pK values (see text). The error bars correspond to the experimental errors in reading differential band intensities. Reproduced from S. N. Timasheff and J. A. Rupley, *Arch. Biochem. Biophys.* **150,** 318 (1972).

equation, within the approximation of spherical symmetry and a smeared charge distribution[4,7,25]

$$\log \frac{\alpha_i}{1 - \alpha_i} = \mathrm{pH} - \mathrm{p}K_i + 0.868 w \bar{Z}$$

$$w = \frac{e^2}{2DkT} \left(\frac{1}{b} - \frac{\kappa}{1 + \kappa a} \right)$$

(2)

where $\bar{Z}$ is the average net charge on the protein molecule, e is the electronic charge in electrostatic units, D is the dielectric constant of the medium, k is Boltzmann's constant, T is the thermodynamic temperature, b is the radius of the protein molecule, a is the distance of closest approach of the center of a small ion to the center of the protein molecule, usually taken as $b + 2.5$ Å, and κ is the reciprocal thickness of the ionic atmosphere. For the calculation of the electrostatic work term, the value of the protein average net charge, $\bar{Z}$, as a function of pH should be taken from a potentiometric acid-base titration curve. The effective radius of the protein, b, can be set equal to the Stokes' radius calculated either from a hydrodynamic parameter, such as the sedimentation coefficient, or from the radius of gyration of the protein measured by small-angle X-ray scattering.[7,26] Alternatively, the parameter w may be taken directly from a potentiometric titration curve, if such a curve gives a straight line in a Linderstrøm-Lang plot. In the case when the protein is known to associate in the pH (pD) range of interest, the electrostatic interaction parameters of the various associated species, w_A, can be estimated by assuming that each protein A-mer is a sphere with A times the volume of the monomer and A times its charge, using the relation[13]

$$w_A Z_A = A^{2/3} w_M Z_M \frac{1 + \kappa a_M}{1 + \kappa[(A^{1/3} - 1)b_M + a_M]}$$

(3)

where the subscripts M and A stand for monomer and aggregate, respectively, and A is the degree of association. The term, $w\bar{Z}$, is taken then as the average of the contributions of all the aggregated species in proportions calculated from association data. While this calculation is highly approximate, any uncertainties resulting from it normally should not lead to significant errors, since the small variations in w introduce only second-order errors in the final assignment of $\mathrm{p}K$ values.

In this manner the experimental results are fitted to a set of $\mathrm{p}K$ values, varying the values of n_i and α_i until a good fit is obtained for the experimental curve, the total number of groups, Σn_i, being always

[26] H. Pessen, T. F. Kumosinski, and S. N. Timasheff, this volume [9].

set equal to the number of carboxyls in the protein. Such a comparison between theory and experiment requires that a value be assigned to the absorbance change developed through the ionization of one group. As is the normal practice in other types of spectrophotometric titrations, this is taken to be the same for all groups and is set equal to $1/n$ of the total absorbance change between extreme acid and alkaline pD. Since differences in environment between protein surface and interior normally do not affect the absorptivities of chromophoric residues by more than 10%, this simplification introduces an uncertainty of not more than 0.1–0.2 groups in the titration curve and in the assignment of pK values from the infrared spectrophotometric titrations of carboxylic residues.

[24] Differential Conductimetry

By Alkis J. Sophianopoulos

The method described here provides mainly two pieces of information, the number of ions bound by a protein and the isoelectric point of the protein. In addition, conductivity measurements give accurately the concentration of an electrolyte in the presence of a protein or a nonelectrolyte such as a sugar or a polysaccharide. Nonelectrolytes affect the conductance of an electrolyte, and the corrections for the effect of nonelectrolyte described here can be used to give accurate values of electrolyte concentration. The main application for the study of enzymes is the use of the method along with pH titration studies[1] because it gives Z, the net charge of the protein as a function of pH and the isoelectric point of the protein. Thus, by the use of the method, all information necessary for pH titration studies is obtained as a function of pH, namely the number of protons bound, and the number of ions other than protons. The method cannot be used for the separation or detection of proteins as electrophoresis can, as it determines the properties of the sum of the components. It is well known that interactions of enzymes may involve alteration of their tertiary or quarterary structure accompanied in many instances by a change in their protonic charge, Z_H, and/or net charge Z. Such interactions can be followed nicely by the use of conductimetry.

In describing the method we shall proceed from the simple measurements of conductance in the absence of ion binding, to the more complex

---

[1] pH titration studies have been described by Y. Nozaki and C. Tanford, see Vol. 11, p. 715. The reader is referred to this article for details on pH titrations.

measurement of ion binding at constant pH and finally to the determination of the isoelectric point and ion binding as a function of pH.

Conductance in the Presence of Nonelectrolytes

Conductance in the Absence of Ion Binding

The measured specific conductance $\bar{K}$, in ohm^{-1} cm^{-1}, is related to the electrolyte concentration (M) or (N), in equivalents per liter of solution by:

$$\bar{K} = \Lambda \cdot (M) \cdot 10^{-3} \tag{1}$$

where Λ is the equivalent conductance of the particular electrolyte in ohm^{-1} cm^{-1} N^{-1}. Λ is ionic strength dependent:

$$\Lambda^\circ = \Lambda + B_3 M^{1/2} \tag{2}$$

where Λ° is the limiting equivalent conductance at zero ionic strength and B_3 is given by:

$$B_3 = \frac{(B_1\Lambda^\circ + B_2)}{1 + B\alpha M^{1/2}} \tag{3}$$

where for aqueous solutions:

$$B = 50.29 \times 10^8 \times (\epsilon T)^{1/2}$$
$$B_1 = 8.204 \times 10^5 \times (\epsilon T)^{-3/2}$$
$$B_2 = 82.51 \times \eta(\epsilon T)^{-1/2}$$

ϵ is the dielectric constant of the solvent (78.30 for water at 25°C), T is the absolute temperature, η is the viscosity of the solvent in poises, and α is the ion size parameter in angstroms. Values for common electrolytes can be found conveniently[2,3] as well as in other standard sources. Equation (2), due to Robinson and Stokes,[2] is accurate up to at least 0.1 M for 1:1 electrolytes. If in addition to the electrolyte there is also a nonelectrolyte present, such as sucrose, the measured equivalent conductance is reduced. The experimental proof as well as other details for the effect of nonelectrolytes is given elsewhere,[4] and so here we shall give the basic relations for calculating specific conductance in the presence of nonelectrolytes. Proteins can be considered to be nonelectrolytes. Later we shall see that we can take into account the small contribution

[2] R. A. Robinson and R. H. Stokes, "Electrolyte Solutions," rev. 2nd ed. Butterworth, London, 1965.

[3] H. S. Harned and B. B. Owen, "The Physical Chemistry of Electrolytic Solutions," 3rd ed. Van Nostrand-Reinhold, Princeton, New Jersey, 1958.

[4] E. A. Sasse, Ph.D. Thesis, University of Tennessee Medical Units, Memphis, Tennessee, 1968.

to conductance of highly charged proteins. For measurements in the presence of a nonelectrolyte we use primed symbols, and we shall use subscripts 1, 2, etc., to distinguish the various concentrations of the electrolyte. Thus, the equivalent conductance of an electrolyte such as KCl of concentration M_1 in the presence of a nonelectrolyte such as sucrose is given by:

$$\Lambda_1 = \Lambda'_1 + Q_1 C_p \qquad (4)$$

where C_p is the concentration of nonelectrolyte in grams per 100 ml of solution and Q_1 has the units of equivalent conductance per nonelectrolyte concentration. In other words, this new quantity Q is the quantity by which the equivalent conductance of an electrolyte is reduced by a particular nonelectrolyte at a concentration of 1 g per 100 ml. This Q is unique for each nonelectrolyte and must be determined for each one. Like equivalent conductance, Q is also ionic strength dependent. It would be extremely tedious to determine Q not only for each nonelectrolyte, but also for each concentration of nonelectrolyte. Fortunately, it turns out that once Q is determined for a nonelectrolyte at a single concentration of an electrolyte, Q can be calculated for every other concentration of the same electrolyte. Moreover, the value of Q for a different kind of electrolyte can also be calculated. It suffices then to determine accurately Q for a particular nonelectrolyte at one concentration of one electrolyte. One way to do so is given in Fig. 1. To calculate Q at any other concentration of the same electrolyte, Eq. (5) is used:

$$\frac{Q_1}{Q_2} = \frac{\Lambda_1}{\Lambda_2} \quad \text{or} \quad \frac{Q_1}{Q_0} = \frac{\Lambda_1}{\Lambda_0} \qquad (5)$$

where the subscripts refer again to a particular electrolyte concentration. Since the Λ_i values refer to equivalent conductances in the absence of the nonelectrolyte, they can be calculated using Eqs. (2) and (3) or be experimentally determined. Another way of expressing the relationships in Eq. (5) is

$$\frac{\Lambda_1}{\Lambda_2} = \frac{\Lambda'_1}{\Lambda'_2} \quad \text{or} \quad \frac{\Lambda_1}{\Lambda^\circ} = \frac{\Lambda'_1}{\Lambda^{\circ\prime}} \qquad (5\text{-}1)$$

If Q is known for an electrolyte "a", Q for another electrolyte "b" of the same concentration can be calculated by:

$$\frac{Q_a}{Q_b} = \frac{\Lambda_a}{\Lambda_b} \quad \text{or} \quad \frac{Q^\circ_a}{Q^\circ_b} = \frac{\Lambda^\circ_a}{\Lambda^\circ_b} \qquad (6)$$

The equations apply to 1:1 electrolytes. Although a limited amount of information indicates that they hold for other than 1:1 electrolytes, at present it would be wise for one to determine Q experimentally for other

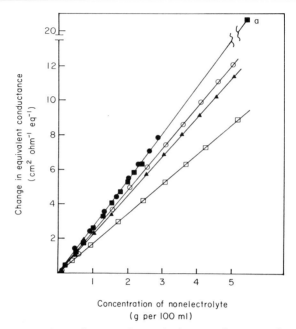

Fig. 1. Nonelectrolyte effect on the equivalent conductance of potassium chloride; 25°C. ○—○, Sucrose in $0.02\,M$ KCl; ▲—▲, sucrose in $0.1\,M$ KCl; ■—■, dextran 250 in $0.063\,M$ KCl ($a = 7.59$ g per 100 ml of dextran 250; ●—●, dextran 20 in $0.05\,M$ KCl; □—□, isoionic hemoglobin in $0.055\,M$ KCl. From E. A. Sasse, Ph.D. Thesis, University of Tennessee, 1968.

than 1:1 electrolytes. For 1:1 electrolytes, Eqs. (5) and (6) give results accurate to 0.1–0.2% of equivalent conductance. The nonelectrolyte concentration may be as high as 10%, and experiments with sucrose even up to 20% show no deviations from the above equations. Table I gives Q values for some nonelectrolytes.

Determination of Q in the Presence of Ion Binding—Basic Equations and Definition of Solutions

When ions bind to a protein and are therefore removed from solution, the conductance is reduced. The question is, then, how can one distinguish between the reduction in conductance due to ion binding and that due to the nonelectrolyte effect described by the quantity Q? Equilibrium dialysis gives unequivocal answers; however, before describing it, we must define some useful relations.

It is convenient to define two states for concentration. The initial concentration of protein and of salt before interaction occurred, and their respective concentrations afterward. This is particularly useful in

TABLE I
Q VALUES FOR COMMON NONELECTROLYTES

Nonelectrolyte and electrolyte at the ionic strength shown, 25°C	Λ^a	Q^b
Sucrose, KCl 0.02 M	138.34	2.53
Sucrose, KCl 0.05 M	133.37	2.44
Sucrose, KCl 0.10 M	128.96	2.33
Dextran 20, KCl 0.02 M	138.34	2.75
Dextran 20, KCl 0.05 M	133.37	2.70
Dextran 250, KCl 0.02 M	138.34	2.80
Dextran 250, KCl 0.05 M	133.37	2.57
Carboxyhemoglobin, human isoionic KCl 0.02 M	138.34	1.87
Carboxyhemoglobin, human isoionic KCl 0.05 M	133.37	1.78
Lysozyme pH 6.4–6.6, KCl about 0.022 M	138.34	2.10
Lysozyme pH 6.5–6.7, KCl about 0.042 M	135.00	2.00
Mannitol, KCl 0.029 M	136.68	2.70
Mannitol, HCl 0.004–0.006 M in 0.029 M KCl	415.8[c]; 136.68	7.20[c]
Mannitol, HCl 0.03 M in 0.025 M KCl	404.52[c]; 137.42	6.77[c]

[a] Equivalent conductances are for the specified electrolytes in the absence of nonelectrolytes; the units of Λ are cm^2 ohm^{-1} eq^{-1}.
[b] Q values for the specified nonelectrolyte at the concentration of electrolyte given. Q units are cm^2 ohm^{-1} eq^{-1} per gram per 100 ml.
[c] These are Λ and Q values for HCl at the specified solution conditions.

this method, because continuous titrations are carried out and one must define changes in charge, pH, etc. in terms of the previous state of the solution. In all situations, we use an asterisk to distinguish the initial conditions of an experiment. For example, if the protein carried positive charge Z^* and its molar concentration was M^*_p, the concentration of counterions would be $Z^*M^*_p$. If upon addition of salt of concentration M^*, ν anions bind per mole of protein, the concentrations, etc., of the various species are as follows: cation "molarity" = $M = M^*$; anion "molarity" = $(M^* - \nu M_p + Z^*M_p) = M^* + (Z^* - \nu)M_p$; final protein charge = $Z^* - \nu = Z$.

In general, in this method ν is used to designate bound ions. The above description may be disturbing to those accustomed to more orthodox definitions, but our purpose is to keep track of ions, and although, for example, the chloride ions that might originally be the counterions of the protein are indistinguishable from the chloride ions in a potassium chloride solution, it is useful for our bookkeeping to describe their respective concentrations clearly. Another problem with regard to the conductance of the solution is that the concentrations of cations and anions (excluding the macroion) are usually unequal, as in the above example. To describe the concentration of free or unbound electrolyte, it is convenient to define a "salt concentration," M_s, in terms of equivalent ionic conductances so that we deal with "one" kind of salt. For example, if in the above example potassium and chloride were the ions involved, we can describe M_s as:

$$M_s = M_{KCl} + (Z^* - \nu)M_p \frac{\lambda^\circ_{Cl}}{\Lambda^\circ_{KCl}} \tag{7}$$

and the specific conductance of the above electrolyte would be described by:

$$\bar{K}'_s = \Lambda'_{KCl}M_s \tag{7-1}$$

Later on we shall describe how the charge and the number of ions bound can be calculated.

Another complication arises when the protein contributes to the conductance, which is the case when it carries appreciable charge and it has a relatively low molecular weight. Two general methods are used. In the first, the conductances of two different solutions are measured and the conductances are subtracted from each other to cancel out the protein conductance. In the second, the actual value of the equivalent conductance of the protein is used. We shall give here the general equations that are used. The specific conductance of the protein (exclusive of its counterions) is given by:

$$K'_p = \lambda'_p Z M_p \tag{8}$$

So that:

$$\bar{K}' = \bar{K}'_p + \bar{K}'_s \tag{9}$$

where $\bar{K}'$ is the experimentally determined specific conductance and $\bar{K}'_p$ and $\bar{K}'_s$ are given by Eqs. (8) and (7-1), respectively.

It is assumed that λ obeys Eq. (5-1), so that

$$\lambda'_{p_2} = \lambda'_{p_1} \frac{\Lambda_{ps_2}}{\Lambda_{ps_1}} \tag{10}$$

Λ_{ps_1} and Λ_{ps_2} are the known equivalent conductances of *any* electrolyte, determined at ionic strengths $(Z_1 p_1)$ and $(Z_2 p_2)$, and *not* at the ionic strengths of the supporting electrolyte. This is emphasized by the use of subscripts ps_1 and ps_2, whereas the equivalent conductances of the supporting electrolyte carry only numbers. An example where the protein conductance, if applicable, must be canceled out is in the determination of Q by equilibrium dialysis at two ionic strengths and at about the same protein concentration. Using subscripts 1 and 2 for the two different experiments, Eq. (9) can be written in detail as:

$$\bar{K'}_1 = \lambda'_{p_1} Z_1 M_{p_1} + \Lambda'_1 \cdot M_{s_1} \tag{9-1}$$

and also:

$$\bar{K'}_2 = \lambda'_{p_2} Z_2 M_{p_2} + \Lambda'_2 M_{s_2} \tag{9-2}$$

In order to subtract out the conductance component due to protein, Eqs. (10) and (5) are utilized to rewrite Eq. (9-2) as:

$$\bar{K}_2 \frac{(Z_1 M_p)_1}{(Z_2 M_{p_2})} \frac{\Lambda_{ps_1}}{\Lambda_{ps_2}} = \lambda'_{p_1} Z_1 M_{p_1} + \bar{K'}_{s_2} \frac{(Z_1 M_{p_1}) \Lambda_{ps_1}}{(Z_2 M_{p_2}) \Lambda_{ps_2}} \tag{9-3}$$

Letting:

$$f = \frac{Z_1 M_{p_1} \Lambda_{ps_1}}{Z_2 M_{p_2} \Lambda_{ps_2}} \tag{9-4}$$

and subtracting Eq. (9-1) from Eq. (9-2), there results:

$$\bar{K'}_2 f - \bar{K'}_1 = \Lambda'_2 \left(M_{s_2} f - \frac{\Lambda_1}{\Lambda_2} M_{s_1} \right) \tag{11}$$

In the case of ion binding, the only unknowns in Eq. (11) are Z_1, Z_2, M_{s_1}, M_{s_2}, and Q. The values Z_1, Z_2, M_{s_1}, and M_{s_2} can be determined by equilibrium dialysis, and therefore Q can be determined by Eq. (11) even in the presence of ion binding and appreciable protein conductance. Of course, Q is related to Λ' by Eq. (4).

Determination of Q by Equilibrium Dialysis

Equilibrium dialysis is the most unequivocal and most informative way to determine Q. Specific situations may not require equilibrium dialysis, and these will be discussed later, but the judgment as to whether dialysis is to be used or not is, in fact, based on knowing the information equilibrium dialysis can provide.

As employed here, dialysis demands the highest degree of accuracy in preparation of solutions and experience in manipulating equipment. If the worker is inexperienced, he or she should practice with known and inexpensive polymers such as dextran. The experimental design is described here briefly to facilitate understanding the calculations. Details will be given in the experimental section.

It is assumed that the protein along with its counterions is initially in water. A known weight of the solution is placed in a dialysis sac, and it is dialyzed against a known weight of an electrolyte solution. Preferably the weights of the two solutions should be about equal although this is mainly a matter of convenience in manipulation. After equilibration the weights of the two solutions are determined and their conductance is measured. The protein concentration after dialysis is also determined. The following symbols are used to describe the solutions. The quantities of the free ions inside the sac carry as superscript the Roman numeral I. The quantities of the solvent outside the sac carry no superscript. In addition, the ions inside the sac exclusive of the counterions carry the superscript II.

We may use as a specific example lysozyme chloride dialyzed against potassium chloride at neutral pH. Under these conditions, lysozyme binds chloride ions, it has a positive charge, and also it has appreciable conductance.

Outside the sac:

$$m_{K^+} = m_{Cl^-} = m_{KCl} \tag{12}$$

Inside the sac the concentration of the free potassium ions can be represented in two ways:

$$m_{K^+}{}^I = m_{K^+}{}^{II} \tag{12-1}$$

However, the concentration of the free chloride ions inside the sac, m_-^I, is given by:

$$m_-^I = m_{Cl}{}^{II} + m_z \tag{12-2}$$

where m_z is the molality of the counterions. The number of moles of the total electrolyte added is known and $m_+ = m_-$; the electrolyte outside the sac can be measured extremely accurately by conductance. Thus the quantity $m_-{}^{II} = m_+{}^{II}$ can be calculated subtractively. The remaining quantity m_-^I can therefore be calculated from the Donnan relationship:

$$m_-^I m_+^I = m_{KCl}{}^2 \tag{13}$$

It must be emphasized that the required accuracy demands the use of molalities in the Donnan relationship (Eq. 13). Molarities must be used in the equations involving conductance.

A complication arises if the ions that bind to the protein carry the same charge as the protein. In this case the quantity m_+^I for example, i.e., the molality of the ions having the same charge as the protein is not equal to the quantity calculated subtractively, as described above. Thus we must use a specific symbol for the *quantity calculated by subtraction*, so it carries the primed superscript I'. If the protein does bind ions of opposite charge, $m_+^I = m_+^{I'}$. However, equilibrium dialysis reveals whether a protein binds ions of like or unlike charge, and testing the above relationship reveals the kind of ion bound, as we shall see. Table II gives the basic calculations for a protein which does not bind ions at its isoelectric point, human carboxyhemoglobin.

Having defined all necessary quantities, we proceed to ask the following questions: (a) What is the net charge Z of the protein? (b) Does it bind ions of like or unlike charge? (c) Does the protein contribute to the conductance? (d) What is the Q of the protein?

In the case where a protein binds ions of opposite charge, the quantity calculated, subtractively, $m_+^{I'}$, is equal to m_+^I, so that the remainly unknown, m_-^I is calculated using the Donnan equation (13). Since the protein concentration has been measured, the net charge is simply:

$$Z = \frac{m_-^I - m_-^{II}}{m_p} \tag{14}$$

To calculate the ions bound under these dialysis equilibrium conditions, one needs to know Z_H, the "protonic" charge, which can be calculated from the pH titration curve. To calculate Z_H one needs to know the isoelectric point. Thus, the reader sees how all the experiments described here fit together; the calculations for one experiment depend on the results of the others.

In the above case where the concentration of the free ions inside the sac is known, the electrolyte concentration can be calculated for use in the conductance equations. The total "salt" concentration M_s is calculated by Eq. (7) and M_s then is used in Eq. (7-1) to calculate the equivalent conductance of salt Λ'_s. If the protein contributes to the conductance, two rather than one equilibrium dialyses must be performed. These are performed at approximately the same protein concentration but different equilibrium ionic strengths, for example $0.02\,M$ and $0.04\,M$. The Eqs. (8) through (11) are then used.

A word of caution: Λ' quantities should be calculated rather than Q values, as shown in Eq. (8) through (11) because the numbers used are derived by subtracting quantities nearly equal to each other, and significant figures would be lost if not carried to enough significant places. If a computer is used, such precaution is unnecessary. It should be emphasized that one should retain significant figures in all calculations in this method since in most instances the derived number is a small difference between two relatively large numbers.

It is interesting that once Q has been determined, the equivalent ionic conductance of the protein, λ'_p, can be calculated using Eq. (8). This method of calculating λ'_p was described by Sasse for the first time.[4] The value determined for lysozyme at $25°$ and 0.02–0.04 ionic strength was $\lambda'_p \approx 20$, which is nearly twice as large as the value expected from a simplified relation of electrophoretic mobility and diffusion. The exact theoretical significance of such a calculated value cannot be speculated upon until further experiments provide adequate evidence. At any rate,

TABLE II

(1) *Initial measurements*

Tare (glass-stoppered cylinder)	68.01530 g
Protein solution in bag + tare	80.71536 g
Protein solution + electrolyte + tare	95.83744 g
Normality of KCl (determined previously)	0.070087 N
Density of KCl solution (determined from graph)	$\rho = 1.0003$ g/ml

(2) *Initial calculations*

Protein concentration (spectrophotometrically)	2.375 g/100 ml
The protein initially was dissolved in water.	
Density of protein was determined by formula.[a]	
At 25°C, $\rho_p = 0.99707 + (1 - 0.749 \times$	
$0.99707) \times 0.02375 =$	1.003083 g/ml
Weight of protein + bag (in air)	12.70006 g
Weight of wet dialysis tubing (determined from tubing of equal size)	0.67568 g
Weight of protein solution (in air)	12.02438
Weight of dry dialysis tubing[b] $= \dfrac{0.67568}{1.403} =$	0.4816 g
Weight of protein solution (vacuum)[c] $=$	
12.0243 g $+ \left(\dfrac{12.02}{1} - \dfrac{12.02}{8.0}\right) \times 0.0012 =$	12.03696 g
Volume of protein solution $= \dfrac{12.03696}{1.0031} = 11.9998 =$	12.000 ml
Weight of electrolyte (air) $= 95.83744 - 80.71536 =$	15.12208 g
Weight of electrolyte (vacuum)	15.13796 g
Volume of electrolyte added $= \dfrac{15.13796}{1.0003} =$	15.13346 ml
Millequivalents total of KCl $=$	
15.13346×0.070087 $N =$	1.06066 mg

(3) *Measurements and calculations at equilibrium*

Protein concentration (spectrophotometrically)	2.350 g/100 ml
Total weight of cylinder $=$	95.83740
	(i.e., same as original)
Weight after removing bag with protein	83.04306 g
Weight of bag with protein (by subtraction)	12.79434 g
Subtract 10% of bag weight for wetness outside	0.048 g
Net weight of protein bag	12.74634 g
Weight of bag-bound water	0.0906 g
Net weight of protein solution (in air) $=$	
$12.74634 - 0.0906 - 0.4816 =$	12.17414 g
Weight of protein solution (vacuum)	12.18686 g
Density of protein solution (assume 0.04 M KCl)	$\rho = 1.00482$
Volume of protein solution $\dfrac{12.18686}{1.00482} =$	12.12842 ml
Specific conductance of protein solution $\bar{K}' =$	5.05188 ohm^{-1} cm^{-1}
Weight of solvent outside $=$	
$83.04306 + 0.048 - 68.01530 =$	15.07576 g
Weight of outside solvent (vacuum) $=$	15.09160 g

TABLE II (*Continued*)

Specific conductance of outside solvent	$\bar{K} = 5.29491$ ohm^{-1} cm^{-1}
Normality of KCl (from graph of $\bar{K}$ vs. N)	$0.039251\ N$
Density of KCl outside (from graph)	0.9989
Volume of KCl solution outside	15.1082 ml
Total meq of KCl outside	0.593012
Total meq of KCl inside $= 1.06066 - 0.59301 =$	0.46765 meq KCl

(4) *Calculation of molalities and Q*

Grams of KCl outside $= 0.59301 \times 10^{-3} \times 74.56 =$	0.044215 g
Grams of H$_2$O outside $=_{,} 15.09160 - 0.044215 =$	15.04738 g
m_{KCl} outside $= \dfrac{0.59301}{15.0474} =$	$0.0394095\ m$
Grams of protein $= 0.0235 \times 12.128 =$	0.28501 g
Grams of H$_2$O $= 12.18686 - 0.28501 - 0.03487 =$	11.8670 g
$m^{\text{II}'} = \dfrac{0.46765}{11.8670} =$	$0.0394076\ m$

Assuming $0.0394076\ m$ is "molality" of one of the kind of ions inside, use Donnan relationship to get the other:

$$m^{\text{I}} = \frac{(0.0394095)^2}{0.0394076} = 0.0394114\ m$$

Difference $(0.0394114 - 0.394076) = 3.8 \times 10^{-6}\ m$, which is insignificant, thus mK$^+$ = mCl$^-$ inside. Molarity of KCl inside $= 0.46765/12.1284 = 0.03856\ M$ of KCl. At this KCl molarity from graph of $\bar{K}$ vs. M KCl, was obtained $\bar{K} = 5.20495$ at $0.03856\ M$ KCl, so that $\Lambda = 134.98$.

From actual measurement: $\quad \Lambda' = \dfrac{5.05188}{0.03856} = 131.01$

$$Q = \frac{(134.98 - 131.01)}{2.35} = 1.69$$

Alternatively, the protein solution volume at equilibrium inside the bag is calculated from the initial volume added and the ratio of absorbance values before and after equilibrium:

At equilibrium: $\quad$ Volume $= 12.000 \times \dfrac{2.375}{2.350} = 12.128$ ml

Also, when ions do not seem to bind, the molarity of electrolyte inside can be calculated from the millimoles of electrolyte and volume of solution:

$$M_{\text{KCl}} = \frac{0.46765}{12.128} = 0.038559 = 0.03856\ M$$

---

[a] $\rho = \rho_0 + (1 - V\rho_0)C_p$, where ρ is density of protein solution, V is the partial specific volume of the protein, C_p is the protein concentration in grams per milliliter, and ρ_0 is the density of the solvent.

[b] 1.430 is the ratio of wet:dry tubing at the specified conditions.

[c] The weight in vacuum is determined by:

$$w_v = w_a + \left(\frac{w_a}{d_s} - \frac{w_e}{d_{wt}}\right) \times d_a$$

where w_v is weight of substance in vacuum, w_a is weight of substance in air, d_2, d_{wt}, and d_a are the densities of the substance, of the weights used and of air, respectively.

the experimentally determined λ'_p can still be used to adequately correct for protein conductance in the experiments described here. The only remaining question is whether the protein binds ions of like or unlike charge. We have not studied a protein which binds ions of like charge, but the equations to be used are similar to the other ones which have been tested. If the ions that bind are of like charge to that of the protein they are given a special designation by using double charge symbols such as ν_{++} or ν_{--}. Assuming again a positively charged protein, the protein net charge at equilibrium would be $Z = Z^* + \nu_{++}$, and thus $m_+^{\mathrm{I}} = M_+^{\mathrm{I}'} - \nu_{++} \cdot m_p$. The way to write the Donnan equilibrium becomes

$$(m^{\mathrm{I}'} - \nu_{++} \cdot m_p)(m^{\mathrm{I}'} + Z^* m_p) = m^2 \tag{15}$$

In this case two dialysis experiments are again necessary. However, here in the second experiment the ionic strength inside the dialysis sac is kept as nearly equal to the first as possible, and the protein concentration is changed by 30–50%. The reason for keeping the ionic strength constant is to keep the protein charge constant. It is easy to show that under these conditions:

$$\nu_{++} = \frac{(m^{\mathrm{I}'})^2 - m^2 + Z^* \cdot m_p \cdot m^{\mathrm{I}'}}{m_p(m^{\mathrm{I}'} + Z^* \cdot m_p)} \tag{16}$$

$$Z^* = \frac{m^2 - (m^{\mathrm{I}'})^2 + \nu_{++} \cdot m_p \cdot m^{\mathrm{I}'}}{m_p(m^{\mathrm{I}'} - \nu_{++} m_p)} \tag{16-1}$$

Any of the Eqs. (15) through (16-1) is solved simultaneously at two protein molalities, and thus the "salt" concentration M_s can be calculated and therefore the value of Q obtained.

The reader has come to the conclusion by now that if the protein also contributes to conductance, a total of three equilibrium experiments are necessary. Two dialyses are carried out at constant, rather high, protein concentration and two different ionic strengths, such as 0.02 and 0.04 M. The third dialysis experiment is carried out preferably at the lower ionic strength, such as 0.02 M, used and lower protein concentration. Donnan calculations are carried out by assuming first that anions, and also that cations, bind. One set of calculations will give consistent results. Perhaps the best way to realize the difference is through use of actual numbers, and Table III was constructed to show such differences. The trends of numbers in Table III are easy to see, but some major points need elaboration. Wrong calculations tend to show a much smaller net charge; if true, this means that the protein conductance should be small. Moreover, M_s also tends to be less than the true value. Both of these miscalculations reinforce each other and would give not only a small Q value, but in many instances a negative one. This should

TABLE III

A SIMULATED DIALYSIS EQUILIBRIUM EXPERIMENT

Quantities measured or calculated[a]	6 Anions bind conc × 10³ $m_p =$ 0.5 × 10⁻³ Column 1	6 Cations bind conc × 10³ $m_p =$ 0.5 × 10⁻³ Column 2	6 Cations bind conc × 10³ $m_p =$ 1 × 10⁻³ Column 3	6 Cations bind conc × 10³ $m_p =$ 1 × 10⁻³ Column 4
m added KCl (initially)	40.00	34.00	40.00	20.00
m_{KCl} out (equil.)	20.488	17.271	20.238	9.5454
m_I calc. (wrong)	—	16.729	19.762	10.455
$m'_{k+} = M$ (true)	19.512	13.729	13.762	4.4546
m'_{Cl} (wrong)	—	17.831	20.725	8.7149
m'_{Cl} (correct)	21.512	21.727	29.762	20.454
Z (true)	4.0	16.0	16.0	16.0
Z (false)	—	2.2	0.963	−1.74
m_s (true)	20.531	17.804	21.913	—
m_s (false)	—	17.290	20.253	? (neg.)
Λ	138.2	138.9	137.9	—
$\bar{K}'$ meas.	2.8082	2.4477	2.9596	—
Q (true)	2.0	2.0	2.0	2.0
Q (false)	—	−3.75	−5.8	—

[a] The quantities in this table were calculated by assuming that a protein had an initial charge $Z = 10$. The values (wrong) in columns 2, 3, and 4 were calculated by assuming that the protein bound anions rather than cations. Columns 2 and 3 would correspond to experiments designed to calculate Q as well as the true charge Z. Notice the values M'_{K+} of 13.729 × 10⁻³ M and 13.762 × 10⁻³ M which are nearly equal, so that the number of cations bound per mole remains unchanged. Column 4 shows how much more pronounced the inconsistencies at lower ionic strength become by making the wrong assumption that anions rather than cations bind.

Since these are illustrative rather than actual values, some simplifications have been used. It has been assumed that the volumes of the solutions inside and outside the bag are equal, and that they remain equal after equilibrium. Also, the conversions from molality to molarity and vice versa are omitted so that one type of concentration unit is used. The same concentration unit used to calculate the Donnan relationship is also used to calculate equivalent conductance. Table II illustrates a real example where the differences between molarity and molality are shown.

become apparent when the experiments to cancel out protein conductance are carried out, i.e., at constant protein concentration and two ionic strengths. As Table III shows, many results are quite absurd; to point one out, the concentration inside the bag of the ion of like charge to that of protein is higher than its concentration outside the bag. All this may seem rather involved, but it becomes clear when actual data are used. The information collected is more than adequate to determine both the kind of ions bound and the extent of binding. In the following section a value of Q will also be calculated from a titration with an electrolyte. In that case if the ion binding is neglected, the calculated value of Q is larger than the true one. This effect is opposite to that given by wrong calculations of dialysis equilibrium data. With all these checks, it is rather difficult to arrive at the wrong conclusions. Dialysis experiments should be carried out between pH 5 and pH 9 so that the concentration of hydrogen and hydroxyl ions can be neglected. When it is necessary, the Q values for protons and hydroxyl ions are calculated as for the other ions, and their effect is additive.

One final complication might arise in the case where a protein associates appreciably with concomitant binding or release of ions. This binding or release of ions which is protein concentration dependent would cause ν_{++} not to be constant at constant ionic strength. A solution may be to choose a pH at which the protein self association is not so ionic strength dependent. Another approach might be to use other means to determine the *kind* of ion bound, such as ion-specific electrodes. In the next section the measurement of ions bound as a function of ionic strength will be discussed, and this provides an additional way to determine the. change in ions bound at the two ionic strengths used for equilibrium dialysis.

Ion Binding with Varying Ionic Strength

A protein can be titrated continuously with concentrated electrolyte, at constant pH. The concentration of the titrant should be such that the total volume change by the end of the titration does not exceed 10–15%; using two different titrants the useful range of ionic strength can be covered adequately, as shown in Fig. 2 with bovine serum albumin. Each protein is in essence an individual case as far as net charge, kind of ions bound and related parameters are concerned, and explicit equations must be written in each instance to describe the particular situation. To illustrate the method we use a protein positively charged which binds chloride ions and is titrated with potassium chloride. The initial state is:

$$M_{p^{*}}^{z^{*}} + (Z^{*} \cdot M_{p^{*}})_{\text{Cl}^-} + M^{*}_{\text{KCl}} \tag{17}$$

A solution of a protein from a dialysis equilibrium experiment is an excellent one to use since all quantities in Eq. (17) are known. After an aliquot of a concentrated electrolyte is added to the above solution, it becomes:

$$M_{p_1}^{(Z^{*}-\nu_1-j_1)} + M_{\text{K}_1^+} + (M_1 - \nu M_{p_1})_{\text{Cl}_1} + (Z^{*}M_p)_{\text{Cl}^-} \tag{17-1}$$

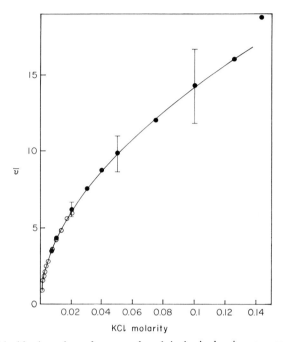

Fig. 2. Chloride ions bound per mole of isoionic bovine serum albumin as a function of KCl molarity. The limits describe the maximum overall deviation; such deviation was calculated by using maximum deviation values in every step of the procedure. ○—○ = Conductance titration up to 0.02 M KCl. ●—● = Conductance titration from 0.0075 M to 0.144 M KCl. From E. A. Sasse, Ph.D. Thesis, University of Tennessee, 1968.

M_1 is the new predetermined, calculated concentration of electrolyte, and v_1 are the chloride ions bound in going from state (*) to state (1). The actual concentration of unbound or free electrolyte is not known *a priori*, and conductance is used to determine this. The conductance equations for solutions (17) and (17-1) are written as:

$$K'_* = \lambda_{p*}(ZM_p)_* + (\Lambda_* - Q_*C_{p*})\left[M_{*\mathrm{KCl}} + \frac{\lambda^\circ}{\Lambda^\circ}(ZM_p)_* \right] \qquad (18)$$

$$K'_1 = \lambda_{p*}\frac{\Lambda_{ps_1}}{\Lambda_{p*}} \cdot \frac{V_*}{V_1} \cdot M_{p*}|Z_1|$$
$$+ \frac{\Lambda_1}{\Lambda_*} \cdot \left(\Lambda_* - Q_* \frac{V_*}{V_1} \cdot C_{p*}\right) \cdot \left[M_{1,\mathrm{KCl}} + \frac{\lambda^\circ}{\Lambda^\circ} \cdot \frac{V_*}{V_1} \cdot (ZM_{p_1})_{\mathrm{Cl}} \right] \qquad (18\text{-}1)$$

The ratio V_*/V_1 is used to correct for dilution differences in states (*) and (1), due to addition of titrant. The equivalent conductance ratios are used to account for ionic strength changes between two states. The correct values of Λ_1 and Λ_{ps_1} are obtained iteratively since they depend on the exact value of ionic strength. The quantity in brackets in (18) and (18-1) is the quantity of M_s of Eq. (7). Equations (18) and (18-1) are simply a detailed representation of Eq. (9-1). The quantity Z_i is an absolute value, and it emphasizes that protein contributes to conductance regardless of the kind of charge it carries. It should be emphasized here that a protein may contribute to conductance as its net charge is increased, something likely to happen as the number of ions bound increases. One way to get an idea of the extent of conductance of the protein is to assume that its equivalent conductance is directly proportional to its diffusion coefficient, and then use the values for lysozyme in the proportionality calculations. These values for lysozyme are: $\lambda_p = 20\text{--}25$ ohm^{-1} cm^{-1} equiv^{-1} and $D = 1 \times 10^{-6}$ at 20°C. Although the reliability of this assumption may be still questionable since it has not been substantiated by a large number of tests with proteins, the estimated conductance is probably close to the upper limit, a conservative estimate. From a simplified direct relationship of mobility and conductance, the value of equivalent conductance would in all likelihood be one half or less:

$$\lambda_i = F \cdot U_i \qquad (19)$$

where F is the Faraday and U_i is the mobility in cm^2 volt^{-1} sec^{-1}. Another point that must not be overlooked is that, as ions bind, they are removed from solution so the term (νM_p) is always subtracted from the added electrolyte. Closely related is the meaning of ν as it is used here. This describes the difference in the number of ions bound in going from state (*) to state (1), a relative number, unique to the experimental conditions. Its relationship to Z^* has already been elaborated upon.

Finally, such a binding study at constant pH may be an exploratory one and precede a dialysis experiment. If the experimenter intends to study ion binding as a function of ionic strength, it is advisable to carry one such binding study prior to equilibrium dialysis. It will provide an upper limit for a value of Q, as such direct titration data will give in case of ion binding. The slope of a graph of Eq. (18-1) will give an "apparent" value of Q. If the number of ions bound does not change appreciably with ionic strength (i.e., in cases of very strong or very weak ion binding) the slope will give a value of Q^* which is close to the true one. In any case, the value of Q^* from a slope of a plot of Eq. (18-1) should be equal or *less* than the Q values that can be determined from single point determinations—excluding protein conductance—obtained from data of the same experiment. Such discrepancies in Q values are unmistakable indications of ion binding. Thus, a very simple experiment such as described here, when carefully analyzed, provides information to plan the whole ion binding study.

Determination of Isoelectric Point and pH Dependent Ion Binding

Proteins That Do Not Bind Ions

The method is outlined by using a simple example, that of a positively charged protein that binds no ions in the pH range under investigation. The supporting electrolyte in the example is potassium chloride. In order to avoid cumbersome notation, small letters are used to signify concentrations of the various species, with the understanding that appropriate expressions which have been given above are used to convert specific conductance to concentrations, equivalent conductances, and so forth.

To a protein of initial charge Z_*^+ accompanied by its counterions and in supporting electrolyte such as KCl of concentration (g) is added potassium hydroxide and the initial pH_* is changed to a new pH_i:

$$P^{z*} + Z^*Cl + gKCl + (j + i)KOH \rightarrow P^{(z*-i)} + Z^*Cl^- + jK^+ + gKCl + iK^+ + iOH^- \quad (20)$$

where j is the number of moles of protons removed per mole of protein and i is the number of moles of free OH^- to maintain pH_i. If to an aliquot of solvent of $gKCl$ concentration and of the same volume as the protein solution are added the following, to make up the reference solution:

$$jK^+ + jCl + iK^+ + iOH^- + gKCl \quad (20-1)$$

and the concentrations of the reference solution are subtracted from the right-hand side of the expression (20), there results:

$$P^{(z*-i)} + (Z^* - j)Cl^- \quad (20-2)$$

when $j = Z^*$, the net charge of the protein is zero, and $(Z^* - j)Cl^- = 0$, which is the isoelectric point of the protein. The main difference between the protein and reference solutions is that instead of adding $(j + i)$ moles of KOH to the reference, j moles of KCl and i moles of KOH are added. The following is a brief description of how the experiment is carried out. Figure 3 gives a schematic summary of the process. A continuous pH titration of the protein is carried out in a pH range of about $+2$ and -2 pH units about the expected isoelectric point. This determines the quantity of $(j + i)KOH$ as a function of pH. A blank pH titration of the solvent $(gKCl)$ determines the quantity $iKOH$, or this quantity is determined from a previously constructed curve. Thus, the quantity $jKOH$ and of course $jKCl$ are determined, at various chosen pH values that the experimenter considers appropriate. Next, the conductance of another

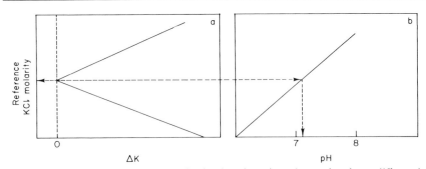

Fig. 3. Graphical evaluation of the isoelectric point of proteins from differential conductimetry and pH titration data. (a) Molarity of KCl in the reference solution versus the absolute difference in specific conductance between the protein and reference solution to determine the molarity of KCl at $\Delta \bar{K} = 0$. (b) Reference of KCl molarity to pH titration data. Value at $\Delta \bar{K} = 0$ gives the isoelectric pH. From E. A. Sasse, Ph.D. Thesis, University of Tennessee, 1968.

aliquot of a protein solution of the same volume is measured continuously as the $(j + i)$ moles of KOH are added, to correspond to the pH values chosen. Also, the procedure is repeated with the reference solution to which jKCl and iKOH are added. The conductance of the protein solution needs to be corrected to a value that would represent the true conductance of the electrolyte present, as if the protein were absent. That is, it is corrected for the protein nonelectrolyte effect, protein conductance, etc. The equation used to get K'_c, the corrected specific conductance at pH_i is:

$$\bar{K}'_c = \bar{K}' + Q_s \cdot C_p \cdot M_{si} + Q_{\mathrm{KOH}} \cdot C_p \cdot M_{\mathrm{KOH}(i)} - \lambda' M_{pi} \cdot Z_i \quad (21)$$

The Q correction for acid or base is used when significant, and the correction for protein conductance becomes insignificant as the isoelectric point is approached.

The difference $(\bar{K}'_c - \bar{K}_{\mathrm{ref}}) = \Delta \bar{K}$ is thus a direct measure of the concentration of the ions in the expression (20). Divided by the protein concentration, this in fact is the net charge of the protein at any pH_i. If ions of charge like that of the protein bind, an additional correction is necessary for this to be true, as discussed below. When $\Delta \bar{K} = 0$, the value of $(M_{j_{\mathrm{KCl}}})$ is read off a graph such as in Fig. 3a, the same value in $M_{j_{\mathrm{KCl}}}$ is located from a graph as in Fig. 3b, and the corresponding pH is the isoelectric point of the protein.

It is important to point out that, in principle, it is not necessary to perform experiments with the reference solution because the concentration of the supporting electrolyte $(g + j)$ and free acid or base are known, and so are the equivalent conductance values at their respective concentrations. In practice, it is not only more convenient but far more accurate to obtain the conductance values with identical volumes, settings of the microsyringe, etc., to those used for the protein solution. The important fact is that one can obtain the concentration of ions related to the protein in some manner, apart from the supporting electrolyte. This is the basis for calculating not only the isoelectric point in the more realistic cases discussed below, where ions bind or are released as the pH changes, but also calculating the number of ions bound or released as a function of pH.

Isoelectric Point of Proteins Binding Ions

With proteins which bind ions, it is necessary to know what kind of ions they bind at the pH at which the titration is started. Thus, there are only two distinct cases to be discussed, first the proteins which bind ions of opposite charge and the second is those which bind ions of like charge. Table IV gives in detail all the cases possible, and it states whether a true or false isoelectric point will be obtained *if* the criterion $\Delta \bar{K} = 0$ is used. On the basis of this criterion, all cases where the ion bound is of opposite charge will give the true pI. Another fact needs to be established, and this is whether or not the bound ions can be removed by pretreatment such as exhaustive dialysis, deionization on a column, etc. The reason for this distinction is to know whether the protein, besides its counterions, brings with it into the solution other ions. In any case, this fact does not invalidate the determination of the isoelectric point, but it bears on the conclusions with regard to ion binding.

We shall discuss only two cases in order to illustrate the calculations involved. The results of individual cases will indicate the category in Table IV in which a particular protein belongs.

Case 1. In case 1a a cationic protein binds anions; such ions are removable by pretreatment. When bound at a particular electrolyte concentration, the ions are not released up to the value of pI. The titration, reference concentration, and subtraction of the two are as follows:

Initial: $\text{P}^{z*} + Z^*\text{Cl}^- + g\text{KCl} \rightarrow \text{P}^{(z*-b)} + (Z^* - b)\text{Cl}^- + g\text{KCl}$

Base addition: $\text{P}^{(z*-b)} + (Z^* - b)\text{Cl}^- + g\text{KCl} + (j + i)\text{KOH} \rightarrow$

$\qquad\qquad\qquad \text{P}^{(z*-b-j)} + (Z^* - b - j)\text{Cl}^- + j\text{KCl} + g\text{KCl} + i\text{K}^+ + i\text{OH}^-$

Reference: $g\text{KCl} + j\text{KCl} + j\text{KOH}$

$\Delta \bar{K}$: $\text{P}^{(z*-b-j)^+} + (Z^* - b - j)\text{Cl}^-$ (22)

The reader may have noticed the artificiality of having two quantities

TABLE IV

Effect of Ion Binding on pI Determination[a]

		Type binding			
Protein charge at beginning pH of pI determination	Ion bound	pH independent	pH dependent binding increase or decrease	Bound ions removable by pretreatment	pI given by method at $\Delta\bar{K} = 0$
1 a. Cationic (+)	Anion, Cl⁻	x		Yes	True
b. Cationic (+)	Anion, Cl⁻	x		No	True
c. Cationic (+)	Anion, Cl⁻		x	Yes	True
d. Cationic (+)	Anion, Cl⁻		x	No	True
2 a. Anionic (−)	Cation, K⁺	x		Yes	True
b. Anionic (−)	Cation, K⁺	x		No	True
c. Anionic (−)	Cation, K⁺		x	Yes	True
d. Anionic (−)	Cation, K⁺		x	No	True
3 a. Cationic (+)	Cation, K⁺	x		Yes	False
b. Cationic (+)	Cation, K⁺	x		No	True
c. Cationic (+)	Cation, K⁺		x	Yes	False[b]
d. Cationic (+)	Cation, K⁺		x	No	False
4 a. Anionic (−)	Anion, Cl⁻	x		Yes	False
b. Anionic (−)	Anion, Cl⁻	x		No	True
c. Anionic (−)	Anion, Cl⁻		x	Yes	False[a]
d. Anionic (−)	Anion, Cl⁻		x	No	False

[a] In this table it is assumed that it is not known what kind of ion binds and that a pI titration is carried out as described in the text. Note that in the cases where a false pI is given by the regular calculations, calculations which take into account that a cationic protein binds a cation or an anionic protein binds an anion will give the true pI. The manner for carrying out such calculations is described in the text.

[b] This situation will give a true pI for the special case where the ion binding decreases as the pH → pI and at the pI all the original ions bound are released.

for KCl, but again, this is done for illustrative purposes to emphasize that it is essential to account for all ions present.

In Eq. (22), the quantity $\Delta\bar{K}$ always gives the concentration of ions related in some manner to the protein charge and the ions bound or released with change in pH. In the specific case above and in all cases where a protein binds ions of opposite charge, this quantity gives the net charge of protein at any pH. When $j = (z - b)$ above, this pH is the isoelectric point. Let us now look at a case identical to the above in all respects except that at the isoionic point, all bound ions are released. The equations are:

$$P^{(z*-b)+} + (z* - b)Cl^- + gKCl + (j + i)KOH \rightarrow$$
$$P^{(z*-i)+} = bCl^- + (z* - b)Cl^- + gKCl + jK^+ + iKOH$$

Reference: $gKCl + jKCl + iKOH$

Δ: $\qquad P^{(z-i)+} + (z - b + b - j)Cl^-$ or $P^{(z-i)+} + (z - j)Cl^-$ $\qquad$ (23)

Since the pI is at the pH where $z = j$, $\Delta\bar{K} = 0$ at that pH also. The true isoelectric point is obtained.

Let us look now at one of the ways to determine pH-dependent ion binding or release. The initial concentration difference is known; it is labeled Δ^*. In the cases 1a and 1c of Table IV, this is $\Delta^* = (z - b)Cl^-$. Also the quantity jK^+ or jCl^- is known since this is the amount added. The quantity Δ_{concj} is defined so that:

$$\Delta_{concj} = \Delta^* - jCl^- \qquad (24)$$

and, with some corrections for dilution, it can be calculated at any pH. The quantity $\Delta\bar{K}$ obtained by conductance may or may not be identical to Δ_{concj}. In case 1a it is the same, but in case 1c it is:

$$\Delta\bar{K} - \Delta_{concj} = (z - b + b - j)Cl^- - (z - b - j)Cl = bCl^- \qquad (25)$$

Thus, the difference in Eq. (25) gives the concentration of ions released or bound by the protein. This is one of the ways to obtain this quantity with change in pH. Expression (24) becomes more important in the cases where protein binds ions of like charge, which is discussed below.

Proteins That Bind Ions of Like Charge

An additional requirement in this case is that the number of ions bound be known at the starting pH and ionic strength of the titration. This number is obtained by equilibrium dialysis, and one can dialyze enough protein to use for the pI titration. Also, a study of ions bound at constant pH, such as the pH used for equilibrium dialysis, would give the number of ions bound at different ionic strengths. With such information available, the pI titration can be conducted at an ionic strength different from that of the equilibrium dialysis. Let us examine first case 3a of Table IV, which is the same as case 1a except that ions of like charge bind. The equations are:

$$P^{*z+} + z*Cl^- + gKCl^- \rightarrow P^{(z+b)} + z*Cl^- + (g - b)K^+ + gCl^-$$

Add base: $P^{(z*+b)} + z*Cl^- + (g - b)K^+ + gCl^- + (j + i)KOH \rightarrow$
$$P^{(z+b-i)^+} + z*Cl^- + jK^+ + (g - b)K^+ + gCl^- + iOH^-$$

Reference: $\qquad jKCl + iKOH + gKCl$

$\Delta\bar{K}$: $\qquad P^{(z*+b-i)} + (z* - j)Cl^- + (-b)K^+$ $\qquad$ (26)

Since the ions are not released with change in pH, the isoelectric point should be in fact at the pH at which $j = Z$, however, at the pH at which $\Delta \bar{K} = 0$, $j = Z^*Cl^- - bK^+$, that is if the pH at which the conductance difference is zero were taken as the isoelectric point, it would be false. If the initial value of b is known, however, it is easy to determine the isoelectric point by the following procedure. The titrations are carried out exactly as before, only the calculations are somewhat different. The initial concentration of the reference electrolyte could still be $gKCl$, however, the calculations are neater if a concentration $(g - b)KCl$ is used. A more important reason is that in such cases, one would often use protein solution from a dialysis equilibrium experiment, in which case the supporting electrolyte concentration is indeed $(g - b)KCl$. It is easy to show that in the case 3a discussed above, if $(g - b)KCl$ were used the true isoelectric point would be obtained. This is a trivial case, and a more illustrative example is case (3c), which is like case (3a) except that the "b" cations bound initially are released before or at the isoelectric point. The initial states are:

Protein: $P^{(z^*+b)} + (z^* + b)Cl^- + (g - b)KCl$

Reference $(g - b)KCl$

$\Delta\bar{K}$: $(z^* + b)Cl^-$ (27)

If upon adding $(j + i)KOH$ the b cations are released, the equations are:

Protein: $P^{(z^*-i)} + (z^* + b)Cl^- + bK^+ + jK^+ + (g - b)KCl + iKOH$

Reference: $(g - b)KCl + jKCl + iKOH$

$\Delta\bar{K}$: $P^{(z^*-i)} + (z^* - j)Cl^- + bKCl$ (28)

However, writing out $\Delta_{conc}j$ we find that:

$$\Delta_{conc}j = (z^* + b)Cl^- - jCl^-$$

and so:

$$\Delta b\Delta = \bar{K} - \Delta_{conc}j = [(z^* - j)Cl + bKCl] - [(z^* + b)Cl^- - jCl^-]; \quad \Delta b = +bK^+ \quad (29$$

Therefore, the way for correcting the conductance difference to give the true isoelectric point is as follows: (1) Use reference electrolyte $(g - b)KCl$. (2) Measure $\Delta\bar{K}$ as previously. (3) Calculate Δb to determine the concentration of ions bound or released in the course of titration. (4) Calculate what would be the specific conductance of a concentration "b" of supporting electrolyte (such as KCl), using the equivalent conductance of the salt at the *total* ionic strength at pH_i. Note that although bK^+ ions are released, the equivalent conductance to be used is for the salt (KCl) and *not*, in this case, just the ionic conductance of the cation (K^+). (5) If cations are released, *subtract* the specific conductance calculated in step (4) from the experimental $\Delta \bar{K}$; if more cations are bound, that is $\Delta b = -bK^+$, *add* the calculated specific conductance to the experimental $\Delta \bar{K}$. (6) Use this corrected $\Delta \bar{K}$ in constructing the graph for the determination of the pI. In the case (3c) we just considered, it is seen that subtracting the conductance of bKCl from the experimental $\Delta \bar{K}$ would give:

$$\Delta \bar{K}_{\text{corr}} = (z - j)\text{Cl}^- \tag{30}$$

so that the isoelectric point is again at $j = z$, which is correct. In the previous case (3a) it is obvious that $\Delta b = 0$ would be calculated throughout the titration, and therefore no corrections were necessary. Thus, the difficulties encountered with the binding of ions of like charge are mainly a matter of additional arithmetic.

In examining Table IV, it should be noted that if titration with acid would give a false pI (if the corrections suggested are not made) because ions of like charge bind, a titration of the same protein with base would give the correct pI. In other words, a discrepancy between the pI values obtained by titration with acid and with base would indicate that ions of like charge bind during one of the titrations.

Examination of equations (26) through (29) will tell the experimenter whether the "wrong pI" obtained is less or more than the true pI value. To repeat, it is very wise to write out the relations involved, since what is "intuitively obvious," in this case, is often wrong; at least that has been our experience.

Ion Binding as a Function of pH

The data from the titration for the determination of pI can be utilized to determine ion binding as a function of pH. The difference in expression (24a) and (29) gives the overall change of ions bound from the initial pH(*) up to pH(j). The change with pH of ions bound per mole of protein is obtained by dividing the overall change by the molarity of protein. A way of obtaining the actual number of ions bound at any pH is to subtract the average protein charge Z from the protonic charge Z_{H}. The protonic charge is known from the pH titration data and the isoelectric point, plus the knowledge of ions bound at the isoelectric point. The net charge Z is given by equations such as (20), (22), and (29). The change therefore in going from pH(i) to pH(k) can be represented in general:

$$\Delta \nu_{ik} = \Delta \bar{r}_{ik} - \Delta Z_{ik} \tag{31}$$

where $\bar{r}$ is the number of protons removed or bound per mole of protein.

Another way of representing the data for the determination of the isoelectric point is as follows. At any pH$_j$ the conductance difference is:

$$\Delta \bar{K}_j = \lambda_{j\text{Cl}}(M^*_{\text{Cl}} - M_{bj\text{Cl}} - M_{j\text{Cl}}) \tag{32}$$

Equation (32) applies when a protein binds or releases anions. The right-hand side of Eq. (32) can be directly proportional to the number of moles of the component if Eq. (32) is multiplied through by the total solution volume V_j:

$$V_j \Delta K_j = \lambda_{j\text{Cl}}(m^*\text{Cl} - m_{bj\text{Cl}} - m_{j\text{Cl}^-}) \tag{32a}$$

In Eq. (32a), m stands for the number of *moles* of the particular component, not molality. Subtracting Eq. (32a) at two pH values, pH_i and pH_k, there results:

$$V_k \Delta \bar{K}_k - V_i \Delta \bar{K}_i = -\lambda_{ik\text{Cl}}(m_{k(\text{Cl}^-)} - m_{i(\text{Cl}^-)}) - \lambda_{ik\text{Cl}}(m_{bk\text{Cl}} - m_{bi\text{Cl}}) \tag{32b}$$

The conductance λ_{ik} represents an average of the values at the ionic strengths in solutions of pH, and this is adequate in most cases where i and k are fairly close. The change in moles of ions bound can be calculated directly by:

$$\Delta m_{bik(\text{Cl}^-)} = m_{bk} - m_{bi\text{Cl}} = -(m_{k(\text{Cl}^-)} - m_{i(\text{Cl}^-)}) - \frac{(V_k \Delta \bar{K}_k - V_i \Delta \bar{K}_i)}{\lambda_{ik(\text{Cl}^-)}} \tag{33}$$

Equation (33) has the advantage that a graph of $V_i \Delta \bar{K}_i$ versus $m_{(\text{KCl})}$ gives readily the change in moles of ions bound, and of course also the isoelectric point.

We have given here the basic relations for a model study correlating pH titration, measurement of ions bound as a function of pH at (relatively) constant ionic strength, ions bound as a function of ionic strength at constant pH, and the isoelectric point at one or more ionic strengths. The model equations may be adapted easily to fit specific situations.

As a final account before description of the experimental techniques, the cases of carbomonoxy human hemoglobin and hen's egg white lysozyme will be presented to illustrate some basic calculations. A more complete description of the studies with these two proteins has been given elsewhere.[4] Only select data are given here for calculations which are more likely to be ambiguous or complicated.

Hemoglobin. Hemoglobin is simple to handle with this technique. At its isoionic point it binds no ions, and around pH 7 there is no need for corrections of Q for acid or base used. Yet hemoglobin provides a wealth of information, some of which is presented here to provide experience in handling data. First, the isoionic point of human carbomonoxy hemoglobin in the ionic strength range 0.02–0.1 KCl, is about 7.2 at 25°, but it is about 7.5–7.6 at 7°. This simple fact shows why isoelectric points obtained at low temperatures are different from those at room temperature. This is necessarily so for all proteins with isoelectric points in the neutral and basic regions, because the groups which ionize in this pH range have an enthalpy of ionization which is fairly high. As a matter of fact, this is the main reason which led to the establishment of the method described here.

Examination of Fig. 4 shows that not all slopes of the lines are the same. At $0.05\,M$ KCl the ionic equivalent conductance for chloride is about 68 whereas that of the potassium ion is about 66. The slope of the titration with potassium hydroxide beginning at pH 5.47 is only 50. Also, the curve produced by titration with acid has two different slopes: the slope from pH 9.43 to the isoelectric point is normal, but the slope of the portion below the isoelectric point is about 78. Both these slopes in the acid region tell us that: ions are released as the pI is approached from the acid region, or (the same statement in other words) ions are bound as the pH is decreased below the isoelectric point. Calculations show that between pH 7.2 and pH 5.47, about four ions bind. The question of which ion binds remains to be answered. To carry out dialysis below the isoelectric point is unwise because of the instability of hemoglobin. However, the fact that the isoionic point increases slightly with increasing ionic strength suggests that potassium ions bind. It should be noted that no ions bind above the isoelectric point, which of course coincides with the isoionic pH of carbomonoxyhemoglobin of the same protein concentration and ionic strength. Ion binding by hemoglobin

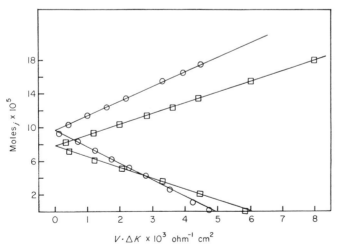

FIG. 4. Hemoglobin pI determination at $0.05\,M$ KCl ionic strength and 25°. ○—○ = pI titration with KOH beginning at pH 5.47; pI = 7.26. □—□ = pI titration with HCl beginning at pH 9.43; pI = 7.26. From E. A. Sasse, Ph.D. Thesis, University of Tennessee, 1968.

below its isoelectric point is reported here for the first time, and more experiments are needed to establish the facts.

Lysozyme. Lysozyme on the other hand presents the most complex example we have studied. The isoelectric titration is shown in Fig. 5. It should be noted that it is a complex curve and the fact that it is not a straight line shows why a good number of points are needed near the pI region. Lysozyme also releases chloride ions as its pI is approached. At about pH 6.7 it binds 2.5–3.0 chloride ions which are gradually released as the pH increases. Lysozyme was positively charged during dialysis because its isoelectric point is near pH 11, and therefore dialysis experiments established that ions of charge opposite to that of the protein were bound. The equilibrium dialysis data for lysozyme were analyzed in the manner exemplified by Table III. Because of the highly basic pH necessary, it is difficult to obtain reliable data above the isoelectric point. However, these pieces of evidence point out that no ions bind to lysozyme at its isoelectric point. First the slope of Fig. 5 above the pH of the isoelectric point is close to the equivalent ionic conductance of chloride ions. The pH of an isoionic solution of lysozyme shows that the pH of $Z_H = 0$ is the same as the isoelectric point, assuming no ion binding at about pH 11. The Z of lysozyme at pH 6.6–6.7 obtained by

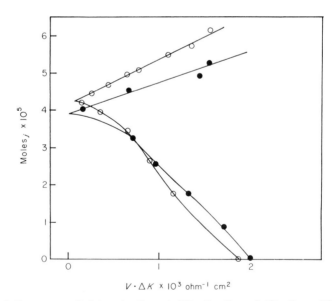

Fig. 5. Lysozyme pI determination at 25°. ●—● = pI titration with KOH beginning at pH 9.51 and 0.02 M KCl ionic strength; pI = 10.98. ○—○ = pI titration with KOH beginning at pH 9.50 and 0.05 M KCl ionic strength; pI = 10.94. From E. A. Sasse, Ph.D. Thesis, University of Tennessee, 1968.

equilibrium dialysis is the same as the Z obtained from the pI conductimetric titration, assuming no ions bound at the isoelectric point. The examples given here show the major difficulties one is likely to encounter.

Materials and Methods

The methods of pH titration and conductivity measurement are standard enough so that only the experimental details which are pertinent here will be discussed. The article of Nozaki and Tanford[1] describes pH titrations, and the article of Craig[5] provides information for conducting dialysis. The kind of dialysis described[5] is of a more sophisticated nature than the one used here; however, the precautions and methods of preparation apply here equally well. A wide variety of instruments is available, and the instruments we used are mentioned simply to indicate the level of accuracy adequate enough for the experiments.

The water was passed through Amberlite MB 1 mixed-bed resin and then distilled from an all-glass apparatus; it had a minimum specific conductivity of 1×10^{-6} int. ohm^{-1} cm^{-1}.

All solutions, whenever possible, were prepared by weight, and the volume was calculated by the formula for density in Table II or, if available, another relationship, more exactly fitting the case. For potassium chloride, the densities agreed within 0.01% of those taken from a graph of literature values.[6] The agreement of equivalent conductances was within 0.1% of those in the literature.[7,8] All weight values used were corrected to vacuum. Standard acid was prepared from constant-boiling hydrochloric acid.[9] Carbonate-free potassium hydroxide was prepared from the $1 N$ potassium hydroxide solution available from Fisher Scientific, and it was standardized against the standard acid or potassium acid phthalate.

It is convenient to classify titrations as one of four types. Types A and B are those for the determination of the isoelectric point. Types C and D are used for the determination of Q and ion binding at constant pH. All titrations are of the continuous type, either automatic by using a titrigraph as in type A or stepwise. Type A is a pH titration of the protein solution, and of the reference electrolyte solution to determine the concentration of free protons. Type B is a conductimetric titration

[5] L. C. Craig, see Vol. 11, p. 870.
[6] International Critical Tables (E. W. Washburn, ed.), Vol. 1, p. 87. McGraw-Hill, New York, 1928.
[7] T. Shedlovsky, J. Amer. Chem. Soc. 54, 1411 (1932).
[8] J. F. Chambers, J. M. Stokes, and R. H. Stokes, J. Phys. Chem. 60, 985 (1956).
[9] G. A. Hulett and W. D. Bonner, J. Amer. Chem. Soc. 31, 390 (1909).

of a protein and a reference solution at pH values determined and selected from a type A titration. In type C, a protein or reference solution is titrated with a fairly concentrated standard solution of an electrolyte and the conductance is measured after each addition of the electrolyte. Type D is also a conductimetric titration, but the titrant here is a fairly concentrated protein or other nonelectrolyte solution at the same electrolyte concentration as the electrolyte solution which is titrated. Type D gives the conductance of the electrolyte at constant ionic strength as a function of the nonelectrolyte concentration.

Titration Type A. Only the practices that are different somewhat from those of Nozaki and Tanford[1] will be described here. The reference calomel electrode assembly is constructed as follows. The calomel electrode is an open type and is immersed in saturated KCl in a short test tube with a side arm. The test tube is stoppered loosely with a one-hole, sulfur-free grooved rubber stopper. The side arm is connected with a piece of tygon tubing to the glass tubing that is immersed in the solution to be titrated. The glass tubing, 2–4 mm in diameter and 10–15 cm long, is prepared as follows. Asbestos fibers, which have been cleaned with aqua regia, are inserted into glass tubing and the tubing is drawn out over a flame and cut. Thus, the asbestos forms a plug, a fraction of a millimeter in diameter. Many such tubes are prepared, and from these a few are chosen that conform to the leakage requirements of the experimenter. The leakage is tested by filling the tubes with water and applying pressure from a compressed air line. The amount of leakage can be tested by measuring the increase in conductance of water in which such a bridge filled with electrolyte is immersed. There are two main advantages of this arrangement. First, the calomel electrode and tygon tubing are filled with KCl at room temperature, whereas the bridge glass tubing is filled with KCl saturated at the desired temperature of the titration. Second, KCl leakage is reduced to a minimum by mounting the test tube with side arm on a stand, so that the level of the saturated KCl solution can be adjusted to only a fraction of an inch higher than the level of the solution titrated.

For pH titrations a Radiometer type SBR2 Titrigraph was used, combined with a type SBU 1 syringe burette, a type TTT 1 titrator and Radiometer electrodes. A variety of thermostated vessels were used. The solution to be titrated was delivered to the vessel from a weighing burette. Inert atmosphere over the solution was maintained by passing Matheson prepurified nitrogen, which was washed through carbon dioxide, base, and acid traps and finally water. Constant temperature for pH titrations was maintained with a type FSe Haake thermostat. Blank electrolyte titrations were carried out with every pH titration.

Titration Type B. This is a conductimetric titration and its end product is pictured in Figs. 4 and 5. From a titration type A, 6 to 12 pH values are chosen about the expected isoelectric point. A protein as well as a reference solution is titrated. Acid or base and/or electrolyte are added to produce a chosen pH and ionic strength, and the conductance is measured. Table V shows from titrations of type A and type B.

In addition to a thermostated vessel as in type A, a peristaltic pump, a conductivity cell, and a good thermostat bath are needed. This is because the titrants are added to the solution in a vessel as in type A, but the protein solution is then circulated with the pump through the thermostated conductivity cell. After mixing, circulation is stopped and after temperature equilibration the conductance is measured. More titrant is added, and the cycle is repeated. A low heat producing peristaltic Sigmamotor kinetic clamp pump was used. A small-bore polyethylene tubing was used for the connections between the titrating vessel and the conductivity cell except for the portion used with the peristaltic pump, which was flexible and of appropriate bore. The time of circulation of solution and waiting to reach temperature are best determined with a blank titration. However, during every titration the conductance is read a number of times until readings, random within the precision sought, are obtained.

For conductance measurements an LKB 3216B precision conductivity bridge was used with a Wagner ground and an external, precision, calibrated variable 1 ohm resistor, or with an external decade resistor, 10^4 to 10^5 ohms in steps of 10,000 ohms. The resistance readings were obtained at 2000 cycles per second and were accurate to within ±0.1 ohm over the range 0.1–11,110 ohms with the external 1 ohm resistor and to within ±0.1% over the range 100–111,110 ohms with the external 100,000 ohm decade resistor. The LKB conductivity cells of the Shedlovski type were calibrated with the Jones and Bradshaw 0.01 demal potassium chloride solution.[10] For cells with very low cell constant, 0.001 N potassium chloride was used. The use of an oil bath was avoided by insulating the conductivity cells themselves. The cells were coated with three to four layers of clear epoxy glue. Properly coated cells gave no or negligible difference in resistance readings at the two frequencies of 1000 and 2000 cycles per second.

A thermostated bath that could be controlled to ±0.002° was built as follows. A 10-gallon aquarium tank was inserted into a larger, 15-gallon tank, and the gap between them closed at the top with 1 inch thick pieces of plastic, thus maintaining air insulation between them.

[10] G. Jones and B. C. Bradshaw, *J. Amer. Chem. Soc.* **55**, 1780 (1933).

TABLE V

ISOELECTRIC POINT DETERMINATION OF LYSOZYME, 25°

Part 1. Conductimetric Titration of Protein

pH[a] 1	KOH used $N = 0.20150$ (ml) 2	Total volume of the protein solution (ml) 3	Resistance (R) (ohms) 4	$\bar{K}'$ specific conductance protein solution[b] 5	Molarity of free (OH^-) 6
9.50	0.00	10.00	1314.33	6.79776	5.888×10^{-5}
10.41	0.150	10.15	1287.36	6.94017	3.981×10^{-4}
10.85	0.250	10.25	1259.34	7.0708	1.047×10^{-3}
10.94	0.275	10.275	1250.85	7.10874	1.259×10^{-3}
11.03	0.30	10.30	1241.83	7.16462	1.549×10^{-3}

Part 2. Conductimetric Titration of Solvent

pH 1	KOH to produce pH (ml) 7	0.2015 N KCl: columns 2-7 (ml) 8	M_{KCl} 9	Resistance (R) of KCl solution (ohms) 10	$\bar{K}$ of column 10 11	$K'p$ due to protein[c] 12
9.50	—	—	0.049722	1344.34	6.64601	0.09158
10.41	2.005	0.12995	0.051567	1285.27	6.95145	0.0721
10.85	5.325	0.19675	0.052377	1240.61	7.20169	0.01787
10.94	6.422	0.21080	0.052525	1229.28	7.26807	0.00091
11.03	7.916	0.22085	0.0525933	1216.76	7.34286	0.00444

Part 3. Calculations with Data from Parts 1 and 2

pH 1	Protein concentration (C_p) (g/100 ml) 13	KCl correction $Q \times M_{KCl} \times C_p$ 14	KOH correction $Q \times M_{KOH} \times C_p$ 15	K' correction, columns $5 + 14 + 15 - 12$ 16	ΔK columns $16-11$ 17	$V \times \Delta K \times 10^3$ columns 3×17 V in liters 18
9.50	1.381	0.14887	3.19800	6.85537	0.20936	2.0936
10.41	1.360	0.15204	0.00213	7.02224	0.07079	0.71852
10.85	1.347	0.15296	0.00554	7.21143	0.00974	0.099835
10.94	1.344	0.15305	0.00665	7.26753	−0.00054	−0.005548
11.03	1.341	0.152903	0.00817	7.32125	−0.02161	−0.22258

TABLE V (Continued)

[a] Initially a pH titration with base, starting at pH 9.50, was carried out. A set of pH values was chosen from the pH titration, and they were used for the conductimetric titrations. A few pH points are reproduced here as examples. The same pH values are repeated in all three sets to facilitate reading the table. Part 1 contains data from the conductimetric titration of the protein solution with 0.20150 N KOH. Part 2 is the conductimetric titration of "blank" 0.049722 M KCl at pH 9.50 with both 0.20150 N KOH (to produce the desired pH) and electrolyte of the same normality, i.e., 0.20150 N KCl. All the pH values used in this titration were: 9.50, 10.18, 10.41, 10.65, 10.85, 10.94, 11.03, 11.10, 11.17, 11.23, 11.35, 11.45, 11.57, and 11.66.

[b] The units of specific conductance are in millimhos per centimeter. The cell constant in this case was 8.9345 cm^{-1}. The specific conductance of column 5 was obtained by dividing the cell constant by the resistance in column 4. The protein specific conductance is then corrected as it appears in column 16 by using the following Q values: for 0.0527 M KCl, $Q = 2.17$; for KOH, $Q_{KOH} = 3.93$. The low base concentrations do not warrant correcting the Q of the base for concentration changes by Eq. (5-1) unless the base concentration becomes high. Here, the Q_{KOH} was obtained indirectly, by use of Eq. 6.

[c] The equivalent conductance of lysozyme under these conditions was 23.9, and the net charge Z at pH 9.0 was $Z = 4.5$. The following Z values were used for the pH values in the table: pH 9.50, $Z = 4$; pH 10.41, $Z = 3.2$; pH 10.85, $Z = 0.8$; pH 10.94, $Z = 0.4$; pH 11.03, $Z = 0.2$. The Z values were calculated by iterating twice the Z values obtained as explained in the text.

Most of the top was covered with wood board, which served as a support for the various immersed parts. The cooling, heating, and stirring was located at one end of the bath and the thermostat probe and conductivity cell at the other end. The temperature was controlled with an Arthur H. Thomas microset thermoregulator Model 9655F in conjunction with a Princo Model T681 transistorized relay which operated a 125 W Thomas knife blade immersion heater Model 6147-E2. The bath was cooled by tap water flowing through a Thomas Model 9927-C cooling tube. The temperature was measured with a Brooklyn calorimeter thermometer 22214 which had been tested against a National Bureau of Standards certified thermometer to read 25.00 at 25°. Such a constant-temperature bath is also extremely well suited for viscosity measurements.

Titrations Types C and D. These were conductimetric titrations carried out with the apparatus described for type B. Figure 2 describes a titration type C and Fig. 1 describes a titration type D.

The concentration of standard electrolytes should be known to ±0.00001 M. One should keep in mind that the protein molarity is usually in the range 10^{-3} M to 10^{-4} M, and the uncertainty of standard solutions is critical in determining the uncertainty of the fraction of titrable groups per protein molecule.

Equilibrium Dialysis. The procedure used is fairly standard. The only important difference is that quantitative accounting must be made of all solution components involved. The reader is referred to another article[5]; this deals with a different approach to dialysis, but the details and discussion of handling of the membranes and related materials is relevant and highly recommended. The reason for complete quantitation stems from the need to calculate one of the two ions of the supporting electrolyte inside the bag by subtraction from the total electrolyte added. The second kind of ion can then be calculated from the Donnan equation (13). We found that the commercial dialysis cells we tested were unsuitable because there was significant loss either through leakage or evaporation from the unsealed edge of the membrane. We used, therefore, the simple approach of placing the protein solution inside a membrane sausage and suspending it in solvent in a glass-stoppered cylinder. The solution volumes were in the range of 12–15 ml. The procedure involves weighing in every step of the process and Table II gives in detail the various quantities weighed during an equilibrium dialysis experiment. One comment should be included here. The process of equilibrium dialysis as described here is indeed tedious, and a cell of some sort that could be used more conveniently, without sacrificing quantitation, is needed. As soon as a satisfactory cell becomes available, it will be reported.

Usually, all the electrolyte used in dialysis is contained in the solution against which the protein is dialyzed. This is done to avoid error accumulating in the calculation of total electrolyte from two sources rather than one.

Table III shows a simplified set of calculations; these are used to determine whether cations or anions bind to the protein—if any. The main simplification is assuming that the protein and dialyzate solution volumes are equal so that the reader can tell directly the differences between the various possibilities, without resort to additional calculations.

Details for manipulation of tubing and calculations of water associated with it are as follows. The tubing is thoroughly soaked and washed with $1 \times 10^{-3}\ M$ disodium ethylenediamine tetraacetate (EDTA) and rinsed with conductivity water. Because of the small molecular weight lysozyme (14,400), the size 18/32 Visking Nojax casing was used, which holds back lysozyme. It is important to know the weight of the tubing used at a well defined state of "wetness" because its weight has to be taken into consideration in the calculations, as shown in Table II. Also, some water "binds" to the tubing in the sense that it is unavailable either to the protein solution or the solvent used in dialysis; this has to be calculated also, as shown below. The weight of the tubing used in dialysis is determined from the weight of another piece of tubing of exactly equal size. Two pieces of tubing are cut simultaneously, both are cleaned and rinsed, and one is used for dialysis while the other is further treated as follows: The weight of one of the pieces may be taken before cleaning, "as is," to double check on the weight of it obtained after cleaning. The clean duplicate piece is dried in a jar under aspirator vacuum for 3 hours and then weighed. Knowing this weight (and also the weight under condition "as received"), one can calculate the dry weight of the tubing by correlating it to separate determinations of its weight under specified conditions described below. The conditions we found convenient to define as states of "wetness," and average weights, were as follows: (a) washed and oven dried at 110° for 2 hours, 1.0000 g; (b) washed and dried under aspirator vacuum for 3 hours, 1.0364 g; (c) untreated, as received, 1.2720 g. Of course oven-dried tubing was not used for dialysis. Thus, the "dry" weight of the tubing used can be calculated from its duplicate. The protein solution is added to the tubing using a weight burette and thus its weight is known. The weight of "wet tubing plus protein solution" is determined. Thus, the water that came along with the dialysis tubing can be determined by difference. The last correction that is necessary is to determine the portion of water actually bound to the tubing. The bound water depends on ionic strength and must be determined by separate experiment as follows: Pieces of dialysis tubing of known weight, which are aspirator dried, are added to an electrolyte solution. The weight of tubing is approximately 10–15% of the weight of the solution to which it is added, although good results are obtained with as low as 5% tubing. Upon addition of the tubing, the conductance of the electrolyte solution is increased because water is removed by binding to the tubing. Knowing the conductance of the electrolyte before and after addition of tubing and therefore the change in concentration of electrolyte, the water removed (or bound to the tubing) is calculated. Our results, expressed per gram of "oven-dried" tubing are as follows: for 0.01 M KCl, 0.586 g water per gram of "oven-dried" tubing; and for 0.04 M KCl, 0.195 g of water per gram of "oven-dried" tubing. In Table II is given the grams of water bound to the tubing used and also the grams of water adhering due to "wetness" of the tubing. To emphasize again, the water adhering due to wetness is obtained by subtracting the bound water from the total water that came along with the tubing calculated by weight difference. The important difference is that the water bound to the dialysis tubing is unavailable to both protein solution and dialysis electrolyte; the remaining water of "wetness" is available to the total system.

Protein Solutions and Deionization. The way of preparation of the protein solution varies widely and depends largely on the nature of the protein. Three comments need be made. First, in dialyzing a protein at a pH other than its isoelectric point, instead of using plain water we used extremely dilute electrolyte such as $10^{-6} M$ KCl to ensure that the counterions were of the desired kind. Second, in deionizing a protein solution we found the mixed bed resin Amberlite MB-1 to be entirely satisfactory. Third, a protein may precipitate out on complete deionization with a mixed bed resin especially if it is concentrated. This is the case with lysozyme. In such a case, the protein solution is diluted, and after passing it through the resin, the solution is partly lyophilized under nitrogen to reconcentrate it.

[25] Thermal Titrimetry

By MARIO A. MARINI and CHARLES J. MARTIN

In almost every physical or chemical process, there is an accompanying evolution or absorption of heat. For many years, the measurement of the heat change of a reaction has been used to determine the extent of the reaction. That is, in the process

$$xA + yB \rightleftharpoons nP \pm Q$$

the formation of n moles of P releases $(-)$ or absorbs $(+)$ a quantity of heat, Q, expressed in calories. This caloric value may be related to the heat of the reaction, ΔH, expressed as calories per mole P by:

$$\Delta H = Q/nP \tag{1}$$

The caloric value of a reaction is most easily measured as a temperature rise of a system with a known and near constant heat capacity.

Although a number of procedures for the measurement of the heats of reaction are known (see Vol. 26 [11]), we have found that thermal titrimetry offers the most convenient mode to obtain information on a variety of molecular processes. For thermal titrimetry, a concentrated reagent, B, is added to the material, A, at a constant rate; the heat change generated by the reaction is recorded as a function of time or concentration of B added. This principle was first described by Bell and Cowell,[1] who used the heat output to indicate the completion of the titration of citric acid with ammonium hydroxide. Linde et al.[2] replaced

[1] J. M. Bell and C. F. Cowell, J. Amer. Chem. Soc. 35, 49 (1913).
[2] H. W. Linde, L. B. Rogers, and D. N. Hume, Anal. Chem. 25, 40 (1953).

the high heat capacity thermometer with a highly sensitive, low heat capacity thermistor to detect the heat change and introduced the use of a constant flow of titrant in order to display the heat change as a function of reagent added on a recorder. These improvements gave impetus to further refinements that are available today in modern thermal titrimeters.[3,4]

Measurements of heat changes have certain fundamental advantages over other analytical methods. For example, the clear solutions required for optical measurements are not necessary. Indeed the heat of a reaction may be obtained on solutions that are turbid or become turbid during the course of a reaction, and inferences from the heat changes during reaction and precipitation may be made.

Measurement of heat during a reaction leads directly to the evaluation of the enthalpy (ΔH) of the reaction which previously had been derived by the tedious measurement of some parameter as a function of temperature and evaluated by the procedure of van't Hoff. This latter method presents experimental and theoretical problems.[5] In this respect, thermal titrimetry not only measures the enthalpy of the reaction, but can be used to obtain the free energy, ΔG, from the equilibrium constant and thus to determine the entropy, ΔS.[6]

A further advantage of thermal titrimetry or any calorimetric method is the nonselectivity of the procedure. Biological materials have their own intrinsic specificities, and by judicious selection of reagent and reactant one can analytically measure the heat change of a number of reactions with a single heat sensor system. In addition, the measurement of the heat change is dependent on the concentration of reactants per unit volume, not on the total volumes involved. Thus the lower limit of the volumes to be analyzed is a function of the design of the reaction vessel.

Although thermal titrimetry is in its infancy, the power and scope of the technique has already been demonstrated. A great number of standard analytical methods have been adapted to thermal titrimetry, and many more will be in the future. In addition, a number of estimations which cannot now be made may be studied by this technique. Essentially all that is required is that a reagent be added to a sample and that the reaction be reasonably rapid and accompanied by a measurable change in temperature. With the equipment currently available, these tempera-

[3] S. T. Zenchelsky, J. Periale, and J. C. Cobb, *Anal. Chem.* **28**, 67 (1956).
[4] J. Jordan and T. G. Alleman, *Anal. Chem.* **29**, 9 (1957).
[5] D. S. Reid, J. J. Christensen, and F. Franks, *Nature (London)* **224**, 1293 (1969).
[6] R. M. Izatt, D. Eatough, J. J. Christensen, and C. H. Bartholomew, *J. Chem. Soc. A* 47 (1969).

ture changes can be as small as 10^{-5} to 10^{-7} degrees; since these temperature changes are dependent on the molar enthalpy, not on the volume, extremely small samples may be examined. For further discussion on applications and the theory of thermal titrimetry see Tyrrell and Beezer,[7] Bark and Bark,[8] and the review by Jordan.[9] Because of the relative newness of thermal measurements of biological material in solution, molar enthalpy changes are not as well known as might be desired although two impressive lists of thermodynamic data for biological materials have been compiled.[10,11]

The development of thermal titrimetry will do much to extend our knowledge of these thermodynamic parameters. The precise and accurate measurement of heat changes normally associated with reaction calorimetry has now been combined with the high speed necessary to analyze unstable biological materials in concentrations compatible with biological systems.

Methodology

Several companies presently market thermal titrimeters.[12] In this work we have used a prototype model made by the American Instrument Company. The essential features are those introduced by Linde et al.[2] as modified by Jordan and Alleman.[4] The total system is composed of a thermal titrimeter, recorder, and constant-temperature water bath (20°) with provision for circulation to the reaction vessel.

The thermostated cell is composed of two Plexiglas chambers (Fig. 1). The upper chamber has inlet holes for the thermistor, heater, and stirrer and provides access to the reaction cup proper. Inlet and outlet tips for the thermostated titrant solutions are also provided. The lower double-walled chamber contains the reaction cup within an air space that is capable of rapid thermal equilibration by means of a "thermo jet" (cf. Fig. 1). Prior to and during a reaction, the thermo jet is

[7] H. J. V. Tyrrell and A. E. Beezer, "Thermometric Titrimetry." Chapman & Hall, London, 1968.
[8] L. S. Bark and S. M. Bark, "Thermometric Titrimetry." Pergamon, New York, 1969.
[9] J. Jordan, Thermometric enthalpy titrations, in "Treatise on Analytical Chemistry" (I. M. Kolthoff and P. J. Elving, eds.), Vol. 1, Part 1, p. 5175. Wiley (Interscience), New York, 1968.
[10] R. M. Izatt and J. J. Christensen, in "Handbook of Biochemistry" (H. A. Sober, ed.), J-49, Chem. Rubber Publ. Co., Cleveland, Ohio, 1968.
[11] G. C. Kreschek, in "Handbook of Biochemistry" (H. A. Sober, ed.), J-140. Chem. Rubber Publ. Co., Cleveland, Ohio, 1968.
[12] Tronac, 1804 S. Columbia Lane, Orem, Utah; SKC, Inc., P.O. Box 8538, Pittsburgh, Pennsylvania; Luminon, 120 Coit Street, Irvington, New Jersey; American Instrument Co., 8030 Georgia Ave., Silver Spring, Maryland.

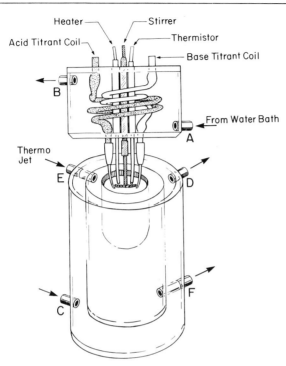

Heater — Stirrer

Acid Titrant Coil — Thermistor

Base Titrant Coil

B

From Water Bath

A

Thermo Jet

E D

C F

Fig. 1. Thermal titration vessel. The upper section contains shielded inlets for the constant-speed stirrer, the thermistor, and the heater probe. Two inlets for the acid and base are led through glass coils immersed in the thermostating solution. The reaction vessel (12 ml total capacity) is contained in the lower chamber and is removable. Sealing is effected with a rubber O ring. Water from the constant-temperature bath enters at inlet port A, exits at B, and then can go simultaneously to inlet ports E (controlled by a valve) and C in the lower chamber. Outlet ports D and F return the thermostating solution to the water bath. Equilibration of the reaction vessel contents is achieved by opening the valve leading to port E (thermo jet) which directs an envelope of water around the surface of the cup. A layer of water is maintained at the bottom of the inner chamber. When equilibration is achieved, the thermo jet valve is closed and the reaction cup is then surrounded by only an air envelope to give pseudoadiabatic conditions.

closed to provide an approach to adiabatic conditions. Just prior to initiation of a reaction, and just after its completion, essentially isothermal conditions prevail.

The recorder we have used is an Esterline-Angus, Model E1101S, with variable voltage spans and chart speeds. Sensitivity below 1.0 mV full scale is obtained through use of a Keithley Microvoltmeter, Model 155.

Good temperature control of the water bath is important. Using the Tronac Model 1040 temperature controller in combination with its special heater/cooling coil, the bath can be maintained within ±0.0001°. Within the reaction chamber and with the thermo jet open, the temperature variation is sinusoidal with a period of about 25 minutes and an amplitude of ±0.0020°. With the thermo jet closed, the temperature variation (in the absence of a reaction) is essentially zero using the 1.0 mV (equivalent to 0.132°) span during the time period generally required for observation of a reaction. At smaller voltage spans, temperature variation from the heat of stirring can be observed but can in part be electrically compensated by a device similar to that described by Arnett *et al.*[13] The computer-matched twin thermistor probe has a resistance of 17 Kohms with a time constant in stirred water in the millisecond range. These may be obtained from the American Instrument Company. The heater has a resistance of 25 ohms.

The titrant drive systems are driven by Sage variable speed pumps, Model 345. For many purposes, it has been found convenient to select a combination of drive speed and syringe size that will deliver about 10 μl of titrant in about 7–8 seconds. The change in temperature generated by the titration of 40 μmoles of sample can then be spread over a chart distance of about 4 inches.

Operating Procedure and Standardization

The syringes are filled with titrant, e.g., fresh 1 *M* HCl and carbonate-free 1 *M* KOH, and pumped into the titrant coils for thermal equilibration. In actual practice, the titrant coils are filled from the exit tips so as to remove air bubbles. The sample (generally 4.0 ml, containing up to 80 μmoles of material) is pipetted into the clean, dry reaction cup and the two chambers joined together. Thermal equilibration is then rapidly achieved by opening the thermo jet. If the sample had previously been equilibrated at the bath temperature, this requires less than 5 minutes. Operationally, this is determined by observing the rate of change of temperature as displayed by a change in the voltage using the 1.0 mV range. When this has become close to zero, the thermo jet is closed and, after about 30 seconds for additional thermal stabilization (in practice this time period will vary somewhat), the recorder chart drive is actuated. This is shown as point A in Fig. 2 which initiates the recording of the initial temperature baseline $\overline{AB}$. The titrant syringe motor drive is switched on at point B. If only water is in the reaction cup and if both isothermal and adiabatic conditions prevail, the trace segment $\overline{BC}$ will be observed since the heat of mixing of either

[13] E. M. Arnett, W. G. Bentrude, J. J. Burke, and P. McC. Duggleby, *J. Amer. Chem. Soc.* **87**, 1541 (1965).

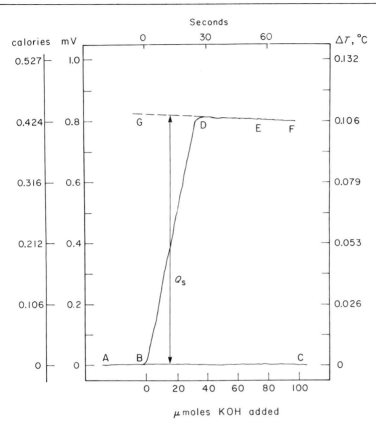

FIG. 2. Thermogram for neutralization of 32 μmoles of HCl in 4.0 ml water by 1 M KOH. See text for discussion.

1 M HCl or KOH is negligible. When titrant delivery is discontinued (point C) the temperature remains constant for a short period of time. This is probably due to the fortuitous compensation effects of self-heating from the thermistor and stirrer and conductive heat losses.

With a sample such as 32 μmoles of HCl (the material used for routine standardization) in a 4.0 ml volume in the reaction cup, titrant (KOH) delivery initiated at point B causes a temperature rise to point D which is linear with time and due to the exothermic reaction of H$^+$ and OH$^-$ to form 32 μmoles of water. Base addition is then continued to point E to ensure that the reaction is complete and to establish the "high" temperature baseline $\overline{DE}$. The trace $\overline{EF}$ represents the cooling curve of the sample. In the titration described, 32 μmoles of base are required to neutralize the 32 μmoles of HCl present. With 1 M KOH as titrant, this corresponds to 0.032 ml of titrant added to a 4.0 ml volume or a volume increase of less than 1%. For most purposes this increase in volume may be ignored.

Reference to Fig. 2 also shows that heat from the reaction is liberated at a constant rate. Since the titrant delivery rate is constant and since the fraction of H^+ neutralized is proportional to the equivalents of base added, it follows that the heat generated is proportional to the base added throughout the course of the reaction.

It will also be noticed that appreciable lag periods are not observed in the recorded trace. This reflects an adequate thermistor response (ca. 25 msec), efficient stirring, and a brisk on/off response of titrant delivery.

The information displayed in Fig. 2 is used to standardize the response of the instrument to enable the calculation of enthalpy changes of other reactions. The heat, Q_s, liberated in the reaction to form 32 μmoles HOH is proportional to the height between the extension of $\overline{AB}$ (in this case $\overline{BC}$) and $\overline{DG}$ (representing the extrapolation of $\overline{DE}$) at the point of half-reaction, i.e., when 16 μmoles of base have been added. For any weakly ionizable monoprotic acid, the point of half-reaction occurs at the pK'. For the case wherein the lower and upper temperature baselines are of positive or negative slope and are either parallel or non-parallel to each other, an approximation to the observed heat can be obtained by measurement of the height between temperature baseline extensions at a point that satisfies the condition of equal areas for the inscribed resultant triangles.[14]

In the apparatus described, a recorder response equal to 0.82 mV is generated in the formation of 32 μmoles HOH in a 4.0 ml reaction volume. Any alteration in the heat capacity of the system, such as a change in the reaction volume, will change this value. Since ΔH_F, the enthalpy of formation of HOH from H^+ and OH^-, is -13.5 ± 0.05 kcal/mole,[15] then 1.00 mV is equivalent to an output of 0.527 cal [(32 $\times$ 10^{-6} mole) (13.5 $\times$ 10^3 cal)/0.82 mV]. Since 1 cal is required to raise 1 ml of water 1°, 1.00 mV is also equivalent to a temperature rise of 0.132° for the 4.0 ml volume used. This information can now be used to calculate changes in the enthalpy accompanying other reactions.

Standardization can also be effected by the reverse of the titration described, i.e., with KOH in the reaction cup and HCl as the titrant. In our experience, however, the heat liberated by this procedure is slightly less than that obtained with KOH as the titrant. Most likely this is due to energy losses from the reaction solution occasioned by the loss of dissolved CO_2 in going from a high to a low pH.

[14] H. C. Dickinson, Nat. Bur. Stand. (U.S.) Bull. 11, 189 (1915).

[15] D. H. Everett and W. F. K. Wynne-Jones, Trans. Faraday Soc. 35, 1380 (1939);
H. S. Harned and B. B. Owen, Chem. Rev. 25, 31 (1939).

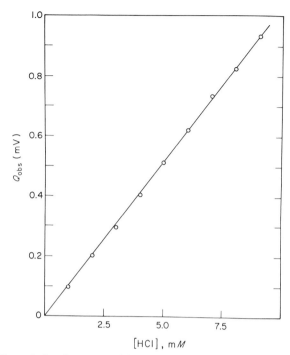

FIG. 3. The relation between millivolt response (Q_{obs}) and HCl concentration in a constant reaction volume (4.0 ml).

The accuracy of the standardization can be confirmed by the titration of an analytically prepared Tris solution. This has been used by a number of workers as a primary standard, and the heat of ionization is given at 11.35 to 11.39 kcal/mole at 25°.[16] Using the thermal titration technique, we have obtained -11.4 kcal/mole at 20° for the protonation of Tris with HCl.

As an alternate method of standardization, one can calibrate the recorder scale by passage of a known current, I, through a heater of resistance R for a time t to give a desired mV response. The heat generated electrically in joules, may be expressed as calories by

$$Q_e = I^2 Rt / 4.184 \tag{2}$$

Provided that one can measure I per period t accurately, this method gives reasonably close agreement with the calibration procedure involving the neutralization reaction of acid with base. However, we have preferred to use the latter method since it closely approximates that

---

[16] R. G. Bates and H. B. Hetzer, *J. Phys. Chem.* **65**, 667 (1961); J. J. Christensen, D. P. Wrathall, and R. M. Izatt, *Anal. Chem.* **40**, 175 (1968).

actually used in the determination of the heat change of unknown reactions.

It is important in these measurements to ensure that the mV response (which is equivalent to temperature rise and calories produced) reflects that which occurred in the reaction. This can be verified in two ways. The mV response should be directly proportional to varying concentrations of HCl in a constant volume if no appreciable heat losses occur during the titration. This is found to be the case (Fig. 3).

Alternatively, one should obtain a constant millivolt response using different reaction volumes of the same HCl concentration. This also was found to be true.[17]

Applications

Heats of Ionization

Thermal Titration of a Weak Acid

In the titration of a weak acid, HA, the fraction, α, of the anion produced is logarithmically related to pH but directly proportional to the equivalents of base added during the neutralization reaction. Therefore,

$$\alpha \approx \text{equivalents OH}^- \approx Q_{\text{obs}}$$

where Q_{obs} is the mV response related to any observed temperature change.

A typical example is shown in Fig. 4 for the titration of 40 μmoles of imidazole (pK' 7.2) in a 4.0 ml volume at an initial pH of 5.0. The heat liberated in the ionization is less than that seen in the neutralization reaction of 40 μmoles of H^+ by OH^- by the amount of energy required to ionize the proton from the imidazole ring. The pertinent reactions are:

$$\text{ImH}^+ \rightleftharpoons \text{Im}^0 + \text{H}^+ \qquad Q_{\text{Im}}, \text{ endothermic}$$
$$\text{OH}^- + \text{H}^+ \rightleftharpoons \text{HOH} \qquad Q_s, \text{ exothermic}$$

$$\text{Sum: OH}^- + \text{ImH}^+ \rightleftharpoons \text{Im}^0 + \text{HOH} \qquad Q_\alpha, \text{ exothermic}$$

with

$$Q_{\text{Im}} = Q_s - Q_\alpha \tag{3}$$

when the same mole equivalents of standard and sample in a constant volume are involved.

[17] M. A. Marini, R. L. Berger, D. P. Lam, and C. J. Martin, *Anal. Biochem.* 43, 188 (1971).

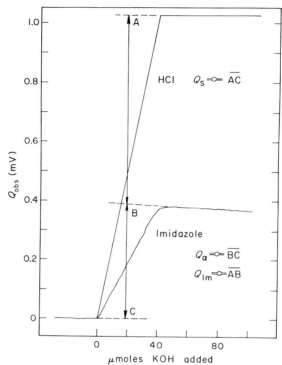

Fɪɢ. 4. Thermogram for titration of 40 μmoles imidazole in 4.0 ml water with initial pH 5.00. The thermogram for the neutralization of 40 μmoles HCl in 4.0 ml water by KOH is schematic.

Accordingly, the heat of ionization, ΔH_i, for the ionization of imidazole can be calculated simply from

$$\Delta H_i = \Delta H_F(1 - Q_\alpha/Q_s) \qquad (4)$$

or the equivalent expression

$$\Delta H_i = C_s Q_{\text{Im}}/n_1 \qquad (5)$$

where C_s is equal to the calibration constant of the voltage scale used, i.e., 0.527 cal/mV, n_1 equals the moles of imidazole, and ΔH_F is the enthalpy of the formation of water (13.5 kcal/mole). Thus, for the data shown in Fig. 4, $Q_s = 1.025$ mV and $Q_\alpha = 0.374$ mV, so that ΔH_i equals 8.58 kcal/mole.

Equations (4) and (5) are of general applicability for the same mole equivalents of standard or sample in a constant volume. A more general equation which is suitable for the case of n moles standard in volume V_s and of n_1 moles sample in volume V_1 is

$$\Delta H_i = \Delta H_F - (V_1 C_s Q_\alpha / V_s n_1) \tag{6}$$

In any such ionization reaction, Q_α will always represent an increase in temperature (overall reaction exothermic) as long as the energy required for the ionization is not equal to or greater than Q_s under the conditions of equal molar concentrations and volume. It will also be noticed that data of the type displayed Fig. 4 can be used as a check on the analytical concentration of the sample. If, for example, a time span of t seconds is required to neutralize n moles of proton by n moles of OH$^-$, then a proportional time, t_1, will be required to neutralize n_1 moles of sample. For equal amounts of standard or sample, the time axis segment for complete reaction should be identical irrespective of volume differences, provided that the chart speed of the recorder, the rate of titrant delivery, and titrant concentration are kept constant.

The heat of ionization of imidazole can be more directly obtained by titration of the free base form with acid. From the resultant thermogram, the heat of ionization is obtained using Eq. (5) and the height that would be equivalent to Q_{Im} in Fig. 4.

Thermal Titration of a Bifunctional Weak Acid or of a Mixture of Two Weak Acids

Consider the acids HA at concentration n_1 moles and HB at n_2 moles with pK's of pK'_1 and pK'_2, respectively, with p$K'_2 >$ pK'_1. Let α equal the fraction of HA in the anionic form and β the fraction of HB in the anionic form at any stage of the titration in the reactions

$$HA + OH^+ \rightleftharpoons A^- + HOH$$
$$HB + OH^- \rightleftharpoons B^- + HOH$$

If the ionization constants of the two groups are sufficiently separated, i.e.,

$$\Delta pK'_{12} = pK'_1 - pK'_2 \geq 2 \text{ pH units} \tag{7}$$

the heat of ionization of each group can be obtained directly from a single thermal titration. The thermogram will exhibit two distinct and separated slopes spread over a time axis equal to the sum of their mole equivalents with the first slope representing the heat liberated as the result of the ionization of the acid with pK'_1.

As an example of this case, the upscale titration of 40 μmoles of histidine in a 4.0 ml volume from an initial pH of 4.0 is shown in Fig. 5. The pertinent ionizable groups are the imidazoyl group (pK'_2 6.23) and the amino group (pK'_3 9.38). Two slopes are obtained with each slope segment spanning a 40 μmole length on the time axis. From such data,

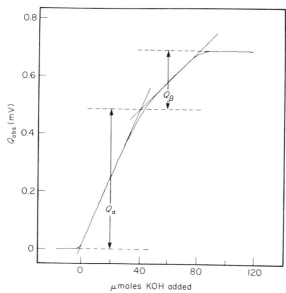

FIG. 5. Thermogram for titration of 40 μmoles histidine in a 4.0 ml volume with initial pH 4.00. The imidazoyl group with pK'_2 6.23 is the first group titrated with resultant slope height Q_α. When it is almost completely titrated the amino group with pK'_3 begins to ionize to yield the second slope height Q_β.

Q_α and Q_β are obtained as indicated at the point of half-neutralization of each group, and ΔH_i values are calculated readily from Eq. (4).

In the above example, the difference in the pK's of the two groups amounts to 3.15 units. It can be shown that a difference of this magnitude will introduce a negligible error (less than 0.001%) in the calculation of ΔH_i values of the two groups arising from the slight degree of overlapping ionizations, irrespective of the difference in magnitude of their heats of ionization. The experimental errors of measurement exceed this uncertainty.

When, however, the ionization of the two groups overlap appreciably, heats of ionization determined as discussed will be in error and dependent on ΔpK'_{12} and the actual heat of ionization of each group. In this case, any point on the resulting thermogram (with the exception of the very beginning and end of the titration provided that ΔpK'_{12} does not equal zero) will contain contributions from both heats of ionization. Furthermore, dependent on the value of ΔpK'_{12} and the ratio of the two heats of ionization, Q_{obs} at any point on the thermogram will be either smaller or larger than that which would be obtained for the titration of each group separately and the identity of the two slope segments will merge. In the extreme case of $\Delta pK'_{12}$ equal to zero, only a single slope will be observed and Q_{obs} will represent the average of the two heats of ionization.

At any stage of the titration, then,

$$Q_{obs} = \alpha Q_\alpha + \beta Q_\beta \tag{8}$$

where Q_α represents the heat liberated from the complete ionization of n_1 moles HA and Q_β the heat liberated from the complete ionization of n_2 moles HB. The total heat liberated will be given by

$$Q_T = Q_\alpha + Q_\beta \tag{9}$$

when α and β equal unity. From Eqs. (8) and (9) one obtains

$$Q_\alpha = \frac{Q_{obs} - \beta Q_T}{(\alpha - \beta)} \tag{10}$$

and Q_β from Eq. (9). Heats of ionization for both HA and HB can then be calculated from Eq. (6). Thus, the heats of ionization for two acids with overlapping ionizations can be calculated from a knowledge of pK'_1 and pK'_2, Q_T, and a single point measurement of Q_{obs} on the thermogram equal to some value of $\alpha + \beta$. For this, it is convenient to select a point on the time axis equal to half-completion of the titration, i.e., at $\alpha + \beta$ equal to 1.0. Then, since at any pH, $\Delta pK'_{12}$ can be expressed as

$$\Delta pK'_{12} = \log \frac{\beta}{1 - \beta} - \log \frac{\alpha}{1 - \alpha} \tag{11}$$

from the Henderson-Hasselbalch equation it follows that

$$\alpha = \frac{1}{1 + \text{antilog} \left(\frac{1}{2} \Delta pK'_{12}\right)} \tag{12}$$

and

$$\beta = \alpha - 1 \tag{13}$$

An example of the distortion produced in thermograms for two ionizable groups with $\Delta pK'_{12}$ equal to -1.0 is shown in Fig. 6 for the case of equal and unequal concentrations of the two groups assuming that the actual ΔH_i of the group with pK'_1 is 6.75 kcal/mole and of the group with pK'_2 is 10.8 kcal/mole. Figure 6 also shows schematically the trace that would be obtained if $\Delta pK'_{12}$ was greater than two pH units. For the conditions given, Q_{obs} is smaller throughout most of the titration than would be expected if overlap of the ionizations did not occur. Therefore, estimations of Q_α and Q_β from subjective slope height

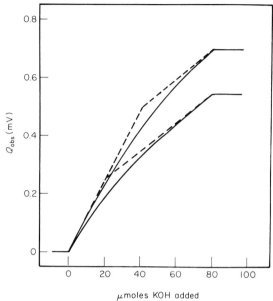

μmoles KOH added

Fig. 6. Theoretical thermograms for a mixture of two ionizable groups with different heats of ionization and with pK's differing by 1 pH unit at equal (upper set) and unequal (lower set) concentrations. The dashed curves show schematically the thermograms that would be obtained if ionization overlap did not occur. See text for further discussion.

measurements will yield a value for ΔH_i of the group with pK'_1 *higher* than the correct value and a ΔH_i for the group with pK'_2 *lower* than the actual value. This discrepancy is further magnified for the case when one group is in the lesser concentration.

As a variant of the above, an opposite situation prevails when the group with lower pK' has the higher heat of ionization. An example of this is shown in Fig. 7A for the schematic titration of an equimolar mixture of serine (pK' 9.20) and phenol (pK' 10.0). Comparison of the calculated concave curve with the slopes that would be obtained if their ionizations did not overlap reveals that straightforward slope height estimations from the schematic titration curve would yield a heat of ionization for serine that is lower and for phenol that is higher than their true values.

The experimental titration curve for an equimolar mixture of serine and phenol is shown in Fig. 7B. From measurements of Q_{obs} (0.398 mV at $\alpha = 0.715$; $\beta = 0.285$) and Q_T (0.933 mV) and using Eqs. (9), (10), and (6), the values of $(\Delta H_i)_{\mathrm{serine}} = 9.7$ kcal/mole and $(\Delta H_i)_{\mathrm{phenol}} = 5.7$ kcal/mole are obtained. These agree fairly closely with those ob-

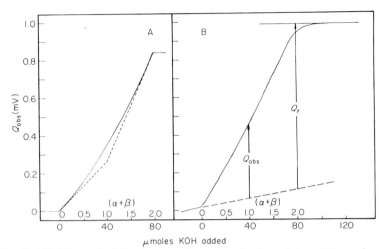

Fig. 7. (A) Theoretical thermogram for an equimolar mixture (40 μmoles each) of serine (pK'_2 9.20) and phenol (pK' 10.0); the dashed curve representing that which would be obtained if the ionization of the two groups did not overlap. Volume, 4.0 ml. (B) Experimental thermogram for the titration of 40 μmoles of serine and 40 μmoles of phenol in a 4.0 ml volume. The apparent disparity in slope heights relative to that shown in (A) is due to the positive slope of the low temperature baseline and a difference in the calibration constant of the ordinate.

tained by the titration of each group separately; 9.9 and 5.7 kcal/mole.[17] In actual practice, the accuracy of the two heats of ionization can be improved by using Q_{obs} values at other points on the thermogram where $1 < \alpha + \beta > 1$ and taking an average of the resulting calculated values.

With the techniques and treatment discussed, the heats of ionization of the commonly occurring amino acids as well as a number of imidazole derivatives have been reported.[17] The nucleotides and a variety of their derivatives have also been studied by thermal titrimetry.[18]

This treatment may be extended to obtain the heats of ionization of a three group mixture in which all three ionizations overlap. The only requirements are that the respective pK's be known exactly and that a minimum of two points be taken from the thermogram, in addition to knowing the total heat, Q_T.

Thermal Titration of Chymotrypsinogen

Heats of ionization of the groups of proteins were first obtained by Wyman,[19] who titrated hemoglobin at three different temperatures to derive the apparent heat of ionization from the observed changes in the

[18] J. J. Christensen, J. H. Rytting, and R. M. Izatt, *Biochemistry* 9, 4907 (1970).
[19] J. Wyman, *J. Biol. Chem.* 127, 1 (1939).

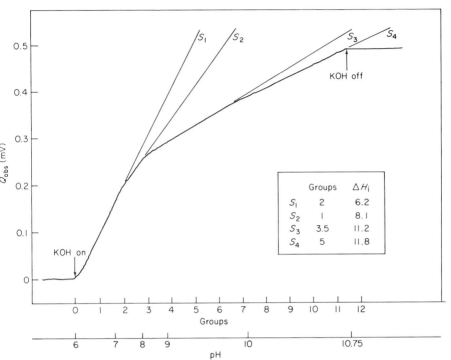

FIG. 8. Thermal titration of 7.55 μmoles of chymotrypsinogen in 4.0 ml 0.15 *M* KCl at an initial pH 6.0. Titration was conducted to pH 11 with 1 *M* KOH.

ionization constants of each class of groups. More recently, heats of ionization of the ionic residues of proteins have been obtained directly by calorimetry.[20,21]

Using thermal titrimetry, a number of proteins have been analyzed by this procedure.[22-24] The procedure involves the addition of acid or base at a constant rate to a known concentration of protein at a known pH. The resulting thermogram is a composite of a series of slopes, each of which is characteristic of the heat of ionization of a given set of groups.

A portion of the thermal titration of chymotrypsinogen is shown in Fig. 8. From tangents drawn to the curve at the points indicated and from the number of groups titrated obtained from the ordinate, the heats of ionization of the various sets of groups may be calculated. Although

[20] J. Hermans, Jr. and G. Rialdi, *Biochemistry* **4**, 1277 (1965).
[21] G. C. Kresheck and H. A. Scheraga, *J. Amer. Chem. Soc.* **88**, 4588 (1966).
[22] M. A. Marini, *Aminco Lab News* **25**, 8 (1969).
[23] N. D. Jespersen and J. Jordan, *Anal. Lett.* **3**, 323 (1970).
[24] C. Phelps, L. Forlani, and E. Antonini, *Biochem. J.* **124**, 605 (1971).

no attempt has been made to correct for the effect of overlapping ionizations on the slope heights or the "thermogram stretch" effect (see below), one can assign (cf. Fig. 8) slope S_1 to the two imidazoyl group ionizations, slope S_2 to the single α-amino group ionization and slopes S_3 and S_4 to a composite of the tyrosine hydroxyl and ϵ-amino group ionizations.

The Thermogram Stretch Effect

Groups that titrate at the extremes of the pH scale will yield distorted thermograms due to the buffer capacity of water. This becomes significant below pH 4 and above pH 10. However, in systems of biological interest the alkaline pH region will more usually be encountered, and discussion will be restricted to this case. Any thermogram which extends above pH 10 will exhibit a "stretching" of the curve along the base addition or time axis which decreases the slope of the thermogram. This can be seen to some extent in the upper portion of the curve in Fig. 7B corresponding to the region of near completion of the titration of the phenolic group. In the absence of the buffer capacity of water, the change in slope that coincides with complete titration of the group would have been much more abrupt (compare with Fig. 4).

Figure 9 illustrates the "thermogram stretch" effect superimposed upon the titration of a group with an assigned pK' 11.0 and ΔH_i of 10.8 kcal/mole. The dashed line represents the idealized thermogram that would be obtained in the absence of any solvent buffering capacity. The solid line represents the "experimental" curve that would be obtained upon titration of 40 μmoles of the group in 4.0 ml 0.15 M KCl. For all practical purposes, complete titration of the group will be achieved at pH 13 but, as can be seen, the approach to this pH by the concentration

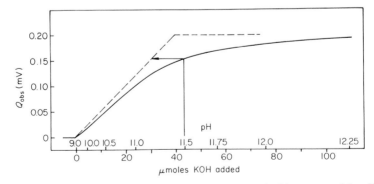

FIG. 9. Theoretical curves for the titration of an ionizable group with pK' 11.0 and heat of ionization of 10.8 kcal/mole. The dashed line represents the thermogram that would be obtained in the absence of the buffering capacity of water. The solid curve represents the thermogram that would be obtained experimentally due to the thermogram stretch effect. See text for discussion.

of base used (1 M) results not only in an increase in volume which decreases Q_{obs}, but also in considerable distortion of the thermogram. As a consequence, a slope height estimation based on a line drawn tangent to the initial segment of the thermogram and intersecting the upper temperature baseline at a point corresponding to the addition of 40 μmoles of base will lead to a calculated ΔH_i that is *larger* than the true value by about 20%. The error will be even larger when the pK' of the group is greater than 11.

Correction of the "thermogram stretch" effect can be made if one knows the pH corresponding to any point on the thermogram in conjunction with independent solvent titration data such as are given in the table. For example, 43.6 μmoles of KOH are required to raise 4.0

BASE REQUIREMENT FOR THE TITRATION OF 4.0 ML 0.15 M KCl

pH	μMoles KOH
9.00	0.2
10.00	1.2
10.25	1.8
10.50	2.4
10.75	3.4
11.00	4.2
11.25	7.8
11.50	13.2
11.75	22.8
12.00	39.8
12.25	71.4
12.50	128 ± 5

ml of 0.15 M KCl containing 40 μmoles of the group to pH 11.5 (cf. Fig. 9). Of this amount, 13.2 μmoles of KOH were required (cf. the table) for the titration of the solvent. The difference, 30.4 μmoles of KOH, thus represents the amount of base required to titrate the group to pH 11.5. With the group pK' equal to 11.0, this represents 76% completion of the reaction. Since the heat of dilution of the titrant can be ignored, this procedure defines a point that is corrected for the titration of the solvent at the same height on the ordinate. This is illustrated in Fig. 9. Although only a correction at one pH value is theoretically required, it is advisable to repeat this procedure at several pH values to better define the corrected slope.

Kinetics and Heats of Reaction

Thermometric titrimetry readily permits observation of the time course of an enzyme-catalyzed reaction with the condition that the heat change is of sufficient magnitude. Even when this does not obtain, however, one can resort to chemical amplification techniques in special cases and obtain data that reflects the reaction rate curve.

An example of this is shown in Fig. 10 for the chymotrypsin-catalyzed hydrolysis of acetyl-L-tyrosine ethyl ester in the presence of Tris buffer. In the absence of Tris, the heat change accompanying this reaction at the substrate concentration used was not detectable using the 1 mV scale. However, as has been mentioned earlier, the enthalpy change for the addition of a proton to the unprotonated species of a Tris buffer is -11.4 kcal/mole. Since, in the hydrolysis of the ester substrate a proton is liberated, one can follow the time course of the reaction by coupling it with the relatively high protonation reaction heat of Tris. Adequate tracking requires that the reactive buffer component be in excess as acceptor as well as for the maintenance of a near-constant pH. For maximum sensitivity, the pH used should represent an optimal balance between the pK' of both the generated carboxyl group and the pK' of

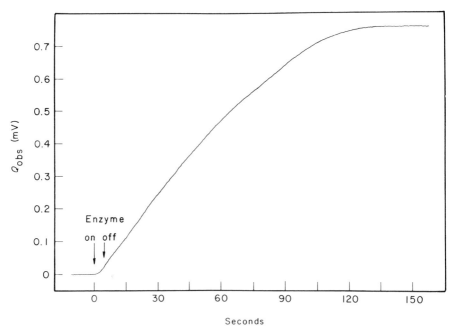

Fig. 10. Reaction rate curve for the chymotrypsin-catalyzed hydrolysis of acetyl-L-tyrosine ethyl ester. The reaction volume was 4.0 ml containing 40 μmoles of the ester and was 0.04 M in Tris buffer at pH 8.0 and at 25°. The chymotrypsin solution (ca. 5 g/l) was in the glass coils of the upper chamber and was added for 6 seconds (ca. 30 μg chymotrypsin) to initiate the hydrolysis at the point indicated. The heat evolved with time represents the enthalpy change associated with the addition of a proton to the unprotonatel Tris buffer component. This curve has been corrected for slope deviations of the upper and lower temperature base lines.

the buffer. In the example discussed, pH 8 fulfills this condition. It is of course obvious that the use of this particular stratagem precludes the determination of the enthalpy change accompanying the ester hydrolysis reaction.

In other reactions, both the enthalpy change and the reaction rate curve may be obtained in a single run and the procedural details are straightforward. As an example, Fig. 11 shows the kinetics of the catalase-catalyzed decomposition of hydrogen peroxide, a reaction which is exothermic to a considerable degree. Several precautions, however, must be observed. It is most desirable that the enzyme concentration be high so that the reaction is completed in a relatively short period of time. This minimizes the uncertainties of temperature baseline extrapolations and, because of the small volume addition, volume corrections are unnecessary. Under such conditions though it becomes imperative to prevent enzyme leakage from the titrant tip into the substrate solution during the equilibration period. This can generally be achieved by ap-

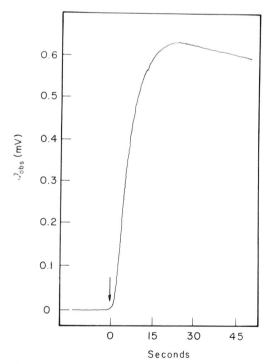

Fig. 11. Reaction rate curve for the decomposition of hydrogen peroxide by catalase. The volume was 4.0 ml and contained 20 μmoles of peroxidase in 0.06 M phosphate buffer, pH 7.0. Catalase (20 μl, ca. 600 IU) was added quickly at zero time.

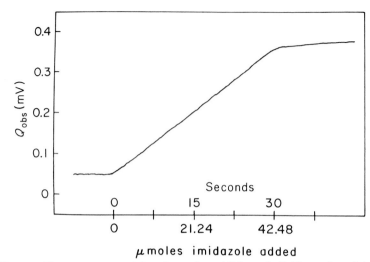

Fig. 12. Thermogram resulting from the addition of 42.48 μmoles of imidazole (titrant: $1\,M$ imidazole, pH 7.0) to 4.0 ml of 2.22 M formaldehyde containing 0.15 M KCl, pH 7.0.

propriate tip design and the occlusion of a small air bubble at the interface junction of enzyme and substrate. Heats of dilution errors must be assessed independently.

Data of the type shown in Fig. 11 also permits calculation of the enthalpy of reaction.[25] Since, from Eq. (1), the observed heat change, Q_{obs}, is directly proportional to the moles of reacting species, one can apply such information to the analytical determination of peroxide. As a practical application consider the conversion of uric acid to allantoin with hydrogen peroxide as an additional product of the reaction in the reaction catalyzed by uricase. This reaction evidently proceeds without much heat change since it could not be detected using the 1.0 mV scale with 5 μmoles of uric acid in a 4.00 ml volume. However, the reaction can be easily followed by coupling it with the catalase reaction and, from the total heat change, applied to the analytical determination of uric acid. An analogous procedure can also be used to amplify the glucose oxidase reaction although this reaction can be readily observed without coupling to the catalase reaction.

As another example of the application of thermal titrimetry to the measurement of the heat of reaction, Fig. 12 shows the heat change

[25] From the data of Fig. 11, the enthalpy of the reaction is calculated to be -17.5 kcal/mole. However, the moles of peroxide present were calculated from the stated concentration of the reagent, and the actual concentration could be much lower. The enthalpy term would, therefore, be higher than the above value.

generated upon the addition of a concentrated imidazole solution (1 M) to a solution of formaldehyde (2.22 M). The shape of the curve indicates that the reaction went to completion. Upon correction for a rather pronounced heat of dilution, the enthalpy of the reaction is -3.30 kcal/mole. In experiments of this type, the time axis was calibrated in terms of volume titrant delivered per chart division during the standardization reaction using HCl and 1 M KOH. This permits a precise knowledge of the moles of any other titrant added from the molar concentration of the titrant and the time axis segment of the slope height.

Conformational Changes in Proteins

Chymotrypsin. The enthalpy change accompanying the state A (folded) to state B (partially unfolded) transition of α-chymotrypsin at pH 2.0 appears to follow two-state transition theory and has been evaluated using such procedures as monitoring changes in absorbance or optical rotation as a function of temperature.[26] The transition temperature at this pH in 0.01 M Cl⁻ is about 31°, and ΔH for the transition has been evaluated from van't Hoff plots to be 120 kcal/mole at 40°.[27] Recently, the enthalpy of this transition has been evaluated calorimetrically using both batch and flow microcalorimeters.[28] For this determination advantage was taken of the fact that at 40° and near pH 4, α-chymotrypsin is completely in state A and at pH 2 and 40°, completely in state B. The data obtained led to a value for the enthalpy of transition of 110 kcal/mole after correction for the heat of protonation of the carboxyl groups. However, through necessity this correction term was evaluated at 25°, where the protein is completely in state A.

It is relatively simple to execute a similar type of experiment using thermal titrimetry. As an example, Fig. 13 shows the trace obtained for the titration of 1.84 μmoles α-chymotrypsin in a 4.0 ml volume (pH 3.5, 40°) with 1 M HCl to a pH less than 2.0. The reaction is endothermic and yields a value for the enthalpy of the state A to state B transition, after correction for the exothermic heat of protonation of the titratable carboxyl groups, of 105 kcal/mole. This can be compared with the value of 110 kcal/mole reported by Biltonen et al.[28]

That the curve shown in Fig. 13 represents adequate tracking of the transition was confirmed by independent kinetic measurements. In a

[26] R. Lumry and R. Biltonen, *in* "Structure and Stability of Biological Macromolecules" (S. N. Timasheff and G. D. Fasman, eds.), Vol. 2, Chapter 2. Dekker, New York, 1969.

[27] R. Biltonen and R. Lumry, *J. Amer. Chem. Soc.* **91**, 4251 (1969).

[28] R. Biltonen, A. T. Schwartz, and I. Wadsö, *Biochemistry* **10**, 3417 (1971).

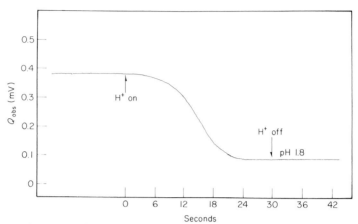

Fig. 13. The state A to state B transition for α-chymotrypsin at 40°. The 4.0 ml volume contained 1.84 μmoles of enzyme at a pH of 3.5 and the phase change was initiated by adding 1 M HCl at the point indicated. The acid was added at a constant rate to the final pH 1.8 at which time the addition of acid was stopped.

separate experiment, the change in absorbance at 230 nm with time was monitored upon changing the pH from 3.5 to 2 at 40°.

Ligand Binding

A number of papers have recently appeared in which calorimetry has been utilized to determine the molar heat of binding, ΔH_B, of molecules, exclusive of protons, to macromolecules of biological interest. For example, to mention but a few, Shiao and Sturtevant[29] and Shiao[30] have investigated the binding of various inhibitors to α-chymotrypsin and Hinz et al.,[31] the binding of inhibitors to aldolase. Lovrien and Sturtevant[32] have measured the enthalpy of binding of ions to bovine serum albumin, and Hearn et al.[33] have obtained the thermodynamic parameters for the binding of S-peptide to S-protein to form RNase-S'. The reconstitution of tobacco mosaic virus[34] and the quaternary structure of hemerythrin[35] have also been studied calorimetrically. In all the

[29] D. D. F. Shiao and J. M. Sturtevant, *Biochemistry* **8**, 4910 (1969).

[30] D. D. F. Shiao, *Biochemistry* **9**, 1083 (1970).

[31] H. J. Hinz, D. D. F. Shiao, and J. M. Sturtevant, *Biochemistry* **10**, 1347 (1971).

[32] R. Lovrien and J. M. Sturtevant, *Biochemistry* **10**, 3811 (1971).

[33] R. P. Hearn, F. M. Richards, J. M. Sturtevant, and G. D. Watt, *Biochemistry* **10**, 806 (1971).

[34] S. Srinivasan and M. A. Lauffer, *Biochemistry* **9**, 2173 (1970).

[35] N. Langerman and J. M. Sturtevant, *Biochemistry* **10**, 2809 (1971).

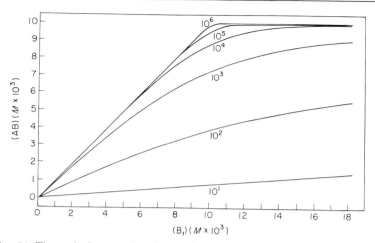

Fig. 14. Theoretical curves for the reaction of A plus B to form AB with various values for the association constant. The volume was assumed to remain constant.

above-reported experiments, flow or batch calorimeters were employed.

Using thermal titrimetry, Benjamin[36] has determined the effect of guanidine, urea, and alcohols on the micellization of dimethyldecylamine oxide. Forlani[37] has investigated the binding of ferricyanide to hemoglobin. Heats of metal binding have been thoroughly studied by the same technique,[38] and Kreschek[39] has measured the interaction of sodium dodecyl sulfate with β-lactoglobulin.

In the application of thermal titrimetry to an investigation of ligand binding, a concentrated solution of reagent B (titrant) is added to the titrand containing component A. Provided that the stoichiometry of the reaction is known, e.g., AB is formed, both the association constant and the enthalpy of binding may be determined from a single thermogram. This approach, however, depends markedly on the value of the association constant (Fig. 14). As the association constant increases, the thermogram will change in shape from an extremely flat curve ($K_{assoc} \leqq 1$) to one which is linear throughout to completion of the reaction. Above a K_{assoc} of about $10^5 M^{-1}$, only ΔH_B can be obtained since the reaction is essentially complete or complete at equimolar concentrations of A and B, e.g., the neutralization of HCl by KOH. At lower values of K_{assoc} (10^4 to $10^2 M^{-1}$), where B will be in excess of A, procedures have been

[36] L. Benjamin, J. Colloid Interface Sci. 22, 386 (1966).
[37] L. Forlani, private communication (1971).
[38] J. J. Christensen, R. M. Izatt, and D. Eatough, Inorg. Chem. 4, 1278 (1965).
[39] G. C. Kresheck, Wenner-Gren Center, Int. Symp. Ser. to be published.

derived to obtain both ΔH_B and K_{assoc} from a single thermogram.[7,40-43] The enthalpy of binding is derived from the initial slope, and K_{assoc} from approximation analysis using the experimental data from the curved portion of the thermogram.

In the execution of such experiments, the idealized curves of Fig. 14 will differ somewhat from those actually obtained. In addition to heat of dilution effects, there will be an increasing temperature difference between the reaction volume and the titrant, and magnified the smaller is K_{assoc} and the more dilute the concentration of B. There will also be a continuous change in the concentration of both reactants and a change in the reaction volume. All these effects must be corrected for and procedures to this end have been published.[7]

The problem becomes more complex when consecutive reactions are involved, as in the stepwise formation of complex species with stability constants that do not differ greatly from each other. For such cases, computer program evaluation of the constants can be achieved.[7]

As a very approximate approach to the evaluation of K_{assoc} for those cases where (a) titrant B is in great excess of A; (b) the stoichiometry of the reaction is known; and (c) the maximum value of Q_{obs} can be measured, i.e., all of A has reacted, it follows that

$$B_t = \left[\frac{1}{K_{assoc}} \right]^{1/n} \tag{14}$$

at the point on the thermogram (corrected for volume changes, concentration of reactants, etc.) where Q_{obs} is half its maximal value and B_t is the total concentration of B.

When conditions are appropriate, i.e., K_{assoc} is in the range in which excess of B is required to drive reaction to completion, both ΔH_B and K_{assoc} can often be more simply evaluated using a different experimental approach and treatment of data.

Consider the reaction

$$A + nB \rightleftharpoons AB_n$$

where the concentration of B is always large in comparison to the concentration of A. Then one can add as the titrant a small volume of A, at concentration A_0, to a solution of B, at concentration B_0, which now becomes the titrand, and measure the resultant Q_{obs} proportional to the amount of AB_n formed. By keeping A_0 constant and repeating the titra-

[40] M. H. Dilke and D. D. Eley, *J. Chem. Soc.* **2601**, (1949).

[41] J. J. Christensen, J. Ruckman, and R. M. Izatt, *Thermochimica Acta* **3**, 203 (1972).

[42] D. J. Eatough, J. J. Christensen, and R. M. Izatt, *Thermochimica Acta* **3**, 219 (1972).

[43] D. J. Eatough, R. M. Izatt, and J. J. Christensen, *Thermochimica Acta* **3**, 233 (1972).

tion at various levels of B_0, one can obtain a series of Q_{obs} values vs. B_0 levels over the range of Q_{obs} close to zero to $(Q_{obs})_{max}$. If the volume of A added as titrant is small (ca. 1% of total reaction volume) volume corrections can be ignored. After correction of the Q_{obs} values for any heat of dilution effects and using the expression

$$K_{assoc} = \frac{(AB_n)}{(A_0 - AB_n)(B_0)^n} \tag{15}$$

it follows that

$$\log K_{app} = \log \frac{(AB_n)}{(A_0 - AB_n)} = n \log B_0 + \log K_{assoc} \tag{16}$$

where K_{app} equals the fraction of A in state AB_n. Since Q_{obs} should be proportional to the amount of AB_n formed, K_{app} can be calculated directly from the data since

$$K_{app} = \frac{Q_{obs}}{(Q_{obs})_{max} - Q_{obs}} \tag{17}$$

One then calculates a series of K_{app} values as a $f(B_0)$ and from a plot of $\log K_{app}$ against $\log B_0$, obtain K_{assoc} from the intercept and n from the slope. The value ΔH_B for the reaction can then be obtained from the concentration of A and the $(Q_{obs})_{max}$ value.

The above technique has been successfully used for the reaction between imidazole and formaldehyde, and Fig. 12 shows one thermogram out of a series run to evaluate K_{assoc} for the reaction. Since, for this reaction a concentration of close to $2\,M$ formaldehyde is required to completely convert ca. 40 μmoles of imidazole to its N-hydroxymethyl derivative, the use of even a $12.34\,M$ formaldehyde solution as titrant into a 4.0 ml reaction volume will result in a volume increase (to reach $2\,M$ formaldehyde) of 0.96 ml. Concomitant with this there will be deviations from the true Q_{obs} values due to the resultant factors of a heat of dilution, temperature differences between titrant and titrand solution in which the titrant will act as coolant, and a cooling effect on the heat evolved with the increase in heat capacity of the system. All effects considered then, the previous outlined procedure has inherent advantages even though a number of thermograms must be obtained. The rapidity in execution of the thermal titration procedure, however, does not make this a serious problem.

Acknowledgment

This work was supported, in part, by Public Health Research Grant HE 14803 of the National Heart and Lung Institute and GM 10902 of the Institute of General Medical Sciences, National Institutes of Health.

Note Added in Proof

By an extention of the principles outlined here, we have been able to calculate the individual heats of ionization of the sets of overlapping ionizable groups in proteins. In certain cases, the heats of ionization of individual groups have been determined in the chymotrypsin family of enzymes and in hemoglobin.[44,45] More recently we have been able to analyze the potentiometric titration curves and the thermal titration curves simultaneously by means of a computer assisted program to obtain both the ionization constants and the heats of ionization. This should make it possible to characterize a protein by its ionic properties which play such an important role in a host of biologically important phenomena.

[44] C. J. Martin, M. A. Marini, and L. Forlani, Abstracts, 27th Ann. Calorimetry Conference, Park City, Utah (1972).
[45] L. Forlani, M. A. Marini, R. L. Berger, and C. J. Martin, Abstracts, 27th Ann. Calorimetry Conference, Park City, Utah (1972).

[26] Negative Stain Electron Microscopy of Protein Macromolecules

By ROBERT M. OLIVER

The purpose of microscopy is the production of images of sufficient magnification, and of such quality and resolution, that relevant details of the object examined are revealed. In recent years, the development of negative stain microscopy has permitted, at least in part, the achievement of that end with respect to protein macromolecules.

Image Formation at Atomic and Molecular Levels

Image formation involves the physical processes of diffraction, a scattering of radiation by the object, and the collection, or focusing, of the scattered radiation by an optical system to form the image. Object detail will not be resolved in the image by magnification to any degree if the wavelength of the radiation employed is larger than the dimension of the detail.

X-rays satisfy the wavelength requirement for studies at the atomic and molecular level and have been extensively so used. Unfortunately, X-rays cannot be focused by any known system, and images cannot be experimentally observed. Information about an object present in the scattered X-rays may be recovered by interpretive procedures; the sophisticated methods of analysis of diffraction data thus are substituted for the optical system in the image formation process.

Electrons of moderate energies also satisfy the wavelength requirement as radiation suitable for the study of matter at the atomic and molecular levels. The diffractions produced by electrons and X-rays contain similar information. X-rays are scattered by the orbital electrons of atoms comprising the object; Fourier treatment of the experimental X-ray data reveals the distribution of electrons within the object. Electrons are scattered by the electrostatic potentials of the atomic nuclei as well as by the orbital electrons. A similar treatment of electron diffraction data reveals the potential distribution within the object. In general, the two kinds of data may be analyzed, or focused, by the same mathematical methods with equivalent results.

Since electrons are charged and, when in motion, generate a magnetic field, they may be focused by means of suitable electrostatic or electromagnetic lenses to directly form an image of the scattering object. In passage through the lens the diffracted radiation is recombined in such a way as to restore in the image plane the amplitude and phase relationships which existed at the object. Although the diffracted electron radiation contains the necessary information, the production of images with atomic resolution is not experimentally realized in most electron-optical systems due to lens aberrations and other instrumental inadequacies. However, resolutions sufficient to image protein molecules are routinely achieved.

Information about the object examined exists in the image only as a result of the particular way in which the object modifies the incident illumination. In the visible portion of the spectrum, such modifications involve the familiar light–matter interaction phenomena of absorption, reflection, diffraction, etc., and combinations of these effects. In an analogous way electron–matter interactions must modify the incident electron beam if useful electron images are to be formed. If the magnitude of the interaction is insufficient to produce observable effects under a given set of experimental conditions, the object is, in effect, invisible. This condition is frequently encountered in transmission microscopy with both light and electrons. In that event, useful information about the object can be collected in the image only by modifying the instrument—as for phase microscopy, or by modifying the object in a systematic way—that is, the object is "stained." Many objects of biological interest suitable for examination by electron microscopy are so electron-transparent that staining is necessary, particularly if detail at the molecular level is to be observed by bright-field transmission microscopy.

If the staining involves the combination of an electron-dense material with the object of interest in such a way that the stained object appears in the micrograph as an electron-opaque area against a relatively light background, the object is said to be positively stained. Alternately, the electron-transparent object may be more or less completely embedded in a matrix of electron-dense material, in which event it will appear in reversed contrast—an electron-transparent area against a dark background. This object is said to be negatively stained.

The effects of reverse contrast in electron microscopy were first reported by Farrant in studies of ferritin molecules.[1] Hall observed similar effects when studying virus particles originally prepared by positive staining techniques.[2] Conditions for the routine use of negative stain microscopy were described by Brenner and Horne.[3]

For the study of protein macromolecules the negative stain technique offers several advantages over other techniques that have been used to enhance the visibility of structures of biological interest. When a specimen is imaged, variations of electron intensity, or contrast, are produced at the image plane in proportion to the mass distribution, or structure, in the specimen plane. Under the conditions of observation usually established, the minimum thickness of unstained protein which will result in noticeable contrast is of the order of 100 Å. With this limit of resolution, features of large complex structures which are of the greatest interest will be lost, and smaller protein molecules will be invisible.

Contrast enhancement by positive staining presumably involves a direct interaction of a stain material with the protein. A spherical protein molecule of a molecular weight of 20,000 will have a diameter of about 25 Å. For such a mass to be visualized under ordinary conditions with noticeable contrast, its density must be increased by a factor of three or four. Since most potential positive stains have densities < 8 g cm^{-3}, the probable binding of sufficient stain material to reveal features as small as 25 Å appears unlikely. Such intense staining, if achieved, would almost certainly distort the structure under study. Negative staining does not require a stain–protein interaction, and denaturation of the subject molecule by staining is minimized, yet visualization to a limit of resolution of about 25 Å is easily realized.

Molecules with diameters < 100 Å may be studied by various shadowing and replication techniques. In general, these techniques require the removal of nonvolatile salts and water from the specimen protein before application of the contrast-enhancing material. These operations are laborious and necessitate the use of freeze drying or critical-point drying to avoid possible distortions of the subject material. A relatively thin cap

[1] J. L. Farrant, *Biochim. Biophys. Acta* **13**, 569 (1954).
[2] C. E. Hall, *J. Biophys. Biochem. Cytol.* **1**, 1 (1955).
[3] S. Brenner and R. W. Horne, *Biochim. Biophys. Acta* **34**, 103 (1959).

of high density material is imposed on the molecular surface, usually from one side and in such a way as to cast a "shadow" of the object molecule on the support film surface. The resultant image of only part of the molecular surface is difficult to interpret. Measurements of molecular dimensions are indirect; particle heights are inferred from the shadow lengths, and width measurements must be corrected for the thickness of the contrast-enhancing material.

Most of the limitations of the shadowing technique are avoided by the use of negative stain. Nonvolatile salts need not be removed, within limits described below, and the preparation of the specimen is simplified. When the protein molecules are dried within a matrix of negative stain, surface tension forces are dissipated against the stain bed surface, thereby minimizing distortion of the specimen ultrastructure. Ideally the immersion of the specimen molecules in the stain is complete; thus ultrastructural detail in the image is easier to interpret, and dimensions may be estimated directly from the micrographs.

Significance of the Negative Stain Image

The ultimate objective of delineation of protein structure at the atomic level is presently accomplished only by means of X-ray diffraction analysis, an approach that is laborious and beset with experimental difficulties. To achieve atomic resolution, suitable crystals of the protein, and possibly of several metal derivatives, must be examined, and a substantial body of data must be collected and evaluated. X-ray investigations to a limit of resolution at the molecular level are less demanding, but not free of difficulty. The potential value of negative stain microscopy derives from the relative ease with which proteins may be imaged at the molecular level of resolution. The validity of the technique may be established by a comparison of molecular models based on electron and X-ray observations.

The general model that emerges from X-ray studies of protein crystals depicts the protein molecule as an entity with a well defined, three-dimensional structure.[4] The polypeptide chain is so folded as to form a distinct boundary that encloses a nonpolar inner environment. The polar amino acid residues are almost always exposed to the solvent, except occasionally when they perform a special function. The interior atoms are located with high precision and are tightly packed so that the molecule is compact. In general, the only side chains that do not assume fixed positions are the lysine, arginine, and glutamic acid residues on the surface of the molecule.

[4] Lubert Stryer, *Annu. Rev. Biochem.* **37**, 25 (1968).

Several observations suggest a high stability of the fundamental internal arrangement of the molecule: for example, proteins have the same structure in different crystal lattices. If proteins were readily deformable, the forces between molecules in crystals would be expected to distort their structure in a manner dependent on the intermolecular contacts. This has not been observed. In the several proteins studied the active site consists of residues that are located in at least three distinct segments of the polypeptide chain. Thus, precise and unique conformations are required to maintain the integrity of the site. Only minor conformational changes are induced when ligands are bound or allosteric changes occur, the tertiary structures of the chains hardly being affected. Finally, the presence of symmetry axes in the architecture of multichain assemblages suggests a high degree of similarity of conformation between the repeated asymmetric units from which the structures are formed.

Negative stain microscopy studies, particularly those of virus particles, oligomeric assemblages, and multienzyme complexes, lead to similar generalizations as to the physical nature of protein molecules. Molecules containing but a single polypeptide chain usually appear as compact entities with well defined boundaries and morphologies. In the absence of denaturation, artifacts, or distortions, the images are consistent in appearance and dimensionally reproducible. When an adequate knowledge of morphology of the folded protein mass has been established, molecular volumes may be calculated from dimensional measurements performed directly on the magnified images. Molecular weight estimates based on these volumes and an average packing density are often comparable to hydrodynamic values.

Multimolecular assemblages, either aggregates of identical chains, such as glutamine synthetase, or structures incorporating several components, such as pyruvate dehydrogenase complex from *Escherichia coli*, appear as particles of definite size and morphology. These entities are not necessarily closely packed at the molecular level, however their subunits apparently are as compact as single-chain molecules. Chemical observations suggest that the noncovalent, intermolecular contacts which join the subunits are highly specific; the morphological and architectural consistency of these complex particles revealed by their negative stain images is further evidence of a precise and relatively rigid conformation of the several component subunits of which they are comprised.

In particular views, the subunits of the multichain proteins frequently appear to be organized about an axis of symmetry, and the more complex assemblies apparently involve more than one symmetry axis in their design. Such obvious consistency would not be observed unless the asymmetric units on which the designs are based possess atomic arrangements of high stability.

A rationale for the interpretation of negative stain images of protein macromolecules may be based on the following relationships: illumination of a specimen object results in a scattering of the radiation, or diffraction. The distribution of intensities in the diffraction can be shown experimentally to be the Fourier transform of the distribution of mass in the object; thus Fourier synthesis from the intensities observed in the diffraction yields the mass distribution in the object. When electrons are used to irradiate a specimen in the electron microscope, the Fourier synthesis is accomplished by means of a lens, the image being the Fourier transform of the scattered radiation. The intensity distribution in the image, represented by the optical density distribution on the micrograph, is then proportional to the mass distribution in the object.

The proportionality of the recorded image intensities and the mass distribution is illustrated by Fig. 1. Comparison of the X-ray diffraction of the object, in this case a crystal of dihydrolipoyl transsuccinylase from *E. coli*, with the optical transform of an image of the negatively stained crystal reveals the fidelity with which information about the mass distribution of the object is preserved in the production of the micrograph. The intensity distributions of the diffraction patterns are very similar[5]; their essential difference results from a loss of the high resolution components of the diffraction in the stain-electron imaging system. Since the Fourier transform of the optical pattern is the electron image of the negatively stained crystal, and the Fourier transform of the X-ray pattern is the electron density distribution (proportional to mass) in the crystal, the similarity of the diffraction patterns demonstrates the proportionality of the optical density distribution on the micrograph to the mass distribution in the specimen.

A photograph is a projection in two dimensions onto the plane of the film of the object details which scatter light into the aperture of the camera lens. If the object is opaque, the recorded detail is related to only one side of the object, viz., that side viewed by the camera. Photographs are quite similar to images formed at the retina, and, ordinarily, their interpretation is uncomplicated.

A usefully thin negative stain specimen for transmission electron microscopy is essentially translucent to the incident beam, and the information in the scattered radiation is relevant to the mass distribution in three dimensions in the specimen. Due to the limited aperture of the

[5] D. J. DeRosier, R. M. Oliver, and L. J. Reed, *Proc. Nat. Acad. Sci. U.S.* **68**, 1135 (1971).

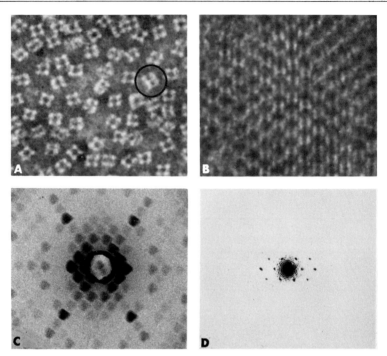

FIG. 1. (A) An electron micrograph of dihydrolipoyl transsuccinylase (LTS) from *Escherichia coli* negatively stained with equal parts of methylamine tungstate and sodium methyl phosphotungstate. LTS has 432 symmetry; its 24 subunits are clustered into 8 morphological units near the vertices of a cube. In negative stain micrographs the molecules appear hollow; the square, 4-fold view and the 2-fold view, which resembles the capital letter "H" (circle), can be seen in the micrograph. (B) A micrograph of a crystal of LTS negatively stained with sodium phosphotungstate viewed down a 2-fold axis of symmetry; the characteristic 2-fold appearance of the individual molecules can be seen in the crystal. (C) A [110] axis, zero-layer precession, X-ray diffraction pattern from a crystal of LTS grown in ammonium sulfate. The space group of the crystal is F432. (D) An optical diffraction pattern from micrograph (B). On the diffraction patterns, a spacing of 1 mm corresponds to $1/402 \text{ Å}^{-1}$. The dimensions of the LTS molecule estimated from measurements of optical and X-ray transforms of stained and unstained crystals, respectively, are identical, within experimental error, and are the same as those based on measurements of individual molecules as seen in micrograph (A). (A) and (B) ×300,000.

objective lens, the depth of field of the electron microscope is several hundreds of angstroms, usually greater than the thickness of the specimen. As a consequence, all levels of the specimen are focused simultaneously, and specimen details from all levels are superposed at the two-dimensional image plane. Thus, the micrograph image is a projection of the mass distribution of the specimen in a direction parallel to the beam onto a plane perpendicular to the beam; the densities recorded do not simply display details from one surface of the object. For this reason the interpretation of a micrograph may not be as straightforward as the interpretation of a photograph, particularly for complex structures.

When a population of identical protein molecules is embedded in negative stain, the individual molecules may be randomly oriented with respect to the beam, and the micrograph will record a variety of views of the molecule. The deduction of three-dimensional structure from such a set of two-dimensional projections of the molecule requires the discovery of the relationships between the views. Only if a molecule is essentially spherical is the projection of its mass from any view identical with any other, and its orientation in the specimen of no consequence to the interpretation. Since most protein molecules are not spherical, the micrograph may exhibit such a large variety of views as to confound simple attempts to correlate the views and derive a model of the molecule. The superposition of component masses of complex structures may significantly complicate the problem in that structural detail may be obscured in a large portion of the views.

When a heterogeneous population of protein molecules is present in the specimen, each unique protein structure will be represented on the micrograph by a set of orientation-related views; thus the collection of images on the micrograph will be comprised of two or more sets of views. Since images from different sets may be confused, serious errors of interpretation can occur. If an individual component of the specimen material differs appreciably in morphology or size from other components, its composite set of images may be easily recognized; by contrast, sorting the images of molecules which are of similar size and shape cannot always be accomplished with certainty.

The sorting of a collection of images from a heterogeneous population of protein molecules into sets of views which represent the several molecular species demands a careful scrutiny of the detailed features of the individual images. Fortunately, the presence of axes of symmetry in the design of many macromolecules tends to simplify the problem by reducing the number of possible views; for example, the presence of helical symmetry reduces all views normal to the helical axis to essentially the same view. If the specimen molecule is so oriented that the view is projected along an axis of symmetry, the micrographic image will reveal the presence of the symmetry operator. Frequently the systematic clustering of asymmetric protein masses about the symmetry axis imparts a distinct and easily recognizable morphology to the molecule. Therefore, the discovery of those views that exhibit symmetry may simplify the postulation of a model by which the whole set of views may be rationalized.

Any pair of images from a set of orientation-related views of the same structure must exhibit a common projected dimension, even though the features of the images differ significantly (Fig. 2). On this basis, pairs of images with a common dimension may be selected from the collection for consideration; the features of the two images may suggest possible three-dimensional distributions of mass in the specimen molecule. If the selected images are of the same set, one view may be transformed to the other by a simple rotation of the envisioned structure about an axis, the projection of the axis being common to the images; thus the relationship of the views may be conjectured. The dimensional features of the images, measured parallel to the rotation axis, are common to the two images. Obviously, each selected view may be paired with any other view of the set, and the correlations of features and dimensions of the images extended. Tentative ideas of the structure thus are tested and refined by an appropriate modification of detail as needed to account for all the views considered. By correlating several combinations of image pairs in this fashion, indirect evidence may be deduced that a certain gallery of views are of the same set; by the same study, a hypothesis as to the structure of the entity represented by the set will be developed.

When the collection of images on the micrograph is actually comprised of a single set of views, that is, the molecules in the specimen population are identical, an appropriate conjecture of structure based on correlations of image pairs must, with particular views, account for all images appearing in the collection. If, in the course of specimen preparation, some of the protein molecules of an identical population are denatured, dissociated, aggregated, distorted, or incompletely engulfed by the stain, the specimen population is, of course, heterogeneous, and sorting of the resulting images is essential to a complete evaluation of the collection.

Some of the uncertainties inherent in an interpretation based on images which are paired solely because of a common projected dimension may be obviated by the study of tilted pairs of images. When, after a field of images are recorded, the specimen is tilted a few degrees with respect to the beam and a second micrograph of the same field is taken, the two images (one from each micrograph) of the same specimen entity constitute a tilted pair of views. When the field of images

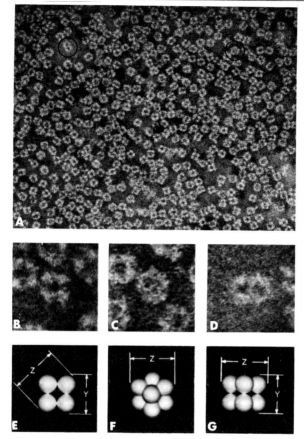

Fig. 2. (A) An electron micrograph of dihydrolipoyl transsuccinylase (LTS) from *Escherichia coli* which was treated with glutaraldehyde before negative staining with sodium methyl phosphotungstate. A random orientation of the LTS molecules in the specimen produces a variety of orientation-related views of the molecules on the micrograph. (B), (C), and (D) are images selected from (A), (circles), which are interpreted as views down a 4-fold axis, a 3-fold axis, and a 2-fold axis, respectively (see legend of Fig. 1). (E), (F), and (G) are photographs of a model of LTS consisting of 8 spheres at the vertices of a cube viewed down the same symmetry axes as (B), (C), and (D), respectively. Dimensions marked Y and Z illustrate the common projected dimensions between pairs of views. (A) ×200,000; (B)–(D) ×700,000.

The specimen was prepared by the addition of 25 μl of 10% glutaraldehyde to 540 μg of LTS contained in 70 μl of 0.05 M potassium phosphate buffer, pH 7.0, at 4°. After 10 minutes, the solution was warmed to room temperature. After 30 minutes, 20 μl of the treated LTS solution were added to 1 ml of 0.25% stain solution, pH 7.0. Sucrose (100 μl of an 0.03% solution) was added to the preparation to enhance wetting of the support film of Butvar B-76. The preparation was sprayed onto the support. Treatment of LTS by glutaraldehyde appears to inhibit the penetration of the molecule by the stain; thus the contrast is noticeably enhanced.

is viewed stereoscopically, study of the detailed features of individual molecules is facilitated, and the general morphology of the molecules may be evident. The views depicted by the individual images of a tilted pair are known to be of the same set of orientation-related views; their relationship also is known from the direction and angle the specimen was rotated about the tilt axis. For these several reasons, inferences as to the possible structure of the molecule may be drawn with greater confidence. Since each image of a pair must also exhibit a common projected dimension with the individual images of other tilted pairs of the same specimen entity, the sorting of images into sets of views and the correlations of views within the set is accomplished with more assurance. However, when the component masses of a complex structure are relatively closely packed, the superpositioning of those masses in the images may so obscure their arrangement that ambiguities in the interpretation cannot be avoided.

As suggested above, the correlation of many views within a set involves concomitant hypotheses as to the distribution of mass in the molecule and the relationships between the views; therefore, corroboration of the structural conjecture necessarily is dependent on an experimental determination that the position of the views have been correctly postulated. When a molecule or complex assemblage possesses distinctive features, a modest rotation of the molecule, or its model, may so change the superpositioning of masses in the direction of the projection as to significantly alter the features of the image. A comparison of the alteration of image features which results from a given tilt with that expected when the conjectured model is subjected to the same tilt will demonstrate the consistency of the interpretation of the gallery of views.

Crowther, DeRosier, and Klug have proposed a method for the unambiguous reconstruction of three-dimensional models from negative stain images when the relationships of the views can be established.[6] Their method depends on the fact that the Fourier transform of a two-dimensional projection of an object is identical with the central section of the three-dimensional transform of the object from the same point of view. The three-dimensional transform can, therefore, be built up plane by plane using the transforms of different projected views of the object. The transforms of the projected views may be calculated from density measurements on negative stain micrographs. In effect this amounts to reversing the focusing process mathematically to obtain data which might have been observed in the diffraction. The three-

[6] R. A. Crowther, D. J. DeRosier, and A. Klug, *Proc. Roy. Soc. Ser. A* **317**, 319–340 (1970).

dimensional image, or model, may be synthesized from an appropriate set of these data.

Applications of Negative Stain Microscopy

Negative stain microscopy has been firmly established as an effective technique for examination of a variety of materials of biological interest, even though inherent limitations preclude the delineation of structure at the atomic level. Information about structure at lower resolutions can be of particular value in studies of protein macromolecules. The observational data from microscopy relate directly to the size and shape of protein masses. Frequently, protein studies by other techniques require the expenditure of considerable effort in the collection of data relevant to these same properties. Measurements from negative stain images and a direct assessment of compactness and morphology of a specimen protein may significantly aid the evaluation of such physiochemical data. The converse is also true; physiochemical data are helpful in the interpretation of micrographs.

The fate of materials subjected to purification procedures and the homogeneity of components which have been isolated are matters of continuing concern. If the components can be distinguished on the basis of size or appearance, negative stain microscopy may be used to good advantage to monitor the purity of preparations and to appraise the efficiency of preparative procedures. Since, in effect, the components are examined particle by particle, minor constituents of a preparation which may escape observation by sedimentation or electrophoretic techniques may be readily detected. Once the conditions for preparation of the negative stain specimen are established, application of the technique to these uses is rapid and straightforward.

Under appropriate conditions, protein molecules tend to aggregate in an ordered way to form more complex structures. These larger assemblies may involve more than one kind of polypeptide chain. Further, many native complex structures of biological interest are ordered aggregates which, under appropriate conditions, may be dissociated. Use of negative stain microscopy in the study of aggregation or dissociation of a multichain system may simplify discovery of the design of the complex structure and disclose the relationship of the components. The effect of specific chemical or physical treatments on the structure of such systems may be directly evaluated. Resolution of a multienzyme system and characterization of the constituent proteins by chemical and physiochemical techniques may not unambiguously demonstrate the stoichiometry of the complex and the structural relationships of the constituents. Microscopic studies may provide an objective basis for the selection among various possible models postulated from chemical and physiochemical observations.

Micrographs of negative stain images of ordered aggregates and small crystals of protein are appropriate objects for study by optical transform methods. The analysis of optical transforms of aggregate images may provide more precise estimates of the dimensions of the aggregating unit than might be obtained from measurements on images of randomly oriented, unaggregated molecules. In the case of crystals, knowledge of the unit cell dimensions and of the symmetry operators present in the crystal can be deduced by such analysis. Information of this nature can be used with X-ray diffraction data to analyze the packing of molecules in the crystal. In addition, micrographs of crystals can also be used to provide images of the structural detail within a single molecule. The procedure is essentially identical to the mapping of electron density from X-ray diffraction patterns with the exception that the phases are determined directly from the micrographs.[7] The three-dimensional map which results in a map of the distribution of negative stain, the structural detail being visible as an absence of stain.

Materials and Techniques for Negative Stain Microscopy

Negative stain microscopy is an art. The production of micrographs of reasonable quality involves the simultaneous control of a considerable number of instrumental and procedural variables that have a direct affect on the contrast and the level of noise in the image. The ease with which structure is detected in the specimen is dependent on the image quality, or contrast-to-noise ratio. Artistry is expressed in the skill of the microscopist in optimizing image quality. In large measure his success is dependent on attention to seemingly small but actually critical details.

Image defects introduced by the nonideality of the instrument, such as lens aberrations, mechanical and electrical instabilities, can seriously degrade the image quality. Competent operation of microscopes of modern design will reduce noise from these sources to such a degree that the limit of resolution imposed by the instrument lies well beyond that imposed by the specimen. The operating practices necessary to achieve an acceptable performance of the instrument are commonly recognized and will not be elaborated here since negative stain specimens impose no unique difficulties in this regard.

[7] D. J. DeRosier and P. B. Moore, *J. Mol. Biol.* **52**, 355 (1970).

Negative Stains

As negative stain specimens are ordinarily prepared, the protein macromolecules and surrounding stain are supported by a thin film, usually of carbon or plastic. Ideally, the stain should be of uniform thickness, and just sufficiently deep to completely engulf the specimen molecules. The whole specimen should be of limited thickness, a few hundred angstroms at most.

When the specimen is examined in the microscope, electrons incident on the specimen usually have been collimated into a beam of limited divergence and cross section. The energy distribution of the electrons is sufficiently narrow that, practically, the beam may be considered monochromatic with a fair degree of coherence. Under typical conditions the probability of a complete transfer of the energy of an incident electron to the specimen is relatively low, and absorption is of no consequence in image formation. The probability of any electron interaction is such that many incident electrons, usually about 40%, traverse the specimen without scatter and constitute a relatively high background intensity against which the image is formed.

A major portion of the electrons incident on the specimen are elastically scattered by interaction with the nuclei of specimen atoms. Since no energy is transferred to the specimen, and since the mass of the electron is small compared to that of the nucleus, a large angle of deflection may result from a single scattering event. As a result, a substantial portion of the elastically scattered electrons are removed from the beam by loss to the walls of the instrument or to an appropriately placed aperture. This subtraction of electrons from the beam is the principal process by which contrast, called "amplitude" contrast, is produced in the image. Phase contrast is of minor importance in the formation of near-focus images of negatively stained protein molecules.

The likelihood that an electron will interact with a specimen is dependent on the scatter cross section of the specimen atoms and the number of atoms encountered along the trajectory of the electron. The scatter cross section, or the probability that a scattering process will occur, is proportional to the energy of the electron and the nuclear charge, which, for the light elements at least, is proportional to the mass of the atom. The specific cross section, S (also called the mass-scattering cross section, cm^2 g^{-1}) for a given species of atoms of atomic weight A is defined,

$$S = \frac{N_A}{A} \sigma_t \tag{1}$$

where N_A is Avogadro's number, and σ_t is the total cross section for scattering to angles greater than the aperture angle of the objective lens under a given set of instrument conditions. Measurements of S have shown it to be remarkably constant over a wide range of atomic weights and specimen densities; it is essentially independent of the chemical composition of the specimen material. The magnitude of S is dependent on the operating parameters of the microscope, principally the objective aperture angle and the accelerating potential of the electron source.[8] The number of atoms encountered along a trajectory of an electron through a specimen composed of atoms of atomic weight A is proportional to the packing density (atoms cm^{-3}) and the length of the trajectory, t. Since the density of the material, ρ, is the product of the packing density and the mass per atom, it follows that the number of target atoms is proportional to the mass thickness, ρt. For a given set of instrument conditions, the electron intensity, I, emerging from any path through the specimen is given by

$$I = I_0 \exp(-S\rho t) \qquad (2)$$

where I_0 is the intensity of the incident beam, and the ratio of intensities emerging from any two paths is proportional to the difference in mass thickness along the paths. Thus if the densities of the stain and protein are ρ_s and ρ_p, respectively, the intensities emerging from a path through the protein and one through the surrounding stain, I_p and I_s, are related by the expression

$$\ln \frac{I_p}{I_s} = S(\rho_s - \rho_p)t_0 \qquad (3)$$

where t_0 is the thickness of protein in the selected path, provided that the specimen molecule is completely immersed in a stain bed of uniform thickness. Obviously, the essential property of a material useful for negative staining is a high density relative to that of the protein.

For structural detail to be shown on a micrograph, it must be faithfully preserved as the specimen is prepared and subsequently exposed to the vacuum and electron beam. The selection of a stain that will render reasonable image quality requires, therefore, a consideration of several other properties of the material. The specimen preparation

---

[8] When viewing a positively stained specimen, an object whose scattering cross section is higher than that of the materials imaged in the background, higher contrast in the image is achieved by severely limiting the aperture angle of the objective lens. This technique is ineffective when the specimen object is negatively stained and is of lower scattering cross section than the background material. A reduction of the accelerating potential will result in contrast enhancement from specimens of either type. R. C. Valentine and R. W. Horne, *in* "The Interpretation of Ultrastructure" (R. J. C. Harris, ed.), p. 263. Academic Press, New York, 1962.

methods in general use involve the commingling of the stain and protein in an aqueous medium. As the specimen dries, the stain presumably replaces water in the interstices and at the boundaries of the protein molecules until, ideally, all hydrated regions are filled. The enveloping stain then supports the protein structure, if the deposit of stain is thicker than the embedded protein. Thus the protein escapes exposure to surface tension forces and, hopefully, retains its native morphology. A material of limited solubility may begin to precipitate before the last stages of drying and fail to engulf and protect the specimen protein. A relatively insoluble material will form deposits from which the specimen protein is excluded. In general, those materials which are satisfactory stains are very soluble.

The drying together of the stain and the protein with its supporting buffer may produce artifacts, or contrast which is unrelated to the structure under study. Such noise degrades the quality of the image and may be of sufficient magnitude to obscure structural detail. To be most effective, the stain must deposit in an amorphous bed of uniform thickness and density to produce an artifact-free background against which the images of the specimen protein are viewed. A tendency to form microcrystals or other anomalies of structure or density in an otherwise homogeneous deposit will result in excessive diffractive or refractive grain effects in the slightly underfocused micrograph. Contrast from these effects may obscure the ultrastructural detail of the specimen protein (Fig. 3).

The practical success of negative staining depends, to a large extent, on control of the stain bed thickness. During passage through a typically thin specimen, a small percentage of the beam electrons are inelastically scattered by the orbital electrons of the specimen atoms. As a result of this interaction, the beam electron is deviated from its trajectory, but for most cases the angle of deviation is not sufficiently large that the electron is lost from the beam. Most of the inelastically scattered electrons are confined to a cone near the optical axis. Since their energies differ appreciably from those of the unscattered electrons, they are not focused in the same plane, and image quality necessarily suffers. The probability of an inelastic scattering event is proportional directly to the thickness of the specimen and inversely to the accelerating potential. Since electron lenses cannot be corrected for chromatic aberrations, and the inelastically scattered electrons cannot be removed by conventional aperturing, noise in the image arising from this effect can be minimized only by working with the thinnest specimens possible. If wetting of the support film surface by the stain is inadequate, the stain solution may not spread well on the surface; as a result much of the

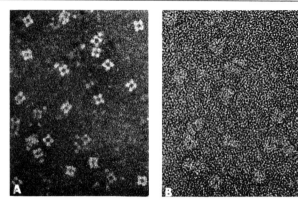

Fig. 3. Micrographs of dihydrolipoyl transsuccinylase from *Escherichia coli* (LTS) negatively stained with phosphotungstate (A) and "Uranium·EDTA" (B). The reticulated appearance of (B) results from excessive phase dependent contrast in the image. This effect is attributed to inhomogeneities in the stain bed. The images recorded were underfocused by the same amount, about 900 Å. ×200,000; electron magnification was ×87,000.

Each specimen was prepared by the addition of 50 μg of LTS contained in 12 μl of 0.02 M potassium phosphate buffer to 1 ml of stain. The solutions were sprayed on support films of carbon. The "Uranium·EDTA" stain contained 0.5% uranyl acetate, 0.25% sodium ethylenediamine tetraacetate, and 0.05% ammonium acetate; its pH was adjusted to 7.0 with ammonium hydroxide. The phosphotungstate solution contained 0.25% of the potassium salt at pH 7.0.

stain bed may be much thicker than is necessary to engulf the specimen protein, and the image quality will suffer. The surface-active properties of the stain material, if any, usually may be exploited to good advantage in the control of the stain bed thickness.

Excellent and uniform wetting of the support film surface is necessary to consistently achieve micrographs of good quality. The effect of both the stain and the protein on the tendency of the solution to uniformly wet the support film surface is direct and immediate; poor wetting often gives rise to artifacts at the stain bed-support interface. Such artifacts are more commonly observed on micrographs from specimens prepared on carbon films (Fig. 4A and B). The effect apparently results from an uneven deposition of the stain at the interface. In extreme cases, small hydrophobic areas on the film surface may not wet at all; no stain will be deposited over these areas, and the micrograph will have a spotted appearance (Fig. 4C and D).

A most desirable, sometimes essential, property of the stain is that it be chemically inert. Stain–protein interactions may involve a simple binding of the stain without a significant alteration of the protein structure or effect on its catalytic activity. In other cases, specific interactions

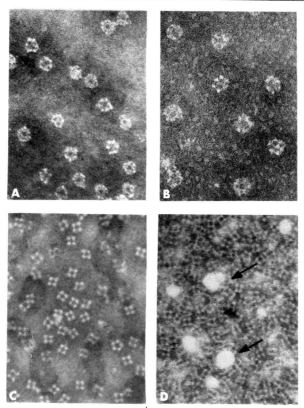

Fig. 4. (A) and (B) Micrographs of dihydrolipoyl transacetylase from beef kidney negatively stained with potassium phosphotungstate prepared on support films of Butvar B-98 (A) and carbon (B). To prepare the specimens, 40 μg of the protein contained in 30 μl of 0.02 M potassium phosphate buffer were added to 1 ml of 0.25% stain solution; the two supports were sprayed simultaneously with the protein-stain solution. The small, gray blotches on (B) result from nonuniform wetting of the carbon film surface. This effect is not ordinarily observed when plastic films are used as specimen supports. (C) and (D). Micrographs of dihydrolipoyl transsuccinylase from *Escherichia coli* (LTS) showing artifacts resulting from poor wetting of the support film by the stain solution; (C), uneven distribution of stain deposit; (D), spots barren of stain (arrows) where wetting did not occur. The specimens were prepared by floating grid-mounted carbon films for 30 seconds on solutions of LTS in 0.02 M potassium phosphate buffer. The solutions contained 40 μg/ml and 400 μg/ml of protein for grids (C) and (D), respectively. After adsorption of the protein, the film surfaces were rinsed with water and sprayed with a stain solution containing 0.125% sodium methyl phosphotungstate and 0.125% methylamine tungstate. (A)–(D) ×200,000.

may induce a consequent denaturation, dissociation of a multichain structure, or the aggregation or precipitation of the protein. An ideal stain would be inert to the protein as well as to the buffers, salts, metal ions, cofactors or other reagents commonly used to maintain the integrity of the protein structure; it would be soluble and chemically stable over a wide range of pH values.

In the course of drying, the specimen protein is briefly, but necessarily, exposed to saturation concentrations of stain, buffers or other salts with the concomitant possibility of protein alteration due to the high ionic strength of the solution or a shift in pH. If the stain is nonionic, or only weakly ionic, and exhibits no buffer capacity, the maintainence of a suitable environment for the protein throughout the preparative procedure may be simpler. If the molecular dimensions of the hydrated stain are the same or larger than the ultrastructure of the specimen protein, the visualization of that structural detail may be prohibited. Thus, some advantages may accrue from the use of stains of low molecular weight.

Finally, an obviously essential property of the stain material is that it be reasonably stable under the conditions to which the specimen is exposed in the microscope. The energy transferred by the inelastic event produces radiation damage in the specimen. The rupture of chemical bonds with the formation of new structures and the ejection of materials from the irradiated area will eventually degrade the specimen. The consequences of radiation damage can be minimized only by providing that the operations of scanning, focusing, and photography and accomplished with a minimum exposure of the specimen to the beam. The transfer of energy to the specimen may also result in a thermal degradation of the specimen and produce image drift due to thermal stresses in the specimen. Heating is reduced by minimizing the specimen thickness and restricting the incident beam to that area under immediate examination. Some materials, e.g., sodium chloride or ammonium sulfate, are readily volatilized under the impact of the beam *in vacuo;* useful stains are highly resistant to such attack. Since the ultrastructural detail observable by negative staining is replicated by the stain, any derangement or degradation of the stain deposit during exposure to the beam will result in the loss of that detail in the image.

In view of the stringent requirements, it is not surprising that comparatively few materials have been found useful as negative stains. The sodium or potassium salts of the complex heteropoly acids derived from phosphate, tungstate, silicate and molybdate, in their several combinations, have been extensively employed; the lithium, ammonium, calcium, magnesium, and tris(hydroxymethyl)aminomethane salts are also useful stains. Phospho- and silicotungstates are the most popular stains of this kind. The uranyl salts of the organic acids, formic, acetic, and oxalic, are also commonly employed as stains. Other materials that are useful, but less widely used, are sodium tungstate, ammonium molybdate, uranium nitrate, and cadmium iodide.

Densities of the materials listed range from about 3.8 to 5.7 g cm^{-3}. Although some contrast advantage results from use of the denser materials, significantly finer detail may not be observed in many applications. The uranyl compounds generally give higher contrast than the complex heteropoly acid salts. All the materials are sufficiently soluble; under appropriate conditions, each forms deposits of acceptable uniformity. The uranyl salts generally produce an underfocus refractive grain of higher contrast than the other materials. Any one of the stains may be found unsuitable for use with a particular protein preparation by reason of its reactivity with the protein or its supporting buffer. The heteropoly acids tend to positively stain proteins at low pH values, and are generally used at pH values >4. The uranyl salts are unstable at pH values >6, and are generally applied below that pH. No one of the materials exhibits superior properties with respect to surface activity or stability under electron bombardment *in vacuo*. Each of the stains fails in some respect to meet the "ideal" specifications set out above. The question of which is better is moot, each having been useful in many applications.

Methyl phosphotungstates and methylamine tungstate may be used as negative stains, at times to good advantage.[9] The methyl phosphotungstates are heteropoly complexes of the methyl esters of phosphoric acid with tungstate. The trimethyl complex is relatively insoluble, and appears to have no value as a negative stain. The mono- and dimethyl complexes are strong acids, and are extremely soluble ($\sim$300 g/100 ml). Their compositions have not been established. Preliminary data suggest their molecular weights are much lower than that of phosphotungstic acid. The density and stability of their sodium salts are comparable to sodium phosphotungstate. They are inert to some proteins which are denatured by phosphotungstate, and probably vice versa, though the converse situation has not been observed. The primary advantage of the methyl phosphotungstate stains is their greater tendency to wet the support film surface and spread sufficiently to deposit large areas of relatively thin stain bed of good uniformity. They appear to be stable over a wide range of pH values, and have been used from pH 4 to pH 9.5.

[9] A. C. Fabergé and R. M. Oliver, unpublished observations.

Methylamine tungstate is a complex product of the reaction of methylamine with freshly precipitated tungstic acid. Elemental analysis and a series of physiochemical measurements suggest its empirical formula to be $CH_3NH_2 \cdot WO_3$. Molecular weight values of the order of 400 are observed by osmometry. The solubility of the material is ~ 100 g/100 ml, and its anhydrous density is 3.9 g cm^{-3}. Solutions of methylamine tungstate are neutral and weakly ionic, but exhibit essentially no buffer capacity. The few proteins which have been critically examined in the presence of the stain did not appear to have been hydrodynamically altered. When used as a negative stain, methylamine tungstate deposits in thin beds of acceptable uniformity, which are stable under electron bombardment *in vacuo*. Its effect as a wetting agent is far superior to either the phosphotungstates or the methyl phosphotungstates.

Phosphotungstates, methyl phosphotungstates, and methylamine tungstate are compatible in any proportions and may be used simultaneously to control the stain bed thickness. Wetting of the support film surface is dependent on the properties of the surface, and on the nature and concentration of proteins and other substances (salts, cofactors, fixatives, etc.) as well as the stain in the preparation. In most situations, a considerable measure of control over the stain bed thickness can be effected by adjustment of the proportions and concentrations of methyl phosphotungstate and methylamine tungstate in the stain medium.

Benzylamine tungstate, a more effective wetting agent than methylamine tungstate, is unsatisfactory as a negative stain due to its low density. It is compatible with methylamine tungstate and the methyl phosphotungstates and may be added in low concentrations to the stain medium to improve wetting in difficult cases.

Preparation of Methyl Phosphotungstates

Phosphotungstate complexes derived from purified mono- or dimethyl phosphates and from unpurified, mixed preparations of those esters have been used as negative stains.[10]

Mixtures of the methyl esters are prepared by alcoholysis of phosphorus pentoxide as described by Hochwaldt *et al.*[11] Phosphorus pentoxide is weighed into a Pyrex crystallizing dish covered by a watchglass with a center hole. After weighing, the dish is half-packed in crushed ice so that the considerable heat generated by the vigorous alcoholytic reaction is quickly dissipated. Methanol is cautiously added, a few

[10] Monomethyl phosphate is available from K & K Laboratories, Inc., Plainview, New York.
[11] C. A. Hochwaldt, J. H. Lum, J. E. Malowam, and C. P. Dryer, *Ind. Eng. Chem.* **34**, 20 (1942).

drops at a time, through the hole in the watchglass. The methanol additions are spaced over a sufficient time that the temperature of the reactants does not rise excessively. When the addition of the methanol is complete, the watchglass is removed, and the syrupy products are stirred slowly until the solids have completely dissolved. Finally, the product is diluted with an equal volume of water.

The proportion of primary to secondary ester formed, which is dependent on the amount of water present or formed during the reaction, may be varied by adjusting the molar ratio of the reactants. When the ratio of alcohol to phosphorus pentoxide is 4:1, the product will be about two-thirds dimethyl phosphate; halving the ratio increases the yield of monomethyl phosphate to about two-thirds of the product. Essentially no trimethyl phosphate is formed by the reaction, but all the products will contain small amounts of phosphoric acid and esters of pyrophosphate and metaphosphate. The temperature of the reactants effects the speed of the reaction, but has little effect on the proportion of the products formed.

Methyl phosphotungstates are prepared by the method of Rosenhein and Jaenicke, as adapted by Bailar, for the preparation of phosphotungstic acid.[12] Typically the yield from preparations of the mixed methyl phosphate esters is about 35% of the weight of $Na_2WO_4 \cdot 2H_2O$ used. Thus to prepare about 10 g of the stain, 1.8 g of P_2O_5 and about 3 ml of methanol are used to prepare the esters, which then are added to 30 g of $Na_2WO_4 \cdot 2H_2O$ dissolved in 40 ml of water. The solution is brought to boiling, and 100 ml of concentrated hydrochloric acid are added, slowly, with constant stirring. The methyl phosphotungstic acids will begin to separate as a white precipitate after about two-thirds of the acid has been added. When the addition of the acid is complete, heating is continued while an additional 50 ml of water is stirred into the solution. After cooling, the solution and the precipitated complex acids are transferred to a separatory funnel.

The methyl phosphotungstic acids form water and ether insoluble complexes with ethyl ether and may be almost completely recovered from the reaction products by ether extraction. Sufficient ether is added to the funnel to form, after gentle agitation, a three-phase system. The heavy methyl phosphotungstic–ether complexes separate below the aqueous phase. The lower phase is removed and retained. The aqueous phase is reagitated, adding more ether as needed to maintain the upper phase; more water may be needed to redissolve any salt which separates.

[12] J. C. Bailar, Jr., *in* "Inorganic Synthesis" (H. Booth, ed.) Vol. 1, p. 132, McGraw-Hill, New York, 1939.

The lower phase is removed, and the aqueous phase is extracted a third time and finally discarded. The combined complex extracts are returned to the funnel and washed with about three volumes of ether-saturated water. Ether is added as needed to maintain the upper phase. The ether complexes are washed three or more times and finally are transferred to a shallow crystallizing dish. The methyl phosphotungstic acids are recovered from the ether complexes merely by permitting the ether to evaporate. The solid, heteropoly acids, which are deposited as a soft, amorphous cake, may be powdered and stored indefinitely. The powder is not hygroscopic, and requires no special handling. The solubility of the product material in water is about ~300 g/100 ml. A 1% (w/v) solution of the mixed ester complexes is strongly acidic (pH <3); salts of the complexes are good buffers in the pH range from 6 to 9. In these properties the product mixture is very similar to phosphotungstic acid. The sodium, potassium, lithium, magnesium, calcium, ammonium, and tris(hydroxymethyl)aminomethane salts of the mixed methyl phosphotungstates have been used as negative stains in the pH range from 4 to 9.5.

For use in the preparation of negative stain specimens, stock solutions (0.25–1% w/v) of the methyl phosphotungstates are prepared. The methyl phosphotungstic acid is dissolved in water and titrated to the desired pH by the addition of fairly concentrated solutions of the appropriate base. Powdered MgO or CaO may be dissolved directly in the methyl phosphotungstic acid solution when stains of these cations are prepared. After standing overnight at room temperature, the stain solutions are passed through a 10 nm Millipore filter and stored in glass containers at 4°.

Preparation of Methylamine Tungstate

The complex of methylamine with tungstic acid is easily prepared by the solution of the freshly precipitated acid in an aqueous solution of the amine. The light yellow dihydrate, $WO_3 \cdot 2H_2O$, is prepared by the method of Biehler.[13] Cold (~4°), molar sodium tungstate solution is added slowly, with stirring, to a cold acid solution composed of one part concentrated nitric acid, three parts concentrated hydrochloric acid, and four parts of water. Sufficient acid is used to provide a 5% excess over the stoichiometric requirement for formation of the tungstic acid. The slurry, which forms immediately on combination of the reagents, is allowed to stand in the cold for about 2 hours.

The precipitated tungstic acid is recovered by centrifugation from

---

[13] G. Biehler, *Ann. Chim.* [12] **2**, 489 (1947).

the slurry and washed by resuspension in deionized water followed by centrifugation. The washing procedure is repeated, usually five or six times, until the supernatant is free of chloride. Following the last wash, the tungstic acid is resuspended in an equal volume of water.

A 5% excess of methylamine should be added without delay to the slurry of washed tungstic acid. The requisite volume of a 40% aqueous solution of the amine is stirred into the slurry. The precipitated acid will quickly dissolve. The product solution is evaporated to dryness on a steam bath. Finally, the product cake is powdered and dried *in vacuo* over phosphorus pentoxide. The yield of $CH_3NH_2 \cdot WO_3$ is about 90% based on the tungstate.

When methylamine tungstate is dissolved to prepare stock solutions (0.25–1% w/v), the pH of the solvent is essentially unaffected. Thus very dilute buffer solutions may be used as solvents to control pH in the range from 4 to 9 when methylamine tungstate is the only negative stain reagent used in the specimen preparation. When methylamine tungstate is used with salts of methyl phosphotungstate, the latter reagent controls the pH. A stock solution of methylamine tungstate in deionized water is prepared and passed through a 10-nm Millipore filter. Aliquots of this solution and stock solutions of the methyl phosphotungstates are mixed to yield stain solutions of the desired concentration and pH.

Specimen Support Films

The techniques of thin film preparation are important components in the art of negative stain microscopy, since the performance of the specimen support is crucial to the quality of the micrograph. Unless the support fabrication procedures are well controlled, a dependable production of supports with desirable characteristics is unlikely; the recording of a micrograph of good quality then will become a matter of chance. As a result, routine studies of specimen proteins will be frustrating, if not frustrated. The specimen support is essential to mechanically position the specimen in the microscope; it also provides a suitable surface over which the stain and specimen proteins are distributed in adequately thin deposits. However, the film must be regarded as a necessary evil; it adds no contrast relating to protein structure, yet it generates noise in the image.

The physical specifications of an ideal support film are self-contradictory. Since the film scatters beam electrons, both elastically and inelastically, image quality is degraded in proportion to the mass thickness of the film; to minimize noise from this source, the thickness of the film must be limited. However, the film must be strong enough to survive the manipulations of the preparative procedures and to rigidly support the specimen under the impact of electron bombardment. Since strength and rigidity usually increase with film thickness, the thickness of the practical support film is a compromise, being acceptably thin yet sufficiently tough.

The granular appearance of most negative stain micrographs is derived from noise which overlays the image and obscures the finer image detail. This noise is generated by inhomogeneities in the specimen, both in the stain bed, as mentioned above (Fig. 3), and in the support film. Since the potential inside a substance is higher than in free space, the refractive index is greater than unity, and the wavelength of radiation in the substance is shorter than in a vacuum. Thus a wave passing through a thin film will suffer a phase delay with respect to parts of the wave passing through an opening in the film. This disturbance in the wave front results in the formation of the familiar diffraction fringes which appear in the defocused image of the edge of a hole in a film. The hole may be considered to be an inhomogeneity in the film, a zone of maximum possible variance of potential in the path of the wave through the film. Variations of lesser magnitude create similar disturbances in the wave which appear at the image plane as the familiar, phase-dependent background grain. To achieve useful specimen resolutions below ~50 Å, the noise introduced by refraction in the specimen from both the film and the stain must be limited.

Since the wave disturbance is dependent on the internal structure of the film, the magnitude of the effect is determined by both the intrinsic nature of the material and the way it is processed to produce the film. For materials now in popular use for film fabrication, carbon, collodion, and other plastics, simply reducing the film thickness to the minimum practical limit does not obviate the need for further control of the grain. Huxley and Zubay have described a technique for preparing specimens over holes in a supporting carbon substrate, thus eliminating the film over limited areas of the specimen.[14] Micrographs taken from these areas are potentially of superior quality since inelastic electron scatter from the specimen is reduced simultaneously with avoidance of the grain effect. A consistent production of useful preparations of this type is difficult to achieve, and use of the technique has been limited.

Amorphous films of aluminum oxide prepared by anodization are of sufficient structural homogeneity that the grain effect is substantially less than that introduced by evaporated carbon films. Films composed

[14] H. E. Huxley and G. Zubay, in "European Regional Conference on Electron Microscopy" (A. L. Heuwink and B. J. Spit, eds.), Vol. II, p. 703. De Nederlandse Vereniging Voor Electronenmicroscopie, Delft, 1960.

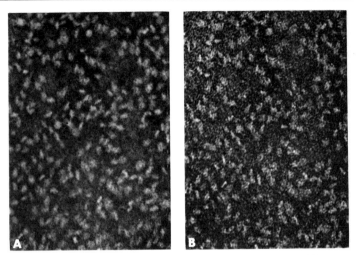

Fig. 5. Views of a field of fructose-1,6-diphosphatase from swine kidney negatively stained with methylamine tungstate and recorded near focus (A) and about 2700 Å underfocus (B). The pebbled appearance accentuated in (B) results from the focus-sensitive, phase-dependent contrast generated by structural inhomogeneities in the stain and supporting film. To prepare the specimen 50 μg of protein contained in 50 μl of 0.01 M phosphate buffer were added to 1 ml of 0.25% stain solution; the stain-protein solution was sprayed onto a support film of Butvar B-98. Micrographs were recorded at an electron magnification of ×105,000; (B) was recorded after the specimen had been tilted about 20° from its position for (A). The specimen protein was prepared by Dr. Joseph Mendicino, Department of Biochemistry, University of Georgia, Athens, Georgia. (A) and (B) ×300,000.

of crystalline materials, e.g., mica and graphite, also generate relatively little background grain.[15] However, films of these materials are comparatively difficult to fabricate and have not been widely used.

Adjustment of focus is the easiest way to routinely control background grain (Fig. 5). Phase-dependent contrast produced by refraction is an interference effect which is maximized by some degree of defocus due to path differences in the divergent and axial rays. As a result, electron images appear to be more sharply focused when actually somewhat underfocused. The background grain from a film of normal thickness appears in highest contrast at 3000–5000 Å underfocus, focus being the plane of the film. At about the same degree of overfocus the virtual image of the grain appears in reverse contrast. At true focus the grain

[15] George H. N. Riddle and Benjamin M. Siegel, in "Proceedings, Twenty-ninth Annual Meeting, Electron Microscope Society of America" (C. J. Arceneaux, ed.), p. 226. Claitors Publ. Division, Baton Rouge, Louisiana, 1971.

contrast is negligible, the film being a plane of uniform intensity.[16] When generated by a specimen protein, phase contrast contributes useful information to the image, particularly at resolutions higher than about 30 Å. This contribution is largely lost if the image is recorded at true focus on the supporting film. However, recent studies suggest this loss to be a moderate price to pay for control of the background grain of micrographs of negatively stained proteins since the resolution limit of such specimens usually is 25–30 Å.[17]

The chemical properties of the support film surface are of prime importance to the satisfactory distribution of protein and stain on the surface. Some methods of preparation involve, as a first step, the adsorption of specimen protein molecules on the support film; subsequently rinsing and stain solutions are applied. The suitability of a particular film for use in these procedures must be determined by experiment. In some cases the protein may be so avidly adsorbed that a monomolecular layer is rapidly built up on the film even from very dilute protein solutions; other specimen protein molecules may not be adsorbed in satisfactory numbers, even after a lengthy exposure of the surface to the protein solution (see Figs. 4C, and 4D, and 6). The quantity or distribution of protein in the specimen when prepared by alternative methods is not dependent on protein adsorption by the film. In these procedures the stain and protein are simultaneously applied to the film surface, usually from a premixed solution. Whichever preparative technique is used, the surface of the film must be receptive to aqueous media. That the role of the stain in wetting the surface was important to the avoidance of artifacts and the control of stain bed thickness was emphasized above. A hydrophilic film surface will further enhance the tendency of the stain solution to spread and deposit the stain in an acceptably thin bed, free of artifacts (see Figs. 4 and 6).

The aesthetic quality of the micrograph is, to a large measure, dependent on the absence of artifacts. Therefore, the maintenance of standards of acceptable quality from routinely prepared specimens requires that the causes of artifacts be at once identified and eliminated.

[16] It is difficult for the novice at electron microscopy to acquire skill at critically focusing the instrument. Fortunately the background grain in the image is an excellent focusing guide. The grain is visible in the image on the final screen when the electron magnification is greater than about 60,000 times, provided the screen luminosity is adequate. Under these conditions the "null" position of focus, that is a minimum grain position as focus is shifted from over- to underfocus (or vice versa), is easily located.

[17] H. P. Erikson and A. Klug, in "Proceedings, Twenty-eighth Annual Meeting, Electron Microscope Society of America" (C. J. Arceneaux, ed.), p. 248. Claitors Publ. Division, Baton Rouge, Louisiana, 1970.

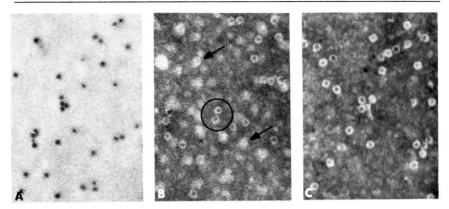

Fig. 6. Micrographs of ferritin from horse spleen which demonstrate the influence of the support film surface on the generation of artifacts which result from poor wetting. Each specimen was prepared by adsorption of the ferritin from a phosphate-buffered solution (0.02 M, pH 6.8) containing 1 mg/ml of the protein. (A) and (B) were prepared on untreated carbon films; the film used for (C) had been floated for 5 minutes on dimethyl sulfoxide and rinsed with water before exposure to the ferritin preparation. After floating for 10 minutes on the ferritin solution, each film was rinsed with water; the excess water was removed by wicking. Before the films were dried, (B) and (C) were sprayed with stain (0.25% sodium methyl phospho-tungstate, pH 6.8). (A) was examined unstained; the dense cores of the ferritin molecules show in positive contrast against the light background. The cores serve to distinguish some of the images of the negatively stained ferritin (circle) from the artifacts (arrows) on micrograph (B). The reduction of the prevalence of such artifacts observed on (C) is attributed to the preliminary treatment of the support film. (A)–(C) ×200,000.

By carefully selecting areas for recording during studies at high magnification, the distracting presence of major artifacts caused by film topography or areas barren of stain usually can be avoided. However, gentle undulations of the film surface, which affect focus, are usually undetected, and small artifacts arising from minor topographic defects of the surface, from poor wetting at the stain–film interface, and from the degradation of materials in the stain bed, may not be noticed under the viewing conditions at the final screen and may be inadvertently recorded; occasionally minor defects are so numerous as to be unavoidable. The improvement of micrograph quality, therefore, usually entails an interpretation of the artifacts recorded.

Aside from the aesthetic aspects, a practical, and sometimes more compelling, justification for the avoidance of major variations from an ideal distribution of the stain derives from a limitation of the photographic processes usually employed to produce positive prints of the image, which is recorded as a negative. The average electron intensity

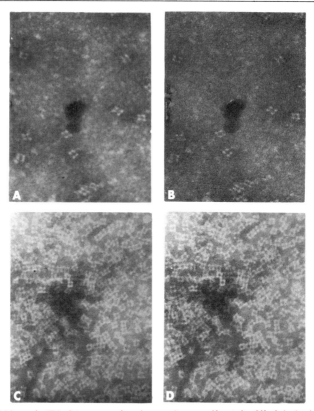

Fig. 7. (A) and (B) Stereoscopic views of a small, stain-filled hole in a carbon film, an artifact in a field of dihydrolipoyl transsuccinylase from *Escherichia coli* (LTS) which was negatively stained with sodium methyl phosphotungstate. (C) and (D) Stereoscopic views of a relatively large, stain-filled pit in a Butvar B-76 film, an artifact in a field of LTS prepared in a similar way. The interpretations of the artifacts are obvious when the micrographs are viewed stereoscopically. As the views are presented, the aspect is through the stain bed to the underlying support film. (B) and (C) were recorded after the specimens had been tilted about 10° from their positions for (A) and (D), respectively. Note the presence on (A) and (B) of wetting artifacts usually observed on carbon films. The diameter of the hole is about 250 Å; the pit is about 1000 Å across. Several molecules of the protein are embedded in the stain which fills the pit. (A)–(D) ×200,000.

The specimen for (A) and (B) was prepared by adding 40 µg of the protein contained in 30 µl of 0.01 M phosphate buffer to 1.0 ml of the stain solution (0.25%, pH 7.0); the specimen for (C) and (D) was prepared by the addition of 115 µg of the protein contained in 15 µl of 0.05 M phosphate buffer to 1.0 ml of the same stain. Sucrose (50 µl of 0.03% solution) was added to each preparation to enhance wetting of the supports. The preparations were sprayed onto the supports.

in the image of a given limited portion of a negative stain specimen is determined largely by the mass thickness of the stain in that portion. Information about objects embedded in the stain is expressed as localized variations of intensity about the average intensity for the area. If the average mass thickness of the specimen in its several portions varies widely, a wide range of electron intensities are presented to the recording photoplate. Since all parts of the image are exposed simultaneously, the range of optical densities produced on the negative photoplate may exceed 2 or more optical density units. This condition is commonly encountered in negative stain microscopy. The gray-scale range of most photographic printing papers is limited to about 1.6 optical density units; thus the whole field of view of a specimen which produces such excessive variations of electron intensity cannot finally be rendered with acceptable quality on a single print.

The topography of the film surface will influence the deposition of the stain bed and create anomalous variations in the mass thickness of the specimen. Ideally the surface should be smooth and free of ripples, rifts, pits, or holes. Small-scale irregularities of the surface are replicated by the stain and will be evident on the micrograph if their dimensions parallel to the beam are larger than 30 Å. Ripples, which project from the surface into the stain bed, locally reduce the mass thickness of the specimen; the resulting artifacts are lighter than the average background contrast. Pits and small holes, which usually will fill with stain, have the opposite effect and appear as dark artifacts (Figs. 7 and 8). If the surface is well wetted by the stain solution, the holes may conduct the solution to the opposite surface, and stain may be deposited on both surfaces of the film. Spreading of the stain will be uneven when influenced by the larger topographic features such as wrinkles or folds, and the stain deposits will be thicker in troughs (Fig. 8). When wetting is good, the stain may spread with acceptable uniformity over gentle undulations of the film, but simultaneous critical focusing of both the troughs and crests may not be possible (Fig. 9).

The nature of a defect which produces a certain artifact may not be obvious from a casual inspection of the micrograph. Local variations in thickness of the stain deposit frequently result from wetting or drying processes imposed by the surface and staining conditions, though the effect may not be easily distinguished from thinning due to ripples in the surface or other effects related to film topography. When the cause of an artifact is topographic, stereoscopic views of the specimen may straightaway suggest a proper interpretation. Pits and undulations are dramatically evident when viewed in three dimensions.

Potentially useful thin films can be prepared from a variety of ma-

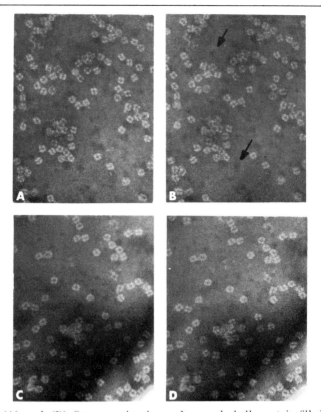

FIG. 8. (A) and (B) Stereoscopic views of several shallow stain-filled pits (arrows), artifacts in a field of negatively stained dihydrolipoyl transsuccinylase from *Escherichia coli* (LTS). (C) and (D) Stereoscopic views of a stain-filled wrinkle in the film supporting a field of the same preparation. When the micrographs are viewed stereoscopically, the nature of the artifacts is revealed. As the views are presented, the aspect is through the stain bed to the underlying support film of Butvar B-76. (B) and (C) were recorded after the specimen had been tilted about 10° from its positions for (A) and (D) respectively. Diameters of the pits range from about 50 to 150 Å. (A)–(D) ×200,000.

The protein was treated with glutaraldehyde at 4° for 30 minutes before exposure to the stain; 25 μl of 10% glutaraldehyde was added to 500 μg of the protein contained in 70 μl of 0.05 M potassium phosphate buffer (pH 7.0); 20 μl of the fixed protein solution were added to 1.0 ml of a sodium methyl phosphotungstate (0.25%, pH 7.0) stain solution which also contained 0.003% sucrose. The stain-protein preparation was sprayed onto the support film.

terials, including carbon, silicon, silicon monoxide, silicon carbide, beryllium, mica, aluminum oxide, and several kinds of plastics. Ideally, the best combination of inherent film properties—density, chemical nature, stability, internal and surface structure, etc.—should determine the choice

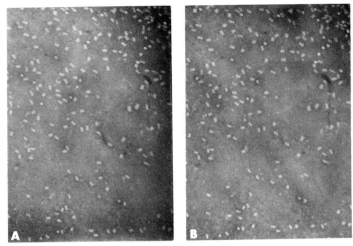

Fig. 9. Stereoscopic views of a field of aminoacyl-tRNA synthetase complex molecules (from rabbit reticulocytes) negatively stained with methylamine tungstate. The micrographs show the uniformity with which stain may be deposited over an undulation in the supporting film when wetting of the film by the stain is good. When viewed stereoscopically, the undulations of the support film are dramatically evident. As the views are presented, the aspect is through the stain to the underlying film. (A) was recorded after the specimen had been tilted about 20° from its position for (B). (A) and (B) ×100,000.

To prepare the specimen, 37 μg of protein contained in 50 μl of Tris-sulfate buffer were added to 1 ml of 0.25% stain solution; the stain-protein solution was sprayed on a support film of Butvar B-98. The specimen protein was prepared by Dr. Boyd Hardesty, Clayton Foundation Biochemical Institute, University of Texas at Austin, Austin, Texas.

of material; usually for routine work the choice is a compromise between the desired properties and the difficulty of fabrication, the latter being of greater influence. Commonly, supports for negative stain specimens are prepared from carbon, collodion (nitrocellulose), or the polyvinyl acetals.

Collodion films are easier to make than films of other materials. They are prepared from amyl or butyl acetate solutions of Parlodion,[18] usually by allowing a drop of the solution to spread on a water surface; alternately, the films may be formed on glass. Films of Parlodion offer the advantages of excellent wettability and a high propensity to adsorb proteins; their principal handicaps are mechanical instability and a tendency to disruption when exposed to electron radiation. Specimens

[18] Parlodion is the trade name of a nitrocellulose plastic manufactured by the Mallinckrodt Chemical Works, St. Louis, Missouri.

supported by collodion tend to move when first exposed to the beam and subsequently to drift during the recording of the micrograph. The film may be stabilized against movement by coating with 10–20 Å of carbon. Since collodion is volatilized by exposure to the beam, its use promotes higher contamination rates in the microscope.

Formvar has been the most widely used of the polyvinyl acetal plastics for the preparation of thin films.[19] Like collodion, Formvar films may be formed either on water or glass, but stronger and thinner films are made on glass and subsequently recovered by flotation on water. The technique of stripping films of Formvar 15/95 E from glass was described by Schaefer and Harker.[20] As usually prepared, Formvar films are inferior to collodion with respect to uniformity of thickness and wettability; they do not adsorb protein as avidly, nor are their surfaces generally as smooth. Many microscopists consider the preparation of Formvar films to be the more difficult procedure owing to the uncertainty that the stripping operation will be successful. Because of the relative toughness of Formvar, films made of Formvar are mechanically superior to those fabricated from collodion. A slight movement of a specimen supported on Formvar is induced by exposure to the beam; however, when firmly attached to a supporting net the film will quickly stabilize so that image drift during the recording of the micrograph is negligible. Because Formvar is less volatile than collodion under the impact of the beam, contamination of the instrument by the plastic is not a serious problem.

Films of the polyvinyl butyral resins Butvar B-76 and B-98 possess, in large measure, the desirable properties of both collodion and Formvar films.[19] These resins contain more hydroxyl ($>10\%$ vs. $\sim5\%$ expressed as polyvinyl alcohol) than do the Formvars; consequently, the Butvar films are the more hydrophilic and are practically as wettable as collodion. When prepared as described below, Butvar films are more tacky and perhaps a bit stronger than Formvar films. A firm attachment of the film to its support is easily formed because of the adhesiveness of Butvar. Although some shrinkage of the film on first exposure to the beam can be observed, a specimen supported by a Butvar film which is well anchored to a supporting net will drift <1 Å per second after a few seconds in the beam. The smoothness and uniformity of Butvar film compares favorably with collodion film, and its volatility under electron bombardment is similar to that of Formvar film. A reliable, routine production of Butvar films of consistent quality may be achieved by the

[19] Formvar and Butvar are the trade names of polyvinyl formal and polyvinyl butyral resins manufactured by Monsanto Company, 5051 Westheimer, Houston, Texas 77027.

[20] V. J. Schaefer and D. Harker, *J. Appl. Phys.* **13**, 427 (1942).

procedure set out below. The degree of difficulty is very little more than for production of collodion films. Butvar films offer a bonus for their use in that routinely prepared specimen supports have an occasional, well formed hole of appropriate size for checking microscope performance at high magnification.[21]

Support films of excellent uniformity and stability under instrument conditions can be prepared from carbon. Ordinarily, graphitic carbon is deposited *in vacuo* onto surfaces of mica, glass, detergent-coated glass, or plastic films by a technique described by Bradley.[22] Carbon deposits on split mica surfaces, which are atomically smooth, are easily stripped by flotation on water. The stripping of carbon from glass surfaces is much more difficult. Films produced either way are fragile, tend to break apart when stripped, and are easily ruptured in the subsequent mounting and drying operations. A firm bond of the film to the supporting grid is difficult to secure, and the yield of dependably useful specimen supports from water-stripped films is usually unsatisfactory. A more reliable method is to deposit the carbon on a plastic substrate previously mounted on a supporting grid. Subsequently the plastic is partially removed by the action of an appropriate solvent; a residue of plastic left sandwiched between the carbon and the grid provides a firm attachment of the film to the grid. Control of the solvent treatment is necessary to achieve the desired result. Rupture of the film over a substantial portion of the grid openings may result from the solvent treatment if a Parlodion substrate is used. This result, which has been attributed to the swelling of the plastic before solution, may be avoided by the use of other nitrocellulose plastics or Butvar for substrates and a procedure described below for the removal of the plastic. Since the surface topography of the substrate is replicated by the technique, carbon films deposited on plastic substrates are not as uniform or smooth as films formed on mica.

A most efficient and reliable procedure for the routine production of grid-mounted carbon films was described by Fabergé.[23] The carbon is deposited on the surface of an organic glass, benzylamine tartrate, into

---

[21] "Holey" specimens of excellent quality for checking instrument performance are easily prepared from the Butvars simply by choosing not to control the humidity when drying the film as described in the preparation procedure. "Holey" specimens may be stabilized with carbon if desired, but usually a minute or so of exposure to the beam will sufficiently stabilize the film surrounding the hole so that specimen drift values <1 Å per second are observed.

[22] D. E. Bradley, *Brit. J. Appl. Phys.* **5**, 65 (1954).

[23] A. C. Fabergé, *in* "Proceedings, Twenty-eighth Annual Meeting, Electron Microscope Society of America" (C. J. Arceneaux, ed.), p. 484. Claitors Publ. Division, Baton Rouge, Louisiana, 1970.

which copper support grids have been inlaid. The grid surface is fully exposed to the carbon which is deposited in a continuous film over both the copper and benzylamine tartrate surfaces. Thus the firmest possible attachment of the film to the grid is assured. Subsequently, the benzylamine tartrate is dissolved away by a simple procedure which does not subject the carbon film to high stress. Details of the preparation of carbon specimen supports by this method are given below.

As ordinarily fabricated, carbon films do not generate significantly less noise in the image than do plastic films even though the carbon film may be substantially thinner. Since the density of carbon is about three times that of the plastics, a carbon film of the same mass thickness can be only about one-third as thick as a plastic film. Fortunately a carbon film of adequate strength need be only 20–50 Å thick even when a supporting net is not used. Thus inelastic scattering of electrons by the film may be reduced by the use of carbon. However, this advantage is largely offset by the relatively stronger background grain generated by inhomogeneities in the carbon film. Carbon supports thinner than 20–50 Å may be expected to generate less noise in proportion to their thickness. Williams, Glaeser, and Richards have developed a technique for the fabrication of extraordinarily thin films of carbon (4–6 Å).[24]

The principal handicap arising from the use of carbon films deposited by the Bradley method is the hydrophobic nature of the carbon surface which inhibits the spreading of the stain and tends to generate wetting artifacts. Unless the carbon surface is treated to render it more hydrophilic, the superior stability of specimens prepared on carbon will not result in a general improvement of specimen quality over specimens prepared on plastic films. A variety of surface treatments have been devised. Surface active materials have been adsorbed on the surface or added to the stain or protein media used in the preparation of the specimen; proteins, notably serum albumin, have been adsorbed on the surfaces prior to the application of the specimen protein and stain. Carbon supports prepared on plastic substrates are usually more wettable than films formed on mica. Apparently during its removal some of the plastic is adsorbed on the carbon surface.

Preparation of Specimen Support Films of Butvar B-76 or Butvar B-98

The polyvinyl butyral resins are nontoxic, free-flowing powders which require no special handling. When stored in moisture-proof con-

[24] Robley C. Williams, R. M. Glaeser, and K. E. Richards, in "Proceedings, Twenty-ninth Annual Meeting, Electron Microscope Society of America" (C. J. Arceneaux, ed.), p. 482. Claitors Publ. Division, Baton Rouge, Louisiana, 1971.

tainers in the dark, their shelf life extends to years. Because the hydroxyl content of Butvar B-98 (20% polyvinyl alcohol) is about twice that of Butvar B-76, films of Butvar B-98 are the more hydrophilic.

The resins are of limited solubility in chlorinated hydrocarbons, but it is from these solvents that the best films are cast. Reagent grades of 1,2-dichloroethane (ethylene chloride) or chloroform (which contains about 0.75% ethanol as a preservative) are preferred. Resin solutions are prepared and stored in glass Erlenmeyer flasks with aluminum-wrapped stoppers. (The resins are adhesives, and will cement glass stoppers in place. Plastic containers are unsatisfactory since the solution may be contaminated by plasticizers from the container.) The resins disperse slowly in the solvent, and solutions should be allowed to stand, with occasional agitation, for several hours before use. The resin solutions are unstable. Films cast from aged solutions may be excessively "holey," are less stable in the beam, and are not as wettable. The aging process is accelerated by light. For best results, film-making solutions should be stored in the dark, and renewed after 3 or 4 weeks.

The molecular weight of Butvar B-98 is somewhat lower than that of Butvar B-76 (~32,000 vs. ~50,000), and its solutions are of lower viscosity. Therefore, Butvar B-98 films are thinner than those from Butvar B-76 when cast the same way from solutions of the same resin content. The thickness of the film may be controlled by adjusting the resin content of the solution. The interference color of films prepared by the procedure given below from a 1% (w/v) chloroform solution of Butvar B-76 indicates the film thickness to be about 1,000 Å.[25] Reducing the resin content to 0.25% produces films that are gray (thickness <500 Å). Films that are transparent and almost invisible may be routinely cast from 0.15% solutions. Films cast from a 1% chloroform solution of Butvar B-98 have a bluish silver color indicating a thickness of 600–900 Å. Films of Butvar B-98 also may be routinely cast from chloroform solutions containing as little as 0.15% resin.

Practical success in Butvar film preparation depends on the rigorous control of several steps in the procedure. The films are formed on standard 1 × 3 inch glass microscope slides. Consistent results are obtained only if the slide surfaces are carefully prepared. As obtained from suppliers, "precleaned" slides are usually coated with residues and are not uniformly wettable. These residues usually affect the quality of a Butvar film formed on the slide. If the slides used in making films are not routinely cleaned, the effect of these materials must be considered when poor results are obtained. Some batches of slides are not usable;

[25] L. D. Peachey, J. Biophys. Biochem. Cytol. 4, 233 (1958).

even after extensive cleaning treatment, films cannot be recovered from their surfaces.

Ordinarily the residues may be removed by soaking in aqua regia. Batches of slides are immersed in freshly prepared aqua regia, and left until needed. Usually after 24 hours the slides are sufficiently clean, but for some batches extended treatment may be necessary. The uniformity of wetting with water is a useful criterion for evaluating the surfaces of the slide after cleaning. While the slide is held with its long edge vertical (with forceps clamped at the lower edge), the surface is rinsed with deionized water. If the surface is sufficiently clean, the residual water will drain without beading, and a film will be retained over the whole surface until drying begins. Any tendency of the residual film to draw into beads (particularly along the edges) indicates that further cleaning is necessary. If the slide is clean, excess moisture is wicked off at its lower end, and it is air-dried in a dust-protected environment.

To assure that the film which will be formed on the surface of the slide can be removed with ease, the clean, air-dried slide surface is coated with a detergent. Brij 35, a polyoxyethylene lauryl ether preparation, is useful for this purpose.[26] About half of one side and the edges of the slide are waxed with the pasty detergent. After lightly polishing off the excess with a lint-free cloth, an even surface film of detergent remains on the glass.

The resin solution from which the film is to be made is contained in a test tube of sufficient diameter to receive the slide end-on, and of sufficient length that the whole slide may be held out of the solution but still be confined in the test tube. Fill the test tube to a depth of about 1 inch with the solution. Holding the slide with forceps (or a hemostat), dip the prepared end into the resin solution for 10–15 seconds. The time of exposure of the slide to the solution is important; less than 10 seconds' exposure will not consistently produce useful films.

Withdraw the slide into the vapor-saturated space just above the fluid, and drain the excess solution from its surfaces by maintaining physical contact between a lower corner of the slide and the wall of the test tube for about 1 minute. Draining in this way in the vapor atmosphere retards drying; thus more of the solution will be removed, and a thinner, more uniform film will result. After draining, the slide is air-dried for a minute or two.

The condensation of water droplets on the film during drying will

<hr/>

[26] Brij 35 is manufactured by Atlas Chemical Industries of Wilmington, Delaware. Highly purified Brij 35 is available from Pierce Chemical Company, Rockford, Illinois.

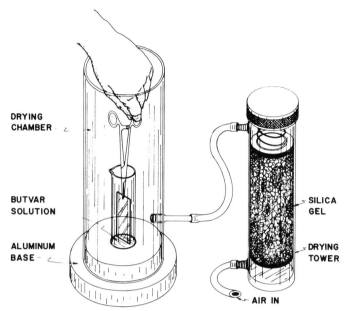

Fig. 10. Sketch of a simple apparatus for control of the humidity of the atmosphere in which Butvar films are dried. The drying chamber, made from a piece of acrylic tubing 4 inches in diameter and 12 inches high, is held upright by a loose-fitted base turned from 6-inch aluminum stock. The base is center-bored to a depth of 1.5 inches to receive a large test tube, the container for the Butvar solution. In use, air is supplied continuously at about 2 psi to the column of silica gel. After the slide has been drained within the test tube, as described in the text, it is dried in the chamber just above the test tube.

produce pits and holes in the film. Under humid conditions the population of holes will be so high that the film is ruined. A simple apparatus useful for controlling humidity during the drying step is shown in Fig. 10.

After it has dried, the film is severed along the edges of the Brij 35-coated surface by scraping the edges with a knife blade or the edge of a spatula; the film is stripped from the slide by flotation on water. A staining dish (Fig. 11) is a convenient container for the water on which the film is stripped. Precautions should be taken to assure that the water surface is free of floating particles. Sweeping the surface is most easily accomplished by the following procedure: the precleaned dish is filled with deionized water to the level of the sides of the dish, then more water is cautiously added until, with the aid of surface tension, the water surface is above the level of the sides. An 8 × 10 inch sheet of lens tissue is gently laid onto the surface. After wetting of the tissue over the whole surface of the dish is complete, the tissue is removed by dragging it

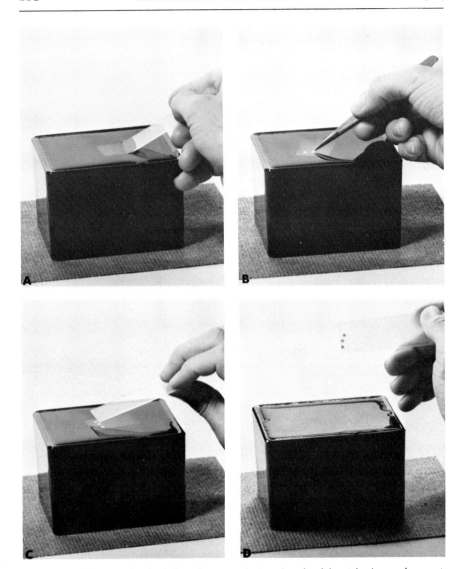

FIG. 11. Photographs depicting the several steps involved in stripping and mounting Butvar specimen support films. The film, formed on a glass microscope slide, is stripped from the slide by flotation on water. Films of useful thickness are hardly visible while floating on the water surface and are most easily viewed by light reflected from that surface; the sides of the water container, a glass staining dish, have been painted black to improve viewing conditions. In (A) a film (extraordinarily thick to improve photographic contrast) had been completely stripped from a slide; another slide (sand-blasted to improve photographic visibility) was used to demonstrate the angle at which the film and slide are presented to the water

horizontally over the surface. Particles on the surface of the water are swept off with the tissue.

To strip the film, hold the slide at a low angle (15°–20°) to the surface of the water and gently lower the end covered by the film to, and then slightly below, the surface. Surface tension forces will lift the outer edges of the film; water will be drawn into the film–glass interface, thus parting the film from the glass. As the slide is slowly lowered into the water, the freed film will be left floating on the surface. The slide may be retained for use in the retrieval of the grid-mounted supports from the water surface.

Firm attachment of the film to its support is essential to the mechanical stability of the specimen in the beam. When mounting the film on a bare grid, a strong bonding of the film to the grid is assured if the grid surface is precoated with an adhesive. Butvars B-76 or B-98 are useful for this purpose. The coating is easily accomplished by spacing clean grids about 5 mm apart on a pad of bibulous paper (2 or 3 thicknesses). When a drop of 0.15% (w/v) chloroform solution of Butvar is dropped onto a grid, the excess solution will quickly move into the paper, leaving a thin, tacky coating on the grid surfaces. The grids should be protected from dust and allowed to dry about 5 minutes before mounting. When the film is mounted on a carbon- or silicon carbide-stabilized net (prepared as described below), adhesive coating of the net is not necessary to achieve a good bond, provided the resin solution from which the film is cast is reasonably fresh.

Support grids, or grid-mounted nets, are laid onto the film as it floats on the water surface. The thickness of the film will gradually vary along its length, being thickest at the end that was lowest during the draining–drying operation. Usually only the mid-portion of the film is mounted. Grids are arranged on the film in any pattern which will leave about 3 mm between grids. No grid should be nearer than about 1 cm to the thickest end of the film. Up to a dozen grids may be applied to each film. After the grids are in place, each in turn is pressed onto the film by gently pushing downward with the tips of the forceps with sufficient pressure to depress the grid and film about 1 mm below the surface of the water. The reaction of surface tension forces to this maneuver

surface. Specimen grids precoated with Butvar, or grids with premounted nets, are placed on the floating film (B). To retrieve the grids, a microscope slide is presented at a low angle to the film (C), and the film is depressed slightly below the water surface. Surface tension will press the film against the underside of the slide. Adherence of the film over its whole surface to the slide assures the application of minimal stress to the grid–film bond during removal from the water surface (D).

will stretch the film and press it onto the grid, thus assuring maximum adherence.

The specimen supports are retrieved from the water surface with the same slide on which the film was formed. Again holding the slide at a low angle, the side of the slide which was not coated with Brij 35, which is now coated with Butvar, is presented to the film. Place the end of the slide on the film about 5 mm from its thickest end, and depress the film a millimeter or two below the surface. Surface tension will press the film and grids onto the lower surface of the slide to which the film will adhere. The slide with film attached is completely immersed, turned edge-up, and retrieved edge-first back through the surface. When the retrieval operation is successful, a small pocket of air is entrapped with each grid, the film adhering continuously to the slide between the grids. If inspection of the grids after retrieval shows the grids to be wet, the film has been ruptured and the specimen mount may not be usable. Excess water is wicked off the edges of the slide, and the grid–film–slide assembly is placed in a dust-free environment to dry at least half an hour before use. When the film has dried, the specimen mounts are easily picked off the slide with forceps.

After several days of aging the film surfaces are not as wettable as when freshly prepared. It is advisable, therefore, to prepare the specimen mounts as needed. When large grid openings are required and the use of stabilized nets is prohibited, carbon may be coated on the Butvar film to achieve improved mechanical stability. Ordinarily a 10–20 Å thickness of carbon is sufficient. The stabilizing carbon may be evaporated through the grid openings onto the back of the film; thus the more wettable plastic surface is preserved, even though its wettability is impaired by exposure to the vacuum of the evaporator. Alternately, the carbon may be evaporated onto the face of the film and the Butvar subsequently partially dissolved away to produce an ultrathin, grid-mounted carbon film.

Preparation of Specimen Support Films of Carbon

Grid-mounted support films of carbon may be routinely prepared by the deposition of the carbon *in vacuo* on substrates of either Butvar or benzylamine tartrate as mentioned above. Butvar B-76 film is prepared as described for specimen supports from a chloroform solution (0.25% w/v) and stripped on water. A batch of adhesive-coated 200–400-mesh grids are laid on the floating film; the whole assembly is retrieved on a microscope slide. After thorough drying, the slide with the entire batch of grids attached is placed in the evaporator, and a 20–50 Å thickness of carbon is deposited on the film. Afterward the grids are individually transferred to an extractor (described below), carbon face up, and most of the plastic is removed with solvent. The carbon film which remains is firmly attached to the grid. The tenacity of film-grid bond assures the survival of the film over a major fraction of the grid openings as the support subsequently is subjected to the stresses of specimen preparation.

Carbon films prepared by this technique are generally suitable for the support of negative stain specimens subject only to the handicap of the films' limited wettability. Since the film replicates the surface irregularities and undulations of the plastic substrate on which it was formed, it will not necessarily be of uniform thickness, nor will its surface be plane. Ordinarily these defects are of minor importance if the Butvar substrate is of reasonable quality.

Benzylamine tartrate is a superior substrate on which to form the carbon film. After melting, the cooled substrate sets to a glass which preserves the smoothness of the liquid surface. Grids will float on, and partially sink into, the surface of the molten substrate; after cooling, the grids are neatly inlaid in the substrate surface, and the metal surfaces of the grids are left exposed. Deposition of the carbon directly on the exposed grid surfaces, as the film is formed on the substrate which fills the grid openings, results in the firmest possible attachment of the film to the grid. Except for zones along the edges of the grid openings where the contact of the substrate and grid is replicated, the films generally are smooth and free of undulations and minor irregularities. A consistent production of grid-mounted films of uniformly good quality is easily attained by this procedure.

A small, shallow sieve, 1 inch to 2 inches in diameter, of 200-mesh woven stainless steel wire cloth is a convenient receptacle for handling a batch of grids during the preparation (Fig. 12). Benzylamine tartrate is melted on a spatula held against a hot plate and applied to the sieve which has also been heated.[27] The mesh of the sieve is exposed to radiant heat from the hot plate until the molten substrate spreads into a reasonably uniform layer about 0.5 mm thick embedding the wires of the sieve. Heating is continued until the small bubbles which appear in the melt

---

[27] Benzylamine tartrate is not commercially available; it is easily prepared from a solution of 10 g of d-tartaric acid in about 70 ml of tetrahydrofuran. The acid dissolves slowly when the solvent is heated to boiling. After the solution has cooled, 15 ml of benzylamine (a slight excess) is added. The benzylamine tartrate which precipitates is collected by filtration on paper and washed with petroleum ether. The product is finally spread on the unfolded paper under the gentle heat of an electric light for removal of the solvent. Benzylamine tartrate crystals melt at 145°–147°. The amorphous material melts at a much lower temperature, about 63°. This was not explained in the original description of the technique.

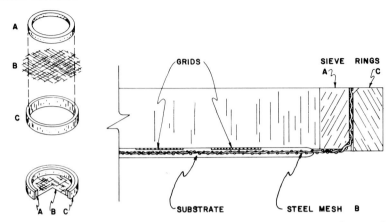

Fig. 12. Sketches (left) illustrating the assembly of a small stainless steel sieve which is used in the preparation of carbon film supports on a substrate of benzylamine tartrate. Stainless steel rings A and C, about 4 mm wide and an inch or more in diameter, form the sides of the sieve. The inside face of C and/or the outside face of A are machined so that clearance between these surfaces, when the rings are concentric, is about 5 mils, a bit less than the thickness of the 200-mesh woven stainless steel wire cloth, B. With the cloth between the rings, the rings are pressed together; the cloth thus is drawn taut around ring A to form the bottom of the sieve. The partial cross section of the sieve (right) depicts the substrate enveloping the wires of the cloth with grids inlaid in its upper surface inside the sieve.

have disappeared. Overheating, which will decompose the substrate, should be avoided.

The sieve is allowed to cool, and grids are placed on the surface of the solidified substrate on the inside of the sieve. The sieve is now held a millimeter or two above the hot plate, and the substrate remelted. The grids will float on the molten surface, and the molten substrate will penetrate the grid openings until its surface is flush with the upper surfaces of the grids. When this has occurred for all the grids, the sieve is allowed to cool. Wetting of the upper surface of the grids by the substrate must be avoided. The substrate and grids are now ready to receive the carbon. After the deposition of the carbon, the entire sieve–grid assembly is placed on the extractor with the carbon uppermost and the underside of the sieve mesh in contact with the paper of the extractor. After a few hours the substrate is entirely removed.

The extractor used in these procedures is a simple device of general utility for gently washing preparations with solvents (Fig. 13). It was devised by H. Studer as reported by Fabergé.[28] Solvent placed in the

[28] A. C. Fabergé, *J. Microsc.* (*Paris*) **6**, 157 (1968).

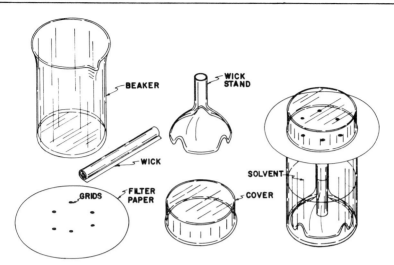

Fig. 13. Sketches of the several components of a simple extractor, shown assembled at far right. The extractor was devised by H. Studer. The solvent receptacle is a 150-ml beaker; the wick, a roll of bibulous paper, is held upright at the center of the beaker by the stand which is made of glass. The length of the wick is adjusted so that its upper end extends slightly above the plane of the beaker opening. The beaker is closed with a circular sheet of filter paper of such diameter as to extend well out from the beaker. Grids, or other objects, placed on the paper are protected from dust by the cover, a 2-inch petri dish cover.

beaker moves by capillarity up the wick and along the paper holding the objects to be washed. Substances removed by solution from the objects are deposited toward the periphery of the paper where the solvent evaporates. The extent of washing is easily controlled by the quantity of solvent placed in the beaker. Fifty to 75 ml of solvent is sufficient to remove substrates from the carbon films. The solvent for Butvar is ethylene dichloride and for benzylamine tartrate is methanol.

Carbon vapor for the deposition of thin films by the Bradley method is generated by the resistive heating of two graphite electrodes at their point of contact.[29] Usually the diameter of one electrode is reduced near the junction as illustrated in Fig. 14; as a result electrical power into the system is dissipated largely at the junction. When the power is raised to several hundred watts, the sublimation temperature of carbon ($\sim2100°$) is exceeded at points of contact of the electrode faces; the smaller electrode, having a lower thermal capacity, becomes the hotter and is burned away. Because the temperature coefficient of resistivity of

[29] Spectrographic grades of graphite rods suitable for this procedure are available from Ladd Research Industries, Inc., P. O. Box 901, Burlington, Vermont.

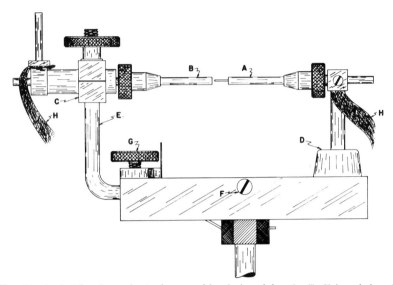

Fig. 14. A sketch of an electrode assembly designed by A. C. Fabergé for the vaporization of carbon by the Bradley procedure. The design provides, by means of clamp C, for the easy alignment of the electrodes, and for a precise adjustment (thumbscrew G), which permits only a predetermined length of the small electrode A to be vaporized. Electrodes A and B are clamped to a fixed, insulated holder, D, and a movable holder, E, respectively. E pivots on pin F and is springloaded to press against thumbscrew G. When the electrodes have been placed in contact, G is turned out by a measured distance thus placing the electrode contact under a positive loading from the spring. When the electrodes are heated by current supplied through cables, H, electrode A burns away; contact of the electrodes is maintained by the pivoting of B inward until E again presses against G.

carbon is negative, electrical control of the sublimation process is difficult. A mechanical control of the quantity of carbon vaporized is feasible; the length of rod which is fed into the junction may be limited. Several commercially available electrode assemblies have been designed to do this. If the rod is reduced to a standard, uniform diameter, say 1 mm, a consistent production of films of a given thickness becomes a simple procedure.

The optical density of a film of carbon deposited on a glass surface is directly dependent on its thickness. Moor has examined this relationship for film thicknesses up to $100 \, \text{Å}$[30]; his data indicate a linear dependency over that range of 0.002 optical density units per angstrom of thickness (measured at a wavelength of $5000 \, \text{Å}$). Using this factor, the thickness of film deposited at a given distance from the source by the

[30] H. Moor, J. Ultrastruct. Res. 2, 393 (1959).

vaporization of a unit length of standard rod can be estimated. In one series of experiments of this kind the thickness of film deposited under typical conditions on a glass surface placed 7 cm from the source was shown to be 44 Å per millimeter of electrode vaporized.[31] The thickness of film deposited when a given quantity of carbon is vaporized varies as the inverse square of the distance from the source to the surface receiving the deposit; thus by altering the source-to-surface distance a range of film thicknesses can be deposited without varying the electrode adjustments or the conditions of the vaporization.

The consistent deposition of films of an intended thickness is dependent, in part, on the design of the electrode assembly and the care with which it is used. Variations in thickness from film to film also may result from the explosive ejection of relatively large graphite particles from the source, so that the intended quantity of carbon is not vaporized, and from an inadequate control of pressure during the deposition. Ordinarily the ejection of material from the source is not subject to control. Due to this uncertainty the thickness of a particular film, as indicated by its optical density, may vary as much as 25% from its intended thickness. Control of the pressure in the region between the source and the surface receiving the deposit is not critical provided the mean free path of vaporized atoms is long compared to the source-to-surface distance; otherwise vaporized atoms are scattered by collisions with gas molecules in this region and may not reach the surface. To a fair approximation the mean free path (MFP) in air is inversely related to pressure by the expression

$$MFP = \frac{5 \times 10^{-3}}{p} \text{ cm} \qquad (4)$$

where p is the pressure in Torr. Thus at pressures $<10^{-4}$ Torr the influence of the atmosphere in the vacuum chamber on the deposition of the film is negligible. Deposition of the carbon at an ultrahigh vacuum ($<10^{-6}$ Torr) does not significantly alter the properties of the film or improve its quality.[32] Unless the electrodes are adequately degassed, the sudden elevation of temperature at the instant of sublimation may result in the evolution of such quantities of adsorbed gases from the electrodes as to raise the pressure above 10^{-4} Torr. This can be avoided by preheating the electrodes for several minutes to a cherry-red glow as the chamber is evacuated.

[31] R. M. Oliver, unpublished observations. The evaporator used was an Elion DV-502. Conditions were: pressure, 8×10^{-5} Torr; maximum electrode current, 18 A; electrode diameter, 1 mm.
[32] J. T. Stasny and R. M. Oliver, unpublished observations.

Carbon vapor for the deposition of thin films may be generated by an arc powered by direct current rather than the alternating power source used by Bradley. The properties of films deposited from the dc arc differ substantially from those of films deposited by the Bradley method. The most striking difference is the wettability of their surfaces; films from the dc source compare favorably with Butvar films in this property.[33]

Unfortunately films of carbon from the dc arc are unsatisfactory supports for negative stain specimens. When exposed to the stain solution, the films appear to expand and sag in the grid openings producing an undulating surface; consequently the thickness of the stain deposit is usually not uniform. However, carbon supports prepared by the Bradley method can be overcoated with a 10–15 Å thickness of carbon from the dc arc; the resulting film, being strong, rigid, and hydrophilic, is an excellent support for negative stain specimens.

An apparatus for the controlled production of carbon vapor from a dc arc is illustrated by Fig. 15. The prime advantage of this apparatus for evaporative depositions derives from the relative ease with which the process is controlled. Usually the power input to the arc need not be more than 40–80 W; consequently, deposition rates of the order of 10 Å per minute are realized. The deposition can be interrupted, and resumed, at will.

Preparation of Fenestrated Film Supports

The minimum practical thickness for specimen support films when mounted on metal grids with openings of conventional size (100–500 mesh) is dependent on the mechanical properties of the film. Unless the film is sufficiently strong and elastic as to be self-supporting over the width of the openings, it will not survive the film mounting procedures. In general, thicker films must be used with the larger openings. A film just rugged enough to withstand the preparative manipulations may still be so flexible over the relatively large span of the grid openings that it fails to firmly position the specimen under the impact of the beam. Thus the routine use of very thin specimen support films requires that the film be rigidly fixed to a support with openings of but a few microns at most. This is accomplished by first mounting a fenestrated film, or net, on the conventional grid; the specimen support film then is mounted on the net.

Usually nets are prepared from plastic materials and subsequently are strengthened by a heavy overcoat of carbon. Several techniques

---

[33] F. C. Maseles and R. M. Oliver, unpublished observations.

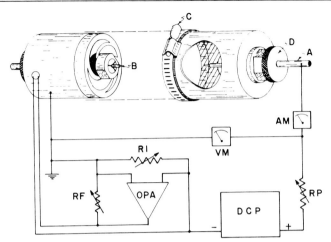

Fig. 15. A diagrammatic sketch of an apparatus developed by F. C. Maseles for the *in vacuo* deposition of carbon, silicon carbide, and other conductive, refractory materials from a direct current arc. Electrodes of the material to be vaporized are fixed to the positioning rod, A, and the spring-loaded armature of a solenoid, B. To complete the assembly, the control unit housings are joined and clamped together by tightening C. The electrodes are placed in contact and the armature is displaced by 2–4 mm from its relaxed position by pushing rod A inward against the restoring action of springs in the solenoid assembly; the electrodes are fixed in this position by tightening the collet clamping nut, D. Current passed through the coil of the solenoid generates a force tending to separate the electrodes against the force of the springs which act to maintain the electrodes in contact. Thus control of the coil current provides a means for controlling the gap between the electrodes when the arc is struck.

Power for the arc from the supply DCP (50 V dc, 5 A) is controlled by rheostat RP. Arc current and voltage are monitored by meters AM and VM, respectively. Current for the operation of the solenoid is provided by an operational power amplifier, OPA, which is controlled by potentiometer RF. Current through the electrodes returns to the supply by way of potentiometer RI; the voltage appearing across RI is fed to the input of OPA causing the solenoid to open the electrodes as necessary to strike the arc and maintain a proper gap width. The gap width is controlled by adjusting RI. As the electrodes are shortened by vaporization in the arc, the solenoid armature compensates by shifting toward its relaxed position to maintain the arc.

have been devised for net production which are more or less satisfactory; a common feature of these methods is the uncertainty of their routine, successful application.

Fenestrated films may be prepared from collodion or Butvar by condensing moisture on the surface of the plastic as it is formed from solution. Butvar nets are formed on microscope slides which have been prepared as for the fabrication of support films described above except that Ethosperse LA-23 is used to coat the slide.[34] The prepared slide is dipped into a solution of Butvar B-76 in ethylene dichloride (0.125% w/v). After 15 seconds the slide is withdrawn and immediately inserted in the vapor over water warmed to 90°. The plastic film will turn translucent in a few seconds owing to the formation of a network of holes. After about 30 seconds in the water vapor, the slide is set aside to dry for about 5 minutes. At this stage the nets should be examined under an optical microscope to locate those areas which are adequately populated with holes of the desired size.

Procedures for stripping, mounting, and retrieving the grid-mounted nets are similar to those employed in making Butvar films. Grids which are precoated with Butvar B-76 as an adhesive are laid onto the floating net over those areas observed to contain suitable populations of holes. Recovery of the grids from the water surface is facilitated by the use of absorbent paper rather than a microscope slide; a sheet of bibulous paper, folded twice for added stiffness, is gently lowered to the water surface on top of the floating net; the net will adhere to the paper which is removed directly without immersion.

Nets of collodion may be prepared in a similar way. The microscope slide without a detergent coating is dipped into a solution of Parlodion in amyl acetate (0.5% w/v) and transferred to the water vapor atmosphere while the film is forming. The relatively thick collodion film which forms will be reticulated with pits and a few holes. After the film has been transferred to grids, the pits are opened into holes by holding the grid for about 5 seconds in saturated amyl acetate vapor.

After thorough drying, the fragile and flexible plastic nets are strengthened by the deposition *in vacuo* of an overcoat of carbon or silicon carbide. A thickness of 200–300 Å of carbon deposited by the Bradley method is needed to make the net sufficiently rigid. Silicon carbide deposited from a dc arc (Fig. 15) is an excellent material for net stabilization. Silicon carbide films are less flexible than carbon films of equal thickness, and a deposit only 50–100 Å thick is needed to stabilize the net. The surface of the silicon carbide deposit tends to granular roughness, and short whiskers (20–40 Å) grow from the surface during deposition. As a consequence the adherence of support films to the net is improved. Butvar films adhere poorly to the surface of carbon deposited

---

[34] Ethosperse LA-23 is available from Glyco Chemicals, Inc., 417 Fifth Avenue, New York, New York 10016. For making nets, a drop of 1–2% aqueous solution of the detergent is applied to one side of the clean microscope slide and spread into a uniform film by wiping with lens tissue. As soon as the film appears to have dried, the slide is ready for use.

from a dc arc; thus carbon from this source is not a useful material for stabilizing nets.

Preparation of Negative Stain Specimens

Techniques for the preparation of negative stain specimens of protein macromolecules are simple and direct. The essential aim of the procedure is to embed the specimen protein in a usefully thin deposit of stain; however, resolution of molecular features is accomplished only by the deposition of the stain immediately at the stain-protein boundary to produce a maximum density contrast. This result is achieved only if the procedures employed severely limit the deposition at that boundary of buffer salts or other materials with densities less than the stain; otherwise the specimen protein molecules will be imaged at low resolution and will appear as nondescript blobs. Commonly the specimen protein preparation is applied directly to the surface of the support film, which has previously been mounted on its supporting grid, and a population of protein molecules is adsorbed on the film surface. Attachment of the molecules is usually so secure that they are not removed by subsequent rinsing and staining operations which remove most of the buffer salts. The chemical natures of the protein and film surface under the environmental conditions established by the buffer largely determine the quantity of protein adsorbed (compare Figs. 4C, 4D, and 6). At times the protein may adsorb so avidly that a monolayer is very quickly deposited, even from very dilute solutions. Such dense packing is usually undesirable, for discernment of the features of individual molecules is difficult under these circumstances. At the other extreme, adsorption of the portein may not occur. Some measure of control may be effected by adjustment of the protein and buffer concentrations, adsorption time, etc., and appropriate conditions must be determined by experiment for each specimen material. Typically the protein concentration may range from 10 to 100 μg/ml and adsorption times from a few seconds to a few minutes.

Exposure of the film surface to the protein is accomplished by a variety of techniques. Usually the grid is held at its edge by clamped forceps so that its surface is nearly horizontal; a droplet of the protein solution is placed on its surface as illustrated by Fig. 16. Care should be taken to limit the size of the droplet so that the fluid is completely retained on the surface of the film by surface tension. After an appropriate interval of time the excess of fluid is wicked away by touching the edge of a sheet of absorbent paper, such as bibulous paper, to the edge of the grid.

Immediately following the removal of the excess specimen, a droplet

Fig. 16. A photomicrograph illustrating a technique for the adsorption of specimen protein molecules on the grid-mounted support film. A droplet of the specimen protein preparation has been placed on the support which is held at its edge by the forcep tips. The edge of a piece of absorbent paper is positioned near the edge of the grid preparatory to the removal of excess specimen by wicking.

of rinse or stain solution is applied to the grid so that the film surface does not begin to dry. If the protein preparation contained sufficiently high concentrations of buffer salts or other solutes as to interfere with deposition of the stain, rinsing is necessary. The nature of an appropriate rinse is, of course, dependent on the conditions that the protein will tolerate. After a minute or so the rinse is wicked away, and the rinsing procedure is repeated if necessary. If the stain preparation is itself a suitable rinse, rinsing and staining may be accomplished simultaneously; otherwise, a droplet of stain is finally applied to the grid. After a minute or two the excess stain is also wicked away, leaving but a thin aqueous film on the surface of the grid. The specimen is thoroughly dried in air, over a desiccant if necessary, before exposure to the vacuum of the microscope.

An alternative procedure for adsorption of the specimen molecules is to float the grid-mounted film on the protein preparation. Conveniently about 0.1 ml of the preparation is placed in a spot plate depression, and the grid is gently lowered to and floated on its surface. After an appropriate interval of time, the grid is transferred to the surface of rinsing and staining solutions contained in other depressions of the plate. Finally, the grid is removed from the stain, wicked to remove the excess fluid, and dried. Accidental immersion of the grid must be avoided; fluids on the grid side of the support film cannot be wicked away. In a variation of this technique, small individual puddles of the protein, rinse, and stain solutions are placed on a flat, nonwetting surface of a material such as Teflon or Parafilm. The grid is floated on each puddle in turn. Since manipulations of a grid on the puddles are easy to perform, immersion of the grid is less likely.

These direct protein-adsorption techniques offer an appealingly simple way to avoid the deposition of materials other than stain and protein in the stain bed; the rinsing operation usually assures the removal of such materials. As a result of preliminary isolation procedures, some specimen protein may be available only at low protein (<1 mg/ml) and moderate to high buffer or salt concentrations, and the protein's inherent tendency to aggregate or dissociate may preclude the removal of electrolytes or water by conventional procedures. Other proteins may aggregate or precipitate in the presence of the stain. The adsorption techniques of specimen preparation may offer distinct advantages for surmounting experimental difficulties of this kind; it is unlikely, however, that adsorption of the protein on the film surface will stabilize it against an adverse tendency to denature or dissociate in the presence of the stain or in the absence of a supporting buffer or reagent.

The quantity of stain solution remaining on the film surface after removal of the excess is dependent on several factors; the wettability of the surface, the quantity of protein which has been adsorbed, and the properties of the stain. Thus the stain bed thickness is difficult to control. Stains usually are applied in the concentration range from 0.25% to 4%, and adjustment of the concentration provides a measure of control over the thickness of the deposit. Frequently in practice a large portion of specimen will be too thick, and only limited areas will yield micrographs of useful quality. Alternatively, if the stain is diluted and removal of the excess is efficient, the specimen structures may not be completely engulfed in the stain and the bed may not continuously cover the support. Therefore, the preparation of a negative stain specimen of adequate and uniform thickness requires a judicious variation of the procedure based on a familiarity with the particular materials and problems involved.

Nonuniform wetting of the support film surface by the stain solution will simultaneously complicate control of the stain bed thickness and foster the production of artifacts. Utilization of the more hydrophilic film materials generally will improve specimen quality; however, exposure of the film to reagents contained in the protein preparation may significantly and adversely alter the properties of the surface. For example, the aldehydes commonly added as fixatives to protein preparations will substantially reduce the wettability of Butvar surfaces. Some proteins appear to exhibit a similar effect when adsorbed on the film surface; by contrast, bovine serum albumin has been used to enhance the wettability of the support surface.

When the concentration of salts may be reduced or interfering substances removed by a preliminary treatment of the protein preparation, the protein may be commingled with the stain before application to the support film. This procedure may be particularly useful when the subject protein is so poorly adsorbed on the support film that application and removal of rinsing and stain solutions also removes the protein. Appropriate volumes of the protein and stain solutions are mixed to finally yield a protein concentration of 20–200 μg/ml in a 0.2–0.5% (w/v) solution of stain. A drop of the mixed preparation may be applied to the grid and the excess removed as described above leaving a suitably thin aqueous film on the support to dry. Adjustment of the stain bed to a proper thickness is effected by variation of the component concentrations and by skillful wicking to leave just the proper quantity of material on the support; owing to uncertainties in the latter, consistent specimen quality is difficult to achieve.

If the stain–protein solution is sprayed in small droplets onto a wettable support surface, and if the solution itself has a propensity to wet and spread over the surface, uniformly thin deposits of negative stain specimen are the natural result. When the action of interfacial forces has caused the droplet to spread to the limit permitted by its surface tension and volume, the resultant aqueous film will tend to be of uniform depth, except for feathering at its edges. Consequently the mass of stain deposited per unit area of support film tends to be constant. By a proper choice of support film and stain materials, the extent of spreading of the droplets can be controlled; thus the capricious effect of the wicking procedure is avoided, and adjustment of the stain bed to a proper thickness is simplified. When the experimental conditions have been adjusted to produce an optimum stain bed thickness, the production of wetting artifacts also will be minimized.

A fine mist of droplets of the stain–protein solution may be generated with the simple spraying apparatus depicted by Fig. 17. When a drop of the solution is aspirated through the capillary, a cloud of droplets of assorted sizes is formed; the cloud will expand as it is carried downward until the whole cross section of the skirt is filled. Most of the droplets are large enough to settle on grids placed below the skirt; the very small droplets will be carried away on the air stream. Adjustment of the diameter of the orifice and of the air velocity at the tip of the capillary allows a measure of control over the average size of the droplets formed. If the droplets are too small, drying on the film may proceed too rapidly to permit optimum spreading of the stain, and the stain deposits will tend to be too thick.

To prepare a specimen, 0.2–0.3 ml of the stain-protein solution is aspirated. Specimen supports placed below the skirt intercept such a limited number of droplets that, ordinarily, less than half of the support film surface is covered by stain bed. Thus the droplets have ample room to spread along the surface before drying. When too much preparation is sprayed, many of the droplets will coalesce and much of the stain deposit will be thicker than desired. Since most of the solution sprayed is lost, spraying is not the most efficient method of specimen preparation; however, several dozen specimen supports can be sprayed simultaneously at a cost of but 20–40 μg of the specimen protein. The specimens can be preserved indefinitely *in vacuo*.

When the average droplet size is appropriate for an optimum result, the fluid deposited on the support surface will not be visible to the naked eye unless several droplets coalesce. For this reason, the drying time of the typical droplet is difficult to judge, but coalesced droplets, which are always produced, appear to dry in a minute or two. When the specimen appears to have dried, some water may be still present in the stain bed, and a derangement of the stain deposit could result from its rapid vaporization if the specimen is suddenly placed in the vacuum of the microscope. The specimen should be dried over a desiccant for at least 10 minutes as a precaution.

Drying of the aqueous film proceeds from the edges, the central area covered by the droplet being last to dry. Minor solutes tend to be held in solution until the last stages of drying and are deposited in highest concentrations in the central area. As a result, the specimen in this area ordinarily is of inferior quality. If the concentrations of the minor solutes in the preparation have been limited, their segregation as the stain is deposited results in a decided improvement in contrast and resolution of micrographs taken from areas near the periphery of the stain bed.

The principal disadvantage of the spray technique derives from the need to severely limit the amounts of substances other than stain and protein in the sprayed solution; to obtain satisfactory specimens a preliminary treatment of the protein preparation to remove such materials is often necessary. When phosphotungstate, methyl phosphotungstate, or methylamine tungstate are used, the concentration of buffer in the stain–protein solution should not exceed 2 mM. (Phosphate or tris(hydroxymethyl)aminomethane sulfate buffers are preferred with these stains.) Even at very low concentrations, materials of limited solubility and low density may deposit around the specimen molecules and

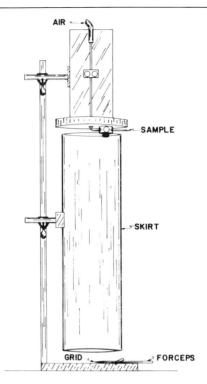

Fig. 17. Sketch of an apparatus used to prepare negative stain specimens by the spraying of a mist of droplets of the stain-protein preparation onto grid-mounted support films. Air at low pressure (2–3 psi) is introduced through flexible tubing to a 6-inch length of glass tubing (2.5 mm i.d.) which is securely clamped to the upper assembly of the apparatus and projects through a hole in the base piece of that part. The specimen solution is introduced through a glass capillary (2.5 inches × 1 mm i.d.) which has been tapered to a fine tip (0.5 mm i.d.). The capillary is secured by an adjustable clamp to the base of the upper assembly and is so positioned that its tip is centered on the bore of the air induction tube about 1 mm below the tube. A drop of the specimen solution presented to the outer end of the capillary will be forcefully aspirated through the capillary to form a cloud of fine droplets at the tip. The cloud will be blown downward through the skirt to emerge through the base-to-skirt gap. Droplets of the mist will impinge the base or be intercepted by specimen supports inserted in the gap. The sketch depicts a single support grid held by clamped forceps, but several grids may be sprayed simultaneously.

The apparatus is assembled in an upright position by mounting on a common laboratory support, the skirt being fixed about 1 inch above the base of the support, and the base of the upper assembly about 0.5 inch above the skirt. The parts are so positioned that the air induction tube is centered on the axis of the skirt. The skirt is a 15-inch length of 4-inch (o.d.) acrylic tubing. Its supporting arm of aluminum rod (0.5 × 4 inches) is threaded into a block of acrylic cemented to its side. The upper assembly is fabricated from 0.5 inch-thick acrylic sheet. The reduce the density contrast so that fine detail in the images is lost. Some ionic materials, such as sodium chloride and ammonium sulfate, are readily volatilized by the electron beam and when present should be completely removed, if possible; when these materials are deposited at or near a protein molecule, the derangement of the stain which results from their degradation may completely obliterate the image of the protein. Other common reagents—sucrose, substrates, reducing agents, detergents, etc.—almost certainly will produce artifacts if present in sufficient amounts, but concentrations <2 mM of many of these substances can be tolerated. If the specimen protein can be obtained at concentrations >10 mg/ml, the dilution of the protein by the stain solution to a concentration suitable for spraying (usually <100 μg/ml) may so reduce the concentration of other components in the protein preparation as to negate their effect, particularly near the periphery of the stain deposited from a single droplet.

That the properties of the film and stain materials are important to the routine production of thin and uniform deposits of negative stain has been emphasized; spreading of droplets of the stain also is significantly influenced by the specimen protein. The admixture of the protein with the stain lowers the surface tension of the stain solution. As a consequence, the area of support film surface wet by droplets of a given volume will be extended, and the stain deposit which results will be thinner. However, within rather wide limits, adequate specimen quality is not critically dependent on the protein concentration in the sprayed preparation. A concentration of 15–20 μg/ml of small protein molecules ($\sim$25,000 molecular weight) when sprayed in a solution containing 0.2–0.5% of the stain material onto a Butvar surface will usually result

---

circular base piece is 4.5 inches in diameter with a 0.5-inch hole at its center. The upright rectangular piece (3 × 6 inches), to which the supporting sidearm is fixed, is cemented to the base piece in an off-center position so that the air induction tube is centered in the hole of the base piece. The vertical position of the air induction tube can be adjusted and firmly fixed by means of the clamp cemented to the center of the rectangular piece. The capillary clamp is a circular block 1 inch in diameter, which is attached to the base piece with a thumbscrew off-centered to the block. By adjusting the position of the capillary in the clamp and rotating the clamp through an arc around the thumbscrew, the tip is easily adjusted to the center of the air induction tube.

The size distribution of droplets in the mist is dependent on the air velocity at the capillary and the diameter of the orifice of the capillary tip; the optimum velocity for a particular orifice to produce the desired droplet size must be determined by experiment. Care must be taken that the end of the capillary at its tip is cut off straight and even, otherwise the cloud of mist will not distribute evenly over the cross section of the skirt as it expands downward.

in sufficient spreading of the droplets as to produce simultaneously the desired distributions of stain and protein. With protein concentrations <10 μg/ml, the stain beds will be noticeably thicker and sparsely populated with protein molecules; with concentrations >50 μg/ml, the stain bed may be so thin that the protein molecules are incompletely embedded, and the images of the specimen will be displeasingly over-crowded on the micrograph. Usually the highest concentration that may be used is limited by the population density of protein molecules in the stain bed. The molecules may become so crowded that images of individual molecules are not easily distinguished on the micrograph.

Presumably the effect of a protein on the spreading of the droplets is proportional to its molarity in the solution; thus for the same weight/volume concentration, the smaller protein molecules should induce comparatively thinner deposits of the stain, a result generally observed. This consideration suggests that for consistent results between preparations the protein concentrations should be adjusted in direct proportion to the molecular weights of the specimen molecules; however, usually this is not desirable. Several factors combine to limit the maximum useful concentrations for the heavier molecules. The protein molecules of greater mass are, of course, dimensionally larger, and somewhat thicker stain beds are needed to assure that the specimen molecules will be completely embedded. Further, proportionally fewer of the larger molecules can be accommodated on the micrograph before a displeasing overcrowding of images occurs. Due to the packing arrangements of their subunits, a significant fraction of the volume occupied by the very large molecules and complex assemblies is void of protein; consequently, the images of these entities are considerably larger than might be anticipated by a consideration of their masses. As a practical rule, 3-fold difference in protein concentrations is a suitable adjustment between specimen preparations of molecules which differ by a factor of ten in their masses.

Acknowledgment

The assistance of L. J. Reed, D. DeRosier, A. Jackson, and F. Hoffman in the preparation of the manuscript is gratefully acknowledged.

Section IV

Conformation: Optical Spectroscopy

[27] Circular Dichroism and Optical Rotatory Dispersion of Proteins and Polypeptides

By Alice J. Adler, Norma J. Greenfield, and Gerald D. Fasman[1]

Introduction

Considerable information concerning the structure of proteins in solution can be obtained from measurement of their optical activity. The great asymmetry of protein molecules is responsible for the large signals they display in the interrelated methods of optical rotatory dispersion (ORD) and circular dichroism (CD). ORD is the measurement, as a function of wavelength, of a molecule's ability to rotate the plane of linearly polarized light; CD is similar data evaluating the molecule's unequal absorption of right- and left-handed circularly polarized light. Although all the amino acids except glycine contain at least one asymmetric carbon atom (the L or D configuration), most amino acids display only small ORD and CD bands.[2] It is the conformation of the protein, that is, the asymmetric and periodic arrangement of peptide units in space, which gives rise to their characteristic ORD and CD spectra.

In recent years X-ray diffraction analysis has lead to the complete mapping of the peptide backbone and side-chain positions of lysozyme,[3,4] several other enzymes,[5] and quite a few other proteins[6,7] in the solid state. Newer techniques such as neutron diffraction (for an example, see Schoenborn[8]) and high resolution nuclear magnetic resonance (for a

[1] Contribution No. 869 of the Graduate Department of Biochemistry, Brandeis University, Waltham, Massachusetts 02154. This work was supported in part by grants from the U.S. Public Health Service (GM 17533), National Science Foundation (GB 29204X), American Heart Association (71-111), and the American Cancer Society (P-577). N. J. G. is at Merck, Sharp, and Dohme Research Laboratories, Rahway, New Jersey.
[2] C. Toniolo, *J. Phys. Chem.* **74**, 1390 (1970).
[3] C. C. F. Blake, D. F. Koenig, G. A. Mair, A. C. T. North, D. C. Phillips, and V. R. Sarma, *Nature (London)* **206**, 757 (1965).
[4] D. C. Phillips, *Sci. Amer.* **215**, 78 (1966); *Proc. Nat. Acad. Sci. U.S.* **57**, 484 (1967).
[5] D. M. Blow and T. A. Steitz, *Annu. Rev. Biochem.* **39**, 63 (1970).
[6] D. R. Davies, *Annu. Rev. Biochem.* **36**, 321 (1967).
[7] R. E. Dickerson and I. Geis, "The Structure and Action of Proteins." Harper, New York, 1969.
[8] B. P. Schoenborn, *Nature (London)* **224**, 143 (1969).

review see Roberts and Jardetzky[9]) have yielded information about the hydrogen atoms of proteins; this is beyond the resolution of X-ray analysis. ORD and CD are techniques which lack the capacity for exact structural determination possible with the methods mentioned above; however, they have the advantage that one can rapidly approximate the percentages of the conformations present in dilute solutions of proteins. Because of the small volume and low concentration required, a protein sample of less than 0.1 mg is often sufficient for a CD and/or ORD determination.

For studies of the structure of synthetic polypeptides, macromolecules that often exhibit regular repeating conformations, ORD and CD may be unsurpassed. Moreover, ORD and CD are very sensitive spectral tools and thus may be of use in studying any reactions involving changes in optical activity. Thus they are excellent for measuring protein denaturation and helix–coil transitions of polypeptides. They can also be used in measurements of enzyme interactions with substrates, inhibitors, or coenzymes and of the binding of metal ions and dyes to proteins and polypeptides.

In 1963 a chapter on the ORD of proteins was written by Fasman[10] for this series. At that time, because of instrumentational limitations, most studies of protein ORD involved measurements in the wavelength range above 240 nm and data analysis by means of the Drude[11] or Moffitt[12–14] equations. Examination of the Cotton effects derived directly from asymmetric peptide chromophore transitions (the 233 nm ORD trough)[15] was just beginning. In the intervening years two types of improvements in recording spectropolarimeters have revolutionized measurements of protein rotatory properties, causing a proliferation of polypeptide and protein investigations, and more than justifying a new review in "Methods in Enzymology." The first advance was development of commercial instruments permitting routine ORD measurements in the spectral range 185–600 nm, thus including Cotton effects arising from peptide chromophores. The second, even more useful, development was that of circular dichroism spectrophotometers capable of operating in this same spectral range. The advantages of CD over ORD for protein conformational studies are that (1) each optically active electronic transition gives

[9] G. C. K. Roberts and O. Jardetzky, *Advan. Protein Chem.* **24**, 448 (1970).
[10] G. D. Fasman, see Vol. 6 [126] p. 928.
[11] P. Drude, "Lehrbuch der Optik," 2nd ed. Hirzel, Leipzig, 1906.
[12] W. Moffitt, *J. Chem. Phys.* **25**, 467 (1956).
[13] W. Moffitt, *Proc. Nat. Acad. Sci. U.S.* **42**, 736 (1956).
[14] W. Moffitt and J. T. Yang, *Proc. Nat. Acad. Sci. U.S.* **42**, 596 (1956).
[15] N. S. Simmons and E. R. Blout, *Biophys. J.* **1**, 55 (1960).

rise to only one CD band instead of both positive and negative ORD signals (see Fig. 1), and the bands are thus more easily resolved and assigned, and that (2) CD, unlike ORD, bands are of finite width and, therefore, CD spectra contain no contribution from transitions outside of the measured spectral range. For these reasons most investigations of proteins utilized CD instead of ORD beginning about 1968, and the present chapter will concentrate on CD.

Another important advance was the discovery of the water–soluble β conformation of heated poly-L-lysine.[16-19] Therefore, the contribution of β forms, as well as α-helical and random structures, to protein conformation in solution could be considered. This made it possible to analyze protein ORD[20] and CD[21] spectra for mixtures of these three structures by using the synthetic polypeptide, poly-L-lysine, as a model. These methods seem to be a useful approximation for the conformation of highly structured proteins. However, unresolved problems in interpretation arise from the choice of model structures, the contribution of optically active nonpeptide chomophores, and the effects of light scattering.

Thus, although excellent CD and ORD data on proteins can now be obtained, the interpretation of these data must be approached with caution. The purpose of this article is to bring up to date the methodology of CD and ORD data gathering and analysis in the rapidly changing field of protein structure, making use of synthetic polypeptide studies when necessary. A systematic review of the theory of optical activity[22-34] or the history of rotational analysis of proteins and other bipolymers will not be attempted, as these can be found elsewhere.[10,26,30,32,33,35-46] Nor will the structure of collagen and polyproline,[37(b,d)47-49] proteins with prosthetic groups like heme, or the new fields of infrared ORD[50] and magnetic CD[51-53] be discussed.

[16] B. Davidson, N. M. Tooney, and G. D. Fasman, *Biochem. Biophys. Res. Commun.* **23**, 156 (1966).

[17] R. Townend, T. F. Kumosinski, S. N. Timasheff, G. D. Fasman, and B. Davidson, *Biochem. Biophys. Res. Commun.* **23**, 163 (1966).

[18] P. Sarkar and P. Doty, *Proc. Nat. Acad. Sci. U.S.* **55**, 981 (1966).

[19] B. Davidson and G. D. Fasman, *Biochemistry* **6**, 1616 (1967).

[20] N. J. Greenfield, B. Davidson, and G. D. Fasman, *Biochemistry* **6**, 1630 (1967).

[21] N. J. Greenfield and G. D. Fasman, *Biochemistry* **8**, 4108 (1969).

[22] T. M. Lowry, "Optical Rotatory Power." Longmans, Green, New York, 1935.

[23] E. U. Condon, *Rev. Mod. Phys.* **9**, 432 (1937).

[24] W. Kauzmann, J. E. Walter, and H. Eyring, *Chem. Rev.* **26**, 339 (1940).

[25] W. Kuhn, *Annu. Rev. Phys. Chem.* **9**, 417 (1958).

[26] C. Djerassi, "Optical Rotatory Dispersion." McGraw-Hill, New York, 1960; (a) Chapter by A. Moscowitz, p. 150.

[27] I. Tinoco, Jr., *Advan. Chem. Phys.* **4**, 67 (1962).

[28] S. F. Mason, *Quart. Rev.* **17**, 20 (1963).

The Phenomena of CD and ORD

A beam of linearly polarized light of wavelength λ can be considered as the sum of two components: beams of right- and left-circularly polarized light, with electric vectors E_R and E_L, respectively. When such light interacts with an asymmetric molecule (such as most biological macromolecules) two phenomena, CD and ORD, are observed, and the molecule

[29] L. Velluz, M. Legrand, and M. Grossjean, "Optical Circular Dichroism." Academic Press, New York, 1965.

[30] N. J. Greenfield and G. D. Fasman, in "Encyclopedia of Polymer Science and Technology." Vol. 15, p. 410. Wiley, New York, 1971.

[31] W. F. H. M. Mommaerts, see Vol. 12B, p. 302.

[32] D. W. Urry, Annu. Rev. Phys. Chem. 19, 477 (1968).

[33] S. Beychok, Science 154, 1288 (1966).

[34] H. Eyring, H.-C. Liu, and D. Caldwell, Chem. Rev. 68, 525 (1968).

[35] S. Beychok, Annu. Rev. Biochem. 37, 437 (1968).

[36] W. B. Gratzer and D. A. Cowburn, Nature (London) 222, 426 (1969).

[37] G. D. Fasman, ed., "Poly-α-Amino Acids: Protein Models for Conformational Studies." Marcel Dekker, New York, 1967. See in particular, chapters by (a) W. B. Gratzer, p. 177; (b) A. Elliot, p. 1; (c) S. Beychok, p. 293; (d) L. Mandelkern, p. 675.

[38] P. J. Urnes and P. Doty, Advan. Protein Chem. 16, 401 (1961).

[39] J. A. Schellman and C. G. Schellman, in "The Proteins" (H. Neurath, ed.), 2nd ed., Vol. 2. Academic Press, New York, 1964.

[40] S. N. Timasheff and M. J. Gorbunoff, Annu. Rev. Biochem. 36, 13 (1967).

[41] J. T. Yang, in "A Laboratory Manual of Analytical Methods of Protein Chemistry" (P. Alexander and H. L. Lundgreen, eds.), Vol. 5, p. 23. Pergamon, Oxford, 1969.

[42] G. N. Ramachandran, "Conformation of Biopolymers." Academic Press, New York, 1967.

[43] C. W. Deutsche, D. A. Lightner, R. W. Woody, and A. Moscowitz, Annu. Rev. Phys. Chem. 20, 407 (1969).

[44] I. Tinoco, Jr. and C. R. Cantor, in "Methods of Biochemical Analysis" (D. Glick, ed.), Vol. 18, p. 81. Wiley, New York, 1970.

[45] B. Jirgensons, "Optical Rotatory Dispersion of Proteins and Other Macromolecules." Springer-Verlag, Berlin and New York, 1969.

[46] I. Tinoco, Jr., in "Molecular Biophysics" (B. Pullman and M. Weissbluth, eds.), p. 269. Academic Press, New York, 1965.

[47] F. A. Bovey and F. P. Hood, J. Amer. Chem. Soc. 88, 2326 (1966).

[48] F. A. Bovey and F. P. Hood, Biopolymers 5, 325 (1967).

[49] W. L. Mattice and L. Mandelkern, Biochemistry 10, 1926 (1971).

[50] Y. N. Chirgadze, S. Y. Venyaminov, and V. M. Lobachev, Biopolymers 10, 809 (1971).

[51] G. Barth, R. Records, E. Bunnenberg, C. Djerassi, and W. Voelter, J. Amer. Chem. Soc. 93, 2545 (1971); E. Bayer, A. Bacher, P. Krauss, W. Voelter, G. Barth, E. Bunnenberg, and C. Djerassi, Eur. J. Biochem. 22, 580 (1971).

[52] D. D. Ulmer, Fed. Proc., Fed. Amer. Soc. Exp. Biol. 30, 1179 (1971).

[53] B. Holmquist, Fed. Proc., Fed. Amer. Soc. Exp. Biol. 30, 1179 (1971).

is said to be optically active. These phenomena arise from the following events:

1. E_R and E_L travel at different speeds through the molecule. This difference in refractive index leads to optical rotation, the rotation of the plane of polarization, measured in degrees of rotation, α_λ. ORD is the dependence of this rotation upon wavelength. In a region where the molecule does not absorb light, the rotation plotted against wavelength yields a plain curve. In the region of light absorption, however, the dispersion is anomalous. The rotation first increases sharply in one direction, falls to zero at the absorption maximum, and then rises sharply in the opposite direction. This anomalous dispersion is called a Cotton effect.

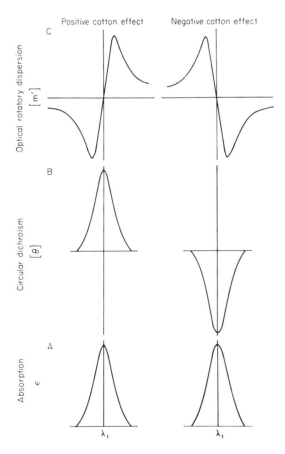

FIG. 1. A typical electronic absorption band (A) with its associated circular dichroism (B) and optical rotatory dispersion (C) curves.

2. In the region of its Cotton effect, an asymmetric molecule which exhibits ORD will also show unequal absorption of left- and right-handed circularly polarized light; this difference in extinction coefficient, $(\epsilon_L - \epsilon_R)$ is known as circular dichroism, and can be measured directly in some instruments as a differential absorbance. When CD occurs the emerging light beam is no longer linearly polarized, but instead is elliptically polarized. Thus, the ellipticity of the resulting light, θ_λ, is another measure of CD, and is proportional to $(\epsilon_L - \epsilon_R)$. A typical absorption band with its associated ORD and CD Cotton effects is shown in Fig. 1. Both dispersive and absorption phenomena are caused by the same charge displacements in a molecule; therefore, ORD and CD are closely related to each other. By means of the Kronig–Kramers transform equations developed by Moscowitz,[26(a)] ORD curves can theoretically be computed from CD data and vice versa. This calculation is sometimes useful for evaluation of CD bands at very low wavelength.

Experimental Methods

General Considerations in Measurement

The practical considerations involved in recording CD or ORD data in a modern circular dichrometer or polarimeter are basically the same as those already indicated[31,54] with respect to nucleic acids. Briefly, sufficient light must pass through the optically active sample and strike the photomultiplier to allow a meaningful rotation or CD measurement with tolerably high signal-to-noise ratio. This requirement becomes important to consider (1) below 225 nm (in the interesting spectral region of backbone peptide chromophores) where light sources begin to lose intensity, and where some light is absorbed by the optics and by solvents, and (2) in the region of aromatic side-chain absorption where molecular optical activity is relatively small. In such cases the parameters of protein concentration, cell path length, and solvent must be chosen with care. Under conditions of high absorbance and low CD or ORD signal, it is desirable to reduce path length and/or concentration.[54]

Our previous article[54] in this series should be consulted for considerations affecting choice of spectral band width, scan speed, and instrumental time constant or pen period, and details affecting resolution and measurement, including tests for rotational artifacts. For tests of CD artifacts d-10-camphorsulfonic acid (or any compound or polymer displaying a CD band in the wavelength region of interest) should be used instead of sucrose, which is utilized in ORD. In a given instrument, the maximum allowable scan speed for CD and for ORD might not be

[54] A. J. Adler and G. D. Fasman, see Vol. 12B, p. 268.

identical. For example, in a Cary 60 spectropolarimeter with 6001 CD attachment, a time constant of 3 seconds in the CD mode corresponds to a pen period of 10 seconds in ORD operation, in that both would require a maximum scan speed of 4 nm per minute and a maximum photomultiplier dynode voltage of 500 V if the spectral band width is 1.5 nm (see Table I of the previous article[54]).

Instruments

At present three commercial instruments, the Cary, the Jasco, and the Jouan, are capable of yielding high resolution CD and/or ORD spectra with relatively low noise levels down to about 185 nm, and are suitable for protein studies.

1. Cary 60 recording spectropolarimeter with 6001 CD accessory, and Cary 61 circular dichrometer.[55] The Cary instruments utilize a double-prism monochromator, a Faraday cell for rotational modulation of the light beam, Rochon prisms for polarizer and analyzer, and a Pockels cell as CD modulator and quarter-wave plate. A description of the polarimeter in the ORD mode has been given[54]; operation and precision of the Cary 60 or 61 as a CD instrument are very similar. Data are reproducible to within 0.001° of rotation (ORD) or ellipticity (CD), and optical densities of ≤ 2 can usually be tolerated. The noise level is 0.0005° rotation or ellipticity under conditions of small absorbance at $\lambda > 220$ nm, and increases to about 0.005° for a protein sample of absorbance ~ 2 at 190 nm. Spectra are recorded on a continuous roll of chart paper.

2. Durrum-Jasco Model J-20 CD (ORD optional).[56] The Jasco J-10 (CD and spectrophotometer) and J-5 (ORD and spectrophotometer) instruments also are similar to the J-20 in specifications. All these instruments have been considerably improved over the Jasco Model 5,[54] and performance is now similar to that of the Cary 60/61. One improvement is that the baseline can now be multipotted. A possible disadvantage is the preprinted chart. The Jasco instruments differ from the Cary in not utilizing the Faraday effect, and in measuring CD directly in differential dichroic absorbance $(A_L - A_R)$ rather than in ellipticity.

3. Roussel-Jouan Dichrograph CD 185 Model II.[57] This instrument is similar to the Jouan previously described,[29,31] but has recently been

---

[55] Manufactured by Varian Instrument Division, 611 Hansen Way, Palo Alto, California 94303.

[56] Distributed by Durrum Instrument Corp., 3950 Fabian Way, Palo Alto California 94303.

[57] Manufactured by Jouan Société, 113, Bd. St. Germain, Paris 6e; distributed in the United States by Bio-cal Instrument Co., 2400 Wright Ave., Richmond, California 94804.

modified to measure CD spectra, as $(A_L - A_R)$, down to 185 nm. It has not yet been extensively employed for proteins or polypeptides. However, data taken on Jouan instruments in some European laboratories[58] are of good quality with respect to noise level.

4. Bendix-Ericsson Polarimatic 62 Recording Spectropolarimeter.[59] The Bendix is capable of precise ORD measurement only at wavelengths above about 220 nm and is, therefore, not useful for peptide chromophores.

Calibration

The absolute accuracy of any quantitative measurement depends upon the standardization of the method. It is easy to check the calibration of ORD instruments: specific rotation values (at several wavelengths from 250 to 589 nm) of a 0.25% solution of National Bureau of Standards sucrose can be compared to literature values.[60] If necessary the polarimeter can be recalibrated.

However, there is at present no CD standard available with the consistent purity of N.B.S. sucrose. The compound commonly used for calibration of circular dichrometers is d-10-camphorsulfonic acid in 0.1% aqueous solution, which displays a large ellipticity band at 290 nm. But 10-camphorsulfonic acid forms a hydrate containing about 7% water under normal laboratory conditions,[61,62] so that weight may not be an accurate measure of concentration. Furthermore, yellow impurities were found[63] in some batches of reagent grade (Eastman Kodak) 10-camphorsulfonic acid. The acid may be purified by recrystallization from acetic acid,[61] followed by vacuum sublimation, drying at 80° under vacuum, and storage in a desiccator.[63] It is then suitable as a CD standard by means of which the signal gain adjustment controlling the magnitude of the observed CD signal on an instrument can be manipulated.

The exact value of the peak molecular ellipticity, $[\theta]_{290}$, of 10-camphorsulfonic acid is not known with certainty, partially because of impurity problems. Fortunately, the $[\theta]_{290}$ value can be calculated,[61,62,64]

[58] See footnotes 219–221, 242, and 262 for articles giving examples on data taken on Jouan dichrographs.

[59] Manufactured by Bendix-Ericsson U.K. Ltd., and distributed in the U.S.A. by Bendix Corp., Cincinnati, Ohio.

[60] E. Brand, E. Washburn, B. F. Erlanger, E. Ellenbogen, J. Daniel, F. Lippmann, and M. Scheu, *J. Amer. Chem. Soc.* **76**, 3037 (1954).

[61] D. F. DeTar, *Anal. Chem.* **41**, 1406 (1969).

[62] W. C. Krueger and L. M. Pschigoda, *Anal. Chem.* **43**, 675 (1971).

[63] G. D. Fasman and P. Lituri, unpublished data.

[64] J. Y. Cassim and J. T. Yang, *Biochemistry* **8**, 1947 (1969).

by means of the Kronig–Kramers transform, from accurate ORD data on the same 10-camphorsulfonic acid sample obtained on a well-calibrated polarimeter. For this calculation to be valid the sample must not contain optically active impurities, although small amounts of water are tolerable. A simple way to obtain an absolute CD value for a rotationally pure aqueous solution of 10-camphorsulfonic acid standard is to use Cassim and Yang's[64] calculated ratios of peak molecular ellipticity to peak and trough molecular rotations: $[\theta]_{290}/[M]_{306} = 1.76$ and $[\theta]_{290}/[M]_{270} = -1.37$. For example, one dry, purified sample of 10-camphorsulfonic acid[63] yielded measured rotations of $[M]_{306} = +4480$ and $[M]_{270} = -5700$, from which $[\theta]_{290}$ equals the average of $4480 \times 1.76 = 7880$ and $-5700 \times -1.37 = 7800$, or $[\theta]_{290} = 7840$ (corresponding to $\Delta\epsilon_{L-R} = 2.37$). The resulting $[\theta]_{290}$ magnitude can then be used to calibrate the circular dichrometer, even though the 10-camphorsulfonic acid sample may contain some water. Note that the calibration setting recommended in the Cary 60 operating manual corresponds to $[\theta]_{290} = 7150$, and should not be used with 10-camphorsulfonic acid standards of unknown purity and dryness. The value of 7840 agrees well with that for another[62] pure sample of 10-camphorsulfonic acid, but not with a third[61] value, possibly because of different handling of the Kronig–Kramers relationships.

Solutions

The conditions of solvent, protein concentration and cell path length must be chosen so that the solution remains relatively transparent in the wavelength region of interest (optical density below 2 in most cases), but so that enough solute is present to register an easily measurable rotation or ellipticity with high signal-to-noise ratio.

The first consideration is to use a solvent which transmits sufficient light. Water is the usual solvent for protein and polypeptide spectral studies. Tris and acetate buffers, in moderate concentration, may be used at $\lambda \geq 200$ nm, but farther in the ultraviolet there is no buffer of suitably low absorbance. Fluoride and perchlorate salts may be employed down to the spectral limit of instrumentation to maintain desired ionic strengths, but most common salts, including chlorides, hydroxides, and phosphates, are to be avoided in the far ultraviolet.[37(a)] Many organic solvents in which some proteins and polypeptides are soluble, such as dimethyl formamide and dimethyl sulfoxide, are not suitable for rotatory studies at $\lambda < 250$ nm unless cells of extremely thin path length are used.[65] There are a number of protein solvents of sufficient transparency,

[65] J. Engel, E. Liehl, and C. Sorg, *Eur. J. Biochem.* **21**, 22 (1971).

for example, trifluoroethanol, hexafluoroisopropanol, trimethyl phosphate, and methanesulfonic acid. Addition of any organic solvent, as well as any change of pH or temperature, may affect the protein's conformation, and must be used with caution.

The next consideration is that of sample concentration. For protein and polypeptide studies in the 185–240 nm spectral range, solution concentrations of 0.01–0.1%, with CD or ORD measured in cells of 0.1–1 mm path length, usually result in data showing a good compromise between sufficient signal and adequate light intensity. An absorbance of about 0.7 is usually optimal. Near-ultraviolet experiments ($\lambda >$ 240 nm) usually require larger concentrations and/or longer cells. It may be necessary to vary the concentration and/or path length during the experiment in order to measure accurately each CD or ORD band of interest. In circumstances of low rotatory strength or low light intensity, or if aggregation is suspected, it is a good practice to take measurements at different concentrations (or path lengths), as a test for artifacts.

The solutions to be measured should be filtered[54] (Millipore filters[66] are useful) or centrifuged to remove suspended material.

The concentrations of samples must be known accurately in order to calculate molar or residue ellipticities or rotations. The measured weight of the sample is not a sufficient concentration determination, since most proteins and polypeptides retain some water even after drying under vacuum at moderate temperatures. Therefore, the concentration of the solution used for CD or ORD must be assayed (after filtration). For many proteins, accurate values for aromatic side–chain extinction coefficients are known[67,68]; in these cases concentration can be obtained by a simple optical density determination. For all proteins and polypeptides, the Nessler micro-Kjeldahl analysis[69] for total nitrogen can be employed; however, in order to obtain concentration in terms of peptide residues per liter the amino acid composition must be known. In addition biuret,[70] Lowry,[71] or ninhydrin[72] colorimetric assays can be used, but for accurate results each method should be standardized with the protein to be measured.

[66] Manufactured by Millipore Filter Corp., Bedford, Massachusetts.

[67] D. M. Kirschenbaum, *Int. J. Protein Res.* 3, 109, 157, 237 (1971).

[68] D. M. Kirschenbaum, *in* "The Handbook of Biochemistry." (H. Sober, ed.), 2nd ed. Chem. Rubber Publ. Co., Cleveland, Ohio, 1970.

[69] C. A. Lang, *Anal. Chem.* 30, 1692 (1958).

[70] S. Zamenhof, see Vol. 3, p. 696.

[71] O. H. Lowry, N. J. Rosebrough, A. G. Farr, and R. J. Randall, *J. Biol. Chem.* 193, 265 (1951).

[72] S. Moore and W. H. Stein, *J. Biol. Chem.* 211, 893 (1954).

Cells

Only fused quartz, circular cells should be used for ORD and CD. Such cells are available[73] in a great variety of path lengths and special adaptations (such as water jacketed), and should be tested for birefringence.[54] For path lengths of 1 mm or less, face-filling cells and double-necked cells are recommended because they are relatively easy to fill and clean. In any case, syringes fitted with thin Teflon tubing will be needed for filling thin cells. The exact path length of thin cells (1 mm or less) may be measured by counting interference fringes in the near infrared[74]; this procedure is recommended for cells thinner than 1 mm.

Cell holders should be designed so that cells can be positioned reproducibly and firmly; V-blocks are not usually suitable. A cell holder assembly for ORD measurements in the Cary 60 has been described.[54] A very similar assembly can be machined for use in the Cary CD compartment. However, because of space limitations, the brass block with the groove (part A, Fig. 3, of reference cited in footnote 54) must be removable; it can be equipped with two holes on its underside which fit onto pins on a small brass plate permanently fixed to the floor of the sample compartment. Similar assemblies can be designed for other instruments.

Measurements

Most polarimeters and circular dichrometers now in use can be adjusted to yield a flat baseline (for a given solvent in a given cell). The previous discussion of ORD data collection[54] (the need to take frequent air blanks, etc.) applies also to CD. In CD measurements the sample signal should coincide with the solvent baseline in spectral regions where there is no absorption; if it does not, then the cell may not be properly positioned. (There is no such simple test for ORD data.)

Some instruments can be equipped with modified sample cell holders to permit direct measurement of difference ORD[75,76] and difference CD.[76] These methods involve compensation for artifacts, but they may be useful for analysis of small changes in protein conformation, such as may occur upon binding of substrates, inhibitors, or coenzymes. (The same information can be obtained indirectly, for example, by subtraction of independently determined enzyme and inhibitor CD spectra from the

[73] Manufactured by Opticell Co., 10792 Tucker St., Beltsville, Maryland 20705.
[74] W. J. Potts, "Chemical Infrared Spectroscopy," Vol. I, p. 119. Wiley, New York, 1963.
[75] B. J. Adkins and J. T. Yang, *Biochemistry* **7**, 266 (1968).
[76] J. T. Yang and K. H. Chan, this volume [28].

measured CD of enzyme–inhibitor complex under the same conditions, e.g., see Simons[77]).

Most instruments are thermostated so that measurements can be performed at a constant temperature (usually near 25°). Sample temperatures between about −20° to 100° can be obtained by means of circulating thermostat baths[78] connected to jacketed sample cells or to hollow cell holders. Electrical heating and cooling may also be employed.[44,79] It is necessary to monitor the sample temperature during heating or cooling experiments.[80] At present extensive apparatus modification is necessary to achieve very low sample temperatures.[81]

Calculations

Treatment of Raw Data

The recorder chart tracing of any CD or ORD spectrum will contain significant noise, which must usually be averaged by eye. If the spectrum has been scanned slowly enough, this presents no problem. Alternatively, if digitizing accessories are available,[44] the ORD or CD curve (a single spectral scan) can be smoothed by computer. This technique improves data precision by reducing noise. In addition, a computer of average transients (CAT) can be utilized to increase resolution through repetitive scans.[44,82,83] This method is particularly useful in cases of small rotational strength, for example, with L-tryptophan.[83]

The noise-averaged baseline (for the same cell, solvent, and sensitivity range) then has to be subtracted from the sample spectrum at each wavelength of interest, and the chart difference multiplied by the instrumental sensitivity. The result for ORD is the observed rotation, α_{obs}, in degrees; for CD, depending upon the instrument, it is either the observed ellipticity, θ_{obs}, in degrees, or the differential circular dichroic absorbance,

[77] E. R. Simons, *Biochim. Biophys. Acta* **251**, 126 (1971).
[78] Circulating thermostat baths with built-in heating and cooling units, suitable for measurements in the −10 or −20 to 100° range, are manufactured by Tamson, Lauda, and Haake (among others). They are distributed, respectively, by Neslab Instruments (Durham, New Hampshire), Brinkmann Instruments (Westbury, New York), and Greiner Scientific (22 N. Moore St., New York, New York). Mixtures of ethylene glycol and water can be used as the circulating fluid.
[79] C. Formoso and I. Tinoco, Jr., *Biopolymers* **10**, 531 (1971).
[80] The plastic tip of a YSI thermistor probe (Yellow Springs Instrument Co., Yellow Springs, Ohio) can be inserted into the cell jacket and plugged into a YSI Tele-thermometer. A small flat, YSI surface temperature probe may be clipped to the Pockels cell of a Cary 60 to monitor its temperature during a heating experiment.
[81] E. H. Strickland, J. Horwitz, and C. Billups, *Biochemistry* **8**, 3205 (1969).
[82] J. Horwitz, E. H. Strickland, and E. Kay, *Anal. Biochem.* **23**, 363 (1968).
[83] Y. B. Myer and L. H. MacDonald, *J. Amer. Chem. Soc.* **89**, 7142 (1967).

$(A_L - A_R)$, in absorbance units. The calculations which follow convert these data into quantitative measures of rotation or CD, and are amenable to computerization with most desk-top calculators.

Optical Rotatory Dispersion

Optical activity is usually reported in terms of molar rotation $[M]$ (for low molecular weight substances such as amino acid derivatives) and of residue rotation $[m]$ (for macromolecules). The residue rotation gives an indication of the optical activity of a single chromophoric peptide unit in a polypeptide or protein chain. The specific rotation at fixed wavelength, $[\alpha]_\lambda$, is of use mainly as a criterion of purity, for comparison to the older literature, and for calculation of Drude plots.

1. Specific Rotation. The optical activity is defined in terms of the specific rotation, $[\alpha]_\lambda$:

$$[\alpha]_\lambda^T = \frac{\alpha_{obs} \times 100}{l \times c}$$

where T = temperature; λ = wavelength; α_{obs} = observed rotation in degrees (solution minus solvent blank); l = path length of cell in decimeters; c = concentration in grams per 100 ml. For films, c = concentration in grams/cm^2, and l is unnecessary.

2. Molar Rotation. For low molecular weight substances $[M]_\lambda$ is defined as:

$$[M]_\lambda = \frac{MW}{100} [\alpha]_\lambda$$

where MW = molecular weight of the solute.

3. Residue Rotation. For comparison of rotations of proteins, oligopeptides and polypeptides, which differ greatly in molecular weights, a more meaningful unit, the mean residue rotation, is used:

$$[m] = \frac{MRW}{100} [\alpha]_\lambda$$

where MRW = mean residue weight of the repeating unit. The physical units for $[m]$ (as well as for $[M]$ and for these quantities reduced to vacuum) are degree centimeters^2 per decimole. The repeating molecular unit for proteins is the amino acid residue, so that the residue weight is the sum of the atomic weights in the unit—[C(=O)—CHR—NH]—. The MRW's of synthetic polypeptides and of proteins with known compositions can be calculated. The MRW for a large group of proteins is approximately 115, and this value can be used for comparative purposes when the exact amino acid composition is not known.

A convenient equation for calculation of $[m]$ when the molar residue concentration, c' (moles peptide residue/liter), is known is:

$$[m] = \frac{\alpha_{obs} \times 10}{l \times c'}$$

where the other symbols have been defined above.

4. Reduced Molar and Residue Rotations. The optical rotatory power is dependent upon the refractive index of the medium. Therefore, to compare observed rotations in a variety of solvents, the rotations are reduced to the value they would have in a vacuum, by means of the Lorentz correction factor, $3/(n^2 + 2)$, where n is the refractive index of the solvent at wavelength λ. Values of n and of $3/(n^2 + 2)$ for many solvents have been tabulated,[10,54] and methods for measuring n have been given.[54] For example, $3/(n^2 + 2)$ values for water are 0.77 at 250 nm, 0.76 at 220 nm, and 0.74 at 195 nm. The reduced mean residue rotation, $[m']$, at wavelength λ incorporates the refractive index correction in the following manner:

$$[m'] = [m]\frac{3}{(n^2 + 2)} = \frac{MRW}{100} \times \frac{3}{(n^2 + 2)}[\alpha]_\lambda$$

A similar expression can be written for $[M']$, the reduced molar rotation, by substitution of molecular weight, MW, for residue weight.

5. Drude and Moffitt Equations. The most informative method of ORD data analysis in recent studies is to present dispersion plots ($[M']$ or $[m']$ vs. λ), and to obtain the characteristic parameters (positions and magnitudes) for the Cotton effects resulting from peptide and side-chain absorption in the ultraviolet. However, before ultraviolet ORD studies became feasible, much use was made of the Drude[11] and, especially, the Moffitt[12-14] equations for protein ORD analysis. The application of these equations to protein studies, and the graphical methods employed for data reduction (including examples of calculation) have been extensively discussed.[10] The Moffitt calculation is still performed occasionally.

Very briefly, the Drude equation, which describes the ORD in spectral regions far from optically active absorption bands, is:

$$[\alpha]_\lambda = \frac{k}{\lambda^2 - \lambda_c^2}$$

where k is a constant, and λ_c represents the mean wavelength of optically active electronic transitions.

The Moffitt equation, developed for synthetic polypeptides, is:

$$[m']_\lambda = \frac{a_0\lambda_0^2}{\lambda^2 - \lambda_0^2} + \frac{b_0\lambda_0^4}{(\lambda^2 - \lambda_0^2)^2}$$

where a_0, b_0, and λ_0 are constants. When λ_0 is taken to be 212 nm, b_0 is about -630 for polypeptides in a totally right-handed α-helical conformation and zero for random coils; thus b_0 for a protein can be used as a measure of its helical content, if it is devoid of any beta structure.

Circular Dichroism

Circular dichroism data are reported either as $[\theta]$, the molar or residue ellipticity, or as $(\epsilon_L - \epsilon_R)$, the differential molar circular dichroic extinction coefficient. The two measurements are proportional: $[\theta] = 3300(\epsilon_L - \epsilon_R)$. The rotational strength, R_K, of each optically active absorption band is sometimes calculated, provided that the experimental CD bands can be resolved; a du Pont 310 Curve Resolver[84] is useful for this purpose. There is no method of phenomenological CD analysis comparable to the Drude or Moffitt equations.

1. Molar or Residue Ellipticity, $[\theta]$. The molar ellipticity (for small molecules) or mean residue ellipticity (for proteins and polypeptides) is defined as:

$$[\theta]_\lambda = \frac{\theta_{\text{obs}} \times \text{MW (or MRW)}}{10 \times d \times c''}$$

where λ = wavelength; θ_{obs} = observed ellipticity, in degrees; MW = molecular weight; MRW = mean residue weight (see ORD); c'' = concentration in grams per milliliter; d = path length in centimeters. If the molar concentration of peptide residues, c', is known directly, then $[\theta]$ may be calculated from:

$$[\theta] = \theta_{\text{obs}} \times \frac{10}{l \times c'}$$

where l = path length in decimeters c' = concentration in moles residue/liter. The units for $[\theta]$ are degrees per square centimeter per decimole.

2. Reduced Molar or Residue Ellipticity, $[\theta']$. The Lorentz refractive index correction (see ORD) is not usually applied to CD data. However, this correction is occasionally useful for literature comparisons:

$$[\theta'] = [\theta] \times \frac{3}{n^2 + 2}$$

3. Differential Molar CD Extinction Coefficient. In some instruments the difference in absorbance between left- and right-handed circularly polarized light, $(A_L - A_R)$ is measured directly. In such cases, the molar

---

[84] Manufactured by du Pont Instrument Products Division, Wilmington, Delaware 19898.

circular dichroic extinction coefficient (also called the molar dichroic absorption), $\epsilon_L - \epsilon_R$, is obtained from:

$$(\epsilon_L - \epsilon_R) = \frac{(A_L - A_R)}{c' \times d}$$

where d = path length in centimeters; c' = concentration in moles of residue per liter; $(\epsilon_L - \epsilon_R)$ has units of liters per mole centimeter.

4. Relation between Ellipticity and Differential Absorption. The proportionality, $[\theta] = 3300(\epsilon_L - \epsilon_R)$, has already been given. Another useful relationship, for comparison of raw CD data, is: $\theta_{obs} = 33(A_L - A_R)$. Thus, an observed ellipticity of 0.001 degree corresponds to an observed differential dichroic absorbance of 3×10^{-5} absorbance units.

5. Rotational Strength. The rotational strength R_K of the Kth optically active absorption band is defined by an integral which can be found elsewhere.[26(a),29,31] If the CD band is nearly Gaussian in shape, then:

$$R_K \approx 1.23 \times 10^{-42} [\theta_{max}]_K \frac{\Delta}{\lambda_{max}}$$

where λ_{max} = wavelength of the Kth transition; $[\theta_{max}]_K = [\theta]$ at λ_{max}; Δ = half-width of the band.

Relations between CD and ORD

CD and optical rotation are related, for the Kth optically active transition (Cotton effect), by the Kronig–Kramers relations.[85,86] By means of these integral transforms[26(a),44] information contained in a complete ORD spectrum may be deduced, in principle, from a CD curve, and vice versa. Computer programs are available for calculation.[61,62,64] Various methods of data manipulation can be compared in studies[61,62,64] on d-10-camphorsulfonic acid. The transforms can be used to search for optically active transitions beyond the observable UV range.[87] A generalization useful for qualitative calculation of band magnitude is that, if a CD band is approximately Gaussian in shape then:

$$([M]_{peak} - [M]_{trough})_{ORD} \approx 1.2([\theta]_{maximum})_{CD}$$

Analysis of Data

The interpretation of protein CD and ORD spectra in terms of backbone conformation, side-chain interactions or active site geometry, is not a routine or straightforward matter, even with precise data collected for pure material under the best of experimental conditions. The fault

[85] R. de L. Kronig, *J. Opt. Soc. Amer.* **12**, 547 (1926).
[86] H. A. Kramers, *Atti Congr. Int. Fis. Como* **2**, 545 (1927).
[87] J. Y. Cassim and J. T. Yang, *Biopolymers* **9**, 1475 (1970).

lies in the methods of data analysis currently available; these are usually based upon synthetic polypeptides as models for protein structure, and cannot cope with the complexities of structure found in natural proteins. The model polypeptides can assume only three fundamental conformations in solution: α-helix, extended β structure, and random coil. Even so, the CD and ORD spectra for each standard conformational form vary with the polypeptide chosen (see Figs. 4–6 and 9–11). Furthermore, the unique tertiary structure of any protein in solution may contain distorted or extremely short segments of α or β chains (which have no counterpart in synthetic polypeptides), portions which are "random" in the sense that there is no regular, repeated structure (but which may have rotational properties very different from ionized poly-L-glutamic acid or poly-L-lysine), side-chain interactions which may contribute to the CD and ORD, and optically active disulfide and aromatic groups (which may cause rotational bands in the far- as well as the near-ultraviolet region). Additional complications in the CD and ORD of real protein systems may be caused by aggregation and light scattering. It is at present impossible to adequately calculate the rotational effects of all these complicating factors, although increasingly sophisticated and successful attempts are being made. A recent paper on insulin[88] demonstrates some of the problems involved.

Nevertheless, as shown below in the table, fair to excellent approximations to real protein structures determined by X-ray diffraction can be obtained from chiroptical data, especially if the protein contains large amounts of conventional secondary structure. It should be emphasized that CD and ORD techniques are now applied to proteins mainly to yield information on gross secondary structure; however, these methods are beginning to be used for study of individual peptide residues in proteins.[89-92] Considerable information concerning enzyme action can be deduced by measuring the small conformational changes that occur upon complex formation with inhibitors[77,90-95] or coenzymes.[96,97]

[88] M. J. Ettinger and S. N. Timasheff, *Biochemistry* **10**, 824 (1971).

[89] M. Z. Atassi, M. T. Perlstein, and A. F. S. A. Habeeb, *J. Biol. Chem.* **246**, 3291 (1971).

[90] V. I. Teichberg, C. M. Kay, and N. Sharon, *Eur. J. Biochem.* **16**, 55 (1970).

[91] W. D. McCubbin, K. Oikawa, and C. M. Kay, *Biochem. Biophys. Res. Commun.* **43**, 666 (1971).

[92] J. P. Halper, N. Latovitzki, H. Bernstein, and S. Beychok, *Proc. Nat. Acad. Sci. U.S.* **68**, 517 (1971).

[93] L. Fretto and E. H. Strickland, *Biochim. Biophys. Acta* **235**, 489 (1971).

[94] E. Breslow, *Proc. Nat. Acad. Sci. U.S.* **67**, 493 (1970).

[95] P. Cuatrecasas and C. A. Anfinsen, *Proc. Nat. Acad. Sci. U.S.* **64**, 923 (1969).

[96] D. D. Ulmer and B. L. Vallee, *Advan. Enzymol.* **27**, 37 (1965).

[97] R. Koberstein and H. Sund, *FEBS (Fed. Eur. Biochem. Soc.) Lett.* **19**, 149 (1971).

The use of ORD and CD for the evaluation of polypeptide and protein structure has been the subject of several extensive reviews.[10,32-46] Since the 1950's great progress has been made in the elucidation of such structure, largely through X-ray diffraction analysis. Pauling and Corey[98] showed that polypeptide chains could assume only a limited number of stable structures, among them the α-helix (with 3.6 peptides per turn) and the interchain hydrogen-bonded β-structure (a fully extended parallel or antiparallel arrangement of peptide chains). These conformations were then found in various synthetic polypeptides by means of X-ray studies (see A. Elliot[37(b)] for a review).

As the various conformations of polypeptides were elucidated they were correlated with the optical activity of the polymers, and the application of CD and ORD spectra to characterize polypeptide and protein structure was initiated. At first only ORD in the spectral region $\lambda > 250$ nm was available; much use was made of the Drude and (especially) the Moffitt equations. The early work has been reviewed.[10,38] Then, as instrumentation improved, Cotton effects in the far-ultraviolet peptide region (185–240 nm) could be examined directly in both ORD and CD. The great majority of studies since about 1963 have utilized these Cotton effects; the near-ultraviolet region above 240 nm is now examined primarily for side-chain chromophore Cotton effects, not for backbone conformation.

Evaluation of Protein and Polypeptide Conformation from ORD in the Visible and Near-Ultraviolet Region

In 1955 Cohen suggested that the change in optical rotation that was noted upon protein denaturation may be due to a helix to random coil conformational transition.[99] The first measurements of optical activity of polypeptides and proteins were performed in the visible and near-ultraviolet regions ($\lambda > 250$ nm). Empirical equations were found to correlate ORD with polypeptide structure. The Drude equation (see calculations) could be satisfied for polypeptides of low helical content and for some proteins, but the ORD data of completely helical polypeptides and many other proteins[12,13,100] could not be fitted. For polypeptides λ_c was found to be proportional to the percentage of helical content[100] up to 40% helix.

The Moffitt equation[12,13] (see calculations) describes the rotation of helical polypeptides (such as un-ionized poly-L-glutamic acid) in the visible spectral region. The constant, b_0, is proportional to backbone

[98] L. Pauling and R. B. Corey, *Proc. Nat. Acad. Sci. U.S.* **37**, 729 (1951); *ibid.* **39**, 247, 253 (1953).

[99] C. Cohen, *Nature (London)* **175**, 129 (1955).

[100] J. T. Yang and P. Doty, *J. Amer. Chem. Soc.* **79**, 761 (1957).

helical content and independent of side chain and environment, while a_0 is a function of side-chain and solvent. The ORD of many fully helical polypeptides yielded $b_0 = -630$, while fully random polypeptides gave a value of $b_0 = 0$. Polypeptides of mixed conformation gave a b_0 value which was proportional to helicity. Thus the value of b_0 was used as a measure of helical content, $b_0/-630 = \%$ α-helix.[14,101-103] However, the contribution of any β structure was ignored. The use of the Moffitt equation has been reviewed.[10,38] Moffitt's equation was originally derived theoretically. However, it was later shown to have neglected some important terms.[104,105] The equation, nonetheless, gave reasonable results for proteins and peptides whose helical content was then known, e.g., myoglobin.[106,107]

Evaluation of Protein Structure from ORD in the Far Ultraviolet: Use of Cotton Effects

In 1960 Simmons and Blout[15] were the first to obtain ORD measurements in the absorption region of the polypeptide backbone. They detected the trough of a Cotton effect at 233 nm for tobacco mosaic virus. Following this, instrumentation rapidly improved and measurements of the entire ORD and CD spectra from 185 to 300 nm were soon achieved.

The first attempts to utilize ORD Cotton effects as a measure of helical content involved correlation of the magnitude of the 233 nm trough of proteins and polypeptides with helicity.[108] Later Yamaoka[109] and Blout and co-workers[110-114] proposed a modified four-term Drude equation to account for the near ultraviolet and visible ORD. This Schechter–Blout equation is essentially a summation of the dispersion

[101] C. Cohen and A. Szent-Györgyi, *J. Amer. Chem. Soc.* **79**, 248 (1957).
[102] K. Imahori, E. Klemperer, and P. Doty, *Abstr. 131 Meet. Amer. Chem. Soc.*, Miami, April, 1957.
[103] J. T. Yang, *Tetrahedron* **13**, 143 (1961).
[104] W. Moffitt, D. D. Fitts, and J. G. Kirkwood, *Proc. Nat. Acad. Sci. U.S.* **43**, 723 (1957).
[105] J. A. Schellman and P. Oriel, *J. Chem. Phys.* **37**, 2114 (1962).
[106] S. Beychok and E. R. Blout, *J. Mol. Biol.* **3**, 769 (1961).
[107] P. J. Urnes, K. Imahori, and P. Doty, *Proc. Nat. Acad. Sci. U.S.* **47**, 1637 (1961).
[108] N. S. Simmons, C. Cohen, A. G. Szent-Györgyi, D. B. Wetlaufer, and E. R. Blout, *J. Amer. Chem. Soc.* **83**, 4766 (1961).
[109] K. K. Yamaoka, *Biopolymers* **2**, 219 (1964).
[110] E. Shechter and E. R. Blout, *Proc. Nat. Acad. Sci. U.S.* **51**, 695 (1964).
[111] E. Shechter and E. R. Blout, *Proc. Nat. Acad. Sci. U.S.* **51**, 794 (1964).
[112] E. Shechter, J. P. Carver, and E. R. Blout, *Proc. Nat. Acad. Sci. U.S.* **51**, 1029 (1964).
[113] J. P. Carver, E. Shechter, and E. R. Blout, *J. Amer. Chem. Soc.* **88**, 2550 (1966).
[114] J. P. Carver, E. Shechter, and E. R. Blout, *J. Amer. Chem. Soc.* **88**, 2562 (1966).

of the α-helical and random coil forms of polypeptides, and attempts to account for the rotation due to the Cotton effect of each electronic transition of both polypeptide forms. The Shechter-Blout equation actually does not take into account all the terms present and combines many terms. Thus it is empirical, like the Moffitt equation, can be put into the same form as the Moffitt equation,[115] and thus gives similar results for the estimation of helical content as the Moffitt equation.

The methods of helical content estimation considered so far involved measurements of optical rotation in the visible and near UV ($\lambda > 220$ nm) region. They generally ignored regular polypeptide conformations other than α-helix and random, although other conformations were known to exist in proteins. In particular, the β-form (that is the pleated sheet interchain hydrogen bonded form) was neglected. The β-form had not been studied extensively because of the difficulty of obtaining water-soluble model polypeptides in this conformation. Nevertheless, some attempts were made to include the β-form in the estimation of protein structure by means of the Moffitt equation.[116-119] This extension was difficult because precise values for the contribution of the β structure to a_0 and b_0 were hard to determine due to uncertainties in the conformation of model polypeptides. Estimates of the Moffitt parameters for β polymers have resulted in a wide range of values for a_0 and b_0.[16,18,116,120-128]

In 1966 two groups[16-19] obtained good ORD (and CD) measurements of the β form of poly-L-lysine in aqueous solution. (Similar data were obtained for the β form of silk fibroin in organic solvents.[42,129]) ORD curves for the three fundamental conformations of poly-L-lysine (α-

---

[115] J. T. Yang, *Proc. Nat. Acad. Sci. U.S.* **53**, 438 (1965).
[116] A. Wada, M. Tsuboi, and E. Konishi, *J. Phys. Chem.* **65**, 1119 (1961).
[117] J. A. Schellman and C. G. Schellman, *J. Polym. Sci.* **49**, 129 (1961).
[118] G. V. Troitski, *Biofizika* **10**, 895 (1965).
[119] S. N. Timasheff, R. Townend, and L. Mescanti, *J. Biol. Chem.* **241**, 1863 (1966).
[120] K. Imahori, *Biochim. Biophys. Acta* **37**, 336 (1960).
[121] G. D. Fasman and E. R. Blout, *J. Amer. Chem. Soc.* **82**, 2262 (1960).
[122] E. M. Bradbury, A. Elliott, and W. E. Hanby, *J. Mol. Biol.* **5**, 487 (1962).
[123] B. S. Farrap and I. W. Stapleton, *Biochim. Biophys. Acta* **75**, 31 (1963).
[124] S. Ikeda, H. Maeda, and T. Isemura, *J. Mol. Biol.* **10**, 223 (1964).
[125] K. Imahori and J. Yahara, *Biopolym. Symp.* **1**, 421 (1964).
[126] E. V. Anufrieva, I. A. Bolotina, B. Z. Volchek, N. G. Illarionova, V. I. Kalikhevich, O. Z. Korotkina, Y. V. Mitin, O. B. Ptitsyn, A. V. Purkina, and V. E. Eskin, *Biofizika* **10**, 918 (1965).
[127] E. V. Anufrieva, N. G. Illarionova, V. I. Kalikhevich, O. Z. Korotkina, Y. V. Mitin, O. B. Ptitsyn, A. V. Purkina, and B. Z. Volchek, *Biofizika* **10**, 346 (1965).
[128] L. Velluz and M. Legrand, *Angew. Chem. Int. Ed. Engl.* **4**, 838 (1965).
[129] E. Iizuka and J. T. Yang, *Biochemistry* **7**, 2218 (1968).

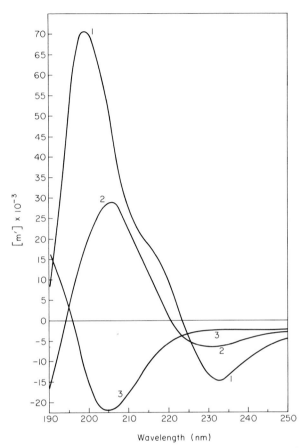

FIG. 2. The optical rotatory dispersion of poly-L-lysine in the α-helical (curve 1), β (curve 2), and random (curve 3) conformations. [From N. J. Greenfield, B. Davidson, and G. D. Fasman, *Biochemistry* **6**, 1630 (1967).]

helix, β, and random) are shown in Fig. 2. At this point in time, realistic calculations of protein conformation, based upon polypeptide ORD Cotton effects, could be attempted. Greenfield *et al.*[20] used these data for poly-L-lysine to calculate ORD curves expected for various mixtures of the three basic conformations, and compared these calculated curves with measured ORD spectra for several proteins. (The proteins chosen for this study had conformations determined by X-ray diffraction.) The ORD comparison for myoglobin is shown in Fig. 3. The estimation of conformation for both myoglobin and lysozyme tended to overestimate the amount of β structure and to underestimate the amount of α-helix and random coil as found by X-ray diffraction studies. The

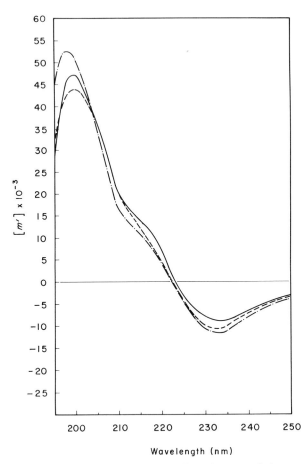

Fig. 3. The measured optical rotatory dispersion of sperm whale myoglobin (———) [from S. C. Harrison and E. R. Blout, *J. Biol. Chem.* **240**, 294 (1965)] compared with the calculated optical rotatory dispersion for 77% α-helix and 23% random coil (–·–·–·) and for 54% α-helix, 36% β structure, and 10% random coil (- - - - - -) [from N. J. Greenfield, B. Davidson, and G. D. Fasman, *Biochemistry* **6**, 1630 (1967)].

differences between the calculated and X-ray determined structures were attributed to aromatic side chain chromophores, disulfide bridge contributions, prosthetic groups contributions, and possible contributions from conformations of the amide groups other than those in the three reference conformations. It can also be seen from Figs. 4–6 that the absolute rotatory values for the three reference conformations are in doubt, since they vary with polypeptide side chain and solvent, adding further

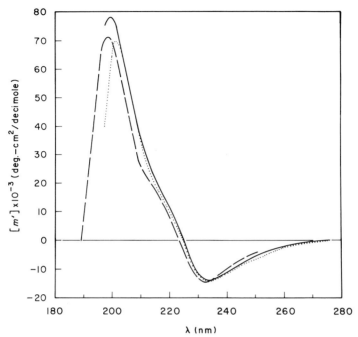

FIG. 4. The optical rotatory dispersion of the α-helix in various polypeptides: Poly-L-glutamic acid in water:dioxane, 1:1, —— [from E. Iizuka and J. T. Yang, *Biochemistry* **4**, 1249 (1965)]. Poly-L-lysine in water, pH 11.0, – – – – – – [from N. J. Greenfield, B. Davidson, and G. D. Fasman, *Biochemistry* **6**, 1630 (1967)]. Poly-[N^5-(2-hydroxyethyl)-L-glutamine] in methanol:water, 8:2, · · · · · · [from A. J. Adler, R. Hoving, J. Potter, M. Wells, and G. D. Fasman, *J. Amer. Chem. Soc.* **90**, 4736 (1968).]

difficulty in estimating protein conformation. Magar[130] performed similar calculations using a more precise method of minimizing the variance between the calculated and experimental ORD curves, but he reached essentially the same conclusions as Greenfield et al.[20]

Evaluation of Protein Conformation from CD Measurements in the Ultraviolet

The preceding methods of estimating protein structure all used optical rotatory dispersion because instrumentation was not available to study circular dichroism in the far ultraviolet. The first measurements of the circular dichroism of polypeptides were made by Holzwarth

[130] M. E. Magar, *Biochemistry* **7**, 617 (1968).

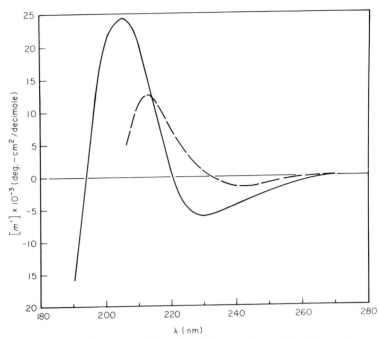

FIG. 5. The optical rotatory dispersion of the β form of various polypeptides: Poly-L-lysine in water, pH 11, ——— [from N. J. Greenfield, B. Davidson, and G. D. Fasman, *Biochemistry* 6, 1630 (1967)]. Poly-S-carboxymethyl-L-cysteine in water, pH 4.25, - - - - - [from S. Ikeda and G. D. Fasman, *J. Mol. Biol.* 30, 491 (1967)].

et al.,[131,132] who examined the Cotton effects associated with peptide electronic transitions in α-helical and randomly coiled synthetic polypeptides and in myoglobin (which is largely helical). For helical polymers they assigned the negative CD band at 222 nm to the $n \rightarrow \pi^*$ amide transition, and the negative 208-nm band and positive 190-nm band to the parallel and perpendicularly polarized, respectively, $\pi \rightarrow \pi^*$ amide transitions. (This exciton splitting had been predicted by Moffitt.[12,13]) They also made tentative assignments of the random coil CD bands, and showed (by means of the Kronig–Kramers transform) that the ORD helical and random-coil spectra were consistent with the CD bands.

After this major breakthrough, improved CD instruments became

[131] G. M. Holzwarth, W. B. Gratzer, and P. Doty, *J. Amer. Chem. Soc.*, 84, 3194 (1962).
[132] G. M. Holzwarth and P. Doty, *J. Amer. Chem. Soc.* 87, 218 (1965).

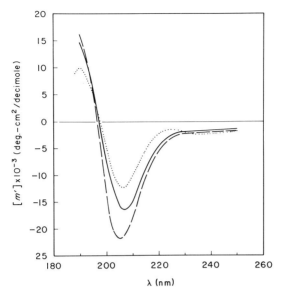

λ (nm)

Fig. 6. The optical rotatory dispersion of the random coil form of various poly-peptides: Poly-L-glutamic acid in water, pH 7.3, —— [from E. Iizuka and J. T. Yang, *Biochemistry* **4**, 1249 (1965)]. Poly-L-lysine in water, pH 4.7–5.0, - - - - - [from N. J. Greenfield, B. Davidson, and G. D. Fasman, *Biochemistry* **6**, 1630 (1967)]. Poly-[N^5-(2-hydroxyethyl)-L-glutamine] in water, · · · · · · [from A. J. Adler, R. Hoving, J. Potter, M. Wells, and G. D. Fasman, *J. Amer. Chem. Soc.* **90**, 4736 (1968)].

widely available, CD spectra for β-form poly-L-lysine were measured,[16–19] and it became feasible to interpret CD data on proteins in terms of polypeptide conformations, as was done previously for ORD.[20] CD is now being used extensively to study the conformation of polypeptides and proteins in solution. Circular dichroism has an advantage over ORD in that there is less overlap between optically active transitions, and usually one can separate transitions due to the polypeptide backbone from other chromophores.

CD spectra for the three reference conformations of poly-L-lysine (α-helix, β-form, and random coil) are shown in Fig. 7. (Note that the β structure exhibits only one negative CD band in the 220-nm region.) Greenfield and Fasman[21] utilized these curves for poly-L-lysine to calculate CD spectra expected for various mixtures of the three funda-mental conformations. They used these calculated CD curves (in a manner analogous to their ORD study[20]) to compare with experimental CD spectra of several proteins of known conformation (from X-ray studies). The results of this analysis, based on data from 208 to 250 nm,

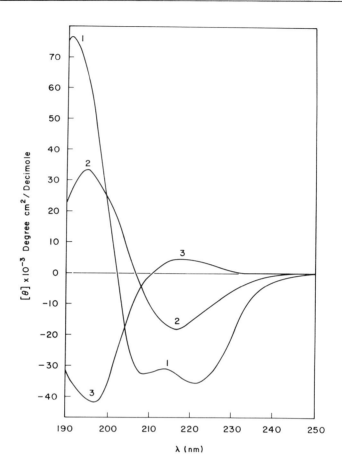

FIG. 7. Circular dichroism spectra of poly-L-lysine in the 100% α-helical (curve 1), β (curve 2), and random coil (curve 3) conformations. [From N. J. Greenfield and G. D. Fasman, *Biochemistry* **8**, 4108 (1969).]

are shown in Fig. 8 and in the table. Greenfield and Fasman[21] found, essentially, that when a protein is highly ordered, with either α-helix or β structure predominating, the results are within 5% of the X-ray data. Thus the estimates obtained from myoglobin, lysozyme, and RNase are quite good. When the proteins studied lacked a large amount of regularity, the deviations were much larger. In the latter case the fit of the estimated structure to the actual conformation obtained was not as good; however, these results did give an approximate idea of the secondary structure of the proteins and were more informative than all previous ORD methods of estimating protein conformation based on such param-

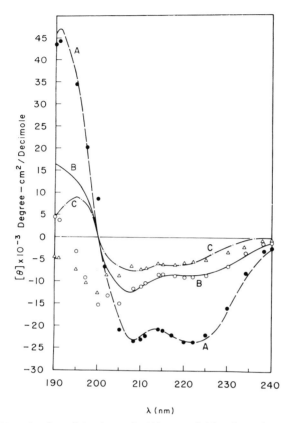

λ (nm)

FIG. 8. The circular dichroism of (A) myoglobin (experimental curve) and 68.3% α-helix, 4.7% β structure, and 27.0% random coil, calculated from poly-L-lysine reference spectra in water (●●●●). (B) Lysozyme (curve) and 28.5% α-helix, 11.1% β structure, and 60.4% random coil (○○○), calculated as in part A. (C) RNase (curve) and 9.3% α-helix, 32.6% β structure, and 58.1% random coil, calculated as for A (△ △ △). [From N. J. Greenfield and G. D. Fasman, *Biochemistry* 8, 4108 (1969).]

eters as the Moffitt equation, the Shechter–Blout equation or the magnitude of the 233 nm trough of the ORD Cotton effect.

Refinements of the Analysis of Cotton Effects

As in the case of the ORD analysis,[20] the differences between calculated and experimental protein CD curves[21] can be attributed largely to the imperfect nature of synthetic polypeptides as CD standards for protein chain conformations. The problems involved will be discussed in detail later. It can be mentioned here that (1) CD band parameters for α, β, and random-coil polypeptides vary with side chain and solvent (Figs. 9–11),

Comparison of Conformation Obtained by Circular Dichroism (CD) and X-Ray[a,b]

Protein	X-Ray structure: fraction				CD calculated structure: fraction[c]					
	α	β	Random coil	Reference	α		β		Random coil	
					b	c	b	c	b	c
Myoglobin	65–72, 77[d]	0	32–23	[h]	68.3	68.2	4.7	7.9	27.0	23.9
Lysozyme	28–42[d]	10	62–48	[i,j]	28.5	29.9	11.1	9.3	60.4	61.0
RNase	6–18[d]	36	58–46	[k]	9.3	12.0	32.6	43.4	58.1	44.5
RNase S	15[e]	31	54	[l]						
Carboxypeptidase A	23–30[d]	18	59–52	[m]	13.0	15.9	30.6	39.9	56.4	44.2
α-Chymotrypsin[f]	3[g]	—		[n]	11.8	13.4	22.8	31.9	65.5	54.8
Chymotrypsinogen					13.8	17.3	25.2	28.9	60.9	53.8

[a] From N. J. Greenfield and G. D. Fasman, *Biochemistry* **8**, 4108 (1969).

[b] The best fit to the experimental CD curve was found by minimizing the variance between the experimental CD curve and a linear combination of CD curves for the α-helix, β structure, and random coil from 208 to 240 nm.

[c] Columns b: Values used in calculations are for poly-L-lysine in H₂O. Columns c: Values used in calculations same as for columns b with data for β-poly-L-lysine in SDS substituted for β poly-L-lysine in H₂O.

[d] Lower value represents true regular α-helix; upper value, total helix including 3₁₀ and distorted helices.

[e] Helical type not distinguished by authors.

[f] α-Tosyl chymotrypsin used for X-ray work.

[g] Only one short-chain section α-helix included; isolated helical turns and β structure not reported, although they may be present.

[h] J. C. Kendrew, R. E. Dickerson, B. E. Strandberg, R. G. Hart, D. R. Davies, D. C. Phillips, and V. C. Shore, *Nature (London)* **185**, 422 (1960).

[i] C. C. F. Blake, D. F. Koenig, G. A. Mair, A. C. T. North, D. C. Phillips, and V. R. Sarma, *Nature (London)* **206**, 757 (1965).

[j] D. C. Phillips, *Sci. Amer.* **215**, 78 (1966); D. C. Phillips, *Proc. Nat. Acad. Sci. U.S.* **57**, 484 (1967).

[k] G. Kartha, J. Bello, and D. Harker, *Nature (London)* **213**, 862 (1967).

[l] H. W. Wyckoff, K. D. Hardman, N. M. Allemwell, T. Inagami, L. N. Johnson, and M. Richards, *J. Biol. Chem.* **242**, 3984 (1967).

[m] G. N. Reeke, J. A. Hartsuck, M. L. Ludwig, F. A. Quiocho, J. A. Steitz, and W. N. Lipscomb, *Proc. Nat. Acad. Sci. U.S.* **58**, 2220 (1967).

[n] P. B. Sigler, D. M. Blow, B. W. Matthews, and R. Henderson, *J. Mol. Biol.* **35**, 143 (1968).

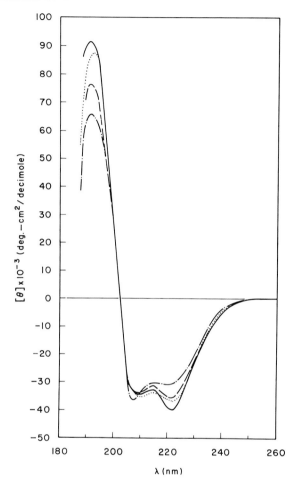

FIG. 9. The circular dichroism of the α-helix in various polypeptides: Poly-L-glutamic acid in water, pH 4.4——— [from G. M. Holzwarth and P. Doty, *J. Amer. Chem. Soc.* **87**, 218 (1965)]. Poly-L-lysine in water, pH 11.0, ------ [from R. Townend, T. F. Kumosinski, S. N. Timasheff, G. D. Fasman, and B. Davidson, *Biochem. Biophys. Res. Commun.* **23**, 163 (1966)]. Poly-[N^5-(2-hydroxyethyl)-L-glutamine] in methanol:water, 8:2, · · · · (G. D. Fasman, unpublished data). Poly-L-alanine in trifluoroethanol:trifluoroacetic acid, 98.5:1.5, –·–·–· [from F. Quadrifoglio and D. W. Urry, *J. Amer. Chem. Soc.* **90**, 2755 (1968).]

(2) CD curves of unordered proteins bear only a qualitative relation to those for random-coil poly-L-lysine or poly-L-glutamic acid, (3) other regular conformations (such as 3_{10}-helix) and very short or distorted segments of the three standard forms undoubtedly contribute to protein CD in a different manner, and (4) optically active aromatic, cystine disulfide,

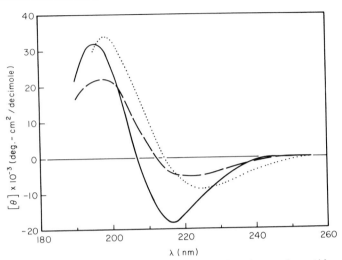

Fig. 10. The circular dichroism of the β form of various polypeptides: Poly-L-lysine in water, pH 11, ———— [from N. J. Greenfield and G. D. Fasman, *Biochemistry* **8**, 4108 (1969)]. Poly-L-serine in water, - - - - [from F. Quadrifoglio and D. W. Urry, *J. Amer. Chem. Soc.* **90**, 2760 (1968)]. Poly-S-carboxymethyl-L-cysteine in water, pH 4.3, · · · · · · (G. D. Fasman, unpublished data).

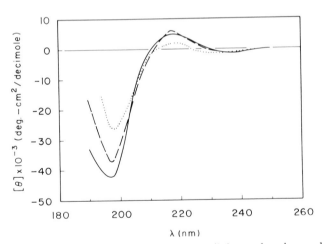

Fig. 11. The circular dichroism of the random coil form of various polypeptides: Poly-L-glutamic acid in water pH 7.5, ———— [from A. J. Adler, R. Hoving, J. Potter, M. Wells, and G. D. Fasman, *J. Amer. Chem. Soc.* **90**, 4736 (1968)]. Poly-L-lysine in water, pH 5.7, - - - - [from N. J. Greenfield and G. D. Fasman, *Biochemistry* **8**, 4108 (1969)]. Poly-[N^5-(2-hydroxyethyl)-L-glutamine in water, · · · · · · [from A. J. Adler, R. Hoving, J. Potter, M. Wells, and G. D. Fasman, *J. Amer. Chem. Soc.* **90**, 4736 (1968)].

and prosthetic group chromophores may well influence protein CD spectra in the far ultraviolet. Although many data have been collected on these subjects, there has been little attempt as yet to apply this mass of information to protein studies.

Several recent papers represent variations and refinements of the methods of CD[21] and ORD[20] analysis presented above. Straus et al.[133] made similar calculations of percentage of α, β, and random forms for several proteins, utilizing various polypeptides (not only poly-L-lysine) as standards, and allowing CD band parameters to vary for the calculation. Myer[134] utilized isodichroic points at 198, 204, and 208 nm for poly-L-lysine in its various conformations for analysis of CD spectra of myoglobin, lysozyme, RNase, and cytochrome c. Rosenkranz and Scholtan[135] used poly-L-serine at high salt concentration as their reference for the unordered form and calculated CD curves for myoglobin, lysozyme, and RNase. They obtained a better fit to the experimental data at $\lambda = 195$–208 nm; but a worse fit at $\lambda > 208$ nm than did Greenfield and Fasman.[21]

Saxena and Wetlaufer[136] avoided the problems inherent in polypeptide standards; they used native proteins of known conformation to obtain reference CD curves for α, β, and unordered conformations. These standard spectra, obtained from crystallographic and CD data for myoglobin, RNase, and lysozyme [and considerably different from those for poly-L-lysine (Fig. 7), especially in the random form], were then used for computation of CD curves for other proteins. Saxena and Wetlaufer[136] obtained very good agreement (better than did Greenfield and Fasman[21]) for carboxypeptidase, but only a fair fit for α-chymotrypsin. Thus, even total avoidance of polypeptide standards does not ensure a perfect method of CD conformational analysis of proteins. Each protein probably contains regions of unique structure which cannot be analyzed by any set of α, β, and random standard curves; furthermore, side-chain chromophore CD contributions are likely to be different for each protein. However, this type of CD interpretation,[136] based on protein data, is a step in the right direction. Chen and Yang[137] made use of crystallographically determined proteins to obtain b_0, $[m]_{233}$ and $[\theta]_{222}$ parameters for α-helix and random-coil forms. An additional study bearing on CD analysis[138] shows that ORD data in the visible and near UV region may

[133] J. H. Straus, A. S. Gordon, and D. F. H. Wallach, Eur. J. Biochem. **11**, 201 (1969).
[134] Y. P. Myer, Res. Commun. Chem. Pathol. Pharm. **1**, 607 (1970).
[135] H. Rosenkranz and W. Scholtan, Hoppe-Seyler's Z. Physiol. Chem. **352**, 896 (1971).
[136] V. P. Saxena and D. B. Wetlaufer, Proc. Nat. Acad. Sci. U.S. **68**, 969 (1971).
[137] Y.-H. Chen and J. T. Yang, Biochem. Biophys. Res. Commun. **44**, 1285 (1971).
[138] S. Sugai, K. Nitta, and M. Ishikawa, Biophysik **7**, 8 (1970).

be used to calculate CD and ORD peptide Cotton effects for polypeptides. There are many investigations utilizing CD or ORD of polypeptides (for example, the references cited in footnotes 139–143) whose main concern is stability or thermodynamics of the polymers, and which will not be discussed here.

Critique of the Analysis: Polypeptide Backbone

Long-chain synthetic polypeptides are an imperfect choice as models for protein ORD and CD structure determination. Conformational reference parameters for α, β, and random forms vary with the polypeptide, and a variety of possible protein structures are excluded in these simple polymers. The use of synthetic polypeptides as standards for the protein backbone, and the effects of chain length and of aggregation will be discussed. This section will also include a consideration of light scattering (which is particularly applicable to membrane studies), and the final section will discuss aromatic and cystine side-chain chromophore contributions.

α-Helix

Long-chain acidic poly-L-glutamic acid and alkaline poly-L-lysine have long been used as the canonical models for the α-helix in proteins. Although the chiroptical spectra are qualitatively similar for polypeptides with different side chains in various solvents, as illustrated in Figs. 4 (ORD) and 9 (CD), the peak values vary considerably. For example, the CD data for α-helical poly-L-glutamic acid,[132] poly-L-lysine,[17,20] poly-[N^5-(2-hydroxyethyl)-L-glutamine],[144] poly-L-alanine,[145] poly-γ-methyl-L-glutamate,[87,146] poly-L-methionine,[146] and poly-L-homoserine[146] (in various solvents) are far from coincident. This change in α-helix parameters is not due only to solvent effects.[87,146] Calculations[147] have predicted that nonaromatic side-chain effects will contribute to the rotatory strength of α-helical peptide bands, due to changes in the geom-

---

[139] H. A. Scheraga, *Chem. Rev.* **71**, 195 (1971).

[140] M. Goodman, A. S. Verdini, N. S. Choi, and Y. Masuda, in "Topics in Stereochemistry" (E. L. Eliel and N. L. Allinger, eds.), Vol. 5, p. 69. Wiley, New York, 1970.

[141] G. Conio, E. Patrone, and S. Brighetti, *J. Biol. Chem.* **245**, 3335 (1970).

[142] J. Steigman and A. Cosani, *Biopolymers* **10**, 357 (1971).

[143] D. Puett and A. Ciferri, *Biopolymers* **10**, 547 (1971).

[144] A. J. Adler, R. Hoving, J. Potter, M. Wells, and G. D. Fasman, *J. Amer. Chem. Soc.* **90**, 4736 (1968).

[145] F. Quadrifoglio and D. W. Urry, *J. Amer. Chem. Soc.* **90**, 2755 (1968).

[146] J. R. Parrish and E. R. Blout, *Biopolymers* **10**, 1491 (1971).

[147] J. N. Vournakis, J. F. Yan, and H. A. Scheraga, *Biopolymers* **6**, 1531 (1968).

etry of the helical backbone. This change in rotatory strength has been shown experimentally with copolymers of L-leucine and L-lysine in aqueous solution,[148] for which $[\theta]_{208} = -34{,}000$ remains constant, but $[\theta]_{222}$ varies with the leucine content. Furthermore, the CD spectrum of the α-helix constructed from data on proteins[136] differs from those based on synthetic polypeptides. Aggregation is known to change the ORD of helical poly-L-glutamic acid,[149] and care should be taken to avoid comparable situations in proteins.

Another respect in which long-chain polypeptides may not be a good model is that of chain length. Segments of α-helix in proteins are short, ranging from 3 to about 20 peptide units. ORD data[150] show that $[m']_{233}$ values for oligo-L-lysines, even at $n = 22$, do not approach that for poly-L-lysine because of incomplete helix formation. No helical models for short peptide chains are available. Theoretical calculations indicate that the rotatory strength of the $\pi - \pi^*$ transition of the α-helix 208 and 191 nm CD bands should be greatly dependent upon chain length,[151,152] but vary in predictions about the $n - \pi^*$ transition (222 nm CD band).[147,153,154]

β-Forms

Poly-L-lysine at high pH, after heating and recooling, is usually used as a reference for the β form. However, it has been shown that the β form of poly-L-lysine produced at neutral pH with sodium dodecyl sulfate has slightly different ORD[18] and CD[155] spectra than those found in water alone at pH 11. The magnitude of the CD ellipticity band at 218 nm for this β form has only one-half the magnitude of the β form produced by heating poly-L-lysine in water at pH 11, as found by Townend et al.[17] and by Sarkar and Doty.[18] Li and Spector[155] stated that β poly-L-lysine in water alone may form an intermolecular "infinite" pleated sheet and may not be a good model for the short sections of β structure found in proteins. Furthermore, depending upon the concentration and chain length, poly-L-lysine is able to form either inter- or intramolecular β pleated sheets (or, in some cases, mixtures).[156]

[148] C. R. Snell and G. D. Fasman, *Biopolymers* **11**, 1723 (1972).
[149] J. Y. Cassim and J. T. Yang, *Biochem. Biophys. Res. Commun.* **26**, 58 (1967).
[150] A. Yaron, G. D. Fasman, E. Katchalski, H. A. Sober, and A. Berger, *Biopolymers* **10**, 1170 (1971).
[151] R. W. Woody and I. Tinoco, Jr., *J. Chem. Phys.* **46**, 4927 (1967).
[152] R. W. Woody, *J. Chem. Phys.* **49**, 4797 (1968).
[153] I. Tinoco, Jr., R. W. Woody, and D. F. Bradley, *J. Chem. Phys.* **38**, 1317 (1963).
[154] D. W. Urry, *Proc. Nat. Acad. Sci. U.S.* **60**, 394 (1968).
[155] L. K. Li and A. Spector, *J. Amer. Chem. Soc.* **91**, 220 (1969).
[156] S.-Y. C. Wooley and G. Holzwarth, *Biochemistry* **9**, 3604 (1970).

Different poly-α-amino acids in the β form display different CD (Fig. 10) and ORD (Fig. 5) spectra depending on side-chain and solvent.[157-161] For example, the CD curves for poly-L-lysine,[21] poly-L-serine,[157] and poly-S-carboxyethyl-L-cysteine[161] in aqueous solution bear little more than qualitative relationship to one another or to a β-form CD curve constructed for globular proteins.[136] Fasman and Potter[158] have examined films cast from several β-forming polypeptides, and have grouped their ORD spectra into two classes; these films have also shown two classes of CD spectra.[160] Poly-L-lysine (Figs. 5 and 10) and poly-L-valine are found in form I, while poly-L-serine (Fig. 10) is found in form II.[158,160]

Another difficulty is that there are three types of β structure possible in polypeptides[98]: intermolecular parallel and antiparallel hydrogen-bonded sheets, and intramolecular antiparallel cross-β structure. It has been impossible to assign CD or ORD spectra unequivocally to one of these forms or another; this problem has been discussed experimentally[156,158,160,] and theoretically.[162-164] An exception is poly-[L-ala-L-glu-(OEt)-gly],[165] which appears to be in a cross-β conformation[166] and has the optical properties of form I-β.

Urry[167] calculate an extremely large theoretical chain-length dependence for the $n - \pi^*$ transition in antiparallel pleated β sheets. However, Goodman et al.[168] found that oligo-isoleucines, $n = 7$ and 8, in organic solvents had CD spectra similar to those of long-chain β polypeptides.

Random Coil

The use of synthetic polypeptides, usually fully charged poly-L-lysine·HCl, sodium poly-L-glutamate, as models for "unordered" seg-

[157] F. Quadrifoglio and D. W. Urry, J. Amer. Chem. Soc. 90, 2760 (1968).
[158] G. D. Fasman and J. Potter, Biochem. Biophys. Res. Commun. 27, 209 (1967).
[159] N. M. Tooney and G. D. Fasman, J. Mol. Biol. 36, 355 (1968).
[160] L. Stevens, R. Townend, S. N. Timasheff, G. D. Fasman, and J. Potter, Biochemistry 7, 3717 (1968).
[161] H. Maeda and S. Ikeda, Biopolymers 10, 1635 (1971).
[162] E. S. Pysh, Proc. Nat. Acad. Sci. U.S. 56, 825 (1966).
[163] K. Rosenheck and B. Sommer, J. Chem. Phys. 46, 532 (1967).
[164] R. W. Woody, Biopolymers 8, 669 (1969).
[165] J. M. Anderson, W. B. Rippon, and A. G. Walton, Biochem. Biophys. Res. Commun. 39, 802 (1970).
[166] W. B. Rippon, J. M. Anderson, and A. G. Walton, J. Mol. Biol. 56, 507 (1971); J. C. Andries and A. G. Walton, J. Mol. Biol. 56, 515 (1971).
[167] D. W. Urry, Proc. Nat. Acad. Sci. U.S. 60, 114 (1968).
[168] M. Goodman, F. Naider, and C. Toniolo, Biopolymers 10, 1719 (1971).

ments of proteins (which have well defined although not regularly repeating asymmetric structures) is open to criticism. First, use of such polyelectrolytes has been questioned by Krimm and co-workers,[169-172] who stated that in water at low salt concentration charged sodium poly-L-glutamate forms an extended helical structure due to charge repulsion. (Such a structure exists in the solid state.[173]) Because of this, the suggestion has been made that polypeptides (sodium poly-L-glutamate,[170] poly-L-proline,[170] or poly-L-serine[135]) in very concentrated salt solutions may be better standards for unordered protein chains. However, this suggestion is disputed by Fasman et al.,[174] who cite work[175] showing that in high salt it is possible that PGA collapses into a compact structure which cannot be termed a true statistical coil. Moreover, it was found[176] that salt may interact with the peptide carbonyl groups and consequently may shift transition moments. There is evidence against any extended helical form for protonated poly-L-lysine at low salt concentration.[177] Furthermore, the random coil ORD (Fig. 6) and CD (Fig. 11) curves for uncharged poly-$[N^5$-(2-hydroxyethyl)-L-glutamine]^{144} in water and for several un-ionized polyamino acids in various solvents[178] are qualitatively similar to those of charged polypeptides at low ionic strength; all exhibit negative CD bands at about 198 nm and positive bands at about 218 nm, although the band magnitudes are greater for the polyelectrolytes. It may be mentioned here that strong sulfuric acid (sometimes employed to produce random coil forms) may protonate the peptide backbone.[179,180] The origin of the small 238-nm CD band is uncertain,[40,113,181] and the assignment of other bands has been theoretically discussed.[182]

A second, more serious, criticism of the use of synthetic polypeptides

---

[169] M. L. Tiffany and S. Krimm, *Biopolymers* **6**, 1379 (1968).
[170] M. L. Tiffany and S. Krimm, *Biopolymers* **8**, 347 (1969).
[171] S. Krimm, J. E. Mark, and M. L. Tiffany, *Biopolymers* **8**, 695 (1969).
[172] S. Krimm and J. E. Mark, *Proc. Nat. Acad. Sci. U.S.* **60**, 1122 (1968).
[173] H. D. Keith, *Biopolymers* **10**, 1099 (1971).
[174] G. D. Fasman, H. Hoving, and S. N. Timasheff, *Biochemistry* **9**, 3316 (1970).
[175] E. Iizuka and J. T. Yang, *Biochemistry* **4**, 1249 (1965).
[176] J. Bello and H. R. Bello, *Nature (London)* **194**, 681 (1962).
[177] D. G. Dearborn and D. B. Wetlaufer, *Biochem. Biophys. Res. Commun.* **39**, 314 (1970).
[178] D. Balasubramanian and R. S. Roche, *Chem. Commun.* **1970**, 862 (1970).
[179] E. Peggion, A. Cosani, M. Terbojevich, and A. S. Verdini, *Macromolecules* **3**, 318 (1970).
[180] K. Rosenheck, *in* "Molecular Associations in Biology" (B. Pullman, ed.), p. 517. Academic Press (1968).
[181] Y. P. Myer, *Macromolecules* **2**, 624 (1969).
[182] D. Aebersold and E. S. Pysh, *J. Chem. Phys.* **53**, 2156 (1970).

as model random coils is that their CD patterns in solution are quite different from those of denatured (presumably unordered) proteins. These proteins, denatured by a variety of methods, may still have polypeptide chains subject to some conformational constraints; they exhibit CD spectra characterized by a negative shoulder at about 220 nm and a negative band at about 200 nm.[170,174,177] This type of CD curve is also shown by films of unordered synthetic polypeptides,[174] in which the peptide chains are presumably restricted although they do not have definite regular, repeating asymmetry. Thus, synthetic polypeptides in solution can be seriously questioned as suitable models for unordered proteins.

Other Backbone Conformations

The applicability of long-chain polypeptides as models for α-helix, β-sheet, and random-coil segments in proteins is even more tenuous if these segments are distorted or very short. In addition, backbone structures other than α, β, and unordered are known to occur in proteins, and each of these structures may contribute its own chiroptical properties to the ORD or CD data. For example, calculations[151,152] indicate that the rotatory strength of the 3_{10}-helix, found in proteins,[3] differs from that of the α-helix and displays its own chain-length dependence.

The situation may occur where an arbitrary backbone structure may mimic the CD of one of the reference conformations (α, β, random) and thus seriously interfere with interpretation of CD data. A good example of this is gramicidin S. This molecule has ten amino acids arranged in a cross-β-like structure.[183] The ORD and CD patterns obtained[184-187] are somewhat similar to those of an α-helix. The ORD curve[184,186,187] is particularly misleading,[187] containing a trough, $[m]_{233} = -18,000$. CD data[185,187] include negative bands at 208 and 217 (not 222) nm of $-30,000$ to $-40,000$, and a positive band,[187] $[\theta]_{186} = 28,000$ (not $[\theta]_{192} = 80,000$ which is the value for an α-helix). (Another CD study[186] found only one broad negative band.) Therefore, the similarity of gramicidin CD spectra to α-helix is only superficial. Pysh[188] calculated that the gramicidin S structure could indeed yield optical parameters

[183] A. Stern, W. Gibbons, and L. C. Craig, *Proc. Nat. Acad. Sci. U.S.* **61**, 734 (1968); W. A. Gibbons, J. A. Sogn, A. Stern, L. C. Craig, and L. F. Johnson, *Nature (London)* **227**, 840 (1970).
[184] L. C. Craig, *Proc. Nat. Acad. Sci. U.S.* **61**, 152 (1968).
[185] S. Laiken, M. Printz, and L. C. Craig, *J. Biol. Chem.* **244**, 4454 (1969).
[186] D. Balasubramanian, *J. Amer. Chem. Soc.* **89**, 5445 (1967).
[187] F. Quadrifoglio and D. W. Urry, *Biochem. Biophys. Res. Commun.* **29**, 785 (1967).
[188] E. Pysh, *Science* **167**, 290 (1970).

resembling those of the α-helix, the resemblance being coincidental. Another example is the simple amide, L-5-methylpyrrolid-2-one, whose CD curve in cyclohexane[189] is similar to that of a right-handed α-helix (although the bands are shifted and are much smaller in magnitude).

The variety of CD and ORD spectra obtained for other small model compounds containing one,[190,191] two, [191,192] and six[193] amide groups illustrates the great variability in rotatory properties conceivable for peptide residues in regular conformations. In addition, there is evidence[194] that some polypeptides can change conformation upon aggregation with other polypeptides; this finding may have relevance to proteins under conditions of inter- or intramolecular aggregation.

Light Scattering

A final factor that can distort the CD and ORD spectra of proteins is light scattering. There have recently been several studies[195-206] showing that rotational bands can be red-shifted and reduced in magnitude for turbid suspensions (for example, for proteins in membrane preparations). Urry and Ji[195] were the first to attempt to correlate the known distortions in α-helical CD bands of membrane proteins with calculations of expected differential light scattering and absorption flattening in particulate systems. Such calculations have since been refined.[196-199] CD data have been obtained for suspensions[195,200] and scattering films[201] of polyamino acids and for membrane preparations[198,199,202-206] The tentative conclusion (based largely on studies of

[189] D. W. Urry, J. Phys. Chem. **72**, 3035 (1968).
[190] N. J. Greenfield and G. D. Fasman, J. Amer. Chem. Soc. **92**, 177 (1970).
[191] N. J. Greenfield and G. D. Fasman, Biopolymers **7**, 595 (1969).
[192] E. B. Nielsen and J. A. Schellman, Biopolymers **10**, 1559 (1971).
[193] S. M. Ziegler and C. A. Bush, Biochemistry **10**, 1330 (1971).
[194] G. G. Hammes and S. E. Schullery, Biochemistry **7**, 3882 (1968).
[195] D. W. Urry and T. H. Ji, Arch. Biochem. Biophys. **128**, 802 (1968).
[196] C. A. Ottaway and D. B. Wetlaufer, Arch. Biochem. Biophys. **139**, 257 (1970).
[197] D. J. Gordon and G. Holzwarth, Arch. Biochem. Biophys. **142**, 481 (1971).
[198] D. J. Gordon and G. Holzwarth, Proc. Nat. Acad. Sci. U.S. **68**, 2365 (1971).
[199] D. W. Urry, L. Masotti, and J. R. Krivacic, Biochim. Biophys. Acta **241**, 600 (1971).
[200] D. W. Urry, T. A. Hinners, and L. Masotti, Arch. Biochem. Biophys. **137**, 214 (1970).
[201] D. W. Urry, T. A. Hinners, and J. Krivacic, Anal. Biochem. **37**, 85 (1970).
[202] T. H. Ji and D. W. Urry, Biochem. Biophys. Res. Commun. **34**, 404 (1969).
[203] A. S. Schneider, M.-J. T. Schneider, and K. Rosenheck, Proc. Nat. Acad. Sci. U.S. **66**, 793 (1970).
[204] M. Glaser and S. J. Singer, Biochemistry **10**, 1780 (1971).
[205] G. L. Choules and R. F. Bjorklund, Biochemistry **9**, 4759 (1970).
[206] J. C. Reinert and J. L. Davis, Biochim. Biophys. Acta **241**, 921 (1971).

fragmented, nonscattering red blood cell membranes[198,203,204] and of scattering suspensions of helical proteins[203]) is that particulate distortion effects (mainly absorption flattening[196,197]) and not special protein conformations, are responsible for the characteristic CD spectra of membranes. Because of this distortion there is considerable ambiguity in interpretation of CD spectra of particulate systems: films of poly-L-alanine known to be α-helical from IR spectroscopy, display β-type CD spectra when scattering is present,[201] and analysis shows that *Mycoplasma* membrane proteins are largely in the β conformation, although their CD curve appears similar to that of an α-helix with scattering distortion.[205] Therefore, great caution should be exercised in the interpretation of CD spectra of turbid systems, and independent physical methods of conformation determination should be used whenever possible. Recently a cell has been devised which can correct scattering artifacts and shows great promise.[206a]

Contribution of Nonamide Chromophores

General Comments

Two types of nonbackbone chromophores, aromatic amino acid residues and disulfide groups, may cause complications in the interpretation of CD data. The absorption spectra[207] of the aromatics (phenylalanine, tyrosine, and tryptophan) and cystine include bands in the near-UV region (240–300 nm). In addition, these amino acids and histidine have absorption bands[207] in the 185–240 nm "peptide" region. Any of these bands may be optically active in a protein, especially if the side chain is held in an asymmetric environment. The near-UV Cotton effects, when properly assigned to specific side chains,[89,92] can be helpful in determining conformational interactions in proteins, for example, the tertiary structure of active sites. On the other hand, the lower wavelength side-chain CD and ORD bands can interfere with estimation of secondary structure from analysis of peptide Cotton effects. Studies of absorption and chiroptical properties of model compounds (amino acid derivatives and polypeptides) and of proteins have shown that in the near-UV range although the wavelengths of Cotton effects due to a given side-chain chromophore are relatively constant, the amplitude and even the sign of these bands can vary with the composition and geometry of the molecule. In the far-UV region ($\lambda < 240$) interaction between side-chain and peptide chromophores can even cause the position of CD and

---

[206a] B. P. Dorman, J. E. Hearst, and M. F. Maestre, this volume [30].
[207] D. B. Wetlaufer, *Advan. Protein Chem.* **17**, 303 (1962).

ORD bands to be variable. The purpose of this section is to summarize methods of utilizing side-chain Cotton effects to gain structural information, as well as to point out the dangers inherent in trying to interpret CD or ORD spectra containing overlapping peptide and side-chain bands.

Goodman and co-workers have written two very useful review articles on aromatic Cotton effects in proteins (and model compounds)[208] and in polyamino acids,[209] which cover the important literature up to 1968. An interesting series of articles from Strickland's laboratory[81,82,93,210–217] deals with the resolution and analysis of near-UV aromatic CD bands in proteins (and models), usually at 77° K where characteristic fine structure becomes apparent. Another method often used for classification of side-chain bands is the red-shift upon ionization (at high pH) of tyrosine absorption and Cotton effects (see Fig. 13 for an example).

Many studies have been concerned with aromatic poly-α-amino acids as models for these chromophores locked into fixed orientations such as may exist in proteins. Two problems are common in these studies: overlapping peptide and aromatic bands, and low solubility of the polypeptides in water. To overcome the second problem, extensive use is made of nonaqueous solvents,[65,218–221] and of copolymers (random or block) with water-soluble residues[218,221–227] and modified side chains.[209,228–232]

[208] M. Goodman and C. Toniolo, *Biopolymers* **6**, 1673 (1968).
[209] M. Goodman, G. W. Davis, and E. Benedetti, *Accounts Chem. Res.* **1**, 275 (1968).
[210] J. Horwitz, E. H. Strickland, and C. Billups, *J. Amer. Chem. Soc.* **91**, 184 (1969).
[211] E. H. Strickland, M. Wilchek, J. Horwitz, and C. Billups, *J. Biol. Chem.* **245**, 4168 (1970).
[212] J. Horwitz, E. H. Strickland, and C. Billups, *J. Amer. Chem. Soc.* **92**, 2119 (1970).
[213] E. H. Strickland, J. Horwitz, E. Kay, L. M. Shannon, M. Wilchek, and C. Billups, *Biochemistry* **10**, 2631 (1971).
[214] E. H. Strickland, J. Horwitz, and C. Billups, *Biochemistry* **9**, 4914 (1970).
[215] E. H. Strickland, E. Kay, and L. M. Shannon, *J. Biol. Chem.* **245**, 1233 (1970).
[216] J. Horwitz and E. H. Strickland, *J. Biol. Chem.* **246**, 3749 (1971).
[217] L. Fretto and E. H. Strickland, *Biochim. Biophys. Acta* **235**, 473 (1971).
[218] E. Peggion, A. S. Verdini, A. Cosani, and E. Scoffone, *Macromolecules* **2**, 170 (1969).
[219] E. Peggion, L. Strasorier, and A. Cosani, *J. Amer. Chem. Soc.* **92**, 381 (1970).
[220] V. N. Damle, *Biopolymers* **9**, 937 (1970).
[221] A. Cosani, E. Peggion, A. S. Verdini, and M. Terbojevich, *Biopolymers* **6**, 963 (1968).
[222] H. J. Sage and G. D. Fasman, *Biochemistry* **5**, 286 (1966).
[223] H. E. Auer and P. Doty, *Biochemistry* **5**, 1708 (1966).
[224] G. D. Fasman, M. Landsberg, and M. Buchwald, *Can. J. Chem.* **43**, 1588 (1965).
[225] A. Ohnishi, K. Hayashi, and J. Noguchi, *Bull. Chem. Soc. Jap.* **42**, 1113 (1969).
[226] B. Shechter, I. Shechter, J. Ramachandran, A. Conway-Jacobs, and M. Sela, *Eur. J. Biochem.* **20**, 301 (1971).

The optical activity of cystine residues has been discussed by Beychok.[33,37(c)] Examples of proteins displaying near-UV CD bands, at 250–
280 nm, attributed to S-S bridges are insulin,[233] ribonuclease,[234] and the
neurophysins[94]

In general, the presence of aromatic residues may result in too low
an estimate for the α-helical (or β-form) content of a protein. This conclusion is based upon many studies of far-UV CD and ORD for synthetic
aromatic containing polypeptides in ordered conformations. (Little can
be predicted from smaller aromatic amino acid derivatives about the
sign and magnitude of Cotton effects in proteins.) Positive side-chain
Cotton effects partially (or completely) cancel the negative peptide
Cotton effects in the range $\lambda = 200$–240 nm for polymers and copolymers
of L-phenylalanine,[219] L-tyrosine,[65,220,226,227,235–241] L-tryptophan,[241,242] and
L-histidine.[243] This may be true under a variety of solvent conditions,
including many where the first three polymers are thought to be α-helical,
and where poly-L-histidine may be in an α or a β structure. Thus, overlap
of peptide with aromatic CD bands may lead to apparently weak negative ellipticity values for the peptide wavelength region in structured
proteins, and result in an underestimation of the amount of secondary
structure actually present. For example, one study on poly-L-tyrosine[65]
notes that "in unfavorable cases one tyrosine side chain may compensate
for the contribution of about one peptide bond in an α-helical or β conformation." A striking case of far-UV aromatic bands in proteins is

[227] J. Ramachandran, A. Berger, and E. Katchalski, *Biopolymers* **10**, 1829 (1971).
[228] M. Goodman, A. M. Felix, C. M. Deber, A. R. Brouse, and G. Schwartz, *Biopolymers* **1**, 371 (1963).
[229] M. Goodman, A. M. Felix, C. M. Deber, and A. R. Brouse, *Biopolym. Symp.*
1, 409 (1964).
[230] C. Toniolo, M. L. Falxa, and M. Goodman, *Biopolymers* **6**, 1579 (1968).
[231] M. Goodman, C. Toniolo, and E. Peggion, *Biopolymers* **6**, 1691 (1968).
[232] M. Goodman and E. Peggion, *Biochemistry* **6**, 1533 (1967).
[233] S. Beychok, *Proc. Nat. Acad. Sci. U.S.* **53**, 999 (1965).
[234] M. N. Pflumm and S. Beychok, *J. Biol. Chem.* **244**, 3982 (1969).
[235] G. D. Fasman, E. Bodenheimer, and C. Lindblow, *Biochemistry* **3**, 1665 (1964).
[236] S. Beychok and G. D. Fasman, *Biochemistry* **3**, 1675 (1964).
[237] E. Patrone, G. Conio, and S. Brighetti, *Biopolymers* **9**, 897 (1970).
[238] F. Quadrifoglio, A. Ius, and V. Crescenzi, *Makromol. Chem.* **136**, 241 (1970).
[239] S. Friedman and P. O. P. Ts'o, *Biochem. Biophys. Res. Commun.* **42**, 510 (1971).
[240] M. Shiraki and K. Imahori, *Sci. Pap. Coll. Gen. Educ. Univ. Tokyo* **16**, 215
(1966).
[241] G. D. Fasman, R. McKinnon, and R. Hoving, unpublished results (1969).
[242] E. Peggion, A. Cosani, A. S. Verdini, A. Del Pra, and M. Mammi, *Biopolymers*
6, 1477 (1968).
[243] Y. P. Myer and E. A. Barnard, *Arch. Biochem. Biophys.* **143**, 116 (1971).

that of avidin,[244] whose CD spectrum contains a positive tryptophan band at 228 nm; this protein has an ORD peak near 233 nm instead of the trough at this wavelength characteristic of the α-helix.

There is little information on the contribution of asymmetric disulfide chromophores to the CD of proteins or polypeptides in the wavelength region below 240 nm. However, the presence of CD bands below 240 nm for cystine itself (in mulls[245] and in KBr disks[246,247]), and the existence of a positive CD band at 230 nm, for gramicidin S, attributed (at least partially) to an S-S transition[248] may serve as cautionary notes. Several other amino acids (cysteine, methionine, asparagine, and glutamic and aspartic acids and amides) also absorb light in the 185–240 nm region,[207] although they do not have absorption maxima in this region. Their contribution to protein rotatory properties has not been extensively investigated, but it is not expected to be large. The remainder of this section will summarize the CD and ORD properties of individual aromatic and S-S chromophores in model compounds and in proteins.

Amino Acid Derivatives and Poly-L-Amino Acids

(i) *Phenylalanine.* Phenylalanine can be taken as typical of the aromatic chromophores, although somewhat fewer studies have been performed with it than with tyrosine or tryptophan. All aromatic studies have in common the problems of choice of model compounds and resolution of aromatic rotatory bands from peptide bands; the same reviews[208,209] can be consulted for all aromatic residues. The ultraviolet absorption spectrum[207] of phenylalanine is characteristic of a monosubstituted benzene ring.[208] The weak band, with vibrational fine structure, in the 260-nm region gives rise to several weak Cotton effects; not all of the transitions are optically active in any given compound. In addition, there are strong, optically active, electronic absorption bands further in the UV.

ORD studies of L-phenylalanine,[249-252] its small peptides,[251] and N-acetyl-L-phenylalanine amide[253] (a more realistic model than the free amino acid for the side-chain chromophore in proteins) showed very

[244] N. M. Green and M. D. Melamed, *Biochem. J.* **100**, 614 (1966).

[245] P. C. Kahn and S. Beychok, *J. Amer. Chem. Soc.* **90**, 4168 (1968).

[246] A. Imanishi and T. Isemura, *J. Biochem.* (*Tokyo*) **65**, 309 (1969).

[247] N. Ito and T. Takagi, *Biochim. Biophys. Acta* **211**, 430 (1970).

[248] U. Ludescher and R. Schwyzer, *Helv. Chim. Acta* **54**, 1637 (1971).

[249] A. Moscowitz, A. Rosenberg, and A. E. Hansen, *J. Amer. Chem. Soc.* **87**, 1813 (1965).

[250] A. Rosenberg, *J. Biol. Chem.* **241**, 5119 (1966).

[251] E. W. Gill, *Biochim. Biophys. Acta* **133**, 381 (1967).

[252] N. Sakota, K. Okita, and Y. Matsui, *Bull. Chem. Soc. Jap.* **43**, 1138 (1970).

[253] M. Shiraki, *Sci. Pap. Coll. Gen. Educ. Univ. Tokyo* **19**, 151 (1969).

weak multiple Cotton effects near 260 nm superimposed on a large rotation at 220–230 nm due to the amide (or carboxyl) chromophore. The development of CD instrumentation permitted much better resolution of the optically active transitions for L-phenylalanine,[210,252,254,255] N-acetyl-L-phenylalanine amide,[201,253,256] N-acetyl-L-phenylalanine esters,[208,210] and peptides (linear[257] and cyclic[258,259]). Simmons et al.[256] were the first to achieve good resolution of the near-UV CD bands (all negative in the case of N-acetyl-L-phenylalanine amide) and to measure the positive CD bands at 195 and 217 nm; they attributed a small, negative, band at 240 nm to the primary amide group (by analogy to N-acetyl-L-alanine amide). The CD spectra of N-acetyl-L-phenylalanine amide (and the analogous L-tyrosine and L-tryptophan derivatives) in water are shown in Fig. 12.

As pointed out by Horwitz et al.,[210] the CD vibrational fine structure varies with the compound: in L-phenylalanine itself only the prominent absorption bands at 264 and 258 nm are dichroic[254] (giving rise to positive Cotton effects), but, on the other hand, in the amide[256] and in proteins[260] only the weak absorption bands of the phenylalanine chromophore at 268 and 262 nm are optically active. The resolution of the near-UV CD bands can be increased by working at 77° K[210]; band assignments useful in characterizing phenylalanine in proteins can be made. Organic solvents can greatly affect the entire CD spectrum of N-acetyl-L-phenylalanine amide (as well as those of the other aromatics).[253]

Studies of the phenylalanine chromophore in polypeptides are complicated by the insolubility of poly-L-phenylalanine in water (necessitating the use of copolymers or organic solvents) and by the overlapping of side-chain and peptide bands (making conformational determination very difficult). Early ORD studies in aqueous solution of block copolymers of L-phenylalanine with DL-glutamic acid[222,223] revealed the presence of near-UV aromatic bands and of a weak 228-nm trough, not typical of normal polypeptide conformations, but suggestive of an α-helix. Chiroptical investigations of phenylalanine containing polymers performed before 1968 have been summarized.[209]

[254] M. Legrand and R. Viennet, Bull. Soc. Chim. Fr. 1966, 2798 (1966).
[255] L. Verbit and P. J. Heffron, Tetrahedron 23, 3865 (1967).
[256] N. S. Simmons, A. O. Barel, and A. N. Glazer, Biopolymers 7, 275 (1969).
[257] I. Weinryb and R. F. Steiner, Arch. Biochem. Biophys. 131, 263 (1969).
[258] K. Bláha and I. Frič, Collect. Czech. Chem. Commun. 35, 619 (1970).
[259] K. Bláha, I. Frič, and J. Rudinger, Collect. Czech. Chem. Commun. 34, 3497 (1969).
[260] E. H. Strickland, E. Kay, L. M. Shannon, and J. Horwitz, J. Biol. Chem. 243, 3560 (1968).

Later CD studies[32,218,219,261] of such polymers showed positive or weakly negative dichroism in the peptide region, and it was suggested that an ordered structure (possibly α-helical) is formed under certain solvent conditions. Urry[32] found that poly-L-phenylalanine forms such a structure in 99% ethylene dichloride plus 1% trifluoroacetic acid (TFA). The CD displays finely structured bands of $\sim$100 deg.-cm²/dmole near 260 nm, plus a band at 227 nm ($[\theta]_{227} = -9000$) and a positive shoulder at 215 nm; this spectrum is destroyed by additional TFA. Peggion and co-workers[218,219] have concluded, from two lines of evidence, that the ordered structure may be a right-handed α-helix, even though its CD properties are unlike a typical α-helix. First,[218] the CD spectra of copolymers of L-phenylalanine with ϵ-carbobenzoxy-L-lysine in tetrahydrofuran showed a gradual perturbation of the α-helical pattern of poly-ϵ-carbobenzoxy-L-lysine as additional phenylalanine was introduced, but no change in helical sense. Second,[219] poly-L-phenylalanine undergoes a transition in water–methanesulfonic acid mixtures, in a manner analogous to that of the α-helix-to-coil transition of poly-L-cyclohexylalanine; near UV bands were also observed in this study. Thus, the CD patterns of poly-L-phenylalanine may be attributable to an overlap of conformation-dependent side-chain chromophore bands with normal random-coil and α-helical peptide bands. The same can be said (with somewhat more assurance) for poly-L-tyrosine and poly-L-tryptophan. (A recent CD study by Peggion et al.[261] on copolymers of lysine and phenylalanine in water at various pH values suggests that the presence of phenylalanine in this system induces β-structure formation.)

(ii) Tyrosine. The absorption spectrum[207] of un-ionized tyrosine (the form present in proteins at neutral pH) contains a weak band at 275 nm (with a shoulder at 282 nm) and two stronger peaks at 224 and 194 nm. ORD studies of this amino acid[249,250] showed small near-UV Cotton effects. The CD spectrum of un-ionized L-tyrosine[236,255,262] has positive bands corresponding to the 275 and 225 nm transitions (see Fig. 13); the latter CD band interferes with conformational determination of tyrosine polypeptides. The CD signal is positive also at $\lambda < 200$ nm. Theoretical calculations of tyrosine optical activity have been made.[263] Vibrational fine structure becomes apparent in the CD of tyrosine (and its derivatives) upon cooling.[212] When the phenolic group is ionized (at pH above about 12) all absorption and CD bands are shifted to longer

[261] E. Peggion, A. S. Verdini, A. Cosani, and E. Scoffone, Macromolecules 3, 194 (1970).

[262] M. Legrand and R. Viennet, Bull. Soc. Chim. Fr. 1965, 679 (1965).

[263] T. M. Hooker, Jr. and J. A. Schellman, Biopolymers 9, 1319 (1970).

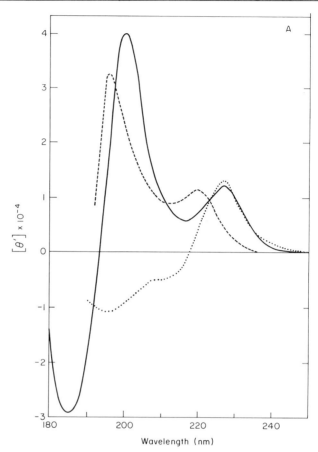

FIG. 12. Circular dichroism spectra of aromatic amino acid derivatives in water in the far ultraviolet (A) and near ultraviolet (B) regions: N-acetyl-L-tyrosine amide, ————; N-acetyl-L-phenylalanine amide, - - - - - - -; N-acetyl-L-tryptophan amide, · · · · · · [from M. Shiraki, *Sci. Pop. Coll. Gen. Educ. Univ. Tokyo* **19**, 151 (1969).] Note that molar ellipticities reduced to vacuo, [θ'], are given; these can be multiplied by $(n^2 + 2)/3 \approx 1.3$ to yield [θ] values.

wavelength.[262] (This red-shift is useful to diagnose tyrosine residues in proteins.)

The CD pattern of N-acetyl-L-tyrosine amide[212,253] is shown in Fig. 12; the 275-nm band for this compound in water is negative. L-Tyrosine ethyl ester and N-acetyl-L-tyrosine ethyl ester also have been examined at different temperatures[212] and pH values.[33] The near-UV CD spectra of N-acetyl-L-tyrosine amide[253] and ethyl ester[212] are extremely solvent dependent; the ester 275-nm CD band is positive in dioxane and negative in methanol. Horwitz *et al.*[212] suggested that it should be possible to

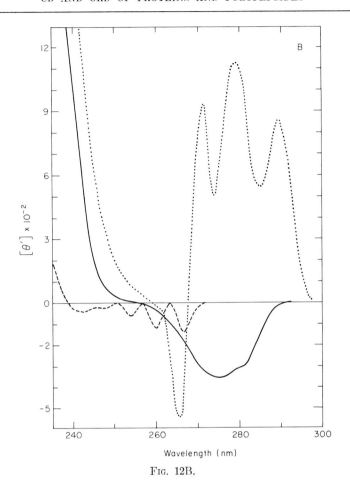

Fig. 12B.

identify the tyrosine residues in proteins from their CD vibrational structure between 275 and 290 nm. The bands between 250 and 270 nm were deemed not as useful due to overlap with phenylalanine residues and disulfide bridge contributions.

The near-UV CD properties of diketopiperazines (cyclic dipeptides) containing aromatic chromophores have been examined at room temperature[264] and below.[211] Ellipticity values for L-tyrosine-containing cyclic dipeptides are much larger than those for the analogous linear peptides, apparently because of the rigid conformation of the diketopiperazines.[264] Red shifts are caused by tyrosine ionization and by organic solvents.[264] Cyclic-L-tyr-L-tyr shows a small exciton contribution in

[264] H. Edelhoch, R. E. Lippoldt, and M. Wilchek, *J. Biol. Chem.* **243**, 4799 (1968).

its CD.[211] Ziegler and Bush[193] compared the CD spectrum of cyclo-(glycine)$_5$-L-tyrosine with those of the analogous linear hexapeptide and cyclo-(Gly)$_5$-L-leucine. They concluded, with the help of nuclear magnetic resonance data,[265] that the observed enhancement of the amide $\pi - \pi^*$ CD bands is the result of coupling between the tyrosyl and amide chromophores.

As with the cyclic hexapeptides, the main concern of studies on poly-peptides containing tyrosine (and other aromatics) is to resolve peptide from side-chain CD bands and to determine conformation. The estima-tion of secondary structure of these polymers is important if they are to be used as models for protein structural determination, since the con-tribution of aromatic chromophores to the CD spectra is conformation dependent. The polymer CD spectrum is not the simple sum of that for a monomeric amino acid (e.g., L-tyrosine) derivative and that for a known polypeptide conformation (e.g., α-helix)[236,241]; side-chain chro-mophores interact with one another and with peptide chromophores. The problem, as illustrated by poly-L-tyrosine, is a difficult one: the CD curves for this polymer (Fig. 13) obviously reflect a mixture of peptide and tyrosine bands. The resultant spectra do not correspond, even qualitatively, to polypeptide CD data for the known α, β, and random forms. If it can be shown that a given poly-L-tyrosine CD spectrum corresponds to the polymer in any known backbone conforma-tion (for example, α-helix), then the normal CD pattern for this con-formation can be subtracted, the side-chain CD contribution can be estimated, and these results can be applied to correct the CD of proteins containing tyrosine residues in segments with this conformation. However, if the poly-L-tyrosine conformation is not known, then backbone and side-chain CD contributions cannot be separated, (because the tyrosine Cotton effects are conformation dependent) and the polymer studies are much less useful for protein applications.

After an enormous number of studies on the ORD,[235,237,240,266] CD,[32,65,220,236,238-240] and other physical properties[220,238,267,268] of poly-L-tyrosine and copolymers[226,227,241] in various solvents, plus theoretical CD calculations,[269-271] all that can be concluded is that poly-L-tyrosine

[265] K. D. Kopple, M. Ohnishi, and A. Go, *J. Amer. Soc.* **91**, 4264 (1969); *Biochem-istry* **8**, 4087 (1969).

[266] J. D. Coombes, E. Katchalski, and P. Doty, *Nature (London)* **185**, 534 (1960).

[267] J. Applequist and T. G. Mahr, *J. Amer. Chem. Soc.* **88**, 5419 (1966).

[268] M. B. Senior, S. L. H. Gorrell, and E. Hamori, *Biopolymers* **10**, 2387 (1971).

[269] Y. H. Pao, R. Longworth, and R. L. Kornegay, *Biopolymers* **3**, 537 (1965).

[270] T. Ooi, R. A. Scott, G. Vanderkooi, and H. A. Scheraga, *J. Chem. Phys.* **46**, 4410 (1967).

[271] A. K. Chen and R. W. Woody, *J. Amer. Chem. Soc.* **93**, 29 (1971).

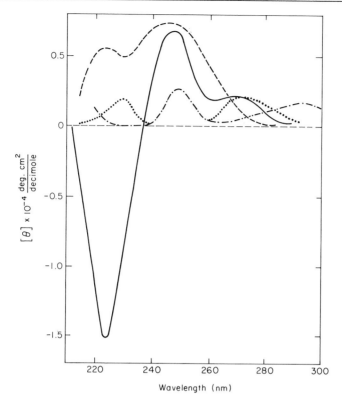

Fig. 13. Circular dichroism of helical poly-L-tyrosine (——), random poly-L-tyrosine (– – –) (both at pH 11.2), and L-tyrosine at pH 8 (· · · ·) and at pH 12 (–·–·) [from S. Beychok and G. D. Fasman, *Biochemistry* **3**, 1675 (1964)]. The helical form of the polymer was prepared by direct dissolution into water at pH 11.2; the random-coil form was first brought to pH > 12 and then back-titrated to pH 11.2.

is probably a right-handed α-helix in aqueous solution under certain conditions. There is also evidence for left-handed helix[267] (see Goodman et al.[209]) and β-sheet[237,268] formation. However, one generalization useful for protein CD interpretation emerges from the chiroptical data: all ORD and CD spectra of poly-L-tyrosine (no matter what the corresponding structure is) have ellipticity values in the 210–240 nm range that are much more positive than those of normal polypeptides in known, ordered conformations. Therefore, tyrosine (and other aromatic) residues in proteins are likely to cause too low an estimation of any ordered conformation from CD data.

The first sign of peculiarity for poly-L-tyrosine was its positive b_0

value.[235] Beychok and Fasman[236] examined its CD and found the spectra shown in Fig. 13. Since poly-L-tyrosine is not soluble in water in its totally un-ionized form, the polymer had to be examined at high pH. In the structured ("helical") form there are CD bands corresponding to the un-ionized tyrosine absorption bands at 275 and 225 nm, plus a CD band at 248 nm attributable to ionized tyrosine. All of these are superimposed on the peptide CD bands. Later work showed the presence of a positive CD band at about 200 nm in trimethyl phosphate[220,238] and in 0.1 M NaClO$_4$ at pH 10.8 [238]; this band, too, is probably a mixture of the 194 nm tyrosine absorption plus $\pi - \pi^*$ peptide absorption bands. Somewhat different CD curves found for poly-L-tyrosine,[65,220,239] in which the negative 225-nm band is missing, may arise from variations in ionization.

Some recent studies of copolymers containing tyrosine are of interest. Shechter et al.[226] found that for a copolymer of sequence (L-Tyr-L-Ala-L-Glu)$_{200}$ the 275-nm CD band is negative in the helical conformation (although the 225-nm band retains the same sign as in the homopolymer). This result indicates the sensitivity of the tyrosine chromophore to its asymmetric environment and shows that its ability to interact with other tyrosines may change its rotatory contribution. Ramachandran et al.[227] studied the same copolymer by means of ORD. They compared it to a random sequence copolymer (Tyr, Ala, Glu)$_n$ in both α-helical and random-coil conformations, and gave X-ray diffraction evidence for the helical form of the sequence copolymer. They concluded that, in the α-helical form, the ORD contribution of tyrosine residues in the 200–250 nm region (a peak at 233 nm and a trough below 200 nm) nearly cancels the contribution of the α-helix. The tyrosine contribution of the random copolymer in the α-helical conformation is similar in shape but has only one-third the amplitude.

Polymers of hydroxyethyl-L-glutamine incorporating small amounts of L-tyrosine display negative Cotton effects in the near-UV region[241] in 80% aqueous methanol (where the polymers are α-helical) as shown in Fig. 14. These CD bands (calculated per mole of tyrosine residue) grow disproportionately as the tyrosine content increases, again showing specific interaction. Furthermore, block copolymers of L-tyrosine with DL-glutamate have positive 275-nm CD bands, similar to poly-L-tyrosine. It is not possible in any of these cases[241] to calculate the copolymer CD spectra at any wavelength by adding the CD curves of model tyrosine derivatives to those of helical or random polypeptides, even though these copolymers have normal conformations. This result thus shows that aromatic CD contributions are conformation dependent, or that the side chains become immobilized due to the backbone structure and

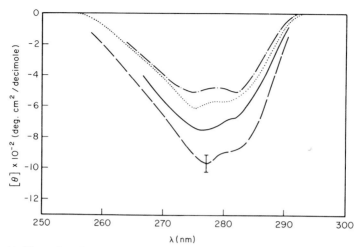

Fig. 14. Near-ultraviolet circular dichroism of copolymers of N^5-(2-hydroxyethyl)-L-glutamine with various amounts of L-tyrosine (randomly incorporated) (from G. D. Fasman, R. McKinnon, and H. Hoving, unpublished results). The solvent is 80% aqueous methanol, in which the copolymers are α-helical. [θ] values are given per mole of tyrosine residues. Mole percent tyrosine in the copolymers: 4.5, –·–·–·; 8, ———; 17, – – – – –. The CD spectrum of N-acetyl-L-tyrosine amide in methanol, · · · · ·, is shown for comparison. Error bars indicate agreement of duplicate experiments.

are therefore locked in an asymmetric configuration. Examination of CD curves in the peptide wavelength region for these helical copolymers[241] shows an apparent decrease in magnitude of the peptide CD bands as the tyrosine content is increased; this finding is very similar to the case of tryptophan copolymers (see Fig. 15, below).

(*iii*) *Tryptophan.* This amino acid absorbs light at 277 nm (with vibrational structure), 218, and 196 nm.[207] The ORD of L-tryptophan,[250,251] N-acetyl-L-tryptophan amide,[253] and small peptides[251] plus the CD spectra of the amino acid,[33,81,83,213,255,262] its esters,[81,213] and N-acetyl-L-tryptophan amide[81,253] (see Fig. 12) have been examined. Several positive CD bands are present between 270 and 300 nm for all these compounds. Strickland *et al.* have assigned these vibronic bands[81,213] at room temperature and 77°K (where they are greatly enhanced) with the assistance of indole absorption spectra.[214] Two electronic transitions, ^1L_a and ^1L_b, contribute to the CD of model tryptophan compounds in this region. ^1L_b vibronic CD bands are located at 290 nm for the 0–0 transition and at 283 nm for the 0 + 850 cm^{-1} transition; these positions are not shifted by different solvents. Several ^1L_a bands, located between 265 and 297 nm are solvent dependent.

Several laboratories[33,83,255,262] agree that in the low wavelength region L-tryptophan at neutral pH has CD bands at 222 nm (positive) and at about 195 nm (negative), which correspond to its absorption peaks. In addition, small bands at 240 and about 200 nm are noted in some reports. The CD spectrum of L-tryptophan is pH dependent,[33,255,262] and is very similar to that of N-acetyl-L-tryptophan amide.[253] Studies of diketopiperazines[211,264] show that the near-UV CD bands of cyclic-Gly-L-Trp and cyclic-(L-Trp)$_2$ are several times as large as those for the corresponding, less rigid, linear dipeptides. Dipeptide solvent effects were demonstrated[264]: dioxane appears to destroy the interaction between the indole chromophores and the dipeptide ring, permitting greater rotational freedom. The tryptophan CD bands were correlated[264] with bands in β-lactoglobulin and carbonic anhydrase.

The chiroptical properties of poly-L-tryptophan and its copolymers show overlap of peptide and side-chain bands, similar to the case of poly-L-tyrosine. Much of the work has been summarized.[209] Fasman et al.[224] examined the ORD of a poly-L-tryptophan film and found a trough at 233 nm plus small near-UV Cotton effects. They concluded that poly-L-tryptophan, despite its positive b_0 value, is a right-handed α-helix, since a series of copolymers with γ-benzyl-L-glutamate yielded a linear relationship between b_0 and percentage of tryptophan. Stevens et al.[160] obtained CD spectra of poly-L-Trp films. Cosani et al. studied the ORD[221,272] and CD[221] spectra of a block copolymer of tryptophan with γ-ethyl-DL-glutamate in trifluoroethanol, in which solvent the polymer appears structured. CD bands were found at 290 nm ($\Delta\epsilon = -0.63$), 286 and 280 (positive shoulders), 272 ($\Delta\epsilon = 2.65$), 226 ($\Delta\epsilon = 42.7$), 210 ($\Delta\epsilon = -26.4$), and 190 ($\Delta\epsilon = 14.3$). The two dichroic bands at 210 and 190 nm are of the same sign and position as those observed for polypeptides in a right-handed α-helical conformation. However, the 226 nm band shows that optically active indole transitions are overlapping peptide transitions in this polymer, and the conformational assignment required confirmation.

More evidence from Peggion's group (X-ray analysis of poly-L-Trp films,[242] CD spectra in ethylene glycol monomethyl ether of poly-L-Trp and several copolymers with γ-ethyl-L-glutamate,[242] and a CD study of poly-2,2-nitrophenyl-sulfenyl-L-tryptophan[273]) plus work on a L-Trp, L-Glu copolymer[225] substantiate that poly-L-tryptophan is a right-handed α-helix in some solvents. Therefore, in the case of tryptophan, the polypeptide results can be applied to proteins in a known conforma-

[272] A. Cosani, E. Peggion, M. Terbojevich, and A. Portolan, *Chem. Commun.* **1967**, 930 (1967).

[273] E. Peggion, A. Fontana, and A. Cosani, *Biopolymers* **7**, 517 (1969).

tion. Peggion et al.[242] showed how the CD of an α-helix is distorted by increasing incorporation of L-tryptophan, with the 208 and 222 nm bands of the α-helix being replaced by a band at 220 nm and with a positive band appearing at 230 nm. This change is gradual but not linear with mole fraction tryptophan, perhaps because of exciton interactions between aromatic residues at high tryptophan content.

The effect upon peptide-region CD spectra of incorporation of small amounts (0–15 mole percent) of L-tryptophan into poly-hydroxyethyl-L-glutamine is shown in Fig. 15. These polymers[241] are α-helical in 80% aqueous methanol (Fig. 15A) and mainly unordered in water (pH 7, Fig. 15B). In Fig. 15A is evidence that the presence of aromatic residues in proteins can result in misleadingly low estimates of the amount of helical structure. It can be seen in Fig. 15B that tryptophan residues may initiate and stabilize helical structures in proteins.

(iv) Histidine. The imidazole group of histidine[207] has at 210 nm an absorption peak that could interfere with protein CD measurements. CD spectra of the L-amino acid display a positive band at 213 nm[208,243,255,262] (which is slightly red-shifted in acid[255]) and a negative band at 193 nm.[208] A negative CD band at 217 nm in cyclo-Gly$_2$-L-Tyr-Gly$_2$-L-His was shown[193] to have contributions from $n - \pi^*$ amide and imidazole electronic transitions.

Poly-L-histidine has been studied by means of ORD,[274,275] CD,[32,243,275,276] and infrared spectroscopy.[243,277] There is general agreement that at pH below 4, where the imidazole moiety carries a positive charge and where the CD spectrum is characterized by a positive band at 222 nm,[32,243,275,276] the polymer is a random coil (or perhaps an extended chain[32]). However, in aqueous solutions above the pK of the imidazole transition (pH 5–6), there is controversy about the ordered form (or forms) of poly-L-histidine. At pH $\sim$ 6 the CD contains bands at 221 nm ($[\theta] \sim -5000$)[32,243,275] and at 203 nm ($[\theta] \sim 16,000$)[32,243]; very complicated CD changes (involving more than two species) were noticed as the pH was varied from 4 to 6.[243] It is not clear whether poly-L-histidine at neutral pH is a left-handed helix,[32,274] a right-handed helix (with side-chain contributions confusing the chiroptical properties),[275] a β-sheet,[243] or perhaps some other conformation[276] or a mixture.[243] In any case (as appears to be the rule for aromatic residues) the

[274] K. S. Norland, G. D. Fasman, E. Katchalski, and E. R. Blout, Biopolymers 1, 277 (1963).

[275] S. Beychok, M. N. Pflumm, and J. E. Lehmann, J. Amer. Chem. Soc. 87, 3990 (1965).

[276] E. Peggion, A. Cosani, M. Terbojevich, and E. Scoffone, Macromolecules 4, 725 (1971).

[277] J. Muelinghaus and G. Zundel, Biopolymers 10, 711 (1971).

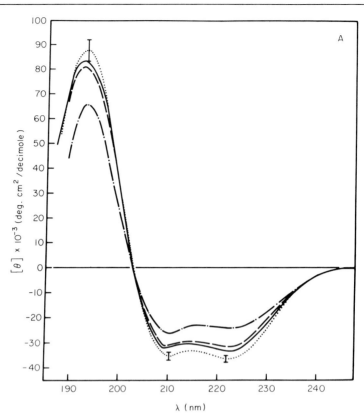

Fig. 15. Far-ultraviolet circular dichroism of copolymers of N^5-(2-hydroxyethyl)-L-glutamine with various amounts of L-tryptophan (randomly incorporated) [from G. D. Fasman, R. McKinnon, and R. Hoving, unpublished results]. Mole percent tryptophan in the copolymers: 0, · · · · (poly-hydroxyethyl-L-glutamine); 28, ———; 8.8, - - -; 14.8, –·–· [θ] values are given per mole of peptide residues. (A) α-helical polypeptides in 80% aqueous methanol. (B) Random-coil conformations in water.

presence of histidine results in a CD spectrum much less negative in the 195–240 nm region than a normal α or β curve.

Peggion *et al.*[276] examined the CD spectra of a series of random copolymers of L-histidine with L-lysine, and concluded that the ordered structure assumed by nonprotonated poly-L-histidine in water cannot be an α-helix. A copolymer of sequence L-(His-Ala-Glu)$_n$, which is water-soluble and a good model for histidine in proteins, has a CD spectrum in water[278] which indicates an essentially random-coil conformation at all pH values, with little sign of interference from side-chain optical ac-

[278] H. J. Goren, M. Fridkin, and E. Katchalski, personal communication (1971).

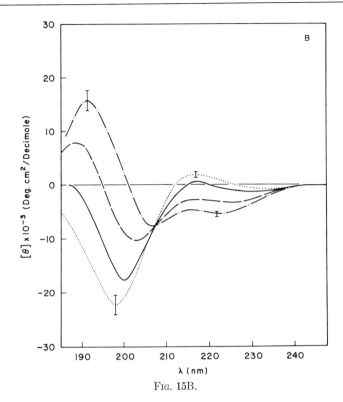

Fɪɢ. 15B.

tivity. The ORD[274] and CD[279] of poly-(1-benzyl)-ʟ-histidine and its copolymers with N-ε-carbobenzoxyl-ʟ-lysine[279] in organic solvents show complexities. The CD spectra of poly-(1-benzyl)-ʟ-histidine show bands similar to those of a right-handed α-helix but with diminished amplitudes.[279]

(v) *Sulfur-containing residues.* In addition to the aromatic residues the contribution of sulfur containing residues, primarily cystine (S-S), must be considered. Cysteine (SH), methionine (SCH₃), and cystine all have long absorption tails in the 195–230 nm region[207]; cystine has an additional band at about 250 nm.

The sulfur chromophore itself does not appear to be an important source of optical activity. ORD studies of ʟ-cysteine,[280,281] its methyl ester,[280] and N-acetyl-ʟ-cysteine methyl amide[281] could all be interpreted in terms of known amide and carboxyl Cotton effects.

[279] M. Terbojevich, M. Acampora, A. Cosani, E. Peggion, and E. Scoffone, *Macromolecules* **3**, 618 (1970).

[280] D. W. Urry, D. Miles, D. J. Caldwell, and H. Eyring, *J. Phys. Chem.* **69**, 1603 (1965).

[281] D. L. Coleman and E. R. Blout, *J. Amer. Chem. Soc.* **90**, 2405 (1968).

The cystine disulfide chromophore on the other hand, presents a very significant source of optical activity. A disulfide grouping is rigidly fixed; the barrier to rotation about the S-S bond is high, and the disulfide can exist in either of two rotomers having either a right- or left-handed screw sense. Thus, a new center of asymmetry is created in the peptide chain, causing Cotton effects in the near- and far-ultraviolet regions. The dihedral angle (defined by C_1-S-S and S-S-C_2) is close to 90° in cystine itself, but can vary in proteins. The absorption band is significantly red-shifted as the dihedral angle decreases,[233] so that near-UV S-S CD bands could appear anywhere between 250 and about 270 nm. Furthermore, these bands have no vibrational fine structure[212]; this fact can be an aid to their diagnosis in proteins.

Beychok[37(c)] and Coleman and Blout[282] found that L-cystine has a broad pH-dependent negative CD band at about 255 nm ($[\theta] \sim -2500$), which is shared by N-acetyl derivatives and by homocystine, and which is unequivocally associated with the disulfide transition. Another pH-dependent band (positive, near 220 nm)[282] is also present in these compounds. ORD data[281] agree well. The optical activity of cystine is modified when the molecules are locked into a fixed configuration within a crystal.[245-247] This has important implications, as the disulfide groupings in a protein are likely also to be specifically constrained. L-Cystine crystals in mulls[245] and in KBr disks[246] show a positive CD band near 300 nm and a negative band near 240 nm. Under the same conditions L-cystine dihydrochloride has a negative band at 270 nm and a positive one near 230 nm. These inversions can be correlated with X-ray evidence (see references in Kahn and Beychok[245]) showing that L-cystine crystallizes with the disulfide as a left-handed screw, while its dihydrochloride crystals have the opposite chirality. Another study[247] of crystalline L-cystine in KBr disks considers the possibility of exciton splitting.

Recent experimental[283-286] and theoretical[287] work has attempted the determination of the sign of the highest wavelength CD band for disulfide derivatives of known screw sense and dihedral angle. A quadrant rule for correlation of CD with geometry was formulated,[287] based largely

[282] D. L. Coleman and E. R. Blout, *in* "Conformation of Biopolymers" (G. N. Ramachandran, ed.), Vol. 1, p. 123. Academic Press, New York, 1967.
[283] M. Carmack and L. A. Neubert, *J. Amer. Chem. Soc.* **89**, 7134 (1967).
[284] G. Claeson, *Acta Chem. Scand.* **22**, 2429 (1968).
[285] R. M. Dodson and V. C. Nelson, *J. Org. Chem.* **33**, 3966 (1968).
[286] P. D. Hensen and K. Mislow, *Chem. Commun.* **1969**, 413 (1969).
[287] J. Linderberg and J. Michl, *J. Amer. Chem. Soc.* **92**, 2619 (1970).

on studies of dithiane rings,[283-285] and confirmed by examination of S-allyl-L-cysteine-S-oxides[286] and [2,7-cystine]-gramicidin S.[248]

Copolymers of L-cystine with L-glutamate, containing inter- and intra-chain S-S bridges, have been studied[282] by CD and ORD spectroscopy. All these polymers, charged or uncharged, showed Cotton effects near 260 nm. In only one case did there appear to be a small disulfide contribution at 200 nm. The same authors[282] described the cyclic oligopeptides, arginine vasotocin and 8-L-ornithine vasopressin, which have optically active S-S transitions near 200 nm, similar to those of model compounds, in addition to near-UV bands. Beychok[233] has examined the CD curves of cystine disulfoxide and oxidized glutathione, which are similar to, but less intense than that of cystine. In conclusion, the CD spectra of cystine[245-247,282] and its peptides[248,282] contain bands in the 210–240 nm region and near 200 nm[282] (which may confuse interpretation of protein peptide CD contributions) as well as a near-UV band of variable sign, wavelength (250–300 nm) and intensity[37(c),245-248,282-286] (which may yield information about the geometry of disulfide bridges and adjacent residues in proteins).

Nonpeptide Contributions in Proteins

Summary. Examination of near-UV chiroptical properties (mainly circular dichroism) of proteins has recently, become a widespread technique; it can yield useful information about specific aromatic and cystine residues and their environments, for example at the active site of an enzyme. On the other hand, very little is known about side-chain contributions to the CD spectra of proteins in the far-UV "peptide" region, or about general methods of conformational interpretation in the presence of side-chain CD bands. Goodman and Toniolo[208] have reviewed work done through 1968 on aromatic optical activity in proteins. A paper by Beychok[233] constitutes a good introduction to CD of disulfides in proteins. A recent study of insulin[88] illustrates the problems of overlapping peptide and side-chain bands.

A brief survey of recent CD investigations of side chains in proteins follows. There have been no CD studies specifically on histidine. Few studies have been centered on the role of phenylalanine in proteins because of the weakness of its CD bands; exceptions are investigations of peroxidase,[215,260] and carboxypeptidase.[93] Studies of tyrosine include ribonuclease,[212,216,288-293] staphylococcal nuclease,[95] cytochrome c,[294] lysozyme,[89,90,92] insulin,[88] concanavalin,[91] and carboxypeptidase.[93] Tryptophan CD or ORD has been examined in cytochrome c,[213,294-296] lyso-

[288] A. N. Glazer and N. S. Simmons, *J. Amer. Chem. Soc.* **87**, 3991 (1965).
[289] R. T. Simpson and B. Vallee, *Biochemistry* **5**, 2531 (1966).

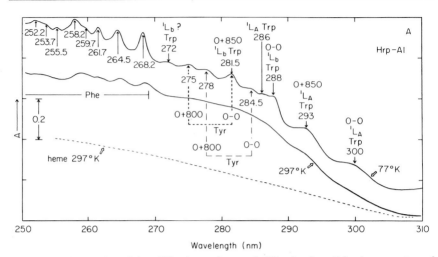

Fig. 16. Near-ultraviolet (A) absorption and (B) circular dichroism spectra of horse radish peroxidase A1 [from E. H. Strickland, J. Horwitz, E. Kay, L. M. Shannon, M. Wilchek, and C. Billups, *Biochemistry* **10**, 2631 (1971)]. Spectra are recorded for 297 and 77°K (———); the 77°K curves have been offset to facilitate viewing. Peroxidase (1.9 mM) was dissolved in water–glycerol (1:1, v/v) containing 50 mM sodium phosphate (pH 7). Notes: (A, absorption): Approximate absorption of heme moiety at 297°K is shown by dashed line. Vibronic assignments are given for aromatic bands at 77°K. B, Circular Dichroism: Dichroism is recorded as $(A_L - A_R)$. Cell path length 0.12 mm. Data are not shown at $\lambda < 270$ nm because this investigation was concerned primarily with tryptophan CD.

zyme,[89,90,92] avidin,[244] hemoglobin,[297] peroxidase,[213] chymotrypsinogen,[81] and carboxypeptidase.[217] Cystine bridges in ribonuclease,[212,233,234] lysozyme,[92,233] insulin,[233] the neurophysins,[94] and other proteins (see Beychok[33,37(c)] for a review) have been investigated.

Strickland's laboratory has been able to differentiate CD bands arising from various side-chain residues by means of their vibronic fine structure. Much of the work was performed at low temperature, where band resolution is better. An example (horseradish peroxidase A1)

[290] N. S. Simmons and A. N. Glazer, *J. Amer. Chem. Soc.* **89**, 5040 (1967).

[291] H. Hashizume, M. Shiraki, and K. Imahori, *J. Biochem. (Tokyo)* **62**, 543 (1967).

[292] E. R. Simons, E. G. Schneider, and E. R. Blout, *J. Biol. Chem.* **244**, 4023 (1969).

[293] J. Bello, *Biochemistry* **8**, 4535 (1969).

[294] D. D. Ulmer, *Biochemistry* **5**, 1886 (1966).

[295] Y. P. Myer, *J. Biol. Chem.* **243**, 2115 (1968).

[296] T. Flatmark, *J. Biol. Chem.* **242**, 2454 (1967).

[297] A. Wollmer and G. Buse, *FEBS (Fed. Eur. Biochem. Soc.) Lett.* **16**, 307 (1971).

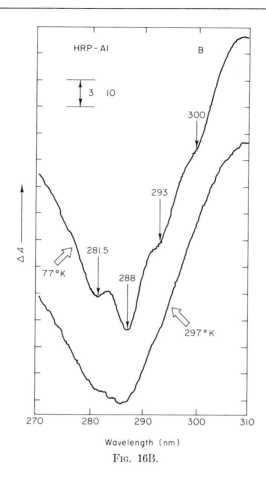

Fig. 16B.

is shown in Fig. 16.[213] The CD bands due to a given type of side chain can then in favorable cases, be correlated with specific residues (see discussion of ribonuclease,[212] which will follow) and yields information about their environment.

Chiroptical properties of prosthetic groups are sometimes studied, for example in hemoglobin. These bands can be resolved from protein bands; this has been done for the heme groups of peroxidase,[213,260,298] hemoglobin,[297] and cytochrome c.[213] Information about stoichiometry and manner of coenzyme and metal ion binding to enzymes can be obtained

---

[298] K. Hamaguchi, K. Ikeda, C. Yoshida, and Y. Morita, *J. Biochem.* (*Tokyo*) **66**, 191 (1969).

by monitoring the extrinsic Cotton effects (in the region above 300 nm) formed upon enzyme–cofactor complex formation in the absence or in the presence of substrate. The ORD of several such systems has been reviewed.[96] An example of a similar CD investigation (of beef liver glutamate dehydrogenase and NADH) can be seen in Koberstein and Sund.[97]

Many investigations employ changes of pH to distinguish between side-chain bands; tyrosine residues show a particularly large red shift upon ionization (see Fig. 13). Protein denaturation [by heat, extremes of pH, or addition of a denaturant (e.g., guanidine hydrochloride)] is another common tool; in this way the protein environment of given residues can be varied, and the effect upon CD or ORD studied. Most of the papers catalogued above measure only near-UV CD ($\lambda > 240$ nm); a few deal also with the shorter wavelength region.

Chemical modification of specific residues shows promise of becoming an important CD technique; tyrosine in insulin[88] and in lysozyme,[89] and tryptophan in lysozyme,[89,90] have been studied in this manner. The effects of substrate or inhibitor binding upon side-chain CD bands have been measured for carboxypeptidase,[93] staphylococcal nuclease,[95] the neurophysins,[94] concanavalin,[91] lysozyme,[90,92] and ribonuclease.[77]

Specific Examples. (i) PANCREATIC RIBONUCLEASE. Several types of information can be obtained from careful ORD and CD studies of a protein. Pancreatic ribonuclease has been a favorite enzyme for study because the molecule does not contain tryptophan residues or much α-helix; thus the assignment of Cotton effects arising from tyrosine and cystine are more certain than in most proteins. Some of the RNase investigations have been summarized.[299] In 1965, Glazer and Simmons[288] first studied the ORD spectrum of ribonuclease. A band was found near 278 nm at pH 6.2 which shifted to 292 nm at pH 11, showing that the band was associated with tyrosine residues exposed to solvent and therefore susceptible to ionization. Sodium dodecyl sulfate removed these bands; thus the asymmetry arose from specific interactions of the tyrosine residues with their native environment. Simpson and Vallee[289] found that these external tyrosine bands could be modified without affecting the rest of the ORD spectrum. Simmons and Glazer[290] continued their experiments using CD and found negative CD band at 273 nm at neutral pH which shifted to 285 nm at pH 11. Hashizume et al.[291] compared the CD of ribonuclease A with that of poly-L-tyrosine and found that the ellipticity of the tyrosine residues in the protein was much higher than

[299] F. M. Richards and H. W. Wyckoff, *in* "The Enzymes" (P. D. Boyer, ed.), 3rd ed., Vol. 4, p. 647. Academic Press, New York, 1971.

would be expected from the ellipticity of poly-L-tyrosine. They felt, therefore, that the residues buried in the protein, because they were constrained, had a greater contribution to the CD than did the external residues. Upon heating of RNase[291] the ellipticity at 278 nm decreased in two steps, suggesting that the buried tyrosine residues (which stabilize the tertiary structure of the enzyme) were not unfolded simultaneously with the exposed ones. Simons et al.[292] and Bello[293] also followed the thermal denaturation of ribonuclease and reached similar conclusions.

Horwitz et al.[212] examined the near-UV CD spectrum of ribonuclease at 77°K. Their results are shown in Fig. 17. By correlating the CD with the vibrational fine structure displayed by the protein and by model compounds in absorption spectra, they were able to differentiate the CD contributions of buried and exposed tyrosine residues and of S-S bridges. The same group studied RNase-S,[216] and found that the environment of a single tyrosine residue was changed upon cleavage.

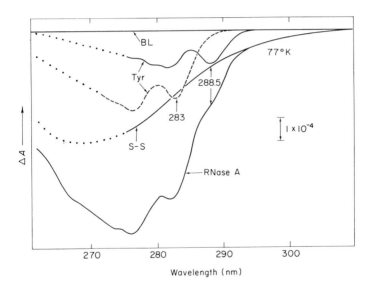

FIG. 17. Near-ultraviolet circular dichroism (CD) spectrum of ribonuclease A at 77°K [from J. Horwitz, E. H. Strickland, and C. Billups, J. Amer. Chem. Soc. 92, 219 (1970)]. Protein concentration 3.6 mM; path length 0.2 mm; solvent water–glycerol (1:1, v:v) with 25 mM sodium phosphate. BL is the baseline. The spectrum has been resolved into components corresponding to CD contributions of various side chains; dotted lines represent regions of extrapolation. S-S designates disulfide contribution; the dashed tyrosine curve results from three exposed tyrosine residues; the small, solid tyrosine curve is caused by one buried tyrosine. Two additional buried tyrosine residues contribute negligibly (<5%) to the total CD spectrum, and were not considered in this analysis.

Pflumm and Beychok[234] studied the reduction and reoxidation of pancreatic ribonuclease and found that the tyrosine CD band at 240 nm appeared to be associated with formation of the correct S-S bridges of the intact enzyme. Simons[77] examined the effect of binding an inhibitor, 3′-CMP, upon this environmentally sensitive 240-nm tyrosine band. She found that a tyrosine side chain, far removed from the active site of RNase and previously buried, became exposed to solvent as the enzyme–inhibitor complex formed.

The far-ultraviolet CD spectrum of RNase[21,37(c)] shows little evidence of the characteristic α-helix 222-nm band (see Fig. 8). Since the CD conformational determination[21] is in good agreement with the X-ray data,[300] RNase is probably not a case (like avidin[244]) in which far-ultraviolet side-chain CD bands seriously obscure the peptide Cotton effects.

(ii) LYSOZYME. An aromatic (probably tryptophan) Cotton effect near 280 nm was found in the ORD spectrum[301] of hen egg-white lysozyme. Exposure of the enzyme to sodium dodecyl sulfate destroys this Cotton effect, but does not change the 233-nm trough characteristic of the α-helical conformation. The near-ultraviolet CD spectrum was resolved[302] into three positive bands between 280 and 300 nm and a negative band at 262 nm; the latter band has been attributed to disulfide.[233] In the presence of the competitive inhibitor, N-acetyl-D-glucosamine, the aromatic CD bands are intensified,[302] which indicates that exposed aromatic residues at the substrate-binding site are optically active. By studying the effect of pH, Ikeda et al.[303] found that some of the optically active tyrosine residues are exposed to solvent, can be ionized at alkaline pH, but become optically inactive when lysozyme is denatured by exposure to pH > 12.

More recently, Halper et al.[92] have distinguished between tyrosine, tryptophan, and disulfide CD bands by means of pH dependence. They compared human with hen egg-white lysozyme and found differences in the near-UV CD region, although the enzymes displayed similar far-UV CD spectra, indicating analogous secondary structure. Two recent articles[89,90] deal with chemical modification of hen lysozyme: Teichberg and co-workers[90] selectively oxidized tryptophan-108, and determined its CD contribution by means of CD difference spectra. They found that this residue is responsible for most of the aromatic CD region of

[300] G. Kartha, J. Bello, and D. Harker, Nature (London) 213, 862 (1967).
[301] A. N. Glazer and N. S. Simmons, J. Amer. Chem. Soc. 87, 2287 (1965).
[302] A. N. Glazer and N. S. Simmons, J. Amer. Chem. Soc. 88, 2335 (1966).
[303] K. Ikeda, K. Hamaguchi, M. Imanishi, and T. Amano, J. Biochem. (Tokyo) 62, 315 (1967).

lysozyme, that its oxidation does not significantly affect the conformation-dependent ellipticity bands[21] at 209 and 192 nm, and that there is a change in orientation of Trp-108 when oligosaccharide inhibitors are bound to the active site. Atassi et al.[89] made derivatives of two tyrosines and six tryptophans of lysozyme; examination of ORD and CD data showed various amounts of unfolding.

Conclusion

In conclusion, circular dichroism and optical rotatory dispersion can yield useful estimates of protein secondary structure. In some cases the agreement with X-ray diffraction results is excellent. Furthermore, for synthetic polypeptides CD and ORD are invaluable tools for structural analysis in solution; each basic conformation can be closely correlated with a specific type of CD or ORD spectrum.

However, the mixture of backbone conformations found in proteins can be more complex and varied than those of simple poly-α-amino acids. The model systems used for calculation of protein conformation are far from perfect. Moreover, for proteins there are factors besides the peptide backbone conformation which influence the optical activity. Light scattering in membrane systems is one such case; the contribution of asymmetric side-chain chromophores is another.

The contributions of aromatic residues and disulfide bonds to the optical activity of proteins are complicated, but very interesting. Recent advances in instrumentation and techniques have made it possible to begin to sort out the chiroptical properties of individual amino acid residues and the effects of inhibitor binding at active sites. Such results can be used to gain insight into the location and function of specific residues within a protein. On the other hand, the complexity of the spectra of polypeptides containing aromatic and cystine residues reemphasizes the pitfalls inherent in trying to estimate protein and polypeptide structure from optical activity alone. There is a need to correlate variations in conformation and in side-chain orientation with their contributions to CD and ORD, and to apply this knowledge systematically to proteins.

[28] Difference Optical Rotatory Dispersion and Circular Dichroism

By Jen Tsi Yang and Kue Hung Chau

Difference measurement is defined in this section as the simultaneous optical subtraction of one measurement from another. The technique has been widely used in many analytical methods, such as difference absorption spectroscopy. In a double-beam spectrophotometer difference absorption spectrum is accomplished by suppressing the zero point of the instrument to allow scale expansion for monitoring small differences in absorbance between two solutions, one placed in the sample beam and the other in the reference beam. Originally, difference measurement was used to increase analytical precision by measuring, for instance, the absorbance of a solution against a moderately high absorbance of another (reference) solution of known concentration.[1,2] In recent years the technique has been developed into a useful tool for detecting small differences or changes in the conformation or configuration of biopolymers in solution (for reviews, see, for example, Donovan[3]). What is true for difference absorption spectroscopy is in principle also applicable to difference optical rotatory dispersion (ORD) and circular dichroism (CD).

The currently available spectropolarimeters and circular dichrometers utilize a single-beam optical system. In the Cary 60 spectropolarimeter, however, the folded optical beam is capable of making difference ORD measurement by means of a specially designed cell holder. The same principle is now applied to the Cary 61 circular dichrometer with slight modifications for difference CD measurement. The attachment for difference CD in Durrum-Jasco J-10, J-15, and J-20 uses two Fresnel rhombs as a half wavelength retardation plate between the sample and reference solutions. This section describes in brief the principles, applications, and limitations of difference ORD and CD techniques. For definitions of ORD and CD, see references cited in footnotes 4–6.

Difference ORD (Cary 60)[7,8]

Method. Difference optical rotation (at any wavelength) or difference ORD can be measured with a Cary 60 spectropolarimeter equipped with

[1] R. Bastian, *Anal. Chem.* **21**, 972 (1949).
[2] C. F. Hiskey, *Anal. Chem.* **21**, 1440 (1949).
[3] J. W. Donovan, this volume [21].
[4] A. J. Adler and G. D. Fasman, see Vol. 12 [122].
[5] W. F. H. M. Mommaerts, see Vol. 12 [123].

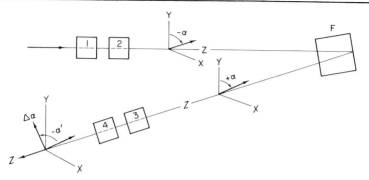

FIG. 1. Relationship of the incident and reflected beam to the mirrored Faraday cell (F), the solution-containing cells 1–4, and the beam-oriented coordinate systems of the Cary 60 spectropolarimeter [B. J. Adkins and J. T. Yang, *Biochemistry* **7**, 266 (1968)].

a difference cell holder by the manufacturer (Cary Instruments). Figure 1 shows a sketch of the optical layout of the light beam through the sample compartment. The incident polarized beam (i-beam) passes through one or more cells (e.g., cells 1 and 2) before reaching a mirrored Faraday cell that modulates the plane of polarization and reflects the beam (r-beam) into the photomultiplier tube. In normal operation there is nothing to interrupt the r-beam, but in difference ORD the r-beam passes through another set of cells (e.g., cells 3 and 4).

The principle is simple. It is based on the inversion of the beam-oriented coordinate system due to reflection of the i-beam by the mirror of the Faraday cell. As the r-beam reaches the reference solution(s), its plane of polarization has in effect been prerotated in a direction opposite to that imparted by the sample solution(s). To put it another way, if an optically active sample rotates the plane of polarization, its mirror image will rotate this plane in an opposite direction. If cells of equal light path (e.g., cells 1 and 3) are filled with the same nonabsorbing but optically active solution (see discussion on artifacts), the rotation in the i-beam is exactly canceled by that in the r-beam and only the baseline of the instrument is recorded. The difference measurement is thus capable of detecting small differences in rotation between two solutions because any large rotations they have in common cancel each other.

The experimental procedure is as follows: As in difference absorption measurements, we select four cells of equal path length that are free of

[6] A. J. Adler, N. J. Greenfield, and G. D. Fasman, this volume [27].
[7] Instruction Manual of Cary 60, Cary Instruments, Monrovia, California.
[8] B. J. Adkins and J. T. Yang, *Biochemistry* **7**, 266 (1968).

birefringence and arrange them in pairs as illustrated in Fig. 1. For example, cell 1 can be filled with a protein solution plus a perturbant, cell 2 with the solvent, cell 3 with the protein solution without perturbant, and cell 4 with the perturbant. Then we record either the difference optical rotation at a chosen wavelength or difference ORD. Ideally, all we need is an additional measurement of the baseline for calculating the difference measurement. This procedure, however, is permissible only for essentially nonabsorbing solutions, since artifacts are generated when an absorbing solution is placed in the r-beam.

Calculation of $\Delta\alpha$. The net measured rotation of two or more cells is equal to the sum of the rotations of the solutions at any wavelength:

$$\alpha_{12} = \alpha_1 + \alpha_2 \tag{1}$$

and

$$\alpha_{34} = \alpha_3 + \alpha_4 \tag{2}$$

The correct difference rotation, $\Delta\alpha$, is then

$$\begin{aligned}\Delta\alpha &= \Delta\alpha_{\text{ex}} - b \\ &= (\alpha_{12} - \alpha_{34}) \\ &= (\alpha_1 + \alpha_2)_i - (\alpha_3 + \alpha_4)_r\end{aligned} \tag{3}$$

where $\Delta\alpha_{\text{ex}}$ indicates the reading on the recording chart, subscripts i and and r refer to the i- and r-beams, b represents the baseline of the instrument, and the minus sign in front of α_{34} or $(\alpha_3 + \alpha_4)_r$ accounts for the reflection and inversion of the beam-oriented coordinate system. For instance, a dextrorotatory solution gives a negative reading on the recording chart when the solution is placed in the r-beam instead of the i-beam.

Artifact in the Reflected Beam.[8] An absorbing solution placed in the r-beam produces false levorotation, even when the solution is optically inactive. For example, a potassium dichromate solution placed in the r-beam gives apparent levorotation (Fig. 2). The profile of the artifact as a function of wavelength parallels the corresponding absorption spectrum. Figure 3 further shows that the magnitude of the false rotation is small below 0.5 OD unit but increases sharply at higher optical densities. The magnitude of artifact also varies with the instruments used and even with the same instrument after each servicing. The vertical bars in Fig. 3 represent the average variation of the false rotations observed with two Cary instruments on five different occasions. A satisfactory explanation for the false levorotation is still lacking. No artifact is observed for an absorbing solution placed in the i-beam as in the normal operation.[9]

[9] T. Samejima and J. T. Yang, *Biochemistry* 3, 613 (1964).

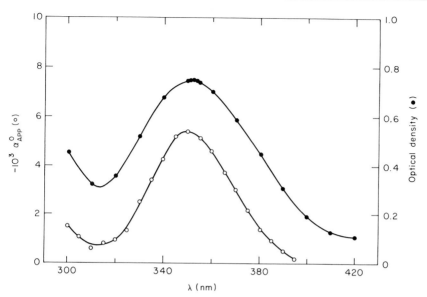

Fig. 2. Apparent rotation at an absorption band of potassium dichromate solution (68.6 mg/l in a 1-cm cell) placed in the reflected beam [B. J. Adkins and J. T. Yang, *Biochemistry* **7**, 266 (1968)].

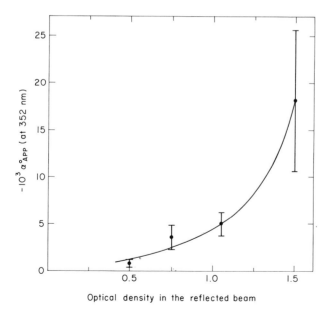

Fig. 3. Dependence of the apparent rotation on the optical density of potassium dichromate solutions placed in the reflected beam. $K_2Cr_2O_7$ concentration: 68.6 mg/l. Path lengths: 5, 10, 15, and 20 mm [B. J. Adkins and J. T. Yang, *Biochemistry* **7**, 266 (1968)].

Because of the artifact Eq. (3) must be modified when measurements are extended into the region of absorption bands.

$$\Delta\alpha' = \Delta\alpha'_{ex} - b$$
$$= (\alpha_1 + \alpha_2)_i - (\alpha_3 + \alpha_4)_r - \alpha_{a34}$$
$$= (\alpha_{12})_i - (\alpha_{34})_r - \alpha_{a34} \qquad (4)$$

where α_{a34} represents the rotational artifact in the r-beam. To circumvent this complication, we can perform a second measurement of difference optical rotation or difference ORD by interchanging cells 1 and 2 with cells 3 and 4. Thus, we have in a similar manner

$$\Delta\alpha'' = \Delta\alpha''_{ex} - b$$
$$= (\alpha_3 + \alpha_4)_i - (\alpha_1 + \alpha_2)_r - \alpha_{a12}$$
$$= (\alpha_{34})_i - (\alpha_{12})_r - \alpha_{a12} \qquad (5)$$

If the total absorbance of cells 1 and 2 and that of cells 3 and 4 are approximately equal, then $\alpha_{a12} \cong \alpha_{a34}$. Furthermore, $(\alpha_1 + \alpha_2)_i = (\alpha_1 + \alpha_2)_r$ and $(\alpha_3 + \alpha_4)_i = (\alpha_3 + \alpha_4)_r$. Subtraction of the second measurement [Eq. (5)] from the first [Eq. (4)] cancels almost all the artifact. The net result is twice the actual difference optical rotation, that is,

$$\Delta\alpha = (\Delta\alpha' - \Delta\alpha'')/2$$
$$= (\Delta\alpha'_{ex} - \Delta\alpha''_{ex})/2 \qquad (6)$$

provided, of course, that the instrument baseline remains unchanged during the two measurements. From Eqs. (4) and (5) we may also obtain the apparent baseline in the presence of artifact, b_a, i.e.,

$$b_a = (\Delta\alpha'_{ex} + \Delta\alpha''_{ex})/2$$
$$= -(\alpha_{a12} + \alpha_{a34})/2 + b$$
$$= -\alpha_{a12} \text{ (or } - \alpha_{a34}) + b \qquad (7)$$

If $\alpha_{a12} \neq \alpha_{a34}$, Eq. (6) should be written as

$$\Delta\alpha = (\alpha_{12})_i - (\alpha_{34})_r + (\alpha_{a12} - \alpha_{a34})/2 \qquad (6a)$$

that is, the last term on the right side does not vanish, and $\Delta\alpha$ in Eq. (6a) does not represent the correct difference rotation. The term α_{a12} can be determined by measuring the apparent rotation when cells 3 and 4 are also filled with the solutions of cells 1 and 2, respectively. Similarly we can determine the term α_{a34}. However, the advantage of difference measurement over conventional separate measurements of solutions is offset by the number of measurements required under these conditions.

Difference measurements can be obtained from precise conventional measurements of two ORD spectra separately, followed by manual subtraction of one measurement from the other, wavelength by wavelength. Such a procedure, however, often produces much larger experimental errors than optical subtraction.[2] In this respect we may mention the method of Chignell and Gratzer[10] for measuring small differences in rotation between two large rotations. Their method increases the sensitivity of direct ORD measurement by increasing the concentration of the sample solution and placing in series with the sample solution, a solution of known ORD of opposite sign to cancel the "background" rotation, for instance, a dextrorotatory sucrose solution versus a levorotatory protein solution.

Test of the Method. To verify the performance of the difference ORD technique, we can place a nonabsorbing sample such as a sucrose solution in cell 1 and water in cell 3 (cells 2 and 4 are not used in this experiment) and record the rotations (curve 1, Fig. 4). After interchanging cells 1 and 3, we record the rotations again (curve 2, Fig. 4). As can be seen, curves 1 and 2 are mirror images with respect to the instrument baseline (curve 7, Fig. 4). The baseline can be calculated from Eq. (7) without the artifact terms and is identical with that measured experimentally. Substitution of the data of curves 1 and 2 into Eq. (6) yields the rotations of sucrose (curve 6, Fig. 4). In this example $\Delta\alpha$ is simply the rotation of sucrose solution.

During normal operation of the Cary 60 spectropolarimeter, when the sample rotates the plane of polarization, the polarizer is rotated in an equal and opposite direction to cancel out the rotation of the sample. Thus the orientation of the plane of polarization reflecting off the back of the Faraday cell remains nearly constant. In difference measurement, the sample rotation is canceled out by the rotation in the reference cell. Since the polarizer is not rotated to maintain the plane of polarization reflecting off the Faraday cell at a constant value, the light becomes slightly depolarized, resulting in an apparent additional rotation. According to the manufacturer's manual,[10] however, rotations up to four or five degrees have been canceled out with only about two millidegrees of error introduced by the reflection. Rotation of such magnitude is not encountered in normal experiments and the effect is therefore negligible.

When an absorbing potassium dichromate solution (cells 2 and 4), which is optically inactive is placed in series with sucrose solution and water in the i and r-beams, respectively, the measured ORD curves 3 (sucrose vs. water) and 4 (water vs. sucrose) (Fig. 4) do not coincide with curves 1 and 2, nor are they mirror images of each other. Instead, both curves are more levorotatory in the absorption region of potassium dichromate than curves 1 and 2, with maximum deviation at the absorp-

[10] D. A. Chignell and W. B. Gratzer, *Nature (London)* **210**, 262 (1966).

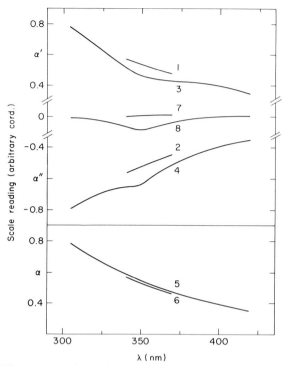

Fig. 4. Difference rotation of sucrose in water with and without potassium dichromate solution in series. Curves: 1, sucrose vs. water; 2, water vs. sucrose; 3, sucrose plus dichromate vs. dichromate plus water; 4, dichromate plus water vs. sucrose plus dichromate; 5, calculated from 3 and 4 with Eq. (6); 6, calculated from 1 and 2 with Eq. (6); 7, baseline calculated from 1 and 2 with Eq. (7); and 8, baseline calculated from 3 and 4 with Eq. (7). Concentrations: sucrose, 0.101 g/dl in water and potassium dichromate, 68.6 mg/liter in water. Cells: 1-cm path length. See text for details. [B. J. Adkins and J. T. Yang, *Biochemistry* **7**, 266 (1968).]

tion maximum of potassium dichromate (about 352 nm). Substitution of the data of curves 3 and 4 into Eq. (6), however, gives the calculated ORD for sucrose (curve 5, Fig. 4), which is close to the direct measurement of sucrose (curve 6) and free of any noticeable artifact. Thus the difference ORD technique can still be applied to a protein solution in its absorption region, provided that the artifact is taken into consideration and the necessary correction is made. Equation (6) is valid as long as the sample and reference solutions have nearly identical total absorbance.

To illustrate the application of difference ORD, we show in Fig. 5 the change in specific rotation as a result of the complex formation of

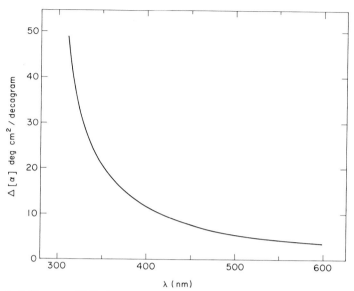

FIG. 5. Difference ORD for lysozyme in the presence and absence of N-acetyl-D-glucosamine (NAG). Concentrations: lysozyme, 0.857 g/dl in 0.1 M NaCl, and NAG, 0.156 M in 0.1 M NaCl. Light path of the cells: 1 cm [B. J. Adkins and J. T. Yang, *Biochemistry* **7**, 266 (1968)].

an enzyme molecule and its inhibitor. X-ray diffraction studies of lyso-zyme from egg white[11,12] indicate that the enzyme molecule strongly binds one molecule of the inhibitor, N-acetyl-D-glucosamine, in the cleft of the protein molecule. Clearly, the enzyme–inhibitor complex in this case is less levorotatory in the visible region than the sum of the rotations of the enzyme and inhibitor measured separately. The precision of the data can easily be further increased by increasing the concentration of the protein and inhibitor solutions or the light path of the cells, or both, since the "background" rotations in the i- and r-beam have been can-celed out. The curve in Fig. 5 was reported to reach a maximum at about 297 nm, below which the difference ORD was of small magnitude (using 1-mm cells to reduce absorption) and almost within the experimental error.[8]

To summarize, the difference ORD technique can be a useful tool for measuring small changes in the optical rotation of protein solutions under various conditions, such as changes in pH and temperature and

[11] C. C. F. Blake, D. F. Koenig, G. A. Mair, A. C. T. North, D. C. Phillips, and V. R. Sarma, *Nature (London)* **206**, 757 (1965).
[12] L. N. Johnson and D. C. Phillips, *Nature (London)* **206**, 761 (1965).

addition of denaturing agents, even though we may not know precisely the mechanism of the conformational or configurational changes. With the reference solution that cancels out the "background" rotation, we can easily increase the concentration or light path, or both, of the protein solution, thereby increasing the magnitude of its difference rotation. The results with this technique will be more precise than heretofore obtainable, especially when the difference rotation by manual subtraction is of marginal signal-to-noise ratio (commercial recording spectropolarimeters have a range of only ±2 degrees of rotation). In addition, manual subtraction involves three direct measurements (for a sample solution, a reference solution, and one solvent baseline), whereas the difference technique requires only two measurements [Eq. (3) plus a baseline or (6)]. Thus the difference technique has the advantages of being more precise, sensitive, and rapid, as well as less laborious, than the normal operation (manual subtraction).

The difference ORD method is particularly useful for measurements in the wavelength ranges where the protein solutions are transparent to the light beam (e.g., the visible region for most protein solutions). But an artifact is produced in the absorption region and must be corrected. Furthermore, the concentration, or light path, or both, of a protein solution is limited by the general rule that the total absorbance of the sample and reference solutions must be less than two (i.e., with more than 1% transmission of the light beam). In other words, the sample solution is allowed to have only one half the absorbance allowed in a normal operation; the reference solution accounts for the other half of the absorbance. Whether the difference measurement in this case has definite advantages over manual subtraction has to be determined for each system studied.

Difference CD (Cary 61 and Jasco J 10)[13,14]

Method. Difference CD can be measured in either a Cary 61 (Fig. 6) or Durrum-Jasco J-10, J-15, or J-20 (Fig. 7) spectropolarimeter, provided that the instrument is equipped with the necessary optical accessories and special dual cell holders. (The dual cell holder of Cary 6101 is similar to the holder for difference ORD; the Durrum-Jasco instruments have the special cell holders incorporated in the optical accessory for measuring difference CD.) As in difference ORD, this method is also based on the principle of optical subtraction. In normal operation,

---

[13] Instruction Manual of Cary 6101, Difference CD Accessory for Model 61 CD Spectropolarimeter, Cary Instruments, Monrovia, California, 1970.

[14] D. P. Sproul, Pittsburgh Conference on Analytical Chemistry and Applied Spectroscopy, Cleveland, Ohio, March 1969.

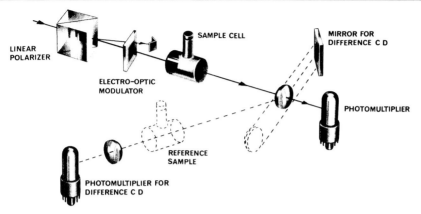

FIG. 6. Optical components for ordinary CD and difference CD in a Cary 61 spectropolarimeter. For difference CD, the mirror replaces the lens in the light path. See Instruction Manual of Cary 6101, Difference CD Accessory for Model 61 CD Spectropolarimeter, Cary Instruments Monrovia, California, 1970.

an electrooptical modulator (Pockels cell) resolves the linearly polarized light into two circularly polarized components. After passing through an optically active solution, the two components differ in magnitude, thus giving rise to circular dichroism. They then reach the optical device for measuring difference CD, which consists of either a mirror (Fig. 6) or two Fresnel rhombs (Fig. 7). The device effects a phase retardation of 180° (half wave) on the two components. The original left circularly polarized component thus becomes right circularly polarized and vice versa; consequently, the sign of $\epsilon_L - \epsilon_R$ is reversed. The components next pass through the reference solution, and finally reach the photomultiplier. If the solutions in the sample cell(s) and the reference cell(s) are identical in optical activity, ideally the resultant CD

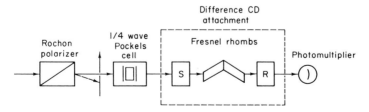

FIG. 7. Optical components for ordinary CD and difference CD in a Durrum-Jasco spectropolarimeter. For difference CD, the attachment is placed in the light path. S and R in the attachment represent positions of the sample cell(s) and the reference cells(s). [D. P. Sproul, Pittsburgh Conference on Analytical Chemistry and Applied Spectroscopy, Cleveland, Ohio, March 1969.]

signal is zero. Since the difference CD device reverses the phase of the two components and the sign of $\epsilon_L - \epsilon_R$, the CD signal from the sample cell(s) is canceled exactly by that of the reference cell(s). If the solution in the reference cell(s) is optically inactive, the resultant CD spectrum is identical to the regular CD spectrum of the sample. Furthermore, the spectra become mirror images of each other if the positions of the reference cell(s) and the sample cell(s) are interchanged. As an example, the CD spectra of d-10-camphorsulfonic acid are shown in Fig. 8. If the solutions in the reference cell(s) and the sample cell(s) differ in optical activity, the result is a difference CD spectrum which is equivalent to subtracting the reference CD from the sample CD. It should be corrected by measuring the baseline with two cells containing either the sample or the reference material. Alternatively, we can use an average method (see below) to calculate the difference CD. In principle, the difference CD technique can resolve and measure small differences between two CD spectra; therefore, it can be used for studying small changes in the conformation or configuration of biopolymers. For instance, the two sample cells can contain a solution of enzyme plus inhibitor and a solvent and the two reference cells can contain the enzyme and the inhibitor separately.

Calculation of $\Delta\theta$. The calculation of difference ellipticity, $\Delta\theta$, is similar to that of difference ORD. If one sample cell and one reference

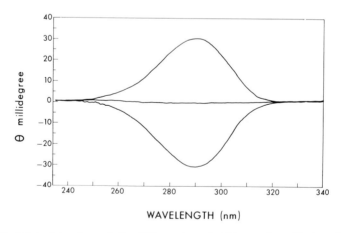

WAVELENGTH (nm)

Fig. 8. Original tracings of the CD spectra of d-10-camphorsulfonic acid (0.01% w/v in water) measured with the difference CD attachment in a Jasco J 10 spectropolarimeter. Both the sample cell and the reference cell were 1.0 mm in path length. The upper spectrum was measured with the solution in the sample cell and water in the reference cell; the lower spectrum was measured after interchanging the positions of the cells; the middle line was measured with water in both cells.

cell are used, and their respective ellipticity is θ_1 and θ_2; then the correct difference ellipticity is

$$\Delta\theta = \theta_1 - \theta_2$$
$$= \Delta\theta_{\text{ex}} - b \qquad (8)$$

where $\Delta\theta_{\text{ex}}$ is the recorded ellipticity of a difference CD spectrum, and b is the baseline.

An alternative method requiring no baseline determination consists of measuring another difference spectrum after exchanging the positions of the cells. The correct difference ellipticity takes the sign of the first spectrum and its magnitude is equal to one half the absolute magnitude between the two difference spectra, which ideally should be mirror images of each other (or one half the value obtained by subtracting the data of the second spectrum from the first one). Thus, we have

$$\Delta\theta = (\Delta\theta_{\text{ex}} - \Delta\theta'_{\text{ex}})/2 \qquad (9)$$

where $\Delta\theta_{\text{ex}}$ and $\Delta\theta'_{\text{ex}}$ are the difference ellipticity of sample versus reference and reference versus sample, respectively [cf. Eq. (6) for difference ORD]. When the material has fairly large absorption but small CD in the wavelength region under investigation, we must take in consideration the artifacts in baseline. Knowing the correct difference ellipticity, one may express the final result in conventional units of molar or mean residue ellipticity [θ].

Limitation and Artifact. Difference CD is a method complementary to difference ORD and shares its advantages over the indirect method of subtracting regular spectra manually. Optical absorbance, however, imposes more limitation on difference CD than on difference ORD, because CD, unlike ORD, is always measured in the absorption region of the solution. The general rule of maintaining less than two units of total light absorbance in the light path permits not more than one absorbance unit in either the sample beam or the reference beam; thus this rule limits the measurable magnitude of the difference CD. This limitation originates primarily from the fact that spectropolarimeters are single-beam instruments, whereas with a double-beam spectrophotometer the absorbance in each beam may be as great as two units. For difference between weak CD bands of strong absorbance, this limitation becomes serious. An example is the system of lysozyme and its interaction with the inhibitor, N-acetyl-D-glucosamine.[8,15,16] Figure 9 shows the CD spectra of lysozyme and lysozyme plus inhibitor; their actual recordings differed by approximately 1 millidegree for the protein concen-

[15] K. H. Chau and J. T. Yang, *Anal. Biochem.* **46**, 616 (1972).
[16] A. N. Glazer and N. S. Simmons, *J. Amer. Chem. Soc.* **88**, 2335 (1966).

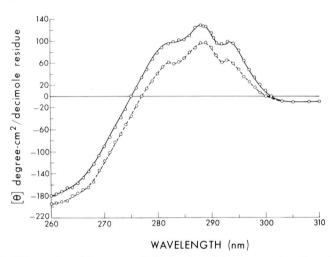

WAVELENGTH (nm)

FIG. 9. CD spectra of lysozyme (LS) (- - -) and lysozyme plus N-acetyl-D-gluco-samine (LIS) (——) measured in a Durrum-Jasco J-10 spectropolarimeter. The points (○) represent the corresponding spectra measured in a Cary 6001 (CD attachment to a Cary 60). Identical concentration of lysozyme ($A^{1\,cm}_{280\,nm}$ = 0.88) was used in each spectrum; the concentration of N-acetyl-D-glucosamine (NAG) in the mixture was 0.015 M; the molar ratio of lysozyme to NAG was approximately 700:1; the salt concentration was 0.1 M NaCl in all solutions.

tration used ($A^{1\,cm}_{280\,nm}$ = 0.88). Figure 10 shows the difference CD spectrum obtained by manual subtraction of the spectra in Fig. 9 and the direct difference CD spectra measured in a Jasco J-10 and a Cary 6101 with the same concentration of each solution as in Fig. 9. The total absorbance ($A^{2\,cm}_{280\,nm}$ = 1.8) in the light path of the direct method approached the upper limit for accurate CD measurement, but the signal-to-noise ratio was far less than unity, which is the minimum measurable quantity. The noise problem is not improved by decreasing the concentration of each solution by half, for the difference CD signal decreases simultaneously.

Light absorption also creates small but significant artifact of false ellipticity in the direct difference CD. In the Durrum-Jasco J-10 which we have tested the artifact occurs even when an optically inactive solution is placed in the sample beam or the reference beam or in both. The magnitude of the artifact increases with absorbance; optically active solutions appear to augment it. For both Durrum-Jasco J-10 and Cary 6101 the effects of such an artifact are: (1) curvature in the baseline measured with the same solution (e.g., lysozyme or lysozyme plus inhibitor) in both the sample and the reference cells, and (2) the reversed difference CD spectrum which is measured with reference solution against

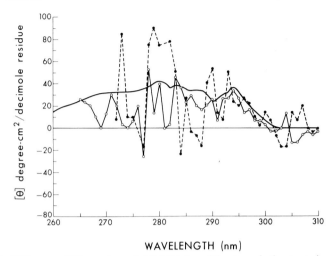

WAVELENGTH (nm)

Fig. 10. Difference CD spectra of LS-LIS (see Fig. 9 and the text for detail). ——, Spectrum from subtraction of spectra in Fig. 9; · - - ·, spectrum from Durrum-Jasco difference CD; ○ - - ○, spectrum from Cary 6101 difference CD (courtesy of the Cary Instruments).

sample solution is not a mirror image of the regular difference spectrum with respect to the instrument baseline. If the artifact in both beams is equal, Eq. (9) is still valid for $\Delta\theta$. If it is not equal (for instance, when the sample and reference solutions differ slightly in absorbance), then the artifact should be corrected by a procedure similar to that for difference ORD [cf. Eq. (6a)]. Because of these problems, it is difficult to use the direct method of difference CD for systems having strong absorption and small difference among weak CD bands. The indirect method always gives spectra of better signal-to-noise ratio than the direct method for the same magnitude of difference CD, and it allows the use of higher concentrations (within the limit of two absorbance units). For systems having weak absorbance and strong CD, errors with the direct method are relatively small, but one may question the necessity of using the direct method for such systems since the indirect method can easily resolve their difference CD.

Comparing the instruments more specifically on the feature of difference CD, we find that the Cary 61 spectropolarimeter has certain advantages over the Durrum-Jasco instrument. Since the Cary 61 uses a mirror for phase retardation, the loss in light intensity is negligible. In contrast, the Fresnel rhombs in a Durrum-Jasco show light absorption and require a narrower slit program to assure a relatively flat baseline; therefore the dynode voltage of the photomultiplier increases between 30 and 50%, as compared with regular CD measurement with the same instrument. The rhombs also reduce the space in the sample compartment, leaving room on each side for a cell-holder which can accommodate only two 1-cm cells. The cell-holder for difference CD in the Cary 61 is more spacious and is capable of holding two cells longer than 1 cm each in either light beam; this feature is especially convenient when cells of thin path length are used to decrease absorbance without diluting the solutions. (Each ORD/CD cell shorter than 1 cm in path length has a jacket 1.5 cm in length; therefore neither of the two holders in the difference CD device of the Durrum-Jasco instruments can accommodate two such cells.) For measurements of difference CD at a fixed wavelength or at only a few wavelengths, such as in kinetics studies of enzyme reactions or melting curves of nucleic acids, another procedure of indirect difference CD may be more sensitive. This is accomplished by utilizing the baseline offset control of the Durrum-Jasco instruments or, for more accurate results, the calibrated zero suppression and pen center controls of the Cary 61 to cancel (suppress) the CD of the reference solution; an expanded scale is then used to measure the difference CD of the sample solution without the difference CD attachment. Obviously this method is inconvenient to use for continuous difference CD spectra. The baseline, judged by air baseline, in the difference CD mode in the Durrum-Jasco takes a few hours to stabilize at a given temperature, although the regular CD baseline is stable after about 30 minutes of warmup and shows small dependence on temperature. Information on the temperature dependence of the Cary 61 is not available.

In conclusion, the direct method of measuring difference CD is potentially a useful technique for studying small changes or differences in the structure of macromolecules in solution, but the present problems associated with use of single-beam instruments limit its application at present.

Acknowledgments

We acknowledge the courtesy of Cary Instruments for the use of their Cary 6101 spectropolarimeter. We also thank Dr. J. J. Duffield and Mr. W. N. Mitchell of Cary Instruments and Mr. D. P. Sproul of Durrum Instrument Corporation for their helpful suggestions.

This work was aided by grants from the U.S. Public Health Service (HE-06285, GM-10880, and K3-GM-3441.

[29] Analysis of Optical Activity Spectra of Turbid Biological Suspensions

By ALLAN S. SCHNEIDER

Ultraviolet circular dichroism (CD) and optical rotatory dispersion (ORD) are among the most sensitive measures of biopolymer structure in solution. However, when these methods are applied to suspensions of intact biological structures such as membranes,[1-4] red blood cells,[1] and polypeptide aggregates,[2] two phenomena are observed: (a) high scattering and turbidity and (b) significant distortions in the shape and magnitude of the circular dichroism and optical rotatory dispersion bands. The resulting distorted spectra were at first unrecognizable and difficult to interpret. It has been demonstrated experimentally that these spectral distortions arise from the scattering properties and particulate nature of the above turbid suspensions,[1,2,5] and it is likely that the phenomenon is quite general. The question that now faces us is how to interpret such anomalous optical activity spectra, and it is this question with which the present article is concerned.

That scattering should influence the circular dichroism spectrum of turbid biological suspensions is quite reasonable, since (a) CD measures the difference in extinction (absorption + scattering) between left and right circularly polarized light and (b) the optically active particles making up biological suspensions will scatter different amounts of left and right polarizations by virtue of their different refractive indices for these polarization states. Consequently a scattering contribution to optical activity will result in addition to the normal intrinsic optical activity of the macromolecules that one observes in solution.

In addition to scattering effects, particulate suspensions also exhibit what has come to be known as absorption flattening[6] effects in their spectra. This effect may be thought of as a light sieve effect due to the regions between suspended particles which contain no chromophore and are thus transparent. The transmission will accordingly increase and the absorption and circular dichroism decrease relative to the equivalent

[1] A. S. Schneider, M.-J. T. Schneider, and K. Rosenheck, *Proc. Nat. Acad. Sci. U.S.* **66**, 793 (1970).

[2] T. H. Ji and D. W. Urry, *Biochem. Biophys. Res. Commun.* **34**, 404 (1969).

[3] J. M. Steim and S. Fleischer, *Proc. Nat. Acad. Sci. U.S.* **58**, 1292 (1967).

[4] J. Lenard and S. J. Singer, *Proc. Nat. Acad. Sci. U.S.* **56**, 1828 (1966).

[5] M. Glaser and S. J. Singer, *Biochemistry* **10**, 1780 (1971).

[6] L. M. N. Duysens, *Biochim. Biophys. Acta* **19**, 1 (1956).

molecularly dispersed solution. The effect is maximal at the peak in absorption and gives absorption bands that appear flattened. This absorption flattening is accounted for in the usual equations relating the optical properties of a suspension to the total cross section (scattering + absorption) of the scattering particle, and its influence on optical activity spectra may easily be separated from that of the scattering per se.

In the present article we outline a method of analyzing the optical activity spectra of turbid biological suspensions. We begin with a brief description of the theoretical principles and equations relating the optical activity (CD and ORD) of a scattering suspension to the intrinsic optical activity of the molecules composing the scattering particle, as for example the measured spectra of a membrane suspension and the corresponding solution spectra of the proteins making up the membranes. We show the separation of scattering and absorption flattening effects, which will be useful in recognizing the various contributions to the spectra as well as in the interpretation of partially corrected experimental spectra measured with a spectropolarimeter capable of collecting the scattered light.[7] This is followed by a section on the practical application of the relevant equations with illustrative examples for turbid suspensions of proteins and cell membranes. We conclude with a discussion of the present state of the art, limitations of the method and outlook for extension of optical activity spectra to the more highly organized and interesting biological structures.

Principles

General. There are two general types of optical activity experiments which one may wish to interpret: (a) those where the biopolymer structure and solution spectra are known and an understanding of the turbid suspension spectra is desired, as for example the hemoglobin spectra in suspensions of sphered red blood cells, and (b) those where only the experimental suspension spectra are known and one wishes to determine the intrinsic particle or solution optical activity and thus learn about the macromolecular structure, as in cell membrane suspensions. In both cases a functional relation between suspension and particle optical activity is required.

Our general approach is to start with known scattering functions (extinction cross sections) for absorbing particles, such as the Mie cross-sections for large spheres, and extend these to optical activity spectra. This approach has the advantage of using tried and tested scattering functions that contain the full dependence on the real and imaginary

---

[7] B. P. Dorman, M. F. Maestre, and J. E. Hearst, this volume [30].

parts of the particle refractive index, and yields both absorption and scattering effects in a consistent way. A large number of such scattering functions are readily available for a variety of particle shapes and sizes, and can be found in the excellent books of Van de Hulst[8] and Kerker.[9]

The extension of the normal scattering functions to optical activity requires a consideration of the circular polarization dependence of the scattered light.[10] This leads to two complementary lines of analysis resulting in functional relationships between the optical activity of a turbid suspension and the intrinsic CD and ORD of the macromolecules making up the scattering particles. One approach is to make separate calculations of the extinction for left and right circularly polarized light, the difference yielding the suspension circular dichroism as a function of the particle CD and ORD. The second method is formulated in terms of a series expansion of the suspension optical activity to first order in the intrinsic particle CD and ORD, and provides direct insight into the effect of particle optical activity on the circularly polarized scattering and absorption properties of the suspension.

In what follows we will be interested in optically active particles whose dimensions are comparable to or larger than the wavelength of incident light. In referring to suspension optical properties we will mean the results of a measurement on a suspension of particles; by intrinsic, particle or solution optical properties we mean the properties of the molecules making up the scattering particle, i.e., the equivalent molecularly dispersed solution. Refractive index m, is considered as complex, $m = n - in'$, where the real part, n, is the usual refractive index, and the imaginary part, n', is directly related to the extinction coefficient. By extinction we mean the total loss of light including both absorptive and scattering losses. The subscripts susp and p will refer to suspension and intrinsic particle properties, respectively.

Suspension Optical Activity via Separate Scattering Calculations for Left and Right Circularly Polarized Light. The optical activity (OA) of a substance is defined by the difference of its complex refractive indices for left and right circularly polarized light. Thus for a scattering suspension we have:

$$OA_{susp} = \frac{\pi}{\lambda} (m_l - m_r)_{susp} \tag{1}$$

Separating into real and imaginary parts gives the ORD and CD:

[8] H. C. Van de Hulst, "Light Scattering by Small Particles." Wiley, New York, 1957.
[9] M. Kerker, "The Scattering of Light and Other Electromagnetic Radiation." Academic Press, New York, 1969.
[10] A. S. Schneider, *Chem. Phys. Lett.* **8**, 604 (1971).

$$\text{ORD}_{\text{susp}} = \frac{\pi}{\lambda} (n_l - n_r)_{\text{susp}} \quad \text{and} \quad \text{CD}_{\text{susp}} = \frac{\pi}{\lambda} (n'_l - n'_r)_{\text{susp}} \qquad (2)$$

where λ is the wavelength of incident light, and the subscripts l and r refer to left and right circular polarizations. As we have previously shown,[10] the suspension refractive index is a function of the complex particle refractive index, which we indicate by

$$m_{\text{susp}} = F(m_p) = F(n_p, n'_p) \qquad (3)$$

The explicit functional dependence on particle refractive index as well as on wavelength and particle size, will be dictated by the appropriate scattering function for the particle under consideration, e.g., the Mie equations for the case of large absorbing spheres. For left and right circular polarizations we may write:

$$m_{l\ \text{susp}} = F(m_{lp}) \quad \text{and} \quad m_{r\ \text{susp}} = F(m_{rp}) \qquad (4)$$

The intrinsic particle optical activity may now be introduced by noting the relation between circular refractive indices and optical activity:

$$m_{lp} = \bar{m}_p + k^{-1}\text{OA}_p; \; m_{rp} = \bar{m}_p - k^{-1}\text{OA}_p; \quad \text{and} \quad \bar{m}_p = \tfrac{1}{2}(m_{lp} + m_{rp}) \qquad (5)$$

where the wave vector $k = 2\pi/\lambda$ and $\bar{m}_p$ is the mean particle refractive index measured with unpolarized light. In terms of real and imaginary parts Eq. (5) separates:

$$n_{lp} = \bar{n}_p + k^{-1}\text{ORD}_p; \; n_{rp} = \bar{n}_p - k^{-1}\text{ORD}_p; \quad \text{and} \quad \bar{n}_p = \tfrac{1}{2}(n_{lp} + n_{rp}) \qquad (6)$$

and

$$n'_{lp} = \bar{n}'_p + k^{-1}\text{CD}_p; \; n'_{rp} = \bar{n}'_p - k^{-1}\text{CD}_p; \quad \text{and} \quad \bar{n}'_p = \tfrac{1}{2}(n'_{lp} + n'_{rp}) \qquad (7)$$

Substituting Eqs. (3)–(5) into Eq. (1) gives a general functional relation between suspension and particle optical activity:

$$\text{OA}_{\text{susp}} = \frac{\pi}{\lambda} [F(m_{lp}) - F(m_{rp})]$$

or

$$\text{OA}_{\text{susp}} = \frac{\pi}{\lambda} [F(\bar{m}_p + k^{-1}\,\text{OA}_p) - F(\bar{m}_p - k^{-1}\,\text{OA}_p)] \qquad (8)$$

In Eq. (8) the optical activities of suspension and particle may easily be written in terms of the respective CD's and ORD's via Eqs. (2), (6), and (7).

There are several important points to be made about Eq. (8). First it suggests a calculational approach for the analysis of turbid suspension spectra by making separate scattering calculations for left and right circularly polarized light using the corresponding particle circular refractive indices. Second we note that the suspension CD and ORD (optical activity) will each depend on both the real and imaginary parts of the particle optical properties, namely, on particle real refractive index, absorption coefficient, CD_p, and ORD_p. Finally, the scattering function, $F(n_p, n'_p)$, contains both absorption and scattering effects, and these can be easily separated for many systems of interest. Such a separation can give insight into the origins of the spectral distortions, and will be useful for the interpretation of spectra measured with a spectropolarimeter modified to collect the scattered light.[7] Such instrumentation can remove part or all of the scattering contribution to circular dichroism, but cannot remove the absorption flattening distortions which are an inherent property of particulate suspensions.

The separation of scattering and absorption contributions to anomalous suspension CD spectra is most easily considered in terms of a total extinction cross section, σ_T, which is linearly related to the imaginary part of the scattering function, $F(n_p, n'_p)$. The total cross section, σ_T, is defined as the geometrical cross-sectional area of the particle multiplied by the fraction of light incident on it which is removed from the forward beam by both scattering and absorption. Accordingly, the total cross-section σ_T, can be written as a sum of scattering (scat) and absorption (abs) cross sections[8]:

$$\sigma_T = \sigma_{\text{scat}} + \sigma_{\text{abs}} \tag{9}$$

and formulas for calculating the individual cross sections for an assortment of particle geometries can be found in standard scattering texts.[8,9] Writing the suspension circular dichroism in terms of total, scattering, and absorption cross sections for circularly polarized light thus provides the desired resolution of the spectra:

$$CD_{\text{susp}} = \frac{N}{4}(\sigma_{lT} - \sigma_{rT}) \tag{10}$$

$$CD_{\text{scat}} = \frac{N}{4}(\sigma_{l\ \text{scat}} - \sigma_{r\ \text{scat}}) \tag{11}$$

and

$$CD_{\text{abs}} = \frac{N}{4}(\sigma_{l\ \text{abs}} - \sigma_{r\ \text{abs}}) \tag{12}$$

where N is the particle concentration. CD_{susp} represents the total measured suspension circular dichroism and is obviously the sum of CD_{scat}

plus CD_{abs}. The absorption contribution, CD_{abs}, includes the flattening effects, and is what would be measured by a spectropolarimeter capable of collecting all the scattered light.[7]

Suspension Optical Activity via Series Expansion.[10] An alternative way to derive the suspension optical activity is by a series expansion to first order in the intrinsic particle CD and ORD.[10] The result is a simple and illuminating expression which can give direct insight into the effects of turbidity on ORD and CD spectra. For suspension circular dichroism this is:

$$CD_{susp} = \frac{1}{2} Nk^{-1} \left[\left(\frac{\partial \sigma_T}{\partial n_p} \right)_{\bar{n}_p, \bar{n}'_p} ORD_p + \left(\frac{\partial \sigma_T}{\partial n'_p} \right)_{\bar{n}_p, \bar{n}'_p} CD_p \right] \qquad (13)$$

and the scattering and absorption contributions are:

$$CD_{scat} = \frac{1}{2} Nk^{-1} \left[\left(\frac{\partial \sigma_{scat}}{\partial n_p} \right)_{\bar{n}_p, \bar{n}'_p} ORD_p + \left(\frac{\partial \sigma_{scat}}{\partial n'_p} \right)_{\bar{n}_p, \bar{n}'_p} CD_p \right] \qquad (14)$$

and

$$CD_{abs} = \frac{1}{2} Nk^{-1} \left[\left(\frac{\partial \sigma_{abs}}{\partial n_p} \right)_{\bar{n}_p, \bar{n}'_p} ORD_p + \left(\frac{\partial \sigma_{abs}}{\partial n'_p} \right)_{\bar{n}_p, \bar{n}'_p} CD_p \right] \qquad (15)$$

where the subscripts $\bar{n}_p, \bar{n}'_p$ indicate that the differentials of cross section with respect to real and imaginary parts of particle refractive index are evaluated at the average particle refractive index.

The beauty and utility of Eqs. (13)–(15) lie in their simple and explicit functional dependence of suspension circular dichroism on particle CD and ORD. Using these equations, it will be possible to predict the general nature of the spectral distortions, i.e., the direction of band shifts and intensity changes, from simply a scattering curve (cross section vs. particle refractive index) and the signs of the intrinsic particle CD and ORD. It will be possible to recognize particle CD versus ORD effects in the suspension spectra and their respective roles in scattering and absorption flattening. Such application of these equations will be outlined in the next section.

Practical Application

General. The application of the two approaches sketched above to the interpretation of circular dichroism spectra of scattering systems follows in a straightforward manner. In the first approach one simply calculates the extinction (scattering plus absorption) for left and right circular polarizations separately, using the intrinsic particle optical properties, and the difference gives the distorted suspension CD. This method is conveniently applied to detailed numerical computation of suspension optical activity spectra. The second approach via Eq. (13) requires a knowledge of the first derivative of the total extinction cross section with particle refractive index, which will be difficult to evaluate analytically for some of the more complicated large particle scattering functions, such as the Mie equations. Thus, although these derivatives can be quantitatively determined either graphically or by numerical methods, the advantage of the series expansion approach is the ease with which it may be applied to the qualitative analysis of the spectral distortions.

For the application of either method there are several things which must be known. First one needs to define the scattering particle by its real refractive index, absorption coefficient (imaginary refractive index), shape, and ratio of its size to wavelength. For the visible region of the spectrum, the above data can usually be determined. However, in the ultraviolet region, which is of interest for proteins and nucleic acids, experimental values of real refractive index are generally not available. One can, however, make reasonable estimates by extrapolation of refractive index dispersion data from the visible to the UV or by applying the Kramers-Kronig transforms to the experimental absorption spectra. One can also try a reasonable range of values for real refractive index and see whether the experimentally observed suspension spectra can be accounted for.

Next one needs the appropriate scattering functions for the given system, i.e., the explicit dependence of the cross sections on particle size, complex refractive index, and wavelength. Such scattering functions are readily available for a variety of particle geometries and refractive indices[8,9] including the large, absorbing particles one is likely to encounter in biological suspensions. Indeed, Latimer and co-workers[11] have pioneered in the application of scattering and extinction calculations to an assortment of turbid biological systems including cellular suspensions (albeit in the visible region of the spectrum). Their excellent agreement with experimental data is highly encouraging for the present extension to optical activity spectra.

Finally, for the calculation of suspension optical activity, one needs the intrinsic CD and ORD of the particle, i.e., the solution CD and ORD. For many proteins and nucleic acids, these spectra are available in the experimental literature. Thus for hemoglobin in a suspension of sphered red blood cells, the solution CD and ORD are known and a calculation of the spectral distortions is straightforward. However, there

[11] F. D. Bryant, B. A. Sieber, and P. Latimer, *Arch. Biochem. Biophys.* **135,** 97 (1969) and references therein.

are many cases of interest where it is the purpose of the experiment to determine the intrinsic molecular spectra for proteins or nucleic acids in some larger biological structure where their conformation is unknown. For these cases, only the measured turbid suspension spectra will be available and an iterative trial procedure can be used to determine the intrinsic ORD and CD. Thus one would assume, for example, a given protein secondary structure (% helix, β-sheet and random coil) which would specify the intrinsic optical activity spectra.[12,13] This would then be used for the calculation of the turbid suspension spectrum. When a reasonable fit to the anomalous suspension spectra is achieved, a good approximation to intrinsic solution spectra will have been obtained.

In some experiments one may wish to know whether a new and unusually looking spectrum is due to some new macromolecular structure or is simply the effect of turbidity. For such cases one can try existing solution spectra for known biopolymer structures in the above scattering calculations and see whether the anomalous suspension spectrum can be obtained. If it can, then one might reasonably suspect the scattering and particulate nature of the suspension to be the source of the unusual spectrum rather than some new macromolecular conformation. However, before such conclusions can be reached, one must be sure that the model scattering functions and refractive index used in the calculation are appropriate to the specific biological particle under consideration.

In the remainder of this section illustrative examples are given of both methods of analysis outlined above. In addition it will be instructive to show a typical scattering curve (cross section vs. refractive index) for the large particles one is likely to encounter in biological suspensions. The separation of scattering and absorption contributions to the spectra will be shown.

Application of Series Expansion Method to a Suspension of Hemoglobin Inside of Sphered Red Blood Cells. We have mentioned that the series expansion method is capable of predicting the nature of the spectral shifts and intensity changes in scattering suspensions. We now demonstrate this capability through the application of Eqs. (13)–(15) to a suspension of spherically symmetric particles. We note in these equations that the expansion coefficients are the derivatives of the cross-sections with respect to particle refractive index. These can be evaluated either by analytical or numerical differentiation of the scattering functions or more simply by taking the slopes from a plot of cross section

[12] B. Greenfield and G. D. Fasman, *Biochemistry* 8, 4108 (1969).
[13] Y.-H. Chen, J. T. Yang, and H. Martinez, *Biochemistry* 11, 4120 (1972).

versus refractive index. Although the series expansion approach can be used for numerical calculations of suspension optical activity spectra, such calculations are best reserved for cases where the scattering functions may be easily differentiated. The use of the method here will be to illustrate how the qualitative nature of the spectral distortions can be resolved.

We consider a turbid suspension of sphered and partially hemolyzed red blood cells. Circular dichroism measurements have been made on such a suspension at various degrees of hemolysis[1] and provide a clear illustration of the effects of turbidity on the spectrum of a well characterized alpha-helical protein. The experimental spectra are shown in Fig. 1, including a solution spectrum (curve A) for comparison. The question now is whether one can account for the characteristic distortions in the suspension spectra, namely the red shift in crossover wavelength near 203 nm and the decreased intensity of the 210 and 222 negative troughs, with the 210 trough showing the greater effect.

A suspension of large absorbing spheres with refractive index near that of the medium is taken as a crude model of our sphered and partially hemolyzed red blood cells. Specifically we consider a suspension of aqueous hemoglobin droplets of diameter 6.6 μ and hemoglobin concen-

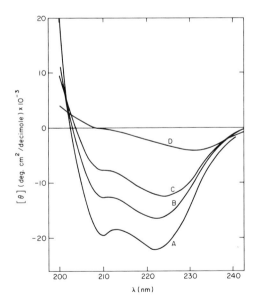

Fig. 1. Circular dichroism of hemoglobin in aqueous solution (curve A) versus inside of sphered red blood cells at different degrees of hemolysis (curves B–D). Curve D represents the least hemolyzed cells.

tration 0.3 mole of peptide/per liter. The size of the spheres and the fact that the particle refractive index is near that of the aqueous medium permits the use of a relatively simple limiting case of the Mie equations, namely the Van de Hulst anomalous diffraction equations for calculating the cross sections.[8]

The plot of total, scattering, and absorption cross-sections vs. refractive index of the particle relative to that of the medium is shown in Fig. 2. These curves illustrate the unusual scattering properties of large particles (compared to the familiar small-particle Rayleigh scattering). Because of the interference of light diffracted from the edges of the particle with that transmitted, the amount of light lost due to scattering will oscillate with increasing real refractive index, and this is seen in the behavior of the scattering and extinction cross sections in Fig. 2. We also note that the absorption cross section is independent of real refractive index.

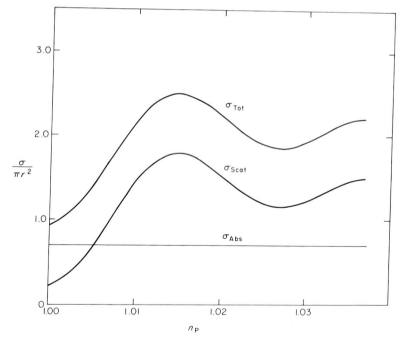

FIG. 2. Normalized cross sections versus real refractive index of the particle relative to that of the medium for aqueous hemoglobin droplet. The anomalous diffraction equations of H. C. Van de Hulst ("Light Scattering by Small Particles," Wiley, New York, 1957) were used for a sphere of $3.3\,\mu$ radius, wavelength of 202.5 nm, imaginary refractive index (relative to the medium) of 0.0036, and hemoglobin concentration of 0.3 mole of peptide per liter.

For the purpose of determining the effects of turbidity on circular dichroism spectra, it is necessary to determine the location of a given biological scattering particle on curves such as those in Fig. 2 and then to use the slope at that point in Eqs. (13)–(15). For our aqueous hemoglobin spheres we estimate the relative refractive index to be less than 1.01, which would be in the initial rising portion of scattering and extinction cross-section curves. Thus at 202.5 nm, where the solution CD spectrum has a crossover point ($CD_p = 0$), the differentials of the cross sections with real particle refractive index are:

$$\frac{\partial \sigma_T}{\partial n_p} = +; \quad \frac{\partial \sigma_{scat}}{\partial n_p} = +; \quad \frac{\partial \sigma_{abs}}{\partial n_p} = 0$$

Since the intrinsic particle rotation, ORD_p, is known to be positive at this wavelength, we would now predict from Eqs. (13)–(15) that CD_{susp} would be positive due to a scattering contribution, whereas the solution circular dichroism, CD_p, at this wavelength is zero. Consequently the increase in positive CD intensity will result in a red shift in crossover wavelength, a prediction readily verified by the exprimental red cell spectra in Fig. 1. Since the absorption cross section is independent of real refractive index and, at the crossover point, $CD_p = 0$, both terms in Eq. (15) will be zero. Thus the crossover red shift and increase in CD intensity can be resolved as a pure scattering contribution deriving from the particle ORD in Eq. (14). Similar analyses may be performed at other wavelengths and might especially be suitable where $ORD_p = 0$, and at peaks and troughs in the intrinsic CD spectrum. Thus one can show that the negative 210 nm CD trough in the hemoglobin solution spectrum will be decreased by both absorption flattening effects and a positive scattering contribution, the latter being determined by the signs and relative magnitudes of the terms in Eq. (14). At 222 nm, however, the loss of CD intensity will not be as great, owing in part to the opposite sign of the CD_{scat} contribution at this wavelength. Consequently, a distortion in the shape of the characteristic double trough in the CD spectrum of alpha-helical proteins should be observed with the 210 nm trough showing a greater loss of intensity than the 222 nm trough. These distortions can be seen in the experimental spectra of Fig. 1.

For cases where the absorption cross section is approximately independent of real refractive index, the first term of Eq. (15) will be very small or zero. The absorption flattening effect on circular dichroism, which can be described in terms of a flattening coefficient, CD_{abs}/CD_p, will then be directly related to the derivative of the absorption cross section with respect to the imaginary part of refractive index, $\partial \sigma_{abs}/\partial n'_p$.

Since the slope of the curve of σ_{abs} vs. n'_p (not shown here, see Gordon[14]) decreases with increasing n'_p, the flattening coefficient, CD_{abs}/CD_p, will decrease with increasing particle absorption. This is just the absorption flattening effect on the suspension circular dichroism.

Another case of interest arises when the scattering cross section is approximately independent of the imaginary part of refractive index. This will make the second term of Eq. (14) small and the scattering contribution to circular dichroism, CD_{scat}, might then be expected to follow the ORD_p, i.e., to resemble a solution ORD spectrum. Gordon[14] has demonstrated such a case for cell membrane suspensions using a spherical shell scattering model and Eqs. (13)–(15) above.

Separate Scattering Calculations for Left and Right Circular Polarizations: Application to Cell Membrane Suspension. The method of separate scattering calculations for left and right circularly polarized light is an approach suitable for detailed numerical computations of suspension optical activity spectra including the resolution of scattering and absorption effects. In the interest of clarity and for those not familiar with the details of such calculations, we list a simple stepwise procedure for computing the circular dichroism of a turbid suspension:

1. Define the geometry and mean complex refractive index of the scattering particle relative to that of the medium.

2. Select the appropriate model scattering function according to the particle geometry and relative refactive index.[8,9,11]

3. (a) If the intrinsic solution CD_p and ORD_p are known, as for hemoglobin inside of sphered red blood cells, use these to determine the circularly polarized refractive indices of the particle as in Eqs. (6) and (7). (b) If the CD_p, and ORD_p, are unknown, as for cell membranes, assume trial spectra and calculate the circularly polarized refractive indices as in 3(a).

4. Perform separate calculations of the extinction for the opposite states of circular polarization, using the scattering function from step 2, and take the difference of the two results to obtain the suspension circular dichroism, as in Eqs. (8) and (10).

5. For cases where the intrinsic solution optical activity, CD_p and ORD_p, are unknown, repeat step 3(b) and 4 until a good fit to the experimental suspension CD spectrum is obtained. The trial CD_p and ORD_p may then be assumed to be a reasonable estimate of the solution optical activity spectra.

6. If a breakdown of the total extinction cross section into scattering and absorption cross sections is available for the particulr scat-

[14] D. J. Gordon, *Biochemistry* 11, 413 (1972).

tering model under consideration, then a further resolution of the suspension CD spectra will be possible, via Eqs. (11) and (12).

In what is probably the most advanced numerical calculation to date of turbid suspension optical activity, Gordon and Holzwarth[15] and Gordon[14] have applied the Mie scattering functions to suspensions of spherically symmetric scattering particles using the above procedure (excluding step 5). Their applications include a spherical shell model for cell membrane suspensions and a solid sphere model for polyglutamic acid aggregrates. Their results are highly encouraging and serve to illustrate the theory[10] and interpretational methods discussed above. We will describe their calculation for a suspension of red cell membranes.

The spherical shell used as a model of the red cell ghost has a radius of 3.5 μ and a shell thickness of 70 Å. The real refractive index of the shell material (membrane) was estimated to be 1.2 times that of the aqueous medium which was inside as well as outside the shell; refractive index dispersion was neglected. The intrinsic optical properties of the shell material, i.e., n'_p, ORD_p, and CD_p, were assumed to be those measured for membranes solubilized in 0.1% sodium dodecyl sulfate solutions, an assumption we will have more to say about later. The membrane protein in such detergent solution had an alpha-helix content of about 40%. The scattering functions appropriate to the geometry and intrinsic optical properties of such a scattering model were derived by Aden and Kerker[16] and are discussed in Kerker's book.[9] Using the intrinsic particle optical rotation and circular dichroism, ORD_p and CD_p, the complex refractive indices of the particle for right and left circularly polarized light were calculated. These were then employed in the separate calculation of the extinction (absorption + scattering) for the two states of circular polarization. The difference then gave the suspension circular dichroism spectrum.

The results of the above calculation are compared in Fig. 3 with the experimental suspension and solubilized membrane solution CD spectra. The characteristic distortions in the membrane suspension spectrum relative to the solution spectrum are seen to be reasonably well accounted for by the calculated suspension CD: the 3–5 nm red shift in crossover and trough wavelengths and the disproportionately decreased amplitude of the 208 nm trough relative to the 222 nm trough. Quantitative agreement is achieved for the size of the red shifts and approximate agreement for the change in CD intensity. The intensity agreement is poorest near the 224 nm trough, and this may reflect inaccuracies in

[15] D. J. Gordon and G. Holzwarth, *Proc. Nat. Acad. Sci. U.S.* **68**, 2365 (1971).
[16] A. L. Aden and M. Kerker, *J. Appl. Phys.* **22**, 1242 (1951).

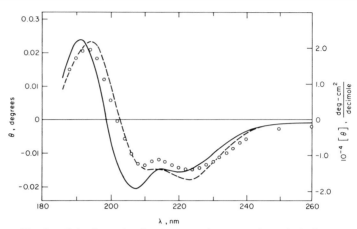

Fig. 3. Circular dichroism of red cell ghosts in suspension (dashed curve) versus solubilized in 0.1% sodium dodecyl sulfate solution (solid curve). Solution spectrum was used to obtain calculated suspension spectrum (open circles). From D. J. Gordon and G. Holzwarth, *Proc. Nat. Acad. Sci. U.S.* **68**, 2365 (1971).

the assumption that the intrinsic membrane protein optical properties could be represented by those of detergent solubilized membrane proteins with a 40% helix content. It might have been interesting to repeat the calculation with several trial optical activity spectra for the particle (step 5 of the above procedure), including higher helix contents, to see whether closer fits to the membrane suspension spectra could be obtained. In general, however, the overall calculated results simulate the main features of the experimental membrane suspension spectrum and adequately serve to illustrate the above interpretational approach.

The same calculation[15] was also applied to the suspension optical rotatory dispersion, and the results are shown in Fig. 4. An impressive quantitative agreement is obtained between calculated and experimental suspension ORD spectra which further confirms the present method of analysis.

The calculated suspension CD spectrum has been resolved into scattering and absorption contributions, and the results are shown in Fig. 5. The scattering contribution is seen to resemble an ORD spectrum, a result readily understood with the aid of the first term of our series expansion, Eq. (14) above, and further discussed by Gordon.[14]

Present State of the Art

The general approach taken here for the interpretation of the anomalous optical activity spectra of scattering suspensions has been

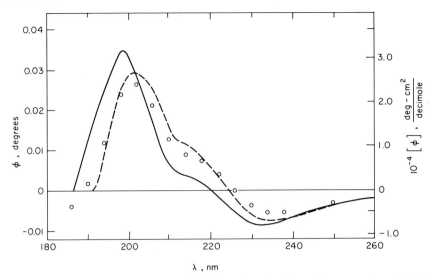

FIG. 4. Optical rotatory dispersion of red cell ghosts in suspension (dashed curve) versus solubilized in 0.1% sodium dodecyl sulfate solution (solid curve). Calculated suspension spectra (open circles). From D. J. Gordon and G. Holzwarth, *Proc. Nat. Acad. Sci. U.S.* **68**, 2365 (1971).

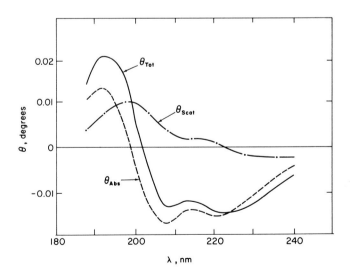

FIG. 5. Resolution of calculated circular dichroism spectrum of red cell ghost suspension into scattering (θ_{scat}) and absorption (θ_{abs}) contributions. From D. J. Gordon and G. Holzwarth, *Proc. Nat. Acad. Sci. U.S.* **68**, 2365 (1971).

to treat scattering and absorption as part of a single total extinction cross section (scattering function) for absorbing particles according to the standard methods of classical scattering theory.[8,17] Such a total cross section is dependent on the complex particle refractive index, i.e., on real refractive index and absorption, and gives both the scattering and absorption flattening effects via its relation to suspension optical properties. The extension of classical scattering theory to CD and ORD spectra requires a consideration of the dependence of scattering on circular polarization which we have previously described in terms of the symmetry properties of the scattering amplitude matrices.[10] Such considerations then formed the basis for the calculational methods described in this chapter.

There have been several recent calculational attempts to interpret the distorted CD and ORD of scattering systems, and these have been aimed primarily at the alpha-helical proteins of biological membranes and polypeptide suspensions. Some workers have calculated only the absorption flattening effects,[5,18] others have used a somewhat involved semi-empirical procedure,[19] and calculations have also appeared which use the methods described in this chapter.[14,15] The most advanced of these are the recent calculations of Gordon and Holzwarth[15] and Gordon.[14] These authors have used the Mie scattering functions with their full dependence on the complex particle optical properties and have thus accounted for the absorption flattening and scattering in a consistent manner. Their results have a firm basis in classical scattering theory and good agreement is achieved with experimental membrane and polypeptide suspension spectra.

We thus conclude that one may reasonably hope to resolve the characteristic spectral shifts, intensity changes and distorted band shapes due to scattering and absorption flattening in the CD and ORD of turbid biological suspensions. For cases where the intrinsic CD_p and ORD_p are unknown, existing solution spectra can be tried in an iterative type calculation until the best fit to the experimental suspension spectra is obtained. It may still be premature, however, to expect the calculations to yield precise quantitative simulation of experimental suspension spectra. One reason for this is the current lack of ultraviolet refractive index data for most biological systems of interest. Another is that physically realistic model scattering functions may not always be avail-

[17] R. G. Newton, "Scattering Theory of Waves and Particles," Part I. McGraw-Hill, New York, 1966.
[18] D. J. Gordon and G. M. Holzwarth, Arch. Biochem. Biophys. 142, 481 (1971).
[19] D. W. Urry, Biochim. Biophys. Acta 265, 115 (1972).

able for the more unusual geometries of some biological particles. Nevertheless, good approximations can often be made as demonstrated by the excellent results of Latimer and co-workers[11] in their scattering calculations for some highly complex turbid suspensions (yeast cells, chloroplasts, bacteria, etc.). The meaningful extension of circular dichroism and optical rotatory dispersion methods to a host of interesting biological structures now looks promising.

[30] UV Absorption and Circular Dichroism Measurements on Light Scattering Biological Specimens; Fluorescent Cell and Related Large-Angle Light Detection Techniques

By BURTON P. DORMAN, JOHN E. HEARST, and
MARCOS F. MAESTRE

This chapter describes a set of new instrumental concepts designed to obtain ultraviolet absorption and circular dichroism (CD) spectra of intensely scattering specimens, such as suspensions of virus, chromatin, chromosomes, nuclei, cells, and numerous other biological structures. The need for special techniques arises in connection with sample particles whose characteristic dimensions are within about one order of magnitude of the measuring light wavelength. Suspensions of such large particles will be likely to scatter light outside the detection region of a conventional spectrophotometer or circular dichrograph. In absorption measurements the scattering will add to the true absorption of the sample. The scattering intensity and its angular distribution can be expected to vary with wavelength. The strength of this wavelength dependence will be a complicated function of the scattering particle geometry and may exhibit substantial anomalies within an absorption band. Because optically active particles may scatter left and right circularly polarized light with different efficiency, serious distortions may also occur in the differential absorption measurement which constitutes CD.

The most common instrumental means of eliminating scattering distortion in absorption measurements has been to enlarge the solid angle of detection for light-scattering samples. For example, Shibata, Benson, and Calvin[1] placed a thin sheet of opalescent glass just after the sample cuvette to diffuse transmitted light and to collect the (diffuse)

---

[1] K. Shibata, A. A. Benson, and M. Calvin, *Biochim. Biophys. Acta* **15**, 461 (1954).

light scattered by the sample through angles approaching 90°. This technique was applicable in the visible, but not in the ultraviolet below 315 nm, where opal glass absorbs strongly. Amesz, Duysens, and Brandt,[2] by replacing opal glass with a thin vessel of fluorescing solution, extended the diffuser method to the ultraviolet region between 350 and 220 nm. An earlier but related device, the diffusely reflecting integrating sphere, has in recent years been used only occasionally or for special applications.[3]

The diffuser techniques are now largely supplanted by commercial instruments adapted to use end-window photomultiplier detectors close behind the sample holders. Such instruments are intended to collect all light scattered through angles up to 90°, and they may substantially reduce the effects of scatter. However, the maximum scatter angle collected by a diffuser or photomultiplier located behind the sample container must fall below 90° for scatter events occurring at any distance from the detector. A two-inch diameter photocathode subtends only about 68° from a point a centimeter away on its axis. Moreover, the scattered light intensity which misses such a detector increases with the scattering coefficient, because a larger fraction of scattering events will occur farther from the detector. Thus despite the use of such devices, many biologically interesting samples scatter light so intensely that details of their absorption spectra have remained obscure. Numerous theoretical and experimental treatments of this problem are detailed in reviews by Beaven,[4] Butler,[5] and Kratohvil.[6]

The new approach developed in our laboratory involves the use of a fluorescent scattering (fluorscat) cell. It is in principle an extension of the fluorescent method of Amesz et al.[2] to the case of a fluorescent solution which surrounds the sample cell. In particular, a standard strain-free fused silica cylindrical cuvette is mounted within a larger cylindrical Pyrex cell so that the axes of the two cylinders are coincident. The Pyrex cell holds a scintillator solution which surrounds the sample cuvette completely, except that the incident light beam has access to the cuvette entrance window without traversing the scintillator solution. The cell design is such that *all* incident beam intensity which is either transmitted or scattered through the sample cell, except light scattered back out the entrance path, is absorbed in the scintillator solution. The

[2] J. Amesz, L. N. M. Duysens, and D. C. Brandt, *J. Theor. Biol.* **1**, 59 (1961).
[3] E. Gratton, *Biopolymers* **10**, 2629 (1971).
[4] G. H. Beaven, *in* "Advances in Spectroscopy," Vol. II (H. W. Thompson, ed.), p. 339, Interscience, New York.
[5] W. L. Butler, *Annu. Rev. Plant Physiol.* **15**, 451 (1964).
[6] J. P. Kratohvil, *Anal. Chem.* **38**, 517R (1966).

scintillator molecule, tumbling rapidly in solution, may emit its fluorescence in any direction with equal probability. As a result, the fluorescent emission exhibits equal intensity in all directions. Thus a photodetector will receive a fluorescence intensity proportional to the total number of measuring beam photons transmitted or scattered into the scintillator solution. If the photodetector is at a large distance compared to the cell dimensions, the fluorescence intensity it receives will be, to a first approximation, independent of the angular scattering distribution produced by particles in the sample cell.

As all other devices for experimental light scattering corrections, the fluorescent scattering (fluorscat) cell has the effect of increasing the solid angle of detection so that scattered light will not miss the detector and register as absorption. In so doing, the scattered photons are weighted identically with the transmitted photons, even though the effective optical path for the scattered light may differ significantly from the geometrical sample path length traversed by transmitted photons. Without a correction for this effect, the spectra of scattering and nonscattering samples are not strictly comparable. But in practice the fluorescent cell calibration spectra we have measured for an absorber in the presence of nonabsorbing scattering particles show only small variation in the effective optical path. And the spectra of sample particles which both absorb and scatter generally exhibit considerably more structure than those measured with other instrumental techniques. The spectra of nonscattering samples (e.g., DNA) are accurately reproduced in our device.

Extension of the fluorescent cell concept to circular dichroism is in principle straightforward. The CD photomultiplier ideally measures only light intensity and is insensitive to its polarization. Therefore, if a scattering sample produces a different CD spectrum in the fluorescence cell than in the standard cuvette, the difference necessarily implies that the fluor has detected light of different intensity for the two circular polarizations. Thus the difference spectrum is a direct measure of the differential scattering[7,8] into a region of space around the sample cuvette where measuring photons are detectable by the fluorescence cell but not by the conventional CD instrument.

UV absorption and circular dichroism spectra for typical light scattering samples in both standard cuvettes and fluorescent scattering cells are shown by way of example in Figs 1A and 1B.

The remainder of this discussion treats the construction, calibration

[7] D. W. Urry and T. H. Ji, *Arch. Biochem. Biophys.* **128**, 802 (1968).
[8] D. J. Gordon, *Biochemistry* **11**, 413 (1972).

and use of the fluorescent scattering cells in absorption measurements, the instrumental considerations influencing their use in circular dichroism including some complementary large-angle detection techniques, and examples of typical CD applications.

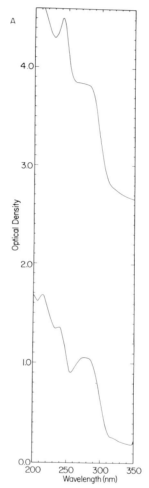

FIG. 1. (A) Cary 15 UV absorption spectrum of cow's milk diluted 25 times by water. Sample path length of 1 cm for both standard cuvette (upper curve) and fluorscat cell (lower curve). The fluorscat cell has reduced light scattering contribution to the optical density by almost an order of magnitude. (B) Intact T2 bacteriophage circular dichroism spectra recorded in Cary 60 Spectropolarimeter with CD accessory using a standard cuvette (------) and a fluorscat cell (———). The fluorscat cell has corrected for the differential light scattering, which contributes a long wavelength positive tail with a maximum at 290 nm to the standard cell spectrum.

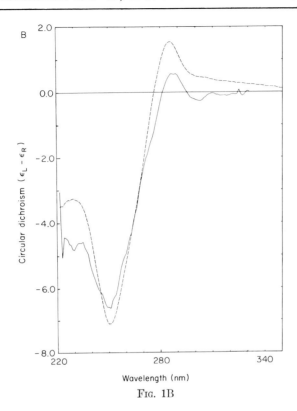

FIG. 1B

Cell Details and Construction

The fluorscat cell and its components are diagrammed in Fig. 2. Three different fluorscat cell configurations are depicted. The small-necked FS cell used in the Cary 1462 scattered transmission accessory is shown assembled (Fig. 2a) and in an exploded view (Fig. 2b). Also shown are the neckless PMF cell used primarily in the Cary 15 absorption spectrophotometer (Fig. 2c) and the large-necked CDF cell used for Cary 60 CD measurements (Fig. 2d). The two latter models incorporate small but significant geometrical modifications necessary for the specific applications to be discussed below. The term "fluorscat" will be applied interchangeably to fluorescent scattering cells of all three types, and the designation FS, PMF, or CDF will be used when it is desired to call attention to a specific cell configuration.

The sample cuvette built into all fluorscat cells is a commercially available 1-cm optical path length cylindrical cell with windows and body of low-fluorescence fused silica, such as Suprasil (Amersil, Inc.). Cuvettes of different path lengths are equally feasible. All other cell components are of Pyrex (Corning Glass Works). The cylindrical body is

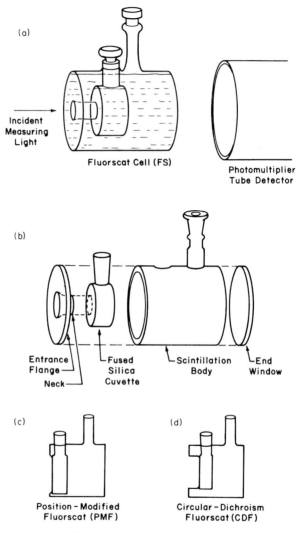

Fig. 2. Fluorscat cells. Three distinct design configurations are shown; FS cell with 8 mm diameter entrance neck in assembled view (a) and exploded view (b); position modified fluorscat (PMF) cell (c) with sample cuvette bonded to entrance flange but no intervening entrance neck; circular dichroism fluorscat (CDF) cell (d) with 19 mm diameter entrance neck. All cells contain 1 cm path length sample cuvette.

38 mm o.d. tubing cut to 38 mm length. The end window is cut from 1/16-inch plate stock with a 38-mm circle cutter and fused onto the body. A standard taper 5/12 outer ground joint is fused around a fill hole located ½-inch from the end window. A 13 mm diameter hole for

the cuvette neck is cut next to the fill hole so that its center is 12 mm from the entrance neck end of the body. The entrance neck is made from tubing just large enough to pass the measuring beam at maximum monochromator slit settings. It is feasible to use 8 mm i.d. tubing for FS cells, but 19 mm i.d. tubing is employed for the entrance neck of CDF cells. The entrance neck flange is formed in a lathe by flaring the entrance neck tubing with a carbon rod. The flared flange section is brought to 37 mm diameter or a dimension just smaller than the body diameter. The cylindrical neck is then cut to length, ground square and brought flush against one window of the cuvette. The cuvette neck must line up with the appropriate hole in the body and the cuvette windows must be normal to the axis of the fluorscat cell body. With conventional 1-cm path length cuvettes, the entrance neck will extend 5.5 mm into the cylindrical body. The *cuvette* neck may have to be shortened before it will pass through the hole in the body, depending on the cuvette manufacturer and model.

At this point all components are cleaned scrupulously. The remaining assembly sequence involves (1) bonding the entrance neck to the cuvette, (2) inserting the cuvette neck through the hole in the body, bringing the flange flush against the body, and aligning the cuvette windows normal to the axis of the body, and (3) bonding the flange to the body. The bonding is done with Silastic 892 RTV silicon polymer (Dow Corning), which has been found to withstand prolonged exposure to ethanol better than other agents. It is convenient to apply the polymer from a 5-ml disposable syringe to the outside of the entrance neck so as not to reduce the clearance for passage of the light beam. The alignment has been found to be rather uncritical for absorption applications but very critical for reproducible CD performance. Cells to be used in circular dichroism studies are aligned with the use of a special jig, optical bench, and ruby laser. When satisfactory alignment is evidenced by coincidence of the laser beam and its reflection off the cuvette entrance window, silicon polymer is applied to the perimeter of the flange where it contacts the cell body.

The fluorescent solution of choice is $0.1 M$ reagent grade sodium salicylate (Allied Chemical) in 100% ethanol (Commercial Solvents Corporation). This system absorbs strongly at wavelengths between 70 nm and 350 nm, fluoresces at 90–95% quantum efficiency with an emission maximum above 4000 Å in the blue visible,[9] and exhibits essentially no self absorption due to an unusually large separation between the absorption and fluorescence bands (Stokes loss equals 4890 cm^{-1}).[10]

[9] R. Allison, J. Burns, and A. J. Tuzzolino, *J. Opt. Soc. Amer.* **54**, 747 (1964).
[10] I. B. Berlman, "Handbook of Fluorescence Spectra of Aromatic Molecules." Academic Press, New York, 1965.

No improvement in fluorescence output intensity has been discerned after deaeration of the solution with oxygen-free nitrogen. At the 0.1 M concentration used, absorptivity of the scintillator solution is greater than or equal to 3 OD per millimeter (!) at all wavelengths below 330 nm.[10] Thus effectively all of the incident UV measuring beam is absorbed after a short penetration into the scintillator solution. The Pyrex cell body acts as an additional UV filter at wavelengths below 280 nm. Elaborate testing of the salicylate–ethanol system has confirmed that only visible fluorescence reaches the photomultiplier tube (PMT) detector, and that the fluorescence output is isotropic.[11] It should be mentioned that this scintillator system is restricted to wavelengths below 350 nm, corresponding to the absorption region of sodium salicylate. Although scintillator–solvent systems suitable for extending this technique to the measurement of visible light absorptions may very well exist, all candidate systems considered to date exhibit significant overlap of their emission and absorption bands. Inasmuch as the emission spectrum changes somewhat with the excitation wavelength, use of such scintillators would presumably necessitate a correction for the amount of self absorption at each excitation wavelength. All fluorescent cell work reported herein has utilized sodium salicylate in ethanol and has been done in the ultraviolet region below 350 nm.

UV Absorption Applications

Principles

The specific procedures to be described below for utilizing fluorscat cells must be understood in the context of relevant design concepts. The surrounding of the sample cuvette with scintillator solution is done as a means of capturing as much scattered intensity as possible. Scattered light or transmitted light entering the scintillator fluid will be absorbed soon after leaving the sample cuvette. Although the absorption event will require an appropriately oriented scintillator molecule, rapid tumbling of the scintillator in solution prior to emission will effectively randomize the fluorescence output. The fluorescence may be emitted in any direction with equal probability. Thus the probability that a fluorescent photon released from a given point in the scintillator solution will strike the detector is proportional to the solid angle which the detector subtends from that point. In order to obtain a detector output which is independent of the angular distribution of light scattered by

[11] B. P. Dorman, Ph.D. Thesis, University of California, Berkeley, 1972.

the sample, the detector must subtend essentially the same angle from all points serving as sources of fluorescent emission around the sample cuvette.

To achieve this condition rigorously requires an infinitely distant detector. In practice, observed spectra do not change significantly when the fluorscat cell to detector separation distance is increased beyond 2 to 3 times the detector diameter. Thus separations of 5 cm to 10 cm are found adequate for typical commercial photomultiplier tube detectors. This is a fortunate result because the more distant the photomultiplier, the smaller the fluorescent intensity it receives and the smaller the signal output it generates. Consequently, it may become necessary to operate with larger monochromator slits, resulting in poorer spectral resolution and the introduction of more stray light. Even at relatively small separations of a few centimeters, much of the fluorescence output misses the detector and typical double monochromator slit readings may range 3 to 4 times normal operating values. Such values correspond to detecting only 5% to 10% of the normal signal intensity. This situation is partially compensated by the fact most photomultipliers exhibit more sensitivity for blue visible fluorescence than for UV measuring light. Where additional improvement is required, it may be obtained by operating the photomultiplier detector at higher signal amplification settings, but only at the expense of significantly reduced signal to noise ratios.

As a further means of obtaining a signal intensity which is independent of angular scattering distribution, we have elected whenever possible to omit optics of any sort between the fluorscat cell and the photomultiplier. In this way it is hoped to minimize the probability that light emitted from certain points in the scintillator might be preferentially detected. Thus, the Cary 14 spectrophotometer affords a relatively unfavorable instrumental environment for fluorscat due to the presence of mirrors between the sample compartment and photomultiplier. In contrast, the Cary 15 spectrophotometer, the Cary 14 fitted with a model 1462 scattered transmission accessory, and the Cary 60 spectropolarimeter with CD accessory all provide the line of sight detection deemed favorable to fluorscat. These applications will be described in the ensuing discussion.

During the preliminary stages of our work, the Cary Model 1462 scattered transmission accessory was modified to test the fluorescent cell concept under most extreme conditions. The modifications permitted a photomultiplier tube detector to be positioned interchangeably either on the incident beam axis or 90° off the beam axis. The photomultiplier tube was the same distance from the center of the sample cell in both positions, and a hand-blown spherical cell was used to maximize sym-

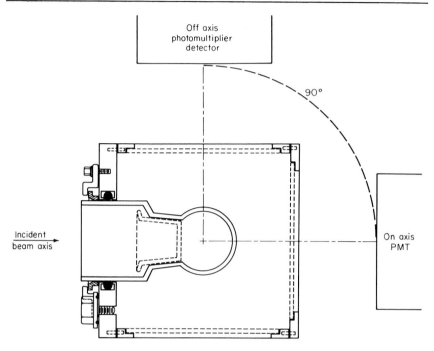

Spherical sample cell in scintillator tank

FIG. 3. Schematic side view of scintillator tank with spherical sample cell mounted through a Viton O-ring seal. Spectrophotometer measuring beam enters cell from left through an extractable transmission window. Fluorescence light exits through Pyrex windows in top, right, and bottom tank walls. The photomultiplier tube detector may be located either on the photomultiplier beam axis (right) or 90° off axis (top) at the same distance from the center of the sample cell. The handblown spherical cell was useful for testing the fluorescence cell concept (see text), but its use was discontinued because of strong internal reflections that produced deviations from Beer–Lambert law behavior. Standard commercial sample cuvettes with planar optical windows were employed for all subsequent work.

metry. The spherical cell fitted through an O-ring seal into a tank of sodium salicylate scintillator solution with Pyrex windows leading to both photomultiplier positions (Fig. 3).

Spectra obtained using the spherical fluorescence cell with either an on axis or 90° off-axis detector may be compared to the spectra obtained in a conventional Cary 14 spectrophotometer (Fig. 4). We see that the optical density measured for a light scattering suspension of alumina particles is reduced an order of magnitude by the fluorescence cell. If the alumina suspension contains an absorbing substance such as potassium dichromate, the dichromate absorption spectrum is badly distorted by the

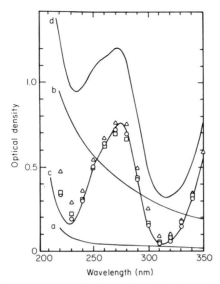

Fig. 4. Comparative spectra made in a Cary 14 absorption spectrophotometer using either a spherical fluorescence cell or a standard sample cuvette. The optical density observed for a light-scattering suspension of alumina particles is shown for the fluorescence cell (curve a) and standard cuvette (curve b). Alumina does not absorb at these wavelengths, so all nonzero values are artifacts attributable to scattering. The scattering artifacts are reduced about an order of magnitude by use of the fluorescence cell, although a part of the scattered light intensity remains uncompensated (curve a). When the alumina suspension is made 0.06 mg/ml in $K_2Cr_2O_7$, the normal dichromate absorption spectrum (curve c) is badly distorted in the standard cuvette (curve d). More accurate spectra are obtained using the fluorescent cell with the photomultiplier located either on the incident light axis (triangles) or 90° off-axis (squares). The difference (circles) between on-axis values and the uncompensated on-axis scattering (curve a) closely resembles the true dichromate spectrum (curve c).

alumina scattering in the conventional instrument. A substantially more accurate dichromate spectrum results with the fluorescence cell, and the fluorescence cell spectrum is nearly independent of whether the photomultiplier is on the axis or 90° off the axis of the measuring beam. This may be taken as dramatic confirmation that the fluorescence output is spherically symmetric and that the spectra obtained using the fluorescence cell concept will be highly insensitive to the angular distribution of light scattered from large particles in the sample suspension. Having established this point, it should be emphasized that handblown spherical cells offer far less satisfactory optical performance than precision-made commercial sample cuvettes of the sort we have employed for all subsequent work.

Cary 15

Use of fluorscat in the Cary 15 spectrophotometer requires every effort to maximize the fluorescent intensity detected. Specifically, the fluorscat cell must be placed to the far right of the sample compartment, or as near the photomultiplier compartment as possible. A matching fluorescent cell must be similarly located in the reference compartment to obtain comparable intensity for both beams. Inasmuch as the beams are diverging and occupy a greater cross-sectional area as they exit from the sample space, it is necessary to use fluorscat cells with large-diameter entrance windows. It has been found convenient to mount the sample cuvette flush with a hole in the entrance flange end of the scintillator body, effectively eliminating the entrance neck. This feature is incorporated in the "position-modified fluorscat" (PMF) cell (Fig. 2c), in which the sample cuvette is located tangent to the bottom of the scintillation cell body. Such cells may be mounted in the Cary 15 beam path by

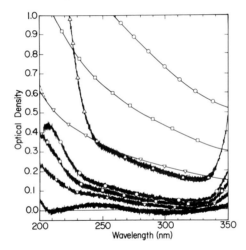

FIG. 5. Cary 15 spectra recorded for various concentrations of alumina particles in $10^{-2} N$ KOH. Both standard cuvette (———) and PMF fluorscat cell (▬▬▬) are used. Baselines are shown without plot symbols. Fluorscat spectra represent tracing of actual instrumental noise level resulting from high photomultiplier dynode setting (see text). Since alumina does not absorb in the near UV, all non-zero optical density readings are due to light scattering. Standard cell spectrum of highest concentration suspension ($\triangle$) gave $OD_{340} = 2.4$ and $OD_{250} = 3.0$ and is off scale of figure. In the fluorscat cell, the same suspension measures $OD_{340} = 0.2$ and $OD_{250} = 0.3$, a reduction of 10-fold. Spectra obtained from a diluted alumina suspension (—○—) and subsequent (approximately) 2-fold series dilutions (—□— and —▽—) are shown for both cell types. Steep curvature in the fluorscat spectra at short wavelengths reflects increased scattered light intensity. Curvature between 350 nm and 335 nm is due to long wavelength tail of the scintillator absorption band (see text).

simply removing the end mask and adding a ¾-inch extension to the standard cell mount. All Cary 15 fluorscat runs were made with PMF cells using 1 second pen constant, sensitivity = 4.7, scan rate of 5–10 Å per second, dynode = 5. For Cary 15 runs with conventional cuvettes, all instrumental settings were the same except dynode = 3.

The ability of PMF cells to correct for scattering effects in the Cary 15 was tested using alumina suspensions (Vitro Labs) of nearly spherical particles, fractionated to contain particles ranging from 0.01 μ to 0.2 μ diameter, in 10⁻² N potassium hydroxide. Results obtained for a series of alumina particle concentrations in PMF cells and standard cuvettes are shown in Fig. 5. The alumina particles do not absorb in the near UV, so all optical density is attributable to scattering. It will be seen that optical densities observed in standard cuvettes are reduced about an order of magnitude in PMF cells. One also may observe the instrumental pen

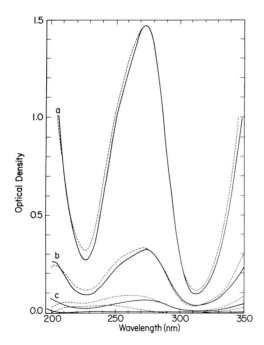

Fig. 6. Absorption Spectra. Cary 14 concentration series in potassium dichromate using standard cuvette (———) and PMF fluorscat cell (------). Most concentrated solution was 0.12 mg K₂Cr₂O₇ per milliliter of 10⁻² N KOH (curves a). Serial dilution by factor of approximately 5 (curves b and c). Two lowest curves are baselines obtained with 10⁻² N KOH in both sample and reference cells. When fluorscat spectra are corrected for curvature in the fluorscat baseline, excellent agreement is obtained between the standard cuvette and fluorscat cell spectra of a pure absorber.

noise envelope is extremely large, on the order of 0.025 optical density unit, due to the high dynode settings used with PMF cells. The PMF baseline curvature is due to imperfectly matched fluorescent cells in the sample and references beam and is typical of what may be expected when the instrument has been adjusted for a flat baseline with standard cuvettes. This amount of curvature can be corrected readily with Cary 15 multipots. The fall in optical density measured in fluorscat between 350 nm and 335 nm corresponds to the long wavelength tail of the sodium salicylate absorption band. The scintillator solution is transparent above 350 nm.

Beer–Lambert law behavior for a nonscattering sample was tested on a concentration series of potassium dichromate in 10^{-2} N potassium hydroxide (Fig. 6). The fluorscat data presented here are the mean of the observed peak-to-peak noise envelope. PMF and standard cuvette spectra are shown for a stock solution containing 0.12 mg $K_2Cr_2O_7$ per milliliter of 10^{-2} N KOH, and two successive (nominally) 5-fold dilutions. It is seen that, subject to a correction for curvature in the fluorscat baseline, excellent agreement is obtained between fluorscat and standard cuvette data.

The ability of fluorscat to retrieve the potassium dichromate absorption spectrum in the presence of alumina particle scattering is depicted in Fig. 7. The instrumental baseline and data curves for a standard cuvette are shown, and the fluorscat data after baseline subtraction are plotted relative to the standard cuvette baseline to permit close comparison. Fluorscat and standard cuvette data are presented for a 0.12 mg/ml dichromate stock solution after dilution to 0.06 mg/ml 10^{-2} N KOH. The scattering densities observed for a similarly diluted stock alumina suspension are also shown for sample cells of both types. The ability of fluorscat to correct for most of the scattering is again observed. Moreover, it also may be seen that mixing the dichromate and alumina stocks in equal parts produces significant distortion of the dichromate absorption when the mixture is measured in the standard cuvette, and a considerably more accurate dichromate spectrum in fluorscat. Inasmuch as the scattering and absorption originate from independent particles, the effects are seen to be additive. In either the fluorscat cell or the standard cuvette, the mixture of dichromate and alumina produces a spectrum which is approximately the sum of the observed optical density for the pure scatterer and the pure absorber.

In summary we have observed that the position modified fluorscat (PMF) cell used in the Cary 15 affords reasonably faithful spectra for pure absorbers, and substantially more accurate spectra than are obtained using standard cuvettes for absorbing specimens in the presence of light-scattering particles. The PMF cell is able to collect and register

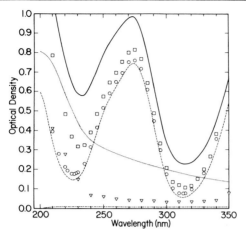

FIG. 7. Cary 14 spectra. The optical density observed for a light-scattering suspension of alumina particles is shown for a standard cuvette (......) and PMF fluorscat cell (▽). The standard cuvette absorption spectrum of 0.06 mg of potassium dichromate per milliliter of $10^{-2} N$ KOH (- - - - -) is displaced upward when alumina particles are introduced to the standard cuvette (———). A considerably more accurate dichromate spectrum is obtained with fluorscat despite the presence of the alumina (□). When the fluorscat spectrum of the scattering–absorbing mixture is corrected by the uncompensated scattering in fluorscat, the correct dichromate spectrum is nearly returned (○). All data are relative to the standard cuvette baseline (—·—·—).

about 90% of the scattered light intensity which was missed with a standard cuvette. In view of the simplicity of the PMF-Cary 15 application, this correction represents an excellent return on a very modest instrumental investment.

Cary Model 1462 Scattered Transmission Accessory

The Cary 1462 Scattered transmission accessory contains a 2-inch diameter end-window photomultiplier tube detector which can be translated along the optical axis over a distance of several centimeters. With standard sample cuvettes, a series of spectra recorded at various tube positions successively closer to the sample cell display a monotonic reduction and even elimination of spectral features due to scattering, such as the long wavelength extinction tails extending outside the absorption band. Optical densities inside the band are also reduced.

When a 1462 accessory is available for use with fluorscat, it offers some substantial advantages over the Cary 15 application. The 2-inch diameter photocathode subtends a larger angle from the fluorscat cell and collects more fluorescence. More conventional dynode settings are therefore feasible, with lower noise amplitudes as a result. In our laboratory, the Model 1462 preamplifier has been relocated to the front of the scattering accessory compartment so the photomultiplier may be positioned farther from the fluorescent cell. Data are customarily recorded with a fluorscat to detector clearance distance of 12 cm. The conventional Model 1462 rectangular cuvette holder has been replaced by a removable 2⅞ × 3¼ inches mounting platform with an outrigger leg support. Independently movable v-block fluorscat mounts separated by a vertical light mask are attached to the platform. Sample and reference beams are adjusted to parallel alignment with the conventional scatter accessory rotating mirror and lens combination. Spherical fused silica lenses (Oriel Optics Corp., Stamford, Connecticut, Cat. No. A-11-641-20; diameter = 2.5 cm; f = 5 cm) have been used to obtain a smaller than normal beam cross section. With the beam focused on the sample cuvette entrance window, it becomes feasible to utilize FS fluorscat cells with an 8 mm i.d. entrance neck, as depicted in Fig. 2a. Cells of this design in principle should be able to detect photons scattered back out the entrance window between the entrance neck and the cuvette wall.

Data recorded using fluorscat (FS) cells in the Model 1462 accessory are presented in Fig. 8 for two alumina particle concentrations, along

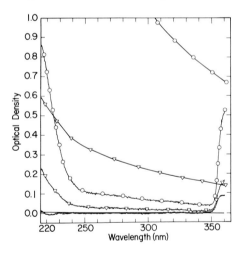

FIG. 8. Optical density due to scattering recorded for a concentrated suspension (○) of alumina particles in $10^{-2} N$ KOH and for a 5-fold dilution (▽); Cary 15 standard cuvette data (solid curves) and fluorscat spectra obtained in Cary 1462 scattered transmission accessory (noisy curves). Baselines from $10^{-2} N$ KOH in sample and reference path are shown for both cases. Cary 1462 instrumental settings for fluorscat runs were slit control = 25, dynode = 3, pen period = 1 second, scan rate = 10 Å per second. Fluorscat baseline was adjusted flat with instrument multipots.

WAVELENGTH CALIBRATION ON BENZENE VAPOR SPECTRA

Sample cell	Wavelength of absorption maxima (nm)									
Standard cuvette	236.4	237.4	241.6	242.6	247.1	248.1	252.7	253.7	259.3	260.1
Fluorscat cell	236.4	237.4	241.8	242.8	247.4	248.4	253.2	254.2	259.3	260.3

Spectral band width of monochrometer beam in Cary 14 for standard cuvettes ≅0.05 nm; for fluorscat cell ≅0.3 nm @230 nm, ≅0.4 nm @260 nm.

with Cary 15 standard cuvette spectra for the same suspensions. Baselines are obtained using 10^{-2} N KOH in both sample and reference cells. The fluorscat baseline has been straightened with Cary 14 multipots. The fluorscat data was obtained with slit control = 25, dynode = 3, pen period = 1, scan rate = 10 Å per second. The signal noise is considerably smaller than was obtained for fluorscat runs in the Cary 15. The characteristic drop of fluorscat optical density values when the salicylate begins to absorb at 350 nm is seen in Fig. 8, as it was in Fig. 5. The Cary 1462 fluorscat cell scattering density is seen to be more than an order of magnitude lower than standard cell Cary 15 data.

As a final calibration step, the ability of fluorscat to resolve the spectral location of sharp absorption maxima was calibrated on benzene vapor spectra. Wavelengths of five double maxima measured in the standard Cary 14 were reproduced by fluorscat in the Model 1462 accessory to within the spectral band width of the spectrophotometer beam (see the table).

To summarize, use of fluorscat with the end-window detector of the Cary 1462 accessory permits (1) the use of cells with smaller entrance necks allowing improved detection of scatter through angles greater than 90°, and (2) collection of more fluorescent light, leading to a low-noise output signal. Interest in these advantages may be offset by the likelihood that the scattered transmission accessory is unavailable in most laboratories.

Light Scattering Effects in Circular Dichroism

Principles

Circular dichroism constitutes the difference in absorption coefficient for left and right circularly polarized light. The CD can be obtained by measuring the ratio of the difference in transmission to the average transmission of a sample, $(T_R - T_L)/(T_R + T_L)/2$, for the two circular polarizations. The difference, if any, is presumed to arise from preferential absorption of one polarization or the other. All commercially available instruments determine circular dichroism by exposing the sample to a measuring beam which alternates in time from one sense of circular polarization to the other. The beam intensity and the photomultiplier response are presumed to be independent of the polarization state. Thus, if sample transmission is the same for both polarizations, a constant (dc) photomultiplier output current will be obtained whose magnitude is proportional to the sample transmission. If sample transmission is slightly different for the two polarizations, the intensity of the detected beam will vary with the polarization, and a small ac component of magnitude proportional to the difference in sample transmission will be superimposed upon the dc photomultiplier output current. The CD is taken to be proportional to the ratio of the ac component to the average dc signal current. Rather standard electronic circuitry enables measurement of ac to dc ratios on the order of 10^{-4}, corresponding to an absorbance difference, $A_L - A_R$, of 5×10^{-5} and to ellipticity of 1.5 millidegrees.

Should a specimen scatter light in equal intensity for both circular polarizations at angles large enough to miss the detector, that light will appear to have been absorbed, the transmission will drop, but T_L and T_R will be changed in the same proportion and the ratio which constitutes CD will be unchanged. However, for a scattering particle which is optically active, the angular distribution of scattered light, i.e., its relative intensity as a function of scattering angle, may not be the same for left and right circularly polarized measuring light.[7,8,12] Thus the light intensity scattered outside of the solid angle subtended by the detector and the scattered intensity which is detected may also vary with the sense of circular polarization. This phenomenon has been termed differential scattering.[13] The difference in detected scattered intensity will be added to the difference in transmitted intensity and will distort the correct circular dichroism value. Inasmuch as the measured intensity of scattered radiation depends upon the solid angle of detection, the observed circular dichroism of optically active light scattering suspensions may be found to vary with the experimental geometry employed—the optical path length and the relative dimensions of beam cross section, cell diameter, and the position and size of the light detection apparatus. Under these circumstances, structural conclusions based upon circular dichroism data obtained from particulate suspensions in conventional CD spectropolarimeters may be unreliable or meaningless. At the very least, the possibility of light scattering contributions must complicate the interpretation of CD data from scattering specimens and the comparison with spectra from molecularly dispersed samples of the same chemical composition.

The use of fluorscat represents one way of changing the instrumental detection geometry. If, as a result of collecting light scattered through large angles, the fluorscat spectrum is different from that obtained with a standard cuvette, the difference constitutes a straightforward assay for the presence of differential light scattering. Our experience with unpolarized light absorption spectra encourages us to believe that fluorscat would produce a more accurate spectrum since it treats the scattering

[12] A. S. Schneider, *Chem. Phys. Lett.* **8**, 604 (1971).
[13] D. W. Urry and J. Krivacic, *Proc. Nat. Acad. Sci. U.S.* **65**, 845 (1970).

as transmission rather than absorption. But the fluorscat cell and the standard cuvette represent only two of many possible detection geometries. A better understanding of the differential light scattering phenomenon should result from studying the dependence of the CD upon systematic variation of the solid angle of detection. Variable detection geometry can be attained by employing an end-window photomultiplier tube detector which can be positioned on the optical axis of a CD instrument at various distances from a standard sample cell.

The advantages of using a large end-window photomultiplier with fluorscat cells have been discussed above. The large photocathode surface collects more fluorescent light and can be operated at low gain, low noise conditions. But end-window photomultipliers possess another important advantage independent of fluorescent cell applications. The large photocathode can be expected to offer much more homogeneous response than corresponding side-window models. That is, the output current per unit of incident intensity is substantially independent of position on the photocathode of an end-window tube, but is often a sensitive function of position for a side-window tube. Homogeneous cathode response permits a far less critical optical alignment procedure for an end-window tube, advantage of which has been taken by installing end-window units in the latest commercial circular dichrographs, such as the Cary Model 61 and the Jasco Model J-20. Homogeneous response becomes an essential requirement for measurements on light-scattering samples which distribute significant intensity across large portions of the photocathode. And it becomes a crucial characteristic if we wish to compare spectra obtained by moving the photomultiplier to various positions along the optical axis, where each position exposes the photocathode to a beam cross section of different dimensions.

To implement and test these concepts, a Cary 60 spectropolarimeter with Model 6001 CD accessory has been modified to accommodate a movable end-window photomultiplier tube detector. Installation, calibration, and use of the modified instrumentation to measure the circular dichroism of light-scattering specimens are detailed in the following sections.

Cary 6001 Modification

1. The ORD modulator, focusing lens, side-window photomultiplier, and the light mask separating the sample elevator and ORD modulator compartments were removed. The opening between these two compartments was widened by 1 inch. The photomultiplier preamplifier was remounted to the underside of the top plate of the elevator.

2. An end-window photomultiplier tube (PMT) was installed on a variable-position roller-bearing mount (Fig. 9) in the ORD modulator compartment. The mount is supported by a plate which can be translated horizontally and which rests in turn on a facsimile of the ORD modulator adjustable-height kinematic leveling table.

3. A positioning rod extending outside the instrument via a light lock allows the photomultiplier to be moved along the optical axis to locations 0.04 to 5.0 inches from the sample cell. The leveling provisions described above enable alignment such that the photocathode remains normal to the measuring beam and centered on the optical axis. The positioning rod is grooved at 1.5-inch intervals to allow several photomultiplier positions to be located reproducibly with negligible error. Whenever the photomultiplier extends into the sample space, the sample elevator is automatically locked in place to preclude tube damage or misalignment.

4. Shielded wiring connects the cathode, anode, and final dynode pin positions of the PMT socket to the preamplifier. The photomultiplier is magnetically shielded, and electrically shielded at cathode potential.

5. The modulator compartment cover panel has been divided to provide access to the photomultiplier without removal of the sample elevator.

6. A special v-block holder was fabricated to hold fluorscat CDF cells (Fig. 2d) in the beam path. A supplementary 8-mm thick v-block which rests on the fluorscat holder is used for standard cuvettes.

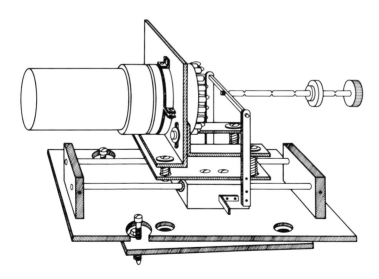

Fig. 9. End-window photomultiplier tube detector mount installed in modified Cary 60 spectropolarimeter with Model 6001 CD accessory. End-window photocathode diameter is 2 inches. See text for further description.

7. The photomultiplier tube is positioned to receive the beam image at the center of the tube face and adjusted so the beam reflection is centered around the incident beam while the detector is moved throughout its range of travel.

Calibration

Large-angle detection techniques have been employed due to our interest in assessing differential light scattering contributions from large, particulate biological specimens. Before dealing with such specimens it was necessary to persuade ourselves that fluorscat cells and/or large area photomultiplier detectors would yield correct CD spectra for nonscattering samples and would register zero CD for scattering samples which are optically inactive. Therefore, an unmodified Cary 6003 accessory on a second Cary 60 spectropolarimeter has been employed to cross check results obtained with the Cary 6001 instrument modified to use a large area end window detector. It has been possible to compare results for standard cuvettes and fluorscat cells in both instruments. All combinations of instruments, standard cuvettes, fluorscat cells, and photomultipliers were calibrated with d-10-camphorsulfonic acid (Eastman Organic), concentration 1 mg per milliliter of H_2O. In every case calibration readings of 0.304 ± 0.004 degree ellipticity at 290 nm were obtained and rechecked periodically.

Calibration tests were also performed on optically inactive samples providing extinction due to: (1) absorption: potassium dichromate solution (Mallinckrodt Chem.), 120 mg/liter $10^{-2} N$ potassium hydroxide. (2) scattering: alumina suspensions (Vitro Labs) of nearly spherical particles in the range of 0.01 μ to 0.2 μ in diameter in $10^{-2} N$ potassium hydroxide. Candidate photomultipliers were rejected if they gave readings, so-called inactive sample artifacts, greater than 1 millidegree per optical density unit for the above samples.

Inactive Sample Artifacts

Calibration testing of the modified Cary 6001 has introduced us to a troublesome measurement anomaly, the inactive sample artifact. It is manifest as an ac component in the photomultiplier tube output which produces a nonzero circular dichroism signal even in the absence of a circularly dichroic sample. When an optically inactive absorber is placed in the beam, the average beam intensity and dc current level falls but the artifactual ac current remains. Thus the ratio of the ac to dc output increases with the sample absorbance, producing a CD resembling the absorption spectrum of the sample. For this reason the anomaly may be characterized as an absorption artifact, although the same result will be observed for extinction due to light scattering. Extinction arising from a combination of scattering and absorption will lead to a cumulative artifact. It may be noted the artifact is a much larger problem for standard sample cuvettes than for fluorscat cells, and we shall return to this point below.

The inactive sample artifacts observed using standard cuvettes with a number of photomultiplier tubes are depicted in Fig. 10. All differences between the solvent baselines and the CD curves obtained for optically inactive samples in the same solvent are artifactual. Even the most favorable photomultiplier tube may be seen to exhibit an anomalous CD signal on the order of 1 millidegree per unit of sample optical density. CD spectra of samples whose actual circular dichroism is of the same order will be significantly distorted. Less distortion will be observed when sample absorbance is low or sample dichroism is high, or both.

Whereas the inactive sample artifact may be detected to various extents for all tubes, its magnitude ranges over an order of magnitude from one tube to another. The magnitude of the artifact observed for a given tube is approximately the same in the modified Cary 6001 as in an unmodified Jasco J-20 spectropolarimeter with CD accessory. Thus the photomultiplier tube is strongly implicated as the source of the artifact. Further important characteristics of the artifact may be listed: (1) An artifact of some magnitude may be generated at any wavelength throughout the range of the CD instrument by use of appropriate absorbing samples. (2) Removing the Pockel's cell modulator from the beam may reduce, but does not entirely eliminate, the artifact. (3) Exposing the photomultiplier to room lights can increase the artifact. (4) Artifact magnitude is roughly proportional to photomultiplier tube gain. All high-gain tubes generate large artifacts; all tested models of the low-gain Dumont KM 2703 photomultiplier generate smaller artifacts. (5) Artifact magnitude is a sensitive function of the rotational and azimuthal orientation of the photomultiplier tube. (6) When a tube is found to generate a small artifact, its performance is not sensitive to orientation.

Item (1) indicates that the source of the problem is probably not due to secondary emission excited by the measuring beam, since suitable excitation wavelengths would not likely range through the visible and ultraviolet. Item (2) indicates that light intensity fluctuations due to beam modulation are not the exclusive cause of the artifactual ac signal, although item (3) suggests that some form of optical or electrical pickup may be involved. This suggestion is further reinforced by the correlation between artifact magnitude and tube gain, item (4), and the observed sensitivity to tube orientation, items (5) and (6).

Further characteristics seem to require a different explanation: (7)

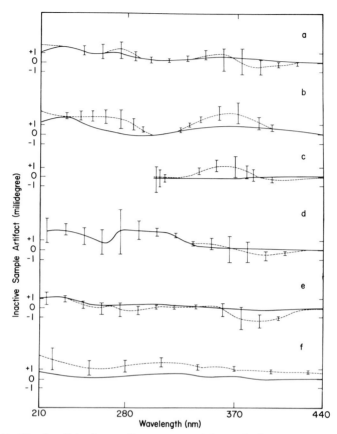

Fig. 10. Circular dichroism artifacts measured for optically inactive samples using various end-window photomultiplier tube detectors. Artifacts were measured on the 40-millidegree scale of a modified Cary 60, using a standard 1-cm path length cuvette. Samples were either 0.12 mg/ml potassium dichromate, a through e, or a highly scattering suspension of alumina particles, f, in $10^{-2} N$ KOH. Solvent baselines were run (————), and all deviations from the baselines produced by absorbing or scattering samples (------) are artifactual. Magnitude of the deviations is proportional to the sample optical density indicated by vertical bars. The potassium dichromate solution exhibits absorption maxima at 275 nm, OD $\cong$ 1.6 cm, and at 372.5 nm, OD $\cong$ 2.1. Optical density of the alumina was 1.1 at 350 nm, 1.55 at 260 nm, and 1.8 at 220 nm. Photomultiplier tubes represented are: Hamamatsu No. R375, for two different positions of the photomultiplier tube detector (a and b); Dumont No. 6292, used without (c) and with (d) a solid coating of sodium salicylate; and a Dumont KM 2703 (e and f). Notice the variability in baseline shape with different photomultipliers.

The magnitude of the artifact is wavelength dependent; it tends to be greater in the visible than in the ultraviolet. (8) The magnitude decreases when the measuring light is depolarized before striking the photomultiplier, as it is when a fluorscat cell is used. (9) The magnitude varies with changes in the beam cross section and its position of incidence on the photocathode. (10) For a symmetrically scattering sample, the artifact varies with changes in the sample cell position, even though the photomultiplier position is held constant.

Item (7) suggests the presence of a refractive index effect, and item (8) suggests there may be some asymmetry in instrumental response to the state of polarization of the incident beam. Items (9) and (10) indicate there is some heterogeneity to the output response of the photomultiplier as a function of position on the end-window photocathode, even though the response is probably much less sensitive to position than would be observed for a side-window tube. Taken together, these latter characteristics lead us to conclude that there is probably a small amount of birefringence which is intrinsic to the tube envelope or the photocathode coating and which varies in magnitude across the tube face.

Photomultiplier Tube Selection

Based upon the preceding analysis, two approaches may be used to obtain an end-window photomultiplier tube exhibiting acceptably small inactive sample artifacts. First, a low-gain photomultiplier may be used, but at the expense of higher dynode voltages and attendant noise levels. If higher gain response is needed, as for measurements at short wavelength, more elaborate precautions against optical and electrical pickup may be required. Second, the measuring beam may be depolarized between the sample cell and the photomultiplier to minimize polarization effects on response characteristics. Depolarization is automatically accomplished when fluorscat cells are used to convert measuring light to visible fluorescence. Similar results may be obtained for standard cuvette spectra by covering the detector face with a solid coating of sodium salicylate. We have obtained very satisfactory, higher gain detector performance from a salicylate-coated 2-inch diameter end-window Dumont 6292 photomultiplier. The coating scatters and depolarizes but still transmits visible and near UV light. The onset of the salicylate absorption at 350 nm produces about a 50% increase in Cary 60 dynode voltage and some change in baseline shape, but it is otherwise undetectable. The coating may be conveniently applied by spraying a concentrated ethanol solution from an inexpensive aerosol nebulizer.[14] Whereas the lime glass

[14] R. A. Knapp, *J. Appl. Optics* **2**, 1334 (1963).

tube face of a Dumont 6292 becomes opaque to UV light below 300 nm, the salicylate coating converts the UV to visible fluorescence and extends the useful range down to 190 nm or even farther into the vacuum UV. Its soft glass envelope makes an S-11 response photomultiplier such as the No. 6292 five to ten times less expensive than the same tube built into a quartz envelope to obtain an S-13 response. Thus a sodium salicylate-coated, high-gain, S-11 photomultiplier deserves serious consideration for large-solid-angle light detection applications in circular dichroism.

Circular Dichroism Applications

An example of large solid-angle detection geometry used to assay and correct for differential light scattering effects in circular dichroism is presented below. The circular dichroism of T2 bacteriophage is obtained using standard cuvettes and varying the detection angle by means of a movable end-window detector in a modified Cary 6001. The circular dichroism measured using standard and fluorscat cells in a conventional, unmodified Cary 6003 is shown for comparison.

An uncoated Dumont KM 2703 photomultiplier was selected for the work to be described below because it generated the flattest instrumental baseline and the smallest inactive sample artifacts and was least sensitive to orientation of any tube we had tested. All sample cells were 1 cm path length. The zero ellipticity line in all spectrograms was obtained from solvent baselines, and no instrumental adjustments were made between solvent and sample runs. All spectra were run without multipots but with the 14 Å spectral band width automatic slit program, unless otherwise noted. Data were run with instrumental time constant of 1 second and digitally recorded by on-line computer using available pen-averaging, baseline-subtracting, and data-smoothing programs.[15] Subsequent processing to obtain difference curves and plotted output was also accomplished by means of CDC 6400, 6600, or 7600 computer. Extinction per mole phosphate at 260 nm was assumed to be 6440 for T2 DNA and 10,206 for intact T2 virus.[16]

Three types of data are reported. (1) "Conventional spectra" obtained using standard sample cuvettes in an unmodified Cary 60 spectropolarimeter with CD accessory. (2) "Large-angle planar detector spectra" obtained in a Cary 6001 CD instrument modified to use a movable end-window photomultiplier detector at a specified position relative to the

[15] B. L. Tomlinson, Ph.D. Thesis, Univ. of California, Berkeley, 1968.
[16] M. F. Maestre, D. M. Gray, and R. B. Cook, *Biopolymers* **10**, 2537 (1971).

sample cell, e.g., distant, intermediate, close. (3) "Fluorscat spectra" obtained using a PMF fluorscat cell in an unmodified Cary 6003.

T2 Bacteriophage

The observation that the protein coat of T2 bacteriophage makes very little contribution to the phage circular dichroism above 245 nm leads to the expectation that the CD spectrum at wavelengths above 245 nm should be representative of the secondary conformation of the phage DNA inside the coat protein.[16] The conventional CD spectrum of intact T2 bacteriophage (Fig. 11) exhibits a negative maximum at 251 nm, $\Delta\epsilon = -6.9$, is zero at 276.5 nm, displays a positive maximum at 285 nm, $\Delta\epsilon = 1.6$, and exhibits a positive tail at long wavelength extending well into the visible. This tail admits no theoretical explanation in terms of a DNA structural modification. However, a nonzero CD signal outside the absorption band is a feature of interest, because it might conceivably be due to differential light scattering from a highly compacted DNA organization inside the phage head. To investigate this possibility, and

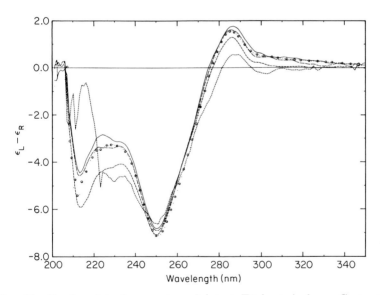

Fig. 11. Circular dichroism spectra of intact T2 bacteriophage. Conventional spectrum (......) made using standard cuvette in unmodified Cary 6003 CD instrument is compared with fluorscat cell spectrum in same instrument (------). Large-angle planar detector spectra measured using standard cuvette in a Cary 6001 CD instrument modified to use an end-window photomultiplier detector are shown for close (-----), intermediate (— — —), and distant (———) photomultiplier positions. Fluorscat spectrum is average of three runs with monochromator slits set at 2.4 mm.

the extent to which other features of the intact T2 CD spectrum might derive from differential light scattering, the CD spectrum of intact T2 phage was measured in a conventional Cary 6003 using both a standard cuvette and a (PMF) fluorscat cell and in a Cary 6001 spectropolarimeter modified to accommodate an end-window photomultiplier tube with 2-inch diameter photocathode, using standard cuvettes. The end-window PMT may be positioned on the optical axis of the instrument at various distances from the sample cuvette.

For the most part, the conventional T2 spectrum exhibits values midway between the large angle planar detector spectra (Fig. 11). However, as the solid angle of detection is increased by moving the photomultiplier to the intermediate and close positions, the 285 nm peak and the long-wavelength tail are seen to be reduced. Indeed, the tail is no longer evident in the close PMT spectrum.

The fluorescent scattering (fluorscat) cell CD spectrum of intact T2 is shown in the lowest curve of Fig. 11. The long wavelength positive tail is completely gone, and the positive maximum, slightly red-shifted, is further reduced to about half the value obtained in the close PMT spectrum. Inasmuch as the fluorscat cell is capable of detecting light scattered through angles up to and even larger than 90 degrees, these data demonstrate that the more scattered light detected the greater is the depression of the positive 285 nm ellipticity band in the intact T2 phage spectrum. It may be concluded that differential light scattering is contributing significantly to the 285 nm band and is the sole source of the positive tail observed in the conventional CD spectrum of intact T2. It is clear that in using circular dichroism data to study secondary conformation of T2 DNA inside the virus coat one must take into account this differential scattering contribution. A detailed analysis of the internal T2 DNA conformation has been made elsewhere.[17]

In order to obtain information about the nature and source of the differential light scattering from intact T2 phage, we have also measured standard cell, large-angle planar detector spectra on purified T2 DNA and disrupted T2 virus. Disruption of the phage has been induced by cyclic freeze thawing leading to release of DNA through a rent in the phage coat.[16] Except for this rent the disrupted T2 coats are thought to be morphologically intact after the DNA is released.[18] The phage coats, it should be noted, scatter sufficient light to register $OD_{350} = 0.04$ per OD_{260} in a Cary 14 absorption spectrophotometer. The shape of the differential light scattering contribution detected for intact T2 phage, dis-

[17] B. P. Dorman and M. F. Maestre, *Proc. Nat. Acad. Sci. U.S.* **70**, 255 (1973).
[18] R. M. Herriott and J. L. Barlow, *J. Gen. Physiol.* **40**, 809 (1957).

rupted T2 phage, and purified T2 DNA has been obtained by subtracting the close PMT spectrum from the distant PMT spectrum. The difference spectra presented in Fig. 12 show that no significant CD spectral changes are observed for the disrupted phage or for purified DNA upon changing the solid angle of detection with the movable end-window photomultiplier. Therefore, it may be concluded that differential light scattering contributions for these specimens, if any, are beneath the sensitivity level of the present measurements.

The intact phage difference spectrum displays a long-wavelength tail, a maximum at 290 nm, goes to zero at 265 nm, and exhibits a larger maximum at 220 nm. The detailed significance of this scattering curve is beyond our present understanding. But inasmuch as the observed differential scattering disappears completely upon release of the viral DNA into solution, we may clearly attribute the differential light scattering from intact T2 virus to the DNA packing organization inside the viral coat protein.

It is an important result of our work that disrupted phage suspensions do not exhibit geometry-dependent CD spectra. In the first place, this result demonstrates the absence of a systematic position dependence to the large-angle planar detector CD spectra even in the presence of a light scattering specimen such as the disrupted phage. But more im-

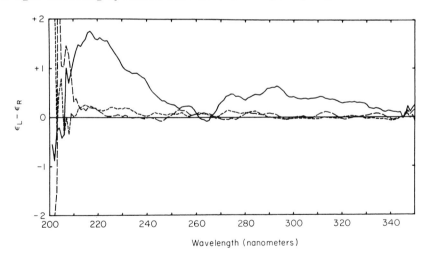

Fig. 12. CD difference spectra computed for large-angle planar detector data by subtracting close PMT spectrum from distant PMT spectrum of intact T2 bacteriophage (——), disrupted T2 bacteriophage (— — — —), and purified T2 DNA (- - - - - -). Only the intact virus exhibits a nonzero difference spectrum indicative of differential light scattering. Nonzero values below 210 nm are probably noise artifacts.

portant, it demonstrates that the presence of large, optically active biological particles in a sample suspension is not in itself sufficient to generate differential light scattering effects in circular dichroism.

Conclusions

Our work to date suggests that differential light scattering will influence the CD of some, not all, particulate suspensions of optically active particles. Whether or not differential light scattering arises may depend on the specific organization of the aggregated chromophores. We infer that where differential light scattering exists, the organization must exhibit ordered asymmetry. If this is the case, it is conceivable the shape of the scattering contribution curves may be used to assess details of the relevant ordered structure. As such, the differential light scattering phenomenon might constitute a uniquely powerful probe for molecular organization at the level of tertiary or quaternary conformation.

The data presented above reveal a large light-scattering effect on the CD of intact T2 virus. We anticipate that differential light-scattering contributions may be influencing the conventional CD of other virus, chromosomes, chromatin, and membrane structures. Light scattering from large, asymmetric biological structures such as these is not presently amenable to rigorous theoretical treatment. Therefore the CD spectra of such structures ought to be remeasured under conditions permitting empirical analysis of possible differential light-scattering contributions. The large-angle detection techniques reported here promise to be quite useful for that purpose.

Acknowledgments

We are indebted to Dr. K. D. Philipson and Professors K. Sauer and I. Tinoco, Jr., for use of essential instrumentation. We thank Mrs. K. Sieux for T2 virus. This work was supported by NIH grants GM-11180 and AI-08427-04 and by NASA grant 05-003-020.

[31] Magneto Optical Rotation Spectroscopy

By VICTOR E. SHASHOUA

Magneto optical rotation (MOR) spectroscopy is based on Faraday's observation[1] that any substance will rotate the plane of polarized light when a magnetic field is applied parallel to the light beam. MOR spectra

[1] M. Faraday, *Phil. Trans. Roy. Soc. London* 3, 1 (1846).

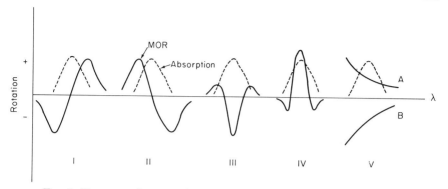

FIG. 1. The general types of magneto optical rotation (MOR) spectra.

can be obtained for both naturally optically active and optically inactive molecules. Verdet[2] showed that the induced magnetic rotation θ, at a fixed wavelength, was related to the magnetic field strength H and the path length of the sample L by the following equation.

$$\theta = VHL \tag{1}$$

where V is a constant known as the Verdet constant. The induced magnetic rotation is a linear function of the magnetic field strength for measurements within and outside the absorption band region of a molecule.[3] In addition, the sign of the induced magnetic rotation is dependent upon the direction of the magnetic field and is linearly additive to the natural optical activity of a substance.[4]

MOR spectra show the same type of dispersion features as those observed for optical rotatory dispersion (ORD) measurements[3] within the absorption band region of molecules. Figure 1 illustrates the five types of MOR spectra obtained. Types I and II have positive and negative sigmoid curves with inflection points coincident with the absorption band maxima. Types III and IV MOR spectra have negative and positive magnetic rotation peaks at the position of the absorption band maxima. Type V shows no anomalous dispersion features and may have a positive or negative sign depending on the magnitude of the magnetic rotation for neighboring absorption bands. The apparent similarity of MOR spectral types to those obtained in ORD measurements, however, has not led to any simple correlations of the sign and shape of a MOR spectrum to stereochemical features. Thus applications of MOR spec-

[2] E. Verdet, *C. R. Acad. Sci.* **39**, 548 (1854).
[3] V. E. Shashoua, *J. Amer. Chem. Soc.* **86**, 2109 (1964).
[4] V. E. Shashoua, *Symp. Faraday Soc.* **3**, 61 (1969).

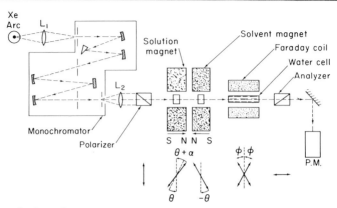

Fig. 2. Optical diagram of the magneto optical rotation (MOR) spectropolarimeter illustrating operation of the instrument: θ, solvent plus sample cell rotation; α, solute rotation; and ϕ, modulation angle in the Faraday coil. From J. G. Forsythe, R. Kieselbach, and V. E. Shashoua, *Appl. Opt.* **6**, 699 (1967).

troscopy, as an analytical technique, have been limited to investigations in theoretical spectroscopy and to empirical structure–property correlations for certain biologically active molecules[4] and complex organic compounds.[3-6] In theoretical studies, MOR spectra have been used for precise assignments of the position of excited states of molecules[3,6] and measurements of magnetic moments of excited states.[7] In structure–property studies MOR spectroscopy has been applied to investigations of the ligand binding of ferri- and ferrohemoglobins,[8] the aggregation of chlorophylls,[8,9] and studies of the kinetics of oxidation and reduction of cytochrome c.[10,11]

Instrumentation

MOR spectra are determined in an automatic recording spectropolarimeter equipped with a system for applying a magnetic field parallel to the light beam at the sample compartment. Figures 2 and 3 show a

[5] B. Briat, D. A. Schooley, R. Records, E. Bunnenberg, and C. Djerassi, *J. Amer. Chem. Soc.* **89**, 7062 (1967).

[6] V. E. Shashoua, *J. Amer. Chem. Soc.* **87**, 4044 (1965).

[7] A. J. McCaffery, P. J. Stephens, and P N. Schatz, *Inorg. Chem.* **6**, 1614 (1967).

[8] V. E. Shashoua, *in* "Hemes and Hemoproteins" (B. Chance, R. W. Estabrook, and T. Yonetani, eds.), p. 93. Academic Press, New York, 1966.

[9] E. A. Dratz, Ph.D. Thesis, University of California, No. UCRL-17200 Berkeley (1966).

[10] V. E. Shashoua, *Nature (London)* **203**, 972 (1964).

[11] V. E. Shashoua, *Arch. Biochem. Biophys.* **111**, 550 (1965).

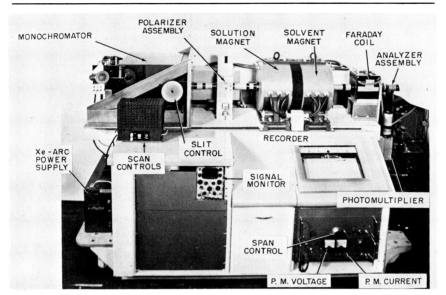

FIG. 3. Magneto optical rotation (MOR) spectropolarimeter. From J. G. Forsythe, R. Kieselbach, and V. E. Shashoua, *Appl. Opt.* **6**, 699 (1967).

diagram and photograph of such an instrument.[12] The MOR spectropolarimeter measures the magnetic rotations induced by a 5000 G field as a function of wavelength with an angular sensitivity of $\pm 0.001°$ and $\pm 0.003°$ at the visible and UV spectral regions, respectively. In the operation of the instrument a plane-polarized light beam is rotated by an angle of $(\theta + \alpha)$ in the first solenoid, where θ is the angle of rotation of the solvent plus sample cell and α is the rotation of the solute. The light beam then passes through a second solenoid with a magnetic field polarized in the opposite direction. This contains an identical sample cell with pure solvent to give a rotation of $(-\theta)$. The two magnetic fields thus produce a net rotation of the polarized light beam of α, the solute rotation. The detection of the magnetic rotation is achieved by the use of a Faraday modulator. This consists of an alternating magnetic field which induces an oscillating 60-cycle rotation $(\pm \phi)$ in a water sample 20 cm long. This provides the carrier frequency for activating the angle-sensing and recording system of the spectropolarimeter. The value of ϕ_{max} varies from $\pm 0.5°$ at red region of the spectrum to $\pm 16°$ at the UV region. The light beam next passes through an analyzer set at the crossed position relative to the quartz polarizer and then onto a mirror which directs it to the photomultiplier. If the angle of polarization of the light reaching

[12] J. G. Forsythe, R. Kieselbach, and V. E. Shashoua, *Appl. Opt.* **6**, 699 (1967).

the analyzer does not correspond to the condition of minimum light transmission (null), then there is a 60 cps component in the photo-multiplier current resulting from the action of the Faraday modulator. The phase of this component is sensitive to the direction of the analyzer position from null. This serves as a signal for operating a servo motor to reposition the analyzer to null. The angle of rotation of the polarizer to achieve a null position is measured by the use of a linear voltage differential transformer.[12] This transduces the angular rotation of the polarizer into an electrical signal which is recorded as a function of wave-length to give a MOR spectrum. The light source for the spectropolar-imeter is a 150-W Xe arc lamp. The light is focused by quartz optics onto the entrance slit of a Cary Model 14 double monochromator.[13] This provides a spectral purity of wavelength selection with a stray light level of less than 1 part in 10^6. The sample temperature is maintained to within 0.1° and the slit width of the monochrometer must be controlled to maintain a band pass equivalent to about one-tenth of the half-width of the absorption band under study.

The magnetic field within the core of the solenoids must be uniform so that the same magnetic field (within 0.5 G) can be reproducibly obtained. The temperature of the air column within the core of the solenoid should be kept constant by some external cooling method to avoid eddy currents of air which can give significant changes in the observed magnetic rotation due to the air. In addition the solenoid magnets should be enclosed in a soft iron shield to eliminate stray magnetic field effects on the optics and electronics of the instrument. Satisfactory shielding is obtained by covering the solenoid magnets with a half-inch layer of soft iron to give a stray magnetic field of less than 2 G at a distance of 2.5 cm away from the solenoids.

Method of Measurement

The MOR spectrum of a substance is obtained by measuring the in-duced magnetic rotation as a function of wavelength. To eliminate any possible stray light effects, the concentration of the solute should be adjusted to give solutions with a maximum optical density of 2.5. Suitable solvents are those which have absorption bands far from the spectral region under investigation, such as water, chloroform, ethanol. In a typi-cal measurement the solvent is placed in the quartz sample cells (1-cm path length) and mounted into the water-cooled cell holders of the instrument. These are inserted into the central cores of the solenoid magnets, and a zero magnetic field baseline of rotation versus wavelength

[13] H. Cary, R. C. Hawes, P. B. Hooper, J. J. Duffield, and K. P. George, *Appl. Opt.* **3**, 329 (1964).

is obtained. The absence of a rotation in such a measurement indicates that there are no strains in the quartz cells. The magnetic fields are then turned on, and a second baseline is obtained. This allows for matching the magnetic field of the solenoid of the sample compartment with that of the solvent compartment by adjustment of the current flow through one of the solenoid magnets. Next the solution is placed in the sample compartment and MOR spectrum is obtained. For substances that are naturally optically active, a fourth wavelength scan is obtained at zero magnetic field to record the ORD data.

Under experimental conditions of measurements, the light beam travels along the north-south direction of the first magnetic field. Measurements of a clockwise rotation of the plane of polarized light for observations opposite to the direction of travel of the light beam are designated as positive rotations. This is in the same sense as the Faraday rotation for a pure solvent outside its absorption band region. The specific magnetic rotation $[\alpha]_{sp}$ is defined by

$$[\alpha]_{sp} = \frac{10,000\theta}{lcH} \tag{2}$$

where θ is the angle of rotation in degrees, l the path length in decimeters, c the concentration in grams per milliliter, and H is the magnetic field in gauss.

The specific molar magnetic rotation $[\alpha]_M$ is defined by

$$[\alpha]_M = \frac{[\alpha]_{sp}M}{100} \tag{3}$$

Where M is the solute molecular weight. A magnetic field strength of 10,000 G is used as a standard for comparison of $[\alpha]_{sp}$ and $[\alpha]_M$ values. Figure 4 shows a typical MOR spectrum obtained for potassium ferricyanide. The compound has a type II and a type I MOR spectrum for its 420 nm and 300 nm absorption bands, respectively.

Magnetic Rotation Measurements Outside Absorption Bands

The induced magnetic rotation for measurements at wavelengths far from the absorption band regions of molecules increases from low values at the visible regions to large rotations at the UV region of the spectrum. For substances such as water, chloroform, and dioxane the results can be fitted to a simplified form of the Drude equation[14]

$$[\alpha]_{sp} = \frac{A}{\lambda^2 - \lambda_0^2} \tag{4}$$

$$[\alpha]_{sp}\lambda^2 = [\alpha]_{sp}\lambda_0^2 + A \tag{5}$$

[14] P. Drude, "Lehrbuch der Optik" (S. Hirgel, ed.), 2nd ed. Verlag, Leipzig, 1906.

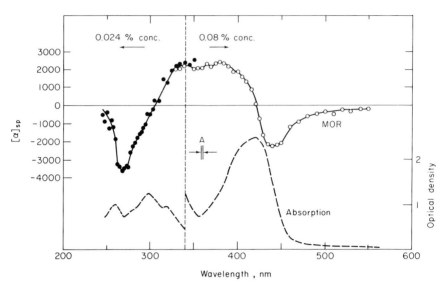

Fig. 4. Magneto optical rotation (MOR) spectrum of potassium ferricyanide; A, monochromator band pass; 1-cm cell; 10°, 9350 G. From V. E. Shashoua, *J. Amer. Chem. Soc.* **86**, 2109 (1964).

where λ is the wavelength in nanometers and A and λ_0 are constants. Figure 5 shows its straight-line plots obtained for $[\alpha]_{\mathrm{sp}}\lambda^2$ as a function $[\alpha]_{\mathrm{sp}}$ for three solvents.[15] Table I lists the λ_0 and A values for the MOR measurements.

MOR Spectra of Porphyrins and Phthalocyanines

Porphyrins and phthalocyanines as a group (see Table II and Figs. 6 and 7) have large magnetic rotations with several anomalous dispersion features in their MOR spectra.[6] For example, zinc phthalocyanine and magnesium phthalocyanine have specific magnetic rotations of $-8.3 \times$

TABLE I

APPLICATION OF THE DRUDE EQUATION TO VARIOUS SOLVENTS

Substance	A^a	λ_0
Water	4.7×10^6	143
Chloroform	5.6×10^6	143
p-Dioxane	8.4×10^6	145

[a] V. E. Shashoua, unpublished data.

[15] V. E. Shashoua, unpublished data.

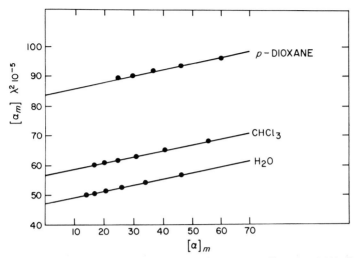

FIG. 5. Applications of the Drude equation; 1-cm cell; 30°; 10,000 G.

10^5 degrees cm^2 per gram per 10,000 G, at their 670 and 675 nm absorption bands, respectively. In addition the MOR spectra of the porphyrins have spectral features that are related to the direction of polarization of their $\mathrm{II} \rightarrow \mathrm{II}^*$ transitions. Fluorescence studies[16] have indicated that for

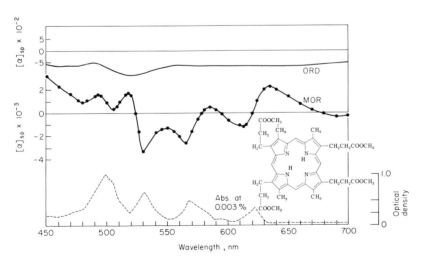

FIG. 6. Magneto optical rotation (MOR) and optical rotatory dispersion (ORD) spectra of coproporphyrin II (tetramethyl ester) in N-methylpyrrolidone. From V. E. Shashoua, *J. Amer. Chem. Soc.* **87**, 4044 (1965).

[16] M. Gouterman and L. Stryer, *J. Chem. Phys.* **37**, 2260 (1962).

TABLE II

Magneto Optical Rotation (MOR) Spectral Data for Porphyrins
and Phthalocyanines[a]

Compound	Solvent[b]	λ (nm)	$[\alpha]_{sp}{}^c$	Type[d]
Coproporphyrin-II (tetramethyl ester)	NMP	622(−)	3.6×10^3	I
		518(−)	2.6×10^3	I
		530(−)	3.3×10^3	II
		498(+)	1.4×10^3	II
Hematoporphyrin	NMP	624(−)	700	I
		570(−)	350	I
		535(−)	420	II
		500(+)	220	II
Zinc hematoporphyrin	NMP	578 −	4.4×10^4	III
		−	1.1×10^{4f}	III
		538 −	1.35×10^4	III
		−	0.45×10^{4f}	III
Phthalocyanine	ClN	698(−)	150×10^4	I
		662 −	66×10^4	II(?)
		635 −	15×10^4	III
		602(+)	7.5×10^4	II
Copper phthalocyanine	NMP	670 −	21×10^4	III
		638 −	0.9×10^4	III
		608(−)	1.9×10^4	II
Magnesium phthalocyanine	NMP	670 −	83×10^4	III
		640 −	5×10^4	III
		605(−)	8.5×10^4	II
Zinc phthalocyanine	NMP	671 −	88×10^4	III
		640 −	5×10^4	III
		605(−)	8.2×10^4	II
Nickel phthalocyanine	ClN	670(−)	66×10^{4e}	III

[a] From V. E. Shashoua, *J. Amer. Chem. Soc.* **87**, 4044 (1965).

[b] NMP, *N*-methylpyrrolidone, ClN, chloronaphthalene.

[c] (+) and (−) represent the sign of the magnetic rotation at the inflection points of the types I and II MOR spectra.

[d] See Fig. 1 for definitions of MOR spectral types.

[e] Owing to the extreme insolubility of the compound, this result was estimated by considering that the extinction coefficient for the 670 nm band was about the same as for magnesium phthalocyanine.

[f] MOR bands associated with the 540 nm absorption region.

coproporphyrin II (see Fig. 6) the two long wavelength absorption bands of the molecule at 622 nm and 568 nm are polarized in the plane of the pophyrin ring parallel to the two central hydrogen atoms of the molecule while the two bands at 530 nm and 498 nm are in the plane of the ring and polarized perpendicular to the longer wavelength bands. The MOR spectrum of the molecule also indicates this difference in polarization of

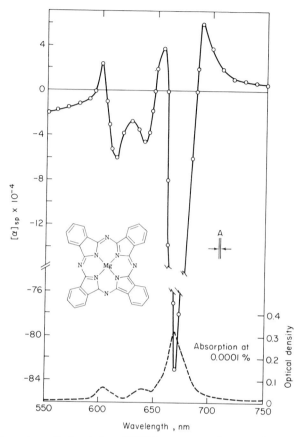

Fɪɢ. 7. Magneto optical rotation (MOR) spectrum of magnesium phthalocyanine in N-methylpyrrolidone. A, monochromator band pass. Data from V. E. Shashoua, *J. Amer. Chem. Soc.* **87**, 4044 (1965).

the transitions, i.e., type I MOR spectra are observed for the 622 nm and 568 nm bands and type II MOR spectra for the 530 and 498 nm bands, respectively. This observation suggests that the shape of the MOR spectrum is related to the polarization of excited states. Additional experimental evidence, however, is required to substantiate this postulate.

MOR Spectra of Hemoglobin, Methemoglobin, and Myoglobin

Figure 8 and Tables III and IV illustrate the MOR spectral changes that can be observed for hemoglobin and methemoglobin as a function of the substituents at the sixth position of porphyrin nucleus of the mole-

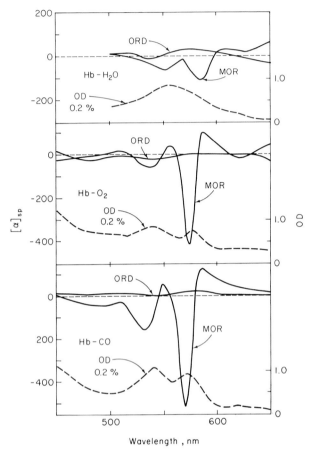

Fig. 8. Magneto optical rotation (MOR) and optical rotatory dispersion (ORD) spectra of bovine ferrohemoglobin with H_2O, O_2, and CO as the ligands. Data from V. E. Shashoua, *Symp. Faraday Soc.* **3**, 61 (1969).

cules.[6] The magnitude of the specific magnetic rotation at the 570 nm band of bovine ferrohemoglobin increases about 4-fold from 150 to 650 for replacement of H_2O with CO as the ligand. Thus substituents with a large ligand field give rise to large magnetic rotations. For methemoglobin (Table IV) the four-absorption-band system characteristic of the high spin state of the molecule, with water as the ligand, changes to a two absorption band system with cyanide as the ligand. The high spin state gives comparatively low magnetic rotations whereas with cyanide as the ligand, the low spin state is characterized by a large magnetic rotation. These results suggest that measurements of the magnitude of

TABLE III

MAGNETO OPTICAL ROTATION (MOR) SPECTRAL DATA FOR HEMOGLOBIN
pH = 6.8, 10°C[a]

Ligand	Absorption maxima (nm)	λ_1/nm	$[\alpha]_{sp}$	λ_2/nm	$[\alpha]_{sp}$	Magnetic moment B.M.[e]	Number of unpaired electrons
			MOR data[b]				
H_2O[c]	554	550	45	585	150	4.9	4
O_2[c]	540	538	100	574	510	0	0
CO[c]	540	530	210	—	—	0	0
	572	—	—	570	650	—	—
O_2[d]	542	535	118	—	—	—	—
	577	—	—	576	785	—	0

[a] From V. E. Shashoua, *Symp. Faraday Soc.* **3**, 61 (1969).
[b] The $[\alpha]_{sp}$ values of the MOR data are the sum of the positive and negative components of the type III MOR spectra.
[c] Commercial grade bovine hemoglobin.
[d] Horse hemoglobin data (see Fig. 8).
[e] Data obtained from C. D. Coryell, F. Slitt, and L. Pauling, *J. Amer. Chem. Soc.* **198**, 33 (1952).

the magnetic rotations may be useful to detect changes in the spin states of certain coordination compounds.[8]

Myoglobin[17] has also been studied by MOR spectroscopy. The specific magnetic rotation of the molecule at 580 nm changes from −200 to −1250 when myoglobin binds oxygen. This change in the induced magnetic rotation may be a useful parameter for studying the kinetics of oxygen binding.[17]

MOR Spectral Studies of Cytochrome c

The MOR spectrum of cytochrome c is very sensitive to the oxidation state of the molecule.[10] The specific magnetic rotation at 549 nm changes from −8000 to −150 when the reduced molecule is oxidized (see Fig. 9). This feature of the MOR spectrum was used for studies of the kinetics of oxidation of cytochrome c with hydrogen peroxide and reduction with formamidine sulfinic acid (see Fig. 10). Another aspect of the MOR spectra of cytochrome c is illustrated in Table V. The α and β absorption bands of the molecule at 550 and 520 nm, respectively, have an optical density ratio of about 1.7 for fully reduced ferrocytochrome c. An analysis of the magnetic rotation per optical density unit at 549 nm for the fully reduced molecule from different sources shows that yeast ferrocytochrome c has over twice the magnetic rotation of cytochrome c isolated from

[17] M. V. Volkenstein, J. A. Sharonov, and A. K. Shemelin, *Nature (London)* **209**, 709 (1966).

TABLE IV

MAGNETO OPTICAL ROTATION (MOR) SPECTRAL DATA FOR METHEMOGLOBIN pH 6.8, 10°C[a]

Ligand	Band I		Band II		Band III		Band IV		Magnetic moment[b]	Unpaired electrons
	λ_1/nm	$[\alpha]_{sp}$	λ_2/nm	$[\alpha]_{sp}$	λ_3/nm	$[\alpha]_{sp}$	λ_4/nm	$[\alpha]_{sp}$		
H_2O	630	50	567	36	540	20	500	28	5.20	(4)
F	606	95	—	—	550	30	485	38	5.92	4
OH[c]	607	32	575	95	538	38	—	—	4.47	4
CN	—	—	565	470	540	129	—	—	2.5	1

[a] From V. E. Shashoua, *Symp. Faraday Soc.* **3**, 61 (1969).
[b] The magnetic moment data were obtained from C. D. Coryell, F. Slitt, and L. Pauling, *J. Amer. Chem. Soc.* **198**, 33 (1952).
[c] This was studied at pH 10.

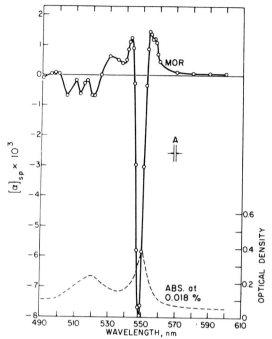

Fig. 9. Magneto optical rotation (MOR) spectrum of ferrocytochrome c; A, monochromator band pass. From V. E. Shashoua, *Arch. Biochem. Biophys.* **111**, 550 (1965).

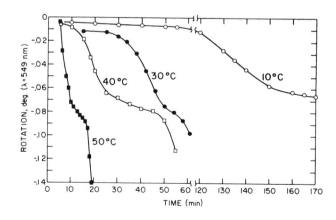

Fig. 10. Kinetics of the reduction of ferricytochrome c to ferrocytochrome c with formamidine sulfinic acid at various temperatures. From V. E. Shashoua, *Arch. Biophys. Biochem.* **111**, 550 (1965).

TABLE V

MAGNETO OPTICAL ROTATION (MOR) SPECTRAL DATA—CYTOCHROME $c^{a,b}$

Source	pH = 7		pH = 10	
	A	B	A	B
Yeast	1.77	0.22	1.70	0.37
Tuna	1.75	0.21	1.80	0.21
Horse hemoglobin	1.79	0.23	1.71	0.15
Pigeon breast	1.75	0.23	1.71	0.15

[a] From V. E. Shashoua, Symp. Faraday Soc. **3**, 61 (1969).
[b] A is the ratio of optical densities at 550–520 nm. B is the Δ MOR divided by the optical density.

pigeon breast. Moreover, this occurs only at pH 10, near the isoelectric point of the molecule. Such changes in the magnetic rotational strengths[18] have been attributed to differences in the polypeptide chain sequences which form ligands with the porphyrin nucleus of ferrocytochrome c.

Theories of MOR Spectroscopy

A number of theories have been proposed for MOR spectroscopy.[19-21] Classical theories have been largely based on the Zeeman effect.[19] The magnetic field essentially splits the excited state of a given electronic transition into two components which have different indices of refraction for the left and right circularly polarized light components. The induced magnetic rotation is related to the refractive indices by Eq. (6):

$$\alpha = \frac{\pi}{\lambda} \left(\eta_L - \eta_R \right) \tag{6}$$

Where η_L and η_R are the indices of refraction for left and right circularly polarized light and λ is the wavelength.

More rigorous quantum mechanical theories have been recently proposed for MOR spectroscopy.[21] However, there has been as yet no quantitative correlations with experimental results. It is not yet possible to predict what type of MOR spectrum, or what magnetic rotational strength, will be obtained for a given transition.

[18] V. E. Shashoua and R. W. Estabrook, in "Hemes and Hemoproteins" (B. Chance, R. W. Estabrook, and T. Yonatani, eds.), p. 427. Academic Press, New York, 1966.
[19] For a general review, see J. R. Partington, "Advanced Treatise on Physical Chemistry" Vol. IV, pp. 592–632. Longmans, Green, New York, 1954.
[20] A. D. Buckingham and P. J. Stephens, Annu. Rev. Phys. Chem. **17**, 399 (1966).
[21] A. D. Buckingham, Symp. Faraday Soc. **3**, 7 (1969).

[32] Differential Spectrofluorometry

By GERALD D. FASMAN and BARKEV BABLOUZIAN[1]

The determination of conformation and conformational changes in biological systems has become of central interest in relation to biological function. Many physicochemical techniques have been applied to such determinations,[2] e.g., optical rotatory dispersion, circular dichroism, nuclear magnetic resonance. The presence of the aromatic side chain of the amino acids tyrosine, tryptophan, and phenylalanine has made the ultraviolet spectral region particularly advantageous for such determinations.

As the ultraviolet absorption spectra of tryptophan and tyrosine are related to the environment about them, a change in the environment about either amino acid will cause a shift in the absorption spectrum.[3,4] Consequently, a conformational change of a protein containing these residues will result in a spectral shift if the environment about either residue is altered, or solvent perturbation of exposed residues will likewise cause spectral shifts. Thus intrinsic tryptophan and tyrosine moieties can be used as probes for detecting conformational changes in proteins. A particularly useful way to follow these changes is by difference spectroscopy.[3,4] In this method the difference between a perturbed and unperturbed protein molecule is measured. This difference technique has allowed investigators to determine, among other things, the fraction of exposed tyrosine and tryptophan residues in a native protein, the effect of substrate binding, and to study the kinetics and thermodynamics of denaturation. In particular, it has allowed the observation of very small differences.

Another technique for studying changes in protein conformation, which is related to absorption, is fluorescence. This technique allows the use of much smaller samples because of the greater sensitivity of the method. Changes in the environment (e.g., solvent perturbation, substrate binding, conformational changes) about a tyrosine or tryptophan

[1] Contribution No. 864 of the Graduate Department of Biochemistry, Brandeis University, Waltham, Massachusetts 02154. This work was supported in part by grants from the U.S. Public Health Service (GM 17533), National Science Foundation (GB 29204X), American Heart Association (71-111), and the American Cancer Society (P-577).
[2] G. Fasman (Ed.), "Poly-α-amino acids." Dekker, New York, 1967.
[3] T. T. Herskovits, see Vol. 11, p. 748.
[4] D. B. Wetlaufer, *Advan. Protein Chem.* **17**, 303 (1962).

residue, either incorporated in a protein molecule or free in solution, have been shown to lead to changes in the fluorescence emission spectrum of these amino acids.[5-8] In addition to these causes of fluorescence alteration, specific quenching effects induced by environmental factors can cause large fluorescent changes. The correlation of the fluorescence spectrum with the absolute immediate medium surrounding these moieties is difficult. However, this method is excellent for observing small changes occurring about these residues. In analogy with difference absorption spectroscopy difference fluorescence spectroscopy offers a highly sensitive method to measure conformational changes in proteins, or to probe the surface to locate aromatic residues. Any system which possesses a fluorescent moiety is amenable to study. The usefulness of difference fluorometry was first shown by the work of Lehrer and Fasman[9] in their study of the effect of pH and inhibitors on the fluorescence of lysozyme. These workers were able to separate the fluorescence of three different tryptophan residues, and to show the effect of binding on their fluorescence. Lehrer[10,11] has elegantly extended this technique to illustrate the effect of iodide quenching in differentiating between tryptophyl side chains exposed to solvent from those buried in the interior of the molecule, and to demonstrate that substrate binding can protect the binding site from iodide quenching. More recently this technique has been used to investigate the mechanism of ATP hydrolysis by heavy meromyosin.[12]

This technique, although powerful in principle, had not found widespread use because of the lack of availability of a difference spectrofluorometer. In 1970 a difference spectrofluorometer was designed and built by Bablouzian, Grourke, and Fasman.[13] This instrument is described and examples of its use are discussed in this article.

Optics

In Fig. 1 is seen an optical schematic of the instrument. The light source is a 150-W dc operated xenon lamp (Osram XBO 150 w/1)

[5] L. Brand and B. Witholt, see Vol. 11, p. 776.
[6] R. F. Chen, in "Fluorescence" (G. G. Guilbault, ed.), p. 443. Dekker, New York, 1967.
[7] G. Weber and F. W. J. Teale, in "The Proteins" (H. Neurath, ed.), Vol. 3, p. 445. Academic Press, New York, 1965.
[8] R. F. Chen, H. Edelhoch, and R. F. Steiner, in "Physical Principals and Techniques of Protein Chemistry," (S. J. Leach, ed.), Part A, p. 171. Academic Press, New York, 1969.
[9] S. S. Lehrer and G. D. Fasman, J. Biol. Chem. 242, 4644 (1967).
[10] S. S. Lehrer, Biochem. Biophys. Res. Commun. 29, 767 (1967).
[11] S. S. Lehrer, Biochemistry 10, 3254 (1971).
[12] M. Werber, A. St. Gyorgi, and G. D. Fasman, Biochemistry 11, 2872 (1972).
[13] B. Bablouzian, M. Grourke, and G. D. Fasman, J. Biol. Chem. 245, 2081 (1970).

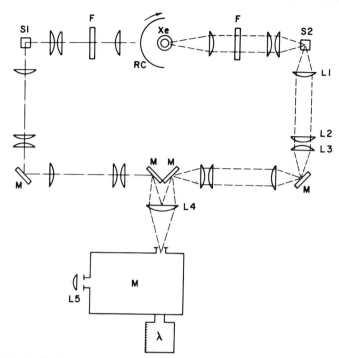

Fig. 1. Optical schematic of the differential fluorometer. Xe, 150 W xenon lamp; RC, rotating light chopper; F, 280 nm interference filters; L1, L2, L3, fused quartz condensing and focusing lenses; S1, S2, samples; M, plane front surface mirrors; L4, focusing lens; L5, field lens; λ, wavelength drive of monochromator.

powered by a regulated power supply (G. W. Gates and Company, Long Island, New York, Model P150D). Additional inductor and capacitor filter networks were installed between the power supply and the lamp, these reduced lamp ripple to less than 1%.

Samples S1 and S2 time share the xenon light source by the "can"-type rotating chopper. This is a metal cylinder with a 180° section milled out and fitted over the xenon lamp. Some difficulties were encountered in igniting the lamp. The high voltage high frequency ignition pulse delivered by the power supply arced over to the inside of the chopper preventing lamp ignition. Rounding of all sharp edges on the lamp holder, increasing the diameter of the chopper, and reducing stray capacitances solved this problem.

The excitation wavelength is selected by a pair of matched interference filters (Thin Film Products Division of I-R Industries, Waltham, Massachusetts) which have a transmission of 20% at 280 nm and a half-band width of 100 Å. The filters were matched to within 2% for band width and peak transmission.

The excitation light is focused at the center of the cuvettes by the fused quartz condensing and focusing lenses supplied with the monochromators (Bausch and Lomb, Rochester, New York). Six sets of the same lenses are used throughout the instrument.

The sample holders are thermostated and accommodate cuvettes of 1-cm internal light paths or smaller. An adjustable shutter between the light source and the samples allows initial adjustment of the exciting light.

Fluorescence emission is observed at 90° to the direction of the exciting light. This fluorescence is focused on a plain front surface mirror which in turn reflects the light to a second mirror and is subsequently collimated at the entrance slit of the monochromator. The monochromator is a Bausch and Lomb 250-mm focal length grating type with a constant resolution of 16 Å per millimeter of slit width. It was coupled to the instrument on its "side." That is, the entrance slit was in the direction of the two fluorescent beams so that each beam illuminated a section of the grating along the rulings. Placing the monochromator in its normal way would have produced spectral shifts caused by the displacement of the two beams on the grating along a line perpendicular to the rulings. A field lens was used at the exit slit which focuses and superimposes the two slit images on the same spot of the photomultiplier photocathode.

Detection and Recording

The chopper was rotated by an 1800-rpm synchronous motor (Bodine type NSY-12) and therefore the output of the photomultiplier (EMI 6255S) is a 30 Hz square wave,[13a] which is preamplified by an operational amplifier (Philbrick SP2A). In Fig. 2 is shown the oscilloscope photograph of the various wave forms. A disc with the same mechanical cut out as the chopper is mounted on the shaft of the chopper with a small tungsten lamp (6 V) and a phototransistor (G. E. L14A502) on either side of the disc. The collector loads of the phototransistors are high speed mercury-wetted contact relays (C. P. Clare HGS 5003) which have operate and drop-out times of 1 msec with no contact bounce. The contacts of the relays switch the alternate half-cycles of the photomultiplier output into three-stage R-C boxcar integrators. Time constants of these integrators are variable from 100 msec to 3 sec. For most of the spectra 100-msec time constants were sufficient at the scanning speed of 0.5 nm

[13a] B. Chance, V. Legallais, and B. Schoener, *Rev. Sci. Instrum.* **34**, 1307 (1963).

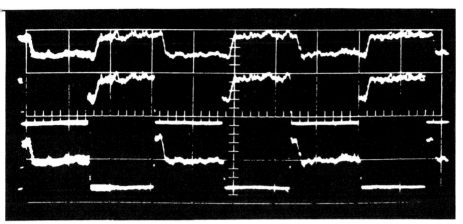

FIG. 2. Oscilloscope photograph of various waveforms in the instrument. Top, the output of the photomultiplier preamplifier. Middle, output waveform of the sample channel demodulating relay. Bottom, output waveform of the reference channel demodulating relay. Vertical 5 V/cm, horizontal 10 msec/cm.

per second. During the dark half-cycles, the contacts of the demodulating relays are grounded.

Several lock-in techniques could be used including commercial tuned amplifiers. However, it was found that the relatively inexpensive photo-transistors and mercury relays formed a good demodulation circuit. The filtered direct current voltages are substracted by a Keithley model 153 differential voltmeter and its output is recorded on a Hewlett-Packard 7590B (RS) X-Y recorder. The X input voltage of this recorder is derived from a 10-turn potentiometer and a voltage source. The potentiometer is coupled to the wavelength drum of the monochromator by a precision gear. The spectrum is scanned by a stepping motor, each step corresponding to a 0.5-nm increment. The speed of scanning is controlled by a low frequency oscillator.

In typical operation the photomultiplier is operated at 1200 V and the outputs of the filters at the peak of the emission band are around 8 V. To amplify small differences the sensitivity of the differential voltmeter is switched to scales which are 10 or 30 times more sensitive.

Performance

For the work discussed herein, the excitation wavelength was always 280 nm while the emission was scanned from 300 to 430 nm. Light output of the lamp decreases appreciably at this excitation wavelength and even though the monochromator grating was blazed for 3000 Å, the overall efficiency of the system is quite low. With these conditions a 10% difference could be measured with a 1% accuracy.

Experimental Procedure

N-Acetyl-L-tryptophanamide (Mann, mp 192–194°) was used as purchased.

Lysozyme (Worthington) was used without further purification. Indole (Matheson Coleman and Bell, lot 384224) was recrystallized from methanol in the dark.

The solutions of N-acetyl-L-tryptophanamide were made by first preparing a stock solution in twice-distilled water and then diluting with either twice-distilled water or reagent grade methanol to equal volumes. The OD_{279} was 0.152 for N-acetyl-L-tryptophanamide in methanol (1% H_2O) and 0.143 in water. The solutions of indole were made by preparing a stock solution in ethanol (USP-NF grade) and then diluting with twice-distilled water or reagent grade methanol. The OD_{270} was 0.157 for indole in methanol (1% ethanol) and 0.151 in water (1% ethanol). The solutions

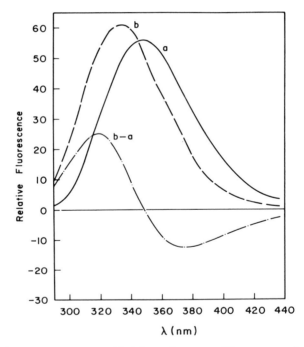

Fig. 3. Fluorescence spectra of indole: a, in water, OD_{270} = 0.151; b, in methanol, OD_{270} = 0.157; b − a, the difference spectrum. Instrumental conditions: photomultiplier tube voltage, 1120 V; full scale of the Keithley voltmeter, 10 V; Y range of the recorder, 50 mV/inch.

for lysozyme were prepared by making stock solutions of twice the desired concentration in 0.2 M NaCl, and diluting in half with either 0.2 M NaCl or tri-N-acetyl-D-glucosamine dissolved in 0.2 M NaCl at the appropriate pH. The final lysozyme solutions had an OD_{279} of about 0.2 (when using the split cells) or 0.1 when using standard fluorescence cells, and the final concentration of inhibitor was slightly greater than 0.2%.

For all but one of the set of spectra described, standard fluorescence cells (40 × 10 × 10 mm) were used. For the set of spectra of lysozyme plus tri-N-acetyl-D-glucosamine versus lysozyme, split cuvettes were used (40 × 10 × 10 mm was divided with a 1-mm quartz plate giving each compartment a 4.5-mm light path). One split cell contained protein solution in one compartment and solvent plus inhibitor in the other; the second split cell contained protein plus inhibitor in one compartment and solvent in the other. The blank solutions were placed facing the excitation source. In this way the absorbing or fluorescing effect of a perturbant is canceled out. Both sets of fluorescence cells were purchased from the Optical Cell Company, Beltsville, Maryland.

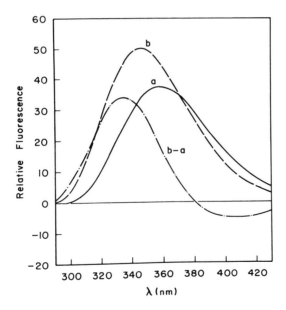

Fig. 4. Fluorescence spectra of N-acetyl-L-tryptophanamide: a, in water, OD_{279} = 0.143; b, in methanol, OD_{279} = 0.143; b – a, the difference spectrum. Instrumental conditions: the sensitivity of the difference spectrum is 1.7 times that of the absolute spectrum; difference spectrum: photomultiplier tube voltage, 1120 V; full scale of the Keithley voltmeter, 3 V; Y range of the recorder, 100 mV/div.; absolute spectrum: photomultiplier tube voltage, 1120 V, full scale of the Keithley voltmeter, 10 V; Y range of the recorder, 50 mV/inch.

CORRECTED PEAK WAVELENGTH FLUORESCENCE VALUES

	λ_{max} (nm)	λ of difference spectra (nm)
Lysozyme, pH 7.0	340	328[a]
Lysozyme, pH 11.4	345	
Lysozyme + tri-NAG[c] at pH 5.6	325	
		310[a]
		365[b]
Lysozyme at pH 5.6	338	
N-Acetyl-L-tryptophanamide in H_2O	350	
		331[a]
		417[b]
N-Acetyl-L-tryptophanamide in methanol	341	
Indole in H_2O	340	
		309[a]
		370[b]
Indole in methanol	318	

[a] Maximum.
[b] Minimum.
[c] Tri-N-acetyl-D-glucosamine.

It is necessary to adjust for small differences in the total optical paths of the two components. Mechanical shutters, attached directly to the two-cell compartments, enable the operator to equalize the fluorescence intensity of a control solution placed in each compartment at a single wavelength. The difference spectrum between two such control solutions constitutes the baseline of the instrument. This baseline is subtracted from the difference between the control solution and a sample solution to give the true difference.

The spectra obtained from the difference fluorometer are not corrected for variation in sensitivity of the photomultiplier tube, the xenon lamp, or the efficiency of the monochromator. It is possible to obtain a correction curve over a limited spectral range by measurement of the corrected fluorescence intensity of, say, indole dissolved in diethyl ether, in methanol, and in water, on a Zeiss ZFM 4C spectrofluorometer.[14] This corrected value is divided by the fluorescence intensity of the same solution measured on the difference fluorometer, when the appropriate solvent is used as

[14] S. S. Lehrer and G. D. Fasman, J. Amer. Chem. Soc. 87, 4687 (1965).

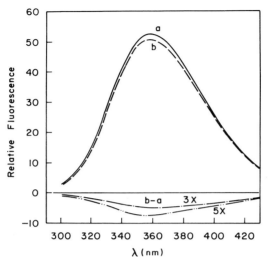

Fig. 5. Fluorescence spectra of N-acetyl-L-tryptophanamide in water at two different concentrations: a, $OD_{279} = 0.143$; b, $OD_{279} = 0.140$; b − a, the difference spectrum at the sensitivities indicated relative to the absolute spectra. Instrumental conditions for the 3 X difference spectrum are: photomultiplier tube voltage, 1150 V; full scale of the Keithley voltmeter, 3 V; Y range of the recorder, 50 mV/div. For the 5 X difference spectrum: photomultiplier tube voltage, 1150 V; full scale of the Keithley voltmeter, 10 V; Y range of the recorder, 10 mV/inch. The conditions for the absolute spectrum are: photomultiplier tube voltage, 1150 V; full scale of the Keithley voltmeter, 10 V; Y range of the recorder, 50 mV/inch.

the reference solution. This ratio (f), which varies with wavelength, constitutes our correction curve. To obtain the corrected fluorescence value of a sample at a particular wavelength one uses the relationship

$$f_c = (f)(f_d)$$

where f is the correction factor, f_d is the fluorescence intensity of the sample measured on the difference fluorometer, and f_c is the corrected fluorescence intensity of the sample. The correctness of this method was shown by use of a solution such as L-tryptophan in water. The peak value of the corrected fluorescence curve (f_c) was 351 nm.[15,16]

For all the spectra measured, the following variables were constant: cell compartment temperature, 24°; monochromator entrance and exit slit widths, 4 mm; slit height, open. Other instrumental variables are given in the figure legends.

[15] F. W. J. Teale and G. Weber, *Biochem. J.* **65**, 476 (1957).
[16] G. M. Bhatnagar, L. C. Gruen, and J. A. MacLaren, *Aust. J. Chem.* **21**, 3005 (1968).

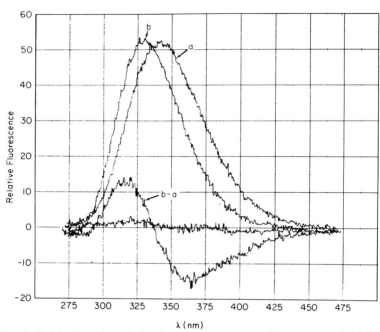

Fig. 6. Original tracing of the fluorescence spectra of lysozyme and inhibitor in water at pH 5.1: a, pure enzyme; b, with tri-N-acetyl-D-glucosamine; b − a, the difference spectrum. Instrumental conditions for the difference and absolute spectra are: photomultiplier tube voltage, 1300 V; full scale of the Keithley voltmeter, 3 V; Y range of the recorder, 100 mV/inch.

Application and Discussion

It is known from previous work that a change in the solvent medium can cause a corresponding change in the fluorescence intensity as well as a shift in the spectrum of a fluorescence solute.[17] In general a decrease in the polarity of the solvent causes an increase in fluorescence and a corresponding blue shift in the fluorescence spectrum. Utilizing this fact, the usefulness of the difference technique in fluorescence can be easily shown. In Fig. 3 is shown the fluorescence of indole in methanol and in water as well as the difference fluorescence spectrum. Lowering the polarity, e.g., water to methanol, of the media causes a blue shift and increase in fluorescence. The difference curve yields a long wavelength negative band and a short wavelength positive band. The corrected peak values for this and following figures are found in the table. The fluorescence spectra of N-acetyl-L-tryptophanamide in methanol, in water, and their difference

[17] B. L. Van Durren, J. Org. Chem. 26, 2954 (1961).

fluorescence spectra is shown in Fig. 4. These spectra are similar to those of the indole curves; the peak shift is not as large, while the increase in fluorescence in methanol is relatively larger. The difference spectrum is composed of a small negative long wavelength band and a large positive short wavelength band. The spectra in Fig. 5 are of N-acetyl-L-trypto-phanamide in water at two concentrations. Here the negative difference curve is the equivalent of that expected for quenching without a shift in emission. The utility of the instrument may be viewed from the small difference between the absolute spectra, which may be expanded in the difference mode by switching to a higher sensitivity. In this particular case there is no spectral shift, but a small difference may be accompanied by a shift, which could be easily evaluated by enlarging the difference spectrum.

It has been shown from the work of Lehrer and Fasman[9] that 3 of the

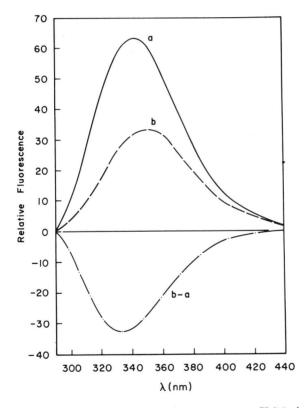

Fig. 7. Fluorescence spectra of lysozyme in water: a, at pH 7.0; b, at pH 11.4; the difference spectrum b − a. Instrumental conditions for the difference and absolute spectra are: photomultiplier tube voltage, 1300 V; full scale of the Keithley volt-meter, 3 V; Y range of the recorder, 100 mV/inch.

6 tryptophan residues in lysozyme are affected by changes which occur in or near the active site of the enzyme. In Fig. 6, which is a photograph of the original tracing, the effect of inhibitor on the enzyme at pH 5.1 is seen. In this pH region a carboxyl group is protonated, and causes a fluorescence decrease in one of the tryptophan residues when inhibitor is added. This can be observed from the negative portion of the difference spectrum. In addition, the binding of inhibitor causes the environment about one or two tryptophan residues to become less polar, and is seen as the positive band of the difference spectrum which is shifted toward the blue relative to the absolute spectra. In the free enzyme a change in pH from 7.0 to 11.4 causes a quenching of one of the tryptophan residues without a corresponding change in the conformation of the enzyme.[9] The fluorescence curve obtained for lysozyme at pH 7.0 is seen in Fig. 7, curve a, and at pH 11.4 as curve b. The negative difference curve (b − a) is that expected for quenching of a tryptophan with emission maximum at 330 nm, namely a tryptophan moiety in a nonpolar environment. This quenching is most likely the result of energy transfer to an ionized tyrosine residue.[18,19] It should be stressed here that only through use of the difference technique could the fluorescence contributions of these three tryptophan residues be separated in lysozyme.

Difference fluorescence spectroscopy offers a sensitive method to detect protein conformational changes involving the aromatic residues and to reveal small emission differences in aromatic molecules. Similar studies involving attached fluorescent probes are feasible. Used concurrently with solvent perturbation or quenching species it can be used to probe the topology of the surface of proteins to evaluate the accessibility of aromatic residues.

[18] R. F. Steiner and H. Edelhoch, *Nature (London)* **192**, 873 (1961).
[19] R. W. Cowgill, *Arch. Biochem. Biophys.* **100**, 36 (1963).

Section V

Resonance Techniques: Conformation and Interactions

[33] Nuclear Magnetic Resonance
Spectroscopy of Proteins

By W. D. Phillips

Over the past ten years, X-ray crystallography has provided us with unprecedented insight into the structures and modes of action of a range of proteins. The pioneering work of Kendrew[1] on myoglobin showed for the first time the folding of a protein into its three-dimensional structure, and formed the basis for detailed deductions regarding relationships between structure and biological function. The first determination of the X-ray structure of an enzyme was that of lysozyme, and the results permitted Phillips and collaborators[2] to draw general conclusions regarding the nature of enzyme action that went far beyond the activity of lysozyme. Similarly, the work of the groups of Perutz on hemoglobin,[3] Richards on ribonuclease,[4] Dickerson on cytochrome c,[5] and Jensen[6] and Kraut[7] on the iron-sulfur proteins has established quantitative structural bases for useful applications to proteins of other physical techniques that at best are capable of providing only limited quantitative structural information. Nuclear magnetic resonance (NMR) spectroscopy is one of these techniques, whose value in applications to protein structure and function has been greatly enhanced by the availability of the results of X-ray crystallography. In every sense, however, X-ray crystallographic and NMR techniques, when applied to proteins, are complementary rather than competitive. The X-ray method is inapplicable to molecules in solution, a state which is required for observation of so-called high resolution spectra and a state which more often than not resembles that of the protein in its physiological environment. However, without the

[1] J. C. Kendrew, H. C. Watson, B. E. Strendberg, R. E. Dickerson, D. C. Phillips, and V. C. Shore, *Nature (London)* **190**, 665 (1961).
[2] C. C. F. Blake, D. F. Koenig, G. A. Mair, A. C. T. North, D. C. Phillips, and V. R. Sarma, *Nature (London)* **206**, 757 (1965).
[3] M. F. Perutz, H. Muirhead, J. M. Cox, and L. G. C. Goaman, *Nature (London)* **219**, 131 (1968).
[4] H. W. Wyckoff, K. D. Hardman, N. M. Allewell, T. Inagami, L. N. Johnson, and F. M. Richards, *J. Biol. Chem.* **242**, 3984 (1967).
[5] R. E. Dickerson, T. Takano, D. Eisenberg, O. B. Kallai, L. Samson, A. Cooper, and E. Margoliash, *J. Biol. Chem.* **246**, 1511 (1971).
[6] K. D. Watenpaugh, L. C. Sieker, J. R. Herriott, and L. H. Jensen, *Cold Spring Harbor Symp. Quant. Biol.* **36**, 359 (1971).
[7] C. W. Carter, Jr., S. T. Freer, N. H. Xuong, R. A. Alden, and J. Kraut, *Cold Spring Harbor Symp. Quant. Biol.* **36**, 381 (1971).

prior results of the X-ray crystallographers, NMR spectroscopists on occasion have been guilty of making claims of the potential of their technique to elicit structural information that have not been realized in practice.

With this caveat in mind, the promise of the NMR approach to study of the structures and interactions of proteins in solution is very bright. It is the one spectroscopic technique that we possess that is capable, in principle at least, of providing spectral responses that are identifiable with specific nuclei throughout the three-dimensional structure of a protein in solution. NMR is without doubt the tool most widely employed by chemists for the elucidation of molecular structure. Useful application of NMR to proteins was, however, impossible until fairly recently because of the size and complexity of proteins. Proteins with their many nuclei of more or less continuously varying chemical shift environments produced, prior to about 1966, proton magnetic resonance (PMR) spectra that were virtually absorption envelopes with few resolved features. This situation was altered with the introduction of the Varian PMR spectrometer that operated at a resonance frequency of 220 MHz and a polarizing field, furnished by a superconducting solenoid, of 52,000 G. Resolution at 220 MHz, although only quantitatively superior to, say, that of a 100 MHz PMR spectrometer, was sufficient that work could begin on assignment of resolved resonances of a few of the component protons of a protein. Today, high resolution PMR spectrometers exist that operate at resonance frequencies of 250 MHz, 270 MHz, and 300 MHz, and there are prospects that PMR spectrometers capable of the higher dispersion that even larger polarizing fields would confer will be designed and introduced.

Another difficulty that has plagued NMR spectroscopists who have attempted to apply their technique to proteins is that of sensitivity. NMR is intrinsically one of the least sensitive spectroscopic techniques available. The strength of an NMR response of an ensemble of nuclei depends on the extent of nuclear spin polarization produced by a magnetic field and on the rate at which the nuclear spin polarization is reestablished once the system has been saturated by the absorption of radiofrequency energy. This latter process is characterized by T_1, the longitudinal relaxation time. Nuclear spin relaxation times characteristically are long, of the order of 10^{-3} seconds or greater, and the extent of nuclear spin polarization is small (of the order of one part in 10^6) because of the very small differences in energy that separate the various spin states under the influence of laboratory magnetic fields. The problem of low signal-to-noise ratios is severe in the proton magnetic resonance spectroscopy of proteins, but effectively precludes in the absence of signal enhancement techniques application of ^{13}C and ^{15}N magnetic resonance spectroscopies. The natural abundances of ^{13}C and ^{15}N are only 1.1% and 0.4%, respectively. In addition, the small nuclear moments of ^{13}C and ^{15}N lead to nuclear spin polarizations considerably less than that for ^{1}H at the same temperature and in the same strength magnetic field.

The problem of low signal-to-noise ratios in NMR has been greatly ameliorated by the introduction, first, of computer averaging techniques and, more recently, of Fourier transform spectroscopy. Signal-to-noise ratios increase as $n^{1/2}$, where n is the number of repetitive scans of a spectrum that are stored and processed by an associated computer. Thus, in principle, the signal-to-noise ratios of ^{1}H, ^{13}C, or ^{15}N spectra of a protein can be increased to any desired value, depending on the temporal stability of the system under study and the patience of the investigator. The introduction of Fourier transform techniques into spectroscopy in general and NMR in particular must be accounted one of the most important advances in scientific instrumentation of this century, and certainly will be in large measure responsible for whatever future advances in our knowledge of proteins with which NMR is credited.[8]

In this brief essay, no attempt will be made to review exhaustively applications of NMR to proteins that have been completed or are in progress. Rather, the philosophy, as the present author sees it, of the application of NMR to protein conformation will be discussed. The discussion will be further restricted to observations of resonances that derive from nuclei that are components of the protein; resonances of substrates or inhibitors interacting with enzymes will not be dealt with here.

It is trite, but nonetheless true, to point out that, in order for NMR as applied to proteins to be useful in anything other than empirical fashion, resonances attributable to single nuclei or structurally equivalent groups of nuclei in a protein must be resolvable and assignable. It was anticipated early that the resolution of resonances from single nuclei or structurally equivalent groups of nuclei in a protein would be a formidable task, and this indeed has proved to be the case. Other things being equal, protons would be the nuclei of choice in NMR studies of proteins because of their essentially 100% natural abundance, their relatively large intrinsic sensitivity in NMR absorption, their ubiquitous distribution throughout the three-dimensional structure of the protein, and their involvement in such important structural and dynamic aspects of proteins as hydrogen bonding, ionization, and proton mobility. Unfortunately, the proton exhibits the smallest range of chemical shifts in

[8] T. C. Farrar and E. D. Becker, "Pulse and Fourier Transform NMR." Academic Press, New York, 1971.

nuclear magnetic resonance of any nucleus possessed by proteins. Since, however, the chemical shift is linearly dependent on the polarizing field or, equivalently, the radiofrequency at which resonance absorption is observed, resolution of the PMR spectra of proteins has been vastly improved by the introduction of spectrometers based on the superconducting solenoid. One can only encourage those interested in instrumentation in their efforts in the difficult task of developing spectrometers that operate beyond the highest presently available PMR resonance frequency of 300 MHz. It is perhaps appropriate to point out here that the present cost of a PMR spectrometer that is equipped with Fourier transform capability and operates in the 220–300 MHz range is upward of $250,000. This formidable figure represents a commitment that cannot be entered upon lightly by investigator, institution, or granting agency. And unless there are presently unforeseen technological advances in radiofrequency and, particularly, magnet technology, PMR spectrometers that operate at resonance frequencies higher than 300 MHz are inevitably going to require investments substantially above this level. Much thought should be given to selection of appropriate sites for the location of a limited number of such spectrometers and to making the instruments broadly available to investigators with significant problems that require their use.

Much more progress has been made in the resolution and assignment of protein PMR resonances that derive from single protons or equivalent groups of protons that might have been anticipated. First, it was found that to the resolution achievable on PMR spectrometers presently in operation, chemical shifts of hydrogen atoms of side chains of denatured proteins are very similar to those of the component amino acids under the same environmental conditions, i.e., temperature, pH, salt concentration. The peptide linkage, while definitely affecting chemical shifts of α-CH protons has little effect on the chemical shifts of carbon-bound, side-chain protons of denatured proteins. And in fact it has been shown that the PMR spectra of carbon-bound, side-chain protons of denatured proteins (Fig. 1b) can be well simulated from the summation of the spectra of the component amino acids, suitably weighted by the amino acid composition of the protein (Fig. 1a).[9] In retrospect, this perhaps is not so surprising since the side-chain protons are considerably removed from perturbations of the peptide linkage and their solvation environments are probably quite similar to those of the free amino acids.

The PMR spectra of proteins in their folded, biologically active forms (Fig. 1c) generally are quite different from those of the denatured protein

[9] C. C. McDonald and W. D. Phillips, J. Amer. Chem. Soc. 91, 1513 (1969).

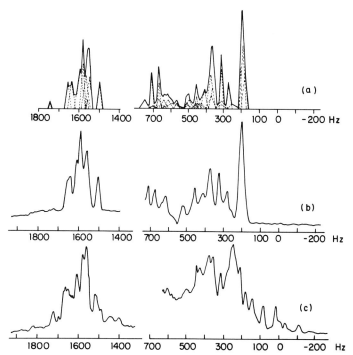

FIG. 1. Paramagnetic resonance spectra (220 MHz) of hen egg white lysozyme. (a) Schematic representation, pD 7.0, 40°C. Low-field intensity ×4; (b) and (c) 10% lysozyme in D_2O, pD 5.0; (b) random-coil spectrum 80°C, low-field amplified ×4.1 relative to high-field; (c) native spectrum 65°C, low-field amplified ×2.3 relative to high-field, preheated to remove NH proton resonances. From C. C. McDonald and W. D. Phillips, *J. Amer. Chem. Soc.* **91**, 1513 (1969).

(Fig. 1b). As the preceding paragraph would imply, such differences would result from alteration in the local shielding environments of the side-chain protons in proceeding from a situation where intramolecular perturbations are virtually absent to one in which such perturbations are important chemical shift determinants. Such perturbations that exist in the folded protein frequently remove coincidences of resonances that exist in the denatured protein and have led to resolution of resonances of single protons or equivalent groups of protons that are identifiable with structurally unique hydrogen atoms of the protein. The magnitude of the problem of PMR resolution in proteins is indicated by the fact that a relatively small protein of 15,000 molecular weight contains about 1200 hydrogen atoms. Some examples where resolution has been achieved and assignments made are indicated below.

A major intramolecular perturbation that is manifested by rather large chemical shift displacements in the PMR spectra of folded forms of proteins derives from the ring-current fields that are associated with aromatic π-electron ring systems. Examples of protein constituents with which are associated such ring-current fields are the phenyl rings of phenylalanine and tyrosine residues, the indole rings of tryptophan residues, and the porphyrin rings of the heme proteins. Ring-current fields exhibit very sensitive angular dependences and fall off as d^{-3}, where d is the distance between the origin of the field and the perturbed nucleus. For example, resonances of a group positioned above or below a phenyl ring will be displaced to high field by the ring-current field, while resonances of groups positioned in the plane of the phenyl ring will be displaced to low field. Frequently, resonances of single protons or equivalent groups of protons, e.g., a methyl group, are displaced to high field by ring-current field effects beyond the complex, poorly resolved bulk of resonances of the majority of the side-chain protons of a folded protein. Such resonances are particularly prominent in the PMR spectrum of native lysozyme, which, in addition to possessing phenylalanine and tyrosine residues, contains six tryptophan residues with their extended π-electron systems. In thermally denatured lysozyme, the 220 MHz PMR absorption is terminated at the high-field end of the spectrum by the intense resonance at 220 Hz (Fig. 1b). From knowledge of the PMR spectra of the component amino acids, this resonance can be assigned to the methyl protons of the component leucine, isoleucine, and valine residues. Upon renaturation, the intense 200 Hz resonance splits, as a result of intramolecular perturbation, into a series of resolved resonances, a number of which are displaced to high field and are to be found in the 180 Hz to -200 Hz region of resonance absorption of Fig. 1c. Ring-current field effects, largely deriving from the six tryptophan residues of lysozyme, are primarily responsible for these displacements.[10,11] As might be expected, ring-current field effects deriving from the extensively de-localized π-electron systems of the component porphyrin rings figure prominently in the PMR spectra of cytochrome c, myoglobin, and hemoglobin.[12,13]

Resonances shifted by ring-current field effects to the extreme high-field region of resonance absorption of native proteins, thus, often can be separately resolved and identified with individual protons or equivalent groups of protons of the protein. Continuing from high-field to low-field in the PMR absorption of native proteins, a broad envelope of

[10] C. C. McDonald and W. D. Phillips, in "Fine Structure of Proteins and Nucleic Acids" (G. D. Fasman and S. N. Timasheff, eds.), p. 1. Dekker, N.Y., 1970.
[11] H. Sternlicht and D. Wilson, *Biochemistry* 6, 2881 (1967).
[12] K. Wüthrich, *Struct. Bonding (Berlin)* 8, 53 (1970).
[13] R. G. Shulman, S. H. Glarum, and M. Karplus, *J. Mol. Biol.* 57, 93 (1971).

poorly resolved resonances is encountered that derive, first, from saturated CH protons (200 Hz to 1000 Hz, Fig. 1c) and then from aromatic CH protons (1400 Hz to 1900 Hz, Fig. 1c). In D_2O as solvent where exchangeable protons (NH, OH, SH) are replaced by deuterium, the low-field region of resonance absorption of diamagnetic proteins is terminated by resonances of the C-2 protons of any histidine residues the protein may possess. Ribonuclease contains four histidine residues, and resonances from the C-2 protons are separately resolved in the PMR spectrum of the native protein. Jardetzky and co-workers have assigned these resolved resonances to specific residues and have carried out important active site studies of ribonuclease utilizing principally these assigned C-2 histidine proton resonances.[14]

When H_2O rather than D_2O is employed as solvent, resonances attributable to at least part of the exchangeable protons of the protein (NH, SH, and OH) can be observed in the PMR spectrum. A very strong resonance that derives from the protons of H_2O inevitably is introduced for this choice of solvent, but aside from the α-CH proton resonances, most of the CH proton resonances still are observable. Not all resonances of exchangeable protons are resolvable, even in principle, for H_2O solutions of proteins. Depending on environmental conditions, i.e., temperature, pH, many exchangeable protons are undergoing sufficiently rapid exchange with the protons of water that their resonances are coalesced with that of the strong water resonance. Reductions of temperature and/or alterations of pH frequently result in reductions of rates of proton exchange to the point where the resonances of such protons are differentiable from that of H_2O.

Most, if not all, of the amide NH protons of proteins undergo sufficiently slow exchange with those of H_2O that, even in the denatured form of the protein, resonances of amide protons are separately resolvable from those of water. In the folded forms of the proteins, features of the secondary and tertiary structures of the protein such as intramolecular hydrogen bonding and accessibility to solvent in the three-dimensional structure exert profound influences on the rates of exchange of the amide NH protons.[15] This potentially profitable area has been little explored by NMR.

Component tryptophan residues of enzymes often are involved at the active sites. This certainly is the case for hen egg white lysozyme, where three of the six such residues are located at the active site, and the indole NH protons of two of these three residues actually are utilized in binding

[14] D. H. Meadows, O. Jardetzky, R. M. Epand, H. H. Ruterjans, and H. A. Scheraga, *Proc. Nat. Acad. Sci. U.S.* **60**, 766 (1968).

[15] A. Hvidt and S. O. Nielsen, *Advan. Protein Chem.* **21**, 287 (1966).

substrate and inhibitor.[2] The PMR absorption of the indole CH protons of the tryptophan residues is sufficiently complex, particularly in the folded form of the protein, that resolution of resonances separately attributable to the component tryptophan residues is out of the question. In H_2O as solvent, however, a single resonance to extreme low field is detected in the PMR spectrum of denatured lysozyme whose intensity corresponds to six protons per molecule of lysozyme. This resonance derives from the indole NH protons of the six tryptophan residues, and the coincidence of chemical shifts in thermally denatured lysozyme is attributed to the absence of tertiary structure. Upon renaturation, this resonance splits into six resonances, each of intensity corresponding to a single proton. Through inhibitor perturbation, deuterium exchange kinetics, and chemical modification of the protein, five of the six were uniquely identified with specific tryptophan residues.[16] Involvement of exchangeable protons in the coupling of subunits to form quaternary structures as in hemoglobin has been evidenced by low-field displacements and reduced rates of proton exchange with water.[17]

It was early recognized that the ideal way, albeit difficult and expensive, to simplify the complex and generally badly overlapped PMR spectra of proteins was to replace with deuterium substituted amino acids most of the amino acids of the protein. In this way, only a few of the twenty different types of amino acids would continue to possess CH protons, thus greatly simplifying the spectrum of the partially deuterated protein and permitting resolution and identification of the ¹H resonances of the remaining nondeuterated residues. The feasibility of this approach has been demonstrated by Jardetzky and co-workers on a nuclease from *Staphylococcus aureus*[18] and by Crespi, Katz, and co-workers on a number of partially deuterated molecules derived from a variety of microorganisms grown on D_2O.[19] The expense and effort required to produce selectively deuterated protein is great, but the effort is more than repaid in the information that becomes available in the well-resolved PMR spectra of such enzymes. It can be confidently predicted that selective deuteration, as well as selective introduction of ¹³C and ¹⁵N for NMR purposes, will become an increasingly popular and profitable activity.

[16] J. D. Glickson, W. D. Phillips, and J. A. Rupley, *J. Amer. Chem. Soc.* **93**, 4031 (1971).

[17] D. J. Patel, L. Kampa, R. G. Shulman, T. Yamane, and M. Fujiwara, *Biochem. Biophys. Res. Commun.* **40**, 1224 (1970).

[18] I. Putter, J. L. Markley, and O. Jardetzky, *Proc. Nat. Acad. Sci. U.S.* **65**, 395 (1970).

[19] H. L. Crespi, H. F. Daboll, and J. J. Katz, *Biochim. Biophys. Acta* **200**, 26 (1970).

To reiterate, the principal objective of NMR spectroscopists who have employed PMR spectroscopy to study proteins has been the resolution and identification of resonances that derive from single protons or equivalent groups of protons. This has proved to be possible at the high-field end of the PMR spectrum by virtue of intramolecular ring-current field effects. In the low-field region of resonance absorption of proteins dissolved in D_2O, certain aromatic CH proton resonances have been useful for purposes of conformational analysis, in particular those of the C-2 protons of histidine residues. With H_2O as solvent, certain exchangeable protons exhibit resolved resonances to low field; resonances deriving from indole NH protons of tryptophan residues have been particularly useful. Selective isotope replacement and enrichment clearly is the ideal way to simplify the more or less unresolved, broad middle portion of PMR absorption of proteins.

The previous discussion has concerned diamagnetic proteins. The presence of paramagnetic centers such as occurs in the heme proteins and in the iron-sulfur proteins introduces a strong perturbation on the positions of resonances of proximal nuclei. This perturbation can be either of the pseudocontact variety or derive from hyperfine contact interaction. In the former, the contact interaction is essentially dipolar in nature and falls off as d^{-3} from the paramagnetic center and generally exhibits a strong angular dependence. The hyperfine contact contribution derives from the electron-nucleus interaction first described by Fermi that gives rise, also, to the hyperfine splittings often observed in electron paramagnetic resonance spectra.

Independent of the origin of the contact interaction, contact shifts frequently displace the resonances of affected protons sufficiently far to low-field or high-field that they are easily resolved from the body of proton resonances of the protein. In fact, it is not uncommon for contact interaction perturbations to be so large as to displace resonances outside the sweep ranges of some PMR spectrometers. Concomitant with the displacement is the effect of the paramagnetic center on the width, or transverse relaxation time, of the contact-shifted resonance. Instead of line widths of 5 Hz to 50 Hz commonly encountered for diamagnetic proteins, widths of contact-shifted resonances of paramagnetic proteins can vary from the above quoted range to widths of thousands of Hertz. Thus, contact-shifted resonances can be undetectable, in a practical sense, on conventional high-resolution PMR spectrometers. In addition to their displacement and widths, a criterion for contact shifted resonances is that they exhibit strong temperature dependences that frequently parallel the temperature dependence of the paramagnetic component of the magnetic susceptibility of the protein. This latter feature frequently is absent in the iron-sulfur proteins, where strong antiferromagnetic spin exchange coupling exists between component iron atoms; local contributions to the overall magnetic susceptibility of individual iron-sulfur proteins are reflected in temperature dependences of contact-shifted resonances that may vary not only in magnitude but in sign as well.

The contact-shifted resonances that have been observed in the heme proteins (cytochrome c, myoglobin, hemoglobin)[12] and the iron-sulfur proteins (the two-iron ferredoxins from plants, the four-iron high potential iron proteins from photosynthetic bacteria, and the eight-iron ferredoxins from anerobic bacteria)[20] have been of great value in elucidating the geometric and electronic structures of these proteins. Since the results have been reviewed elsewhere, we only reemphasize the utility of contact-shifted resonances in the classes of proteins for which they have been and will be observed. Because of the rather long electronic relaxation times of the paramagnetic centers in proteins that contain $Cu(II)$ and in proteins for which the paramagnetic $Mn(II)$ can replace $Mg(II)$, contact-shifted resonances have not been observed, nor is it likely that they will be. An interesting exclusion of observability of EPR signal and contact-shifted resonances in NMR exists here.

Contact shifts observed in the PMR spectra of the heme proteins and the iron-sulfur proteins arise from the paramagnetism of the paramagnetic iron atoms that are integral parts of the protein. Because of the opening up of the PMR spectrum of a protein that results from contact interaction perturbation, it clearly would be advantageous for purposes of resolution and assignment if contact interaction displacements could be induced in the PMR spectrum of a protein that does not contain an endogenous paramagnetic center by the addition of extrinsic paramagnetic ions. For lysozyme this has proved to be possible by addition of $Co(II)$[21] and $La(III)$[22] to aqueous solutions of the enzyme. In both instances it appears that the paramagnetic ions bind loosely to a single site near the surface of the enzyme. Coordination to the enzyme is by way of an aspartic acid and a glutamic acid, and the rate of binding and release of the ion is sufficiently rapid that the observed proton resonances of lysozyme are time-averages of the spectra of the enzyme in the presence and in the absence of the paramagnetic ion. In practice, what is observed is a monotonic increase with increase in concentration

[20] W. D. Phillips and M. Poe, in "The Iron-Sulfur Proteins" (W. Lovenberg, ed.). Academic Press, in press.

[21] C. C. McDonald and W. D. Phillips, Biochem. Biophys. Res. Commun. **35**, 43 (1969).

[22] I. D. Campbell, C. M. Dobson, R. J. P. Williams, and A. V. Xavier, Ann. N.Y. Acad. Sci. (1973), in press.

of the paramagnetic ion in the magnitudes of contact displacements for protons subject to contact interaction perturbation. In the study involving the La(III) ion and lysozyme, a spectral subtraction scheme employing a computer was developed that yielded the resonances with associated nuclear spin couplings of those resonances that were affected by contact interaction. The La(III) approach to contact displacement offers promise as a means of determining partial three-dimensional structures of proteins in instances where the site of interaction can be independently determined and where something is known about the angular dependence of the contact interaction perturbation.

In principle the contact-shift perturbations on the PMR spectrum of lysozyme that result from introduction of the paramagnetic Co(II) and Pr(III) ions are translatable into distance and angular distributions of the perturbed nuclei throughout the three-dimensional structure of the protein. In practice this proves difficult because of the problems associated with identification of resolved resonances with specific nuclei and the coupled nature of the distance and angular dependences of the pseudocontact interaction.

While we are principally concerned in this brief review with factors that affect the positions of resonances of proteins, effect of paramagnetic centers on the breadths of nuclear resonances appears to be the most promising approach that exists for obtaining intramolecular distance parameters of molecules in solution. Electron-nucleus dipolar relaxation falls off as d^{-6}. A number of workers have used this effect to study enzyme kinetics and geometry in systems where a paramagnetic ion such as Mn(II) is involved at the active site.[23] In other instances stable organic free radicals have been covalently bound to specific sites of proteins and intranuclear distances have been derived from the resulting broadening of nuclear resonances. Recently the potent ability of Gd(III) to relax nuclei and the fact that lysozyme binds metals most strongly at a single unique site have been combined with computer spectral subtraction and deconvolution techniques to study the solution structure of lysozyme.[22] Clearly, these NMR approaches to quantitative determination of protein structure in solution, while most promising and certain of extension, are going to be of greatest value when combined with prior X-ray analysis of the protein in the crystalline state.

In conclusion, it can be asserted that of all spectroscopic techniques in hand and on the horizon, NMR offers the greatest promise for elucidation of protein structure and conformation in solution. PMR spectroscopy

[23] E. L. Packer, H. Sternlicht, and J. C. Rabinowitz, *Proc. Nat. Acad. Sci. U.S.* **69**, 3278 (1972).

of proteins is an established approach, and sufficient background information has been obtained over the past few years to permit assessment of its virtues and limitations as a protein probe. With only the recent introduction of appropriate spectrometer and signal enhancing techniques, the breadth of the applicability of ^{13}C magnetic resonance spectroscopy to proteins remains to be defined. The potential, however, is suggested by the recent ^{13}C studies on electron exchange between the oxidized and reduced forms of ferredoxin from *Peptococcus aerogenes*[23] and the detection of separate ^{13}C signals for CO bound to the α- and β-chains of hemoglobin.[24]

[24] R. B. Moon and J. H. Richards, *J. Amer. Chem. Soc.* **94**, 5093 (1972).

[34] The Prospects for Carbon-13 Nuclear Magnetic Resonance Studies in Enzymology

By FRANK R. N. GURD and PHILIP KEIM

Introduction

The development of pulsed Fourier transform methods for the observation of magnetic resonance properties of the ^{13}C nucleus promises to have important applications in enzymology and protein chemistry in general. The following essay is intended to set out some of the more obvious possibilities for application that are apparent at this early stage.

It is convenient and helpful to develop the discussion of ^{13}C nuclear magnetic resonance (NMR) with reference to ^{1}H NMR, which has been applied far more widely. In either case we have first to consider the types of information that may be obtained, such as chemical shifts, spin–spin coupling constants, and rates of relaxation processes. A qualitative presentation of these concepts is introduced first to provide a general basis for appreciating the more detailed theory and the potential applications of ^{13}C NMR spectroscopy to proteins.

Glossary of Frequently Used Symbols

$G(\tau)$ The autocorrelation function
H_0 Applied external magnetic field
H_1 Various radio frequency fields as designated in the text
$\mathcal{H}'_i(t)$ Time-dependent interaction Hamiltonian for relaxation mechanism "i"

h	Planck's constant $= 1.05443 \times 10^{-27}$ erg · seconds
$\hbar$	Planck's constant divided by 2π
I	Spin quantum number
$\bar{I}, \bar{S}$	Spin angular momentum vectors
$J(\omega)$	Spectral density for frequency ω obtained by Fourier transformation of the autocorrelation function
k	Boltzmann's constant $= 6.62377 \times 10^{-16}$ erg · deg^{-1}
$\bar{M}$	Net macroscopic magnetization; subscripts (without the bar) refer to components along the coordinate axes
T	Absolute temperature
T_{1_i}	Spin-lattice relaxation time (seconds); subscript "i" refers to specific relaxation mechanisms
$T^*_{2_i}$	Spin-spin relaxation time; subscript refers to specific relaxation mechanism and superscript * (if present) refers to inclusion of field inhomogeneity and spectrometer instability
W_{ij}	Transition probability per unit time for exchange between spin states i and j
γ_N	Magnetogyric ratio of nucleus N
μ_N	Magnetic moment of nucleus N
ν	Frequency in sec^{-1}; subscript "o" refers to the resonance or Larmor frequency and all other subscripts are defined in the text
τ_c	Generalized correlation time
τ_e	Correlation time for electron–nuclear scalar interactions
τ_{eff}	Effective correlation time; "effective" includes anisotropic or internal motion or other motional factors not explicitly separable by analysis
τ_{en}	Electron–nuclear dipolar correlation time
τ_g	Correlation time for internal motion of a side-chain methine carbon
τ_m	Residence time of nuclear species in the first coordination sphere of a paramagnetic ion
τ_r	Rotational correlation time for overall molecular motion
τ_s	Electron spin relaxation time
τ_{SC}	Correlation time for scalar coupling
τ_t	Correlation time for translational motion
ω_i	Frequency in rad sec^{-1}; subscript "o" refers to the resonance or Larmor frequency and all other subscripts are defined in the text
—	The barred symbol refers to vectoral quantities; symbol without overbar refers to scalar quantities

General Qualitative Basis

An NMR experiment measures the absorption of radio frequency radiation at a given frequency by an assembly of nuclei exposed to an externally applied magnetic field. The responsive nuclei are those with a nuclear magnetic moment, such as ^1H, ^{13}C, ^{15}N, ^{19}F, and so on. A given nucleus in a compound will be sensitive to magnetic interaction not only with the externally applied magnetic field but also with the local magnetic fields of the surrounding electrons and the neighboring magnetic nuclei. For diamagnetic molecules, in which paramagnetic effects due to unpaired electrons are not prominent, the various magnetic contributions within the compound are quite local, and extend at most a few atoms' distance away. These local magnetic contributions moderate the effective magnetic field at a given nucleus so that it differs from the applied magnetic field.

Chemical Shift. The fact that nuclei in nonidentical chemical environments will experience slightly different local magnetic fields, even in a homogeneous externally applied magnetic field, forms the direct basis of the first type of distinction between nuclei that this method allows. The absorption of the radio frequency radiation coincides with the transition of the nucleus in question between two spin energy levels. For ^1H and ^{13}C nuclei, with spin $\frac{1}{2}$, there are two energy levels and a single transition (promoted in either direction by the radiation) between them. The separation between the energy levels is controlled by the local, or effective, magnetic field at the nucleus in question. Usually the externally applied magnetic field is held constant and the radio frequency is varied so as to offset the local magnetic effects by accurately measured amounts, thus bringing each nucleus in turn into resonance. The resulting NMR spectrum usually shows peaks separated along an abscissa calibrated in frequency units. However, resonances are usually expressed in terms of a dimensionless quantity called the chemical shift. In practice the measured resonance is compared to the resonance position for a standard compound measured concurrently. The ratio of the frequency difference (in Hz) between the two resonance positions to the frequency of a "naked" nucleus (in MHz) defines the chemical shift parameter for the measured resonance. The chemical shift is said to be at the calculated number of parts per million (ppm) relative to the chemical shift of the reference compound. The resonances of ^{13}C nuclei are much more difficult to detect than those of ^1H nuclei because of both the much smaller magnetogyric ratio and the low natural abundance of only 1.1%. ^{13}C chemical shifts reported in this essay are often referenced to external carbon disulfide or ethylene glycol or to internal dioxane.

Splitting Patterns. The quantization of the nuclear moments on which the basic experiment depends has a consequence in causing splitting of signals, when the time-scale of interaction is appropriate. If the effective magnetic field at one nucleus is affected by a nearby nucleus with a spin angular momentum of its own, it follows that the alternative states of the latter will make different contributions, additive and subtractive, to the field at the first nucleus. This is the origin of the multiplet patterns that are so valuable in the interpretation of ^{1}H NMR spectra. In large molecules spectral overlap is a major problem so that the spin-spin splitting is relatively difficult to exploit. The natural abundance of ^{13}C is so low that such coupling between two ^{13}C nuclei is hard to observe. Splitting of ^{13}C by neighboring ^{1}H nuclei is observed, and can be used to help identify resonances. Such splitting represents a second type of distinction between nuclei that ^{13}C NMR allows.

Relaxation Processes. The absorption of the radio frequency radiation facilitates passage between the spin energy states that have been separated by the applied magnetic field. In the absence of the radio frequency radiation, the lower energy state is slightly the more populous. With the application of intense radio frequency radiation at resonance, interconversion is rapid in both directions, the population difference is decreased, and further absorption is thus decreased in turn. The signal is attenuated and is said to be saturated. After the radiation is turned off, the original population difference in the applied magnetic field is reestablished. Since the radiation is now absent, however, other mechanisms are required to promote conversions between spin states. The mechanisms responsible for the decay of the perturbed spin population distribution back to the equilibrium distribution of spin states are referred to as relaxation processes.

Relaxation processes concern us in several ways. First, they limit the rate at which observations can be made in accumulating repetitive measurements. Second, they are involved in the factors controlling the breath of signal, for example by influencing the "lifetime" of a spin state. Third, with pulse techniques it is easy to measure directly one of the two relaxation times that define the behavior of a given ^{13}C nucleus. This relaxation time, T_1, the spin-lattice relaxation time, measures the rate of change of spin state by way of mechanisms in which the spin energy change is accommodated by energy dissipation into thermal motion of nearby structures. The nucleus undergoing the change experiences fluctuations in local magnetic fields with components of the appropriate frequency to allow the required quantal change. For ^{13}C the directly bonded ^{1}H nuclei are often the dominant source of the local fluctuating magnetic fields that underlie the effect. The experimental measurement of T_1 thus provides a means of sampling the spectrum of thermal motion in the region of a ^{13}C nucleus, and adds a very promising technique to those available to the enzymologist. The second relaxation time, T_2, the spin-spin relaxation time, measures the rate of change of spin state by way of mechanisms in which the change of spin state of one nucleus is accompanied by a change of spin state of another nucleus. Again, fluctuations in local magnetic fields are required for the interaction and have their origin in molecular motion. However, unlike T_1, T_2 measurements are affected significantly by inhomogeneities in the applied magnetic field and by instrumental instability. This feature limits detailed evaluation of T_2 in terms of molecular motion. Measurements of T_2 are less our concern than of T_1, but the two parameters are related. A qualitative awareness of the sensitivity of T_2 to molecular motion is useful in assessing the meaning of the width of an observed resonance line.

Proton Decoupling. Interactions between nuclei, in which relaxation processes inevitably enter, underlie the use of the proton decoupling procedure, which is commonly employed in ^{13}C NMR. Strong irradiation over a range of frequencies at which ^1H resonance occurs is maintained during the ^{13}C NMR experiments. The very rapid interconversions of spin states of nearby protons average out their contribution to the fields at each ^{13}C nucleus so that the splitting of the ^{13}C resonances disappears. This simplifies the spectrum and obviously tends to help enhance the magnitude of each resonance line with respect to noise. In addition, the accompanying nuclear Overhauser effect causes an enhancement of signal strength by a mechanism that involves terms related to molecular motion. This enhancement can be, and often is, nearly 3-fold. The magnitude of the signal-to-noise improvement for each carbon caused by nuclear Overhauser effects will depend on the motional features of the individual carbon nuclei and the mechanisms contributing to ^{13}C relaxation. Consequently, analysis of integrated intensities in terms of the number of contributing carbons is not as straightforward for ^{13}C NMR as it is for ^1H NMR.

Outline of Applications of Proton Nuclear Magnetic Resonance

The preceding discussion has sketched the general basis of the types of information to be derived from NMR experiments, with emphasis on ^{13}C. As further background to the more specific development of the theoretical treatment, the description of techniques, and the presentation of results which will follow, it is useful at this point to refer to applications of ^1H NMR to proteins and enzymes.[1-3] In this setting it will be easier to foresee the prospects for applications of ^{13}C NMR.

[1] G. C. K. Roberts and O. Jardetzky, *Advan. Protein Chem.* **24**, 447 (1970).

[2] A. S. Mildvan and M. Cohn, *Advan. Enzymol.* **33**, 1 (1970).

[3] A. Allerhand and E. A. Trull, *Annu. Rev. Phys. Chem.* **21**, 317 (1970).

The parameters that describe NMR spectra provide, in principle, a wealth of information on each magnetic nucleus. The chemical shift is sensitive to the electronic and molecular environment. The area under each resonance is proportional to the number of nuclei involved, allowing for exceptions where nuclear Overhauser enhancement occurs to unequal degrees with different nuclei or where relaxation processes are slow enough to lead to selective saturation. The nature of splitting, characterized by coupling constants, describes interactions with neighboring magnetic nuclei. Furthermore, the nuclear spin relaxation times reflect magnetic interactions with neighboring structures, and thereby reflect molecular motions. In practice the potential information content of a spectrum of a protein is not completely extracted.

Sensitivity of the method is limited by the inherently weak signals. Even with signal averaging the presence of solvent protons often interferes seriously. Resolution is a particular problem. Especially with ^{1}H NMR the chemical shifts of many nuclei in the protein are nearly equivalent. The problem of overlap is enhanced by large line widths which are generally more apparent as the size of the protein increases.[8]

Denatured Proteins. Chemical shift assignments are generally made more readily for the spectra of denatured proteins[9-11] in which line broadening is reduced by the increased segmental mobility. Also contributing to the narrowing of the spectral lines is the lessening of the distinctions between the environments of given chemical types of groups in the denatured proteins. Under such conditions it is possible to approximate the overall ^{1}H NMR spectrum as a summation of the resonances due to the constituent amino acids[10] providing that care is taken to adjust the α-carbon chemical shifts for incorporation in the peptide chain.[12,13]

Native Proteins: Ring Currents. To interpret the spectrum of a native

[4] C. C. McDonald and W. D. Phillips, *in* "Biological Macromolecules" (G. D. Fasman and S. N. Timasheff, eds.), Vol. 4, pp. 1–45. Dekker, New York, 1970.

[5] J. C. Metcalfe, *in* "Principles and Techniques of Protein Chemistry" (S. J. Leach, ed.), Part B, Chapter 14. Academic Press, New York, 1970.

[6] H. A. O. Hill, *in* "NMR: Basic Principles and Progress" (P. Diehl, ed.), Vol. 4, pp. 167–179. Springer-Verlag, Berlin and New York, 1971.

[7] B. Sheard and E. M. Bradbury, *in* "Progress in Biophysics and Molecular Biology" (J. A. V. Butler and D. Noble, eds.), Vol. 20, pp. 187–246. Pergamon, Oxford, 1970.

[8] J. H. Bradbury, B. E. Chapman, and N. L. R. King, *Int. J. Protein Res.* 3, 351 (1971).

[9] C. C. McDonald and W. D. Phillips, *J. Amer. Chem. Soc.* 89, 6332 (1967).

[10] C. C. McDonald and W. D. Phillips, *J. Amer. Chem. Soc.* 91, 1513 (1969).

[11] C. C. McDonald, W. D. Phillips, and J. D. Glickson, *J. Amer. Chem. Soc.* 93, 235 (1971).

[12] A. Nakamura and O. Jardetzky, *Proc. Nat. Acad. Sci. U.S.* 58, 2212 (1967).

[13] A. Nakamura and O. Jardetzky, *Biochemistry* 7, 1226 (1968).

protein it is necessary to allow for specific factors in the environment of each magnetic nucleus. The most prominent of these are the ring current effects.[14-17] The aromatic side chains of phenylalanine, tyrosine, and tryptophan residues, or aromatic rings such as heme, in the external applied magnetic field, undergo an induced electron flow in the π-electron system. This ring current produces an anisotropic magnetic field in its neighborhood. Protons lying "outside" the aromatic rings are shifted downfield, those "inside," and more nearly perpendicular to the plane of the ring, are shifted upfield. The appropriate intepretations have been applied to the NMR spectrum of lysozyme.[9,18] Clearly a stringent test of a particular assignment or correction for structural relationships is hardly possible for the majority of the ¹H nuclei in a native protein at this time.

Diamagnetic Perturbations. Very resourceful work in a number of laboratories has made possible the interpretation of certain individual resonances.[19-22] The region of the spectrum in which the aromatic protons are seen has yielded considerable information. The proton on the C^ϵ of the imidazole ring of histidine residues has been observed in several cases. The state of protonation of the neighboring nitrogen atoms in the ring affects the chemical shift of the C^ϵ proton. It has been possible to observe the pH dependence of this signal from each of the 4 histidine residues in bovine pancreatic ribonuclease A[19] and to determine the individual pK values.[23] The technique has been extended to observe perturbations caused by binding of small molecules in the neighborhood of certain of these histidine side chains,[24] and a catalytic mechanism has been proposed.[25]

[14] J. A. Pople, *J. Chem. Phys.* **24**, 1111 (1956).

[15] K. Wüthrich, R. G. Shulman, and J. Peisach, *Proc. Nat. Acad. Sci. U.S.* **60**, 373 (1968).

[16] R. G. Shulman, K. Wüthrich, T. Yamane, D. J. Patel, and W. E. Blumberg, *J. Mol. Biol.* **53**, 143 (1970).

[17] T. C. Farrar, S. J. Druck, R. R. Shoup, and E. D. Becker, *J. Amer. Chem. Soc.* **94**, 699 (1972).

[18] H. Sternlicht and D. Wilson, *Biochemistry* **6**, 2881 (1967).

[19] D. H. Meadows, O. Jardetzky, R. M. Epand, H. H. Rüterjans, and H. A. Scheraga, *Proc. Nat. Acad. Sci. U.S.* **60**, 766 (1968).

[20] J. D. Glickson, C. C. McDonald, and W. D. Phillips, *Biochem. Biophys. Res. Commun.* **35**, 492 (1969).

[21] I. Putter, J. L. Markley, and O. Jardetzky, *Proc. Nat. Acad. Sci. U.S.* **65**, 395 (1970).

[22] J. L. Markley, M. N. Williams, and O. Jardetzky, *Proc. Nat. Acad. Sci. U.S.* **65**, 645 (1970).

[23] G. C. K. Roberts, D. H. Meadows, and O. Jardetzky, *Biochemistry* **8**, 2053 (1969).

[24] D. H. Meadows, G. C. K. Roberts, and O. Jardetzky, *J. Mol. Biol.* **45**, 491 (1969).

[25] G. C. K. Roberts, E. A. Dennis, D. H. Meadows, J. S. Cohen, and O. Jardetzky, *Proc. Nat. Acad. Sci. U.S.* **62**, 1151 (1969).

Intrinsic Paramagnetic Perturbations. Certain proteins contain intrinsically a paramagnetic center such as the iron in ferrimyoglobin. This protein has been studied in the low-spin cyanide complex in which one electron is unpaired. The magnetic moment of an electron is much greater than that of a ^{1}H or ^{13}C nucleus, so that strong magnetic effects of considerable range are seen. Protons of the porphyrin moiety are shifted both upfield and downfield from the protein envelope, by contact or pseudo-contact interactions[26-29] allowing close analysis of their behavior. In this case the large conjugated ring system of the porphyrin provides ring current effects.[15,16] Some chemical shifts are dominated by the paramagnetic effect, which involves electron delocalization, so that the unpaired spin is found to some extent to reside on the porphin skeleton. Such paramagnetic effects can in principle be distinguished from ring current effects on the basis of temperature dependence.[15] Care must be taken to recognize possible direct effects of temperature on the nature of the metal ion coordination and on the protein moiety.

Extrinsic Paramagnetic Perturbations. Paramagnetic effects may be exploited by adding an ion such as Co(II)[30] to produce shifts in ^{1}H resonances originating in groups that are accessible to the ion. This is a promising method for both identification of resonances and for understanding the interaction pattern of the ion in question, in addition to its possible application to determining whether a given structure is exposed to the solvent or not.

Selective Deuteration. Because of the large absorption by solvent protons, ^{1}H NMR is normally done in D_2O solution in which residual HDO may still be troublesome. Ionizable groups normally exchange out the ^{1}H nuclei rapidly. Exchange of amide H will occur rapidly with respect to most measurements unless the amide (peptide) groups are "buried" within the protein. Other relatively acidic protons such as that on C^ϵ of histidine exchange over a period of several days.

The foregoing observations contain the germ of the very successful strategy of selective deuteration. Largely deuterated proteins may be obtained by growing a microorganism in D_2O with one or more amino acids supplied in excess in the nondeuterated form. The groups of Jar-

[26] R. G. Shulman, S. H. Glarum, and M. Karplus, *J. Mol. Biol.* **57**, 93 (1971).
[27] R. G. Shulman, K. Wüthrich, T. Yamane, E. Antonini, and M. Brunori, *Proc. Nat. Acad. Sci. U.S.* **63**, 623 (1969).
[28] H. M. McConnell and D. B. Chesnut, *J. Chem. Phys.* **28**, 107 (1958).
[29] K. Wüthrich, R. G. Shulman, T. Yamane, B. J. Wyluda, T. E. Hugli, and F. R. N. Gurd, *J. Biol. Chem.* **245**, 1947 (1970).
[30] C. C. McDonald and W. D. Phillips, *Biochem. Biophys. Res. Commun.* **35**, 43 (1969).

detzky[19,21] and Katz[31] have had great success with this method of simplifying the ^{1}H NMR spectrum.

Small Molecule Interactions. A small molecule in dynamic equilibrium with a protein, such as an inhibitor with an enzyme, may be observed both in the free and bound forms. The magnetic environment of the small molecule is often sufficiently different in the bound and free states that substantial changes in the chemical shifts occur. If the rate of exchange of the small molecule between the two environments is relatively slow, the spectra of both forms are observed simultaneously. If the exchange is more rapid, the spectra tend to merge. In the limit of fast exchange a single spectrum is observed, with chemical shift positions that represent the averages for the two spectra with a weighting according to the relative abundances of the forms. From such measurements binding constants can be derived. This type of work is well exemplified by the elegant work of Raftery and his colleagues.[32]

The relaxation behavior of a small molecule bound to a protein is very different from that of the free molecule, since the tumbling rate of the protein is much slower. Various ways of exploiting the relaxation effects have been set out by Jardetzky,[33] and specific instances are reported by Jardetzky and co-workers[34,35] and by Metcalfe *et al.*[36]

Relaxation studies of water protons in protein solutions have been useful in enzymology. Relaxation here is promoted both by the slowing of molecular motion when the water is bound to the large molecule, and by changes in spin state of a paramagnetic ion if such is present and in a position to exchange bound water with the bulk solvent. This field has been well reviewed in terms of applications to enzymology.[2]

Points of Comparison of ^{13}C with ^{1}H NMR

The inherently weaker signals of the ^{13}C nuclei mean that the application of ^{13}C NMR to proteins has depended on the development of pulsed Fourier transform techniques. In practice, therefore, the signal accumulation and processing is done in a different way from conventional ^{1}H NMR. Another obvious point is that ^{1}H NMR is almost always done in deuterated solvent, in D_2O solutions of deuterated buffers, for example.

[31] H. L. Crespi, R. M. Rosenberg, and J. J. Katz, *Science* **161**, 795 (1968).

[32] M. A. Raftery, F. W. Dahlquist, S. M. Parsons, and R. G. Wolcott, *Proc. Nat. Acad. Sci. U.S.* **62**, 44 (1969).

[33] O. Jardetzky, *Advan. Chem. Phys.* **7**, 499 (1964).

[34] O. Jardetzky and N. G. Wade-Jardetzky, *Mol. Pharmacol.* **1**, 214 (1965).

[35] J. J. Fischer and O. Jardetzky, *J. Amer. Chem. Soc.* **87**, 3237 (1965).

[36] J. C. Metcalfe, A. S. V. Burgen, and O. Jardetzky, *in* "Molecular Associations in Biology" (B. Pullman, ed.), pp. 487–497. Academic Press, New York, 1968.

Exchange of ^{2}H for ^{1}H is not of direct value for ^{13}C NMR of proteins (see, however, Sternlicht *et al.*[37]) and could have an adverse effect due to the loss of nuclear Overhauser enhancement. With the partial exception of systems in which small molecule–protein interactions are observed, both NMR methods use high concentrations of relatively small proteins. Typically 1–20 mM protein solutions are used, in volumes of as much as 2 ml.

Apart from the foregoing points of similarity and contrast in the experimental conditions and technique, there are four principal ways in which the approach to the ^{13}C studies differs from that for ^{1}H. First, the ^{13}C NMR chemical shifts are greater than those for ^{1}H NMR by almost an order of magnitude.[38-40] Resolution generally benefits, although overlap of resonances remains a major difficulty with the method. Second, the proton decoupling procedure applied concurrently with the measurement of the relatively rare isotope simplifies the spectrum appreciably by reducing the ^{13}C resonances to singlets.[40] Third, the relaxation mechanisms for ^{13}C nuclei are simpler to interpret in many cases.[41,42] The pulse technique[43] on which the Fourier transform method depends is readily adapted to direct measurement of T_1, from which certain inferences may be drawn about the mobility of the groups or segments of the protein. Fourth, the low natural abundance of ^{13}C provides ample room for specific enrichment, calling for strategies that are not quite the obverse of the selective replacement of ^{1}H with ^{2}H in the deuteration work that has been mentioned. By incorporation of residues enriched in ^{13}C at specific positions, the corresponding resonances can be intensified 80-fold or more.

Perhaps too self-evident for inclusion in the list of differences is the fact that ^{13}C resonances report on groups such as the carbonyl and carboxyl that are not reflected directly in the ^{1}H spectrum. Although neither method "sees" an amino group, the ^{13}C resonances of neighboring nuclei show considerable sensitivity to its presence and state of protonation.[38-40,44]

[37] H. Sternlicht, G. L. Kenyon, E. L. Packer, and J. Sinclair, *J. Amer. Chem. Soc.* **93**, 199 (1971).

[38] W. J. Horsley and H. Sternlicht, *J. Amer. Chem. Soc.* **90**, 3738 (1968).

[39] W. J. Horsley, H. Sternlicht, and J. S. Cohen, *Biochem. Biophys. Res. Commun.* **37**, 47 (1969).

[40] W. J. Horsley, H. Sternlicht, and J. S. Cohen, *J. Amer. Chem. Soc.* **92**, 680 (1970).

[41] J. E. Anderson, K. J. Liu, and R. Ullman, *Disc. Faraday Soc.* **49**, 257 (1970).

[42] A. Allerhand and R. K. Hailstone, *J. Chem. Phys.* **56**, 3718 (1972).

[43] R. L. Vold, J. S. Waugh, M. P. Klein, and D. E. Phelps, *J. Chem. Phys.* **48**, 3831 (1968).

[44] F. R. N. Gurd, P. J. Lawson, D. W. Cochran, and E. Wenkert, *J. Biol. Chem.* **246**, 3725 (1971).

Theory

Basic Concepts

A rigorous mathematical treatment of the theory of nuclear magnetic resonance and relaxation phenomena[45-50] is beyond the scope of this chapter. Our intention is to present a conceptual outline punctuated with equations whenever formal representation of ideas is more economical and informative. Such a treatment suffers from loss of exactness, but has the advantage of focusing on fundamental issues basic to the interpretation of ^{13}C NMR of proteins.

The nuclei of isotopes such as ^{1}H, ^{15}N, ^{19}F, ^{33}S, ^{35}Cl, and ^{13}C possess an intrinsic spin defined by a nuclear spin quantum number I. The associated angular momentum and the nuclear charge confer on the nucleus a magnetic moment $\bar{\mu}_N$ given by

$$\bar{\mu}_N = \gamma_N \hbar \bar{I} \tag{1}$$

Here γ_N is the characteristic magnetogyric ratio of the nucleus, and $\hbar$ is Planck's constant divided by 2π. In the presence of a static external field, $\bar{H}_0$, the nuclear magnetic moment interacts with the field and can assume $2I + 1$ spin states characterized by energy levels which are equally spaced with an energy of separation

$$\Delta E = \frac{\mu_N H_0}{I} \tag{2}$$

Transitions between energy levels are restricted by selection rules to adjacent levels and are induced by radio frequency radiation of frequency (in Hz)

$$\nu_0 = \frac{\gamma_N H_0}{2\pi} \tag{3}$$

In contrast with the previous quantum mechanical treatment, the classical mechanical description of the resonance condition provides a readily interpretable model to describe the experimental aspects of resonance and relaxation. The classical model defines the spinning nucleus precessing about $\bar{H}_0$ at an angle θ to the direction of the applied magnetic field (Fig. 1A). This orientation, which corresponds to an energy state

[45] F. Bloch, W. W. Hansen, and M. Packard, *Phys. Rev.* **70**, 474 (1946).
[46] N. Bloembergen, E. M. Purcell, and R. V. Pound, *Phys. Rev.* **73**, 679 (1948).
[47] G. E. Pake, *Amer. J. Phys.* **18**, 438 (1950).
[48] A. Abragam, "The Principles of Nuclear Magnetism." Oxford Univ. Press, London and New York, 1961.
[49] H. G. Hertz, *in* "Progress in Nuclear Magnetic Resonance Spectroscopy" (J. W. Emsley, J. Feeney, and L. H. Sutcliffe, eds.), Vol. 3, Chapter 5. London, Pergamon, Oxford, 1967.
[50] A. G. Redfield, *Advan. Magn. Resonance* **1**, 1 (1965).

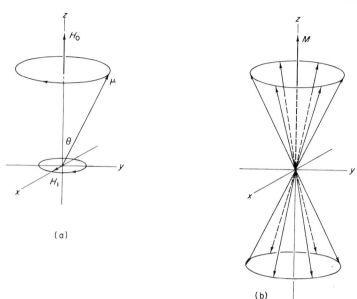

Fig. 1. (A) Precession of a magnetic moment $\bar{\mu}$ about a fixed magnetic field $\bar{H}_0$. The radio frequency field $\bar{H}_1$ rotates in the xy plane. (B) Precession of an ensemble of identical magnetic moments of nuclei with $I = \frac{1}{2}$. The net macroscopic magnetization $\bar{M}$ is oriented along $\bar{H}_0$ and has the equilibrium value M_0.

described in the quantum mechanical treatment, is a consequence of the interaction of the applied field and the spin angular momentum. The resultant torque exerted on the magnetic moment produces a precession analogous to that of a gyroscope in a gravitational field. The angular velocity or Larmor frequency, ω_0, (rad·sec^{-1}) of this nuclear precessional motion is given by

$$\omega_0 = \gamma_N H_0 \tag{4}$$

where $\omega_0 = 2\pi\nu_0$. An alteration of this orientational angle, but not of the angular velocity, can be achieved only if a second magnetic field $\bar{H}_1$ (the radio frequency mentioned previously) continuously rotating with angular velocity ω_0 in phase with the precessing nucleus, is applied in a plane perpendicular to the static field. The change in orientation induced by energy absorption corresponds to the transition between nuclear energy levels. In practice, these conditions obtain for the NMR experiment, and the energy absorption must necessarily be monitored in the plane of $\bar{H}_1$.

In the absence of any other interactions the absorption of energy will be characterized by an observed resonance line width consistent with the Heisenberg Uncertainty Principle:

$$\Delta E \Delta t \geq \hbar \qquad (5)$$

where ΔE and Δt are the uncertainties in measurement of energy and time, respectively. Spin states are characterized by specific lifetimes, that are reflected in resonance line widths. In addition static field inhomogeneities will contribute to the line width presenting difficulties in interpretation.

The characteristic properties for some nuclei of biological interest are listed in Table I. The tabulated physical constants and derived quantities are of general value since they reflect fundamental properties common to nuclei of specific type in any system of biological importance. However, the nuclear magnetic resonance properties of greatest value to protein chemists arise from the modulating effects produced by the electronic nature of the molecule containing the studied nucleus as well as from the properties of the entire sample. Intra- and intermolecular interactions influence both the position of the resonance line (chemical shift) and the line width (relaxation) in a manner which can be understood in terms of chemical structure and solution dynamics. In the sections that follow the basis for interpretation of chemical shift and relaxation phenomena will be discussed with special emphasis on relating ^{13}C spin-lattice relaxation times to the analysis of the structural dynamics of proteins in solution.

Relaxation

Thermal Equilibrium and Spin Relaxation. We will begin our discussion of relaxation with a conceptual analysis of events which prevail when a protein sample is subjected to both a static magnetic field, $\bar{H}_0$, and a time-dependent field, $\bar{H}_1$, rotating in the plane perpendicular to the static

TABLE I
PHYSICAL CONSTANTS OF SOME TYPICAL MAGNETIC NUCLEI

Nucleus	I	γ_N (radians · sec^{-1} · gauss^{-1})	Q ($\times 10^{-24}$ cm^2)	μ (erg · gauss^{-1})	Relative sensitivity at constant field	Frequency (MHz) for 10 kilogauss field
^{1}H	$\frac{1}{2}$	26,753	—	2.79268	1.000	42.276
^{2}H	1	4,107	2.74×10^{-2}	0.85738	9.64×10^{-3}	6.5357
^{13}C	$\frac{1}{2}$	6,728	—	0.70220	1.59×10^{-2}	10.705
^{14}N	1	1,934	7.1×10^{-2}	0.40358	1.01×10^{-3}	3.076
^{15}N	$\frac{1}{2}$	$-2,712$	—	-0.28304	1.04×10^{-3}	4.315
^{19}F	$\frac{1}{2}$	25,179	—	2.6273	8.34×10^{-1}	40.055
^{33}S	$\frac{3}{2}$	2,054	5.3×10^{-2}	0.64274	2.26×10^{-3}	3.266
^{35}Cl	$\frac{3}{2}$	2,624	-7.9×10^{-2}	0.82091	4.71×10^{-3}	4.172

field. If we initially focus our attention on nuclear transitions for one type of nucleus, induced by the rotating field, we can predict system characteristics based on radiation theory. This theory predicts that the probabilities of radiation-*induced* emission and absorption are equal. In addition, the associated probability of spontaneous emission, which is dependent on the magnitude of the nuclear energy difference, is negligible for radio frequency energies. A straightforward analysis based on the above considerations shows that energy absorption induced by $\bar{H}_1$ occurs only if there exists a population difference between the respective spin states. However, based on these considerations alone, any population difference will decay exponentially to zero with time upon application of the resonant radio frequency with a concomitant attenuation in signal strength. This type of behavior is sometimes observed. More generally, the measured absorption signal approaches a constant value invariant with time. Clearly, our description of the system is incomplete and must include consideration of processes which contribute to the maintenance of a spin population difference. Our description will be complete if we recognize that there are interactions between the nuclei and their surroundings which mediate the transfer of spin energy to other degrees of freedom of the system. The consequence of this assumption is that the probabilities of *spontaneous* spin transitions are *not* equal. Mathematical analysis based on this consideration shows that at thermal equilibrium there will be a population difference whose magnitude is determined by the ratio of the spontaneous transition probabilities according to

$$\frac{N_\beta{}^0}{N_\alpha{}^0} = \frac{W_{\alpha\beta}}{W_{\beta\alpha}} \tag{6}$$

Here $N_\beta{}^0$ and $N_\alpha{}^0$ refer to the equilibrium populations for the upper and lower spin states, respectively, and $W_{\alpha\beta}$ and $W_{\beta\alpha}$ denote the upward and downward spontaneous transition probabilities. The sample population is distributed according to the Boltzmann law, where the molecular interactions characteristic of the sample provide the mechanism for distributing the nuclei unequally between the energy levels at thermal equilibrium. As we have seen, an important experimental consequence of this Boltzmann distribution is a small excess of spin states in the lower energy levels sufficient to result in a net observable absorption of radio frequency radiation. Following processes that disrupt the equilibrium distribution, such as absorption of radio frequency energy or alteration of the static field, the spin system must return to thermal equilibrium. In this regard, the relaxation processes responsible for the establishment of equilibrium control the lifetime of a given spin state.

The Bloch Equations. If we consider the hypothetical case of a large number of identical nuclei with $I = \frac{1}{2}$ in the presence of a static magnetic field, we can see that all the magnetic moments will precess at the same frequency. The direction of the field provides a directional reference within the system. Since the Boltzmann factors slightly favor the lower energy state, at equilibrium more nuclei are aligned with the field than against it. This condition results in a net macroscopic magnetization, $\bar{M}$, directed along the z axis, illustrated in Fig. 1B. In the subsequent development, the behavior of the macroscopic magnetization and effects due to relaxation are readily understood in terms of this model and a mathematical formalism developed by Bloch.

The complete Bloch equations are a set of phenomenological differential equations[51] that describe the motion of $\bar{M}$ in the presence of a static magnetic field directed along the z axis, $\bar{H}_0$, and a field constantly rotating in the xy plane, $\bar{H}_1$. Since free spins in magnetic fields possess a nuclear spin angular momentum, the time rate of change of this angular momentum, related to the Larmor frequency, can be evaluated according to classical mechanics,

$$\frac{d\bar{M}}{dt} = \gamma_N \bar{M} \times \bar{H} \tag{7}$$

where $\bar{H} = \bar{H}_0 + \bar{H}_1$. In a static field directed along the z axis the approach of the z component of magnetization, M_z, toward its equilibrium value, M_0, can often be described by a first-order process given by

$$\frac{dM_z}{dt} = \frac{(M_z - M_0)}{T_1} \tag{8}$$

where T_1 is the characteristic spin-lattice relaxation time. Furthermore, if a perturbation of the magnetization produces components M_x and M_y at right angles to the static field, the decay of these components to zero at equilibrium can be described by another first-order rate process given by

$$\frac{dM_z}{dt} = \frac{-M_z}{T_2} \quad , \quad \frac{dM_y}{dt} = \frac{-M_y}{T_2} \tag{9}$$

where T_2 is the characteristic spin-spin relaxation time. Often T_1 and T_2 are referred to as the longitudinal and transverse relaxation times, respectively, with reference to time constants for the decay of magnetization components either parallel or perpendicular to the static field. Bloch combined the above features in a unique way by assuming that the motion of the magnetization due to relaxation could be superimposed on

[51] F. Bloch, *Phys. Rev.* **70**, 460 (1946).

the motion of the free spins. He formalized the assumption with the equation

$$\frac{d\bar{M}}{dt} = \gamma_N \bar{M} \times \bar{H} - \frac{M_x \bar{i} + M_y \bar{j}}{T_2} - \frac{(M_z - M_0)\bar{k}}{T_1} \tag{10}$$

where $\bar{i}$, $\bar{j}$, and $\bar{k}$ are the unit vectors directed along the x, y, and z axes, respectively. Expressed in terms of a fixed coordinate system, the respective components of magnetization assume a time-dependence described by the complete Bloch equations:

$$\frac{dM_x}{dt} = \gamma_N (M_y H_z - M_z H_y) - \frac{M_x}{T_2}$$

$$\frac{dM_y}{dt} = \gamma_N (M_z H_x - M_x H_z) - \frac{M_y}{T_2}$$

$$\frac{dM_z}{dt} = -\gamma_N (M_y H_x - M_x H_y) - \frac{(M_z - M_0)}{T_1} \tag{11}$$

Here ω is the angular frequency of $\bar{H}_1$. The Bloch equations contain terms that refer to the directional components of the static and rotating fields. These components are defined by:

$$H_x = H_1 \cos \omega t$$
$$H_y = -H_1 \sin \omega t$$
$$H_z = H_0 \tag{12}$$

The previous relationships were developed using a fixed coordinate system called the laboratory frame of reference. Further analysis of these equations[45,51] provides a reasonable macroscopic explanation of magnetic resonance absorption and predicts resonance line shapes. Pulsed NMR methods, which make possible the observation of ^{13}C magnetic resonance of proteins, can best be understood by referring to a coordination system which *rotates* about $\bar{H}_0$ in the direction of the precession of nuclei. This coordinate system is referred to as the rotating frame of reference and results in great simplification for subsequent considerations. If the new coordinate system rotates about the z axis with an angular velocity ω', it can be shown[52] that

$$\left(\frac{d\bar{M}}{dt}\right)_{rot} = \gamma_N \bar{M} \times \bar{H}_{eff} \tag{13}$$

where

$$\bar{H}_{eff} = \bar{H} + \bar{\omega}'/\gamma_N \tag{14}$$

[52] T. C. Farrar and E. D. Becker, "Pulse and Fourier Transform NMR." Academic Press, New York, 1971.

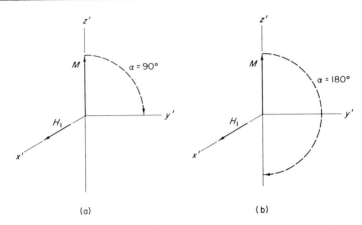

Fig. 2. Precession of $\bar{M}$ about $\bar{H}_1$ in the rotating frame by (A) 90° pulse and (B) 180° pulse.

These equations indicate that equations of motion previously derived for a fixed coordinate system can be applied to the rotating frame provided we visualize that the magnetic field is modified according to the frequency of rotation of the coordinate system. In other words, the magnetization precesses about $\bar{H}_{\text{eff}}$ in the rotating frame and not about $\bar{H}$. In the rotating frame the moving axes are denoted x', y', and z', where the z' axis is parallel to the static field. If the frame rotates with angular frequency ω_0 (the Larmor frequency), we can easily see that if $\bar{H} = \bar{H}_0$, the magnetization is invariant with time. However, if in addition to H_0 we have a second field $\bar{H}_1$, directed in the xy plane (of the laboratory frame) with rotational frequency ω_0, at resonance $\bar{H}_1$ is the only field with which the magnetization interacts. Since $\bar{H}_1$ *rotates with the same frequency as the rotating frame*, we can assign the direction of $\bar{H}_1$ along the rotating x' axis, and we can immediately see that $\bar{M}$ will precess about x' in the rotating frame, as illustrated in Fig. 2. The angle α, in radians, through which the magnetization precesses, is dependent on the time duration of the pulse, t_p, and is given by

$$\alpha = \gamma_N H_1 t_p \tag{15}$$

At resonance the magnetization $\bar{M}$ interacts with the field $\bar{H}_1$ by tending to precess about $\bar{H}_1$. After $\bar{H}_1$ is turned off the components of magnetization will return to their original equilibrium positions. Because of natural processes that promote spin energy exchange between the various nuclei in the sample, the moments in the $x'y'$-plane of the rotating frame begin to spread out (see Fig. 3) or lose phase coherence. The effect of field inhomogeneities is to enhance the loss of phase coherence since the

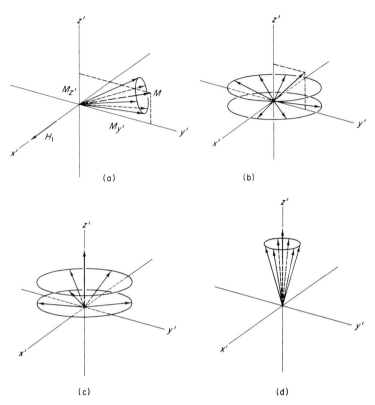

FIG. 3. (A) Tipping of nuclear moments and macroscopic magnetization through an angle α and establishment of $M_{y'}$. (B) Dephasing of nuclear moments by spin-spin relaxation and/or magnetic field inhomogeneity, with reduction of $M_{y'}$. (C) Reduction of $M_{y'}$ to approximately zero. (D) Reestablishment of $M_{z'}$ at its equilibrium value, M_0.

resultant effect is to modulate the field experienced by nuclei in different locations in the magnetic field. Thus $M_{x'}$ and $M_{y'}$ decay to zero with a time constant T_2 or T_2^*, with reference to a homogeneous or inhomogeneous field, respectively. (The origins of the difference between T_2 and T_2^* will be discussed below in the section "Frequency Contributions to T_1 and T_2".) The measured decay of transverse magnetization is referred to as free induction decay (FID). The return of $M_{z'}$ to M_0 along the z' axis follows the processes which dissipate spin energy to the surroundings, or lattice, and is described by T_1. For the situation where the various ^{13}C nuclei of a protein experience slightly different effective fields as a result of the unique magnetic environments of the individual nuclei, the return of the components of magnetization to equilibrium will be characterized by several relaxation times. As we shall see later, the complexity of this situation can be unraveled by means of Fourier analysis.

Fluctuations and Intensities of Local Magnetic Fields. As we have seen, an intense pulse of radio frequency energy, applied for a few microseconds, can displace the macroscopic magnetization of the sample from equilibrium. After removal of the pulse, the magnetization returns to its equilibrium value. This return is controlled by the torque of $\bar{H}_{\text{eff}}$ on the magnetization and by interactions between the nuclear moments of the spin system. Nuclei interact with local fluctuating magnetic fields generated by the thermal motion of spin nuclei comprising the sample. Since the local magnetic fields fluctuate over a broad spectrum of frequencies, but only relatively intense fields fluctuating at the Larmor frequency can cause magnetic relaxation, there must be a means of characterizing both the frequency distribution and associated field intensities.[48,53] The statistical time profile of a magnetic field that fluctuates with time, $H_{\text{loc}}(t)$, is described by the autocorrelation function $G(\tau)$,

$$G(\tau) = H^*_{\text{loc}}(t)H_{\text{loc}}(t + \tau) \simeq \overline{H^*_{\text{loc}}(t)H_{\text{loc}}(t)} \; e^{-|\tau|/\tau_c} \tag{16}$$

in which the persistence of the random fluctuations is assumed to decay exponentially with a time constant τ_c, referred to as the correlation time of the motion, and $H^*_{\text{loc}}(t)$ is the complex conjugate of $H_{\text{loc}}(t)$. The bar indicates a time average over the assembly of spins, and the absolute value of τ refers to the irreversible nature of the motional processes. $H_{\text{loc}}(t)$ can be treated as a correlated fluctuation which averages to zero. Fourier transformation of the autocorrelation function produces the spectral density $J(\omega)$,

$$J(\omega) = \int_{-\infty}^{\infty} G(\tau)e^{i\omega\tau}d\tau = \int_{-\infty}^{\infty} \overline{H^*_{\text{loc}}(t)H_{\text{loc}}(t)} \; e^{-|\tau|/\tau_c}e^{i\omega\tau}d\tau \tag{17}$$

The Fourier inverse is a transformation from the time domain to the frequency domain. The Fourier coefficients at each frequency yield the intensity or spectral density of H_{loc} at that frequency. The rate of approach of the magnetization to its equilibrium value increases with increase in the intensity of the spectral density at the resonance frequency. The frequency distribution of the spectral density can be obtained by evaluating the above integral, whereby

$$J(\omega) = \overline{H^*_{\text{loc}}(t)H_{\text{loc}}(t)} \cdot \frac{2\tau_c}{1 + \omega^2\tau_c^2} \tag{18}$$

and plotting $J(\omega)$ against ω for various values of τ_c (Fig. 4). The area under each curve is identical, so that different τ_c values only alter the

[53] G. B. Savitsky, K. Namikawa, and G. Zweifel, *J. Phys. Chem.* **69**, 3105 (1965).

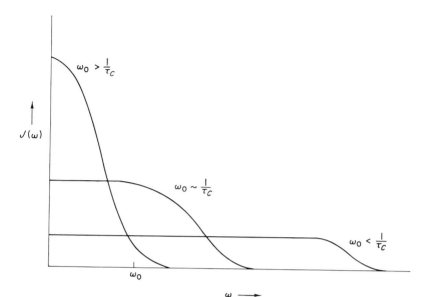

FIG. 4. Spectral density curves plotted against frequency for various correlation times, τ_c. The Larmor frequency is denoted by ω_0.

frequency distribution of spectral density. Clearly, the spectral density at the Larmor frequency, ω_0, is maximized when the correlation time of the motion, τ_c, is comparable to ω_0^{-1}.

The specific relationship between spin-lattice relaxation and $J(\omega)$ is obtained by quantum mechanical treatment using time-dependent perturbation theory. Since T_1 processes induce transitions between nuclear spin states, the quantum mechanical formulation treats $H_{loc}(t)$, fluctuating with the frequency of the transition, as a time-dependent perturbation $\bar{H}$of $_0$ and evaluates the transition probability per unit time, W_{ij}, for the transition between two spin states i and j, coupled to $H_{loc}(t)$.

Time dependence, using this theory, is handled by separating the energy operator (Hamiltonian) into a time-independent part, $\mathcal{3C}_0$, and a time-dependent part $\mathcal{3C}'(t)$. The transition probabilities are defined by evaluating the time-dependence of the coefficients for solutions to the time-dependent Schrödinger equation. These coefficients are related to the time-dependent interaction Hamiltonian. Any time-dependent interaction perturbation, $V(t)$, can be expressed as a product of a static

$$V(t) = V \cdot f(t) \tag{19}$$

interaction term, V, and a term expressing the time dependence of this interaction, $f(t)$. V can be evaluated in terms of the static interaction energy for spins according to the postulated relaxation mechanism. The time-dependence of this interaction, in the laboratory frame, must describe Brownian motion. The solutions to the diffusion equation for Brownian motion are expressed in the form of an expansion in spherical harmonics, and are modified to include the assumed exponential decay of motional persistence. The resultant time-dependent interaction perturbation is transformed into suitable operator notation and can be evaluated in terms of the spin $\bar{I}$ (e.g., ^{13}C) coupled to $H_{loc}(t)$, or

$$\mathcal{H}'(t) = \hbar \bar{I} \cdot H_{loc}(t) \tag{20}$$

However, the interaction operator is usually expressed by an equivalent form

$$\mathcal{H}'(t) = -h \bar{I} \cdot \widetilde{T} \cdot \bar{X} \tag{21}$$

This expression tells us that spin $\bar{I}$ (in our case ^{13}C) and a physical quantity $\bar{X}$ (another spin or field, determined by the specific relaxation mechanism being evaluated) are interacting and are coupled according to the form of the coupling interaction tensor $\widetilde{T}$. The form of $\widetilde{T}$ is determined by the type of interaction being investigated (see section "Mechanisms of Relaxation" below).

The transition probability for a single spin is related to $J(\omega_{ij})$ by

$$W_{ij} = \gamma_N{}^2 J(\omega_{ij}) \tag{22}$$

If the exponential form of the autocorrelation function is used, we can easily see that $T_1{}^{-1}$ exhibits the same time dependence on τ_c as does $J(\omega)$ in Eq. (18),

$$\frac{1}{T_1} = \gamma_N{}^2 \overline{H^*{}_{loc}(t) H_{loc}(t)} \frac{\tau_c}{1 + \omega_{ij}{}^2 \tau_c{}^2} \tag{23}$$

since

$$\frac{1}{T_1} = W_{ij} \tag{24}$$

The manner in which the previously described theoretical treatment is applied to a specific type of relaxation is illustrated in the following qualitative description of dipolar relaxation of a two-spin $\frac{1}{2}$ system.[54] The time-dependent interaction Hamiltonian for two spins ($\bar{I} = {}^{13}C$, $\bar{S} = {}^{1}H$) is formulated for the dipolar interaction (see section "Mechanisms of Relaxation"), and transition probabilities are evaluated in terms of integrals containing this Hamiltonian. The time-rate of change of the various spin-state populations can be related to combinations of

[54] I. Solomon, *Phys. Rev.* **99**, 559 (1955).

the transition probabilities. Since population differences can be shown to result in a macroscopic magnetization, the various components of magnetization (expressed as expectation values) can be defined in terms of these differences. The macroscopic Bloch equations are then used to define the equations of motion of the components of magnetization in terms of the transition probabilities. This set of equations has been modified to include the spin-spin interaction and correctly represents the motion of the macroscopic magnetization. Complete proton decoupling modifies these equations to the extent that equalization of the proton spin state populations allows I_z to decay exponentially to its equilibrium value with a single time constant $1/T_1$. The transverse components of $\bar{I}$ can be described by a simple exponential decay with time constant $1/T_2$. Both time constants are explicitly defined by a set of transition probabilities, which are functions of the appropriate Fourier coefficients derived from the autocorrelation function describing the motion of the system. Since carbons in proteins may experience several types of motion, the autocorrelation function must include several contributing correlation times. The included τ_c values will modify the functional form of the spectral densities and thus be reflected in the equations for T_1 and T_2. Specific results of this approach will be discussed later.

The Relationship between T_1 and T_2 and the Correlation Time. An important limiting condition for small molecules arises for short molecular correlation times, $\tau_c \leq 10^{-10}$ second at fields of 14.1 or 23.5 kG where $\omega \simeq 10^7$ to 10^8 rad/second. The term $\omega^2\tau_c^2$ is much less than unity, and Eq. (23) reduces to

$$\frac{1}{T_1} = \gamma_N{}^2\overline{H^*{}_{\text{loc}}(t)H_{\text{loc}}(t)}\tau_c \tag{25}$$

This is the condition of "motional narrowing" or the "extreme narrowing limit" where T_1^{-1} depends directly on τ_c. A treatment similar to the above can be used to evaluate T_2.[48] Figure 5 illustrates a generalized plot of T_1 and T_2 against τ_c for three values of ω.

Several features of this type of plot, important in subsequent discussion of protein ^{13}C NMR, must be emphasized. For short correlation times, T_1 decreases with increasing τ_c, according to the extreme narrowing condition. When this condition is violated, near $\omega = \omega_0 = 1/\tau_c$, T_1 passes through a minimum and then increases with increasing τ_c, because $\omega^2\tau_c^2 \geq 1$. Since T_1 is a double-valued function of τ_c, one must determine which side of the minimum applies to a particular carbon under particular conditions. This can be determined by measurements at different static fields, since T_1 is independent of $\bar{H}_0$ on the left side of the minimum, but increases with increasing $\bar{H}_0$ on the right side (Fig. 5).

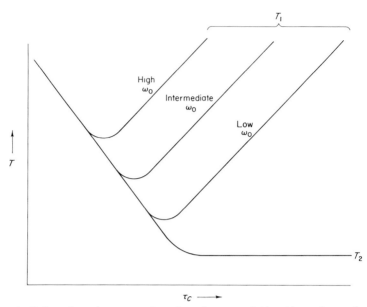

Fɪɢ. 5. Relaxation times are plotted against correlation times for various field strengths expressed in terms of resonant frequencies.

The selection of one τ_c associated with a specific T_1 can be made using qualitative arguments. In one case, the choice may include a τ_c which is much too long based on physical intuition, and this value can be excluded. An alternative method for defining the τ_c associated with a specific T_1 is evaluation of T_2. T_2 is a monotonically decreasing function of τ_c, which approaches a limiting value with increasing τ_c controlled by the dominance of frequency independent terms for long correlation times. Although in principle T_2 can be determined from natural line widths, w, (in Hz) (when relaxation mechanisms are defined) according to

$$w = \frac{1}{\pi T_2} \qquad (26)$$

in practice measured line widths for proteins are difficult to interpret non-aggregating because of two complicating conditions. The inherent spectral overlap of chemically nonequivalent carbons makes individual line widths difficult to observe. Even for exceptional cases of isolated resonances, contribution to line width from magnetic field inhomogeneity and instrumental instability can dominate the contribution from relaxation. In practice line widths may be useful as qualitative indicators for selection of τ_c from a T_1 if the associated T_2 values are different by large amounts.

Obtaining the explicit functional form relating measured T_1 values and correlation times characteristic of the motional features of proteins is a complex problem. Reasonable selection of models describing the various types of motion in terms of multiple correlation times and estimates of the relative magnitudes of contributions from a number of potential relaxation mechanisms must be made. The latter contributions can best be understood by considering, in more detail, the motional frequency contributions to T_1 and T_2 and the nature of various relaxation mechanisms.

Frequency Contributions to T_1 and T_2. If a molecule persists in some state of motion for 10^{-x} second, the frequency distribution of this motion will cover a range of $0\text{--}10^x \text{Hz}$. The directional components of the macroscopic magnetization return to their equilibrium values, after external perturbation, according to relaxation mechanisms that are described by interactions of microscopic spin magnetic moments with resonant frequency contributions of microscopic, fluctuating magnetic fields. These fluctuating fields, $\bar{H}_{\text{loc}}$, generated by other spin nuclei subject to molecular motion, behave in a manner similar to the fluctuating radio frequency pulse, $\bar{H}_1$. Consequently the interaction of the magnetization with the fluctuating field may be described by evaluating the interaction of $\bar{M}$ with $\bar{H}_{\text{loc}}$ in the usual way according to the relationship[52]

$$(\bar{H}_{\text{loc}} \times \bar{M})_{\text{rot}} = \bar{i}(H_{y',\text{loc}}M_{z'} - H_{z',\text{loc}}M_{y'})$$
$$+ \bar{j}(H_{z',\text{loc}}M_{x'} - H_{x',\text{loc}}M_{z'}) + \bar{k}(H_{x',\text{loc}}M_{y'} - H_{y',\text{loc}}M_{x'}) \quad (27)$$

We see that in the rotating frame the longitudinal components of magnetization are relaxed by components of the fluctuating fields along the x' and y' axes, and the transverse components are relaxed by field components along all three coordinate axes. Since the static z'-field component in the rotating frame is static in the laboratory frame, but the static x'- and y'-field components in the rotating frame correspond to high frequency components in the laboratory frame, we conclude that T_1 is influenced by high frequency processes and T_2 is affected by both high and low frequency processes. For small molecules with short correlation times, low frequency components are negligible, and $T_2 = T_1$. However, in the case of proteins, the longer correlation times contain significant low frequency components, which shorten T_2 values in many cases, and consequently $T_2 \leq T_1$.

As mentioned in the preceding section, static field inhomogeneity and instrumental instability lead to broadening of resonances. The corresponding T_2^* for proteins, see Eq. (26), is related to T_2 and T_1 by the inequality

$$T_2^* \leq T_2 \leq T_1 \tag{28}$$

For isolated resonances of proteins the relationships between T_2 and T_1 can be determined if the static broadening contributions to T_2^* are experimentally evaluated. This type of analysis has direct bearing on selection of correlation times, as mentioned previously, and may be very useful in detecting low frequency motional contributions for small molecule binding to proteins.

Mechanisms of Relaxation. As we have seen, ^{13}C relaxation processes originate from the interaction of the magnetic moment of the carbon nucleus with local magnetic fields fluctuating at the resonance frequency. In this section we will describe some important features of the most significant relaxation mechanisms contributing to T_1 processes for ^{13}C nuclei. In the case of spin-lattice relaxation the fluctuating fields originate from magnetic moments of other nuclei, magnetic moments of unpaired electrons, and asymmetric distributions of electronic charges subject to random molecular motion. Unlike most molecular relaxation processes in liquids, the time scale of T_1 processes is relatively long because the nuclear spin system is only weakly coupled to the motions of the system. An essential requirement for relaxation is that molecular motion must be on a suitable time scale compared with that for NMR. Interactions that change sign with a frequency much greater than the NMR resonance frequency have little effect. Thus electron motions and molecular vibrations are relatively unimportant. Other magnetic nuclei interact with ^{13}C nuclei through direct coupling of magnetic dipoles or indirect coupling (scalar, electron transmitted). Nuclei possessing an electric quadrupole are relaxed by fluctuating electric fields and can indirectly influence ^{13}C relaxation through scalar coupling mechanisms. Our treatment[48,52,55] will include a brief description of the origin, the form of the perturbing Hamiltonian and the explicit contribution to T_1 of the following mechanisms: (1) magnetic dipole-dipole; (2) spin-rotation; (3) chemical shift anisotropy; (4) quadrupole; (5) scalar coupling. In addition, we will attempt to estimate the contributions of these mechanisms to ^{13}C relaxation in proteins.

1. MAGNETIC DIPOLE-DIPOLE. This mechanism originates from instantaneous local magnetic fields produced by spin-spin dipolar interactions.

[55] J. R. Lyerla and D. M. Grant, *in* M. T. P. International Review of Science, Physical Chemistry, Series One, Vol. 4 (C. A. McDowell, ed.). University Park Press, State College, Pa., in press, 1973.

The strength of the interaction between two magnetic moments depends on the magnitude of the magnetic moments, their separation, and their relative orientation. The magnitude of the local field at the ^{13}C nucleus (c) generated by nucleus j is given by

$$H_{loc} = \pm \frac{\mu_j}{r_{cj}^3} (3 \cos^2\theta_{cj} - 1) \tag{29}$$

where r_{cj} is the nuclear separation between the carbon and nucleus j and θ_{cj} the angle of r_{cj} relative to the static field $\bar{H}_0$. The $\pm$ sign reflects the directional dependence of H_{loc} on the spin state of nucleus j. Interactions of this kind can be quite large, especially for carbons directly bonded to hydrogen nuclei. In fact the r^{-3} dependence suggests that dipolar relaxation for protonated carbons will be dominated by intramolecular contributions.

The time-dependence of H_{loc} is introduced as a consequence of the motion of molecules to which the dipoles are attached. θ_{cj} becomes time-dependent for intramolecular dipoles experiencing molecular rotation while both θ_{cj} and r_{cj} vary with time for intermolecular dipolar interactions as a result of translational and rotational diffusion. The form of the interaction Hamiltonian is

$$\mathcal{H}'_D(t) = -h\bar{I} \cdot \tilde{D} \cdot \bar{S} \tag{30}$$

where $\bar{I}$ and $\bar{S}$ refer to ^{13}C and ^1H, respectively, and $\tilde{D}$ is the dipolar-coupling tensor. This tensor contains the time-dependence of the interaction associated with geometrical alterations due to Brownian motion.

Using as a model for intramolecular carbon-hydrogen dipolar coupling a CH group in a rigid sphere undergoing isotropic rotational reorientation characterized by a single correlation time, τ_r, the equations for T_1 and T_2 with proton decoupling become

$$\frac{1}{T_{1D}} = \frac{1}{10} \frac{\hbar^2 \gamma_I^2 \gamma_S^2}{r_{IS}^6} \left[\frac{\tau_r}{1 + (\omega_I - \omega_S)^2\tau_r^2} + \frac{3\tau_r}{1 + \omega_I^2\tau_r^2} + \frac{6\tau_r}{1 + (\omega_S + \omega_I)^2\tau_r^2} \right] \tag{31}$$

$$\frac{1}{T_{2D}} = \frac{1}{20} \frac{\hbar^2 \gamma_I^2 \gamma_S^2}{r_{IS}^6} \left[4\tau_r + \frac{\tau_r}{1 + (\omega_I - \omega_S)^2\tau_r^2} + \frac{3\tau_r}{1 + \omega_I^2\tau_r^2} \right.$$
$$\left. + \frac{6\tau_r}{1 + \omega_S^2\tau_r^2} + \frac{6\tau_r}{1 + (\omega_I + \omega_S)^2\tau_r^2} \right] \tag{32}$$

For heteronuclear relaxation we can see from comparison of individual terms of Eqs. (31) and (32) with Eq. (18) that T_2 depends on a low (zero) frequency component while T_1 does not, thus explaining the general functional dependences of T_1 and T_2 on τ_c illustrated in Fig. 5. The extreme narrowing approximation reduces Eqs. (31) and (32) to

$$\frac{1}{T_{1D}} = \frac{1}{T_{2D}} = \frac{\hbar^2\gamma_I^2\gamma_S^2}{r_{IS}^6}\tau_r \tag{33}$$

where there are no frequency-dependent terms.

Multinuclear dipolar relaxation (such as ^{13}C bonded to more than one ^1H) is generally expressed as a simple sum for N nuclei (^1H), which can be simplified, for the extreme narrowing case, to

$$\frac{1}{T_{1D}} = \frac{N\hbar^2\gamma_I^2\gamma_S^2}{r_{IS}^6}\tau_{\mathrm{eff}} \tag{34}$$

due to the rapid attenuation with distance implied by r_{IS}^{-6}. Equations (31) and (32) can also be corrected to include more than one bonded ^1H by introducing N in the same manner. Equation (34) is valid only if all r_{IS} are constant and all coupled spins are governed by the same effective correlation time. In essence, this formalism allows a rough quantitative estimate of correlated motions for anisotropic reorientation, where the motional details are obscured, by use of τ_{eff}. For ^{13}C–^1H dipolar mechanisms, this equation is valid only for complete proton decoupling[14] and only when $1/\tau_{\mathrm{eff}}$ is much larger than the ^1H resonance frequency. Obviously overevaluation of results based on this equation must be avoided.

Compared with ^{13}C–^1H dipolar interactions, dipolar relaxation of ^{13}C by other nuclei is very inefficient. This results from the small r_{CH} and large γ_H compared with the respective parameters for other nuclei. On a scale where ^{13}C–^1H intramolecular coupling is unity, the competing ^{13}C–^{14}N coupling is 0.0021. Moreover, the low natural abundance of ^{13}C eliminates the need to consider ^{13}C–^{13}C dipolar coupling. In addition, intermolecular dipolar relaxation, formulated in terms of translational motion alone, can be described by an expression

$$\frac{1}{T_{1D\,\mathrm{inter}}} = \frac{2\pi\hbar^2\gamma_I^2C\tau_t}{a^2}\left[4\gamma_I^2I(I+1)\sum_S\frac{1}{d_{IS}} + \frac{8}{3}\sum_F\gamma_F^2F(F+1)\frac{1}{d_{IF}}\right] \tag{35}$$

where S refers to spins of the same type as I, F to all other spins, d_{IS} and d_{IF} are distances of closest approach, a the effective radius of the molecule, τ_t the translational correlation time, C the molecular concentration. The distances of closest approach and the effective molecular radius for proteins are so large that the contribution of the intermolecular term to dipolar relaxation in proteins should be small, whenever the carbon in question is directly bonded to hydrogen. This suggests that viscosity effects for ^{13}C NMR should not be as pronounced as they are for ^1H NMR. For proteins, dipolar relaxation of protonated carbons will be dominated by the directly bonded hydrogens, and as a result dipolar relaxation behavior is substantially simplified.

The previous discussion of heteronuclear dipolar relaxation does not provide information about the effects of overall anisotropic reorientation[56] and/or effects due to significant contributions from internal motion.[57,58] Since proteins are likely to possess these motional properties, it is necessary not only to understand that internal motion invalidates the equations usually applied to small molecule analysis [see Eqs. (31), (33), and (34)], but also to develop suitable models for predicting and quantitating these contributions. Analysis of nonspherically symmetric systems with and without internal rotation has been very limited.[58,59] Of particular interest is a recent analysis by Doddrell et al.[60] in which expressions for T_1 and T_2 have been obtained for a methine carbon, with one degree of internal motion defined by τ_g, attached to a rigid, isotropically reorienting sphere with an overall correlation time τ_r. The methine ^{13}C nucleus is assumed to undergo relaxation solely by way of dipolar interaction with the covalently attached, decoupled proton. This model predicts that T_2 will be a monotonically decreasing function of τ_r and τ_g. If the correlation time for overall molecular motion is short, increasing internal motion leads to an increased value of T_1. However, for slow overall reorientation (e.g., the case expected for many native proteins), onset of internal rotation signals an initial decrease in T_1 which passes through a minimum and then increases as internal rotation increases. In terms of this model the T_1 value for a methine carbon will reflect either one or both correlation times depending on whether τ_r and τ_g differ by a few orders of magnitude or are quite similar, respectively. In a general sense the correlation times dominating a particular T_1 process are selected from the shortest of all contributing values according to

$$1/\tau = \sum_i 1/\tau_i \tag{36}$$

where i refers to the independently contributing motions. A tetrahedral methine carbon with one degree of internal motion is predicted to have a T_1 value up to nine times that for a methine carbon which is part of a rigid backbone in a large molecule undergoing isotropic reorientation. The corresponding τ_g for the rotating carbon must be several orders of magnitude shorter than the τ_r for protein molecules.

Several qualitative remarks can be made based on the above discussion provided we assume that the average T_1 for all alpha-carbons in

[56] W. T. Huntress, Advan. Magn. Resonance 4, 1 (1970).

[57] D. E. Woessner, J. Chem. Phys. 36, 1 (1962).

[58] D. Wallach, J. Chem. Phys. 47, 5258 (1967).

[59] D. E. Woessner, J. Chem. Phys. 37, 647 (1962).

[60] D. Doddrell, V. G. Glushko, and A. Allerhand, J. Chem. Phys. 56, 3683 (1972).

native globular proteins reflects the τ_r for the protein. First, if the T_1 of a side chain methine (or the NT_1 for a carbon bonded to N protons) directly bonded to the alpha-carbon backbone of a native globular protein is greater than the T_1 for the "average alpha-carbon," but less than the predicted upper limit, τ_g is not necessarily less than τ_r. This complication is likely to prevail for τ_r values greater than 10^{-8} second because T_1 is a double-valued function of τ_g over this range of τ_r. Second, if the NT_1 for a carbon attached directly to the alpha-carbon backbone is approximately nine times the T_1 of the "average alpha-carbon," only an upper limit for τ_g can be obtained. Third, if the NT_1 for a carbon is approximately equal to the T_1 for a backbone methine, only a lower limit for τ_g can be estimated. Fourth, for carbons beyond the first attached carbon on a side chain, NT_1 can be less than, equal to, or greater than the T_1 of the backbone methines. The possibility of several degrees of internal motion must be considered for these side-chain carbons, which will invalidate predictions based on one degree of internal motion. Fifth, if the NT_1 for a distant side-chain carbon is much greater than nine times the T_1 for the backbone methine carbons, the presence of several degrees of motional freedom is strongly suggested.

The effects of anisotropic molecular reorientation and of a large number of degrees of internal motional have not been analyzed in detail. The functional forms of T_2 and especially of T_1 are likely to be extremely complex. Consequently, the present interpretations of segmental motion in proteins for a dominant $^{13}C-^{1}H$ dipolar relaxation mechanism, reflected in a τ_{eff} obtained from Eq. (31), must be treated with extreme caution since various motional contributions cannot be explicitly quantitated at this time. The use of NT_1 data to provide τ_{eff} values has the advantage of qualitatively describing altered motional behavior in protein systems.

2. SPIN-ROTATION. Spin-rotation effects result from the interaction of nuclear magnetic moments with fluctuating local fields generated by the motion of molecular magnetic moments. The local fields arise from electron and nuclear currents associated with molecular rotation. Fluctuations in the local fields result from alterations in the magnitude and direction of the angular momentum of the rotating molecule caused by strongly perturbing intermolecular interactions. Although currents associated with symmetrical charge distributions should average to zero with molecular motion, the angular momentum of electrons at a nucleus will, on the average, lead to a net local magnetic field at that nucleus. The local magnetic field is proportional to the rotational velocity of the molecular motion and inversely proportional to the moment of inertia of the molecule. In general only small, symmetric molecules can assume angular velocities large enough to generate large associated fields. The frequency

of molecular collisions determines the spin-rotation correlation time, τ_{SR}; therefore, small molecules with little or no intermolecular interactions will be most affected, since the power density at the resonance frequency will be maximized.

The interaction Hamiltonian for spin rotation is described by

$$\mathcal{K}'_{SR}(t) = -h\bar{I} \cdot \tilde{C} \cdot \bar{J}(t) \tag{37}$$

which indicates that the spin angular momentum, $\bar{I}$ (^{13}C) interacts with the time-dependent angular momentum associated with overall molecular rotation, $\bar{J}(t)$. This interaction is defined by the spin-rotation interaction tensor $\tilde{C}$.

Explicit formulation of $T_{1_{SR}}$ has been developed for the following systems: (1) magnetic nuclei at the center of symmetry of spherically symmetric molecules; (2) magnetic nuclei outside of the molecular center of gravity in a cylindrically symmetric electronic environment; (3) magnetic nuclei in totally asymmetric electronic environments of spherical molecules.

For all these cases T_1 is inversely proportional to temperature. This feature results from increased rotational angular momentum at higher temperatures and leads to more efficient spin-rotation relaxation processes. This behavior is opposite to that for other relaxation mechanisms that depend on overall molecular reorientation. Consequently the spin-rotation contribution to overall relaxation is usually signalled by T_1 passing through a maximum with increasing temperature, and $T_{1_{SR}}$ can be extracted from the measured T_1 by careful analysis of temperature dependence.

Previous considerations strongly suggest that the spin-rotation mechanism will not be an important feature of ^{13}C relaxation in proteins, since these interactions should prevail only for small, relatively symmetric molecules.

3. CHEMICAL-SHIFT ANISOTROPY. The magnetic field experienced by a carbon nucleus in a molecule subject to an external magnetic field is not precisely equal to the applied field because of magnetic fields generated by the molecular electrons. In the absence of unpaired electrons, the motions of electrons, in magnetic fields, responsible for the local field at the nucleus differ from those in a field-free system by a common precession frequency. This field-dependent motion is equivalent to a closed electric current loop and, therefore, has an associated magnetic field. This effect is usually expressed by the equation

$$\bar{H}_{loc} = \bar{H}_0(1 - \tilde{\sigma}) \tag{38}$$

where $\tilde{\sigma}$ is the shielding factor for the extranuclear electrons. Chemical bonding can restrict the electronic circulation directly at the nucleus in question through alteration of both the electron density at the nucleus and the overall symmetry of electronic distribution. In addition to local effects, long-range shielding effects arise from the electrons at other atoms or groups in the molecule. The consequence of *anisotropic* screening will be reflected in directional components of $\tilde{\sigma}$, which will fluctuate with molecular motion relative to the axis of the external magnetic field. The carbon nucleus will experience a local field whose fluctuation is determined by the same rotational reorientation time important to the dipolar relaxation mechanism. The interaction Hamiltonian can be described by

$$\mathfrak{FC}'_A(t) = -\bar{H}_0 \cdot \hbar\tilde{\sigma} \cdot \bar{I} \tag{39}$$

In the case of chemical shift anisotropy the interaction of ^{13}C, $\bar{I}$, with the external magnetic field, $\bar{H}_0$, is defined by shielding effects contained in the chemical shift tensor $\tilde{\sigma}$.

Spin-lattice relaxation times for shift anisotropy are inversely proportional to the square of the applied field for molecules to which the motional narrowing limit is applicable. This strong field dependence can be exploited to determine the contribution of this mechanism to relaxation in small molecules.

Chemical shift anisotropy can be an important relaxation mechanism for nonprotonated carbons in large asymmetric molecules, since the dipolar ^{13}C–^{1}H mechanism will not necessarily dominate whenever another mechanism dependent on the same correlation time is available for relaxation. This is the basis of interpretation of the relaxation of nonprotonated carbons in adenosine 5'-monophosphate.[61] Unfortunately, in the case of proteins the relative contribution from chemical shift anisotropy for nonprotonated carbons may be difficult to evaluate, since for slow molecular reorientation this type of relaxation becomes field invariant and, in addition, any other contributing mechanisms exhibit a field dependence due to terms containing $(1 + \omega^2\tau^2)^{-1}$ [cf. Eq. (31)]. Measurement of nuclear Overhauser enhancement, discussed later, may be useful in recognizing contributions from this mechanism.

4. QUADRUPOLE. Nuclei with spin ≥ 1 possess electric quadrupole moments due to nonspherical distribution of nuclear charge. Quadrupoles can interact with electric field gradients produced by the surrounding electronic environment. Molecular motion modulates the local field gradient at the nucleus thus producing a fluctuating electric field capable of inducing transitions between nuclear quadrupole levels. Thus time-dependent field gradients generated by molecular motion produce relaxa-

[61] A. Allerhand, D. Doddrell, and R. Komoroski, *J. Chem. Phys.* **55**, 189 (1971).

tion for quadrupolar nuclei. The interaction Hamiltonian for the quadrupolar nucleus, $\bar{I}$, is given by

$$\mathcal{3C}'_Q(t) = \bar{I} \cdot \tilde{Q}(t) \cdot \bar{I} \tag{40}$$

where the quadrupole coupling tensor, $\tilde{Q}(t)$, defines the quadrupole interaction. In the motional narrowing limit, T_{1Q} is inversely proportional to both the quadrupole coupling constant, J_Q, and to the same correlation time of overall molecular reorientation as that for dipolar relaxation.

Obviously, this mechanism does not directly contribute to ^{13}C relaxation since ^{13}C does not possess a quadrupole moment. This relaxation mechanism is of some interest to protein ^{13}C NMR, however, since quadrupolar relaxation of ^{14}N and ^{2}H can contribute to ^{13}C relaxation when scalar-coupling mechanisms are important.

5. SCALAR COUPLING. Scalar spin-spin coupling mechanisms, distinct from dipolar coupling, are a second-order effect originating from Fermi hyperfine coupling of spin nuclei mediated by molecular electrons. In protein ^{13}C NMR we will generally not be concerned with ^{13}C–^{1}H coupling since the protons are purposefully decoupled by the experimenter. However, we must investigate the potential effects of ^{13}C–^{2}H and especially ^{13}C–^{14}N coupling on ^{13}C relaxation times. The interaction Hamiltonian for scalar coupling is given by

$$\mathcal{3C}'_{SC}(t) = 2\pi\hbar A \cdot \bar{I} \cdot \bar{S} \tag{41}$$

where ^{13}C, $(\bar{I})$, and ^{14}N, $(\bar{S})$, interact with a scalar coupling constant A. The resulting relationships for T_{1sc} and T_{2sc} are given below:

$$\frac{1}{T_{1sc}} = \frac{8\pi^2 A^2}{3} S(S+1) \frac{\tau_{SC}}{1 + (\omega_I - \omega_S)^2 \tau_{SC}^2} \tag{42}$$

$$\frac{1}{T_{2sc}} = \frac{1}{2T_{1sc}} + \frac{4\pi^2 A^2}{3} S(S+1)\tau_{SC} \tag{43}$$

where $\tau_{SC} = T_{1Q}$ for the scalar coupling mechanism of interest.[48] The field dependence of $(\Delta\omega)^2$ terms may be manifest in the relaxation since $T_{1Q}^2 (\Delta\omega)^2 \simeq 1$. Analysis of the functional relationships represented in Eqs. (42) and (43) and the discussion of quadrupole relaxation provides insight into the effects related to scalar coupling of ^{13}C and ^{14}N for peptides and proteins. To a first approximation, T_{1Q} is inversely proportional to τ_{eff}, and τ_{eff} (peptides) $< \tau_{eff}$ (proteins). Since $(\omega_C - \omega_N)^2$ is quite large the quadrupolar spin-lattice relaxation of ^{13}C nuclei bonded to ^{14}N in proteins and peptides will be very inefficient. But since T_{2sc} is inversely proportional to T_{1Q}, the longer spin-lattice relaxation time for ^{14}N in peptides, compared with proteins, should be reflected in smaller values of T_{2sc} in peptides, and correspondingly broader line widths for the directly bonded ^{13}C nuclei. In practice, this prediction is borne out, especially in the case of histidine, where line widths for the protonated ring carbons in pentapeptides[62] are considerably broadened. Less noticeable broadening is observed for equivalent resonances in proteins since the incorporation of ^{14}N into the slowly reorienting macromolecule decreases $T_{1\varrho}$ to the extent that the second term is Eq. (43) should be very small. The identification of associated correlation times will allow direct extraction of the magnitude of such broadening effects for both ^{14}N and ^2H.

The state of protonation and the nature of the bonded substituents of ^{14}N will affect both the ^{14}N quadrupole coupling constant and the $^{13}C-^{14}N$ spin-spin coupling constant. The lack of data for these constants limits further discussion of comparisons of expected line broadening in ^{13}C NMR of proteins.

Nuclear Overhauser Enhancement. Complete proton decoupling has been shown to simplify the description of spin-lattice relaxation of protonated ^{13}C nuclei. Proton decoupling also provides simplification of ^{13}C spectra through the collapse of multiplets derived from $^{13}C-^1H$ scaler coupling, and in addition provides further signal-to-noise improvement referred to as the nuclear Overhauser enhancement effect (NOE). The NOE is a manifestation of the non-Boltzmann distribution of spin states at equilibrium resulting from ^1H saturation. The steady-state spin populations depend on the various transition probabilities of relaxation for the two-spin system, and thus the NOE is responsive to molecular motion. Grant and co-workers[63] have shown that unlike T_1, the NOE is independent of the number of protons involved in the dipolar relaxation mechanism. They also show that the enhancement factor, X, is proportional to the respective magnetogyric ratios, $\gamma_S : \gamma_I$ and the ratio of the various transition probabilities describing the two-spin system. For carbon and hydrogen this factor becomes

$$X = 1 + \frac{\gamma_{^1H}}{\gamma_{^{13}C}} \frac{W_2 - W_0}{W_2 + 2W_1 + W_0 + 2W_1^*} \tag{44}$$

where W_1^* represents transition probabilities related to nondipolar relaxation mechanisms and the other W terms refer to the dipolar-related transition probabilities. Since the dipolar mechanism contributes to all but one of these probabilities, the effect of other relaxation contributions will be to reduce the enhancement factor. For a purely $^{13}C-^1H$ dipolar mechanism, in the narrowing limit, $X = 2.988$, if the mechanism is very

[62] V. G. Glushko, F. R. N. Gurd, P. Keim, P. J. Lawson, R. C. Marshall, A. M. Nigen, and R. A. Vigna, unpublished results.

[63] K. F. Kuhlmann, D. M. Grant, and R. K. Harris, *J. Chem. Phys.* **52**, 3439 (1970).

efficient. If the dipolar mechanism becomes less efficient (long T_1 values) or if other relaxation mechanisms begin to dominate the dipolar effect, the Overhauser enhancement will decrease. Thus, the measurement of NOE by integrated intensities in conjunction with field or temperature variations[17,55,64] is a useful method for determining the relative contributions of various relaxation mechanisms to spin-lattice relaxation for small molecules.

If relaxation is overwhelmingly dominated by $^{13}C-^1H$ dipolar interactions, and if molecular reorientation is very slow, the NOE drops to 1.153.[63] Since the overall correlation time of proteins falls somewhere between this limit and the extreme narrowing condition, expressions describing the associated NOE are useful to protein chemists. It has been shown in a paper cited previously[60] that for a rigid CH group attached to a rigid, isotropically reorienting molecule the NOE varies between the previously stated limits with a decrease in NOE for increasing molecular correlation time (τ_r) beyond the motional narrowing limit. For a CH group with one degree of motional freedom located on a rigid, isotropically reorienting sphere, Doddrell et al.[60] found that for the extreme narrowing limit $[(\omega_C + \omega_H)^2\tau_r^2 \ll 1]$ the NOE is 2.988 regardless of the magnitude of the correlation time for internal motion, τ_g. For slow overall reorientation $[\omega_C + \omega_H)^2\tau_r^2 \geq 1]$ the NOE increases with the onset of internal motion, passes through a maximum, and asymtotically approaches the NOE predicted in the absence of internal motion as τ_g approaches zero. For both models, the NOE is a function of field strength.

Quite clearly the NOE for protonated carbons in proteins will vary according to their motional properties. Thus, direct integrated intensities of ^{13}C resonances will probably not provide an *accurate* residue count for all amino acids in proteins even in the absence of competing relaxation mechanisms. The inaccuracies will prevail largely for those residues where the magnitudes of several correlation times are comparable.

The Chemical Shift and Spin-Spin Coupling

The previous discussion of relaxation processes focused attention on interactions of magnetic nuclei with fluctuating magnetic fields generated by electrons and nuclei subject to random thermal motion. Components of these fields, fluctuating at the resonance frequency, induced transitions between nuclear energy levels, giving rise to relaxation. The local fields were treated as correlated fluctuations with a mean value of zero, consequently they do not contribute to considerations of those factors which determine the transition energies of spin nuclei in liquids. On the

[64] J. R. Lyerla, D. M. Grant, and R. K. Harris, *J. Phys. Chem.* **75**, 585 (1971).

other hand, we will be very much interested in the response of magnetic nuclei to nonzero contributions to local fields and interactions, i.e., chemical shifts and spin-spin coupling constants.

Two of the most informative features of the magnetic resonance experiment are the chemical shifts and coupling constants. The disposition of resonance positions of carbon nuclei is the consequence of the extreme sensitivity of these magnetic nuclei to their environment. This sensitivity reflects the response of the spin nuclei to electronic effects, which modulate the applied magnetic field at the nucleus, and to interactions between two nuclei coupled through the intervening electrons.

The surrounding electrons in a molecule produce effects which modify the Zeeman energy, describing the interaction of the nucleus with the applied magnetic field. The molecular electron currents alter the local field felt at the nucleus, H'_{eff} by shielding effects given by

$$H'_{eff} = H_0(1 - \sigma) \tag{45}$$

where σ is the shielding constant. The Zeeman interaction term of the Hamiltonian is correspondingly modified for each nucleus.[65] Since the effective field varies according to the chemical environment of the nucleus, the resonance condition, at fixed radio frequencies, for chemically nonequivalent nuclei will be achieved at slightly different applied fields.

The shielding of nuclei in diamagnetic molecules can be treated as arising from four separate contributions. The first, diamagnetic screening due to directly bonded nuclei, is proportional to electron density of the electronic ground state at the nucleus. The second, or paramagnetic, contribution, depends on electronic distributions in both ground and excited states. In effect this term takes into account the restriction of electronic circulation by chemical bonding. This contribution can be very large for asymmetric distribution of p and d electrons close to the nucleus, which have low-lying excited states, but vanishes for s orbitals. The third contribution includes the diamagnetic and paramagnetic contributions from other distant nuclei and is especially large for molecules with large and anisotropic magnetic susceptibilities. The fourth contribution refers to ring currents arising from electronic currents in closed rings such as delocalized π-electrons in aromatic compounds. This is a diamagnetic effect in which the shift, $\Delta\nu_{rc}$ (in Hz), is described by[14]

$$\Delta\nu_{rc} = \frac{\gamma_N N e^2 H_0 a^2}{8\pi^2 m c^2 d^3} (3\cos^2\theta - 1) \tag{46}$$

where d is the distance of the nucleus from the center of the ring, N is the

[65] A. Carrington and A. D. McLachlan, "Introduction To Magnetic Resonance." Harper, New York, 1967.

number of delocalized π-electrons with charge e and mass m, c is the velocity of light, θ is the angle between d and the magnetic field $\bar{H}_0$, and a is the radius of the aromatic ring.

Intermolecular contributions to chemical shifts must be recognized for strong solvation or whenever magnetically anisotropic solvents are used. In the case of proteins many of the ^{13}C nuclei are less likely to be influenced by solvent effects due to their high solvent inaccessibility. On the other hand, chemical shifts of amino acid side chains at the protein surface and small bound molecules may be subject to solvent perturbations.

Distant contributions to ^{13}C chemical shifts seem to be relatively small,[66] and paramagnetic factors seem to dominate diamagnetic terms. This suggests that ring current effects should be of the same magnitude for ^{13}C as for ^1H, but that ^{13}C chemical shifts should be very responsive to asymmetric p or d orbital electronic distributions. By the same token, the dominance of paramagnetic effects will tend to complicate correlations based on electron densities such as those calculated for electronegative substituents in heteroaromatic compounds. Electric field effects also contribute to ^{13}C chemical shifts and can be treated as perturbations of the diamagnetic and paramagnetic factors. Electric field effects propagated through the medium will compete with induction effects through chemical bonds and should be prominent over larger distances.

Chemical shifts are also determined by nuclear spin–spin coupling of which the strongest is dipolar coupling. Although this interaction vanishes for small molecules in solution, it is notable that ^1H NMR measurements of proteins are complicated by proton–proton dipolar coupling due to motional restriction of the large protein molecules.[8,67] The analogous ^{13}C–^{13}C dipolar interaction is highly improbable due to the low natural abundance of the isotope. The most important interaction to be considered is contact coupling (scalar coupling) between ^{13}C and ^1H in which one nucleus tends to align electron spins antiparallel to its nuclear spin. The electron spins are coupled in chemical bonds such that spin polarization at the one nucleus is accompanied by an excess of electron spins, parallel to the first nucleus, at the second nucleus. The result is that the field experienced by the second nucleus is influenced by the spin state of other coupled nuclei in neighboring bonded groups. In a more formalistic sense, the spin–spin coupling interaction is another term added on to the Zeeman term to describe the resonance positions of carbons. ^{13}C–^1H scalar coupling

[66] J. Mason, *J. Chem. Soc. (A)* 1038 (1971).
[67] M. Cohn, A. Kowalsky, J. S. Leigh, and S. Maričić, *in* "Magnetic Resonance in Biological Systems" (A. Ehrenberg, B. G. Malmström, and T. Vänngård, eds.), p. 45. Pergamon, Oxford, 1967.

patterns and constants are useful for chemical shift and stereochemical assignments in small molecules. The multiplet structures resulting from various coupling or decoupling techniques,[68] tend to complicate natural abundance ^{13}C NMR of proteins to the extent that such techniques provide serious drawbacks compared with total proton decoupling. Measurements of ^{13}C–^{1}H and ^{13}C–^{13}C coupling constants should be useful for ^{13}C enriched proteins and enzyme substrates. ^{13}C–^{13}C coupling will provide considerable analytical difficulties for high percentage isotopic enrichment. However, limited enrichment for carbons adjacent to or very near each other should produce simple splitting patterns similar to proton splittings. In principle the associated ^{13}C–^{13}C and ^{13}C–^{1}H coupling constants should reflect both substituent and stereochemical information similar to that from ^{1}H–^{1}H coupling in amino acids.[1,69] Selective ^{13}C enrichment in peptides and proteins, sufficient to produce limited coupling, could provide the means for establishing similar relationships.

Paramagnetic Phenomena

The ability of paramagnetic transition metals to modify chemical shift and relaxation properties of diamagnetic compounds is a consequence of unpaired electron interactions with spin nuclei. The unpaired electrons can alter nuclear resonance positions through isotropic shifts. The term "isotropic shift" includes phenomena arising from Fermi hyperfine contact interactions (scalar) or from electron-nuclear pseudocontact interactions (dipolar). The scalar interaction results from electron spin transfer between metal and ligand, mediated through σ and π molecular orbitals. The dipolar interaction is most simply described as a through-space coupling between unpaired electrons with anisotropic g values, and ligand nuclei. The expected contact shift, $\Delta \nu_c$ (in Hz), is given by[70]

$$\Delta \nu_c = \frac{A \gamma_e S(S+1)\nu}{\gamma_I 3kT} \tag{47}$$

where A is the contact interaction constant proportional to γ_I, S is the total electronic spin, γ_e is the magnetogyric ratio for the electron, ν is the resonance frequency of nucleus I and T is the absolute temperature. The pseudocontact shift, $\Delta \nu_{pc}$ (in ppm), for hemes and heme proteins, where rotational tumbling is slow and electronic relaxation is fast, is expressed by[26]

---

[68] E. Wenkert, A. O. Clouse, D. W. Cochran, and D. Doddrell, *J. Amer. Chem. Soc.* **91**, 6879 (1969).

[69] M. Mandel, *J. Biol. Chem.* **240**, 1586 (1965).

[70] N. Bloembergen, *J. Chem. Phys.* **27**, 572, 595 (1957).

$$\Delta\nu_{pc} = CK \left[g_z{}^2 - \frac{1}{2} (g_x{}^2 + g_y{}^2) \right] (1\text{-}3 \cos^2\Omega) + \frac{3}{2} (g_y{}^2 - g_x{}^2) \sin^2\Omega \cos 2\Psi$$

$$(48)$$

where Ω is the polar angle of radius r from the electron spin to the nucleus and Ψ is the angle between the projection of r in the xy plane and the x axis. The temperature-dependence for this shift is contained in C according to the equation

$$C = \frac{\beta^2 S(S + 1)}{9kTr^3}$$

$$(49)$$

where β is the Bohr magneton. K is a reduction factor for the unpaired spin in the transition metal orbitals. Contact and pseudocontact shifts are both inversely proportional to temperature, and this dependence is used for their identification. The pseudocontact term should attenuate rapidly outside the immediate vicinity of the metal due to the r^{-3} dependence. Shifts expected for ^{13}C nuclei should be of the same magnitude as those for ^1H. In this respect ^{13}C shifts may not be as valuable as ^1H shifts have been for proteins,[15,27] since the range of shifts for ^{13}C is much greater than that for ^1H. Isotropic shifts in enriched proteins should provide useful information similar to that obtained in ^1H studies of intrinsically paramagnetic proteins such as the heme proteins[15,27] as well as for those diamagnetic proteins which bind paramagnetic species.[30]

Another important feature of paramagnetic interactions with proteins is related to relaxation phenomena.[2] The large magnetic moment of an unpaired electron provides an intense, fluctuating field which usually dominates other local fluctuating fields. Thus bound paramagnetic metals usually control the relaxation of nuclei in their immediate vicinity. The utility of the isotropic shift is restricted to those paramagnetic metals which do not relax nuclei too quickly. If nuclear relaxation is very rapid, resonance signals are broadened too severely to be seen. The most successful NMR studies have been done using metals whose unpaired electron has a very *short* relaxation time. Such short relaxation times prelude EPR measurements, so as a rule of thumb one selects metals for NMR studies which yield poor EPR signals. Eaton[71] and others have demonstrated the usefulness of several metals for NMR studies, including V(III), Mn(III), Ru(III), Ni(II), Co(II), Cr(II), Fe(III), Eu(III), and Pr(III).

The equations describing longitudinal and transverse relaxation times for nuclei in paramagnetic molecules are given below[54,70]:

[71] D. R. Eaton, *J. Amer. Chem. Soc.* **87**, 3097 (1965).

$$\frac{1}{T_{1M}} = \frac{2}{15} \frac{S(S+1)\gamma_I^2 g^2 \beta^2}{r_{IS}^6} \left[\frac{3\tau_{en}}{1 + \omega_I^2 \tau_{en}^2} + \frac{7\tau_{en}}{1 + \omega_S^2 \tau_{en}^2} \right]$$
$$+ \frac{2}{3} \frac{S(S+1)A^2}{\hbar^2} \left[\frac{\tau_e}{1 + \omega_S^2 \tau_e^2} \right] \quad (50)$$

$$\frac{1}{T_{2M}} = \frac{1}{15} \frac{S(S+1)\gamma_I^2 g^2 \beta^2}{r_{IS}^6} \left[4\tau_{en} + \frac{3\tau_{en}}{1 + \omega_I^2 \tau_{en}^2} + \frac{13\tau_{en}}{1 + \omega_S^2 \tau_{en}^2} \right]$$
$$+ \frac{1}{3} \frac{S(S+1)A^2}{\hbar^2} \left[\tau_e + \frac{\tau_e}{1 + \omega_S^2 \tau_e^2} \right] \quad (51)$$

In the equations τ_{en} is the electron-nuclear dipolar correlation time, τ_e is the correlation time for the scalar interactions, and S refers to the electron. The dipolar correlation time, τ_{en}, will be determined by the fastest rate process selected from the contributions from τ_r, the correlation time for the metal–protein complex, τ_s, the electron spin relaxation time, and τ_m, the residence time of the nuclear species in the first coordination sphere of the paramagnetic ion according to

$$\frac{1}{\tau_{en}} = \frac{1}{\tau_r} + \frac{1}{\tau_s} + \frac{1}{\tau_m} \quad (52)$$

Similarly, the correlation time for the scalar interaction will be determined by the shorter of either τ_s or τ_m. For inert metal–protein complexes τ_{en} may include contributions from τ_r and τ_s for most paramagnetic species of interest. In general, electron relaxation times are poorly understood.[72] The strength of ligand field and covalent bonding can profoundly influence the value of τ_s in rather unpredictable ways. The magnitude of τ_e will generally be dominated by the contribution from τ_s.

In Eqs. (50) and (51) the first term represents the dipolar contribution and the second corresponds to the scalar contribution to paramagnetic relaxation of nuclei. The scalar term for $1/T_2$ will be much larger than that for $1/T_1$ for relatively long electron relaxation times, which explains the substantial broadening mentioned previously for certain paramagnetic species. The small magnetogyric ratio and large ion-nucleus distance for carbon should be sufficiently favorable to offset effects due to long correlation times, so that paramagnetic metal–protein interactions can be studied by ^{13}C NMR.

The effect of the paramagnetic metal on ^{13}C relaxation rates is a function of the distance between the interacting sites and of the motional freedom in the region of the paramagnetic species. Consequently, paramagnetic studies using ^{13}C should provide information on the dynamic

[72] D. R. Eaton and W. D. Phillips, *Advan. Magn. Resonance* **1**, 119 (1965).

properties of enzyme active sites *at least* as useful as that obtained by
^{1}H techniques.[2]

Experimental

Instrumentation and Pulse Techniques

The study of proteins by ^{13}C NMR[73-77] has been realized by the
development of high resolution pulsed Fourier transform techniques.[78-80]
Prior to these developments standard continuous wave (CW) methods
were used with very limited success.[74] The main disadvantage of the
latter technique lies in the fact that the sampling of each resonance in
turn requires a considerable amount of time. Since protein concentrations
are very dilute compared with those for small molecules, thousands of
scans must be computer averaged to provide sufficient signal-to-noise for
accurate analysis. Application of a coherent radio frequency to decouple
all protons in the spectrum is generally employed in both CW and pulsed
modes to improve the sensitivity. Even so, the time required to obtain
sufficient signal-to-noise by CW spectroscopy is measured in days, thus
instrumental and sample stability become significant limiting factors. On
the other hand the application of a short, intense, homogeneous pulse of
an appropriate radio frequency samples *all* ^{13}C resonances simultaneously
over a broad frequency range. However, certain experimental requirements must be met to ensure that all ^{13}C nuclei are sampled. These requirements can best be understood by considering the macroscopic
magnetization associated with all spin states in the sample, in the rotating
frame. For a given static field and a frame rotating with the frequency
of the H_1 field, ω, very near the respective carbon resonance frequencies,
ω_i, the effective field, H_{eff}, in the absence of H_1 will be small and directed
along the z' axis,[52] according to

$$|H_{\mathrm{eff}}| = \frac{1}{\gamma_c} [(\omega_i - \omega)^2 + (\gamma_c H_1)^2]^{1/2} \tag{53}$$

If H_1 is chosen such that

[73] A. Allerhand, D. W. Cochran, and D. Doddrell, *Proc. Nat. Acad. Sci. U.S.* **67**,
1093 (1970).
[74] P. C. Lauterbur, *Appl. Spectrosc.* **24**, 450 (1970).
[75] J. C. W. Chien and J. F. Brandts, *Nature (London)* **230**, 209 (1971).
[76] A. Allerhand, D. Doddrell, V. G. Glushko, D. W. Cochran, E. Wenkert, P. J.
Lawson, and F. R. N. Gurd, *J. Amer. Chem. Soc.* **93**, 544 (1971).
[77] V. G. Glushko, P. J. Lawson, and F. R. N. Gurd, *J. Biol. Chem.* **247**, 3176 (1972).
[78] E. L. Hahn, *Phys. Rev.* **80**, 580 (1950).
[79] R. R. Ernst and W. A. Anderson, *Rev. Sci. Instrum.* **37**, 93 (1966).
[80] R. R. Ernst, *Advan. Mag. Resonance* **2**, 1 (1966).

$$\gamma_c H_1 \gg 2\pi\Delta \qquad (54)$$

where Δ is the entire range of chemical shifts (in Hz) to be studied, H_{eff} will lie along the axis of H_1, the x' axis. As we have seen, when these conditions obtain, the magnetization will tip through a particular angle dictated by the duration of the pulse (usually only a few microseconds). Typically the pulse duration is fixed to produce a 90° tip in the magnetization in less than 25 μsec. The pulse is then shut off and a time-averaging computer is triggered and collects the free induction decay signal, in digital form, in a matter of seconds. After a waiting period (recycle time) 3–5 times the longest T_1 in the sample (typically no longer than about 1 second for proteins), the sequence can be repeated under control of a pulse programmer. Thousands of accumulations improve the signal-to-noise by $(N)^{1/2}$, where N is the total number of accumulations. The time saving by this technique, compared with CW methods, is sufficient to provide comparable natural abundance spectra for 10–15 mM solutions of small proteins in 4–6 hours. The accumulated FID signal is a time domain spectrum describing the decay of the magnitude of the magnetization with time. Data processing is required to convert the FID to the familiar frequency domain spectrum. The transformation of the time domain signal to the frequency domain[81] is accomplished by fast Fourier transformation.[82] Programs suitable for this transformation[77] contain the basic Fourier transform routine, an apodization correction to counter the large transient signal from the radio frequency pulse by equating the signal in the first few channels of memory to a specific signal, and a phase correction to compensate for the combination of absorption and dispersion mode introduced by the spectrometer electronics.[52,65] The resultant Fourier transformed signal called a "normal" spectrum, can now be stored on magnetic tape or "read-out" in the usual manner.

The main thrust of pulsed Fourier transform ^{13}C NMR of proteins is directed toward analysis of dynamical molecular behavior reflected in spin-lattice relaxation times. Several variations of pulse sequences are applicable to T_1 measurements,[52] although most methods have virtues primarily for small molecule studies. The method of interest for proteins is the inversion recovery method in which a 180°-t-90° pulse sequence is employed.[43] Initially the magnetization is inverted by a radio frequency pulse, which is short compared to measured T_1 values. After removal of the pulse the z' component of the magnetization will begin to return to its original equilibrium position and magnitude. A delay time, t, is

[81] I. J. Lowe and R. E. Norberg, *Phys. Rev.* **107**, 46 (1957).
[82] J. W. Cooley and J. W. Tukey, *Math. Comput.* **19**, 297 (1965).

selected when a 90° observing pulse is applied to rotate the z' component into the $x'y'$ plane where its FID is observed. A waiting period of 3–5 T_1 follows to allow the magnetization to return to equilibrium, and the entire pulse sequence is repeated. The pulse sequence is again controlled by the pulse programmer. The resultant partially relaxed Fourier transform (PRFT) spectrum in the frequency domain, for a particular t, is obtained as previously described. The typical spectrum contains resonance peaks whose amplitudes can be negative, null, or positive depending on whether $t < T_1 \ln 2$, $t = T_1 \ln 2$, or $t > T_1 \ln 2$, respectively. For progressively longer delay times, a set of amplitudes for an isolated resonance are obtained whose intensities vary from fully negative to fully positive. Values of T_1, are computed using the equation

$$A = A_0(1-2 \exp[-t/T_1]) \tag{55}$$

where A_0 is the equilibrium amplitude of the peak obtained by reiteration of a single 90° pulse and A is the amplitude at a particular instrumental t setting. Data analysis is accomplished by computer fitting programs (methods in references cited in footnotes 83 and 84 as applied in 77) or by plotting $\ln(A_0 - A)$ *versus* t using the relationship

$$\ln(A_0 - A) = \ln 2A_0 - t/T_1 \tag{56}$$

Although pulsed ^{13}C spectrometers are becoming commercially available, the majority of instruments in present operation are of the "home-built" variety. The complexity of pulsed instrumentation is apparent from the previous discussion. The successful operation of pulsed equipment requires the combined efforts of qualified scientists and versatile electronics personnel. In this regard we have benefited from the expertise of Dr. Adam Allerhand, who provided the conceptual basis for the spectrometer, and of Mr. Arthur Clouse and his associates, who perfected the instrumental design. A brief description of our instrument will serve as an example of the types of components required for protein measurements.

Our experiments are carried out on a high resolution spectrometer equipped with a Varian 14.1 kG electromagnet operating at 15.1 MHz. A Fabri-Tek 1074 computer is used for signal averaging and a PDP-8/I computer for data processing. The apparatus also includes an Ortec pulse programer, a Butterworth filter, a homemade probe which accommodates 11.6 mm (i.d.) sample tubes, and various fabricated amplification systems. The probe is constructed to accommodate the spinner, a proton decoupling coil, the radio frequency pulse coil (which also serves as

[83] P. R. Bevington, "Data Reduction and Error Analysis for the Physical Sciences," p. 104. McGraw-Hill, New York, 1969.
[84] M. J. D. Powell, *Computer J.* **7**, 155 (1964).

the detector) and a temperature control tube connected to an externally cooled dry air source. The equipment is also provided with an external frequency modulated ^{19}F lock, a noise-modulated proton decoupler, and a phase-sensitive detector. Our present spectral resolution is limited to just over 0.1 ppm. Homogeneity adjustments are made using neat ethylene glycol, and spectral assignments are referenced to CS_2. Internal dioxane is used for small molecule experiments.

Potential improvements in instrumentation include the use of larger static fields, automatic shimming systems, wide-range temperature control, larger capacity and faster computers, improved blocking circuits necessary to improve the response characteristics of the detector, and improved computer processing techniques.[85] Since the major fraction of the signal is contained in the first few microseconds following removal of the pulse, the development of improved electronic circuitry to reduce "dead-time" is of critical importance for protein work. Working at higher fields, however, may not significantly improve signal-to-noise for proteins.[60] Although higher field strength provides an increase in basic sensitivity, this improvement can be offset by a decrease in the NOE and T_2/T_1 ratio,[79] which are strongly dependent on the long protein correlation times and the magnetic field strength. Increased spectral resolution at higher fields, will undoubtedly be needed for extended chemical shift analysis of closely spaced resonances. Unfortunately spectral resolution at high field strength is offset by large tube diameters. Since signal is initially determined by the total number of nuclei in the probe, the diameter of the sample tubes at higher field strengths can be a limiting factor even if NOE and T_2/T_1 are favorable.

Sample Preparation

Natural abundance spectra can be obtained in 6 hours from concentrated protein solutions (10–15 mM) in sample tubes of approximately 12 mm diameter. Such concentrations can be obtained by a variety of techniques of which ultrafiltration seems the most gentle. Samples of diamagnetic proteins must be scrupulously devoid of paramagnetic materials which will tend to dominate T_1 values.

Systematic Errors for T_1 Measurements

The most significant contribution to error in T_1 measurements is due to radio frequency inhomogeneity. Inhomogeneity can result from the pulse source itself or from characteristics of the sample and introduces a systematic error in an experiment which is reflected in T_1 values being

---

[85] H. M. Pickett and H. L. Strauss, *Anal. Chem.* **44**, 265 (1972).

too short. Source inhomogeneity is minimized by using the most homogeneous source available within financial reason. Problems due to sample characteristics can be minimized by reducing the extension of sample volume beyond the geometrical confines of the pulse coil. The duration of the pulse and the function of the pulse programmer must be set to minimize systematic contributions caused by large pulse widths, unsynchronized or variable pulse durations, and programmer instability. In addition, the receiver must be tuned to maintain linear response characteristics over the expected range of signal intensities. A test sample, of known T_1 values, should always be used to check the spectrometer prior to experiments on unknown samples.

Applications

Small Molecules

The systematics of chemical shifts with substitution on and near a given carbon nucleus have been worked out by Grant and co-workers[86-88] and by others (see Mooney and Winson[89] and Cochran[90] and references therein). Numerous applications of ^{13}C NMR have been made to solve structural questions in organic and natural products chemistry.[91-95] The technique has been quick to assimilate recent refinements[95a-97] such as the use of rare earth ions to alter chemical shifts or to affect relaxation behavior.[95a,97] To introduce the applications of ^{13}C NMR to amino acids, peptides, and proteins, a concise sketch of the trends of chemical shifts with the nature of substitution is helpful. The relationships observed in

[86] E. G. Paul and D. M. Grant, *J. Amer. Chem. Soc.* **85**, 1701 (1963).

[87] D. M. Grant and E. G. Paul, *J. Amer. Chem. Soc.* **86**, 2984 (1964).

[88] T. D. Brown, Ph.D. Thesis, University of Utah, Salt Lake City, Utah, 1967.

[89] E. F. Mooney and P. H. Winson, *Annu. Rev. NMR Spectros.* **2**, 153 (1969).

[90] D. W. Cochran, Ph.D. Thesis, Indiana University, Bloomington, Indiana, 1971.

[91] A. Rabaron, M. Koch, M. Plat, J. Peyrouk, E. Wenkert, and D. W. Cochran, *J. Amer. Chem. Soc.* **93**, 6270 (1971).

[92] M. Jautelat, J. B. Grutzner, and J. D. Roberts, *Proc. Nat. Acad. Sci. U.S.* **65**, 288 (1970).

[93] E. Wenkert, C.-J. Chang, A. O. Clouse, and D. W. Cochran, *Chem. Commun.* 961 (1970).

[94] R. Neuss, C. H. Nash, P. A. Lemke, and J. B. Grutzner, *J. Amer. Chem. Soc.* **93**, 2337 (1971).

[95] M. Tanabe, T. Hamasaki, D. Thomas, and L. F. Johnson, *J. Amer. Chem. Soc.* **93**, 273 (1971).

[95a] P. G. Gassman and G. A. Campbell, *J. Amer. Chem. Soc.* **93**, 2566 (1971).

[96] O. A. Gansow and W. Schittenhelm, *J. Amer. Chem. Soc.* **93**, 4294 (1971).

[97] O. A. Gansow, M. R. Willcott, and R. E. Lenkinski, *J. Amer. Chem. Soc.* **93**, 4295 (1971).

these studies between the chemical shift and molecular structure form the basis for interpreting chemical shifts of proteins in terms of primary, secondary, tertiary, and quaternary structure.

Aliphatic Carbons. Grant and co-workers[86,87] have demonstrated that chemical shifts for alkanes can be calculated accurately using four empirically determined substituent effects and several branching corrections. Roberts *et al.* have determined substituent parameters for alcohols[98] and carboxylic acids,[99] and Brown[88] has identified the corresponding parameters for amino groups. These substituent parameters are similar to those for methyl groups but are of different magnitude. The influence of substituents at a carbon gamma to the substitution site has been the subject of considerable interest. Grant and Cheney[100,101] interpreted this effect in terms of a force component which polarizes the CH bond at the γ position through nonbonded interactions with the alpha substituent. These interactions have pronounced angle and distance dependence. Correlations have also been reported between the chemical shift and bond angle for X-C-X structures.[102]

[13]C chemical shifts are responsive to conformational constraints in closed rings and to various types of conformational equilibria. Chemical shifts for axial carbons (and the carbons to which they are attached) have been found to occur upfield of their equatorial counterparts.[103,104] Tautomeric equilibria have been studied,[105] as have conformational equilibria at low temperatures.[106] In addition chemical shift anisotropies have been measured for methyl halides in nematic liquid crystals.[107-110] Attempts at calculating ^{13}C chemical shifts for aliphatic systems from first principles have met with mixed success.[66]

Ionization of protonated carboxyl and amino groups is signaled by downfield shifts of several parts per million in covalently attached

[98] J. D. Roberts, F. J. Weigert, J. I. Kroschwitz, and H. J. Reich, *J. Amer. Chem. Soc.* **92**, 1338 (1970).

[99] R. Hagen and J. D. Roberts, *J. Amer. Chem. Soc.* **91**, 4504 (1969).

[100] B. V. Cheney and D. M. Grant, *J. Amer. Chem. Soc.* **89**, 5319 (1967).

[101] B. V. Cheney, *J. Amer. Chem. Soc.* **90**, 5386 (1968).

[102] D. Purdela, *J. Magn. Resonance* **5**, 37 (1971).

[103] G. W. Buchanan and J. B. Stothers, *Can. J. Chem.* **47**, 3605 (1969).

[104] D. K. Dalling and D. M. Grant, *J. Amer. Chem. Soc.* **89**, 6612 (1967).

[105] J. Feeney, G. A. Newman, and P. J. S. Pauwels, *J. Chem. Soc. C* 1842 (1970).

[106] H. J. Schneider, R. Price, and T. Keller, *Angew. Chem. Int. Ed. Engl.* **10**, 730 (1971).

[107] C. S. Yannoni and E. B. Whipple, *J. Chem. Phys.* **47**, 2508 (1967).

[108] T. Yonemoto, *J. Chem. Phys.* **54**, 3234 (1971).

[109] E. Breitmaier, G. Jung, and W. Voelter, *Angew. Chem. Int. Ed. Engl.* **10**, 673 (1971).

[110] G. R. Luckhurst, *Oesterr. Chem. Ztg.* **68**, 113 (1967).

aliphatic carbons.[99] This effect is attenuated considerably three to four carbons away. The direction of these shifts is opposite to that for corresponding ionizations monitored by ^{1}H NMR. This suggests that electric field effects,[111-115] oppositely directed to inductive effects, are dominant for these ^{13}C chemical shifts.[38-40,99] In fact, the electric field effects for ^{13}C can be more than an order of magnitude greater than inductive effects, although the two effects are about the same order of magnitude for [1]H.

The studies discussed in the preceding paragraphs suggest the great potential for interpretation of ^{13}C chemical shifts of aliphatic carbons in terms of protein conformation, in general, and local protein environments, in particular. The sensitivity of these shifts to steric constraints, inductive effects, and especially to field effects should, in principle, provide detailed information about neighboring residues related to amino acid substitution, steric constraints of conformation, and electrostatic interactions.

Aromatic Carbons. Empirical substituent effects appear to be additive for aromatic carbocycles.[116] Deviations at the *ortho* position reflect steric inhibition of resonance.[117] Chemical shifts at the *para* position correlate with Hammett and Taft σ and σ^+ constants.[118] Correlations between measured shifts and those calculated on the basis of electron density depend on the methods of calculation.[119-121] In general, calculated chemical shifts provide fair-to-poor agreement with experimental values.[122,123]

Chemical shifts for various substituted aromatic heterocycles have been obtained (see footnotes 90, 124, 125 and references in works cited therein). Chemical shift calculations based on electron density[126,127] sug-

[111] A. D. Buckingham, *Can. J. Chem.* **38**, 300 (1960).
[112] W. T. Raynes and T. A. Sutherley, *Mol. Phys.* **17**, 547 (1969).
[113] J. C. Hammel and J. A. S. Smith, *J. Chem. Soc. A* 2883 (1969).
[114] W. McFarlane, *Chem. Commun.* 418 (1970).
[115] J. C. Hammel and J. A. S. Smith, *J. Chem. Soc. A* 1852 (1970).
[116] G. B. Savitsky, *J. Phys. Chem.* **67**, 2723 (1963).
[117] K. S. Dhami and J. B. Stothers, *Can. J. Chem.* **45**, 233 (1967).
[118] C. H. Yoder, R. H. Tuck, and R. E. Hess, *J. Amer. Chem. Soc.* **91**, 539 (1969).
[119] P. C. Lauterbur, *J. Amer. Chem. Soc.* **83**, 1838 (1961).
[120] H. Spiesecke and W. G. Schneider, *J. Chem. Phys.* **35**, 731 (1961).
[121] A. J. Jones, P. D. Gardner, D. M. Grant, W. M. Litchman, and V. Boekelheide, *J. Amer. Chem. Soc.* **92**, 2395 (1970).
[122] T. Tokuhiro and G. Fraenkel, *J. Chem. Phys.* **51**, 3626 (1969).
[123] A. Velenik and R. M. Lynden-Bell, *Mol. Phys.* **19**, 371 (1970).
[124] Y. Sasaki and M. Suzuki, *Chem. Pharm. Bull.* **17**, 1778 (1969).
[125] R. J. Pugmire and D. M. Grant, *J. Amer. Chem. Soc.* **93**, 1880 (1971).
[126] T. F. Page, T. D. Alger, and D. M. Grant, *J. Amer. Chem. Soc.* **87**, 5333 (1965).
[127] F. J. Weigert and J. D. Roberts, *J. Amer. Chem. Soc.* **90**, 3543 (1968).

gest that σ-polarization and low energy $n \to \pi^*$ transitions contribute to these chemical shifts, and therefore predictions developed from π electron densities, alone, are usually inaccurate.[122,128,129]

Numerous studies on charged heteroaromatic compounds[130,131] are of some interest. Protonation of imidazole produces upfield shifts at all three carbons,[132] a result unexpected from electron density calculations.[129] In general, however, chemical shifts of carbons not directly bonded to heteroatoms correlate with calculated electron densities.[132] Usually, contributions from bond order[130,132] and excitation energies for low-lying excited states[133] must be included with estimates of electron densities to provide understanding of the sensitivity of aromatic carbons to molecular structure.

The nature of the response of heteroaromatic carbons to molecular structure is quite complex. Presumably, aromatic amino acid side chains should be sensitive to local charge perturbation. Further theoretical and experimental work is necessary before useful structural information can be extracted from the response of aromatic ^{13}C chemical shifts in proteins. The magnitude of the responses of aliphatic carbons to aromatic ring currents, exploited so elegantly in ^1H NMR studies of lysozyme[9,18] has not been assessed for ^{13}C nuclei. Studies on some porphyrins[134] indicate that ^{13}C shifts due to ring currents are of the same order of magnitude as those for ^1H. Thus, relatively speaking, ring current effects should not be as valuable for ^{13}C studies of proteins as they have been for ^1H studies.

Carbonyls. The ^{13}C chemical shifts of carbonyl carbons are extremely sensitive to external perturbations. Carbonyls conjugated with double bonds or aromatic rings are shifted several parts per million upfield.[135] These shifts are also sensitive to the twist angle in π-systems.[136] Good correlations exist between chemical shifts and $n \to \pi^*$ transition energies,[53,135] π-bond polarities,[137] and electron densities.[138] These shifts also

[128] W. Adam, A. Grimison, and G. Rodriguez, *J. Chem. Phys.* **50**, 645 (1969).
[129] G. Del Re, B. Pullman, and T. Yonezawa, *Biochim. Biophys. Acta* **75**, 153 (1963).
[130] A. Mathias and V. M. S. Gil, *Tetrahedron Lett.* 3163 (1965).
[131] R. J. Pugmire and D. M. Grant, *J. Amer. Chem. Soc.* **90**, 697 (1968).
[132] R. J. Pugmire and D. M. Grant, *J. Amer. Chem. Soc.* **90**, 4232 (1968).
[133] A. J. Jones, D. M. Grant, J. G. Russell, and G. Fraenkel, *J. Phys. Chem.* **73**, 1624 (1969).
[134] D. Doddrell and W. S. Caughey, *J. Amer. Chem. Soc.* **94**, 2510 (1972).
[135] J. B. Stothers and P. C. Lauterbur, *Can. J. Chem.* **42**, 1563 (1964).
[136] K. S. Dhami and J. B. Stothers, *Can. J. Chem.* **43**, 479 (1965).
[137] G. E. Maciel, *J. Chem. Phys.* **42**, 2746 (1965).
[138] G. E. Maciel, *J. Phys. Chem.* **69**, 1947 (1965).

respond to solvent[139] or intramolecular hydrogen bonding[140] by moving downfield.

Carbonyl chemical shifts of proteins should be very responsive to bonded substituents and nonbonded perturbations of the polar, charged, or steric types. In addition, these nuclei should be sensitive probes of secondary structure where hydrogen-bonding and electric field effects would be cumulative. The accessibility of certain carbonyls to aqueous solution may provide characteristic resonances. However, the presence of a large number of carbonyl carbons in the protein is reflected in an extended range of associated resonances. Therefore the information concerning structural perturbations monitored by the carbonyls will be restricted to those situations where a class of carbonyl resonances is substantially shifted away from the main resonance band.

Metal Complexes. Reports on paramagnetic metal interactions with organic molecules have been limited. Studies of iron cyanide complexes indicate that large contact shifts are observed.[141-143] These shifts are responsive to σ- and π-delocalization of electrons. Other complexes with Ni(II)[144,145] and Co(II)[144] have been studied. For Ni(II)–amino acid complexes,[146] upfield shifts were observed for those carbons bonded to the ligating heteroatoms, and downfield shifts for all other carbons. Large pseudocontact shifts in small molecules have also been observed.[95a,97] The potential for exploiting these shifts appears to be limited by the already large range of ^{13}C resonances in the absence of paramagnetic ions.

Solvent Effects. Studies of solvent effects on ^{13}C chemical shifts primarily relate to hydrogen-bonding with carbonyls.[139,147,148] Responses of aliphatic or aromatic carbons to water can be substantial, suggesting that "exposed" carbons may be shifted relative to their "buried" counterparts, in some cases.

Relaxation Behavior. Molecules the size of many substrates and most coenzymes have been of particular interest from the point of view of T_1 measurements. Except for some very small (see Lyerla and Grant[55] and references therein) molecules (which could include the smallest sub-

[139] G. E. Maciel and D. D. Traficants, *J. Amer. Chem. Soc.* **88**, 220 (1966).

[140] G. E. Maciel and G. B. Savitsky, *J. Phys. Chem.* **68**, 437 (1964).

[141] M. Shporer, G. Ron, A. Loewenstein, and G. Navon, *Inorg. Chem.* **4**, 358 (1965).

[142] D. G. Davis and R. J. Kurland, *J. Chem. Phys.* **46**, 388 (1967).

[143] J. Conrad, *C. R. Acad. Sci. Ser. B* **266**, 975 (1968).

[144] D. Doddrell and J. D. Roberts, *J. Amer. Chem. Soc.* **92**, 6839 (1970).

[145] D. Doddrell and J. D. Roberts, *J. Amer. Chem. Soc.* **92**, 5255 (1970).

[146] C. E. Strouse and N. A. Matwiyoff, *Chem. Commun.* 439 (1970).

[147] G. E. Maciel and G. C. Ruben, *J. Amer. Chem. Soc.* **85**, 3903 (1963).

[148] G. E. Maciel and J. Natterstad, *J. Chem. Phys.* **42**, 2752 (1965).

strates), the $^{13}C-^1H$ dipole–dipole interactions are dominant for proton-ated carbons relative to the other relaxation mechanisms such as chemical-shift anisotropy,[52,149-151] scalar coupling,[149] and spin rotation.[152-154]

Kuhlmann, Grant, and Harris[63] studied the highly symmetrical molecule adamantane, containing 4 CH carbons and 6 CH_2 carbons. The Overhauser enhancement was the same for both types of carbons. So long as the extreme narrowing limit holds, the theory predicts that the Overhauser effect should be the same independent of the number of directly bonded hydrogens for all carbons in a molecule whenever $^{13}C-^1H$ dipolar interaction is the major relaxation mechanism. The value of T_1 for the CH_2 carbons was 11.4 ± 1 seconds, and for the CH carbons it was 20.5 ± 2 seconds. The deviation from the simple directly bonded ratio of 1 to 2 fits with a more elaborate calculation taking vicinal hydrogens into account.[63] This molecule represents an almost ideal case because of the isotropic rotational diffusion which allows the correlational time to be strictly the same for each C-H vector.

Allerhand, Doddrell, and Komoroski[61] have made elegant use of the partially relaxed Fourier transform technique to study sucrose, cholesteryl chloride, and adenosine 5'-monophosphate (AMP). Under several conditions it was shown that the nuclear Overhauser enhancement for the carbon nuclei in sucrose is the same for all types of carbons. In 2 M sucrose the enhancement for the nonprotonated carbon was determined to be within experimental error of the theoretical maximum of 2.988.[63] The conclusion was drawn that the $^{13}C-^1H$ dipolar relaxation was operative for all types of carbons in sucrose. Measurements in D_2O had little effect, indicating that intermolecular dipolar relaxation was negligible. The overall motion of the sucrose molecules in water was judged to be essentially isotropic, with τ_r values of 7×10^{-11} seconds and 3×10^{-10} seconds for 0.5 M and 2 M solutions, respectively. In 2 M sucrose the carbons of the primary alcohol groups showed relaxation times measurably longer than one-half T_1 of the protonated carbons in the rings. This is evidence for separate or "internal" motion of the side chains, and it was estimated that τ_g might be in the range 9×10^{-11} to 3×10^{-10} second.

The T_1 measurements on the protonated backbone carbons of cholesteryl chloride in carbon tetrachloride were inversely proportional to

[149] J. R. Lyerla, D. M. Grant, and R. D. Bertrand, *J. Phys. Chem.* **75**, 3967 (1971).
[150] R. R. Shoup and D. L. VanderHart, *J. Amer. Chem. Soc.* **93**, 2053 (1971).
[151] H. W. Spiess, D. Schweitzer, U. Haeberlen, and K. H. Hausser, *J. Magn. Resonance* **5**, 101 (1971).
[152] J. R. Lyerla, Ph.D. Thesis, University of Utah, Salt Lake City, Utah, 1971.
[153] S. W. Collins, Ph.D. Thesis, University of Utah, Salt Lake City, Utah, 1971.
[154] A. Olivson and E. Lippmaa, *Chem. Phys. Lett.* **11**, 241 (1971).

the number of attached hydrogens. T_1 for CH_2 carbons was approximately 0.25 second and for CH carbons 0.50 second. Again, therefore, the rotational motion was essentially isotropic, with τ_r of 9×10^{-11} second. C-20 showed the corresponding value of 0.49 second, so that there was no evidence of internal motion on its part. The value of 1.5 seconds for T_1 for the methyl group carbons C-18, C-19, and C-21 could be rather directly interpreted since they are attached to carbons that do not show individual motion, and τ_g for these methyl carbons was taken to be no longer than 5×10^{-12} second. The motion involved is the rotation of the methyl groups. Similar reasoning led to the conclusion that internal motion is greatest for carbons near the free end of the hydrocarbon side chain.

In the case of AMP two of the nonprotonated carbons, C-4 and C-5, in the base showed less intensity than the other carbons in the proton-decoupled spectra. All three of the nonprotonated carbons of the base showed long T_1 values.

Doddrell and Allerhand[155] measured T_1 values of resonances in neat n-decanol-1 that had been assigned by Roberts and co-workers.[98] There was a steady trend towards longer T_1 values in going from the alcoholic group to the methyl group at the other end. These results were interpreted in terms of there being enough restriction in the reorientation of the molecule as a whole to allow the correlation times for internal motion to contribute detectably. In n-decanol-1, τ_r and the various τ_g values are of the same order of magnitude, with τ_g decreasing with increasing distance from the hydroxyl group. In the straight-chain hydrocarbons the effect is not apparent.[156]

Scalar Coupling. A considerable amount of information is available concerning $^{13}C-^1H$ coupling constants.[3,89] Interest in this area is due to the known relationship between the magnitude of the coupling constant and the nature of the bonding carbon orbitals.[157] The value of these studies is limited by the theoretical foundations for analysis[158-160]; however, coupling constants for selectively enriched proteins should be structurally informative.

Although $^{13}C-^{13}C$ coupling effects are insignificant features of natural abundance studies of proteins, studies of proteins enriched at several loci certainly will include consideration of short and long range $^{13}C-^{13}C$

[155] D. Doddrell and A. Allerhand, *J. Amer. Chem. Soc.* **93**, 1558 (1971).
[156] R. Freeman and H. D. W. Hill, *J. Chem. Phys.* **53**, 4103 (1970).
[157] W. McFarlane, *Quart. Rev. Chem. Soc.* **23**, 187 (1969).
[158] J. A. Pople and D. P. Santry, *Mol. Phys.* **8**, 1 (1964).
[159] J. A. Pople and D. P. Santry, *Mol. Phys.* **9**, 311 (1965).
[160] M. Barfield and D. M. Grant, *Advan. Magn. Resonance* **1**, 149 (1965).

coupling. Limited ^{13}C enrichment of proteins should provide additional information about the bonding features of the enriched carbons provided sufficient information is available establishing such correlations in small molecules. Very few reports devoted to the magnitudes of $^{13}C-^{13}C$ coupling constants and theoretical correlations are available (see Bartuska and Maciel[161] and references therein). In general, constants for directly bonded carbons increase with increasing "s" character of the bond.[162-165] Typical values for nonaromatic systems range from 30–40 Hz for sp^3 hybrids, 60–70 Hz for sp^2 hybrids and 150 Hz for sp hybrids. This sensitivity to bond hybridization seems to dominate polarization effects.[166] However, rough correlations do exist between the magnitude of $^{13}C-^{13}C$ coupling constants and the electronegativity of the atom directly attached to one of the carbons.[161,164-166] Increases in coupling constants are observed for increased electronegativity of the bonded substituent. Multiply bonded carbons (carbonyls) are more sensitive to this effect than are sp^3 hybrids.[165] The magnitude of the couplings decreases dramatically beyond the first bond,[165,167] although hybridization and substituent effects are still notable for substituted benzenes.[165] Modulation of $^{13}C-^{13}C$ scalar coupling due to steric features has been reported for cyclopropane.[164] Ring strain is interpreted in terms of unusual hybridization of the ring carbons. Similar studies related to strained regions in protein sequences or repeating helical structures are desirable. At present, availability of coupling data is still a limiting factor. As more data become available, the applicability of existing theoretical treatment (see Bartuska and Maciel[161] and references therein) to carbon–carbon coupling can be thoroughly tested.

Amino Acids and Peptides

Chemical Shifts. The initial systematic work by Sternlicht and co-workers[38-40] on ^{13}C chemical shifts in amino acids and peptides revealed a pattern consonant with the principles outlined in the preceding section. Moving upfield from CS_2 toward increased shielding, the pattern shows first carbonyl carbons (amide and free carboxyl), other nonprotonated carbons (e.g., C^ζ of arginine), aromatic carbons, and finally aliphatic carbons. In general, the aliphatic β carbons lie upfield of the α carbons

[161] V. J. Bartuska and G. E. Maciel, *J. Magn. Resonance* **5**, 211 (1971).
[162] R. M. Lynden-Bell and N. Sheppard, *Proc. Roy. Soc. Ser. A* **269**, 385 (1962).
[163] K. Frei and H. J. Bernstein, *J. Chem. Phys.* **38**, 1216 (1963).
[164] F. J. Weigert and J. D. Roberts, *J. Amer. Chem. Soc.* **89**, 5962 (1967).
[165] A. M. Ihrig and J. L. Marshall, *J. Amer. Chem. Soc.* **94**, 1756 (1972).
[166] W. M. Litchman and D. M. Grant, *J. Amer. Chem. Soc.* **89**, 6775 (1967).
[167] F. J. Weigert and J. D. Roberts, *J. Amer. Chem. Soc.* **89**, 2967 (1967).

followed by γ and δ carbons. The influences of electronegative atoms on the aliphatic side chains are pointed out below.

In the following section dealing with proteins a rather complete listing of amino acid resonances is given for reference. Here it is perhaps more instructive to make a number of comparisons of the chemical shifts of the free amino acids with their residues in peptide linkage. The most representative location for an amino acid residue in a protein being one considerably removed from a chain end,[44] the plan in this laboratory is to complete a set of comparisons in which the given residue is central in a pentapeptide.[168,169] In the pentapeptide the residue under study is flanked on either side by a pair of glycine residues.

In Table II are listed the chemical shifts for each carbon nucleus in ten amino acids in both the peptide and free forms at neutral pH. In all cases the carboxyl carbon lies much the most downfield. On incorporation into the peptide form this resonance shows an upfield shift in most cases of the order of 0.5 ppm. Except inside a peptide chain this resonance is quite sensitive to pH, as shown below. The same observations apply to the α carbons where a similar chemical shift difference of about 1 ppm is seen. Again, the pH sensitivity in a shorter peptide or amino acid is marked. The β-carbon nuclei generally lie considerably upfield from the α-carbons with the exception of serine and threonine which show the marked deshielding influence of the hydroxyl group.

As might be expected the aromatic nuclei of phenylalanine, tyrosine and histidine show both a downfield chemical shift range and a great sensitivity to substituents. This sensitivity is clearly seen for phenylalanine and tyrosine in the peptide form. The substitution of the hydroxyl group on C^ζ in tyrosine not only deshields that carbon itself, but also has marked effects on the ortho C^ϵ nuclei and the para C^γ nucleus. For histidine the protonated imidazole form is represented in parentheses. Both C^γ and C^ϵ are sensitive to ionization, and indeed the effect is not limited to the heterocyclic moiety.

The pattern of chemical shifts in the other side chains shown in Table II is generally toward higher values as the influence of the electronegative atoms diminishes. For lysine the assignments have been aided by T_1 data as described below. It is noteworthy that the gem-dimethyls of valine and leucine are not magnetically equivalent. The collection of data to extend Table II to cover all the protein amino acids is in progress.[62]

[168] F. R. N. Gurd, A. Allerhand, D. Doddrell, V. G. Glushko, P. J. Lawson, A. M. Nigen, and P. Keim, *Fed. Proc., Fed. Amer. Soc. Exp. Biol.* **30**, 1046 (1971).

[169] F. R. N. Gurd, P. Keim, V. G. Glushko, P. J. Lawson, R. C. Marshall, A. M. Nigen, and R. A. Vigna, 3rd American Peptide Symposium, Boston, Massachusetts, 1972.

TABLE II

CHEMICAL SHIFTS OF ZWITTERIONIC AMINO ACIDS AND CENTRAL AMINO ACIDS IN GLYGLY-X-GLYGLY PENTAPEPTIDES[a]

| | Chemical shifts | | | | | | | | | | | | | | | | | | | |
| | Ser | | Thr | | His | | Phe | | Tyr | | Lys | | Met | | Ala | | Val | | Leu | |
Carbon	Peptide[d]	Free	Peptide[d]	Free	Peptide[d]	Free	Peptide[d]	Free	Peptide[d]	Free	Peptide[d]	Free	Peptide[d]	Free	Peptide[d]	Free	Peptide[d]	Free	Peptide[d]	Free
CO	20.6	20.6[c]	20.4	19.6[b]	19.4 (20.9)[e]	18.7[b]	19.2	18.8[b]	19.3	18.6[b]	18.6	18.3[b]	18.8	18.3[b]	17.5	16.8[b]	18.8	18.8[c]	17.8	17.4[c]
$C\alpha$	137.1	136.1[c]	133.7	132.1[b]	138.8 (140.2)[e]	139.2[b]	137.6	136.3[b]	137.4	136.4[b]	139.0	138.4[b]	139.9	138.4[b]	142.9	142.0[b]	132.9	132.1[c]	140.1	138.9[c]
$C\beta$	131.7	132.4[c]	125.9	126.5[b]	164.2 (166.6)[e]	166.2[b]	156.0	156.1[b]	156.8	156.2[b]	162.7	163.0[b]	162.8	162.6[b]	176.3	176.3[b]	163.0	163.6[b]	153.2	152.8[c]
$C\gamma$			174.1	173.1[b]	60.0 (64.4)[e]	65.5[b]	56.4	56.4[b]	64.9		170.9	171.3[b]	163.6	163.5[b]			174.4 / 175.3	174.7[c] / 176.0[c]	168.6	168.4[c]
$C\delta$					75.4 (75.4)[e]	75.4[b]	64.0	65.5[b]	62.4	63.2[b]	166.6	166.5[b]	178.7	178.4[b]					170.6 / 172.1	170.6[c] / 171.6[c]
$C\epsilon$					56.6 (59.2)[e]	59.2[b]	63.5	59.2[b]	77.4	75.7[b]	153.4	153.7[b]								
$C\zeta$							65.6	64.1	38.2	36.7[b]										

[a] All chemical shifts are expressed as parts per million upfield of CS_2. Amino acid data were obtained in neutral D_2O or H_2O. Pentapeptide data were obtained in neutral H_2O.
[b] W. J. Horsley, H. Sternlicht, and J. S. Cohen, J. Amer. Chem. Soc. 92, 680 (1970).
[c] F. R. N. Gurd, P. J. Lawson, D. W. Cochran, and E. Wenkert, J. Biol. Chem. 246, 3725 (1971).
[d] V. G. Glushko, F. R. N. Gurd, P. J. Keim, P. J. Lawson, R. C. Marshall, A. M. Nigen, and R. A. Vigna, in preparation.
[e] Protonated imidazole form.

The results to date confirm the pattern of strong terminal effects reported by Gurd and co-workers.[44] Incorporation of an amino acid residue into an amino-terminal position in a peptide at neutral pH results in an upfield shift for the carbonyl carbon and α-carbon of 4 or 5 ppm and 1 or 2 ppm, respectively. These shifts are much greater than for the internal incorporations discussed above. They are paralleled to a large extent simply by the change in the free amino acid when dissociation of the carboxyl group is suppressed.[170] Conversely, removing the charge on the amino group causes downfield shifts in the carbonyl and α-carbon resonances no matter what set of comparisons is considered. It is reasonable, therefore, to recognize large effects of nearby charges and to look upon the chemical shift positions of carbonyl and α-carbons in the free dipolar amino acids partly as the resultant of canceling charge effects.[111-115,133] Electric field effects appear to be dominant, at least in certain loci in amino acids and peptides, over inductive effects. The result is the observed tendency of the chemical shift with ionization in ^{13}C NMR to be opposite in direction to that in ^{1}H NMR.[38]

The effects of changes in protonation extend beyond the immediate backbone carbon nuclei. Figure 6 shows the dependence on pH of chemical shifts for individual carbon atoms in three peptides: (A) L-valyl-L-leucyl-L-seryl-L-glutamylglycine; (B) L-leucyl-L-seryl-L-glutamic acid; (C) L-seryl-L-glutamylglycine. Theoretical titration curves link the shifts for a given carbon atom according to reasonable pK_a values. For each of the three peptides the largest changes on titration are seen with the titration of the α-amino group. For the carbonyl carbon of the amino-terminal residue the changes in chemical shift on titration of this group are about 8 ppm. The effect of this titration on the vicinal α-carbon is less than that for the carbonyl carbon, and is often exceeded by that for the β-carbon. It can be seen from parts (A) and (B), respectively, of Fig. 6 that the terminal amino group ionization affects even the end of the valine and leucine side chains. In the tripeptides the middle residue is consistently affected by ionization processes of the end residues, an observation that forms the basis of the decision to emphasize pentapeptides as models for polypeptides.

In the pentapeptide glycylglycyl-L-tyrosylglycylglycine C$^\alpha$ and C$^\beta$ of the tyrosine residue show no sensitivity to pH. There is a hint of a pH dependence (0.5 ppm span) centered around pH 9.9 for the carbonyl carbon of this residue. The span of the shifts between acid and basic extremes for C$^\zeta$, C$^\epsilon$, and C$^\gamma$ are 10.5, 3.3, and 6.8 ppm, respectively, with

[170] W. A. Gibbons, J. A. Sogn, A. Stern, L. C. Craig, and L. F. Johnson, *Nature* (*London*) **227**, 840 (1970).

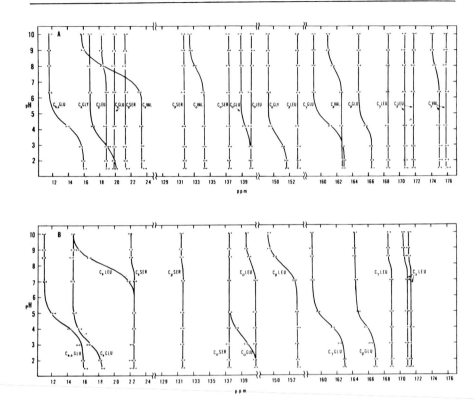

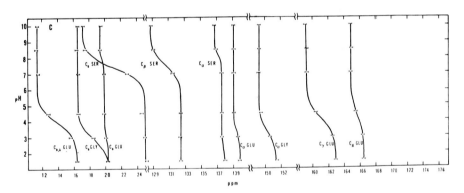

Fig. 6. Dependence on pH of chemical shifts for individual carbon nuclei in parts per million relative to CS_2. (A) L-valyl-L-leucyl-L-seryl-L-glutamylglycine; (B) L-leucyl-L-seryl-L-glutamic acid; (C) L-seryl-L-glutamylglycine. C_0 represents the α-carboxyl carbon atom. The temperature was 36°. Taken from F. R. N. Gurd, P. J. Lawson, D. W. Cochran, and E. Wenkert, *J. Biol. Chem.* **246**, 3725 (1971).

pK values of 10.02, 9.95, and 10.01. For C$^\gamma$ the shift is upfield with deprotonation whereas for the others it is downfield. No change of chemical shift is observed with C$^\delta$.

Apart from the seminal work of Sternlicht and co-workers,[38-40] the work of Freedman, Cohen, and Chaiken,[171] of Christl and Roberts,[172] and of Jung, Breitmaier, and Voelter,[173] should be consulted for other examples of pH dependence of ^{13}C NMR spectra of peptides. Voelter et al.[109,173-175] have published a large collection of data on peptides, amino acids, and related compounds in which pH references are usually lacking, however. The method holds promise for determining microscopic dissociation constants.[44]

Gibbons and co-workers[170] reported the proton-decoupled natural abundance ^{13}C NMR spectrum of gramicidin S-A, the antibiotic cyclo(-Phe-Pro-Val-Orn-Leu-)$_2$. Their reported spectra are reproduced in Fig. 7. Because of the C$_2$ symmetry the spectra amount to those of a pentapeptide without terminal effects. Partial assignments were based on the parameters of Grant and Paul[86,87] and Horsley et al.[38-40] that have already been mentioned, and also on N-acetyl amino acid methyl esters in dimethyl sulfoxide, a solvent used with the decapeptide. The general appearance of the spectra gives an idea of what to expect in a protein spectrum. Part B shows the carbonyl carbons grouped together but clearly resolved except for the apparent overlay of two resonances. Upfield are resonances matching the phenylalanine chemical shifts (cf. Table II). Part A shows the separated α-carbon resonances with nearby the δ carbon resonance of proline, a carbon which is also directly attached to the nitrogen. Comparison of parts A and C shows some differences that may possibly represent solvent effects.[170,173] These are most striking in the γ carbons and gem-dimethyl groups. The downfield shift of the δ carbon of ornithine relative to the corresponding carbon of leucine is similar to the results for lysine in Table II. These results also show that, at least in the gramicidin S-A geometry, the carbons directly bonded to nitrogen do not experience significant broadening effects. Not all resonances appeared with the same intensity.

Several special points about this work stand out. First, the continuous

[171] M. H. Freedman, J. S. Cohen, and I. M. Chaiken, *Biochem. Biophys. Res. Commun.* **42**, 1148 (1971).

[172] M. Christl and J. D. Roberts, *J. Amer. Chem. Soc.* **94**, 4565 (1970).

[173] G. Jung, E. Breitmaier, and W. Voelter, *Eur. J. Biochem.* **24**, 438 (1972).

[174] G. Jung, E. Breitmaier, and W. Voelter, *Hoppe-Seyler's Z. Physiol. Chem.* **352**, 16 (1971).

[175] G. Jung, E. Breitmaier, W. Voelter, T. Keller, and C. Tänzer, *Angew. Chem. Int. Ed. Engl.* **9**, 894 (1970).

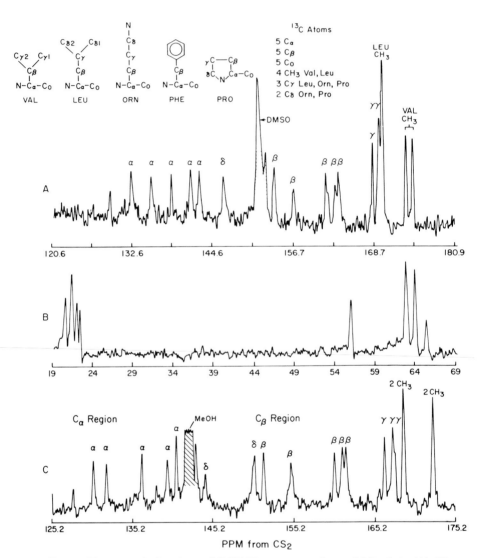

FIG. 7. The natural abundance ¹³C NMR spectrum of gramicidin S-A. (A) The high-field part of the spectrum in DMSO; (B) the low-field part in DMSO; (C) the high-field part in methanol. All chemical shifts are in parts per million upfield from CS₂. Taken from W. A. Gibbons, J. A. Sogn, A. Stern, L. C. Craig, and L. F. Johnson, *Nature* (*London*) **227**, 840 (1970).

wave technique was used, yielding generally broader resonances than are obtained by the pulsed Fourier transform technique that displaces it. Second, the structure as proposed[170,176] contains hydrogen bonds between certain specific backbone components. Third, the geometry is possibly rigid enough to show only coherent tumbling of the backbone carbons. Alternatively, if flexing motions contribute to the relaxation processes in an observable way it may be possible to differentiate such motions at different points in the molecule by the specific enrichments with ^{13}C that we are promised.[170] Fourth, it must be borne in mind that the molecular weight of 1120 puts this peptide well above the "small molecule" size class and well below the protein size class. Because of the symmetry of the molecule the resolution of the spectrum is twice as good as the molecular weight would usually imply. Fifth, the molecule contains some D-amino acid residues.

Relaxation Behavior. Up to now very little has been published on T_1 measurements of amino acids and peptides. Amino acids are small enough that they may fall within the category mentioned under "small molecules" in which T_1 values are characteristically long and may not be dominated by the ^{13}C–^{1}H dipole–dipole relaxation mechanism. A consequence of the long values of T_1 is that the resonances are easily attenuated unless recycling is appropriately delayed. This effect may be a problem when pulsed Fourier transform NMR techniques are first adopted in a laboratory. In this regard the T_1 value can be shortened by a paramagnetic ion, in many cases without any change in chemical shift, and one may foresee examples where a signal will be recognized only after some such manipulation of the system. An interesting example of improving signal acquisition by reducing molecular motility is in the report by Sternlicht and co-workers[37] on ^{13}C NMR of amino acid derivatives bound to polystyrene resin beads.

The pentapeptides described at the beginning of this section, in which a given residue is flanked on either side by a glycylglycine unit, are being studied with respect to the T_1 values of the various carbon nuclei. Figure 8 shows examples of values for the three pentapeptides glycylglycyl-L-alanylglycylglycine, glycylglycyl-L-lysylglycylglycine, and glycylglycyl-L-tyrosylglycylglycine. The results are schematized to show the α-carbons and the side chains. The observed T_1 values are converted to NT_1 by multiplying by N, the number of directly bonded protons. The T_1 values of the carbonyl carbons were too long for accurate measurement with the recycle times used. The same applied for the nonprotonated C^γ and C^ζ in

[176] A. Stern, W. A. Gibbons, and L. C. Craig, *Proc. Nat. Acad. Sci. U.S.* **61,** 734 (1968).

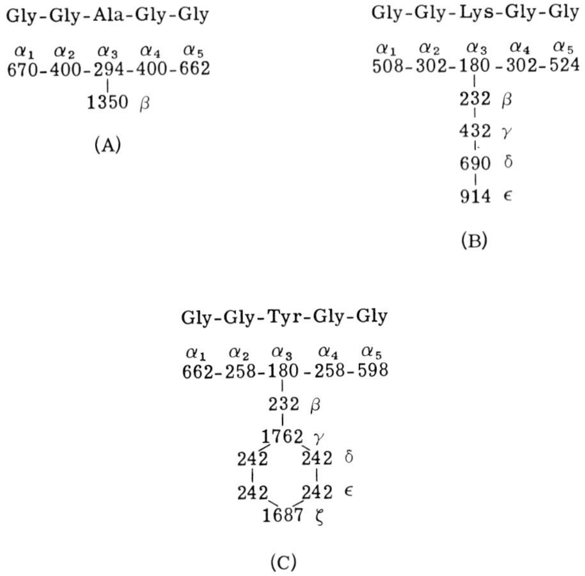

Gly-Gly-Ala-Gly-Gly

α_1 α_2 α_3 α_4 α_5
670-400-294-400-662
|
1350 β

(A)

Gly-Gly-Lys-Gly-Gly

α_1 α_2 α_3 α_4 α_5
508-302-180-302-524
|
232 β
|
432 γ
|
690 δ
|
914 ϵ

(B)

Gly-Gly-Tyr-Gly-Gly

α_1 α_2 α_3 α_4 α_5
662-258-180-258-598
|
232 β
|
1762 γ
242   242 δ
| |
242   242 ϵ
1687 ζ

(C)

FIG. 8. Schematic representation of NT_1 values for the three pentapeptides in neutral solution: (A) glycylglycyl-L-alanylglycylglycine; (B) glycylglycyl-L-lysyl-glycylglycine; (C) glycylglycyl-L-tyrosylglycylglycine. The NT_1 values for the five C^α in each peptide are shown in horizontal succession, and the side-chain values are labeled pendantly. The temperature was 27°. V. G. Glushko, F. R. N. Gurd, P. Keim, P. J. Lawson, R. C. Marshall, A. M. Nigen, and R. A. Vigna, unpublished results.

the tyrosine residue, but the results are given to emphasize the striking difference between them and their neighbors. The measurements were made at neutral pH and 27° at 15.1 MHz.

Several points come out of Fig. 8. First, in each case the α-carbons of residues 2 and 4 are indistinguishable with respect to NT_1 as well as with respect to chemical shift. Second, the trend along the backbone in each case shows the shortest NT_1 values for the central residue with longer ones toward the ends. Third, the C^β of alanine in the first pentapeptide shows a long NT_1 relative to the other protonated carbons. Fourth, the side chain of lysine in the second pentapeptide shows a steady trend in NT_1 as one looks farther out from the backbone. Fifth, the C^β, C^δ, and C^ϵ carbons in the tyrosine in the third peptide all have the same value of NT_1 within experimental error.

For molecules in this molecular weight class one expects the $^{13}C-^1H$ dipole–dipole relaxation mechanism to dominate for the protonated carbons.[61] Bearing in mind Eq. (34), the trends of NT_1 mentioned above indicate the following. First, rotational reorientation occurs most freely at the ends of the backbone and less so in the center. Second, the methyl group of alanine undergoes rapid reorientation qualitatively in keeping with theoretical predictions.[60,61] Third, the lysine side chain shows increasingly free rotational reorientation in passing stepwise from C^β to C^ϵ. The effect is more pronounced than that discussed above for n-decanol-1.[155] Fourth, if the dipolar relaxation mechanism remains dominant in the aromatic ring, it could be argued that the most effective motion of the side chain of the tyrosine is the rotation about the axis determined by C^β, C^γ, and C^ϵ. Allerhand and Hailstone[42] have presented evidence for such an interpretation of results on polystyrene solutions.

Proteins

The preceding sections have brought out most of the points to bear in mind when considering a study of a protein by ^{13}C NMR. Compared to the established ^{1}H NMR methods, ^{13}C NMR offers some advantage in the wider range of chemical shifts, perhaps a somewhat more complete reporting of reactive groups, an absence of multiplet signals in the proton-decoupled mode, and generally a more interpretable relaxation behavior. The method shares with ^{1}H NMR the potentiality of providing a mass of detailed information, perhaps second only to the quantity of information provided for the crystalline state by X-ray diffraction analysis.

Spectral Resolution. The very inclusiveness of the method makes it difficult to exploit. The spectral resolution of the resonance of a given carbon nucleus is limited by the appreciable number of chemically identical carbons with which one must expect to deal, as well as by the frequently very small differences in chemical shift between chemically distinct carbons. In other words the individual peaks or bands actually seen in the spectrum are made up of the sum of signals from several carbon atoms, either in the sense that several carbon atoms of the same chemical type behave sufficiently similarly to overlie each other, or in the sense that signals from different carbons overlap too closely for separate peaks to be distinguished.

Motional Behavior. The large size of proteins has the further consequence that overall molecular motion is relatively slow. Any parts of the structure lacking internal motion will show longer τ_r values than will be encountered with the much smaller molecules that have been discussed in such terms above. The most significant implication is that the product $\omega_c^2 \tau_r^2$ will not be so much less than 1 that the motional narrowing limit holds, and equations of the form of Eqs. (31) and (32) are required even to describe the simplest isotropic case. To signal that the assumption of isotropic motion will generally be inappropriate for a protein, the notation τ_{eff} will be used here (see, for instance the discussion of Eq. 34, above).

On the other hand, it is known that even in protein crystals some atoms such as those on the extremity of lysine side chains occupy no preferred location and are presumably in thermal motion.[177,178] Hence it is to be expected that no matter how stable the secondary structure of the protein, some carbons will undergo relatively rapid internal motion and yield relatively short τ_{eff} values. In this case τ_{eff} arises from a number of terms including rotational anisotropy, internal motion of side chains, and segmental motion of the polypeptide backbone, that cannot at present be evaluated separately.[60,61]

Sensitivity to Conformational Change. It is well known that ^{1}H NMR spectra are very sensitive to conformational change. McDonald and Phillips in particular have shown several series of spectra obtained at temperatures passing through the range of thermal transition from native to denatured or "melted" states.[9-11] The spectra of the denatured state, in which the structure is typically more mobile and less compact than in the native state, are much more sharply defined. The basis for the spectral narrowing lies in two factors. First, more rapid segmental motion lengthens T_2, and thence leads to a narrower line width [Eqs. (26) and (32)]. Second, the distinctions between the various microenvironments of the individual carbon nuclei of a given chemical nature are lessened. This effect may be thought of both in terms of greater exposure to solvent and of rapid time averaging of environments. The overlie of signals from identical nuclei is improved, and overlap with others is reduced.

The effects on T_1 of the more rapid segmental motion are generally toward increased values. However, as shown in Fig. 5, T_1 passes through a minimum with τ_r so that the possibility of observing a decrease in T_1 with a (limited ?) increase in segmental mobility must be kept in mind. In this case the effects of internal motion on T_1 could be pronounced[60] whenever $\tau_r \simeq \tau_g$.

Normal Ribonuclease A Spectra. Every protein spectrum published up to now has represented a new undertaking in the particular laboratory. Lauterbur[74] and Chien and Brandts[75] have published spectra of hen egg white lysozyme. Breitmaier, Jung, and Voelter,[109] Conti and Paci,[179] and an advertising circular produced by Jeol Laboratories show spectra of heme proteins. In this department Professor A. Allerhand has provided the insight, technique, and impetus for studies of ribonuclease A[73,76,77]

[177] G. Kartha, J. Bello, and D. Harker, *Nature (London)* **213**, 862 (1967).
[178] H. W. Wyckoff, D. Tsernoglou, A. W. Hanson, J. R. Knox, B. Lee, and F. M. Richards, *J. Biol. Chem.* **245**, 305 (1970).
[179] F. Conti and M. Paci, *FEBS (Fed. Eur. Biochem. Soc.) Lett.* **17**, 149 (1971).

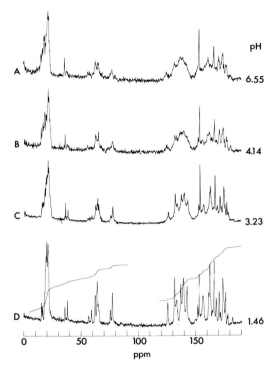

FIG. 9. Natural abundance ^{13}C NMR spectra of ribonuclease A. Resonance positions are given in parts per million upfield from CS$_2$. (A) pH 6.55; (B) pH 4.14; (C) pH 3.23; (D) pH 1.46; the integrated intensity in arbitrary units is shown in the superimposed curves. The temperature was 45°. Taken from V. G. Glushko, P. J. Lawson, and F. R. N. Gurd, *J. Biol. Chem.* **247**, 3176 (1972).

and the other proteins mentioned above.[168] We shall deal first with the more fully studied ribonuclease and then return to the other proteins.

Spectra of ribonuclease A at several pH values are shown in Fig. 9.[77] The spectra at pH 6.55 and 4.14 are very similar overall and reflect mainly some changes in the aromatic region, 50–80 ppm, where histidine resonances should respond to the pH difference (cf. Table II). Much more striking changes are observed at the lower pH values, in a range where conformational changes are recognized[180] to occur at the temperature of this work, 45°. At pH 1.46 the spectrum has much narrower and less overlapping resonances. This is a pH value at which denaturation is complete at the temperature of the experiment, just as the transition is incomplete

[180] J. F. Brandts, *J. Amer. Chem. Soc.* **87**, 2759 (1965).

at pH 3.23 where the spectrum has some of the properties of both the native and acid-denatured spectra.

The relatively well resolved resonances of the spectrum in Fig. 9D of the acid-denatured protein are suitable for making tentative assignments, which are listed in Table III. The listing is identical with that published,[77] with some additions taking account of more recent unpublished work described previously.[62] In considering the spectra in Fig. 9, it is well to bear in mind that the intensity of the resonance from a single carbon nucleus in the protein molecule should be barely observable. Indeed, a broad line would be easily overlooked. Furthermore, these experiments were conducted with a recycle time of 1.36 seconds, which biases the intensities against the nuclei with the longest T_1 values such as (see below) the nonprotonated carbons. While considerable attention has been paid to the reproducibility of the relative intensities in these spectra, a compromise has been made in favor of the most rapid recycling that would yield reliable intensities for the upfield resonances. Nevertheless, in a large number of cases to be discussed, quite good proportionality with composition is observed.

Carbonyl Region. The carbonyl region between 10 and 25 ppm upfield from CS_2 becomes appreciably sharper on denaturation. The two small peaks near 15.5 and 16.5 ppm may represent carboxyl carbon atoms.[40] Most of the absorption in this region occurs between 18 and 22 ppm upfield of CS_2, an observation matched by the values in Table III. By adjusting the observing "window" Glushko has expanded this region and brought out some details.[181] This region of the spectrum would benefit from study at a higher frequency.

Predominantly Aromatic Region. Upfield to about 80 ppm from CS_2 is the region in which unsaturated carbons are found. The first peak at 36 ppm in Fig. 9D represents C^ζ of the four arginine residues. This peak is seen with comparable sharpness and intensity in all the spectra in the figure. It is a clear example of a resonance that is not sensitive to the conformational state of the protein, presumably because its environment in every instance is primarily that of the solvent. Forming a striking and pleasing contrast is the immediately adjacent resonance at 38.2 ppm of C^ζ of the six tyrosine residues. This is a clear peak in Fig. 9D, with an intensity relative to C^ζ of arginine in keeping with the six-to-four numerical proportion in the enzyme. The observation of such a proportion calls for further study under improved recycling conditions to establish the nuclear Overhauser enhancement for these carbon nuclei; it is quite possible from present indications that they are experiencing maximum

---

[181] V. G. Glushko, Ph.D. Thesis, Indiana University, Bloomington, Indiana, 1972.

enhancement as such. The other intriguing aspect of the peak for C^ζ of tyrosine in Fig. 9 is that it becomes broadened in the native spectra (A and B) in a most striking way. Clearly the established nonequivalence of the tyrosines[182] is reflected in these spectra. In the intermediate spectrum in Fig. 9C the native and denatured pictures appear to be combined.

Between 56.5 and 77.3 ppm in Fig. 9D are a group of purely aromatic resonances. The first can be assigned to the nonprotonated C^γ of the three phenylalanine residues, in rough proportion to the neighboring resonance at 59.2 ppm assigned to the four $C^{\epsilon 1}$ of histidine. These resonances are much broadened in the other spectra. That assigned to histidine should shift upfield in part at pH 6.55 under the neighboring complex band of resonances.

The band of partially overlapping resonances in Fig. 9D immediately upfield is described with tentative assignments in Table III. The first nearby separated peak at 62.5 ppm represents the twelve nuclei of $C^{\delta 1}$ and $C^{\delta 2}$ of the six tyrosine residues. Table III shows that the remainder of the overlapping resonances represent four carbons in the three phenylalanine residues that probably dominate the outstanding peak, flanked upfield by C^γ of histidine and tyrosine and C^ζ of phenylalanine. The appearance of this region changes markedly with pH. Part of the change presumably depends on effects of protonation of the imidazole rings,[62] but may depend on shifts[62,172] and disproportionate broadening of other resonances.

The remaining aromatic resonances at 75.5 ppm, histidine $C^{\delta 2}$ and at 77.3 ppm, tyrosine $C^{\epsilon 1}$ and $C^{\epsilon 2}$, are again sharp in Fig. 9D. By inspection, their areas are in keeping with the relative composition of the enzyme. These two resonances broaden in the other spectra in Fig. 9 and show interesting differences between pH 4.14 and 6.55.

Predominantly Methine Region. The nuclei represented between 126 and 143 ppm upfield of CS_2 are almost all assignable to methine carbons. The well defined resonance in Fig. 9D at 125.8 ppm, which broadens but retains its definition in the other spectra, is recognizable as C^β of the ten threonine residues. Table III shows that, with the considerable exception of the contribution of the fourteen serine residues at 131.8 ppm, all carbons resonating below 150 ppm are of the methine type and are α-carbons. These group reasonably in the spectrum at pH 1.46 and merge in the other spectra. Attention should be drawn to the small band at 150.0 ppm assignable to the C^α of the three glycine residues.

Upfield Region. Above 150 ppm, the spectrum in Fig. 9D presents a great deal of detail. The pattern of consideration can be extended by the reader by reference to Table III and the other spectra in Fig. 9. Special

[182] C. Tanford and J. D. Hauenstein, *J. Amer. Chem. Soc.* **78**, 5287 (1956).

TABLE III
Tentative Assignment of Certain Resonance Bands[a]

| Ribonuclease A at pH 1.46 | Tentative assignment | Correlation based on: | |
		Peptide	Amino acid
15–23	Carbonyls	$16.1–25.6^b$	$14.7–25.2^c$
36.0	Arginine C^ζ		36.1^c
38.2	Tyrosine C^ζ	38.2^b	
56.5	Phenylalanine C^γ	56.4^b	
59.2	Histidine C^{ϵ_1}	59.2^b	59.2^b
62.5	Tyrosine C^{δ_1}, C^{δ_2}	62.4^b	
64.3	Phenylalanine C^{ϵ_1}, C^{ϵ_2}	63.5^b	
	Phenylalanine C^{δ_1}, C^{δ_2}	64.0^b	
	Histidine C^γ	64.3^b	65.5^b
64.9(s)	Tyrosine C^γ	64.9^b	
65.7(s)	Phenylalanine C^ζ	65.6^b	
75.5	Histidine C^{δ_2}	75.4^b	75.4^b
77.3	Tyrosine C^{ϵ_1}, C^{ϵ_2}	77.4^b	
125.8	Threonine C^β	125.8^b	126.5^c
131.8	Serine C^β	131.6^b	132.3^c, 132.4^d
	Proline C^α	131.9^b	132.0^c
	Valine C^α	132.9^b, 134.1^d	132.0^c, 132.1^d
	Isoleucine C^α		132.7^c
133.6	Threonine C^α	133.7^b	132.1^c
137.3	Serine C^α	137.1^b, 137.3^d	136.2^c, 136.1^d
	Tyrosine C^α	137.4^b	
	Phenylalanine C^α	137.6^b	
	Glutamine C^α		138.2^c
	Methionine C^α	139.9^b	138.4^c
	Arginine C^α		138.5^c
	Glutamic acid C^α	138.9^d	137.9^c, 137.8^d
	Lysine C^α	138.9^b	138.3^b, 138.3^c
	Leucine C^α	$140.0^{d,b}$	138.9^c, 138.9^d
	Histidine C^α	140.2^b	139.2^b
	Aspartic acid C^α		140.4^c
	Cystine C^α		140.7^c
139.8	Asparagine C^α		140.9^c
142.8	Alanine C^α	142.9^b	142.0^c
150.0	Glycine C^α	150.3^b	151.1^c, 151.2^d
152.6	Arginine C^δ		152.1^c
153.4	Leucine C^β	153.1^d	152.6^c, 152.8^d
	Lysine C^ϵ	153.4^b	153.3^b, 153.6^c
156.7	Phenylalanine C^β	156.0^b	
	Aspartic acid C^β		156.0^c
	Cystine C^β		156.2^c
	Isoleucine C^β		156.5^c
	Tyrosine C^β	156.8^b	156.8^b
	Asparagine C^β		157.6^c

TABLE III (Continued)

Ribonuclease A at pH 1.46	Tentative assignment	Correlation based on:	
		Peptide	Amino acid
161.7	Glutamic acid C^γ	159.1[d]	159.1[c], 159.1[d]
	Glutamine C^γ		161.6[c]
	Methionine C^β	162.8[b]	162.6[c]
	Valine C^β	163.0[b], 162.6[d]	163.4[c], 163.6[d]
	Lysine C^β	162.7[b]	162.7[b]
	Methionine C^γ	163.6[b]	163.5[c]
162.8	Proline C^β	163.6[b]	163.9[c]
165.0(s)	Glutamic acid C^β	164.8[d]	165.5[c], 165.7[d]
	Arginine C^β		165.1[c]
166.6	Glutamine C^β		166.2[c]
	Histidine C^β	166.6[b]	166.2[b]
	Lysine C^δ	166.6[b]	166.6[b]
168.4	Isoleucine C^{γ_1}		167.9[c]
	Leucine C^γ	168.5[d]	168.2[c], 168.4[d]
	Proline C^γ	168.5[b]	169.2[c]
	Arginine C^γ		168.7[c]
170.8	Leucine C^{δ_1}	170.6[d]	170.4[c], 170.6[d]
	Lysine C^γ	170.9[b]	171.2[c]
172.0(s)	Leucine C^{δ_2}	171.7[d]	171.5[c], 171.6[d]
174.1	Threonine C^γ	174.1[b]	173.1[c]
	Valine C^{γ_1}	175.0[d]	174.5[c], 174.7[d]
	Valine C^{γ_2}	176.0[d]	175.7[c], 176.0[d]
176.5	Alanine C^β	176.3[b]	176.3[c]
177.9	Isoleucine C^{γ_2}		177.7[c]
178.4	Methionine C^δ	178.7[b]	178.4[c]
182.5	Isoleucine C^δ		181.3[c]

[a] All chemical shifts are expressed as parts per million upfield of CS_2. Carbon types listed under tentative assignments use the IUPAC–IUB biochemical nomenclature. (s) indicates a shoulder on a larger peak. When separation among peaks is poor, the entire region is bracketed and the intense peaks are used for reference.

[b] V. G. Glushko, F. R. N. Gurd, P. Keim, P. J. Lawson, R. C. Marshall, A. M. Nigen, and R. A. Vigna, in preparation.

[c] W. J. Horsley, H. Sternlicht, and J. S. Cohen, J. Amer. Chem. Soc. 92, 680 (1970).

[d] F. R. N. Gurd, P. J. Lawson, D. W. Cochran, and E. Wenkert, J. Biol. Chem. 246, 3725 (1971).

attention should be drawn to certain features as a preliminary to the following discussion of relaxation phenomena. The small sharp peak at 152.6 ppm is assigned as C^δ of the 4 arginine residues. It is of particular interest because of its narrowness and persistence even at pH 4.14, and because at pH 1.46 its intensity appears to be compatible with that of C^ζ of arginine discussed previously.

The side-chain carbons of the ten lysine residues are represented by four prominent resonances to which they individually contribute heavily. At 153.4 ppm C^ϵ appears as a prominent peak in Fig. 9D, and indeed this is the dominant narrow peak in the other spectra. The other members of this side chain are assigned as follows: C^β at 162.7, C^δ at 166.6, and C^γ at 170.9 ppm.

Table III shows that in the upfield region considerable overlap of classes of carbons is probably encountered. Of special interest are the resonances at 174.1 and 176.5 ppm made up in the first place of the methyl carbon, C^γ, of the ten threonine residues and the two magnetically nonequivalent methyl carbons, $C^{\gamma 1}$ and $C^{\gamma 2}$ of the nine valine residues. Largely separated from that peak is the one at 176.5 ppm representing the methyl C^β of the twelve alanine residues.

Spin-Lattice Relaxation Times. Figure 10 shows a series of partially

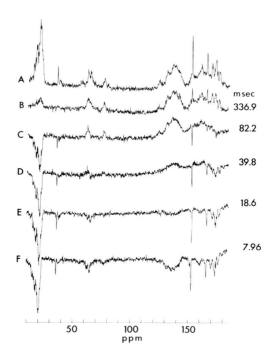

FIG. 10. Set of partially relaxed natural abundance ^{13}C NMR spectra of ribonuclease A, pH 6.55, 23 mM. The delay times, t, were 7.96, 18.6, 39.8, 82.2, and 336.9 msec, respectively in spectra (F) through (B). The normal spectrum is shown in (A). In all cases 16,384 scans were averaged. The temperature was 45°. Taken from V. G. Glushko, P. J. Lawson, and F. R. N. Gurd, *J. Biol. Chem.* **247**, 3176 (1972).

relaxed spectra[77] of ribonuclease A at pH 6.55. The enzyme was 23 mM. Spectra F through B were obtained with delay times, t, set for 7.96, 18.6, 39.8, 82.2, and 336.9 msec, respectively. The normal spectrum, identical with Fig. 9A, is shown in (A). In all cases 16,384 scans were accumulated. The recycle time was 1.36 seconds. Table IV shows the spin-lattice relaxation times, T_1, in milliseconds, for a number of resonances assigned as already discussed. The table covers results at pH 6.55, pH 4.14, and pH 2.12, the latter almost the equivalent of pH 1.46 which has been considered in some detail. The original publication contained more details[77] including results with oxidized ribonuclease. Those results serve to give more emphasis to the trend of change in T_1 values as the native structure is disrupted. To facilitate interpretation in terms of Eq. (31), values of NT_1 are included in Table IV.

Table IV shows several general trends. First, T_1 generally lengthens on passing from the native conditions at pH 6.55 and 4.14 to the acid-denatured state at pH 2.12. This trend is accompanied by a sharpening of the spectrum in the last case approximating that seen in Fig. 9D, and shows that the change in T_1 occurs along with that attributed to T_2 in the previous discussion. Second, the most striking example of the change is illustrated by the C^α region. At pH 6.55 and 4.14 the T_1 values

TABLE IV

T_1 and NT_1 Values for Ribonuclease A Resonances at Several pH Values[a]

Chemical shift	Tentative assignment	pH 6.55		pH 4.14		pH 2.12	
		T_1	NT_1	T_1	NT_1	T_1	NT_1
77.3	Tyrosine C^{ϵ_1}, C^{ϵ_2}	44	44	30	30	57	57
125.8	Threonine C^β	38	38			59	59
132						61	61
137						59	59
	C^α	35	35	30	30		
140						55	55
142						67	67
153.4	Lysine C^ϵ	278	556	291	582	295	590
162.7	Lysine C^β	66	132			93	186
166.6	Lysine C^δ	188	376	166	332	151	302
170.9	Lysine C^γ	100	200	82	164	121	242
174	Threonine C^γ	148	444	148	444	204	612
	Valine C^γ						
176.5	Alanine C^β	118	354	119	357	221	663

[a] All chemical shifts are expressed as parts per million upfield of CS_2. T_1 values are in milliseconds. The table lists the primary carbon types responsible for the resonance. For a more complete list, see Table III.

are comparable when derived by reading off amplitudes (see Eq. 55) anywhere on the envelope. The separate peaks in the acid-denatured protein show some variations in T_1 but are uniformly appreciably longer. The more specifically identifiable resonances of $C^{\epsilon 1}$ and $C^{\epsilon 2}$ of tyrosine residues and C^β of threonine residues both show striking parallels with the C^α resonances. Third, the methyl carbons at the bottom of the table show a similar trend, although T_1 and NT_1 are much longer. Fourth, the resonances dominated by the four side chain carbons of the lysine residues are less sensitive to the transition to the denatured form, with the probably significant exception of C^β.

Components of Relatively Low Mobility. It is a safe assumption that the change on denaturation in T_1 of the C^α resonances, and also those of C^β of threonine residues and of C^ϵ of tyrosine residues, corresponds to an increase in segmental mobility. The simple expression in Eq. (34) cannot be taken as generally appropriate for the protein because of the possibility that longer rotational correlation times apply, of the order of $1/\omega_c$. In this case again Eq. (31) is more appropriate for evaluating τ_r, or providing a working value of τ_{eff}. In the present case it is extremely probable that the NT_1 values of 55 msec and more fall on the side of Fig. 5 where T_1 increases with decreasing τ_r, or its equivalent for our purposes, τ_{eff}. It is interesting that all the nuclei represented in this group show great similarities of behavior because it is quite reasonable to expect them to undergo relatively restricted internal motion in the native protein.

For the isotropic case in the absence of internal motion Eq. (31) should give a good measure of τ_{eff} which would correspond to τ_r. This equation has two roots which have been computed for most cases in Table IV and are shown in Table V.[76,77] In the case of the C^α resonances and the C^β of threonine the two roots of τ_{eff} for the native protein experiments are of the order of 2 and 20 nsec respectively. A secure choice between these solutions is not possible at present. Because of the breadth of the bands in Fig. 9A and B it is not possible to rule out either result as implying an unreasonably large line width (Eq. 32). The longer value of τ_{eff} is of a reasonable order of magnitude for the motion of the molecule as a whole in the absence of internal motion. However, the experimental basis for this statement rests on measurements in much less concentrated solutions.[183] The shorter value of 2 nsec for τ_{eff} would imply some internal motion, i.e., segmental mobility. Reference to Fig. 5 shows that the choice between the two solutions should follow quite easily if T_1 were determined

[183] J. Yguerabide, H. F. Epstein, and L. Stryer, *J. Mol. Biol.* **51**, 573 (1970).

TABLE V
VALUES OF τ_{eff} FOR CERTAIN RESONANCES[a]

Chemical shift	Tentative assignment	pH 6.55 τ_{eff}	pH 4.14 τ_{eff}	pH 2.12 τ_{eff}
125.8	Threonine C^β	1.63		0.87*
		26.8		45.1
132				0.84*
	C^α			46.8
137		1.90	2.70	0.87*
		24.0	18.9	45.1
140				0.95*
				41.7
142				0.75*
				51.8
153.4	Lysine C^ϵ	0.08*	0.08*	0.08*
		446	467	473
162.7	Lysine C^β	0.36*		0.25*
		105		149
166.6	Lysine C^δ	0.12*	0.14*	0.15*
		301	265	242
170.9	Lysine C^γ	0.23*	0.29*	0.19*
		160	131	194
174	Threonine C^γ	0.11*	0.11*	0.08*
	Valine	356	356	491
176.5	Alanine C^β	0.13*	0.13*	0.07*
		284	286	703

[a] All chemical shifts are expressed as parts per million upfield of CS_2. Each pair of τ_{eff} values is calculated for the isotropic case [see Eq. (31)] and corresponds to the appropriate T_1 value in Table IV. The preferred τ_{eff} value, in nanoseconds, is starred (see text).

at another frequency. In either case it is clear that the nuclei discussed here have relatively low mobility in the native enzymes.

Table V also lists the τ_{eff} values corresponding to the NT_1 values for the four C^α resonances of the acid-denatured protein. The longer correlation times do not fit well with the observed line widths, so that the set between 0.84 and 0.95 nsec are taken as correct.

The Lysine Side Chain: Graded Motional Freedom. In Table IV the values of NT_1 for the C^β, C^γ, C^δ, and C^ϵ resonances of the lysine residues show the gradation of 132, 200, 376, and 556 msec, respectively, in the list for pH 6.55. No great sensitivity to pH is apparent, except that in the acid-denatured state the value for C^β goes to 186 when that for C^α probably falls near 60 msec. A similar trend was noted for the lysine-containing pentapeptide discussed previously.[62] The shorter value for

τ_{eff} in Table V is approximately 0.3 nsec for C^β and 0.2 nsec for C^γ, 0.15 nsec for C^δ, and 0.08 nsec for C^ϵ. The longer values fall between 100 and 500 nsec, in each case quite out of keeping with the observed line widths and certainly not correct. Clearly the motional freedom in the lysine side chains is graded toward an upper limit at C^ϵ which is not entirely out of keeping with the behavior of a free small molecule. It is interesting that beyond C^β or C^γ the change in C^α mobility accompanying denaturation has no clearly discernible effect. Although none of the τ_{eff} values quoted here describes a single motional mode (see the section on $^{13}C-^1H$ dipolar relaxation mechanisms), it is entirely understandable that the contributions of the slower motions should be attenuated at the point of C^ϵ.[42] These results fit nicely with those of Doddrell and Allerhand[155] for n-decanol-1.

Spinning Motion of Methyl Group. The C^β of the alanine residues shows an NT_1 value of approximately ten times that of its C^α. The theoretical ratio is nine for a spinning methyl group attached to a relatively slowly isotropically reorienting methine carbon.[60,61] The values of τ_{eff} for the methyl carbon in the native case were calculated as 0.13 nsec, and in the acid denatured case as 0.07 nsec. These numbers are at least an order of magnitude smaller than those for C^α of the alanine residues. In all cases, again, the alternative solution for τ_{eff} was too long to be acceptable. Quite similar results are given in Table V for the C^γ resonances of threonine and valine residues. Although the overlap of the two types of residues weakens the argument, in principle the match with the clearer results for the methyl group of alanine is very good. It is interesting that in all these cases the methyl group in question is attached to a carbon of restricted mobility. The present work supports that view for the alanine and threonine residues, and model building or direct experiments[184] generally point to C^β of valine as being very restricted indeed.

Relaxation of Nuclei in Aromatic Groups. Although relaxation in the aromatic groups has been approached with caution,[42,55] resonances such as those of $C^{\epsilon 1}$ and $C^{\epsilon 2}$ of the tyrosine residues do behave very much as those of the other carbons bearing a single proton (Table IV). In the native X-ray crystallographic structure the six tyrosine residues appear to have some considerable limitations on their freedom of motion relative to the backbone.[177,178] It would not be surprising to find that the relaxation properties of these resonances can be interpreted in the same terms as the aliphatic resonances[42] at least under certain conditions. It is in-

---

[184] E. R. Blout, *in* "Polyamino Acids, Polypeptides and Proteins" (M. A. Stahman, ed.), pp. 275–279. Univ. of Wisconsin Press, Madison, 1961.

teresting that the relative intensities of a number of the better defined aromatic and aliphatic carbon resonances in Fig. 9D are roughly in keeping with the relative composition. For example, the signal for the twelve carbons discussed above fits quite well with the signal for the ten C^{β} of threonine at 125.8 ppm, and so on. Detailed measurements are still required, but it does seem likely that the nuclear Overhauser enhancement in such cases is quite comparable.

Other Proteins. The general features of the ribonuclease spectra are also shown by lysozyme[74,75] and the heme proteins.[109,168,179] A detailed study of myoglobins is in progress in this laboratory,[185] and the work is being extended to hemoglobins. None of the reported spectra conflicts with the general picture of the assignments made for ribonuclease, although a number of discrepancies have been unavoidable without systematic information on peptides in the literature. Simple comparison of the spectra of different proteins has been quite helpful in tracking down assignments. For example, lysozyme differs from ribonuclease strikingly in the relative content of arginine and lysine residues, a fact that helped early with the assignments in this laboratory.[77] The comparison of closely related myoglobins has also been helpful. Clearly a great deal will come out when the relaxation behavior of these proteins is reported.

Enrichment

Positive enrichment with ^{13}C could be exploited in protein systems in three general ways. First, the sensitivity could be enhanced by biosynthetic enrichment so that less concentrated protein solutions would be suitable for study. This type of procedure can be somewhat selective in the case of certain amino acids[37] if the preparation of the enriched amino acids is economically feasible. Biosynthesis on a medium enriched in $^{13}CO_2$ would give a rather general enrichment which could be very useful.[40,171] The limitation might be $^{13}C-^{13}C$ coupling, but at least an order of magnitude enrichment could be made without appreciable difficulty of that sort.

Second, an interacting small molecule such as a substrate or inhibitor could be enriched and its relatively very intense signal could be studied under the influence of the enzyme. It would be feasible to label different parts of the small molecule to different degrees of enrichment. The general approach has been very resourcefully developed for ^1H NMR.[33] Two points need to be emphasized here. The first point is that below 130 ppm in the simple protein spectrum there is a considerable expanse of open

---

[185] V. G. Glushko, F. R. N. Gurd, P. Keim, R. C. Marshall, A. M. Nigen, unpublished results.

territory in which the small molecule signals can be studied in the absence of protein background absorption. This is particularly true of carbohydrate moieties, but in many cases an aromatic resonance may escape interference. The second point is that the relaxation mode can be exploited most elegantly in a small molecule–protein system. The full study with T_1 measurements has many obvious advantages and ramifications. To gain a preliminary understanding of a system, much use can be made of varying the recycle times, since the more rapid tumbling of the unbound small molecule makes dipole–dipole relaxation generally much less efficient. For example, Allerhand et al.[61] found that most of the carbons in sucrose (0.5 M) are near the null point at about 0.3 second with T_1 values of the order of 0.5 second. On the other hand the C^α region of ribonuclease, for example, is near a null at 0.02 second (Fig. 10) with T_1 near 0.03 second. It is not difficult to find circumstances in which a bound molecule of the size of sucrose gives a strong signal whereas the signal of the unbound molecule is greatly attenuated. The choice of recycle time is most easily made directly from a relaxation study which is itself very valuable.

The third general way of exploiting enrichment is through the specific incorporation of highly enriched nuclei into the protein. Up to now the scope of this kind of work would be limited to adding groups to the protein as small adducts under the heading of protein modification or to substituting a prosthetic group with its enriched form. An example of a useful and simple adduct for this purpose is the carboxymethyl group introduced through reaction with [2-^{13}C]bromoacetate. Nigen and co-workers[186] have made observations on carboxymethyl derivatives of three proteins, sperm whale myoglobin, harbor seal myoglobin, and ribonuclease A.

Results for the two myoglobins were very similar. Figure 11 shows (A) the untreated seal myoglobin, (B) the seal myoglobin treated with bromoacetate at natural abundance with respect to isotopes, and (C) the seal myoglobin treated under the same conditions with 30% enriched [2-^{13}C]bromoacetate. The compositions of the preparations in (B) and (C) were extremely similar. The enriched preparation contained 6.5 out of its 13 histidine residues per molecule as unmodified histidine, and the modified histidine residues were distributed as 4.7 residues of the dicarboxymethyl derivative, 1.2 residues of the N^ϵ-carboxymethyl (3-Cm in the old nomenclature) and 0.7 residue of the N^δ-carboxymethyl derivative. The amino-terminal glycine residue was modified, and approxi-

[186] A. M. Nigen, P. Keim, R. C. Marshall, J. S. Morrow, and F. R. N. Gurd, J. Biol. Chem. 247, 4100 (1972).

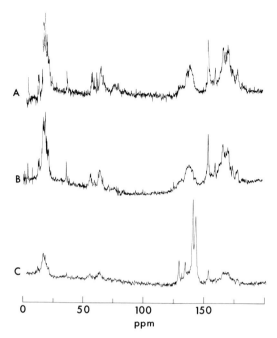

Fig. 11. Carbon-13 NMR spectra of (A) harbor seal myoglobin, (B) harbor seal myoglobin carboxymethylated with natural abundance bromoacetate, (C) harbor seal myoglobin treated as for (B) but with 30% enrichment of the methylene carbon of the bromoacetate. The temperature was 27°. Taken from A. M. Nigen, P. Keim, R. C. Marshall, J. S. Morrow, and F. R. N. Gurd, *J. Biol. Chem.* **247**, 4100 (1972).

mately one residue of lysine was also modified. The individual histidine residues have been identified in such a preparation.[187,188] Comparison of the three spectra in Fig. 11 shows four resonance bands attributable to the enriched loci. The two most prominent, which appear in the unenriched carboxymethylated preparation as two delightfully modest details, are readily attributable to the N^ϵ- and N^δ-derivatives mainly in the form of the dicarboxymethyl histidine residues. The other two resonance bands at 129.4 and 134.4 ppm upfield from CS_2 appear to represent, respectively, a glycolate ester derivative[cf. 189] and a modification of the amino terminus. The last two peaks represent approximately one or two adducts per molecule of protein.

Measurements of T_1 for the two large peaks yielded values of 38 and 39 msec. These values correspond to about 80 msec for NT_1, and suggest

[187] T. E. Hugli and F. R. N. Gurd, *J. Biol. Chem.* **245**, 1939 (1970).

[188] A. M. Nigen, Ph.D. Thesis, Indiana University, Bloomington, Indiana, 1972.

[189] K. Takahashi, W. H. Stein, and S. Moore, *J. Biol. Chem.* **242**, 4682 (1967).

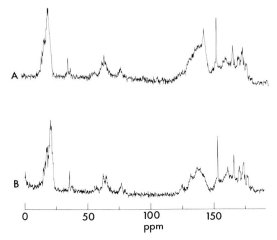

FIG. 12. Carbon 13 NMR spectra of (A) pancreatic ribonuclease A bearing 60% [^{13}C]H$_2$COO$^-$ on N$^\delta$ of histidine residue 119, (B) the untreated control ribonuclease A. The temperature was 27°. Taken from A. M. Nigen, P. Keim, R. C. Marshall, J. S. Morrow, and F. R. N. Gurd, *J. Biol. Chem.* **247**, 4100 (1972).

some degree of motional freedom. This very tentative suggestion is supported to some extent by the modest line widths.

Figure 12 shows (A) the spectrum of pancreatic ribonuclease A bearing the -[^{13}C]H$_2$COO$^-$ group on N$^\delta$ of histidine residue 119.[186] Approximately 0.85 residue of the N$^\delta$-carboxymethyl histidine per molecule was obtained on amino acid analysis, along with mere traces of other modified derivatives. Close comparison with the unenriched spectrum in Fig. 12 (B) shows that the adduct resonance is centered at 143.7 ppm upfield of CS$_2$. The value of T_1 was determined as 29 msec. While the chemical shift in this case is not as convenient for observation against the protein background as, for example, the two downfield adduct resonances in Fig. 11C, it is clear that a great deal of information is potentially available with the introduction of a single enriched locus in a small protein. In this case measurements of the chemical shift, T_1, NOE, and ^{13}C–^{1}H coupling constant under a variety of conditions will provide the necessary information for structural analysis.

Some Future Prospects

The fullest exploitation of ^{13}C NMR in enzymology and protein chemistry will come with the introduction of specifically enriched residues in the polypeptide chain. The necessary basis for the peptide chemistry is being developed, particularly by Offord.[190,191] If one or more strategically located residues can be removed and replaced in appropriately enriched form, there will be a prospect of correlating changes in their properties, e.g., T_1 values, with other manipulations. A simple example of such an application would be to compare the segmental motion of a given C^α in a helical region with that in a nonhelical, possibly even a terminal, location. As background to the interpretation of such experiments it is desirable to measure the effective rotational correlation time of the whole molecule. Specifically enriched heme would allow such measurements for myoglobin or hemoglobin in much the way that fluorescence depolarization measurements on artificial fluorophors have been used.[183] An important question is whether the overall rate of rotational reorientation of the protein is sensitive to bringing the concentration up to the range of the NMR experiments. Nigen et al.[186] have made a start by observing no concentration dependence of T_1 values over a rather small concentration range.

The interpretation of T_1 values in terms of motional aspects is limited to certain components a few orders of magnitude on either side of the resonance frequency used. This range nevertheless can be very useful for enzymology. For example, it is comparable with the range of phenomena studied by temperature and pressure perturbation techniques.[192,193] Appropriate enrichment of the enzyme and of a substrate or inhibitor might open up questions of variations in freedom of motion of the components in an interaction of that type. In this connection it is helpful to draw attention to the resourceful experiments by Sternlicht et al.[37] on amino acid derivatives attached to resin particles. Measurements were made of T_1, and also lower frequency components were assessed by T_2 measurements. The study is also interesting in that selective deuteration with $^{13}C-^2H$ double resonance was exploited to provide narrow line widths. This technique may be used to reduce spectral overlap and provide a key method for resolving individual resonances in narrow frequency ranges where the more efficient $^{13}C-^1H$ dipolar interactions lead to large line widths and lack of spectral resolution. In conjunction with calculated τ_{eff} values, this procedure can provide detailed information about the components of T_2.

[190] R. E. Offord, Nature (London) 221, 37 (1969).
[191] F. Borrás and R. E. Offord, Nature (London) 227, 716 (1970).
[192] M. Eigen, Z. Electrochem. 64, 115 (1960).
[193] M. Eigen and L. DeMaeyer, in "Technique of Organic Chemistry" (A. Weissberger, ed.), Vol. 8, pp. 895–1054. Wiley (Interscience), New York, 1963.

[35] Mössbauer Spectroscopy

By Thomas H. Moss

I. Introduction

For the experimentalist unfamiliar with the application of the Mössbauer effect in biological research, a starting point in judging its usefulness is the fundamental similarity with other spectroscopic techniques. In common with all ranges of optical spectroscopy, Mössbauer spectroscopy monitors energy absorption in a sample of interest as an irradiating frequency is scanned over an attainable frequency range. Absorption peaks occur at frequencies characteristic of the sample and its chemical or physical state, and the experimentalist infers from the positions of the peaks, using either empirical or more basic theoretical principles, something about the structure of the sample. An important practical point common to many forms of spectroscopy is implied in the words "empirical" and "theoretical" above: the method can be used effectively by workers with widely varying backgrounds and frames of reference. With judiciousness it may be possible to do good research using the positions of absorption peaks as simple empirical markers for particular materials or conditions, without the need to understand details of the interactions causing the particular energies and intensities of the lines. At the other end of the spectrum of approaches, one can obtain exact structural information by fitting data carefully to detailed molecular models.

This article will first outline in simple fashion the basic principles of Mössbauer spectroscopy, emphasizing features distinct from other spectroscopic techniques, type of information obtainable, and special problems. Experimental problems will then be treated, from the point of view of the needs of the researcher starting an effort in this area. For those whose interest is more detailed, the following section will discuss in greater depth the conditions for applicability and the physical basis of the absorption spectral details. Last, a number of typical applications will be reviewed as examples of what can be done and what limitations must be faced. Recent reviews pertaining to Mössbauer applications in biochemistry are indicated in footnotes 1-5.

[1] P. G. Debrunner, in "Spectroscopic Approaches to Biomolecular Conformation" (D. W. Urry, ed.), p. 209. Amer. Med. Ass., Chicago, Illinois, 1971.
[2] A. J. Bearden and W. R. Dunham, *Struct. Bonding* (*Berlin*) 8, 1 (1970).
[3] G. Lang, *Quart. Rev. Biophys.* 3, 1 (1970).
[4] C. E. Johnson, *Phys. Today*, p. 35, February, 1971.

II. Basic Principles

A. Distinctive Features of Mössbauer Spectroscopy

The most distinctive feature of Mössbauer spectroscopy is that the absorptions observed correspond to highly energetic transitions between ground and excited nuclear energy levels—as opposed to the transitions between nuclear spin orientations observed in NMR or electronic transitions observed in optical spectroscopy. That the effect is relevant to chemistry or biochemistry at all is due to the fact that there are easily observable shifts and splittings of the nuclear energy levels caused by interactions of the nucleus with the surrounding chemically important electrons. It is distinct, too, in that it is a *resonant* effect in the sense that the source of radiation must be a radioactive nucleus of the same isotope as that in which the absorption occurs. The source decays from an excited to a ground nuclear energy level, emits radiation, and excites an identical nucleus in the absorbing material to the same excited state (Fig. 1). Since nuclear energy levels are involved, the radiation emitted is much higher in energy than that used in spectroscopy based on electronic levels— depending on the nucleus, the gamma rays used span energies from the X-ray region (6 keV $\approx$ 1.9 Å) up to 100 keV.

One simple but very important advantage of the resonance relation between source and absorber is that it makes the chemical origin of the absorption lines unequivocal. If an excited iron isotope is the source, only iron nuclei in the sample will absorb as a sharp function of source fre-

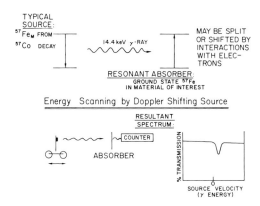

FIG. 1. Schematic of essential features in Mössbauer spectroscopy.

---

[5] Y. S. Moshkovskii, *in* "Chemical Applications of Mössbauer Spectroscopy" (V. I. Goldanskii and R. H. Herber, eds.). Academic Press, New York, 1968.

quency, and absorption lines cannot be attributed to any other site. If the iron ion is the active center, its properties will be perfectly resolved from the rest of the molecule.

Another unique feature is that the radiation from the source is scanned in frequency or energy not by adjusting a prism or grating as in optical spectroscopy, but by *moving* the source. The fact that the source is a long-lived excited nuclear state makes the source radiation highly monochromatic. By moving the source one uses the electromagnetic Doppler effect (analogous to the acoustical Doppler effect which changes the apparent pitch of moving acoustic emitters) to shift the energy of the gamma radiation up or down as the source is moved toward or away from the absorber (Fig. 1). The Doppler shift of the source energy is just

$$\delta E = \frac{v}{c} E_0 \tag{1}$$

where v is the source velocity, E_0 the unshifted source energy, and c the speed of light. The line widths and electronic splitting of the useful nuclear levels are very small, so that the energy range scanned need not be large—source movements of a few centimeters per second are usually adequate to cover the entire spectrum of interest (1 cm/second = 4.8 × 10^{-8} electron volt).

A last distinctive feature of Mössbauer spectroscopy is also its most serious limitation: permitted line widths of the nuclear transitions are determined on the one hand by the need to resolve energy levels spaced closely by hyperfine interactions with surrounding electrons, and in the extreme of narrowness by the need to have absorption peaks correspond to experimentally obtainable ranges of source velocities. These and other factors discussed further in Section IV combine to exclude most nuclei as possible Mössbauer isotopes. Many of the suitable nuclei are rare-earth group elements, of little natural interest biologically, though they may be used as chemically added probes or substitutions.[6] Iron-57 is by far the easiest isotope to use, and the overwhelming majority of Mössbauer applications in biology have been based on studies of iron in proteins and other biologically relevant systems. Of the other conceivable isotopes, only ^{127}I and ^{129}I and ^{40}K seem to show much promise for biological work, and these are experimentally much less convenient to use than iron. Figure 2 illustrates the distribution of Mössbauer isotopes in the periodic table. A resonance in ^{67}Zn has been reported,[7] but observation alone is so difficult that few applied studies are likely. The one redeeming feature

[6] R. J. P. Williams, *Quart. Rev. Chem. Soc.* **24**, 331 (1970).
[7] H. de Waard and G. J. Perlow, *Phys. Rev. Lett.* **24**, 566 (1970).

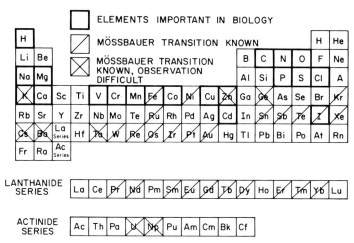

FIG. 2. Distribution of Mössbauer isotopes throughout the periodic table.

of the nearly total restriction to work with iron-containing systems is the fact that of all the transition metals iron seems to be by far the most ubiquitous in the biological makeup of most organisms. Unless otherwise stated, many of the details of this article will refer primarily to work with ^{57}Fe.

B. Type of Information Obtained

In basic terms, a Mössbauer spectrum contains information concerning the configuration of electrons and other charges within a few angstroms of the absorbing nucleus—as monitored by the perturbations of the energy levels of that nucleus by a particular configuration. For example, the number and arrangement of the electrons belonging to the absorbing atom are reflected in the absorption peaks—hence these are an indication of valence and the number of unpaired electrons, or "spin state." Information about bond type is provided indirectly to the extent that bonding is correlated with spin state or valence. The symmetry and charge of surrounding ligands may also cause characteristic splitting or shifts of the absorption lines and the extent of covalent sharing of electrons with ligands further dictates some spectral features. Covalency effects, especially where the absorbing nucleus is part of a large conjugated ring system, sometimes enable the Mössbauer spectroscoper to "see" rather large (the size of the ring) distances from the absorbing nucleus.

The parameters usually cited in biological Mössbauer literature are *isomer shift, quadrupole splitting,* and parameters related to *magnetic hyperfine splitting*. The *isomer shift* is simply the energy of the center of the split or unsplit absorption spectrum with respect to a given standard. As with all the parameters, this will be discussed further in the more detailed analysis of Section IV; briefly, the isomer shift is a measure of the s-electron density at the nucleus. This density is determined both by covalent sharing of s electrons with neighbors, and by interactions with particular arrangements of atomic electrons belonging to the nucleus. Thus this parameter alone cannot be associated with any one feature of the nuclear site, and its interpretation must be considered in the light of other information. The *quadrupole splitting* similarly has multiple origins. If the excited nuclear state has an electric quadrupole moment, it will be split by an electric field gradient at the nucleus. The quadrupole splitting is then measured from the separation of the absorption lines representing transitions to the split state. The field gradient can arise from asymmetries in the arrangement of the atomic electrons, or equally well in the arrangement of surrounding ligand fields. The *magnetic hyperfine parameters* are measures of the further splitting observed in the absorption spectrum when a magnetic field is applied. The excited and ground states may have magnetic moments as well as quadrupole moments, and if so a magnetic field will split the levels observed in zero field into a more complex pattern. The effective magnetic field may be external, or can originate from interactions of the nucleus with nearby unpaired electrons. As with the other measurements, both the number and configuration of the atomic electrons, and their covalent interaction with nearby ligands, act together to determine the precise magnetic hyperfine splitting. Each of the commonly measured Mössbauer parameters is determined, then, by a combination of chemically significant variables and can be completely interpreted only within the framework of the others and any other available experimental data.

As a way of comparing the efficacy of Mössbauer experiments to other techniques, one can note that Mössbauer spectroscopy is closest to electron spin resonance (ESR) in the information provided about an iron site in a biological molecule. Both techniques provide data about short-range influences on the iron electron configuration, although ESR is not limited to studies involving iron. Unlike ESR, however, Mössbauer spectroscopy is not restricted to paramagnetic configurations. Although some of the ambiguities in Mössbauer data interpretation have already been pointed out, it is probably fair to say that even in the paramagnetic forms more complete information about an iron ion in a biological molecule can be obtained from Mössbauer data than from ESR. Though the reader will be warned of even more limitations in the detailed discussion below, it is hoped that he will conclude that Mössbauer spectroscopy is an important addition to the set of physical techniques applicable to studies of iron-containing systems.

C. Special Problems and Limitations

In addition to its fundamental limitation to a few elements, other special problems associated with Mössbauer spectroscopy are its restriction to measurements in the solid state, and sensitivity problems that usually require use of cryogenic absorber temperatures and highly concentrated samples. Enrichment with the Mössbauer isotope may also be necessary. The requirement for solid absorbers occurs because of the resonance relation between source and absorber. In order that the source radiation be sufficiently energetic to excite the absorber nucleus, the source or absorber must not lose an appreciable amount of energy as the γ-ray is emitted or absorbed. This requires that the momentum of the γ-ray photon be taken up by the source or absorber matrix as a whole, without exciting any energetic vibrational modes (phonons) in the lattice. The factors that determine the probability of a zero-phonon nuclear transition will be discussed further in Section IV, but in general terms it means that the absorber and source must be bound in a fairly rigid structure, which can be only a solid or a very viscous liquid. Frozen solutions, crystals, or lyophilized materials must therefore be used in the biological studies.

In a solid, the only way that the nucleus can exchange energy with the lattice vibrations is by creating or destroying phonons, so that the zero-phonon processes are unbroadened by any thermal processes. This is the basis for the extremely narrow Mössbauer line widths ($\Delta E/E < 1/10^{12}$ for ^{57}Fe) which permit observation of the closely spaced hyperfine lines. The zero-phonon line widths are given just by the uncertainty principle for τ, the excited state lifetime:

$$\Delta E = \frac{\lambda}{\tau} \tag{2}$$

The cryogenic requirement arises because simple solidification of the absorber (by freezing, drying, or crystallizing a solution) is usually not sufficient to bring about a large Mössbauer absorption. The percentage of absorption processes that do not excite a phonon increases exponentially with descending temperature so that for many materials absorber temperatures of liquid nitrogen or lower are necessary for a strong effect. The requirements of solid state and low temperature combine to ensure that Mössbauer measurements can never be made under true physiological conditions, and the measurements must then bear the same burden as X-ray crystallography and much electron spin resonance work in showing that biologically important structures are maintained even under non-biological physical conditions.

The sensitivity problem also requires that, for iron, on the order of 1 μmole of the necessary ^{57}Fe isotope must be present in the sample. This is given simply by the cross section for nuclear excitation and the need to see absorption peaks of several percent for accurate interpretation. Unfortunately ^{57}Fe is only 2.2% abundant among the iron isotopes, so nearly 45 μmoles of the natural abundant iron are needed. For a protein with one iron per 10,000 MW, this represents roughly half a gram of material—possibly more than is obtainable in some cases, or more than can be maintained in small volumes of solution in others. For this reason it is often imperative to enrich the native material with ^{57}Fe. At best this is expensive and time consuming: it can be done by growing microorganisms or cell cultures on media made with enriched ^{57}Fe isotope,[8] or injecting the enriched isotope in animals made anemic.[9] This means that the iron will be taken up along biochemical pathways, but requires large amounts of the isotope at several dollars per milligram. The alternative to these procedures is to remove the iron chemically from the purified biological sample, and reconstitute it with enriched isotope.[10-12] Because there is generally much less dilution of the ^{57}Fe by this means, this is a less expensive procedure. However, it leaves another large burden of proof to the experimenter to show that his reconstituted material is native.

III. Experimental Considerations

Although the range of application is much more limited, the experimental difficulties and costs involved in beginning an effort in Mössbauer spectroscopy are much less than those involved in using physical techniques such as electron spin or nuclear magnetic resonance. There is no requirement for a large homogeneous magnet; further, the electronics involved are considerably simpler in the Mössbauer case. Maintenance and "tuning" is much more straightforward. One simply requires a source of radiation with a device to move it in a regular way in order to scan the required energy. Usually a means to cool the sample is also required to

[8] T. H. Moss, A. J. Bearden, R. A. Bartsch, M. A. Cusanovich, and A. San Pietro, *Biochemistry* **7**, 1591 (1968).

[9] G. Lang and W. Marshall, *Proc. Phys. Soc.* **87**, 3 (1966).

[10] W. S. Caughey, W. Y. Fujimoto, A. J. Bearden, and T. H. Moss, *Biochemistry* **5**, 1255 (1966).

[11] G. Lang, T. Asakura, and T. Yonetani, *Proc. Phys. Soc. London (Solid State Phys.)* **2**, 2246 (1969).

[12] T. H. Moss, A. Ehrenberg, and A. J. Bearden, *Biochemistry* **8**, 4159 (1969).

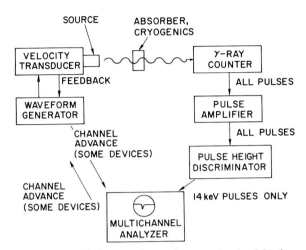

Fɪɢ. 3. Block diagram of the experimental apparatus for Mössbauer studies.

increase absorption peak intensity and sensitivity. A block diagram is shown in Fig. 3.

A. Source Movers

The most important part of the Mössbauer experiment is the device which moves the source and thereby scans the energy range of interest. The most rigorous requirement of the source motion is that it be repeatable to high accuracy through many cycles. This is because all Mössbauer spectra are obtained by superimposing many (typically ≈ 1000) repeated energy scans. Source decay rate is a process with an inherent uncertainty of $\sqrt{N}$ where N is the total number of counts detected in an interval of time. Since absorption peaks are detected by changes in count rate as a function of source velocity, peaks of absorption on the order of 5% of transmission require $\sqrt{N}/N \sim 1/20$ or $N \sim 400$ counts per velocity interval for detection even at signal-to-noise ratio of one. In practice, to see detailed features of complex spectra signal-to-noise should be at least 10 so $\sqrt{N}/N = 1/200$, $N \sim 40,000$ counts per velocity interval. A count rate of a few hundred counts per velocity interval over a normal velocity sweep is the most that can be expected even for a strong source so that more than 10^2 sweeps are almost always needed for an intelligible spectrum. Unless the 10^2 sweeps are exactly superimposable, the energy axis of the absorption vs. energy will be blurred. For the same reason, the source mover must be vibration free.

Source movers are usually designed to provide a nearly constant acceleration to the source, so that the velocity or energy of the gamma rays is a linear function of time, and the total counting time is equivalent for each velocity interval observed. If the count rate as a function of time is then plotted, one has a display of absorption vs. a linear energy scale, which is easy to comprehend visually. In theory, practically any velocity function (sine wave, etc.) could be used to drive the mover, and a computer analysis is subsequently used to correct the count rate and energy scale for convenient display. Such an approach could be preferable if computational facilities were easily available, as a sine wave velocity drive is often easier to achieve either mechanically or electromechanically than a constant acceleration mode.

A rather fantastic array of gadgetry has been used to provide the necessary source motion. Many odd-appearing mechanical devices have been proposed, but now electromechanical movers, such as loudspeaker cones, are overwhelmingly prevalent. The mechanical devices include shafts driven by eccentric cams,[13] model airplane engines,[14] rotating disks,[15] lead screw devices,[16] and many others. The mechanical devices are often less expensive and involve less electronic design, but in general they suffer from distortions due to mechanical wear and vibrations, and are more difficult to design for automatic control electronics. Only where it is necessary to move large masses of material along with the source are they likely to be optimal.

The electromechanical devices operate by driving the voice coil of a loudspeaker (or similar electromagnetic device), with a sawtooth voltage.[17] The source is mounted on the voice coil. The sawtooth drive gives the voice coil and source a constant acceleration, to a first-order approximation, over a considerable duty cycle. Correction of deviations from linear response of the voice coil are made by comparing the output voltage of a second, rigidly connected, transducer with the driving signal, and using the difference signal as a feedback voltage to the original source bearing coil. Problems of long-term drift of the speaker coil from the linear response region after many cycles of operation have been alleviated in some designs by building in a position-sensitive photocell activated reset system which operates after each cycle of motion.[18] The

[13] A. J. Bearden, M. G. Hauser, and P. L. Mattern, in "Mössbauer Effect Methodology" (I. J. Gruverman, ed.), Vol. 1, p. 67. Plenum, New York, 1965.

[14] J. J. Tamul, L. A. York, and E. J. Seykora, Nucl. Instrum. Methods 84, 317 (1970).

[15] A. J. Bearden, P. L. Mattern, and P. S. Nobel, Amer. J. Phys. 32, 109 (1964).

[16] R. H. Nussbaum, F. Gerstenfeld, and J. K. Richardson, Amer. J. Phys. 34, 45 (1966).

[17] G. K. Wertheim and R. L. Cohen, in "Applications of the Mössbauer Effect in Chemistry and Solid State Physics," Tech. Rep. Ser., No. 50. International Atomic Energy Agency, Vienna (1966).

[18] F. G. Ruegg, J. J. Spijkerman, and J. R. DeVoe, Rev. Sci. Instrum. 36, 356 (1965).

motion of the source mover is correlated with the counting system by using the sawtooth voltage to generate or trigger the address advancing pulses in a multichannel analyzer used in multiscalar mode.[17] In this way the numbers of γ-rays transmitted through the sample are counted in each channel for fixed intervals of time, and each interval of time corresponds to a particular velocity or energy range. Alternatively, the analog voltage of the address scalars of the multichannel analyzer itself can be used as the input sawtooth wave form driving the transducer.[18] In either case the channel-velocity relationship can be made reproducible over many cycles, with equal time spent in each channel. This yields the desired display of absorption as a function of a nearly linear velocity or energy scale. In addition to the constant acceleration mode of operation, the electro-mechanical devices can also be used, with different driving signals, as constant velocity source movers.[18]

Reliable and accurate electromechanical velocity transducers are available commercially for less than 1000 dollars and complete descriptions of the mechanical parts and circuitry for similar devices are available in the literature.[17-20] The U.S. National Bureau of Standards has invested considerable effort in Mössbauer design and may be able to provide drawings and consultation.

B. Sources

For work with iron absorbers, a source of radioactive ^{57}Co is used. The radioactive ^{57}Co isotope decays by the scheme shown in Fig. 4, with the resonant γ-ray being the 14.4 keV transition. In general these sources are prepared by diffusing a radioactive cobalt salt into a metal foil matrix.

There are several criteria which determine the effectiveness of the source. The activity should be plated on a small area of foil, so that the γ-rays come from an approximate point source, and any necessary collimation does not result in a substantial loss of intensity. The radioactive cobalt should diffuse only a short distance from the surface of the host lattice, so that the loss of intensity due to absorption within the host lattice is minimal. The matrix material should form a very rigid lattice so that there is a high probability (f value) of resonant emission even without cooling the source. Just as important as these considerations relating to intensity is the requirement that the source cobalt nuclei be in equivalent sites in a homogeneous lattice, so that the source line width is not broadened to an envelope of overlapping inequivalent lines. Palla-

[19] E. Kankeleit, *Rev. Sci. Instrum.* **35**, 194 (1964).
[20] J. J. Spijkerman, *in* "Techniques of Inorganic Chemistry" Vol. 7 (H. B. Jonassen, ed.), p. 71. Wiley, New York, 1968.

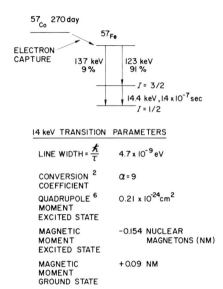

FIG. 4. Decay scheme of ^{57}Co producing the ^{57}Fe 14.4 keV Mössbauer transition.

dium and chromium have proved to be good sources in this sense, though chromium has some difficulties related to problems of diffusion into the host, and palladium has an X-ray emission line at 21 keV which can interfere with the 14 keV γ-ray of interest if low resolution counters are used. It should be emphasized that broadening of the source emission line causes a corresponding broadening of the absorption lines; if high resolution is necessary, as in cases where there are many overlapping magnetic hyperfine lines, much information can be lost by the "smearing" effect of broad source lines.

^{57}Co sources can be "homemade" from ^{57}CoCl$_2$, available commercially, following one of the references on this subject.[21,22] They are also available commercially, at a cost of roughly 50 dollars per millicurie in the 5–10 mCi range and 30 dollars per millicurie in the 10–30 mCi range. Although source strengths of 1–5 mCi are usually adequate for reasonably rapid data accumulation with nonbiological materials, stronger sources (10–30 mCi) are usually needed for biological studies. This is because these are often examined in frozen aqueous solution, and much of the incident γ-ray intensity is absorbed in nonresonant processes (Compton scattering or photoelectric absorption) by the ice matrix instead of by the specific Mössbauer process involving the iron. A sample about 0.5

[21] J. Stephen, *Nucl. Instrum. Methods* **26**, 219 (1964).
[22] T. A. Kitchens, W. A. Steyert, and R. D. Taylor, *Phys. Rev.* **138**, A467 (1965).

cm thick of ice will absorb about 50% of the incident 14 keV γ-rays. Given protein solubilities, this thickness is usually required to contain enough iron nuclei to obtain substantial Mössbauer absorption. In addition to these considerations, a strong source is usually a good investment in that it will remain adequate longer even as the 270-day half-life attenuates its strength.

A variant approach to the standard technique is to use the biological material as the *source* host and a simple known compound as absorber. In the $^{57}Co-^{57}Fe$ system this would possibly permit the study of many of the cobalt enzymes and cofactors. However, as the decay scheme (Fig. 4) indicates, the Mössbauer γ-ray is emitted *after* the cobalt has undergone electron capture and is transmuted to iron. One must face the problem that the structure indicated by the Mössbauer spectrum may not be simply related to the structure that existed before the transmutation. A few studies of this sort have been attempted,[23] but a firm basis for believing that they will be generally useful has not been established.

Much of the difficulty in using iodine and potassium Mössbauer resonances arises from source preparation problems. The iodine sources can be prepared by irradiation in a reactor, but the most favorable parent state has a half-life of only 70 minutes[24] (compared to the 270-day ^{57}Co half-life). Strong sources, high count rates, and good counting statistics are therefore troublesome to maintain. The potassium source problem is far more difficult, for the excited state is produced only by interaction with a continuous beam of bombarding charged particles or neutrons.[25] There is no long-lived parent nucleus "feeding" the Mössbauer excited state. The experiments can therefore be done only in tandem with a charged particle accelerator or high flux neutron source.

C. γ-Ray Detectors

Both scintillation and proportional counters have been used. Crystals of NaI doped with thallium give good results for the scintillation technique, but at the energy of the iron γ-rays, high efficiency is available from the proportional counters and the resolution is much superior. In a proportional counter, a γ-ray photon ionizes the fill gas in a cylindrical tube, and the resultant charged particles are collected at a high voltage center wire yielding a current pulse. The magnitude of the pulse is pro-

---

[23] A. Nath, M. Klein, W. Kundig, and D. Lichtenstein, *in* "Mössbauer Effect Methodology" (I. J. Gruverman, ed.), Vol. 5, p. 163. Plenum, New York, 1970.

[24] D. W. Hafemeister, *in* "The Mössbauer Effect and Its Application to Chemistry," *Advan. Chem. Ser.* **68**. Amer. Chem. Soc., Washington, D.C., 1967.

[25] P. K. Tseng, S. L. Ruby, and D. H. Vincent, *Phys. Rev.* **172**, 249 (1968); S. L. Ruby and R. E. Holland, *Phys. Rev. Lett.* **14**, 591 (1965).

portional to the number of ionized particles, and thus the initial energy of the γ-ray. The high energy resolution of the counter is useful because the ^{57}Co source emits radiation at 136, 122, and 6 keV in addition to the resonant 14 keV Mössbauer γ-ray (Fig. 4). If the counter can resolve the 14 keV γ-ray pulses completely from the pulses due to this additional radiation, the electronics can be set so that only the transmission of the resonant γ-ray is monitored and the sensitivity-decreasing background can be reduced.

Important design criteria for the proportional counter are its efficiency as a function of energy and ability to handle a high count rate. The 122 keV photons can be 30 times or more as intense as the 14 keV Mössbauer radiation, and the large pulses they generate when depositing their full energy in the counter can overload the electronics and confuse the apparent energy of any pulses in near coincidence. The optimum geometric design and fill gas for ^{57}Fe work thus minimizes probability of stopping 122 keV photons while maintaining high efficiency for the 14 keV radiation. Argon-filled, 4-inch diameter end-window counters, with 1 ft γ-ray path length, are $\approx$60% efficient for the 14 keV γ-rays and minimize the overloading effect of the 122 keV photons.[3] Argon is superior to xenon as a fill gas for this purpose. Compton scattering, in which the high energy γ-rays deposit only a part of their energy in the counter, can increase background seriously by generating many nonresonant pulses in the 0–40 keV region. Xenon presents fewer problems than argon in this respect. For low count rates, when pileup of charge in the counter is not a problem, xenon might well be the better choice. Xenon also has the advantage of a higher cross section for 14 keV γ-rays so that detection can be made with 90% efficiency in less bulky 2-inch diameter tubes using transverse windows. Krypton is not used because of the strong possibility of generating a krypton X-ray which can escape and shift the apparent energy of any incoming pulse away from its true value. A small amount (10%) of "quench gas" is always added to ensure that very large voltage breakdowns are not generated in the tube. Methane or nitrogen have proved most effective for this purpose. Proportional counters are also commercially available, although some researchers prefer to build their own for optimum design. Resonant counters have been designed,[26] in which the transmitted radiation is absorbed by ^{57}Fe foil and the subsequent decay of the excited nuclei is detected. These avoid the background problems mentioned above, but their efficiencies are so low that their overall usefulness for biological studies does not compare with conventional counters.

[26] J. Fenger, *Nucl. Instrum. Methods* **69**, 268 (1970).

D. Other Electronics

Other aspects of the counting system are the electronics shown in the block diagram (Fig. 3). A preamplifer mounted on the proportional counter tube is best to minimize noise, and this generally feeds the detected pulses into a differential discriminator. The discriminator can be set to generate a large ($\sim$10 V), square pulse for only those incoming signals in a preset voltage range. This enables one to count transmission of only the resonant 14 keV γ-rays, eliminating the background at other energies. The standard discriminator output pulse can then be counted on an ordinary scalar as a measure of transmission through the sample. Using a multichannel analyzer in multiscalar mode, with the channel used for storage determined by time or source mover drive voltage, this count rate can then be stored as a function of time and/or source mover velocity.

E. Cryogenics

As with other aspects of the Mössbauer apparatus, it is possible to begin effective research with a very simple system, although some special applications may require considerable sophistication. Generally, for a strong probability of Mössbauer absorption, it is usually necessary to cool biological samples, either crystalline, lyophilized, or in ice matrices, to $\leq$150°K. A simple liquid nitrogen Dewar will suffice for this. The sample should be mounted in a small-diameter tail piece, because the γ-ray beam is diverging, not parallel, and maximum count rate is achieved by having the counter a minimal distance from the source. For powder or lyophilized samples it may be necessary to have the sample surrounded by a suitable heat exchange gas such as helium to ensure full cooling. Moreover, if the sample is in a gastight chamber it can be heated to any desired intermediate temperature above that of liquid N_2 by removing most of the exchange gas and applying electrical heating to the sample mount. The materials for the windows through which the γ-rays pass must be of low atomic weight to avoid loss of γ-ray intensity. Beryllium and Mylar are generally used. Beryllium has the disadvantage of being difficult and hazardous to machine, sometimes given to sudden dramatic failure, and not always free of iron impurity. Mylar can also fail suddenly and certainly does fatigue over long use. It can also cause vacuum problems because it is permeable to helium at room temperature. Mylar is much less expensive, however, and probably superior for most applications.

Magnetic hyperfine splitting can generally be observed only well below liquid nitrogen temperature, so for these observations one must have a helium Dewar. The small diameter tail piece design must again be preserved. Small Dewars (<1 foot long) which cool by evaporating helium at a given rate in the sample mount are available commercially and can be much more economical in helium use and space than a conventional reservoir Dewar. For applying small magnetic fields to paramagnetic samples, external permanent magnets or helmholtz coils can be used. Equally well, magnetic flux can be trapped in lead washers mounted near the sample in the Dewar by cooling then in a magnetic field and then withdrawing the magnet.[3]

To resolve magnetic hyperfine structure in diamagnetic samples, large fields of more than 30 kilogauss are needed. A possible Dewar design is shown in Fig. 5. High field homogeneity is not necessary, but compactness is, to minimize tail piece path length. It is desirable to be able to apply the field either parallel or perpendicular to the γ-ray beam, as the selection rules for the appearance of certain lines in the two cases can help

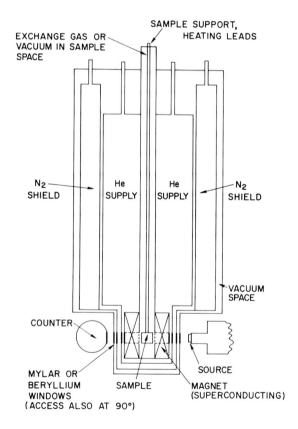

Fig. 5. A possible cryogenic arrangement with superconducting magnet for Mössbauer studies.

make the interpretation much more straightforward. The magnet shown has γ-ray access either parallel or perpendicular to the magnetic axis for this reason.

F. Computing Techniques

Although simple contemplation of derived spectra is a possible avenue to interpretation, there are a number of ways that computing techniques can be applied. There are some corrections to the raw data that usually need to be made. It is often the case that small nonlinearities occur in the velocity response of the drive system, for instance. In addition, a small curvature may be imposed on the transmission baseline because of the fact that the source subtends, at the two extrema of its motion, a slightly different solid angle at the counter. These simple effects can be estimated from the characteristics of the spectrometer or calibration runs. If there is any impurity iron absorption in the cryostat or counter windows, this can also be subtracted from the data. More sophisticated programs have been aimed at least-square fitting of spectra to a pre-determined number of Lorentzian lines, in order to get accurate values for peak positions and splittings, line widths, and intensities.[27,28] These generally use iterative procedures to obtain the best values for the above parameters.

The most difficult but most fruitful approach is to simulate the Möss-bauer spectrum from a theoretical model of the coordination center studied, and make comparisons with the data. These programs range widely in complexity, with the simplest cases dealing with diamagnetic configurations and combining the effects of quadrupole splitting and external magnetic field.[29] Where the absorber is itself paramagnetic, the interaction between any applied field, the unpaired electrons, and the nuclear moment can be quite complicated (see Section IV, D). In biological cases where frozen solutions are used as samples, averages over all orientations of absorber molecules with respect to external fields and the γ-ray beam must be used in computing magnetic effects and transition probabilities. Several programs[30-32] have been described which make these calculations. At least one has been adapted to oscilloscope display[30] so

[27] D. Agresti, M. Bent, and B. Persson, *Nucl. Instrum. Methods* **72**, 235 (1969).

[28] W. Kundig, *Nucl. Instrum. Methods* **75**, 336 (1969).

[29] R. L. Collins and J. C. Travis, *in* "Mössbauer Effect Methodology" (I. J. Gruverman, ed.), Vol. 3, p. 123. Plenum, New York, 1968.

[30] E. Munck, P. G. Debrunner, J. C. M. Tsibris, and I. C. Gunsalus, *Biochemistry* in press (1973).

[31] W. R. Dunham, A. J. Bearden, I. T. Salmeen, G. Palmer, R. H. Sands, W. Orme-Johnson and H. Beinert, *Biochim. Biophys. Acta* **253**, 134 (1971).

[32] G. Lang, T. Asakura, and T. Yonetani, *Proc. Phys. Soc. London* (*Solid State Phys.*) **2**, 2246 (1969).

that results of various combinations of parameters can be quickly compared with data in seeking an optimum fit. In addition, families of computed curves with typical sets of parameters have been published[33] for comparison with actual spectra when computational facilities are not readily available.

G. Sample Preparation

As has been mentioned above, samples should contain on the order of 1 μmole of ^{57}Fe, often attained by isotopic enrichment, and must be in solid form. All high molecular weight buffer or salt ions should be minimized, as these attenuate the γ-ray beam by non-Mössbauer processes. A path length of 0.5–1.0 cm through material with the average atomic composition of water is the maximum practical in order to avoid overattenuating the γ-beam. This means that frozen solutions must generally be quite concentrated. The necessity of the solid and concentrated form of the samples nearly always presents some disadvantages. There is some evidence that the ice or crystal matrix may cause some changes in the protein as samples are cooled.[34] This does not seem to be true for lyophilized materials, but in that case the dehydration process itself may cause alterations in native structure.[10] All possible correlative checks on the state of the sample under conditions of the Mössbauer measurement should be made.

IV. Details of Principles and Parameters Measured

A. Probability of Mössbauer Absorption

The probability of γ-ray emission or absorption without exciting an internal vibration (zero-phonon process) and destroying the resonant relationship of source and absorber, depends on the γ-ray energy, the elasticity of the lattice bonds, and the temperature. For harmonic lattice forces the probability of a zero-phonon process is given by

$$f = \exp - \left[\frac{4\pi^2 \langle x^2 \rangle}{\lambda^2} \right] \tag{3}$$

where $\langle x^2 \rangle$ is the mean square amplitude of nuclear vibration and λ is the γ-ray wavelength. For higher γ-ray energies, or shorter wavelengths, this probability is small, limiting suitable isotopes to those with fairly low energy excited states. Mean square vibration amplitudes increase with temperature, so the probability can be small except when the source

[33] G. Lang and W. T. Oosterhuis, *J. Chem. Phys.* **51**, 3608 (1967).
[34] A. Tasaki, *J. Appl. Phys.* **41**, 1000 (1970).

or absorber is cold. However, source lattices are usually chosen to be so rigid that $f \approx 1$ even at ordinary room temperature. Analysis of vibrational modes of complex absorber molecules in frozen ice matrices, or lyophilized materials, is hopelessly complex and the term lattice itself is not really appropriate. Empirically, however, many biological absorbers in these conditions seem to have f values on the order of 0.8–0.6 at 4.2 or 77°K, decreasing to less than 0.1 at 273°K. The explanation for the relatively high f values in these nonrigid lattices probably lies in their atomic heterogeneity which restricts the kind of internal modes allowed.[35] Because $\langle x^2 \rangle$ depends on the elasticity of chemical bonds, f can be anisotropic with respect to the angle between the incident γ-ray and the molecular axes. In principle then, f values can give information about anisotropic structural features. Even where polycrystalline samples or frozen solutions are used, the effect of anisotropic f values may be detectable, when combined with quadrupole splitting, as unequal line intensities (Gol'danski-Karyagin effect).[36] However, this is not a very sensitive measure of bonding and usually single crystal measurements would be necessary for accuracy. Large (0.01 cm³) crystals are not often available, and for this reason this line of research has been largely stillborn.

B. Isomer Shift

Because the isomer shift measures a directly "chemical" parameter, s-electron density at the nuclear position, much of the early hope for information on chemical bonding from the Mössbauer effect was based on it. The mathematical formulation of the shift and its effect on the Mössbauer spectrum are illustrated in Fig. 6. The excited and ground states of the iron nucleus have slightly different radii, R, and thus their energy levels are shifted differently by interactions with s-electrons which have a finite density, $|\psi(0)|^2$, at the nuclear position. As the s-electron density varies, the energy between ground and excited state will change, and thus the Mössbauer transitions will have slightly different energies for different absorbers relative to the same source. The source velocities at which they absorb will thus be different. Most promising for elucidating electronic structure is the fact that the s-electrons are shielded from the nucleus by interactions with the d-shell. In this sense the isomer shift is a potential measure of valence in transition metals, as the *number* of d-electrons determines the amount of shielding and thus the nuclear s-electron density. It was in fact observed that isomer shifts correlated

[35] Y. Kagen, *Zh. Eksp. Teor. Fiz.* **41**, 659 (1961).
[36] S. V. Karyagin, *Dokl. Phys. Chem.* **148**, 110 (1964).

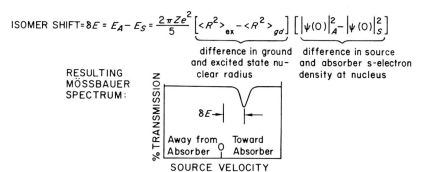

ISOMER SHIFT $= \delta E = E_A - E_S = \dfrac{2\pi Z e^2}{5} \underbrace{\left[<R^2>_{ex} - <R^2>_{gd} \right]}_{\substack{\text{difference in ground} \\ \text{and excited state nu-} \\ \text{clear radius}}} \underbrace{\left[|\psi(0)|^2_A - |\psi(0)|^2_S \right]}_{\substack{\text{difference in source} \\ \text{and absorber s-electron} \\ \text{density at nucleus}}}$

RESULTING MÖSSBAUER SPECTRUM:

Fig. 6. Essential features of the isomer shift as measured in Mössbauer spectroscopy.

well with valence for ionic compounds.[37] An ambiguity in this parameter arises, however, because the s-electron density can also be changed by covalent interactions of the s-orbital with neighboring ligands. Thus the isomer shift has the extra value of measuring covalency to some extent, although uncertainty is introduced into the measurements of d-shell interaction. More important for iron are covalent interactions involving d-orbitals. These can be so extensive that in a reduced iron, nominally $3d^6$ configurations, the isomer shift is indistinguishable from that of an oxidized $3d^5$ state. Electron delocalization from the d-shell is so great in these cases that s-electrons see only the shielding of an effective $3d^5$ configuration. Table I gives some typical isomer shifts for various states of iron. The overlap due to covalency is obvious. Only the high-spin ferrous state can usually be identified by isomer shift alone. This is simply because the stable high spin state implies weak ligand field and little covalency to confuse the isomer shift interpretation. For other states, direct interpretation of the isomer shift alone is possible only within a series of closely analogous compounds where the variables of symmetry and bond type are strongly constrained.

Any temperature dependence of the isomer shift can be revealing,

[37] L. R. Walker, G. K. Wertheim, and V. Jaccarino, *Phys. Rev. Lett.* **6**, 98 (1961).

TABLE I

Some Typical Isomer Shifts of Iron in Inorganic and
Biological Compounds

Compound			Isomer shift[a] (mm/sec)
Fe(II) $3d^6$	High spin	FeF_2	0.9
		$FeSO_4 \cdot 7H_2O$	0.9
		Deoxyhemoglobin	1.3
	Low spin	$K_4Fe(CN)_6$	0.4
		Oxyhemoglobin	0.6
Fe(II) $3d^5$	High spin	FeF_3	0.1
		$Fe_2(SO_4)_3 \cdot xH_2O$	0.1
		Acid methemoglobin	0.6
	Low spin	$K_3Fe(CN)$	0.5
		Methemoglobin cyanide	0.5

[a] All isomer shifts at 195°K relative to a source of ^{57}Co in platinum. Much higher accuracies are attainable but are not indicated here as somewhat differing experimental conditions were used in obtaining the various data.

for it may indicate the presence of excited electronic states within thermal energies of the ground state. If relaxation between two closely spaced states is rapid compared to $h/(\delta E_1 - \delta E_2)$, where the δE's are the isomer shifts of the two states, a single "average" line may be observed with an isomer shift which changes with temperature as the Boltzmann populations of the two states. If relaxation between states is slow, lines representing each state will be observed with intensities varying as the Boltzmann populations. This interpretation of "thermal mixtures" applies, of course, to all of the Mössbauer parameters of an electronic configuration. Special caution must be used in analyzing the isomer shift in this way, however, because of the temperature dependence due to the "second-order Doppler shift."[37a] This effect can be thought to result from a relativistic mass change of the emitting or absorbing nucleus due to its energy change. The shift in energy caused by this effect is given by:

$$\frac{\delta E}{E} = \frac{\langle v^2 \rangle}{2c^2} \tag{4}$$

where $\langle v^2 \rangle$ is the root mean square velocity of the vibrating emitter or absorber. In a harmonic approximation of crystal forces the temperature dependence of $\langle v^2 \rangle$ over a range ΔT can be simply expressed so that:

$$\frac{\delta E}{E} = \frac{C\Delta T}{2Mc^2} \tag{5}$$

[37a] B. D. Josephson, *Phys. Rev. Lett.* **4**, 341 (1960).

where C is the specific heat and M the mass of the emitter or absorber. This temperature-dependent term will cause an apparent change in the isomer shift independent of any effects of excited states and should be subtracted from the raw data in advance of interpretation of the temperature dependence of the data. Even then there can be small ambiguities as real biological samples are not harmonic crystals, and it may not be possible to subtract the second-order Doppler shift simply by using Eq. (5).

C. Quadrupole Splitting

Because of the quadrupole moment, Q, of the excited state of the ^{57}Fe isotope, any asymmetric arrangement of valence electrons or neighboring ligands which causes an electric field gradient at the iron nucleus will split the excited state energy level and the corresponding Mössbauer transition. The interaction is formulated and illustrated in Fig. 7. The effect of the valence d-electrons on the quadrupole splitting is, to first approximation, straightforward. Each of the occupied d-orbitals contributes a term to the axial and nonaxial parts of the field gradient, and from a given arrangement of orbital occupancy one can simply sum the effects to predict a total interaction. Table II gives the contributions of a single electron in each d-orbital. A completely filled d-shell is spher-

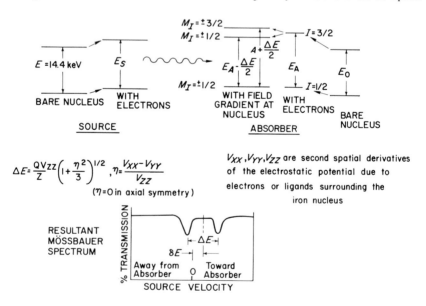

FIG. 7. Essential features of the quadrupole splitting as measured in Mössbauer spectroscopy.

TABLE II

CONTRIBUTIONS OF A SINGLE ELECTRON IN A d-ORBITAL TO THE EXPECTATION
VALUE OF THE FIELD GRADIENT AT THE NUCLEUS

Orbital	V_{zz}	$V_{xx} - V_{yy}$
$\lvert 3z^2 - r^2 \rangle$	$-\dfrac{4}{7}e\langle r^{-3}\rangle$	0
$\lvert x^2 - y^2 \rangle$	$+\dfrac{4}{7}e\langle r^{-3}\rangle$	0
$\lvert xz \rangle$	$-\dfrac{2}{7}e\langle r^{-3}\rangle$	$-\dfrac{6}{7}e\langle r^{-3}\rangle$
$\lvert yz \rangle$	$-\dfrac{2}{7}e\langle r^{-3}\rangle$	$-\dfrac{6}{7}e\langle r^{-3}\rangle$
$\lvert xy \rangle$	$+\dfrac{4}{7}e\langle r^{-3}\rangle$	0

ically symmetric, and the contributions from all the orbitals sum to zero. This is why we consider only the valence shell in general—the contributions of electrons in the complete inner shells also sum to zero.

As in the case of the isomer shift, thermally accessible excited states will cause a temperature-dependent shift or change in intensity of quadrupole lines, depending on relaxation times between states. Because quadrupole splittings vary more widely among the accessible configurations, these temperature shifts of quadrupole splitting are likely to be more dramatic. Several studies have based considerable interpretation on these temperature effects.[38,39]

Although, in principle, it appears that the exact electron arrangement could be determined from the magnitude of the quadrupole splitting, a variety of factors introduce ambiguities here as well. Charged ligand atoms can also introduce field gradients at the iron site, and these are difficult to evaluate in all but the most ionic cases because it is not easy to judge how the charge is localized. The magnitude of the ligand field effect can be a substantial fraction of the gradient introduced by the valence electrons and confuse interpretation based on the Table II.

Covalent interactions, where extra charge is transferred in or out of orbitals along particular molecular axes can have very large effects on the magnitude of the quadrupole splitting, for these amount to adding or subtracting a fractional orbital in the sum over the valence electrons. An alternative way to look at the same effect is to note, in Fig. 7, the dependence of the field gradient terms on $\langle r^3 \rangle$. Covalent effects are often

[38] H. Eicher and A. Trautwein, *J. Chem. Phys.* **52,** 932 (1970).

[39] T. H. Moss, A. J. Bearden, R. G. Bartsch, and M. A. Cusanovich, *Biochemistry* **7,** 1591 (1968).

estimated by scaling this expectation value of the radial part of the wave function by factors of up to 20%. With bond types varying along the different molecular axes, there is no reason that the scaling should be the same for each of the d-orbitals. Certain low-spin $3d^6$ Fe(II) compounds, with a nominally spherically symmetric electron arrangement, have quadrupole splittings as large as the inherently asymmetric compounds because of their extensive covalency.[40] The sensitivity of the field gradient to small perturbing electrostatic fields arises from the fact that it is the result of a fine balance of electron density near the molecular z axis compared to that in the x-y plane. These two regions of charge contribute terms opposite in sign to the field gradient so that the final result is a small difference between several large terms. This sensitivity is a great advantage of Mössbauer spectroscopy in measuring molecular bond asymmetry, but it makes prediction of field gradients from approximate models extremely difficult. A small error in evaluating any single term determining the field gradient can cause even the sign to be predicted incorrectly. Detailed molecular orbital calculations on iron porphyrins,[41] for instance, which have been successful in accounting for many properties of the molecules, have yielded orbital occupancies which give a field gradient opposite in sign to that observed.[42] Necessary constraints which make such calculations computationally feasible, such as restricting the radial parts of all iron d-orbitals to be identical, probably introduce small changes in electron distribution which can make large errors in the finely balanced field gradient contributions.

A further ambiguity in interpreting and predicting quadrupole splittings is given by the fact that the symmetric inner shell electrons distort in response to field gradients from either the valence shell or the ligand neighbors, and shield or antishield the nucleus from these fields by large and difficult to predict factors. For effects of the lattice, there is an antishielding factor[43] of about 12, whereas for the valence electrons this figure is estimated[44] to be about 0.8. These are of uncertain accuracy, even when calculated for the free ion, and are certainly not well known for iron in coordination complexes. Moreover, in highly covalent compounds the distinction between valence electrons and ligand field becomes blurred, and obviously the choice of 12 or 0.8 for the antishielding can introduce substantial variation. Further, the quadrupole moment of the

[40] T. H. Moss and A. B. Robinson, *Inorg. Chem.* **7**, 1692 (1968).
[41] M. Zerner, M. Gouterman, and H. Kobayashi, *Theor. Chim. Acta* **6**, 366 (1967).
[42] T. H. Moss, A. J. Bearden, and W. S. Caughey, *J. Chem. Phys.* **51**, 2624 (1969).
[43] R. Ingalls, *Phys. Rev.* **133**, A787 (1964).
[44] R. Ingalls, *Phys. Rev.* **128**, 1155 (1962).

excited nucleus itself is not well known,[45,46] so that even if the field gradients were thought to be accurately determined the response of the nucleus to them would be uncertain.

As in the case of the isomer shift, then, interpretation of quadrupole splitting can best be made relative to a series of similar compounds, rather than as a parameter with absolute meaning. It is usually possible to assume that at least some of the ambiguous factors are constant throughout such a series and interpret differences in terms of real structural or electronic differences.

D. Magnetic Hyperfine Splitting

Magnetic hyperfine splitting might appear on the surface to contain less chemically important information than the isomer shift or quadrupole splitting which relate directly to electron distributions. However, since the ease of observation of these splittings in biological systems was first appreciated, it has developed that careful fitting of magnetic hyperfine data has become the most powerful tool for analysis of Mössbauer results.

The underlying reason for the usefulness of the magnetic data is that unpaired electrons in the iron configuration can create very large fields at the nuclear position, enough to split the energy levels widely and clearly resolve many of the possible information-containing transitions. Also important is the fact that biological systems are usually very dilute, magnetically, so that the chief mechanism relaxing electronic spins is the temperature-dependent spin–lattice, not spin–spin, interaction. At low temperatures this relaxation can be sufficiently slow so that the electronic spin maintains its direction long enough that the time average of its interaction with the nucleus is nonzero over the nuclear spin precession time ($\sim 10^{-8}$ second). It is then that the magnetic hyperfine effects can be observed. In nonbiological preparations it is often difficult to dilute paramagnetic centers enough to make the temperature independent spin–spin relaxation rate slow enough to fulfill this condition.

The onset of magnetic hyperfine effects with temperature is itself a convenient diagnostic for the presence of unpaired electrons. However, the analysis of the interaction is very difficult because it involves coupled nuclear and electronic coordinates.[47] The interactions are indicated in Fig. 8. It is for this reason that the application of a small external mag-

[45] C. E. Johnson, in "Hyperfine Structure and Nuclear Radiation" (E. Matthies and D. A. Shirley, eds.). North-Holland Publ., Amsterdam, 1968.

[46] J. Chappert, R. B. Frankel, and A. Misetich, Phys. Lett. 28B, 406 (1969).

[47] G. Lang, in "Conference on the Physical Properties of Iron Proteins," Biophysics Laboratory (Stanford) Report 208 (1968).

$$\mathfrak{K}_{mag} = 2g_N \beta \beta_N \sum_k \frac{1}{r_k^3} \left\{ l_k \cdot I + \underbrace{3(r_k \cdot s_k)(r_k \cdot I) - (s_k \cdot I)}_{\substack{\text{with} \\ \text{spins as} \\ \text{dipoles}}} \right.$$

with orbital moment

$$\left. -k(s_k \cdot I) \right\}$$

Fermi contact
(from core polarization,
S-electron admixture in $3d$)

$= \underline{I} \cdot \underline{A} \cdot \underline{S}$ with no applied field

$= \underline{I} \cdot \underline{A} \cdot \langle \underline{S} \rangle$ with small (≈ 100 gauss) applied field

FIG. 8. Magnetic interactions of the nucleus with electron spins and external magnetic fields.

netic field can have dramatic effects. The magnetic moment of the ^{57}Fe nucleus makes an effective field of the order of only 10 G at the iron d-shell radius, so that external fields of a few hundred gauss can easily break the coupling between the nucleus and the electrons. The electronic spin will then couple to the external applied field in a simple fashion, and the entire interaction of the magnetic nucleus with the unpaired electrons can be simplified to the interaction of the nucleus with an effective external field. This is illustrated in Fig. 8, where the operator S is replaced by its expectation value in the presence of an applied field. The effect on the spectrum in a simple case of an effective internal field is shown in Fig. 9. It should be emphasized that the direct effect of the applied field in further splitting the hyperfine levels is usually not important, as the 100-G fields typically applied compare to fields of hundreds of thousands of gauss generated by the unpaired electrons. The effect of *alignment* of the electronic spins is the essential point.

Examples of how the magnetic data can be used to extract chemical information are shown in Figs. 10–12, originally presented by G. Lang.[47] Figure 10 shows the quite fundamental difference between the effects of an unpaired electron in an iron d_{z^2} and in a $d_{x^2-y^2}$ orbital. The "Fermi contact interaction" is a direct spin–spin interaction between the nuclear moment and any unpaired s-electron density at the nucleus. This can arise when unpaired s-electron character is mixed into the valence electron wave functions.[48] More important, it can also occur when the amount of spin up and spin down density of the inner shell s-electron at the nuclear position is made unequal by differential interactions with unpaired electrons in the d-shell (core polarization).[49] This interaction

[48] A. Abragam and M. H. C. Pryce, *Proc. Roy. Soc. Ser. A* **205**, 135 (1950).
[49] R. E. Watson and A. J. Freeman, *Phys. Rev.* **123**, 2027 (1961).

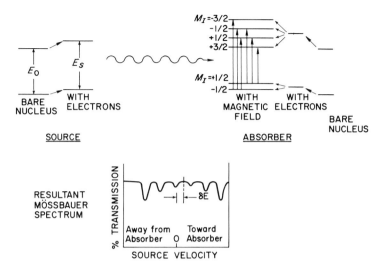

FIG. 9. Essential features of the magnetic hyperfine splitting as measured in Mössbauer spectroscopy.

always produces an effective field opposite in direction to the net electron spin. However, the effective field due to the dipole moment of the electron can either add to or subtract from this, depending on whether the electron is in an orbital close to the magnetic axis or in the plane perpendicular to it (Fig. 10). Figure 11 presents calculated spectra based on these two cases.[47] The important point is that even after folding in reasonable line widths and averaging transition probabilities over all possible molecular

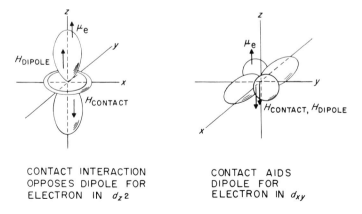

FIG. 10. Effective magnetic fields at the nucleus due to electrons in different d-orbitals.

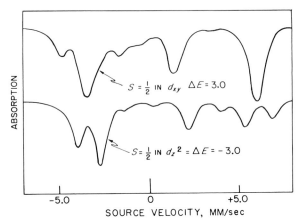

FIG. 11. Typical hyperfine split Mössbauer spectra for the two cases of Fig. 9.

orientations in a frozen solution sample, the very complexity of the spectrum from a given configuration almost guarantees its uniqueness. An example of a fit to real data is given in Fig. 12.[47] The fitting procedure generated g values for the electronic state in this case which were previously unknown, but subsequently verified by electron spin resonance.[47]

For configurations with a number of magnetic states spaced by energies comparable to thermal energies, there can also be complex but revealing temperature dependence. As in determining other parameters, relaxation rates relative to the nuclear precession times determine whether superpositions of individual states or thermal averages are observed.

Although the efficacy of small applied magnetic fields has been emphasized thus far, there are definite uses for fields of 50 kG or stronger.

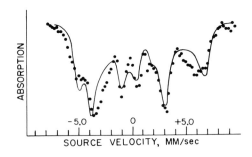

FIG. 12. Theoretical fit and data for methemoglobin cyanide in a 700-G applied magnetic field at 4.2°K.

In the paramagnetic cases these fields are big enough perturbations of the ≈ 100 kG per spin effective field of the unpaired electrons to cause measurable changes in the observed spectra. The precise way the external field couples to the internal one can provide additional information concerning the electronic state.[32] For diamagnetic metal configurations the direct effect of an applied magnetic field can also be important. Applied fields of 50 kG are not really sufficient to resolve many transitions of the magnetic hyperfine levels, but they are sufficient to change in distinctive ways the shapes of the two lines corresponding to transitions to the $M_I = \pm 1/2$ and $\pm 3/2$ levels of a quadrupole split pair. In axial symmetry, the sign of the electric field gradient determines which of these two lines lies highest in energy, so that an analysis of the higher and lower energy line shapes can determine this sign.[29] Such a determination can sometimes be the key to judging the molecular symmetry, and it is only this combination of quadrupole splitting and strong applied magnetic field which can give the information for diamagnetic configurations.

Although the complexity of the magnetic hyperfine interactions makes them much less equivocal indicators of electronic state than are the isomer shift or quadrupole splitting, there are still considerable difficulties. For the heme proteins, where many details of the geometry and electronic configuration are well known, enough parameters can be fixed in advance of Mössbauer data analysis so that the extra insight given by the Mössbauer experiment can be decisive. In unknown structures, however, there may be so many combinations of possible interactions that a unique fit to the data is hopeless. One must then retreat to much less detailed, though often still useful, discussion. Covalent interactions can also be troublesome in interpreting the magnetic data. As in the case of quadrupole interactions the effective fields often scale as $\langle r^3 \rangle^{-1}$. As long as this quantity remains ambiguous through covalent electron delocalization or other effects, interpretation of magnetic data will have to rely heavily on empirically determined factors.

V. Applications

Almost all the applications of Mössbauer spectroscopy in biology have dealt with either heme proteins or the iron–sulfur class of nonheme iron proteins.[50] The work in the former area has concentrated on delineating rather detailed features of electronic structure,[3] as much other information was previously available. In the latter class more general interpretations of the data were made until very recently, when supporting information began to make possible more exact discussion. Work on iron outside

[50] D. O. Hall and M. C. W. Evans, *Nature (London)* **223**, 1342 (1969).

of these two classes has been limited to a few isolated cases, and no biological study has been reported on other isotopes.

Heme Proteins

One area where Mössbauer spectroscopy had a special promise was in analysis of the classic problem of the oxygen–iron bond in hemoglobin. Considerable speculation had gone into rationalizing the basis of the ability of the iron to carry oxygen without losing an electron to the oxygen molecule—and becoming irreversibly oxidized.[51] The ability of the isomer shift to indicate (Section IV, B), in conjunction with other Mössbauer parameters, the electron density in the d-shell, threw an important new light on this discussion. Analysis of the data showed it to be consistent with a model where almost an entire electron actually was delocalized from the iron d-shell to oxygen π-orbitals.[52] The expected oxidation of the iron did in fact take place, and a more appropriate question was thus shown to focus on why the electron was "returned" to the iron as the oxygen became dissociated.

Many more detailed aspects of the iron configuration in heme proteins have been delineated for the first time through Mössbauer spectroscopy. By using the differences in magnetic hyperfine effects of electrons in orbitals of different symmetries (Section IV, D) the unpaired spin in the nitroxide derivative of hemoglobin was localized.[52] The zero-field splittings and relative mixtures of high and low spin states of many heme proteins and derivatives have been accurately measured by careful fitting of the temperature dependence of both quadrupole and magnetic hyperfine splitting.[32,53]

In the peroxidase heme enzymes, the long-standing question[54] of the existence of iron valence states higher than 3^+ was also a natural problem for Mössbauer spectroscopy (Section IV, B). The existence of a true Fe(IV) state was inferred by consideration of how the isomer shift changed in going from resting enzyme to compound II of horse radish or Japanese radish peroxidase,[12,55] or complex ES of cytochrome c peroxidase.[56] The lack of magnetic hyperfine splitting in complex ES in the presence of a small applied field, which indicated a nonmagnetic ground state and hence precluded an Fe^{3+} configuration, confirmed this

[51] J. H. Wang, in "Oxygenases" (O. Hayaishi, ed.). Academic Press, New York, 1962.

[52] G. Lang and W. Marshall, Proc. Phys. Soc. 87, 3 (1966).

[53] G. Lang, T. Asakura, and T. Yonetani, in "Proceedings of the Conference on Applications of the Mössbauer Effect," Tihany, Hungary, to be published.

[54] P. George and D. H. Irvine, Nature (London) 168, 164 (1957).

[55] Y. Maeda, Nippon Seirigaku Zasshi 24, 151 (1968).

[56] G. Lang, T. Asakura, and T. Yonetani, to be published.

interpretation.[56] On the other hand, the lack of change in Mössbauer parameters in going from complex I to complex II of the horse radish or Japanese radish peroxidases established that the postulated Fe(V) to Fe(IV) valence change did not exist, and that the extra oxidizing equivalent of these enzymes had to be localized elsewhere on the protein.

Early work on the iron–sulfur proteins focused on broad questions such as the similarity of iron sites in proteins from different organisms,[57] and used the observation of magnetic hyperfine splitting in all the iron absorption lines of a polynuclear iron complex to infer electron delocalization over a large system.[8] The lack of magnetic hyperfine splitting in applied field for the oxidized derivatives of the two-iron, two-sulfur proteins was taken as evidence for a diamagnetic, antiferromagnetically coupled, state of the two irons.[58] The large number of cases where similar results were obtained helped establish the prevalence of the closely coupled polynuclear iron center as a design principle in metal proteins. Recently ENDOR measurements[59] have delineated many of the components of the hyperfine tensor, **A,** which couples the electronic and nuclear spins in these proteins. Using this as supporting data, nearly complete models of the structure about the iron center have been derived from the Mossbauer data in several cases.[30,31]

These are but a few representative examples of Mössbauer experiments on biological systems, which bibliographic compilations[60,61] indicate total well over 200 published papers at this writing.

[57] C. E. Johnson, E. Elstner, J. F. Gibson, G. Benfield, M. C. W. Evans, and D. O. Hall, *Proc. Nat. Acad. Sci. U.S.* **63,** 1234 (91969).

[58] C. E. Johnson, E. Elstner, J. F. Gibson, G. Benfield, M. C. W. Evans, and D. O. Hall, *Nature (London)* **220,** 1291 (1968).

[59] J. Fritz, R. Anderson, J. Fee, G. Palmer, R. H. Sands, J. C. M. Tsibris, I. C. Gunsalus, W. H. Orme-Johnson, and H. Beinert, *Biochim. Biophys. Acta* **253,** 110 (1971).

[60] L. May, *in* "Index of Publications in Mössbauer Spectroscopy of Biological Materials," Department of Chemistry, Catholic University of America, Washington, D.C., 1971.

[61] J. Stevens and V. E. Stevens, *in* "Mössbauer Data Index, 1969" IFI/Plenum, New York, 1970.

Addendum. *Enzyme Structure, Part B–Sequence Determination*

[36] Automated Edman Degradation: The Protein Sequenator

By HUGH D. NIALL

I. Introduction

The most significant recent development in techniques for sequence analysis of proteins and peptides is the advent of automated procedures for degradation by the phenylisothiocyanate method. Although the first detailed description[1] of this approach appeared only in 1967, automated equipment based on the design of Edman and Begg is at the time of writing already being used in about 100 laboratories around the world. This major advance was based on elegant and painstaking work on the detailed chemistry of the isothiocyanate method, carried out in the laboratory of Pehr Edman over the last two decades, and described in a classic series of papers.[2-5] Particular attention should be drawn to a recent comprehensive review by Edman.[6]

The purpose of the present article is to describe the automated procedures in some detail from a practical point of view. For a more theoretical treatment of the subject the reader is referred elsewhere.[6,7] Some of the important principles, however, must be summarized here since they are essential to a proper understanding of the methodology. No attempt will be made to review the already large number of applications of the sequenator to specific proteins. Illustrations mainly from work carried out in the author's laboratory will be briefly discussed to illustrate the scope and present limitations of the methodology.

Abbreviation

The following abbreviations and proprietary names are used in this chapter: PITC, phenylisothiocyanate; PTC, phenylthiocarbamyl; PTFE, polytetrafluoroethylene (Registered Trademark, Teflon, Dupont Chemical Company); Kel-F, polytrifluorochloroethylene (Registered Trademark, 3M Company); FEP, fluorinated ethylene-propylene copolymer; Quadrol,

[1] P. Edman and G. Begg, *Eur. J. Biochem.* **1**, 80 (1967).

[2] P. Edman, *Acta Chem. Scand.* **4**, 283 (1950).

[3] P. Edman, *Acta Chem. Scand.* **10**, 761 (1956).

[4] P. Edman, *Proc. Roy. Aust. Chem. Inst.* **24**, 434 (1957).

[5] P. Edman, *Ann. N.Y. Acad. Sci.* **88**, 602 (1960).

[6] P. Edman, *in* "Protein Sequence Determination" (S. B. Needleman, ed.), p. 211. Springer-Verlag, Berlin and New York, 1970.

[7] H. D. Niall, *J. Agr. Food Chem.* **19**, 638 (1971).

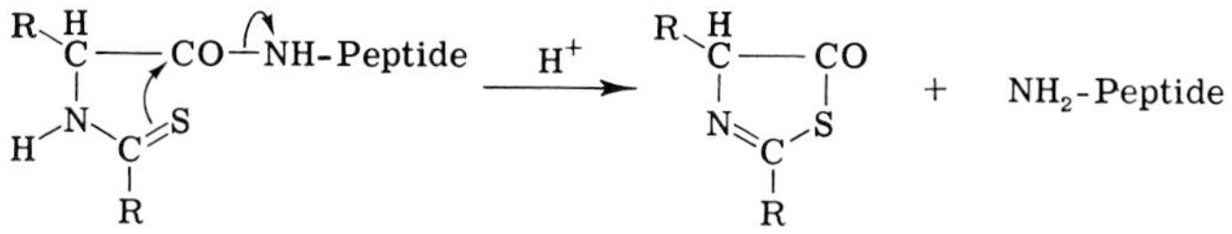

FIG. 1. Mechanism of the Edman degradation, showing cleavage of the terminal amino acid as its thiazolinone.

N,N,N^1,N^1-tetrakis(2 hydroxypropyl)ethylenediamine (Registered Trademark, Wyandotte Chemicals Corp.); HFBA, heptafluorobutyric acid; PTH, phenylthiohydantoin; GLC, gas-liquid chromatography; TLC, thin-layer chromatography; BTMSA, bis(trimethylsilyl)acetamide; DMAA, dimethylallylamine; DEAA, diethylallylamine.

II. Chemistry of the Edman Degradation

In the Edman procedure, the reagent phenylisothiocyanate (PITC) couples with the terminal alpha amino group of a peptide or protein to form a phenylthiocarbamyl (PTC) adduct. Under anhydrous acidic conditions the N-terminal amino acid residue is selectively cleaved from the peptide chain as a heterocyclic derivative (an anilino-thiazolinone) through the attack of the sulfur of the PTC adduct on the carbonyl component of the first peptide bond (Fig. 1). The cleaved amino acid derivative is separated from the residual peptide by extraction with an organic solvent and then converted to a more stable isomer (a phenylthiohydantoin) prior to identification by one of several possible procedures. The shortened peptide chain may then be subjected to further cycles of coupling and cleavage, with identification of each successively removed amino acid, thus establishing the amino-terminal sequence.

Under appropriate conditions[1,5] the coupling and cleavage reactions take place with close to 100% efficiency and a minimum of side reactions. Hence automation of what had previously been a manual "test tube" procedure became a realistic possibility.

III. Automation of the Edman Degradation: The Sequenator

Edman and Begg immensely simplified the task of automation by introducing the novel concept of carrying out the degradation with the protein spread as a thin film on the inside wall of a continuously spinning cylindrical glass cup (Fig. 2).

The cup is spun on its vertical axis by a motor; it is situated in a sealed thermostated chamber which may be subjected to vacuum or filled with an inert gas (Fig. 2, II). The protein is added in solution while the cup spins so that it is distributed over the lower half of the inside wall.

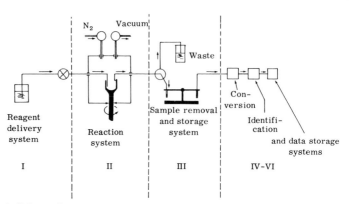

Fig. 2. Schematic representation of the sequenator shown in modular fashion. The present automated instrument includes modules I, II, and III. Modules IV–VI have not yet been automated. See text for details.

After application of a vacuum to dry the protein film, the degradation is initiated by the delivery of the reagents for the coupling reaction.

Reagent and solvent additions are made via a thin Teflon tube which delivers the liquids in a continuous stream at the bottom of the cup (Figure 2, I and II). As the liquid emerges from the tube it is spun by centrifugal force onto the wall and spread out as a thin film over the surface of the protein. Volumes of reagents (coupling buffer or acid) are metered so that the level is sufficient just to cover the protein film, dissolving and reacting with it. After coupling and cleavage reactions, a vacuum is applied to remove the bulk of the liquid. The dried or semidried protein film is then washed with a stream of solvent to remove nonvolatile side products and excess reagents (after coupling) or to extract the thiazolinone from the residual protein (after cleavage). The protein itself, being insoluble in these organic solvents, is not washed out but precipitates at its original site. The ascending layer of solvent after washing over the protein leaves the cup via another Teflon line the tip of which is situated tangentially in a circumferential groove at the top of the cup (Fig. 2, II). The outflowing solvent stream is directed by a valve to a waste container or a fraction collector (Fig. 2, III). The samples may be removed from the fraction collector at intervals for conversion from the thiazolinone to the thiohydantoin form, followed by identification (Fig. 2, IV–VI). A description of this system may be found in the original publication.[1] Further details are discussed in this article.[8]

[8] Protein sequenators are available from several commercial sources. Those listed in the "Guide to Scientific Instruments" (*Science* **174A**, No. 4010A, 133, November, 1971) are as follows: Beckman Spinco, 1117 California Ave., Palo Alto, California

IV. General Design Features

The original design of the protein sequenator was based upon a very detailed examination of many important parameters. A large number of problems, some obvious, some extremely subtle, were encountered and solved or minimized by Edman and Begg. Most of these problems are described in their paper,[1] which should be read extremely carefully. Failure to appreciate the importance of some of these design features led to serious problems when arbitrary design changes were made by workers attempting to build a modified or "improved" sequenator. Some of the features critical for the design and operation of a sequenator will be reviewed here. Some repetition of material already described in the paper of Edman and Begg is necessary for the sake of clarity.

Materials. Only a restricted range of materials may be used in contact with the corrosive liquids and vapors used in the degradation. Borosilicate glass, gold, and certain chemically inert, highly resistant synthetic polymers such as PTFE, FEP, and Kel-F may be used in contact with liquids. Stainless steel may be used where contact is solely with vapors. These facts, pointed out by Edman and Begg, are repeated here because they were ignored, with disastrous results, in the construction of more than one instrument.

There is at least a theoretical danger that condensates from metal parts situated above the cup may drip into it, bringing trace amounts of metallic ions, which may interfere with the degradation reaction. Hence metal should probably be avoided in the design of this area.

Two additional comments may be made. Commercially available stainless steel varies considerably in quality. Some grades undergo severe corrosion after exposure to the vapors for only a few months. Hence only a high quality grade should be used. Plastic materials such as PTFE and Kel-F, while chemically inert, adsorb organic vapors, which may

---

94304; Bio-cal Instrument, 2400 Wright Ave., Richmond, California 94804; Illinois Tool Works, 2501 N. Keeler Ave., Chicago, Illinois 60639; Scientific Products, 1430 Waukegan Road, McGaw Park, Illinois 60085; Sondell Scientific, 870 San Antonio Road, Palo Alto, California 94303. Instruments are manufactured outside the United States by the Jeolco Instrument Company, in Japan, and the Socosi Instrument Company in France. In addition a number of individual laboratories have built their own instruments. See, for example, M. D. Waterfield, C. Corbett, and E. Haber, *Anal. Biochem.* **38**, 475 (1970); J. D. Lynn and J. C. Bennett, *Anal. Biochem.* **45**, 498 (1972).

Wherever possible the discussion in this article has been written so as to be applicable to any sequenator based upon the spinning cup principle of Edman and Begg, except where otherwise specified.

"out-gas" only very slowly even under high vacuum. Successive exposure to acidic and basic vapors during the degradation with incomplete removal leads to accumulation within the system of poorly volatile organic salts which may interfere with the function of valves, vacuum gauges, or bearings. They also interfere with the maintenance of pH control during the coupling reaction when volatile buffers are being used for peptide degradation, as will be described more fully below.

For these reasons, the surface area of these polymers within the reaction cell and its surroundings should be kept to a minimum.

The other cogent reason for minimizing the use of plastic materials throughout the instrument is their permeability to oxygen, which may interfere with the degradation through oxidative desulfuration of the phenylthiocarbamyl group. The protection afforded by tight sealing of connections and the use of an inert gas such as nitrogen or argon within the system is quite illusory if plastic tubing is employed for the delivery line, as it usually is. PTFE is particularly permeable to oxygen, FEP and polyethylene somewhat less so. Kel-F annealed at high temperature has a relatively low oxygen permeability, but this tubing has less suitable mechanical properties. It can be calculated that a single 1-foot length of PTFE tubing exposed to the atmosphere will admit enough oxygen (on a molar basis) to completely desulfurate the protein sample in the cup several times over at each cycle. Oxygen also leaks in through Teflon valves and through plastic diaphragms on pressure regulators. Hence though nitrogen containing less than 10 parts per million of oxygen may be used for purging the system, the true oxygen level within the cell is probably several hundred parts per million unless special efforts are made to minimize diffusion.

At present the practical significance of these facts is not clear. At times the presence of a leak into the vacuum system is associated with poor repetitive yields; with elimination of the leak, the yields have improved. This kind of result has usually been attributed to amino-terminal blocking of the protein chain through oxidative desulfuration of the PTC group by atmospheric oxygen. However in some experiments in which leaks are known to have been present, repetitive yields have been apparently unaffected. As has been pointed out, plastic delivery lines which are known to be oxygen permeable are present in most instruments so far built. Hence oxygen must contact and react with the PTC-protein very inefficiently. One factor may help to explain these puzzling observations. A leak may introduce deleterious substances other than oxygen— for example, volatile oxidants present as atmospheric pollutants. The actual presence or absence of these compounds and their concentration might vary from one laboratory to another. In this connection, Schroeder[9] has pointed out the particular difficulty of carrying out successful manual Edman degradations in the Los Angeles area and has attributed this to atmospheric pollutants. It should also be pointed out that a severe leak could lead to excessive drying of the sample during coupling and cleavage reactions, and thus cause incomplete reaction. Even if oxygen per se does not severely affect the degradation, since even a small increase in yield would lead to much longer degradations, the design and testing of a totally "oxygen-free" system (in practice, one with less than say 10 parts per million O_2) would seem well worth while. For these reasons the quantity and length of plastic tubing should be minimized. Choice of material for tubing, valves, and regulator diaphragms should take into account oxygen permeability as well as other mechanical and chemical properties. The possible use of reducing agents in the reagents to scavenge oxygen is discussed below.

The Programmer Unit. All operations of the sequenator are controlled by a single programmer unit. Although the programming of the degradation could be made as complex as required, in practice a single system is adequate for most purposes and subprogramming routines have not been utilized. There would, however, be some advantage in repeat degradations in optimizing conditions at particular cycles. For example, at cycles where serine was known or suspected to be present, special cleavage conditions could be used to improve the yield (as described in the section on automated degradation of short peptides). In degradations on peptides, the extent of solvent extraction could be decreased stepwise or continuously as the residual peptide chain became shorter. However, there is insufficient need for these maneuvers in the present state of the art to warrant the use of an elaborate programming unit. In the Beckman instrument a single punch-tape programmer is used. This has proved reliable and has the advantage that a library of prepunched programs can be built up. This is particularly useful for the automated degradations on shorter peptides, where several different programs may need to be used in the course of a single degradation.

It is obvious that in the general design of the sequenator, the programming unit should be so situated that it is in no danger of its becoming overheated by proximity to the reaction chamber. It should also be well away from the areas where the reagents are being stored and handled to minimize the danger of spillage.

Safety Features. A detailed discussion of the essential and desirable

[9] W. A. Schroeder, see Vol. 11, p. 445.

safety features is outside the scope of this article. However, the safety of the operator, the instrument and the sample must all be considered. Since flammable solvents are used, all electrical contacts must be sealed to avoid sparking. The fumes from the pump must be vented off in a way which avoids exposure of laboratory personnel and which conforms to applicable antipollution regulations. A reliable temperature cutoff device should be installed to prevent overheating of the reaction chamber; temperature "runaway" can cause serious damage to the instrument. With the magnetic drive system, it is important to incorporate features which will prevent the drive from becoming "uncoupled." This is particularly likely to happen during sudden changes in cup speed. When the nitrogen pressure type of delivery system is used, if a delivery valve opens and then fails to close, the entire contents of the reservoir can be emptied into the cup. This can be the result of either a mechanical failure of the valve or to an error in the programmer unit. It can be prevented or at least rendered extremely unlikely by suitable safety features.

A number of safety features are aimed at protecting the sample, which at times will be more valuable than the instrument itself. Thus an automatic shutdown of the degradation can be made to come into operation if certain events happen—for example, if the pressure in the nitrogen cylinder falls below a certain value or if the fraction collector advances beyond a certain point.

V. Specific Design Features

A. Delivery System

There are several basic requirements for this system:

1. As discussed above, the materials used for the reagent bottles, seals and delivery lines and valves should consist only of borosilicate glass or highly resistant polymers (Teflon, Kel-F, FEP). Plastic delivery lines should be kept as short as possible to minimize oxygen diffusion. There should be a minimum dead space within the delivery line and valve. This again minimizes the volume of reagent exposed directly to oxygen diffusion, and allows the system to be flushed through economically with fresh reagent prior to starting a new degradation. It also lessens the opportunity for formation of nitrogen or air bubbles which may mechanically interfere with accurate delivery of reagents. Avoidance of unnecessary dead space within the actual delivery valve is particularly important since inefficient flushing may lead to local accumulation of insoluble deposits, derived, for example, from the Quadrol or the PITC, interfering with valve function and contaminating the reagents as they are delivered.

2. Reagent reservoirs must be of a volume adequate to allow at least 50 degradation cycles without need for replenishing. This permits the machine to be operated untended over the weekend.

3. Reagents must be able to be purged with nitrogen and sealed to exclude atmospheric contamination.

4. Liquid cross-contamination of one reagent with another must be avoided. However, the small amount of cross-contamination resulting from diffusion in the vapor phase may be harmless. For example, the three solvent bottles (for benzene, ethyl acetate, and 1-chlorobutane) may be pressurized from a common nitrogen manifold, since the small degree of cross diffusion does not appreciably alter the physical or chemical behavior of the solvents. However, any diffusion of vapor between the reagent reservoirs should be prevented. Diffusion of acid may alter the pH of the coupling buffer. Diffusion of water vapor from the coupling buffer into the HFBA could predispose to the occurrence of nonspecific hydrolysis of the protein during cleavage.

5. Stopcock grease or other lubricants should not be used to seal the solvent or reagent bottles, since it is almost impossible to avoid contamination. Kel-F grease, even in trace amounts, gives rise to multiple artifactual peaks on gas chromatography interfering with the identification of PTH amino acids.

6. Accurate reagent metering is of the utmost importance. To achieve quantitative coupling and cleavage at each cycle, the levels in the cup of coupling buffer and acid must be absolutely reproducible. Any inequality in delivery results in an upper rim of protein remaining uncoupled (or uncleaved), and serious asynchrony (overlap) develops in the degradation.

Design of the Delivery System

The Edman–Begg delivery system was based on fluid displacement by positive pressure of nitrogen. The reagent and solvent bottles are pressurized from a nitrogen cylinder, through pressure regulators. The reaction cell is maintained through a separate regulator at a lower pressure. Since the pressure differential between the reagent or solvent bottle and the cell is kept constant, when the intervening valve is opened, the liquid will flow at a constant rate. Valve opening and closing should be rapid so as to allow accurate control of the volume delivered.[10]

The Beckman delivery system is based on the same principle, with minor differences. Four separate double delivery valves replace the single

---

[10] The height to which the liquid reservoir is filled has a minor but perceptible influence on the volumes delivered, because of gravitational effects.

multiport valve of Edman and Begg, and the single delivery line into the cup is replaced by four separate lines. Each double-port valve and its corresponding delivery line is used to deliver a reagent and a solvent (e.g., PITC and benzene; Quadrol and ethyl acetate; HFBA and 1-chlorobutane). The solvent may be used to flush out residual reagent left in the delivery line. This is particularly important with the Quadrol buffer. If it is left in the delivery line, the volatile components (propanol and water) tend to evaporate, plugging the line with the viscous, nonvolatile Quadrol. Suitable program alterations, discussed below, can prevent this complication. In the Edman–Begg delivery system, no reagent is left in the line between the delivery valve and the cup since a nitrogen stream pushes the reagent into the cup. This is desirable since reagent left in the line contributes nothing to the reaction and requires additional solvent extraction or application of vacuum for its removal. A potential problem with the more volatile reagents and solvents is that a rise in ambient temperature can lead to the build up of vapor pressure in the reservoir. If the pressure reached is higher than the pressure being used to deliver the reagent, then as soon as the valve between the pressure regulator and the reservoir opens, vapor and perhaps liquid may back up into the valve and into the pressure regulator itself. This can lead to damage to the components in the pressurizing system and to contamination of the reagent in the reservoir.

The Edman–Begg instrument is designed for use in a constant temperature, which virtually eliminates the problem. The Beckman design incorporates provision for venting the reservoir to the atmosphere, thus relieving the pressure before each delivery. The Beckman instrument may, therefore, be used in laboratories that vary widely in ambient temperature.

Overall, the nitrogen-pressure delivery system has proved to be extremely reliable and effective. Some specific problems are discussed below.

An alternative delivery system based on positive displacement by motor-driven syringes has also been used.[11] There seems to be no reason why such a system could not be an effective one; however, the author has no experience with its use, and to date such a system has not been widely evaluated.

B. Reaction System

The general plan of the reaction vessel and chamber is illustrated in Fig. 2. Some of the design features important for proper functioning of the system will be discussed in detail.

---

[11] M. D. Waterfield, C. Corbett, and E. Haber, *Anal. Biochem.* **38**, 475 (1970).

The Cup. A suitable design for the cup has been described. It should be pointed out that the tolerances required for the inside cylindrical surface, though stringent by some standards, are quite coarse in the light of modern technology used, for example, in the construction of optical lens systems. The bottom "corner" of the cup (where the side meets the floor) should not be sharp but be rounded to prevent accumulation of protein, to aid cleaning and to provide a more uniform protein film. It is extremely important that the surface of the cup be highly polished. On prolonged use, the surface tends to become etched by the chemicals, particularly if alkaline cleaning solutions are used injudiciously.[12] Scratches on the surface may also be produced during cleaning. When this occurs, inefficient reaction and extraction and poor results are observed. The surface may be repolished through gentle abrasion with solid ceric oxide.

Little work has been done on alternative cup designs.[13] Several unsuccessful attempts have been made to use cups composed of Kel-F or Teflon to simplify the problem of manufacture. The overwhelming disadvantage results from the inability of the operator to see what is happening in the cup during reagent additions and reactions, making proper monitoring of the degradation impossible.

The Drive. The Edman–Begg and the Illinois Tool instruments used a direct-drive system for spinning the cup; i.e., the cup was mounted directly on an extension of the motor shaft. In other instruments a magnetic drive system has been used. If one ignores possible differences in cost or ease of construction, neither system seems clearly superior to the other. With the direct drive the vacuum seals around the shaft gradually become ineffective due to contact with organic vapors. Hence, vacuum leaks are likely to develop in time, and it has been found necessary to replace the seals routinely about every 6 months.

With the magnetic drive there is no problem with seals or vacuum leaks; however, the bearings necessarily used in this type of drive are also exposed to organic vapors. Over a period of months with constant use of the instrument the bearings gradually become coated with viscous organic deposits. The resulting increased frictional resistance to rotation

---

[12] The use of sodium hydroxide solution for cleaning the cup is not recommended. An effective procedure is to use in succession Quadrol coupling buffer, ethanol/water and acetone. The cup is then washed through with highly purified ethyl acetate from the sequenator reservoir, and dried by vacuum. A cotton-wool tipped wood (not metal) applicator is used to remove adherent protein.

[13] The most recent model of the Beckman sequenator has a cup which has been slightly undercut halfway up its inside surface. As the liquids rise during delivery, they encounter temporary resistance to flow from the undercut edge of the glass. This provides a margin of error in equalizing the levels reached by the coupling buffer and the HFBA.

is manifested by noise and by vibration of the cup assembly. The increased resistance also increases the likelihood that the magnetic drive may become "uncoupled" and the cup stop during a run. For these reasons the bearings must be inspected and cleaned or replaced routinely every 6–12 months, whether the drive is noticeably noisy or not.

Cup Speed. In the Edman–Begg design the cup spins at a fixed speed throughout the degradation (1400 rpm). Since results obtained with protein degradations on this instrument are excellent, it is clear that a multiple speed drive is not necessary. However, the provision of different cup speeds (1200 and 1800 rpm) in the Beckman instrument has been useful. A change from high to low speed during coupling or cleavage will cause a fall in the level of the protein film. This can be useful in helping to keep the film low in the cup and in minimizing the effects of slight inequalities of deliveries of coupling buffer and HFBA. Use of the higher cup speed is helpful during degradation of shorter peptides (see below) since the layer of extracting solvent is thinner than at low speed. Smaller volumes can be used for the solvent extractions, and there is therefore less extractive loss of peptide. It should be remembered that unless precautions are taken the magnetic drive system may become "uncoupled" during sudden changes in cup speed.

The Vacuum System. There has been some misinterpretation of the purpose of Edman and Begg in introducing the vacuum stages following coupling and cleavage. The aim was only to dry off the bulk of the volatile components of the coupling buffer (propanol and water) and the bulk of the acid. This is necessary to facilitate precipitation of the protein when the subsequent extraction with an organic solvent takes place. If the film is not sufficiently dry, the protein may be washed up the wall of the cup by the solvent. However, it is difficult, unnecessary, and even disadvantageous to dry the sample completely.

The Quadrol in the coupling buffer is completely nonvolatile and acts as a "sink" from which more volatile reagents outgas only very slowly. Hence the vacuum reading achieved after coupling may appear to be "poor" (100–200 μ). This has led workers to introduce very prolonged vacuum stages at this point. This is unnecessary since both the Quadrol and its dissolved volatile components will be removed anyway by subsequent extractions. It is harmful since it increases the time during which the coupled PTC-protein is at risk of oxidative desulfuration by molecular oxygen or trace oxidants in the reagents. Excessive drying after the acid cleavage step is also disadvantageous, since the over-dried protein is very poorly penetrated by 1-chlorobutane. Hence, the extraction of the cleaved anilinothiazolinone is extremely inefficient, giving low apparent yields.

Hence, a low reading on the vacuum gauge attached to the cell should not be made the prime objective, nor need an excessively powerful vacuum pump be used. The important requirements for the vacuum system are (1) that it be well sealed, free from leaks; (2) that the pump itself be sturdy and capable of continuous running for prolonged periods (1–2 years); (3) that it be resistant to corrosion by the organic reagents used, in particular the HFBA.

In some designs two vacuum pumps are used. One pump (the rough pump) is used first to dry off most of the liquid. The oil in this pump rapidly becomes contaminated during the course of a run, and it achieves only a moderate vacuum. However, a second pump is programmed in after most of the liquid has been removed. It dries off the final traces of liquid (for the reasons outlined above it is not necessary or desirable to attempt complete drying at certain stages) and achieves a higher vacuum.

Nitrogen System. The reaction chamber must be filled with an inert gas at all times during the cycle except during vacuum stages. The main objective in this is to prevent oxidative desulfuration reactions resulting from atmospheric oxygen. However, the inert gas (usually nitrogen, though argon or in theory any inert nonflammable gas would do as well) also serves other functions. The nitrogen can be used in conjunction with appropriate pressure or flow regulators to open or close valves, to establish a pressure difference between the reagent bottles and the reaction cell in the delivery system described above; to dry off volatile reagents or solvents from the cup; dry the 1-chlorobutane samples in the fraction collector; to flush residual liquid out of delivery lines or from the exit lines.

An additional, less obvious, function is in heat transfer. Considerable cooling goes on in the cup during evaporation of volatile reagents and solvents. Reheating occurs more quickly in a nitrogen atmosphere, because of conduction and convection, than it does when a vacuum is maintained in the reaction cell.

The nitrogen or argon used should be highly purified, if necessary through the use of appropriate "scrubbers" in the inflow line to remove oxygen, moisture, or dirt from the cylinder. As already discussed, the lines and pressure regulators, particularly the diaphragms of the latter, must be relatively impermeable to oxygen. All lines should be as short as possible with the minimum of joints where leaks might occur.

Heating System. The reaction cell must be kept at a controlled if not constant temperature during the cycle. This is because a number of parameters in the system are temperature dependent. The coupling and cleavage reactions obey the rules for most organic chemical reactions and proceed more rapidly at higher temperatures. Competing side reactions are also accelerated *pari passu*. It has been suggested that nonspecific peptide bond cleavage is much more prominent at higher temperatures, as is the decomposition of the somewhat unstable PITC. Higher temperatures may aid in dissolving poorly soluble proteins in the coupling or cleavage media. Viscosity and surface tension of the reagents, in particular Quadrol and HFBA, vary with temperature. They can lead to irregularities in delivery volumes and in the height of liquid levels in the cup. The solubility of the apolar PITC in the coupling medium is limited by the presence in it of water and of polar hydroxyl groups in the Quadrol and propanol. Hence a temperature above ambient has been found to be necessary to dissolve enough PITC for coupling to occur in a reasonably short time.

The useful range of temperature is roughly 45° to 65°. Below 45° the reactions proceed too slowly and solubility problems are prominent. The author has carried out identical degradation runs on a standard protein (sperm whale myoglobin) at 50°, 55° and 60°, using the Beckman instrument. These runs were all satisfactory, and no significant differences in repetitive yield could be detected. No definitive examination of the practical upper limit of temperature has yet been made. For routine use a temperature, at equilibrium, of 55° is recommended. Whatever temperature is chosen, fluctuation due to evaporative cooling occurs during the cycle, and adequate time has to be allowed for reheating to occur. This is particularly important in peptide degradations using volatile tertiary amines in the coupling buffer, as described below.

The two methods of heating used have been (1) use of a transparent metallic electroconductive heating layer on the glass surface of the reaction chamber and (2) use of circulating hot air within an outer Perspex chamber which surrounds the reaction chamber. Both these methods are adequate.

Size of Reaction Cell. The reaction cell should be constructed so as to have the smallest possible vapor space, as pointed out by Edman and Begg. This minimizes evaporation of volatile reagents. Particular attention to this point was applied in the design of the Beckman instrument, facilitating automated degradation of peptides, as described below.

Openings into Reaction Cell. The number of openings into the reaction cell should be kept to a minimum to reduce the risk of leaks. A port into which a vacuum gauge or a pressure gauge can be plugged is useful, however, since this may be used to check for vacuum leaks or pressure leaks. It is doubtful whether a vacuum gauge permanently in communication with the cell is of much value, since contamination by organic vapors soon renders it inaccurate.

Inflow Line. The end of the Teflon inflow line should be cut at 45° so that the opening faces outward toward the wall of the cup. The tip of the line should be almost but not quite touching the floor of the cup. This ensures that the emerging liquid stream makes contact with the floor of the cup and is spun smoothly across the floor into the cup wall. If the line is too high, droplets of liquid appear to "explode" from the end of the line and are flung directly onto the wall, causing turbulence and resulting in splashing. If the line is too low or not cut obliquely, there is some danger that the small gap between the top of the line and the floor may become obstructed by dried nonvolatile components of the reagents. If multiple inflow lines are used, as in the Beckman instrument, then the tips of the lines should be cut so that the emerging stream of liquid from one line does not strike another line, causing splashing.

C. Sample Removal and Storage System

Outflow Line. This Teflon line projects obliquely into the groove at the top of the inside cup wall. It points in the direction opposite to the direction of rotation of the cup at a slight angle, so that the opening faces the oncoming stream of liquid. Positioning of the line in the groove is still more of an art than a science. The aim is to adjust it so that as the advancing liquid front reaches the cup wall and flows into the groove, it is at once removed smoothly through the outflow line. There should be no turbulence and little perceptible accumulation of liquid in the groove. If the line is badly positioned, turbulence is produced in the groove; if this is severe, an obvious wave pattern develops throughout the liquid in the cup, extraction becomes inefficient, and there is danger of disruption of the protein film. It is helpful if the line is held by a pivoting arm which can be reproducibly returned to the same position relative to the cup wall and clamped there. This obviates the necessity for manual realignment of the line every time it is moved, for example, whenever the cup is cleaned. If the line is inadvertently placed in contact with the floor of the groove, then its tip is abraded by friction from the spinning cup, and small Teflon shavings are generated. These may obstruct the lumen of the tube and may be mistaken for flakes of protein, which can at times also be swept up into the outflow line.

The exact angle of projection of the outflow line into the groove and the angle at which its tip should be cut varies somewhat with the diameter of the tube and the size of the groove, which may differ from one model of sequenator to another and must be found by trial and error. Hence no attempt will be made to describe or illustrate this in detail.

Outflow Valve. This two-way valve directs the stream of solvent to waste or to a fraction collector. The requirements for this valve are similar to those for delivery valves, i.e., Teflon composition, low dead space, rapid opening and closing, ability to hold a vacuum. The common passage and the passage through which solvent is directed to the waste container are kept reasonably clean since an excess of a polar organic solvent, ethyl acetate, flushes through at every cycle. However, this is not true of the passage leading to the fraction collector. This is flushed only with a small volume (2–6 ml) of 1-chlorobutane at each cycle, and deposits are prone to form obstructing the outflow and providing a nidus for contamination of samples. Low molecular weight peptide material dissolving in the chlorobutane and precipitating in the valve contribute heavily to such deposits. Hence, the valve and line to the fraction collector should be periodically flushed through with a solvent capable of dissolving such peptide deposits. A flush with ethanol/water, propanol/water, or 50% acetic acid, followed by one with ethyl acetate, is a suitable procedure. The valve may occasionally need to be opened and directly cleaned.

Waste Container. This should be able to be emptied easily without undue exposure of laboratory personnel to the organic vapor.

Fraction Collector. As has been pointed out, the materials used should be resistant to acid vapors. Since the thiazolinone amino acid derivatives are not stable, it is highly desirable that the collector be refrigerated, since samples may be left there for 48 hours or longer, prior to conversion and identification. It is also desirable that the samples be stored under a nitrogen atmosphere to minimize oxidative breakdown. No conclusive studies are yet available on the relative merits of refrigeration (alone), storage under nitrogen (alone), or a combination of both measures for sample preservation. Ideally both measures should be adopted. The use of reducing agents for improved sample preservation is discussed below. It is helpful if the nitrogen system used for providing an inert atmosphere in the fraction collector is also used to evaporate the samples to dryness. This saves time by carrying out automatically a step which otherwise must be done manually. Dried samples of thiazolinone also appear to be more stable than those kept in solution.

The fraction collector lid should be sealed adequately. A leak causes excessive loss of nitrogen and during vacuum stages causes air to be drawn through the chamber promoting oxidation of the samples.

D. Conversion, Identification, and Data Storage Systems

In the sequenators currently available the mechanization stops with the delivery of the cleaved thiazolinone derivatives into the fraction collector. Clearly the subsequent steps of conversion of the derivatives to the thiohydantoin form, and their identification could, and ideally should, be automated. There is no technical reason why this cannot be done. However, since many samples can be manually converted and identified each day by a single worker, the substantial additional cost necessary for a fully automated instrument has so far inhibited this development. An additional consideration is that no really satisfactory identification procedure has as yet been devised.

1. Conversion

Methods based on direct identification of phenylthiohydantoin derivatives require the conversion step. This is performed under conditions well established by Ilse and Edman.[14] The 1-chlorobutane is removed from the samples by evaporation, and 0.2–0.3 ml of 1 N hydrochloric acid is added to the dried thiazolinones. The sample tubes are flushed with a stream of nitrogen to remove dissolved oxygen from the hydrochloric acid, stoppered rapidly, and heated for 10 minutes at 80° in a temperature-controlled oil bath. A commercial frypan (e.g., Sunbeam) is relatively cheap and just as effective for this purpose as more elaborate pieces of equipment. At the end of 10 minutes, the samples are cooled and extracted twice with 1 ml of highly purified ethyl acetate (of purity comparable to that used in the sequenator itself). The aqueous and organic layers are thoroughly mixed using a "Vortex" type mixer and then separated by centrifugation. The organic layer is transferred to a separate tube, and evaporated to dryness. It contains the PTH derivatives of all the amino acids except for arginine, histidine, and cysteic acid. These are ionized at the pH used for conversion and are thus too polar to dissolve except to a limited extent in ethyl acetate. Hence, they remain in the aqueous phase.

2. Identification

The procedures to identify the cleaved amino acids differ from laboratory to laboratory. Here only the procedures used in our laboratory will be described in detail, with the reasons for their use (see Section VI, E). Other approaches, briefly reviewed here, can also be used. In the absence of a single really superior method, the choice of an identification system to some extent depends upon personal preference, prior experience, and availability of equipment.

a. Gas–Liquid Chromatography (GLC) of PTH Amino Acids

This is probably the best single identification procedure available. Most PTH amino acids can be identified quantitatively at the level of

---

[14] D. Ilse and P. Edman, *Aust. J. Chem.* **16**, 411 (1963).

a few nanomoles. The system used in our laboratory is described in detail below.

b. Thin-Layer Chromatography (TLC) of PTH Amino Acids

This approach is that adopted by Edman and Begg.[1] It is less sensitive than gas chromatography and is nonquantitative. However, it is rapid and simple, and is used in our laboratory to supplement gas chromatography, as described below.

c. Mass Spectrometry of PTH Amino Acids

Some effort is presently being expended in this area, particularly using chemical ionization and field ionization techniques. Lack of widespread availability of the rather expensive equipment required and lack of experience in the technique by biochemists generally has so far limited the evaluation and application of mass spectrometry for this purpose.

d. Back Hydrolysis of Thiazolinones or Thiohydantoins to the Free Amino Acids with Amino Acid Analysis

This approach is being used in a number of laboratories either as the sole procedure or in conjunction with gas chromatography or thin-layer chromatography.

Advantages

i. The equipment (an amino acid analyzer) is available in almost all protein chemistry laboratories.

ii. Many samples may be hydrolyzed at a time. If available, use of automatic sample loading equipment and automatic data reduction greatly increases the convenience of the method.

iii. A single column system can be used to separate all amino acids in 2 hours or less.

iv. The method can with some qualification be regarded as quantitative.

v. The method is less susceptible than GLC or TLC to interference by impurities or breakdown products derived from the reagents, since only ninhydrin-positive components are detected. GLC with the usual hydrogen flame detector will register almost any volatile organic molecule. Any UV-absorbing impurity can interfere with TLC as usually carried out. Hence, the stringent requirements for reagent purity can be to some extent relaxed when hydrolysis and amino acid analysis is to be used for identification, provided that any impurities present do not interfere with coupling or cleavage reactions.

Disadvantages

i. There is an obligatory delay of at least one day, more usually two or more days, before any results can be known. This means that it is impossible to monitor a sequenator run as it proceeds. If the run is proceeding badly, this may not be realized until 30 or 40 more degradation cycles have been completed. This useless prolongation of a degradation is wasteful of expensive chemicals and of time on an expensive instrument. With TLC or GLC, identifications can be completed within 1–2 hours of the completion of a particular degradation cycle.

ii. Although the actual measurement of the generated free amino acids is quantitative, the back hydrolysis step is not. Yields of the free amino acids are variable and some labile residues are totally destroyed (e.g., serine, tryptophan) when acid hydrolysis is used. Determination of the presence of amide groups on aspartyl or glutamyl residues is indirect, depending on the measurement of ammonia generated during the hydrolysis.

iii. When smaller peptides are being degraded, some of the peptide material is inevitably extracted from the cup with the cleaved thiazolinone amino acid. This peptide is therefore hydrolyzed together with the thiazolinone and generates free amino acids which can obscure the identification.

Smithies and his co-workers[15] have put considerable effort into developing a workable system based on hydrolysis by two different procedures (hydrogen iodide and NaOH/dithionite). They have succeeded in overcoming many of the difficulties in identification and quantitation and have presented impressive evidence as to the practicability of the procedure in their hands. Since they have recently published their methods,[15] no further comment need be made here.

e. CONCLUSIONS

i. There are only three approaches to identification which are practical at the present time. These are GLC, TLC, and back-hydrolysis with amino acid analysis. Other approaches (mass spectrometry, high-pressure liquid chromatography, back-hydrolysis with dansylation, infrared spectroscopy, and a number of others) either have obvious drawbacks or are still in an early stage of development.

ii. No one of these three identification methods is adequate alone. The deficiencies of GLC and TLC, respectively, are discussed below. The

---

[15] O. Smithies, D. Gibson, E. M. Fanning, R. M. Goodfliesch, J. G. Gilman, and D. L. Ballantyne, *Biochemistry* **10**, 4912 (1971).

difficulties with back-hydrolysis techniques have already been pointed out. Reliance on only one method can lead to serious risks of error.

iii. Any two of these three methods could be chosen as the basis of a workable system. At least one of the two quantitative procedures is mandatory for accurate determination of repetitive yields, and to facilitate identification at the later stages of a degradation when overlap and nonspecific cleavage may obscure identifications based solely on thin-layer chromatography. Quantitation is also essential when peptide or protein mixtures are being sequenced.

Which two methods are chosen is partly a matter of personal preference and availability of equipment. We prefer the combination of GLC and TLC, and only very occasionally use back-hydrolysis and amino acid analysis, usually on isolated samples (e.g., PTH cysteic acid, arginine, or histidine) where the direct identification procedures have for one reason or another given ambiguous results. In general we strongly recommend the use of procedures that depend on direct identification of the cleaved residue. The case for the direct as opposed to the indirect procedures (back-hydrolysis, dansylation) for identification has been very forcefully and convincingly stated by Edman.[6]

f. DETAILED PROCEDURES FOR GAS CHROMATOGRAPHY

The general approaches used are those devised by Horning and collaborators[16] several years ago for gas chromatography of steroids and amino acid derivatives using inert supports thinly coated with thermostable stationary phases. Workers in several laboratories have further refined the procedures for PTH amino acids,[17] and most of their findings are also applicable to other similar derivations, e.g., methyl (MTH) and pentafluorophenyl (PFPTH) thiohydantoins.[18,19] Since gas chromatography has been the most widely used single identification method used in conjunction with the protein sequenator, there is now a considerable cumulative experience in its use, and many minor variations in the technique have been introduced. The details of the method currently used in the author's laboratory are given here.

Instrument. A Beckman GC45 gas chromatograph[20] is used. It is

---

[16] E. C. Horning, W. J. A. Vanden Heuvel, and B. G. Creech, *in* "Methods of Biochemical Analysis" (D. Glick, ed.), Vol. XI. Wiley (Interscience), New York, 1963.

[17] J. J. Pisano and T. J. Bronzert, *J. Biol. Chem.* **244**, 5597 (1969).

[18] M. Waterfield and E. Haber, *Biochemistry* **9**, 832 (1970).

[19] R. M. Lequin and H. D. Niall, *Biochim. Biophys. Acta* **257**, 76 (1972).

[20] Gas chromatographs incorporating similar features are available from other commercial sources. See "Guide to Scientific Instruments" (*Science* **174A**, No. 4010A, November, 1971).

equipped with a single-column oven holding two U columns, a linear temperature programmer, twin on-column injection ports, and twin hydrogen flame ionization detectors. The columns are 2 mm i.d. × 4 foot glass; each is connected to its own detector, electrometer, and recorder. This allows simultaneous injections to be made on each column.

It is important that the injections be made "on column," the needle of the syringe entering the top of the glass column, and the injected sample being swept on by the stream of inert carrier gas. This is because any contact with metal in the injection area by the PTH amino acids leads to their decomposition. For the same reason metal columns must of course be avoided. We have observed that certain syringes used for injection appear to give inferior results with poor yields and decomposition. This is probably due to variability in the quality of stainless steel used for the needle; the PTH amino acids decompose during their brief exposure to the metal of the needle as it sits in the hot injection port. This problem can be avoided (a) by use of "Hamilton" syringes (10 μl capacity; a guide for the plunger is recommended to reduce the chance of bending it during injection) and (b) by "sandwiching" the sample to be injected between 0.5–1.0 μl aliquots of pure solvent (ethyl acetate). This means that the needle is filled only with solvent as it is inserted, and thus with a rapid injection the contact of the PTH amino acids in the sample with hot metal is reduced to a minimum.

Column Packing. PTH amino acids tend to be adsorbed to a variable degree to "bare" areas on the support. Hence it is important that the particles of support be completely inactivated by silanization prior to coating with the stationary phase, and that the stationary phase be applied so as to form a complete and uniform coating. The procedures used to achieve this in our laboratory are those devised by Horning et al.[16] as modified by Pisano and co-workers.[17,21]

Gas Chrom P or Chromosorb W (100–120 mesh; Supelco, Inc.) should be used as the starting material. Occasional batches are too friable and give poor results.

REMOVAL OF FINES. Suspend 30 g of support in distilled water in a large beaker and decant off the "fines" (any particles not settling within 5–10 minutes). Repeat the suspension and decanting twice more with water, twice with methanol and once again with water.

ACID TREATMENT. Suspend in concentrated HCl in a large beaker. Leave 12–18 hours with occasional stirring. Decant acid. Do three 1-hour acid washes, decanting supernatant. Wash exhaustively with glass-

---

[21] "Mixed phase" columns as advocated by Pisano and Bronzert[17] have in our hands proved too unstable for routine use with the sequenator.

distilled water (until pH is above 5) decanting each time. Then wash with acetone to remove water. Dry the support at room temperature for several hours, then in an oven at 80–110°C.

DEACTIVATION OF SUPPORT. Take the support from the oven and *at once* place in a solution (100 ml) of 5% dimethyldichlorosilane in toluene in a sidearm filter flask. The flask is stoppered and its contents subjected to reduced pressure by connecting the sidearm to a water aspirator for a few minutes. The flask is shaken to dislodge bubbles from the surface of the support, and the pressure is returned to atmospheric. Repeat this twice. Then filter the support on a Büchner funnel and wash it with toluene and then with methanol. Resuspend it in methanol (anhydrous) and leave for 2 hours. Decant, filter, and wash with acetone. Dry the packing in air for 2 hours (spread it out on a flat glass surface); then dry it in an oven at 80–110°.

COATING BY FILTRATION. For the 10% DC560 column, 10 g of the stationary phase is weighed into a measuring cylinder and the volume is made up to 100 ml with acetone. The solution is mixed thoroughly and immediately used to coat the support. For the OV-25 column, 2 g of the stationary phase is dissolved in chloroform, the volume again being made up to 100 ml. (The 1.5% AN600 column packing was not prepared in this way by us but has so far been obtained directly from Beckman Instruments, Fullerton, California.)

The solution of stationary phase is added to 10 g of the support in a sidearm filter flask attached to a water aspirator. The flask is stoppered, and the contents are maintained under reduced pressure for a few minutes, the flask being swirled to aid the removal of air bubbles. The vacuum should not be too powerful nor applied for long since the solvent tends to boil easily, and undue evaporation can concentrate the stationary phase. Leave at atmospheric pressure for 5 minutes. Place the slurry on a Büchner funnel on a sidearm filter flask and allow the solution to drain freely through the support. Apply vacuum to flask for 5 minutes. The coating should be damp but not wet. Spread the coated support on a smooth glass surface. Dry at room temperature for 2 hours, then in an oven at 80–100°. The packing should flow freely as a powder and is now ready for filling the column.

COLUMN. The columns themselves and the glass wool used to plug the outlet end must be carefully silanized. They are exposed for 5 minutes to a solution of 5% dichlorodimethylsilane in toluene, then washed with toluene and anhydrous methanol, and dried.

PACKING. The columns can now be filled with the packing. The procedure should be carried out using extreme care not to damage or crush the packing. Any damage will produce new "bare" surfaces not covered by the stationary phase, which will irreversibly adsorb the samples.

OMISSION OF INLET PLUG. We routinely do not put a glass wool plug in the inlet end of the column. Black deposits of carbonaceous material tend to accumulate on such a plug and on the top of the column with prolonged use. These deposits seem to adsorb PTH amino acids or catalyze their decomposition. It is more satisfactory to omit the plug. Deposits still form on the top of the column packing and on the glass near the injection site. These are readily removed by aspiration of the top 2 cm of column packing. The glass is then cleaned with a pipe cleaner dipped in acetone. Use of these procedures prolongs useful column life to several months. Care must be taken to release the pressure head of carrier gas from the column before removing it from the instrument. Otherwise, in the absence of a retaining glass wool plug the packing is blown out of the column.

CONDITIONING. The column is conditioned overnight at 275°.

COMMENTS. Further information can be found in references 16 and 17 cited above. The above-described procedure seems obsessive but is essential for reproducibly good results. Some commercially available column packings are usually satisfactory. Column packing coated with SP400 (Beckman Instruments, Fullerton, California) gives results comparable to those obtained with DC560. One reason why we prefer to make our own column packing is that the fragile supports are readily damaged by shaking or rough handling during shipping. Column packing once prepared is perfectly stable for several years, if kept in a sealed container.

Gas Chromatographic Procedure. The basic procedure we have used depends on injection of aliquots of each sample isothermally at three different temperatures, on 4-foot long, 2-mm internal diameter glass columns filled with Chromosorb W coated with 10% DC560. The carrier gas (nitrogen) is adjusted to a flow rate of 50 ml/minute on each column, and the temperature is raised to 180–200°. The exact temperature is chosen so that the first group (Group I) of PTH amino acids elute in 10–15 minutes. The order of elution is alanine, serine/SCMC (eluting together), glycine, valine, dehydrothreonine/proline (eluting together), isoleucine and leucine (eluting together), norleucine internal standard (see Scheme 1). If it is found that standards injected simultaneously on both columns elute at different rates, it is convenient to modify the carrier gas flow rates on either or both columns within the range 30–50 ml/minute to "balance" them. If this cannot be done, then the slower column can be made to give more rapid elutions by removal of a few centimeters of column packing from the inlet side. This can usually be done without significant loss of resolution.

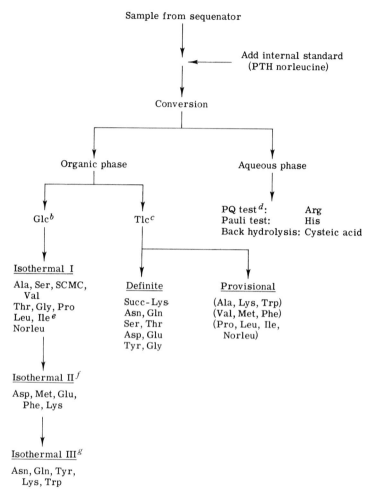

SCHEME 1. Identification for PTH Amino Acids (a) see text for further details, (b) gas-liquid chromatography injections were carried out on DC560 or SP400 columns; (c) silica gel plates; ethylene dichloride:glacial acetic acid, 30:7 v/v see text; (d) PQ:phenanthrenequinone; (e) leucine and isoleucine were distinguished on AN600 column; (f) this injection is carried out after the sample has been mixed with bis(trimethylsilyl)acetamide; (g) these derivatives may be identified at greater sensitivity on an OV25 column.

The second isothermal injection is made at 210–230°. PTH aspartic acid (trimethylsilyl derivative), methionine, glutamic acid (trimethylsilyl derivative), and phenylalanine (Group II) elute in that order in 10–15 minutes. PTH lysine (trimethylsilyl derivative) elutes just after PTH phenylalanine.

The third isothermal injection is made at 230–250°. PTH asparagine, glutamine, lysine/tyrosine (eluting together), and tryptophan (Group III) elute in that order. These derivatives elute within 10 minutes except for PTH tryptophan, which at these temperatures takes about 30 minutes to appear.

This basic procedure is supplemented by further injections on OV-25 and AN-600 columns, by TLC and by other maneuvers as required to confirm the identification. Since the details are described below (see Section VI, E), only a few more general comments are made here.

In the interest of efficiency as many identifications as possible are made on the DC560 column. We have found that with careful attention to preparation of the column packing, even polar and less volatile Group III derivations can be identified at the nanomolar level on this column. Where smaller amounts of material are available, the OV-25 column is recommended.

Trimethylsilylation of PTH amino acids for GC analysis is employed as sparingly as possible. It is used routinely only for PTH aspartic acid and glutamic acid and occasionally for confirmation of other residues.

Isothermal injections are preferred to temperature programming. We find that the reproducibility of elution times with temperature programming leaves something to be desired, and much time is wasted on recooling and reequilibration of the columns. Recovery of asparagine and glutamine PTH derivatives is reduced when their transit time through the column is prolonged. Hence rapid isothermal elution of these derivatives at high temperature gives better yields than when a temperature program is used. The total time taken per sample for the 3 isothermal injections is less than 45 minutes, and the quantity of material needed is less than 5 nmoles. These values compare favorably with the corresponding figures when a temperature program is used.

g. DETAILED PROCEDURES FOR THIN-LAYER CHROMATOGRAPHY

TLC is carried out using precoated silica gel plates (glass, 20 × 20 cm) obtained from Analtech (Newark, Delaware). These plates, which incorporate a fluorescent indicator, are superior to those from other sources tested and give results identical to those reported by Edman.[6] Only one solvent system is used. The mobile phase consists of ethylene dichloride:glacial acetic acid (30:7, v/v), as devised by Edman and Kluh.[6] R_f values for PTH standards are illustrated in Fig. 3. Definite identifications are possible using this system alone for most PTH amino acids with R_f values from zero to 0.5. Asparagine is distinguished from glutamine because some deamidation of these derivatives invariably occurs during conversion. Hence a minor spot corresponding to aspartic

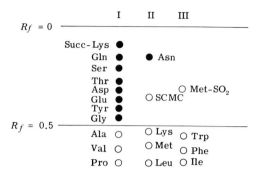

Fig. 3. Identification of PTH amino acids by thin-layer chromatography (TLC) using system H of Edman and Kluh [for details, see text and see P. Edman, *in* "Protein Sequence Determination" (S. B. Needleman, ed.), p. 211. Springer-Verlag, Berlin and New York, 1970]. Derivatives shown as closed circles ($R_f < 0.5$) can usually be definitely identified by TLC. If derivatives shown as open circles are present, gas chromatography is also required.

or glutamic acid is always seen. In performic acid-oxidized proteins or those alkylated with iodoacetic acid, the PTH derivatives of methionine sulfone and S-carboxymethylcysteine may be present. They are distinguished from PTH aspartic and glutamic acids by GLC. The PTH derivative of ε-maleyl lysine coelutes with that of ε-succinyl lysine. Both are stable under the conditions of Edman degradation.

For derivatives with R_f values between 0.5 and 1, only provisional or "group" identifications may be made. Spots seen in this region may be assigned to one of the three groups: Ala, Lys, Trp; Val, Met, Phe; and Pro, Leu, Ile. Identifications within each of these groups are made by GLC.

This TLC system is used routinely in conjunction with GLC as described below. Less than 1 μg of derivative can be readily identified (about 4 nmoles).

3. Data Storage System

The GC tracings are stored in their original form and can at any point be reexamined. Since standards are injected routinely, quantitative yields can be calculated. Variable destruction of several PTH amino acids takes place during cleavage, conversion, and/or identification procedures. Hence quantitative yields vary. However, the progress of the degradation can be followed by measuring the yield of more stable derivatives (e.g., alanine, valine, leucine) which occur commonly in most proteins.

The TLC plates are viewed under ultraviolet light (254 nm). The derivatives appear as dark spots on a bright green background. Reasonably adequate photographs can be taken of these plates, using a Polaroid camera and film, for a permanent record.

The advent of automated sequence procedures has greatly expanded the volume of sequence data submitted for publication in the biochemical literature. Particularly when mainly qualitative or indirect methods have been used for identification, this situation has posed editorial problems in deciding what constitutes "reasonable" proof of structure. The very rapidity of the procedure introduces new possibilities of error in mislabeling or mislocating samples, since so many are handled at one time. The efficiency of the procedure means that a sequence can be proposed on the basis of only a limited number of observations on any one residue. Although the automated Edman degradation is obviously a major advance and overall greatly increases the accuracy as well as the speed of sequence analysis, the technique must be used thoughtfully and the results be interpreted cautiously.

VI. Procedures

A. Preparation and Maintenance of Instrument

The procedures necessary for the general maintenance of a sequenator and the specific preparations necessary before starting a run will naturally vary from one instrument to another. What is outlined here can be taken as applicable mainly to the Beckman instrument. It is important to distinguish between procedures that can and should be carried out by the person in charge of the instrument and those that should be left to a trained serviceman. In the second category should be placed (a) any repairs or adjustments to the programmer unit; this requires training in electronics and a proper understanding of circuitry not often possessed by the average worker; (b) adjustments to the delivery valves; these should not normally be opened, since the Teflon seat (in Beckman valves) may not be returned to its original site and a leak may be introduced; (c) adjustments to the cup mounting; (d) cleaning or replacement of bearings in the drive assembly.

Procedures which should be routinely carried out by the worker himself include (a) regular changing of oil in the vacuum pumps; this should be done routinely not less often than every 2 weeks of machine operation; (b) cleaning of the cup and reaction chamber area; this should be done before each run. Even with the use of Quadrol, which is allegedly completely nonvolatile, sticky deposits form in the whole reaction cell area after a few days' operation. (These deposits are markedly reduced in extent if a nitrogen bleed is introduced into the reaction chamber while the vacuum pump is operating. The flow through of pure nitrogen seems to make the removal of deposits by the vacuum pump much more efficient.) Keeping the reaction cell area clean is particularly important for successful automated degradation of shorter peptides using volatile reagents.

It should be remembered that any space or any portion of the instrument in contact with the reaction cell area can and will become contaminated to some extent with oily or sticky deposits derived from the reagents. For this reason *cul de sacs* should be avoided in the design of this area. All parts of the instrument in communication with the reaction cell vapor space should also be heated. Otherwise much local condensation will occur. Solenoid valves and vacuum gauges, as well as any other moving part in communication with the reaction cell area, may require regular cleaning to avoid malfunction.

Two other points should be made. First, it is important that, even if several different workers or groups are using a sequenator, a single person be designated to be in charge of it. He should be responsible for its maintenance and should perform all sample and reagent additions to the instrument. Ideally he should also be responsible for identification of the PTH samples obtained in the degradations. Instruments with many masters have very poor performance records. Second, a program of preventive maintenance is essential. Do not wait for trouble to develop. Prevent it by regular cleaning and servicing of the instrument. Otherwise the results will be bad and the machine may suffer permanent damage.

B. Preparation of Chemicals

Particular attention has been directed by Edman and Begg to the need for highly purified reagents and solvents for use in the sequenator. Only a few comments need be added here.

Source. Despite the commercial availability of chemicals in allegedly highly purified form (e.g., "spectral" grade benzene, ethyl acetate), such chemicals are not necessarily of adequate purity for carrying out prolonged automated degradations.

A particular batch of such chemicals may of course be found to be perfectly good for this purpose. However, considerable variability may be encountered from batch to batch, and one simply cannot rely upon this grade of reagent. Chemicals used in the sequenator should, therefore, without exception be specifically purified for that purpose, either by a reliable commercial chemical supplier or by the individual worker in his own laboratory. Reagents specifically purified for the sequenator are offered by several firms (Beckman Instruments, Palo Alto, California; Pierce Chemical Company, Rockford, Illinois; Eastman Organic Chemicals, Rochester, New York; Merck Chemical Company, Germany). We have used chemicals supplied by Beckman Instruments with generally very good results. We have not attempted to evaluate other sources. However, Smithies and co-workers[15] have carried out prolonged and successful degradations using "Sequenal" grade chemicals obtained from the Pierce Chemical Company.

Since the available sources, the quality, and the cost of sequenator reagents are all subject to change, the current experience of workers in the field, ascertained by word of mouth, is probably the best guide.

The purification procedures are difficult, particularly when carried out on a large scale. Necessary quality control measures are also difficult and time consuming. Despite the best efforts of the supplier, occasional bad batches of chemicals are encountered. Even if the original purification is adequate, the chemicals may become contaminated very easily in the laboratory once the bottles are opened. Hence, before any degradation is carried out on an important sample, a trial run on a model protein such as sperm whale myoglobin is recommended strongly, as described below.

The cost of commercially available reagents is quite high owing to the complexity of the purification and quality control procedures. At present a year's supply of reagents can cost between $5,000 and $10,000, depending on the source used and the amount of time the instrument is in operation. A reasonable alternative to consider is to purify the reagents oneself. In the United States this is a doubtful economic proposition, since it involves the use of one technician on a full time basis with a substantial commitment of space, supplies, and glassware. In parts of the world where technical assistance is less expensive, it may be worthwhile. If the reagents are to be purified, it is recommended that the purification procedures described by Edman and Begg[6] be scrupulously followed. Shortcuts cause more trouble than they are worth.

Several additional chemicals are used for automated degradation of peptides. Quadrol (for making up dilute coupling buffer), n-propanol, dimethylallylamine, and pyridine are obtained from the Pierce Chemical company (Rockford, Illinois) or purified according to previously described procedures.[22]

---

[22] H. D. Niall and J. T. Potts, Jr., in "Peptides, Chemistry and Biochemistry" (S. Lande and B. Weinstein, eds.), p. 125. Dekker, New York, 1970.

Storage and Handling. Reagents (phenylisothiocyanate, Quadrol coupling buffer, heptafluorobutyric acid) should be stored at 4°. Solvents (benzene, ethyl acetate, heptane) may be stored at room temperature in the absence of light. Under these conditions reagents, if pure initially, are stable for at least 6 months. Bottles and vials should be opened only immediately prior to use and should preferably be used in entirety. In some areas, atmospheric pollution is a serious problem, and reagents once opened to the laboratory air quickly deteriorate owing to the accumulation of oxidants. However, elaborate systems for transferring chemicals from their original sealed containers into the sequenator reservoirs while continuously under nitrogen are unnecessary. The reagents can simply be poured in. If the atmosphere is so bad that momentary exposure to it makes a difference, it is doubtful if a successful degradation is possible. Even in the instrument some atmospheric exposure takes place, as discussed above with particular reference to oxygen. The need for avoiding any contact of the chemicals with stopcock grease must be once again underlined.

Use of Reducing Agents. Addition of reducing agents to chemicals used in the sequenator has been advocated[15,23] to eliminate oxidative side reactions. In theory, this sounds like an excellent idea, and it has certainly proved to be of some practical value. Our own views and experience can be summarized as follows: (1) Addition of reductants with the aim of converting "bad" (i.e., unpurified or incompletely purified) reagents into "good" reagents cannot be recommended. Both the original impurities and the added reductants can interfere with identification procedures. Some impurities are still capable of causing desulfuration of the PTC protein even in the presence of excess reducing agents. (2) Given purified reagents, a case can still be made for the limited use of reductants, since they appear to have a specific stabilizing effect on PTH serine, which is very susceptible to a side chain β-elimination reaction with subsequent polymerization of the dehydroserine form. Our evidence suggests that reductants stabilize the derivative at the dehydroserine stage, preventing or at least slowing further decomposition. The derivative can be readily identified as the dehydroform by GLC. (3) Volatile reductants are preferable. Nonvolatile ones (e.g., dithiothreitol) may precipitate in the instrument, obstructing delivery lines and valves. Our current practice is to add 1,4-butanedithiol (50 μl/liter) to the 1-chlorobutane solution. No interference with the identification system has been noted. Addition of

---

[23] M. A. Hermodson, L. H. Ericsson, and K. A. Walsh, *Fed. Proc. Fed. Amer. Soc. Exp. Biol.* **29**, 728 (1970) (Abstract).

ethanethiol to the 1 N hydrochloric acid used for conversion has also been advocated.[24]

C. Preparation of Sample

The protein sample (0.1–0.5 μmole) is dissolved or suspended in a small (0.5 ml) volume of a suitable solvent. The requirements for choice of solvent are that it be reasonably volatile and available in a purified form free of contaminants that might interfere with the degradation. It is not necessary that the protein be completely in solution, since it may be transferred into the sequenator cup as a suspension. It can even be added to the spinning cup in solid form. However, this is not recommended since it is hard to achieve an evenly dispersed uniform protein layer. If necessary, the sample can be dissolved in a larger volume and added in several portions over a period of time, with application of vacuum between each addition to dry off the liquid. The protein film should be restricted to the lower one-third to one-half of the cup. Solvents we have found useful for the transfer include water, 20% acetic acid, water containing 1% (w/v) triethylamine, and various mixtures of pyridine and water or of ethanol and water. Very insoluble proteins can often be dissolved in anhydrous heptafluorobutyric acid or trifluoroacetic acid. This may denature globular proteins to some extent, thus facilitating their solution in the coupling buffer when the degradation begins. Successful degradations may be carried out even on very insoluble proteins, as discussed below.

Protein samples should be free of salts. Volatile ammonium salts (e.g., bicarbonate, acetate) should be removed by repeated lyophilization, since residual ammonia can react with PITC to form phenylthiourea. Guanidine, urea, and sodium dodecyl sulfate should also be removed as thoroughly as possible by appropriate procedures prior to automated sequence analysis. Although not directly interfering with the chemistry of the degradation, their presence in more than trace quantities interferes with the formation of a proper film of protein. Heavy metals interfere with the degradation and must be excluded. (Some gradient development chambers used for column chromatography contain metal parts, and contamination of proteins by metallic ions can therefore result when acidic solutions are used for elution.)

Calibration Run. It is a good routine procedure to carry out a calibration run, using a protein of known sequence, prior to a run on an important sample or whenever the instrument has been idle for more than a

[24] M. A. Hermodson, L. H. Ericsson, and K. A. Walsh, personal communication.

few days. Sperm whale apomyoglobin is a suitable substrate, although there is now growing concern about using material obtained from a species threatened with extinction. Oxidized ribonuclease or oxidized hen egg lysozyme are possible alternatives. A run of 10–15 cycles is usually sufficient. A repetitive yield of 92% or better is desirable for a moderate length run to be feasible (20 cycles). Yields of 94–96% will allow degradations of 50–60 cycles on suitable proteins. The calibration run also provides the opportunity for close observation of reagent deliveries, function of valves and vacuum pumps, and other parameters of instrument function. Any aberrations can readily be corrected before the crucial degradation on an unknown sample is initiated.

D. Operation of Instrument

An outline of the main events to be observed during a degradation cycle has been given by Edman and Begg,[1] together with details of a program suitable for use in their instrument. No program details will be given here, since they vary from one instrument to another.[25] However, a few general comments will be added to the earlier description. It is advisable that the operator start the run early enough in the day to observe at least the first cycle or two cycles in entirety and preferably convert and identify at least the first residue. This ensures that he is in a position to detect any mechanical problems, any omissions or errors in the setting of the instrument or disposition of reagents, or any insolubility or other observable problems with the protein. These problems can often be corrected and a run salvaged which would otherwise have been lost. A stroboscope is useful for observing events within the cup.[1]

Addition of Sample. The sample is dissolved or suspended as indicated above and transferred into the cup. A vacuum is applied to dry the protein. Before the degradation is started, adequate time must be left to ensure that temperature equilibrium has been achieved, since cooling occurs during the sample addition and evaporation steps. If the sample has been added in acidic solution, it is advisable to wash the protein film with ethyl acetate to remove excess acid, which might otherwise cause a fall in pH in the coupling buffer.

Addition of Coupling Reagents. The solution of PITC (5% or 2.5%, v/v) in heptane is added first. The exact level reached in the cup is not critical, but the reagent should at least cover the protein film, which it does not wet or penetrate. Most of the heptane is evaporated by a combination of nitrogen drying and application of vacuum. It does not appear

---

[25] Programs suitable for both protein and peptide degradations in the Beckman instrument may be obtained from Beckman Instruments, 1117 California Ave., Palo Alto, California 94304.

to make much difference whether or not the heptane is completely evaporated. If it is, then the PITC is deposited as multiple droplets over the surface of the protein and part of the surface of the cup above it. The Quadrol coupling buffer is then pumped in. Usually the white protein film becomes transparent as the advancing front of the buffer rises in the cup. If the heptane has been completely removed, the droplets of PITC dissolve in the coupling buffer. If there is still a continuous layer of heptane, the Quadrol solution, being denser, will be disposed against the glass wall while the lighter heptane layer floats on top of it. The heptane continues to evaporate to some extent and mixes with the buffer sufficiently to transfer the dissolved PITC into the buffer layer. Solution of the PITC (and the protein) in the coupling buffer is aided by the elevated temperature. Though one might imagine that the liquid layer, being spun smoothly, would not become mixed, in fact there are at least two sources of turbulence. The delivery line or lines projecting into the cup somewhat off center set up currents in the nitrogen atmosphere as the cup spins. These currents are transmitted to the surface of the liquid. In addition there is some vibrational energy transmitted to the liquid layer from the motor, particularly when the direct-drive system is used.

Coupling. During coupling there is a fall of a few millimeters in the liquid level due to partial evaporation of the volatile constituents of the coupling buffer (propanol and water). In the Beckman instrument the reagent additions are often carried out at high cup speed (1800 rpm), and the speed dropped back to 1200 rpm after a few minutes coupling. This causes a fall in the liquid level, intended to keep the protein lower in the cup and reduce the risk of incomplete reaction. However, when the speed of rotation of the cup is suddenly reduced, the Quadrol layer does not fall cleanly, but a thin film clings to the glass. Hence the speed change maneuver, though helpful, by no means obviates the need for exact adjustment of delivery levels.

The usual duration of coupling is 30 minutes when carried out at a temperature of 50°. However the reaction is probably completed within 5–10 minutes provided the protein dissolves at once. The extra coupling time allows for some delay in dissolving the protein film.

Drying. After coupling, a vacuum is applied to remove most of the volatile components of the coupling buffer. As stated already, it is unnecessary and disadvantageous to attempt complete drying. A problem can arise at this point with the Beckman instrument. Some Quadrol buffer is left in the delivery line between the valve and the cup. When a restricted vacuum is applied at the end of coupling, this buffer may dry in the line leaving its nonvolatile component (Quadrol) as a sticky plug which may partly or completely obstruct the lumen. This prevents the ethyl acetate (which shares the same delivery line with the Quadrol) from entering the cup. Extraction is incomplete and the degradation comes to an abrupt halt. This problem can be overcome by carrying out a very brief (4-second) delivery of ethyl acetate at the end of coupling before applying the vacuum. This dislodges the Quadrol from the line into the cup and prevents plugging.

Solvent Extraction. Removal of nonvolatile reagent (mainly Quadrol) and side products is effected by extraction first with benzene and then with ethyl acetate, as described by Edman and Begg.[1] There is no need for a delay between the two extractions; the ethyl acetate delivery can be started without drying off the benzene left in the cup. At the end of the ethyl acetate extraction the protein film should be white and appear dry. A glassy appearance suggests that Quadrol removal has been incomplete.

Cleavage. After removal of residual ethyl acetate by vacuum, hepta-fluorobutyric acid is added for the cleavage reaction. The delivery time should be carefully adjusted so that the acid rises to a height just less (by 3 mm) than the upper level reached by the coupling buffer. The acid layer will then "creep" upward by capillarity to cover the upper extremity of the protein film. It is quite difficult at times to see the upper limit of the acid layer even with a stroboscope, since acid vapors precede the liquid and may appear to dissolve protein higher up on the wall than the liquid itself ever reaches. The best opportunity to observe the exact height of the acid is as it is actually being delivered. After this, the protein is in solution and the entire cup interior has a glassy appearance so that the upper level of the acid layer is obscured. The usual duration of the cleavage is 3 minutes. The cleavage reaction itself is probably almost instantaneous even at room temperature. However, the additional time is necessary to make sure the protein has dissolved completely.

Extraction of Thiazolinone. At the end of the cleavage a vacuum is applied briefly to dry off most of the acid. It is undesirable to dry the acid completely, since the protein then tends to form a very compact glassy film which is poorly penetrated by the 1-chlorobutane. A prolonged drying time has the additional disadvantage that it increases the breakdown of the serine derivative, since the anhydrous acidic conditions favor the β-elimination reaction. A vacuum time of 1–2 minutes is usually adequate. Smithies and co-workers[15] have pointed out that the efficiency of extraction of PTH arginine varies greatly with the amount of residual acid and the cup temperature at which the extraction takes place. The extraction is carried out with 3–5 ml of 1-chlorobutane to which the reducing agent 1,4-butanedithiol has been added to improve serine recoveries.

The cleavage is usually repeated to achieve complete reaction as described by Edman and Begg.[1] This repetition is unnecessary if a shorter degradation (20 cycles or less) is to be carried out, since the degree of cumulative overlap does not become a significant problem.

At the end of the cycle of degradation the protein should again appear white and dry. The reagent additions at the beginning of the next cycle should be observed to make sure the volumes delivered are reproducible.

E. Identification of Samples

1. General GLC and TLC Procedures

Samples from a sequenator run are handled in the following way.

a. Before starting the run, aliquots of PTH norleucine are added to each tube in the sequenator fraction collector as an internal standard. The amount to be added is calculated on the basis of the amount of protein to be degraded, so that both internal standard and sample peaks remain on scale on the GC tracings. It is best to add the internal standard in about one-third the molar amount of end group yield expected from the protein sample at the first cycle; e.g., if 300 nmoles of sample are being degraded, 100 nmoles of PTH norleucine should be put into each tube. Then the sample and standard peaks will both remain on scale through at least a 10-fold decrease in yield of PTH amino acids in the course of a long degradation. Beyond that point increasing nonspecific cleavage and overlap usually halt the degradation anyway. Although there are some minor inconveniences in the use of PTH norleucine as an internal standard, it is most useful in improving the accuracy of quantitation of the degradation, particularly since it allows correction for handling losses during the conversion and extraction procedures.

b. As samples accumulate in the fraction collector, they are subjected to the conversion and ethyl acetate extraction procedures. Usually it is convenient to handle up to 10 samples simultaneously. Extreme care must be taken to avoid mixing or mislabeling samples. This is so important that it is discussed in Section VI, F (Logistics).

c. After evaporation of the ethyl acetate the samples are redissolved in a small volume (20–100 μl) of ethyl acetate. Glass-stoppered centrifuge tubes (Kimax, 3-ml capacity) are suitable. The tapered end of the tube aids withdrawal of aliquots from the small volumes of ethyl acetate used. A suitable concentration range to aim for in the final sample is 0.5–2.0 nmoles per microliter of ethyl acetate. The conversion, extraction, evaporation, and redissolving of 10 samples takes about an hour.

d. While the conversion and solvent extraction procedures are being carried out, the gas chromatograph oven is heated up to the temperature for the first isothermal run (usually 180–190°) and the nitrogen flows are adjusted as described above. Two standard mixtures of PTH amino acids are injected on each column. One mixture contains 2 nmoles each of PTH alanine, glycine, valine, proline, and leucine. The second mixture contains 2 nmoles each of PTH serine and threonine. These injections of standards should be repeated at the end of the series of sample injections to detect any minor variations in GC or column behavior. The concentration of the standard solution is 1 nmole per microliter. The injection volume is therefore 2 μliters. The "sandwich" technique described above is used for all injections, 0.5 μl of solvent on either side of the standard or sample. These volumes can be accurately measured in the 10-μl Hamilton syringe. Total injection volumes greater than 3 μl tend to give too broad a solvent front on the GC.

e. Samples are then serially injected. Residues 1–5 are injected on one GC column while residues 6–10 are simultaneously injected on the other column. Each injection takes about 15 minutes until the last peak of Group I (the PTH norleucine internal standard) elutes. If for some reason the yields from the degradation are much higher or lower than expected, it is better to evaporate the small volumes of ethyl acetate in the tubes and redissolve them in a more appropriate volume so that the same volume range (1–3 μl) can be used for injection. It is convenient to carry a pocket timer (Endura Memo Minder, Laboratory Supplies Company, Inc., Hicksville, New York). If this is set with each injection, the worker can carry out the thin-layer chromatography while the GC samples are running. This group of 7–8 injections per column, including standards, takes less than 2 hours to complete. At the end of this time the thin-layer plate will also be ready for inspection.

At the end of 3 hours a considerable amount of information about the run is available. Usually definite identifications have been achieved for alanine, glycine, valine, proline, leucine/isoleucine, serine/SCMC, and threonine by gas chromatography, and additionally for asparagine, glutamine, aspartic acid, glutamic acid, tyrosine, lysine, and methionine sulfone by thin-layer chromatography. Comparison of the GC and TLC results may also allow provisional identification of tryptophan and of methionine/phenylalanine as described in detail below. If no residue is seen with either the initial isothermal GC injections or by TLC, then arginine, histidine, or cysteic acid must be searched for in the aqueous phase.

Since most of the residues identified by the first isothermal GC analysis are those occurring very frequently in proteins, it is likely that an accurate repetitive yield can be at once calculated. This allows a decision to be made about continuing the run or not. When shorter peptides are being degraded, the information on rate of fall in yield may indicate that program alterations are advisable.

f. Further identification and quantitation is carried out as follows. The temperature of the GC column oven is raised to that for the second isothermal injection (usually 220°). A standard mixture of 2 nmoles each of PTH aspartic acid, methionine, glutamic acid, and phenylalanine is drawn up into the Hamilton syringe together with 2 μl of bis(trimethylsilyl)acetamide (Pierce Chemical Company), and the mixture is at once injected into the gas chromatograph. Silylation of all four derivatives takes place in the heated injection area of the GC. The standards elute in the order given in 10–15 minutes. The 10 samples are then injected, again half on each column. This phase of the identification takes 1–2 hours.

g. The temperature of the columns is then raised to that established for the third isothermal injection (usually 240°) and a standard mixture consisting of 2 nmoles each of PTH asparagine, glutamine, and tyrosine is injected. They elute in that order in 10 minutes. Again the samples are injected as before. On this column at this temperature PTH lysine coelutes with PTH tyrosine, and PTH tryptophan elutes very late, at 30 minutes. Usually unless the presence of tryptophan is suspected from the thin-layer chromatography it is unnecessary to wait until the elution position of this residue has been passed before injecting the next sample. This set of injections therefore usually only takes 1–2 hours. If the conversion and identification procedures are started at 9 AM after an overnight run, by 4 PM the 3 sets of isothermal injections on the DC500 column above and the thin-layer chromatography have been completed. On occasion, special difficulties or uncertainties may appear with particular residues; these are discussed below, with the means of overcoming them. However, as a general rule at this stage, positive identification and quantitation will have been completed for 16 amino acids: i.e., alanine, serine, S-carboxymethylcysteine, glycine, valine, proline, threonine, aspartic acid, glutamic acid, methionine, phenylalanine, asparagine, glutamine, tyrosine, lysine, and tryptophan. Leucine and isoleucine coelute on DC500 and require special procedures for distinguishing one from the other. Arginine, histidine, and cysteic acid also require the use of special techniques since they are too polar for ready identification by TLC or GLC.

2. Leucine and Isoleucine Identification by GLC

Distinction between these residues is achieved by isothermal injection of a 2-nmole aliquot on a 1.5% AN600 column at 140° with a carrier gas flow rate of 20 ml/minute. Isoleucine and leucine elute, in that order, in 20 minutes. The peaks are only partly resolved, forming a double-humped pattern. However, both the position of takeoff of the isoleucine peak from the baseline and its summit clearly precede the takeoff point and summit of the leucine peak. The *absolute amount* of sample injected is quite critical. Using a standard mixture of PTH isoleucine and leucine, the amount giving the best resolution is first established. As stated, this is about $2 \pm .05$ nmoles for a 4 foot, 2 mm-i.d. column, with a 1.5% coating of AN600. Much less than this amount is difficult to detect. Increasing the amount of sample causes the peaks to become asymmetric; the takeoff point from the baseline is unaltered, but the summit is delayed. Serious error can result if this is not realized. For example, if too great a load of a sample which happens to be isoleucine is injected, the summit of the peak may be delayed sufficiently that it may coincide with the position of the leucine summit. An incorrect identification would then be made. If there is any doubt it is best to make two further injections of the unknown (2 nmoles) mixed with 2 nmoles of each standard separately. Appearance of a double-humped peak when the unknown is mixed with the other (i.e., the nonidentical) standard is the most reliable basis for identification. It can be seen that this whole procedure is by no means optimal. However, we believe it to be reliable provided the above precautions are taken. It is convenient to keep all the leucine/isoleucine samples (which are quite stable at $-20°$ in a freezer) until the end of a run, and identify them all in a single session.

3. Other Methods for Leucine-Isoleucine Identification

Thin-Layer Chromatography. Partial resolution of PTH leucine from isoleucine is achieved using system D as described by Edman and Begg.[1] The distinction can be quite difficult to make, however, particularly late in a run where there is a significant background of the derivatives arising from nonspecific cleavage. Again the device of mixing the unknown with each standard in turn can be helpful.

Back-Hydrolysis and Amino Acid Analysis. This is a reliable method since the free amino acids are completely resolved on the analyzer column.

Silylation and Injection on DC560. Here the main silylated peaks formed do not resolve. However, PTH leucine gives a single peak on silylation and PTH isoleucine a double peak, the second component probably arising from the formation of some allo-isoleucine.

Presence of a Post-Leucine Peak on DC560. On direct injection of the PTH derivatives (i.e., without silylation) on DC560, with leucine samples a small peak eluting just after the main leucine peak is frequently seen, the "postleucine peak."

Mass spectrometry shows it to be an oxidation product of PTH leucine, with a parent ion 2 mass units less than that of leucine, presumably owing to the loss of 2 hydrogens.[26] The product may be generated from PTH leucine standards by mild treatment with hydrogen peroxide. It is more prominent in sequenator runs when "bad," i.e., contaminated, chemicals are used. No such degradation product is seen with PTH isoleucine. This peak, however, coelutes with the PTH norleucine standard and hence cannot normally be observed.

Of these four additional procedures, only back-hydrolysis and amino acid analysis (see above) can be recommended. The others, while providing useful indications as to which residue is present, are not felt to be sufficiently reliable to use as primary evidence for identification.

4. Arginine, Histidine, and Cysteic Acid

If none of the 18 PTH amino acids detectable by the above scheme is found on examination of the organic phase after conversion and ethyl acetate extraction, then arginine and histidine derivatives must be sought in the aqueous phase. If the protein sample has been subjected to performic acid oxidation, then PTH cysteic acid must also be looked for. The first step is to evaporate the aqueous phases to dryness, either in a stream of nitrogen or by lyophilization. The latter is more convenient (and incidentally cheaper) as it can be proceeding while the organic phase identifications are being made without monopolizing the nitrogen blowdown equipment. When the samples are dry, they are redissolved in a small volume (20–50 μl) of absolute ethanol. An aliquot is taken and further diluted in absolute ethanol for UV spectra determination. Measurements are made from 220 to 340 nm, and the extinctions are plotted. If one of these three PTH derivatives is present in the aqueous phase, then the typical thiohydantoin spectrum with a maximum at 270 nm and a minimum at 245 nm should be observable. The absolute amount present can be estimated using the known extinction coefficients of these derivatives. Occasionally UV-absorbing impurities present in the samples may obscure the spectrum. As controls, spectra should be read on the sample from the degradation step preceding and that following the residue in question. A small amount of thiohydantoin is always detected due to incomplete extraction of the aqueous phase by the ethyl acetate. However, a clear elevation in the amount present by spectral analysis is expected when one of these three PTH amino acids is present. The absolute yields are usually lower than one would expect from the amount of protein being degraded. One explanation is that the efficiency of extraction of the polar derivatives from the sequenator cup following the cleavage

---

[26] D. Raulais and H. D. Niall, unpublished work.

step is variable. Also the ethyl acetate extraction of the aqueous phase after conversion removes some of these derivatives, PTH arginine in particular. The presence of Quadrol salts in the aqueous phase seems to promote this extraction.

These three derivatives are not readily identified by gas chromatography or thin-layer chromatography. At present, the best two approaches, neither of which is very good, involve the use of (1) specific spray tests (for arginine and histidine) and (2) back-hydrolysis and amino acid analysis (for all three).

a. Identification of PTH Arginine[27]

Reagents

Solution A: 0.02% (w/v) phenanthrenequinone in absolute ethanol. The phenanthrenequinone is conveniently stored as multiple pre-weighed 10-mg aliquots in small Erlenmeyer flasks. One of these is dissolved in ethanol (50 ml) immediately prior to use.

Solution B: 10% (w/v) NaOH in 60% ethanol. The sodium hydroxide pellets are conveniently stored as multiple 2.5-g aliquots in small Erlenmeyer flasks sealed with Parafilm to exclude atmospheric moisture. One of these is dissolved first in 10 ml distilled water. When the solution is clear, 15 ml of absolute ethanol is added.

Procedure. The aqueous phase samples from the conversion mixture (see above), which have been dried, are redissolved in 30–100 μl of water or ethanol/water. Aliquots (10 μl) are applied to a strip of Whatman No. 1 filter paper. A PTH arginine standard (5 μl of a 0.5 mg/ml solution in water) is also spotted. It is essential that the aqueous phase from the preceding cycle be used as a negative control. Equal volumes of A and B are mixed. The strip is immersed briefly, then dried in air, and viewed under long-range UV light (340 nm) to detect the white-blue fluorescence which indicates PTH arginine.

b. Identification of PTH Histidine[28]

Reagents

Solution I: 4.5 g sulfanilic acid; 5 ml concentrated HCl. Make up to 500 ml with water. Warm slightly to dissolve. This solution

[27] C. Easley, B. J. M. Zegers, and M. deVijlder, *Biochim. Biophys. Acta* **175**, 211 (1969).
[28] C. Easley, *Biochim. Biophys. Acta* **107**, 386 (1965).

can be stored for months at 4°C, in 10-ml aliquots in test tubes.
Solution II: 5% (w/v) sodium nitrite in water. This solution must
be made up freshly each time. However, 0.25-g aliquots of solid
sodium nitrite can be stored in small stoppered tubes. One of
these is dissolved in 5 ml water when needed.

Solution III: 10% (w/v) Na_2CO_3 in water. This solution can be
stored at 4° for months, as 10-ml aliquots.

Procedure. The samples (10 μl of the same solution used for the
arginine procedure) are applied in the same way to a strip of Whatman
No. 1 filter paper, together with a PTH histidine standard. Solutions I
(10 ml), II (5 ml), and III (10 ml) are chilled separately in an ice bath.
I is mixed with II, and the paper is sprayed at once with the mixture
until it is moist but not wet. The paper is dried briefly in the air and
sprayed with III. A pink or orange spot indicates PTH histidine. PTH
tyrosine can give a weak yellow or orange reaction; however, this derivative is extracted well into ethyl acetate and is therefore not present in
the aqueous phases being examined.

Comment. These reactions are fairly reliable if there is adequate
sample in the spot applied (4 μg or more). Although standards may be
detected at higher sensitivity, sequenator samples often give equivocal
reactions due to interference by other substances. We have seen rare
false positives in the phenanthrenequinone reaction, the reason for which
is unknown. Neither test is very reliable in the late stages of a long
degradation, since the buildup of "background" from nonspecific cleavage
makes every sample give a weakly positive reaction, rendering it difficult
to detect an increase.

c. IDENTIFICATION OF PTH CYSTEIC ACID

Back-hydrolysis and amino acid analysis is recommended for this
derivative. It can with difficulty be identified by paper or thin-layer
electrophoresis. However, Quadrol salts in the sample (probably Quadrol
heptafluorobutyrate) may interfere with the analysis by these procedures.

5. Special Problems in Identification

Sensitivity. The present system of identification starts to become
inadequate when the total amount of sample available is reduced to 10
nmoles or less. Although the more stable PTH amino acids can be detected readily in amounts less than 1 nmole by gas chromatography,
others (serine, lysine, arginine, and histidine in particular) are much
more difficult to identify. Hence, "gaps" start to appear in the sequence
at which no derivative can be identified with certainty. It is usually

worth prolonging the degradation for 10 cycles or so beyond the first gaps, since the provisional information obtained on the more stable residues may be useful in identifying subfragments or in planning cleavage procedures. A search for a methionine residue may for example allow the limits of a cyanogen bromide fragment to be defined. However, information obtained beyond a gap should not be regarded as definitive or used as primary evidence for the sequence.

Quantitation: Criteria for Identification. The present system leaves much to be desired. The values obtained by gas chromatography are reasonably accurate (probably ±10%) for the stable PTH amino acids. Recovery of the others, however, is variable; for example, values obtained for serine, lysine, proline, and tryptophan are quite unpredictable. Hence the use of "correction factors" does not seem justified. Asparagine and glutamine derivatives obtained from the sequenator appear to be more subject to breakdown and loss during the gas chromatographic analysis than are the crystalline standards injected under the same conditions. Presumably impurities present in the sample are catalyzing their decomposition. A thoughtful discussion of the quantitative aspects of identification has been published by Smithies and co-workers.[15] We have not attempted to analyze our data in such detail, but have taken the view that an identification which is not obvious on inspection must in any case be independently verified. Our usual procedure is to measure the total yield of the particular residue (n) in question, subtract "background" calculated from the average yield of that residue at the preceding 2–3 cycles, and add on a correction for overlap calculated from the (total yield minus background) at the next or $n + 1$ cycle. This gives a figure for specific yield which should be at least twice as great as the background for that residue for an identification to be made. No other residue should show an increase of greater than 30% above its background at that cycle. All figures should be corrected with regard to recovery of the added PTH norleucine internal standard. All identifications must be made in at least two independent automated degradations. These criteria are, of course, quite arbitrary, but with occasional exceptions they have proved to be reliable.

Mixture Analysis. Qualitative identification of the several amino acids split off at each cycle of a degradation on a protein mixture presents no problems provided no more than 3 or 4 different components are present at significant concentrations. However the inherent variability in the quantitation procedure makes it difficult to assign particular residues to a particular sequence unless the situation is a very simple one. For example, three sequences present in the quantitative proportions 4:2:1 can probably be sorted out on the basis of the yields obtained.

However it is not hard to see that, for the PTH amino acids which are subject to variable recovery, an incorrect assignment might be made. At present the accuracy of quantitation is not good enough for the kind of mixture analysis proposed on theoretical grounds by Gray.[29] Richards and co-workers[30] have also described an approach to mixture analysis of thiohydantoins based upon mass spectrometry and have applied it with some success.

Problems in Identifying Specific Residues. Any identification system has its own peculiarities and problems, not necessarily applicable to any other identification system. The following comments apply to our own combined GLC–TLC procedure. Where recovery yields are referred to, these are expressed as a percentage of the recovery of "stable" residues (e.g., PTH valine) which do not appreciably decompose during extraction from the sequenator, conversion, or identification. Although some handling losses of these stable residues occur, the use of the internal standard (PTH norleucine) allows a correction to be made.

ALANINE. Occasionally a breakdown product of Quadrol gives a peak which coelutes with PTH alanine on the DC-560 column, and thus a false positive identification might be made. The Quadrol peak is often broader than the alanine peak, however, and on thin-layer chromatography the Quadrol derivative stays at the origin, while the PTH alanine migrates with an R_f of 0.6.

SERINE AND S-CARBOXYMETHYLCYSTEINE. These PTH derivatives undergo a variable degree of decomposition due to β-elimination. The main product from serine, which is to some extent stabilized by reducing agents, is the dehydro derivative. This elutes as a sharp peak between PTH alanine and PTH glycine on DC-560, and is the only significant serine peak seen. The PTH derivatives of S-carboxymethylcysteine and S-benzylcysteine (and probably other cysteine derivatives) give a peak identical to that seen with serine, since they also undergo sidechain β-elimination to give an identical product (i.e., PTH dehydroserine). These cysteine derivatives can be readily distinguished from serine by thin-layer chromatography or by reinjection on the DC-560 column after silylation. Since the recovery of serine is variable (10–50% of the yield of "stable" PTH derivatives such as valine and leucine), its identification late in a degradation should be made with caution.

GLYCINE. Yields of glycine are frequently low, since its conversion is incomplete (about 70%) under the usual conditions. The glycine derivative tends to be absorbed to the GLC column packing, particularly if the

[29] W. R. Gray, *Nature (London)* **220**, 1300 (1968).
[30] T. Fairwell, W. T. Barnes, F. F. Richards, and R. E. Lovins, *Biochemistry* **9**, 2260 (1970).

column is old; a fall in the peak height for the glycine standard relative to the alanine or valine standards is an early sign of column ill health. On GLC analysis there is occasional difficulty in distinguishing glycine from threonine (see below). Phenylthiourea has the same R_f as PTH glycine on TLC.

PROLINE. Yields of proline are at times very low with a wide range (10–80% of the yield of "stable" residues). Since the progress of the degradation does not seem to be affected, the coupling and cleavage reactions must proceed quantitatively. The fate of the "missing" proline is uncertain, although one might speculate that the thiazolinone of proline might have different properties to the corresponding derivatives of the other amino acids due to the presence of the imino nitrogen in the ring. Hence, there may be an alternative pathway which competes with the conversion reaction.

LEUCINE AND ISOLEUCINE. Methods of distinguishing these two derivatives have already been discussed. A residual difficulty is quantitation of each separately when both are present in a mixture analysis. This cannot be done with accuracy, since the peaks resolve only partially on the AN-600 column.

THREONINE. The yield of this derivative is slightly lower than yields of stable PTH amino acids (about 70%). It has the same tendency to β-elimination as PTH serine, forming a dehydrothreonine derivative, which is reasonably stable. On gas chromatography, threonine gives rise to two peaks. The first coelutes with PTH glycine and has been shown by combined gas chromatography–mass spectrometry to have a mass number identical to that of PTH glycine. It must represent a breakdown product of threonine in which the side chain has been completely eliminated, presumably in the high temperature of the injection port of the gas chromatograph. The second peak elutes very slightly earlier than PTH proline; it is the dehydrothreonine form. The proportions of these two peaks vary depending in part on the injector port temperature. When a crystalline standard is injected, the first (glycine) peak is usually the dominant one. Sequenator samples, however, may give rise to predominantly the first peak, predominantly the second peak or a combination of both. Confusion with PTH glycine or proline may result when one or the other threonine peak predominates. When the main peak is in the glycine position, the results of thin-layer chromatography are conclusive, since the glycine and threonine spots are easily distinguished. When the main peak is in the proline position there is more difficulty. However, threonine should be suspected if the peak is broader than the PTH proline standard and if it elutes slightly earlier. Usually there is a slight elevation of the PTH glycine peak above background levels. On thin-layer chromatography there is usually no visible spot in the threonine position, since essentially complete conversion to the dehydrothreonine form has taken place. The dehydrothreonine spot may be seen; it runs in or near the PTH alanine position. The PTH proline spot is obscured by the presence of the PTH norleucine internal standard. A useful confirmatory test is to examine the UV spectrum. The dehydrothreonine derivative gives a major peak at 320 nm. PTH proline has no peak at this wavelength; only the usual thiohydantoin peak at about 270 nm is seen. Reinjection of the sample on the gas chromatograph after silylation has been recommended for identification of PTH threonine. We have found this procedure gives very variable results and is not especially helpful.

ASPARTIC ACID AND GLUTAMIC ACID. Occasionally these derivatives when obtained from sequenator degradations fail to silylate under conditions where the standards silylate well. PTH aspartic acid is particularly prone to this behavior. Sometimes spurious peaks on the gas chromatograph are seen in the PTH aspartic acid position. These derive from a breakdown product of the bis(trimethylsilyl)acetamide (BTMSA) used for silylation. It is wise to check the BTMSA periodically, to store it at −20°, and to replace it when any extraneous peaks appear. These difficulties are circumvented by routine use of thin-layer chromatography.

TYROSINE. Yields may be low, possibly due to oxidative side reactions. Occasionally tyrosine and glutamic acid derivatives do not separate well on thin-layer chromatography, but they are readily distinguished by gas chromatography.

ASPARAGINE AND GLUTAMINE. Partial deamidation of these derivatives occurs during the conversion reaction, usually to the extent of 20–30%. The presence of the deamidated derivative (i.e., PTH aspartic acid or glutamic acid) helps the identification by thin-layer chromatography since PTH asparagine and glutamine themselves are not resolved. Yields measured by gas chromatography are often considerably lower than expected. This may be due in part to breakdown on the GC column catalyzed by impurities in the sample. Quantitation of PTH aspartic acid in the presence of PTH asparagine may be rendered inaccurate, since their silylated derivatives coelute. Usually however PTH asparagine is not derivatized when the on-column silylation technique is used. The use of the OV-25 column for identification and quantitation of asparagine and glutamine (as well as lysine, tyrosine, and tryptophan) is recommended when only small amounts of sample are available.

LYSINE. Identification by gas chromatography is not very satisfactory. PTH lysine coelutes with tyrosine, and at least 10 nmoles are required for detection. Better sensitivity is achieved after trimethylsilylation, and the silylated derivatives of lysine and tyrosine are well separated. Silylated

PTH lysine elutes after PTH phenylalanine in the second isothermal injection on DC-560. The PTH derivatives of ϵ-succinyllysine and ϵ-maleyllysine decompose on gas chromatography and cannot be identified by this means. However, they are separated from all other PTH derivatives by thin-layer chromatography (Fig. 3).

TRYPTOPHAN. Yields are low because the indole ring of tryptophan is attacked by the anhydrous heptafluorobutyric acid used in the degradation. Tryptophan is progressively destroyed, therefore, as the degradation continues. We have successfully identified tryptophan at position 23 in degradations on bovine and porcine parathyroid hormone. However, in degradations on proteins where tryptophan was known to be present at residue 35 from the amino terminus, it could not be identified. It seems likely that the addition of reducing agent to the HFBA would minimize tryptophan destruction. This approach has been effective in solid phase peptide synthesis for protection against tryptophan breakdown during repetitive exposure to trifluoroacetic acid. Derivatization of the tryptophan with Koshland's reagent[31] prior to the degradation might also be effective.

METHIONINE. PTH methionine itself is stable and presents no difficulty in identification. However, PTH methionine sulfoxide is unstable on gas chromatography. The derivative of methionine sulfone may be identified by thin-layer chromatography; however, it does not resolve from PTH aspartic acid. On gas chromatography PTH methionine sulfone elutes between PTH tyrosine and PTH tryptophan on the DC-560 or OV-25 columns.

PHENYLALANINE. An impurity derived from dimethylallylamine coupling buffer elutes very close to the phenylalanine position on gas chromatography and may be a cause of confusion when shorter peptides are being degraded.

VALINE. This residue seems to be the only one which presents no particular identification problem.

ARGININE, HISTIDINE, AND CYSTEIC ACID. These derivatives have been discussed fully already.

Despite this long list of potential difficulties, the combined TLC–GLC system used has been found to be reliable and reasonably efficient.

F. Logistics

There are two logistic problems created by the speed with which the sequenator generates samples for identification. The first involves the capacity of the identification system to keep up with the sequenator. The

---

[31] Koshland's reagent: 2-hydroxy-5-nitrobenzyl bromide.

second involves the danger of errors resulting from handling many samples simultaneously.

Logistics of Identification. The combined GLC–TLC system described here represents a compromise. Ideally one should obtain a quantitative value for each one of the twenty amino acids at each cycle, analyze the data statistically, and obtain confidence limits for each identification. The nearest approach to such a system has been described by Smithies and co-workers.[15] For the reasons discussed, we prefer the direct identification procedures. We have not routinely attempted complete and quantitative analyses at each cycle because of the cost in terms of technician time and in data handling equipment. The major rate-limiting step is in gas chromatography, in spite of the use of a double system that will allow two simultaneous injections. Hence, we use thin-layer chromatography as a screening procedure. Since a definite or at least provisional identification can be made, it is often possible to omit at least one of the three isothermal injections. This allows the selective use of gas chromatography to verify the identifications and for quantitation. In doubtful situations, when, for example, the background and overlap become significant, a complete gas chromatographic analysis is essential.

It is helpful to make up multiple aliquots of the solutions of PTH standards used for gas chromatography. These can be dried down and redissolved as required. New standards are used each day so that the problems of decomposition during storage in solution are avoided. The reagents used for the phenanthrenequinone and Pauli tests are also prepared in a form ready for use.

Sample Handling. Procedures for handling up to 10 samples simultaneously with minimal chance of error have been described by Edman.[6] The consequences of mixing up samples are quite serious, since there is no way of detecting and correcting the error short of repeating the whole degradation at least two more times. It is important to be continuously aware of the danger and to take careful routine measures to avoid it. At the risk of being tedious, the precautions which should be taken will be described.

The tubes to be used for the degradation should be numbered legibly and placed one at a time in order in the fraction collector. Each automated degradation is given a code number; this is written on each tube as well as the cycle number. When all the tubes are in the fraction collector, the numbering should be checked once again without removing the tubes, before the run starts. A further check should be made as each tube is removed for conversion and identification. The next point at which an error is liable to occur is at the extraction step following conversion. Since centrifugation is necessary to separate the organic and aqueous phases, the tubes are necessarily removed from the rack. Hence the numbering must be checked once again after the centrifugation before transfer of the organic phases to a second set of numbered tubes.

Edman has described several devices which greatly facilitate handling of samples and reduce the possibility of error.[6] An "evaporator manifold" allows 10 samples to be simultaneously evaporated to dryness under a stream of nitrogen. An "extractor manifold" allows the aqueous phases after conversion to be extracted by a countercurrent system in which a constant stream of ethyl acetate droplets is released at the bottom of the hydrochloric acid layer. The droplets rise through the aqueous layer, extracting the PTH amino acids *en route,* and a layer of ethyl acetate gradually accumulates on top of the acid. Since the process is gentle, the layers remain separate and no centrifugation is required. Hence, the need to remove the tubes from the rack is avoided. An "applicator manifold" is used to apply 10 samples simultaneously to a thin-layer plate. Since these manifolds are clearly ilustrated in the review by Edman, they will not be further described here. However, their use is very strongly recommended from the viewpoint both of convenience and of reduction in handling errors. An applicator manifold based on Edman's design is commercially available ("The Tenspot," Pierce Chemical Company, Rockford, Illinois). Suitable extractor and evaporator manifolds are not yet available.

One useful feature of the overlap which tends to develop in the course of longer degradations is that it provides a means of detecting inadvertent reversal of samples.

G. Special Problems in Automated Degradation

1. Problems Observable during the Degradation Cycle

A considerable variety of problems may come to light during the actual degradation. As already pointed out, it is highly desirable that at least the first cycle be closely observed in entirety. Subsequently, the worker should try to monitor the progress of the degradation at frequent intervals, particularly during the several reagent addition steps. Relatively common malfunctions which can usually be readily corrected include minor vacuum leaks, unequal delivery levels, failure of solenoid valves, and poor positioning of the scoop. Three other problems deserve more detailed comment.

a. Insolubility

Certain proteins dissolve poorly in the Quadrol coupling buffer, and in the HFBA. This does not necessarily mean that the degradation cannot proceed, since at times the coupling and cleaving reactions appear to go to completion without the protein ever dissolving. Presumably the protein is sufficiently swollen by the reagents to allow diffusion of PITC and HFBA into its matrix, and what amounts to a heterogeneous reaction in the solid phase takes place. However, this does not always happen. Some proteins plainly do not react at all. Others couple and cleave but form such an impenetrable gel that it is extremely difficult to extract out the cleaved thiazolinone derivative. The solubility properties depend partly on the past history of the protein. Proteins which have been exposed to denaturing agents may be particularly troublesome (e.g., some immunoglobulin heavy chains which have been subjected to reduction and alkylation in urea or guanidine). The agent used for sulfhydryl group modification may make a difference. For example, a particular protein may be soluble after performic acid oxidation but totally insoluble after reduction and alkylation with iodoacetic acid, even though the charge on the cysteine derivative is the same in both cases.

The best procedure to adopt if the protein does not appear to dissolve in the Quadrol buffer is to allow the coupling to go to completion, extract out the Quadrol with benzene and ethyl acetate, dry the protein film, and then repeat the whole coupling. Whether or not the protein appears to dissolve on the second coupling, the degradation should be continued since the solid-phase type of reaction mentioned above may well have taken place. If not, there is little to lose since the protein is now useless for other purposes. Often the solubility properties of a protein will improve after the completion of the first cycle.

If the protein does not yield to this approach, then it is worth trying to degrade another sample which may have been handled differently, either in terms of exposure to denaturants or in terms of modification of sulfhydryl groups. If this is unsuccessful, it may be worth trying coupling buffers other than Quadrol. Quadrol is a highly associated viscous fluid which can form a gel with some proteins. Certain immunoglobulin heavy chains which were poorly soluble in Quadrol were degraded quite successfully in a buffer containing 1 M dimethylallylamine, in pyridine/water. If this does not help, then it is not worthwhile to pursue the sequenator approach. Attempts can be made to degrade the protein manually in the presence of urea or sodium dodecyl sulfate, by procedures that cannot be carried out in the present kind of sequenator. Alternatively, cleavage of the protein into smaller fragments can be attempted; these may well be susceptible to automated degradation.

b. Contraction of Protein Film

Some proteins undergo a physical contraction in the film formed after coupling and solvent extraction. Usually it is noted that when a vacuum is applied after the ethyl acetate extraction, the protein layer peels away from the glass wall and collapses into the center of the cup. In the Beckman instrument the protein film may become entangled with the plug which projects into the cup or with the delivery lines. This phenomenon appears to be a peculiarity of certain proteins. We have observed it in the case of the cyclic AMP receptor protein obtained from *Escherichia coli*.[32] Here the film contraction occurred with native, but not with performic acid-oxidized, material. In other proteins, contraction has occurred despite modification of disulfides. The approaches to be used are essentially those already discussed under (a) above for handling poorly soluble proteins. Use of an inert filler (possibly cellulose) to disperse the protein might help; this has not, to my knowledge, been tried.

c. Quadrol Retention

Failure to remove Quadrol adequately by the solvent extractions after coupling is a relatively common problem. Small amounts of Quadrol probably always escape extraction because of binding to acidic groups on the protein. When the HFBA is added, it dispaces the Quadrol from the protein and renders it available for extraction by the 1-chlorobutane. Hence, some Quadrol finds its way into the sample, and this is distributed between the organic and aqueous layers after conversion. In the amounts usually present, no interference with identification is noted. A moderate increase in the amount of Quadrol in the sample causes some interference with identification by GLC, since it gives rise to one or more broad peaks that overlap or coelute with several PTH amino acids. The most common Quadrol peak coelutes with PTH alanine. Accumulation of Quadrol on the GLC columns may interfere with their function. Usually there is less interference with thin-layer chromatography since the Quadrol is highly polar and adsorbs strongly to the silica gel plate at the point of application. Occasionally, however, PTH aspartic acid and glutamic acid show minor abnormalities in R_f values, presumably due to salt formation between the basic Quadrol and the acidic side-chain groups.

If more severe degrees of Quadrol retention occur, identification becomes impossible. Ultimately the cleavage reaction is compromised because much of the added HFBA is neutralized by the Quadrol bound to the protein film.

Diagnosis of Quadrol Retention. Quadrol retention should be suspected if at the end of the ethyl acetate extraction after coupling, the protein is present as a clear glassy film instead of a dry white powder. One or even two repetitions of the 10-minute ethyl acetate wash should be carried out.

[32] H. D. Niall, W. Anderson, and I. Pastan, unpublished work.

This is often enough to remove the Quadrol from the protein regardless of the cause of its retention.

If the Quadrol retention is not noticed at the stage of the ethyl acetate extraction, then it will become obvious when the 1-chlorobutane samples are dried down. Normally, the sample should dry completely, leaving only a thin film which adheres to the bottom of the tube. When there is significant Quadrol retention, clear viscous droplets are seen which resist further drying. The volume of these may be equivalent to several hundred microliters of liquid. It is helpful to convert such samples using 3–5 times the regular volume of 1 N HCl. This increased volume of acid minimizes the amount of Quadrol extracted into the ethyl acetate layer, since the four polar hydroxyl groups on each Quadrol molecule tend to keep it in the aqueous phase.

Causes and Prevention of Quadrol Retention. EXCESSIVE QUADROL DELIVERY. The volume of Quadrol delivered should be the minimum necessary to reach approximately half way up the cup wall while the cup spins at its highest speed (1800 rpm in the Beckman instrument). The surface tension properties of Quadrol are such that it is easy to deliver too much; a thick film is formed which is slow to find its equilibrium level in the cup. Hence, when calibrating the delivery time of Quadrol several minutes should be allowed to lapse after the delivery before noting the height reached. If too much Quadrol is added initially, extremely long and wasteful ethyl acetate extractions are necessary to remove it.

INADEQUATE ETHYL ACETATE EXTRACTION. Quadrol retention can result when the ethyl acetate extraction is inefficient. This is often due to improper adjustment of the scoop, so that turbulence and stationary wave patterns develop in the cup, and the extraction is incomplete. In the Beckman instrument, a single delivery line is used for both Quadrol and ethyl acetate. Unless special precautions are taken, as described elsewhere (see "Operation of Instrument"), the line may become plugged by Quadrol. The delivery pressure of ethyl acetate is too low to readily force the Quadrol plug out of the line. Hence, the ethyl acetate delivery is either slowed down or completely prevented, resulting in severe Quadrol retention. This mechanism can operate intermittently, resulting in the irregular appearance of Quadrol in the degradation samples.

PROPERTIES OF INDIVIDUAL PROTEINS. Some proteins appear to interact with the Quadrol to form a gel. As already noted, this tendency may be sufficiently marked to completely prevent successful degradation. However, in a less serious form the interaction merely interferes with Quadrol extraction. Some immunoglobulin heavy chains have shown this property. Measures that may be taken include: (a) prolongation of the ethyl acetate extraction, (b) addition of 0.1% (v/v) acetic acid to the ethyl acetate. This addition, which was routinely carried out by Edman and Begg, does not appear to be essential for the majority of proteins; however it is helpful in promoting more efficient Quadrol extraction.

If Quadrol retention persists after attention to the various factors listed above, then an alternate coupling buffer (e.g., dimethylallylamine: pyridine:water) should be tried. As with the solubility properties of a protein, tendency to Quadrol retention may vary with disulfide status and past history. We have noted some tendency for Quadrol retention in glycoproteins, but the presence of carbohydrate is certainly not an essential prerequisite.

In a discussion of specific problems associated with the use of Quadrol (which include the considerable difficulty in its purification), it should not be forgotten that the chemical properties which cause the difficulties are also those which make Quadrol an excellent protein solvent. Quadrol contains four hydroxyl groups and two tertiary amine functions per molecule. The tertiary amine groups make it a good buffer in the pH range required for coupling. The hydroxyl groups are responsible for its property of self-association, which makes it nonvolatile and viscous. They also allow it to associate strongly with the protein through hydrogen bonding to side-chain groups and to the polypeptide backbone. This property makes it a good protein solvent, but also leads to difficulty in its extraction.

2. Problems in Interpretation of Results

No Amino-Terminal Residue Seen. The protein probably has a blocked α-amino group, either through acetylation or pyroglutamic acid formation. However, it is worth running a few cycles in case there is an amino-terminal residue such as cysteine or serine which gives rise to a very labile PTH derivative.

A completely blank tracing on the gas chromatograph is rather suspicious of a mechanical or chemical mishap with the degradation, since there are always trace quantities of PTH amino acids from contaminating polypeptides. Detergent inadvertently left in glassware can completely destroy PTH amino acids.[33]

Enzymes are now available for the specific removal of amino terminal pyroglutamyl residues.[34,35]

---

[33] All glassware which is used in conjunction with the sequenator procedures should be washed successively with aqueous acetic acid, distilled water, and acetone; detergents and oxidant-containing washing solutions must be avoided.

[34] R. F. Doolittle and R. W. Armentrout, *Biochemistry* **7**, 516 (1968).

[35] R. E. Fellows, Jr., personal communication.

Amino Terminal Heterogeneity. Most probably this is due to an unsuccessful or incomplete purification of the protein in question. However, it is worth remembering that there are frequent instances of amino-terminal "fraying" of proteins which are otherwise quite homogeneous. This is most likely due to partial enzymatic digestion by tissue or blood aminopeptidases in the course of isolation of the protein. This phenomenon has been described with bovine growth hormone,[36] human placental lactogen,[37] and bovine luteinizing hormone.[38] If the fraying is not severe, it may still be possible to discern the basic sequence.

Low Yield of Amino-Terminal Residue. It is rare to obtain more than 50–70% of the yield of the amino-terminal residue expected from the weight of protein being degraded. A combination of factors may be involved in this discrepancy. Partial amino-terminal blocking may occur during purification, by impurities in the solvents or reagents used. Carbamylation of the α-amino group by cyanate present as an impurity in urea is the most obvious example. The protein usually contains some residual salt and at least 15% water, unless special attempts are made to remove it. Even if the coupling and cleavage reactions go to completion, losses of the cleaved amino acid residue occur due to incomplete extraction, and further chemical and mechanical losses occur during the conversion and identification procedures. The PTH derivatives of different amino acids vary considerably in their lability, as discussed elsewhere. Overdrying after the HFBA step causes very inefficient extraction of the thiazolinones from the cup and low absolute yields are seen, though the repetitive yields may be relatively unaffected. Contamination of the ethyl acetate used for extraction after conversion, usually by oxidants, can give rise to very low yields and the appearance of extraneous peaks on the gas chromatograph. It is important to remember that, unless the ethyl acetate in the sequenator is also contaminated, which obviously need not be so, the degradation may be proceeding perfectly normally. Hence, before abandoning the degradation it may be worth checking the effect of replacing the ethyl acetate (and the hydrochloric acid) used for the conversion and extraction procedures.

Low Repetitive Yields. Degradations of reasonable length (20 or more cycles) are not possible unless the repetitive yields are greater than 92%. Long degradations (40–60 cycles) require repetitive yields of 95–96%. There may be obvious problems either with protein behavior in the

---

[36] R. E. Fellows, Jr., and A. D. Rogol, *J. Biol. Chem.* **243**, 1567 (1969).

[37] K. J. Catt, B. Moffat, and H. D. Niall, *Science* **157**, 321 (1967).

[38] D. N. Ward, L. E. Reichert, Jr., W.-K. Liu, H. S. Nahm, and W. M. Lamkin, *in* "Proceedings of the International Symposium on Gonadotrophins." Excerpta Med. Found., Amsterdam, 1972.

cup or with mechanical performance of the instrument. This may cause a rapid falloff in yields; however, usually a severe degree of overlap develops as an index of incomplete coupling or incomplete cleavage. Where yields fall rapidly without the appearance of marked overlap, there are two common causes. A leak may be present into the reaction cell. This may be either a pressure leak, a vacuum leak, or a combination of the two. As discussed earlier (see Section V, B, Reaction System), the presence of a leak need not be deleterious. However, sometimes it is clearly harmful, and removal of the leak leads to correction of the yield problem. The mechanics of leak detection vary with the specific instrument design and cannot be discussed in detail. However, a clear description of the steps to take to detect and correct a leak should be part of the instruction manual.

The second common cause of a falloff in yields is contamination of the chemicals due to inadequate purification or to contamination after addition to the sequenator reservoirs. The usual culprits are the ethyl acetate, which may contain oxidants, and the HFBA, which may contain anhydrides due to excessive exposure to drying agents during purification.

A further cause of rapidly falling yields is sometimes forgotten. The polypeptide chain being degraded may be much shorter than appreciated, leading to excessive extractive losses during the cycle. Such extractive losses can become severe when the residual peptide chain is less than about 50 amino acids, though there is considerable variation from one peptide to another, depending on the amino acid composition.

Sudden Fall in Yield. If a sudden rather than a gradual fall in yield occurs during the course of a degradation, the appearance of an instrumental malfunction must be suspected. However, there are other possibilities. If the reagents have been replenished during the run, bad chemicals may have been added to good ones. A sudden fall in yield has been observed by us at residue 29 in bovine parathyroid hormone. This residue is glutamine, and is followed by an aspartic acid at position 30. Though we have no direct evidence, it seems quite likely that cyclization of the glutamine takes place during the cleavage of the preceding residue (leucine). The sequence Gln-Asp- has been found to be especially favorable for promotion of the cyclization reaction, possibly due to an intramolecular participation by the neighboring β-carboxylic acid function. In subsequent degradations on parathyroid hormone only a single, shortened (2-minute) cleavage was carried out at the cycle preceding the glutamine residue. These cleavage conditions proved adequate for quantitative removal of the leucine residue, and no fall in yield was noted at the glutamine cycle. However, the apparent success of this maneuver is not conclusive since, among other possible explanations, trace quantities of water in the HFBA used for cleavage in the first degradation may have catalyzed the ring closure. Usually, no appreciable fall in yield is seen during sequenator degradations at glutamine residues. The speed with which the next cycle is started (in comparison with the manual degradation) may help to avoid this danger. A further possibility for a sudden fall in yield exists at aspartyl or asparaginyl residues due to an $\alpha - \beta$ rearrangement through the intermediate cyclic imide form.

Excessive Overlap. Some of the protein inevitably fails either to couple or to cleave at every cycle. This leads to the appearance of "overlap," i.e., a portion of the sample lags one cycle behind in the degradation. As Edman and Begg have pointed out, repetition of the cleavage step at each cycle is most helpful in reducing the extent of overlap. According to Smithies,[15] elevation of the reaction temperature within limits is also helpful, though it has the side effect of increasing the rate of nonspecific peptide bond cleavage. An acceptable amount of incomplete reaction is 1% (or less) of the total protein coupled and cleaved at that cycle. If overlap is reduced to this extent, very prolonged degradations are possible.

The most common cause of excessive overlap is improper adjustment of the relative heights of acid and coupling buffer, leading to the formation of an upper rim of protein which either fails to couple or fails to cleave. Smithies has pointed out that the fraction of the protein which is already one step out of phase may have a greater chance of (further) incomplete reaction than does the protein which remains in phase. The physical distribution in the cup of the out-of-phase protein would presumably favor this phenomenon.

Insolubility of the protein is another obvious cause of overlap. A third possibility for incomplete reaction leading to overlap is the temporary blocking of the α-amino group by a group which is labile to the degradation conditions and is later removed. Carbon dioxide could represent such a reversible blocking group.[7]

Excessive Internal Peptide Bond Cleavage. The gradual accumulation of a "background" of PTH amino acids due to a small degree of internal peptide bond cleavage at every cycle was noted by Edman and Begg.[1] Whether this is due to truly "nonspecific" cleavage or whether certain bonds are particularly labile is at present uncertain. Whatever the explanation, the background gradually increases in magnitude, and together with the fall in yield eventually brings the degradation to a halt, as illustrated diagrammatically in Fig. 4. Prolongation of the cleavage time and elevation of the temperature increase the rate of accumulation of the background. The longer the polypeptide chain being degraded, the more rapidly the background becomes a problem, since there are cor-

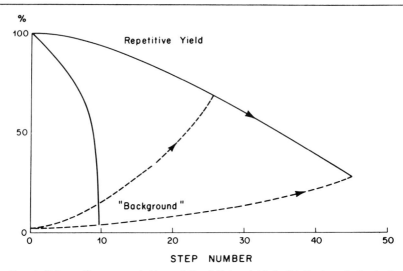

Fig. 4. Schematic representation of the fall in yield (solid line) and rise in background (dashed line) which occur during a degradation and sooner or later make identification impossible. See text.

respondingly more chances for it to occur in any individual molecule. For proteins of the size of macroglobulin heavy chains, for example, it is difficult to identify residues beyond 20–25 cycles. Some proteins seem more susceptible to internal peptide bond cleavage than others; this would suggest that the cleavages are not truly random. It should be noted that the cleavages leading to the appearance of background also contribute to the fall in yield seen with the "in-step" protein, since many of the cleaved fragments would be small enough to be lost in the solvent extractions.

Short of discovering a totally different kind of degradation requiring only very mild cleavage conditions, very little can be done to reduce the background. Reduction in temperature of reaction or in time of exposure to HFBA can be tried. However, overlap tends to increase so there is little overall gain. An unusually rapid appearance of background could be due to overheating the reaction chamber, to the existence of labile bonds in the protein, or to underestimates of the correct molecular weight of the protein chain. It can also be due to the presence of a substantial proportion of amino-terminal blocked chains in the sample being degraded, since this material contributes to background but not to the observed yield of the degradation.

Gaps in the Sequence. Sometimes no residue can be found at a particular cycle, though the sequence up to that point and beyond it can be identified. Such blank cycles are most often due to the presence of a labile residue such as serine, cysteine, or tryptophan, which has been completely destroyed. At times a mechanical failure, such as failure to add any PITC or to collect the 1-chlorobutane extraction in the fraction collector, can be responsible. It is helpful to look for the presence of the "background" and of the "out of step" residue, the identity of which is known from the previous cycle. If these are seen, then obviously the coupling and cleavage reactions at that cycle have been completed normally, and the presence of a labile residue is likely.

Premature Cleavage Reactions. Schroeder,[9] and Blombäck *et al.*[39] have described degradations in which amino-terminal histidine apparently undergoes a cleavage reaction under the alkaline conditions of coupling. Since PITC is still present, the next residue in the chain becomes coupled and later cleaves in the usual fashion. Hence the histidine can be completely overlooked. We have encountered this reaction in a manual degradation on bovine parathyroid hormone using a dimethylallylamine: pyridine:water buffer. The sequence Met^8-Asn^9-Leu^{10} was found in the initial manual degradation; however, in several subsequent automated degradations using the Quadrol system the sequence was established as Met^8-His^9-Asn^{10}-Leu^{11}. The mechanism for the premature cleavage of histidine, which may be partial or complete, is unknown. It has not so far been encountered in hundreds of automated degradations on histidine-containing proteins by ourselves and others.[6,7,15]

3. Degradations on Valuable Proteins

Occasionally degradations must be carried out on samples of proteins that are either unique or extremely difficult or expensive to isolate. One might cite the example of a myeloma protein from an individual patient where no further material can ever be obtained, or a protein on whose isolation many man-years of work have been expended. Special precautions can be taken to minimize the risk of failure in the degradation. If there is enough material it is prudent to undertake a preliminary run on a small amount (a milligram or less) to ascertain whether there are any special solubility problems, before committing the entire sample. After a thorough examination of the mechanical performance of the instrument, the solvent and reagent reservoirs should be filled, and a short calibration run carried out on a known protein such as myoglobin. If the yields are satisfactory, one has only to clean out the cup and start the degradation on the unknown sample. The reservoirs will contain

---

[39] B. Blombäck, M. Blombäck, B. Hessel, and S. Iwanaga, *Nature (London)* **215**, 1445 (1967).

enough reagents for at least 40–50 more cycles so that one does not have the problem of adding further untested chemicals during the course of the degradation. The run should be observed throughout by someone with enough experience to detect and at once correct any machine malfunction. The degradation can if necessary be interrupted at night usually without deleterious effects. Before it is restarted, however, the reagents which have been sitting in the oxygen-permeable delivery lines during the delay must be flushed out and replaced by fresh, nitrogen-purged reagents. If the degradation is to be stopped and restarted, the interruption should be timed to fall at the end of a cycle, not during it.

H. Automated Degradation of Short Peptides

For sequence determination on proteins and larger polypeptides, there are at present no good alternative approaches that can compete with the sequenator. However, the advantage is much less marked for short peptides since, as pointed out by Edman and Begg,[1] these tend to be lost from the cup in the solvent extractions used during the degradation. Despite the availability of effective manual procedures for peptide degradation, the sequenator approach remains a most attractive one because of its speed and ease. Hence, we and others have put considerable effort into extending and adapting the basic sequenator procedures for use with small peptides. The theoretical aspects of automated peptide degradation have been recently reviewed[7] and will not be discussed in any detail here. However, some practical approaches, mostly those used in our laboratory, will be described. The two general approaches to minimizing the extractive losses of peptide are (1) the use of techniques that modify the properties of the peptide to make it less soluble in the extracting solvents; (2) the use of techniques that limit the volume and the polarity of the solvents used for extraction.

1. Modification of the Peptide

There is a marked difference in the rate of loss of peptide depending on its overall polarity, which in turn depends on its relative content of amino acids with hydrophilic and hydrophobic amino acids. This observation is illustrated in Fig. 5. Bovine calcitonin (a 32-residue peptide which contains 2 trypsin-sensitive bonds and an amidated COOH-terminal α-carboxyl group) was digested with trypsin, and the mixture of three fragments was subjected to automated Edman degradation. As shown in the figure, the carboxyl-terminal fragment (an undecapeptide) which was the most hydrophobic was extracted out most rapidly. Next followed the amino-terminal fragment (a tetradecapeptide) which had COOH-termi-

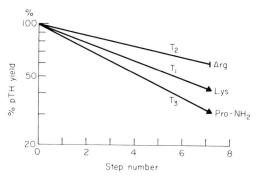

Fig. 5. Effect of amino acid composition on rate of extractive loss of peptides during Edman degradation. T_1, T_2, T_3: amino-terminal, middle, and carboxyl-terminal tryptic peptides of bovine calcitonin. Arg, Lys, Pro-NH₂ represent the carboxyl-terminal residues of the peptides, as shown. The polar guanidinium group decreases the extractive losses of T_2 relative to T_1 and T_3. See text.

nal lysine. The bulky hydrophobic ϵ-PTC group which is added to the lysine side chain at the first coupling step greatly increases its solubility in the extracting organic solvents. The third peptide containing arginine as its COOH-terminal residue, though the smallest (7 residues), had the lowest rate of loss. This is attributable to the presence of the highly polar guanidinium group.

These observations prompted us to attach polar substituents co-valently to peptides to reduce their tendency to be lost in the solvent extractions. Preliminary results[26] show that it is possible to couple arginine methyl ester to the free side-chain and alpha carboxyl groups of a peptide by use of a water soluble carbodiimide. Yields obtained in the sequenator when the derivatized peptide was degraded were sub-stantially higher than with the native peptide.

Procedure. The peptide (0.2–0.8 μmole) is dissolved in 0.5 ml of a 1 N solution of arginine methyl ester dihydrochloride in water, the pH of which has been adjusted to 4.75 with 10% NaOH. An equal volume of an 0.1 N solution of 1-ethyl-3(3-dimethylaminopropyl)-carbodiimide hydrochloride is added, and the pH of the solution is maintained at 4.75 by additions of 0.1 N HCl as required until the pH becomes stable (2–4 hours). The solution is then left overnight. The procedure is carried out at room temperature. The next day the peptide is separated from salts and excess reagent by gel filtration on a column of Biogel P2 equilibrated with 0.1 N acetic acid. It is then ready for automated Edman degradation using the approaches described below. Amino acid analysis shows that addition of the arginine derivative to α, β, and γ carboxylic acid groups takes place to the extent of about 70%. The method is adapted from that of Brew and co-workers[40] for the attachment of glycine methyl ester to carboxylic acid groups on proteins.

An alternative approach to the modification of peptides to reduce extractive losses has been devised by Braunitzer,[41] who has used a series of hydrophilic isothiocyanates incorporating one or more polar sulfonic acid groups. A reagent of this kind is used for the first degradation step on peptides containing lysine. The ε-amino group is then derivatized with a substituent much more polar than the phenylthiocarbamyl group. At degradation steps subsequent to the first, phenylisothiocyanate can be used. Though limited to lysine-containing peptides, this approach is a good one, and initial results using the sodium salt of isothiocyanobenzene sulfonic acid (Pierce Chemical Company, Rockford, Illinois) have been most promising.

The ultimate procedure for making peptides less soluble is to couple them to a solid support. The solid-phase degradation method has been reviewed recently, and the reader is referred to this article for further details.[42] Here it need only be pointed out that the solid-phase approach and the spinning-cup type of sequenator are not mutually incompatible. We have carried out successful degradations in the Beckman Sequenator on peptides covalently coupled both to Merrifield-type resins (polystyrene-divinylbenzene) and to the graft copolymer supports developed by Tregear and co-workers.[43]

The use of solid supports to reduce peptide losses through noncovalent bonding has so far not been evaluated in the sequenator. However, a manual degradation procedure in which silica gel was used effectively for this purpose has been described[44] and it would seem worth trying to extend this approach to the automated Edman degradation.

2. Reduction in Solvent Extraction

When the sequenator method as developed for proteins is applied without modification to short peptides, most of the sample is washed out of the cup in the first few cycles. When the solvents used for extraction are collected and examined by quantitative amino acid analysis, it is found that the 1-chlorobutane used to extract the thiazolinones after

[40] K. Brew, F. J. Castellino, T. C. Vanaman, and R. L. Hill, *J. Biol. Chem.* **245**, 4570 (1970).

[41] G. Braunitzer, B. Schrank, and A. Ruhfus, *Hoppe-Seyler's Z. Physiol. Chem.* **351**, 1589 (1970).

[42] G. R. Stark, *Advan. Protein Chem.* **24**, 261 (1970).

[43] H. D. Niall, J. Jacobs, and G. W. Tregear, *in* "Proceedings 3rd American Peptide Symposium" (J. Meienhofer, ed.). Ann Arbor Sci. Publ., Ann Arbor, Michigan, 1972.

[44] T. Wieland and U. Gebert, *Anal. Biochem.* **6**, 201 (1963).

cleavage contains most of the peptide, the ethyl acetate contains substantial amounts, but the benzene contains very little. On the basis of these and other observations a series of programs have been developed in our laboratory for use in the Beckman sequenator. As previously described, the design of this instrument is such that it is possible to use volatile as well as nonvolatile reagents. Reduction in the volume of the reaction cell is the most important factor in this. The use of a volatile tertiary amine in the coupling buffer is then possible, and this allows great reduction in the solvent extractions.

Before discussing the details of the peptide programs, it should be pointed out that there are disadvantages in reducing the extent of solvent extraction, whether this is done by reducing the total volume or by extracting with less polar solvents. Polar impurities are incompletely removed by the extractions after coupling. Since the extractions after cleavage have to be thorough enough to remove the thiazolinones of the polar amino acids, the polar impurities end up in the sample and may interfere with identification. Another problem is that of overlap, which for several reasons is more severe with the peptide programs. Here one must distinguish between true overlap, due to the presence of a proportion of peptide which lags one step behind in the degradation, and what might be described as "pseudo-overlap." The latter refers to the phenomenon whereby a proportion of the cleaved thiazolinone derived from peptide which is still "in step" in the degradation fails to be removed by the reduced 1-chlorobutane extraction. Since the extractions after the coupling at the next cycle have been reduced, it is not removed at that point either, but is extracted after the following cleavage. Hence, it causes an apparent increase in the "out of phase" peptide. Pseudo-overlap is more marked with polar amino acids such as aspartic acid or arginine, since these are less efficiently removed by the modified conditions for extraction with 1-chlorobutane. We have described[19] the use of two different isothiocyanate derivatives in an alternating fashion (for example, phenylisothiocyanate at cycles 1, 3, 5, etc., of a degradation and pentafluorophenylisothiocyanate (Pierce Chemical Company, Rockford, Illinois) at cycles 2, 4, 6, etc.). Since the thiohydantoins obtained from the two coupling agents can be separated and independently quantitated at each cycle, the extent of pseudo-overlap can be measured separately from true overlap. This approach is not really suitable for routine use because of the additional complexity of identification. However, it has proved very useful in methodological studies during the development of the peptide programs and in certain other special circumstances.

In order to minimize interference from the accumulation of polar impurities and the increased overlap associated with peptide programs, it is advisable to use conditions that are as close to those used for proteins as the particular situation will allow. Hence, for reasonably hydrophilic peptides with chain length of 40–50 amino acids, only minor modifications on the basic protein program are necessary. For peptides of 10 residues or less, substantial reduction in solvent extraction is essential. Between these extremes, it is a matter of judgment to what extent the conditions must be altered to give the maximum protection from extractive losses with the minimum interference with identification from impurities and from overlap.

Modified Cleavage Conditions for Peptide Degradation. Most of the peptide losses occur in the extractions after cleavage, since the presence of residual fluoroacid in the 1-chlorobutane makes it an excellent peptide solvent. The second cleavage can be routinely omitted, with some consequent increase in overlap. This by itself, however, does not usually cause problems in identification for degradations of up to about 30 cycles. The volume of acid used should be kept as small as possible (about 0.15 ml) to facilitate its speedy removal at the end of cleavage. Brunfeldt and Thomsen have used an effective maneuver in which the minimum volume of HFBA required for cleavage is displaced into the delivery line.[45] A short delivery of 1-chlorobutane (which shares the same line in the Beckman instrument) is then used to push the HFBA into the cup. This avoids the need to deliver a larger volume of acid to fill up the dead space in the delivery line between the delivery valve and the cup. Depending on the sequenator design and on the disposition of reagents, it may also be possible to use nitrogen pressure to blow the HFBA into the cup, and achieve the same result. At the end of cleavage the film must be more thoroughly dried than with proteins by 2–3 minutes application of high vacuum. If there is no HFBA left in the delivery line, use of a restricted vacuum is not necessary.

The 1-chlorobutane is then delivered, but initially only enough to cover the peptide film. The small amount of acid remaining helps the 1-chlorobutane penetrate the film, extracting the thiazolinone. After a delay of 30 seconds to ensure that the peptide has precipitated on the wall of the cup, the 1-chlorobutane delivery is continued, and sweeps the thiazolinone into the fraction collector. This is the most crucial part of the program. If the acid is not dried sufficiently, there is excessive loss of peptide, and in extreme cases the peptide film may be washed up the cup by the advancing acid-rich front of butyl chloride. If the acid is dried off too thoroughly before the butyl chloride is brought in, the peptide dries as a dense transparent film which is very poorly penetrated

[45] K. Brunfeldt and J. Thomsen, personal communication.

by the solvent. Thiazolinone extraction is very inefficient and severe pseudo-overlap results. The procedure given here works well, but the delivery times for the HFBA and the 1-chlorobutane and the vacuum times have to be carefully calibrated. One problem with this cleavage procedure is that the additional delays necessary for more thorough drying of the HFBA and the initial "precipitating" delivery of 1-chlorobutane lead to increased destruction of serine thiazolinone, which is very labile under these acidic anhydrous conditions. Hence an auxiliary program designed to give optimum recovery of serine is employed at cycles suspected to be serine. A double cleavage system is used. The first cleavage (0.05 ml of HFBA; 30 seconds duration) allows most serine to be cleaved and extracted before destruction of the labile residue occurs. A second cleavage (0.15 ml of HFBA; 3 minutes duration) completes the cleavage reaction and prevents uncyclized phenylthiocarbonyl-peptides from proceeding through to the next cycle. Since this program employs two 1-chlorobutane extractions, extractive peptide loss is greater than for the standard program and thus it is used only during repeated degradations to confirm the locations of serine residues.

Modified Coupling Conditions for Peptide Degradation. As already indicated, for longer and more hydrophilic peptides only minor changes are necessary. The Quadrol buffer may be used but reduction in the molarity of the Quadrol and thus in the amount delivered is recommended. A $0.1 M$ solution of Quadrol in propanol/water (1:1, v/v), pH 9.0 is suitable. With this reduction in the amount of Quadrol, the extraction times after coupling can be reduced to 100 seconds for benzene and to 120 seconds for ethyl acetate, at flow rate of 2 ml per minute for both solvents.

Further reduction in the ethyl acetate extractions is not possible with Quadrol since it cannot be effectively removed by benzene. However, nonvolatile tertiary amines which are benzene soluble can be used. Dimethyl benzylamine (Pierce Chemical Company, Rockford, Illinois) has been introduced for this purpose by Hermodson, Ericsson, and Walsh with considerable success.[24] We have evaluated[7] several other amines, including N,N'-dimethylpiperazine and 4-dimethylaminopyridine. A third amine, 1,4-diazabicyclo-[2,2,2]-octane, though possessing excellent buffering properties and benzene solubility, has too great a tendency to sublimation at the temperatures usually used for coupling.

Despite the benzene solubility of the amines, their removal by that solvent is not highly efficient since benzene penetrates the peptide film rather poorly. A short ethyl acetate extraction is therefore desirable although not absolutely essential.

A great reduction in solvent extractions with complete omission of the ethyl acetate step is possible when volatile tertiary amines are used in the coupling buffer, in a sequenator with a suitable design for the reaction cell area.

Degradations Using Volatile Reagents. Many automated degradations of shorter peptides have been performed on the Beckman "Sequencer" using this approach. The coupling buffer is a solution of diethylallylamine (0.3 M) in n-propanol:water (60:40, v/v), or a solution of dimethyl-allylamine (1.0 M) in n-propanol:water (60:40, v/v).

In earlier work the pH of these solutions was not adjusted; the apparent pH is then in the range 11.0–11.5. However, variable PITC destruction takes place under such alkaline conditions, so we adjust the pH to 9.5 with trifluoroacetic acid.

The basic program used is as follows: The peptide (0.1–0.5 μM) is transferred into the cup, usually in a small volume of coupling buffer, and dried under vacuum. Phenylisothiocyanate (5% v/v in heptane; 0.4 ml) is delivered to the reaction cup and the mixture is dried with a restricted vacuum (20 mm Hg). Coupling buffer (0.55 ml) is then delivered, and the coupling reaction is allowed to proceed for 20 minutes. The reaction mixture is then dried with restricted, rough (1 mm Hg) and fine (0.1 mm Hg) vacuums. A single 2.5-ml benzene extraction is performed, and residual benzene is dried with nitrogen, followed by restricted and fine vacuums. A single cleavage (duration 3 minutes) is performed by delivery of 0.15 ml of HFBA to the reaction cup, as described above. The acid is dried by sequential application of rough and fine vacuums, and a single 2.0 ml 1-chlorobutane extraction is performed as described above to remove the cleaved amino acid thiazolinone. Residual 1-chlorobutane is dried with nitrogen and with restricted and fine vacuums. This completes the cycle of degradation.

The temperature of the degradation is 55°. All deliveries of reagents and solvents are carried out at a cup speed of 1800 rpm, except for the delivery of coupling buffer. Here the lower speed (1200 rpm) is used to allow an increased volume of buffer to be added. Evaporation can then take place to saturate the vapor space of the reaction cell without compromising the coupling reaction.

The main practical problem with this program (in fact, the main residual problem in the peptide methodology) is overlap or "lag." This amounts to an average of 2% per cycle. After two cycles the lag is 4%, after 10 cycles the lag is 20% and so on. The range of variation in lag is 1–3% per cycle. To put this in perspective, a 1% lag is in our experience about what one has with a good protein program, and is consistent with long degradations. With quantitative identification by GLC, there is usually no difficulty in reading out the correct sequence while the overlap is less than 50% of the true yield. This would allow a 50-cycle degradation. In looking at the results with Quadrol on myoglobin and other proteins, it can be seen that a 1% lag is quite acceptable. Certainly the lag may be much less than 1% per cycle, being 0.5% or less in very good runs.

With peptides, therefore, a 1% lag figure can be regarded as an excellent result. After 30 cycles the overlap still does not cause significant difficulty in identification. A lag of 2% we regard as acceptable, and we have achieved this figure regularly using the program described here. Values of 2.5–3% overlap per cycle are not usually acceptable since identification becomes confusing after 20 cycles or less. However, with shorter peptides the sequence is often completely established before the lag becomes a problem.

The cause of lag or overlap is probably multifactorial. All single cleavage programs (as pointed out by Edman and Begg) carry the risk of incomplete cleavage. With peptides it is better usually to accept this degree of incomplete reaction rather to add a second cleavage and extraction which would increase extractive losses of peptide. Our experiments in coupling indicate that with the procedures described here, pH is maintained very well in the alkaline range during coupling so that incomplete coupling due to excessive loss of base is not a problem. Prolongation of coupling does not alter the percentage lag. At times incomplete extraction of thiazolinone increases the apparent overlap. Lag due to this cause does not tend to increase from cycle to cycle, however, since about the same proportion of thiazolinone tends to get left behind each time around.

In general, initial repetitive yields on a peptide of moderate size (e.g., calcitonin, insulin chains) are in the range 90–94%. This is lower than the yields seen with the Quadrol system on proteins, but quite adequate for peptide degradations. If initial yields are much below 90% the reagents, particularly the coupling buffer, must be held suspect.

Clearly, lower repetitive yields and higher overlap figures which would be quite unacceptable for a long peptide may be reasonable for a shorter peptide available in smaller quantities. There is always a "trade-off" between overlap and repetitive yields. Because of the short chain length of the peptides being degraded, background due to nonspecific cleavage is not usually a problem.

Problems in the Use of Volatile Reagents. Successful degradations on short peptides can be readily carried out in the Beckman instrument. Some examples are given below. However, in sequenators with a different design of the reaction cell module, difficulties will be encountered using this approach, for reasons that will now be described.

If the vapor space surrounding the cup is large, considerable evaporation of the volatile base occurs during coupling. This has two immediate effects. The pH of the coupling buffer drops, owing to loss of buffering capacity. If it drops appreciably the rate of coupling slows, since the α-amino group of the peptide tends to become protonated. The second effect is that the coupling buffer tends to evaporate owing to loss of volatile components. As it becomes more concentrated the phenylisothiocyanate (and at times the peptide) comes out of solution. Both of these factors lead to incomplete coupling. A further deleterious effect results from condensation of the volatile base outside the cup, in other parts of the system. This condensation is accentuated by any tendency toward uneven heating of the vapor space since the base tends to condense in cooler areas, in effect "distilling" out of the cup. This condensation can occur anywhere, but is particularly marked on the outside of the cup which becomes cool due to the evaporation of its contents. Since some of the base becomes adsorbed onto surfaces (particularly to Kel-F and Teflon) it is usually not removed completely during the vacuum stages subsequent to the coupling. During the cleavage step vapors of acid saturate the area, and, since some base is still present, salts are formed. These become immediately evident as oily droplets on the outside of the cup and elsewhere, as soon as the acid enters the system. Accumulation of these salts increases from cycle to cycle, since they have very low volatility and are not appreciably removed by vacuum. The physical presence of the droplets accentuates the problem, since they act as a "sink" for further uptake of volatile base, during coupling, and acid, during cleavage. Since the salts are formed by a very strong acid (e.g., heptafluorobutyric or trifluoroacetic acids) and a relatively weaker base (e.g., a volatile tertiary amine) they have a low pH. (If some of the salt is dissolved in a few drops of water, the pH of the solution is usually between 1 and 2.) To the extent that some of the salt accumulates in or drips into the cup, it interferes with maintenance of a sufficiently high alkaline pH for subsequent coupling with phenylisothiocyanate. It can be readily seen that the two major effects (i.e., an accumulation of acid salts within the system and an ever increasing tendency for evaporation of volatile base) cooperate to make control of pH, and thus efficient coupling, impossible. The accumulation of salt deposits may also adversely affect the functioning of valves, vacuum gauges, or bearings.

In the Beckman Sequenator, the reduction in vapor space together with other design features minimize these problems and allow effective peptide degradation. Obviously the same factors still operate, but the tendency of amine to condense in the system is much reduced. The nitrogen flushing system has proved very effective in reducing salt accumulations during longer degradations.

Choice of Program. The decision to change from a protein program to a peptide program is partly empirical. A discussion of the factors involved has already been published.[7] As a rough guide it can be stated that the unmodified protein program should not be used with peptides shorter than 30–40 residues, and programs which incorporate any ethyl acetate extraction at all should not be used with peptides shorter than 20–30 residues. Our own practice is to use either the unmodified Quadrol or the dilute Quadrol program for peptides in the range of 25–50 residues, and the DMAA–propanol–water system for peptides in the range of 2–25 residues. When the peptide is hydrophobic and less than 10 residues long or when less than 100 nmoles of peptide is available, we use a micromodification[46] of the three-stage manual degradation rather than the sequenator.

Results of Automated Peptide Degradation. A series of automated degradations have been carried out, both on model peptides and on unknown peptides whose sequence we were investigating. Pure peptides suitable for developmental work and available in quantity are hard to locate. Insulin chains A and B from commercial sources often are not pure and are amino-terminal blocked to a greater or lesser extent. Performic acid-oxidized insulin A chain (Mann) is reasonably good, however. We have degraded about 30 different peptides in the Sequencer, ranging in size from 4 to 32 residues. These include both insulin chains, calcitonins from various species, subfragments of calcitonin and of parathyroid hormone, and peptides from collagen and human lysozyme. Synthetic peptides produced by solid-phase procedures have also been successfully degraded. However, much developmental work has been done with calcitonin, which is extremely hydrophobic and represents one of the worst possible peptides to degrade because of severe extractive losses.

The sequence of bovine calcitonin, which we have used most regularly, is as follows:

```
         1              5
   Cys-Ser-Asn-Leu-Ser-Thr-Cys-Val-Leu-
        10             15
   Ser-Ala-Tyr-Trp-Lys-Asp-Leu-Asn-Asn-
           20             25
   Tyr-His-Arg-Phe-Ser-Gly-Met-Gly-Phe
              30
   Gly-Pro-Glu-Thr-Pro-NH₂
```

Repetitive yields were routinely measured, Leu⁴-Leu⁹-Leu¹⁶.

Overlap was quantitated by measuring the yield of valine at step 9 as a percentage of the yield of leucine at that step. The range of results for 15 longer runs on bovine calcitonin may be summarized thus:

[46] H. D. Niall and J. T. Potts, Jr., in preparation.

Amount of peptide used: 0.1–0.5 μmole
Length of degradation: 18–28 cycles; usually 22–25
Repetitive yields: 84–94%
Overlap: 1–3%; averages about 2% per cycle (i.e., average Val^9/Leu^9 = 16%). The usual overlap at step 22 was in the range of 30–50%.

These results indicate that the major portion of the structure of a hydrophobic 32-amino acid peptide may easily be determined in the sequenator. The rest of the structure can be determined by automated degradation of a carboxyl-terminal fragment; this has been accomplished in the sequence analysis of natural bovine and ovine calcitonins.[47]

Peptides with a carboxyl-terminal arginine residue are particularly suitable and can often be completely degraded. This has been accomplished with a tryptic peptide from porcine calcitonin with the sequence Asn-Leu-Asn-Asn-Phe-His-Arg, and on a tryptic peptide from bovine parathyroid hormone with the sequence Asp-Gly-Ser-Ser-Gln-Arg-Pro-Arg. The terminal arginines were identified by amino acid analysis. A 31-cycle automated degradation has been carried out on 0.17 μmole of a 32-residue peptide from bovine parathyroid hormone. The first 6 cycles were carried out using the unmodified Quadrol protein program. The DEAA peptide program was substituted, and a further 25 cycles were completed. All amino acids were identified and quantitated.

VII. Applications and Limitations of Method

No attempt will be made here to discuss the numerous applications of the sequenator technique that have so far been published. The automated Edman procedure is particularly suitable for obtaining long degradations on polypeptides of 60–150 residues (Fig. 6). With longer polypeptides, the increased background limits the length of the degradation. With shorter peptides extractive losses become a problem, though these can be minimized by appropriate procedures discussed above. In certain instances useful information has been obtained from degradations carried out on mixtures of proteins or peptides, though problems with the identification system have limited this kind of application.

There are several obvious limitations in the present technique. Very long degradations are limited mostly by the need to use vigorous conditions for the cleavage reaction, i.e., exposure to fluoroacids. For progress in this direction a degradation employing a much more gentle cleavage procedure must be devised. Degradations on short peptides are limited

[47] R. Sauer, H. D. Niall, and J. T. Potts, Jr., Fed. Proc., Fed. Amer. Soc. Exp. Biol. 29, 728 (1970).

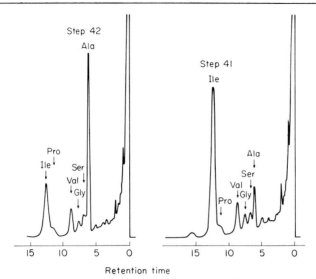

Fig. 6. Gas chromatographic tracings from steps 41 (isoleucine) and 42 (alanine) of a degradation on bovine parathyroid hormone, obtained on a DC560 column using conditions described in the text. This illustrates the ease of identification despite the presence of a background of other residues and the presence of overlap (elevated Ile in step 42).

by the problem of extractive losses of material in the organic solvents. Coupling the peptide to either a high molecular weight hydrophilic carrier in solution or to a solid support would seem to be the best approach to this problem. Although progress is being made in this direction, much more work remains to be done. Limitations in the sensitivity of the procedure are partly dependent on the capacity of the sequenator itself to handle small samples and partly on the identification system. It seems likely that the gas chromatographic method of identification of PTH amino acids has reached its practical limits of sensitivity. New approaches possibly involving the use of radioactive or fluorescent coupling reagents will be necessary for further improvement.

The advent of the sequenator has dramatically altered the strategy of protein sequence determination. The emphasis is now on the isolation of a small number of larger fragments of the protein, produced by suitable cleavage procedures. Cyanogen bromide cleavage, and tryptic cleavage limited to either lysine or arginine residues are the two most useful approaches. We have found that cleavage limited to tryptophan also provides fragments very suitable for degradation. A detailed discussion of sequencing strategy is, however, beyond the scope of this article.

VIII. Alternative Approaches

Though alternative approaches to protein and peptide sequencing are under investigation (for example sequencing by direct mass spectrometry) these so far are not developed to the point of competing with the sequenator method. Minor changes in the degradation (such as the substitution of thiobenzoyl derivatives for isothiocyanates as the coupling agent) will not alter its basic character, and much of the discussion in this article would still be applicable.

Acknowledgments

The author is most grateful to Dr. Pehr Edman for many pleasant hours of discussion and instruction spent in his laboratory, initially as a medical student and later as a graduate. Others too numerous to list provided helpful information. However, particular thanks are due to Mr. Geoffrey Begg and Dr. Francis Morgan, and to Mr. Harry Penhasi, Dr. Jack Ohms, and Dr. Fulvio Perini of Beckman Instruments, Palo Alto, California. The experiments on automated peptide degradation were carried out in collaboration with Mr. Robert Sauer. The author, however, takes responsibility for the accuracy of the facts and for the validity, or lack of it, of the opinions expressed in this article.

Author Index

Numbers in parentheses are reference numbers and indicate that an author's work is referred to although his name is not cited in the text.

Baierlein, R., 309, 316(24), 317, 320, 325, 330

Bailar, J. C., Jr., 637

Bailey, J. E., 498, 504

Baker, T. W., 170

Balasubramanian, D., 709, 710

Baldwin, R. L., 7, 20(22), 31, 61, 66, 118, 122, 340, 360

Ballantyne, D. L., 959, 969(15), 970(15), 974(15), 982(15), 987(15), 995(15), 997(15)

Banerjee, K., 266

Bareiss, R., 395, 406(56)

Barel, A. O., 716

Barfield, M., 885

Bark, L. S., 592

Bark, S. M., 592

Barker, R., 262, 263, 264, 509, 511, 512

Barlow, G. H., 96

Barlow, J. L., 794

Barnard, E. A., 714

Barnes, R. B., 549

Barnes, W. T., 983

Barnett, L. B., 473

Baro, R., 151, 177(11), 179(11), 180(11), 182(11), 188(11)

Barrand, P., 118

Barth, G., 677(51), 678

Bartholomew, C. H., 591

Barton, J. S., 328

Bartsch, R. A., 918, 941(8)

Bartsch, R. G., 933

Bartuska, V. J., 886

Basch, J. J., 549

Basinger, S. F., 96

Bastian, R., 736

Batchelder, A. C., 265

Bates, R. G., 542, 543(35), 597

Batterman, B. W., 177, 188(75)

Bauer, N., 83

Bauer, W., 46, 49(79), 119, 120(22), 121

Baum, F. J., 229

Bayer, E., 677(51), 678

Bayliss, N. S., 500, 504, 509(19)

Bayzer, H., 525

Beams, J. W., 83, 191, 310, 323(28)

Bear, R. S., 180

Beard, J. W., 94

Bearden, A. J., 912, 918, 920, 927, 928 (10), 933, 934, 940(12)

Beaven, G. H., 498, 504(11), 525, 529, 547, 768

Becker, E. D., 827, 842, 869(17)

Becker, R. R., 309, 324, 325

Beeman, W. W., 151, 171, 178, 181

Beers, W. H., 97

Beezer, A. E., 592, 614(7)

Begg, G., 942, 943(1), 944(1), 958, 972, 974, 975, 978, 998

Behnke, W. D., 355, 414, 431(19)

Beinert, H., 97, 927, 941

Bell, J. M., 590

Bell, R. P., 550

Bellamy, B. A., 170

Bello, H. R., 709

Bello, J., 204, 508, 509, 510, 511(37, 42), 702, 709, 896, 906(177)

Benedetti, E., 713, 715(209), 716(209), 721(209)

Benesch, R., 541

Benesch, R. E., 541

Benfield, G., 941

Benjamin, L., 613

Bennett, J. C., 945

Benoit, H., 165, 210, 211(22, 23), 221

Benson, A. A., 767

Bent, M., 927

Bentrude, W. G., 594

Berger, A., 707, 713(227), 714, 720(227), 722(227)

Berger, R. L., 598, 604(17), 616

Berlman, I. B., 773

Berman, M., 171

Bernardi, G., 242, 471, 473, 474, 475, 476, 477, 478

Bernstein, H., 691, 712(92)

Bernstein, H. J., 886

Berry, G. C., 486, 487(20), 488(20)

Bertrand, R. D., 884

Bessman, M. J., 490

Bethune, J. L., 60, 62, 66, 273, 274, 316, 318(34), 327, 331, 332, 333(55), 334 (34, 55), 337(34, 55), 338, 339, 397, 456, 457(1, 2), 459, 460(2), 461(7), 462(1)

Bevington, P. R., 877

Beychok, S., 677(33, 35, 37c), 678, 691, 692(33, 35, 37c), 693, 712(92), 714, 715, 717(236), 718(33), 720(236), 721, 722, 723(33)

Subject Index

D

S